Aerospace Encyclopedia of
WORLD
AIR FORCES

Aerospace Encyclopedia of

WORLD AIR FORCES

Editor: David Willis

Aerospace Publishing London
AIRtime Publishing USA

Published by
Aerospace Publishing Ltd
179 Dalling Road
London W6 0ES
England

Published under licence in USA and
Canada by
AIRtime Publishing Inc.
10 Bay Street
Westport, CT 06880
USA

Aerospace **ISBN: 1-86184-045-4**
AIRtime **ISBN: 1-880588-30-7**

Distributed in the UK,
Commonwealth and Europe by
Airlife Publishing Ltd
101 Longden Road
Shrewsbury SY3 9EB
England
Telephone: 01743 235651
Fax: 01743 232944

Publisher: Stan Morse

Managing Editor:
        David Donald

Editor: David Willis

Sub Editor:
        Karen Leverington

Design: Steve Horton

**Origination by
Chroma Graphics Ltd**

**Printed in Singapore by
Kim Hup Lee**

Distributed to retail bookstores in the
USA and Canada by
AIRtime Publishing Inc.
10 Bay Street
Westport, CT 06880
USA
Telephone: (203) 226-3580
Fax: (203) 221-0779

US readers wishing to order by mail,
please contact
AIRtime Publishing Inc. toll-free at
1 800 359-3003

WORLD AIR POWER JOURNAL
**is published quarterly and provides an in-depth analysis of
contemporary military aircraft and their worldwide operators.
Superbly produced and filled with extensive color photography,
WORLD AIR POWER JOURNAL is available by subscription from:**

**UK, Europe and
Commonwealth:
Aerospace Publishing Ltd
FREEPOST
PO Box 2822
London, W6 0BR
UK
Telephone: 0181-735 1200
Fax: 0181-746 2556
(no stamp required if posted in
the UK)**

**USA and Canada:
AIRtime Publishing Inc.
Subscription Dept
10 Bay Street
Westport, CT 06880
USA
Telephone: (203) 226-3580
Toll-free number in USA:
1 800 359-3003**

Note on sources

Where possible, the information in this volume has been acquired from
official sources, or gathered during officially sanctioned visits. The
editorial team would like to thank the many service personnel and
government officials who have assisted them, and their contributors,
with the compilation of this book. In the cases where such official help
has not been forthcoming, the information has been compiled from a
wide variety of open sources. Although every effort has been made to
verify this material, it is by its very nature subject to varying degrees of
speculation, especially where it concerns those air arms which maintain
heavy security around their operations and inventories. In such cases
this book claims to be no more than a guide to those air arms, and may
not represent their official position.

Authors and acknowledgments

**The editor would like to extend his thanks to
the following authors and correspondents for
their written contributions to this
encyclopedia:**

Yoshitomo Aoki Japan
Bob Archer USAF
Rick Burgess US Navy, USMC
Luigino Caliaro Italy
David Donald Asia Introduction, Korea (South),
Singapore
Robert F. Dorr USAF, USCG
Dylan Eklund United Kingdom
Adrian J. English Belize, Chile, Costa Rica, Cuba,
Dominica, El Salvador, Guatemala, Haiti, Honduras,
Jamaica, Mexico, Nicaragua, Panama, Trinidad &
Tobago
Tieme Festner Belgium, Croatia, Cyprus, Lithuania,
Malta, Slovenia, Uzbekistan
Peter R. Foster Bolivia, Colombia, Ecuador, Guyana,
Paraguay, Peru, Surinam, Uruguay, Venezuela
Richard Fisher China
John Fricker Afghanistan, Algeria, Angola, Armenia,
Azerbaijan, Bahamas, Benin, Bhutan, Botswana, Brazil,
Burkino Faso, Burundi, Cambodia, Cameroon, Cape

Verde, Central African Republic, Chad, Comores, Congo,
Congo (Zaïre), Côte d'Ivoire, Djibouti, Egypt, Equatorial
Guinea, Eritrea, Ethiopia, Gabon, Gambia, Georgia,
Ghana, Guinea-Bissau, Guinea Republic, Iraq, Korea
(North), Laos, Lebanon, Lesotho, Liberia, Libya,
Madagascar, Malawi, Maldives, Mali, Mauritania,
Mauritius, Mozambique, Myanmar, Namibia, Niger,
Nigeria, Oman, Palestine, Qatar, Rwanda, Saudi Arabia,
Senegal, Seychelles, Sierra Leone, Somalia, Sudan,
Swaziland, Syria, Tanzania, Togo, Tunisia, Uganda,
United Arab Emirates, Yemen, Zambia
Sqn Ldr Jon Hancock NATO
Robert Hewson Ireland
Salvador Mafé Huertas Portugal, Spain
Jan Jørgensen Denmark, Estonia, Latvia, Norway,
Sweden
George Kamp Albania
Gert Kromhout Germany, Netherlands, Poland
Jon Lake Israel, Jordan, Taiwan
Frederic Lert France
Georg Mader Austria, Czech Republic, Hungary,
Slovakia
Alexander Mladenov Bulgaria, Romania
Nigel Pittaway Australia
Jorge Felix Nuñez Padin Argentina
Chris Pocock Brunei, Indonesia, Malaysia, Philippines
Jeff Rankin-Lowe Canada
Tim Ripley Bosnia-Herzegovina, SFOR, UN,
Yugoslavia
Tom Ring US Army

Frank G. Rozendaal Belarus, Kazakhstan, Russia,
Ukraine, Vietnam
Pushpindar Singh Bangladesh, India, Nepal,
Pakistan, Sri Lanka
Tulio Soto Central America & Caribbean Introduction
Baldur Sveinsson Iceland
David Willis African Introduction, Australasia
Introduction, Bahrain, Barbados, Europe Introduction,
Indian Sub-continent Introduction, Iran, Kenya, Kuwait,
Malaysia, Middle East Introduction, Moldova, Mongolia,
Multi-National Introduction, North American
Introduction, Philippines, South Africa, South American
Introduction, Tajikistan, Turkmenistan, Zimbabwe
Jim Winchester Fiji, New Zealand, Papua New
Guinea, Tonga
René van Woezik Finland, Greece, Macedonia,
Switzerland, Thailand, Turkey

Thanks are also due to the following for their additional
assistance: BARG, MAR, Sqn Ldr Paul Harrison RNZAF,
Mrs I.M. Winchester, Mrs M.V. Willis, Achille Vigna,
Chris Knott, Carlos Lorch, Tulio Soto, Robert F. Dorr
and those who wish to remain anonymous.

Photographic credits appear on page 320.

*Frank Rozendaal would like to dedicate the accounts of the air forces
of Belarus, Kazakhstan, Ukraine and, in particular, that of the
Russian Federation, to the late Frits von Münching, whose gracious
assistance was instrumental in compiling these features.*

CONTENTS

MULTINATIONAL FORCES

The destruction and suffering experienced during World War II led to the desire to create an international organisation to prevent – with force if required – armed aggression and to promote diplomacy and understanding as the way forward for mankind. Based on the failed League of Nations, the United Nations (UN) was created, and first met in 1946.

Its early challenges included North Korea's invasion of its southern neighbour in 1950 and the formation of a mandate to protect South Korea. This was led by the United States, which poured large amounts of military hardware and personnel into the conflict. Wars of independence from the late 1940s onwards

also involved the UN, which throughout the years used member nations' aircraft and the personnel of their armed services, or hired aircraft from civil operators. Traditionally painted overall white, the United Nations' fleet of aircraft can be found operating at many of the world's trouble spots today.

Of the many military alliances that have existed throughout the history of armed conflict, few have survived as long as the North Atlantic Treaty Organisation (NATO), forged by the Western powers following the uneasy political settlements in Europe at the conclusion of World War II. NATO grew from the France-Britain Dunkirk Treaty of

1947 and the Brussels Treaty of 1948, which was signed by most countries of Western Europe. Subsequent inclusion of Canada and the US in mutual defence negotiations – following Soviet activities in the Berlin Blockade from March 1948 – led to the creation of NATO. For nearly 50 years, the sole purpose of the organisation had been to protect each member state through a system of mutual defence, prompted initially by the threat of war with or attacks from the USSR. The collapse of the Soviet Union in the late 1980s brought an end of this traditional threat. New opportunities and roles have since arisen for NATO, including peacekeeping in Bosnia-Herzegovina and expansion to include some of its former adversaries in Eastern Europe, the latter through the Partnership for Peace programme.

Wearing large 'United Nations' titles, this King Air 200 carries an American civil registration. The organisation has been allocated '4U' as its international civil aircraft markings code.

UNITED NATIONS

In December 1997 some 14,879 UN peacekeeping personnel were operating around the world as part of 15 missions. The 1990s have seen a massive growth in UN peacekeeping operations, with some 78,000 'blue helmets' in the field at the height of UN operations in Somalia and the former Yugoslavia in 1993. To support these operations, the UN has had to field a wide range of aircraft and helicopters. Its white-painted air force patrols disputed borders, shuttles diplomats to peace talks, flies aid to refugees and resupplies remote observation posts.

UN air assets are either provided by troop-contributing nation's armed forces or hired from commercial air operators. Military helicopters are paid for from the UN budget, with nations reimbursed for their costs. Since the end of the Cold War, the UN has been a keen employer of chartered East European aircraft, including Ilyushin Il-76s, Antonov An-26s, An-32s and An-124s, Mil Mi-8s and Mi-26s.

UNFIL (UNITED NATIONS INTERIM FORCE IN LEBANON)

One of the longest-running UN missions has been the difficult job of separating Israeli and Islamic guerrilla forces in Israel's self-proclaimed 'security zone'.

The Italian Army provides a detachment of Agusta-Bell AB 205 Hueys for air transport and casualty evacuation for the 4,564 strong force.

UNPREDEP (UNITED NATIONS PREVENTATIVE DEPLOYMENT FORCE)

Established in 1992 to monitor the sensitive northern borders of the Former Yugoslav Republic of Macedonia (FYROM), this UN mission was the first designed to pre-empt the outbreak of a conflict. It is currently supported by a single contract Eurocopter Squirrel and a Ukrainian Army Aviation Mil Mi-17. A pair of US Army Sikorsky UH-60L Blackhawks supports the US contingent under national command.

UNMIBH (UNITED NATIONS MISSION IN BOSNIA AND HERZEGOVINA)

This follow-on to the three-year UN mission in Bosnia concentrates on running the International Police Task Force and aid operations. Ukrainian Army Aviation Mi-17s support the operation, moving key personnel around the mountainous country.

UNSCOM (UNITED NATIONS SPECIAL COMMISSION TO DISARM IRAQ)

The operation to monitor compliance with UN Security Council Resolutions calling for the Iraqi regime to dismantle its weapons of mass destruction

was set up immediately after the Gulf War in 1991. Air transport is vital to the success of the mission, moving international inspection teams into Iraq and then flying them to suspect locations. Germany has been a keen supporter of UNSCOM, providing Luftwaffe C.160 Transall airlifters to provide communications between the mission's regional headquarters in Bahrain and Baghdad. In the first years of the mission, the German Heeresflieger also deployed three Sikorsky CH-53G Stallions to Al Rasheed Airport in Iraq; they have since been replaced by five Chilean Air Force UH-1H Hueys. In their cat-and-mouse game with the Iraqi authorities, the UNSCOM helicopters are often shadowed by Iraqi helicopters. On one occasion an Iraqi official flying in a Chilean helicopter tried to grab the controls of the machine when it strayed close to a 'sensitive' site.

MINUGUA (UNITED NATIONS VERIFICATION MISSION IN GUATEMALA)

Just under 200 military observers have been deployed to the Central American country since January 1997 on a ceasefire monitoring mission. It is supported by chartered Bell 212s.

UNFICYP (UNITED NATIONS PEACEKEEPING FORCE IN CYPRUS)

The divided island of Cyprus has been patrolled by UN peacekeepers since 1964. British Royal Air Force Westland Wessex HC.Mk 2 helicopters of No. 84 Squadron based at the UK Sovereign Base area are available, under national command, to support the 1,200-strong UN force, which includes a strong Argentine contingent.

UNIKOM (UNITED NATIONS IRAQ-KUWAIT OBSERVATION MISSION)

This mission to monitor the sensitive border between Iraq and Kuwait began after the Gulf War in 1991. The Chilean Air Force provided a detachment of Bell UH-1H Hueys to provide communications and casualty evacuation support to the UN's border observation posts. The Chileans now also support UNSCOM.

Right: Painted in overall white, this Puma is seen firing infra-red decoy flares. The nature of United Nations missions mean its members' aircraft can be required to operate in hostile environments.

Below: The United Nations has used a vast range of types for different missions over the years. This French air force C.160 Transall is seen loading troops at an African airfield.

The great expense of individual nations acquiring and operating the E-3A Sentry was overcome by the creation of the NATO Airborne Early Warning Force, sharing data and costs among the treaty members.

NORTH ATLANTIC TREATY ORGANISATION

FORMATION AND MEMBERSHIP

The North Atlantic Treaty Organisation (NATO) was established by the 1949 North Atlantic Treaty, commonly referred to as the Treaty of Washington. NATO's 16 member states are Belgium, Canada, Denmark, France, Germany (since 1955), Greece (since 1952), Iceland, Italy, Luxembourg, The Netherlands, Norway, Portugal, Spain (since 1982), Turkey (since 1952), United Kingdom and the United States. France and Spain do not participate in the Integrated Military Command Structure and the Collective Defence Planning, while Iceland has no defence force.

Article 51 of the United Nations Charter established the North Atlantic Alliance as a defensive one, based on political and military co-operation among independent member nations. Articles 4 and 5 of the North Atlantic Treaty are absolutely fundamental to NATO's foundation. Article 4 provides for consultations among the member states whenever any of them believes that their territorial integrity, security or political independence is under threat. Article 5 commits the member states to the defence of one another; an attack against any one member nation in Europe or North America is an attack against all member states. It must also be noted that NATO decisions are taken on the basis of consensus, and a course of action can only be undertaken if all members are in agreement.

POLITICAL AND MILITARY STRUCTURE

The North Atlantic Council: The NAC is the most important decision-making component of NATO, having political authority and powers of decision for the Alliance. Permanent Representatives of the 16 member states meet weekly, and from time to time the council meets at higher levels involving foreign ministers or heads of government.

Defence Planning Committee: The DPC is composed of Permanent Representatives, but meets at defence minister level twice a year, dealing with issues in defence policy and planning.

Nuclear Planning Group: The NPG handles NATO's nuclear issues, meeting twice a year at defence minister level.

NATO Committees: The NATO Committee structure supports NATO's consultation and decision making processes, ensuring fair representation for all states at every level. The committees are supported by the International Staff.

The Secretary General: The Secretary General of NATO is a senior international statesman who chairs the NAC, DPC, NPG and other senior committees. He also acts as NATO's principal spokesman.

Military Committee: The Military Committee has the responsibility for recommending to NATO's political authorities those measures considered necessary for the common defence of NATO's area and for the provision of guidance on military matters to the major NATO commanders – the Supreme Allied Commander Europe (SACEUR) and the Supreme Allied Commander Atlantic (SACLANT). The Military Committee is the highest military authority in the Alliance, and is supported by the International Military Staff.

PARTNERSHIP FOR PEACE

The January 1994 NATO Summit Meeting saw the introduction of the Partnership for Peace (PfP) programme, an initiative mainly aimed at creating a robust co-operation between the Alliance and the Eastern and Central European states. So far, 28 states have joined the PfP, and invitations have now been extended to the Czech Republic, Hungary and Poland to become fully integrated members of NATO.

AIRBORNE EARLY WARNING FORCE

The NATO Airborne Early Warning Force (NAEWF) was procured following a NATO Defence Planning Committee decision in December 1978 to acquire a NATO-owned capability to provide air surveillance and command and control functions for all NATO commands. The NAEWF is the largest single financial expenditure by the Alliance and consists of two components. The E-3A Component comprises 17 airframes (one example having been lost), located at the main operating base at Geilenkirchen in Germany. The E-3D Component is UK-owned and -operated, with seven airframes equipping Nos 8 and 23 Squadrons at RAF Waddington in Lincolnshire.

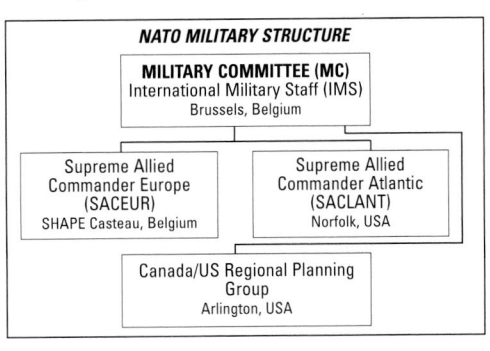

NATO MILITARY STRUCTURE

MILITARY COMMITTEE (MC)
International Military Staff (IMS)
Brussels, Belgium

Supreme Allied Commander Europe (SACEUR) SHAPE Casteau, Belgium	Supreme Allied Commander Atlantic (SACLANT) Norfolk, USA

Canada/US Regional Planning Group Arlington, USA

NATO'S COMMAND STRUCTURE FOR EUROPE (ACE)

Supreme Allied Commander Europe (SACEUR) SHAPE Casteau, Belgium

Allied Forces Southern Europe (CINCSOUTH) Naples, Italy	Allied Forces Central Europe (CINCENT) Brunssum, Netherlands	Allied Forces Northwest Europe (CINC-NORTHWEST) High Wycombe, UK
LANDCENT Heidelberg, Germany	BALTAP Karup, Denmark	AIRNORTHWEST High Wycombe, UK
AIRCENT Ramstein, Germany	NAVNORTHWEST Northwood, UK	NORTH Jatta, Norway

NATO Stabilisation Force (SFOR)

SFOR was established in December 1996 to continue the NATO mission in Bosnia-Herzegovina, i.e., the implementation of the Dayton Peace Accords. Its ground component is drawn from all the NATO countries and 22 non-NATO countries. Fixed-wing air support is provided by NATO's 5th Allied Tactical Air Force (5 ATAF) based in Italy, and occasionally augmented by aircraft-carriers in the Adriatic Sea.

A significant helicopter force is integrated into the SFOR ground component as follows:

NATO SFOR

Commander SFOR Flight (US)	Sarajevo Airport
2 x UH-60L Blackhawk, 1 x C-12	

Multi-National Division North East

Headquarters Eagle Base Tuzla	
4th Aviation Brigade (US)	Comanche Base, Tuzla
501st Aviation Regiment (US)	
1st Battalion 24 x AH-64A	
2nd Battalion 24 x UH-60A/L, 4 x EH-60C	
Medical Company (US)	'Blue Factory', Tuzla
15 x UH-60A/L	

Multi-National Division South West

Headquarters Metal Factory, Banja Luka	
British Aviation Squadron	Banja Luka, Gornji Vakuf, Split
663 Squadron, AAC (UK) 11 x Lynx AH.Mk 7, 4 x Gazelle AH.Mk 1	
Support Helicopter Force (UK)	Banja Luka, Gornji Vakuf, Split
1310 Flight, RAF	4 x Chinook HC.Mk 2
845 Squadron, FAA	2 x Sea King HC.Mk 4
322 Squadron (Czech)	2 x Mil Mi-8	Ljubija
Viper Flight, RNl AF	2 x MBB BO 105	Sisava

Multi-National Division South East

Headquarters Ortijes Airfield, Mostar	
Detachment ALAT (French)	Ploce, Croatia
AS 532UL Cougar, SA 341 Gazelle	
Heeresfliegerregiment 15 (German)	Rajlovac
CH-53G/GS Sea Stallion	
Italian Army Aviation Detachment	Rajlovac
AB 205A-1/AB 212	

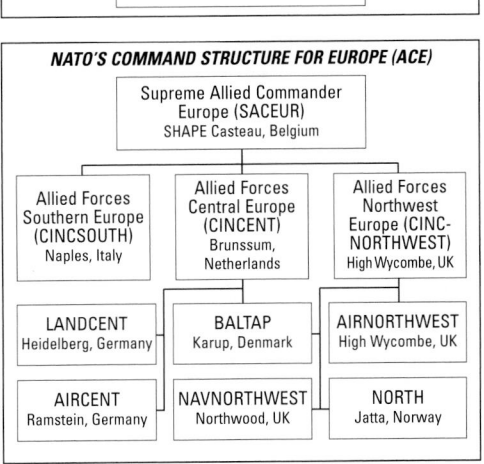

NATO'S CIVIL AND MILITARY STRUCTURE

National Authorities

Permanent Representatives (Ambassadors to NATO)   Military Representatives to NATO

DEFENCE PLANNING COMMITTEE	NORTH ATLANTIC COUNCIL	NUCLEAR PLANNING GROUP
OTHER COMMITTEES	SECRETARY GENERAL	MILITARY COMMITTEE
	International Staff	International Military Staff

Major NATO Commands

ALLIED COMMAND EUROPE	ALLIED COMMAND ATLANTIC

Integrated Military Command Structure

North America

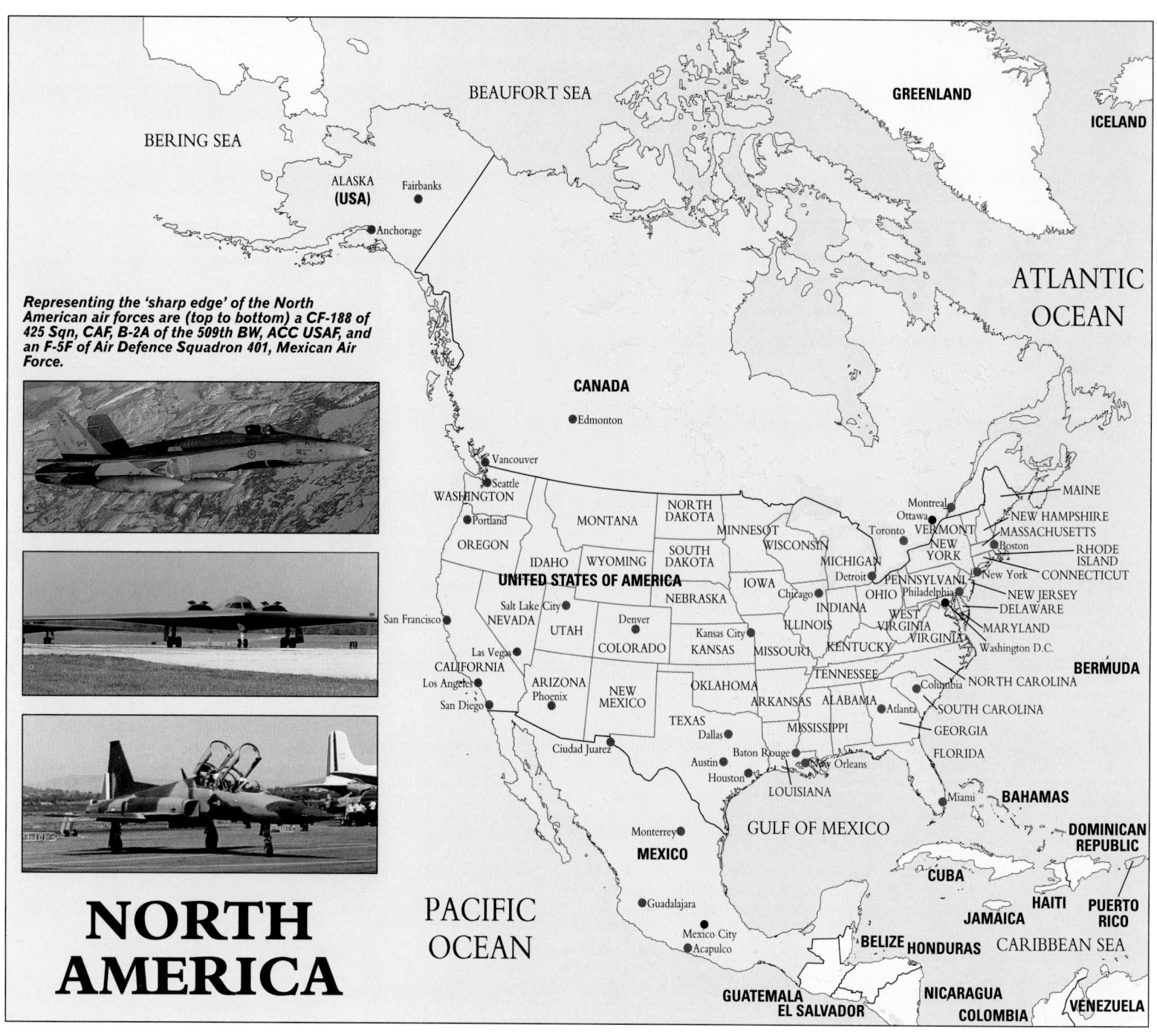

Representing the 'sharp edge' of the North American air forces are (top to bottom) a CF-188 of 425 Sqn, CAF, B-2A of the 509th BW, ACC USAF, and an F-5F of Air Defence Squadron 401, Mexican Air Force.

NORTH AMERICA

North America consists of the countries of Canada, Mexico and the United States of America. The region is dominated, militarily and economically, by the United States, which is the world's last military superpower.

The Unites States Air Force is one of the very few air forces with the full range of offensive and defensive aircraft, operating types as diverse as nuclear bombers and motorised gliders, piston-engined trainers and Mach 2 fighters. Supported by a large and diverse military-industrial complex, the USAF has striven to maintain its technological edge, as well as adequate quantities. Supported by its reserve organisations, Air Force Reserve Command and the so-called 'weekend warriors' of the Air National Guard, the air force has undergone dramatic changes over the last decade. The United States Navy and Marine Corps also control air forces larger than most countries' primary air arms. Large aircraft-carriers capable of projecting firepower from the sea onto the land provide mobile assets that allow the USA to respond quickly to changing military situations. Independent from, but associated with, the Navy, the Coast Guard provides coastal and inland SAR and would report to the Navy in times of mobilisation. Often overlooked as a major air arm, the US Army is

the world's largest rotorcraft operator; its aviation assets are integrated within its overall structure and thus are not as readily apparent.

The end of the Cold War, which had helped guarantee a continuous flow of funds, has left the US military with no 'natural' perceived threat and thus made it harder to justify spending huge sums of money on new hardware and maintaining large standing forces. Post-Cold War cuts bit deeply into the military, leaving only the core forces. On the international scene, the country was also adjusting to the new political landscape. Still committed to the North Atlantic Treaty Organisation (NATO), US force levels in Europe were greatly cut back as tensions decreased. The same cannot be said for Pacific Rim commitments, where deployed assets are only slightly smaller than before the end of the Cold War. Aircraft are still deployed to Japan, and the Korean peninsula remains a potential flash-point. America's relations with Latin America have included support for countries in the continuing war against the narcotics trade in the form of hardware and personnel. The USA shares with Canada the defence of the North American continent through NORAD (North American Aerospace Defence Command).

Much of Canada – the second-largest country in the world, after Russia – is empty wilderness with small pockets of habitation; most of the modest population lives near the Canada-US border, and there are relatively few large conurbations to rival those of its southern neighbour. This is reflected in the structure of its air force, in that squadrons are based throughout the country; about half of the units have an overtly military tasking, and the others are devoted more to providing aid to the civil powers, such as search and rescue. The Canadian Armed Forces today maintain a balanced core of defensive fighters and support types. Great emphasis is placed on the ability of the force to provide assistance to civil authorities during times of emergency, as aviation is the quickest way of get around the vast expanses of the country. The end of the Cold War also negated Canada's only external threat – danger of attack by the USSR – and removed the impetus to maintain large air defence forces and forward-deployed assets in Europe.

Mexico, placed between the United States and the Central America, is in the position of being much weaker that its northern neighbours and much stronger than its southern. As such, its air force is concerned more with internal actions than external.

UNITED STATES OF AMERICA

Capital: Washington, D.C.
Population: 248.7 million
Land area: 9.363 million km² (3,614,170 sq miles)
Major cities: Anchorage, Albuquerque, Atlanta, Baltimore, Boston, Buffalo, Chicago, Cincinnati, Cleveland, Columbus, Dallas, Denver, Detroit, Honolulu, Houston, Jacksonville, Kansas City, Indianapolis, Las Vegas, Los Angeles, Memphis, Miami, Milwaukee, Minneapolis, Nashville, Newark, New Orleans, New York, Oklahoma City, Omaha,

One of a pair of Boeing VC-25As of the Executive Flight of the 89th Airlift Wing passes over Mount Rushmore, South Dakota. Known as 'Air Force One' when the President of the United States is aboard, the VC-25 is used for high-level government transport, frequently accompanied by Boeing E-4s.

Orlando, Philadelphia, Phoenix, Pittsburgh, Portland, Salt Lake City, San Antonio, San Diego, San Francisco, Seattle, St Louis

United States Air Force

The United States Air Force will enter the 21st century as the world's most powerful air arm. The B-2 Spirit 'Stealth Bomber', F-22 Raptor fighter and C-17 Globemaster III airlifter form the vanguard of the USAF's futuristic, high-tech flying fleet. No other air force can boast warplanes as advanced as these and, following the end of the Cold War, none can seriously challenge the USAF in terms of technology.

Still, the US Air Force is challenged to maintain combat readiness in its new role as self-declared world policeman in a relentless series of overseas calamities from Somalia to Haiti to Bosnia. In the late 1990s, downsized and stripped of many of its overseas bases, the USAF is a 'smart' but very lean fighting service, down 36 per cent in size since 1990 – yet committed to military operations all over the world at the highest pace in its history. The end of the Cold War reduced the service's need for nuclear-capable aircraft and missiles, but the

propensity of US leaders to commit forces to trouble spots around the world has heightened the need for conventional warplanes and weaponry.

THE POST-COLD WAR AIR FORCE

Gone forever is the leviathan air armada that once confronted the Soviet Union and maintained bases around the world, ready for a nuclear holocaust or an armoured assault on the plains of Europe. The Cold War, and with it the prospect of an atomic World War III, has vanished from the warfighting plans and targeting codes of the US Air Force. This change is obvious to most observers. Less apparent is the harsh truth that the US Air Force which fought so well during Operation Desert Storm no longer exists, either. The USAF of today is in many respects newer and smarter, as well as smaller, and it has become an expeditionary force rather than an overseas garrison. Most USAF combat units are now located in the United States, not offshore, and warfighting doctrine now relies heavily on rapid deployment to a trouble spot In 1998, the USAF was deploying far too much,

with about 20 per cent of its personnel exceeding the recommended maximum of 155 days per year away from home base; morale was sagging, and retention of skilled personnel – especially pilots – was at the lowest point in the service's history. The USAF's deputy chief of staff said plainly that the exodus of skilled pilots, abandoning their military careers for the lure of the airlines, "has reached crisis proportions." This problem is, of course, shared by many other air arms as airlines press ahead with the biggest hiring round in aviation history.

Ahead, in place of nuclear war with Moscow or even conventional war with Saddam Hussein, lies an era of smaller, sharper flash points in a world not quite at war but not yet at peace – and with USAF people and aircraft operating in a constant crisis mode. Just one military effort, the policing of the 'No-Fly Zone' in southern Iraq known as Operation Southern Watch, has now lasted longer than US participation in World War II and has caused some air crews and maintenance people to be deployed abroad a dozen times in half as many years.

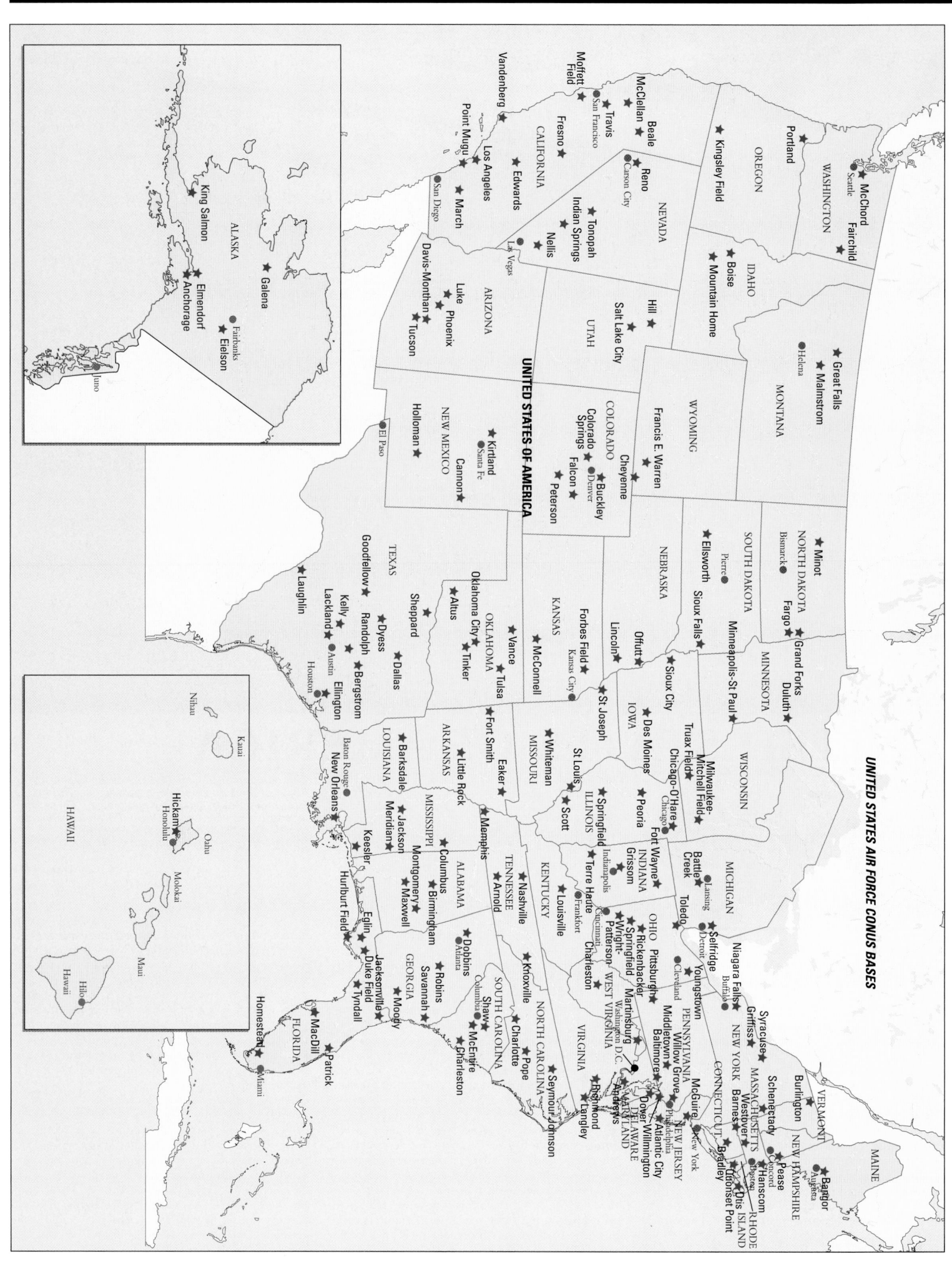

UNITED STATES AIR FORCE CONUS BASES

The aircraft in service with the USAF today are constantly upgraded to be survivable, adding equipment to counter perceived threats. An example is the fitting of flares to helicopters such as this MH-53J.

Few air forces have the full range of combat aircraft – from large bombers to tactical fighters – that the USAF is able to field today. This formation contains a B-1B, F-15Cs, F-15Es, F-16C and a KC-135R.

AN INDEPENDENT FORCE

The United States Air Force (USAF) traces its history to Army components that flew fabric-covered biplanes in World War I and a giant armada of combat aircraft in World War II. The USAF became separate from the US Army on 18 September 1947 following the agreement by Congress of a unification bill which had been signed by President Truman on 26 July. The National Security Act of 1947 created a tripartite Defense Department, with sub-departments of the Army, Navy (including the Marine Corps) and Air Force each presided over by a civilian Secretary. The new military organisation was commanded by the President, who delegated his responsibility to the Department of Defense. The Secretary of the Air Force was directly responsible for the Chief of Staff, USAF, who in turn was the Commander-in-Chief of the vast organisation divided into major commands, numbered air forces, air divisions, wings and squadrons. The new independent air arm inherited a personnel of slightly over 300,000 men and women, together with the equipment and infrastructure that had been part of the Army Air Forces.

JET AIRCRAFT

At the time of its independence, the USAF was on the verge of introducing swept-wing jet aircraft like the F-86 Sabre and the B-47 Stratojet. These resembled combat aircraft of today but were packed with vacuum tubes instead of microchips, equipped with analog instruments instead of digital ones, and powered by early jet engines that were not yet fully reliable. The advance of USAF air power over five decades is more than the story of jet warplanes – arguably, the service's most important aircraft are transports, which fly a real-world mission every day – but pointy jets serve as symbols of progress in all areas. Early jets like the F-86 were dubbed 'flying blow-torches' with good reason: to start the engines, pilots had to simultaneously manipulate fuel pump and throttle, a tedious and time-consuming process which, when rushed, led to a 20- to 30-ft tongue of flame shooting back from the engine. The USAF built up its long-range bomber force in the 1940s and 1950s, and fielded hundreds of radar-equipped interceptors assigned to halt a Soviet bomber attack.

EARLY TASKS

The Air Force was barely a few months old when it was required to respond to the first of many human-

itarian operations, following the closure by the Russians of all ground links between West Germany and Berlin. Operation Vittles involved numerous resupply flights into the city for 11 months. Slightly over a year later, the USAF was part of the United Nations force which began combat operations against the North Korean army that invaded the South. The war seesawed back and forth as both sides gained and then lost the advantage, until eventually a ceasefire was brokered in July 1953. Early jets battled each other while ageing F-51 Mustangs, B-26 Invaders, and C-47 Skytrains – all relics of an earlier time, and all pulled through the sky by propellers – did much of the hard work. By the end, improvements to the F-86, including redesign of its wing, made the Sabre vastly superior to its arch enemy, the Soviet MiG-15.

The next 10 years saw the USAF increase the quantity and quality of its weaponry as the United States and the Soviet Union faced each other over a divided Europe. The Cold War began in the late 1940s and lasted until the late 1980s when the Berlin Wall was breached, signalling the end of the Warsaw Pact. On several occasions – including the Berlin crisis of 1961 and the Cuban crisis of 1962 – the Cold War threatened to get 'hot', but the horror of a nuclear exchange was avoided.

The Vietnam conflict (1961-75) produced an odd disparity where fighters like the F-105 Thunderchief attacked strategic targets near Hanoi while B-52 bombers spent most of the war dropping ordnance into the jungle in search of elusive guerrillas. Aircraft types thought obsolete over the modern battlefield were reinstated into front-line use, and the service life of aircraft was extended to cover the drain of resources sent to the Far East. Vietnam was a long, drawn-out, bloody conflict conducted in the full glare of the world's media. It left a scar on the psyche of the American services which did not heal until the early 1980s.

POST-VIETNAM

The 1970s and 1980s were preoccupied with operations against small-scale regimes in the Caribbean, Central America, Middle East and North Africa. Combat missions were flown over several nations to rid them of their unpopular or undemocratic leaders, or to stem the growing terrorist menace. However, it was the Gulf War against Iraq in 1991 which proved to the world how effective airpower could be if co-ordinated in a cohesive manner. Operation Desert Storm was highly successful and enabled the failures of the Vietnam War to be finally exorcised. Desert Storm involved the large-scale use of the latest technological advances in weaponry, including stealthy aircraft and smart munitions. The F-117 Nighthawk

'Stealth Fighter', developed in an ultra-secret 'black' programme, was the star of the Persian Gulf campaign, but rumours of other extremely secret, 'black' warplanes are probably exaggerated.

The Gulf War also demonstrated a change in tactics, with the previously clear divisions between strategic and tactical operations overlapping; this enabled many of the commands which had been in existence since the Air Force was created in 1947 to be reorganised. In June 1992 the heavyweights of Strategic Air Command (SAC), Tactical Air Command (TAC) and Military Airlift Command (MAC) all ceased to exist, having been superseded by Air Combat Command (ACC) and Air Mobility Command (AMC). Other commands were realigned, consolidated or renamed soon afterwards, as the Air Force prepared itself for a whole new method of operation born from the ashes of the Cold War.

The most expensive warplane in history, the B-2 has always been controversial. A planned buy of 132 was cut to 21 in the light of post-Cold War economics.

Above: Used to strike heavily defended targets during Desert Storm, the F-117A Nighthawks of the 49th FW are the high-technology edge of the air force's tactical strike force.

Left: The most public unit in the air force are the F-16Cs of the Thunderbirds aerial demonstration unit. The team has operated the F-84G, F-84F, F-100C, F-105B, F-100D, F-4E, T-38A, F-16A before the F-16C.

CURRENT USAF AIRCRAFT

The current United States Air Forces is much smaller than at any time in its history. The massive reorganisation which was implemented in June 1992, combined with the effects of the drawdown, have left the USAF with a structure considerably leaner and far more compact. Much of the overseas commitment performed by in-theatre forces has been replaced, particularly in Europe, by assets deployed on a regular basis from the continental USA and elsewhere. The need to police the air exclusion zones above north and south Iraq, and over Bosnia-Herzegovina, has resulted in numerous front-line and reservist crews sharing the commitment, with the added realism of performing missions under actual combat conditions. Such activities have proved to be a double-edged sword in as much as aircrew have gained an invaluable wealth of experience of operations within a hostile environment, but this has for the most part been at the expense of being away from home for long periods of time. For many personnel, particularly those of the E-3 and RC-135 community, it is impossible to adhere to the requirement to spend not more than 120 duty days away from home.

BOMBER FORCE

At the end of the 20th century, as in the past, the primary mission of the USAF is to strike at deep and distant targets in the heart of an adversary's homeland. In addition to its fleet of ICBMs (intercontinental ballistic missiles), the USAF maintains about 210 long-range bombers divided among three types. The roster includes 21 B-2 Spirits and almost 100 each of B-1B Lancers and B-52 Stratofortresses.

The black, bat-like B-2 is the 'silver bullet' of US policy, reserved for use against targets of the highest priority. This small force of 'flying wings' is garrisoned at a single base in the centre of the US and is charged with flying nuclear and conventional missions to targets as far as 6,000 miles (9655 km) away. The B-2 was designed to use LO (low observables) or 'stealth' technology to evade radar detection. It carries the heaviest bombload of any operational bomber now in service, but is relatively slow and can be vulnerable when detected. The USAF is working to increase the flexibility of the B-2 with a new family of precision-guided conventional weapons, and has ambitious plans to operate the aircraft temporarily from forward bases like Guam and Diego Garcia. The B-2 is the costliest warplane ever built (around $900 million per copy), is difficult to maintain and is prone to trouble with the coating that provides much of its 'stealth'. On the type's first forward deployment to Andersen AFB, Guam in April 1998, a pair of B-2 bombers had to be supported by more than 200 personnel brought to the scene by transports. The much-publicised 'global reach' of the B-2 is very real, but has yet to be employed in a crisis situation.

The backbone of the USAF's bomber force remains the 50-year-old B-52 Stratofortress which is charged with both nuclear and conventional bombing missions. The 94 B-52H models now in service are the only American bombers that routinely carry all types of nuclear and conventional bombs, and missiles, in inventory. The B-52 is slow and unarmed (tail guns having been removed in 1992) and relies heavily on electronic deception to defeat an adversary's defences. The USAF's force of ICBMs, B-2s and B-52s, coupled with the US Navy's SLBM (submarine-launched ballistic missile) force, makes up the strategic forces available to carry out the SIOP (Single Integrated Operations Plan), the American targeting plan for the first hours of a nuclear conflict.

The USAF's fleet of 93 B-1B Lancer bombers, removed from the SIOP in October 1997, now has a conventional bombing mission only. The swing-wing, barely supersonic B-1B is slowly – and with difficulty – being adapted to a wider range of bombs and munitions, and has been forward-deployed to overseas bases with some success. More than a decade after entering service, the B-1B has a lower 'in-commission' rate than the USAF wants. Still, teething troubles with the B-1B's electronic warfare suite have been resolved and the bomber makes a valuable contribution to the overall USAF force.

AIR-TO-AIR COMBAT

The USAF's top priority for weapons procurement is the F-22 Raptor fighter. Two YF-22A prototypes began a test-flight programme in September 1990, and the first pre-production F-22 was rolled out with much fanfare on 9 April 1997. Flight testing has proceeded since, albeit with frequent delays. The F-22 is the first fighter to combine stealth, 'super-cruise' (the ability to sustain supersonic flight while proceeding to a distant target) and integrated avionics (a unified cockpit display giving the pilot information from all of the aircraft's sensors). Carrying its weapons internally, the F-22 is nearly as fast as, and far more manoeuvrable than, the F-15C Eagle it is meant to replace. The USAF's original requirement was for 750 F-22s, reduced to 648 in April 1991 because of budget constraints. In February 1994, the number of aircraft to be purchased was reduced further to 422, and in May 1997 to 339. The changes reflect both the downsizing of the USAF since the end of the Cold War and the difficulty the higher echelons are facing 'selling' the F-22 to Congress, the media and the public, with its cost of $158 million per copy. Flight test of the pre-production aircraft is said to be back on track following a six-month delay caused by technical problems. Current plans are for the first F-22 to be operational in 2004.

Above left: The B-1B has relinquished the nuclear role for that of conventional bombing. Able to fly fast and low, the aircraft is expanding the range of munitions it can carry to increase its versatility.

Left: The B-52 Stratofortress has seen over 40 years of service in its various versions. Current plans call for the B-52H to be in service up to the year 2040!

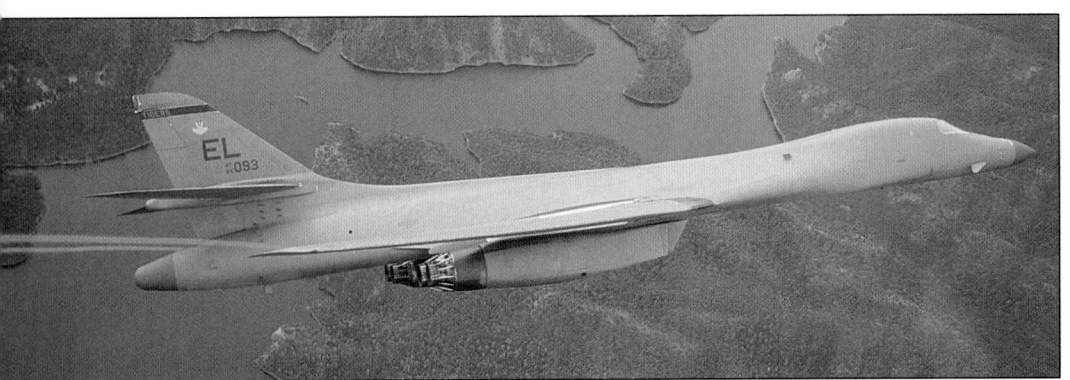

The F-15C/D Eagle remains the USAF's premier air-to-air fighting machine, with about 500 flying in active-duty, ANG (Air National Guard), and Air Force Reserve squadrons. (The Guard and Reserve are integral to American warfighting plans and participate in virtually all exercises and deployments.) With improved radar, AIM-120 AMRAAM missiles, and updated engines, the 20-year-old Eagle remains a formidable warplane, as attested by its claim to 36 of the 39 USAF aerial victories in Operation Desert Storm, without a single combat loss. The two-seat F-15D is fully combat-capable and serves alongside the F-15C. Upgrades of avionics and electronics equipment are expected to improve the Eagle and there are no plans to take it out of the inventory, even after the F-22 enters service.

A few units operate F-15A/B MSIP (Multi-Stage Improvement Program) Eagles with improved avionics and provision for AMRAAM and other missiles carried by the C/D models. As with other USAF tactical aircraft, a few units are pioneering the use of night-vision goggles (NVGs) and NVG-compatible instrument displays with the F-15A/B MSIP Eagle.

STRIKE AIRCRAFT

In the future, the USAF will join other US service branches in acquiring the JSF (Joint Strike Fighter) for the air-to-ground combat role. JSF is an ambitious programme that could ultimately produce 3,000 combat aircraft for the USAF, US Navy, US Marine Corps and Britain's Royal Navy. The USAF version is expected to take off and land conventionally, but other services will employ a short take-off and vertical landing (S/TOVL) variant. JSF is expected to fly in 2002 and to enter service in 2010.

The F-15E Strike Eagle is the USAF's two-seat, dual-role strike aircraft for a variety of air-to-ground missions ranging from close air support to deep penetration. With the recent retirement of the F-111F 'Aardvark', the F-15E force is heavily tasked to deploy to world trouble spots. In contrast to earlier Eagles, the Strike Eagle has improved radar, a wide field of view HUD, numerous internal changes, a fully-equipped station for the rear-seat weapon systems officer, and the capacity to carry up to 24,500 lb (11112 kg) of ordnance. The USAF had 132 F-15Es in combat-ready status at the start of 1998, and the manufacturer was expected to deliver a handful of additional aircraft early in the new century as attrition replacements.

The F-16 Fighting Falcon is the most numerous combat aircraft in USAF service, and has seen active-force, Reserve, and ANG duty. Known as the 'Viper' to its proponents and the 'Lawn Dart' to its detractors, the ubiquitous F-16 has evolved over two decades into the current Block 50/52 F-16C/D model with advanced cockpit displays, a wide-angle HUD, AMRAAM capability and avionics improve-

The heavily tasked Rivet Joint fleet of RC-135V/Ws is based at Offutt AFB, Nebraska. Current planning calls for the aircraft to have their TF33-P-9 engines replaced by CFM56s.

ments. The F-16 is billed as a dual-role warplane that can engage in aerial combat and attack ground targets, but its primary mission in the force is as an air-to-ground 'shooter'. A few F-16C/D aircraft use the AN/ASQ-213 HARM Targeting System (HTS) in the 'Wild Weasel' role against enemy defensive radars. Advanced versions are proposed, but the F-16 is likely to lose the limelight to the coming JSF.

The F-117 Nighthawk, better known as the 'Black Jet' or the 'Stealth Fighter', was the original stealth aircraft in USAF inventory and is remembered for flying 1,270 sorties against top-priority targets, mostly in Baghdad, during the Gulf War. The F-117 continues to operate with one fighter wing at a single base and is likely to be deployed in any overseas situation requiring its ability to carry 2,000-lb (907-kg) precision-guided bombs to a heavily-defended target. But the single-seat F-117 is ageing and a programme to upgrade its FLIR and DLIR sensing equipment has been seriously delayed.

The USAF's air-to-ground inventory is rounded out by the A-10 Thunderbolt II, or 'Warthog', and the AC-130 Spectre gunship – two slow-movers that defy the adage 'speed is life'. The A-10, deemed vulnerable to man-portable SAMs but irreplaceable as a tank-killer and a combat rescue aircraft, is slated to be out of service by 2005. The AC-130H and AC-130U, which bring enormous firepower to bear on a ground target while in a pylon turn, are too slow to be useful where the adversary has modern air defences, but they will remain in service for Third World contingencies.

RECONNAISSANCE

The USAF depends heavily on its mixed fleet of photo- and electronic reconnaissance aircraft, all of which are old and slow. Media reports of a hyper-advanced and super-secret reconnaissance aircraft, called the Aurora in popular reporting, appear to be without foundation. No evidence has emerged that Aurora, or a more prosaic tactical reconnaissance aircraft dubbed the TR-3 Manta, are anything but

The USAF's premier air-superiority aircraft since the early 1970s has been the F-15 Eagle. Since late 1990, when the F-15E first entered service with an operational wing, the type has been responsible for the all-weather attack role as well. This pair of F-15Es is from the 57th Wing F-15E Weapons School.

figments of overworked imagination. Although the USAF still operates a base at Groom Lake, Nevada (in a region popularly called 'Area 51') that is kept secret from the outside world, it appears likely that the only reconnaissance craft being tested there are UCAVs (uninhabited combat air vehicles).

The USAF's most famous reconnaissance aircraft, the SR-71 Blackbird, has been retired after a brief revival in the mid-1990s.

The U-2 is the USAF's principal piloted reconnaissance aircraft, and the service's fleet of about 35 aircraft is thinly spread to meet worldwide contingencies. The U-2 is viewed as both a 'national' and a 'tactical' asset, providing critical intelligence to political decision makers and theatre commanders. It is capable of collecting multi-sensor photo, EO, IR and radar imagery, as well as Sigint. In 1998, the USAF completed the task of upgrading the U-2 fleet to U-2S standard with new engines and avionics. Progress has been slow in exploring an unmanned replacement for this familiar spyplane, earlier versions of which were in service since the 1950s.

The USAF is expanding its RC-135 Rivet Joint fleet, converting two additional aircraft to augment this heavily-tasked fleet of nine aircraft that includes RC-135S, RC-135U, RC-135V and RC-135W models. These reconnaissance versions of the familiar Stratotanker carry out electronic, telemetry and other intelligence duties. RC-135 Rivet Joints have operated in the Persian Gulf since 1990 and have a new focus on loitering near and monitoring radar defences. Another variant of this ubiquitous airframe is the OC-135B, which uses an IRLS, synthetic aperture radar, and video cameras to monitor the 1992 Open Skies Treaty. Three OC-135Bs will be in operation by 2001.

North America

Right: The Boeing (McDonnell Douglas) C-17 Globemaster III was designed to be able to carry all the loads a C-5 Galaxy could, but be able to deliver them to very austere airfields near battle fronts. Given the day-to-day operational profile of the aircraft, budget holders have questioned the need for the costs such an ability incurs.

Below: The original version of the C-130 Hercules first took to the air in 1954 and the type is still in production for the USAF, albeit in a very different form. The Hercules in USAF service performs more than the dozen tasks of its legendary namesake.

Below right: The McDonnell Douglas KC-10A Extender undertakes dual tanker/transport tasks with AMC units. It is also operated by Air Force Reserve Command under the 'associate' programme. The Extender is an excellent vehicle for deployment support as it can transport personnel and supplies at the same time as providing fuel.

ELECTRONIC WARRIORS

After the retirement of the EF-111A Raven (in April 1998), the USAF achieves the tactical radar-jamming and Comint mission through its reliance on the US Navy's EA-6B Prowler. Under a joint-service agreement, four of the Navy's dozen Prowler squadrons are manned by both Navy and USAF flight crews and perform electronic warfare and reconnaissance duty for both services, although the arrangement has suffered delays and has been criticised as inadequate. Many believe that the Air Force was premature in putting to pasture an EW aircraft that was fast enough to stay with strike packages, something the EA-6B is not, while having no replacement planned.

The wide range of EW and command-and-control functions performed by USAF aircraft has become ever more important as warfare increasingly shifts to the electronic spectrum. Critics argue that the USAF may not be keeping up, since much of its electronic capabilities reside in the interiors of ageing airframes that have accumulated high flight hours. The E-3 Sentry AWACS aircraft has become a common sight around the world, especially in the troubled Middle East, and represents another community of fliers who

are frequently deployed and worked hard. In its role as a flying radar station – as well as a key communications and intelligence platform – this variant of the Boeing 707-320B has been constantly upgraded, although the effort to keep up with changing technology remains a challenge. All 32 USAF E-3s are eventually to be upgraded to 'full' E-3C standard with JTIDS (Joint Tactical Information Distribution System) for anti-jam digital communications and with improved ESM equipment. In continuing to field the E-3 Sentry on a worldwide basis, the USAF is in the position of using an old American aircraft when others have a newer example – the Boeing E-767s which perform the same mission for Japan were manufactured as recently as 1997.

The Boeing E-8 J-STARS (Joint Surveillance and Target Attack Radar System) is the other derivative of the Boeing 707-320B wearing US Air Force colours. The current E-8C evolved from earlier versions that first saw action in the 1991 Persian Gulf fighting. The E-8C J-STARS's job, in effect, is to look at the battlefield situation on the ground in the same way that AWACS monitors the situation in the air. J-STARS is now responsible for ground surveillance, targeting, attack and battlefield management, and bomb-damage

assessment. The communications and electronics systems also have a role in SEAD and the detection of elusive ground targets such as mobile missile launchers. Unlike AWACS aircraft, which are long in the tooth but were new when delivered, J-STARS aircraft are previously-owned 707 airframes that have been extensively refurbished. When Britain's Royal Air Force selects an aircraft for a similar mission in 1999, the RAF will probably have a far newer J-STARS equivalent than the American air force.

For airborne command duties the USAF employs the Boeing E-4B NAOC (National Airborne Operations Center), and the EC-130E ABCCC (Airborne Battlefield Command and Control Center) for very different strategic and tactical purposes. The E-4B, a version of the Boeing 747-200B (like the VC-25A presidential aircraft described below), provides an aerial command centre for US leaders in the event of nuclear war. Dubbed the 'Doomsday Plane', the aircraft is kept in readiness to carry the president, or others in the chain of leadership, during the first hours or days of a general conflict. Its mission is supported by the C-20C version of the Grumman Gulfstream which is expected to whisk the president and other key leaders of the National Command Authority away from a wartime target to a location where they can transfer to the E-4B. The very different job of handling battle operations in a combat theatre falls to the EC-130E ABCCC. In the strategic realm, many of the airborne command duties once performed by USAF EC-135 Stratotankers are now the province of the US Navy's E-6B Mercury – itself yet another Boeing 707 variant. This includes the Looking Glass mission of command and control of the US ICBM force.

The 'E' for 'electronic' prefix also applies to the EC-130E Commando Solo/Rivet Rider psychological operations broadcasting aircraft – able to intercept or even displace an enemy's TV transmissions – and the EC-130H Compass Call communications jammer. These vital assets are on the service's 'most deployed' roster, their crews often being away from home base for up to 200 days per year. Never leaving home station at all, in contrast, are the USAF's two E-9As, highly-modified de Havilland Canada (Boeing

The C-5 Galaxy has had a chequered career in service. The original C-5As had to be rewinged due to fatigue cracks, and the type suffers from poor maintainability, giving rise to several derogatory nicknames.

(Canada) DHC-8 Dash 8M-100s used for sea surveillance and weapons range policing duties in Florida.

AIRLIFT OPERATIONS

In the view of many, the USAF's most vital flying machines are not its glitzy fighters or its weapons-laden bombers but, instead, the transports that span the globe every day, year in and year out. Crews and maintainers in the airlift community argue, with justification, that they fly a 'real' mission every day while others merely "go out and bore holes in the sky," as one put it. The C-17A Globemaster III is both the reality and the symbol of a revitalised airlift force. The Desert Shield airlift of 1990-91 – by far the greatest air transport effort in history – might be slightly outside the capability of today's downsized force, but there can be no doubt that C-17s, C-5s and other transports give the USAF global reach. Airlifters are used primarily to carry equipment and freight: most routine movement of military personnel is handled by contract carriers who operate civil Boeing 747s, Lockheed L-1011s and other types.

The C-17A Globemaster III has settled into everyday use following a period of high visibility and after emerging from a developmental programme that was both delayed and mismanaged, but has been reshaped into a model for industry. This transport is unique in being able to fulfil a typical mission with a crew of just three, comprising two pilots and a loadmaster. The high-tech, high-ticket C-17 has set 22 world records for payload-to-altitude, time-to-climb, and short take-off and landing with payload, including one record effort where a C-17 ('Bubba' in the slang of some crews) landed in just 1,400 ft (426 m) with a cargo of 44,000 lb (19960 kg). Although the USAF rarely uses the short take-off and landing capability that drove up the C-17's sticker price, the 'outsized' capacity of this heavy lifter has been a godsend in contingencies from Somalia to Bosnia. Serious consideration is being given to increasing the currently-planned, long-time buy of 120 Globemaster IIIs.

The C-5A/B Galaxy is the largest aircraft ever

This Boeing 727-35 joined the USAF as a C-22B in 1985 after service with Pan American World Airways. Today it flies with the 201st AS of the 113th FW, DC ANG, which also flies F-16s of the 121st FS.

operated routinely by US forces. About 125 Galaxies now in service do much of the USAF's heavy lifting and, until the arrival of the C-17, were alone in being able to transport main battle tanks and other outsized items. The Galaxy is likely to be employed to support virtually any US combat situation or humanitarian relief effort. The C-5A/B fleet has received some, but not nearly enough, attention to bring it up to date, including installation of a revised MADAR II (Malfunction Detection Analysis and Recording) instrumentation package. In recent years, the C-5A/B has suffered from serious reliability problems, and consideration has been given to an extensive and costly programme of major modifications to keep the fleet in service until 2030. A pair of C-5As was modified for cargoes of even larger size than those routinely accommodated and is designated C-5C. Because of its low 'mission capable rate', crews commonly refer to this portly transport not as the Galaxy but as 'Fred', an acronym for 'F—king Ridiculous Economic Disaster'.

The Boeing KC-135 Stratotanker is now part of the same command structure that supports airlift operations. Re-engined KC-135s still have many years ahead of them acting as the principal air refuellers for all US and Allied military aircraft.

The KC-135 is joined in its 'flying gas station' role by a handful of HC-130/MC-130 Combat Shadow variants of the Hercules (below) and by the USAF's 59 KC-10A Extender (military Douglas DC-10 Series 30CF) dual-role tanker/transports. The Extender was brought into service in the 1980s to enable tactical assets to deploy accompanied by an aircraft that could both carry and refuel. Both the KC-10 and KC-135 can employ the flying boom and the probe-and-drogue refuelling methods – the former favoured by the USAF, the latter by the US Navy. A proposal to add additional drogue refuelling stations on the KC-10 wing, to allow more aircraft to be refuelled at once, died after one aircraft was converted.

The venerable C-130 Hercules is the bulwark of USAF tactical air transportation in addition to its other mission duties as a gunship, command post and electronic platform, and remains the workhorse of the tactical air transport fleet. Fully 65 per cent of the USAF's C-130 force is assigned to AFRC or ANG units, and 'trash-hauler' C-130 squadrons represent

Three C-20A Gulfstream IIIs are based at Ramstein AFB, Germany with USAF Europe for operational support airlift. The Gulfstream family is well represented within the USAF, with GIIIs (C-20A, B and C), a GIV (C-20H) and GVs (C-37A – above left) in service.

another community that is frequently deployed. The familiar sight of a Hercules making an 'assault landing' on an unpaved surface in 3,000 ft (914 m) or less has become a staple of American diplomacy in the late 1990s. A priority of the USAF is to reduce the number of C-130 transport variants, since the fleet is currently the service's most disparate: according to one measure, there are 26 versions of the C-130 pilot's instrument panel. Ageing C-130E models are expected to follow the now-departed C-130A/B variants to the 'boneyard', concurrent with a major programme to standardise instruments and equipment in the C-130H fleet. Plans to introduce the C-130J Hercules II variant are proceeding slowly.

While the USAF works toward an early retirement of its fleet of C-141B StarLifters, many of which are approaching the type's 35,000-hour flight limitation, the service continues to fly a variety of other transport types.

The C-9A Nightingale, a version of the civil Douglas DC-9, soldiers on as the USAF's only dedicated aeromedical aircraft (while the C-9C model does VIP transport). The service is slowly reducing its number of shorter-range C-12C/F Huron and C-26A/B Merlin transports, but has no plans to cut its reliance on the C-20A/B/C/H Gulfstream III/IV or C-21A Learjet for operational support airlift (OSA) duty. Four C-22B transports operated by the Air National Guard for OSA and medium-range transport are the service's only examples of the Boeing 727.

VIP TRANSPORT

Flagship of the USAF's VIP fleet is the Boeing VC-25A, two of which are much-modified versions of the Boeing 747-200B. To many people, the beautiful and expensive VC-25A is 'Air Force One'. In fact, that term is used as the radio callsign for any aircraft carrying the US president, but the VC-25A is the aircraft normally used and is operated with expensive and elaborate support facilities for no other purpose than as personal transport for the occupant of the White House. The VC-25A boasts a Bendix Aerospace EFIS-10 electronic flight instrument system and a lavish suite of state-of-the-art communications equipment, an executive conference room and a private bedroom, but does not have, contrary to myth, a handball court or swimming pool. A pair of self-contained air stairs is located on the port side and a baggage loader on the right. Together with an auxiliary power unit, they allow the president's aircraft to be practically self-sufficient and to reduce the need for ground support equipment. The VC-25A is also hardened against EMP, although in wartime the president would be aboard the E-4B NAOC instead.

The USAF's fleet of VC-137B/C Stratoliner (Boeing 707-120) executive transports – the 'Air Force One' prior to introduction of the VC-25A – numbered seven aircraft as recently as 1993, and is to be fully retired by the end of 1999. To replace the much-loved and long-serving C-137s, the USAF has ordered a 'package' of four C-32As (Boeing 757-

North America

200s) and three C-37As (Gulfstream Vs). Production and delivery of the C-32A suffered delays associated with Boeing's long-standing backlog on its 757 production line, but the first C-32A was flying and scheduled for delivery in June 1998, about a year behind original schedule.

The VIP fleet is also home to the C-20A/B/C/H Gulfstream III/IV transports (but no longer C-21A Learjets, which have now been scattered among squadrons throughout the US and abroad). The VIP unit operates some of the USAF's last Bell VH-1N/UH-1N Iroquois (Huey) helicopters.

ROTARY WING

The US Air Force can lay no particular claim to having early recognised the value of the helicopter – it was far behind the Coast Guard in pioneering the development of rotary-wing flying machines – but the helicopter has become indispensable today. Among other things, helicopters have totally replaced the liaison aircraft of the past for utility transportation and miscellaneous duties. And, while the USAF's combat search and rescue force (CSAR) has been trimmed to the bone by the budget axe, the helicopter remains the principal tool of those who fly rescue missions.

The H-60 Hawk series of helicopters represents the only rotary-wing aircraft used by all five branches of the US armed forces. So far, the USAF has operated the type principally for combat rescue and special operations duties. About 100 MH/HH-60G Pave Hawk models provide a wide variety of special-operations capabilities, including infiltration/exfiltration and personnel recovery. The HH-60G, flown by active-duty, Reserve and ANG squadrons, is the dedicated rescue version. In contrast to the rather prosaic UH-60L Black Hawks which abound in the US Army, the MH/HH-60Gs often have integrated navigation systems using GPS (global positioning system), INS and Doppler. A weather-mapping radar is also a prominent feature. Pararescue jumpers, or PJs, find the Pave Hawk helicopter nimble and versatile but complain that it has too little interior space: when carrying two 180-US gal (680-litre) internal fuel tanks for increased range, the Pave Hawk has barely enough room to handle two PJs and two survivors. High on the service's 'wish list', although not part of official planning, is a far larger and longer-legged helicopter for the dedicated rescue mission.

The MH-53J Enhanced Pave Low III, or 'Super Jolly Green', offers greater size and range for the special-operations mission, and can be made available for dedicated rescue duty. This is one of the few Vietnam-era aircraft remaining in inventory, and despite extensive modifications it suffers from a near-terminal case of old age. An MH-53J plucked to safety the only US airman rescued during the 1991 Gulf War (of more than three dozen shot down) but the artificial barrier between special-operations and air-rescue users has hampered the most effective use of this popular helicopter, and fatigue life is a looming problem. Both special-operations and combat-rescue people would like to have a version of the three-engined CH-53E model that has performed well in the US Navy and Marine Corps, but no plans are on the horizon for a purchase – in part because of the Pentagon's massive joint commitment to the V-22 Osprey tilt-rotor aircraft.

The USAF's fleet of about 60 UH-1N Hueys is expected to be retired soon, with the exception of the handful used for VIP transport. The V-22 Osprey tilt-rotor aircraft fills out the USAF's current plans for rotary-wing aircraft, although the USAF is far less heavily committed to the Osprey than the Marine Corps and Navy. Within the Air Staff in the Pentagon, the Osprey is regarded as something of a diversion and is distrusted by the rescue community

because of its cramped interior, noise level, and rotor downwash.

The USAF's CV-22 variant of the Osprey (known in Marine Corps parlance as the MV-22) will give special-operations forces a high-speed, long-range, V/STOL combat aircraft capable of penetration and extraction behind enemy lines in secrecy and poor weather conditions. Combining key features of the aircraft and the helicopter, the tilt-rotor CV-22 is expected to carry 18 troops over a 575-mile (925-km) combat radius at up to 265 mph (426 km/h). (Normally the 'M' prefix would denote special operations and the 'C' prefix a troop- or cargo-carrying model, but in the case of the Osprey the 'M' stands for Marines.) The CV-22 is expected to make use of an integrated navigation package, FLIR and NVGs. Fanciful ads for the special-operations CV-22 depict the aircraft performing all manner of snatch-and-grab operations deep in an enemy's heartland, even landing on rooftops in the adversary's capital. In fact, tactics being developed for the CV-22 will rely heavily on carrying out most missions without detection. The USAF is expected to receive its first operational CV-22 in 2003 and to become operational in 2005, with an eventual purchase of 50 aircraft.

TRAINING

The training of pilots, known as SUPT (specialised undergraduate pilot training), gains new importance because of the unprecedented departure of experienced pilots from military ranks. The USAF is employing new equipment and methods in the instruction of pilots and other aircrew members. Ironically, the training community in 1998 boasted the aircraft types with the best (T-1A Jayhawk) and worst (T-3A Firefly) safety records in all of American aviation, military or civil.

The pilot trainee begins with EFS (Enhanced Flight Screening), a process intended to quickly and cheaply rule out those individuals who lack the aptitude to handle the controls of an aircraft in flight. Beginning in 1993, the USAF purchased 113 T-3A Firefly trainers (its version of the Slingsby T67M-260) for the EFS mission and for training at the Air Force Academy, replacing the T-41A/C (Cessna 172), known as the Mescalero in its Army version. Unfortunately, the T-3A experienced three crashes that produced six fatalities in a matter of months, and in mid-1997 was grounded while a review was conducted of possible ways to resolve a suspected fuel-vapour problem and other alleged control difficulties of an essentially simple and supposedly fool-proof aircraft. A year after the grounding, it was being questioned whether the

T-3A would ever take to the skies again, and in the meantime the USAF had suspended both EFS and Academy flying.

The initial flight trainer for pilot candidates is the Cessna T-37B 'Tweet', which also has the distinction of being the oldest airframe in USAF service. (The B-52 and C-130 flew earlier but the examples in service today are slightly younger than the mid-1950s 'Tweet'.) The T-37 is the standard two-seat primary trainer and has a side-by-side seating arrangement which the USAF has subsequently decided is undesirable. The T-37 is both simple and versatile, but is near the end of its life span. Flying the T-37B at training bases in the American Southwest in the summer months can be especially gruelling because the 'Tweet' has the least effective air conditioner of any aircraft in the inventory. While about 450 remain on duty, the service is eager to move to its planned replacement, the T-6A.

The new primary trainer, the Raytheon T-6A Texan II, also known as the JPATS (Joint Primary Aircraft Training System) and as the Beech/Pilatus PC-9 Mk II, is a tandem, two-seat turboprop chosen recently after a competition which also included turbojet and turbofan candidates. The T-6A has a modern 'glass' cockpit (ironically, newer than many of the operational warplanes to which student pilots will graduate) and in comparison to the T-37B offers visibility, quietness, and many of the same handling qualities as a pure jet – its propeller being virtually unnoticeable to the student in the front seat. The USAF expects to benefit from placing the student pilot on the centreline and from the fuel-economy afforded by the T-6A. A forward canopy bow added to the basic design to improve birdstrike characteristics slightly reduces student visibility in comparison with variants of the Swiss-designed PC-9 used by some other air forces. The first of about 375 operational T-6As was scheduled for delivery in mid-1999, and the aircraft will later be employed by the US Navy as well.

Following *ab initio* training in the T-37B (or, soon, the T-6A), the student pilot is assigned to BAFT (bomber, attack, fighter) or multi-engined (tanker, transport) flight training. For the first time in history, a preponderance of student pilots who are given a choice go into the multi-engined track, many with an eye on the current airline hiring of aircrew. Those

The T-37Bs are due to be replaced in the near future by the Swiss-designed, American-built Raytheon T-6A Texan II, which is being procured under the JPATS (Joint Primary Aircraft Training System) programme with the US Navy.

Seen on its maiden flight, the Ryan RQ-4A Global Hawk high-altitude reconnaissance unmanned aerial vehicle (UAV) illustrates the growing importance of UAVs to the USAF.

The future air dominance fighter for the USAF is the Lockheed Martin F-22 Raptor, combining stealthy qualities with supercruise ability in excess of Mach 1.5. The aircraft is due to enter service in 2004. This is the first Engineering and Manufacturing Development (EMD) F-22A, painted in a scheme called 'Mod Eagle'.

chosen for combat aircraft proceed into the T-38A Talon advanced trainer. Once the world's only supersonic trainer, the ageing T-38 is not scheduled for replacement for decades. About 400 remain in service, accompanied by a handful of T-38Bs that are mostly used to teach fighter fundamentals. An upgraded version of the Talon with improved structure and new avionics, the T-38C, made its first flight on 8 July 1998.

Students in the multi-engine track fly the Raytheon T-1A Jayhawk, based on the Beechjet 400T. Barring change, as of April 1998 the T-1A was the safest aircraft in the history of aviation. One hundred and eighty were built and after five years of operational service no one has been killed or injured in one, and none has been involved in a major or minor mishap. The T-1A flight deck is configured for a student in the left seat, an instructor in the right seat, and a second student to the rear. The aircraft is equipped with a Rockwell Collins avionics package, EFIS, turbulence detection radar, digital autopilot and a central diagnostics and maintenance system. Proponents say the T-1A is the ideal aircraft to introduce the student to a multi-engine, or crew environment. A minor criticism is that the instrument panel, like that on many business jets, is flimsy. The appearance of the T-1A in the early 1990s marked the first time since the Korean War era that USAF pilots pursued separate tracks.

The USAF operates the T-43A (Boeing 737-200) as the navigator trainer for both USAF and Navy personnel. A handful of T-43As, plus civil 737s, are also used by the civil contractor which serves the remote and secretive air base at Groom Lake, Nevada. The USAF operates two UV-18B Twin Otters for parachute jump training at the Air Force Academy and a small number of TG-3A (Schweizer 1-26E) and TG-4A (2-33A) sailplanes, also at the Academy. The service has a TG-9A (Schleicher ASK.21) sailplane and a TG-11A (Stemme S-10) motorised glider.

The USAF operates numerous uninhabited aerial vehicles (UAVs), drones, and decoys. The last QF-106A Delta Dart has been retired and the principal target drone today is the QF-4E/G Phantom. The Air Force's weather-reconnaissance aircraft, the WC-130H Hercules, is being augmented, then replaced, by the WC-130J Hercules II.

FUTURE FLIGHTS

With its array of fast fighters, big bombers, tiny trainers and everything in between, the United States Air Force is generally in good shape, although there are glaring weaknesses in parts, supplies and necessary upgrades. Despite its overtaxed mood caused by unprecedented deployments and its personnel problems caused by unprecedented departures of trained personnel, the USAF can legitimately claim to be the only air force fully prepared, today, to fly and fight in any part of the world. For tomorrow, the emphasis will focus on newer equipment and better ways of achieving more with less.

The air force has retired from service several types since the Gulf War, most notably the F-4 Phantom and F-111, while older versions of other aircraft, such as the B-52G, are no longer in service. They have been replaced by the new 'smart' aircraft and weapons systems, which can perform operations that were only dreamed about two decades ago. However, their acquisition has been a trade-off

between the cost of development and purchase against the quantity required to fulfil the role. In most cases the number actually obtained has been considerably smaller than that originally envisioned – the B-2A Spirit being a prime example, with only 21 funded instead of the 132 originally anticipated.

A force the size of the USAF always has an ongoing 'wish list' for which contractors have been keen to develop new or upgraded equipment featuring the latest technology, while modifications to existing types can enhance their capabilities significantly. There is also an ongoing process of retirement for aircraft which have come to the end of their useful lives, with most being placed in open storage to be cannibalised for spare parts, transferred to friendly nations under mutual aid programmes, or sold for scrap. Some non-combat types find a second lease of life with government agencies or in civilian hands. Collectively, these factors ensure that the US Air Force is a constantly-changing entity.

The Air Force embarked on a programme to introduce into service in the next few years a range of new aircraft types which will serve the organisation well into the new millennium. The B-2A 'Stealth Bomber' will be the flagship of the strategic bombing role, while the F-22 air superiority fighter (which Lockheed Martin prefers to call an air dominance fighter) with its stealthy qualities will add a new dimension to the air combat arena. The planned Joint Strike Fighter, also enjoying the next generation of stealth technology, will come into service a few years after the F-22 begins to be delivered to operational units during the next decade. The service may receive as many as 1,200 of more than 3,000 Joint Strike Fighters that are seen as a 21st century replacement for the A-10, F-16 and other current types. In the near-term, the USAF will begin flying its first C-32A and C-37A VIP transports, as well as a small number of ANG C-38 Astras.

Not yet part of official plans, but certain to materialise, will be versions of the Boeing 767 to replace current AWACS, J-STARS, and possibly tanker aircraft.

'Star Wars' will become a reality during the next decade when the Boeing YAL-1A, which is a version of its 747-440F fitted with a giant laser system dedicated as a missile defence system to destroy theatre ballistic missiles, takes to the air. Following much test and development work, the first live demonstration of its capabilities could be carried out as early as 2002, prior to the Air Force acquiring seven AL-1As. Production could commence a year or two later, with the first examples in service by 2006.

For the foreseeable future, though, most USAF aircraft will be familiar types that have already given good service for some time. Finding ways to use them better will be a priority challenge.

The airlift role has proved to be a pivotal factor in the success of rapidly deploying large numbers of forces over vast distances. The motto 'Global Reach, Global Power' can only become a reality if each of the major commands is capable of performing its missions effectively. The combat elements of Air Combat Command are ineffective if the tanker and airlift units of Air Mobility Command are unable to provide the

necessary support in the numbers required to speed the deployment of personnel, equipment and munitions to their destinations. Similarly, the back-up and resupply of spare parts and replacement commodities must be sufficient to ensure a campaign is not hampered by a break in the logistics chain. Through the regular rotation of squadrons to participate in the air exclusion zones and in realistic exercises such as Red Flag, the Air Force has built up a cadre of combat-experienced personnel capable of deploying anywhere in the world and having the ability to commence operations immediately.

COMMAND CONSOLIDATION

The reorganisation which began in 1992 brought together various Commands whose roles had ceased to be as clearly defined as when they were originally formed. The distinction between strategic and tactical bombardment, for example, was gradually eroded, and disappeared completely during the Gulf War when F-15E and F-111F tactical fighter/bombers attacked strategic targets. The massive overhaul of the USAF structure also served to eliminate some of the duplication of missions by bringing together certain organisations fulfilling similar duties. The combination of Air Force Systems Command (AFSC) with Air Force Logistics Command (AFLC) into Air Force Materiel Command (AFMC) consolidated the functions of development, acquisition, overhaul, support and disposal under a single manager – in effect providing a watchdog to oversee all USAF aircraft and equipment from the cradle to the grave. Furthermore, the creation of Air Education & Training Command (AETC) to replace Air Training Command (ATC) highlighted the emphasis to not only train but also to educate forces to a high degree of professionalism, thereby ensuring they were on a par with the private sector.

EXPEDITIONARY AEROSPACE FORCES

In August 1998 the USAF announced that it was to establish 10 permanent air expeditionary forces by January 2000, called Expeditionary Aerospace Forces (EAF). To be named after heroes of the air force, each will contain about 175 aircraft of a variety of types from units of the active and reserve component of the air force. The EAFs will be highly mobile and able to engage in the full spectrum of conflict, and on deployment overseas during three months of a 15-month cycle. Most combat and transport units in the USAF will be allocated to an EAF, with the exception of some high-value assets such as the E-3s of the 552nd ACW. While helping to regularise the deployment cycle for the majority of air force personnel, it is these assets that are currently most in demand for overseas deployments. Overall, this organisational change will have little impact on the current order of battle.

Air Combat Command (ACC)

Air Combat Command, the largest and most visible of the operational major commands, was established on 1 June 1992, with headquarters at Langley AFB, Virginia. Its mission is to operate the fleet of active-duty strategic bombers, combat fighter, attack and rescue aircraft and helicopters, as well as the small fleet of battlefield management and command and control aircraft. The command was also the control-

ling authority for the theatre airlift role until 1 April 1997, when this function was transferred to Air Mobility Command. ACC provides combat-capable air power to support other branches of the Defense Department as well as to NATO and other allies. In addition, the command maintains one-third of the nuclear triangle with manned strategic bombers. Although ACC has the majority of its forces stationed within the Continental United States, its area of responsibility is global, with certain squadrons required to be ready to deploy rapidly to meet challenges of peacetime air sovereignty and wartime defence. The command has approximately 900 aircraft and helicopters and a personnel active-duty strength of more than 113,000 officers and enlisted. ACC has a primary aircraft inventory of 100-plus strategic bombers, almost 400 fighters, and approximately 200 ground attack aircraft. This hard-core

The B-2A represents ACC's technological edge; they are operated by the 509th Bombardment Wing.

Air Combat Command

USAF Air Warfare Center		Nellis AFB, Nev.
57th Wing		**Nellis AFB, Nev.**
57th OG		
Combat Rescue School		
	HH-60G 'WA'	Nellis AFB
11th RS	RQ-1A UAV 'WA'	Indian Springs AAF, Nev.
15th RS	RQ-1A UAV 'WA'	Indian Springs AAF, Nev.
64th AgS	F-16C/D	Nellis AFB
– Red Flag exercises (Adversary Tactics Division)		
66th RQS	HH-60G 'WA'	Nellis AFB
'Thunderbirds'	F-16C/D	Nellis AFB
A-10 Weapons School [A-10 Weapons School Thunderbolt]		
	A-10A 'WA'	Nellis AFB
F-15 Weapons School [F-15 Weapons School Eagle]		
	F-15C/D 'WA'	Nellis AFB
F-16 Weapons School [F-16 Weapons School Falcon]		
	F-16C/D 'WA'	Nellis AFB
F-15E Weapons School [F-15E Weapons School Night Eagle]		
	F-15E 'WA'	Nellis AFB
Air Force Air-Ground Operations School		**Nellis AFB, Nev.**
548th CTS	no aircraft assigned	Barksdale AFB, La.
(hosts Air Warrior II exercises at Fort Polk, La.)		
549th CTS	no aircraft assigned	Nellis AFB
(hosts Air Warrior exercises at Fort Irwin, Calif)		
53rd Wing		**Eglin AFB, Fla.**
53rd OG		
84th TES	aircraft on loan from 85th TES	Tyndall AFB, Fla.
85th TES	F-15A/B/C/D, F-15E, F-16B/C/D 'OT'	Eglin AFB, Fla.
422nd TES	A-10A, F-15C, F-15E, F-16C/D 'OT'	Nellis AFB, Nev.
Det 1	F-117A 'OT'	Holloman AFB, N.M.
475th WEG		**Tyndall AFB, Fla.**
82nd ATRS	E-9A 'WE', QF-4E/G, BQM-34A, MQM-107D	Tyndall AFB, Fla.
Det 1	QF-4E/G, C-12J	Holloman AFB, N.M.
475th TSS	QF-4E/G, QRF-4C	Tyndall AFB, Fla.

1st Air Force	HQ Tyndall AFB, Fla.
[part of ACC but manned by ANG from mid-1997]	
Air Defense Sectors	**no aircraft assigned**
Northeast ADS	Griffiss AFB, N.Y.
Western ADS	McChord AFB, Wash.
Southeast ADS	Tyndall AFB, Fla.

Air Warfare Center Aircraft

The USAF Air Warfare Center's A-10s are assigned to the A-10 Weapons School of the 57th Wing, based at Nellis AFB, Nevada.

Three units of the 57th Wing operate the F-16C – the 'Thunderbirds' display team, the 64th Aggressor Squadron and the F-16 Weapons School (above).

Based at Tyndall AFB, Fla., the 475th WEG uses two E-9As, equipped with side-looking radar and telemetry equipment, as range support aircraft.

This Phantom left AMARC on 12 February 1997 for TFSI at Mojave for conversion to a QF-4G drone. It last served with the 561st FS of the 57th Wing.

8th Air Force	HQ Barksdale AFB, La.
2nd Bombardment Wing	**Barksdale AFB, La.**
2nd OG	
11th BS	B-52H 'LA'	Barksdale AFB, La.
20th BS	B-52H 'LA'	Barksdale AFB, La.
96th BS	B-52H 'LA'	Barksdale AFB, La.
25th TS	no aircraft assigned	Barksdale AFB, La.
346th TES	no aircraft assigned	Barksdale AFB, La.
5th Bombardment Wing	**Minot AFB, N.D.**
5th OG	
23rd BS	B-52H 'MT'	Minot AFB, N.D.
7th Bombardment Wing	**Dyess AFB, Texas**
7th OG	
9th BS	B-1B 'DY'	Dyess AFB, Texas
13th BS	B-1B 'DY'	Dyess AFB, Texas
28th BS	B-1B 'DY'	Dyess AFB, Texas
27th Fighter Wing	**Cannon AFB, N.M.**
27th OG	
428th FS	F-16C/D	Cannon AFB, N.M.
(trains crew for Singapore's air force)	
522nd FS	F-16C/D 'CC'	Cannon AFB, N.M.
523rd FS	F-16C/D 'CC'	Cannon AFB, N.M.
524th FS	F-16C/D 'CC'	Cannon AFB, N.M.
28th Bombardment Wing	**Ellsworth AFB, S.D.**
28th OG	
37th BS	B-1B 'EL'	Ellsworth AFB, S.D.
77th BS	B-1B 'EL'	Ellsworth AFB, S.D.
85th Group	**NS Keflavik, Iceland**
56th RQS	HH-60G 'IS'	NS Keflavik
509th Bombardment Wing	**Whiteman AFB, Mo.**
509th OG	
325th BS	B-2A 'WM'	Whiteman AFB, Mo.
393rd BS	B-2A, T-38A 'WM'	Whiteman AFB, Mo.
394th CTS	borrows B-2As	Whiteman AFB, Mo.

9th Air Force	HQ Shaw AFB, S.C.
1st Fighter Wing	**Langley AFB, Va.**
1st OG	
27th FS	F-15C/D 'FF'	Langley AFB, Va.
71st FS	F-15C/D 'FF'	Langley AFB, Va.
94th FS	F-15C/D 'FF'	Langley AFB, Va.
4th Fighter Wing	**Seymour Johnson AFB, N.C.**
4th OG	
333rd FS	F-15E 'SJ'	Seymour Johnson AFB, N.C.
334th FS	F-15E 'SJ'	Seymour Johnson AFB, N.C.
335th FS	F-15E 'SJ'	Seymour Johnson AFB, N.C.
336th FS	F-15E 'SJ'	Seymour Johnson AFB, N.C.
20th Fighter Wing	**Shaw AFB, S.C.**
20th OG	
55th FS	F-16C/D 'SW'	Shaw AFB, S.C.
77th FS	F-16C/D 'SW'	Shaw AFB, S.C.
78th FS	F-16C/D 'SW'	Shaw AFB, S.C.
79th FS	F-16C/D 'SW'	Shaw AFB, S.C.

23rd Fighter Group	Pope AFB, N.C.
23rd OG	
74th FS	OA/A-10A 'FT'	Pope AFB, N.C.
75th FS	OA/A-10A 'FT'	Pope AFB, N.C.
33rd Fighter Wing	**Eglin AFB, Fla.**
33rd OG	
58th FS	F-15C/D 'EG'	Eglin AFB, Fla.
60th FS	F-15C/D 'EG'	Eglin AFB, Fla.
93rd Airborne Control Wing	**Robins AFB, Ga.**
93rd OG	
12th ACCS	E-8C 'WR'	Robins AFB, Ga.
16th ACCS	no aircraft 'yet'	Robins AFB, Ga.
93rd TS	aircraft borrowed as required	Robins AFB, Ga.
347th Wing	**Moody AFB, Ga.**
347th OG	
68th FS	F-16C/D 'MY'	Moody AFB, Ga.
69th FS	F-16C/D 'MY'	Moody AFB, Ga.
70th FS	OA/A-10A 'MY'	Moody AFB, Ga.
41st RQS	HH-60G 'MY'	Moody AFB, Ga.
71st RQS	HC-130N/P 'MY'	Moody AFB, Ga.

combat element is backed up by electronic warfare, aerial tankers, reconnaissance, UAV, combat rescue and mixed types numbering almost 200.

While most ACC units are assigned to one of four numbered air forces, the Air Warfare Center at Nellis AFB, Nevada is a direct reporting unit whose role is to evaluate and develop all manner of weapons and tactics for employment by operational forces in wartime. The 1st Air Force at Tyndall AFB, Florida is responsible for three Air Defense Sectors covering the Southeast, Northeast and Western United States, but with no aircraft directly assigned. Despite being part of ACC, the 1st AF and its subordinate sectors are all manned by Air National Guard personnel.

The 8th Air Force at Barksdale AFB, Louisiana has responsibility for many of the ACC units located

8th Air Force Bombers

Dyess AFB, Texas, is home to three squadrons of 8th Air Force B-1Bs. 85-0073 belongs to the 28th BS – 'Mohawk Warriors'. B-1s no longer carry nuclear stores.

Barksdale AFB, Louisiana, houses three squadrons with B-52Hs assigned, the last version in service with the air force. Ninety-six remain active.

9th Air Force Assets

Shaw AFB, South Carolina, is home to the 20th Fighter Wing, which flies four squadrons of F-16C/Ds. The wing is trained in the ground attack role.

The 9th Air Force's 93rd ACW controls the airborne element of the USAF/US Army J-STARS (Joint Surveillance Target Attack Radar System), the E-8C.

within the central United States. This includes all the strategic bomber assets, as they have been positioned at inland bases to provide a natural geographic defensive barrier. Two are equipped with the B-1B and two with the B-52H, while a fifth wing operates the B-2A. The 27th FW at Cannon AFB, New Mexico is also part of the 8th AF, which curiously has assigned the two ACC overseas facilities, the 65th ABW at Lajes AB in the Azores, and the 85th Group at NAS Keflavik, Iceland. The Pentagon is currently considering the possibility of axing the 8th AF to reduce administrative costs, with their assets being redistributed between the 9th and 12th Air Forces. The 9th Air Force at Shaw AFB, South Carolina, encompassing the eastern region, is predominantly equipped for the tactical fighter role, with two wings operating the F-15C/D in air defence, two flying the F-16C/D and one the F-15E for ground attack. The remainder is composed of an A-10 wing for close air support, and the unique Joint-STARS unit performing the ground surveillance and battlefield management role. One of the F-16 units has a composite structure with a squadron of A-10s as well as two combat rescue squadrons. The 12th Air Force at Davis-Monthan AFB, Arizona has units located within the western part of the central USA, with a mixed complement of aircraft types and duties. One F-16C/D unit, one F-117 wing and an A-10 unit are dedicated to the ground attack role, while the 366th Wing at Mountain Home AFB, Idaho is the sole air intervention wing designed for rapid reaction. The latter has strategic bomber, air superiority, ground attack and air refuelling aircraft operating together as a cohesive team. The remainder constitute the USAF's strategic reconnaissance capability, consisting of the U-2S and its trainers, the recently retired pair of SR-71As, and the highly valued RC-135 fleet. The latter are assigned to the 55th Wing at Offutt AFB, Nebraska which also operates the E-4B National Airborne Operations Centers, two OC-135B Open Skies photographic platforms, and a small number of other

12th Air Force		HQ Davis-Monthan AFB, Ariz.
9th Reconnaissance Wing		**Beale AFB, Ca.**
9th OG		
1st RS	T-38A, U-2S/S(T) 'BB'	Beale AFB, Ca.
5th RS	U-2S 'BB'	Osan AB, RoK
99th RS	U-2S 'BB'	Beale AFB, Ca.
Det 1	U-2S 'BB'	RAF Akrotiri, Cyprus
OL CH	U-2S 'BB'	Al Kharj AB, Saudi Arabia
OL FR	U-2S 'BB'	Istres AB, France
49th Fighter Wing		**Holloman AFB, N.M.**
49th OG		
7th FS	F-117A, T-38A 'HO'	Holloman AFB, N.M.
8th FS	F-117A 'HO'	Holloman AFB, N.M.
9th FS	F-117A 'HO'	Holloman AFB, N.M.
48th RQS	HH-60G 'HO'	Holloman AFB, N.M.
433rd FS	T-38A, AT-38B 'HO'	Holloman AFB, N.M.
435th FS	T-38A, AT-38B 'HO'	Holloman AFB, N.M.
55th Wing		**Offutt AFB, Neb.**
55th OG		
1st ACCS	E-4B	Offutt AFB, Neb.
7th ACCS	KC-135E 'OF'	Offutt AFB, Neb.
45th RS	TC-135B, OC-135B, RC-135S, TC-135S 'OF'	Offutt AFB, Neb.
38th RS	aircrew of RC-135U/V/W and TC-135W 'OF'	Offutt AFB, Neb.
343rd RS	sensor operators	Offutt AFB, Neb.
82nd RS	dets from USA	Kadena AB, Okinawa
95th RS	dets from USA	RAF Mildenhall, UK

355th Wing		Davis-Monthan AFB, Ariz.
355th OG		
41st ECS	EC-130H 'DM'	Davis-Monthan AFB
42nd ACCS	EC-130E/H 'DM'	Davis-Monthan AFB
43rd ECS	EC-130H 'DM'	Davis-Monthan AFB
354th FS	OA/A-10A 'DM'	Davis-Monthan AFB
357th FS	OA/A-10A 'DM'	Davis-Monthan AFB
358th FS	OA/A-10A 'DM'	Davis-Monthan AFB
366th Wing		**Mountain Home AFB, Idaho**
366th OG		
22nd ARS	KC-135R 'MO'	Mountain Home AFB
34th BS	B-1B 'MO'	Mountain Home AFB
389th FS	F-16C/D 'MO'	Mountain Home AFB
390th FS	F-15C/D 'MO'	Mountain Home AFB
391st FS	F-15E 'MO'	Mountain Home AFB
Det 1	nil	NAS Whidbey Island, WA
(admin for USN/USAF EA-6B sqns)		
388th Fighter Wing		**Hill AFB, Utah**
388th OG		
4th FS	F-16C/D 'HL'	Hill AFB, Utah
34th FS	F-16C/D 'HL'	Hill AFB, Utah
421st FS	F-16C/D 'HL'	Hill AFB, Utah
Det 1	nil	Utah Test & Training Range
552nd Airborne Control Wing		**Tinker AFB, Okla.**
552nd OG		
963rd AACS	E-3B/C 'OK'	Tinker AFB, Okla.
964th AACS	E-3B/C 'OK'	Tinker AFB, Okla.
965th AACS	E-3B/C 'OK'	Tinker AFB, Okla.
966th AACTS	E-3B/C, TC-18E 'OK'	Tinker AFB, Okla.

specialised versions of the C-135. All operational E-3 Sentries are operated by the 552nd ACW to perform the command and control of hostile skies, with aircraft allocated to two PACAF units at any one time. Finally, two specialised electronic EC-130E and H models of the Hercules are assigned to the 355th Wing, both based at Davis-Monthan but regularly deployed to operational theatres.

ACC performs a limited training role, notably that for the B-1, B-2, B-52, F-15E, F-117 and U-2, as it

would be uneconomic for AETC to have squadrons established expressly for this task. ACC has combined the strategic assets of SAC with the tactical capability of TAC into a global force capable of intervening to prevent war through deterrence, as its first step. Should this fail, and war become the only alternative, then ACC has contingencies to fight a swift and devastating campaign with as little collateral damage as possible, thereby minimising innocent civilian casualties.

Tinker AFB, Oklahoma is home base to the E-3 fleet of the 552nd ACW.

The 355th Wing at Davis-Monthan AFB, Ariz. has three A-10A squadrons.

12th Air Force Aircraft

RC-135V Block VIC 64-14842 is the first aircraft of the 55th Wing to carry the High Game satellite communications antennas along the top of the fuselage.

All the USAF's U-2s are controlled by the 9th Reconnaissance Wing at Beale AFB, Calif.

Air Mobility Command (AMC)

Air Mobility Command (AMC) was formed on 1 June 1992, to operate the majority of assets previously assigned to Military Airlift Command (MAC), with the addition of the vast majority of SAC's KC-10 and KC-135 tankers. Headquarters AMC is housed in the same complex at Scott AFB, Illinois which accommodated HQ MAC. The new command has taken some time to settle its structure, as the transfer of tankers was fragmented with several units being initially assigned to Air Combat Command and only gradually being taken on charge by AMC. Similarly, the Stateside-based theatre airlift role performed by the C-130 Hercules was a function

of AMC, then transferred to ACC before returning to AMC in the spring of 1997. The US-based active-duty C-21 Learjets, which were assigned to various commands for VIP and communications duties, were also centralised under AMC in April 1997.

From the outset, AMC set about rectifying some of the airlift shortcomings which became prevalent during the early stages of Operation Desert Shield in the late summer and autumn of 1990. The build-up of forces in the Middle East taxed MAC's operational capabilities, with crews flying long hours, often without the statutory rest periods, and some aircraft not receiving their routine servicing at the required intervals. To prevent a repetition of the command being over-stretched to satisfy the sudden requirement for a massive additional airlift capability, the Tanker Airlift Control Center was formed at Scott AFB, Illinois to more effectively co-ordinate the worldwide flow of traffic with the volume of cargo and the number of passengers. AMC also reduced significantly the number of tanker units by concentrating them at just

The C-17A was the winner of the C-X competition for a wide-body transport. After a long gestation, Air Mobility Command expects to gain 120 aircraft.

Air Mobility Command

15th Air Force		HQ Travis AFB, Calif.
22nd Air Refueling Wing		**McConnell AFB, Kans.**
22nd OG		
344th ARS	KC-135R/T	McConnell AFB, Kans.
349th ARS	KC-135R	McConnell AFB, Kans.
350th ARS	KC-135R	McConnell AFB, Kans.
384th ARS	KC-135R	McConnell AFB, Kans.
60th Air Mobility Wing		**Travis AFB, Calif.**
60th OG		
6th ARS	KC-10A	Travis AFB, Calif.
9th ARS	KC-10A	Travis AFB, Calif.
19th AS	C-141B	Travis AFB, Calif.
21st AS	C-5A/B/C	Travis AFB, Calif.
22nd AS	C-5A/B	Travis AFB, Calif.
62nd Airlift Wing		**McChord AFB, Wash.**
62nd OG		
4th AS	C-141B	McChord AFB, Wash.
7th AS	C-141B	McChord AFB, Wash.
8th AS	C-141B	McChord AFB, Wash.
92nd Air Refueling Wing		**Fairchild AFB, Wash.**
92nd OG		
92nd ARS	KC-135T	Fairchild AFB, Wash.
93rd ARS	KC-135R/T	Fairchild AFB, Wash.
96th ARS	KC-135T	Fairchild AFB, Wash.
97th ARS	KC-135R	Fairchild AFB, Wash.
375th Airlift Wing		**Scott AFB, III.**
375th OG		
11th AS	C-9A	Scott AFB, III.
457th AS	C-21A	Andrews AFB, Md.
12th AF	C-21A	Langley AFB, Va.
47th AF	C-21A	Wright-Patterson AFB, Ohio
54th AF	C-21A	Maxwell AFB, Ala.
458th AS	C-21A	Scott AFB, III.
11th AF	C-21A	Offutt AFB, Neb.
84th AF	C-21A	Peterson AFB, Colo.
332nd AF	C-21A	Randolph AFB, Texas

The USAF's fleet of C-21A Operational Support Aircraft has been concentrated under the control of the 375th AW at Scott AFB, Illinois. The wing's assets are dispersed throughout the United States.

21st Air Force		HQ McGuire AFB, N.J.
6th Air Refueling Wing		**MacDill AFB, Fla.**
6th OG		
91st ARS	KC-135R	MacDill AFB, Fla.
USCentCom	EC-135N, EC-135Y,	MacDill AFB, Fla.
	EC-137D, CT-43A	
19th Air Refueling Group		**Robins AFB, Ga.**
19th OG		
99th ARS	KC-135R	Robins AFB, Ga.
43rd Airlift Wing		**Pope AFB, N.C.**
43rd OG		
2nd AS	C-130E	Pope AFB, N.C.
41st AS	C-130E	Pope AFB, N.C.
89th Airlift Wing		**Andrews AFB, Md.**
89th OG		
1st AS	C-12C/D, C-32A,	Andrews AFB, Md.
	C-135E	
1st HS	UH-1N	Andrews AFB, Md.
99th AS	C-9C, C-20/C/H, C-37A	Andrews AFB, Md.
Executive Flight	VC-25A	Andrews AFB, Md.
AFFSA	C-21A	Andrews AFB, Md.
305th Air Mobility Wing		**McGuire AFB, N.J.**
305th OG		
6th AS	C-141B	McGuire AFB, N.J.
13th AS	C-141B	McGuire AFB, N.J.
32nd ARS	KC-10A	McGuire AFB, N.J.
72nd ARS	KC-10A	McGuire AFB, N.J.
317th Airlift Group		**Dyess AFB, Texas**
317th OG		
39th AS	C-130H	Dyess AFB, Texas
40th AS	C-130H	Dyess AFB, Texas
319th Air Refueling Wing		**Grand Forks AFB, N.D.**
319th OG		
905th ARS	KC-135R	Grand Forks AFB
906th ARS	KC-135R/T	Grand Forks AFB
911th ARS	KC-135R	Grand Forks AFB
912th ARS	KC-135R/T	Grand Forks AFB
436th Airlift Wing		**Dover AFB, Del.**
436th OG		
3rd AS	C-5A/B	Dover AFB, Del.
9th AS	C-5A/B	Dover AFB, Del.
437th Airlift Wing		**Charleston AFB, S.C.**
437th OG		
14th AS	C-17A	Charleston AFB, S.C.
15th AS	C-17A	Charleston AFB, S.C.
16th AS	C-17A, C-141B	Charleston AFB, S.C.
17th AS	C-17A	Charleston AFB, S.C.
fourth squadron converting to the C-17A commencing 1998		
463rd Airlift Group		**Little Rock AFB, Ark.**
463rd OG		
50th AS	C-130H	Little Rock AFB, Ark.
61st AS	C-130E	Little Rock AFB, Ark.

Above: The KC-10A was ordered in the late 1970s to give fighter wings greater mobility, by transporting equipment and refuelling fighters being deployed.

Below: AMC's C-17s are currently concentrated in the 437th AW at Charleston AFB, S.C. The 62nd AW at McChord AFB, S.C. will convert to C-17s from 1999.

The USAF received a total of 81 C-5A Galaxies in the 1970s. During the mid-1980s the type was reinstated into production, and a further 50 C-5Bs were produced. About 125 remain in service.

seven bases. Two air mobility hubs were created at Travis AFB, California and McGuire AFB, New Jersey with a mixed complement of airlifters and tankers, designed to be the gateway for personnel travelling between the United States and the Pacific or European theatres, respectively. The Air Mobility Warfare Center has been established at Fort Dix, New Jersey to co-ordinate airlift requirements for the Army in the event of potential hostilities.

The transfer of the tanker fleet to AMC has enabled KC-10s and KC-135s performing intercontinental air refuelling support for deploying fighter and bomber aircraft to also airlift spares, equipment and personnel. Surprisingly, this was not always possible during the tenure of MAC and SAC due to the overriding needs of each command taking precedent. The KC-10 was designed as a dual tanker/airlifter and had provision for roller pallets to be affixed to the cargo floor. The KC-135 did not possess this capability until shortly after the Gulf War ended, when some KC-135s were modified with the installation of the 463-L cargo roller system (as fitted to the C-5 Galaxy and C-141 StarLifter) to enable pallets of cargo to be loaded and moved into position with ease.

The entrance of the C-17 into operational service has also alleviated the possibility of a recurrence of the Desert Shield predicament. The 437th AW at Charleston AFB has converted three squadrons, with its fourth commencing transition in late 1998. The 62nd AW at McChord AFB will begin conversion in 1999. Both units have AFRC associate programmes in place that enable reservists to man active-duty C-17s in place of front-line crews. The only other new equipment currently entering service is the Boeing C-32A, based on the Model 757-200, which is joining the 89th AW at Andrews AFB to replace the C-137. The C-32s will be used to transport the vice president and other senior government officials.

The AMC structure is similar to most other major commands. Two numbered air forces comprise the

Above: Wearing a civilian-style 'biz-jet' scheme devoid of national insignia, this C-20C flies in support of visiting American VIPs throughout the world.

15th AF at Travis AFB, California to administer forces on the western side of the USA and the 21st AF at McGuire AFB, New Jersey to co-ordinate activities in the east. During the early days of MAC the geographical division was implemented, with units on the western side tending to service their own area and bases across the Pacific Ocean, while the eastern side, the Atlantic region and Europe were supplied by units of the 21st AF. That division is no longer in place, as AMC units have a global responsibility. AMC personnel number more than 48,000 active-duty with a further component of 54,000 reserve who augment the active-duty and would join the command if mobilised. More than 1,400 fixed-wing airlifters and tankers are assigned directly to AMC, plus a dozen C-9s for aeromedical evacuation. The mixed VIP fleet at Andrews consists of 39 aircraft and helicopters.

Various upgrade programmes are underway or are being planned to include the fitting of a 'glass' cockpit to the entire KC-135 fleet, and wingtip refuelling pods to 50 of these tankers, and a possible service life extension upgrade to the C-5. AMC has no plans at present to replace the elderly C-130Es with the C-130J, although this is likely to change once production of the new advanced Hercules is established.

Above: The primary tanker aircraft of the USAF are the KC-135 Stratotankers, such as this KC-135R. In AMC the KC-135 fleet is used for its secondary transport role more frequently than when the fleet reported to SAC.

Responsible for executive transport, the 89th AW at Andrews AFB has a large variety of aircraft types, including the only three C-9Cs in the inventory (above) and a pair of VC-25As which is used for presidential transport on overseas visits. The wing has recently retired its long-serving C-137s and replaced them with Boeing C-32As (757-200 – below, seen at Renton on a pre-delivery flight) under the VC-X Large programme, and acquired Gulfstream C-37As (Gulfstream V) under the VC-X Small programme.

The C-141 StarLifter has been a stalwart of the heavy transport fleet since the early 1960s. The first transport aircraft designed under the weapon system approach, the C-141 is being withdrawn from active service and replaced by the C-17.

Air Education & Training Command (AETC)

Operator of the most aircraft and helicopters in the Air Force, Air Education & Training Command (AETC) was formed on 1 July 1993, and acquired the assets of the former Air Training Command, together with the operational conversion units for various major types including the F-15, F-16, C-5, C-130, KC-135 and C-141. From its headquarters at Randolph AFB, Texas, AETC has the primary mission of recruiting and training all enlisted personnel and officers to a standard acceptable to perform the host of trades required for the Air Force to run effectively. The command is also required to continue with ongoing education programmes to ensure its personnel remain effective, which involves providing basic training for all new entrants into the service. Following this stage, personnel receive initial and advanced flying, technical and professional military training to prepare them for operational duties. Training centres are not restricted to those joining the Air Force, as AETC provides courses to students from other branches of the Department of Defense, as well as for those from overseas air arms. In recent years the emphasis has switched to joint operations in an effort to reduce duplication of duties and thereby save money. Among these efforts have been the assignment of USAF aircraft to Navy bases for training duties, including the detachment of the T-1A Jayhawk to the US Navy facility at Pensacola, Florida. Furthermore, the Air Force has established four and the Navy five fixed-wing training squadrons to teach basic flying or navigational skills to students from all branches of the military. AETC is also responsible for the Air University, with headquarters at Maxwell AFB, Alabama.

AETC has a flying inventory of 1,164 trainer aircraft composed of the T-1A, T-3A, T-37B, T-38A, T-43A and AT-38B. A further 277 F-15 and F-16 fighters train combat pilots at Luke AFB, Arizona and Tyndall AFB, Florida, respectively. Transports, tankers and fixed-wing special operations types account for a further 110 aircraft, while 27 helicopters are mainly employed to train aircrews destined for

The T-1A Jayhawk symbolises the re-emergence of separate training paths for pilots destined to fly either in the transport and tanker fleets or in bomber/fighter types.

special operations or combat rescue duties. Personnel strength numbers almost 45,000 active-duty who oversee more than 375,000 students annually. The command does not perform all flying training, as it would be uneconomic for squadrons to be formed specifically for those types which are small in number, so most of these form part of the major command to which the operational role is assigned. For example, ACC conducts the operational conversion training for F-15E aircrew at Seymour Johnson AFB, North Carolina with the 4th FW, and likewise U-2S pilots receive transition tuition from the 9th RW at Beale AFB, California. The reserves perform training of their own personnel plus a limited number of active-duty and overseas crews. A-10 pilots are trained at Barksdale AFB, Louisiana by the 917th Wing of the Air Force Reserve Command, while hundreds of F-16 pilots including many from overseas air arms are trained by Air National Guard squadrons in Arizona and Oregon.

The Air Force has introduced the T-3A Firefly to replace the T-41 for screening prospective pilots before they embark on flying training. The first T-3s joined AETC in March 1994 and 113 aircraft are currently divided between the 3rd FTS at Hondo

The T-37 has been the primary trainer of the USAF for the best part of four decades. Scheduled to have been replaced by the T-46A, cancelled in 1987, the 'Tweet' has had to soldier on. Its replacement, the T-6A Texan II, will enter service from mid-1999.

Air Education & Training Command

Air University		HQ Maxwell AFB, Ala.
42nd ABW		
nil	no aircraft assigned	Maxwell AFB, Ala.

2nd Air Force		HQ Keesler AFB, Miss.
17th TRW	no aircraft assigned	Goodfellow AFB, Texas
37th TRW	no aircraft assigned	Lackland AFB, Texas
81st TRW		
45th AS	C-21A 'KS'	Keesler AFB, Miss.
82nd TRW	**Sheppard AFB, Texas**	
Sheppard TW GYA-10A, GA-10A, GB-52G, GC-130B/D/E, GNKC-135A, GKC-135A, GEC-135H, GC-141B, GF-4D/E, GF-15A/B, GF-16A/B/C, GF-111A/E, GUH-1N, GCH-53A 'ST'		
381st TRG	**Vandenberg AFB, Calif.**	
	no aircraft assigned	

Airport, Texas and the 557th FTS at the USAF Academy in Colorado. Another recent acquisition is the T-1A Jayhawk specialised undergraduate flying training aircraft, on which personnel who graduate from basic flying training and who are destined to progress to transport or tanker aircraft receive tuition to become proficient with the flight characteristics of multi-engined types. The two jet primary trainer aircraft with AETC are of 1950s' vintage, and despite various upgrade programmes are gradually becoming costly to operate. The Air Force has completed its selection for the Joint Primary Aircraft Training System (JPATS), with Raytheon Aircraft being awarded a contract to build the Beech/Pilatus PC-9 Mk II under licence as the T-6A Texan II. The Air Force is to receive 372 and the Navy 339 T-6s commencing in May 1999. The T-38 supersonic trainer is the subject of an ongoing upgrade programme called Pacer Classic which will integrate full avionics and a major structural refit to extend the service life to 2020.

The transfer to AETC of the additional training requirements from SAC, TAC and MAC has resulted in the new command having a structure divided between two numbered Air Forces. The Air University, which previously reported directly to headquarters USAF, is now subordinate to headquarters AETC. The 2nd Air Force is primarily concerned with technical training and has four training wings and one training group. Whereas most technical training wings are responsible for duties not directly related to aircraft operations, the 82nd Training Wing at Sheppard AFB, Texas undertakes education for numerous technical trades involving 'hands on' tuition on 50 or so aircraft which have been retired from flying duties. They include most of the types likely to be encountered operationally. The 19th Air Force is the primary operator of the flying training wings, with five devoted to the training of aircrew from the time they commence their basic flight education until graduation. The remaining units are devoted to operational conversion training for specific duties such as air superiority, ground attack, airlift, tanker and special operations. One plan recently announced is the intention to transfer the Introduction of Fighter Fundamentals task from Columbus AFB, Mississippi and Randolph AFB, Texas to Moody AFB, Georgia where a new unit will be formed in 1999. A total of 48 T-38Cs will be involved with deliveries commencing in FY 2000. Furthermore, Air Force Reserve Command is to form associate squadrons at AETC bases to enable reservists to conduct flying training in place of active-duty personnel performing this task.

Above: Winner of the Enhanced Flight Screener competition in the early 1990s, the T-3A has been plagued by problems since entering service, resulting in it being grounded for long periods.

Below: The Avionics Upgrade Program is under way to give the Talon fleet a 'glass' cockpit, as the T-38C. This AT-38B of the 435th FTS displays the new scheme to be adopted by the fleet.

Right: The T-43A is used by the 562nd FTS, 12th FTW based at Randolph AFB, Texas as a navigation trainer for both the USAF and the US Navy. Eight aircraft are used in this role.

19th Air Force	HQ	Randolph AFB, Texas
12th Flying Training Wing		**Randolph AFB, Texas**
12th OG		
3rd FTS	T-3A 'RA'	Hondo Apt, Texas
99th FTS	T-1A 'RA'	Randolph AFB, Texas
435th FTS	AT-38B 'RA'	Randolph AFB, Texas
557th FTS	T-3A 'RA'	USAF Academy, Colo
559th FTS	T-37B 'RA'	Randolph AFB, Texas
560th FTS	T-38A 'RA'	Randolph AFB, Texas
562nd FTS	T-43A 'RA'	Randolph AFB, Texas
(joint pilot training)		
14th Flying Training Wing		**Columbus AFB, Miss.**
14th OG		
37th FTS	T-37B 'CB'	Columbus AFB, Miss.
48th FTS	T-1A 'CB'	Columbus AFB, Miss.
49th FTS	AT-38B 'CB'	Columbus AFB, Miss.
50th FTS	T-38A 'CB'	Columbus AFB, Miss.
47th Flying Training Wing		**Laughlin AFB, Texas**
47th OG		
85th FTS	T-37B 'XL'	Laughlin AFB, Texas
86th FTS	T-1A 'XL'	Laughlin AFB, Texas
87th FTS	T-38A 'XL'	Laughlin AFB, Texas
56th Fighter Wing		**Luke AFB, Ariz.**
56th OG		
21st FS	F-16A/B 'LF'	Luke AFB, Ariz.
(Rep of China AF training)		
61st FS	F-16C/D 'LF'	Luke AFB, Ariz.
62nd FS	F-16C/D 'LF'	Luke AFB, Ariz.
63rd FS	F-16C/D 'LF'	Luke AFB, Ariz.
308th FS	F-16C/D 'LF'	Luke AFB, Ariz.
309th FS	F-16C/D 'LF'	Luke AFB, Ariz.
310th FS	F-16C/D 'LF'	Luke AFB, Ariz.
425th FS	F-16C/D 'LF'	Luke AFB, Ariz.
(Singapore AF training)		
58th Special Operations Wing		**Kirtland AFB, N.M.**
58th OG		
512th SOS	UH-1N, HH-60G	Kirtland AFB, N.M.
550th SOS	MC-130H/P	Kirtland AFB, N.M.
551st SOS	TH-53A, NCH-53A, MH-53J	Kirtland AFB, N.M.
23rd FTF	UH-1N	Fort Rucker AAF, Ala.
71st Flying Training Wing		**Vance AFB, Okla.**
71st OG		
8th FTS	T-37B 'VN'	Vance AFB, Okla.
(joint pilot training)		
25th FTS	T-38A 'VN'	Vance AFB, Okla.
32nd FTS	T-1A 'VN'	Vance AFB, Okla.
(joint pilot training)		
80th Flying Training Wing		**Sheppard AFB, Texas**
80th OG		
88th FTS	AT-38B 'EN'	Sheppard AFB
89th FTS	T-37B 'EN'	Sheppard AFB
90th FTS	T-38A 'EN'	Sheppard AFB
97th Flying Training Wing		**Altus AFB, Okla.**
97th OG		
55th ARS	KC-135R	Altus AFB, Okla.
56th AS	C-5A	Altus AFB, Okla.
57th AS	C-141B	Altus AFB, Okla.
58th AS	C-17A	Altus AFB, Okla.
314th Airlift Wing		**Little Rock AFB, Ark.**
314th OG		
53rd AS	C-130E	Little Rock AFB, Ark.
62nd AS	C-130E	Little Rock AFB, Ark.
314th LSS	GC-130E	Little Rock AFB, Ark.
325th Fighter Wing		**Tyndall AFB, Fla.**
325th OG		
1st FS	F-15C/D 'TY'	Tyndall AFB, Fla.
2nd FS	F-15C/D 'TY'	Tyndall AFB, Fla.
95th FS	F-15C/D 'TY'	Tyndall AFB, Fla.
336th Training Group		**Fairchild AFB, Wash.**
36th RQF	UH-1N 'FC'	Fairchild AFB, Wash.

Air Force Materiel Command (AFMC)

Air Force Materiel Command (AFMC) was formed on 1 July 1992, with headquarters at Wright-Patterson AFB, Ohio, to replace both Air Force Systems Command and Air Force Logistics Command. The creation of the new organisation, as stated earlier, effectively brought together the functions of development, acquisition, overhaul, support and disposal, enabling a single command to manage the entire life cycle of all USAF aircraft and equipment. AFMC manages the integration of research, development, test, acquisition and the sustained support of all weapons systems. The command is also required to produce and obtain advanced systems, operate 'super laboratories', major product centres, logistics centres and test facilities, and to operate the USAF Test Pilots School and the School of Aerospace Medicine. An active-duty personnel strength of 34,000 is bolstered by a civilian workforce of more than 73,000.

AFMC has a nominal strength of between 200 and 300 fixed- and rotary-winged types, with the majority performing test and evaluation duties. The 412th Test Wing at Edwards AFB, California is the flying component of the Air Force Flight Test Center whose primary role is to evaluate new and upgraded weapons systems to determine their suitability for possible inclusion into USAF operations. The majority of work involves new or modified aircraft types, with recent programmes focusing on the B-2A, C-17A, MC-130H and AC-130U. At present, the AFFTC is performing the preliminary test work on the F-22A air superiority fighter and will be making plans early in the next decade for the first Joint Strike Fighter test programme. The AFFTC works closely with manufacturers to enable deficiencies highlighted during development to be rectified, where possible, prior to production commencing. The centre also conducts ongoing evaluation of their test airframes after the aircraft concerned has entered service. A fleet of support aircraft – including F-15s, F-16s and T-38s

The oldest Hercules in the USAF is this NC-130A used by the Air Armament Center at Eglin AFB, Florida, for airborne sensor and seeker tests.

Air Force Materiel Command

Development		
Aeronautical Systems Center (ASC)		
		Wright-Patterson AFB, Ohio
88th ABW no aircraft assigned	Wright-Patterson AFB, Ohio	
Electronic Systems Center (ESC)		**Hanscom AFB, Mass.**
66th ABW no aircraft assigned	Hanscom AFB, Mass.	
Space & Missile Systems Center (SMC)		
		Los Angeles AFB, Calif.
61st ABG no aircraft assigned	Los Angeles AFB, Calif.	

AFMC's Air Force Flight Test Center uses a variety of modified aircraft for test duties. EC-135E 61-0374 is a Cruise Missile Mission Support Aircraft, used to track and test cruise missiles over ranges during flight tests.

Air Armament Center (AAC) Eglin AFB, Fla.		
46th Test Wing		**Eglin AFB, Fla.**
46th OG		
39th TS	F-16A/B/C/D 'ET'	Eglin AFB, Fla.
40th TS	F-15A/B/CD, F-15E,	Eglin AFB, Fla.
	UH-1N 'ET', NC-130A	
Det 1	MV-22B on loan	Hurlburt Field, N.M.
46th Test Group		
586th TSS	AT-38B 'HT'	Holloman AFB, N.M.
377th Air Base Wing		
	no aircraft assigned	Kirtland AFB, N.M.

Air Force Flight Test Center (AFFTC) Edwards AFB, Calif.		
412th Test Wing		**Edwards AFB, Calif.**
412th OG		
410th TS	YF-117A, F-117A 'ED'	Palmdale Apt, Calif.
411th TS	YF-22A, F-22A	Edwards AFB, Calif.
412th TS	T-38A, AT-38B 'ED', C-135C	Edwards AFB, Calif.
413th TS	unknown	Edwards AFB, Calif.
415th TS	F-15A/B/C/D, F-15E 'ED'	Edwards AFB, Calif.
416th TS	F-16A/B/C/D, NF-16A/D 'ED'	Edwards AFB, Calif.
417th TS	C-17A 'ED'	Edwards AFB, Calif.
418th TS	no aircraft assigned at present	Edwards AFB, Calif.
419th TS	B-1B, B-52H 'ED', B-2A	Edwards AFB, Calif.
445th TS	UH-1N, U-6A, UV-18A 'ED'	Edwards AFB, Calif.
452nd TS	C-18A, EC-18B/D, C-135E, EC-135E, NC-141A	Edwards AFB, Calif.
453rd TS	T-39B, NT-39A	Edwards AFB, Calif.
Air Force Test Pilots School		Edwards AFB, Calif.
	operates aircraft as required from test squadrons	

Several early production F-15B Eagles are used by the 415th TS/Air Force Flight Test Center (as chase aircraft) and the Air Force Test Pilots School.

AFMC has five Air Logistics Centers which conduct the major overhaul and repair for the majority of aircraft, helicopters, missiles and weapons systems in the USAF inventory, and perform work on the types which have been acquired in substantial numbers. The command also arranges for manufacturers to undertake work on other equipment on a contract basis. Some of the centres have one or two aircraft for limited development work and for proficiency flying by test pilots. Development work is carried out by several major centres, none of which has aircraft directly assigned, although airframes are borrowed from manufacturers and other test organisations as necessary. Sacramento ALC is due to close in July 2001, while the San Antonio ALC at Kelly AFB was taken over by Boeing, becoming the Boeing Aerospace Support Center. It is believed that the 76th ABW and its 313th TS have disbanded. Boeing undertakes KC-10A overhauls at the facility, and won a nine-year KC-135 overhaul contract in late 1998.

A number of other specialised support organisations make up the remainder of the AFMC establishment, including the most famous of all, the Aerospace Maintenance and Regeneration Center at Davis-Monthan AFB, Arizona. AMARC (known more colloquially as the 'boneyard') has more than 5,000 aircraft, helicopters and other items in store, many of which will be reclaimed for usable components, scrapped, sold to civilian owners, transferred to other government agencies or exported. AMARC is one of the few sections of the USAF which actually makes a monetary profit.

Specialised Support

Aerospace Maintenance and Regeneration Center (AMARC)	
no aircraft operated	Davis-Monthan AFB, Ariz.

Direct Reporting Units

645th MTS	EC 130H, NC-130E 'D4'	Palmdale Apt, Calif.
Det 2 645th MTS	WC-135W 'MF'	Majors Field, Greenville, Texas
US Military Training Mission		
	C-12C	Dhahran AB, Saudi Arabia
Embassy Flights:		
	C-12C	Abidjan, Ivory Coast
	C-12C	Ankara, Turkey
	C-12C	Athens, Greece
	C-12C	Bangkok, Thailand
	C-12C	Bogotá, Colombia
	C-12C	Brasilia, Brazil
	C-12D	Budapest, Hungary
	C-12C	Buenos Aires, Argentina
	C-12C	Cairo, Egypt
	C-12C	Canberra, Australia
	C-12C	Djakarta, Indonesia
	C-12C/D	Islamabad, Pakistan
	C-12C	Kinshasa, Zaïre
	C-12D	La Paz, Bolivia
	C-12C	Manila, Philippines
	C-12D	Mexico City, Mexico
	C-12C	Riyadh, Saudi Arabia
	C-12C	Tegucigalpa, Honduras

The aircraft of the 46th Test Wing carry the 'ET' tailcode, standing for 'Eglin Test'. The unit's role is the evaluation of weapon systems over the vast Eglin AFB test ranges.

– is operated as chase planes to film the flight characteristics of craft under evaluation for later analysis. AFFTC also houses a small number of EC-18s and EC-135s which operate as Advanced Range Instrumentation Aircraft (ARIA) to relay telemetry and voice communications for a variety of roles including missiles under test by the US military and NATO, and the space shuttle for NASA.

The second major flying component within AFMC is the Air Armament Center (ex-Air Force Development Test Center) at Eglin AFB, Florida which has the 46th Test Wing as its flight unit. The AAC has a complement of approximately two dozen F-15s and F-16s as it is primarily concerned with weapons evaluation for these two types. All new weapons destined for Air Combat Command are rigorously tested over the vast Eglin ranges to ensure they are capable of performing as advertised. The unit also has the role of evaluating new software for weapons delivery. The centre has an NC-130A on strength, the oldest Hercules in USAF service. Following clearance of a particular weapons system by the AAC, Air Combat Command's 53rd Wing, located on the same flight line at Eglin AFB, commences the operational test and evaluation phase prior to issue to front-line units.

The 412th Test Squadron of the Air Force Flight Test Centre at Edwards AFB, Calif. uses the T-38A Talon as a chase aircraft.

Air Force Special Operations Command (AFSOC)

Despite having been formed as recently as 22 May 1990, Air Force Special Operations Command is one of the oldest in the USAF as it was created as a separate entity from Military Airlift Command prior to the 1992 reorganisation. AFSOC serves as the Air Force component of US Special Operations Command which oversees all special forces sections within the Department of Defense. As its name implies, AFSOC is charged with performing all manner of special operations, within the broad categories of the deployment and assignment to regional unified commands to conduct unconventional warfare, direct action, counter-terrorism, special reconnaissance, foreign internal defence, counter-proliferation, civil affairs, humanitarian assistance, psychological operations, personnel recovery, and counter-narcotics operations. AFSOC has an active-duty personnel strength of almost 10,000 officers and enlisted, with a further 2,400 available from the Air Reserve Component, which includes the Commando Solo EC-130Es of the 193rd SOS Pennsylvania ANG as well as the AFRC 919th SOW at Duke Field, Florida with its mixed complement of MC-130Es and MC-130Ps.

AFSOC's operations within the United States are concentrated at Hurlburt Field (its only facility), although the command works extensively with other USAF units as well as US Army special forces and

Navy Seal teams. Overseas, the command has a commitment within Europe and the Pacific regions, with AFSOC personnel permanently stationed within these two theatres. European operations, in particular, have meant ongoing involvement in United Nations peacekeeping duties in Bosnia, as well as supporting forces engaged in the implementation of the air exclusion zones over parts of Iraq and the Balkans. Europe-based AFSOC forces have also been involved in dozens of humanitarian operations, including the escort of American and allied civilians from nations in Africa during periods of unrest.

AFSOC operates approximately 135 fixed- and rotary-winged types including a mixed complement of Hercules models which have been tailored to meet the needs of the command. The elderly AC-130H Spectre gunships have been superseded by the newer AC-130U Spectre versions. However, instead of the former joining the reserves, they have been maintained in front-line service alongside the 'U-boat'. Likewise, the MC-130E Combat Talon I continues in active-duty service although some have joined AFRC. The newer MC-130H Combat Talon II is now fully operational, enabling some of the MC-130Es to be updated to enhance their capabilities. Despite reports that the Fulton recovery system installed on some of the MC-130Es was no longer used, two aircraft from the 16th SOW were deployed to the Middle East in February 1998 during Operation Desert Thunder with the yokes fitted. The fleet of HC-130N and P models employed for combat rescue duties were redesignated MC-130P Combat Shadow to reflect their special operations capability more accurately. AFSOC also employs a sizeable rotary-wing element, consisting of three dozen MH-53J Pave Low III helicopters alongside seven MH-60G Pave Hawks. In addition, a pair of UH-1Ns is used to train special forces operations for

The USAF's Sikorsky MH-60G Pave Hawk was developed as a dedicated combat rescue and special operations forces support helicopter.

overseas customers. The Air Support Operations Squadron at Fort Bragg, North Carolina has the 21st Special Tactics Squadron stationed at nearby Pope AFB flying the CASA C.212.

AFSOC has completed its first round of re-equipment with the introduction into service of the MC-130H Combat Talon II and the AC-130U Spectre, both of which are fully operational. The helicopter force of the MH-53J and MH-60G is the backbone of special operations in-theatre; they will be joined in the next decade by the tilt-rotor CV-22 Osprey which will offer fixed-wing flight operations combined with the rotary-wing ability to hover and land/take-off vertically. First deliveries will be to Hurlburt Field, but additional aircraft will eventually be stationed in Europe and the Pacific regions. Air Mobility Command has approximately nine C-141B StarLifters which have been modified for the Special Operations low-level mission to delivery special forces and their equipment. The aircraft have been modified with terrain-following radar housed in two small oval bulges on either side of the lower nose section, and a retractable FLIR turret beneath the nosecone. The aircraft also have radar/missile warning receivers positioned on the nose and rear fuselage. The cockpit and cargo hold lighting has been adapted to enable the crew to wear night-vision goggles. At least two AMC C-5A Galaxies have a special forces function.

Above: The Air Force Special Operations Command fixed-wing fleet mostly consists of specialised versions of the C-130 Hercules, including gunships. Heavily armed, they are also fitted with defensive aids such as anti-infra-red missile decoy flares, as demonstrated by this AC-130H.

Left: The AFSOC fleet of Combat Shadow Hercules was redesignated as MC-130Ps during February 1996. 69-5808 was produced as an HC-130N.

The MH-53J 'Pave Low III' equips three squadrons based in the continental United States, the United Kingdom and the Republic of Korea. This example is based at Hurlburt Field, Florida.

Air Force Special Operations Command

16th Special Operations Wing		Hurlburt Field, Fla.
16th SOG		
4th SOS	AC-130U	Hurlburt Field, Fla.
6th SOS	CASA C.212, UH-1N	Hurlburt Field, Fla.
8th SOS	MC-130E, C-130E	Hurlburt Field, Fla.
9th SOS	MC-130P	Eglin AFB, Fla.
15th SOS	MC-130H	Hurlburt Field, Fla.
16th SOS	AC-130H	Hurlburt Field, Fla.
19th SOS	AC-130H/U	Hurlburt Field, Fla.
(borrowed for training from 4/16th SOSs)		
20th SOS	MH-53J, TH-53A, NCH-53A	Hurlburt Field, Fla.
55th SOS	MH-60G	Hurlburt Field, Fla.
Air Support Operations Squadron		Fort Bragg, N.C.
21st STS	CASA C.212	Pope AFB, N.C.
352nd SOG		
7th SOS	MC-130H	RAF Mildenhall, UK
21st SOS	MH-53J	RAF Mildenhall, UK
67th SOS	MC-130P, C-130E	RAF Mildenhall, UK
353rd SOG		
1st SOS	MC-130H, C-130E	Kadena AB, Okinawa
17th SOS	MC-130P	Kadena AB, Okinawa
31st SOS	MH-53J	Osan AB, RoK

Air Force Space Command (AFSPC)

Air Force Space Command (AFSPC) was created on 1 September 1992, to operate the fleet of Intercontinental Ballistic Missiles (ICBMs) formerly assigned to Strategic Air Command, plus the missile warning radars, optical sensors, satellites, space launch facilities and space surveillance radars. In addition, AFSPC provides command and control for Department of Defense satellites and operates the ballistic missile warning systems for the North American Air Defense network and for US Space Command. The command has its headquarters at Peterson AFB, Colorado close to the Cheyenne Mountain Air Station complex buried deep within the Rocky Mountains.

The command has 50 Boeing LGM-118A Peacekeeper and 380 Boeing LGM-30G Minuteman III ICBMs in silos located across the midwestern and northern United States. The Peacekeepers are assigned to the 90th Space Wing at Francis E. Warren AFB, Wyoming which also has a number of Minuteman missiles. The remaining Minutemen are assigned to the 91st SG (Space Group) at Minot AFB, North Dakota and the 341st SW (Malmstrom AFB, Montana). These units are part of the 20th Air Force with headquarters at F. E. Warren AFB. In recent years the number of ICBMs has been reduced significantly through implementation of the Strategic Arms Limitation Treaty (SALT). The treaty has seen the complete withdrawal from service of the LGM-30F Minuteman II missile. A contract awarded to TRW will involve the conversion of missiles from multiple warhead to single warhead capability.

The UH-1N Iroquois serves with Space Command in the missile site support role, the majority of the command's squadrons operating strategic missiles. The flights were redesignated as Helicopter Flights on 1 May 1998, instead of Rescue Flights.

AFSPC has no fixed-wing assets and fewer than three dozen HH-1H and UH-1N versions of the Iroquois helicopter which are employed for missile site support at the four missile bases and at Vandenberg AFB, California, where the 30th Space Wing houses launch and range operations for various DoD, NASA and commercial space programmes. The base also houses testing and support facilities for ICBMs, plus technical training for all AFSPC activities through the 381st Space and Missile Training Group, which is part of AETC. Other equipment in the AFSPC inventory includes all manner of military satellites such as Navstar, Milstar and the Defense Satellite Communications System (DSCS II and III). Various ballistic missile warning radars and space surveillance systems are located at more than 50 strategically convenient facilities around the world to prevent unfriendly nations from launching a surprise attack on the United States and its allies. A personnel strength of almost 22,000 is assigned to AFSPC along with a civilian workforce of 4,700 and 11,400 contractors.

Above: The USAF's CASA 212s are employed by the Air Support Operations Squadron's 21st STS at Pope AFB, N.C. and the 6th SOS at Hurlburt Field, Fla. Unusual for an in-service US military aircraft, it has not received a Mission Design Series designation.

Below: This MC-130H 'Combat Talon II' is one of 24 new-build aircraft, of which few carry the Fulton STAR recovery equipment of the previous MC-130E-Cs. They are equipped with a low-level aerial delivery and container release system.

Air Force Space Command

14th Air Force		HQ Vandenberg AFB, Calif.
21st Space Wing, Peterson AFB, Colo.		
21st OG		
nil	no aircraft assigned	Peterson AFB, Colo.
30th Space Wing, Vandenberg AFB, Calif.		
30th OG		
576th FLTS	LGM-30G, LGM-118A	Vandenberg AFB, Calif.
76th HF	UH-1N 'HV'	Vandenberg AFB, Calif.
45th Space Wing, Patrick AFB, Fla.		
45th OG		
no sqns	no aircraft assigned	Patrick AFB, Fla.
50th Space Wing, Falcon AFB, Colo.		
50th OG		
no sqns	no aircraft assigned	Falcon AFB, Colo.

20th Air Force		HQ Francis E. Warren AFB, Wyo.
90th Space Wing, Francis E. Warren AFB, Wyo.		
90th OG		
319th MIS	LGM-30G	Francis E. Warren AFB
320th MIS	LGM-30G	Francis E. Warren AFB
321st MIS	LGM-30G	Francis E. Warren AFB
400th MIS	LGM-118A	Francis E. Warren AFB
37th HF	UH-1N 'FE'	Francis E. Warren AFB
91st Space Group, Minot AFB, N.D.		
91st OG		
740th MIS	LGM-30G	Minot AFB, N.D.
741st MIS	LGM-30G	Minot AFB, N.D.
742nd MIS	LGM-30G	Minot AFB, N.D.
54th HF	UH-1H 'MT'	Minot AFB, N.D.
341st Space Wing, Malmstrom AFB, Mont.		
341st OG		
564th MIS	LGM-30G	Malmstrom AFB, Mont.
40th HF	UH-1N 'MM'	Malmstrom AFB, Mont.

Air Force Reserve Command

4th Air Force		HQ March ARB, Calif.
349th Air Mobility Wing		**Travis AFB, Calif.**
(Associate unit – no aircraft assigned)		
349th OG		
70th ARS	KC-10A	Travis AFB, Calif.
79th ARS	KC-10A	Travis AFB, Calif.
301st AS	C-5A/B	Travis AFB, Calif.
312th AS	C-5A/B	Travis AFB, Calif.
710th AS	C-141B	Travis AFB, Calif.
433rd Airlift Wing		**Kelly AFB, Texas**
433rd OG		
68th AS	C-5A	Kelly AFB, Texas
446th Airlift Wing		**McChord AFB, Wash.**
(Associate unit – no aircraft assigned)		
446th OG		
97th AS	C-141B	McChord AFB, Wash.
313th AS	C-141B	McChord AFB, Wash.
728th AS	C-141B	McChord AFB, Wash.
452nd Air Mobility Wing		**March AFB, Calif.**
452nd OG		
336th ARS	KC-135E	March AFB, Calif.
729th AS	C-141C	March AFB, Calif.
730th AS	C-141C	March AFB, Calif.
507th Air Refueling Wing		**Tinker AFB, Okla.**
507th OG		
465th ARS	KC-135R	Tinker AFB, Okla.
513th Air Control Group		**Tinker AFB, Okla.**
(Associate unit – no aircraft assigned)		
513th OG		
970th AACS	E-3B/C	Tinker AFB, Okla.
931st Air Refueling Wing		**McConnell AFB, Kans.**
(Associate unit – no aircraft assigned)		
931st OG		
18th ARS	KC-135R/T	McConnell AFB, Kans.
44th ARS	KC-135R/T	McConnell AFB, Kans.
932nd Airlift Wing		**Scott AFB, Ill.**
(Associate unit – no aircraft assigned)		
932nd AW		
73rd AS	C-9A	Scott AFB, Ill.
940th Air Refueling Wing		**Beale AFB, Calif.**
940th OG		
314th ARS	KC-135E	Beale AFB, Calif

Air Force Reserve Command (AFRC)

The newest command in the USAF, Air Force Reserve Command (AFRC) was formed on 17 February 1997 when the former field operating agency called the Air Force Reserve (AFRes) changed status. The new title was authorised primarily in recognition of the vital role played and the experience gained during Operations Desert Shield and Desert Storm. AFRC is one of two elements of the Air Reserve Component, the other being the Air National Guard which remains a field operating agency.

Headquartered at Robins AFB, Georgia, the command has the primary duty of supporting the active-duty Air Force. Earlier in its history AFRes was exclusively an airlift organisation equipped with surplus aircraft types but, gradually, the older airlifters

While slowly being replaced by the C-17A in Air Mobility Command, a total of 64 C-141Cs equipped with a 'glass' cockpit are planned to serve with Air Force Reserve Command. The first aircraft was handed over at the end of October 1997.

were traded in for newer equipment and AFRC now includes strategic bombers, fighters, tankers, special operations, rescue, aeromedical evacuation, aerial fire-fighting, weather reconnaissance, space operations and airborne air control types among its fleet. AFRC mirrors the active-duty Air Force and frequently performs roles to supplement or replace front-line units. The Associate programme under which reservist crews replace their active-duty colleagues operating front-line aircraft types such as the C-5, C-9, KC-10, C-17, C-141 and E-3 has been judged a great success. Instead of being supplied exclusively with surplus aircraft handed down from active-duty units, much of the equipment assigned to AFRC has been purchased especially, in particular numerous

The air force reserve component has recently become responsible for the operation of heavy bombers, but never in the nuclear strike role. This B-52H belongs to the 93rd BS based at Barksdale AFB, La.

10th Air Force		HQ Bergstrom AFB, Texas
94th Airlift Wing		**Dobbins AFB, Ga.**
94th OG		
700th AS	C-130H [ex-'DB']	Dobbins AFB, Ga.
301st Fighter Wing		**NAS Fort Worth, Texas**
301st OG		
457th FS	F-16C/D 'TF'	NAS Fort Worth JRB
302nd Airlift Wing		**Peterson AFB, Colo.**
302nd OG		
731st AS	C-130H [ex-'CR']	Peterson AFB, Colo.
403rd Wing		**Keesler AFB, Miss.**
403rd OG		
53rd WRS	WC-130H	Keesler AFB, Miss.
815th AS	C-130E [ex-'KT']	Keesler AFB, Miss.
419th Fighter Wing		**Hill AFB, Utah**
419th OG		
466th FS	F-16C/D 'HI'	Hill AFB, Utah
440th Airlift Wing		**General Mitchell IAP, Milwaukee, Wisc.**
440th OG		
95th AS	C-130H [ex-'MK']	Mitchell IAP, Wisc.
442nd Fighter Wing		**Whiteman AFB, Mo.**
442nd OG		
303rd FS	OA/A-10A 'KC'	Whiteman AFB, Mo.
482nd Fighter Wing		**Homestead AFB, Fla.**
482nd OG		
93rd FS	F-16C/D 'FM'	Homestead AFB, Fla.
908th Airlift Wing		**Maxwell AFB, Ala.**
908th OG		
357th AS	C-130H [ex-'MX']	Maxwell AFB, Ala.
910th Airlift Wing		**Youngstown MAP, Ohio**
910th OG		
757th AS	C-130H [ex-'YO']	Youngstown MAP
773rd AS	C-130H [ex-'YO']	Youngstown MAP
911th Airlift Wing		**Greater Pittsburgh IAP, Pa.**
911th OG		

758th AS	C-130H [ex-'PI']	Greater Pittsburgh IAP
913th Airlift Wing		**Willow Grove ARS, Pa.**
913th OG		
327th AS	C-130E [ex-'WG']	Willow Grove ARS
914th Airlift Wing		**Niagara Falls IAP, N.Y.**
914th OG		
328th AS	C-130H [ex-'NF']	Niagara Falls IAP
917th Wing		**Barksdale AFB, La.**
917th OG		
47th FS	OA/A-10A 'BD'	Barksdale AFB, La.
93rd BS	B-52H 'BD'	Barksdale AFB, La.
919th Special Operations Wing		**Duke Field, Fla.**
919th OG		
5th SOS	MC-130P	Duke Field, Fla.
711th SOS	MC-130E, C-130E/H	Duke Field, Fla.
926th Fighter Wing		**NAS New Orleans, La.**
926th OG		
706th FS	OA/A-10A 'NO'	NAS New Orleans
934th Airlift Wing		**Minneapolis-St Paul IAP, Minn.**
934th OG		
96th AS	C-130E [ex-'MS']	Minneapolis-St Paul IAP
939th Rescue Wing		**Portland IAP, Ore.**
939th RQG		
303rd RQS	HC-130P 'PD', C-130E	Portland IAP, Ore.
304th RQS	HH-60G 'PD'	Portland IAP, Ore.
305th RQS	HH-60G 'DR'	Davis-Monthan AFB
920th RQG		**Patrick AFB, Fla.**
920th RQG		
39th RQS	HC-130N/P 'FL', C-130E	Patrick AFB, Fla.
301st RQS	HH-60G 'FL'	Patrick AFB, Fla.

Note: the 920th Rescue Group has been formed to co-ordinate activities at Patrick AFB, but is ultimately responsible to the 939th RW

944th Fighter Wing		**Luke AFB, Ariz.**
944th OG		
302nd FS	F-16C/D 'LR'	Luke AFB, Ariz.

examples of the C-130H. The command has also begun receiving the first C-141C models fitted with a 'glass' cockpit to extend their service lives. The final examples of the StarLifter are due to be operated by both AFRC and the ANG until possibly 2008, with 64 examples to be converted to C-141C standard.

Personnel number over 78,000 including some who are full time, although the majority are part time. The command is divided into three numbered air forces. The 10th Air Force at NAS Forth Worth Joint Reserve Base, Texas, operates the fighter, bomber, ground attack and special forces squadrons, equipped with the OA/A-10A, B-52H, MC-130E, HC-130P, F-16C/D and HH-60G. The 4th Air Force at McClellan AFB, California and the 22nd Air Force at Dobbins AFB, Georgia are equipped with the C-5, C-130, KC-135 and C-141 squadrons together with the associate units. Collectively, almost 400 fixed- and rotary-winged types are directly assigned. AFRC is to receive new equipment in the near future with the introduction into service of the WC-130J to perform the weather reconnaissance role in place of the WC-130Hs of the 53rd WRS, most of which are over 30 years old. The C-130J is to join the co-located 815th AS during 1999. Many of the older C-130Es have also seen service for more than 30 years.

The overall reduction in the size of the active-duty Air Force has seen the transfer to the reserves of a significant portion of the USAF's duties, particularly that of airlift and aerial refuelling. At the same time as the reduction has been implemented, the Air Force has been called upon to participate in more humanitarian and peacekeeping duties. Understandably, the Air Force has had to rely increasingly upon the reserves to fulfil some of this commitment, in particular replacing front-line combat units enforcing air exclusion zones over Iraq and Bosnia-Herzegovina. Further rounds of cutbacks may affect the active-duty commands, but AFRC is unlikely to see any wholesale reductions in its overall size.

Above: Leading 'LR' coded Block 32 F-16C/Ds of the 302nd Fighter Squadron/944th Fighter Wing (the first Reserve unit to be issued with factory-fresh F-16C) is a 'FM' coded 93rd Fighter Squadron/482nd Fighter Wing example (previously on the F-16A ADF).

Below: AFRC units use the Hercules more than any other type. The 815th AS/403rd Wing operates the C-130E from Keesler AFB, Miss. and from mid-1998 was involved in tests for a new paintless appliqué film coating as a potential replacement for paint.

22nd Air Force		HQ Dobbins AFB, Ga.
315th Airlift Wing		**Charleston AFB, S.C.**
(Associate unit – no aircraft assigned)		
315th OG		
300th AS	C-17A	Charleston AFB, S.C.
317th AS	C-17A	Charleston AFB, S.C.
701st AS	C-141B	Charleston AFB, S.C.
707th AS	C-17A	Charleston AFB, S.C.
434th Air Refueling Wing		**Grissom AFB, Ind.**
434th OG		
72nd ARS	KC-135R	Grissom AFB, Ind.
74th ARS	KC-135R	Grissom AFB, Ind.
439th Airlift Wing		**Westover ARB, Mass.**
439th OG		
337th AS	C-5A	Westover ARB, Mass.
445th Airlift Wing		**Wright-Patterson AFB, Ohio**
445th OG		
89th AS	C-141B	Wright-Patterson AFB
356th AS	C-141B	Wright-Patterson AFB
459th Airlift Wing		**Andrews AFB, Md.**
459th OG		
756th AS	C-141B	Andrews AFB, Md.
512th Airlift Wing		**Dover AFB, Del.**
(Associate unit – no aircraft assigned)		
512th OG		
326th AS	C-5A/B	Dover AFB, Del.
709th AS	C-5A/B	Dover AFB, Del.
514th Air Mobility Wing		**McGuire AFB, N.J.**
(Associate unit – no aircraft assigned)		
514th OG		
76th ARS	KC-10A	McGuire AFB, N.J.
78th ARS	KC-10A	McGuire AFB, N.J.
702nd AS	C-141B	McGuire AFB, N.J.
732nd AS	C-141B	McGuire AFB, N.J.
916th Air Refueling Wing		**Seymour Johnson AFB, N.C.**
916th OG		
77th ARS	KC-135R	Seymour Johnson AFB
927th Air Refueling Wing		**Selfridge ANGB, Miss.**
927th OG		
63rd ARS	KC-135E	Selfridge ANGB, Miss.

Air National Guard (ANG)

The Air National Guard together with Air Force Reserve Command form the Air Reserve Component. Organised on completely different lines, the two services nonetheless perform similar duties, enabling substantial numbers of former USAF officers and enlisted personnel to continue duty on a part-time basis. The ANG has a personnel strength of more than 110,000, just a small percentage of whom are full time with the Guard. Ninety wings in the ANG are under state government jurisdiction during peacetime, but could be mobilised and assigned to the active-duty Air Force if required. The airlift and tanker units would join AMC; the fighter, attack, air defence, air rescue and bomber elements would be assigned to ACC; the two training units would come under AETC; and the single EC-130E-equipped psychological warfare unit would join AFSOC. In the Pacific region, the C-130, KC-135 and F-15 units in Alaska and Hawaii would come under PACAF.

The image of the 'weekend warrior' has long since been replaced by that of an efficient secondary capability ready to augment the active-duty Air Force in both peacetime and in combat. The ANG has assumed many duties of the front-line Air Force, including 100 per cent of the USAF's fighter interceptor force, almost half of the Air Force's airlift and aerial refuelling capability, and one-third of its fighter potential. Traditionally, the Guard was equipped with aircraft types supplanted from front-line units, which resulted in man hours expended to maintain these

Above: Today's Air National Guard uses earlier examples of the current generation of USAF aircraft. The 122nd FS/Louisiana ANG flies the F-15A Eagle from NAS New Orleans.

Below: Displaying '125th FW' on the inside fin, this early F-15A (75-0039) serves with the 159th Fighter Squadron of the Florida Air National Guard. The Eagles replaced F-16ADFs.

older airframes in flyable condition. Those days have long since been consigned to history as the Guard has received modern equipment to mirror the front-line squadrons. In many cases, aircraft have been obtained new, in particular the A-10A, C-130H and F-16C. Other aircraft types in service are identical to those of front-line squadrons, enabling the Guard to replace the active-duty force performing numerous operations. In particular, the ongoing United Nations air exclusion zones above Bosnia and Iraq are manned for the majority of the time by front-line squadrons, although Air National Guard units frequently provide a necessary replacement to enable the active-duty personnel to take a well-earned break.

New equipment planned for the Air National Guard includes the C-130J which is due to join the 135th AS, Maryland ANG during 1999. Other ANG units will also receive the C-130J in due

course, and the 193rd SOS at Harrisburg IAP, Pennsylvania will exchange its EC-130E Commando Solo psychological warfare airframes for the new EC-130J. Early in the next decade, the 183rd AS at Jackson, Mississippi will replace its C-141Cs with the C-17A, becoming the only reservist unit to have the Globemaster III directly assigned. In the meantime, the two ANG squadrons will send their C-141Bs to Warner Robins Air Logistics Center to be converted to C-141C configuration and have the 'glass' cockpit fitted. The first ANG example was due for completion in January 1999. A small number of squadrons continue to operate the F-16A/B, but are expected to upgrade to the F-16C/D model eventually. Several air refuelling squadrons have received the KC-135R, and 11 still equipped with the KC-135E may yet be funded to convert their aircraft to the R model.

Left: The New York ANG's 139th Airlift Squadron will soon be the sole unit in the US military operating ski-equipped Hercules, after the US Navy transfers its aircraft. This example is an FY83 LC-130H.

Below: Based at NAS Willow Grove, the 103rd FS flies OA-10As and a single C-26A, which it uses in the operational support aircraft role.

The 133rd ARS/157th ARW at Pease AFB, N.H. flies the KC-135R Stratotanker. If mobilised, this unit would be gained by Air Mobility Command. Eleven ANG squadrons fly the CFM56-engined KC-135R.

Displaying large fire-fighting codes, this MAFFS-equipped 115th AS/146th AW C-130E of the California Air National Guard is able, during the forest fire season, to supplement civil assets.

Flying F-16C/Ds, the New Mexico ANG is responsible for the 188th FS and the co-located 150th DSE operating from Kirtland AFB, New Mexico.

Air National Guard

Unit	Aircraft / Base
101st ARW	**Bangor IAP, Me.**
132nd ARS	KC-135E
102nd FW	**Otis ANGB, Mass.**
101st FS	F-15A/B, C-26B
103rd FW	**Bradley ANGB, Conn.**
118th FS	A-10A 'CT'
104th FW	**Barnes MAP, Westfield, Mass.**
131st FS	A-10A 'MA'
105th AW	**Stewart ANGB, N.Y.**
137th AS	C-5A
106nd RQW	**Francis S. Gabreski IAP, Suffolk County, N.Y.**
102nd RQS	HC-130P, HH-60G 'LI'
107th ARW	**Niagara Falls IAP, N.Y.**
136th ARS	KC-135R
108th ARW	**McGuire AFB, N.J.**
141st ARS	KC-135E
150th ARS	KC-135E
109th AW	**Schenectady County Apt, N.Y.**
139th AS	C-130H, LC-130H, C-26B
110th FW	**W.K. Kellogg Apt, Mich.**
172nd FS	OA/A-10A 'BC'
111th FW	**Willow Grove ARS, Pa.**
103rd FS	OA-10A 'PA', C-26A
113th FW	**Andrews AFB, Md.**
121st FS	F-16C/D 'DC'
201st AS	C-22B, C-38A
114th FW	**Joe Foss Fd, Sioux Falls, S.D.**
175th FS	F-16C/D 'SD', C-26B
115th FW	**Dane County RAP, Truax, Wisc.**
176th FS	F-16C/D 'WI', C-26B
116th BW	**Robins AFB, Ga.**
128th BS	B-1B 'GA', C-26B
117th ARW	**Birmingham MAP, Ala.**
106th ARS	KC-135R
118th AW	**Nashville Metro Apt, Tenn.**
105th AS	C-130H
119th FW	**Hector Fd, Fargo, N.D.**
178th FS	F-16A/B, C-26B
120th FW	**Great Falls IAP, Mont.**
186th FS	F-16A/B, C-26B
121st ARW	**Rickenbacker ANGB, Ohio**
145th ARS	KC-135R
166th ARS	KC-135R, C-26A
122nd FW	**Fort Wayne MAP, Ind.**
163rd FS	F-16C/D 'FW', C-26B
123rd AW	**Standiford Fd, Louisville, Ky.**
165th AS	C-130H
124th Wg	**Boise Air Terminal, Idaho**
189th AS	C-130E [ex-'ID']
190th FS	A-10A 'ID', C-26B
125th FW	**Jacksonville IAP, Fla.**
159th FS	F-15A/B, C-26B
126th ARW	**Chicago-O'Hare IAP, Ill.**
108th ARS	KC-135E
127th Wg	**Selfridge ANGB, Mich.**
107th FS	F-16C/D 'MI'
171st AS	C-130E [ex-'MI']
128th ARW	**Gen. Mitchell IAP, Milwaukee, Wisc.**
126th ARS	KC-135R
129th RQW	**NAS Moffett Field, Calif.**
129th RQS	HC-130P, HH-60G 'CA'
130th AW	**Yeager Apt, Charleston, W.Va.**
130th AS	C-130H
131st FW	**Lambert Fd, St Louis, Mo.**
110th FS	F-15A/B, C-26B 'SL'
132nd FW	**Des Moines MAP, Iowa**
124th FS	F-16C/D 'IA', C-26B
133rd AW	**Minneapolis-St Paul IAP, Minn.**
109th AS	C-130H [ex-'MN']
134th ARW	**McGhee Tyson Apt, Knoxville, Tenn.**
151st ARS	KC-135E
136th AW	**NAS Fort Worth JRB, Texas**
181st AS	C-130H [ex-'TX']
137th AW	**Will Rogers Apt, Oklahoma City, Okla.**
185th AS	C-130H [ex-'OK']
138th FW	**Tulsa IAP, Okla.**
125th FS	F-16G/D 'OK'
139th AW	**Rosecrans MAP, St Joseph, Mo.**
180th AS	C-130H [ex-'XP']
140th Wg	**Buckley ANGB, Colo.**
120th FS	F-16C/D 'CO', C-26B
200th AS	C-26B, CT-43A
141st ARW	**Fairchild AFB, Wash.**
116th ARS	KC-135E, C-26B
142nd FW	**Portland IAP, Ore.**
123rd FS	F-15A/B, C-26A
143rd AW	**Quonset Point State Apt, Providence, R.I.**
143rd AS	C-130E [ex-'RI']
144th FW	**Fresno Air Terminal, Calif.**
194th FS	F-16C/D, C-26B
145th AW	**Charlotte/Douglas IAP, N.C.**
156th AS	C-130H
146th AW	**NAS Point Mugu, Calif**
115th AS	C-130E [ex-'CI']
147th FW	**Ellington ANGB, Texas**
111th FS	F-16C/D, C-26A/B 'EF'
148th FW	**Duluth IAP, Minn.**
179th FS	F-16A/B, C-26B
149th FW	**Kelly AFB, Texas**
182nd FS	F-16C/D 'SA'
150th FW	**Kirtland AFB, N.M.**
150th DSE	F-16C/D
188th FS	F-16C/D 'NM', C-26B
151st ARW	**Salt Lake City IAP, Utah**
191st ARS	KC-135E
152nd AW	**Reno Cannon IAP, Nev.**
192nd AS	C-130E [ex-'NV'], C-26B
153rd AW	**Cheyenne MAP, Wyo.**
187th AS	C-130H [ex-'WY']
154th Wg	**Hickam AFB, Hawaii**
199th FS	F-15A/B
203rd ARS	KC-135R
204th AS	C-130H
155th ARW	**Lincoln MAP, Neb.**
173rd ARS	KC-135R, C-26
156th AW	**Muniz ANGB/Puerto Rico IAP, San Juan, PR**
198th AS	C-130E, UC-26C
157th ARW	**Pease AFB, N.H.**
133rd ARS	KC-135R
158th FW	**Burlington IAP, Vt.**
134th FS	F-16C/D, C-26B
159th FW	**NAS New Orleans, La.**
122nd FS	F-15A/B, C-130H 'JZ'
161st ARW	**Sky Harbor IAP, Phoenix, Ariz.**
197th ARS	KC-135E
162nd FW	**Tucson IAP, Ariz.**
148th FS	F-16A/B 'AZ'
152nd FS	F-16A/B 'AZ', C-26B
195th FS	F-16A/B/C/D 'AZ'
163rd ARW	**March ARB, Ca.**
196th ARS	KC-135R
164th AW	**Memphis IAP, Tenn.**
155th AS	C-141B
165th AW	**Savannah IAP, Ga.**
158th AS	C-130H [ex-'GA']
166th AW	**Gtr Wilmington Apt, Del.**
142nd AS	C-130H [ex-'DE']
167th AW	**Eastern W.VA Reg Apt, Martinsburgh, W.Va.**
167th AS	C-130H [ex-'WV']
168th ARW	**Eielson AFB, Ark.**
168th ARS	KC-135R
169th FW	**McEntire ANGB, S.C.**
157th FS	F-16C/D, C-130H

The 'Quiet Professionals' of the 193rd SOS/PA ANG undertake the psychological warfare role using two versions of the EC-130E. This is a 'Rivet Rider' radio/television relay and transmission aircraft.

Above: The 186th Fighter Squadron of the Montana Air National Guard was one of the last units in the USAF to replace its F-106 Delta Daggers. From June 1987 it became the second Guard unit to employ the F-16A in the air defence role.

Below: The 'Happy Hooligans' of the 178th FS/ 119th FW North Dakota ANG operate the Block 15 Air Defense Fighter variant of the F-16A/B.

The 11th Air Force became part of the Pacific Air Forces in 1990. Based at Elmendorf AFB, Alaska, these F-15Cs of the 19th FS 'Fighting Gamecocks' form part of the 3rd Wing.

Air National Guard continued

171st ARW	**Gtr Pittsburgh IAP, Pa.**
146th ARS	KC-135E
172nd AW	**Allen C. Thompson Fd, Jackson, Miss.**
183rd AS	C-141B
173rd FW	**Kingsley Field, Ore.**
114th FS	F-15C/D
174th FW	**Hancock Fd, Syracuse, N.Y.**
138th FS	F-16C/D 'NY'
175th Wg	**Glenn L. Martin State Apt, Baltimore, Md.**
104th FS	OA/A-10A 'MD'
135th AS	C-130E [ex-'MD']
176th Wg	**Kulis ANGB, Anchorage, Ark.**
144th AS	C-130H 'AK'
210th RQS	HC-130N, HH-60G
177th FW	**Atlantic City IAP, N.J.**
119th FS	F-16C/D 'AC'
178th FW	**Springfield-Beckley MAP, Ohio**
162nd FS	F-16C/D 'OH'
179th AW	**Mansfield-Lahm Apt, Ohio**
164th AS	C-130H [ex-'OH']
180th FW	**Toledo Express Apt, Ohio**
112th FS	F-16C/D 'OH'
181st FW	**Hulman Fd, Terre Haute, Ind.**
113th FS	F-16C/D 'TH'
182nd AW	**Gtr Peoria Apt, Ill.**
169th AS	C-130E [ex-'IL']
183rd FW	**Capital Apt, Springfield, Ill.**
170th FS	F-16C/D 'SI'
184th BW	**McConnell AFB, Kans.**
127th BS	B-1B
185th FW	**Sioux Gateway Apt, Sioux City, Iowa**
174th FS	F-16C/D 'HA'
186th ARW	**Key Fd, Meridian, Miss.**
153rd ARS	KC-135R, C-26B
187th FW	**Dannelly Fd, Montgomery, Ala.**
160th FS	F-16C/D 'AL', C-26B
188th FW	**Fort Smith MAP, Ark.**
184th AS	F-16A/B 'FS'
189th AW	**Little Rock AFB, Ark.**
154th AS	C-130E
190th ARW	**Forbes Field, Topeka, Kans.**
117th ARS	KC-135D/E
192nd FW	**Byrd IAP, Richmond, Va.**
149th FS	F-16C/D 'VA', C-26B
193rd SOW	**Harrisburg IAP, Middletown, Pa.**
193rd SOS	EC-130E (CL), EC-130E (RR)

Pacific Air Forces (PACAF)

The Pacific Air Forces (PACAF) is the only major flying command which appears to have survived unscathed the numerous rounds of drawdown and budget cuts in recent years. Its size and structure is largely the same as it was at the end of the last decade, due mainly to the planners having established the basic size of the command after the US involvement in the Vietnam War. From its headquarters at Hickam AFB, Hawaii, PACAF has jurisdiction over probably the largest area in the USAF, as its forces are thinly spread from Alaska to the Indian Ocean.

A personnel level of over 32,000 active-duty and 3,500 reservists administers PACAF, whose mission is to be an effective force ready to conduct offensive and defensive air operations in the Pacific and Asiatic theatres. The command is divided into four numbered air forces which are organised on a geographic basis, including the 13th Air Force at Andersen AFB, Guam whose area of responsibility extends from its headquarters all the way to Singapore, but which has no aircraft assigned. The other three air forces are responsible for forces in Japan, South Korea and Alaska. Equipment consists primarily of fighter aircraft which vary between the OA/A-10A, F-16C/D, F-15C/D and the F-15E and number approximately 260. Since the major reorganisation of Stateside units in 1992, PACAF has had its own tanker force of 15 KC-135Rs permanently assigned at Kadena AB, Okinawa to make up the Pacific Tanker Task Force. In addition, a mixed theatre airlift, communications,

and aeromedical evacuation element is assigned. Four E-3 Sentries are also stationed in the region, operated by the 3rd Wing at Elmendorf AFB, Alaska and the 18th Wing at Kadena AB, Okinawa. The aircraft themselves are drawn from the 552nd ACW at Tinker AFB, Oklahoma, but carry the markings of the local unit assignment while stationed in PACAF.

The command is situated in an area of the world which contains some of the United States' most vociferous opponents, in particular the regime of North Korea whose leadership is volatile and unpredictable. The area also contains many advancing Third World markets of the Southeast Asian region whose importance to world trade depends upon a stable economy. Therefore, PACAF's presence is primarily a forward presence designed to enable swift reaction over vast distances to react to almost any theatre crisis, whether humanitarian or combat. Numerous training exercises held at various PACAF bases offer the chance for other air arms such as Australia, Malaysia, Thailand, South Korea and Singapore to operate alongside the USAF and forge contingency plans. In particular, Cope Thunder in Alaska, Commando Sting in Singapore, and Foal Eagle in South Korea are all joint exercises designed to integrate operations effectively.

Although no announcement has yet been made, it is highly likely the command will receive the F-22 to replace the F-15C/D and later the Joint Strike Fighter in place of the A-10 and F-16.

This formation illustrates the aircraft of the 18th Wing at Kadena AB, Okinawa. The KC-135R (909th ARS), E-3 (961st ACS) and F-15Cs (12th, 44th and 67th FS) all display the 'ZZ' code of the wing.

Pacific Air Forces

Direct reporting		
15th Air Base Wing		**Hickam AFB, Hawaii**
15th OG		
65th AS	C-135E/K, KC-135E	Hickam AFB, Hawaii
5th Air Force		**HQ Yokota AB, Japan**
18th Wing		**Kadena AB, Okinawa**
18th OG		
12th FS	F-15C/D 'ZZ'	Kadena AB, Okinawa
33rd RQS	HH-60G 'ZZ'	Kadena AB, Okinawa
44th FS	F-15C/D 'ZZ'	Kadena AB, Okinawa
67th FS	F-15C/D 'ZZ'	Kadena AB, Okinawa
909th ARS	KC-135R 'ZZ'	Kadena AB, Okinawa
961st AACS	E-3B/C 'ZZ'	Kadena AB, Okinawa
35th Fighter Wing		**Misawa AB, Japan**
35th OG		
13th FS	F-16C/D 'WW'	Misawa AB, Japan
14th FS	F-16C/D 'WW'	Misawa AB, Japan
Det 2, 33rd RQS	HH-60G 'WW'	Misawa AB, Japan
374th Airlift Wing		**Yokota AB, Japan**
374th OG		
36th AS	C-130E/H 'YJ'	Yokota AB, Japan
374th AS	C-9A	Yokota AB, Japan
459th AS	C-21A, UH-1N	Yokota AB, Japan
7th Air Force		**HQ Osan AB. Republic of Korea**
8th Fighter Wing		**Kunsan AB, RoK**
8th OG		
35th FS	F-16C/D 'WP'	Kunsan AB, RoK
80th FS	F-16C/D 'WP'	Kunsan AB, RoK
51st Fighter Wing		**Osan AB, RoK**
51st OG		
25th FS	OA-10A 'OS'	Osan AB, RoK
36th FS	F-16C/D 'OS'	Osan AB, RoK
Det 1, 33rd RQS	HH-60G 'OS'	Osan AB, RoK
55th ALF	C-12J 'OS'	Osan AB, RoK
11th Air Force		**HQ Elmendorf AFB, Alaska**
3rd Wing		**Elmendorf AFB, Ak.**
3rd OG		
19th FS	F-15C/D 'AK'	Elmendorf AFB, Ak.
54th FS	F-15C/D 'AK'	Elmendorf AFB, Ak.
90th FS	F-15E 'AK'	Elmendorf AFB, Ak.
517th AS	C-130H 'AK'	Elmendorf AFB, Ak.
962nd AACS	E-3B/C 'AK'	Elmendorf AFB, Ak.
354th Fighter Wing		**Eielson AFB, Ak.**
354th OG		
18th FS	F-16C/D 'AK'	Eielson AFB, Ak.
353rd FS	no aircraft assigned	Eielson AFB, Ak.
– organises Cope Thunder exercises		
355th FS	OA/A-10A 'AK'	Eielson AFB, Ak.
13th Air Force		**HQ Andersen AB, Guam**
36th ABW	no aircraft assigned	Andersen AFB, Guam
497th FTS	no aircraft assigned	Paya Lebar AB, Singapore

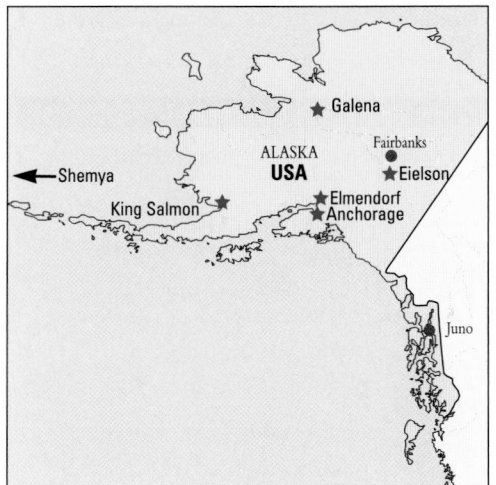

United States Air Forces in Europe (USAFE)

The United States Air Forces in Europe was formed on 7 August 1945 – and therefore is the only current major command which was established prior to the formation of the USAF itself in September 1947 – and has its headquarters at Ramstein Air Base, Germany. The command has slightly over 27,000 officers and enlisted personnel at present, which is a shadow of the 70,000-manpower level assigned during the Cold War. However, the peace dividend afforded by the unification of East and West Germany, the elimination of the Warsaw Pact and the dissolution of the USSR, has enabled USAFE to reduce strength significantly. Much of the former USAFE infrastructure has been returned to the host nations, including many of the bases in England that have closed following their units being withdrawn. In Germany the US presence has also been cut back fairly drastically, whereas in the Netherlands, Spain and Belgium the US has withdrawn altogether. Currently, the 3rd Air Force manages all USAFE assets north of the Alps, while the 16th Air Force is responsible for activities to the south.

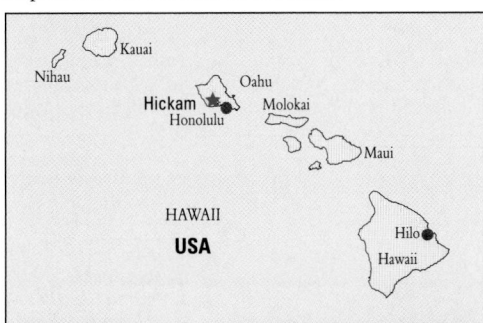

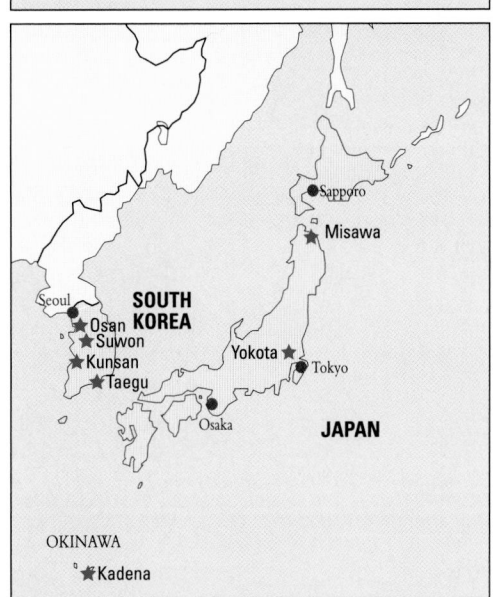

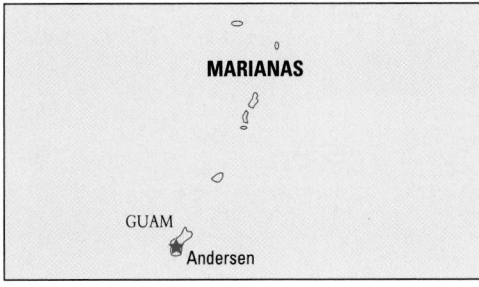

From a whole wing in the 1980s, the A-10 Thunderbolt II today equips a single squadron in USAF Europe – the 81st FS of the 52nd FW at Spangdahlem AB, Germany. Both A-10As and OA-10As are flown by the squadron, tasked with anti-tank/close support and forward air control duties.

USAFE currently operates little more than 125 fighters (F-15C/Ds and F-16C/Ds) along with 75 attack aircraft (OA/A-10A and F-15E), a total that is less than one-third the USAFE offensive capability of 10 years ago. The remainder of the USAFE complement is composed of theatre airlift, VIP, communications and aeromedical evacuation types, and nine KC-135Rs for aerial refuelling. The latter are bolstered by additional tankers deployed from the USA, although the 351st ARS at RAF Mildenhall is currently increasing its assigned complement to 15 KC-135s.

The humanitarian and peacekeeping roles in Bosnia-Herzegovina have been one of the major challenges which have occupied a large portion of the USAFE's flying programme. The civil war which raged throughout the Balkans for almost five years has now ceased, but the US and its allies maintain a peacekeeping role, initially through NATO's Implementation Force (IFOR) and latterly as the Stabilisation Force (SFOR). The air exclusion zone over Bosnia is patrolled by aircraft from various nations including the US, with operations conducted at several bases in eastern Italy as well as aircraft-carriers. Aviano AB is one of the major facilities and has large numbers of US aircraft in temporary residence operating beside the two squadrons of F-16C/Ds of the 31st FW. Sustained combat operations by USAFE squadrons deployed to Aviano have been relieved to a degree by the deployment from the USA of ANG and AFRes/AFRC units. The stabilisation force composed of ground personnel is resupplied by the theatre airlift C-130s based at Ramstein, which are bolstered by an ongoing rotation of additional C-130s from the USA. The US has also stationed personnel in Albania and Hungary to perform operations associated with the Bosnian crisis – a situation which would have been unthinkable just 10 years ago.

Farther east, USAFE has a second theatre of operations in Turkey to implement the 'No-Fly Zone' over northern Iraq under Operation Northern Watch. The participants for the operation are all located at Incirlik Air Base near Adana in Turkey. The composition of Northern Watch varies according to the likely threat from Iraq, but usually consists of six F-16C/Ds, six F-15Cs, three E-3B/Cs, five KC-135Rs, a pair of HC-130Ps and three MH-60Gs. The electronic warfare capability provided by three EF-111As was withdrawn during 1997 and replaced by the EA-6B crewed jointly by Air Force and Navy personnel. The operation was boosted by additional aircraft deployed during February 1998 as the likelihood increased of air strikes being mounted against Iraq; however, the easing of tensions enabled some aircraft to return home.

USAFE completed its current re-equipment programme early in the 1990s as the final F-15Es were delivered to the 48th FW at RAF Lakenheath to replace the F-111F. The present composition will probably remain unchanged until the next decade, when the F-22A Raptor is due to enter service with one squadron of the 48th FW. Later still, the Joint

Strike Fighter is also due to be assigned to Lakenheath. No details concerning any re-equipment for the 52nd FW at Spangdahlem AB have been released. It would seem unlikely under the current circumstances – with USAFE committed to two air exclusion operations – that there would be further reductions in the size of the command. There is a possibility that some units could be relocated, particularly in Germany where the Royal Air Force presence will cease during the next decade. One possibility under consideration is to relocate some of the 52nd FW elements eastward, possibly into Poland.

The 31st Fighter Wing at Aviano Air Base, Italy forms part of the 16th Air Force. The wing is equipped with two squadrons of Lockheed Martin F-16C/Ds, this example being operated by the 555th Fighter Squadron 'Triple Nickel'. The Aviano-based F-16s were the first service aircraft to receive the Improved Data Modem, allowing automatic targeting.

The F-15E replaced the F-111F in two squadrons of the 'Statue of Liberty Wing' – the 48th Fighter Wing – based at RAF Lakenheath, UK. The blue tail band denotes that this aircraft, 91-0305/'LN', belongs to the 492nd FS.

United States Air Forces in Europe

3rd Air Force		HQ RAF Mildenhall, UK
48th Fighter Wing		**RAF Lakenheath, UK**
48th OG		
492nd FS	F-15E 'LN'	RAF Lakenheath, UK
493rd FS	F-15C/D 'LN'	RAF Lakenheath, UK
494th FS	F-15E 'LN'	RAF Lakenheath, UK
52nd Fighter Wing		**Spangdahlem AB, Germany**
52nd OG		
22nd FS	F-16C/D 'SP'	Spangdahlem AB, Germany
23rd FS	F-16C/D 'SP'	Spangdahlem AB, Germany
81st FS	OA/A-10A 'SP'	Spangdahlem AB, Germany
86th Airlift Wing		**Ramstein AB, Germany**
86th OG		
37th AS	C-130E 'RS'	Ramstein AB, Germany
75th AS	C-9A	Ramstein AB, Germany
Det 1	C-9A	Chièvres AB, Belgium
76th AS	C-20A, C-21A	Ramstein AB, Germany
HQ USEUCOM		
	C-21A	Stuttgart-Echterdingen AP, Germany
100th ARW		**RAF Mildenhall, UK**
100th OG		
351st ARS	KC-135R 'D'	RAF Mildenhall, UK

16th Air Force		HQ Aviano AB, Italy
31st Fighter Wing		**Aviano AB, Italy**
31st OG		
510th FS	F-16C/D 'AV'	Aviano AB, Italy
555th FS	F-16C/D 'AV'	Aviano AB, Italy
39th Wing		**Incirlik AB, Turkey**
39th OG		
no sqns	no aircraft assigned	Incirlik AB, Turkey
(tactical support for rotational units for Operation Northern Watch)		
HQ TUSLOG		
no sqns	no aircraft assigned	Ankara AS, Turkey
7217th ABG		
no sqns	no aircraft assigned	Ankara AS, Turkey
7241st ABG		
no sqns	no aircraft assigned	Izmir AS, Turkey

Above: Used by the CinC USAFE and other dignitaries are the C-20As on the strength of the 76th Airlift Squadron of the 86th Airlift Wing. The aircraft have been at Ramstein AB, Germany since 1986/87.

Below: Seven C-9A Nightingales are assigned to USAF Europe in the aeromedical and transport role. One aircraft is available to Det 1 based at Chièvres, Belgium, in support of SHAPE.

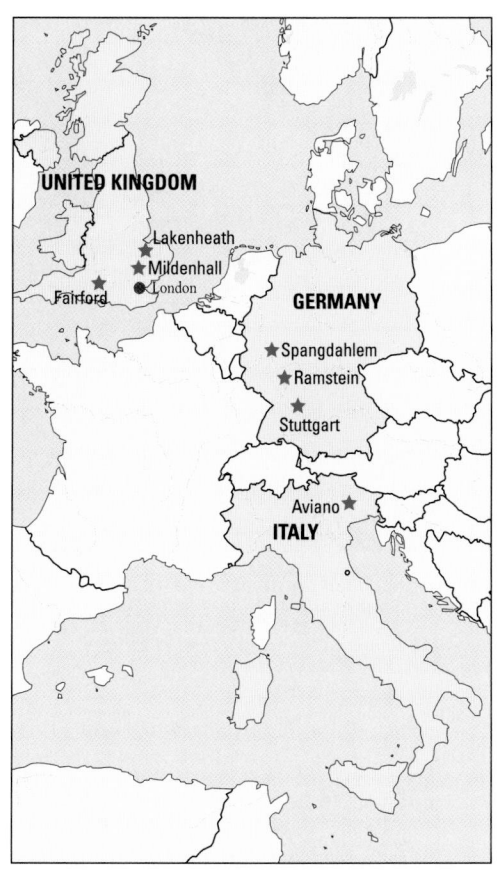

US Air Force Academy		USAF Academy, Colo.
94th ATS	TG-3A, TG-4A, TG-7A, TG-11A, UV-18A/B	USAF Academy
Air Force Operational Test and Evaluation Center		**Kirtland AFB, N.M.**
Det 2	no aircraft assigned	Eglin AFB, Fla.
Det 4	no aircraft assigned	Peterson AFB, Colo.
Det 5	no aircraft assigned	Edwards AFB, Calif.
11th Wing		**Bolling AFB, DC**
	no aircraft assigned	

The USAF Academy uses a variety of light aircraft, motorised gliders and gliders to provide air experience for the cadets. Wearing both military and civil serials, this TG-7A (Schweizer SGM-2-37) is one of 14 examples that have been used.

Direct-Reporting Units

There are a number of direct reporting units (DRU) which are sub-divisions of the Air Force directly subordinate to USAF Headquarters at the Pentagon. They are separate from any major command or field operating agency due to the specialisation of their missions, but they have the same organisational responsibilities as major commands.

Among the DRUs which are relevant to the USAF structure are the Air Force Operational Test and Evaluation Center at Kirtland AFB, New Mexico which was formed on 1 January 1974. The centre's primary role, as its title suggests, is to plan and perform operational test and evaluation to determine the effectiveness and suitability for service of aircraft and weapon systems and their capacity to meet the requirement of their allocated mission. The structure consists of Det 2 at Eglin AFB, Florida; Det 4 at Peterson AFB, Colorado; and Det 5 at Edwards AFB, California. While having no aircraft assigned, the centre is conducting tests related to the B-2 Spirit bomber, C-17A Globemaster III, and upgrades to the radar and detections systems at the Cheyenne Mountains complex.

The Air Force Academy is a DRU, having been established on 1 April 1954 at Colorado Springs, Colorado. The AFA educates and trains cadets, with the average number enrolled at any one time being approximately 4,000. Personnel strength is 2,100 with a further 2,000 civilian employees. The Academy has 94 aircraft in residence, most of which are T-3As of the 557th FTS, which is part of AETC's 12th FTW. A small mixed fleet of aircraft is directly assigned, including TG-3 and TG-4 gliders, TG-7 and TG-11 motorised gliders, ASK-21 sailplanes, and three UV-18A/B Twin Otters for parachute jump training.

The 11th Wing at Bolling AFB, D.C. is the primary unit located at the only Air Force Base within the capital of Washington, and has no aircraft assigned. It has a wide-ranging responsibility to provide administrative and ceremonial support to the Air Force in the nation's capital, all 50 states and in almost 100 countries worldwide.

Finally, United States Central Command in the Middle East is responsible for a rotation of aircraft to implement Operation Southern Watch. The operation is located in Saudi Arabia, Kuwait and the Gulf States to patrol southern Iraq to prevent any military aircraft activity. The participating squadrons are mostly drawn from the continental United States although on occasions squadrons from USAFE and PACAF as well as the air reserve component have been deployed. The 4404th Wing (Provisional) with headquarters at Prince Sultan Air Base, Al Kharj, Saudi Arabia is responsible for more than a dozen provisional squadrons which have been formed to operate fixed- and rotary-winged assets deployed from their home bases to impose and support the operation. These include:

4402nd Reconnaissance Squadron (Provisional) at Al Kharj AB with the U-2S deployed from the 9th RW at Beale AFB, California;
4405th Airborne Command and Control Squadron (Provisional) at Al Kharj operating the E-3B/C from the 552nd ACW at Tinker AFB, Oklahoma;
4406th Operations Group operating from Kuwait with OA/A-10A from active and reserve units;
4408th Air Refueling Squadron (Provisional) at Al Kharj with the KC-135R/T drawn from Stateside-based, active-duty air refuelling wings;
4410th Airlift Squadron (Provisional) at Al Kharj with the C-130E/H deployed from Air Mobility Command and reserve airlift wings;
Det 1 4410th Airlift Squadron (Provisional) at Seeb International Airport, Oman with the C-130E/H drawn from the above squadron;
4412th Rescue Squadron (Provisional) at Kuwait City with the HH-60G from various active-duty and reservist units;
4413th Air Refueling Squadron (Provisional) at Al Dhafra, UAE with the KC-10A from the 60th and 305th AMWs at Travis AFB, California and McGuire AFB, New Jersey, respectively.

In addition, several other provisional squadrons operate the F-15C/D, F-15E, F-16C/D, F-117A and EA-6B from the joint US Navy/USAF force, although their provisional squadron designations are unknown. Most of these aircraft are from Air Combat Command and reservist units in the USA. The A-10s, F-117s and some reservist F-16s are deployed to Al Jaber AB, Kuwait, while the remainder are located at Al Kharj. During the last two years the USAF has deployed Air Expeditionary Wings to various locations in the Middle East formed around a core unit, but with other squadrons assigned for the duration. Among the facilities hosting these have been Sheikh Isa in Bahrain where F-15C/Ds and F-16C/Ds of the 347th AEW were joined by a small number of B-1Bs and KC-135s.

IN CONCLUSION

The US Air Force is now a much smaller fighting force than at almost any time in its history. The active-duty component has at its disposal approximately 4,500 aircraft and helicopters, while Air Force Reserve Command and the Air National Guard have a further 447 and 1,426, respectively. Almost 25 per cent of the ANG complement is more than 25 years old, with this figure rising to 40 per cent for the active-duty and 50 per cent for the AFRC. On the surface these aircraft may seem antique, but many have been the subject of upgrades and conversions to extend their service lives. Furthermore, what the Air Force lacks in quantity it more than adequately makes up for in quality. It is the airframes which carry the weapons load, passengers, cargo or other commodity to their destination, but it is the pilot and aircrew – whose own considerable skills are supported by a host of computer systems – which enhance their capabilities. The advent of computer-aided smart weapons and the development of stealth to cloak combat operations from radar screens have together provided the Air Force with a method of using far fewer aircraft to perform a larger number of successful missions. The next generation of weapons will be the Unmanned Aerial Vehicles (UAV) which are being developed to exclude the human factor in the cockpit. These will also revolutionise the battlefield, as it is likely they will perform much of the reconnaissance over well defended areas; eventually they will conduct jamming of radars, and one day even deliver ordnance.

The Air Force of the next millennium will be radically different from that formed in September 1947.

Due to be supplemented soon by the F-22A Raptor, the F-15 Eagle will remain the USAF's primary air superiority fighter as the century draws to a close. Over 500 of the type remain in service.

US Naval Aviation

The United States Navy has, for a long time, maintained one of the world's largest air arms. US naval aviation has evolved from a supporting arm to the Navy's power projection force, the centrepiece of its conventional offensive warfighting capability. Naval aircraft, operating from a fleet of aircraft-carriers, surface combatants, amphibious warfare ships and shore bases, support US policy objectives worldwide, in war and peace. The primary roles of naval aircraft are power projection, control of the seas, and deterrence. Naval aircraft perform these roles in conjunction with surface ships and submarines, as well as independently or jointly with other services. Specific missions of naval aviation include strike warfare, anti-air warfare, sea control, anti-submarine warfare, electronic warfare, special operations, amphibious assault, mine countermeasures, reconnaissance, communications, search and rescue, and logistics.

THE HIGHEST TRADITIONS

Soon after the US Navy recognised the potential of the aircraft, the Navy began experimenting with aircraft at sea. Eugene Ely made the first shipboard

take-off in November 1910, from the cruiser *Birmingham*, and the first shipboard landing, in January 1911, onto the cruiser *Pennsylvania*. The date on which Captain W. I. Chambers prepared the requisition papers for two Curtiss aircraft – 8 May 1911 – is considered the official birthday of US naval aviation.

US naval aviators were called into action fewer than three years later, flying minehunting and scouting missions at Tampico and Veracruz, Mexico, drawing hostile fire and sustaining some damage to their aircraft. Shortly after this action, naval aviation was formally recognised by the Navy with the establishment of the Office of Naval Aeronautics. In 1916, Congress authorised the formation of a naval flying corps and a reserve counterpart.

When the United States entered World War I in April 1917, naval aviation was small, but it grew rapidly, and in June 1917 deployed the First Aeronautical Detachment to France, from where it flew missions with the Northern Bombing Group. Other naval aircraft deployed to Italy, Panama and the Azores. Naval aircraft had their greatest effect in anti-submarine warfare, damaging several German submarines. Soon after the war, the Navy's NC flying-boats became the first aircraft to fly across the Atlantic.

Despite rapid post-war demobilisation, naval aviation pushed ahead with fleet integration. The Navy's first aircraft-carrier, USS *Langley* (CV 1), entered service in 1922, followed by seven more before World War II. Naval aircraft developed the tactics, including dive-bombing, that would prove successful in World War II. Giant rigid airships enjoyed a brief period of fleet service until they were terminated for safety reasons.

The 1941 Japanese attack on Pearl Harbor in Hawaii destroyed or disabled most of the Navy's battleships, leaving intact a carrier force which raided Japanese-held islands and stopped Japanese carrier

The contraction of the number of F-14 Tomcat squadrons has resulted in there being just one fighter wing, based at NAS Oceana, Virginia. This aircraft belongs to the wing's fleet readiness squadron, VF-101, the 'Grim Reapers'.

forces in the battles of the Coral Sea and Midway. Carrier and land-based naval aircraft – primarily F4Fs, SBDs, TBFs and PBYs – prevailed in the long campaign to retake Guadalcanal and the rest of the Solomon Islands chain, and supported the drive across the central Pacific. Fast carrier task forces formed the main maritime striking arm against the Japanese. Navy aircraft – primarily F6Fs, SBDs, TBMs and SB2Cs – inflicted a decisive blow against Japanese naval aviation in the 1944 Battle of the Philippine Sea, and assisted in the final defeat of the Japanese battle fleet in the Battle of Leyte Gulf. The invasions of Iwo Jima and Okinawa were supported by Navy aircraft, which also defeated the *kamikaze* threat and raided Japanese bases in the western Pacific.

In the Atlantic theatre, carriers supported the landings in North Africa and southern France and raids against German targets in Norway. However, the main contribution of carrier aircraft, patrol aircraft and blimps was toward the decisive defeat of the U-boat menace.

The Navy ended the war with over 100 aircraft-carriers. Naval aviation was cut back in the post-war draw-down, but carriers were forward-deployed in support of US and NATO strategy. Carriers also became launch platforms for aircraft armed with nuclear weapons.

Carrier-based aircraft were instrumental in turning back the 1950 North Korean invasion of South Korea, and for three years carriers launched strike, reconnaissance and close air support sorties in support of UN forces in Korea. Carrier-based jet fighters were used for the first time in the Korean conflict.

The truce in Korea did not end all hostilities, however. Several patrol and reconnaissance aircraft were damaged or shot down by Soviet and Chinese fighters during the 1950s. Naval aircraft patrolled the Formosa Straits to deter a Communist Chinese take-over of Taiwan. The 1950s was a decade of tremendous technological advancement in naval aviation, with the advent of angled-deck carriers, jet attack aircraft and bombers, air-to-air missiles, and improved intercept radars and anti-submarine sensors. The A-3, A-4 and F-8 entered service in the late 1950s.

The early 1960s saw the service entry of the A-5, A-6, F-4, H-2, SH-3 and P-3 aircraft. Carriers stood by for action (but refrained from it) during the 1961 Bay of Pigs invasion of Cuba, but naval aircraft were instrumental in the 1962 showdown during the Cuban Missile Crisis, during which

Each Carrier Air Wing (CVW) has one Fleet Airborne Early Warning Squadron assigned, consisting of four E-2C Hawkeyes. VAW-126, to which this Hawkeye belongs, is assigned to CVW-3.

The ability of the fleet to move large amounts of stores to and from vessels underway is aided by seven squadrons equipped with versions of the Sea King, Sea Knight and Sea Dragon. This 'VR'-coded HH-46D is operated by Det 6 of HC-11.

reconnaissance jets pinpointed missile sites in Cuba and patrol aircraft tracked the Soviet ships importing the nuclear-tipped missiles.

Naval aviation supported South Vietnamese efforts to defeat a Communist insurgency starting in 1960, and in 1964 flew retaliatory strikes against North Vietnam following the Tonkin Gulf incident. A long air campaign against North Vietnam began in April 1965, and for eight years carriers deployed to the South China Sea, launching strikes and reconnaissance missions over Southeast Asia. Navy squadrons downed many enemy aircraft in aerial combat, but suffered a heavy toll in aircraft and crews to North Vietnamese air defences. Patrol aircraft and helicopters also supported the riverine and coastal forces trying to defeat the insurgency. The war saw the introduction of the A-7, E-2 and C-2 and the retirement of the A-1.

The January 1973 truce in Vietnam brought respite, but in April 1975 carriers were back off the coast of Vietnam, covering the evacuation of Americans as South Vietnam collapsed. Carrier aircraft also struck targets in Cambodia during the rescue of the American cargo ship *Mayaguez*.

Further advancements in aircraft during the early 1970s produced the F-14 fighter. The latter half of the 1970s was marked with a return to a peacetime routine, punctuated by occasional responses to crises. Naval aviation was 'hollowed' by budget restraints, causing a decline in material and personnel readiness.

During the 1980s, naval aviation maintained a vigorous pace as it benefited from a general build-up to counter the Soviet threat. The F/A-18 and SH-60 entered service and other aircraft were improved. Carrier battle groups ranged close to the Soviet motherland in a bold demonstration of capability. Clashes with Libya in 1981, 1986 and 1989 over navigation rights and retaliation for terrorist activity resulted in the destruction of several Libyan aircraft, naval vessels and ground targets. Carrier aircraft also struck Cuban targets in Grenada and Syrian targets in Lebanon in 1983. In 1988, carrier aircraft dealt a punishing blow to the Iranian navy in Operation Praying Mantis.

US naval aviation contributed significantly to the American contingent in Operation Desert Shield (the response to the August 1990 Iraqi invasion of Kuwait) and to combat in Operation Desert Storm in January 1991. Six aircraft-carriers, the largest sustained gathering since World War II, launched strikes against Iraqi forces. Naval aircraft performed a variety of missions, including the destruction of the Iraqi navy.

Since Desert Storm, naval aviation – despite the loss of two carrier air wings and the cut of over 100 squadrons – has maintained a high tempo of operations, supporting relief efforts in Kurdish Iraq, Somalia, the Philippines, Liberia, the Congo, Sierra Leone, Albania, Rwanda and Haiti, as well as the UN and NATO peacekeeping efforts in the former Yugoslavia. Carrier aircraft flew retaliatory strikes in Iraq (1993) and Bosnia (1995), and have maintained a vigil over Iraq for most of the decade.

In service since 1972, the EA-6B Prowler is the prime electronic warfare aircraft of the US military services, and as such is constantly in demand to support not only the carrier deployments, but also USAF operations.

The F/A-18C forms the backbone of the US Navy's aerial offensive capability. The Hornet is due to increase its prominence in the service when the F/A-18F replaces the F-14 Tomcat. This example serves with VFA-82, based at Cecil Field, but due to move to MCAS Beaufort in March 2000.

US NAVAL AVIATION ORGANISATION

US Naval Aviation is not a separate organisation, despite its strong identity, but is a fundamental part of the US Navy, and is operationally integrated into the full spectrum of naval operations.

THE CHIEF OF NAVAL OPERATIONS

The Chief of Naval Operations (CNO) is the Department of the Navy's agent for training and equipping naval forces, including Naval Aviation, and making them available for operational control by joint commanders to deploy in time of war or crisis or for routine peacetime operations. The CNO's staff (OPNAV), headquartered in the Pentagon in Arlington, Va., includes a Director of Air Warfare who co-ordinates air warfare and strike warfare policy and requirements.

FLEETS

Most operational Naval Aviation forces are assigned to the US Atlantic Fleet or the US Pacific Fleet, which provide the forces used by joint commanders to execute US defence policy. An echelon below, five numbered fleets have both operational and administrative roles.

The Pacific Fleet includes the US Third Fleet, the US Fifth Fleet and the US Seventh Fleet. The Third Fleet conducts operations in the eastern Pacific, and is available to serve as the naval component commander for a joint operation. The Fifth Fleet, created during the mid-1990s to command the forces that comprised the Middle East Force, operates in the Persian Gulf and northwest Indian Ocean. The Seventh Fleet operates in the Western Pacific and most of the Indian Ocean.

The US Second Fleet (Norfolk, Va.) operates in the Atlantic, with a NATO role as Striking Force Atlantic. An adjunct to the Second Fleet, the Western Hemisphere Group (Mayport, Fla.) is the naval component commander for the US Southern Command, which deploys ships to the Caribbean Sea, particularly in counter-drug operations. The Second Fleet also trains forces for the US Sixth Fleet (Gaeta, Italy), permanently deployed to the Mediterranean Sea, with a NATO role as Striking Force Southern Europe.

Above: With the retirement of the F-14 Tomcat, the fast jets operated from the carriers will be limited to the F/A-18 Hornet and remaining EA-6B Prowlers.

Below: An SH-60F Seahawk is seen over the (now decommissioned) 'Forrestal'-class aircraft-carrier USS Saratoga (CV-60).

TYPE COMMANDERS

Virtually all naval aircraft and units are grouped administratively under a type commander, a flag officer with responsibility for administration, training and maintenance support. Operational aircraft, aircraft-carriers and units assigned to the Atlantic and Pacific Fleets are administered, respectively, by Commander Naval Air Force, US Atlantic Fleet (COMNAVAIRLANT, based at NAS Norfolk, Va.), and Commander Naval Air Force, US Pacific Fleet (COMNAVAIRPAC, based at North Island, Calif.). Training aircraft and units are the responsibility of the Chief of Naval Air Training (CNATRA). Commander Naval Air Reserve Force (CNARF) administers the Naval Air Reserve's aircraft and units (described separately below). The Navy's test aircraft fleet falls under the Commander Naval Air Systems Command (NAVAIR, or NASC).

Additionally, there are commands overseas that act as agents for type commanders which provide similar support (under COMNAVAIRLANT or COMNAVAIRPAC) to aircraft stationed at or deployed to forward bases. Commander Fleet Air, Western Pacific (NAF Atsugi, Japan), Commander Fleet Air, Mediterranean (NSA Naples, Italy), and Commander Fleet Air, Keflavik (NAS Keflavik, Iceland), provide such support, and some have a few units directly assigned.

NAVAL AIR STATIONS AND FACILITIES

The Navy's aviation infrastructure consists of naval air stations (NASs), naval air weapons stations (NAWSs), and naval air facilities (NAFs). Air stations host the wings and squadrons assigned to them and provide airfield services and personnel support, but do not exercise operational control over the tenant units. Most bases have UC-12 station liaison aircraft (a few have other types such as UP-3, VP-3, RC-12 or CT-39), and many have HH-1 or UH-3 search and rescue helicopters assigned. Some air stations operated by the Naval Air Reserve that host units of other services or the National Guard are designated NAS-JRBs (for joint reserve base).

The airfields at Point Mugu and China Lake are NAWSs; they are under the command of the Naval Air Warfare Center and host aircraft involved in weapons testing (see below).

Naval air facilities are in some cases autonomous airfields, but many are tenant activities hosted by airfields operated by other US services or the air arms of other nations.

Some Navy airfields are operated as bases designated naval stations (NSs), such as Mayport, Fla., or Roosevelt Roads, Puerto Rico; or naval support activities (NSAs), such as Naples, Italy or Memphis, Tenn. Some Naval Aviation units are hosted by Air Force bases or federal airfields.

BASE FLIGHTS

Versions of the twin-turboprop Beech C-12 Huron are used extensively for base flight support aircraft, with a few specialised variants assigned the additional duty of range clearance and support. The UC-12B is the most numerous variant, most being assigned to Stateside bases. The UC-12F is used at naval air facilities in Japan, while the UC-12M is used for base flights in Europe (and at NAS Norfolk, Virginia for type training). Two RC-12Fs (converted from UC-12Fs) support the Pacific Missile Range Facility in Hawaii, while two new-build RC-12Ms assigned to NS Roosevelt Roads, Puerto Rico support the Atlantic Fleet Weapons Training Facility. Excess UC-12Bs are being converted to TC-12Bs to augment the T-44A multi-engined training track with Training Air Wing 4.

CARRIER BATTLE GROUPS

The centrepiece of US naval striking power is the aircraft-carrier battle group (CVBG). Each aircraft-

The Naval Air Weapons Station at China Lake assigns its base rescue helicopters to the based Naval Weapons Test Squadron. Three HH-1Ns are used for localised emergency evacuation.

The USN acquired 49 UC-12Bs, 12 UC-12Fs and 12 UC-12Ms for service with various base flights and Naval Air Facilities (NAF). This UC-12M is one of three on the strength of NAF Mildenhall, UK.

carrier is assigned to a battle group, comprised of the carrier, cruisers, destroyers, frigates, submarines and supply ships. The carrier serves as a base for the assigned carrier air wing. Battle groups typically deploy for six months to the Mediterranean, Persian Gulf or Western Pacific. An intervening year is occupied by rest and work-up cycles. Normally, two or three CVBGs are on station overseas at any given time. Battle group commanders are flag officers who are either carrier group commanders or cruiser-destroyer group commanders. The Atlantic and Pacific Fleets each employ four carrier group commanders (three for deployment and one for work-ups) and three cruiser-destroyer group commanders (all deployable). One group, Carrier Group 5, is stationed in Yokosuka, Japan and deploys onboard USS *Kitty Hawk*. Battle groups are named after the carrier around which they are organised.

AIRCRAFT-CARRIERS

The Navy's 12 aircraft-carriers, around which battle groups are deployed, are the capital ships that launch the aircraft of the carrier air wings into combat and other operations. The Navy has staunchly maintained that a minimum force of 12 carriers is necessary to the national defence. Of the 12 carriers in service, nine are nuclear-powered (CVNs, eight 'Nimitz'-class and one 'Enterprise'-class), and three are conventionally-powered (CVs). One of the conventionally-powered carriers, USS *John F. Kennedy* (CV-67) is part of the Naval Reserve Force and is based on the East Coast. This ship was intended to be a dedicated training carrier, replacing USS *Forrestal* (AVT-59) and available for use as a deployable CV, but has in fact been retained as a CV in the overseas deployment cycle. The other two CVs – USS *Kitty Hawk* (CV-63) and USS *Constellation* (CV-64) – are assigned to the Pacific Fleet. In mid-1998, *Kitty Hawk* replaced USS *Independence* (CV-62) as the carrier forward-deployed to Japan. (*Independence* was decommissioned in September 1998.) Four CVNs each are based on the East and West Coasts. USS *Harry S. Truman* (CVN-75) is the newest, having been commissioned in mid-1998. A ninth CVN, *Ronald Reagan* (CVN-76), is under construction, and a follow-on carrier, the CVN-77 (a transitional 'Nimitz'-class ship) has funding. The CVX (a new design) has been abandoned due to financial constraints.

Several carriers have been decommissioned during the 1990s, including *Lexington* (AVT-16), *Midway* (CV-41), *Coral Sea* (CV-43), and the entire 'Forrestal' class comprising *Forrestal* (AVT-59), *Saratoga* (CV-60), *Ranger* (CV-61) and *Independence* (CV-62).

Carriers normally deploy overseas to the Mediterranean, Persian Gulf, and Western Pacific with their battle groups for six months. The home cycles combine work-ups with upkeep periods. Normally, two carriers are undergoing extensive overhauls at any one time.

CARRIER AIR WINGS

The carrier air wing (CVW) projects the major conventional striking power of the Navy. Each wing includes a finely tuned balance of strike and support aircraft designed to maximise the operational effectiveness of the aircraft-carrier. The carrier air wing commander and his staff not only conduct the affairs of the wing but advise the battle group commander on strike warfare doctrine and tactics.

Although the Navy operates 12 aircraft-carriers, the number of CVWs is set at 10, with one reserve CVW (described separately). With two carriers normally in long-term overhaul, the 10 wings are fully occupied with operational commitment, normally deploying for six months in each 18-month cycle.

CVWs are assigned five each to the Atlantic and Pacific Fleets. One Pacific Fleet wing, CVW-5, is permanently based at NAF Atsugi, Japan and deploys onboard USS *Kitty Hawk* (CV-63).

Squadron assignment to CVWs is more standardised now than at any time since before World War II, in terms of types of squadrons, detachments and aircraft assigned. The only significant anomaly is the assignment in two CVWs (7 and 8) of two F-14 fighter squadrons, with one of the VF units in each CVW displacing an F/A-18 Hornet strike fighter squadron. As the F-14 squadrons transition to F/A-18F Super Hornets, this anomaly will disappear. Four CVWs each include one Marine Corps fighter-attack (VMFA) F/A-18 Hornet squadron in place of a Navy strike fighter (VFA) squadron. For several years, until recently, one Marine tactical electronic warfare

F-14A Tomcats of VF-102 are parked on the deck while a VFA-82 F/A-18C approaches the wires. The limited number of Tomcat airframes means only two air wings have two Tomcat-equipped VF squadrons. With this exception, the retirement over the last 10 years of types such as the EA-3B, A-6E, KA-6D, A-7E and RF-4B has significantly reduced the variations in the make-up of each carrier air wing.

(VMAQ) squadron deployed with one CVW in place of a Navy tactical electronic warfare (VAQ, now called electronic attack) squadron.

TYPE WINGS

For administrative convenience, squadrons flying the same type of aircraft in the same role are grouped into type wings and are usually assigned together at the same base. When deployed on aircraft-carriers or to overseas bases, they are assigned to an operational wing, such as a carrier wing. The type wing commander provides administrative, training and maintenance support to the squadrons in the wing. The fleet readiness squadron (FRS) is also normally assigned to the type wings (see below).

Generally, the Atlantic and Pacific Fleets each include one of each category of type wing. For some aircraft types, such as the EA-6B, E-6 and F-14, the

CVN-72 USS Abraham Lincoln, a 'Nimitz'-class nuclear-powered aircraft-carrier, is 'home away from home' for 5,500 sailors and Marines, and carries up to 80 combat and support aircraft. Displacing nearly 100,000 tons and with a flight deck covering 4.5 acres, the warship's home port is currently Everett, Wash.

Typical CVW Composition		
Squadron type	**Aircraft type**	**No. of aircraft**
VF	F-14A/B/D	14
VFA	F/A-18C	12
VFA	F/A-18C	12
VFA or VMFA	F/A-18C	12
VAW	E-2C	4
HS	SH-60F/HH-60H	6/2
VAQ	EA-6B	4
VS	S-3B	8
VQ Detachment	ES-3A	2
VRC Detachment	C-2A	2
(from 1999 no CVWs will have a VQ detachment)		

Left: Aircraft-carriers give the United States the ability to project power throughout the world. The US Navy has eight 'Nimitz'-class carriers. Two more of the class are to be constructed, the final example being a transitional carrier between the 'Nimitz'-class and the (cancelled) CVX.

Carrier Air Wing Assignments

Listed below are the Navy's carrier air wings (CVWs) – 10 active and one reserve – and the assigned squadrons and detachments, as well as their aircraft types. Carrier assignments, which change frequently, are not listed.

US ATLANTIC FLEET

CVW-1

VF-102	F-14B
VMFA-251	F/A-18C
VFA-82	F/A-18C
VFA-86	F/A-18C
VAW-123	E-2C
HS-11	SH-60F, HH-60H
VAQ-137	EA-6B
VS-32	S-3B
VQ-6 Det	ES-3A
VRC-40 Det	C-2A

CVW-3

VF-32	F-14B
VMFA-312	F/A-18C
VFA-37	F/A-18C
VFA-105	F/A-18C
VAW-126	E-2C
HS-7	SH-60F, HH-60H
VAQ-130	EA-6B
VS-22	S-3B
VQ-6 Det	ES-3A
VRC-40 Det	C-2A

CVW-7

VF-11	F-14B
VF-143	F-14B
VFA-131	F/A-18C
VFA-136	F/A-18C
VAW-121	E-2C
HS-5	SH-60F, E-2C
VAQ-140	EA-6B
VS-31	S-3B
VQ-6 Det	ES-3A
VRC-40 Det	C-2A

CVW-8

VF-14	F-14A
VF-41	F-14A
VFA-15	F/A-18C
VFA-87	F/A-18C
VAW-124	E-2C
HS-3	SH-60F, HH-60H
VAQ-141	EA-6B
VS-24	S-3B
VQ-6 Det	ES-3A
VRC-40 Det	C-2A

CVW-17

VF-103	F-14B
VFA-34	F/A-18C
VFA-81	F/A-18C
VFA-83	F/A-18C
VAW-125	E-2C
HS-15	SH-60F, HH-60H
VAQ-132	EA-6B
VS-30	S-3B
VQ-6 Det	ES-3A
VRC-40 Det	C-2A

NAVAL RESERVE FORCE

CVWR-20

VMFA-142	F/A-18A
VFA-201	F/A-18A
VFA-203	F/A-18A
VFA-204	F/A-18A
VAW-78	E-2C
HS-75	SH-3H, UH-3H
VAQ-209	EA-6B

Note: Early in 1999, VF-201 was redesignated VFA-201 upon transition to the F/A-18A Hornet strike fighter from the F-14A . Three shore-based squadrons, VFC-12, VFC-13 and VAW-77, are assigned for administrative purposes only and do not deploy with the wing.

fleets are both supported by only one type wing. In the case of patrol wings, each fleet is structured differently. The Atlantic Fleet has a functional-wing flag-level staff, Commander Patrol Wings, US Atlantic Fleet, who supervises the two type wings, Patrol Wings 5 and 11, as well as the P-3 FRS, VP-30. Commander Patrol Wings, US Pacific Fleet, supervises Patrol Wing 10, but also directly supervises the patrol squadrons based in Hawaii since Patrol Wing 2 was disbanded. Patrol Wing 1 in Japan supervises only squadrons and detachments deployed in its operational area.

US PACIFIC FLEET

CVW-2

VF-2	F-14D
VMFA-323	F/A-18C
VFA-137	F/A-18C
VFA-151	F/A-18C
VAW-116	E-2C
HS-2	SH-60F, HH-60H
VAQ-131	EA-6B
VS-38	S-3B
VQ-5 Det	ES-3A
VRC-30 Det	C-2A

CVW-5

VF-154	F-14A
VFA-27	F/A-18C
VFA-192	F/A-18C
VFA-195	F/A-18C
VAW-115	E-2C
HS-14	SH-60F, HH-60H
VAQ-136	EA-6B
VS-21	S-3B
VQ-5 Det	ES-3A
VRC-30 Det	C-2A

CVW-9

VF-211	F-14A
VMFA-314	F/A-18C
VFA-146	F/A-18C
VFA-147	F/A-18C
VAW-112	E-2C
HS-8	SH-60F, HH-60H
VAQ-138	EA-6B
VS-33	S-3B
VQ-5 Det	ES-3A
VRC-30 Det	C-2A

CVW-11

VF-213	F-14D
VFA-22	F/A-18C
VFA-94	F/A-18C
VFA-97	F/A-18C
VAW-117	E-2C
HS-6	SH-60F, HH-60H
VAQ-135	EA-6B
VS-29	S-3B
VQ-5 Det	ES-3A
VRC-30 Det	C-2A

CVW-14

VF-31	F-14D
VFA-25	F/A-18C
VFA-113	F/A-18C
VFA-115	F/A-18C
VAW-113	E-2C
HS-4	SH-60F, HH-60H
VAQ-139	EA-6B
VS-35	S-3B
VQ-5 Det	ES-3A
VRC-30 Det	C-2A

Controlled by the Commander, Patrol Wings, US Atlantic Fleet, the P-3 Orion fleet readiness squadron is VP-30, based at NAS Jacksonville, Fla.

Naval air test wings perform the functions of type wings for the test aircraft fleet, as training air wings do for training aircraft and units.

FLEET READINESS SQUADRONS

Often called fleet replacement squadrons or referred to by the obsolete replacement air group term, 'RAGs', the fleet readiness squadrons (FRSs) provide training for new aviation and maintenance personnel in the type of aircraft operated by fleet squadrons. The FRS also assists operational squadrons with conducting tactical simulator training during work-up cycles. Many aircraft 'communities' have only one FRS that provides personnel to both fleets; others have one FRS on each coast. In some cases, such as for MH-53E and HH-1N helicopters, Marine training squadrons train Navy personnel; in others, such as for EA-6Bs, Navy FRSs train Marines. In yet other cases, FRS training is conducted by an air station base fight (as NAS Norfolk does for C-12 crews), a training support unit (as for E-6 pilots) or the Air Force (such as for the C-130).

FRSs perform an important function in promoting fleet-wide standardisation of operating procedures and tactics among crews flying a certain aircraft.

STRIKE FIGHTERS

The primary striking power of each Navy carrier battle group lies in its 50 strike aircraft, fighters adept in air-to-ground as well as air superiority missions. The Boeing F/A-18 Hornet strike fighter, in its second decade of service, is numerically the Navy's primary combat aircraft. The Grumman F-14 Tomcat, in its third decade as the Navy's front-line interceptor, has been modified during the 1990s as a capable strike aircraft. Replacements for both of these ageing (but still first-rate) aircraft are in production (Boeing F/A-18E/F Super Hornet) or in the design phase (Joint Strike Fighter).

The F-14 Tomcat, flown by the Navy's fighter (VF) squadrons and designed originally to counter the Soviet bomber and cruise missile threat, remains the most capable naval interceptor in the world, despite the relatively recent addition of strike missions to its roles. Armed with the AIM-54C Phoenix air-to-air missile, the F-14 can track 24 targets and attack six at ranges of more than 100 miles (160 km). The Tomcat also is armed with AIM-9 Sidewinder and AIM-7 Sparrow air-to-air missiles and a 20-mm cannon for shorter ranges. (Plans to arm the F-14 with the AIM-120 AMRAAM have been cancelled.)

The F-14 also performs photo-reconnaissance

Operational conversion to the Hornet is undertaken on courses run by VFA-106 'Gladiators' for Atlantic fleet units, using examples of all the versions of the first-generation Hornet. 'AD'/342 is one of the unit's twin-stick F/A-18Bs.

The F-14 Tomcat has come a long way from the problematic aircraft that first entered squadron service in the early 1970s. Today's aircraft are not just air superiority machines, having acquired the reconnaissance roles as well as a limited strike capability. This example, an F-14D(R), flies with VF-2.

Above: Thundering away from the photographer, this F-14A-135-GR of VF-211 is seen during the squadron's cruise on USS Nimitz as part of CVW-9.

Below: Wearing an Iranian-style camouflage scheme, this F-14A serves with the Naval Strike Air Warfare Center at Fallon, Nevada.

with the TARPS (Tactical Air Reconnaissance Pod System). A digital system – TARPS-DI – entered service in 1996 and has the capability of down-linking near-real-time imagery for immediate threat and battle-damage assessment. A further development – TARPS-CD – began testing in 1998; this improvement includes real-time electro-optical step-framing imagery.

The Navy is accelerating the retirement of the initial long-serving Tomcat version, the F-14A, which is still powered by troublesome TF30 engines. The F-14B (initially designated F-14A+) featured F110 engines that resisted compressor stalls and markedly improved performance, increasing thrust by 30 per cent, thereby eliminating the need for after-burner on launch and greatly increasing engine life and time on station. Some F-14As were converted to F-14Bs to augment the production F-14Bs, while a few others were modified into NF-14As and NF-14Bs for test work. F-14As participated in combat during the 1980s against Libya, downing two Su-22 and two MiG-23 aircraft. F-14As and F-14Bs also participated in Operation Desert Storm; one Iraqi Mi-8 helicopter was downed by an F-14A.

The last production version of the F-14, the F-14D Super Tomcat, featured the F110 engine, the jam-resistant APG-71 radar, the Joint Tactical Information Distribution System, and the dual Television Camera Set/Infra-Red Search and Track sensor. Some F-14As were remanufactured into F-14Ds. The Super Tomcat entered service too late to participate in the Gulf War. A few F-14Ds have been converted to NF-14Ds for test work.

The Navy has steadily upgraded its F-14 fleet during the 1990s to keep it effective until its scheduled retirement in 2008. The most significant change was the addition of an air-to-ground attack capability latent in the F-14 since its initial design. The retire-

VX-4 flew an all-black F-4 Phantom for years, and continued the tradition when the unit received Tomcats. Today the aircraft serves with the VX-9 detachment at NAWS Point Mugu, California. In the distance are seven QF-4 Phantom drones.

ment of the A-6 Intruder, the cancellation of the A-12, and the Navy's requirement for 50 strike aircraft in each carrier air wing led to the fruition of an F-14 strike capability. F-14s were fitted with the capability of dropping 'dumb' bombs; this was expanded to the capability to drop laser-guided bombs. This air-to-ground capability was put into use in air strikes in Bosnia during 1995. Installation of the LANTIRN (low-altitude navigation and targeting, infra-red and night) system, first deployed by VF-103 in June 1996, gave the F-14 the capability to laser-designate targets for its own bombs. LANTIRN-equipped F-14s also received upgrades including the ALR-67 radar warning system, chaff dispensers, night-vision capability, and the GPS. Installation of the GPS began in 1995 on the F-14D and in 1996 on the F-14As and F-14Bs.

The first aircraft to be remanufactured under the F-14A/B Upgrade programme was delivered in 1997. These aircraft received the F110 engine, a major computer upgrade, digital avionics, and structural and survivability enhancements to make them comparable to the F-14D. A Navy plan to mix F-14B Upgrade and F-14D aircraft in the same squadrons was cancelled in 1997.

The F/A-18 Hornet replaced the A-6, A-7 and some F-4s in front-line carrier service. The first production version of the Hornet – the F/A-18A – equips only one active-duty operational squadron, in

The 'World Famous (Pukin) Dogs' (VF-143) fly the F-14B version of the Tomcat, fitted with the F110-GE-400 turbofan. This aircraft carries USS George Washington titles on the wing leading edge.

addition to five reserve squadrons, the two Hornet fleet readiness squadrons (FRSs) and the Naval Flight Demonstration Squadron ('Blue Angels'). The two-seat counterpart of the F/A-18A, the F/A-18B, serves in small numbers in the two Hornet FRSs, one reserve adversary squadron, with test squadrons and with the 'Blue Angels'. The twin-engined F/A-18A, armed with a 20-mm cannon, AIM-9 Sidewinder and AIM-7 Sparrow air-to-air missiles, AGM-84 Harpoon anti-shipping missiles, AGM-88 anti-radiation missiles, and high-explosive and cluster bombs, is designed to be equally adept at fighter and attack missions. Its thrust-to-weight ratio and its digital flight control system give the aircraft exceptional manoeuvrability. The F/A-18A first flew in combat during retaliatory strikes against Libya in 1986.

The F/A-18C is the primary version of the Hornet in front-line carrier service, equipping all but one active-duty strike fighter (VFA) squadron, as well as the two Hornet fleet readiness squadrons (FRSs). The F/A-18C incorporates improved weapons in the form of the AIM-120 AMRAAM air-to-air missile, the AGM-84E SLAM (stand-off land-attack missile) and the infra-red imaging version of the AGM-65 Maverick air-to-ground missile. F/A-18Cs produced from 1989 feature improved night-attack capabilities, including a FLIR pod, a raster HUD, night-vision goggles, cockpit lighting compatible with night-vision goggles, a digital colour moving map and a multi-purpose colour display. Enhanced-performance engines were introduced in production F/A-18Cs during the mid-1990s. The F/A-18C fleet is being upgraded with a tactical FLIR/laser designator pod, ARC-210 radios, GPS, and a cockpit video recorder. Programmed upgrades by 2001 include the AGM-84H SLAM-ER (Stand-off Land-Attack Missile-Expanded Response), AIM-9X version of the Sidewinder, the JDAM (joint direct attack munition),

reconnaissance and FAC roles in addition to its interceptor and strike roles.

The Joint Strike Fighter is envisioned as a 'stealthy' replacement for the Hornet in the second decade of the 21st century. One of four planned variants of the JSF, the Navy's carrier-capable version will be a single-seat, single-engined supersonic fighter with an internal weapons bay and state-of-the-art avionics. Boeing is expected to fly its X-32 concept demonstration aircraft (CDA) in 1999. The Lockheed Martin team, which includes Northrop Grumman and British Aerospace, is producing the X-35 CDA. The winning design will be engineered into a full-scale JSF in the engineering and manufacturing development programme phase to begin in 2001. The JSF is scheduled to enter service in 2008.

FIGHTER SQUADRONS

The number of fighter (VF) squadrons per carrier air wing has changed in most wings during the mid-1990s. Of the 10 CVWs, only two (CVW-7 and CVW-8) retain the traditional pair of F-14 Tomcat-equipped VF squadrons. The remaining eight wings deploy with a single, larger, TARPS-capable, 14-aircraft VF squadron, the second squadron having been replaced by an F/A-18 Hornet-equipped strike fighter squadron. This change is the result of an air wing restructuring that enshrines a 50-strike-aircraft wing, ideally consisting of 36 Hornets and 14 Tomcats (eventually to be replaced by F/A-18F Super Hornets). Eleven active-duty F-14 squadrons (most non-TARPS-equipped) have been disestablished since the end of the Cold War.

The reduction in carrier-based VF squadrons also has resulted in the consolidation of almost all of the Navy's VF squadrons at NAS Oceana, Va. with Fighter Wing, US Atlantic Fleet, commensurate with the transfer of NAS Miramar, Calif., to the Marine Corps. The 11 fleet VF squadrons at Oceana (three with F-14As, five with F-14Bs, and three with F-14Ds) deploy onboard carriers from both coasts. Another squadron, VF-154, is permanently

With 'Fist' displayed on the overwing strake, from VFA-25's 'Fist of the Fleet' name, the colourful tail markings denote this F/A-18C as the 'CAG-bird' (Carrier Air Wing commander's aircraft).

the JSOW (joint stand-off weapon), Link 16, the GPS, an improved tactical FLIR and the ATARS (advanced tactical air reconnaissance system). By 2004, further upgrades will include advanced multi-colour cockpit displays, satellite communications and the Block 6 version of the AGM-88 HARM.

The F/A-18C saw extensive combat in Kuwait and Iraq during 1991 in Operation Desert Storm. On the first day of the campaign, two F/A-18Cs, each carrying four 2,000-lb bombs, shot down two Iraqi MiG-21s and proceeded to deliver their bombs on target.

The two-seat F/A-18D, counterpart to the F/A-18C, equips the two Hornet FRSs and some test squadrons. Unlike the Marine Corps, which uses the F/A-18D as a front-line combat aircraft, the Navy uses the aircraft only for training F/A-18C pilots and for test work.

A nagging shortcoming with the Hornet has been a deficiency in range in certain mission profiles. Expensive range improvements were deferred for higher-priority demands. Further growth potential of the Hornet has been limited by constraints in avionics cooling, electrical capacity, and space. The larger,

longer-range F/A-18E Super Hornet and its two-seat version – the F/A-18F – are expected to rectify the shortcomings of the Hornet.

The F/A-18E/F, which entered low-rate production in September 1997, is the highest-priority programme in US Naval Aviation. The F/A-18E is 4.2 ft (1.28 m) longer than a standard Hornet, and has a wingspan 4.7 ft (1.53 m) greater. The Super Hornet's wing area is 25 per cent larger and its internal fuel capacity is 33 per cent greater – expected to increase range by 41 per cent and endurance by 50 per cent. The Super Hornet features two more wing-mounted stores stations, and can carry the full array of new Navy air-to-ground ordnance, including the JDAM, the JSOW and the AGM-84H SLAM-ER. The Super Hornet's F414 engines give it the ability to land aboard a carrier with a load of expensive 'smart' weapons. The Super Hornet will also be able to carry an aerial refuelling store. The aircraft incorporates low-observable concepts but is not a 'stealth' aircraft.

The Super Hornet is scheduled to become operational in 2001. Navy procurement plans project a production run that will equip each carrier wing with one F/A-18E and one F/A-18F squadron, along with two F/A-18C squadrons that eventually will acquire the Joint Strike Fighter as the Hornets are phased out. The F/A-18F will replace the F-14, and fulfil its

Left: Carrier operations are stressful on both men and machines. At least five Prowlers have been lost due to carrier launch or landing failures.

Below: With the callsign of EAGLE, the 'Vikings' of VAQ-129 are the fleet readiness squadron for the EA-6B community, training not only the Naval squadrons but also Marine and USAF personnel.

Above: A pair of VAQ-140 'Patriots' EA-6Bs peels away from the camera ship. The aircraft display the codes and markings of CVW-7.

forward-deployed to NAF Atsugi, Japan, being the F-14A squadron assigned to CVW-5 onboard USS Kitty Hawk (CV-63). A 13th F-14 squadron, VF-101, is the FRS for the entire F-14 community. This large squadron operates F-14A/B/D versions, and a few T-34C trainers for spotter missions.

Since the end of the Cold War, the VF squadrons have assumed an air-to-ground strike role never before employed by the F-14. Equipped with LANTIRN targeting pods and laser-guided bombs, the F-14s have become a potent strike force. In addition to retaining their TARPS photo-reconnaissance role, the VF units also have assumed a forward air controller role for the carrier air wing. This role will be retained as 10 squadrons transition to the two-seat F/A-18F Super Hornet, a process scheduled to begin in 1999, resulting in the first two F/A-18F squadrons being operational by 2001. The squadrons equipped with the F-14A, including VF-154 in Japan, will be the first to make the transition. Yet to be determined is whether any of the VF squadrons will move to the West Coast after they transition to the F/A-18F, and whether the change to the two-seat Super Hornet will precipitate a redesignation to strike fighter (VFA) squadrons.

Several land-based fighter squadrons (VFs 43, 45 and 126) which performed the air combat manoeuvring adversary role were disbanded during the mid-1990s when this mission was assumed completely by the Naval Air Reserve.

A single fleet readiness squadron performs the training tasks for both the Atlantic and Pacific airborne early warning communities. VAW-120 'Cyclones' also trains crews for the COD role.

STRIKE FIGHTER SQUADRONS

The major striking power of carrier-based Naval Aviation resides in 24 strike fighter (VFA) squadrons, all of which operate the F/A-18C version of the Hornet. The number of VFA squadrons is insufficient to provide three 12-aircraft squadrons in each of the 10 carrier air wings; the deficit is compensated for by two Tomcat-equipped VF squadrons (see above) and four Marine fighter-attack (VMFA) squadrons assigned to carrier duty. Most VFA squadrons are equipped with night-attack-capable F/A-18C(N) versions of the F/A-18C.

The Hornet squadrons have both strike and air superiority roles, using a variety of air-to-air and air-to-ground weapons. The F/A-18E Super Hornet and later the Joint Strike Fighter are scheduled to replace the F/A-18Cs in these squadrons. Each air wing is currently scheduled to operate one F/A-18E squadron.

The VFA squadrons are evenly divided between two strike-fighter wings, one each on the East and West Coasts. Three Pacific Fleet squadrons are stationed at NAF Atsugi, Japan, from where they deploy with CVW-5. The Atlantic Fleet VFA squadrons are scheduled to move to NAS Oceana, Va., and MCAS Beaufort, N.C., starting in late 1998, as NAS Cecil Field is prepared for closure. Each wing includes an FRS for training Hornet pilots and maintenance personnel. VFA-106 and VFA-125 conduct training for the East and West Coasts, respectively, using F/A-18A/B/C/D versions of the Hornet, as well as a few T-34C trainers for spotter missions. These squadrons also train Marine Corps Hornet crews; in reciprocation, the Marine Corps Hornet FRS, VMFAT-101, trains Navy Hornet crews as well.

This 'cross-pollination' promotes high standardisation in the large Hornet community.

The training of F/A-18E/F crews will begin in 1999 after the stand-up of the FRS, VFA-122, at NAS Lemoore, Calif. in January 1999. An FRS is planned eventually for the East Coast at NAS Oceana.

Two VFA squadrons equipped with F/A-18As were disbanded during the last decade and VFA-127, a land-based air combat manoeuvring adversary squadron, was disbanded in 1996.

ELECTRONIC ATTACK

The Navy has operated the Northrop Grumman EA-6B Prowler electronic warfare aircraft since 1972, when the Prowler protected American strike aircraft by jamming North Vietnamese radars. The carrier-capable EA-6B, a variant of the A-6 Intruder attack aircraft and a replacement for the Douglas EKA-3B Skywarrior, is capable of jamming enemy radars and communications, and attacking enemy radars with AGM-88 HARM (high-speed anti-radiation missiles). The EA-6B has been upgraded through the years with the EXCAP (expanded capability) and ICAP (increased capability) I and II upgrades. The ICAP II is now the standard fleet configuration.

The Prowlers are planned for upgrade to the Block 89A configuration, which includes an upgraded AYK-14 mission computer, an inertial navigation system integrated with the GPS and two anti-jam ARC-210 radios. The Block 89A version, which first flew in 1997, is scheduled to enter service in 1999 as the standard configuration for all EA-6Bs. (Wing centre-sections are also being replaced as funds permit to alleviate the airframe limitations caused by stress cracks.) Approval has been given to Northrop Grumman to develop an ICAP III (Increased Capability III) 'Warfighter Upgrade System' configuration, expected to enter service in 2002. ICAP III upgrades include installation of a new receiver to replace the ALQ-99, improved low-band jamming capability, and integration of sensor information from other platforms.

The addition of the AGM-88 HARM during the mid-1980s gave the EA-6B a weapon to supplement its impressive jamming capability, and in 1998 the Navy's tactical electronic warfare squadrons were redesignated electronic attack squadrons. During the mid-1990s, the EA-6B was chosen to be the jamming platform to support Air Force as well as Navy and Marine Corps forces, finally replacing the EF-111A Raven in May 1998. As such, the EA-6B, in addition to its carrier operations, routinely deploys overseas in Navy expeditionary squadrons in support of Air Force operations. Navy EA-6Bs have been deployed to Japan and frequently to sites in Italy, Turkey and Saudi Arabia in support of joint operations.

The EA-6B has been out of production since the early 1990s, and the high level of defence commitments has strained the availability of this ageing

aircraft, which is intended to serve until 2015. No replacement for the EA-6B has been identified; an electronic warfare version of the F/A-18F Super Hornet (EF/A-18G) has been mentioned as a possible replacement.

The Navy operates 14 operational electronic attack (VAQ) squadrons (redesignated in March 1998 from 'tactical electronic warfare squadrons'). These units, all based with one electronic attack wing at NAS Whidbey Island, Wash., fly the EA-6B Prowler carrier-based electronic warfare aircraft, capable of suppressing enemy defences with electronic jamming and with AGM-88 high-speed anti-radiation missiles (HARMs). Ten four-ship squadrons are assigned one each to the 10 carrier air wings. The remaining four squadrons were organised in the mid to late 1990s as expeditionary squadrons to replace the Air Force EF-111A Ravens in USAF expeditionary wings and include some Air Force personnel. They deployed to the Middle East, replacing the EF-111A in April 1998, and to Japan to cover gaps (caused by commitments in Bosnia and the Middle East) in Marine Corps EA-6B deployments to MCAS Iwakuni, Japan.

FRS training of Navy and Marine Corps EA-6B crews and maintenance personnel is conducted by VAQ-129 at Whidbey Island.

Three fleet EA-6B squadrons (VAQs 133, 137, and 142) were disestablished after the end of the Cold War. A fourth, VAQ-134, was reduced to cadre status before standing up as an expeditionary squadron. Four new squadrons (three numbered the same as the three disbanded squadrons and one numbered VAQ-128, after the former A-6 FRS, VA-128) were established, three being expeditionary squadrons. The readiness of the EA-6B force has been severely stressed in the late 1990s by aircraft shortages aggravated by fatigued airframes and aircraft modification programmes.

AIRBORNE EARLY WARNING

Airborne early warning for carrier battle groups is performed by the Northrop Grumman E-2C Hawkeye. The twin-turboprop Hawkeye, with its distinctive rotating dorsal radome, first flew in 1961; the E-2C version first flew in 1971. The current variant's radar can detect targets anywhere in a 6 million-cubic mile surveillance envelope and simultaneously track 2,000 targets. The five-man E-2C crew can control more than 20 airborne intercepts simultaneously. The Hawkeye also is used in the drug-interdiction role to detect drug-running aircraft.

The E-2C has been in service and in production so long that aircraft have been retired at the end of their fatigue lives even as new E-2Cs roll off the production line. The E-2C fleet has been progressively upgraded with improvements in mission avionics and communications equipment, and has been produced in baseline, Group 0, Group I, Group II (with T56-A-427 improved engines), and now Hawkeye 2000 versions. The Hawkeye 2000 development aircraft first flew in early 1998. This version includes a mis-

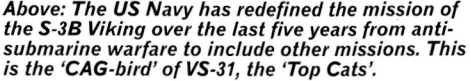

Above: The US Navy has redefined the mission of the S-3B Viking over the last five years from anti-submarine warfare to include other missions. This is the 'CAG-bird' of VS-31, the 'Top Cats'.

Left: A pair of VS-38 'Red Griffins' Vikings is seen on the deck of USS Constellation. The far example displays a bright red unit badge on an otherwise grey aircraft.

Below: The Pacific Fleet Viking FRS is VS-41, the 'Shamrocks', based at NAS North Island, Calif.

sion computer upgrade, advanced workstations, integrated satellite communications capability, and CEC (co-operative engagement capability).

Group I variants in service will be retrofitted to Group II standard and later to the Hawkeye 2000 configuration. By 2010 all production Group II aircraft also will be upgraded to the Hawkeye 2000 standard. All Group 0 aircraft will be retired by 2004.

A few E-2Cs have been stripped of mission equipment and converted to TE-2C pilot trainers. A variant of the Common Support Aircraft (CSA) is programmed to replace the E-2C.

Each carrier air wing includes one carrier airborne early warning (VAW) squadron equipped with four E-2C Hawkeye radar surveillance aircraft. These aircraft provide radar early warning of enemy aircraft, guide interceptors to counter them, and aid the battle group commander in controlling strikes at sea or ashore. Ten fleet VAW squadrons are evenly divided in two wings between the Atlantic and Pacific Fleets. One Pacific Fleet unit, VAW-115, is stationed at NAF Atsugi, Japan with CVW-5; the other four, based at MCAS Miramar, Calif., are scheduled to move, most likely to NAWS Pt Mugu, Calif. The E-2C FRS, VAW-120, conducts crew training at NAS Norfolk, Va., where the Atlantic VAW units are based. VAW-120 operates E-2C and TE-2C aircraft, as well as C-2A Greyhound carrier onboard delivery (COD) aircraft to train C-2A crews.

SEA CONTROL

The carrier-based sea control mission is carried out by the Lockheed S-3B Viking, a twin-engined four-seat jet originally designed in the early 1970s with a sophisticated sensor suite for anti-submarine warfare (ASW). The S-3B, which incorporates anti-surface warfare upgrades such as the APS-137 inverse synthetic aperture radar and the AGM-84 Harpoon missile, replaced the S-3A in the early 1990s. The demise of the Soviet Union and the increasing dominance of littoral warfare led to decreased emphasis on ASW and more emphasis on anti-surface warfare and land-attack missions. S-3s can strike lightly-defended targets with bombs and rockets, and can lay sea mines. The S-3B and ES-3A also are the only carrier-based aircraft configured as aerial tankers, a role vital to carrier operations.

The S-3 fleet, intended to serve through 2015, is going through a service-life assessment programme (SLAP) to determine requirements to formulate an effective service-life extension programme (SLEP). Several upgrades are being installed on Vikings,

Left: The standard carrier-based airborne early warning aircraft is the E-2C. This is a 'Screwtops' (VAW-123) aircraft, based at Norfolk NAS, Va., but deployed with CVW-1 of the Atlantic fleet.

Below: Displaying a large unit badge just behind the cockpit, this E-2C of the Pacific Fleet's 'Sun Kings' (VAW-116) is seen wearing titles received during a cruise on USS Constellation with CVW-2.

Above: The Orion fleet readiness unit is VP-30 'Pros Nest', based at NAS Jacksonville, Florida. The squadron has retired its TP-3As and instead uses the 'older' P-3Cs, such as 158935 'LL/43'.

Right: The end of the Cold War reduced the need to hunt vast fleets of submarines, and other duties have been found for the P-3C, including overland roles.

Land-based long-range electronic reconnaissance is undertaken by the EP-3E Aries. The original 12 aircraft, conversions of P-3As and EP-3Bs, were replaced by 12 converted P-3Cs. This is a second-generation aircraft of VQ-2.

including the GPS, carrier aircraft inertial navigation system II (CAINS II), new tactical displays, computer memory, SATCOM equipment and improved radios.

The S-3B is planned for replacement by a variant of the Common Support Aircraft (CSA). The US-3A carrier onboard delivery (COD) version was withdrawn from service in the early 1990s. The ES-3A is described separately below.

Each carrier air wing includes one sea control (VS) squadron (formerly air anti-submarine squadron) equipped with eight S-3B Vikings. VS squadrons perform anti-submarine, anti-shipping, mine-laying, and surveillance missions for the carrier battle group. An important secondary mission is aerial refuelling; with the 1997 retirement of the KA-6 Intruder, the S-3B and ES-3A are the only US carrier-based tankers available. For this reason, the number of S-3Bs per squadron, which had declined from 10 to six, has been increased to eight.

Fleet VS squadrons are evenly divided in two wings between the Atlantic and Pacific Fleets. The Atlantic fleet squadrons moved from NAS Cecil Field, Fla., to nearby NAS Jacksonville during late 1997. One Pacific fleet unit, VS-21, is based at NAF Atsugi, Japan, for duty with CVW-5. S-3 FRS training is handled by an additional unit, VS-41 at North Island, Calif. (The Atlantic Fleet FRS, VS-27, was disbanded in 1994.) All S-3A versions have been withdrawn from service.

MARITIME PATROL

The Navy has long operated the land-based Lockheed P-3 Orion maritime patrol aircraft for anti-submarine and anti-shipping, as well as for surveillance, reconnaissance, mine-laying, drug-interdiction, and search and rescue missions. Although developed to counter the Soviet submarine threat, the maritime patrol force, greatly reduced since the end of the Cold War, finds itself in great demand in the littoral warfare environment of the 1990s, and even can be found flying missions over Bosnia and deep in the African continent.

The P-3C, powered by four turboprop engines and

operated by a crew of 12, is armed with Mk 46 and Mk 50 torpedoes, mines, rockets, high-explosive and cluster bombs, AGM-84 Harpoon and AGM-65 Maverick anti-shipping missiles, and AGM-84E SLAM land-attack missiles. The aircraft has a sophisticated sensor suite including the UYS-1 acoustic sonobuoy processor and ALR-66 ESM system, plus MAD gear, an infra-red detection set and a search radar. Some aircraft are equipped with the APS-137 inverse synthetic aperture radar, which can display an image of its target.

The current front-line version, the P-3C, first entered in service in 1969 and has been upgraded frequently ever since, through Updates I, II, II.5 and III, and is currently undergoing several programmes to extend the life of the airframe and improve its mission suite and armament. The Boeing Update IV package, with a vastly improved mission suite and new engines, was cancelled in 1992, a victim of post-Cold War budget cutbacks. The Navy is gradually upgrading most P-3Cs to an Update III Common Configuration that will be the fleet standard.

A roll-on counter-drug upgrade (CDU) sensor package, which included Cluster Ranger high-powered optical sensors, became available in the early 1990s and proved useful in NATO peace enforcement over Bosnia. The Anti-surface Improvement Program (AIP), planned for 146 P-3Cs, includes enhancements in anti-surface weapons (Maverick, SLAM and SLAM-ER), command, control, communications and intelligence (C3I), over-the-horizon targeting and survivability. The P-3C AIP development aircraft began testing in December 1996 and entered service in March 1998.

The Sustained Readiness Program (SRP) is designed to keep the P-3C airframe in service until 2015. A SLAP is being conducted to determine requirements for a kit to install in P-3Cs going through a SLEP. The first P-3C to go through SRP returned to service in February 1998.

All active and reserve patrol squadrons are equipped with the P-3C. A few P-3Bs and P-3Cs in service with patrol squadron special projects units are modified with specialised reconnaissance equipment. A single surviving EP-3J is used in fleet exercises to simulate enemy electronic threats. A few UP-3A and UP-3B utility transports remain in service, as do a few VP-3A executive transports. One UP-3A, one

NP-3C, and a dozen NP-3Ds (former RP-3A, RP-3D, and EP-3B versions) are used for RDT&E and oceanographic survey. The TP-3A cockpit crew trainers were withdrawn from service in early 1998. The EP-3E electronic reconnaissance version is described separately below.

The Multipurpose Maritime Aircraft (MMA) programme is in early stages of conception; the MMA is intended to replace the P-3, EP-3 and C-130 in naval service.

The Navy maintains a force of 12 active-duty operational patrol (VP) squadrons, half the number maintained at the height of the Cold War, 12 squadrons having been disestablished since 1990. The squadrons are evenly divided between East and West Coasts. Six squadrons under Commander Patrol Wings, US Atlantic Fleet (Norfolk, Va.), are based, three each, at NAS Brunswick, Maine (Patrol Wing 5) and NAS Jacksonville, Fla. (Patrol Wing 11). The six West Coast squadrons, under Commander Patrol Wings, US Pacific Fleet, are based, three each, at NAS Whidbey Island, Wash. (Patrol Wing 10) and NAS Barbers Point, Hawaii (directly under Patrol Wings, US Pacific Fleet, to move to MCAF Kaneohe Bay, Hawaii by 2 July 1999).

All VP squadrons operate P-3C Orion patrol aircraft. Since the late 1980s, the Navy has been upgrading older P-3Cs to the Update III configuration. As this process is not yet complete, many squadrons operate a mix of Update II, II.5, and III versions, but shuffle aircraft to be equipped with nine Update III aircraft (10, in the case of Hawaii-based squadrons) when deployed.

All MPA crew and maintenance training is conducted by the FRS, VP-30, at Jacksonville. VP-30, which reports directly to Commander Patrol Wings Atlantic, operates P-3Cs, having retired its remaining TP-3A versions in early 1998. The squadron also maintains a VIP transport detachment comprised of three VP-3A aircraft, assigned to support the Chief of Naval Operations and the Commander in Chief, US Atlantic Fleet.

Four VP squadrons are deployed at any given time, two from each fleet. The Pacific squadrons deploy for six months to Misawa AB, Japan, or to Diego Garcia, B.I.O.T. These squadrons both maintain detachments in Kadena AB, Okinawa, Japan, and the Diego Garcia squadron keeps a detachment in the Middle East, normally at Al Masirah, Oman, in support of Fifth Fleet operations in the Persian Gulf. One Atlantic Fleet squadron is deployed to NAS Sigonella, Sicily, recently being engaged in operations in the Adriatic and over Bosnia. A second Atlantic Fleet squadron, responsible for covering areas of the Atlantic, is split between NAS Keflavik, Iceland, and NS Roosevelt Roads, Puerto Rico, from where it supports drug-interdiction operations in the Caribbean. Routine MPA operations are no longer conducted from Bermuda, Lajes (Azores) or Rota (Spain).

Patrol squadrons primarily engage in anti-submarine and anti-shipping missions, as well as reconnaissance, shipping surveillance, rescue, logistics, range support, threat simulation, and communica-

tions relay. Improved optical sensors have made the P-3C useful in overland surveillance in Bosnia and central Africa. Recent advancements, embodied in the Anti-surface Improvement Program (AIP) in the form of long-range optics, improved command and control, and weaponry (such as the SLAM-ER missile), will make the MPA force potent in a stand-off land-attack role.

VP SPECIAL PROJECTS UNITS

For three decades the Navy has fielded a small number of specialised reconnaissance versions of the P-3, which initially were operated by special projects departments of selected VP squadrons. In 1982, however, the departments became separate units, designated Patrol Squadron Special Projects Units (VPUs), and included VPU-1 at NAS Brunswick, Maine and VPU-2 at NAS Barbers Point, Hawaii (to move to MCAF Kaneohe Bay, Hawaii by July 1999). In 1996, these units became full commands, although they remain designated as units instead of squadrons. The VPUs fly modified P-3B and P-3C aircraft for tactical and strategic intelligence collection, and each maintains a P-3C for logistic support and crew training. VPU detachments typically deploy on short notice, frequently for short periods.

ELECTRONIC RECONNAISSANCE

For long-range electronic reconnaissance, the Navy relies on a fleet of 12 land-based Lockheed EP-3E Orions equipped with the Aries II mission avionics suite. During the mid-1990s, these EP-3Es, converted from older P-3Cs, replaced two EP-3B 'Bat Rack' and 10 EP-3E 'Aries I' (all P-3A conversions) in service with fleet air reconnaissance (VQ) squadrons.

The unarmed EP-3E, powered by four turboprop engines, carries a large crew, mostly electronic warfare operators, communications intercept operators and linguists. The aircraft often fly missions off the coast of potentially hostile nations, monitoring their radar and communications activity. Satellite communications enable the EP-3E mission crew to relay mission intelligence to a carrier battle group or to a ground commander. The EP-3E is being upgraded with improved mission avionics to improve command and control connectivity. The Aries III upgrade fitted to one EP-3E is being evaluated by the Naval Air Warfare Center – Aircraft Division. The EP-3E also will act as the Department of Defense prototype for the high-band system of the Joint Sigint Avionics Family (JSAF).

Sixteen Lockheed ES-3A Shadow carrier-based electronic reconnaissance aircraft were converted from S-3A Vikings during the early 1990s. Too late for the Gulf War, the ES-3As filled the void left by the withdrawal in 1989 of the EA-3B Skywarrior from carrier decks. Serving in fleet air reconnaissance squadrons, the unarmed twin-jet ES-3A is equipped

with a variety of electronic surveillance and intercept equipment to locate and identify hostile emitters and communications stations. The aircraft also shares the S-3B's role as a tanker for the carrier air wing.

The number of ES-3As is insufficient to equip two-aircraft detachments with each of the 10 carrier air wings; the aircraft are rotated frequently to keep deployed carrier air wings equipped.

In mid-1998, the Navy made the decision to withdraw the ES-3A from service without replacement. VQ-5 and VQ-6 will be disestablished in July and September 1999, respectively. The aircraft's mission avionics suite, becoming obsolescent in the age of interconnectivity in the 'electronic battlefield', was deemed as too expensive to upgrade. Carrier battle groups in the future will rely on land-based aircraft (such as EP-3Es and RC-135s) and space-based sensors to support signals intelligence requirements.

FLEET AIR RECONNAISSANCE

The Navy operates six active-duty fleet air reconnaissance (VQ) squadrons, two each of three types, one of each assigned to each coast. Two of the squadrons have a strategic communications role, and are described separately.

Two VQ squadrons, VQ-1 and VQ-2, collect Sigint with long-range EP-3E Orions fitted with the Aries II collection suite. Flying reconnaissance missions, typically along the periphery of foreign nations, these aircraft collect Elint and Comint for further analysis. These squadrons also operate UP-3A, UP-3B and P-3C aircraft for crew training and logistics. VQ-1, based at NAS Whidbey Island, Wash. with Patrol Wing 10, routinely deploys aircraft to sites in the Pacific and Indian Oceans and the Persian Gulf. VQ-2, based at NS Rota, Spain, under Commander, Fleet Air Mediterranean, deploys aircraft in the Mediterranean, Europe and the Middle East, and frequently mans a detachment at Souda Bay, Crete.

For carrier-based intelligence collection, the Navy maintains one VQ squadron on each coast. These VQ squadrons were formed in the early 1990s during the gap in carrier-based VQ presence precipitated by the retirement of the EA-3B Skywarriors flown by VQs 1 and 2. VQ-5, at NAS North Island, Calif., deploys two ES-3A Shadows with each carrier in the Pacific Fleet. A permanent VQ-5 detachment at Misawa, Japan, deploys with CVW-5. VQ-6, which moved in January

1998 from NAS Cecil Field, Fla. to nearby NAS Jacksonville, deploys two ES-3As with each Atlantic Fleet carrier. Only 16 ES-3As were modified from S-3As, a number barely sufficient to maintain VQ detachments onboard each carrier on deployment or in training work-ups. VQ-5 also operates two S-3Bs for crew training. The ES-3As have a secondary mission, aerial refuelling, all the more important since the retirement of the A-6 Intruder, which was the aircraft formerly charged with the aerial refuelling mission.

STRATEGIC COMMUNICATIONS

Since the late 1960s the Navy has operated strategic communications aircraft which linked the national command authority with the nuclear deterrent force's fleet ballistic missile submarines. Known by the programme name TACAMO (take charge and move out), 16 Boeing E-6A Mercuries (variants of the Boeing 707 airliner) replaced the Lockheed EC-130G/Q Hercules aircraft by 1992. The Mercury transmits VLF communications with 30,000-ft (9144-m) trailing-wire antennas. The E-6As, able to remain on station for extended periods with aerial refuelling, are being modified into E-6Bs with airborne command post (ABNCP) suites to assume the Looking Glass role from the Air Force EC-135C. The E-6B commenced ABNCP operations in April 1998.

Two VQ squadrons are not reconnaissance squadrons at all but rather strategic communications units, with the mission of maintaining communications between the National Command Authority and the strategic nuclear deterrent forces. VQ-3 and VQ-4, both based with Strategic Communications Wing One at Tinker AFB, Okla., operate the Boeing E-6 Mercury aircraft. The 16 E-6As replaced the EC-130G and EC-130Q Hercules during the early 1990s and are themselves being upgraded to E-6Bs, with the last scheduled for completion in 2001. The E-6B not only supports the ballistic missile submarine fleet, but also began in April 1998 to assume the Looking Glass command post role from Air Force EC-135s in support of the USAF's Minuteman missile force. E-6 aircraft operate from permanent detachments at Travis AFB, Calif., NAS Patuxent River, Md., and, since 1998, Offutt AFB, Neb. Flight crew training for the E-6 fleet is conducted at Tinker AFB by the Naval Training Support Unit, which operates two modified Boeing 707-320s, designated TC-18F.

Above: The carrier onboard delivery (COD) role is undertaken by the C-2A Greyhound. This Greyhound belongs to VAW-120, which is responsible for training crews on the type.

Below: Ready for a carrier catapult launch, this Greyhound is operated by VRC-40 'Rawhides', the Atlantic COD squadron. .

The SH-60F Seahawk, unofficially called the 'Ocean Hawk' or 'CV-Helo', serves as the inner-zone anti-submarine warfare helicopter for the carrier battle group, as well as undertaking plane-guard duty. These examples are from HS-3 (above) and HS-10 (right).

CARRIER ONBOARD DELIVERY

For the specialised mission of delivering personnel, mail, spare parts, and cargo to aircraft-carriers at sea, the Navy relies on a fleet of Grumman C-2A Greyhound carrier onboard delivery (COD) aircraft, derivatives of the E-2 Hawkeye early warning aircraft. The 38 C-2As currently in service, produced during the mid-1980s, replaced an earlier production run of 17 produced in the mid-1960s. A SLEP, delayed from 1994, is awaiting the assessment of a full-scale fatigue-life study. The SLEP will include installation of ARC-210 radios, full-face oxygen masks, an improved pitot-static system, and the CAINS II inertial navigation system. The C-2A is planned for eventual replacement by a version of the Common Support Aircraft (CSA).

The Navy is as dependent as ever on the use of COD aircraft to carry passengers, mail, supplies and spare parts (including aircraft engines) to and from aircraft-carriers. Although the requirement has not decreased even in the post-Cold War draw down, the Navy has consolidated its COD operations from four squadrons to two, and changed its mode of operations from forward-deployed squadrons servicing deployed carriers to that of assigning COD detachments to each deploying carrier air wing.

Each carrier deploys with detachment of two C-2As. When operating on station in areas such as the Mediterranean, Persian Gulf or Western Pacific, the C-2As are normally staged at shore bases, each flying one to three sorties per day to the carrier and back. Atlantic Fleet carriers draw C-2As from Fleet Logistics Support (VRC) Squadron 40. West Coast carriers deploy with C-2As from VRC-30, which also maintains a permanent detachment in Atsugi, Japan to support the USS Kitty Hawk's CVW-5. FRS training for the C-2A is conducted by the E-2 FRS, VAW-120, at NAS Norfolk, Va.

VR-24 and VRC-50, the COD squadrons in the Mediterranean and Western Pacific, respectively, were disbanded in the early 1990s. The long-range US-3A COD aircraft, which proved so useful in supporting carriers in the North Arabian Sea from Diego Garcia, were withdrawn from service shortly before VRC-50 was disbanded.

ASW HELICOPTERS

The vast majority of US Navy helicopters are versions of the Sikorsky H-60 Seahawk. If the Navy's Helicopter Master Plan is executed in full, the Navy's operational helicopter force eventually will be equipped completely with versions of the H-60. Three types currently in service are the SH-60B LAMPS III (Light Airborne Multi-purpose System III), the SH-60F ('CV-Helo') and the HH-60H. Two follow-on types under development are the SH-60R and the CH-60.

Overall low-visibility grey is the standard scheme for the SH-60B LAMPS III light ASW helicopters. The Pacific Helicopter Anti-submarine Wing Light includes the 'TY'-coded HSL-47 (left), while HSL-46 (below) serves with the Atlantic Fleet.

The SH-60B, which replaced the SH-2 Seasprite, functions as an extension of the shipboard weapon system of the cruiser, destroyer or frigate on which it is deployed. With radar, ESM, MAD, infra-red, and sonobuoy sensors, the SH-60B can detect and track submarines and surface ships and attack with ASW torpedoes and anti-shipping missiles. For rescue and special operations missions, the SH-60B also carries a door-mounted 7.62-mm machine-gun. An upgrade programme for 93 SH-60Bs, called Block I, will continue through mid-1999, and includes installation of the ability to use the Mk 50 torpedo, the GPS, and AGM-119 Penguin anti-shipping missiles, as well as the AAS-44 infra-red sensor.

The SH-60F has replaced the SH-3H Sea King in all active-duty HS squadrons as the aircraft-carrier's ASW helicopter. Equipped with dipping sonar and armed with ASW torpedoes, the SH-60F provides inner-zone defence to a carrier battle group. The AQS-22 dipping sonar will replace the current AQS-13F. The SH-60F also serves in plane guard, rescue, and logistics roles.

In HS squadrons, the SH-60F is augmented by the HH-60H strike rescue version that has the primary role of conducting combat SAR, and insertion and extraction of special warfare forces. The HH-60H, which replaced the HH-1K in the Naval Air Reserve, is armed with 7.62-mm machine-guns and AGM-114 Hellfire missiles.

The Navy is currently developing the SH-60R as a replacement for both the SH-60B and SH-60F. Most SH-60B/F and HH-60H airframes will be remanufactured into SH-60Rs, including 38 SH-60Bs originally scheduled for the Block I upgrade. The SH-60R will feature many improvements, including an increase in gross operating weight; two additional stores stations; a 1553 databus; an AYK-14 mission computer; improved cockpit displays; an AQS-22 dipping sonar; a UYS-2 acoustic processor; a multi-mode radar; an upgraded ESM system; an infra-red sensor; and an integrated self-defence system. The MAD system will be deleted, but the SH-60R will retain the 7.62-mm machine-gun, the Penguin and Hellfire missiles, and the Mk 46 and Mk 50 torpedoes as armament. The SH-60R is scheduled to enter service in 2002, and the remanufacture programme is to continue through 2010.

A proposed variant of the SH-60R, the SH-60R(V), is planned for HS squadrons as a less-expensive 'CV-helo'. This version is envisioned to be wired for all SH-60R systems but would not have all systems installed. A final decision on production of this version has not been made.

In late 1997, Sikorsky flew the YCH-60 development helicopter, fabricated from an Army UH-60L

Above: Differing from the CH-53E by the provision of large fuel sponsons and mission-dedicated equipment, such as rear-view mirrors under the nose for watching the towed sled, mine-hunting MH-53Es equip two squadrons – this one is from HM-15.

Left: The US Navy operates a dedicated mine-countermeasures helicopter in the guise of the MH-53E Sea Dragon, this example being from the NAS Norfolk, Va. based HM-14 – the 'Sea Stallions'.

The MH-53E can tow a variety of anti-mine warfare systems including the ALQ-160 acoustic countermeasures system, the ALQ-166 magnetic minesweeping hydrofoil sled and the Airborne Mine-countermeasures System Mk 105 (seen above).

Black Hawk and a Navy SH-60F, and demonstrated it in the vertical replenishment role. Basically a Black Hawk with rotor, engines, tail pylon, gear box, rescue hoist and automatic flight control system of a Seahawk, the CH-60 also has dual large cargo doors, a cabin with reversible floorboards (one side with rollers to handle pallets), and an external cargo hook. The CH-60 (no series suffix designation yet has been chosen) was approved in 1998 for low-rate initial production and is planned for service as a replacement for the CH-46D, HH-46D, and UH-46D (vertical replenishment and amphibious ship rescue detachments); UH-3H and HH-1N (rescue and utility); VH-3A (executive transport); and HH-60H (strike rescue). The CH-60's advantages over the HH-60H include the ability to carry a rigid inflatable boat internally; a Black Hawk-style tailwheel that allows steeper approaches to landing; and the Black Hawk's versatile external stores support system.

The YCH-60 also is undergoing development as a mine-warfare platform and may yet prove to be a suitable replacement for the MH-53E by using remotely operated systems and laser detection systems.

The Kaman SH-2G Seasprite (LAMPS Mk I) remains in service with two reserve squadrons, but is expected to be retired from service by 2000. This twin-turbine helicopter, armed with ASW torpedoes and 7.62-mm door-mounted machine-gun, is the only Navy platform currently using the Magic Lantern laser mine-detection system. SH-2Gs deploy on reserve early-build 'Perry'-class guided-missile frigates (FFGs) not equipped to handle the SH-60B. With the pending retirement of the early 'Perry' FFGs, the SH-2Gs will no longer be required.

Another ASW helicopter serving only in the reserve is the Sikorsky SH-3H Sea King, one squadron of which provides ASW and rescue services for the reserve carrier air wing. That HS squadron and a reserve HC squadron also operate the UH-3H, a utility version of the SH-3H (modified with a cargo hook and redesigned cabin) for torpedo recovery and other range services. The UH-3H is also operated by VC-8, the Pacific Missile Range Facility, and by several air station base flights. A few former presidential transports, VH-3As, provide executive transportation in the Norfolk, Virginia area. A single NVH-3A is used by the

Rotary-Wing Aircraft Test Squadron as a test platform for the presidential fleet of Marine Corps VH-3Ds. The UH-3H and the VH-3A are scheduled for replacement by the CH-60.

CARRIER-BASED ASW HELICOPTERS

Since the early 1970s, each carrier air wing has included one helicopter anti-submarine (HS) squadron for ASW, rescue, plane guard, logistics, radar calibration and other duties. The SH-3H Sea King was phased out in 1995 and all fleet HS squadrons operate a mixture of eight SH-60F and HH-60H Seahawk helicopters. The HH-60Hs were added to give the carrier a more survivable combat SAR capability. The originally intended normal mix for deployment is six SH-60Fs and two HH-60Hs, but this has frequently been modified in favour of the HH-60H in such areas as the Adriatic Sea and the Persian Gulf. Future plans call for the SH-60Fs and HH-60Hs to be remanufactured into an SH-60R configuration, and for replacement of the HH-60H with the CH-60.

The 10 active-duty HS squadrons are divided evenly between the Atlantic and Pacific Fleets, with one Pacific Fleet squadron (HS-14) forward-deployed in Atsugi, Japan, for duty with CVW-5. The HS FRS, HS-10, trains pilots and crews for the SH-60F and HH-60H at North Island, Calif.

LAMPS HSL SQUADRONS

Ten helicopter anti-submarine squadrons light (HSL) operate the SH-60B Seahawk Light Airborne Multi-Purpose System (LAMPS) Mark III helicopter, which is deployed in detachments of one or two helicopters (depending on ship capacity) onboard cruisers, destroyers and frigates. These detachments provide ASW, anti-shipping, rescue, logistics and utility services for surface warships.

The 10 fleet squadrons are divided evenly between the Atlantic and Pacific fleets. One Pacific squadron, HSL-37, is based permanently in Hawaii (at MCAF Kaneohe Bay since 1999, from NAS Barbers Point) to deploy with ships based at Pearl Harbor. A second Pacific squadron, HSL-51, is based at NAF Atsugi, Japan, to operate from destroyers based in Yokosuka, Japan. HSL-51 also operates a VIP detachment (Detachment 11) that flies a UH-3H in support of the Commander, US Seventh Fleet from his flagship, USS *Blue Ridge* (LCC 19).

The LAMPS community employs two FRSs to train its pilots, crewmen and maintenance personnel. The West Coast FRS, HSL-41, is based at North

Island, Calif.; the East Coast FRS, HSL-40, is based at NS Mayport. Fla.

MINE COUNTERMEASURES

The massive three-engined Sikorsky MH-53E Sea Dragon, a derivative of the CH-53E Super Stallion assault transport, is the Navy's sole dedicated mine countermeasures helicopter, equipping two joint active/reserve HM squadrons. MH-53Es also equip an HC squadron in the vertical onboard delivery role. The Sea Dragon tows a variety of mine countermeasures sleds and side-scan sonars. The installation of a GPS has increased their effectiveness. The MH-53E fleet was programmed to begin a service-life extension programme in late 1998; future plans call for the installation of T64-GE-419 (versus -416) engines to allow for one-engine-out operation. If suitable mine-countermeasures systems are developed, the CH-60 may be adapted to replace the MH-53E.

Only two airborne mine countermeasures (AMCM) squadrons are on strength. Both operate the MH-53E Sea Dragon in mine countermeasures duties, and perform vertical on-board delivery duties as well. These squadrons deploy detachments overseas for exercises and deploy onboard the mine countermeasures command ship USS *Inchon* (MCS 12). During the early 1990s HM-14 and HM-15 absorbed their reserve counterpart squadrons, HM-18 and HM-19, respectively, becoming joint active-duty/reserve squadrons. With the closure of NAS Alameda, Calif., HM-15 moved to NAS Corpus Christi, Texas, near the mine warfare forces based at Ingleside. Plans to move HM-14 from Norfolk, Va. to Corpus Christi have been delayed indefinitely by funding constraints.

FRS training for the HM squadrons, conducted until 1994 by now-disbanded HM-12, is handled by Marine Corps H-53 training squadron HMT-302, which operates several MH-53Es at MCAS New River, N.C. Tactical training takes place at the AMCM School at Norfolk.

In order to reduce the variety of helicopters in service, the Navy is considering a version of the H-60 to replace the MH-53E. The probability of such a step would depend upon the success of matching AMCM technology with the H-60 airframe.

VERTICAL REPLENISHMENT

An ageing fleet of Boeing Vertol H-46 Sea Knight twin-engined tandem rotor helicopters sustains the Navy's VertRep capability and provides rescue/utility detachments for the amphibious assault ships and the

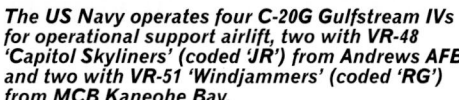

single mine countermeasures command ship. The CH-46D, HH-46D and UH-46D are essentially similar and used interchangeably by the HC squadrons, with the HH-46Ds equipping the rescue detachments. The Sea Knights are going through a dynamic component upgrade (DCU) to extend their operating life and to lift flight restrictions imposed by fatigue. The Navy anticipates a shortage of 48 Sea Knights by 2000. The CH-60 (see above) is the likely replacement for the H-46 in this role.

HELICOPTER COMBAT SUPPORT

Seven helicopter combat support (HC) squadrons provide the Navy with the vertical lift capability necessary to keep its ships at sea properly provisioned. HC squadrons also provide rescue detachments and perform a variety of utility missions.

Four HC squadrons, two each with the Atlantic and Pacific Fleets, provide H-46 Sea Knight vertical replenishment detachments to ships of the Combat Logistics Force and the Military Sealift Command that replenish Navy ships at sea. These helicopters shuttle supplies, spare parts, ordnance and personnel to other ships in a battle group. During the early 1990s, these HC squadrons assigned detachments to assume the rescue and utility role formerly performed by UH-1N and later HH-1N helicopters onboard amphibious assault ships (LHDs, LHAs, and LPDs). An H-46 detachment also deploys onboard the mine countermeasures command ship USS *Inchon* (MCS 12). These HC squadrons operate a mixture of essentially similar CH-46D, HH-46D and UH-46D versions. HC-11, based at North Island, Calif., also operates a single UH-3H Sea King from the command ship USS *Coronado* (AGF 11) to support the commander of the Third Fleet. The Navy has experimented with civil contractor helicopters, such as the Kaman K-Max and Evergreen's Bell 214, but is considering the CH-60 Blackhawk as well to replace the H-46. FRS training for the Navy's H-46 fleet is performed by a fifth H-46 squadron, HC-3 at North Island, Calif.

Vertical onboard delivery missions in the Mediterranean, Europe, Africa and the Middle East are handled by HC-4, based at NAS Sigonella, Sicily. HC-4 operates MH-53E Sea Dragon heavy-lift helicopters, having acquired them during the mid-

1990s when the Navy turned over its CH-53E Sea Stallions to the Marine Corps. HC-4 supports carrier battle groups and amphibious ready groups with migrating detachments staged at shore bases near the ship's operating areas. FRS training for HC-4 personnel is accomplished by the Marine Corps CH-53E training squadron, HMT-302 at MCAS New River, N.C., which has MH-53Es on strength.

HC-2, based at NAS Norfolk, Va., has three assigned roles. The squadron's VH-3As (former presidential support helicopters) provide VIP transportation to flag officers assigned to the Norfolk area, including the Commander-in-Chief, US Atlantic Command and to the Commander-in-Chief US Atlantic Fleet. The unit's UH-3Hs operate in several detachments. One, formed as needed, transports the commander of the Second Fleet, operating from the command ship USS *Mt Whitney* (LCC 20). Detachment 1, based at NSA Naples, Italy, supports the Sixth Fleet commander onboard his flagship, USS *LaSalle* (AGF 3). Detachment 2, the 'Desert Ducks', is based in Bahrain and supports the Fifth Fleet in the Persian Gulf. HC-2 also serves as the FRS (succeeding HS-1 in this role in 1996) for the Navy's H-3 helicopter fleet, training pilots and crews for HC-2 detachments, several air station rescue units, and the Pacific Missile Range Facility in Barking Sands, Hawaii. HC-2's H-3s are planned for replacement by the CH-60 Blackhawk.

FLEET LOGISTICS

The Navy has managed to maintain control over a sizeable logistics aircraft fleet ever since World War II. The current fleet is used to meet rapidly changing requirements in transporting personnel, mail, spare parts, and other cargo to ports for transfer to ships at sea. Aircraft rotate on detachments to forward fleet operating areas such as the Mediterranean, Persian Gulf and the Western Pacific.

The McDonnell Douglas C-9B Skytrain II, a military version of the twin-engined DC-9-32 airliner, has been in naval service since 1973. These aircraft have been augmented by second-hand DC-9-31 airliners and later by DC-9-33 versions, which have retained their civil designations. The C-9B and DC-9 aircraft have been operated by the Naval Air Reserve for almost two decades; they are currently going through an avionics upgrade to remain consistent with commercial aviation standards and FAA requirements. In 1997 the Navy decided to procure a version of the Boeing 737-700, designated C-40A, to begin replacing the C-9 fleet. The C-40A is scheduled to enter service in 2000.

During the early 1990s, as the Navy retired its fleet of Lockheed C-130F and KC-130F forward-deployed transports (the latter borrowed from the Marine Corps), four squadrons of new C-130T transports were procured for the Naval Air Reserve. These aircraft rotate on detachment to forward operating areas along with the C-9s to provide overseas fleet units with logistics support. The modern C-130Ts have since been upgraded with ARC-210 radios and the GPS.

In 1998, the Navy began retiring its ski-equipped LC-130F and LC-130R Hercules transports used to supply the scientific research stations in Antarctica. The LC-130R aircraft, owned by the National Science Foundation and operated by the Navy, are being turned over to the New York Air National Guard, which has assumed the Antarctic support role and operates similar LC-130H aircraft.

A few modern Gulfstream C-20D Gulfstream III and C-20G Gulfstream IV twin-engined long-range jets are operated by the Naval Air Reserve for a variety of transport duties, but mostly for liaison and transportation of high-level Navy and Marine Corps military and civilian officials.

COMPOSITE SQUADRONS

The number of Navy composite (VC) squadrons has steadily dwindled as budget cutbacks reduced squadrons and economies were achieved with a shift to use of contractors to provide aircraft services for fleet training. Only two VC squadrons, distinctly different in composition, remain in service.

VC-8, based at NS Roosevelt Roads, Puerto Rico, provides UH-3H helicopters for rescue, torpedo recovery and logistics for ships exercising in the Puerto Rican Op Area (PROA), including the torpedo range at St Croix. VC-8 is the only operational squadron in the Navy still flying the Skyhawk; its TA-4Js are used to tow targets and to simulate enemy aircraft and cruise missiles. The TA-4Js are due to be retired in 1999; a replacement has not been identified, but turnover of its duties to a contractor has been considered.

UNMANNED AERIAL VEHICLES (UAVS)

VC-6, based at NAS Norfolk Va., has long been an operator of drones for target services. The squadron also operates a number of fast boats to simulate fast patrol craft. VC-6 led the way for the introduction of the RQ-2A Pioneer unmanned aerial vehicle (UAV) during the 1980s. The squadron sent two detachments to the Persian Gulf during Operation Desert Storm, one each deployed onboard the battleships USS *Missouri* and USS *Wisconsin*. (One Pioneer UAV is credited with forcing the surrender of a band of Iraqi soldiers.) VC-6's UAV detachments are based at NAS Patuxent River, Md. and are available for deployment onboard amphibious warfare ships.

A replacement for the RQ-2A has not been selected, but the Navy is considering procurement of a VTOL UAV to augment the Pioneer.

NAVAL AIR FORCE, US PACIFIC FLEET

USN Pacific Fleet

Note: Unless otherwise noted, units are located with their parent wings or other parent commands.

Chief of Naval Operations	Arlington, Va.

Commander-in-Chief. US Pacific Fleet	
	NB Pearl Harbor, Hawaii

Commander, Naval Air Force, US Pacific Fleet	NAS North Island, Hawaii
Commander, Carrier Group 1	NAS North Island, Calif.
Commander, Carrier Group 3	NS Bremerton, Wash.
Commander, Carrier Group 5	NS Yokosuka, Japan
Commander, Carrier Group 7	NAS North Island, Calif.
USS *Kitty Hawk* (CV 63)	NS Yokosuka, Japan
USS *Constellation* (CV 64)	NAS North Island, Calif.
USS *Carl Vinson* (CVN 70)	NS Bremerton, Wash.
USS *Abraham Lincoln* (CVN 72)	NS Everett, Wash.
USS *John C. Stennis* (CVN 74)	NAS North Island, Calif.
Commander, Carrier Air Wing 2	'NE'	NAS Leemore, Calif.
Commander, Carrier Air Wing 5	'NF'	NAF Atsugi, Japan
Commander, Carrier Air Wing 9	'NG'	NAS Lemoore, Calif.
Commander, Carrier Air Wing 11	'NH'	NAS Lemoore, Calif.
Commander, Carrier Air Wing 14	'NK'	NAS Lemoore, Calif.

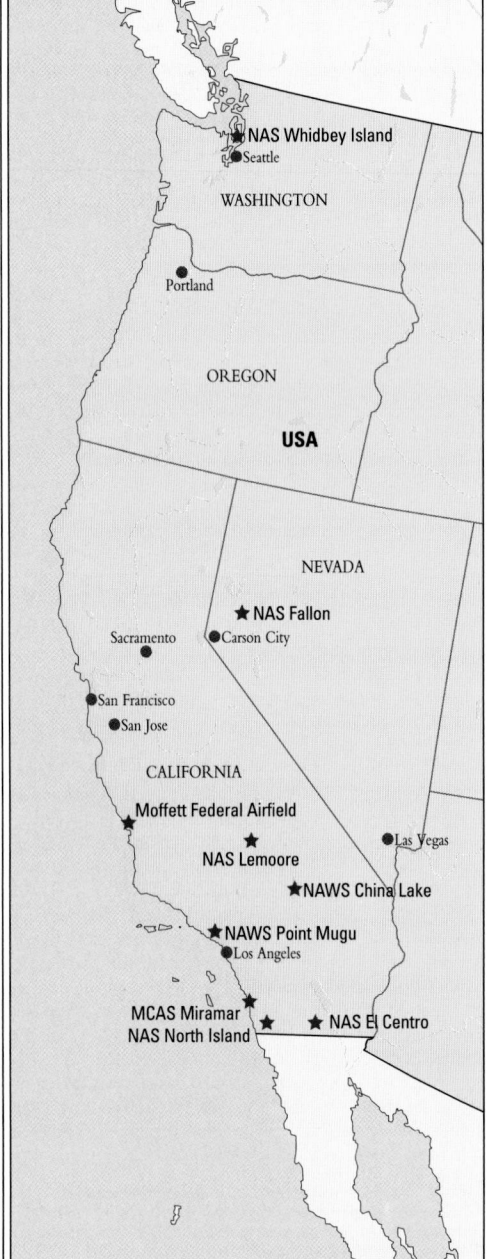

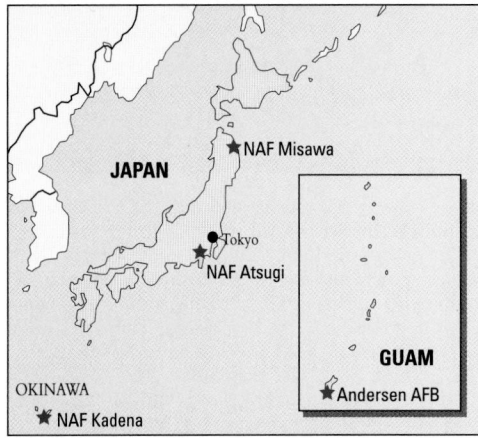

Commander, Patrol Wings, US Pacific Fleet		
		MCAF Kaneohe Bay, Hawaii
VP-4	P-3C	'YD'
VP-9	P-3C	'PD'
VP-47	P-3C	'RD'
VPU-2	P-3C, UP-3A	'SP'

Commander, Patrol Wing 10		NAS Whidbey Island, Wash.
VP-1	P-3C	'YB'
VP-40	P-3C	'QE'
VP-46	P-3C	'RC'
VQ-1	EP-3E, P-3C, UP-3A/B	'PR'

Commander, Electronic Attack Wing, US Pacific Fleet		
		NAS Whidbey Island, Wash.
VAQ-128	EA-6B	'NL'
VAQ-129 (FRS)	EA-6B	'NJ'
VAQ-130	EA-6B	
VAQ-131	EA-6B	
VAQ-132	EA-6B	
VAQ-133	EA-6B	'NL'
VAQ-134	EA-6B	'NL'
VAQ-135	EA-6B	
VAQ-136	EA-6B	NAF Atsugi, Japan
VAQ-137	EA-6B	
VAQ-138	EA-6B	
VAQ-139	EA-6B	
VAQ-140	EA-6B	
VAQ-141	EA-6B	
VAQ-142	EA-6B	'NL'

Note: VAQs 128, 133, 134, 137 and 142 provide electronic support to Air Force expeditionary wings.

Commander, Strike Fighter Wing, US Pacific Fleet		
		NAS Lemoore, Calif
VFA-22	F/A-18C	
VFA-25	F/A-18C	
VFA-27	F/A-18C	NAF Atsugi, Japan
VFA-94	F/A-18C	
VFA-97	F/A-18A	
VFA-113	F/A-18C	
VFA-115	F/A-18C	
VFA-122	F/A-18E/F	'NJ'
VFA-125 (FRS)	F/A-18A/B/C/D, T-34C	'NJ'
VFA-137	F/A-18C	
VFA-146	F/A-18C	
VFA-147	F/A-18C	
VFA-151	F/A-18C	NAF Atsugi, Japan
VFA-192	F/A-18C	NAF Atsugi, Japan
VFA-195	F/A-18C	NAF Atsugi, Japan

Commander, Airborne Early Warning Wing, US Pacific Fleet		
		NAWS Point Mugu, Calif.
VAW-112	E-2C	
VAW-113	E-2C	
VAW-115	E-2C	NAF Atsugi, Japan
VAW-116	E-2C	
VAW-117	E-2C	
VRC-30	C-2A, UC-12B	'RW'	NAS North Island, Calif.

Commander, Sea Control Wing, US Pacific Fleet		
		NAS North Island, Calif.
VS-21	S-3B	NAF Atsugi, Japan
VS-29	S-3B	
VS-33	S-3B	
VS-35	S-3B	
VS-38	S-3B	
VS-41 (FRS)	S-3B	

VQ-5	ES-3A, S-3B	

(VQ-5 is due to be disestablished in July 1999)

Commander, Strategic Communications Wing 1			
			Tinker AFB, Okla.
VQ-3	E-6A/B	'TZ'	
Det Travis	E-6A/B	'TZ'	Travis AFB, Okla.
VQ-4	E-6A/B	'HL'	
Det Patuxent River	E-6A/B	'HL'	NAS Patuxent River, Md.
Det Offutt	E-6B	'TZ/HL'	Offutt AFB, Neb.
NTSU (FRS)	TC-18F		

Commander, Helicopter Anti-submarine Wing, US Pacific Fleet			
			NAS North Island, Calif.
HS-2	SH-60F, HH-60H		
HS-4	SH-60F, HH-60H		
HS-6	SH-60F, HH-60H		
HS-8	SH-60F, HH-60H		
HS-10 (FRS)	SH-60F, HH-60H	'RA'	
HS-14	SH-60F, HH-60H		NAF Atsugi, Japan

Commander, Helicopter Anti-submarine Wing Light, US Pacific Fleet			
			NAS North Island, Calif.
HSL-37	SH-60B	'TH'	
HSL-41 (FRS)	SH-60B	'TS'	
HSL-43	SH-60B	'TT'	
HSL-45	SH-60B	'TZ'	
HSL-47	SH-60B	'TY'	
HSL-49	SH-60B	'TX'	
HSL-51	SH-60B, UH-3H	'TA'	NAF Atsugi, Japan

Commander, Helicopter Tactical Wing, US Pacific Fleet			
			NAS North Island, Calif.
HC-3 (FRS)	CH/HH-46D	'SA'	
HC-5	CH/HH/UH-46D	'RB'	Andersen AFB, Guam
HC-11	CH/HH/UH-46D, UH-3H	'VR'	
VXE-6	LC-130F/R	'XD'	NAWS Point Mugu, Calif.

Note: VXE-6 is scheduled for disestablishment in March 1999.

VX-9	F/A-18A/B/C/D/E, EA-6B, AV-8B, AH-1W	'XE'	NAWS China Lake, Calif.
Det Pt. Mugu	F-14A/B/D	'XF'	NAWS Pt Mugu, Calif.

Naval Strike and Air Warfare Center			NAS Fallon, Nev.
	F/A-18A/B, F-14A, SH-60F		

Naval Fighter Weapons School		
Pacific Missile Range Facility		Barking Sands, Hawaii
	RC-12F, UH-3H	

NAS Barbers Point	UP-3A		NAS Barbers Point, Hawaii
NAS North Island	UC-12B	'7M'	NAS North Island, Calif.
NAS Lemoore	UC-12B, HH-1N	'7S'	NAS Lemoore, Calif.
NAS Fallon	UC-12B, HH-1N	'7H'	NAS Fallon, Nevada
NAS Whidbey Island	UC-12B, UH-3H	'7G'	NAS Whidbey Island, Wash.
NAF El Centro	UC-12B	'8N'	NAF El Centro, Calif.

Commander, Fleet Air Western Pacific NAF Atsugi, Japan

Note: HC-5, part of Helicopter Tactical Wing, US Pacific Fleet, operates under control of Commander Fleet Air, Western Pacific.

Commander, Patrol Wing One			Kamiseya, Japan
VP-XX (Rotation)	P-3C		NAF Misawa, Japan
VP-XX (Rotation)	P-3C		NSA Diego Garcia, B.I.O.T.
VP-XX Det (Rotation)			NAF Kadena, Okinawa, Japan
	P-3C		
VQ-1 Det	EP-3E		NAF Misawa, Japan
NAF Atsugi	UC-12F		NAF Atsugi, Japan
NAF Diego Garcia	None		NSA Diego Garcia
NAF Misawa	UC-12F	'8M'	NAF Misawa, Japan
NAF Kadena	UC-12F	'8H'	NAF Kadena, Okinawa, Japan

NAVAL AIR FORCE, US ATLANTIC FLEET

USN Atlantic Fleet

Note: Unless otherwise noted, units are located with their parent wings or other parent commands.

Chief of Naval Operations	**Arlington, Va.**

Commander-in-Chief, US Atlantic Fleet	**Norfolk, Va.**

Commander, Naval Air Force, US Atlantic Fleet	
	NAS Norfolk, Va.
Commander, Carrier Group 2	NAS Norfolk, Va.
Commander, Carrier Group 4	NAS Norfolk, Va.
Commander, Carrier Group 6	NS Mayport, Fla.
Commander, Carrier Group 8	NAS Norfolk, Va.
USS *Enterprise* (CVN 65)	NS Norfolk, Va.
USS *Nimitz* (CVN 68)	Newport News, Va.
(3-year refuelling and overhaul)	
USS *Dwight D. Eisenhower* (CVN 69)	NS Norfolk, Va.
USS *Theodore Roosevelt* (CVN 71)	NS Norfolk, Va.
USS *George Washington* (CVN 73)	NS Norfolk, Va.
USS *Harry S. Truman* (CVN 75)	NS Norfolk, Va.
Commander, Carrier Air Wing 1 'AB'	NAS Oceana, Va.
Commander, Carrier Air Wing 3 'AC'	NAS Oceana, Va.
Commander, Carrier Air Wing 7 'AG'	NAS Oceana, Va.
Commander, Carrier Air Wing 8 'AJ'	NAS Oceana, Va.
Commander, Carrier Air Wing 17 'AA'	NAS Oceana, Va.

Commander, Patrol Wings, US Atlantic Fleet			
			NAS Norfolk, Va.
VP-30 (FRS)	P-3C, P-3A	'LL'	NAS Jacksonville, Fla.
Commander, Patrol Wing 5			**NAS Brunswick, Maine**
VP-8	P-3C	'LC'	
VP-10	P-3C	'LD'	
VP-26	P-3C	'LK'	
VPU-1	P-3B, P-3C	'OB'	
Commander, Patrol Wing 11			**NAS Jacksonville, Fla.**
VP-5	P-3C	'LA'	
VP-16	P-3C	'LF'	
VP-45	P-3C	'LN'	

Commander, Fighter Wing, US Atlantic Fleet		
		NAS Oceana, Va.
VF-2	F-14D	
VF-11	F-14B	
VF-14	F-14A	
VF-31	F-14D	
VF-32	F-14B	
VF-41	F-14A	
VF-101 (FRS)	F-14A/B/D, T-34C	
VF-102	F-14B	
VF-103	F-14B	
VF-143	F-14B	
VF-154	F-14A	NAS Atsugi, Japan
VF-211	F-14A	
VF-213	F-14D	
VC-8	TA-4J, UH-3H 'GF'	NS Roosevelt Roads, Puerto Rico

Commander, Strike Fighter Wing, US Atlantic Fleet		
		NAS Oceana, Va.
VFA-15	F/A-18C	
VFA-34	F/A-18C	
VFA-37	F/A-18C	
VFA-81	F/A-18C	
VFA-82	F/A-18C	MCAS Beaufort, S.C.
VFA-83	F/A-18C	
VFA-86	F/A-18C	MCAS Beaufort, S.C.
VFA-87	F/A-18C	
VFA-105	F/A-18C	
VFA-106 (FRS)	F/A-18A/B/C/D	
VFA-131	F/A-18C	
VFA-136	F/A-18C	
Strike Fighter Weapons School		
	T-34C	

Commander, Airborne Early Warning Wing, US Atlantic Fleet		
		NAS Norfolk, Va.
VAW-120 (FRS)	E-2C, TE-2C, C-2A 'AD'	
VAW-121	E-2C	
VAW-123	E-2C	
VAW-124	E-2C	
VAW-125	E-2C	
VAW-126	E-2C	
VRC-40	C-2A	

Commander, Sea Control Wing, US Atlantic Fleet,		
		NAS Jacksonville, Fla.
VS-22	S-3B	
VS-24	S-3B	
VS-30	S-3B	
VS-31	S-3B	
VS-32	S-3B	
VQ-6	ES-3A	'ET' *(not usually carried)*

(VQ-6 due to disestablish during September 1999)

Commander, Helicopter Anti-submarine Wing, US Atlantic Fleet	
	NAS Jacksonville, Fla.
HS-3	SH-60F, HH-60H
HS-5	SH-60F, HH-60H
HS-7	SH-60F, HH-60H
HS-11	SH-60F, HH-60H
HS-15	SH-60F, HH-60H

Commander, Helicopter Anti-submarine Wing Light, US Atlantic Fleet		
		NS Mayport, Fla.
HSL-40 (FRS)	SH-60B	'HK'
HSL-42	SH-60B	'HH'
HSL-44	SH-60B	'HP'
HSL-46	SH-60B	'HQ'
HSL-48	SH-60B	'HS'

Commander, Helicopter Tactical Wing, US Atlantic Fleet			
			NAS Norfolk, Va.
HC-2	(FRS)	UH-3H, VH-3A 'HU'	
	Det 1	UH-3H	NSA Naples, Italy
	Det 2	UH-3H	ASU Bahrain
HC-4		MH-53E	'HC'	NAS Sigonella, Sicily
HC-6		CH/HH/UH-46D	'HW'	
HC-8		CH/HH/UH-46D	'BR'	
HM-14		MH-53E	'BJ'	
HM-15		MH-53E	'TB'	NAS Corpus Christi, Texas
VC-6		RQ-2A		NAS Norfolk, Va.

Detachments of VC-6 at NAS Patuxent River, Md.
Note: HM squadrons are joint active/reserve squadrons.
HC-2 serves as both an FRS and an operational squadron.

VX-1	P-3C, S-3B, SH-60B/F	'JA'	NAS Patuxent River, Md.
NAS Brunswick	HH-1N	'7F'	NAS Brunswick, Maine
NAS Oceana	UH-3H	'7R'	NAS Oceana, Va.
NAS Norfolk	UC-12B/M	'7M'	NAS Norfolk, Va.
NAS Jacksonville	UC-12B	'7E'	NAS Jacksonville, Fla.
NAS Cecil Field	None	'7U'	NAS Cecil Field, Fla.
NAS Key West	UC-12B, UH-3H	'7Q'	NAS Key West, Fla.
NS Mayport	None	'8U'	NS Mayport, Fla.
NS Roosevelt Roads	RC-12M	'8E'	NS Roosevelt Roads
NS Guantanamo Bay	UC-12B, HH-1N	'8F'	NS Guantanamo Bay, Cuba

Commander, Fleet Air Keflavik		**NAS Keflavik, Iceland**
VP-XX (Rotation)	P-3C	
NAS Keflavik	UP-3A	

Commander, US Naval Forces Europe		**London, UK**
NAF Mildenhall	UC-12M	'8G'	NAF Mildenhall, UK

Commander, Fleet Air Mediterranean			**NSA Naples, Italy**
VQ-2	EP-3E, P-3C	'JQ'	NS Rota, Spain
VP-XX (Rotation)	P-3C		NAS Sigonella, Sicily

Note: HC-4, based at NAS Sigonella, is part of Helicopter Tactical Wing, US Atlantic Fleet, but operates under control of Commander, Fleet Air Mediterranean.

NAS Sigonella	UC-12M, VP-3A	'8C'	NAS Sigonella, Sicily
NS Rota	UC-12M	'8D'	NS Rota, Spain
NSA Naples	UC-12M		NSA Naples, Italy
NSA Souda Bay	None		NSA Souda Bay, Crete
Admin Support Unit Bahrain	UC-12B		

Naval Air Reserve

United States naval aviation is augmented by a powerful reserve force that provides reinforcement in time of national emergency, and in some cases provides capabilities not found in the active-duty forces. The Naval Air Reserve traces its origins to 1911, when aviation components to naval militias were formed in New York and Massachusetts. Congressional appropriations in 1917 provided for the formation of a naval reserve flying corps and the purchase of 12 aircraft. The first unit of the corps was organised at Yale University as the First Yale Unit, the official beginning of the Naval Air Reserve. Naval air reservists comprised more than 75 per cent of naval aviation forces during World War I, though they were not organised into units distinct from active forces. The reserve almost disappeared in the post-war demobilisation, a trend reversed in 1923 with the formation of a naval aviation reserve establishment. In January 1941, all 16 existing reserve units were called to active duty. During World War II, naval air reservists served alongside regular fliers in units Navy-wide. The large-scale demobilisation following World War II made available plenty of modern piston-engined aircraft and experienced aviators to staff the reserve structure, formed as the Naval Air Reserve

Training Command. Many of these reservists saw action in the Korean conflict, in the course of which over 40 reserve squadrons were activated (including an entire carrier air group), many of which were retained as active-duty squadrons. Reserve equipment after the Korean conflict lagged in modernity behind that of the active-duty forces rapidly entering the jet age. However, 18 reserve squadrons were mobilised for the 1961 Berlin crisis, and reserve transport crews supported US forces waging the Vietnam War. The Naval Air Reserve today reflects the reserve force squadron structure created in 1970 in the wake of the lack of readiness displayed during the failed reserve call-up during the 1968 *Pueblo* crisis off North Korea. Today the Naval Air Reserve, headquartered in New Orleans, Louisiana, consists of five wings (soon to be only four) and 37 squadrons. Since the end of the Cold War, 21 reserve force squadrons (as well as two patrol squadron master augmentation units and most squadron augmentation units) have been disbanded, and only one formed. In certain mission areas, the Naval Air Reserve operates all of the Navy's aircraft assigned those missions. All non-carrier-capable airlift squadrons, strike rescue helicopter squadrons, and air combat manoeuvring adversary squadrons are part of the reserve force structure. The Naval Air Reserve is manned largely by experienced officers and sailors who have been released from active duty following their initial service. A core of reservists on active duty,

The Seasprite has disappeared from the US Navy in all organisations except the Reserve, where two squadrons operate the type. This is an HSL-84 SH-2G.

designated TARs (Training and Administration of Reserves), ensures the day-to-day operation of the reserve force squadrons and wings.

RESERVE CARRIER AVIATION

The Navy's determined effort to maintain two reserve carrier air wings (CVWRs) on strength since 1970 and to continuously upgrade their aircraft common to fleet standard was dashed with the end of the Cold War. Budget reductions forced the disestablishment of the West Coast's CVWR-30 and most of its squadrons, leaving only East Coast-based CVWR-20. The Navy's attempt to modernise the CVWRs was largely successful, with their aircraft and squadron composition mirroring those of the active-duty CVWs for the most part. Their aircraft were usually earlier models of the fleet-standard aircraft. Unlike active-duty CVWs, CVWRs were never equipped

Two squadrons in the Reserve fly the dedicated rescue and special operations version of the SH-60 Seahawk, the HH-60H – unofficially called the 'Rescue Hawk' – and are the only squadrons so-equipped in the Navy.

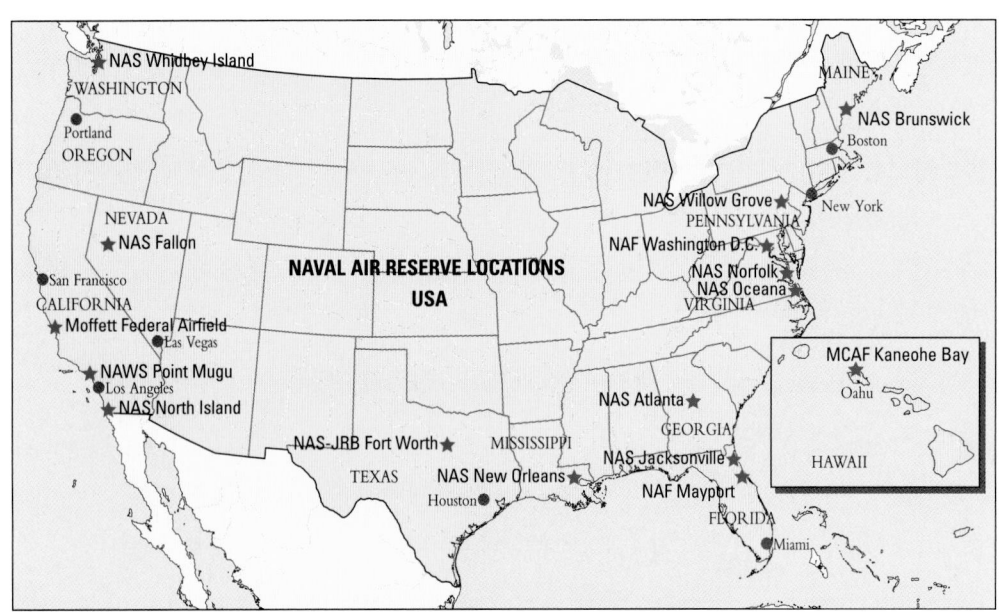

NAVAL AIR RESERVE LOCATIONS
USA

have augmented active-duty VAQ units operating from carriers off Bosnia, and have served in rotation in support of Air Force expeditionary deployments to Turkey in support of Operation Northern Watch over Iraq. A shortage of EA-6Bs in the fleet has produced mounting pressure to relocate VAQ-209 from NAF Washington to the main EA-6B base, NAS Whidbey Island, to allow the squadron's aircraft to be used to support fleet deployments.

AEW SQUADRONS

CVWR-20 has two E-2C Hawkeye squadrons assigned, but only one, Carrier Airborne Early Warning (VAW) Squadron 78, is intended to operate onboard aircraft-carriers. Airborne Early Warning Squadron (also designated VAW) 77 was formed in 1995 to replace active-duty squadron VAW-122 as a dedicated drug-interdiction and fleet exercise support squadron. VAW-77, assigned to CVWR-20 as a matter of administrative convenience, has greatly relieved the operational tempo required of active-duty VAW squadrons.

ADVERSARY SERVICES

The Navy's once-expansive active-duty air combat adversary force (VF-43, VF-45, VF-126 and VFA-127) has been disbanded and adversary services are provided by only two squadrons, both with the Naval Air Reserve. Fighter Composite Squadron (VFC) 12 flies F/A-18A/B Hornets on the East Coast. VFC-13, which moved to Fallon, Nevada from Miramar, California, flies the F-5E/F Tiger II in support of carrier air wing work-ups at Fallon. Both VFC units are administratively assigned to CVWR-20.

PATROL SQUADRONS

For nearly three decades, one-third of the Navy's patrol (VP) squadrons have been operated by the Naval Air Reserve. Reserve VP units routinely augment active-duty VP squadrons at overseas deployment sites and in fleet exercises in what has long been a remarkably successful active-reserve operational integration. The number of reserve VP squadrons, however, has declined at a pace commensurate with that of the active-duty patrol force, from 13 squadrons in 1990 to seven by the beginning of 1999 (with the scheduled disestablishment of VP-91). The drawdown will result in the January 1999 consolidation of Commander Reserve Patrol Wing, US Pacific Fleet (also designated Commander Patrol Wing 4) with Commander Reserve Patrol Wing, US Atlantic Fleet into one organisation, Reserve Patrol Wing. Reserve VP squadrons operate a mixture of Update I, II, II.5, and III variants of the P-3C Orion; eventually, all units will operate a common Update III version.

Currently, the US Navy Reserve has four squadrons of Hornets, having gained a fourth when VF-201 trades in its Tomcats. This example is a F/A-18A belonging to VF-203.

with VS (anti-submarine) squadrons, relying instead on squadron augmentation units that would draw aircraft from the S-3 FRSs. The CVWRs also have to rely on active-duty squadrons, whose aircraft are in short supply, for fleet air reconnaissance and carrier onboard delivery (COD) detachments. With the dissolution of the fleet's three electronic adversary squadrons (VAQs 33, 34 and 35), the CVWRs assumed the role of providing electronic adversary services (using special pods) to Navy ships. In recent years, for convenience, the CVWRs have also assumed administrative command of land-based VFC and VAW units (fighter composite squadron and carrier airborne early warning squadron, respectively) that had no carrier mission. CVWR-20 is nominally assigned to USS *John F. Kennedy* (CV 67), which is part of the Naval Reserve Force. This carrier, however, is still integrated in the overseas deployment cycle with active-duty CVWs.

After operating the less capable E-2B Hawkeye from 1973 until the 1980s, the Reserve's VAW-77 and -78 (seen here) received the E-2C.

FIGHTER SQUADRONS

The retirement of CVWR-30 and the consolidation in most carrier air wings of the two fighter squadrons resulted in the retention of only one reserve fighter squadron, VF-201, which used the F-14A Tomcat in the fleet air defence and strike roles. The Navy's initiative to accelerate retirement of the F-14A led to the decision to transition VF-201 to a VFA squadron with F/A-18A Hornet strike fighters in 1999, which leaves CVWR-20 with four Hornet squadrons and no Tomcat fighters.

STRIKE FIGHTER SQUADRONS

Most of CVWR-20's combat capability resides in the four F/A-18A Hornet strike fighter squadrons. One of the assigned squadrons is Marine Fighter-Attack Squadron (VMFA) 142, which replaced VA-205 in the wing's structure when the A-6 Intruder was retired. The two VFA squadrons have a secondary mission of electronic adversary training support.

ELECTRONIC ATTACK SQUADRONS

CVWR-20 is assigned one electronic attack (formerly tactical electronic warfare) squadron, VAQ-209, which performs suppression of enemy air defences with electronic countermeasures and missile strikes. VAQ-209 crews and EA-6B Prowler aircraft

The Reserve has a single electronic attack squadron, VAQ-209 'Star Warriors', based at NAF Washington.

Flying the F/A-18A, the 'River Rattlers' of VFA-204 use the 'AF' tailcode and '400'-series Modex.

US Naval Air Reserve

Chief of Naval Reserve		NAS New Orleans, La.
USS *John F. Kennedy* (CV 67)		NS Mayport, Fla.

Commander, Naval Air Reserve Force
NAS New Orleans, La.

Commander, Reserve Patrol Wing		**NAS Norfolk, Va.**
VP-62	P-3C	'LT'	NAS Jacksonville, Fla.
VP-64	P-3C	'LU'	NAS Willow Grove, Pa
VP-65	P-3C	'PG'	NAWS Point Mugu, Calif.
VP-66	P-3C	'LV'	NAS Willow Grove, Pa.
VP-69	P-3C	'PJ'	NAS Whidbey Island, Wash.
VP-92	P-3C	'LY'	NAS Brunswick, Maine
VP-94	P-3C	'PZ'	NAS New Orleans, La.
VQ-11	EP-3J, P-3C 'LP'	NAS Brunswick, Maine

(VQ-11 is due to disestablish during 1999)

Commander, Reserve Carrier Air Wing 20
NAS Atlanta, Ga.

VFA-201	F/A-18A	'AF'	NAS Fort Worth, Texas
VFA-203	F/A-18A	'AF'	NAS Atlanta, Ga.
VFA-204	F/A-18A	'AF'	NAS New Orleans, La.
VAQ-209	EA-6B	'AF'	NAF Washington, D.C.
VAW-77	E-2C	'AF'	NAS Atlanta, Ga.
VAW-78	E-2C	'AF'	NAS Norfolk, Va.
VFC-12	F/A-18A/B 'AF'	NAS Oceana, Va.
VFC-13	F-5E/F	'AF'	NAS Fallon, Nev.

(VFC-12 and VFC-13 do not carry the allocated 'AF' code)

Commander, Helicopter Wing Reserve
NAS North Island, Calif.

HC-85	UH-3H	'NW'	NAS North Island, Calif.
HCS-4	HH-60H	'NW'	NAS Norfolk, Va.
HCS-5	HH-60H	'NW'	NAWS Point Mugu, Calif.
HS-75	SH-3H, UH-3H 'NW'	NAS Jacksonville, Fla.
HSL-84	SH-2G	'NW'	NAS North Island, Calif.
HSL-94	SH-2G	'NW'	NAS Willow Grove, Pa.

Commander, Fleet Logistics Support Wing
NAS Fort Worth, Texas

VR-1	C-20D	'JK'	NAF Washington, D.C.
VR-46	C-9B, DC-9 'JS'	NAS Atlanta, Ga.
VR-48	C-20G	'JR'	NAF Washington, D.C.
VR-51	C-20G	'RG'	MCAF Kaneohe Bay, Hawaii
VR-52	DC-9	'JT'	NAS Willow Grove, Pa.
VR-53	C-130T	'WV'	NAF Washington, D.C.
VR-54	C-130T	'CW'	NAS New Orleans, La.
VR-55	C-130T	'RU'	NAWS Point Mugu, Calif.
VR-56	C-9B	'JU'	NAS Norfolk, Va.
VR-57	C-9B, DC-9 'RX'	NAS North Island, Calif.
VR-58	C-9B	'JV'	NAS Jacksonville, Fla.
VR-59	C-9B, DC-9 'RY'	NAS Fort Worth, Texas
VR-61	DC-9	'RS'	NAS Whidbey Island, Wash.
VR-62	C-130T	'JW'	NAS Brunswick, Maine

NAS New Orleans	UC-12B '7X'	NAS New Orleans, La.
NAS Fort Worth	UC-12B '7D'	NAS Fort Worth, Texas
NAS Atlanta	UC-12B '7B'	NAS Atlanta, Ga.
NAS Willow Grove	UC-12B '7W'	NAS Willow Grove, Pa.
NAF Washington	UC-12B '7N'	NAF Washington, D.C.
NARC Santa Clara	UC-12B '7Y'	NARC Santa Clara, Calif.

The US Navy received 20 C-130Ts, to replace the C-130Fs used in the fleet logistic squadrons. This C-130T belongs to VR-55.

and VRC-50. The C-20 aircraft, based in Maryland and Hawaii, are used for rapid-response airlift and for transport of senior Navy Department official and flag officers. Two DC-9 squadrons were disbanded in the early 1990s, but the redistribution of their aircraft to other VR units resulted in no reduction in C-9B/DC-9 aircraft numbers. The replacement for the ageing C-9Bs and DC-9s has been selected: three Boeing C-40As, a version of the Boeing 737-700 IGW (increased gross weight) airliner, are on order.

Helicopter Wing Reserve

All of the Navy's reserve helicopter squadrons are administratively assigned to Helicopter Wing Reserve. One carrier-based helicopter anti-submarine (HS) squadron operates as the helicopter component for CVWR-20 in performing anti-submarine, rescue, and utility missions. HS-75 is the only Navy squadron still operating the SH-3H Sea King.

Light Airborne Multi-purpose System (LAMPS) Units

Since the mid-1990s, the Navy's remaining SH-2G Super Seasprite LAMPS helicopters have been operated by the Naval Air Reserve's helicopter anti-submarine squadron light (HSL) units, two of three such units still being in operation. In recent years, detachments of these squadrons have deployed in support of drug-interdiction operations. Their SH-2Gs are the only aircraft currently configured to operate the Magic Lantern laser mine detection system. These HSL squadrons have been kept in operation to provide LAMPS helicopters to the Naval Reserve's early 'Oliver Hazard Perry'-class guided-missile frigates, which were incapable of operating the SH-60B LAMPS Mk III helicopter flown by active-duty squadrons. The imminent retirement of these ships has resulted in the planned disestablishment of the two reserve HSL units by 2000.

Strike rescue and special operations

The Naval Air Reserve operates the Navy's only two helicopter combat support squadron special (HCS) squadrons, which operate the HH-60H Seahawk for strike rescue missions and support of special warfare forces, such as the Navy's Sea-Air-Land (SEAL) commandos. In recent years, HCS squadron personnel and helicopters have augmented active-duty HS squadrons onboard aircraft-carriers operating off Bosnia. The HH-60Hs are scheduled to be withdrawn for conversion to SH-60Rs and to be replaced by new CH-60 variants.

Mine countermeasures

The Navy's two helicopter mine countermeasures (HM) squadrons are joint active-reserve squadrons, which fly the MH-53E Sea Dragon. (See Mine Countermeasures in the Navy section.) During the mid-1990s the two reserve HM units, HM-18 and HM-19, were disbanded and their aircraft and personnel combined with the two active squadrons to form the joint squadrons.

Utility helicopter operations

Helicopter Wing Reserve operates one helicopter combat support (HC) squadron, HC-85, which provides rescue, utility and torpedo recovery services in the southern California area. HC-85, which operates the Sikorsky UH-3H Sea King, was redesignated from HS-85 when Reserve Carrier Air Wing 30 was disbanded. HC-85 assumed the duties once performed by active-duty squadron HC-1. The Naval Air Reserve is assigned several Beech UC-12B (King Air 200) aircraft for air station liaison support.

Fleet Air Reconnaissance Squadrons

In 1997, the Naval Air Reserve formed its first Fleet Air Reconnaissance (VQ) squadron, VQ-11, to operate the two EP-3J Orion (one having since crashed) electronic adversary aircraft formerly operated by VP-66, plus one P-3C for training and support. The squadron's name does not reflect its true mission; its aircraft do not perform reconnaissance, but simulate enemy electronic threats during work-up training of Navy ships.

Airlift

The Navy's entire organic airlift force resides in the Naval Air Reserve. Drawing on former active-duty aviators, many of whom are airline pilots, Commander Fleet Logistic Support Wing's (CLSW's)

14 fleet logistic support (VR) squadrons transport high-priority cargo and passengers throughout the United States and overseas. A major responsibility of the VR units is to transport carrier air wing personnel and cargo from their bases to ports to meet deploying aircraft-carriers. The logistic support wing's squadrons fly a mixture of C-9B, DC-9, C-130T, C-20D and C-20G aircraft. Normally, one C-130T and one C-9B or DC-9 each are deployed to the Mediterranean and Western Pacific to support airlift requirements in the Sixth and Seventh Fleets, respectively, partially filling the gap created by the disestablishment in the early 1990s of active-duty squadrons VR-22

Naval Air Training Command

The Navy has maintained a world-renowned flight-training programme that produces pilots and navigators proficient in operating aircraft from aircraft-carriers and other ships in any weather conditions. The Naval Air Training Command, headed by the Chief of Naval Air Training (CNATRA) and head-quartered at NAS Corpus Christi, Texas, trains aviators and navigators for the Navy, Marine Corps, Coast Guard and, increasingly, for the Air Force. In turn, the Air Force conducts some phases of pilot and navigator training for the Navy, Marine Corps and Coast Guard. The institution of joint training between the Navy and the Air Force since the mid-1990s has resulted in large-scale consolidation and resource sharing, with a large emphasis on inter-changeability in training. CNATRA also trains a number of pilots and navigators for foreign navies and air forces. In addition to training functions, CNATRA also operates Naval Aviation Schools Command and the Naval Flight Demonstration Squadron (the 'Blue Angels'), and maintains the National Museum of Naval Aviation.

NAVAL AVIATION TRAINING ASSETS

Naval training aviation is currently going through a period of change which started with the introduction of the T-45 Goshawk and will be completed with the entry of the T-6A Texan II into service.

PRIMARY TRAINING AIRCRAFT

The two-seat single-engined turboprop Beech T-34C Turbo Mentor is operated in large numbers by the Naval Air Training Command for basic and inter-mediate training of naval aviators (including helicopter pilots) and flight officers. The T-34C replaced the piston-engined T-34B Mentor and many T-28B/C Trojans in training service during the late 1970s. A few T-34Cs are used by the F/A-18 and F-14 fleet readiness squadrons for target spotting duties. A single NT-34C serves as a development aircraft at the Naval Air Warfare Center – Aircraft Division; it has been used to develop a collision-avoidance system.

The eventual replacement for the T-34C has been selected through the Joint Primary Aircraft Training System (JPATS) competition. A militarised variant of the Beech Mk II, itself a variant of the Pilatus PC-9, is being developed by Raytheon and has been designated T-6A Texan II. The T-6A, equipped with a modern 'glass' cockpit, will be just one part of a total

The most famous unit of Naval Air Training Command is the Naval Flight Demonstration Squadron, otherwise known as the 'Blue Angels'. The team currently flies the F/A-18A Hornet.

interactive training system. The T-6A will enter service initially in the Air Force to replace the T-37, and will begin replacing Navy T-34Cs in 2003.

ADVANCED TRAINING AIRCRAFT

Advanced maritime multi-engine training is conducted in the Beech T-44A Pegasus, a military version of the Beech King Air 90 corporate aircraft. The twin-turboprop T-44A, which entered service in the late 1970s, replaced the TS-2A and US-2B Trackers. The proposed T-44B follow-on version was cancelled before development. The shift of Air Force multi-engine pilot training to the Navy in 1997 increased the demand for training aircraft; some 20 Beech UC-12Bs are being converted to TC-12Bs to meet the demand (see above).

Two training squadrons still operate the two-seat, twin-engined North American T-2C Buckeye for intermediate and advanced strike training of aviators and flight officers. The T-2C has suffered several groundings because of structural problems, but is planned to serve until 2003, at which time sufficient T-45s will be available to replace it. A few T-2Cs are

also operated by the US Naval Test Pilot School.

The two-seat, single-engined TA-4J remains the only variant of the McDonnell Douglas Skyhawk in US naval service; the remaining A-4M, OA-4M, TA-4F, NTA-4F and NTA-4J versions all have been retired since the mid-1990s. The TA-4J is used as an advanced strike trainer in one training squadron, and also serves as an adversary and target-tug in one composite squadron. The T-45C is in production to replace the TA-4J in the strike training role. All TA-4Js are scheduled for phase-out by the end of 1999, but problems with the T-2C fleet may delay the Skyhawk's demise.

The Boeing (formerly McDonnell Douglas) T-45 Goshawk is replacing the T-2C and TA-4J in the strike-training role. A development of the British Aerospace Hawk trainer with redesigned landing gear, a strengthened mid-section, and a tailhook, the T-45 is one component of an entire interactive training system, the T45TS (T-45 Training System). The two-seat, single-engined T-45A, which entered service during the mid-1990s, equips two training squadrons. Deliveries of the T-45A were completed in mid-1998. The proposed T-45B non-carrier-capable version was never developed.

The T-45C, first delivered in November 1997, features Cockpit-21, a 'glass' cockpit compatible with modern military tactical aircraft that includes an iner-tial navigation set integrated with the GPS and the 1553 digital databus. The T-45C began training

Left: After being replaced by the T-47A, the Sabreliner has returned to the aircrew training role in the guise of the T-39N.

Below: Naval basic training will be revitalised with the introduction of the T-6A Texan II. This is PT-4, the first production standard aircraft, at its rollout.

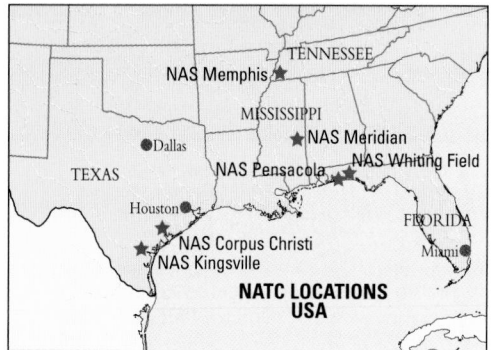

student pilots in one training squadron and will eventually replace the T-2C and TA-4J in two others. All T-45As are planned for eventual conversion to T-45Cs by 2007.

The North American T-39N Sabreliner is used by one training squadron for advanced strike training for flight officers destined to crew the F-14, F/A-18D/F, S-3 and EA-6B. The T-39Ns are modifications of a variety of earlier T-39 and CT-39 models with avionics suited for training in radar navigation and air intercepts. During the mid-1990s, the T-39Ns replaced the Cessna T-47A Citations, which earlier had replaced the T-39D. The T-39Ns were purchased by the Navy (and issued new Bureau numbers) after having been owned and operated under contract by Boeing North American. Raytheon Systems Company won the contract to operate and maintain the Sabreliners beginning in 1998.

A few Northrop T-38A Talon two-seat trainers remain in service with the US Naval Test Pilot School.

Only one type of training helicopter, the Bell TH-57 Sea Ranger, is used to instruct the many helicopter pilots who serve with the Navy, Marine Corps and Coast Guard. The TH-57B and TH-57C are used mainly for visual flight rules training and instrument flight rules training, respectively.

TRAINING AIR WINGS

CNATRA supervises five training air wings (TAWs), each based at a separate airfield in Florida, Mississippi or Texas. Two wings, TAW-1 at NAS Meridian, Miss., and TAW-2 at NAS Kingsville, Texas, conduct strike syllabus training. (TAW-3 at NAS Chase Field, Texas, and its squadrons, Training Squadrons (VT) 24, 25 and 26, were disbanded in the post-Cold War draw-down.) TAW-4 conducts primary training and multi-engined turboprop training at NAS Corpus Christi. TAW-5 trains primary

The days of the TA-4J Skyhawk as the advanced trainer for the US Navy are drawing to a close. In service since 1968, it is now operated as a trainer only by VT-7 at NAS Meridian.

students and conducts all basic and advanced helicopter training at NAS Whiting Field, Fla. All basic and some advanced navigator training is conducted by TAW-6 at NAS Pensacola, Fla. (TAW-6 also includes the German Luftwaffe's 2nd Training Squadron.)

Upon completion of primary training, student naval aviators proceed on one of six different training tracks. Navigators (naval flight officers) and Air Force navigators have five training tracks after primary training.

AVIATOR TRAINING – PRIMARY

After aviation pre-flight indoctrination (API), student aviators undergo 23 weeks of primary flight training at TAW-4 (with VTs 27 or 28 flying the T-34C Turbo Mentor), TAW-5 (with VTs 2, 3 or 6, flying the T-34C), or with the Air Force at Vance AFB, Okla., flying the T-37B. After primary training, students proceed on one of the six pilot training tracks.

STRIKE TRAINING

After primary training, student aviators beginning the strike aviator syllabus are channelled into one of two tracks, determined by the type of training aircraft available. Students assigned to TAW-2 will complete their advanced strike training flying the T-45A with VTs 21 or 22 in a 39-week syllabus. Students assigned to TAW-1 will undergo a 23-week intermediate phase with VT-19 in the T-2C Buckeye, followed by a 25-week advanced phase with VT-7 in the TA-4J Skyhawk, or a 29-week advanced phase with VT-23 in the T-45C. The T-45C, introduced with VT-23 in early 1998, eventually will replace the TA-4J in 1999 and the T-2C in three to four years. At that point, the TAW-1 and TAW-2 syllabi will be identical. TAW-2 T-45As are programmed to be modified as T-45Cs.

Strike training includes carrier qualification and weapons employment. Graduates proceed to FRS training in the F/A-18, F-14, EA-6B and S-3B.

Replacing the T-2C Buckeye and TA-4J Skyhawk for the advanced training phase of flight training is the T-45 Goshawk. This is one of the 'Red Hawks' (VT-21) T-45As based at NAS Kingsville, Texas. T-45As are due to be modified as T-45Cs with a 'glass' cockpit layout known as 'Cockpit 21'.

E-2/C-2 TRACK

Students selected to fly the E-2 or C-2 proceed through a combined turboprop and jet syllabus. These students complete a 15-week intermediate phase in the T-44A and TC-12B with TAW-4's VT-31, followed by a 22-week carrier qualification phase flying the T-2C with TAW-1's VT-19. Upon graduation, aviators undergo FRS training at VAW-120 at NAS Norfolk, Va.

MARITIME TRAINING

Students destined to fly large multi-engined turboprop aircraft undergo a 26-week intermediate syllabus flying the T-34C with the TAW-4's VT-27 or 28 or with TAW-5's VTs 2, 3 or 6. Upon completion, students train in a 20-week syllabus (25 for Air Force students) flying the T-44A Pegasus and TC-12B Huron with TAW-4's VT-31. TAW-4 has assumed all training for future Air Force C-130 pilots. Graduates of the maritime syllabus proceed to FRS training in the P-3, EP-3, HU-25, or C-130.

TACAMO TRAINING

Students selected to fly the Boeing E-6 Mercury TACAMO communications aircraft complete the 26-week intermediate phase with TAW-4's VT-27 or VT-28 or TAW-5's VTs 2, 3 or 6 in the T-34C, or with the Air Force in the T-37B. A 26-week advanced phase follows, conducted by the Air Force in the T-1A Jayhawk. Graduates report to the Navy Training Support Unit at Tinker AFB, Okla., for transition into the E-6 aircraft using the TC-18F for training. The latter is an ex-civil Boeing 707.

Naval Air Training Command

Chief of Naval Education and Training		NAS Pensacola, Fla.
Commander, Naval Air Training Command		NAS Corpus Christi, Tex.
Commander, Training Air Wing 1		**NAS Meridian, Miss.**
VT-7	TA-4J	'A'
VT-9	T-2C	'A'
VT-23	T-45C	'A'

In service since 1969, the T-2C Buckeye is another type to be retired soon. In NATC, only VT-9 (coded 'A') and VT-86 (coded 'F') continue to operate them.

Commander, Training Air Wing 2		**NAS Kingsville, Tex.**
VT-21	T-45A	'B'
VT-22	T-45A	'B'
Commander, Training Air Wing 4		**NAS Corpus Christi, Tex.**
VT-27	T-34C	'G'
VT-28	T-34C	'G'
VT-31	T-44A, TC-12B	'G'

Commander, Training Air Wing 5		**NAS Whiting Field, Milton, Fla.**
HT-8	TH-57B/C	'E'
HT-18	TH-57B/C	'E'
VT-2	T-34C	'E'
VT-3	T-34C	'E'
VT-6	T-34C	'E'
Commander, Training Air Wing 6		**NAS Pensacola, Fla.**
VT-4	T-34C, T-1A	'F'
VT-10	T-34C, T-1A	'F'
VT-86	T-2C, T-39N	'F'

Undergraduate Naval Flight Officers are introduced to aerial navigation in the T-39Ns of VT-86. The aircraft are equipped with the APG-66NT radar.

NFDS 'Blue Angels'	F/A-18B, TC-130G		NAS Pensacola, Fla.
NAS Meridian	HH-1N	'A'	NAS Meridian, Miss.
NAS Corpus Christi	UC-12B, HH-1N	'G'	NAS Corpus Christi, Texas
NAS Kingsville	None	'B'	NAS Kingsville, Texas
NAS Pensacola	UC-12B, UH-3H	'F'	NAS Pensacola, Fla.
NAS Whiting Field	None	'E'	NAS Whiting Field, Fla.
NSA Mid-South	UC-12B	'6M'	NSA Memphis, Tenn.

HELICOPTER TRAINING

Helicopter pilot trainees complete a 26-week T-34C syllabus with TAW-4's VT-27 or VT-28, or TAW-5's VTs 2, 3 or 6. The students are introduced to helicopters with a 21-week syllabus flying the TH-57B and TH-57C Sea Ranger with TAW-5's

Like the US Army, the Navy uses a version of the Bell 206 as its helicopter training vehicle. This TH-57C Sea Ranger belongs to Training Air Wing-5. Two squadrons of the wing operate the type.

helicopter training (HT) squadrons HT-8 or HT-18. Graduates proceed to FRS training in the H-1, H-3, H-46, H-53, H-60, or H-65.

NFO PRIMARY TRAINING

Student naval flight officers (NFOs) and Air Force navigators undergo 14 weeks of primary training after completion of API. Students fly the T-34C at TAW-6 (with VTs 4 or 10). Upon completion, students are funnelled into one of five training tracks.

MARITIME NAVIGATION TRAINING

Upon completion of primary training, NFO students selected for long-range maritime navigator training are sent to Randolph AFB, Texas, for 26 weeks of advanced navigation training in the T-43A aircraft operated by the Air Force's 562nd FTS. Navy graduates proceed to FRS training in the P-3, EP-3 and E-6. EP-3 NFOs will also complete electronic warfare officer school at Corry Field, Pensacola, Fla. Air Force graduates will fly the E-3, E-8, C-5, C-130, C-135 and C-141.

STRIKE SYLLABUS

NFOs and navigators destined to fly non-fighter tactical jets or Air Force bombers will remain with TAW-6 and undergo a 14-week intermediate syllabus flying the T-34C and T-1A with VTs 4 or 10. Strike training consists of 19 weeks flying the T-2C and T-39N Sabreliner. Navy graduates proceed to FRS training in the S-3, ES-3 or EA-6B; NFOs destined for the latter two types will also complete electronic warfare officer school at Corry Field. Air Force graduates will fly the B-1 or B-52 bombers.

STRIKE FIGHTER SYLLABUS

Students bound for fighter aircraft as radar intercept officers (RIOs) or weapon system operators (WSOs) receive 14 weeks of intermediate training with TAW-6 in the T-34C and T-1A operated by VTs 4 and 10. VT-86 conducts 26 weeks of training (including air combat manoeuvring) in the T-2C and T-39N. Navy graduates undergo FRS training to become F-14 RIOs. Marine Corps graduates (and eventually Navy graduates) will become F/A-18 WSOs. Air Force graduates will become F-15E WSOs.

ATDS SYLLABUS

Students destined to become airborne intercept controllers onboard E-2C aircraft undergo 14 weeks of intermediate training with TAW-6 in the T-34C and T-1A. Advanced training in the air tactical data system (ATDS) is provided in 32 weeks of training at VAW-120, the E-2 FRS at NAS Norfolk, Va. During the FRS ATDS training, the students are designated NFOs.

'BLUE ANGELS'

The Naval Flight Demonstration Squadron, popularly known as the 'Blue Angels', is based at NAS Pensacola, Fla. and conducts its winter training at NAF El Centro, Calif. The aerobatic team keeps an intensive air show schedule, performing throughout the United States and occasionally in other nations. The squadron's Navy and Marine Corps pilots fly F/A-18A Hornets (one two-seat F/A-18B is also assigned). A Marine Corps TC-130G Hercules provides logistic support for the team.

Below: The US Navy has recently been made responsible for the multi-engined training for all branches of the US military. To relieve the strain on the T-44A Pegasus of VT-31, a number of TC-12Bs have been delivered to the unit.

Above: The T-34C Turbo-Mentor has been the US Navy's basic trainer since 1976. The type's replacement is the JPATS winner, the T-6A Texan II, which is currently in the flight test phase.

Naval Air Systems Command

RESEARCH, DEVELOPMENT, TEST AND EVALUATION

Development, procurement and depot-level maintenance of Navy and Marine Corps aircraft and aviation weapons and systems is the responsibility of the Naval Air Systems Command (NAVAIR, or NASC), which in 1996 moved its headquarters from Arlington, Va. to NAS Patuxent River, Md. A Navy-wide reorganisation of its research, development, test and evaluation establishment during the early 1990s resulted in a consolidation of its aircraft-operating organisations into fewer sites and into six squadrons, arranged into two test wings, themselves under two divisions of the Naval Air Warfare Center (NAWC).

NAVAL AIR WARFARE CENTER

Much of NAVAIR's structure comprises the Naval Air Warfare Center, also headquartered at Patuxent River. Most of its work is performed by the Center's three main divisions – the NAWC Aircraft Division (NAWC-AD) at Patuxent River, Md., Trenton, N.J., and Lakehurst, N.J., the NAWC Weapons Division (NAWC-WD) at China Lake and Pt Mugu, Calif., and the Training Systems Division in Orlando, Fla.

NAWC-AIRCRAFT DIVISION

The NAWC-AD's aircraft fleet is assigned to Commander Naval Air Test Wing Atlantic. The four squadrons assigned to the wing were organised in the mid-1990s from the directorates of the former Naval Air Test Center at Patuxent River. The Naval Strike Aircraft Test Squadron performs RDT&E for fighter, attack and electronic attack aircraft; the Naval Rotary-Wing Test Squadron does the same for Navy and Marine Corps helicopters. The Naval Force Warfare Test Squadron has test responsibilities for anti-submarine, electronic reconnaissance, early warning, communications, aerial refuelling, and transport aircraft. The US Naval Test Pilot School (USNTPS), now considered a squadron, trains test pilots and flight officers on a variety of fixed-wing aircraft, gliders and helicopters. USNTPS operates some helicopters and fixed-wing aircraft on loan from the US Army.

When new aircraft are being evaluated at NAWC-AD, integrated test teams (ITTs) comprised of naval and industry pilots, engineers, and support personnel are formed to shepherd new aircraft through the development process. Current ITTs are evaluating the F/A-18E/F Super Hornet and the MV-22B Osprey.

The NAWC-AD facilities at Lakehurst, N.J. are used for aircraft-carrier catapult and arresting gear trials. The Trenton, N.J. facility, used for aircraft engine test and development, is scheduled to close as its functions are transferred to NADEP North Island, Calif.

Long retired from the front-line units, F-4 Phantoms continue to find employment as testbeds and target drones with the NWTS Point Mugu (such as this QF-4). The drones are usually only shot down after a long service as testbeds, and only if tests require it. The prototype F-4J has recently found employment as a ejection seat testbed.

NAWC-WEAPONS DIVISION

The NAWC-WD is responsible for development and testing of air-launched weapons and (along with NAWC-AD) integrating them with aircraft weapons systems. NAWC-WD operates two major sites – Naval Air Weapons Stations (NAWS) at China Lake and Pt Mugu, Calif. – with two naval weapons test squadrons (NWTS) under Commander Naval Air Test Wing Pacific. NAWS China Lake, with purview of testing air-to-ground weapons, is the home of NWTS China Lake. NAWS Point Mugu operates the Pacific Missile Range and hosts the NWTS Point Mugu, with responsibility for testing air-to-air weapons. NWTS Point Mugu also operates the QF-4 drones as targets for missile test shots.

NWTS China Lake operates the station's HH-1N rescue helicopters.

TEST RANGES

The Navy maintains an extensive test range off the island of Kauai in the Hawaiian Islands, used for missile shots and exercise torpedo firings. The Pacific Missile Range Facility operates UH-3H Sea King helicopters for torpedo recovery and RC-12Fs for range clearance. The Atlantic Fleet Weapons Test Facility, with ranges in Puerto Rico and at St Croix in the Virgin Islands, is supported by UH-3H helicopters from VC-8 and RC-12Ms from NS Roosevelt Roads, Puerto Rico. The Advanced Underwater Test and Evaluation Center (AUTEC) at Andros Island in the Bahamas is now supported by contractor helicopters.

The US Navy has used the T-38A Talon for adversary training and various test duties. The only unit that still operates the type is the US Naval Test Pilots School at Patuxent River.

The oldest design operated by the US Navy is the U-6A Beaver, active with the USNTPS.

Naval Aviation Depots

NASC operates three major industrial facilities, called Naval Aviation Depots (NADEPs), which overhaul and modify Navy and Marine Corps aircraft. Formerly called Naval Air Rework Facilities (NARFs), these depots now compete with commercial companies for some contracts. The number of depots is down by half since the end of the Cold War; those at NAS Alameda, Calif., Pensacola, Fla., and Norfolk, Va. were closed and their work shifted to the other depots, to similar facilities of the other services, or to commercial contractors. NADEP North Island, Calif. handles work for the F/A-18, S-3, E-2 and C-2; NADEP Jacksonville, Fla. overhauls the F-14, EA-6B and P-3; and NADEP Cherry Point, N.C. conducts work on the AV-8, H-46 and H-53. Many aircraft are overhauled by commercial contractors, some affiliated with aircraft manufacturers.

Air Test and Evaluation Squadrons

The Navy maintains two air test and evaluation (VX) squadrons operationally assigned to the Commander, Operational Test and Evaluation Force (COMOPTEVFOR), headquartered in Norfolk, Va. These squadrons, co-located but separate from the NASC organisation, put new aircraft and systems through operational testing after they have gone through developmental testing with the Naval Air Warfare Center's squadrons and ITTs. The VX squadrons test the new systems under 'real-world' conditions and make recommendations to COMOPTEVFOR regarding the suitability of the system for fleet introduction. VX-1 at Patuxent River is responsible for operational testing of anti-submarine, electronic reconnaissance, and strategic communications aircraft and associated weapons and systems. VX-9, based at China Lake with a detachment in Pt Mugu, conducts operational tests of strike fighter and electronic countermeasures aircraft, helicopter gunships, and associated weapons. (VX-9 was formed in 1994 from elements of VX-4 and VX-5 and assumed their roles.)

Strike Warfare Development

The Naval Strike and Air Warfare Center (NSAWC), established in 1995 at NAS Fallon, Nev., is an expansion of the Naval Strike Warfare Center ('Strike University') that was set up in the mid-1980s to improve strike warfare doctrine and tactics following the flawed December 1983 strike on anti-aircraft sites in Lebanon. The increasing use of cruise missiles in war at sea and in strikes against land targets led to an expanded syllabus for students (aviators, surface warriors, and submariners) attending the NSWC. The Navy decided to consolidate the Navy Fighter Weapons School ('Topgun'), and the Carrier Airborne Early Warning School ('Topdome') into the Fallon centre.

The NSAWC operates a small inventory of fleet aircraft, including F-14As, F/A-18A/Bs and SH-60Fs for tactical development. Fleet E-2Cs are assigned on a temporary basis as needed. The famous 'Topgun' school offers a syllabus in air combat manoeuvring for fleet aviators and flight officers in fighter and strike fighter squadrons.

The US Naval Test Pilot School operates the last U-21Fs in US service, on loan from the US Army.

Antarctic Research Support

The Navy has long maintained an Antarctic development support squadron, which ferried scientists and their support personnel in and out of the Antarctic. Antarctic Development Squadron (VXE) 6, home-based at NAWS Point Mugu, Calif., flies ski-equipped LC-130F/R Hercules transports from Christchurch, New Zealand. Custody of the Antarctic support mission was turned over to the New York ANG's 109th Mobility Air Wing in March 1998; VXE-6 is scheduled for disestablishment in April 1999 and will turn over three of its LC-130Rs to the NY ANG to augment the 109th's LC-130Hs.

Naval Research Laboratory

The Naval Research Laboratory (NRL) in Washington, D.C. maintains a Flight Support Detachment (FSD) at Patuxent River, Md. which operates several NP-3D Orions in support of a variety of scientific and military research projects. NRL-FSD assumed some of the aircraft and oceanographic survey missions formerly assigned to Oceanographic Development Squadron 8 (VXN-8) when that squadron was disbanded in 1993.

Coastal Systems Station

The Naval Surface Warfare Center operates a small facility, Coastal Systems Station, at Panama City, Fla. for the development of undersea warfare systems, particularly mine countermeasures systems. For these purposes, N/MH-53E Sea Dragon mine countermeasures helicopters are assigned, as is an HH-1N for utility support.

Miscellaneous Types

The Navy has long included in its inventory numerous aircraft in small quantities, many on loan from other services, for special purposes. Many of these serve with the US Naval Test Pilot School in order to give students the opportunity to fly in a wide variety of flight conditions. Aircraft in service with the USNTPS include the de Havilland Canada NU-1B Otter and U-6A Beaver, Beech U-21F Ute, Hughes TH-6B Cayuse, and Bell OH-58A/C Kiowa.

McDonnell Douglas QF-4N and QF-4S Phantom II target drones are expended periodically in missile tests. They also serve (while manned) as chase planes for tests. A single YF-4J serves when needed as an ejection-seat test platform.

Naval Air Systems Command

Commander, Naval Air Systems Command, NAS Patuxent River, Md.

Naval Research Laboratory		**NS Washington, D.C.**
Flight Support Detachment		NAS Patuxent River, Md.
NP-3D	'NRL'	
Naval Surface Warfare Center		**Arlington, Va.**
Naval Coastal Systems Station		Panama City, Fla.
MH-53E, NMH-53E, HH-1N		

Commander. Naval Air Warfare Center — NAS Patuxent River, Md.

Commander, Naval Air Warfare Center-Aircraft Division — NAS Patuxent River, Md.
Commander, Naval Test Wing Atlantic — NAS Patuxent River, Md.

Naval Force Warfare Aircraft Test Squadron		
P-3C, UP-3A, NP-3C/D, S-3B, E-2C, C-2A, KC-130F, T-34C, NT-34C, NC-130H		
Naval Rotary-Wing Aircraft Test Squadron		
UH-1N, AH-1W, NVH-3A, CH-46E, CH-53E, NSH-60B, SH-60B/F, YSH-60F, HH-60H, TH-57C		
Naval Strike Aircraft Test Squadron	'SD'	
F/A-18A/C/D, NF/A-18A/C/D, F-14A, NF-14A/D, AV-8B, EA-6B		
US Naval Test Pilot School		
F/A-18B, T-2C, T-38A, NP-3D, NU-1B, U-6A, U-21F, TH-6B, OH-58C, UH-60A, NSH-60B, X-26A		
NAS Patuxent River	UC-12B, UH-3H	'7A'

Commander, Naval Air Warfare Center-Weapons Division — NAWS China Lake, Calif.
Commander, Naval Test Wing Pacific — NAWS Pt Mugu, Calif.

Naval Weapons Test Squadron China Lake			NAWS China Lake, Calif.
F/A-18A/C/D, NF/A-18D, F-14A, AV-8B, NAV-8B, TAV-8B, T-39D, AH-1W, HH-1N			
Naval Weapons Test Squadron Point Mugu			NAWS Point Mugu, Calif.
NF-14A/B/D, NP-3D, QF-4N,QF-4S,YF-4J			
NAWS Point Mugu	none	'7L'	NAWS Pt Mugu, Calif.
NAWS China Lake	none	'7P'	NAWS China Lake, Calif

Some aircraft bearing Navy markings are operated by companies providing services under contract to the Navy. A number of DC-130A Hercules, operated by AVTEL Services, a contractor based at NAWS Pt Mugu, Calif., perform drone launch services for missile shots. Raytheon (formerly Hughes) also operates EA-3B, ERA-3B, NRA-3B, TA-3B and NTA-3B Skywarriors for a variety of test services from Van Nuys, Calif.

A single Douglas EC-24A (a modified DC-8) was operated by Raytheon Systems Company for the Navy as an electronic aggressor for fleet exercises, but was retired to Davis-Monthan AFB storage in January 1999. Two Boeing NKC-135A Stratotankers used for the same purpose were retired in the mid-1990s. Presently the US Navy does not have a dedicated organic electronic aggressor. It is possible that the navy will convert some ex-USAF C-26Bs as electronic aggressor aircraft, but during 1999 this was only in the study phase.

The USN received five DC-130As from the USAF. Three remain, including 560514, operated under contract by Flight Systems International. They are the oldest US military C-130s in service.

US Marine Corps Aviation

The US Marine Corps maintains a large air arm separate from, but closely aligned with, US naval aviation. The primary role of Marine Corps aviation is to support Marine ground forces ashore and during amphibious operations. To fulfil this role, the Marine Corps operates tactical fixed-wing and rotary-wing aircraft (and soon tilt-rotor aircraft) to provide close air support, assault transport, air defence, and electronic warfare support, and has augmented Navy carrier-based aviation in projecting air power at sea or against enemy forces ashore.

Although distinct from the Navy, Marine Corps aviation is supported by the Navy, particularly the Naval Air Systems Command, in procurement, maintenance, logistics support, and training. Marine Corps aviation personnel receive their training from the Naval Air Training Command and the Naval Air Technical Training Center. Marine Corps aviators and flight officers wear the same wings of gold as their Navy counterparts, and Marine Corps jet pilots qualify to operate from aircraft-carriers.

A BRIEF HISTORY

Marine Corps aviation traces its origin to 22 May 1912, the day First Lieutenant Alfred A. Cunningham reported for flight training at the Naval Academy in Annapolis, Md. Cunningham became Marine Corps aviator number 1 and naval aviator number 5. The number of aircraft and fliers grew slowly, but Marine Corps aviators participated in combat during amphibious landings at Tampico and Veracruz, Mexico in April 1914.

US entry into World War I dramatically accelerated the development of Marine Corps aviation. The First

Above: A pair of VMFA-312 F/A-18Cs streaks across the western United States, armed with HARM missiles. Hornets and Harrier IIs provide the Corps with aerial firepower, day or night, in any weather.

Right: In service with the Marines since the early 1960s, the CH-46 Sea Knight is instrumental in the Corps' vertical envelopment doctrine. This armed HMM-266 CH-46E is seen at Mogadishu Airport, Somalia, during the Marine intervention in the civil war in the country in 1994.

Marine Aeronautic Company flew anti-submarine patrols from the Azores, and the First Aviation Force flew combat missions with the Northern Bombing Group in northern France.

Post-war demobilisation resulted in a reorganisation, with squadrons or flights stationed in Guam, Haiti and the Dominican Republic. The inter-war period was rich in the development of aviation combat doctrine, such as close air support, dive-bombing, radio communications, aerial resupply, and suppression and isolation of beach defences. In 1927, these concepts were tested by Marine Corps aviators in combat in Nicaragua, and were refined for successful use in World War II. Marine aviators also gained extensive experience in carrier operations during the 1930s.

During World War II, Marine Corps aviation suffered from early setbacks but recovered to become instrumental in the defeat of the Japanese empire. Many Marine aircraft were destroyed during the attack on Pearl Harbor, Hawaii, and during the valiant but unsuccessful defence of Wake Island. Marine fighters and dive-bombers helped in the successful defence of Midway Island.

Marine Corps fighters, dive-bombers and, later, torpedo-bombers formed the core of the 'Cactus Air Force' that defended the foothold on Guadalcanal in the Solomon Islands from a sustained Japanese onslaught. For the remainder of the war, Marine fliers supported the advance up the Solomon chain and the isolation of the Japanese bastion at Rabaul,

New Britain. Marine squadrons kept pressure on other bypassed Japanese bases in the Central Pacific as well.

During the last year of World War II, Marine Corps SBD dive-bombers provided close air support to Army troops driving the Japanese forces out of the Philippines. The *kamikaze* threat that US forces faced during that final year was in part countered by Marine F4U fighter squadrons deployed onboard fleet and escort carriers. Marine squadrons also provided close air support to US troops during the intense Okinawa campaign.

After the war, Marine squadrons deployed to China to help enforce the Japanese withdrawal and to stall a Communist take-over. However, the post-war period of the late 1940s was characterised by a rapid demobilisation and a wholesale cut of squadron types, leaving mainly fighter and transport squadrons by the end of the decade. Progress was marked by the advent of jet fighter aircraft and the helicopter in Marine Corps service.

The North Korean invasion of South Korea in June 1950 reversed the decline of Marine Corps aviation. Boosted by activated reserve squadrons, the USMC entered combat in August 1950, flying close air support missions from bases in Korea and from Navy carriers offshore. Marine crews also flew night-attack and night interceptor missions, and escorted Air Force B-29 bombers striking enemy forces. Marine helicopters were used to airlift troops into combat.

The Marine Corps, denied a sea-going version of the Apache, opted for the twin-engined AH-1W SuperCobra, acquiring new-build helicopters and converting its AH-1Ts to the new standard.

After the Korean conflict, Marine Corps aviation modernised with more jet aircraft, retiring its piston-engined F4U Corsairs and AD Skyraiders by 1959. Jet attack aircraft and supersonic jet fighters entered service during the late 1950s, followed by turbine-powered helicopters. More investment of resources was made into helicopter vertical envelopment forces, with some aircraft-carriers being converted to amphibious assault ships.

The 1960s brought Marine Corps aviation into more confrontation with Communist forces in various areas of the world. Marines flying RF-8A Crusader reconnaissance jets monitored the Soviet build-up of nuclear missiles during the Cuban Missile Crisis of 1962. The same year, Marine helicopters deployed to South Vietnam to assist that nation's forces in resisting the Communist insurgency from North Vietnam, and a substantial portion of Marine Corps aviation was deployed there for the next decade. Marine attack and fighter aircraft (A-4, A-6, F-4 and F-8, including some deployed onboard Navy carriers offshore, mostly flew close air support and interdiction missions. Helicopter transports (UH-34, CH-37, CH-46 and CH-53) and gunships (UH-1, AH-1) gave Marines the mobility to carry the ground war to enemy strongholds and sanctuaries. Marine F-4s, A-4s and A-6s returned to South Vietnam and Thailand in 1972 to fly strikes in the campaign to defeat the North Vietnamese offensive and support the Linebacker offensive that eventually forced a truce in 1973. Marine Corps helicopters were on hand to evacuate Americans and others from South Vietnam in 1975 when the North Vietnamese completed their conquest of the South.

In the decades since the end of the Vietnam War, Marine Corps aviation continued its modernisation with improved tactical jet aircraft (such as the F/A-18 Hornet) and assault helicopters (CH-53E, UH-1N and AH-1T/W). The Corps also adopted the British-designed AV-8 Harrier V/STOL attack aircraft as a front-line combat aircraft.

Frequently during the 1980s and 1990s, USMC aircrews were called into combat action or crisis response in various areas of the world. Helicopter crews participated in the aborted attempt to rescue American hostages in Iran in 1979, and duelled with Cuban anti-aircraft gunners during the 1983 invasion to thwart a Communist take-over of Grenada. AH-1T gunships countered Iranian boats threatening oil tankers in 1987-88 during the Iran-Iraq War. Helicopters performed numerous evacuations of Americans and other nationals from countries wracked by civil strife (Liberia in 1990; Somalia in 1991 and 1993; Albania, Congo and Sierra Leone in 1997); provided humanitarian aid to areas devastated by war or natural disaster (Kurdish northern Iraq, Bangladesh and the Philippines in 1991); performed search and rescue missions (Bosnia in 1992 and 1995); and supported peacekeeping operations (Bosnia in 1992-1997, Cambodia in 1993 and Haiti in 1994). F/A-18 and EA-6B aircraft also supported United Nation and NATO peacekeeping efforts in Bosnia, including air strikes flown in 1995.

By far the largest Marine Corps aviation deployment since World War II took place after the August 1990 Iraqi invasion of Kuwait. More than 70 per cent of USMC aviation forces were sent to the Arabian peninsula or to amphibious assault ships off the shores of Kuwait as part of Operation Desert Shield. These forces formed part of the attack waves that drove Iraqi forces out of Kuwait in early 1991 in Operation Desert Storm.

The years since Operation Desert Storm have seen a decline in the force levels of Marine Corps fixed-wing

Currently the subject of a remanufacturing order, the AV-8B Harrier II/II+ serves in seven operational squadrons – including this Harrier II with VMA-223 – and one conversion unit.

aircraft and the phase-out of such long-serving types as the A-6 Intruder and the OV-10 Bronco. The Corps has modernised its force with F/A-18C/D Hornets, remanufactured AV-8B Harriers, and upgraded CH-46E Sea Knights, and looks forward to remanufactured UH-1Y and AH-1Z helicopters, as well as the new MV-22B Osprey tilt-rotor assault transport.

CLOSE AIR SUPPORT

To perform its premier mission of close air support, the Marine Corps operates two basic types of tactical jets, both built by Boeing (formerly McDonnell Douglas) – the V/STOL AV-8 Harrier II and the carrier-capable F/A-18 Hornet.

The AV-8B Harrier II entered service in 1984 and eventually replaced the earlier, less capable AV-8A/C Harriers and many A-4M Skyhawks. The Harrier II is designed to operate from primitive airstrips, clearings, and amphibious ships in support of Marines on the ground. For this role, AV-8Bs are armed with cannon, bombs (including laser-guided types), rockets and AGM-65 Maverick missiles. The aircraft also has a limited air-to-air capability with its AIM-9 Sidewinder missiles. AV-8Bs participated extensively in the 1991 Gulf War, and routinely deploy onboard amphibious assault ships.

The Harrier II fleet, which equips seven operational squadrons, has gone through two avionics upgrades to the baseline version, and is currently going through a remanufacturing programme. A night-attack sensor upgrade equips four squadrons. The Harrier II+, which incorporates the night-attack upgrades and the APG-65 multi-mode radar, equips two squadrons. A remanufacture programme for 72 aircraft is underway, involving replacing the Pegasus F402-RR-406 engine with the increased-thrust -408A version. The upgrade includes the APG-65, a navigation infra-red set, NVG-compatible cockpit and exterior lighting, and a moving-map display. (The remanufactured aircraft are even being issued new bureau numbers.) The Marine Corps has requested funds for an additional 26 remanufactured aircraft.

AV-8Bs also are going through an avionics upgrade with the GPS, the common missile approach warning system, frequency-agile digital radios and the digital Advanced Target Hand-off System (ATHS). Weapons enhancements include the ability to deploy the JDAM (Joint Direct Attack Munition). The Marine Corps is trying to fund an advanced targeting infra-red set for the Harrier.

The two-seat TAV-8B, used by a Harrier training squadron, is also scheduled for upgrade with the F402-RR-408A engine.

The F/A-18 Hornet, the Corps' front-line strike fighter, has served since 1982. This versatile supersonic jet, like the F4U Corsair before it, is equally adaptable as an air-superiority fighter and as a ground-attack aircraft. Armed with cannon, Sidewinder, Sparrow and AMRAAM missiles, and a wide variety of air-to-ground ordnance, the Hornet is the most heavily armed aircraft in Marine Corps inventory.

The single-seat F/A-18A, which replaced the F-4 Phantom II and some A-4Ms, remains in service in two active-duty and four reserve fighter-attack (VMFA) squadrons. The single-seat F/A-18C, with improved engines and avionics, serves in six active-duty squadrons, four of which are assigned to Navy carrier air wings. During the 1990s, these carrier-based squadrons have supported enforcement of the 'No-Fly Zone' over Iraq.

The two-seat F/A-18D, which replaced the A-6 Intruder, is assigned the FAC(A)/TAC(A) – forward air controller (airborne)/tactical air controller (airborne) – role in addition to its all-weather strike role. The addition of the ATARS (Advanced Tactical Airborne Reconnaissance System) in 1999 will give an imagery reconnaissance capability superior to that lost with the 1990 retirement of the RF-4B. The ATARS is a self-contained system with three EO and IR sensors that will provide digital real-time or near-real-time tactical reconnaissance imagery via datalink to a ground site. High-resolution vertical or oblique imagery (dawn to dusk) will be provided, and an IRLS will provide high-resolution infra-red imagery. The aircraft's APG-73 radar will be integrated to provide ground maps and radar imagery to the receiving site. Four ATARS systems are planned for each F/A-18D squadron, plus others for training and spares.

During the mid-1990s, F/A-18Ds deployed to Aviano Air Base in Italy in support of NATO combat operations in Bosnia. The Marine Corps is seeking to procure additional F/A-18Ds to ensure adequate force levels until the advent of the Joint Strike Fighter (JSF).

The Hornet training squadron operates a few two-seat F/A-18Bs along with F/A-18A/C/D versions. The older Hornets in service are being upgraded with the GPS, Link-16, JSOW (Joint Stand-off Weapon) and JDAM (Joint Direct Attack Munition) that are standard with later Hornet deliveries.

The Marine Corps made a deliberate budget decision to forgo acquisition of the Navy's F/A-18E/F Super Hornet as a replacement for its Hornets. The Super Hornet production schedule was such that any aircraft for the Marine Corps would be available about the time that the JSF entered production, and that aircraft is intended to replace both the F/A-18 and the AV-8 in Marine Corps service.

The USMC will be procuring a STOVL version of the JSF, which will be chosen from competing entries from Boeing and Lockheed Martin and which is scheduled to enter service in 2010. The JSF will be a stealthy single-engined supersonic strike fighter, capable of air-to-air and air-to-ground combat as well as reconnaissance and SEAD roles. The Marine Corps anticipates receiving 609 JSFs.

North America

FIGHTER-ATTACK SQUADRONS

Much of the Marine Corps' air combat power resides in its eight Marine Fighter-Attack (VMFA) squadrons, equivalent to Navy Strike Fighter (VFA) squadrons. Four squadrons, based at MCAS Beaufort, are assigned to 2nd MAW, two equipped with F/A-18A and two with F/A-18C Hornets. The two F/A-18C units, VMFAs 251 and 312, are assigned to Navy carrier air wings (CVWs). Three squadrons, all equipped with F/A-18Cs, are based at MCAS Miramar with the 3rd MAW; two of these squadrons, VMFAs 314 and 323, are assigned to Navy CVWs. Three VMFA squadrons not assigned to Navy CVWs share a rotation in six-month deployments to MCAS Iwakuni, Japan on UDP assignment to MAG-12 of the 1st MAW. Because of the carrier commitment and the deactivation of four VMFA squadrons (235, 333, 451, and 531) during the early and mid-1990s, the UDP programme came under such strain that the Marine Corps decided to move one unit, VMFA-212, permanently from California to Iwakuni, Japan.

Replacement training of pilots and maintenance personnel for the VMFA units is accomplished by Marine Fighter-Attack Training (VMFAT) Squadron 101 at MCAS Miramar and by two Navy VFA fleet readiness squadrons, VFA-125 at NAS Lemoore, California and VFA-106 at Cecil Field, Florida (soon to move to Oceana, Virginia).

ALL-WEATHER FIGHTER-ATTACK SQUADRONS

Grouped with the VMFA squadrons at MCAS Beaufort and MCAS Miramar are six all-weather fighter-attack (VMFA(AW)) squadrons, three at each base, all equipped with two-seat F/A-18D Hornet strike fighters. In addition to close air support, these squadrons, all former A-6 Intruder squadrons, have the additional roles of forward air controller-airborne (FAC/A), and tactical air controller-airborne (TAC/A), and, with the ATARS (advanced tactical air reconnaissance system), the photo-reconnaissance mission as well. For several years until 1997, VMFA(AW) units were deployed to Aviano AB in northern Italy in support of UN and NATO operations in Bosnia. A VMFA(AW) squadron is rotated to MCAS Iwakuni, Japan every six months under the UDP to support the 1st MAW.

Replacement training for VMFA(AW) pilots, weapon systems officers (WSOs) and maintenance personnel is accomplished by VMFAT-101, VFA-125, and VFA-106, as for VMFA units (see above).

ATTACK SQUADRONS

Seven attack (VMA) squadrons, four at MCAS Yuma, California and three at MCAS Cherry Point, North Carolina, operate the AV-8B Harrier II V/STOL (vertical/short take-off and landing) attack aircraft for close air support missions. The squadrons operate a mixture of baseline, night-attack, and radar-equipped Harrier II Plus versions. Eventually, the VMA units will fly the remanufactured AV-8B incorporating new engines and the latest avionics improvements.

VMA detachments also deploy as part of the MEU's ACE on each amphibious assault ship (LHA or LHD). A detachment of AV-8Bs from VMA-311 is normally rotated to NAF Kadena, Okinawa, Japan every six months under the UDP for deployment with the 31st MEU's ACE onboard USS *Belleau Wood* (LHA 3). On these deployments, the VMA detachment becomes a part of the HMM squadron around which the ACE is formed. HMM squadron markings are applied to the Harriers during these deployments.

Harrier pilot and maintenance training is conducted at MCAS Cherry Point, North Carolina by Marine Attack Training (VMAT) Squadron 203, using AV-8B and two-seat TAV-8B aircraft.

One AV-8B squadron, VMA-331, was deactivated in 1992.

Above: Six all-weather fighter-attack squadrons fly the F/A-18D, VMFA(AW)-533 from MCAS Beaufort.

Below: When operating with a Marine Expeditionary Unit, AV-8Bs form part of a composite reinforced helicopter squadron. AV-8B+ '53' was attached to HMM-365 on USS Nassau early in 1997.

Above: VMFAT-101 flies all operational versions of the Hornet as the Marine conversion training unit. This example is an F/A-18B operational trainer.

Below: VMA-542 'Flying Tigers' (with 'WH' codes and tiger-striped fins) became the first Harrier II Plus squadron, during July 1993.

AERIAL REFUELLING

Lockheed KC-130 Hercules transports have been used by the Marine Corps since 1960 as long-range aerial tankers and secondarily as transports. The KC-130 is also used to refuel helicopters, tactical vehicles and ground support equipment on the ground. As a transport, the KC-130 can carry 92 troops or 64 paratroops, or more than 38,000 lb (17237 kg) of cargo. The 37 KC-130F versions still in service are the oldest aircraft in front-line US naval service, and some have recently been struck from inventory because of corrosion. The remainder of the active force is made up of 14 KC-130Rs that have served for two decades. The Marine Corps Reserve operates 28 modern KC-130T and 'stretched' KC-130T-30 versions, the last of which was delivered in October 1996. An avionics systems improvement programme (ASIP) is updating older KC-130F/R versions to KC-130T standard with the installation of ARC-210 radios and the GPS, among other alterations.

During 1999, the KC-130J, with new engines and six-bladed propellers, a 'digital' cockpit, head-up displays, night-vision lighting compatibility, aerial refuelling capability improvements, a 21-per cent increase in speed and a 35-per cent increase in range, will begin replacing the KC-130F. Five KC-130Js have been ordered, and the Marine Corps hopes eventually to acquire 51 KC-130Js.

The USMC also operates one TC-130G version to provide air show logistic support for the Naval Flight Demonstration Squadron, the 'Blue Angels'.

AERIAL REFUELING SQUADRONS

Each MAW has one Marine Aerial Refueler/Transport (VMGR) Squadron permanently assigned. These squadrons operate a mixture of KC-130F and KC-130R versions of the Lockheed Hercules. Although they are often engaged in routine transport missions, their primary mission is aerial refuelling of Marine Corps tactical jet aircraft and helicopters. This role is extended to include the refuelling of aircraft and ground support equipment on the ground.

VMGR-152 is permanently deployed to MCAS Futemma, Okinawa, Japan. VMGR-352 moved from MCAS El Toro, California in September 1998. VMGR-252 supports Atlantic area operations from MCAS Cherry Point. Replacement training for KC-130 crews is conducted by Marine Aerial Refueler/Transport Training (VMGRT)-253 at Cherry Point, using only the KC-130F version.

The only Marine Hercules unit permanently forward-deployed is VMGR-152 at MCAS Futemma, Okinawa. KC-130F 149812 carries the unit's 'QD' code.

Above: With a large – but toned-down – 'Moon Dog' on the tail, this VMAQ-3 EA-6B has the codes it received during a tour on the USS America.

Right: VMAQ-1 was deployed to Aviano to fly missions in support of NATO in Bosnia in 1997.

ELECTRONIC WARFARE

The Marine Corps operates four squadrons of Northrop Grumman EA-6B Prowler electronic warfare aircraft, which replaced the EA-6A Intruder. The carrier-capable EA-6B is capable of jamming enemy radars and communications, and attacking enemy radars with AGM-88 HARMs. The Prowlers are planned for upgrade to the Block 89A configuration, which includes an upgraded AYK-14 mission computer, an inertial navigation set integrated with the GPS, and two anti-jam ARC-210 radios. The Block 89A version, which first flew in 1997, is scheduled to enter service in 1999 as the standard configuration for all EA-6Bs. (Wing centre-sections are also being replaced as funds permit to alleviate the airframe limitations caused by stress cracks.) Approval has been given to Northrop Grumman to develop an ICAP III (Increased Capability III) 'Warfighter Upgrade System' configuration, expected to enter service in 2002. ICAP III upgrades include installation of a new receiver to replace the ALQ-99, improved low-band jamming capability, and integration of

sensor information from other platforms.

Although not designated expeditionary squadrons, Marine EA-6B units together with Navy EA-6B squadrons assumed the electronic jamming role for the Air Force following the 1998 retirement of the EF-111 Raven. Marine EA-6Bs have been deployed routinely to Japan and frequently to sites in Italy and Turkey in support of joint operations. On occasion, they have been assigned to carrier air wings.

The EA-6B has been out of production since the early 1990s, and the high level of defence commitments has strained the availability of this ageing aircraft, which is intended to serve until 2015. No replacement for the EA-6B has been identified; an electronic warfare version of the F/A-18F Super Hornet (EF/A-18G) has been mentioned as a possible replacement.

TACTICAL ELECTRONIC WARFARE SQUADRONS

The ability of Marine Corps combat aircraft to conduct warfare successfully is aided by the four Marine

tactical electronic warfare (VMAQ) squadrons based at MCAS Cherry Point. These units are each equipped with five EA-6B electronic countermeasures aircraft, able to suppress enemy air defences with jamming and AGM-88 HARM missiles. The Marine Corps EA-6B fleet was operated in detachments by one active-duty squadron, VMAQ-1, and by one reserve squadron, VMAQ-4. A reorganisation in 1992 broke VMAQ-1 into three squadrons (VMAQs 1, 2 and 3) and permanently activated VMAQ-4 as a regular squadron, moving it to Cherry Point. Until recently, these four units rotated every six months under the UDP to MCAS Iwakuni. Deployments to Aviano AB, Italy, in support of UN and NATO operations in Bosnia, and to Incirlik, Turkey, have increased the operational tempo of these squadrons, so now the Iwakuni and Incirlik sites are gapped every six months. Also, for a few years in the mid-1990s, one VMAQ squadron was assigned to a Navy carrier air wing.

EA-6B crew and maintenance training is conducted by the Navy's fleet readiness squadron, VAQ-129, at NAS Whidbey Island, Washington.

HEAVY LIFT

The Sikorsky CH-53D Sea Stallion, a twin-engined heavy-lift helicopter in service since 1968, is used to lift troops, supplies and equipment (including small vehicles) to the battlefield. The CH-53D, armed with two 0.50-in machine-guns for self-defence, can carry 37 troops or 8,000-12,000 lb (3628-5443 kg) of cargo. All CH-53Ds are based at Kaneohe Bay in Hawaii. The earlier CH-53A versions were retired in the early 1990s, and the RH-53Ds (former Navy minesweeping helicopters) have been replaced in Marine Corps Reserve service by CH-53Es.

Succeeding the CH-53D in production at Sikorsky was the CH-53E Super Stallion – a larger, three-engined upgrade of the CH-53D, capable of transporting 55 troops or 16 tons (16.25 tonnes) of cargo. A true heavy-lift helicopter, the CH-53E can externally carry the 26,000-lb (11794-kg) LAV (light armoured vehicle) or any tactical jet aircraft in Marine Corps inventory. It is armed with two 0.50-in machine-guns for self-defence, and can be refuelled in flight from a KC-130. Production of the CH-53E is ending with FY 1998 procurement; the helicopter is expected to remain in service until 2025, when a proposed replacement, the Joint Transport Helicopter, is expected to supplant it.

The Marine Corps plans a two-phase service-life extension programme (SLEP) to keep the CH-53E fleet capable through the first quarter of the 21st century. Phase I includes modifying the airframe in critical structural wear points, improving the tail-rotor drive-shaft components, and replacing older wiring. Phase II is more extensive, including installation of upgrades to avionics (including ARC-210 radios, a GPS and a navigation infra-red system), cockpit instrumentation (including a night-vision-compatible HUD), internal and external cargo systems, dynamic components, and safety and survivability components.

HEAVY HELICOPTER SQUADRONS

Ten Marine Heavy Helicopter (HMH) squadrons are on strength. Six units operate the CH-53E Super Stallion: two units are based at MCAS New River, North Carolina with the 2nd MAW and four are assigned to the 3rd MAW in California. The 3rd MAW squadrons are scheduled to move from MCAS Tustin to MCAS Miramar between September 1998 and May 1999.

Replacement training for the CH-53E is conducted at New River by Marine Helicopter Training Squadron 302. This unit also has MH-53E Sea Dragon versions, in which Navy MH-53E crews are trained.

During the mid-1990s the Marine Corps consolidated all CH-53D Sea Stallion helicopters at MCAS (now MCAF) Kaneohe Bay, Hawaii. Four HMH

squadrons with the 1st MAW are based at Kaneohe. A CH-53D replacement training squadron, HMT-301, formerly a CH-46E training squadron, was reactivated at Kaneohe in 1995.

HMH squadrons routinely assign detachments as part of the MEU's ACE onboard amphibious assault ships (LHAs or LHDs). When deployed, these detachments are assigned to the embarked HMM squadron and are so marked. CH-53E-equipped HMH squadrons also participate in the Unit Deployment Program, deploying to MCAS Futemma, Okinawa, Japan; normally one HMH squadron is so-deployed at any time.

Since the mid-1960s, Marine heavy helicopter airlift has used versions of the Sea (and later Super) Stallion. Current planning has the CH-53E in service for another quarter of a century.

MEDIUM LIFT

At the heart of Marine Corps aviation is the support of ground troops, including providing mobility in the battlefield. A substantial number of Marine aircraft are assault transport helicopters, soon to be augmented by advanced tilt-rotor aircraft.

The most numerous USMC helicopter is the tandem-rotor Boeing CH-46E Sea Knight medium-lift helicopter, many of which are three decades old. The CH-46, which can be armed with two 0.50-in machine-guns for self-defence, is assigned the role of transporting Marines and their equipment and supplies to the battlefield. The CH-46E fleet represents an upgrade from CH-46A/D/F versions, and is currently going through the Dynamic Component Upgrade (DCU) programme, which involves replacement of the flight controls, rotor heads, drive-train systems, hydraulic pump, and other dynamic components, and removal of engine exhaust device plumbing. This upgrade is designed to extend the life of the CH-46E fleet until the advent of the MV-22B Osprey, and to restore performance by allowing the removal of many flight restrictions imposed by limitations caused by fatigue.

A few CH-46Es are used to support the Presidential executive flight detachment of HMX-1. Several HH-46D search and rescue versions are used by air station base flights.

The CH-46E fleet is on track for replacement by the world's first operational tilt-rotor aircraft, the Bell Boeing MV-22 Osprey, which entered low-rate initial production in 1997. The MV-22B, with a crew of two, can carry 24 fully equipped troops or 12 litters to a radius of 200 nm (230 miles; 370 km), or farther with aerial refuelling. Cargo can also be carried externally. The aircraft is stressed to handle 0.50-in machine-guns for self-defence; a proposed nose-mounted Gatling gun has not yet been funded. The four MV-22B EMD (engineering and manufacturing development) aircraft have continued the test and evaluation programme initiated with the heavier MV-22A versions, which now have been retired.

The first production MV-22B is scheduled for delivery in May 1999, and crew training is to begin in mid-2000. As CH-46E and CH-53D squadrons make the transition to the MV-22B, each squadron will take delivery of 12 Ospreys. Production is expected to total 360 MV-22Bs by 2014.

MEDIUM HELICOPTER SQUADRONS

The Marine Corps operates a total of 15 Marine Medium Helicopter (HMM) squadrons, more than any other type. Although the number of tactical jet squadrons was reduced after the Cold War, no helicopter transport units, whose numbers were never adequate, were disbanded. All HMM squadrons are equipped with the CH-46E Sea Knight helicopter. During 1998, deliveries began of the Dynamic Component Upgrade (DCU) version CH-46Es.

HMM squadrons routinely deploy onboard amphibious ships (LHAs or LHDs) for six-month cruises in the Mediterranean, Persian Gulf or Western Pacific. When deployed, the detachments of CH-53s, AH-1s, UH-1s and AV-8s assigned to the MEU become parts of the HMM and are marked accordingly for the duration of the deployment.

The 1st MAW has two HMM squadrons permanently assigned, based at MCAS Futemma, Okinawa, Japan. The squadrons moved to Okinawa from Kaneohe Bay, Hawaii, during the mid-1990s when all HMM squadrons were displaced from Hawaii because of the decision to base all CH-53D squadrons at Kaneohe Bay. The basing of an amphibious assault ship, USS *Belleau Wood* (LHA 3), at Sasebo, Japan, necessitated the permanent assignment of an HMM squadron in Okinawa around which to form the core of an ACE for the assigned MEU.

The 3rd MAW has seven HMM squadrons assigned, all formerly based at MCAS Tustin, California, which is scheduled for closure. The squadrons were moved to MCAS El Toro, California as an interim measure. With the closure of El Toro scheduled for 1999, three units – HMMs 164, 163, and 161 – were scheduled to move to MCAS Miramar in November 1998, January 1999, and February 1999, respectively. The 3rd MAW's other four CH-46E units – HMMs 166, 165, 268 and 364 – moved to MCAS Camp Pendleton, California between January and March 1999.

The 2nd MAW has six HMM squadrons assigned, all based at MCAS New River, North Carolina. Also based at New River is the CH-46E replacement training squadron, Marine Helicopter Training Squadron (HMT) 204. HMT-204 is scheduled to be redesignated VMMT-204 (Marine Medium-Lift Training Squadron 204) when the MV-22B Osprey tilt-rotor assault transport enters service. At that time, it is intended that an operational HMM squadron will be removed from front-line service to assume the role of training CH-46 crews. Current plans call for the transition of four HMM squadrons at New River and four in California to transition to the MV-22B starting in 2000.

HELICOPTER GUNSHIPS

Versions of the Bell AH-1 Cobra helicopter gunship have been in Marine Corps service for three decades. The AH-1J Sea Cobra has been retired, and all remaining AH-1Ts were upgraded to AH-1W Super Cobra configuration. Production of the AH-1W for the Marine Corps ended in mid-1998. The AH-1W remains a formidable battlefield weapons platform, armed with a turret-mounted 20-mm cannon, 2.75-in and 5-in rockets, Hellfire and TOW anti-armour missiles and, for aerial combat, the AIM-9 Sidewinder missile.

All AH-1Ws are scheduled to complete an extensive avionics upgrade (to improve night and poor weather capability) by 1999. One feature of this upgrade is a night targeting system that includes an IR system, a low-light television system, laser-designation range finder, and an auto-track capability for the TOW missile system. Communications and navigation will be enhanced by the ARC-210 agile-frequency radio, an upgraded TACAN, and a GPS linked with an INS.

For battlefield command and control, observation, utility, medevac and gunship roles, the Marine Corps has long used the Bell UH-1N Iroquois ('Huey'). The UH-1N is armed with 7.62-mm and 0.50-in machine-

guns plus 2.75-in rockets. (A few unarmed versions, HH-1Ns, are used for replacement training and air station SAR duties.) The UH-1N fleet is going through a block upgrade programme that includes installation of ARC-210 radios, satellite communication, a miniature GPS, and an improved TACAN. Night-vision enhancements include improved external lighting, NVG-compatible HUD, and an IR navigation system.

A remanufacturing programme (4BW/4BN) for the both the UH-1N and AH-1W has been approved, which will involve standardising the engines and drive trains for both types, and replacing the two-bladed rotor with a four-bladed unit. Plans call for 180 AH-1Ws and 100 UH-1Ns and HH-1Ns to be remanufactured into AH-1Zs and UH-1Ys, respectively. A few HH-1Ns will serve as development aircraft for the programme. The first flight is scheduled for October 2000, and first deliveries of operational aircraft are planned for 2003. Initial operational capability for the UH-1Y and AH-1Z are scheduled for 2004 and 2006, respectively.

The remanufacturing programme will permit a high degree of commonality between the UH-1 and AH-1 fleets, which equip the same squadrons. Both aircraft will be fitted with T700 turboshaft engines (already fitted on the AH-1W), a four-bladed, hingeless, bearingless, composite material main rotor system and pusher tail rotor system, and identical new drive trains, hydraulics and electrical systems. The rotor system will feature a semi-automatic blade-fold system. The new rotors and engines will greatly increase the speeds and operating weights of both aircraft.

An open-architecture avionics system will be installed in both types, including colour multifunction

Displaying SFOR titles on the boom, this Marines AH-1W has also adopted the marking of HMM-365 for its cruise in the Mediterranean early in 1997.

displays, mission and weapon system computers, advanced digital navigation and communication equipment, and ribbonised wiring. The AH-1Z also will have the ATSS (Advanced Target-Sight System), a focal-plane array infra-red system to enhance detection ranges. Installation of a long-range colour television camera is being contemplated.

A proposed Joint Replacement Aircraft (JRA) is being considered to replace both the UH-1 and AH-1 in approximately 2020.

LIGHT ATTACK HELICOPTER SQUADRONS

Six Light Attack Helicopter (HMLA) squadrons provide the Marine Corps with helicopter gunship support, utility, and forward air controller support with a mixture of AH-1W Super Cobra and UH-1N Iroquois ('Huey') helicopters. Two squadrons are based with the 2nd MAW at MCAS New River, while four are assigned to the 3rd MAW at MCAS Camp Pendleton. One HMLA squadron from Camp Pendleton is normally deployed on six-month rotation to Okinawa under the UDP.

HMLA detachments of both AH-1Ws and UH-1Ns routinely deploy onboard amphibious assault ships (LHAs or LHDs) as part of the MEU's ACE. When deployed, the detachments are assigned to the deployed HMM squadron and their aircraft are marked accordingly.

Replacement training is conducted by Marine Helicopter Training Squadron (HMT) 303 at Camp Pendleton, using AH-1W, UH-1N, and HH-1N versions.

Early in the next decade, the four-bladed-rotor AH-1Z and UH-1Y, remanufactured from AH-1Ws, UH-1Ns and HH-1Ns, will enter service with HMLA squadrons.

Assigned to the base flights of the Marine Corps Air Stations are UC-12 Hurons. MCAS Beaufort's base flight is allocated the '5B' code and flies a pair each of UC-12Bs and HH-46Ds.

C-20G 165153 was the sole example operated by the Marines, but was badly damaged on 2 February 1998, during a tornado.

UNMANNED AERIAL VEHICLES (UAVS)

The Marine Corps has used RQ-2A Pioneer unmanned aerial vehicles (UAVs) for battlefield reconnaissance since the late 1980s. A replacement has not been selected, but procurement of a VTOL UAV to augment the Pioneer is an option.

UAV SQUADRONS

The most recent addition to Marine Corps aviation is the Marine Unmanned Aerial Vehicle (UAV) Squadron, designated VMU. Both of the Marine Corps' VMUs, which were activated on 15 January 1996, fly the RQ-2A Pioneer UAV, and have seen extensive service in Bosnia. VMU-1 was formerly designated as the 1st Marine UAV Company; VMU-2 was newly activated.

ADVERSARY SUPPORT

The Marine Corps operates one reserve squadron of Northrop F-5E (single-seat) and F-5F (two-seat) fighters as air combat manoeuvring adversary aircraft to train fighter crews of all services. The F-5s replaced the F-21A Kfirs that were withdrawn from service during the early 1990s. Plans to replace the F-5s have not been announced.

RANGE SUPPORT

A small number of Beech T-34C Turbo Mentor trainers are used by the Marine Corp's F/A-18 replacement training squadron (VMFAT-101) as target-spotting aircraft for the squadron's weapons training programme.

PROGNOSIS

Since before the end of the Cold War, the Marine Corps has been aggressively pursuing a goal of reducing the number of different aircraft types in service, in order to operate at peak efficiency in fiscal austerity. More than any other US air arm, Marine Corps aviation, despite a constricted budget, has pushed farther in streamlining its force and its future procurement. By the end of the first quarter of the 21st century, Marine Corps aviation's front-line force will be based on one type each of strike fighters, aerial tankers, tilt-rotor assault transport, heavy-lift helicopter, and helicopter gunship.

Above right: Helicopter transportation for the President and VIPs is provided by HMX-1 based at MCAF Quantico, Virginia, using VH-60Ns and VH-3Ds. The unit also has CH-46Es and CH-53Es, used for general transport and the development of tactics.

Right: The Marines' most important near-term aviation acquisition is the Bell Boeing MV-22 Osprey tilt-rotor. The type is the replacement for the long-serving CH-46 Sea Knight, and is due to start entering service with the Corps in 2000. A total of eight squadrons is scheduled to gain the type.

LOGISTICS SUPPORT

The Marine Corps operates a small number of logistics and liaison aircraft to support air stations, squadron movements and executive transport. Two C-9B Skytrain logistic aircraft have been in service for over two decades; although no decision has been announced, it is likely that the Boeing C-40A being procured by the Navy to replace its C-9s will replace the Marine Corps C-9s as well. Several Beech UC-12B Hurons are used for air station liaison. A few Rockwell CT-39G Sabreliners remain in service for executive transport. A single Gulfstream-built C-20G Gulfstream IV executive transport operated for Headquarters, USMC was severely damaged by a tornado in early 1998, and it has not yet been determined whether it will be returned to service.

PRESIDENTIAL SUPPORT

The Marine Corps is assigned the role of providing helicopter transportation to the President and Vice President of the United States, as well as to other senior government officials. To carry out this role, the Marine Corps operates two types not found elsewhere in its inventory, the Sikorsky VH-3D Sea King and the Sikorsky VH-60N Black Hawk. The ageing but well-maintained VH-3Ds replaced similar VH-3As in presidential service. The VH-60Ns, based on the Army's UH-60A, are less spacious than the VH-3Ds. No replacement for either type has yet been identified.

MARINE HELICOPTER SQUADRON 1

Marine Helicopter Squadron (HMX) 1, based at MCAF Quantico, Virginia, has two primary missions. One, using CH-46E and CH-53E aircraft, is to develop helicopter assault tactics under sponsorship of the Marine Corps Combat Development Command and the Navy's commander of the Operational Test and Evaluation Force. The other mission, far more famous, is to provide executive transportation (and emergency evacuation) for the President and Vice President of the United States, members of the Cabinet, and other senior government officials. HMX-1 maintains a detachment at Naval Station Washington, D.C. (the site of the former NAS Anacostia) which flies VH-3D Sea King and VH-60N Blackhawk helicopters. The squadron's CH-46Es and CH-53Es also assist in transporting officials and their equipment.

HMX-1 is not part of Marine Forces Atlantic, but reports directly to the Marine Corps' Deputy Chief of Staff for Aviation.

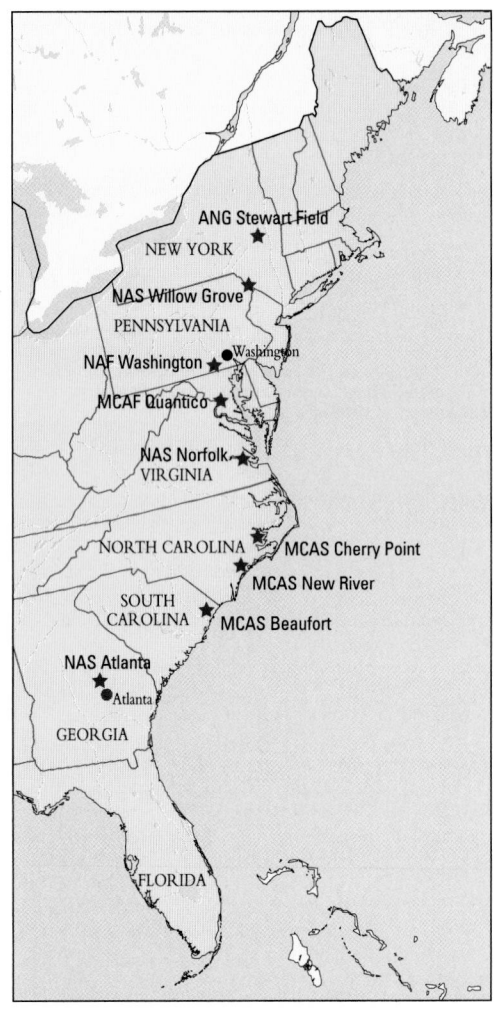

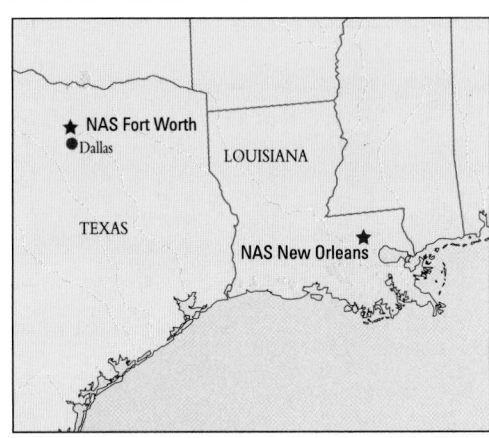

HEADQUARTERS, US MARINE CORPS

Marine Corps Aviation is not a separate organisation within the USMC, but is integrated in every facet of its training and operations. The Commandant of the Marine Corps, a general headquartered at the Pentagon in Arlington, Virginia, has never been an aviator. The Assistant Commandant, also a general, normally is an aviator. The senior officer responsible for aviation matters is a lieutenant general, the Deputy Chief of Staff for Aviation.

MARINE FORCES ATLANTIC/PACIFIC

Marine Forces Atlantic and Marine Forces Pacific are the Marine Corps components equivalent to the Navy's Atlantic and Pacific Fleets, respectively. The three active-duty Marine Aircraft Wings are the aviation components of the Marine Forces commands.

UNITED STATES MARINE CORPS FORCES ATLANTIC

HQ US Marine Corps

Arlington, Va.

Direct Reporting Units		
HMX-1	CH-46E, CH-53E	MCAF Quantico, Va.
Executive Flight Detachment		
	VH-3D, VH-60N	NS Washington, D.C.
MAWTS-1	None	MCAS Yuma, Ariz.

Marine Forces Atlantic

HQ Naval Station Norfolk, Va.

Second Marine Aircraft Wing		MCAS Cherry Point, N.C.
MWHS-2	none	
SOES Cherry Point	C-9B, UC-12B, HH-46D	MCAS Cherry Point
SOS New River	UC-12B	MCAS New River, N.C.
SOS Beaufort	UC-12B, HH-46D	MCAS Beaufort, S.C.

Marine Aircraft Group 14		MCAS Cherry Point, N.C.
MALS-14	none	
VMA-223	AV-8B	
VMA-231	AV-8B	
VMA-542	AV-8B	
VMAT-203	AV-8B, TAV-8B	
VMAQ-1	EA-6B	
VMAQ-2	EA-6B	
VMAQ-3	EA-6B	
VMAQ-4	EA-6B	
VMGR-252	KC-130F/R	
VMGRT-253	KC-130F	
VMU-2	RQ-2A	

Marine Aircraft Group 26		MCAS New River, N.C.
MALS-26	none	
HMM-261	CH-46E	
HMM-264	CH-46E	

HMM-266	CH-46E	
HMT-204	CH-46E	
HMH-461	CH-53E	
HMLA-167	AH-1W, UH-1N	

Marine Aircraft Group 29		MCAS New River, N.C.
MALS-29	none	
HMM-162	CH-46E	
HMM-263	CH-46E	
HMM-365	CH-46E	
HMH-464	CH-53E	
HMT-302	CH-53E, MH-53E	
HMLA-269	AH-1W, UH-1N	

Marine Aircraft Group 31		MCAS Beaufort, S.C.
MALS-31	none	
VMFA-115	F/A-18A	
VMFA-122	F/A-18A	
VMFA-251	F/A-18C	
VMFA-312	F/A-18C	
VMFA(AW)-224	F/A-18D	
VMFA(AW)-332	F/A-18D	
VMFA(AW)-533	F/A-18D	

Marine Wing Support Group 27		MCAS Cherry Point, N.C.
MWSS-271	none	MCAS Cherry Point, N.C.
Detachment	none	MCALF Bogue Field, N.C.
MWSS-272	none	MCAS New River, N.C.
MWSS-273	none	MCAS Beaufort, S.C.
MWSS-274	none	MCAS Cherry Point, N.C.

Marine Air Control Group 28		MCAS Cherry Point, N.C.
MASS-1	none	MCAS Cherry Point, N.C.
MACS-2	none	MCAS Beaufort, S.C.
MACS-6	none	MCAS Cherry Point, N.C.
MWCS-28	none	MCAS Cherry Point, N.C.
MTACS-28	none	MCAS Cherry Point, N.C.
2nd LAAD	none	MCAS Cherry Point, N.C.

Note: Unless otherwise listed, assigned units are based at the location of the parent group.

MARINE AIRCRAFT WINGS

Almost all active-duty aircraft and aviation units are administratively assigned to the three Marine Aircraft Wings (MAWs), each of which is paired with one of the three active-duty Marine infantry divisions. The 1st MAW, headquartered at Camp Smedley D. Butler, Okinawa, Japan, supports the 3rd Marine Division in the Western Pacific from bases in Japan and Hawaii. The 1st MAW has some permanently assigned groups and squadrons, but draws heavily on squadrons and detachments from the other two MAWs for six-month deployments under the Unit Deployment Program (UDP). The 3rd MAW supports the 1st Marine Division from bases in southern California. Both the 1st and 3rd MAWs are controlled by Marine Forces Pacific. The 2nd MAW, under Marine Forces Atlantic, supports the 2nd Marine Division from bases in North and South Carolina.

Each MAW is supported by a Marine Wing Headquarters Squadron (MWHS), an administrative unit with no aircraft assigned. Each MAW is composed of several Marine Air Groups (MAGs), plus one Marine Wing Support Group (MWSG) and one Marine Air Control Group (MACG).

MAWs deploy for operations as the Aviation Combat Element (ACE) of a standing Marine Expeditionary Force (MEF), the Marine Corp's largest Marine Air-Ground Task Force (MAGTF). The Marine Corps maintains three standing MEFs, each of which includes a MAW, a Marine Division, and a Force Service Support Group (FSSG).

Smaller MAGTFs include Marine Expeditionary Units (MEUs), which routinely deploy overseas as the Marine Corps combat force embarked on the ships of an amphibious task force, called an Amphibious Ready Group (ARG). Each MEU includes an ACE, usually a medium helicopter squadron (HMM) with 12 aircraft, reinforced by detachments of six AV-8, four CH-53, three UH-1, and four AH-1 aircraft.

MARINE AIRCRAFT GROUPS

Each MAW is comprised of several Marine Aircraft Groups (MAGs), roughly equivalent to Navy type wings. Typically, each group includes all of the squadrons of a given type in the MAW. MAGs provide administration, training and maintenance support for the assigned squadrons. Provisional MAGs are formed for overseas crisis response, as in the case of Operation Desert Shield/Storm, in which the MAG commander will deploy and assume an operational role as commander of the provisional MAG.

Some MAGs also control the replacement training squadrons assigned to the base at which the MAG is located.

Each MAG has a Marine Air Logistics Squadron (MALS) assigned for intermediate-level maintenance support for aircraft assigned to the group. No aircraft are assigned to the MALS units. The Marine aircraft assigned to Hawaii are supported by a Marine Air Logistics Support Element (MALSE).

UNITED STATES MARINE CORPS FORCES PACIFIC

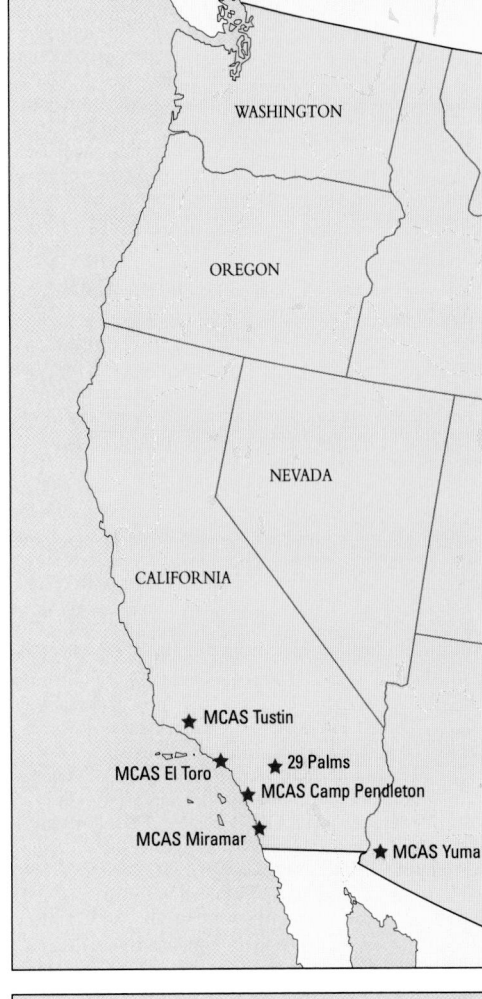

Marine Forces Pacific

HQ Marine Corps Base Hawaii

First Marine Aircraft Wing		Camp S.D. Butler, Okinawa
MWHS-1	none	Camp S.D. Butler, Okinawa
MCAS Iwakuni	UC-12F, HH-46D	MCAS Iwakuni, Japan
MCAS Futemma	UC-12F, CT-39G	MCAS Futemma, Okinawa

Marine Aircraft Group 12		MCAS Iwakuni, Japan
MALS-12	none	
VMFA-212	F/A-18C	
VMFA-XXX	F/A-18A/C	
VMFA(AW)-XXX	F/A-18D	
VMAQ-X	EA-6B	
VMA-311 Det	AV-8B	NAF Kadena, Okinawa

Note: X denotes units from other MAWs rotated under the Unit Deployment Plan.

Marine Aircraft Group 36		MCAS Futemma, Okinawa
MALS-36	none	
VMGR-152	KC-130F/R	
HMM-262	CH-46E	
HMM-265	CH-46E	
HMH-XXX	CH-53D/E	
HMLA-XXX	AH-1W/UH-1N	

First MAW Air Support Element	MCAF Kaneohe Bay, Hawaii
HMT-301	CH-53D
HMH-362	CH-53D
HMH-363	CH-53D
HMH-366	CH-53D
HMH-463	CH-53D
MALSE	none

Marine Wing Support Group 17		Camp Foster, Okinawa
MWHS-171	none	MCAS Iwakuni, Japan
MWSS-172	none	MCAS Futemma, Okinawa
AGSE	none	MCAF Kaneohe Bay, Hawaii

Marine Wing Control Group 18	MCAS Futemma, Okinawa
MASS-2	none
MACS-4	none
MWCS-18	none
MTACS-18	none

Third Marine Aircraft Wing		MCAS Miramar, Calif.
MWHS-3	none	MCAS Miramar, Calif.
SOMS Miramar	UC-12B, CT-39G, HH-1N	MCAS Miramar, Calif.
SOMS Yuma	UC-12B, HH-1N	MCAS Yuma, Calif.

Marine Aircraft Group 11		MCAS Miramar, Calif.
MALS-11	none	
VMFA-232	F/A-18C	
VMFA-314	F/A-18C	

VMFA-323	F/A-18C
VMFA(AW)-121	F/A-18D
VMFA(AW)-225	F/A-18D
VMFA(AW)-242	F/A-18D
VMFAT-101	F/A-18A/B/C/D, T-34C
VMGR-352	KC-130F/R

Marine Aircraft Group 13		MCAS Yuma, Ariz.
MALS-13	none	
VMA-211	AV-8B	
VMA-214	AV-8B	
VMA-311	AV-8B	
VMA-513	AV-8B	
VMU-1	RQ-2A	

Marine Aircraft Group 16		MCAS Miramar, Calif.
MALS-16	none	
HMM-161	CH-46E	
HMM-163	CH-46E	
HMM-164	CH-46E	
HMH-361	CH-53E	
HMH-462	CH-53E	
HMH-465	CH-53E	
HMH-466	CH-53E	
HMM-165	CH-46E	MCAS Camp Pendleton, Calif.
HMM-166	CH-46E	MCAS Camp Pendleton, Calif.
HMM-268	CH-46E	MCAS Camp Pendleton, Calif.
HMM-364	CH-46E	MCAS Camp Pendleton, Calif.

Marine Aircraft Group 39		MCAS Camp Pendleton, Calif.
MALS-39	none	
HMLA-169	AH-1W, UH-1N	
HMLA-267	AH-1W, UH-1N	
HMLA-367	AH-1W, UH-1N	
HMLA-369	AH-1W, UH-1N	
HMT-303	AH-1W, UH-1N, HH-1N	

Marine Wing Support Group 37		MCAS Miramar, Calif.
MWSS-371	none	MCAS Yuma, Calif.
MWSS-372	none	MCAS Camp Pendleton, Calif.
MWSS-373	none	MCAS Miramar, Calif.
MWSS-374	none	MCAS Miramar, Calif.
AGSE	none	MCAGC 29 Palms, Calif.

Marine Air Control Group 38		MCAS Miramar, Calif.
MASS-3	none	MCAS Camp Pendleton, Calif.
MACS-1	none	MCAS Camp Pendleton, Calif.
MACS-7	none	MCAS Yuma, Calif.
MWCS-38	none	MCAS Miramar, Calif.
MTACS-38	none	MCAS Miramar, Calif.
3rd LAAD	none	MCAS Camp Pendleton, Calif.
1st LAAM	none	MCAS Yuma, Ariz.

Note: Squadrons scheduled to move from MCAS El Toro, Calif. to MCAS Miramar, Calif. between publication date and May 1999 have been listed as being based at Miramar. New sites for HMH-769 and HMM-764 have not been determined. HMLA-773 Det A will move to Johnstown, Pa., upon completion of a new hangar.
Note: Unless otherwise listed, assigned units are based at the location of the parent group.

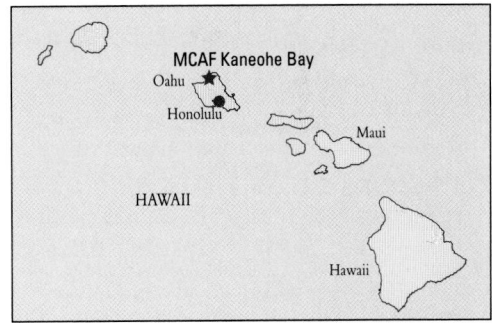

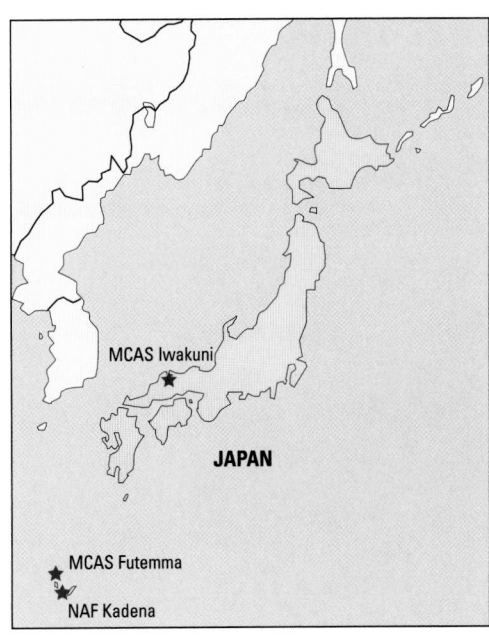

MAWTS-1

Based at MCAS Yuma, Arizona, Marine Air Weapons and Tactics Squadron (MAWTS) 1 develops aviation warfare tactics and provides instruction in close air support and helicopter assault tactics to Marine Corps aircrews and those of other services. The squadron trains WTIs (Weapons and Tactics Instructors) from other squadrons and standardises weapons training throughout Marine Corps aviation. MAWTS-1 has no aircraft assigned, but conducts training using aircraft of hosted units.

MARINE AIR CONTROL GROUPS

Each MAW is assigned a Marine Air Control Group (MACG), to which no aircraft are assigned, for command and control of aerial operations. MACGs include Marine Air Control Squadrons (MACSs) that provide air traffic control and aerial surveillance for anti-air support for Marine combat forces. MACSs maintain air traffic control detachments at one or more other airfields. Marine Tactical Air Control Squadrons (MTACSs) provide planning and co-ordination of air operations for the ACE's Tactical Air Command Center (TACC). Marine Air Support Squadrons (MASSs) control aircraft performing close air support or direct air support missions. Marine Wing Communications Squadrons (MWCSs) provide communications support for the MAW and its elements. MACGs also control the assigned Light Anti-Aircraft Missile Battalions (LAAMs; only one active-duty unit remains, based at MCAS Yuma, but was scheduled for phase-out in 1998), Low-Altitude Air Defense Battalions (LAADs), and Stinger Batteries.

North America

MARINE WING SUPPORT GROUPS

Each MAW is also assigned a Marine Wing Support Group (MWSG), which has no aircraft assigned and control Marine Wing Support Squadrons (MWSSs). MWSSs provide a wide range of logistic and personnel services to the wing's units, including fuel, ground-support equipment, supply, medical services, and weather forecasting. An Aviation Ground Support Element (AGSE) performs the same functions for the airfield at Marine Corps Air-Ground Center Twenty-Nine Palms, California.

MARINE CORPS AIR STATIONS AND FACILITIES

Most active-duty Marine Corps aviation units are based at a small number of air stations (MCASs), air facilities (MCAFs), and a few other installations. These bases and the assigned support squadrons provide the services needed to operate the tenant air groups and squadrons.

Three bases – MCAS Cherry Point and MCAS New River in North Carolina, and MCAS Beaufort in South Carolina – host all squadrons assigned to Marine Forces Atlantic. Bogue Field in North Carolina serves as an auxiliary landing field for East Coast units.

West Coast units (3rd MAW) are assigned to MCAS Miramar and MCAS Camp Pendleton in southern California, and MCAS Yuma, Arizona, as well as a few units at Twenty-Nine Palms, California. MCAS Tustin and MCAS El Toro in California were scheduled to be closed by 1999. Units of the 1st MAW are based at MCAS Iwakuni, MCAS Futemma, and Kadena Air Base, Japan, as well as MCAF Kaneohe Bay, now part of Marine Corps Base (MCB) Hawaii. MCAF Quantico, part of MCB Quantico, Virginia, hosts HMX-1, which maintains a detachment at Naval Station Washington, D.C.

Most air stations have a base flight and/or a SAR helicopter component, operated by Station Operations & Maintenance Squadrons (SOMS) or Station Operations Squadrons (SOS). These units variously operate UC-12B, UC-12F, C-9B, CT-39G, UH-1N, and HH-46D aircraft. The two C-9Bs assigned to SOS Cherry Point, North Carolina are used in much the same manner as their Navy counterparts in providing airlift to the Marine Corps and others. The HH-1Ns assigned to SOMS Miramar, California are due for distribution to other units.

US MARINE CORPS RESERVE

The Marine Corps maintains a small but modern aviation reserve force that can be mobilised in times of national emergency. Marine Corps Reserve aviation (MCRA) began shortly after the establishment of a Marine Corps reserve in 1916, and many reservists saw action during World War I, flying bombing sorties and anti-submarine patrols. Post-war demobilisation reduced MRCA to a tiny force, but in 1928 a few reserve pilots were activated to revitalise the air reserve programme, which was strengthened by legislation during the 1930s that increased pay and training opportunities.

The reserve force maintained before World War II was able to mobilise 13 squadrons for service in the war which fought alongside their active-duty coun-

Above: The US Marine Corp Reserve has a pair of KC-130T-30 Hercules – the only stretched Hercules in US military service. They fly with VMGR-452 from Stewart ANGB, N.Y.

Below: The 'Snipers' of VMFAT-401 are the Marines' only aggressor unit, using F-5E/Fs. F-5E 741572/'12' has an ACMI pod on its port wingtip station and a Sidewinder acquisition round on its starboard.

terparts in regular squadrons. Marine Corps Reserve aviation units, which operated aircraft shared with the Naval Air Reserve Training Command, were organised during the post-war demobilisation as part of Marine Air Reserve Training Command. The number of squadrons grew to over 40 by the time war broke out in Korea, during which few of the 11 reserve squadrons activated saw combat although many reservists flew in combat with active-duty squadrons.

In 1962, all reserve aviation units were grouped under the 4th Marine Aircraft Wing. Though no units were activated, Marine Corps reservists flew with active-duty squadrons during the Vietnam War.

The Navy's 1970 reorganisation of the Naval Air Reserve extended benefits to MCRA in the form of revised force structure and more modern equipment, a trend that accelerated during the mid-1980s as reserve equipment closed the gap to near-fleet standard.

After the 1990 Iraqi invasion of Kuwait, several reserve units deployed to the battlefront as part of

Operations Desert Shield/Storm, and several back-filled for active-duty units taken from forward sites.

4TH MARINE AIRCRAFT WING

All MRCA units are organised under the 4th Marine Aircraft Wing, headquartered at New Orleans, La. Four Marine Aircraft Groups (MAGs), one Marine Air Control Group (MACG), and one Marine Wing Support Group (MWSG) located around the country support the squadrons assigned to each location and, in some cases, in other locations. The assigned aircraft are compatible with fleet-standard aircraft. Some aircraft types, such as the two-seat F/A-18D, EA-6B, AV-8B and RQ-2A, are not operated by the Marine Corps Reserve. The reserve is manned mostly by experienced aviators, naval flight officers and Marines. A cadre of active-duty reservists in each unit runs the day-to-day operations of each unit.

FIGHTER-ATTACK SQUADRONS

The striking power of the Marine Corps Reserve resides in its four fighter-attack (VMFA) squadrons, which operate the F/A-18A Hornet strike fighter. If mobilised, these units would perform strike, close air support, and air defence missions. One unit, VMFA-142, is nominally assigned to the Naval Air Reserve's Carrier Air Wing 20 (CVWR-20) and would deploy onboard an aircraft-carrier if CVWR-20 were activated.

For convenience, VMFA-321 at NAF Washington has custody of the C-20G that supports Headquarters Marine Corps. This aircraft was severely damaged by a tornado in early 1998, and no decision has yet been made to repair it.

Two squadrons of CH-46Es are assigned to the USMC Reserve, including HMM-774 based at NAS Norfolk, Va. HMM-774 was activated and flew combat missions during Operation Desert Storm.

Fourth Marine Aircraft Wing	NAS New Orleans, La.	
MWHS-4	none	NAS New Orleans, La.
Marine Aircraft Group 41	**NAS Fort Worth, Texas**	
MALS-41	none	
VMFA-112	F/A-18A	
VMGR-234	KC-130T	
Marine Aircraft Group 42	**NAS Atlanta, Ga.**	
MALS-42	none	
VMFA-142	F/A-18A	
HMLA-773	AH-1W, UH-1N	
HMLA-775 Det A	AH-1W, UH-1N, CT-39G, UC-12B	NAS New Orleans, La.
HMM-774	CH-46E	NAS Norfolk, Va.
Marine Aircraft Group 46	**MCAS Miramar, Calif.**	
MALS-46		
VMFA-134	F/A-18A	
HMM-764	CH-46E	MCAS El Toro, Calif.
HMH-769	CH-53E	MCAS El Toro, Calif.
VMFT-401	F-5E/F	MCAS Yuma, Calif.
HMLA-775	AH-1W, UH-1N	MCAS Camp Pendleton, Calif.
Marine Aircraft Group 49	**NAS Willow Grove, Pa.**	
HMH-772	CH-53E	
HMLA-773 Det A	AH-1W, UH-1N	
VMFA-321	F/A-18A, C-20G, UC-12B	NAF Washington, D.C.
VMGR-452	KC-130T/T-30	Stewart ANGB, N.Y.
MALS-49	none	Stewart ANGB, N.Y.
Marine Wing Support Group 47	**Detroit, Mich.**	
MWSS-471	none	NAS Fort Worth, Texas
Det A	none	Minneapolis, Minn.
Det B	none	Green Bay, Wis.
MWSS-472	none	NAS Atlanta, Ga.
Det A	none	Wyoming, Pa.
Det B	none	Detroit, Mich.
MWSS-473	none	MCAS Miramar, Calif.
Det A	none	Fresno, Calif.
Det B	none	NAS Whidbey Island, Wash.
MWSS-474	none	NAS Willow Grove, Pa.
Det A	none	Johnstown, Pa.
Det B	none	Westover AFB, Mass.
Marine Air Control Group 48	**Fort Sheridan, Ill.**	
MASS-6	none	Westover AFB, Mass.
Det A	none	MCAS Miramar, Calif.
MACS-23	none	Aurora, Colo.
TAOC Det	none	Aurora, Colo.
EW/C Det	none	Cheyenne, Wyo.
MACS-24	none	FCTC Dam Neck, VA.
ATC Det A	none	NAS Fort Worth, Texas
ATC Det B	none	NAS Willow Grove, Pa.
TAOC Det	none	FCTC Dam Neck, VA.
MWCS-48	none	Fort Sheridan, Ill.
Det A	none	Fort Sheridan, Ill.
Det A (Fwd)	none	MCAS Miramar, Calif.
MTACS-48	none	Fort Sheridan, Ill.
4th LAAD Bn	none	Pasadena, Calif.
Det A	none	NAS Atlanta, Ga.

Above: USMC Reserve unit HMLA-775 (Detachment A) serves from NAS New Orleans, Louisiana. The UH-1Ns are scheduled to be upgraded as UH-1Ys.

Below: The other reserve light attack helicopter squadron is HMLA-773 'Cobras'. Marine AH-1Ws are also due for upgrades, becoming AH-1Zs.

VMFA-124, formerly Marine Attack Squadron (VMA) 124 (until its A-4Ms were retired in 1994 and the squadron moved to NAS Dallas, Texas), was deactivated before it received any F/A-18 aircraft. Another A-4M squadron, VMA-131, was deactivated without ever being redesignated VMFA-131.

AERIAL REFUELLING SQUADRONS

The Marine Corps Reserve operates two aerial refuelling (VMGR) squadrons, which are heavily engaged in supporting both active-duty and reserve forces in aerial refuelling and transport roles. The two squadrons fly a mix of KC-130T and KC-130T-30s.

ASSAULT TRANSPORT HELICOPTER SQUADRONS

The 4th MAW operates four assault transport helicopter squadrons with fleet-standard equipment. Two HMM squadrons fly the CH-46E Sea Knight and two HMH units operate the CH-53E Super Stallion, the latter recently having replaced their former Navy minesweeping RH-53Ds.

LIGHT ATTACK HELICOPTER SQUADRONS

Two Marine light attack helicopter (HMLA) squadrons are operated by the reserve wing, each of which also maintains a large detachment at a separate location. HMLA-773's detachment at NAS Willow Grove, Pa., was activated in 1997. Both squadrons operate a fleet-standard mixture of heavily-armed UH-1N 'Huey' and AH-1W Super Cobra gunships. For convenience, HMLA-775's detachment at NAS New Orleans, La., has custody of the UC-12Bs and CT-39G that support Marine Corps Reserve headquarters.

AIR COMBAT ADVERSARY SQUADRON

The 4th MAW operates the Marine Corps' only adversary squadron, Marine Fighter Training Squadron 401 (VMFT-401). This unit flies F-5E/F Tiger II fighters in the air-combat manoeuvring adversary training role for fighter units of all services.

SUPPORT SQUADRONS

Like its active-duty counterparts, the 4th MAW includes a number of support units that are not equipped with aircraft, including Marine Aviation Logistics Squadrons (MALS), Marine Air Control Squadrons (MACS), Marine Wing Support Squadrons (MWSS), and Marine Air Support Squadrons (MASS). The wing also includes a low-altitude air defence battalion (LAAD). The former Hawk-equipped light anti-aircraft missile battalion (LAAM), later changed to Tactical Missile Defense detachments of a MACS, have been disbanded.

The reserve component of the US Marine Corps has four units flying the first-generation F/A-18A. This aircraft belongs to VMFA-112 'Cowboys', based at NAS Fort Worth, Texas.

US Coast Guard Aviation

The United States Coast Guard is the smallest branch of the US armed forces. Its aircraft, adorned with a familiar, blue-bordered orange sash, operate mostly over waters around the USA. Best-known for their aerial campaign against drug smuggling, the Coast Guard's HU-25 Guardians, HC-130 Hercules and HH-60 Jayhawks have numerous other missions – among them, policing oil spills and flying search and rescue – linked to the United States' commitment as a seafaring and maritime power.

Co-operation between the airborne and surface fleets is essential for the SAR mission. A survivor is seen being transferred to Cutter 41498 from a HH-60J Jayhawk during an exercise.

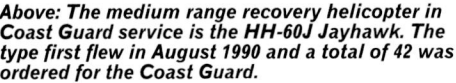

Above: The medium range recovery helicopter in Coast Guard service is the HH-60J Jayhawk. The type first flew in August 1990 and a total of 42 was ordered for the Coast Guard.

The Coast Guard belongs to the US Department of Transportation in peacetime and reverts to control of the US Navy in time of war. Should anyone doubt that this is a military service, Title 14, United States Code says that the Coast Guard is 'at all times an armed force of the United States', as are the other branches – Air Force, Army, Navy and Marine Corps. The service has had an aviation arm since the 1920s when Lieutenant Commander C. G. von Paulsen borrowed a Vought UO-1 seaplane from the Navy to demonstrate how aviation might combat liquor smuggling. Decades later, Coast Guard HU-25B Guardians went to the Persian Gulf to chart the oil spill created by Iraq during the Desert Storm fighting. Today, the Coast Guard has 28 Coast Guard Air Stations (CGASs), 180 aircraft, and 33,000 people.

In the 1990s, the Coast Guard abandoned aircraft with 'single-mission' capability. Familiar types –

Above: An HH-65A Dauphin approaches the USCG Cutter Matagorda, lowering a transfer basket.

HH-65A Dauphins, HH-60J Jayhawks and HU-25 Guardians – took on new and more plentiful tasks.

No aircraft are currently in production for the Coast Guard. After 2000, the service's biggest challenge will be to replace the HH-65A Dauphin, which is short of range and costly to maintain.

HH-65A DAUPHIN

The Eurocopter HH-65A Dauphin is the service's SRR (short-range recovery) SAR helicopter, normally stationed ashore but at times carried onboard cutters. The HH-65A is an 'Americanised' variation of the SA 366G Dauphin and is powered by two 680-shp (510-kW) Lycoming LTS101-750B-2 turboshaft engines driving 38-ft (11.59-m) four-bladed rotors. The 'plastic puppy', as it is called, makes extensive use of composite materials. With a crew of four, the HH-65A can proceed 150 nm (173 miles; 278 km) to a target (which may be a person requiring rescue), engage in a 15-minute search and 15-minute hover, and return to its base or ship after taking aboard three rescuees. The Coast Guard has 94 HH-65As: 80 in use and 14 in maintenance.

HH-60J JAYHAWK

The Sikorsky HH-60J Jayhawk is the Coast Guard's MRR (medium-range recovery) helicopter for the SAR (search and rescue) mission. The Jayhawk cruises at 135 kt (155 mph; 250 km/h) to carry out a six-hour mission over a radius of 300 miles (483 km), remaining on station for 45 minutes to hoist up to six people on board. The Jayhawk employs state-of-the-art radar, radio, and navigational equipment, including the Navstar GPS, and extends the range at which search and rescue missions are possible, making it possible to come to the aid of the distressed at greater distance than before. The Coast Guard has 42 Jayhawks: 35 in use, six in maintenance, and one in storage.

HU-25 GUARDIAN

The Dassault HU-25 Guardian is ideally suited for today's multi-mission doctrine, but expensive to operate. Twenty-five standard HU-25As fly air rescue missions (18 more are in storage). Seven HU-25Bs have an APS-131 SLAR (side-looking airborne radar) pod and an IR/UV line scanner under one wing. Nine HU-25C Night Stalkers or 'HU-25C-Plus' aircraft have Westinghouse AN/APG-66 search

Nine of the HU-25 Guardian fleet – as HU-25C Night Stalkers – are tasked with the interception of drug smugglers' aircraft.

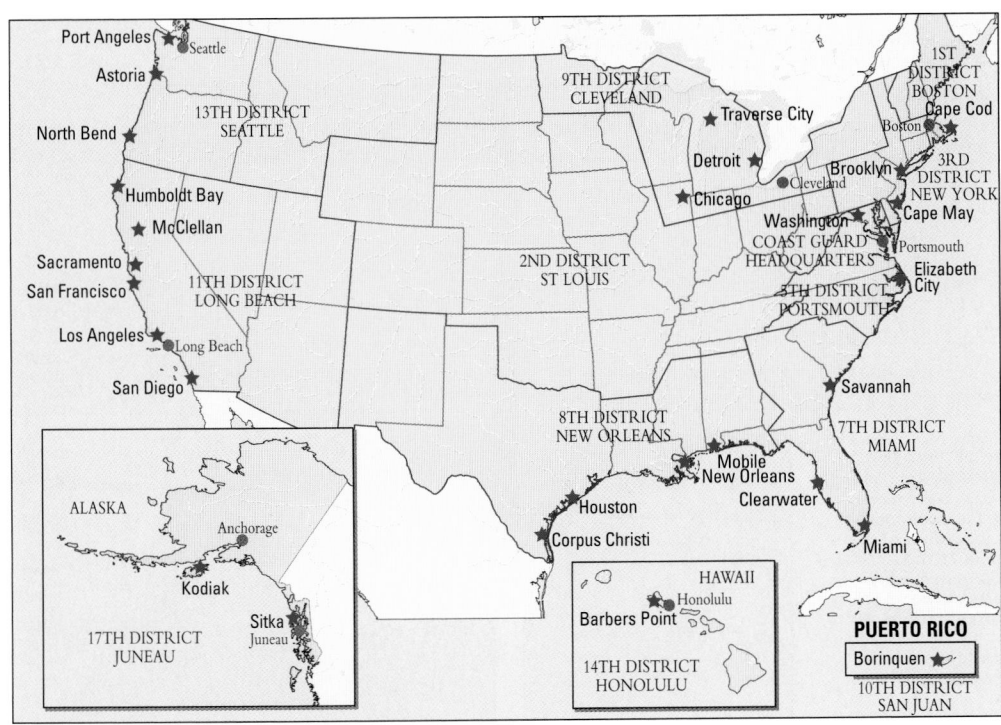

US Coast Guard – Atlantic

(Aviation units only)

Commander Atlantic Area		Governors Island, N.Y.
1st Coast Guard District		**Boston, Mass.**
CGAS Brooklyn	HH-65A	Floyd Bennett Field, N.Y.
CGAS Cape Cod	HH-60J, HU-25A	Otis ANGB, Mass.
5th Coast Guard District		**Portsmouth, Va.**
CGAS Cape May	HH-65A	Cape May, N.J.
CGAS Elizabeth City	HC-130H, HH-60J	Elizabeth City, N.C.
CGAS Washington	C-20B	Washington National AP
Aviation Technical Training Center		Elizabeth City, N.C.
(uses ground instruction airframes)		
Aviation Repair and Supply Center		Elizabeth City, N.C.
(maintenance and repair organisation)		
7th Coast Guard District		**Miami, Fla.**
CGAS Borinquen	HH-65A	Borinquen, PR
CGAS Clearwater	HC-130H, HH-60J	St Petersburg IAP, Fla.
CGAS Miami	VC-4A, HH-65A,	Opa Locka, Fla.
	HU-25A, RU-38A	
CGAS Savannah	HH-65A	Hunter AAF, Ga.
8th Coast Guard District		**New Orleans, La.**
CGAS Corpus Christi	HH-65A, HU-25A	NAS Corpus Christi, Texas
CGAS Houston	HH-65A, HU-25A	Ellington ANGB, Texas
CGAS New Orleans	HH-65A	NAS New Orleans, La.
Aviation Training Center	HH-60J, HH-65A, HU-25A	Mobile, Ala.
9th Coast Guard District		**Cleveland, Ohio**
CGAS Detroit	HH-65A	Selfridge ANGB, Mich.
CGAS Traverse City	HH-65A	Cherry Capitol AP, Mich.

US Coast Guard – Pacific

(Aviation units only)

Commander Pacific Area		Alameda, Calif.
11th Coast Guard District		**Long Beach, Calif.**
CGAS Humboldt Bay	HH-65A	Humboldt Bay, Calif.
CGAS Los Angeles	HH-65A	Los Angeles IAP Calif.
CGAS Sacramento	HC-130H	McClellan AFB, Calif.
CGAS San Diego	HH-60J	Lindbergh Field, Calif.
CGAS San Francisco	HH-65A	San Francisco IAP, Calif.
13th Coast Guard District		**Seattle, Wash.**
CGAS Astoria	HH-60J	Astoria, Ore.
CGAS North Bend	HH-65A	North Bend MAP, Ore.
CGAS Port Angeles	HH-65A	Port Angeles, Wash.
14th Coast Guard District		**Honolulu, Hawaii**
CGAS Barbers Point	HC-130H, HH-65A	Barbers Point, Hawaii
17th Coast Guard District		**Juneau, Alaska**
CGAS Kodiak	HC-130H, HH-65A	Kodiak, Alaska
CGAS Sitka	HH-60J	Sitka, Alaska

radar and a FLIR. Night Stalkers detect, intercept, and lead law enforcement teams to smugglers' aircraft carrying narcotics and other contraband. The active HU-25 fleet totals 51 aircraft.

HC-130H HERCULES

The Coast Guard operates 30 HC-130H Hercules transport/SAR aircraft. While other air arms use the fabled 'Herk' as a transport, to drop paratroops, or for utility missions, the Coast Guard also employs the type for search and rescue, ice patrol and maritime surveillance. The HC-130H is often used for transport, but is a SAR aircraft, able to drop life rafts, food, medicine, and other supplies. An HC-130H can exceed 3,000 miles (4828 km) in low-altitude flight carrying a crew of seven, with an endurance of up to 14 hours. The potential replacement for the HC-130H (of which the oldest was delivered in 1973) is the second-generation HC-130J Hercules II.

OTHER AIRCRAFT

Other Coast Guard aircraft are less well known. A single C-20B serves as the Coast Guard's executive jet and commandant's aircraft. One VC-4A Gulfstream turboprop transport performs miscellaneous duties. Until recently, the Coast Guard oper-

ated two Schweizer RG-8A Condor (civil SA 2-37A) powered gliders, with 'stealth' capability to soar quietly over smugglers and pinpoint their location. One has been modified to twin-engined RU-38A Condor configuration and was undergoing flight-test early in 1999, and a second Condor is due to be built.

FUTURE DEVELOPMENTS

The Coast Guard is keen to reduce the number of helicopter types it uses by finding a replacement for both the HH-65A Dauphin and the HH-60J Jayhawk. The development of the tiltrotor – from Bell's XV-15A through to the V-22 Osprey and the proposed Bell 609 – has interested the Coast Guard as it offers a high cruising speed, making it ideal for

reaching disaster scenes and getting to survivors quickly, and has a good load-carrying ability. Tiltrotor technology has matured to a point where this once revolutionary and experimental concept is on the point of entering service with civil operators and interest in the Bell 609 has led to Bell producing 'photographs' of the tiltrotor in USCG colours, but, as of 1999, the Bell 609 had not been ordered by the Coast Guard.

Over the last 10 years the Coast Guard has taken on new duties, as well as continuing in its traditional role. The next 10 years will see changes in equipment, but the primary role – serving the needs of the seafaring community in US waters – will remain.

In Coast Guard service the HH-65A has been found to be underpowered, but the 1991 evaluation of an HH-65A powered by the LHTEC T800 has not provided the helicopter with a viable alternative powerplant.

The USCG operates 23 standard HC-130Hs and 11 HC-130H-7s as long-range SAR and ice patrol aircraft. These standard HC-130Hs of CGAS Clearwater differ from the H-7 in small details, such as using Allison T56A-7B instead of T56A-7Bs.

US Army Aviation

Today's US Army forces are a mix of those focused on power projection, forward deployment and strategic reserve. Among the most capable elements of the service is one of the largest and most effective aviation forces in the world, yet few people outside the Army recognise its flexibility and lethality, as it conducts most of its operations in dark brown helicopters, at treetop level and at night. The service has methodically updated and modernised its aviation systems over the last two decades while retiring its fleet of relatively unsophisticated, Vietnam-era platforms. Army Aviation has moved to a smaller inventory of sensor-equipped, digitised, highly-lethal airborne platforms, prepared to self-deploy anywhere in the world. The service is having to make choices now that will guide the transition of its force structure from one of massive employment of firepower to one able to simultaneously support focused, precision strike task forces, packaged for rapid and deadly response. The service is moving to a new model, known as dominant manoeuvre warfare (DMW), which synchronises that doctrine across the combined arms of infantry, armour, aviation, field artillery and all other Army branches. Army aviation performs diverse operations across a broad spectrum of military and political conflict scenarios, each and every day.

US Army Aviation has a heritage stretching back to the American Civil War. Fixed-wing operations began in 1913 and expanded until 20 June 1941, when the Army Air Force (AAF) was separated from the Army Ground Force (AGF). The field artillery constituted a force of light spotter aircraft from 6 June 1942 at Fort Sill, Oklahoma. The helicopter was still in its infancy in June 1950 when North Korea invaded South Korea and the Army began a progressive upgrading of its aviation resources under the jealous, watchful eye of the US Air Force, established in 1947. Just prior to the Vietnam War, the Bell HU-1, redesignated as the UH-1, introduced turbine powerplants to battlefield transportation, and the massive build-up of Army forces in-theatre from early 1965 quickly validated the role of the modern helicopter in combat and combat support roles. The modification of the Huey as a gunship led to the development of the Bell AH-1 'Huey-Cobra', the first attack helicopter, proving that airborne platforms could be decisive in land warfare.

With the end of involvement in Vietnam, the Army turned its attention again to the Cold War in Europe. In the 1970s it launched a strategy to develop and field more sophisticated, and survivable, aviation systems. Quietly, but systematically, the Army focused on enhancing its ability to synchronise its forces under cover of darkness and foul weather. Army aviation adapted a doctrine of establishing all-weather/night warfighting dominance through a series of initiatives that combined upgrades with new technology that has extracted extraordinary value and lethality from the meagre slice of budget resources allocated to the Army by the US Congress. The service was rebuilt along this model, concurrently developing the M1 Abrams main battle tank, the M2/M3 Bradley

Above: In service since 1984, the AH-64A Apache is the Army's primary attack helicopter. Upgrading of the fleet to the AH-64D Longbow Apache standard has started, providing the Army with an ability to attack targets at greater distance.

Right: Heavylift transport is provided by about 410 CH-47D Chinooks. The D entered service in 1982, being an upgrade of the previous models. Plans are being formulated to improve the Chinook for further service.

infantry fighting vehicle, the UH-60A Blackhawk, the AH-64A Apache, and upgrading its fleet of Boeing CH-47A/B/C aircraft to the CH-47D variant. The AH-1s were modernised to enhance their survivability and lethality in the face of armour threats from Soviet and Warsaw Pact forces. The service made its combat forces robust and then activated the combat training centres (CTCs) to modernise the way soldiers and complete units trained.

The aviation branch home is at Fort Rucker, Alabama. The Army Aviation Center (USAAVNC) is based there and is responsible for the training of all the Army's aviators and maintainers, as well as the US Air Force helicopter personnel and numerous international military services. The commander of USAAVNC is also the head of the Aviation Branch and, as such, he is the chief proponent of aviation doctrine, systems development, deployments and tactics. The Training and Doctrine Command (TRADOC) oversees the USAAVNC and the other training centres of excellence.

The aviation branch began to adapt night-vision goggle (NVG) systems to their aircraft and, through trial and error, the service became a pioneer in adapting the technology to airborne warfighting tactics. The Army developed and fielded a unique special operations aviation (SOA) organisation that provides an unequalled capability to conduct clandestine missions. OH-58As were modified into the OH-58D that featured the first all-'glass' cockpit and digital weapons systems in armed helicopters, integrating infra-red, laser and television systems in the same compact platform.

By the mid-1980s the Army had adopted the Army of Excellence (AoE) programme that led to the restructuring and reorganisation of its units, their equipment levels and staffing. The programme brought changes in the size and organisation of aviation unit strength levels and the distribution of weapons and equipment, known as the table of organisation and equipment (TOE). The aviation brigade was added to all divisions and corps echelons in 1985-87 to provide commanders with a co-ordinated aviation component. Aviation

brigade inventories varied according to the mission of the assigned division or corps, organised as composite units that operated attack, scout, air assault, general support, combat support, target acquisition, and electronic warfare aviation assets. The composite structure of assigned battalions was created to permit teams of aircraft to be chopped – detached – to the operational command of other manoeuvre brigade commanders, such as infantry or armour units. The AoE structure survived through Operation Desert Shield/Storm. From March 1993 the service implemented the Army Aviation Restructure Initiative (ARI) that has reorganised the battalions in order to essentially pure-fleet their component airframes, and to reduce the number of unmodernised UH-1H and OH-58A/C aircraft in combat and combat support units. Since then the aviation branch has been in constant reorganisation, fulfilling the ARI programme.

The Army Aviation and Missile Command (AMCOM), located at Redstone Arsenal, Huntsville, Alabama was activated in 1996 to manage the acquisition, development, evaluation, deployment and sustainment of aviation and missile systems. AMCOM is a major subordinate command of Army Material Command (USAMC) and is supported in its efforts by other organisations in the command including Communications-Electronics Command (CECOM), Industrial Operations Command (IOC), Simulation, Training & Instrumentation Command (STRICOM) and Test and Evaluation Command (TECOM).

Today's Army aviators and soldiers, male and female, are trained to conduct operations in a wide variety of warfighting scenarios against threats that include armour and mechanised infantry forces, low-intensity conflicts (LICs), operations other than war (OOTWs), peacekeeping, and even disaster relief. Army aviation proves its versatility in each type of operation. Since Operation Desert Shield/Storm, the United States has engaged in a variety of operations in Iraq, Turkey, Somalia, Haiti and Bosnia, while showing the flag in a number of potential conflict locations including the Korean peninsula.

Forces Command (FORSCOM)

The US Army's Forces Command (FORSCOM) is the sustaining command of the Department of the Army and is subordinate to the US Atlantic Command (USACOM); it is tasked to train, mobilise and project land warfare forces in support of world-wide contingency and theatre operations. FORSCOM, located at Fort McPherson, Atlanta, Georgia, draws its strength from active component (AC) and reserve component (RC) force structure based in the continental US (ConUS), the District of Columbia and the Trust Territories. The US Army Reserve Command (USARC) and Army National Guard (ArNG) are key components of FORSCOM, providing over 50 per cent of the combat and combat support forces subordinate to the command. The only ConUS Army forces not assigned are US Special Operations Command (SOCOM or USSOCOM) and its Army component, the US Army Special Operations Command (USASOC) assigned forces. Fiscal Year 1999 force projects show an active component force of 480,000 soldiers and reserve components totalling 565,000, with fully 80 per cent under direct FORSCOM command in peacetime.

FORSCOM is responsible for preparing these forces to deploy to potential battlefields and contingency operations anywhere in the world. The command directly controls three combat corps in peacetime, composed of six active-duty divisions, numerous active and reserve brigades, groups, battalions (squadrons), companies and platoons. The training of reserve component (RC) forces is managed by two active-duty numbered ConUS armies, the First US Army (ONEUSA) at Ft Gillem, Georgia and the Fifth US Army (FIVEUSA) at Ft Sam Houston, Texas, respectively commanding forces east and west of the Mississippi River. During peacetime, FORSCOM commands the Third US Army (THREEUSA), assigned as USCENTCOM's Army land warfare component. Additional task forces (TFs) and separate functional activities are managed by FORSCOM including the Multinational Force and Observers (MFO) in the Sinai Desert, and Joint Task Force SIX, the principal control point for US military forces co-ordinating their efforts in drug interdiction and surveillance along the southern US border.

I Corps is tasked as the Pacific theatre warfighting corps, and over 80 per cent of its warfighting force structure is with RC units. Fully 90 per cent of the 66th Aviation Brigade, corps aviation brigade, is reserve component assigned, as is the brigade headquarters itself, a Washington ARNG command. III Corps, or the Phantom Corps, is the Army's mobile armoured corps, headquartered at Ft Hood, Texas. It sustains the other corps and theatre commands with massive armour, field artillery and mechanised infantry forces. The Army's rapid-response contingency corps is the 18th Airborne Corps, based at Ft Bragg, N.C. It is guaranteed that units of the XVIIIABNCORPS will be among the first Army units to engage in combat, if called to deploy anywhere in the world. FORSCOM supports US Army South (USARSO) with extensive deployment of RC forces in a variety of roles. Ft Buchanan, Puerto Rico, presently a FORSCOM installation, is slated to become the next home for USARSO as the command withdraws from Panama.

Another critical element is unit combat training through a series of combat training centres (CTCs). The National Training Center (NTC), Ft Irwin, California conducts training of brigade-sized, armour and mechanised infantry forces in dominant manoeuvre warfare tactics in desert terrain. The Joint Readiness Training Center (JRTC) at Ft Polk, Louisiana focuses on training light infantry and special operations forces in contingency operations that include a variety of scenarios including those involving civilian and military hostages, disaster and humanitarian relief, military operations in urban terrain (MOUT), low-intensity conflict (LIC), and peacekeeping assistance roles, which fall under the general term of operations other than war (OOTW). Numerous field and command post exercises are held annually, among them Roving Sands, which is a joint air defence artillery exercise involving up to several hundred aircraft.

Highly trained aggressor forces, known as Opposing Force or OPFOR, closely follow the doctrine of potential adversaries at each of the CTCs, providing very realistic scenarios. They operate in conjunction with specialised threat simulation units such as OTSA that operate actual and surrogate threat weapons and support systems including tanks, infantry fighting vehicles, communication systems and vehicles. A small fleet of UH-1Hs and AH-1Ss with visually modified (VISMODs) threat surrogate aircraft are supported by actual threat systems such as the Mil-24/-35 'Hind' attack helicopters to provide added realism.

The standard utility helicopter of the US Army is the UH-60 Blackhawk. Production of the UH-60A has ended; the current model being built is the UH-60L.

FORSCOM

Forces Command HQ (FORSCOM), Ft McPherson, Atlanta, Ga.

USAG Flight Detachment	2 x UH-1H *GS*
Fulton CAP, Atlanta, Georgia	

First US Army, HQ (ONEUSA), Ft Gillem, Georgia

Fifth US Army, HQ (FIVEUSA) Ft Sam Houston, San Antonio, Tex.

National Training Center (NTC), Ft Irwin, Calif.

11th Armored Cavalry Regiment (Opposing Force) (11 ACR/OPFOR)	
NTC Aviation Co.	17 x UH-1H *GS*
Barstow-Daggett AP Calif.	6 x JUH-1H *AGGRESSOR*
247th MED DET (AA)	6 x UH-60A *AA: ICORPS assigned*
Barstow-Daggett AP, Calif.	

Joint Readiness Training Center (JRTC), Ft Polk, La.

Warrior Brigade	
36th MED DET (AA)	6 x UH-1V *AA*
Polk AAF, Ft Polk, La.	*IIICORPS assigned: to gain UH-60A*

Third US Army (3rd ARMY)

Details for Army National Guard (ArNG) and US Army Reserves (USARC) units listed above are provided in separate listing.

Third US Army (3rd ARMY)

The 3rd Army is the US Army theatre command assigned to US Central Command (USCENTCOM), the joint warfighting command responsible for maintaining American military interests in Southwest Asia; this area covers 20 countries, from eastern Africa to western Asia. The command headquarters, located at Ft McPherson, Atlanta, Georgia, spends most of its time preparing for theoretical, future conflict scenarios. They range from large operations with armoured formations, such as was executed during Operation Desert Storm, to limited warfare options of a special operations nature, to peacekeeping and humanitarian missions, and everything in between.

The command was reactivated in 1982 as the Army component headquarters for CENTCOM, composed of Army elements of what had previously been known as the Rapid Deployment Force (RDF). The Third Army is resourced primarily from Forces Command (FORSCOM) units for potential contingency operations with the forces of XVIII Airborne Corps (XVIIIABNCORPS) and US Army Special Operations Command (USASOC) constantly preparing for rapid deployment to the theatre to deter, reinforce and fight, if so directed by the National Command Authority (NCA). Additional warfighting assets are also tasked to reinforce the command from reserve components assigned to FORSCOM. Third Army was resourced to command American and coalition forces deployed to the theatre for what became Operation Desert Shield/Storm. The command built to a peak force of over 350,000 US soldiers, consisting of two US Army corps with seven divisions, numerous brigades, commands and regiments, US Marine Corps forces equal to another corps (with two divisions and an air wing), along with two divisions from France and the United Kingdom. Aviation assets directly controlled by 3rd Army included over 2,080 US Army combat and combat support aircraft, along with several hundred Marine Corps helicopters, plus aviation assets belonging to the coalition countries. VII Corps was resourced with six aviation brigades, the 2nd Regimental Aviation Squadron (4-2 ACR), plus military intelligence and corps support aviation assets, totalling 957 aircraft. XVIII Airborne Corps had five aviation brigades – three assigned to divisions and two corps-level brigades – along with the assets of 4-3 ACR and aviation units assigned to their intelligence brigades and corps support assets, totalling over 1,025 aircraft. Special operations aircraft (SOA) of the 160th SOAR operated at least 50 aircraft in-theatre, including AH-6Gs and MH-60A/Ls operating as gunships plus MH-6Hs, CH/MH-47Ds and MH-60A/Ls flying covert insertion and extraction missions, often behind Iraqi lines.

The 3rd Army's theatre aviation brigade, the 244th Aviation Brigade, a US Army Reserve command, did not deploy to Saudi Arabia. It is the only remaining aviation brigade with a combat capability assigned to the US Army Reserve Command (USARC). The units attached to the command are drawn from the USARC or ARNG components. The lean force structure of the theatre aviation brigade can be boosted with additional units as needed in time of conflict or contingency.

USCENTCOM

US Central Command HQ (USCENTCOM), MacDill AFB, Fla.

3rd US Army/US Army Central Command HQ (THREEUSA/3rd Army/ARCENT), Ft McPherson, Ga.

377th Theater Army Area Command (377 TAACOM), New Orleans, La.

244th Aviation Brigade (Theater) (244 AVN BDE) Ft Sheridan, Ill.	
1-147 AVN (CMD) (WI ARNG)	20 x UH-60A *CS/GS*, 4 x UH-60A/C *C2*
5-159 AVN (HHB) (USARC)	32 x CH-47D *HH*
2-228 AVN (TAB) (USARC)	15 x C-12R *TA*
	1 x UC-35A *TA*

I CORPS

I Corps, pronounced 'eye-corps', is tasked with contingency operations in the Pacific theatre. The corps draws much of its combat and combat support strength from reserve component (RC) units. In peacetime the corps commands only brigade-sized, active-duty units, but could draw from other AC and RC units up to division size. The 40th Infantry Division (Mech.) of the California ARNG is attached to I Corps and has received upgraded equipment and funding in recent years.

III CORPS

The 3rd Corps, represented as III Corps, is supported by brigade-level aviation units including 13th COSCOM, 3rd ACR, 6th CAV BDE, 21st CAV BDE and the 504th MI BDE. These units are deployable combat and combat support units, with the exception of 21st CAV BDE which is the unit combat aviation skills and tactics training unit, performing as a Navy fleet replacement squadron (FRS), or an Air Force replacement training squadron (RTS). 6th CAVBDE(AC), or 6 ACCB, has been deployed to Korea for over two years to support United Nations forces on the peninsula. The 1st Cavalry Division is essentially an armour division that has deployed task forces to Korea and Kuwait from 1996, and in the late summer of 1998 deployed a brigade task force to Bosnia, together with 2-227 AVN, the division's general support aviation battalion (GSAB).

The 4th Infantry Division (Mech.) was designated as an experimental division in the mid-1990s and has been at the vanguard of Force XXI (XXI=21st

Part of the 211th Aviation Group (211 AVNGRP) (Attack), the 1-183th AVN (ATK) of the Idaho Army National Guard reports to I Corps for operational tasks. It operates OH-58As, UH-60As (above) and AH-64As (above left).

Century) test and experimentation of new systems, training and tactics. The division completed a brigade-sized NTC rotation to Ft Irwin, California in March 1997 and performed a division exercise from Ft Hood later that autumn. A variety of improved aviation systems was fielded including two of the prototype AH-64D aircraft with 1-4 ATK; two prototype UH-60C A2C2S, battlefield command and control platform, with 2-4 AVN; eight OH-58D(I)s with upgraded digital communication links with 1-10 CAV; and at least eight CH-47Ds assigned to G Company/3rd BN/149th AVN REGT (TX ARNG) with similar upgrades.

Many RC brigades and groups co-ordinate their training with III Corps including several ARNG divisions such as the 35th Infantry Division (Mech.), Kansas ARNG, and the 49th Armored Division, the Texas ARNG command based at North Ft Hood.

I CORPS

I CORPS HQ (ICORPS), Ft Lewis, Washington
311th Corps Support Command (311 COSCOM),
Los Angeles, Calif. (USARC)

1-109 AVN (AVIM) (IAARNG)	6 x UH-60A *AVIM*

175th Medical Brigade (175 MEDBDE),
Sacramento, California (CA ARNG)

62nd Medical Group (62 MEDGRP), Ft Lewis, Wash.

24th MED CO (AA)(NE ARNG)	15 x UH-60A	*AA*
54th MED CO (AA) (Active)	15 x UH-60A	*AA*
Lewis AAF, Ft Lewis, Washington		
68th MED DET (AA) (Active)	6 x UH-60A	*AA*
Wheeler AAF, Schofield Barracks, Hawaii		
126th MED CO (AA)(CA ARNG)	15 x UH-60A	*AA*
247th MED DET (AA) (Active)	6 x UH-60A	*AA NTC support*
Barstow-Daggett AP, California		
281st MED DET (AA) (Active)	6 x UH-60A	*AA*
Wainwright AAF, Ft Wainwright, Alaska		
717th MED CO (AA) (NM ARNG)	15 x UH-60A	*AA; Det 1, NV ARNG*
1022nd MED CO (AA) (WY ARNG)	15 x UH-60A	*AA Det 1 CO ARNG*
1042nd MED CO (AA) (OR ARNG)	15 x UH-1V	*AA to UH-60A*
1085th MED CO (AA) (SD ARNG)	15 x UH-1V	*AA; to UH-60A*
		Det 1 MT ARNG
1187th MED CO (AA) (IA ARNG)	15 x UH-1V	*AA; to UH-60A*
		Det 1 MN ARNG

66th Aviation Brigade (66 AVNBDE),
Camp Murray, Tacoma, Washington (WA ARNG)

185th Aviation Group (185 AVNGRP) (Lift),
Hawkins Field, Jackson MAP, Mississippi (MS ARNG)

1-108th AVN (ASLT)(KS ARNG)	30 x UH-60A	*ASLT*
1-112th AVN (LUH) (ND ARNG)	32 x UH-1H	*LUH*
3-140th AVN (HHB) (CA ARNG)	48 x CH-47D	*HH*
	16 x CH-47D	*HH; Active duty*
1-185th AVN (CMD)(MS ARNG)	4 x UH-60A *CS*, 4 x UH-60A/C *C2*	
	16 x MH-60A *GS*, 15 x OH-58C *GS*	
	15 x OH-58D(I) *TARC*	
1-189 AVN (CSAB)(MT ARNG)	32 x UH-60A	*CS*

211th Aviation Group (211 AVNGRP) (Attack),
SLC AP #2 W. Jordan, Utah (UT ARNG)

1-183th AVN (ATK) (ID ARNG)	13 x OH-58A *SCT*, 3 x UH-60A *CS*
	18 x AH-64A *ATK*
1-285th AVN (ATK) (AZ ARNG)	13 x OH-58A *SCT*, 3 x UH-60A *CS*
	18 x AH-64A *ATK*

278th Armored Cavalry Regt (278 ACR), Knoxville, Tenn.
(TN ARNG)

D/107 AVN (AVIM) (TN ARNG)	2 x UH-60A *AVIM*
4-278 ACR (RAS) (TN ARNG)	40 x AH-1F *ATK/SCT*
	3 x EH-60C *CEWI*
To replace AH-1F with (18 x UH-60A) ASLT/CS, OH-58D(R) &	
AH-64A, in 1999	

III CORPS

III CORPS HQ (IIICORPS), Ft Hood, Texas
13th Corps Support Command (13 COSCOM), Ft Hood, Texas

3-135th AVN (AVIM) (MO ARNG)	4 x UH-60A *AVIM*; 2 *ARNG*;
	and 2 *USARC*
	2 x UH-60A *AVIM*; *Active duty*

112th Medical Brigade (112 MEDBDE),
Columbus, Ohio (OH ARNG)

1st Medical Group (1 MEDGRP), Ft Hood, Texas

36th MED DET (AA) (Active)	6 x UH-1V *AA*; *JRTC Support*;	
Polk AAF, Ft Polk, La.	*to gain UH-60A*	
82nd MED CO (AA) (Active)	15 x UH-60A *AA*; *Det 1, Ft Riley,*	
Henry Post AAF, Ft Sill, Okla.	*Kansas*	
507th MED CO (AA) (Active)	15 x UH-60A *AA*;	
Hood AAF, Ft Hood Texas	*Det 1 Ft Sam Houston, Texas*	
571st MED CO (AA) (Active)	15 x UH-60A *AA*; *Det 1, Ft Bliss, TX*	
Butts AAF, Ft Carson, Colo.		
172nd MED CO (AA) (AR ARNG)	15 x UH-1V *AA*	
812th MED CO (AA) (LA ARNG)	15 x UH-1V *AA*	
832nd MED CO (AA) (WI ARNG)	15 x UH-1V *AA*	

3rd Armored Cavalry Regt (3 ACR), Ft Carson, Colo.

4-3rd ACR (RAS)	24 x AH-1F(N) *AIR CAV SCT*
Butts AAF, Ft Carson, Colo.	3 x EH-60C *CEWI*
To replace AH-1F(N) with 15 UH-60L ASLT/GS/CS OH-58D(R), 1999	
	16 x AH-64A *ATK*

6th Cavalry Brigade (Air Combat) (6 ACCB),
Hood AAF, Ft Hood, Texas

Brigade forward deployed with Eighth US Army (EUSA) and relocated to Republic of Korea, 1996

63rd Aviation Group (63 AVNGRP) (Lift),
Boone AHP, Frankfort, Kentucky (KY ARNG)

1-106th AVN (ASLT) (IL ARNG)	30 x UH-60A *ASLT*
2-135th AVN (LUH) (CO ARNG)	32 x UH-1H *LUH*
1-137th AVN (CSAB) (OH ARNG)	32 x UH-60A *CS*
3-149th AVN (HHB) (TX ARNG)	48 x CH-47D *HH; ARNG*
	16 x CH-47D *HH; USARC*
1-244th AVN (CMD) (LA ARNG)	15 x OH-58C *TARC*, 16 x UH-60A *GS*
	8 x UH-60A *CMD*;
Active-duty company operates with 21 CAVBDE/IIICORPS below	

385th Aviation Group (385 AVNGRP) (Attack),
Papago AAF, Phoenix, Arizona (AZ ARNG)

3-6th CAV (AC)	24 x AH-64A *SCT/ATK*
(Forward deployed to 8th US Army, Korea)	
4-6th CAV (AC)	24 x AH-64A *SCT/ATK*
(Forward deployed to 8th US Army, Korea)	
7-6 CAV (AC) (USARC)	24 x AH-64A *SCT/ATK*;
	attached to 1st CAV DIV, 1996-99

21st Cavalry Brigade (21 CAVBDE) (Air Combat),
Hood AAF, Ft Hood, Texas

Instructor cadre	6 x OH-58D(I) *AIR CAV CBT TRAINING*
AH-64D since Apr 98	6 x AH-64A *ATK/SCT CBT TRAINING*
	6 x AH-64D *ATK/SCT CBT TRAINING*
B/1-158 AVN (CMD) (Active)	8 x UH-1H *CMD/CS*;
IIICORPS support unit; attached to 1-244 AVN/6ACCB	

504th Military Intelligence Bde (504 MI BDE)
W. Ft Hood, Texas

15th MI BN (AE)	6 x RC-12P/Q *CEWI*;
Robert Gray AAF, W. Ft Hood, Texas	

1st Cavalry Division (1CD), Ft Hood, Texas

Division Support Command (1CD DISCOM), Ft Hood, Texas

615th AVN SPT BN (ASB)	Robert Gray AAF, W. Ft Hood, Texas
C/227th AVN (AVIM)	2 x UH-60A *AVIM*

Aviation Brigade/4th Brigade Robert Gray AAF, Ft Hood, Texas

1-227th AVN (ATK)	24 x AH-64D *ATK/SCT*;
	Delivered from April 1998
2-227th AVN (GSAB)	20 x UH-60L *ASLT/GS*,
	4 x UH-60L(C) *C2*
	4 x EH-60C *CEWI*
1-7th CAV	16 x OH-58D(R) *AIR CAV*
7-6th CAV (AC) (USARC)	24 x AH-64A *ATK/SCT*
Montgomery City AP, Conroe, Texas	

4th Infantry Division (Mech.) (4ID(M)), Ft Hood, Texas

Division Support Command (4ID DISCOM), Ft Hood, Texas

404 AVN SPT BN (ASB) Hood AAF, Ft Hood, Texas	
F/4 AVN (AVIM)	2 x UH-60A *AVIM*

Aviation Brigade (4ID AVNBDE), Hood AAF, Ft Hood, Texas

1-4 AVN (ATK)	24 x AH-64A *ATK/SCT*,
	? x (Y)AH-64D *ATK/SCT*;
	2 a/c operated 1996-98
2-4 AVN (CMD)	20 x UH-60A *GS*, 2 x UH-60A(C) *C2*,
	4 x EH-60C *CEWI* , 4 x UH-60L *CS*
	? x (Y)UH-60C *C2*;
	2 a/c operated 1996-98
1-10 CAV	16 x OH-58D(I) *AIR CAV*

XVIIIABNCORPS

XVIII Airborne Corps is the Army's elite, rapid-reaction corps, configured with a variety of capabilities to control varied conflict scenarios. The corps prepares to operate anywhere in the world and is the service's primary force projection land warfare command. Known as the 'Dragon Corps', it is assigned a number of unique light infantry, mechanised infantry, armour and combat support units which provide operational planners with the ability to customise force packages, task forces and any size combat formation up to and including multiple divisions. These force packages train routinely to converge on an objective from geographically dispersed sites.

The corps is the first to receive the latest equipment and has been among the most active in the US Army since the Vietnam War, performing a crucial role in Operation Desert Shield/Storm. The 2nd Armored Cavalry Regiment (Light) is the only airborne deployable ACR trading heavy firepower of a traditional ACR armour focus for rapid mobility, deployability and manoeuvrability. Its regimental aviation squadron (RAS) operates armed OH-58D(I) Kiowa Warriors and UH-60Ls that can be loaded onto Air Force airlift aircraft in a matter of hours and deployed,

prepared to conduct operations almost immediately. The 82nd Airborne Division (82 ABN) is the best-trained, best-prepared light infantry division in the world, completely deployable by parachute delivery. The 10th Infantry Division (Light), also known as the 10th Mountain Division, operates at almost an equally aggressive training and deployment posture. The 101st Airborne Division (Air Assault) is another light infantry division, configured for heliborne assault

The 'Golden Knights' parachute display team uses two C-31As as jump-ships, based at Simmons AAF, Ft Bragg, N.C., as part of the 18th Aviation Brigade.

with the largest combat aviation brigade inventory assigned to any division-sized unit in the world. The 3rd Infantry Division (Mechanized) is the corps's heavy armour force. III Corps-assigned units such as the 3rd ACR and 1st Cavalry Division have augmented XVIII Airborne Corps in recent years.

XVIIIABNCORPS is assigned many ARNG and USARC combat support units, which must maintain their combat readiness and train for the corps's rapid-deployment contingency roles. These units are all compliant with ARI-specified inventory and equipment levels. The 18th Aviation Brigade gains most of these assets while the 1-111 AVN (ATK), an Apache battalion of the Florida ARNG, is assigned directly to the 3rd ID, the only heavy armour division still resourced with two attack helicopter battalions.

The UH-1V is widely used in the air ambulance role by medical companies. The Huey is not ideal for the role and its replacement is planned to be the HH-60L.

XVIIIABNCORPS

XVIII Airborne Corps, HQ (XVIIIABNCORPS), Ft Bragg, N.C.

1st Corps Support Command (1st COSCOM), Ft Bragg, N.C.

4-159 AVN (AVIM)	8 x UH-60A *AVIM*
Simmons AAF, Ft Bragg, North Carolina	

44th Medical Brigade (44 MEDBDE), Ft Bragg, N.C.

MED CECAT (AA) (TN ARNG)	3 x UH-60A *AA; upgrade to UH-60Q*
	1 x YUH-60A-Q *AA; prototype UH-60Q*
57th MED CO (AA)	15 x UH-60A *AA*
Simmons AAF, Ft Bragg, North Carolina	
104th MED CO (AA) (MD ARNG)	15 x UH-1V *AA*
107th MED CO (AA) (OH ARNG)	15 x UH-1V *AA*
121st MED CO (AA) (DC ARNG)	15 x UH-1V *AA; Det 1 WV ARNG*
148th MED CO (AA) (GA ARNG)	15 x UH-1V *AA*
198th MED CO (AA) (DE ARNG)	15 x UH-1V *AA; Det 1 PA ARNG*
214th MED DET (AA)	6 x UH-60A *AA deployed USARSO*
Ft Kobbe, Howard AFB, Panama	
229th MED DET. (AA)	6 x UH-60A *AA*
Wheeler Sack AAF, Ft Drum, New York	
498th MED CO. (AA)	15 x UH-60A *AA*
Lawson AAF, Ft Benning, Georgia	
681st MED CO (AA) (IN ARNG)	15 x UH-1V *AA*
1059th MED CO (AA) (MA ARNG)	15 x UH-1V *AA*
1133rd MED CO (AA) (AL ARNG)	15 x UH-1V *AA*
1159th MED CO (AA) (NH ARNG)	15 x UH-1V) *AA; Det 1 NJ ARNG*

2nd Armored Cavalry Regiment (Light) (2 ACR)(L), Ft Polk, La.

4-2 ACR(L)	32 x OH-58D(I) *AIR CAV*
Polk AAF, Ft Polk, La.	15 x UH-60L *ASLT/CS/GS*

18th Aviation Brigade (18 AVNBDE), Simmons AAF, Ft Bragg, N.C.

'Golden Knights', US Army Parachute Team	
Simmons AAF, Ft Bragg N.C.	1 x C-12C, 2 x C-31A,
	2 x UV-20A *GS*

159th Aviation Group (159 AVNGRP), Campbell AAF, Ft Campbell, Ky.

1-58 AVN (ATS)	1 x C-12C *OSA*
Simmons AAF, Ft Bragg, North Carolina	
1-126 AVN (LUH) (RI ARNG)	32 x UH-1H *LUH*

1-131 AVN (ASLT) (AL ARNG)	30 x UH-60A *CS*
1-159 AVN (CMD)	4 x UH-60L *CS/GS*
Simmons AAF, Ft Bragg N.C.	4 x UH-60L(C) *C2*
	16 x UH-60L *GS; NC/SC ARNG assigned*
1-169 AVN (HHB) (CT ARNG)	32 x CH-47D *HH*
	32 x CH-47D *HH: Active-duty*
1-171 AVN (CSAB) (GA ARNG)	16 x UH-60L *CS*
	16 x UH-60L *CS; Active-duty*

229th Aviation Group (229 AVNGRP), Simmons AAF, Ft Bragg, North Carolina

1-130 AVN (ATK) (NC ARNG)	24 x AH-64A *ATK/SCT*
1-229 AVN (ATK)	24 x AH-64A *ATK/SCT*
Simmons AAF, Ft Bragg, North Carolina	
3-229 AVN (ATK)	24 x AH-64A *ATK/SCT*
Simmons AAF, Ft Bragg, North Carolina	

525th Military Intelligence Bde (525 MIBDE), Ft Bragg, North Carolina

224th MI BN (AE)	12 x RC-12N *CEWI*
Hunter AAF, Savannah, Georgia	

3rd Infantry Division (Mech.) (3ID(M), Ft Stewart, Georgia

Division Support Command (3ID DISCOM), Ft Stewart, Georgia	
A/603 ASB	2 x UH-60A *AVIM*
Hunter AAF, Savannah, Georgia	
Aviation Brigade (3ID AVNBDE), Hunter AAF, Savannah, Georgia	
1-3 AVN (ATK)	24 x AH-64A *ATK/SCT*
1-111 AVN (ATK) (FL ARNG)	24 x AH-64A *ATK/SCT*
2-3 AVN (GSAB)	4 x EH-60C *CEWI*
	20 x UH-60L *ASLT/CS/GS*
	4 x UH-60A(C) *C2*, 6 x OH-58C *TAR*
3-7 CAV	16 x OH-58D(I) *AIR CAV*

10th Infantry Division (Light) (10ID(L), Ft Drum, New York

Division Support Command (10ID DISCOM), Ft Drum, New York	
C/10th AVN (AVIM)	2 x UH-60L *AVIM*
Wheeler Sack AAF, Ft Drum, New York	
Aviation Brigade (10ID AVNBDE), Wheeler Sack AAF, Ft Drum, New York	

1-10 AVN (ATK)	24 x OH-58D(I) *ATK/SCT*
2-10 AVN (ASLT)	34 x UH-60L *ASLT/CS*
	4 x UH-60L(C) *C2*, 4 x EH-60C *CEWI*
3-17 CAV (RECON)	16 x OH-58D(I) *AIR CAV*

82nd Airborne Division (82AD), Ft Bragg, North Carolina

Division Support Command (82AD DISCOM), Ft Bragg, North Carolina

D/82 AVN (AVIM)	2 x UH-60L *AVIM*
Simmons AAF, Ft Bragg, North Carolina	

Aviation Brigade (82AD AVNBDE), Simmons AAF, Ft Bragg, Texas

1-82 AVN (ATK)	24 x OH-58D(I) *ATK/SCOUT*
2-82 AVN (ASLT)	34 x UH-60L *ASLT/CS*
	4 x UH-60L(C) *C2*, 4 x EH-60C *CEWI*
1-17 CAV (AIR)	24 x OH-58D(I) *AIR CAV*

101st Airborne Division (Air Assault) (101AD(AASLT)), Ft Campbell, Kentucky

Division Support Command (101AD DISCOM), Ft Campbell, Kentucky

8-101 AVN (AVIM)	6 x UH-60A *AVIM*
Campbell AAF, Ft Campbell, Kentucky	

326th Medical Evacuation Bn (326 MEB), Ft Campbell, Kentucky

50th MED CO. (AA)	15 x UH-60A *AA*
Campbell AAF, Ft Campbell, Kentucky	

Aviation Brigade (101AD AVNBDE), Campbell AAF, Ft Campbell, Kentucky

1-101 AVN (ATK)	24 x AH-64A *ATK/SCT*
2-101 AVN (ATK)	24 x AH-64A *ATK/SCT*
8-229 AVN (ATK) (USARC)	24 x AH-64A ATK/SCT
Goodman AAF, Ft Knox, Kentucky	
3-101 AVN (ASLT)	30 x UH-60L *ASLT*
4-101 AVN (ASLT)	30 x UH-60L *ASLT*
5-101 AVN (ASLT)	30 x UH-60L *ASLT*
6-101 AVN (CMD)	20 x UH-60L *GS/CS*
	4 x UH-60L(C) *C2*, 4 x EH-60C *CEWI*
7-101 AVN (HHB)	48 x CH-47D *HH*
2-17 CAV	32 x OH-58D(I) *AIR CAV*
Sabre AHP, Ft Campbell, Kentucky	

Above: The US Army has been retiring the once vast fleet of OH-58A/Cs from Army National Guard service over the last few years. About 300 remain. This OH-58C serves with the Flight Det of the Operations Group of the National Training Center, based at Barstow-Daggett Airport, California.

Right: The UH-1H has been a stalwart of the US Army National Guard for over a quarter of a century. '081' displays the Minuteman insignia on its door.

US Army Reserve Command (USARC)

The US Army Reserve Command (USARC), headquartered at Ft Gillem, Atlanta, Georgia, is essentially tasked with supplying FORSCOM (Forces Command) combat support (CS) and combat service support (CSS) forces. Established in October 1990, the command relinquished most of its aviation force structure of over 600 aircraft during the early 1990s and has fewer than 100 aircraft directly under its control today. What remains are two Apache units, two fixed-wing theatre aviation battalions, and a CH-47D unit, along with less than a handful of UH-1H and UH-60As assigned for combat support tasks. In peacetime, the 244th Aviation Brigade (Theater) supports the aviation assets and personnel, but the two AH-64A units are assigned to other warfighting commands. 244 AVN BDE directly supports Third US Army, or US Army Central Command (3A/ARCENT), forces under US Central Command (USCENTCOM/CENTCOM).

Regional Support Commands (RESCOMs) provide support and training along with newly fielded training support brigades (TSBs), and their assigned battalions. One such organisation is the 166th Aviation Brigade, which conducts lanes training, without any aircraft assigned.

USARC

reports to US Army Forces Command HQ (FORSCOM), Ft McPherson, Atlanta, Georgia

US Army Forces, Central Command/Third US Army (ARCENT/3A), Ft Gillem, Atlanta, Georgia

377th Theater Army Area Command (377 TAACOM), New Orleans, Louisiana (USARC)

244th Aviation Brigade (Theater), Ft Sheridan, Illinois (USARC)

7-6 CAV (AC) (USARC)	24 x AH-64A *ATK/SCT;*
Montgomery CAP, Conroe, Texas	*assigned to 1 CAV DIV*
8-229 AVN (ATK) (USARC)	24 x AH-64A *ATK/SCT;*
Goodman AAF, Ft Knox, Ky.	*assigned to 101 ABN DIV(AASLT)*
6-52 AVN(TAB) (USARC)	12 x C-12F/R *TA/OSA Det 1*
	Robert Gray AAF, Ft Hood, Texas
Los Alamitos AAF, California	4 x UC-35A
2-228 AVN (TAB) (USARC)	15 x C-12R *TA/OSA*
	Det 1 Johnstown AP, Pa.
NAS Willow Grove, Pa.	1 x UC-35A
M/159 AVN (AVIM)	2 x UH-60A *AVIM*
Hood AAF, Ft Hood, Texas	
Army Avn Spt Facility	2 x UH-1H *CS*
Montgomery CAP, Conroe, Texas	

166th Aviation Brigade (Training) Ft Riley, Kansas
no aircraft assigned presently

Army National Guard (ARNG)

Army National Guard (ARNG) aviation units perform combat and combat support roles while operational in all 50 states, along with the territories of Puerto Rico and the Virgin Islands. The citizen-soldier Army is about 10 per cent full-time personnel, with the remainder drilling with their assigned units slightly more than one weekend per month, and for two weeks of co-ordinated unit drills, usually in the summer months. The ARNG is considered a reserve component (RC) force, although it is managed at the state level. The force level for the ARNG is about 350,000 soldiers, and during peacetime they are a strategic reserve to US Forces Command (FORSCOM), the Army component of the US Atlantic Command (USACOM).

The Army Guard, along with the Air National Guard, is represented in the Pentagon hierarchy by the National Guard Bureau (NGB). The NGB performs as the central advocate for the state area commands (STARCs) among military and political spheres of influence in the Washington, D.C. power circles. While much of the ARNG equipment is acquired from or funded by the federal budget, the funding for operations, resources and staffing is administered at the individual state levels. ARNG soldiers perform their duty to the state but they can be called up by the Congress or the President.

The Adjutant General (TAG or AG, as it is sometimes referred to) is usually a one- or two-star general who has prior active-duty experience and has spent several years in the ARNG organisation. Most are appointed but Vermont elects its, and has recently chosen a female for this command billet – the first ever for the ARNG. The State Area Command (STARC) manages the total state resources. Each STARC will command units at division, brigade or group levels.

If there are other units active in the state that report to commands in other states, or with US Army Reserve Command (USARC) or active component (AC) commands, they are usually commanded by a separate numbered troop command for administrative support in the state military structure. The District of Columbia operates as the only District Area Command (DARC) while the three separate US trust territories – Guam, Puerto Rico and the Virgin Islands – are known as TARCs, of which only Guam does not operate aviation units.

ARNG units have undergone tremendous change, reorganisation and restructuring since the early 1990s. Units were inactivated while many of the resources and personnel were reorganised under new unit designations. Most of these changes have affected the aviation units assigned at corps or theatre echelons of command. Nearly 1,000 UH-1s, AH-1s and OH-58s have been retired from ARNG service since 1991, with less than half replaced by modernised types. While some consider these to be the legacy of obso-

lete systems, leftover from Vietnam and Cold War requirements, they perform mundane tasks well and are maintainable, reliable and considerably less costly to operate than replacement types such as the Blackhawk. Some states have postulated reviving combat enhancement capability detachments (CECAT), equipped with UH-1H/V aircraft, to keep the Huey available for non-tactical missions.

The ARNG has attempted to keep attack battalions and cavalry squadrons assigned to a single state, while other units are usually distributed among several states with companies and detachments geographically dispersed to allow more states to utilise their capabilities. This applies to air ambulance (AA), assault (ASLT), command and control (CMD), combat support (CS), heavy helicopter (HH) and light utility helicopter (LUH). The aviation intermediate maintenance (AVIM) battalions likewise deploy their three assigned companies. The fixed-wing, theatre aviation companies deploy two Sherpas in detachments to different states. Several air ambulance units have converted to UH-60As in the last several years.

ARNG units are completing the acceptance of 28 improved, rebuilt C-23B+ Sherpa fixed-wing transports. The AH-64A Apache fleet continues to grow slightly, due to restructuring of aviation units, but no new units will be fielded with the type. The service has absorbed over 350 additional UH-60A Blackhawk variants since the early 1990s, but about 620 UH-1H/Vs remain operational. The fleet of OH-58A/C aircraft was to have been reduced to fewer than 100 aircraft, but political wrangling has allowed about 300 Kiowas to continue in service. Army Guard units have absorbed additional CH-47D airframes since 1996, reassigned from inactivated active-duty units. Fewer than 400 AH-1Fs remain in service with attack and cavalry units, and many are being upgraded with the Night Targeting System (NTS), improvements which permit greater night target acquisition and give the ability to deliver AGM-118 Hellfire missiles.

The ARNG completed fielding of the AH-64A Apache in 1995, and about half of the assigned battalions have undergone transition to the ARI unit structure. The 1-135 AVN of the Missouri ARNG was scheduled to undergo transition from the AH-1F Cobra, but the aircraft were utilised elsewhere. The 40th Infantry Division (M) has gained the 1-211 AVN(ATK) of the Utah ARNG as its resourced attack battalion. The Army National Guard accepted its first Apache in 1987 and now fields seven battalions. The AH-64D Longbow Apache will be fielded to active component (AC) units first, with the first ARNG battalion not expected to transition to the type until at least 2004.

The Operational Support Airlift Command (OSACOM) was redesignated and realigned on 2 October 1995 from the assets of what was formerly known as OSAC, also called the Operational Support Airlift Command. The difference between them was the realignment of the command from a component command of the Military District of Washington (MDW) to a field operating agency (FOA) of the US

US Army

UH-60Ls, such as those of 'D' Company 1-171 CSAB, serve in the Army National Guard in increasing numbers, replacing Hueys and earlier Blackhawks.

Army, assigned to the National Guard Bureau (NGB). The Operational Support Airlift Agency (OSAA) was created in 1997 to be the reporting agency to NGB and to interact with the Joint Operational Support Airlift Command (JOSAC), a subordinate command to the US Transportation Command (USTRANSCOM). From its headquarters at Scott AFB, Illinois, JOSAC administers the OSA fleet that is assigned to the Air Force, Army, Marine Corps, Navy and their reserve component (RC) commands. The function of all these new organisations is supposedly to provide better utilisation of the OSA fleet assigned to the different services, and to provide better customer service and satisfaction. The Joint Airlift Logistics Information System (JALIS)

ARNG

US Atlantic Command, HQ (USACOM), Sewalls Point, Norfolk, Virginia
Forces Command HQ (FORSCOM), Ft McPherson, Atlanta, Georgia

National Guard Bureau (NGB) Pentagon, Washington, D.C.

Counter-Drug Directorate	
State Adjutant Generals (AG/TAG) (50 STARC, 2 TARC, 1 DARC)	
District Area Command (DARC)	2 x UH-60A GS
RAID Detachment	3 x OH-58A(R) RAID
State Area Commands (STARC)	
RAID Detachments	150 x OH-58A(R) RAID
Territory Area Commands (TARC)	4 x UH-1H GS
RAID Detachment	6 x OH-58A(R) RAID

Army National Guard Readiness Center (ARNGRC), Arlington, Virginia

Operational Support Airlift Agency (OSAA)
Davison AAF, Ft Belvoir, Virginia

Operational Support Airlift Cmd (OSACOM), Davison AAF, Ft Belvoir, Virginia

US Army Priority Air Transport Detachment (USAPAT)	
Andrews AFB, Maryland	1 x C-20E, 1 x C-20F, 1 x C-20J
	3 x C-21A OSA/VIP
Alaska Regional Flight Center (AK RFC)	3 x C-12F OSA
Wainwright AAF, Ft Wainwright, Alaska	
Ft Belvoir Regional Flight Center (FB RFC)	2 x C-12C OSA
Davison AAF, Ft Belvoir, Maryland	4 x C-12D OSA
Georgia Regional Flight Center (GA RFC)	2 x C-12C OSA
Brown Field, Fulton CAP, Georgia	4 x C-12F OSA
Hawaii Regional Flight Center (HI RFC)	2 x C-12F OSA
Hickam AFB, Hawaii	1 x C-20E OSA
Lewis Regional Flight Center (WA RFC)	2 x C-12C OSA
Lewis AAF, Ft Lewis, Washington	3 x C-12F OSA
Panama Regional Flight Center (RP RFC)	3 x C-12F OSA
Ft Kobbe, Howards AFB, Panama	
Texas Regional Flight Center (TX RFC)	2 x C-12C OSA
Robert Gray AAF, Ft Hood, Texas	2 x C-12F OSA
State Adjutant Generals (AG/TAG)	
State Area Commands (STARC)	
OSACOM Detachments	42 x C-12D/F OSA;
	41 STARC, 1 TARC
	10 x C-26B OSA;
	9 STARC, 1 DARC

Active-Duty Theater Army assigned units:

USCENTCOM/3rd Army/244th AVN BDE (USARC) assigned:

1-147 AVN (CMD) (WI ARNG)	20 x UH-60A CS/GS,
Truax Field, Madison, Wisconsin	4 x UH-60A(C) C2

USEUCOM/USAREUR/7A assigned:

1-168 AVN (TAB) (WA ARNG) Camp Murray, Tacoma, Washington
(operational control of C-23B/B+ companies in peacetime)

H/171 AVN (TA)	8 x C-23B/B+ TA/OSA;

Dets in Lakeland MAP, Florida, Alabama, Kentucky and Texas USARCENT/3A mission

F/192 AVN (TA)	8 x C-23B/B+ TA/OSA

Dets in IN, PA, Muniz AP, San Juan, P.R. RI and VA USARSO mission, F.L.R. Dominicci AP, San Juan, Puerto Rico

A/249 AVN (TA)	8 x C-23B+ TA/OSA

Dets in OK, SD, McNary Field, Salem, Ore. and WA USAREUR/7A mission

USPACOM/USA Alaska/Arctic Support Brigade assigned:

B/1-207 AVN (TAB) (AK ARNG)	8 x C-23B+ TA/OSA
Bryant AAF, Ft Richardson, Alaska	

Active-Duty Corps assigned aviation units:

I Corps/278th Armored Cavalry Regiment assigned:

D/107 AVN (AVIM) (TN ARNG)	2 x UH-60A AVIM
4-278 ACR (RAS) (TN ARNG)	26 x AH-1F(N) ATK,
	26 x OH-58A SCT
	18 x UH-60A ASLT/CS,
	3 x EH-60C CEWI

I Corps/66th Aviation Brigade assigned:

1-108 AVN (ASLT) (KS ARNG)	30 x UH-60A ASLT
Forbes Field, Topeka, Kansas	
1-109 AVN (AVIM) (IA ARNG)	6 x UH-60A AVIM
Boone MAP, Iowa	
1-112 AVN (LUH) (ND ARNG)	32 x UH-1H LUH
Bismarck RAP, South Dakota	
3-140 AVN (HHB) (CA ARNG)	48 x CH-47D HH:
Stockton MAP, California	ARNG in CA, HI, NV or WA
(16 x CH-47D are assigned to the active-duty 4-123 AVN/USARAK)	
1-183 AVN (ATK) (ID ARNG)	18 x AH-64A ATK
Boise Air Terminal, Idaho	13 x OH-58A SCT,
	3 x UH-60A CS
1-185 AVN (CMD) (WA ARNG)	24 x UH/MH-60A CMD
Camp Murray, Tacoma, Washington	15 x OH-58C GS,
	15 x OH-58D(I) TARC
1-189 AVN (CSAB) (MT ARNG)	32 x UH-60A CS
Helena RAP, Montana	
1-285 AVN (ATK) (AZ ARNG)	18 x AH-64A ATK/SCT
Pinal Airpark, Marana, Arizona	13 x OH-58A SCT,
	3 x UH-60A CS

I Corps/311 COSCOM/175 Medical Brigade assigned:

24th MED CO (AA) (NE ARNG)	15 x UH-60A
Lincoln MAP, Neb.; Det 1 Forbes Field, Topeka, Kans. (KS ARNG)	
126th MED CO (AA) (CA ARNG)	15 x UH-60A AA
Mather AP, Sacramento, California	
717th MED CO (AA) (NM ARNG)	15 x UH-60A AA;
Santa Fe MAP, N.M. Det 1, Reno-Stead AP, Nevada (NV ARNG)	
1022nd MED CO (AA) (WY ARNG)	15 x UH-60A AA;
Cheyenne MAP, Wyo. Det 1, Buckley ANGB, Aurora (CO ARNG)	
1042nd MED CO (AA)	15 x UH-1V AA;
McNary Field, (AA) (OR ARNG) Salem, Ore. to gain UH-60A	
1085th MED CO	15 x UH-1V AA;
Rapid City RAP, S.D. (AA) (SD ARNG)	Det 1, Helena RAP, Mont. (MT ARNG); to UH-60A
1187th MED CO	15 x UH-1V AA;
Boone MAP, Iowa (AA) (IA ARNG)	Det 1, St Paul Downtown AP, Minn.; to UH-60A

III Corps/6th Cavalry Brigade assigned:

1-106 AVN (ASLT) (IL ARNG)	30 x UH-60A ASLT
Decatur MAP, Illinois	
2-135 AVN (LUH) (CO ARNG)	32 x UH-1H LUH
Buckley ANGB, Aurora, Colorado	
3-135 AVN (AVIM) (MO ARNG)	6 x UH-60A AVIM
Springfield MAP, Montana	
1-137 AVN (CSAB) (OH ARNG)	32 x UH-60A CS
Akron-Canton AP, Ohio	
3-149 AVN (HHB) (TX ARNG)	48 x CH-47D HH: ARNG

Grand Prairie AASF, Texas	in IL, IA, MS, OK and TX
(16 x CH-47D are assigned to the F/158 AVN USARC)	
1-244 AVN (CMD) (LA ARNG)	16 x UH-60A* CMD/GS
Lakefront AP, New Orleans, Louisiana	15 x OH-58C TARC

III Corps/13 COSCOM/112 Medical Brigade assigned:

172nd MED CO (AA) (AR ARNG)	15 x UH-1V AA
Robinson AAF, N. Little Rock, Arkansas	
812th MED CO (AA) (LA ARNG)	15 x UH-1V AA
Lakefront AP, New Orleans, Louisiana	
832nd MED CO (AA) (WI ARNG)	15 x UH-1V AA
West Bend AP, Wisconsin	

V Corps/12th Aviation Brigade assigned:

3-126 AVN (LUH) (CT ARNG)	32 x UH-1H LUH maybe replaced by 1-140 AVN Bradley AP, Windor Locks, Conn. (CA ANG)
3-142 AVN (CSAB) (PR ARNG)	32 x UH-60A CS
Muniz AP, San Juan, Puerto Rico	
1-151 AVN (ATK) (SC ARNG)	24 x AH-64A ATK/SCT
McEntire ANGB, Eastover, South Carolina	

XVIII ABN Corps/18th Aviation Brigade assigned:

1-126 AVN (LUH) (RI ARNG)	32 x UH-1H LUH
Quonset State AP, Rhode Island	
1-130 AVN (ATK) (NC ARNG)	24 x AH-64A ATK/SCT
Raleigh-Durham AP, North Carolina	
1-131 AVN (CSAB) (AL ARNG)	32 x UH-60A CS
Dannelly Field, Montgomery, Alabama	
1-169 AVN (HHB) (CT ARNG)	32 x CH-47D HH :ARNG
Enfield, Connecticut	in AL, CT, GA and PA
(32 x CH-47D assigned to active-duty and A & B/2-159 AVN, N.C. and Georgia-based.)	
1-171 AVN (ASLT) (GA ARNG)	30 x UH-60L ASLT
Winder AP, Georgia	

* Plus active component (AC) assigned companies
**Plus USARC assigned company

XVIII ABN Corps/1 COSCOM/44 Medical Brigade assigned:

MED CECAT (TN ARNG)	3 x UH-60A AA; to
Lovell Field, Chatanooga, Tennessee	convert to HH-60L
	1 x YUH-60A(Q) AA prototype HH-60L
104th MED CO (AA) (MD ARNG)	15 x UH-1V AA
Weide AAF, Aberdeen PG, Maryland	
107th MED CO (AA) (OH ARNG)	15 x UH-1V AA
Akron-Canton RAP, Canton, Ohio	
112th MED CO(AA)(ME ARNG)	15 x UH-60A AA
Bangor IAP, Maine	
121st MED CO (AA) (DC ARNG)	15 x UH-1V AA; Det 1;
Davison AAF, Ft Belvoir, Virginia	Wood CAP, Parkersburg, W.Va. (WV ARNG)
148th MED CO (AA) (GA ARNG)	15 x UH-1V AA
Winder AP, Georgia	
198th MED CO (AA) (DE ARNG)	15 x UH-1V AA; Det 1,
Wilmington-New Castle AP, Delaware	Muir AAF, Ft Indiantown Gap, Pa. (PA ARNG)
681st MED CO (AA) (IN ARNG)	15 x UH-1V AA
Shelbyville AP, Indiana	
1059th MED CO (AA) (MA ARNG)	15 x UH-1V AA
Otis ANGB, Falmouth, Massachusetts	
1133rd MED CO (AA) (AL ARNG)	15 x UH-1V AA
Brookley AP, Mobile, Alabama	
1159th MED CO (AA) (NH ARNG)	15 x UH-1V AA;
Concord MAP, N.H Det 1, Mercer CAP, Trenton, N.J. (NJ ARNG)	

Above: Winner of the US Army's C-XX programme, 35 UC-35As are in the process of being delivered. This example is operated by the 78 AVN BN based at NAF Atsugi, Japan.

Left: The C-23B/B+s are used for theatre airlift, but are commonly assigned to operational support airlift missions. This example, the third C-23B ordered, belongs to the Connecticut ARNG.

Below: The operational support airlift fleet of C-12Cs will be retired by 2000. 76-22560 belongs to the Alabama ARNG.

ARNG (continued)

Active-Duty Division assigned aviation units:

3rd Infantry Division (M) Aviation Brigade assigned:
1-111 AVN (ATK) (FL ARNG)	24 x AH-64A *ATK/SCT*	
Craig Field, Jacksonville, Florida		

Army Material Cmd (AMC)/Industrial Operations Cmd (USAIOC) assigned:

Maryland AVCRAD Combat Enhancement (MACE)		
Weide AAF, Aberdeen, Md.	1 x C-23B *GS/TA*	
I/185 AVN (TA) (MS ARNG)	8 x C-23B *GS/TA AVCRAD*	
Gulfport-Biloxi RAP, Miss.	*assigned Dets in California, Connecticut and Missouri*	

Training and Doctrine Command (TRADOC) assigned:
US Army Aviation Center and School (USAAVNC)
Ft Rucker, Alabama
Western Army Aviation Training Site (WAATS) (AZ ARNG)	
Pinal Airpark, Marana, Ariz.	24 x AH-1F, 5 x OH-58A(R) *Training*
Eastern Army Aviation Training Site (EAATS) (PA ARNG)	
Muir AAF, Anneville, Pa.	12 x UH-1H, 4 x CH-47D *Training*
	12 x OH-58A, 5 x UH-60A *Training*
Det 1, Clarkburg AP, W.Va.	4 x C-12C, 1 x C-23B+ *Training*
	1 x C-26B *Training*

28th Armored Div. (28AD),
Ft Indiantown Gap, Anneville, Pennsylvania (PA ARNG)
28AD DISCOM
F/104 AVN (AVIM) (PA ARNG)	2 x UH-1H *AVIM*

28AD AVNBDE Muir AAF, Ft Indiantown Gap, Pa. (PA ARNG)
2-104 AVN (CMD) (PA ARNG)	8 x UH-1H *CMD/CS*
	6 x OH-58C *TAR*, 16 x UH-60A *GS*
1-104 CAV (PA ARNG)	16 x AH-1F(N) *AIR CAV*
1-104 AVN (ATK) (PA ARNG)	24 x AH-1F(N) *ATK/SCT*
Johnstown AP, Pa.	

29th Infantry Div. (Light) (29ID(L),
Ft Belvoir, Virginia (VA ARNG)
29ID DISCOM
F/224 AVN (AVIM) (MD ARNG)	2 x UH-60A *AVIM*

29 AVNBDE, Weide AAF, Aberdeen PG, Md. (MD ARNG)
1-150 AVN (ATK) (NJ ARNG)	24 x AH-1F(N) *ATK/SCT*
Mercer City AP, Trenton, New Jersey	
2-224 AVN (GSAB) (VA ARNG)	15 x UH-1H *ASLT*
Richmond IAP, Sandston, Va.	6 x OH-58C *TAR*
	23 x UH-60A *ASLT/CMD/CS*
1-158 CAV (RECON) (MD ARNG)	16 x AH-1F(N) *AIR CAV*
Weide AAF, Aberdeen, Maryland	

34th Infantry Div. (M)(34ID(M),
Minneapolis, Minnesota (MN ARNG)
34ID DISCOM
F/147 AVN (AVIM) (MN ARNG)	2 x UH-1H *AVIM*

34 AVNBDE St Paul Downtown AP, Minn. (MN ARNG)
1-147 AVN (ATK) (MN ARNG)	24 x AH-1F *ATK/SCT*, ex-3-134 AVN
2-147 AVN (GSAB) (MN ARNG)	38 x UH-1H *CMD/CS/GS*
	6 x OH-58A *TAR*
1-113 CAV (IA ARNG)	16 x AH-1F *SCT ex 1-194 CAV*
Waterloo MAP, Iowa	

35th Infantry Div. (M)(35ID(M),
Ft Leavenworth, Kansas (KS ARNG)
35 DISCOM
E/135 AVN (AVIM) (MO ARNG)	2 x UH-1H *AVIM*

35 AVNBDE Jefferson City MAP, Missouri (MO ARNG)
1-135 AVN AVN (ATK) (MO ARNG)	24 x AH-1F *ATK/SCT*
Whiteman AFB, Missouri	
1-114 AVN (GSAB) (AR ARNG)	8 x UH-1H *CMD/CS*
Robinson AAF,	*replaced 2-135 AVN*
North Little Rock, Arkansas	6 x OH-58A *TAR*, 16 x UH-60A *GS*
1-167 CAV (NE ARNG)	16 x AH-1F *AIR CAV*
Lincoln MAP, Nebraska	

38th Infantry Div. (M)(38ID(M),
Indianapolis, Indiana (IN ARNG)
38 DISCOM
F/238 AVN (AVIM) (IN ARNG)	2 x UH-1H *AVIM*

38 AVNBDE Grand Ledge MAP, Mich. (MI ARNG)
1-238 AVN (ATK) (MI ARNG)	24 x AH-1F *ATK/SCT*
Grand Ledge AP, Michigan	
3-137 AVN (ASLT) (IN ARNG)	38 x UH-1H *CMD/CS/GS*
Shelbyville MAP, Indiana	6 x OH-58C *TAR;*
	replaced 2-238 AVN
2-107 CAV (OH ARNG)	16 x AH-1F *AIR CAV*
Rickenbacker ANGB, Ohio	*replaced 1-238 CAV*

40th Infantry Div. (M)(40ID(M),
AFRC Los Alamitos, California (CA ARNG)
40 DISCOM
F/140 AVN (AVIM) (CA ARNG)	2 x UH-60A *AVIM*

40 AVNBDE Fresno, Calif. (CA ARNG)
1-211 AVN (ATK) (UT ARNG)	24 x AH-64A *ATK/SCT*
SLC AP #2, W. Jordan, Utah	*replaced 1-140 AVN*
2-140 AVN (GSAB) (CA ARNG)	24 x UH-60A *CMD/CS/GD*
Los Alamitos AAF, California	6 x OH-58C *TAR*
1-18 CAV (CA ARNG)	16 x AH-1F(N) *AIR CAV*
Los Alamitos AAF, California	

42nd Armored Div. (42AD), Troy, New York (NY ARNG)
42 DISCOM
F/142 AVN (AVIM) (NY ARNG)	2 x UH-1H *AVIM*

42 AVNBDE Patchogue, New York
1-142 AVN (ATK) (RI ARNG)	24 x AH-1F *ATK/SCT*
Rochester MAP, New York	
2-142 AVN (GSAB) (NY ARNG)	24 x UH-60A *CMD/CS/GS*
Latham, New York	6 x OH-58C *TAR*
5-117 CAV (RI ARNG)	16 x AH-1F(N) *AIR CAV*
Quonset State AP, R.I.	*ex 1-122 AVN*

49th Armored Division (49AD),
North Ft Hood, Texas (TX ARNG)
49 DISCOM
F/149 AVN (AVIM) (TX ARNG)	2 x UH-1H *AVIM*

49 AVN BDE Mueller AP, Austin, Texas.
1-124 CAV (TX ARNG)	16 x AH-1F(N) *AIR CAV*
1-149 AVN (ATK) (TX ARNG)	13 x OH-58A *SCT*
Ellington Field, Houston, Texas	3 x UH-60L *CS*, 18 x AH-64A *ATK*
2-149 AVN (GSAB) (TX ARNG)	8 x UH-1H *CMD/CS*
Martindale AAF,	6 x OH-58C *TAR*
San Antonio, Texas	3 x EH-60C *CEWI*,
	16 x UH-60L *ASLT/GS*

co-ordinates the activity of OSAA/OSACOM and other aircraft through its dispatching network.

OSAA and OSACOM are headquartered at Davison Army Airfield (AAF), Ft Belvoir, Virginia. The rotary-wing assets formerly assigned to OSAC were reassigned to the 12th Aviation Battalion (12 AVN) in 1996 and remained assigned to MDW. All OSA aircraft come under four categories: long-range jets, medium-range jets, short-range turbines and cargo aircraft. These aircraft deploy worldwide, as needed, 24 hours a day. OSAA/OSACOM operates the smallest number of long-range jets with an inventory of just four C-20E/F/J aircraft, although they have a stated requirement for 15. The service has a stated requirement for 35 medium-range jets and, from a single C-21A acquired in 1987, currently operates three of the type. The Cessna UC-35A Citation V fills that requirement, being assigned to theatre aviation (TA) companies primarily, but at least three and perhaps more will be assigned to OSAA/OSACOM if the entire programme is funded in the next few years. The Army's fleet of cargo aircraft has increased with the procurement of remanufactured C-23B+ aircraft, and along with other TA assets including the 16 older C-23B Sherpas, receives control from JOSAC/OSAA/OSACOM when they are utilised for OSA missions, which is common.

The short-range turbine category is the primary asset operated and controlled by OSAA/OSACOM, represented by a fleet of aircraft assigned to regional flight centres (RFCs). In September 1993 OSACOM's predecessor was operating 38 C-12s and 28 U-21s with the 14 RFCs. On 1 September 1994 the OSA aircraft assigned to the Army National Guard (ARNG) state area commands (STARCs) were organised as ARNG OSA state flight detachments and another 18 C-12s, 10 C-26Bs and 24 U-21s were counted on the OSACOM roster. This change made better use of OSAC scheduling tools to incorporate the ARNG aircraft into the system to allow greater availability and flexibility to meet transportation requirements. These state flight dets became affiliated with OSACOM upon its activation. OSACOM has been reducing its fleet through the last few years, eliminating most of the RFCs, while acquiring new ones in Alaska, Hawaii and Panama. The last of the U-21 Ute aircraft in OSA service were retired by early 1997, followed by the remaining C-12L aircraft, to be followed by the remaining fleet of C-12Cs. By 2000, the OSAA/ OSACOM fleet will likely consist of 22 C-12Ds, 43 C-12Fs, five C-12Rs, ten C-26Bs and the fleet of jet aircraft, which may exceed the seven now on strength. The C-12D/F aircraft are being modernised with digital, EFIS flight instrumentation and other improvements designed to reduce support and training costs. Type training for fixed-wing aircraft begins at Ft Rucker, Alabama and is supplemented by Det 1, EAATS with the West Virginia ARNG.

US Army Special Operations Command (USASOC)

The US Army Special Operations Command (USASOC), headquartered at Ft Bragg, North Carolina, is the Army SOF component and has over 35,000 soldiers assigned. USASOC commands a variety of component personnel, systems and units that can be packaged for numerous contingencies. It trains, equips and operationally deploys the SOF and special operations aviation (SOA) assets under its command. Potential conflict scenarios include counter-terrorism (CT), counter-insurgencies (COIN), low-intensity conflict (LIC), operations other than war (OOTW) and conventional warfare scenarios, in which special forces would conduct missions peripherally to, or directly with, large manoeuvre forces of infantry or armour.

USASOC is the Army component of the US Special Operations Command (USSOCOM), which was activated on 16 April 1987, headquartered at MacDill AFB, Tampa, Florida. It is the unified command responsible for the equipment, training and readiness of nearly 50,000 special operations soldiers, sailors and aircrew, assigned to the individual service components, active, reserve and national guard. Most combat positions in SOCOM are not open to female members of the military.

USASOC was activated in 1988, redesignating the assets of what had been previously been known as the 1st Special Operations Command (Airborne), or 1st SOCOM(A), becoming a major command (MACOM) in the process. Its principal combat strength is five Special Forces Groups (SFGs) each tasked for primary assignment to a unified theatre CINC, but cross-trained for worldwide contingency deployment, along with two ARNG SFGs. The 75th Ranger Regiment provides three deployable battalions of highly trained, light infantry forces that perform direct action missions. The command maintains extensive civil affairs and psychological operations forces, most in USARC formations.

Army SOA operations are focused in the 160th Special Operations Aviation Regiment (Airborne), or 160th SOAR(A), from Campbell AAF at Ft Campbell, Kentucky. The unit began as a provisional task force in 1979 assigned with developing tactics and technology to perform with optimal operational stealth while conducting clandestine special operations, usually at night. It advanced to battalion, group and then to brigade command echelon status through 1990, just before Operation Desert Shield/Storm. The unit structure has evolved throughout this time as new missions, systems, tactics and technologies have been inserted into the force structure. Approximately 130 aircraft are resourced to the 160th at the present time.

The brigade, known as the 'Nightstalkers', conducts low-level, clandestine missions that include insertion, resupply and extraction of SOF and other designated personnel, special reconnaissance, raids by Rangers or SOF forces, direct action with airborne weapons or provision of target acquisition and terminal guidance for precision munitions, combat search and rescue (CSAR) missions, and command and control augmentation. The aircrews and maintainers are among the most experienced and dedicated in the world. The aviators routinely employ deck landing qualification (DLQ) skills that permit them to operate from US Navy warships as small as frigates. During the occupation of Haiti in 1994, the 160th SOAR effectively took over flight operations on the USS *America* (CV-66) for a few weeks.

The Army has operated a variety of SOA platforms for its unique roles. Modified variants of the Hughes/McDonnell Douglas OH-6A Cayuse were used from the early 1980s, modified as MH-6B, EH-6B and AH-6C variants for assault, command and control, and as gunships, respectively. The aircraft were modified with NVGs for routine night operations. In the mid-1980s the AH-6F gunship variant was introduced, along with the MH-6E for assault, both adapted from the MD.500 helicopter series. They have been upgraded to the AH-6F and MH-6H variants, respectively, with 'glass' EFIS cockpits, FLIR and improved comm/nav systems, and remain operational today. The service tried to incorporate NOTAR technology for greater stealth, but aircraft range was degraded, leading to the return to conventional tail rotor configurations. The fleet will be upgraded from 1998 to the Mission Enhanced Little Bird (MELB) configuration that offers improved systems and an increase of over 20 per cent in aircraft payload. The term Little Birds and Little Bird Guns are commonly applied to these discrete systems – they are rarely seen, and even more rarely photographed.

From 1981 the 160th Aviation Battalion began operating CH-47Cs and UH-60As for longer-range assault missions. The unit has operated progressively improved variants of both types over the years. The 160th progressed to CH-47Ds optimised for the role, becoming CH-47D Enhanced or CH-47D SOA variants. At least 12 Chinooks were modified with inflight refuelling, radar and FLIR to become MH-47Ds. The MH-47E programme was launched in 1988, designed from the ground up for SOA operations with more fuel, digital EFIS cockpit and other significant combat systems, entering service in late 1993.

The first Blackhawks used by the unit were standard production aircraft, followed by modified and upgraded aircraft in two versions: 42 'enhanced' UH-60A(E)s, and 21 MH-60As with AAQ-16 FLIR, SATCOM, rescue hoist and other improvements. Upgraded UH-60L airframes were brought up to the MH-60L configuration from late 1988, while concurrent development began on the MH-60K with inflight refuelling, radar and an integrated EFIS cockpit, common to the MH-47E. Both variants serve today as the primary SOA assault platforms and a number of the MH-60Ls have been modified into a direct action penetrator, or MH-60 DAP configuration for service as gunships, armed with Hellfire missiles, rockets and a variety of machine-guns and up to 30-mm cannon.

Army SOA assets are heavily supported by the Air Force Special Operations Command (AFSOC), headquartered at Hurlburt Field, Eglin AFB, Florida. AFSOC's 16th Special Operations Wing (16th SOW) operates a variety MC-130 variants for inflight refuelling and other operational support, along with specially tasked airlift aircraft including the C-5, C-17 and C-141. USASOC forces also use other Army fixed-wing OSA/GS/TA aviation assets, such as C-12 and C-23 aircraft, for transport and parachute insertion roles.

Future upgrades are in progress, with the capability to equip its MH-60L fleet with inflight refuelling already demonstrated. The 160th has tested a variety of fixed-wing aircraft for duty and the procurement of some types for medical evacuation, gunship, C2 and other unconventional operations may be imminent.

Seen at Heidelberg, Germany, this C-12R – the Army version of the Beech Model B200C – belongs to A/2-228 AVN. A total of 29 C-12Rs has been ordered.

USASOC

US Army Special Operations Com (USASOC), Ft Bragg, N.C.
(*part of US Special Operations Com (USSOCOM), MacDill AFB, FL*)

USASOC Flight Det.	Simmons AAF, Ft Bragg, N.C.
2 x C-12F, 1 x Pa 31T *GS/OSA*	

160th Special Operations Aviation Regiment (Airborne) (160 SOAR(A)), Campbell AAF, Ft Campbell, Kentucky

SOA Training Company
 12 x MH-6H, 3 x MH-47D, 2 x MH-47E, 2 x MH-60A,
 2 x MH-60L, 2 x MH-60K *all for Type Training*
1-160 SOAR(A)
 18 x AH-6G, 18 x MH-6H, 20 x MH-60K *all SOA ASLT*
 15 x MH-60L *SOA ASLT/ATK/C2*
2-160 SOAR(A)
 24 x MH-47E *SOA ASLT*

3-160 SOAR(A)	Hunter AAF, Savannah, Georgia

 10 x MH-60L (*SOA ASLT/ATK*), 8 x MH-47D (*SOA ASLT*)

D/3-160 SOAR(A)	Ft Kobbe (Howard AFB), Panama

 10 x MH-60L *SOA ASLT/ATK*

US Army Europe/7th Army (USAREUR/7A)

US Army Europe/7th Army is the Army component of the US European Command (USEUCOM or EUCOM), the joint, unified warfighting command assigned to command US forces in the European theatre. Both commands are assigned four-star generals, with EUCOM being headquartered in Stuttgart, Germany and composed of US Air Force and Army units based on the continent and seaborne Navy and Marine Corps forces deployed to the eastern Atlantic Ocean and Mediterranean Sea. Both of these commands represent a fraction of the land warfare, combat capability available to the Supreme Allied Commander, Europe (SACEUR) and its headquarters element, the Supreme Headquarters Allied Powers, Europe (SHAPE), the military command of the North Atlantic Treaty Organisation (NATO).

USAREUR/7A has responsibility for preparing the defence of Europe – from Norway to the Mediterranean Sea, and from the Atlantic to eastern Turkey. The command is headquartered in Heidelberg, Germany, the country where most of its forces have been deployed since the end of World War II. The command's strength has reduced from 213,000 soldiers and civilians at the beginning of FY90 to 65,000 by the end of FY95, a loss of more than half of its combat and combat support forces.

USAREUR/7A underwent a tremendous expansion of capabilities in the early to mid-1980s and then underwent a major contraction, with aviation assets being affected in both situations. The manoeuvre force is focused on V Corps which commands two armoured divisions – 1st Armoured Division and 1st Infantry Division (Mechanized) – along with numerous specialised brigades including the 3rd Corps Support Command, 30th Medical Brigade, 12th Aviation Brigade and the 205th Military Intelligence Brigade, all with aviation units attached. Additional combat forces are assigned to the Southern European Task Force (SETAF), a unit composed of allied military forces, headquartered at Vicenza, Italy.

SACEUR/SHAPE

Supreme Allied Commander, Europe/Supreme Headquarters Allied Powers Europe (SACEUR/SHAPE), Mons, Belgium

USAG Flight Detachment	2 x UH-60A(C) *C2/VIP*
Chièvres AB, Belgium	

US European Command (USEUCOM/EUCOM), Patch Barracks, Stuttgart, Germany

Flight Operations Det	2 x UH-1H *GS/VIP*
Echterdingen AAF, Stuttgart, Germany	

US Army Europe/7th Army (USAREUR/7A), Campbell Barracks, Heidelberg, Germany

USAREUR/7A

US Army Europe/7th Army (USAREUR/7A), Campbell Barracks, Heidelberg, Germany

7th Army Training Command (7 ATC) Grafenwoehr, Germany

Aviation Detachment	1 x C-12F *GS*
Grafenwoehr AAF, Germany	3 x UH-1H *GS*
Combat Maneuver Training Center (CMTC) Hohenfels, Germany	
Aviation Detachment	10 x UH-1H *GS/OC, Threat Spt*
Hohenfels AAF, Germany	

21st Theater Army Area Command (21 TAACOM), Kaiserslautern, Germany

207th AVN CO (TA)	5 x C-12F *TA*; Det 1, 6th AVN Det
Heidelberg AAF, Germany	(2 x C-12R) *TA;*
	USARC, Fwd Deployed
	2 x UC-35A *TA;*
	USARC, Fwd Deployed
	4 x UH-60A *GS*
2-502 AVN BN (AVIM)	2 x UH-1H *AVIM/GS* ex-B/70TRANS
Coleman AAF, Mannheim, Germany	2 x UH-60A *AVIM/GS*

Southern European Task Force (SETAF), Vicenza, Italy
5th Theater Army Area Command (5 TAACOM), Vicenza, Italy

6th AVN Det	2 x C-12F *TA*, 4 x UH-60A *GS*

V Corps (VCORPS) Campbell Barracks, Heidelberg, Germany

3rd Corps Support Command (3 COSCOM), Wiesbaden AB, Germany

7-159 AVN (AVIM)	4 x UH-60A *AVIM*
Illesheim AAF, Ansbach, Germany	

30th Medical Brigade (30 MEDBDE), Heidelberg, Germany

421 Medical Evacuation Bn (421 MEB), Wiesbaden AB, Germany	
45th MED CO (AA)	15 x UH-60A *AA*
Fliegerhorst AAF, Ansbach, Germany	
159th MED CO (AA)	15 x UH-60A *AA*
Darmstadt AHP, Germany	
236th MED CO (AA)	15 x UH-60A *AA*
Landstuhl AHP (Kaiserslautern), Germany	

12th Aviation Brigade (12 AVNBDE), Illesheim AAF, Ansbach, Germany

449th Aviation Group (449 AVNGRP) Kinston, N.C. (NC ARNG)		
3-126th AVN (LUH) (CT ARNG)	32 x UH-1H LUH *maybe*	
Bradley IAP, CT (CA ARNG)	*replaced by 1-140th AVN*	

3-142nd AVN (CSAB) (PR ARNG)	32 x UH-60A *CS:*
	ARNG assigned
F.L.R. Dominicci AP, San Juan, Puerto Rico	
A/5-159 AVN (HHB)	16 x CH-47D *HH*
Giebelstadt AAF, Mannheim, Germany	
C/6-159 AVN (ASLT)	15 x UH-60L *ASLT*
Giebelstadt AAF, Germany	
C/7-158 AVN (ASLT)	15 x UH-60L *ASLT*
Wiesbaden AAF, Germany	
5-158 AVN (CMD)	4 x UH-60A *CS*
Wiesbaden AAF, Germany	4 x UH-60A(C) *C2*

11th Aviation Group (11 AVNGRP), Illesheim AAF, Ansbach, Germany

1-151st AVN (ATK) (SC ARNG)	24 x AH-64A *ATK/SCT:*
McEntire ANGB, Eastover, S.C.	*ARNG assigned*
2-6 CAV (AC)	24 x AH-64A *ATK/SCT*
Illesheim AAF, Ansbach, Germany	
6-6 CAV (AC)	24 x AH-64A *ATK/SCT*
Illesheim AAF, Ansbach, Germany	

205th Military Intelligence Bde (205 MIBDE), Wiesbaden AB, Germany

1st MI BN (AE)	1 x RC-12D *Training*
Wiesbaden AAF, Germany	8 x RC-12K *CEWI*

1st Armored Division (1AD), Bad Kreuznach, Germany

Division Support Command (1AD DISCOM), Bad Kreuznach, Germany

A/127 ASB	2 x UH-60A *AVIM*
Fliegerhorst AAF, Hanau, Germany	

Aviation Bde/4th Bde (1AD AVNBDE)
Fliegerhorst AAF, Hanau, Germany

1-501 AVN (ATK)	24 x AH-64A *ATK/SCT*
	ex-2-227 AVN
2-501 AVN (GSAB)	6 x OH-58C *TAR;*
	ex-7-227AVN(GSAB)
	4 x UH-60A *CS,* 4 x UH-60A(C) *C2*
	4 x EH-60C *CEWI*
	16 x UH-60L *GS/ASLT*
1-1 CAV	16 x OH-58D(I) *AIR CAV*
Armstrong AHP, Budingen, Germany	

1st Infantry Division (Mech.) (1ID(M)), Wuerzburg, Germany

Division Support Command (1ID DISCOM), Kitzingen, Germany

A/603 ASB	2 x UH-60A *AVIM*
Katterbach AHP, Ansbach, Germany	

Aviation Bde/4th Bde (1ID AVNBDE)
Katterbach AHP, Ansbach, Germany

1-1 AVN (ATK)	24 x AH-64A *ATK/SCT*
2-1 AVN (GSAB)	6 x OH-58C *TAR,* 4 x UH-60A *CS*
	4 x UH-60A(C) *C2,* 4 x EH-60C *CEWI*
	16 x UH-60L *GS/ASLT*
1-4 CAV	16 x OH-58D(I) *AIR CAV*
Schweinfurt AAF, Germany	

US Army Land Force Southeast, (USLANDSOUTHEAST/ALFSE), Cigli Airfield, Izmir, Turkey

USAG Flight Detachment	2 x C-12F, 3 x UH-1H *GS*

Turkey US Logistics Group (TUSLOG), Sinop, Turkey

USAG Flight Detachment	1 x C-12F *GS*

SETAF is subordinate to the Allied Land Forces, Southern Europe. The CH-47D unit (E/502 AVN) assigned to SETAF was inactivated in 1997.

The evolution of military and political changes in Europe from the late 1980s has caused planners to reassess the roles and missions of its assigned forces. The force structure of US Army Europe/7th Army remains one of a mobile, armoured strike force that is available for deployment within the theatre or for assignment to other commands to reinforce contin-

The 377th Medical Company based at Seoul, Republic of Korea, is the only UH-60A medical evacuation unit with special markings (for work in the demilitarised zone) in addition to the red cross.

gency operations, since its extensive involvement in Operation Desert Shield/Storm. The US Army deployed a substantial force during December 1995 as a major element of NATO's Implementation Force (IFOR) to Bosnia in order to stabilise a three-way conflict between that state, Croatia and Serbia. The Army has rotated units through there since, redesignating it as the Stabilisation Force (SFOR).

Training of USAREUR/7A forces is managed by the 7th Army Training Command (7 ATC) which operates the Combat Maneuver Training Center (CMTC) at Hohenfels, Germany. This fully instrumented training and evaluation facility allows manoeuvre training over 40,000 acres of Bavarian terrain.

The 3-6th Cavalry, 8th Army, has three companies of AH-64As, each operating eight helicopters. These four examples represent half of Alpha Company.

Eighth US Army (EUSA/8th Army)

The Eighth US Army (EUSA) is the numbered, theatre army command that controls US Army forces deployed to the Republic of Korea (RoK), also known as South Korea. In 1998 the command was designated as the Army component force under US Forces Korea (USFK), the joint US warfighting command in Korea. The commander of USFK is also the commander of the UN Command/Combined Forces Command (UNC/CFC), composed of Republic of Korea, US and some allied military forces, totalling nearly 750,000 soldiers. The forces of UNC/CFC/USFK are maintained at a 'go to war' level of readiness and are directly subordinate to the commander-in-chief, US Pacific Command (USPACOM). EUSA is supported and sustained by the US Army Pacific (USARPAC), Ft Shafter, Hawaii, and its units and commands located primarily in Alaska, Hawaii and Japan. The 8th Army would be reinforced by assets and soldiers commanded by I Corps, the designated contingency corps command for the Pacific theatre, in wartime. The Ft Lewis, Washington-based corps would gain unit strength from a variety of FORSCOM subordinate commands including ARNG (Army National Guard), US Army Reserve Command (USARC) and select units, up to division echelons, from III Corps and XVIII Airborne Corps.

Aviation units of Eighth US Army have been upgraded and reinforced in recent years, as it conducts its mission to deter war and defend South Korea from aggression or invasion. The 6th Cavalry Brigade (Air Combat) (6 ACCB) was forward deployed to the theatre from III Corps in the summer of 1996 and it commands two Apache squadrons tasked for deep strike missions. The 2nd Infantry Division (Mechanized) has gained Apaches and OH-58D(I) Kiowa Warrior helicopters, along with upgraded UH-60L Blackhawks in recent years.

The 17th Aviation Brigade is part of the Combined Aviation Force (CAF), or Ground Component Aviation Force (GCAF) assigned to CFC. This joint RoK/US aviation command is composed of at least 14 aviation battalions with almost 400 aircraft assigned, roughly 150 of them from the US Army. This is an impressive aviation force equivalent in size to that assigned to the 101st Airborne Division (Air Assault). RoK Army aviation forces are organised along the lines of US Army aviation units and there is tremendous symmetry in their doctrine, equipment, organisation and training. South Korean Army aviation units operate the UH-1H, AH-1F/J, CH-47D and UH-60P (an indigenously assembled variant equivalent to the UH-60L) along with large numbers of McDonnell Douglas 500/530 military variants optimised for gunship, observation, scout, target acquisition and special operations roles.

Numerous other combat and combat support units are assigned at the theatre level, or echelons above corps (EAC), with 8th Army. They include the 8th Personnel Command, 8th Military Police Brigade, 18th Medical Command, 19th Support Command, 175th Theater Finance Command, 501st Military Intelligence Brigade and Special Operations Command- Korea (SOC-K), along with numerous other units based in and outside the theatre.

Army heavylift in the Republic of Korea is provided by the 2-52nd Aviation Regiment. This CH-47D operates with the Regiment's Alpha Company.

The UH-60As based at Camp Zama, Japan, are finished in a smart white and black scheme.

EUSA/8th Army

8th US Army (EUSA/8th Army) Yongsan, Seoul, RoK
reports to United Nations Command/Combined Forces Command/US Forces Korea HQ (UNC/CFC/USFK), Seoul, RoK

6th Cavalry Bde (Air Combat) (6 ACCB),
Desiderio AAF, Camp Humphreys, RoK
1-6 CAV (AC)	24 x AH-64A *ATK/SCT*
Camp Eagle, Hoengsong, RoK	
3-6 CAV (AC)	24 x AH-64A *ATK/SCT*
Camp Humphreys, Pyongteak, RoK	

17th Aviation Brigade (Theater) (17 AVNBDE),
Seoul AB, Seoul, RoK
1-52 AVN (CMD)	12 x UH-60A *CS*
Seoul AB, Seoul, RoK	4 x UH-60A(C) *C2*, 16 x UH-60A(E) *GS*
2-52 AVN (HHB)	32 x CH-47D *HH*
Camp Humphreys, Pyongteak, RoK	
A/6-52 AVN (TA)	7 x C-12F *TA; Det 1, Atsugi, Japan*
	1 x UC-35A *TA*
6-52 AVN (TAB) (USARC)	8 x C-12F/R *TA*
Los Alamitos AAF, Calif.	8 x UC-35A *TA*

18th Medical Command (18 MEDCOM), Yongsan, Seoul, RoK
52nd Medical Evacuation Bn (52 MEB) Yongsan, Seoul, RoK	
377th MED CO (AA)	15 x UH-60A *AA*
Seoul AB, RoK	
542nd MED CO (AA)	15 x UH-60A *AA*
Camp Page, Chunchon, RoK	

19th Support Command (19 SUPCOM),
Camp Henry, Taegu, RoK
3-52 AVN (AVIM)	4 x UH-60A *AVIM*
Desiderio AAF, Camp Humphreys, Pyongteak, RoK	

501st Military Intelligence Bde (501 MIBDE),
Camp Humphreys, Pyongteak, RoK
3rd MI BN (AE),	1 x RC-12D *Training,*
Desiderio AAF,	6 x RC-12H *CEWI*
Camp Humphreys, RoK	3 x RC-7B(ARL)/OE-5B *CEWI*

2nd Infantry Division (Mech.) (2ID(M)), Camp Casey, Tongduchon, RoK

Division Support Command (2ID DISCOM),
Camp Casey, Tongduchon, RoK
DASB Camp Stanley, Uijongbu, RoK	
C-2 AVN (AVIM)	2 x UH-60A *AVIM*

Aviation Brigade (2ID AVNBDE),
Camp Stanley, Uijongbu, RoK
1-2 AVN (ATK)	24 x AH-64A *ATK/SCT*
Camp Page, Chunchon, RoK	
2-2 AVN (GSAB)	16 x UH-60L, 4 x UH-60A *GS*
Camp Stanley,	4 x UH-60A(C) *C2*
Uijongbu, RoK	4 x EH-60C *CEWI*, 6 x OH-58C *TAR*
4-7 CAV	16 x OH-58D(I) *AIR CAV*
Camp Edwards, Yono-Re-Ti, RoK	

USARPAC

US Army Pacific (USARPAC) Ft Shafter, Hawaii
reports to US Pacific Command (USPACOM), Camp H.M. Smith, Hawaii

45th Corps Support Grp (Fwd) (45 COSGRP),
Schofield Barracks, Hawaii
(311 TAACOM/I CORPS assigned units)	
C/193rd AVN (HH) (HI ARNG)	16 x CH-47D *HH*
Wheeler AAF,	*subordinate to 3-140 AVN (CA ARNG)*
Schofield Barracks, Hawaii	
E/214th AVN (AVIM)	2 x UH-60A *AVIM*
Wheeler AAF, Schofield Barracks, Hawaii	
68th MED DET (AA)	6 x UH-60A *AA*

US Army, Japan/9th Theater Army Area Cmd (USARJ/9 TAACOM), Camp Zama, Japan

78th AVN BN (P)	5 x UH-1H *GS*, 3 x UH-60A *GS/VIP*
Camp Zama, Japan	3 x UC-35A *GS/TA; Det to A/6-52*
	AVN based at NAF Atsugi, Japan

US Army Alaska (USARAK), Ft Richardson, Alaska

USAG Aviation Det	3 x UH-1H *GS*
Allen AAF, Ft Greely, Alaska	

Arctic Support Brigade (ARCTIC SUPBDE),
Ft Richardson, Alaska
F/228 AVN (AVIM)	1 x UH-60A *AVIM*

Wainwright AAF, Ft Wainwright, Alaska	
4-123 AVN (TA)	15 x UH-60L *ASLT*, 16 x CH-47D *HH*
Wainwright AAF, Ft Wainwright, Alaska	
283rd MED DET (AA)	6 x UH-60A *AA; ICORPS assigned*

207th Infantry Grp (Scout) (207 INFGRP) (AK ARNG),
Anchorage, Alaska
23rd AVN Det (AVIM)	1 x UH-60L *AVIM*
Bryant AHP, Ft Richardson, Alaska	
1-207 AVN (SCT) (AK ARNG)	8 x C-23B+ *TA: USARPAC mission*
Bryant AAF,	15 x UH-1H *GS/CS, 15 x UH-60L ASLT*
Ft Richardson, Alaska	
Dets at Nome, Bethel, Kotzebue and Juneau, Alaska	

25th Infantry Division (Light) (25ID(L)), Schofield Barracks, Hawaii

Division Support Command (25ID DISCOM),
Schofield Barracks, Hawaii
C/25 AVN CO (AVIM)	2 x UH-60A *AVIM*
Wheeler AAF, Schofield Barracks, Hawaii	

Aviation Brigade (25ID AVNBDE),
Wheeler AAF, Schofield Barracks, Hawaii
1-25 AVN (ATK)	24 x AH-1F(N) *ATK/SCT;*
	transition to OH-58D(I) moved to 1999
2-25 AVN (ASLT)	4 x UH-60A *CS*, 4 x UH-60A(C) *C2*
	3 x EH-60C *CEWI*, 30 x UH-60L *ASLT*
3-4 CAV (RECON)	16 x OH-58D(I) *AIR CAV;*
	replaced AH-1F(N), 1998

US Army Pacific (USARPAC)

The US Army Pacific (USARPAC) is the Army component assigned to the US Pacific Command (USPACOM), the unified warfighting command headquartered at Marine Corps Base Camp H.M. Smith, located on a hill overlooking Pearl Harbor, on the island of Oahu, Hawaii. From this command headquarters, the commander-in-chief (CINCPAC) and his staff are assigned the responsibility of maintaining American interests in a territory covering one-third of the earth's surface – from the eastern coast of Africa, northeast to the Southeast Asian land mass, up to the waters bordering Russia, up to the North Pole, including Alaska, along the west coast of the United States, to a line extending due south from the border of Mexico and Guatemala, to the waters of Antarctica, and everything in between. The distances are enormous and key areas of interest to the US, such as Japan and Korea, are halfway around the world from Washington, D.C. The Asia-Pacific theatre of operations includes 100 million sq miles (259 million km²), bordered by over 50 countries, is home to a population of almost 3 billion people (or roughly three-fifths of the world's population) and is where nearly 40 per cent of all US trade is conducted. The seven largest armies in the world reside in this area, collectively with more than 12 million soldiers under arms, 85 per cent being ground forces. Over 300,000 US military personnel are routinely stationed in the theatre, although the Army strength is concentrated first in Korea, then in Hawaii and in smaller numbers of combat and combat support forces positioned in Japan, and Alaska.

US Army Pacific was known as the US Western Command (WESTCOM) until 1990, and remains headquartered at Ft Shafter, Hawaii. The command represents the land warfare component of USPACOM, charged with maintaining a flexible and responsive force posture that can deploy and fight anywhere in the region. The active-duty combat forces directly assigned to USARPAC are primarily light forces, capable of rapidly deploying to trouble spots with all their equipment capable of being loaded onboard US Air Force strategic and tactical airlift aircraft. Most are based in the states of Alaska, Hawaii and Washington and on several of the islands of Japan. In 1995, the Army designated I Corps as the manoeuvre corps for US Army Pacific, replacing IX Corps, which was realigned and became the 9th Theater Army Area Command (9th TAACOM). The commander of 9 TAACOM is 'dual-hatted' as commander of US Army Japan (USARJ), both at Camp Zama, outside the city of Tokyo. Both organisations focus on maintaining the logistics, support and sustaining capabilities vital to deployed combat forces from its forward location. A small aviation element under a provisional (P) battalion command, 78th Aviation Battalion, serves the commands.

I Corps, headquartered at Ft Lewis, Washington, is assigned to Forces Command (FORSCOM) for training and support during peacetime. In a major force structure reorganisation, the commander of USAPACOM delegated operational control of the I Corps headquarters to US Army Pacific in 1994, designating it as the Army's warfighting headquarters in the Pacific theatre. The net effect is a closer working relationship between both commands and affirms the focus of the soldiers assigned to this geographically dispersed command. With the inactivation of IX Corps, I Corps has taken over the added responsibility for the land warfare defence of Japan, closely co-ordinating with the Japanese Ground Self-Defence Force (JGSDF). The corps trains and deploys to locations throughout the Pacific and is unique in that it constitutes most of its warfighting capability from reserve component (RC) units, both Army National Guard (ARNG) and US Army Reserve Command (USARC). In the event of a contingency or major conflict, the command would be rapidly reinforced from units assigned to FORSCOM, primarily from XVIII Airborne Corps, and those assigned to US Army Special Operations Command (USASOC).

The principal active-duty warfighting force available to USARPAC and ICORPS is the 25th Infantry Division (Light), based at Schofield Barracks in Hawaii, in close proximity to Hickam AFB. One of the division's three infantry brigades is based at Ft Lewis, Washington, and the Hawaii ARNG's 29th Infantry Brigade (Light) regularly trains with the unit. The aviation brigade is in process of upgrading from AH-1F NTS aircraft to the OH-58D(R) Kiowa Warrior, with the latest systems upgrades, delayed from its planned fielding date in 1996.

An extensive logistics and support infrastructure is maintained on the island under the command of US Army, Hawaii. It is enhanced by corps support elements of I Corps, the USARC 9th Army Reserve Command (9th ARCOM) and units of the Hawaii ARNG. A Special Operations Theater Support Element supports deployments of SF and SOA assets to the theatre.

The US Army Alaska (USARAK) is a principal component of USARPAC tasked to support training and deployment of warfighting assets throughout the theatre. It provides cold weather training and is sustained by a deployed brigade task force attached to the 10th Infantry Division (Light), Ft Drum, New York, along with the 207th Infantry Group (Scout) of the Alaska ARNG. This unique unit is composed primarily of native Eskimos who are well adapted to living in the harsh and wild terrain of the state's territory. This group is regarded as a first-line unit providing an indigenous defence capability over the barren, remote Alaskan wilderness. An organic aviation battalion replaced its UV-18As with C-23B+s in 1997-98, reorganising them into a theatre aviation (TA) company.

Ft Richardson is adjacent to Elmendorf AFB, which would be a key airlift facility in the event of any contingency movement throughout the theatre. The principal warfighting training centre for Alaska, and allied units in the Pacific, is the Northern Warfare Training Center (NWTC) at Ft Greely. It was scheduled to relocate to Ft Wainwright in 1997.

North America

US Army South (USARSO)

The US Army South (USARSO) is the Army command element of the US Southern Command (USSOUTHCOM or SOUTHCOM), responsible for US military operations in and around the countries of the Americas, specifically those of Central and South America. The area of interest includes a land area measuring 3,000 x 7,000 miles (4830 x 11265 km) and a rapidly growing population of over 350 million people. The command's geographic area of interest begins at the US/Mexico border, continues to the island nations of the Caribbean Sea basin and to the countries of Central and South America; this includes peripheral portions of both the Atlantic and Pacific Oceans. SOUTHCOM has relocated to Miami, Florida and in 1999 US Army South is to begin relocating to Ft Buchanan, Puerto Rico.

US Army South (USARSO) is currently the smallest US Army theatre command, directly commanding about 4,000 soldiers in 1998. The command commonly uses active and reserve component (AC/RC) soldiers to perform temporary duty (TDY) missions of up to 179 days in length, to augment the assigned forces. The command has also been active in performing various counter-drug operations, in assistance to host countries and in interdiction missions across international borders. Combat reinforcement to USARSO would come first from XVIII Airborne Corps and I Corps, under the direction of Forces Command (FORSCOM).

The command theatre aviation brigade, the 128th Aviation Brigade, was inactivated in 1995. At its peak in the late 1980s, the brigade controlled over 130 aircraft, in several battalions and detachments. Most of the remaining combat-coded aircraft operate at Soto Cano AB, Honduras, reporting to 4-228 AVN, but additional assets are still based in Panama and in the United States. The main operating base in Panama until recent years was Ft Kobbe, which is adjacent to Howard AFB and shares ramp space.

The other remaining battalion-sized aviation unit assigned to the theatre is the 204th Military Intelligence Battalion (Low Intensity) (204 MIB(LI)). The unit's numerical designation was classified until at least 1997, when the unit relocated from its main operating base at Orlando International Airport, Florida to Biggs AAF, Ft Bliss, Texas. Battalion assets regularly forward deploy throughout the USARSO geographic area of responsibility. Three four-engined Bombardier/de Havilland Canada DHC-7s were modified specifically for surveillance missions in a programme known as Airborne Reconnaissance Low (ARL). The aircraft are capable of self-deployment and operating at austere, forward locations. Two of the aircraft were originally configured with mission packages optimised for Comint, as ARL-Cs or O-5As, and the third was given additional Imint capability, as ARL-Is (I for imagery) or OE-5Bs. All aircraft were due to be brought up to ARL-M (Multi-mission) RC-7B capability. The sensor data can also be offloaded directly to ground-based units, which receive the data in briefcase-size display workstations. The Army has designated the aircraft as either RC-7Bs or O-5A/OE-5B. The integrated airborne

platform is operated under a mission configuration known as Crazyhawk. The battalion is augmented by the 138th MI Company which has stayed in Orlando flying RC-12G Crazyhorse aircraft, still assigned to US Army Reserve Command (USARC).

USARSO

US Army South (USARSO) Ft Buchanan, Puerto Rico
(relocated from Ft Clayton, Panama in October 1998, reports to US Southern Command (USSOUTHCOM), Miami, Florida

505th Military Intelligence Grp. (505 MIGRP), Ft Gillem, Ga.		
204 Military Intelligence Bn (Low Intensity) (204 MIB(LI))		
Biggs AAF, Ft Bliss, Texas	3 x OE-5B	*CEWI; as RC-7Bs*
	2 x C-12F	*GS*
138 MI Company (USARC)	1 x RC-12D	*Training*
Orlando IAP, Florida	3 x RC-12G	*CEWI*
Special Operations Command, South (SOC-SOUTH), Ft Kobbe, Panama		
D/3-160 SOAR	10 x MH-60L	*SOA ASLT/ATK*
167th Support Command (Theater), Birmingham, Alabama (AL ARNG)		
Theater Support Brigade (SPTBDE) Ft Clayton, Panama		
142nd Medical Evacuation Bn (142 MEB) Ft Clayton, Panama		
214 MED DET (AA)	6 x UH-60A	*AA*
Ft Kobbe, Howard AFB,	*XVIIIABNCORPS assigned*	
Panama		
4-228 AVN (CMD)	24 x UH-60A	*ASLT/GS/AVIM*
Soto Cano AB, Honduras	2 x UH-60A(C)	*C2*
D/1-228 AVN (TA)	8 x C-12D/F	*TA; Dets in FL & KS;*
Simmons AAF, Ft Bragg, N.C. *to gain UC-35A*		
Puerto Rico ARNG San Juan, Puerto Rico (PR ARNG)		
Troop Command Flight Det	1 x C-23B	*TA; assigned to F/192 AVN*
F.L.R. Dominicci AP, P.R.	2 x UH-1H	*GS*
Virgin Island ARNG, St Croix, V.I. (VI ARNG)		
Troop Command Flight Det	1 x C-23B	*TA;attached to F/192 AVN*
Alexander Hamilton AP,	2 x UH-1H	*GS*
St Croix, Virgin Islands		

A total of 66 EH-60Cs was procured, equipped with the ALQ-151 Quickfix II direction-finding, intercept and communications jamming equipment.

US Army Intelligence and Security Command (INSCOM)

Since 1977, the Army Intelligence and Security Command (INSCOM) has maintained worldwide responsibility for Army intelligence collection and production, counter-intelligence and security. This major command (MACOM) is headquartered at Ft Belvoir, Virginia and it operates some of the most unusual aviation assets in the world. Most of INSCOM's activities, budgets and resources remain highly classified owing to its work with other DoD and federal US agencies which include the Central Intelligence Agency (CIA), Defense Intelligence Agency (DIA), and National Security Agency (NSA).

INSCOM uses several types of fixed-wing aircraft for aerial exploitation (AE) military intelligence roles. These systems are usually assigned to corps or theatre echelon units. The most common is the RC-12, which operates in several distinct variants, supporting national communications intelligence (Comint) programmes. The RC-12D/H/K/N/P/Q variants are operated in families of Elint systems under the broad programme name of Guardrail. Each aircraft variant represents an upgraded version of the Sigint systems package that offers greater signal processing, direction-finding and computational capability of an advanced generation. Each of the successive systems has been developed to meet evolving threats with greater flexibility and systems automation. The Guardrail platforms are flown by a crew of two pilots and no processing of data is performed on the aircraft during the missions. A small number of RC-12Gs are assigned to a programme known as Crazyhorse and equip a single USARC MI company (as detailed above). The RU-21 variants, the OV-1D and the RV-1D that were very common in the 1980s and early 1990s have all been withdrawn from service.

The command is also involved with the development and fielding of two other major fixed-wing Sigint systems, the most visible of which is the Joint Surveillance Target Attack Radar System (Joint STARS or J-STARS). This is mounted on a modified, militarised Boeing 707 commercial airline airframe, incorporating the synthetic aperture radar (SAR), data processing systems and workstations. Two development aircraft were built, designated as E-8As, and at least five production E-8Cs have been fielded from a total programme requirement of about 20 aircraft. The aircraft are flown and maintained by the 93rd Air Control Wing (93rd ACW), USAF,

Above left: The Airborne Reconnaissance Low airframes are DHC-7s acquired on the commercial market. All retain civilian schemes and serials with discrete US Army titles under the cockpit.

Left: The RC-12N carries the Guardrail Common Sensor (System 1) Elint equipment. A total of 15 of these aircraft was delivered to the Army.

based at Robins AFB, Georgia, and Army personnel make a significant contribution to the mission crew of the aircraft, operating workstations on every J-STARS sortie.

The other fixed-wing surveillance aircraft currently being fielded is the Airborne Reconnaissance Low (ARL), which evolved from another programme formerly known as Grisly Hunter. This programme also involves using former commercial airliners, DHC-7 aircraft modified for low-intensity conflict (LIC) and operations other than war. The aircraft were known under the programme name of Magic Dragon during the system development and flight test. It was proposed to be a multi-purpose Comint/DF and Imint system, and is outfitted with a sophisticated ESM kit to collect communications and imagery data, process the data and communicate the information in a near-real-time mode. A visual intelligence/imagery reception station that is the size of a briefcase has been developed to permit simultaneous viewing of a target by personnel on the ground, in a boat or in another aircraft. The first two aircraft were equipped only with the Comint collection capability, the third was equipped with an Imint system that included an infra-red line scanner, a retractable FLIR turret, and another turret that mounts a day/night video imaging system. The systems are networked to a workstation in the cabin that can process, transmit and record the intelligence data. The ARL has been designed to take advantage of emerging technologies, and additional sensors could be added in the future. Another capability could be the control of unmanned aerial vehicles (UAVs).

When the aircraft entered service in 1994 they were assigned to the Military Intelligence Battalion (Low Intensity) (MIB(LI)) and were forward-deployed to sites in Central and South America. The aircraft performed so well that they have been employed in Haiti and Bosnia, and from 1997 the type has been deployed to South Korea with the 3rd MIB, replacing the last OV-1D aircraft in Army

Army Material Command's Army Aviation and Missile Command looks after the development of the Army's latest battlefield helicopters, the RAH-66 Comanche (above) and the AH-64D Longbow Apache (the example on the right being from the 1-227 AVN).

service. The aircraft are deployed in three aircraft companies, and presently six are in service, with three more undergoing modification (for XVIIIABN-CORPS); the total requirement is for 15 aircraft. The aircraft operate in low-profile, civilian-style paint schemes. The designation for the initial Comint-only equipped aircraft were E-5As, while the Imint and those aircraft with both capabilities are classified as OE-5Bs, but the Army has also referred to them publicly as RC-7Bs.

The EH-60C is outfitted with a command and control Sigint (Comint) jamming system known in the latest field version as the Quick Fix IIB. This programme began in the late 1970s, intended to provide division-level commanders with a state-of-the-art intelligence asset. The first implementation of the system was on EH-1H aircraft which saw limited service with contingency and Europe-based units. The Quick Fix II was originally to be put into the improved EH-1X aircraft but the decision was made to field the system in the newer, more survivable UH-60A airframe in the early 1980s after operational testing discovered drawbacks to the use of the EH-1X. After a single prototype was converted, the Army procured 66 EH-60Cs which remain in service today. The aircraft are fielded in platoons of three to four aircraft to each division and armoured cavalry regiment, assigned

to either assault, command or headquarters aviation battalions. The Army has funded the development of an upgraded variant of the system under a programme known as Advanced Quick Fix (AQF). Three aircraft were modified by Chrysler Technologies Airborne Systems, Inc. to evaluate the improved system. The aircraft will be designated as EH-60L, and an initial production contract to Loral was let in 1996. Planned fielding of the type is set for 1999.

INSCOM is also trying to field unmanned aerial vehicles (UAVs) for RSTA roles. At present the Army is without an operational UAV system, or at least one abouth which they will publicly talk. The Hunter tactical UAV, developed by TRW and Israel Aircraft Industries, is no longer being fielded after only one unit, the 15th MIB(AE), operated the type in operational service. The Alliant Outrider UAV, with a unique twin-winged design, is the next designated choice for the tactical UAV mission, but it has experienced its own teething problems and the fielding of an operational intelligence UAV remains in the extended future. INSCOM-assigned aircraft are listed under corps and theatre commands.

Army Material Command (AMC)

The Army has consolidated its research and development, acquisition and systems test units within the US Army Material Command (USAMC/AMC). The command directs the entire process of managing material and systems from development, test, procurement, distribution, maintenance and eventual disposition from the service. The command is regarded as a major command (MACOM) and is organised with a variety of research facilities and major subordinate commands that are each focused on a particular combat and combat support systems. AMC is headquartered in Alexandria, Virginia, close to the power corridors of the Pentagon and the funding source in the Congress. On 1 October 1996 the command merged with Army Missile Command (MICOM) and with the Aviation and Troop Support Command (ATCOM) to create the Army Aviation and Missile Command (AMCOM), headquartered at Redstone Arsenal, Huntsville, Alabama. AMCOM is the focal point for the development, maintainability and sustainability of the Army's aviation, missile and UAV systems.

AMCOM operates several centres and directorates that manage the material and information flow regarding new and previously fielded aviation systems. AMCOM facilities and personnel also support the offices of the Program Executive Office, Aviation (PEO AVN). The PEO reports to executives in the Department of the Army, in the Pentagon, on the

progress and problems encountered in new aviation programmes. These include the Comanche, Kiowa Warrior and Longbow Apache which are directly managed by their respective project manager's offices (PMOs) with AMCOM. AMCOM also manages the Army aircraft loaned to other US federal agencies including the US Customs Service (USCS), and those operated by the Navy for test pilot training with the US Naval Test Pilots School (USNTPS).

The US Army Chemical and Biological Defense Command was activated in October 1993 with headquarters at Aberdeen Proving Ground, Maryland with a small fleet of support aircraft assigned. The Communications-Electronics Command (CECOM) is responsible for the development of a variety of Army systems including command and control, communications, computer and intelligence electronic warfare (C4IEW). While AMCOM is responsible for the aviation platforms, CECOM is responsible for many of the systems that are integrated into the airframes and those which integrate those systems into the Army's numerous data and voice networks. CECOM was heavily involved with Joint STARS,

the Guardrail Sigint programme variants for the RC-12s and Airborne Reconnaissance Low (ARL) that operates in DHC-7s designated as either the E-5A/OE-5B or RC-7B for low-intensity conflict (LIC) surveillance missions. The Army's airborne command and control system (A2C2S) is another important new programme, performing as an airborne tactical command post. The aircraft will be designated as the UH-60C and will provide combat commanders with the ability to communicate with joint, theatre and national command centres.

CECOM's Research, Development and Engineering Center (RDEC) manages new electronic systems integration efforts into various aviation platforms through the centre's Command/Control & Systems Integration Directorate (C2SID). Projects are tested and evaluated in aviation systems by the Electronic Systems Division, Airborne Engineering Evaluation Support Branch (AEESB) in its fleet of aircraft. The unit is based at NAWC Lakehurst, New Jersey, with its Night Vision and Electronic Sensors Directorate based at Davison Army Airfield, Ft Belvoir, Virginia, charged with research, development and acquisition of night vision,

Currently being evaluated by the Med CECAT (AA) of the 44th Medical Brigade is the prototype UH-60Q, a dedicated air ambulance version of the Blackhawk. A successful conclusion to the trials will result in over 200 UH-60Ls being modified as HH-60Ls.

AMC

Army Research Laboratory HQ (USARL), Adelphi, Md.
Survivability/Lethality Analysis Directorate, Electronic Warfare Division (SLAD/EWD) White Sands Missile Range (WSMR), N.M. *(supports Big Crow NKC-135A, see TECOM/WSMR below)*

Army Aviation and Missile Command (AMCOM), Redstone Arsenal, Huntsville, Alabama

Program Executive Office, Aviation (PEO-AVN)
Comanche PMO   1 x YRAH-66A *Test*
   Sikorsky-Boeing Team West Palm Beach, Florida
Apache Attack Helicopter PMO 4 x (Y)AH-64D *Test*
   Boeing Helicopter, PMO Falcon Field, Mesa, Arizona
Apache Modernization PMO   2 x UH-1H *Test Support*
   Boeing Helicopter, PMO   2 x AH-64A *Test*
   Mesa, Arizona
Kiowa Warrior PMO   2 x OH-58D(R) *Test*
   Bell Helicopter/Textron, Arlington AP, Texas
Deputy for Systems Acquisition
Fixed Wing PMO (C/RC-12; cargo/transport fleet management)
Raytheon/Beechcraft Aircraft 5 x C-23A, 4 x C-12C *Storage*
   Beech Field, Wichita, Kans. 1 x RC-12K *Test*
Scout/Attack PMO (AH-1F; OH-58A/C fleet management)
Bell Helicopter Textron   1 x AH-1F(N) *Test*
   Arlington AP, Texas
Utility Helicopters PMO (UH-1; CH-47; H-60 fleet management)
Sikorsky Helicopter   2 x UH-60A/L *Test*
   West Palm Beach, Florida
Lockheed Martin   1 x EH-60L *Test*
   Owego, New York
Raytheon E-Systems   1 x EH-60L *Test*
   Connally AP, Waco, Texas

Aviation Research, Development, Engineering Center, (ARDEC) Redstone Arsenal, Alabama
Flight Operations Division   6 x UH-1H *Test Support*

Aeroflightdynamics Directorate (AFDD), NASA Ames Research Center, Moffett Field, Santa Clara, Calif.
Rotorcraft Dynamics Division 1 x UH-1H, 2 x UH-60A *Test*
Simulation and Aircraft   1 x NAH-1S *Test*
Systems Division   2 x JUH-60A *Test*

Aviation Applied Technology Directorate (AATD), Ft Eustis, Va.
Flight Detachment   1 x Beech B100 *GS*
   Felker AAF, Ft Eustis, Va.   3 x UH-1H *Test Support*
Research & Development Programs
    1 x AH-1F, 1 x OH-58D(I) *Test Support*
    1 x UH-60A, 4 x AH-64A *R&D Support*
Air-to-Ground Missile Systems Project Office,

Redstone Arsenal, Alabama
Joint Tactical Unmanned Aerial Vehicles PMO
UAV Flight Operations   8 x JT-UAV *Test*
   Libby AAF, Ft Huachuca, Arizona

US Department of the Treasury Customs Service (USCS), Air Operations Directorate, El Paso, Texas
Aircraft on loan   16 x UH-60A *Air Interdiction*

US Naval Test Pilot School (USNTPS), NAWC Patuxent River, Maryland
Aircraft on loan   4 x U-21F, 4 x OH-58A,
    3 x UH-60A *Test pilot training*

Chemical and Biological Defense Command (CBD COM), Aberdeen PG, Maryland

Chemical R&D Center Aberdeen PG, Maryland
Flight Detachment   1 x C-23A, 2 x UH-1H *GS*
   Phillips AAF, Aberdeen PG 1 x JUH-1H *Test Support*

Communications-Electronics Command (CECOM), Ft Monmouth, New Jersey

Program Executive Office Intelligence, Electronic Warfare and Surveillance (PEO IEWS), Ft. Monmouth N.J.
Program Manager, Joint Target Attack Radar System (PM JSTARS),
Joint STARS Combined Test Force (J-STARS CTF)
   Melbourne IAP, Florida   1 x E-8C *EMD/OT&E: USAF operated*
Program Manager, Signals Intelligence (PM SIGINT)
   California Microwave   3 x RC-7B (ARL-M) *Sys. Integration*
   Hagerstown MAP, Maryland

Research, Development & Engineering Center (RDEC), Ft Monmouth, New Jersey
Command/Control & System Integration Directorate, Electronic Systems Division (C2SID-ESD), Ft Monmouth, New Jersey
Airborne Engineering Evaluation Support Branch (AEESB)
NAS Lakehurst, New Jersey 2 x UH-1H *Test Support*
    4 x JUH-1H, 1 x NUH-1H *Test*
    1 x AH-1F(N), 2 x UH-60A *Test*
    1 x NUH-60A, 1 x AH-64A *Test*
    1 x JAH-64A , 2 x RC-12D *Test*

Night Vision Directorate, Airborne Applications Branch (NVD-AAB), Davison AAF, Ft Belvoir, Virginia
Flight Detachment   2 x UH-1H *Test Support*
    1 x AH-1F, 1 x AH-1S *Test*
    1 x BN-2B, 1 x UV-18B *Test*

Naval Research Laboratory Washington, D.C.
Naval Center for Space Technology, Space Systems Development
Department   1 x (Y)UH-60C *Test*

Industrial Operations Command (USAIOC), Rock Island Arsenal, Illinois

---

Corpus Christi Army Depot (CCAD), NAS Corpus Christi, Texas
Flight Detachment   1 x UH-1H *GS*
Mobilisation Control AVCRAD Element (MACE)
Weide AAF, Aberdeen PG, Md. 1 x C-23B *GS/TA*
I/185 AVN (TAB)   8 x C-23B *GS/TA; AVCRAD assigned;*
   Gulfport-Biloxi RAP, Miss. *Dets in Calif., Conn., and Miss.*
Sierra Army Depot Herlong, California
USAG Flight Detachment   2 x UH-1H *GS*
Aerospace Maintenance & Regeneration Center (AMARC), Davis-Monthan AFB, Arizona
(USAF facility)   miscellaneous aircraft in storage

Army Simulation, Training and Instrumentation Command (STRICOM), Orlando, Florida
Project Manager Instrumentation, Targets and Threat Simulators (PM ITTS): fixed- and rotary- winged aerial targets; *inventory varies*
    X x QUH-1H/M, A/C X x AH-1 *UDS*,
    X x QS-55 *R/W Targets*
    X x MQM-107, X x QF-106A/B,
    X x QF/QRF-4 *F/W Targets*

Test and Evaluation Command (TECOM)   Aberdeen PG, Md.

Aviation Technical Test Center (ATTC), Cairns AAF, Ft Rucker, Alabama
    1 x JC-23A *Test/AMST testbed*
    3 x JAH-1F, 5 x JUH-1H *Test*
    1 x CH-47D, 2 x JCH-47D *Test*
    3 x JOH-58C, 3 x JOH-58D(I) *Test*
    4 x JUH-60A, 1 x JUH-60L *Test*
    2 x AH-64A *Test Support*
    5 x JAH-64A, 1 x JU-21H *Test*
    1 x C-12C *Test*
    3 x T-34C *Pace/Chase*

Dugway Proving Grounds, Utah (DPG)
USAG Flight Detachment   2 x UH-1H *GS*
   Michael AAF, Dugway PG, Utah
White Sands Missile Range (WSMR), White Sands, N.M.
Directorate of Applied Technology, Test and Simulation
Big Crow Program   1 x NKC-135A *Test*
(shared aircraft operated by USAF for Army)
Operations and Support Div.   1 x RC-12D *GS/Test Support*
Holloman AFB, N.M.   10 x JUH-1H *Test Support*
Electronic Proving Grounds (EPG) Ft Huachuca, Ariz.
Flight Detachment   1 x JEH-1H, 1 x JUH-1H *Test Support*
    1 x EH-60C, 2 x O-2A *Test Support*
Yuma Proving Grounds (YPG), Yuma, Arizona
Test Support Division   4 x JUH-1H *Test Support*
Flight Det   1 x NUH-1H *Test*, 2 x OH-58D(I)
Laguna AAF, Arizona   1 x OH-58C *Test Support*

Displaying a large Soviet red star and '03' Bort code, this JUH-1H has also been visually modified to resemble an Mi-24 'Hind'.

HH/MH/SH/UH-60s including Air Force, Coast Guard and Navy assigned aircraft, and the AH-64A/D.

The service provisionally established the Army Simulation, Training and Instrumentation Command (STRICOM) on 16 March 1992, with its headquarters in Orlando, Florida, to co-ordinate disparate computer simulation training efforts, becoming a major subordinate command of AMC on 1 August 1992. STRICOM operates full- and sub-scale aerial target vehicles and its 'customers' include test organisations such as AMCOM and TECOM, training commands and combat units. STRICOM uses numerous rotary-winged aircraft for targets including surplus Army and Navy UH-1C/H/L/M variants since the late 1970s and QS-55 target aircraft modified by Orlando in the late 1980s to perform as Mi-24 'Hind' surrogates. The requirement to simulate the Kamov Ka-50 'Hokum' will be met by upgrading surplus AH-1 airframes for the Universal Drone System (UDS) programme, to be designated as Hokum-X targets. The aircraft receive external VISMODs (visual modifications) and signature replicators that emulate infra-red and radar cross-sections of the type. The Navy-developed QH-50 is used as a sub-scale rotary-wing target (SSRW). Fixed-wing aerial targets include the MQM-107 Variable Speed Training Target (VSTT) and Full-Scale Aerial Targets (FSATs). The service has operated QF-86E, QF-100 and QF-106 aircraft, with the latter two types taken from the US Air Force inventory. The service is currently using USAF QF-4s as needed for the role.

---

electronic warfare and intelligence sensors on its own flight detachment of modified aircraft.

The US Army Industrial Operations Command (USAIOC/IOC) was activated on 1 October 1994 from most of the assets of the Army Armament, Munitions and Chemical Command (AMCCOM) and Army Depot Systems Command (DESCOM). The command is responsible for the service's 'guns and bullets', which includes field artillery, mortars, explosive ordnance,

rocket launchers, aircraft armament, small arms, tank armament and the special tools and systems used to make the systems supportable. The principal aviation focus of IOC is the Corpus Christi Army Depot (CCAD), at NAS Corpus Christi, Texas. This depot performs overhaul, component repair, and crash damage rebuild of AH-1s for the Army and Marine Corps, Huey variants for the Air Force, Army, Marine Corps and Navy, CH/MH-47s, OH-58A/C/D variants, and the

The Test and Evaluation Command (TECOM) is the principal material testing organisation. TECOM manages technical testing, assessments and safety evaluation of a variety of systems including aviation, headquartered at the Aberdeen Proving Ground, Maryland. The command operates the White Sands Missile Range, New Mexico, which is used primarily to conduct tests of missile systems and air defence artillery systems, up to and including impact. The Electronics Proving Ground (EPG) at Ft Huachuca, Arizona is now subordinate to WSMR and it operates its own small fleet of test and test support aircraft in addition to operating facilities for the testing and development of unmanned aerial vehicles (UAV) systems for all US military services. The Yuma Proving Ground, Arizona is a multi-purpose test centre that is used to conduct weapon systems qualification and evaluations.

The Aviation Technical Test Center (ATTC), formerly known as the Aviation Development Test Activity (ADTA), is located at Cairns Army Airfield, Ft Rucker, Alabama and is responsible for testing developing and production aviation systems and components. The organisation is uniquely configured to conduct systems testing, airworthiness testing, lead the fleet, icing testing and assessment of internationally developed helicopters, such as the Soviet-built aircraft used by OPTEC/OTSA and those that are used by other US federal agencies. The Airworthiness Qualification Test Directorate (AQTD) relocated from Edwards AFB, California to Ft Rucker from April to October 1996.

The Army is assigned a single NKC-135A aircraft for electronic warfare assessment of new systems, under a programme known as Big Crow. The aircraft is maintained and flown by the US Air Force, but mission crews are Army contractor and military personnel. It is tasked by the Electronic Warfare Division (EWD) of the Survivability/Lethality Analysis Directorate (SLAD) of ARL, and the Big Crow Program Office of the Directorate of Applied Technology, Test and Simulation, both at White Sands Missile Range (WSMR), New Mexico.

A number of AMC subordinate commands are not known to operate any aircraft directly, although they will utilise aviation resources assigned to other commands such as OSACOM for operational support. They include the US Army Security Assistance Command (USASAC), Soldiers Systems Command (USASSC), and Tank-Automotive and Armaments Command (TACOM).

*Russian-built aircraft are employed by **OPTEC** to provide realistic training during exercises. This Mi-25 is nicknamed Devil', and operates alongside another Mi-25 'Hind-D' and an Mi-24.*

US Military Academy (USMA)

The historic US Military Academy (USMA), located at West Point, on the Hudson River in New York State, is a primary source of officers for the US Army. Its cadets are provided with a full, four-year college education, after which they have a multi-year military service obligation. The USMA has recently reduced its aviation resources assigned to the 2nd Aviation

Military District of Washington (MDW)

The Military District of Washington (MDW) is one of 15 major commands in the US Army. The command supports the Army's units, organisations and facilities in and around the District of Columbia, Maryland, Pennsylvania and Virginia. MDW has been headquartered at Ft Lesley J. McNair since 1966 and is the Army's administrative element, managing most major facilities in the Washington, D.C. area including the Pentagon and its heliport. The command is primarily organised as a service and support command, with the 1st Bn/3rd Infantry Regiment, the 'Old Guard', to provide active defence if needed. The unit mainly performs ceremonial duties throughout the US capital region, including standing guard at the Tomb of the Unknown Soldier at the Arlington National Cemetery.

From the late 1960s until 1992, MDW managed the Davison Aviation Command (DAC), headquartered at Davison Army Airfield, Ft Belvoir, Virginia, which operated fixed- and rotary-winged helicopter assets for operational support airlift (OSA) and priority air transport (PAT), or VIP roles. The DAC operated eight Sikorsky VH-3A helicopters, used to transport the President of the United States, with the Marine Corps, until 1976. UH-1Hs were gained in 1969 and most of the 30 aircraft assigned then remain in service today, some operating as JUH-1Hs. Two UH-60As configured for the VIP role entered service in 1984 along with two UH-60Ls in 1991. The command also operated nearly a dozen C-12C/D and U-21F aircraft and accepted the Army's first operational jet aircraft, a Learjet 35 business jet, in April 1987. Four Gulfstream business jets, in three different models, were acquired from that year, becoming C-20F/H/J variants. The latter did not enter service until after MDW's aviation fleet was reassigned.

On 1 October 1992, the Davison Aviation

Detachment, based at nearby Stewart International Airport. The unit operates no more than two UH-1Hs for general support (GS) and VIP tasks, along with two Cessna 182 aircraft to provide parachute training for students and instructors assigned to the institution.

Command was redesignated as the Operational Support Airlift Command (OSAC) and MDW gave up all its aviation assets to the new organisation. On 1 October 1995, the 12th Aviation Battalion was activated at Davison and all the helicopter assets that had been assigned to OSAC for three years were returned to MDW control, where they remain today.

Operational Test & Evaluation Command (OPTEC)

The Operational Test and Evaluation Command (OPTEC) is headquartered in Alexandria, Virginia and was established on 16 November 1990 from the disparate assets formerly assigned to the several unique test and evaluation activities, including Operational Test and Evaluation Agency, a Field Operating Agency (FOA), and the Test & Experimentation Command (TEXCOM), which was then assigned to TRADOC (Training and Doctrine Command). The

mission statement of OPTEC is to plan and conduct operational tests, evaluations and assessments of material and systems, for the Army and other services. Their perspective is that of an independent activity, reporting directly to the Chief of Staff/Vice Chief of Staff (CSA/VCSA) of the Army. OPTEC and its subordinate commands are very heavily involved in the Army's combat development process, working on numerous, concurrent projects including BattleLab field experiments, advanced technology demonstrations and the Force XXI programme focused on digitising the modern battlefield.

OPTEC is composed of several unique organisation and activities including the OPTEC Threat Support Activity (OTSA), based at Ft Bliss, Texas. The unit is assigned Army aviators and contractor personnel who fly, maintain and present representative samples of threat aircraft, principally those operated by the former Soviet Union and its allies and customers. The OTSA fleet has about a dozen aircraft assigned and they are unmodified variants of aircraft in first-line service with many international military arms. These aircraft perform as threat surrogates during training exercises and test simulations, to provide

realistic potential threats to US forces. The aircraft are equipped with the MILES laser combat training tracking and scoring system, so the aircraft can operate over instrumented ranges. At least two stealthy VariEze aircraft have been brought on strength since 1996 and rumours persist of a BO 105 in the fleet. OTSA is known to require additional fixed-wing threat surrogates, having studied platforms such as the MiG-21 or T-38A supersonic types. A permanent detachment operates at Ft Polk, Louisiana to support the Joint Readiness Training Center (JRTC).

The Army's Test and Experimentation Command (TEXCOM) is the primary activity that tests concepts and training programmes to be sure the user requirements are operationally effective, reliable and maintainable. In 1997 it was spun off from OPTEC and made a separate field operating agency (FOA). TEXCOM prepares and conducts operationally oriented tests through 10 operational test activities and directorates, and at least three directorates operate aircraft in support of their test although they are usually operated by another organisation performing tests under TEXCOM auspices.

The principal TEXCOM activity tasked with

The largest aircraft to wear US Army titles at the moment is the heavily modified prototype Boeing 767-200, the Airborne Surveillance Testbed.

conducting a variety of operational testing and evaluation of aviation material is the Aviation Test Directorate. The directorate relocated from Ft Rucker, Alabama to West Ft Hood, Texas in January 1991. It works very closely with the agencies of TRADOC, particularly those with the US Army Aviation Center (USAAVNC), the other combat arms proponent centres, material development activities, the technical test community and the BattleLabs in order to obtain the highest quality products and services. In recent years the directorate has been a major contributor to the operational evaluation of the OH-58D(I) Kiowa Warrior, AH-64D Longbow Apache and RAH-66A Comanche combat helicopter programmes. The TEXCOM Airborne & Special Operations Test Directorate at Pope AFB, co-located at Ft Bragg, North Carolina, and the Intelligence & Electronic Warfare Directorate at Ft Huachuca, Arizona, have both operated aircraft to support various projects, but none is known to be assigned today.

US Army Space and Missile Defense Command (SMDC)

The Army activated the Space and Missile Defense Command (SMDC) on 1 October 1997 from assets formerly organised under the Army Space and Strategic Defense Command (SSDC). This significantly upgraded the status from a field operating agency to a major command tasked with providing warfighting capability for the Army's space-based assets and systems, theatre missile defence (TMD) and national missile defence (NMD). The three-star general who commands SMDC operates from the Arlington, Virginia command headquarters with facil-

Test and Experimentation Command's (TEXCOM) Aviation Test Directorate was heavily involved in the Kiowa Warrior programme, which produced the armed OH-58D(I) from the OH-58D fleet.

ities dispersed around the world. Among the sites is Redstone Arsenal, Huntsville, Alabama, where most of the command's research, development and testing operations are based. Among the assigned units are the Missile Defense and Space Technology Center, the Space and Missile Defense BattleLab, the Force Development and Integration Center, and the Space and Missile Defense Acquisition Center. The US Space Command (USSPACECOM), a small subordinate command, is based at Peterson AFB, Colorado, and supports a small cadre of Army astronauts based at the NASA Johnson Space Center at Houston, Texas.

The restructured command maintains a small fleet of specialised airframes for sensor tests of space systems including the massive Airborne Surveillance Testbed (AST), formerly the Airborne Optical Adjunct (AOA), with sophisticated sensors and information systems mounted in a specially modified Boeing 767 airframe. The aircraft features a dorsal cupola mounted on the fuselage that is 86 ft (26 m) long, 8 ft (2.4 m) high, and 10 ft (3 m) wide, is streamlined to permit the aircraft to fly at high altitudes with the sensor doors open to the atmosphere. The cupola is essentially a dust-free, clean room, totally environmentally controlled and cooled to sub-freezing temperatures for best viewing. The sensor turrets are mounted facing out the port side of the cupola, mounted on tracking gimbals and protected from the outside environment by a pair of 7-ft (2-m) sliding, viewing port doors. The infra-red sensors can be adapted for observations through a range of visible and invisible light spectrums, according to the mission needs. The aircraft has also been modified with the addition of two ventral strakes located on the aft fuse-

lage. The aircraft carries a crew of up to 15 persons, but additional observers and technicians are routinely embarked. The aircraft completed its systems and sensor integration in 1989, makings its first flight in 1990. In 1995, the aircraft was declared operational and redesignated from AOA to AST to reflect the use of the platform as a mission capable system, employing it for numerous warfighting exercises in recent years.

Another unique platform is a former civilian Gulfstream aircraft that has also been extensively modified for sensor development and space observation roles. The programme is known as the High Altitude Observatory/Infrared Instrumentation

SMDC

US Army Space and Missile Defense Command (SMDC), Arlington, Virginia

US Army Space Command (Forward) (SARSPACE), Colorado Springs, Colorado
Army component of US Space Command (USSPACECOM)

Missile Defense and Space Technology Center (MDSTC), Huntsville, Alabama

Sensors Directorate
Airborne Surveillance Testbed (AST) Program,
Boeing Field, Seattle, Washington   1 x Boeing 767-200 *Test*
High Altitude Observatory/Infrared Instrumentation System (HAO/IIS) Program
Kirtland AFB, New Mexico   1 x Gulfstream II *Test*

Space and Missile Defense Acquisition Center (SAMDAC)
US Army Kwajalein Atoll (USAKA)/Kwajalein Missile Range (KMR)
Kwajalein Atoll, Republic of the Marshall Islands
USAKA Flight Detachment 5 x UH-1H *GS; float equipped*
Dyess AAF and Bucholz AAF 3 x DHC-7 *GS/OSA*

Ft Rucker, Alabama, is the home of US Army Aviation Center (AAVNC) and consists of several army airfields (AAF), including Cairns AAF, Guthrie AAF (out of use), Hanchey AAF, Knox AAF, Lowe AAF and Shell AAF. The fort is the home of all stages of the Army's helicopter training programme, with basic and advanced training being undertaken by the 1-212 AVN (Training) and 1-223 AVN (Training) on the TH-67A Creek (above). Operational conversion training is also undertaken at Ft Rucker. The 1-223 AVN (Training), based at Cairns AAF, trains crew for the UH-60 units (right), while the Hanchey AAF-based 1-14 AVN (Training) provides the same service for the AH-64A community (far right).

Below: Via its TRADOC-assigned units, the Arizona Army National Guard controls the Western Army Aviation Training Site (WAATS) based at Pinal Air Park, Marana, which has 24 AH-1Fs and five OH-58As on charge. With the reduction in use of these types, the need for new aircrew will diminish.

System (HAO/IIS). The aircraft mounts several infrared, laser and visual sensors in the cabin, optimised to assess and measure the unique signatures of missiles and warheads as they traverse space through various flight phases.

The Kwajalein Missile Range (KMR) located southwest of Hawaii is composed of numerous sites that launch, track and record the re-entry of missiles and warheads engaged in TMD/NMD development under the management of the US Army Kwajalein Atoll (USAKA). This facility operates a few aircraft to support nearly 2,700 military, civilian and contractor personnel that maintain the surveillance and ballistic missile defence (BMD) research efforts. Four DHC-7s replaced at least four, and as many as six, Shorts SD3-30s in 1990. The aircraft are used to transport contractor personnel on a daily shuttle between Kwajalein Island and Roi-Namur Island, a distance of 50 miles (50 km) across the world's largest lagoon, encompassing 1,100 sq miles (2848 km²). A platoon of UH-1Hs conducts general support, SAR and utility throughout the 75-mile (120-km) long atoll. These aircraft operate on amphibious floats in place of the skids, in case they need to perform emergency water landings. USAKA operates from Bucholz AAF and Dyess AAF (not to be confused with Dyess AFB), based on separate islands of the archipelago.

Training and Doctrine Command (TRADOC)

US Army Training and Doctrine Command (TRADOC) was activated in the early 1970s to be responsible for determining how the Army fights, how it is organised, and how it is equipped to fight potential conflict scenarios that might erupt. TRADOC co-ordinates every step of soldier training from basic training to technical skills to graduate-level instruction for command and staff officers. The command was activated in 1973 from elements of the Continental Army Command (CONARC). From its headquarters at historic Ft Monroe, Virginia, TRADOC has driven the Force XXI warfare experiments that will shape the Army of the next 20 years, and they are developing the Army After Next (AAN), which is looking to warfighting challenges from 2020-2025.

The command utilises considerable resources to develop and execute combat warfighting doctrine, throughout the scope of potential scenarios. TRADOC was the command that implemented the combat training centre (CTC) programme that led to the simulated battlefields that are created at the National Training Center (NTC), Joint Readiness Training Center (JRTC) and the Combat Maneuver Training Center (CMTC). They are staffed with true-to-life opposing forces (OPFORs) that provide

very realistic adversaries, trained in the doctrine and tactics of the former Soviet Union, or other potential adversaries. TRADOC system managers (TSMs) are fielded to provide doctrinal interfaces for systems including the AH-64D Longbow Apache and the RAH-66A Comanche programmes. Veteran soldiers are tasked as subject matter experts (SMEs), using their operational expertise to develop training requirements from the earliest stages of these aircraft development programmes.

The Army activated the Aviation Branch in 1983, and the US Army Aviation Center (USAAVNC), headquartered at Ft Rucker, Alabama, is the TRADOC centre for aviation training and support of of the branch. All Army aviators, including those who will fly military intelligence (MI), medical evacuation/air ambulance and VIP/special missions, are regarded as members of the Aviation Branch. The centre manages and conducts initial entry rotary-wing (IERW), initial entry fixed-wing (IEFW), type training and advanced combat skills training for the Army, international military forces and US federal agencies including the US Air Force.

Three brigade-sized units are assigned directly to USAAVNC. The Aviation Training Brigade (ATB) conducts initial, type and combat skills training of all Army aviators and air traffic controllers, with four battalions assigned. The training for the AH-64D is conducted from Boeing Helicopter's facility at Falcon Field, Mesa, Arizona, using aircraft straight from the production line before they are fielded to new units, but it will relocate to Ft Rucker in late 1999. Two ARNG facilities are supported by ATB. The Eastern Army Aviation Training Site (EAATS) supplies training for lift helicopters and fixed-wing aircrews through its Det 1 with the West Virginia ARNG. The Western Army Aviation Training Site (WAATS) conducts training for those aircrews being assigned to the AH-1F HueyCobra, attack and air cavalry units, and for OH-58A RAID aircraft.

Below: Army training aircraft, the majority of which are flown from Ft Rucker, Alabama, are marked with high-visibility panels and large alphanumeric codes, as displayed on this OH-58D.

The 1st Aviation Brigade conducts advanced combat skills and leader courses for Army officers assigned to aviation and the other combat arms branches in order to teach them the employment of Army aviation systems. The unit also conducts air assault training, where infantry soldiers learn how to embark and disembark from helicopters in combat conditions.

The centre commands the Army Aviation Logistics School (USAALS), a brigade-sized unit located at Ft Eustis, Virginia which is responsible for training all airframe, electrical, flight systems, propulsion and weapons systems maintenance personnel. The school uses grounded Category B airframes, dedicated to the maintenance instruction role. The aircraft are redesignated with the G- prefix (e.g., GCH-47D, etc.) to indicate their non-flying status. An increasing number of advanced system trainers are utilised as instructional aids. A number were rebuilt from aircraft reworked after major accidents, and these systems negate the need to pull equal numbers of Class A, flyable aircraft out of service as training aids, although a few complete aircraft will always be required.

Several other TRADOC and Forces Command

The current generation of army helicopters is air-deployable without the need for major disassembly, increasing the speed of response and giving the Army global reach. The AH-64A (above) can be carried by the C-5 or C-17, while the OH-58D (left) breaks down to fit inside a C-130 Hercules.

(FORSCOM) centres operate small numbers of aircraft, either to support the training missions or to conduct advanced skills training on specific types of aircraft. Air ambulance, medical evacuation combat skills training is conducted at Ft Benning, Georgia, where the 145th Medical Detachment operates a mixed fleet of UH-1H and UH-60A aircraft in that role. The US Army Intelligence Center and School (USAICS), based at Ft Huachuca, Arizona, utilises 'B' Company/304th Military Intelligence Battalion as the schoolhouse for combat skills of combat electronic warfare/intelligence (CEWI) aviation systems. The company had lost its OV-1Ds by 1995 and it is expected to acquire some EH-60L Advanced Quick Fix (AQF) aircraft by 1999. TRADOC continues to support JRTC and NTC with small numbers of helicopters, used primarily to support observer/controllers (OCs), or judges, who monitor the 'warring' parties engaged in intense warfare simulations and exercises.

GLOSSARY

AA	Airborne Ambulance
AAF	Army Airfield
AFRC	Armed Forces Reserve Center
AIR CAV	Air Cavalry
ASLT	Assault
ATK	Attack
AVIM	Aviation Intermediate Maintenance
CAP	County Airport
CEWI	Combat Electronic Warfare/Intelligence
CMD	Command
CS	Combat Support
C2	Command and Communication
DARC	District Area Command
GS	General Support
HH	Heavy Helicopter Airlift
LUH	Light Utility Helicopter
MAP	Metropolitan Airport
OSA	Operational Support Airlift
RAID	Reconnaissance Air Interdiction Detachments
RAP	Regional Airport
SCT	Scout
SOA	Special Operations Aviation
STARC	State Area Command
TA	Theater Airlift
TARC	Trust Area Command
USAG	US Army Garrison

As the US Army enters the 21st century, the number of Vietnam-era helicopters continues to dwindle, with the OH-6 having been retired and the numbers of UH-1 and AH-1 much reduced, leaving the remanufactured CH-47D and OH-58D fleets to serve into the 21st century. This pair of OH-58Ds is seen with the large temporary codes carried for the duration of exercises undertaken at the National Training Center.

TRADOC

Training and Doctrine Command (TRADOC), Ft Monroe, Va.

US Army Aviation Center (USAAVNC), Ft Rucker, Alabama
Aviation Training Brigade (ATB), Ft Rucker, Alabama

1-11 AVN (ATS)	– no aircraft assigned: air traffic	
Cairns AAF, Ft Rucker, Ala.	services training	
1-14 AVN (Training)	32 x CH-47D, 36 x AH-64A	*Type*
Hanchey AAF, Ft Rucker, Ala.	18 x OH-58C, 29 x OH-58D(I)	*Training*
A Co., Falcon Field, Mesa, Ariz.? x AH-64D *Type Training*		
1-212 AVN (Training)	90 x TH-67A *IERW Training*	
1-223 AVN (Training)	45 x TH-67A *IFR Training*	
Cairns AAF, Ft Rucker, Ala. 18 x UH-1H, 48 x UH-60A, 2 x C-12C,		
	1 x C-12D, 1 x RC-12D,	
	1 x C-12F *Type Training*	
FORSCOM Jet Training Detachment		
Dobbins AFB, Georgia	3 x UC-35A *Type Training*	
Flight Safety International	6 x C-12C *IEFW Training*	
Dothan AP, Alabama	(contractor operation)	
Eastern Army Aviation Training Site (EAATS)		
Muir AAF, Ft Indiantown Gap, Anneville, Pa. (PA ARNG)		
	10 x UH-1H, 3 x CH-47D	
	5 x UH-60A *Type Training*	
Det 1, Benedum Airport, Clarksburg, West Virginia		
	4 x C-12C, 1 x C-23B+	
	1 x C-26B *Type Training*	
Western Army Aviation Training Site (WAATS)		
Pinal Air Park, Marana, Ariz. (AZ ARNG)		
	24 x AH-1F, 3 x OH-58A(R) *Type Training*	

1st Aviation Brigade (Air Assault) (1 AVNBDE) (AASLT), Ft Rucker, Alabama

1-13 AVN (Training)	*no aircraft assigned*
1-145 AVN (Training)	*no aircraft assigned*

1-210 AVN (Training)	*no aircraft assigned*

US Army Aviation Logistics School (USAALS), Ft Eustis, Va.

1-222 AVN (Training)	*no aircraft assigned*
Department of Attack	8 x GAH-1E/F *Instructional airframe*
Helicopter Training	3 x GAH-1S, 10 x GOH-58D(I)
	14 x GAH-64A *Instructional airframe*
Department of Aviation	13 x GUH-1H *Instructional airframe*
Systems Training	9 x GCH-47D, 2 x GYCH-47D
	9 x GOH-58A/C, 21 x GUH-60A
	1 x GYUH-60A, 1 x GYEH-60B
	Instructional airframes

US Army Engineer Center and School (USAENGCS), Ft Leonard Wood, Missouri

USAG Flight Detachment	2 x UH-1H *GS/Support*
Forney AAF, Ft Leonard Wood, Missouri	

US Army Infantry Center and School (USAINFCS), Ft Benning, Ga.

145th MED DET (AA)	5 x UH-1H, 6 x UH-60A AA *Training*
Lawson AAF, Ft Benning, Georgia	

US Army Intelligence Center and School (USAICS), Ft Huachuca, Arizona

111th Military Intelligence Brigade (Training) (111 MIBDE)	
Ft Huachuca, Arizona	
B/304 MI BN (Training)	2 x UH-1H *GS/Support*
Libby AAF,	3 x EH-60C, 2 x RC-12D
Ft Huachuca, Arizona	3 x RC-12N, 8 x UAVs *CEWI Training*

Joint Readiness Training Center (JRTC), Ft Polk, La.

Operations Group (OG)	5 x UH-1H, 5 x OH-58C *GS/Observation*
Flight Det, Polk AAF, Ft Polk, Louisiana	

National Training Center (NTC), Ft Irwin, California

Operations Group (OG)	6 x OH-58C *GS/Observation*
Flight Det., Barstow-Daggett Airport, California	

CANADA

Capital: Ottawa
Population: 27 million
Land area: 9.922 million km² (3.831 million sq miles)
Major cities: Toronto, Montreal, Vancouver, Edmonton, Calgary, Winnipeg, Quebec, Hamilton

Canadian Armed Forces

Canada is the world's second-largest country in terms of area and it has the world's longest coastline. Its military tradition is a proud one and out of all proportion to its population. Its geographic position between the two Cold War superpowers brought it into two strategic alliances as a co-founder of both the North Atlantic Treaty Organization and the North American Aerospace Defence Command (initially named North American Air Defence Command). Canada also has an excellent reputation for its peace-keeping duties with the United Nations, having participated in all but a very few UN peacekeeping missions since its inception.

The sheer scale of the country is difficult to grasp. It ranges 2,879 miles (4634 km) from north to south and 3,426 miles (5514 km) from east to west. The closest piece of land to the North Pole is Canadian. Ellesmere Island reaches to 83° 07'. Most of Canada is extremely sparsely populated, as some 80 per cent of its residents live within 100 miles (160 km) of the Canada-US border. That boundary is the world's longest undefended border, an indication of the close relations between the two countries, each of which is the other's biggest trading partner.

EARLY YEARS

In 1920, a non-permanent militia organisation, the Canadian Air Force, was formed. Its *raison d'être* was the provision of refresher training for some of the thousands of Canadians who had served in the British air forces during World War I. Their successes were notable, for 10 of the 27 leading Imperial aces were Canadian, all of whom had 30 or more kills to their credit.

The refresher training programme ended on 1 April 1922, after which the CAF lay virtually dormant for some time. The Department of National Defence was created on 1 January 1923, combining the Department of Militia and Defence, the Department of Naval Service, and the Air Board.

The CAF was awarded Royal status as the Royal Canadian Air Force on 1 April 1924. It included both permanent and non-permanent components. Under the Ministry of Defence, three branches were formed: the RCAF, Civil Government Air Operations (CGAO), and the Controller of Civil Aviation. Between the two world wars, the Royal Canadian Air Force had a primarily civilian role, including such tasks as mapping, forest-fire fighting, air and ground photography, customs patrols, crop dusting, and 'mercy' flights.

Most of these tasks were transferred to the Department of Transport, which was created in 1936. With a renewed emphasis on military duties, the

The McDonnell Douglas CF-188 Hornet is the sharp end of 1 Canadian Air Division of the Canadian Armed Forces. This CF-188B (fore) and pair of CF-188s belongs to 441 Tactical Fighter Squadron.

417 Sqn operates eight Canadair CT-133 Silver Stars in the combat support role from Cold Lake. Silver Stars first entered service in 1953.

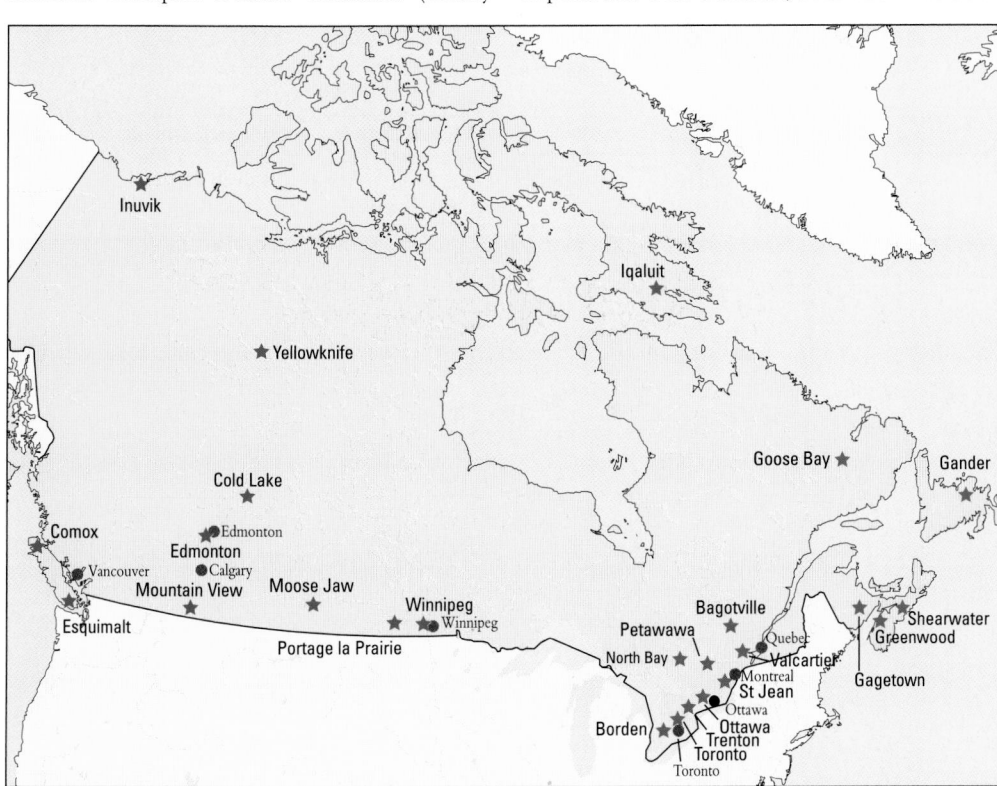

Above: Based at Yellowknife, Northwest Territories, 440 Transport and Rescue Sqn uses the last four CC-138 Twin Otters from a 1971 delivery of nine.

Right: The current workhorse of Canada's air-rescue fleet is the CH-113A Labrador, to be replaced soon by the AW 520 Cormorant.

Left: The Buffalo operates solely with 442 Sqn. Its STOL capabilities make it more suitable than the Hercules for rescue missions in British Columbia.

RCAF commenced a programme of expansion and rearmament. Nevertheless, the RCAF was quite small when World War II began: it consisted of just eight permanent and 12 auxiliary squadrons, and had a grand total of 4,061 officers and men. Three squadrons were sent to Britain in 1940, one of which, No. 1 (later renumbered as No. 401 by the RAF), participated in the later stages of the Battle of Britain.

The other squadrons were initially assigned to home defence duties (including supporting the Americans in the Aleutian Islands) or to the British Commonwealth Air Training Plan. By the end of the war, the BCATP was a massive effort that produced more than 131,000 aircrew, of which some 50,000 were pilots. By the end of 1943, the Home War Establishment had been built up to 39 squadrons. As mentioned, some West Coast units participated in the Aleutian campaign of 1942/43, while East Coast units took part in the struggle against enemy submarines in the North Atlantic.

Nearly a quarter of a million men and women served in the RCAF during World War II, of whom approximately 17,000 were killed. At the peak of its wartime strength, the Royal Canadian Air Force had 46 squadrons overseas, ranging from Britain (and, later, the Continent) to North Africa and the Far East. In addition, many thousands of Canadians, most of them aircrew, served in the Royal Air Force, despite the best efforts of the Canadian government and senior RCAF officers to bring them back into Canadian squadrons. For example, nearly 25 per cent of Bomber Command's aircrew were Canadian – and that figure excludes those serving in RCAF squadrons in 6 (RCAF) Group. At one point, there were more RCAF officers serving as aircrew in the Royal Air Force than there were in all of the RCAF's Regular and Auxiliary squadrons.

At the end of the war, the RCAF was the fourth-largest air force in the world, but it rapidly went through a tremendous reduction in strength immediately afterwards. Several squadrons had been earmarked for the 'Tiger Force', but the surrender of Japan meant they were no longer required and they, too, were disbanded.

The CH-146 Griffon was selected in 1992 to provide tactical support for the Canadian ground forces, a total of 100 being ordered. This example is seen fitted with a rescue hoist.

Canadian pilots served exchange tours with the USAF during the Korean Conflict and 426 (Transport) Squadron was heavily tasked in support of the air bridge from North America. The then-diplomat (later Prime Minister) Lester B. Pearson won the Nobel peace prize for his work in creating peace-keeping forces under the auspices of the United Nations. Canada has also participated in other peace-keeping missions that were not UN operations, such as in the former Yugoslavia. The Canadian air force's major roles have been the transport by air of personnel, support equipment and supplies, and relief supplies. Usually, the aircraft types involved were transport aircraft and helicopters. However, during Operation Friction (the name assigned to Canada's participation in the Gulf War), the CAF provided two squadrons of CF-188s, five CH-124A Sea Kings, a CC-137 tanker/transport, a CC-144A Challenger 600 liaison aircraft, and also flew many support missions with CC-130s and CC-137s. CP-140 Auroras assisted in the work-up of the naval component as it crossed the Atlantic. They also participated in patrols over the Adriatic during the civil war in the former Yugoslavia and, for several months in 1997, six CF-188s were deployed to Aviano AB in Italy, from where they flew armed patrols.

NORAD

In 1958, Canada and the United States formed the North American Air Defence Command ('Defense' in the US). NORAD always has an American commander and a Canadian deputy commander. These positions are reversed in Canadian NORAD Region Headquarters, and the same pattern is followed at other NORAD sites. Renamed North American Aerospace Defence Command in 1981, NORAD continues to watch the skies over the continent, although off-course aircraft and drug smugglers are encountered far more often than the Russian 'Bears' that used to regularly test its defences during the Cold War.

NATO

At its peak, the RCAF's NATO commitment totalled 12 air superiority squadrons in four wings.

Later, one Sabre squadron in each wing was replaced by a CF-100 squadron in the all-weather interceptor role. While the Sabre was built under licence by Canadair (with a Canadian engine), the CF-100 Canuck was a wholly-Canadian creation. Including prototypes and trials aircraft, a total of 692 was produced, including 53 for the Belgian Air Force. The CF-100 was to have been replaced in front-line service by the very advanced CF-105 Arrow, but that programme was cancelled on 'Black Friday', 20 February 1959. Instead, the government purchased BOMARC missiles and CF-101 Voodoos for the squadrons dedicated to NORAD and contracted with Canadair to build CF-104 Starfighters for the NATO-dedicated units. They initially served in the nuclear strike and reconnaissance roles. In 1970, they switched to conventional attack.

On 1 February 1968, the culmination of a process that began as Integration and developed into Unification was marked by the disbandment of the RCAF, Royal Canadian Navy and Canadian Army,

In Canadian service the Dash-8 is used as the CC-142 (fore) and CT-142 navigation trainer.

Canada's aerobatic team, the 'Snowbirds', is officially known as 431 Air Demonstration Squadron. Fifteen CT-114s are assigned.

407 'Demon' squadron participated in the 1997 Fincastle competition with this specially marked CP-140 Aurora.

Canadian Armed Forces

Chief of the Air Staff	National Defence HQ. Ottawa, Ontario

1 Canadian Air Division	CFB Winnipeg, Manitoba
Canadian NORAD Region Headquarters	CFB Winnipeg, Manitoba
Regional Operations Control Centre	CFB Winnipeg, Manitoba
Contingency Capability Centre	CFB Trenton, Ontario

1 Wing	**CFB Kingston, Ontario**
400 Tactical Helicopter Squadron	CFB Borden, Ontario
8 x Bell CH-146 Griffon	
403 Helicopter Operational Training Squadron	
14 x Bell CH-146 Griffon	CFB Gagetown, New Brunswick
408 Tactical Helicopter Squadron	CFB Edmonton, Alberta
24 x Bell CH-146 Griffon	
427 Tactical Helicopter Squadron	CFB Petawawa, Ontario
19 x Bell CH-146 Griffon	
430 Tactical Helicopter Squadron	CFB Valcartier, Québec
16 x Bell CH-146 Griffon	
438 Tactical Helicopter Squadron	CFB Montréal (St-Hubert), Québec
8 x Bell CH-146 Griffon	
3 Wing	**CFB Bagotville, Québec**
425 Tactical Fighter Squadron	CFB Bagotville, Québec
15 x McDonnell Douglas CF-188/CF-188B	
433 Tactical Fighter Squadron	CFB Bagotville, Québec
15 x McDonnell Douglas CF-188/CF-188B	
439 Combat Support Squadron	CFB Bagotville, Québec
3 x Bell CH-146 Griffon, 5 x Canadair CT-133 Silver Star	
3 Air Maintenance Squadron	CFB Bagotville, Québec

(plus QRA det at CFB Goose Bay and FOL at Iqaluit, NWT; neither is permanently manned)

4 Wing	**CFB Cold Lake, Alberta**
410 Tactical Fighter Operational Training Squadron	CFB Cold Lake, Alberta
23 x McDonnell Douglas CF-188/CF-188B	
416 Tactical Fighter Squadron	CFB Cold Lake, Alberta
15 x McDonnell Douglas CF-188/CF-188B	
417 Combat Support Squadron	CFB Cold Lake, Alberta
3 x Bell CH-146 Griffon, 8 x Canadair CT-133 Silver Star	
440 Transport and Rescue Squadron	Yellowknife, Northwest Territories
4 x de Havilland Canada CC-138 Twin Otter	
441 Tactical Fighter Squadron	CFB Cold Lake, Alberta
15 x McDonnell Douglas CF-188/CF-188B	
1 Air Maintenance Squadron	CFB Cold Lake, Alberta

(plus QRA det at CFB Comox and FOLs at Yellowknife, NWT, and Inuvik, NWT)

A flight of four 433 Squadron CF-188s, based at CFB Bagotville, releases infra-red decoy flares.

5 Wing	**CFB Goose Bay, Labrador, Newfoundland**
444 Combat Support Squadron	CFB Goose Bay, Labrador, Newfoundland
3 x Bell CH-146 Griffon	
8 Wing	**CFB Trenton, Ontario**
412 Transport Squadron	Ottawa, Ontario
1 x Canadair CC-144A Challenger-600	
3 x Canadair CC-144B Challenger-601	
424 Transport and Rescue Squadron	CFB Trenton, Ontario
3 x Boeing Vertol CH-113A Labrador (aircraft belong to 8 Wing)	

(to receive 3 x AW 520 Cormorant and retire CH-113A Labradors)
(plus use of two of 20 pooled CC-130E Hercules)

426 Transport Training Squadron	CFB Trenton, Ontario

(uses some of 20 pooled CC-130E and CC-130H Hercules)

429 Transport Squadron	CFB Trenton, Ontario

(uses some of 20 pooled CC-130E and CC-130H Hercules)

436 Transport Squadron	CFB Trenton, Ontario

(uses some of 20 pooled CC-130E and CC-130H Hercules)

437 Transport Squadron	CFB Trenton, Ontario
5 x Airbus CC-150 Polaris (aircraft belong to 8 Wing)	
8 Air Maintenance Squadron	CFB Trenton, Ontario
Canadian Forces Aircrew Selection Centre	CFB Trenton, Ontario
Aerospace Telecommunications and Engineering Support Sqn* (no assigned aircraft)	CFB Trenton, Ontario & CFD Mountain View

* ATESS was formed from the merger of AMDU and TRACS in July 1995.

Note: 8 Wing's CC-130 fleet includes two 'stretched' CC-130H

9 Wing	**CFB Gander, Newfoundland**
103 Search and Rescue Squadron	CFB Gander, Newfoundland
2 x CH-113 Labrador	

(to receive 3 x AW 520 Cormorant and retire CH-113 Labradors)

12 Wing	**CFB Shearwater*, Nova Scotia**
406 Maritime Operational Training Squadron	CFB Shearwater*, Nova Scotia

(uses some of 30 pooled Sikorsky CH-124A/B Sea Kings)

423 Maritime Helicopter Squadron	CFB Shearwater*, Nova Scotia

(uses some of 30 pooled Sikorsky CH-124A/B Sea Kings)

443 Maritime Helicopter Squadron	Victoria IAP, Patricia Bay, British Columbia

(uses some of 30 pooled Sikorsky CH-124A/B Sea Kings)

12 Air Maintenance Squadron	CFB Shearwater*, Nova Scotia

* CFB Shearwater has been merged with CFB Halifax for administrative purposes

14 Wing	**CFB Greenwood, Nova Scotia**
404 Maritime Patrol and Training Squadron	CFB Greenwood, Nova Scotia

(uses some of 13 pooled Lockheed CP-140 Auroras and 3 pooled Lockheed CP-140A Arcturuses)

405 Maritime Patrol Squadron	CFB Greenwood, Nova Scotia

(uses some of 13 pooled Lockheed CP-140 Auroras and 3 pooled Lockheed CP-140A Arcturuses)

413 Transport and Rescue Squadron	CFB Greenwood, Nova Scotia
3 x CH-113 Labrador	

(to receive 4 x AW 520 Cormorant to replace CH-113 Labradors)

4 x Lockheed CC-130E Hercules	
415 Maritime Patrol Squadron	CFB Greenwood, Nova Scotia

(uses some of 13 pooled Lockheed CP-140 Auroras and 3 pooled Lockheed CP-140A Arcturuses)

434 Combat Support Squadron	CFB Greenwood, Nova Scotia
1 x CC-144A Challenger-600	
3 x CE-144A Challenger-EST	

(interim variant; one with airborne sensors for Elint gathering)

3 x CE-144A Challenger-EST *(definitive variant; first is at Canadair for modifications; other two not yet upgraded)*	
2 x CP-144A Challenger-CP *(designation assigned, but*	

programme cancelled; aircraft are nevertheless used in the role)

5 x Canadair CE-133 Silver Star *(designation assigned, but only one aircraft fully modified so far)*	
10 x Canadair CT-133 Silver Star	
Maritime Proving and Experimental Unit	CFB Greenwood, Nova Scotia

(uses some of 13 pooled Lockheed CP-140 Auroras and 3 pooled Lockheed CP-140A Arcturuses)

14 Air Maintenance Squadron	CFB Greenwood, Nova Scotia
14 Software Engineering Squadron	CFB Greenwood, Nova Scotia

(formerly Aurora Software Development Unit)

15 Wing	**CFB Moose Jaw, Saskatchewan**
2 Canadian Forces Flying Training School	
62 x Canadair CT-114 Tutor	CFB Moose Jaw, Saskatchewan
431 Air Demonstration Squadron ('Snowbirds')	
15 x Canadair CT-114 Tutor	CFB Moose Jaw, Saskatchewan
Canadian Forces Flying Instructor School	
11 x Canadair CT-114 Tutor	CFB Moose Jaw, Saskatchewan
16 Wing	**CFB Borden, Ontario**

(various training schools, most notably Canadian Forces School of Aerospace Engineering)

17 Wing	**CFB Winnipeg, Manitoba**
402 Squadron	CFB Winnipeg, Manitoba
2 x CC-142 Dash 8, 4 x CT-142 Dash 8 Nav Trainer	
435 Transport Squadron	CFB Winnipeg, Manitoba
8 x CC-130E Hercules and CC-130H Hercules	

(includes 5 x CC-130H with refuelling pods)

Central Flying School	CFB Winnipeg, Manitoba
3 x Canadair CT-114 Tutor	
Canadian Forces Air Navigation School	CFB Winnipeg, Manitoba

(uses 402 Squadron's CT-142 Dash 8 Nav Trainers)

19 Wing	**CFB Comox, British Columbia**
407 Maritime Patrol Squadron	CFB Comox, British Columbia
5 x Lockheed CP-140 Aurora	
414 Combat Support Squadron	CFB Comox, British Columbia
14 x Canadair CT-133 Silver Star	

(five to be converted to CE-133 Silver Star)

442 Transport and Rescue Squadron	CFB Comox, British Columbia
6 x de Havilland Canada CC-115 Buffalo, 4 x CH-113A Labrador	

(to receive 5 x AW 520 Cormorant and retire CH-113A)

Canadian Forces School of Survival and Rescue	
22 Wing	**CFB North Bay**

(eastern and western Sector Operations Control Centres)

Other Flying Units

Aerospace Engineering Test Establishment	CFB Cold Lake, Alberta
2 x Canadair CT-114 Tutor, 3 x Canadair CT-133 Silver Star	
2 x Bell CH-146 Griffon, 2 x McDonnell Douglas CF-188	
2 x McDonnell Douglas CF-188B	

('borrows' other types as required for specific trials)

3 Canadian Forces Flying Training School	Portage la Prairie, Manitoba

(supervises flying training provided by the private contractor, Canadian Aerospace Training Centre; provides instructors for Basic Helicopter School and Multi-Engine School; Primary Flying School instructors are provided by the contractor; the PFS uses 12 Slingsby T-67C3 Fireflys; the MES uses 8 Beech King Air C90As; the BHS uses 14 CH-139 Jetrangers leased from the CAF, which wear both their military serials and Canadian civil registrations)

•*Note that it is 1 AMS at 4 Wing. In all other wings with an AMS, the number of the AMS matches that of the wing. Cold Lake is an exception in order to perpetuate the heritage of 1 AMS, which had served in Europe.*

Above: The Hercules fleet is varied and includes both CC-130Es, 'Hs and 'H-30s in the transport role, and 'T tankers. 130340 is one of 8 Wing's five CC-130T tankers, equipped with two FRL Mk 32B hose-drum units under the wings.

Above left: Five CC-150 Polaris serve with 437 Sqn, having replaced the CC-137 Husky. Two will gain air-to-air refuelling capabilities in the near future. The fleet is painted in high-gloss grey with a blue cheat line.

Left: The homegrown Canadair Challenger serves as an electronic warfare platform as the CE-144A Challenger EST. 144603 flies with 414 Sqn. The size of the fleet was reduced in late 1998.

and the formation of the Canadian Armed Forces. At that time, the RCAF had 45,000 personnel and 19 types of aircraft. After several reorganisations that saw the merger of some Commands and the creation or disbandment of others, the arrangement settled on Air Command, Maritime Command, Land Forces Command, and Communications Command. 1986 brought a return to separate uniforms, a welcome change from the CAF-wide use of dark green, and the navy readopted its traditional ranks. The air force, however, did not bring back an RAF-style rank structure.

Several recessions in a row hit Canada hard, and the Department of National Defence was a popular target for politicians looking to save money. The 1997/98 DND budget was just CN$9.25 billion, down 23 per cent from the 1993/94 budget of CN$12 billion.

Other than personnel assigned to the NATO Airborne Early Warning Force and NATO headquarters, Canada had withdrawn all of its air and ground assets from Europe by 1994. There have been deep spending and personnel cuts. Following the successful lead of the United States, Canada has adopted the Total Force concept. Almost every unit has both Regular and Reserve personnel, and it is quite common for a unit to have personnel from all three branches.

As the cutbacks continued, Air Command considered several reorganisations and various cutbacks. Although modified aircraft were still being delivered, it was decided to withdraw all of the CF-116 CF-5 aircraft and disband 419 Squadron, which had provided fighter lead-in training. It was also planned to combine Fighter Group, Air Transport Group, and 10 Tactical Air Group (the latter was, essentially, 'army aviation') as Air Combat and Mobility Group, leaving Maritime Air Group as it was. A further review halted those plans and instead saw the disbandment of all four Groups. Air Command itself, along with Maritime Command, Land Forces Command, and Communications Command also stood down. In place of AirCom and the Groups, 1 Canadian Air Division was formed as the tactical operational level. The Commander of Air Command became the Chief of the Air Staff and moved, along with a small staff, from Winnipeg to Ottawa.

Previously in 'the hole' at CFB North Bay,

Canadian NORAD Region Headquarters is now co-located with 1 CAD HQ. Although often written as 1 CAD/CANR HQ, CANR is actually a small cell and not a major component of 1 CAD, at least in terms of the number of personnel. Both the western and eastern Sector Operations Control Centres remain at North Bay. This reduction in the administrative overhead has saved substantial funds and minimised the number of squadrons that had to be disbanded or cut back in size.

Several Reserve squadrons were disbanded, however. All but two were 'twinned' with Regular Force squadrons and the adoption of the Total Force policy meant that it was more economical to combine the units. Previously, there had been two Reserve squadrons each in Toronto and Montreal. In both cases, the units were combined into one, with the other being disbanded. Technically, most of the 'disbanded' Reserve squadrons were actually 'zero-manned' and still exist. Part of the reason for this is to properly accommodate heritage considerations, to ensure the preservation of battle honours.

The decision was also made to severely reduce the number of CF-188s in service. Of the original total of 138 aircraft, only 60 remain in front-line squadrons. The OTU has a further 23 aircraft, mostly two-seaters. Each of the four operational squadrons has a Unit Establishment of 15 aircraft, although they often have a few more on hand. The remainder are stored for varying lengths of time by Canadair, which cycles them back into the active fleet to keep the number of flying hours per aircraft relatively even. However, some of the oldest CF-188s are very close to the end of their airframe hours and are unlikely to fly again.

Consideration had been given to disbanding one of the CF-188 units, but Canadian politics put an end to those thoughts. Western Canada had already lost 419 Squadron when the CF-116 was withdrawn. To disband one of the two CF-188 squadrons in Quebec would have raised an equal storm of protest, and to leave just one squadron at CFB Bagotville would not be cost effective since much of the support infrastructure would still be required. Instead, all four squadrons were cut back in size.

In theory, each of the four front-line squadrons (416 and 441 at Cold Lake, 425 and 433 at Bagotville) is dual-role, but, in practice, each specialises in either air-to-air (and is NORAD-dedicated) or air-to-

ground (and is NATO-dedicated). For a while, each squadron switched roles twice a year, but the cost in time and loss of capability during the transition was too great and the change is now made on an approximately annual basis.

IMPROVING THE FLEETS

Air-to-ground capabilities have received a significant boost with the introduction of precision-guided munitions, this leading directly to the deployment of six aircraft to Aviano AB between August and November 1997. For the first time, Canadian peacekeepers on the ground were supported by their countrymen overhead. During Operation Friction, CF-188s had been limited to dropping 'dumb' bombs when they were authorised to conduct ground-attack missions partway through the war. Canada flew a total of 77 training sorties and 261 operational sorties in support of the Stabilisation Force (SFOR). A typical warload consisted on one 500-lb laser-guided bomb, two AIM-9 Sidewinders, one AIM-7 Sparrow, and one Maverick missile.

Early 1998 saw the announcement of the purchase of 15 EH Industries AW 520 Cormorants for the SAR role; it is a version of the EH101, which made the selection ironic because, immediately upon taking office in November 1993, the Liberal government of Prime Minster Jean Chrétien cancelled the previous Progressive Conservative government's order for 15 EH101s for SAR and 35 more as Sea King replacements. It remains to be seen what will be selected to replace the ageing Sea Kings, some of which have been in service since 1963 (the 'newest' entered service in 1969).

Other major capital expenditures still pending are the Aurora Life Extension Project, which is still being defined (every budget cut causes the ALEP wish list to be cut back) and a life extension programme for the CF-188. The latter has had the pressure eased somewhat by the reduction in the active fleet, but can not be put off indefinitely. Canada has made a CN$10 million 'buy-in' to the Joint Strike Fighter programme so as to be kept fully advised of its progress. There is no plan at all to upgrade from what are essentially F/A-18As and Bs to the new Es and Fs. The previous Chief of the Air Staff has been quoted as saying that Canada could conceivably "cash in" the 126 surviving CF-188s and 188Bs and, with the money available for the CF-188 life extension programme, buy between 20 and 30 Es and Fs. Given the extreme difficulty in getting the government to replace ancient CH-113 and CH-113A Labradors in the politically-correct SAR role, the thought of

approaching its masters for a CF-188 replacement must be nothing short of terrifying to the air force's leadership.

With regard to the Aurora upgrades, the CAF is in close contact with the US Navy, monitoring its P-3 Orion fleet and the work necessary to extend its useful life. 1 CAD hopes to keep the CP-140s in service until 2015. As with the CF-188, the airframe and engines are the easy things to keep going – it is the avionics that are incredibly expensive, but they need to be updated or replaced. Canada is already experiencing problems with interoperability with its allies, a problem that was highlighted during the Aviano deployment. The Canadian aircrews were unable to link with the NATO command and control structure, lacking secure radios and datalinks.

Another ongoing project involves the diverse assortment of Hercules models operated by 1 CAD. Nominally consisting of just CC-130Es and Hs, there are, in fact, seven sub-types, making crew training and maintenance unnecessarily complicated. CAE Aviation (formerly Northwest Industries) has a contract to modernise and standardise the fleet as much as possible. The work performed includes reconfiguring the cockpits to a common standard, replacing the outer wings of the E models (and some of the Hs), installing sensors to detect missile launches, cockpit armour, and some avionics upgrades. Two 'white-tail' L-100-30 'stretched' Hercules were bought from Lockheed and were brought up to military standard by CAE Aviation. Both joined 8 Wing's pool of Hercules aircraft in mid-1997. Because of their higher empty weight, they have a restriction on payload and/or range, but allow the carriage of bulkier items.

TRAINING

Beginning in 1992, the CAF began contracting out some of its flying training. A consortium led by Bombardier (corporate owner of Canadair, de Havilland Canada, Learjet, and Shorts) took over Primary Flying Training using the Slingsby T-67C3 Firefly. Instructors for the PFT are civilians, although the majority are former CAF pilots. The Canadian Aerospace Training Centre (as the consortium is known) took over the operation of the former CFB Portage la Prairie for the operation. CATC also provides the Beech King Air C90A for the Multi-Engine Flying Training course and leases the CAF's 14 CH-139 JetRangers for the Basic Helicopter Training course. (The CH-139s wear their CAF serials and civilian registrations.) Instructors for both the MEFT and BHT are CAF personnel assigned to 3 CFFTS.

The privatised training has been deemed a success and is to be expanded. Another Canadair-led consortium will be responsible for providing training to wings level, which is now provided by 2 CFFTS using CT-114 Tutors. Initially, the consortium was going to use the EMB-312H Super Tucano and the BAe Hawk 100, but the Tucano has been replaced by the new Beech T-6A Texan II (much to the

displeasure of Brazil), to be known as the Harvard II. The aircraft will be commercially owned, but certified by DND as if they were military aircraft, wearing military livery, markings and serials. In order to make the new training economical, it was offered to Canada's allies as the NATO Flying Training in Canada (NFTC) programme. It will encompass three phases. Phase I will continue to be the Primary Flying Training at CATC. Phase II will be basic training, Phase III will be advanced training, and Phase IV will be tactical fighter lead-in training. Students will then move on to operational conversion unit training, followed by combat readiness training at their assigned squadron.

Canada has a long history of providing flying training for its allies, beginning, of course, with the British Commonwealth Air Training Plan (1940–45). It was followed by the NATO Air Training Plan (1950–59), various NATO bilateral air training agreements (1959–83), Military Training Assistance Plan (1964 to the present), and NATO and International aircrew training (1992 to the present).

Currently, students complete Phase I at CATC, then go to 2 CFFTS at CFB Moose Jaw for Phase II. That course was revised to allow for the streaming of students into fighters towards the end of the course after the withdrawal of the CF-116 fighter lead-in trainer. Those students receive different training during the latter stages of Phase II to better prepare them for flying the CF-188. Interestingly, it is felt that the CF-188 is simple to learn to fly, with its actual employment as a combat aircraft being the hard part. Therefore, the lack of a lead-in trainer is not as serious as might be believed at first.

Students chosen for transport and patrol squadrons return to Southport (as CFB Portage has been renamed) for Phase III, then go to either 426 Squadron for CC-130 and CC-150 training, to 442 Squadron for CC-115 Buffalo training, or to 434 Squadron for Challenger training. Those destined for maritime patrol training go to 404 Squadron. Future helicopter pilots also return to Southport for Phase III and then move on to either 403 Squadron for CH-146 Griffon training, to 406 Squadron for Sea King training, or 442 Squadron for Labrador training.

Students heading to the CT/CE-133 complete the same Tutor course as those going on to CF-188s, but then go to 414 Squadron for type conversion.

Under NFTC, Phase II is divided into Phase IIA and IIB. Phase IIA will be common to all students, who are then streamed into either fighter, multi-engine, or helicopter, with Phase IIB specific to each. Phases IIA, IIB, and III will be conducted at CFB Moose Jaw. Some Hawks will be based at CFB Cold Lake for Phase IV. Extensive use will be made of simulators at all levels of the training, which is not surprising since CAE is a member of the consortium.

The CT-114 Tutor will continue in CAF service. It will be employed by the 'Snowbirds' aerobatic team as well as by AETE, CFS and FIS.

NFTC will commence in 1999 with the training of an international corps of instructor pilots to a common standard. A total of 24 Harvard IIs and 18 Hawk 100s will be used for the NFTC programme. The current USAF-hosted NATO Joint Jet Pilot Training programme is due to expire in 2005, but demand will exceed capacity by 2000. Canada will provide both overall direction and day-to-day management, in addition to supplying some of the instructors, and will also set training standards.

The years ahead will be difficult ones as both the CAF in general and 1 CAD in particular are faced with ever-smaller budgets coupled with little or no reduction in taskings. The purchase of four surplus Royal Navy 'Upholder'-class conventional submarines (announced in April 1998) will not help the budgetary situation, for their CN\$750 million price tag is to be covered from the existing DND budget. Some of the cost may be offset in a barter arrangement allowing the RAF and British Army to use training facilities in Canada, but the (necessary) purchase will be felt throughout the CAF. With the airlines hiring again, pilot attrition is a major concern. It was recently addressed with the offer of a cash bonus paid to those who remain in the air force, but personnel also have concerns that come under the 'quality of life' category. DND admits that far too many housing units are substandard and further acknowledges that pay scales (despite an increase in April 1998 with a couple of small raises to follow) are not keeping up with the cost of living in many parts of Canada. One newspaper illustrated this with the story of a Moose Jaw-based lieutenant who 'moonlights' as a security guard at a shopping mall. For several decades, Liberal governments have not exactly been kind to the CAF. The current government has several years left in its mandate, so the situation will likely get worse before it gets better.

However, the professionalism and dedication of the men and women of the Canadian Armed Forces cannot be faulted. They have learned how to do more with less and will have to put those lessons to good use for the foreseeable future.

The Canadian Armed Forces will be reliant on the Hornet well into the future. The type is the focus of a planned life extension programme, probably funded by the sale of some of the fleet.

MEXICO

Capital: Mexico City
Population: 86.4 million
Land area: 1.973 million km² (762,000 sq miles)
Major cities: Guadalajara, Juarez, Leon, Mexicali, Monterrey, Puebla, Tijuana

Above: Air Defence Squadron 401 operates the token fighter force of eight F-5E Tiger IIs (from 10 ordered in 1981), supported by a pair of F-5Fs.

Left: Ten ex-USAF C-130As have been operated over the years, but four were written off and one has been withdrawn. A single L-100-30 acquired in 1993 was sold in 1994. The five C-130As remain the FAM's principal heavy transport, with the 6th Air Group.

Right: Recent deliveries of surplus US Army UH-1Hs supplement the fleet of Bell 212s in service. The 212 is used for search and rescue missions as well as equipping special operations squadrons.

Mexico is the smallest country in North America and has an area equivalent to about 25 per cent that of the United States. It largely entrusts its defence against external military threats to its northern neighbour, with which it would be unable to compete on any realistic level; and, since its own size and resources overwhelmingly exceed those of its southern neighbours, Mexico is relatively free from external threat. The Mexican Armed Forces are therefore primarily orientated towards internal security. In recent years, the activities of drug-traffickers, combined with peasant uprising in the southern part of the country, have occasioned a considerable expansion of the Mexican Armed Forces without any significant alteration in their essentially internal security role.

Mexico was the first country south of the US to appreciate the value of air power: a single aircraft, flown by a US mercenary pilot, Captain Hector Worden, was used on bombing missions by the Maderista revolutionaries as early as 1911. Small numbers of aircraft continued to be used by the various warring factions throughout the revolutionary period (1910-20). Although naval and military air arms were established upon the return to relative political stability, they both remained small.

The sinking of Mexican merchant shipping by German submarines provoked a declaration of war against the Axis in 1942, and Mexican military and naval aircraft patrolled the country's Pacific and Gulf coasts. Mexico also contributed a fighter squadron to the Allied war effort, which saw action in the Pacific during the closing stages of World War II.

When the Air Force achieved semi-autonomous status, in 1944, the Navy retained control of its own vestigial air arm and, to date, the Army has not

re-established any aviation element. Mexican military aviation also remained relatively insignificant even after the 1947 Rio Treaty of Mutual Defence occasioned comparatively generous transfers of defence material by the US to most of its southern neighbours.

The Secretary of National Defence is responsible for both the Army and Air Force, an Under Secretary having direct responsibility for the latter. The Commander of the Air Force is the senior serving Air Force officer and is directly responsible to the Under Secretary of National Defence (Air Force). The Commander-in-Chief of Naval Operations is the senior serving naval officer and reports to the Secretary of the Navy via the Under Secretary of the Navy. The Director of Naval Aviation, who reports to the Commander-in-Chief of Naval Operations via the Chief of the Naval Staff, enjoys equal status to the other naval functional directors and to the officers commanding the Marines, the Pacific and Gulf Fleets and the Naval Zones.

Mexican Air Force

The Mexican Air Force currently numbers approximately 14,000 personnel of all ranks, operating some 300 aircraft. It is organised into two combat wings, with a total of five groups and 10 squadrons plus six independent groups, deployed throughout three air regions and comprising a total of 12 squadrons. They fly from 17 bases and three air stations, not all of which are manned on a permanent basis. The former Air Force Paratroop Battalion has been transferred to the Army.

The Central Air Region includes the principal base of the Fuerza Aérea Méxicana, Base Aérea Militar No. 1 General Alfredo Lezama Alvarez (Santa Lucia, México), which houses the HQ of the 1st Combat Wing, together with its 5th and 7th Air Groups and the 6th and 9th Independent Air Groups. BAM No. 7 Gustavo Leon Gonzalez (Pie de la Cuesta, Guerrero), is home to the 2nd Independent Air Group and BAM No. 11 Teniente Coronel Juan Pablo Aldasoro Suartez (Benito Suarez International Airport, Mexico City) accommodates the 8th Independent Air Group. This region also includes Military Air Bases Nos 13 (Chihuahua), 14

(Monterrey) and 15 (Oaxaca), which have no permanently attached units.

The Western Air Region includes BAM No. 5 Capitán Emilio Carranza Rodreguez (Zapopán, Jalisco), home to the 1st Air Group, which is in turn subordinate to the 1st Combat Wing and the unnamed BAM No. 10 (Culiacán, Sinaloa) which houses the independent 11th Air Group.

The South-Eastern Air Region is operationally the most important and includes BAM No. 2 General Antonio Cardenas Rodriguez (Ixtepec, Oaxaca) with Squadrons 211 and 212 of the 10th Air Group, which is in turn part of the 2nd Fighting Wing; BAM No. 4 General Eduardo Aldasoro Suarez (Cozumel, Quintana Roo) with Squadron 201 of the 4th Air Group; BAM No. 6 General Angel H. Corzo Molina (Tuxtla Gutiérrez, Chiapas), the HQ of the 4th Air Group and home to 205 Squadron; and BAM No. 8 General Roberto Fierro Villalobos (Mérida, Yucatán), HQ of the 10th Air Group, which is subordinate to the 2nd Fighting Wing and home to 210 Squadron. This region also includes Military Air Bases Nos 16 (Copalar, Chiapas) and 17 (Ciudad Pemex, Tabasco), which have no permanently attached units.

Outside the air regions are BAM No. 3 General Alberto L. Salinas Carranza (El Ciprés, Baja California), HQ of the 3rd Air Group and home to 204 Squadron, and BAM No. 9 General Gustavo S. Salinas Camia (La Paz, Baja California Sur) with 203 Squadron of the 3rd Air Group.

BAM No. 12 (Tijuana) houses the third echelon workshops of the Air Force. There is also Logistic Base No. 1 at Mexico City, which is the main depot of the Air Force and houses fourth echelon maintenance. Military Air Bases Nos 2, 5 and 10 also house meteorological stations, and there is another third echelon maintenance facility at Military Air Base No. 5. The remaining bases and stations, most of which are of recent creation, have no permanently attached flying units but are used in accordance with operational requirements.

Officer cadets pursue a four-year course at the Colegio del Aire at Zapopán before commissioning as second lieutenants, and most officers receive some post-graduate training abroad, principally in the United States. The Escuela Militar de Aviación (Air Force Flying School) is also at Zapopán and is

Three Boeing 737s (a Srs 112, 247 and 33A) are used by the Presidential Transport Squadron from Mexico City. This is the 737-112, TP-03/XC-UJJ.

Apart from the Presidential Transport Squadron, the other VIP unit of the 8th Air Group is the Executive Transport Squadron which operates Boeing 727s and Sabreliners, including this Srs 75A.

equipped with three Piper Aztecs, 20 Beech Musketeers, 40 F33C Bonanzas, 20 CAP-10Bs and eight PC.7s. Both the Air Force NCO School and the Escuela Militar de Especialistas de la Fuerza Aérea (Air Force Specialists School), where most enlisted personnel receive their training, are also located at Zapopán. The training schools for most of the other main non-flying elements are at Balbuena. Air Force officers attend the Army's Escuela Superior de Guerra, at Mexico City, for senior post-graduate studies, leading to promotion to field rank or to staff appointments. The combined arms Military School of Maintenance and Supply is also located at Mexico City.

Mexican Air Force

1st Combat Wing		HQ Santa Lucía
1st Air Group		**HQ Zapopán**
SAR Sqn 209	Bell 205, Bell 206, Bell 212, Pilatus PC-6	Zapopán
Special Op Sqn 216	Sikorsky UH-60, MD-530F	Santa Lucía
5th Air Group		**HQ Santa Lucía**
Photo-Recce Sqn 101	Shrike Commander 500	Santa Lucía
7th Jet Fighter Group		**HQ Santa Lucía**
Jet Fighter Sqn 202	Lockheed T-33	Santa Lucía
Air Defence Sqn 401	Northrop F-5E/F	Santa Lucía

2nd Fighter Wing		HQ Ixtepec
4th Air Group		**HQ Tuxtla Gutiérrez**
Fighter Sqn 201	Pilatus PC.7	Cozumel
Fighter Sqn 205	Pilatus PC.7	Tuxtla Gutiérrez
10th Fighter Air Group		**HQ Mérida**
Fighter Sqn 210	Lockheed T-33A	Mérida
Fighter Sqn 211	Lockheed T-33A	Ixtepec
Fighter Sqn 212	Lockheed T-33A	Ixtepec
2nd Air Group		**HQ Pie de la Cuesta**
Fighter Sqn 206	Pilatus PC.7	Pie de la Cuesta
Fighter Sqn 207	Pilatus PC.7	Pie de la Cuesta
3rd Air Group		**HQ El Ciprés**
Fighter Sqn 203	Pilatus PC.7	La Paz
Fighter Sqn 204	Pilatus PC.7	El Ciprés
6th Air Group		**HQ Santa Lucía**
Heavy Transport Sqn 301	Douglas C-118	Santa Lucía
Heavy Transport Sqn 302	Lockheed C-130A	Santa Lucía
8th Air Group		**HQ Mexico City**
Executive Transport Sqn	Boeing 727, Sabreliner	Mexico City
Presidential Transport Sqn	Boeing 757-200, Boeing 737, Boeing 727, Fairchild F27, Bell 212, SA 332L Super Puma	Mexico City
Special Operations Sqn	Beech King Air C-90	Mexico City
9th Air Group		**HQ Santa Lucía**
SAR Sqn 208	IAI Arava, Maule MX-7-80	Santa Lucía
Medium Transport Sqn 311	Douglas C-47	Santa Lucía
Medium Transport Sqn 312	Douglas C-47	Santa Lucía
11th Special Operations Group		**HQ Culiacán**
Special Operations Sqn 214	Bell 212, MD-530F, Pilatus PC-6	Culiacán

NOTE: During 1996/97 the Mexican Air Force received 18 Bell 206 and 73 Bell UH-1H helicopters from the United States plus 12 Mi-8s and four Mi-17s from the Russian Federation. These aircraft are reported to equip new units which may be distributed between the military air bases which hitherto had no permanently attached elements.

Mexican Naval Aviation

Like the Air Force, Mexican Naval Aviation has undergone a recent reorganisation. It currently has a personnel strength of 1,050 operating a total of approximately 80 aircraft, organised in six shore-based and one embarked squadrons, an independent flight and a flying training school. The Mexican Marines, who number 8,650 from a total naval manpower of 54,200, include a Paratroop Regiment of two battalions.

There are naval air bases at Mexico City, Las Bajadas (Vera Cruz), Tulum, Campeche, Chetumal, Puerto Cortés, Isla Mujeres, La Paz, Salina Cruz and Tapachula. Not all house operational units on a permanent basis.

The 1st Naval Air Squadron is based at Chetumal and is primarily a maritime patrol unit equipped with armed CASA 212s. The 2nd Naval Air Squadron, based at Mexico City, is the main transport element of the force, with a mixture of CASA 212s, FH-227s, and a single example of the Beech King Air 90, and a pair of Rockwell Turbo Commander 1000s. The 3rd Naval Air Squadron, based at Las Bajadas, Vera Cruz, is primarily a maritime patrol unit. Las Bajadas is also the shore base of the 1st Embarked Naval Air Squadron, which deploys 11 MBB BO 105Cs aboard the helicopter-capable units of the Mexican fleet; of the un-numbered Search and Rescue Squadron; of the 1st Rotary-Winged Squadron; and of the Escuela de Aviación Naval. The 4th Naval Air Squadron, based at La Paz, Baja California, is a large general-purpose unit; the 5th Naval Air Squadron, at Campeche, is primarily a maritime patrol unit; and the 6th Naval Air Squadron, at Tapachula, is a small unit with a primary communications function. The 2nd and 3rd Rotary-Winged Squadrons are recently-created units, equipped with Russian Mi-8 helicopters and based at

Teacapán, Sinaloa and Chetumal, respectively.

The Naval Academy at Vera Cruz incorporates courses for Naval Aviation officers, the basic duration of which is four years. There is a Naval Staff College at Mexico City and there are schools for Naval Aviation enlisted personnel and petty officers at Vera Cruz. All flying training is carried out at the Escuela de Aviación Naval, at Vera Cruz.

Mexican Naval Aviation

1st Naval Air Sqn, Chetumal
    CASA 212
2nd Naval Air Sqn, Benito Juarez IAP, Mexico City
    CASA 212, FH-227, Beech King Air 90, Turbo Commander 1000
3rd Naval Air Sqn, Las Bajadas, Vera Cruz
    CASA 212, Piper Aztec
4th Naval Air Sqn, La Paz, Baja California
    Beech Baron, Beech Twin Bonanza, CASA 212, Cessna 206, Cessna 337, Cessna 441, Piper Aztec, SA 319 Alouette III
5th Naval Air Sqn, Campeche
    CASA 212, Cessna 402, Cessna 404
6th Naval Air Sqn, Tapachula
    Beech Baron
1st Embarked Naval Air Sqn, Las Bajadas, Vera Cruz
    BO 105
SAR Sqn, Las Bajadas, Vera Cruz
    AS 550
1st Rotary-Winged Sqn, Las Bajadas, Vera Cruz
    Mil Mi-8
2nd Rotary-Winged Sqn, Teacapán, Sinaloa
    Mil Mi-8
3rd Rotary-Winged Sqn, Chetumal
    Mil Mi-8
Tulum Naval Air Flight, Tulum
    Tonatiuh, Maule MX-7
Escuela de Aviación Naval, Las Bajadas, Vera Cruz
    Beech Baron, Beech Bonanza, MD 500E, Cessna 152, Maule MX-7, Valmet L-90

The Mexican Navy operates 11 CASA 212-300s, for a variety of roles including maritime patrol and transport. This example belongs to the 3rd Naval Air Squadron based at Las Bajadas, Vera Cruz.

CENTRAL AMERICAN AND CARIBBEAN MILITARY AVIATION

The official end to armed confrontation in Central America has resulted in cuts to military budgets. As a result, little money is available, in some cases not enough even to cover the day-to-day operation of equipment, much less the acquisition of new aircraft. In Central American countries, infantry get the lion's share of the military budget: the expense of a single helicopter might very well keep an infantry unit supplied and operational for a long time.

Guatemala, El Salvador and Nicaragua are in real need of new helicopters, transport aircraft and some sort of interceptor/ground support aircraft, to replace ageing and obsolescent equipment which, in some cases, has seen more than 25 years of service. Cuba has the largest and most powerful air arm in the region, but it is limited by funding problems in the post-Cold War era and the associated ending of Soviet assistance. The Dominican Republic has become a minor player, after many years of regional prominence. Most other island countries have aircraft for Coast Guard, transportation and SAR. Their politics are more sedate and less violent than those of the Isthmian nations, and thus combat aircraft are not really necessary. Even with obsolete equipment and limited resources, these air arms still survive. Their roles include territorial border protection, SAR, civic protection, and providing transportation to isolated areas.

The interdiction of air, land and sea drug traffic is fast becoming a new mission for the air arms of the region, and one which will remain important for a long time. It is unlikely that these countries will soon introduce large numbers of combat-capable aircraft and helicopters. Acquisitions, if any, will be geared towards transportation and economic zone protection, using rugged aircraft that are relatively cheap to acquire and easy to operate.

Light communications aircraft, such as this Costa Rican Cessna U206G, form a large percentage of many of the air arms in the region.

Few countries in the area operate dedicated combat aircraft. Many use trainers fitted with hardpoints for light strike duties, such as the Guatemalan PC-7s.

The majority of transport aircraft operated are of lesser capacity than this Honduran C-130A, and are usually second-hand commercial designs.

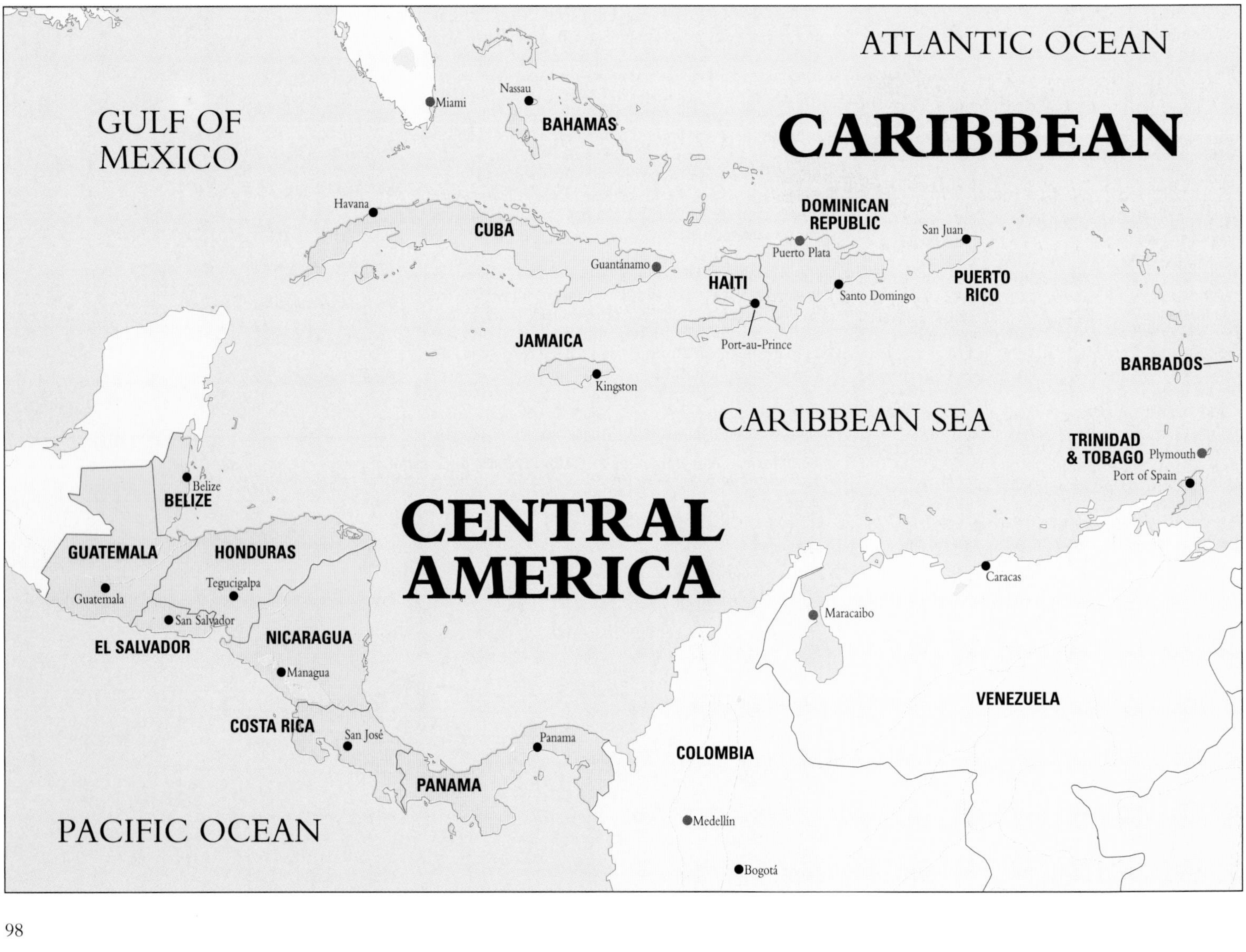

BAHAMAS

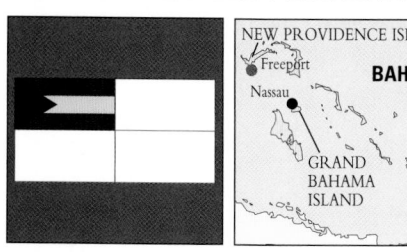

Capital: Nassau
Population: 280,000
Land area: 10070 km² (3,888 sq miles)
Major city: Freeport

Cessna 421C Golden Eagle DF-1001 is one of two aircraft of the Royal Bahamas Defence Force Air Wing, flying from Nassau International.

Air Wing, Royal Bahamas Defence Force

Tasked with mainly drug interdiction, surveillance and internal security roles over the 700 or so islands that make up the Bahamas, an air wing of the Royal Bahamas Defence Force has evolved since initial acquisition of three civil-registered Aero Commander 500 light twins following BDF formation in 1982. By

Royal Bahamas Defence Force		
Air Wing, Nassau International Airport		
Cessna 421C Golden Eagle	1	DF-1001/01
Cessna 404	1	DF-1002

Royal Bahamas Police Force		
Air Element		
Colemill PA-31-325 Panther	2	RBPF-1 & -2

early 1990, they were supplemented by the first military-registered Cessna 404 light twin (DF-1002), soon followed by a slightly larger Cessna 421 (DF-1002), and based at Nassau International Airport. They support the Royal Bahamas Police Force, which also operates a pair of impounded Colemill Panther utility upgrades of the Piper Navajo twin.

BARBADOS

Capital: Bridgetown
Population: 250,000
Land area: 430 km² (166 sq miles)
Major city: Speightstown

Barbados Defence Force

Barbados gained its independence from the United Kingdom in 1966. No regular military forces had been maintained prior to the formation in 1978 of the Defence Force and its territorial militia, the Barbados Regiment.

The Barbados Defence Force contributes to the Eastern Caribbean's Regional Security System, established after the 1983 US-led invasion of Grenada. The

Barbados Defence Force

Air Element of Barbados Defence Force			
Cessna 402C	1	utility	8P-DFA c/n 0427

small air wing is part of the Defence Force and uses a twin-engined Cessna 402C which was delivered in 1981. It is based at the island's only airfield with a permanent-surface runway, the Grantley Adams International Airport.

BELIZE

Capital: Belmopán
Population: 230,000
Land area: 22,800 km² (8,804 sq miles)
Major city: Belize City

Formerly known as British Honduras, Belize became an independent state in 1981. Guatemala claims the entirety of Belize's territory. Until October 1994 Britain guaranteed Belizean independence by the presence of a small military garrison.

Although permanent British forces have withdrawn, the UK still maintains a commitment to protect Belize in the event of aggression.

Belize Defence Force

The Air Wing, which is an integral part of the tiny tri-service Belizean Defence Force, numbers about 50 people and is equipped with two armed Pilatus/Britten-

Belize Defence Force

Air Wing, Belize International Airport		
P/B-N BN-2A Islander	1	transport
P/B-N BN-2B-21 Defender	2	COIN
Slingsby T.67M-200 Firefly	1	trainer

Norman BN-2B Defenders (BDF-01 and -02) and a single Pilatus/Britten-Norman BN-2A Islander transport. A Slingsby T.67M-200 Firefly was acquired for training in 1996. In addition to its own fleet, the Belizean government also effectively has the use of two Ayres S.2RHG-65s Turbo Thrush NEDS aircraft (N3100E and N3100N), a Bell UH-1H helicopter (N81526) and a Cessna 208 (N3090M) of the US Drug Enforcement Administration, which is based in Belize in connection with the campaign against narcotics traffickers. There was an earlier version of the Ayres Thrush, the S.2RT (N-3090M) which first came into use around April 1988, but did not have the 'stealthy' capabilities of the S.2RHGs. A Cessna O-2, registered V3-1KW, is also employed in anti-drug operations.

All aircraft are based at Belize City. Training is undertaken there, with assistance provided by Britain and the US. Airports at Belmopán and Punta Gorda serve as refuelling points for the aircraft of the DFAF, and there are over 30 usable air strips.

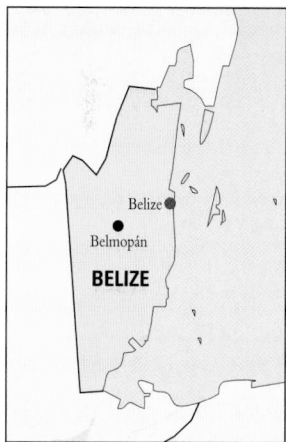

The Belize Defence Force Air Wing uses a single Cessna O-2 Skymaster in the war against drug traffickers. Based at Belize International airport, the aircraft carries a civil registration and may not belong to the Air Wing, which provides its aircraft with an identity prefixed with 'BDF'. The aircraft may have been transferred from USAF stocks.

COSTA RICA

Capital: San José
Population: 3.5 million
Land area: 50660 km² (19,561 sq miles)
Major cities: Alajuela, Cartago, Heredia, Limón, Puntarenas

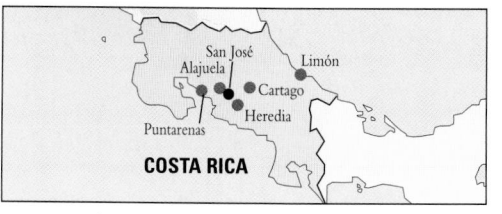

Costa Rica is the most developed country in Central America. Following a bloody civil war in 1948, Costa Rica abolished its armed forces and replaced them with a paramilitary Civil Guard. The Sección Aérea de la Guardia Civil Costaricuense was established, on a *de facto* basis, during the 1955 invasion of Costa Rica by Nicaraguan-supported expatriates.

Sección Aérea de la Guardia Civil Costariquense

Today the Air Section is an integral part of the Civil Guard and is commanded by a Director of colonel or lieutenant colonel rank. It operates from three bases: Base Aérea No 1 (Liberia, Guanacaste), Base Aérea No 2 (San José), and Base Aeronaval (Golfito). Each of the first two bases has a permanent staff of approximately 30, the naval air base

A total of three DHC-4 Caribou has been used by the Costa Rican forces over the years. Only this example is still active.

having a staff of 50. The three Cessna 337s, which are the only combat aircraft operated by the Sección Aérea, are based at the latter installation. Otherwise,

Wearing Fuerza Publica – Vigilancia Aérea titles, this Hughes 369E is one of a pair operated by the Sección Aérea de la Guardia Civil Costariquense. Few, if any, of the aircraft wear the national roundel.

Sección Aérea de la Guardia Civil Costariquense

reports to Ministerio de Seguridad Publica

Base Aérea No 1	**Liberia, Guanacaste**	
Base Aérea No 2	**San José – Juan Santamaria Int'l Airport**	

(aircraft are deployed to the above bases as required)

Colemill Panther PA-31 Navajo	2	communications
Cessna U206G	2	communications
de Havilland Canada C-7A Caribou	1	transport
Hiller FH-1100	1	communications
Grumman American AA-5A	1	communications
Hughes 500D	2	communications
Mi-17MT 'Hip-H'	1	transport
Piper Aztec	1	communications
Piper PA-31T	1	communications
Piper Seneca	1	communications
Piper PA-32-300 Cherokee Six	2	communications
Soloy (Cessna) U206G	2	communications
Base Aeronaval, Golfito		
Cessna 337 (O-2As ?)	3	light attack

flying units are deployed in accordance with operational requirements. The force has recently been expanded, but a CASA C.212 light transport and two Cessna T-41 trainers, reputed to be on order, have not materialised.

All personnel receive their basic military training at the National Police School, at San José. Some flying training is carried out by the Air Section but most flying and maintenance personnel have received the bulk of their training abroad, principally in the United States, Venezuela and Germany.

CUBA

Capital: Havana
Population: 11 million
Land area: 110860 km²
(42,806 sq miles)
Major cities: Camagüey, Holguín, Santiago de Cuba

Establishment of a military air arm was proposed in 1915, with the first squadron of the Cuerpo de Aviación activated in May 1919. The service was subdivided into the Aviación del Ejército (Army Air Corps) and Aviación Naval (Naval Air Arm) after the 1933 revolution. Cuba sided with the Allies in World War II and signed the 1947 Rio Treaty of Mutual Defence. In 1955 another reorganisation combined both army and naval aviation elements in a semi-autonomous force with the title of Fuerza Aérea Ejército de Cuba (Cuban Army Air Force). After the revolution in 1959, the name of the force was changed to Fuerza Aérea Revolucionaria. Following the 'Bay of Pigs' debacle and Castro's unequivocal espousal of Communism, large numbers of Soviet combat, transport and training aircraft were delivered. The air force, now known as the Defensa Anti-Aérea y Fuerza Aérea Revolucionaria/DAAFAR (Revolutionary Air and Anti-Aircraft Defence Force), remained a semi-autonomous adjunct of the Revolutionary Army until 1972, when it was raised to co-equal status with the army and navy.

Defensa Anti-Aérea y Fuerza Aérea Revolucionaria

The Commander of the DAAFAR ranks as a Vice Minister and reports directly to the Chief of the Revolutionary Armed Forces General Staff. The Air Force Commander controls the General Staff of the Revolutionary Air Force and Air Defence Forces, and

a subordinate command structure.

Cuban security is both obsessive and all-pervasive. It is known that the Cuban Revolutionary Air Force divides the country into three territorial commands, known as air zones, and that each of these contains an air brigade made up of a variable number of air regiments and independent squadrons. Each air regiment has a minimum front-line strength of 30 aircraft, the independent squadrons being of variable composition. The Air Force currently numbers approximately 15,000 men and women, operating 530 aircraft. The major bases of the Revolutionary Air Force are San

Antonio de los Baños, Havana, San Julian, Santa Clara, Sancti Spiritus, Cienfuegos, Camagüey and Santiago de Cuba.

Each of the three air zones has an Anti-Aircraft Missile Brigade. There is also an Anti-Aircraft Missile Brigade for the defence of the capital. The Missile Brigades each contain three battalions and comprise about three dozen batteries with SA-3, SA-6, SA-7, SA-9 and SA-13 SAMs. There is also a comprehensive electronic early-warning system.

Operationally, the Interceptor Squadrons are subordinate to Air Defence Command, which also

Above: Cuba has a single squadron of MiG-29s, the 231º Escuadrón de Caza, based at San Julian. The number of aircraft delivered to the air force has been reported as between 12 and 36.

Below: The majority of Cuba's combat fleet consists of nine squadrons of MiG-21s in -21bis (such as 665), -21MF and -21PFM sub-variants.

Left: San Julian is also the home base of the single squadron of MiG-23MF 'Flogger-B' interceptors.

Above: Tactical reconnaissance is conducted by the MiG-21 fleet equipped with pod-mounted cameras fitted onto the central hardpoint.

Left: The Cuban air force operates two-seat conversion trainer versions of all its main combat types. '702' is a MiG-23UB 'Flogger-B'.

Defensa Anti-Aérea y Fuerza Aérea Revolucionaria

Operationally, squadrons are sub-ordinated to Air Defence Command, Tactical Air Command, Logistic Support Command and Air Training Command.

Zona Aérea Oeste
2ª Brigada de Guardia 'Playa Giron'

21° Regimiento de Interception

211° Escuadrón de Interception	MiG-21bis/UM	San Antonio de los Baños
212° Escuadrón de Interception	MiG-21bis/UM	San Antonio de los Baños

22° Regimiento de Interception

221° Escuadrón de Interception	MiG-21bis/UM	Baracoa
222° Escuadrón de Interception	MiG-21bis/UM	Barocoa

23° Regimiento de Caza

231° Escuadrón de Caza	MiG-29/UB	San Julian
232° Escuadrón de Caza	MiG-23MF/UB	San Julian

24° Regimiento de Apoyo Táctico

241° Escuadrón de Apoyo Táctico	MiG-23BN/UB	Guines

25° Regimiento de Transporte

251° Escuadrón de Transporte	Il-76, Yak-40, An-32, An-24, An-2	Jose Martin
252° Escuadrón de Transporte	An-2, An-26	San Antonio de los Baños

26° Regimiento de Helicópteros

261° Escuadrón de Helicópteros de Propósitos Generales	Mi-8	Havana
262° Escuadrón de Helicópteros de Propósitos Generales	Mi-8	Havana

Zona Aérea Central
1ª Brigada de Guardia 'Batalla de Santa Clara'

11° Regimiento de Interception

111° Escuadrón de Interception	MiG-21bis/UM	Santa Clara
112° Escuadrón de Interception	MiG-21bis/UM	Santa Clara

12° Regimiento de Interception

121° Escuadrón de Interception	MiG-21PFM/UM	Sancti Spiritus
122° Escuadrón de Interception	MiG-21PFM/UM	Sancti Spiritus

14° Regimiento de Apoyo Táctico

141° Escuadrón de Apoyo Táctico	MiG-23BN/UB	Santa Clara

15° Regimiento de Transporte

151° Escuadrón de Transporte	An-2, An-26	Cienfuegos

16° Regimiento de Helicópteros

161° Escuadrón de Helicópteros de Guerra Anti-Submarina	Mi-14PL	Cienfuegos
162° Escuadrón de Helicópteros de Propósitos Generales	Mi-8	Cienfuegos
163° Escuadrón de Helicópteros de Propósitos Generales (?)	Mi-17	Cienfuegos

Zona Aérea Oriente
3ª Brigada de Guardia 'Cuartel Moncada'

31° Regimiento de Interception

311° Escuadrón de Interception	MiG-21MF/UM	Camagüey

34° Regimiento de Apoyo Táctico

341° Escuadrón de Apoyo Táctico	MiG-23BN/UB	Holgüín

35° Regimiento de Transporte

351° Escuadrón de Transporte	An-2, An-26	Santiago de Cuba

36° Regimiento de Helicópteros

361° Escuadrón de Helicópteros de Ataque	Mi-25	Santiago de Cuba
362° Escuadrón de Helicópteros de Propósitos Generales	Mi-8	Santiago de Cuba
363° Escuadrón de Helicópteros de Propósitos Generales	Mi-8	Santiago de Cuba

Escuela de Vuelo de la DAAFAR

? Escuadrón de Enseñanza Básica	Zlin 326	San Julian
? Escuadrón de Enseñanza Básica de Vuelo	L-29	San Julian
? Escuadrón de Enseñanza Básica de Vuelo	L-29	San Julian
? Escuadrón de Enseñanza de Vuelo Avanzada	L-39C	San Julian
? Escuadrón de Enseñanza de Vuelo de Combate	MiG-17F/UTI	San Julian

controls the Anti-Aircraft Missile Brigades. The Fighter-Bomber Squadrons are subordinate to Tactical Air Command, which also controls the single Helicopter Attack and the Helicopter ASW Squadrons. Transport units are subordinate to Logistic Support Command, and training units to Air Training Command.

Most training is carried out at the Aviation Cadet School at San Julian, which offers courses in sophisticated aircraft and missile operation, with specialist technical and flying training schools. The period of training for officers varies between four and five years according to specialisation. The Instituto Técnico Militar, at Havana, offers comprehensive specialist training in communications, avionics and aeronautical engineering, and for anti-aircraft troops. All flying personnel formerly received part of their training in the Soviet Union, where combat aircrew also underwent additional advanced training. Officers receive post-graduate training at specific points in their careers at the 'General Maximo Gomez' Academy, such training also formerly being supplemented by advanced courses in the Soviet Union. Following the collapse of the USSR, all foreign training assistance to Cuba was withdrawn. The previous dependence of the Revolutionary Air Force on foreign training aid is reflected in its small inventory of training aircraft relative to its overall size. Lack of fuel has all but put an end to flying training and, even without this handicap, the available number of training aircraft would be inadequate to produce a sufficient output of trained aircrew to replace natural attrition from retirement or other causes.

Like the Revolutionary Army and Navy, the Revolutionary Air Force is well trained and motivated. It is also the best-equipped military air arm in Latin America. By its nature, it must, however, be affected to the greatest degree of all the Revolutionary Armed Forces by the current shortage of replacements and spares and the almost total lack of fuel.

DOMINICAN REPUBLIC

DOMINICAN REPUBLIC

Capital: Santo Domingo
Population: 7.9 million
Land area: 48380 km² (18,681 sq miles)
Major cities: Puerto Plata, San Francisco, Santiago

Fuerza Aérea Dominicana

An Aviation Company of the Dominican army was formed in 1933, equipped only with second-line aircraft until 1942, when the Dominican Republic began to receive aircraft under Lend-Lease. In 1947 a well-equipped group of Dominican exiles threatened to invade the Dominican Republic from Cuba, and President Trujillo's agents managed to obtain a number of war-surplus combat aircraft in Britain. With this equipment, the Compañía de Aviación expanded to become the Cuerpo de Aviación Militar Dominicana. Dominica signed the Rio Treaty of 1947, while further aircraft were obtained from commercial sources. By 1952, the Cuerpo de Aviación

The FAD's Escuadrón de Entrenamiento received 11 ex-US Navy T-34Bs in 1980. Only about five are thought to be operational at any time.

Militar Dominicana had become independent of the army, as the Fuerza Aérea Dominicana, and had approximately 240 aircraft.

Following the assassination of Trujillo in 1961, the brief golden age of the Dominican Armed Forces came to an end. By 1963 the Fuerza Aérea Dominicana had shrunk to 110 aircraft. Nevertheless, 70 of these were combat types and the air force was active in the 1965 civil war. During the years of democratic rule which have followed the civil war, the Dominican Air Force has continued to decline.

The Chief of Staff of the Air Force is effectively both its administrative and operational commander. He reports to the Secretary of State for National Defence through the Under Secretary of State for War.

The FAD divides the country into two Air Zones. The Southern and Northern zones have their respective headquarters at San Isidro, to the east of the capital of Santo Domingo, and Santiago de los Caballeros, in the northwest. The air force currently has 70 aircraft, manned by approximately 4,200 personnel of all ranks. It is organised into an Air Command, a Base Defence Command, a Combat Support Command, a Maintenance Command, a Command of Special Forces, and a

Base Security Command.

Air Command is responsible for all flying operations. There are military air bases at Santiago, Barahona, Puerto Plata, Azua, La Romana, La Vega, Monte Cristi and San Cristóbal, although most aircraft are based at San Isidro (Santo Domingo). Only San Isidro, Santiago and Barahona air bases have permanently deployed operational units. All training is carried out at San Isidro.

Air Force cadets undergo the first two years of their four-year course at the Military Academy, at Haina, before transferring to the Escuela de Aviación Militar at San Isidro air base, for specialist training. Other ranks receive all-through training at the Escuela de Aviación Militar.

Like the Army and Navy, the Dominican Air Force is in urgent need of re-equipment if it is to continue as an effective force.

Below: A single OH-6A Cayuse continues to serve with the FAD, from a total of six delivered.

Fuerza Aérea Dominicana

Northern Air Zone HQ,
Base Aérea Santiago de los Caballeros
Southern Air Zones HQ, Base Aérea San Isidro

Air Command	
Escuadrón de Combate	8 x A-37B Dragonfly
Escuadrón de Transporte	3 x Douglas C-47,
2 x Rockwell Commander 680, 1 x Mitsubishi MU-2	
Escuadrón de Enlace (Liaison)	3 x Beech Queen Air 80,
1 x Cessna 210 L, 5 x Cessna O-2A, 2 x Piper PA-31 Navajo	
Escuadrón de Entrenamiento (Training) 10 x Beech T-34B Mentor,	
4 x Cessna T-41D Mescalero, 2 x N.A T-6G Texan	
Helicópteros	9 x Bell UH-1H,
1 x Bell 47, 1 x Hughes OH-6A Cayuse, 2 x SA 318C,	
1 x SA 3130, 1 x SA 365 Dauphin (VIP)	

Base Defence Command
4 air defence artillery battalions (20-mm)

Combat Support Command
controls all base services

Maintenance Command
maintenance of all aircraft, vehicles and buildings

Dominican Army

A single Cessna 207 is operated by the Dominican Air Force on behalf of the army

Left: The Helicópteros unit operates the FAD's motley collection of helicopters, among which is its sole Aérospatiale (Eurocopter) SA 365C Dauphin.

GUATEMALA

Capital: Guatemala City
Population: 11.7 million
Land area: 108430 km²
(41,868 sq miles)
Major cities: Antigua, Mazatenango, Puerto Barrios, Quezaltenango

Fuerza Aérea Guatemalteca

In 1920 a French military aviation mission opened a flying training school. The Cuerpo de Aviación Militar de Guatemala was established in 1929, and expanded in 1934. The outbreak of World War II hindered any further expansion until 1942, when Guatemala began to receive Lend-Lease military assistance. Guatemala signed the Rio Treaty of Inter-American Reciprocal Assistance in 1947, and the force's title was changed to Fuerza Aérea Guatemalteca in 1948. Following the suspension of US military aid in 1978 due to Guatemala's human rights record, the country turned to Israel and Argentina as suppliers of defence material. The election of Ronald Reagan as President of the United States (and a 1982 coup which brought a more moderate military junta to power) brought with it a mellowing of US foreign policy towards Guatemala; the formal arms embargo was lifted in January 1983.

The Guatemalan Armed Forces constitute a single institution with army, navy and air force elements. The air force has a nominal personnel strength of about 700, a figure which is misleading as it refers only to aircrew and personnel directly involved in the operation of aircraft. Taking into account logistic and other support personnel (who are nominally part of the army), air force strength is probably in the region of at least 1,000. To this could also reasonably be added the army's two paratroop battalions and four anti-aircraft artillery batteries, plus the Agrupamiento Táctica de Seguridad, all of which come under the operational control of the air force. Including these and their logistic support elements, true manpower is

Of the 12 Pilatus PC-7s acquired in the late 1970s/early 1980s, it is believed that only four are operational at any one time. A total of eight of the type was recorded during August 1997 as still being on charge. Like much of the air force, they suffer from a lack of spare parts.

probably of the order of 3,000.

Front-line types are limited to the survivors of 13 A-37Bs delivered in 1974/75 and 12 PC-7s in 1979/80, bought as trainers but fitted with hardpoints. Most existing aircraft require modernisation and the FAG is plagued by an endemic shortage of spares. Limited equipment has dictated that the air force confines its recent operations and deployment to the counter-insurgency role.

Guatemala has never operated Cessna T-37s, Fouga Magisters or Aerotec Uirapurus, as has been mistakenly reported.

The major air bases are BA La Aurora (Guatemala City airport), Base Militar de Tropas Paracaidistas General Felipe Cruz (San José, Escuintla) and BA Teniente Coronel Danilo Eugenio Henry Sánchez (Santa Elena, Petén), with minor bases at Puerto Barrios, Retalhuleu (Base Aérea del Sur) Huehuetenango and Poptún. There are more than a dozen landing strips suitable for the operation of military aircraft available throughout the country,

Fuerza Aérea Guatemalteca

Ala Fija (Fixed-Wing Aircraft Wing)	
Escuadrón de Ataque	4 x A-37B Dragonfly
(3 operate from La Aurora, 1 from Santa Elena)	
Escuadrón de Entrenamiento y Ataque	4 x Pilatus PC-7
(2 operate from La Aurora, 1 from Santa Elena,	
1 (as trainer) from Retalhuleu – a/c rotate bases)	
	5 x T-35 Pillán
Escuadrón de Avionetas	BA La Aurora
1 x Bellanca Decathlon, 4 x Cessna R 172K (ex-Escuela Militar de Aviación), 1 x Cessna T210M, 1 x Cessna 210 Centurion (TG-MAR), 1 x Gulfstream Aerospace Commander 1000, 3 x King Air 200 (TG-CFA, -MDN, -CPG), 3 x PA-31-350 Navajo, 1 x PA-34 Seneca, 1 x WACO VPF-7	
Mainly formed by aircraft seized from drug traffickers, the exact number of aircraft is uncertain, as they are in the custody of the Judiciary System and their numbers change constantly	
Escuadrón de Transporte	4 x Basler Turbo-67
(3 operate from La Aurora, 1 from Santa Elena), 4 x C-47, 1 x F27 Troopship Mk 400M, 2 x F27-200, 4 x IAI Arava Srs 201	

Ala Rotativa (Rotary-Wing Aircraft Wing)	
Escuadrón de Helicópteros	6 x UH-1H Iroquois
(3 operated at La Aurora, 1 at Retalhuleu, 2 at Santa Elena), 4 x Bell 212 (s/n 001 VIP helo used by Grupo Presidencial), 1 x Bell 214B, 2 x Bell 412, 2 x Bell 206B-3 JetRanger (165/TG-WOI at La Aurora, other at Retalhuleu), 2 x Bell 206L-1 LongRanger (130/TG-WOC at La Aurora, other at Santa Elena)	
(The squadron deploys assets across the country.)	
Escalón de Mantenimiento (Maintenance Wing)	
aircraft maintenance	

and most are occupied sporadically.

Air force officers receive their basic military training at the Escuela Politécnica, the national Military Academy at San Juan Sacatepéquez. Contrary to what is customarily reported, Base Aérea Los Cipresales ceased to exist in the late 1960s as the base for the Escuela Militar de Aviación. The area it occupied is now covered by housing. The Escuela Militar de Aviación (EMA) relocated to Base Aérea del Sur, Retalhuleu several years ago, but it is inactive at this time. Most officers also receive advanced training abroad, either in the United States, Mexico or Venezuela. Specialist ground crews and other support personnel receive their training at the Escuela Técnica de Mecánicos de Aviación, located at Aeropuerto La Aurora, but on the FAG's main base.

The Fuerza Aérea Guatemalteca operates four C-47s, and also four examples of the stretched and turboprop-powered Basler BT-67s. The type flies with Escuadrón de Transporte from La Aurora and from Santa Elena.

HAITI

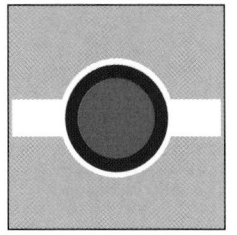
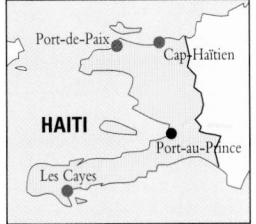

Corps d'Aviation d'Haiti

The Corps d'Aviation d'Haiti was established in 1943, with the help of a US Marine Corps aviation mission. During the years immediately following World War II, small quantities of second-line types were acquired. In 1950 a US Air Force mission arrived in Haiti and

Capital: Port-au-Prince
Population: 6.6 million
Land area: 27560 km² (10,642 sq miles)
Major cities: Cap-Haïtien, Les Cayes, Port-de-Paix

the first combat unit was formed shortly afterwards with four North American F-51D Mustangs.

In 1994, following the US-led OAS invasion and occupation to restore the legitimate President Jean Bertrand Aristide, the Haitian armed forces were formally disbanded. The current status of the former air arm is unknown. The Corps d'Aviation was an integral part of the Haitian armed forces rather than an independent air force. As such, its Commander was subordinate to the Chief of Staff of the Armed Forces in the overall command structure. It consisted of a combat unit, a transport unit and a helicopter unit. Personnel strength was approximately 300 people of all ranks. Bowen Field, Port-au-Prince, was the only permanently occupied base. All officers were trained at the Military Academy at Frères, NCOs being trained at the Camp d'Application. Flying and specialist training were both carried out at Bowen Field.

Corps d'Aviation d'Haiti		
(formally disbanded after 1994)		
The aircraft inventory, most of which was non-operational, consisted of:		
Beech D50 Twin Bonanza	1	communications
Beech B55 Baron	1	communications
Beech King Air 65-90	1	communications
Cessna 150	3	primary training
Cessna 402B	1	communications
Curtiss C-46 Commando	1	transport
de Havilland Canada DHC-2	2	utility transport
de Havilland Canada DHC-6-200	1	transport
Douglas C-47 Dakota	3	transport
Hughes 269C	4	
Hughes 369C	4	
Pilatus/B-N BN-2A Islander	1	transport
Piper PA-34 Seneca	1	communications
SIAI-Marchetti SF.260TP	4	COIN
Sikorsky S-58	4	
Summit Sentry O2/337G	7	COIN

HONDURAS

Capital: Tegucigalpa
Population: 5.8 million
Land area: 111890 km² (43,204 sq miles)
Major cities: La Ceiba, San Pedro Sula, Puerto Cortés

Fuerza Aérea Hondureña

A military flying school was established in 1921 and expanded to become the Military Aviation Service in 1934. In 1954, the Aviación Militar became independent of the army, as the Fuerza Aérea Hondureña. During the brief Honduras-El Salvador 'Football War' of 1969, Honduras gained complete control of the air at an early stage, although its ground forces had to retreat in the face of the superior Salvadorean army. A considerable build-up, including the replacement with jet aircraft of obsolete piston-engined operational material (mainly Vought F4U Corsairs), took place during the 1970s. Since the early 1980s, the Honduran Air Force has also received considerable material and training assistance from the United States.

The Fuerza Aérea Hondureña currently consists of approximately 1,800 military personnel and 400 civilian employees, manning about 130 aircraft. The force is organised into a Fighter Squadron, a Fighter-Bomber Squadron, a Light Strike Squadron, a Transport Squadron, a Communications Squadron and a Helicopter Squadron. The Commander-in-

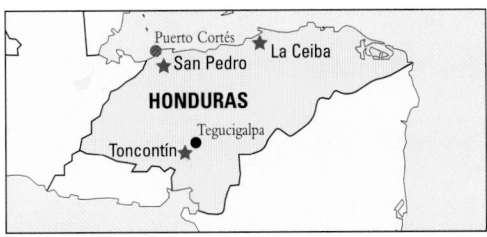

Chief of the Air Force reports to the Commander-in-Chief of the Armed Forces. The air force has its own General Staff which controls three separate divisions dealing with (i) Personnel, Intelligence, Operations and Technical Support, (ii) Air Bases, and (iii) Training.

The major bases are located at Toncontín (Tegucigalpa), Palmerola, Moncada and San Pedro Sula. There are secondary bases at El Aguacate, Cucuyagua, Dursuna, Jamastrán, Trujillo, Puerto Lempira, San Lorenzo, San Marcos, Marcala and Huanpusipi. The air force also maintains radar stations at Amapala and El Lorque, plus a significant presence at each of its four main bases and small detachments at many of the secondary ones, including those at Dursuna, Jamastrán, Puerto Lempira and Marcala.

All training is nominally carried out at Toncontín but the flying elements of the Escuela de Aviación are regularly deployed to each of the other three main air bases and appear to be currently based at 'Enrique Soto Cano' Air Base, Palmerola. Officers receive their initial training at the Escuela Militar, the National Military College, at Tegucigalpa, and their subsequent specialised training at the Escuela de Aviación Militar. Although the flying elements of this institution appear to have moved to Palmerola, various non-flying courses still take place at Toncontín in addition to *ab initio* training for enlisted personnel. Most officer and some selected non-commissioned personnel also

Fuerza Aérea Hondureña

Primer Grupo Táctico, Base Aérea Coronel Hector Caraccioli Moncada (La Ceiba)	
Escuadrón de Apoyo Táctico	A-37B Dragonfly
Escuadrón de Caza	ENAER A-36 Halcón (CASA C.101CC)
Escuadrón de Caza	F-5E/F Tiger II
Escuadrón de Caza	Super Mystère B2
Escuadrón de Helicópteros	6 x Bell UH-1B, 7 x Bell UH-1H, 9 x Bell 412SP, 3 x MD 500D, 1 x Sikorsky S-76C (VIP)
Escuadrón de Transporte	5 x C-47 Dakota, 2 x IAI Arava 201, 1 x C-130A Hercules, 1 x L-188A Electra, 2 x IAI 1121 Westwind
Escuadrón de Enlace (Liaison)	3 x Cessna 172, 3 x PA-31-235 Navajo, 1 x PA-34 Seneca, 3 x Rockwell 114 Commander
Academia Militar de Aviación	4 x C.101BB Aviojet, T-41D Mescalero, 11 x EMB-312 Tucano, 5 x TH-55A Osage
Escuadrón de Comunicaciones	1 x Beech Baron, 2 x Cessna 180, 21 x Cessna 185

receive continuation training abroad.

A mystery surrounds the current status of the Honduran Air Force's F-5s, of which it originally operated 10 examples of the E and two of the F variant. A number were delivered by sea to Chile in 1995, but after being unloaded at Valparaiso disappeared from view. As late as September 1996, at least six were still observed in Honduran service. An unknown number of A-36 Halcóns were obtained from Chile as part of this deal.

During 1998 the decision was taken to return at least 11 Super Mystère B2s to service, possibly to make up the shortfall created by the missing F-5s.

Above: Withdrawn and stored from 27 January 1996, the Fuerza Aérea Hondureña Super Mystère B2s were returned to service in 1998.

Left: As part of the F-5E Tiger II consignment, Honduras received a pair of F-5F conversion trainers. The mystery surrounding the Chilean deal makes it difficult to say if both are still in service.

JAMAICA

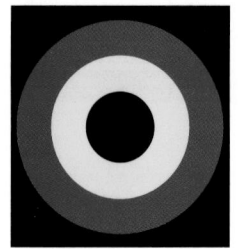

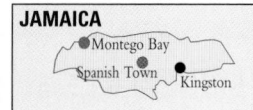

JAMAICA
Montego Bay
Spanish Town
Kingston

Capital: Kingston
Population: 2.6 million
Land area: 10830 km²
(4,182 sq miles)
Major cities: Montego Bay,
Spanish Town

Jamaica Defence Force Air Wing

Jamaica Defence Force

Fixed-Wing Flight, Kingston		
Beech King Air A100	1	VIP
Cessna 210M Centurion	1	communications
Cessna 337G	1	communications
Pilatus/B-N BN-2A Islander	1	maritime patrol
Helicopter Flight, Kingston		
Bell 206A/B JetRanger	4	communications
Bell 212	3	SAR
Bell 222	1	communications
Bell UH-1H	4 (?)	communications
Eurocopter AS 355N Twin Squirrel	4	communications

The island of Jamaica, formerly the largest British possession in the Caribbean, became independent in 1962. It maintains a very small tri-service defence force, with a minuscule air arm. The Jamaican armed forces are known as the Jamaica Defence Force and

It is believed that Jamaica received four UH-1Hs prior to 1995, most likely from surplus US Army stocks.

comprise land, sea and air elements. The Air Wing was formed in 1963. As an integral element of the Defence Force, the Commander of its land element reports to the Governor General, via the Minister for Defence. The two main bases of the JDFAW are both at Kingston. There is also an air strip at Montego Bay, which is not permanently manned by JDF elements. Limited fixed-wing training is undertaken at Manley International Airport, with helicopter training at Up Park Camp. Most flying and technical personnel receive at least part of their training abroad.

NICARAGUA

Puerto Cabezas
NICARAGUA
Estelí
Matagalpa
Bluefields
Leon
Managua
Granada
El Bluff

Capital: Managua
Population: 4.4 million
Land area: 120254 km² (46,434 sq miles)
Major cities: Bluefields, Granada, Leon, Mataglapa

Fuerza Aérea Nicaragüense

Fuerza Aérea Nicaragüense

Much of the current aircraft inventory is of doubtful serviceability

Escuela de Aviación		
Cessna 172	1	communications
Cessna 185	1	trainer
Cessna T-41D Mescalero	2	primary training
PA-18 Super Cub	2	utility
PA-28	2	utility
PZL/Mil Mi-2 'Hoplite'	4	light helicopter
SIAI-Marchetti SF.260W	4	basic trainer/light strike
Escuadrón de Transportes		
Antonov An-2 'Colt'	6	communications
Antonov An-26 'Curl'	3	medium transport
Cessna U-17B	1	communications
Cessna 337	4	light transport
Cessna 421	1	communications
Escuadrón de la Rotativa		
Mil Mi-17 'Hip-H'	15	armed transport helicopter

The Fuerza Aérea de la Guardia Nacional was formed in 1938. From 1942 small numbers of trainers and light transports were acquired from the United States, and by 1945 a total of 20 aircraft was on strength. In 1952 a US aviation mission arrived in Nicaragua and, subsequently, additional quantities of trainers and transports were received, followed by P-38, P-47 and P-51 fighters. For some years to follow, the

Nicaraguan Air Force was the strongest in Central America. Many aircraft fell victim to rebel ground fire during the civil war which finally toppled the Somozas, and others were flown into exile by their pilots. The Sandinista years saw delivery of much Eastern Bloc aircraft and equipment.

Following the electoral defeat of the Sandinistas in 1990, much of the Eastern European equipment was disposed of. Despite the change of government, the

Nicaraguan Air Force retained the title of Fuerza Aérea Sandinista until 1996, when it reverted to Fuerza Aérea Nicaragüense. It currently numbers 1,200 personnel and possesses approximately 30 aircraft organised into a single Transport Squadron, a Helicopter Squadron and a Training/Light-Strike Squadron. Most of its aircraft are unserviceable. Following Soviet and Cuban practice, the air force is also primarily responsible for air defence.

The principal base of the FAN is at 'Augusto César Sandino' International Airport, Managua. Other major bases are at Punta Huete, to the northeast of Managua, and Puerto Cabezas, on the Atlantic coast. In addition, there are military airfields at Estelí, Bluefields, Montelimar, La Rosita, El Bluff and Puerto Sandino. Nicaragua is the only country in Central America to have anything approaching an air early-warning system, boasting AEW and intercept control facilities at Masaya, Toro Blanco, Estelí and El Bluff, plus a coastal surveillance radar station at El Polvón.

Personnel receive their basic training in the central school system of the Nicaraguan Armed Forces. The major elements of this are the 'Carlos Agüero Echeverría' Military Academy; the Centro de Estudios Militares 'Comandante Hilario Sánchez Vázquez', which provides technical and specialised training; and the 'Eduardo Contreras' Basic Military Training School.

PANAMA

Colón
Panama City
Bocas del Toro
PANAMA
David
Chitré
La Palma

Capital: Panama
Population: 2.7 million
Land area: 75990 km²
(29,342 sq miles)
Major city: Colón

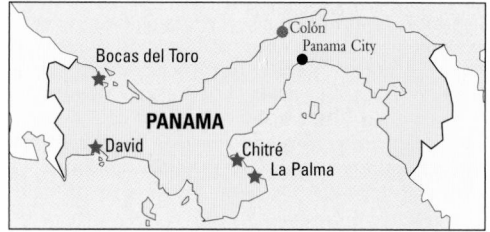

Servicio Aéreo Nacional

Although theoretically independent, Panama has remained a virtual dependency of the US, whose military occupation of the Canal Zone has always been resented by Panamanians. Under the Carter-Torrijos Canal Treaty of 1979, the United States is supposed to hand over both Canal and Canal Zone to Panama at midnight on 31 December 1999. The Treaty, however, allows the United States to intervene if Panama shows itself to be unable or unwilling to defend the Canal and guarantee its availability to shipping.

An air service, established as part of the National Police during the early 1960s, became known as the Panamanian Air Force in 1969. After the destruction

and subsequent formal abolition of the Panamanian Defence Forces after the 1989 United States intervention, the National Air Service, as the former air force is now known, retains a total of only 16 fixed- and rotary-winged aircraft. The Service is an integral part of the new paramilitary police force, known as the Panamanian Public Forces, to the commander of which its commander reports.

The major bases of the Panamanian National Air Service are Tocumen International Airport and Albrook Air Force Base, Panama City. Additional air strips are located at Bocas del Toro, David, Santiago, Chitré and La Palma. Most training is carried out at Albrook Air Force Base. Aircrew are believed to receive their advanced training abroad.

Panamanian Public Forces

Servicio Aéreo Nacional (SAN)		
Bell 205A	2	helicopter transport
Bell 212	3	helicopter transport
CASA C.212-200 Aviocar	2	transport
CASA C.212-300 Aviocar	3	transport
Airtech CN.235-2M	1	transport
ENAER T-35D Pillán	6	primary training
Pilatus/B-N BN-2D Defender	1	transport
Government		
Boeing 727-44	1	(civil reg) VIP
Gulfstream Aerospace Gulfstream II	1	(civil reg) VIP

Above: Training for the Servicio Aéreo Nacional's pilots is undertaken on six ENAER T-35D Pilláns.

Left: A single Gulfstream II is based at Tocumen International Airport as the presidential transport.

EL SALVADOR

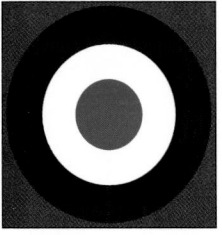 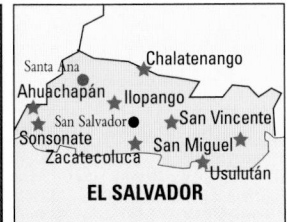

Capital: San Salvador
Population: 5.7 million
Land area: 20720 km² (8,001 sq miles)
Major cities: San Miguel, Santa Ana

Fuerza Aérea de El Salvador (FAES)

El Salvador formed a Military Aviation Service in 1922. Following its signature of the Rio Treaty of 1947 (which provided for mutual defence among American states, including the US), El Salvador benefited from the activities of a US Air Mission and increased transfers of aircraft under the Mutual Defence Assistance Program. The Military Aviation Service was renamed the Fuerza Aérea Salvadoreña and acted as an independent component of the Armed Forces. Unlike the army, which enjoyed a certain regional primacy which it more than vindicated in the 1969 war with Honduras, the Salvadorian Air Force was among the least significant in Central America. Its performance in the 1969 war was mediocre.

Isolated, sporadic guerrilla and terrorist activities which had occurred with increasing frequency from the late 1970s rapidly developed into a full-scale civil war. The anti-government forces soon united under the Marxist-led FMLN (Frente Farabundo Martí para la Liberación Nacional), thus guaranteeing the hostility of the United States, and substantial military backing for El Salvador. US aid to El Salvador had been forthcoming since the 1970s, but by 1984 it had leaped to $196.6 million. The US supplied six UH-1Hs in 1980 and four in 1981, to be used as gunships. From February 1982 onwards US aid to the FAS accelerated with the delivery of eight A-37B Dragonflies, 12 Bell UH-1Hs, four Cessna O-2As and three Fairchild C-123Ks. Subsequent UH-1H deliveries brought the air force's operational fleet to approximately 40. Two AC-47s were delivered via the United States, in December 1984, to supplement the three C-47 transports in use. Although the civil war ended in a truce of mutual exhaustion in 1990, the Salvadorian Armed Forces remain largely orientated towards internal security.

The Salvadorian Air Force, which currently has a personnel strength of approximately 2,000 people of all ranks, operates 138 aircraft. Only a small proportion of these are serviceable, and are divided between two brigades. One comprises an attack squadron with A-37s and a FAC squadron with O-2s, while the other brigade has transports, helicopters and trainers. The air force includes a battalion of paratroops and exercises operational control over the army's anti-aircraft battalion. The Air Force Commander is subordinate to the Chief of the General Staff in the overall chain of command but enjoys almost absolute independence in operational matters.

In addition to the main base at Ilopango, there is a new base at Comalapa and air strips at San Miguel, Ahuachapán, Sonsonate, Zacatecoluca, San Vicente, Chalatenango and Usulután. As the air force contracts, on a peace footing, it may concentrate all its forces at these two locations. All training is carried out at Ilopango. Officer personnel complete the four-year course at the 'Capitán General Gerardo Barrios' National Military Academy, at San Salvador, before commencing flying training at the 'Capitán Reynaldo Cortez' Escuela de Aviación Militar or other specialist training at the Escuela de Especialización de la Fuerza Aérea, both of which are located at Ilopango.

Fuerza Aérea Salvadoreña

Headquarters	**Ilopango**
Cessna 210 (2), Rockwell Commander 114 (1), Merlin IIIB (1 VIP)	

Brigada Aérea 1, Ilopango

Grupo de Transporte
Basler Turbo-67 (4), AC-47 (2), Cessna T-41D (1), Cessna 337 (1)
Grupo Helicoptero
Bell UH-1H (?), Bell UH-1M (?)
approximately 20 Hueys are active in total
Escuela de Aviación Militar
SOCATA 235 Rallye (4), ENAER T-35B Pillán (4), Hughes TH-55 (5)
Fond de Actividades Especiales
MD 500MD (2), Bell UH-1M (1), SOCATA 235 Rallye (1)

Brigada Aérea 2, Comalapa

Attack Squadron
Cessna A-37B (9), Fouga CM.170 (2, potentially airworthy)
Forward Air Control Squadron
Cessna O-2A (8)

There is also an Airborne Forces School at Ilopango. Specialist other ranks receive their training at the Escuela de Especialización. Most officer personnel also pursue additional training abroad.

Above: The FAES has a number of Dakotas, including Basler Turbo 67s and AC-47s. This example is armed with three 0.50-in machine-guns firing out of the door and the last two windows.

Below: Eight Cessna O-2s equip the FAC squadron. Several types have recently been withdrawn from service, including the DC-6, Ouragan (still on strength but grounded), T-34 and C-123.

TRINIDAD & TOBAGO

 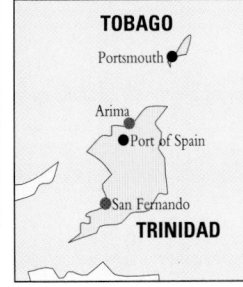

The island republic of Trinidad & Tobago lies off the coast of Venezuela and is geographically part of South America. It gained its independence from Britain in 1962 and a republic was declared in 1976.

Capital: Port of Spain
Population: 1.1 million
Land area: 5130 km² (1,981 sq miles)
Major cities: Arima, San Fernando

Trinidad & Tobago Defence Force

The Trinidad & Tobago Coast Guard formed an air element in 1966. In 1977 the Air Wing, although still very small, became an independent element of the Trinidad & Tobago Defence Force. The President is titular Commander-in-Chief of the Defence Force. The Commander of the Air Wing reports to the Prime Minister, as Minister of National Security, via the Chief of the Defence Staff. There are air bases at Piarco (Trinidad) and Crown Point (Tobago). The current aircraft inventory of the Trinidad & Tobago Defence Force Air Wing comprises three fixed-wing aircraft

Trinidad & Tobago Defence Force

Air Wing			
Cessna 172M	1	communications	9Y-TES
Cessna 310R	1	coastal patrol	9Y-TEW
Cessna 402C	1	coastal patrol	TTDF-201
National Helicopter Services Ltd			
MBB BO 105CBS	2		9Y-THP and 9Y-TIW
MBB BO 105CBS-4	1		9Y-TIC
Sikorsky S-76C	2		9Y-TGW and 9Y-TGX

and fuor helicopters. The helicopters are operated by National Helicopter Services Ltd, a state-owned quasi-commercial organisation which earns revenue by chartering out elements of its fleet to private users when they are not required by the Air Wing.

South America

COSTA RICA

Barranquilla

Maracaibo

Caracas

TRINIDAD
& TOBAGO

PANAMA

VENEZUELA

Medellín

Georgetown

Bogotá

GUYANA

Paramaribo

Cayenne

Cali

SURINAM

COLOMBIA

FRENCH GUIANA

Quito

ECUADOR

Guayaquil

Macapá

Belém

Manaus

Fortaleza

PERU

BRAZIL

Recife

ATLANTIC
OCEAN

Lima

Salvador

La Paz

Brasília

BOLIVIA

Belo Horizonte

PARAGUAY

Rio de Janeiro

Asunción

São Paulo

SOUTH
AMERICA

CHILE

Pôrto Alegre

ARGENTINA

Rosario

URUGUAY

Buenos Aires

Montevideo

Santiago

PACIFIC
OCEAN

FALKLAND ISLANDS/
ISLAS MALVINAS

TIERRA DEL FUEGO

SOUTH GEORGIA

Blocked by the US from acquiring state-of-the-art
American combat aircraft, many countries in the
region operate aircraft of the Mirage family. The
Argentine air force Mirage/IAI Finger fleet, acquired
new from France and second-hand from Israel and
Peru, currently serves with Grupo 6 de Caza.

Military aviation in South America is divided into two leagues: countries that have a claim to regional military leadership; and smaller countries which maintain token combat air forces, with larger support organisations operating aircraft that have a social rather than military role. Militarily, the traditional 'big four' of South America have been Argentina, Brazil, Chile and Peru, each of which maintains defensive and offensive assets and has aerospace industries of some form. The air forces of Bolivia, Paraguay and Uruguay are geared towards maintaining an internal air transport infrastructure, with a limited ability to protect their borders from smuggling. Many of the smaller countries use the confiscated aircraft of convicted smugglers, which usually have a limited service life because a full maintenance history is unavailable. The war against smuggling is also of concern for Colombia, whose international reputation has been tainted by the narcotics cartels within its borders. Colombia has a long history of political unrest and has a number of still-active terrorist factions. Guyana and Surinam have tiny forces, while French Guyana remains a French territory and as such is not covered in this section.

The southern region was home to one of the continent's longest-standing disputes – between Argentina and Chile – but recent agreements have seen a reduction in tension. Another dispute concerns the border between Ecuador and Peru. Venezuela suffered an attempted coup in 1992 in which different factions in the air force attacked air bases. Civil and political unrest on the continent is still a problem, but overall these countries have scaled down regional conflicts, and most budgets for military equipment have been reduced over the last five years. A recent willingness by the United States to supply surplus combat military hardware to certain South American countries, notably Argentina, has only been sanctioned because of the current political climate. The fight against the narcotics traders and producers has also attracted American support.

106

ARGENTINA

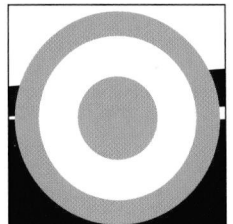

Capital: Buenos Aires
Population: 32.8 million
Land area: 2.778 million km² (1,072,240 sq miles)
Major cities: Córdoba, Rosario, La Plata, Mar del Plata, Mendoza, San Miguel de Tucuman, Santa Fe

Fuerza Aérea Argentina

The Fuerza Aérea Argentina became an independent force on 4 February 1945. Soon, with British help, Argentina had the strongest air arm in the area. When the United States became the primary military source, Argentina's request for more advanced equipment in the late 1960s was rejected. The country turned to European sources, as well as developing aircraft to meet its specific requirements. After a traumatic post-Falklands War period, the Fuerza Aérea is undergoing a period of reform and reorganisation.

The Fuerza Aérea Argentina is subordinate to the Ministerio de Defensa. All operational aircraft are controlled by the Comando de Operaciones Aéreas through the Grupos Aéreos of the Brigadas Aéreas. All training assets are the responsibility of the Comando de Instrucción. Technical support and maintenance facilities are provided by the Comando de Material. Control of civil and commercial aviation is assigned to the Comando de Regiones Aéreas which manages such activities as air traffic control, airport management and security and civil aviation training.

THE COMBAT FORCE

The Mirage family fleet is now consolidated at Base Aérea Militar Tandil, where Grupo Aéreo 6 is based. Escuadrón II operates the surviving 12 Mirage IIIEAs plus two Mirage IIIDAs for the air-to-air role; Escuadrones I and III are the fighter-bomber units with 20 IAI Fingers and nine Mirage 'Maras' (Mirage 5P upgrades). The IIIEA/DA fleet is due to be overhauled to increase its service life by another 15 years, while the Finger fleet is also to be modernised. The FAA has a requirement for 24 next-generation fighters to replace the Mirage fleet from 2005/2006.

Only a mixed fleet of 12 A-4P/C Skyhawks is still flying with Grupo Aéreo 5 at BAM General Pringles. They are the remains of a powerful force of 50 A-4P Skyhawks delivered from 1966 and 25 A-4C Skyhawks purchased in the 1970s. The replacement arrived in the form of another Skyhawk variant. A total of 32 A-4AR Fighting Hawks (rebuilt OA/A-4Ms) and four TA-4ARs is on order, the first aircraft having arrived at El Palomar air base on 23 December 1997 from Lockheed-Martin. All aircraft are due to be with Grupo Aéreo 5 by 1999. The extensive rework performed under the Fighting Hawk programme extended the aircraft's service lives to 2015. After the first 18 aircraft, the others will be produced by Lockheed-Martin Argentina SA at the former FMA at Córdoba. A-4ARs are due to take part in the US Red Flag 2000 exercises.

Two squadrons of Pucarás were equipped primarily for armed reconnaissance and ground strike support with Grupo Aéreo 3 de Ataque, based at Reconquista. Wartime experience gave way to specific roles for the Pucarás, such as helicopter hunter and photo-reconnaissance. For the last task several IA-58A Pucarás were adapted to operate the Halcón del Sur RPV system. The IA-58A is used as a launcher and control platform for the Quimar MQ-2 Bigua, the local designation for the Italian Meteor Andromeda system.

Half a dozen of the venerable BAC Canberra B.Mk 62s/T.Mk 64s are still in flying condition with Grupo Aéreo 2, possibly modified to perform long-range reconnaissance and EW missions.

TRAINING ASSETS

The training of Argentine pilots is undertaken at the Córdoba home of the Escuela de Aviación Militar (EAM) of the Grupo Aéreo Escuela. The young officers must complete the one-year basic training course,

Indigenous aircraft in Fuerza Aérea Argentina service include the IA-63 Pampa advanced trainer (above) and the IA-58A Pucará (left). The Pampa was to have replaced the MS.760 in the FAA, but only 18 are to be built, from a once-planned procurement of 68. Pucarás still equip two squadrons.

Fuerza Aérea Argentina

HQ	Edificio Condor

Comando de Operaciones Aéreas

Iª Brigada Aérea	BAM El Palomar, Buenos Aires
Grupo Aéreo 1 de Transporte	
Escuadrón I	C-130B/H, KC-130H, L-100-30
Escuadrón II	F28-1000C
Escuadrón IV	F27-400/500/600
Escuadrón V	707-365C/372C/387C/389B
Dept de Aviación Presidencial (at Aeroparque Jorge Newbery)	
	757-23A, F28-4000, S-70A, S-76A/B+, Sabreliner75A
IIª Brigada Aérea	BAM General Urquiza, Paraná
Grupo Aéreo 2 de Bombardeo	
Escuadrón I	Canberra B.Mk 62/T.Mk 64
Escuadrón II	Learjet 35A, IA-50A/B Guaraní II
Escuadrón de Servicios	Cessna A 182
IIIª Brigada Aérea	BAM Reconquista
Grupo 3 de Ataque	
Escuadrón I	IA-58 Pucará
Escuadrón II	IA-58 Pucará
Escuadrilla de Servicios	Cessna A 182
IVª Brigada Aérea	BAM El Plumerillo, Mendoza
Grupo 4 de Caza	
Escuadrón I	IA-63 Pampa
Escuadrón II	MS 760 Paris IIR, TA-4J Skyhawk
Escuadrón III	SA 315B Lama
Escuadrilla de Servicios	Cessna A 182
Escuadrilla Acrobática	Sukhoi Su-29AR
Vª Brigada Aérea	BAM General Pringles, Villa Reynolds
Grupo 5 de Caza	
Escuadrón I	A-4AR, TA-4AR
Escuadrón II	A-4C/P (to get A-4AR)
Escuadrilla de Servicios	Cessna A 182
VIª Brigada Aérea	BAM Tandill
Grupo 6 de Caza	
Escuadrón I	IAI Finger A/B
Escuadrón II	Mirage IIIEA/DA
Escuadrón III	Mirage 5PA Mara
Escuadrilla de Servicios	Shrike Commander 500U, Hughes 500
VIIª Brigada Aérea	BAM Mariano Moreno, Buenos Aires
Grupo Aéreo 7 de Helicópteros	
Escuadrón I	MD 500C/D/M, MD 530F, UH-1H
Escuadrón II	UH-1D/H, Bell 212
Escuadrón III	BV 308 Chinook, Shrike Commander 500U, Sikorsky S-61R
IXª Brigada Aérea	BAM Comodoro Rivadavia
Grupo Aéreo 9 de Transporte	
Escuadrón VI	F27-400/600
Escuadrón VII	DHC-6-200 Twin Otter
Xª Brigada Aérea	BAM Rio Gallegos
	Bell 212

Comando de Instrucción

Escuela de Aviación Militar	BAM Córdoba
Grupo Aéreo Escuela	
Escuadrón I	Beech B45 Mentor
Escuadrón II	EMB-312A Tucano
Escuadrilla de Servicios	IA-50B Guaraní, Hughes 500
Escuadrilla Veleros	DINFIA IA-46 Ranquel, Grob G.113 Twin Astir, Standard Astir II, Let L-13 Blanik, SZD-30 Pirat II
Cuerpo de Cadete	BAM Mariano Moreno, B.A.
Escuela de Nolovelismo	Fw 44J

Comando de Material

Centro de Ensayos en Vuelo BAM Córdoba
     Mirage IIICJ, IA-58A, IA-63
Technical support and maintenance undertaken at:-
Aérea Material Córdoba (Lockheed Martin Aircraft Argentina S.A.)
Aérea Material Rio Cuarto
Aérea Material Quilmes

Comando de Regiones Aéreas

Instituto Nacional de Aviación Civil, Morón Airport
    Cessna A182, MD 500D, PA-A-25-235, PA-A-28-236, PA-A-28RT-201, PA-A-34-220T, PA-31-350
Controls civil and commercial aviation, and also looks after air traffic control, airport management and security and civil aviation training

known as Curso de Aviador Militar (CAM), before earning their wings, starting on the FMA-built Beech B.45 Mentor. A total of 30 Mentors has been modernised by LMSA. Thirty EMB-312A Tucanos are used for advanced training. Pilots selected for combat roles continue their training with the Escuela de Caza at BAM El Plumerillo with Grupo Aéreo 4 de Caza. Escuadrón II undertakes the first stage with its armed MS.760A Paris. The Paris is in the process of being replaced by 18 ex-US Navy TA-4Js. Escuadrón I, which concludes the training of fighter pilots, is the sole operator of the FMA IA-63 Pampa. After completion of the CAM (Curso de Aviadores de Combate), combat pilots start their operational career

Above: Three examples of the MD.530F supplement earlier versions of the Hughes 500 in Ecuadrón I of Grupo Aéreo 7 de Helicópteros.

Below: The MS.760 Paris IIR has been in service since 1958 in the armed training role. It flies with Grupo Aéreo 4/Esc II from BAM El Plumerillo.

with the Pucarás of Grupo Aéreo 3 de Ataque. Transport crews are assigned to Escuadrón VII of Grupo Aéreo 9. Its DHC-6 Twin Otters are used for multi-engine training conversion, also known as Curso de Estandarización de Procedimientos para Pilotos de Transporte (CEPTA), which normally takes a year. Rotary-wing crews follow the Curso de Estandarización de Procedimientos Aéreos para Helicópteros (CEPAH), flying the RACA-built Hughes 500Ds of Escuadrón I of Grupo Aéreo 7 de Helicópteros. The Instituto Nacional de Aviación Civil (INAC), based at Morón Airport, is the civil aviation flying school but is under control of the Fuerza Aérea Argentina through its Comando de Regiones Aéreas. All aircraft used by the school fly with Fuerza Aérea markings and serials.

TRANSPORT AND SUPPORT ASSETS

Most transports are concentrated in Grupo Aéreo 1 de Transporte based at BAM El Palomar. Long-range heavy transport and air refuelling is performed by the Hercules fleet of Escuadrón I. Escuadrón II flies four Fokker F28 Mk 1000C Fellowships, mainly as personnel transport and normally under charter contract to commercial operators. Escuadrón IV has a line-up of three Fokker F27 Mk 400M Troopships and three F27 Mk 600 Friendships, which are shared with Lineas Aéreas del Estado (LADE), flying scheduled low-fare under-developed routes. Escuadrón V flies four Boeing 707-372C/3387B/389Bs for long-range transport, photographic reconnaissance and Elint tasks. A 707-365C was acquired and converted into a tanker/transport and the FAA is looking for another example. Grupo Aéreo 9 de Transporte is headquartered at BAM Comodoro Rivadavia, and its flights are closely associated with LADE activities. Escuadrón VI flies two Fokker F27 Mk 400M Troopships and two F27 Mk 600 Friendships. Escuadrón VII complements its role of multi-engine school with passenger transport, SAR and Antarctic operations. Normally, a couple of DHC-6

Twin Otters fitted with skis and extra fuel tanks are detached at BAM Marambio, Antarctica. Escuadrón II of Grupo Aéreo 2 flies the survivors of 20 FMA IA-50 Guaraní IIs. The Comando de Operaciones Aéreas and the Escuela de Aviación Militar each has its own Guaraní as staff transports. Also based at Paraná is Grupo Aéreo 1 Aerofotogrametrico, which uses three Learjet 35As of Escuadrón II of Grupo Aéreo 2 for aerial photography, and another two for navaid checks.

The bulk of the helicopter fleet is concentrated at Grupo Aéreo 7 de Helicópteros. Escuadrón I operates most of the 18-strong fleet of Hughes 500D/Es, and also received eight surplus US Army UH-1Hs from September 1998 onwards. Other Bell UH-1D/Hs and Bell 212s are assigned to Escuadrón II for transport and SAR duties. Heavylift is supplied by the surviving two Boeing Vertol Model 308 Chinooks of Escuadrón III, which also has a single VIP-configured Sikorsky S-61R. Although a helicopter unit, Grupo Aéreo 7 also operates six Rockwell 500U Shrikes for liaison duties. Escuadrón III of Grupo Aéreo 4 is a specialised unit, flying SA 315B Lamas into the rugged and dangerous Andean mountains for SAR missions.

Every Grupo Aéreo maintains its own Escuadrilla de Servicios (base flight) for general duties, the composition of which varies from one unit to another. The Fuerza Aérea also maintains a glider fleet of Let L-13 Blaniks, Grob G.113 Twin Astir, Grob Standard Astir II and SZD-30 Pirat II, distributed between the Escuela de Aviación Militar and Grupos Aéreos 4, 5 and 6.

The Escuadrón de Aviones Presidenciales is based at Aeroparque Jorge Newbery and specialises in VVIP transport duties. Head of the fleet is a Boeing 757-23ER delivered in 1992, with two Fokker F28 Mk 1000s, a Mk 4000, a single Sikorsky S-70A, an S-76 and an S-76B Plus.

Comando de la Aviación Naval Argentina

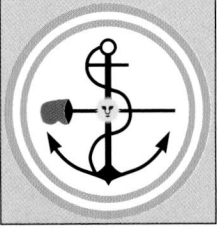

The most capable combat aircraft with the Armada are the Super Etendards delivered from 1981. They are capable of firing the AM39 Exocet ASM.

Flight operations of the Armada Argentina started in 1918. It was the first Latin American navy to operate in the Antarctic, the first to introduce helicopters (1946), the first to use combat jets (F9F Panther and TF-9 Cougar), the first to operate from aircraft-carriers (in 1958), and the first naval aviation in the world to sink a major combat unit with a single ASM, in the 1982 Falklands conflict.

The Comando de la Aviación Naval Argentina (COAN) is in the midst of a restructuring process, consolidating its squadrons at bases Comandante Espora and Almirante Zar. The Aviación Naval Argentina is organised into three Fuerzas Aeronavales, each divided into a variable number of Escuadras Aeronavales. The smallest operational unit is the Escuadrilla Aeronaval, which uses aircraft to accomplish a primary role.

FUERZA AERONAVAL NO 1

The Fuerza Aeronaval No 1 (FAE 1), based at Base Aeronaval Punta Indio (Veronica – Buenos Aires), is responsible for training naval personnel. After flying sailplanes and Boeing N2S Kaydets, student pilots begin basic and elementary training in the Beech T-34C-1 Turbo Mentor. Operational training is accomplished by 1ª Escuadrilla

Aeronaval de Ataque, using 10 EMB-326GB Xavantes.

FUERZA AERONAVAL NO 2

FAE 2 is in charge of all aircraft assets that formerly comprised the aircraft-carrier component. In May 1997 the Armada Argentina made the drastic decision to deactivate the ARA 25 de Mayo (V-2), its only carrier. Carrier qualification is maintained via the use of the Brazilian light carrier Minas Gerais (A-11). The Escuadra Aeronaval No 2 has the primary role of anti-submarine warfare. Under the Tata programme, six Trackers had turboprop engines installed by early 1998, becoming S-2Ts. The Sea Kings of the

2ª Escuadrilla Aeronaval de Helicópteros include five Sikorsky-built S-61Ds, locally designated SH-3, and two Agusta-built AS-61D-4s or PH-3s. Eleven Super Etendards remain with Escuadra Aeronaval No 3's Esc Aeronaval de Caza y Ataque No 2, although normally only six are in flying condition. The 1ª Escuadrilla Aeronaval de Helicópteros flies the SA 316B Alouette III and the AS 555SN.

FUERZA AERONAVAL NO 3

With Escuadra Aeronaval No 6, at Base Aeronaval Almirante Zar (Trelew-Chubut), the Escuadrilla Aeronaval de Exploración re-equipped with six ex-US Navy P-3B 'Super Bs', retiring its five L-188

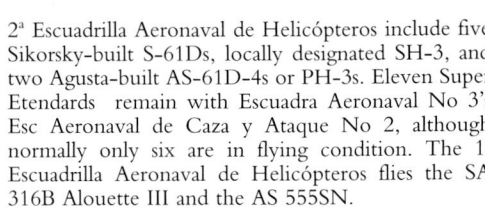

Comando de la Aviación Naval Argentina

Fuerza Aeronaval No 1	
Escuadra Aeronaval 1	**BA Punta Indio**
Escuela de Aviación Naval	T-34C-1, PC-6/B2-H2
Escuadra Aeronaval 4	**BA Comandante Espora**
1ª Escuadrilla Aeronaval de Ataque	MB.326GB, EMB.326GB

Fuerza Aeronaval No 2	
Escuadra Aeronaval 2	**BA Comandante Espora**
Escuadrilla Aeronaval Antisubmarina	S-2E/T
2ª Escuadrilla Aeronaval de Helicópteros	AS-61D-4, S-61D-4
Escuadra Aeronaval 3	**BA Comandante Espora**
1ª Escuadrilla Aeronaval de Helicópteros	
	SA 316B Alouette III, AS 555MN Fennec
2ª Escuadrilla Aeronaval de Caza y Ataque	Super Etendard

Fuerza Aeronaval No 3	
Escuadra Aeronaval 6	**BA Vice Almirante Zar**
1ª Escuadrilla Aeronaval de Exploración	
	P-3B, L-188E Electra WAVE
2ª Escuadrilla Aeronaval de Sostén Logístico Móvil	F28-3000
Escuadrilla Aeronaval de Vigilancia Marítima	King Air 200/200T

Above: Four Beech 200s of the Comando de la Aviación Naval Argentina have been converted to Beech 200T Maritimes under the Comoran project.

Below: Since 1978, the Armada's basic trainer has been the T-34C-1 Turbo-Mentor. Fifteen aircraft were delivered, with five having been lost to date.

Electras (or Electrons) in the process. The Orions have received the Tac/Nav modification, adding advanced mission systems and communications gear. The first arrived in December 1997 and the last was received during 1998. The Sigint Electra WAVE conversion continues to be used by the unit.

The three Fokker F28 Mk 3000s with the 2ª Escuadrilla Aeronaval de Sostén Logístico Móvil fulfil the transport role. Among its secondary roles, the 2nd Transport Squadron undertakes maritime surveillance and Antarctic observation flights.

Under the Cormoran Programme, Arsenal Aeronaval No. 2 is converting four (of seven) Beech 200 King Airs to B.200T Maritime standard. The first example was redelivered to the Escuadrilla Aeronaval de Vigilancia Maritima in March 1997. The final aircraft was expected with the squadron at the end of 1998. Cormoran-modified aircraft are used for coastal patrol, fishery surveillance and SAR. Two other King Airs are used as lights transports, and a third serves as a photo-reconnaissance aircraft.

Comando de Aviación del Ejército

All the operational elements of the Aviación del Ejército are under control of the Agrupación Aviación del Ejército 601 headquartered at Campo de Mayo Army Aviation Airfield. The Escuadrón de Aviación de Exploración y Reconocimiento 601 uses the OV-1D Mohawk for observation; 23 OV-1Ds have been obtained, and possibly seven others.

The Bell UH-1H is the workhorse helicopter of the Argentine army. After receiving 42, another 20 ex-US Army examples arrived in 1998. Most serve with Grupo de Helicópteros de Asalto 601, sharing the transport role with AS 332B Pumas and the last SA 330 Puma. Escuadrón de Aviación de Exploración y Ataque 602 operates UH-1Hs as gunships, in addition to the armed Agusta A 109A Hirundos.

Escuadrón de Aviación de Apoyo 603 is tasked with logistic support and operates a mixed fixed-wing fleet. Transport operations are performed by a trio of Alenia G222s, plus two DHC-6-300 Twin Otters and a single CASA C.212-200 Aviocar. Three Fairchild SA-226AT Merlin IVs are configured as medevac transports, and four Fairchild SA-226T Merlin IIIs are used as staff transports. The squadron's complement includes most of the eight Cessna T-207As and one Cessna U-17A-CE. Every large Army Corps has a flight used mostly for staff transport. Fixed-wing aircraft are on loan from Escuadrón de Aviación de Apoyo 603.

Most army aviation pilots arrive with their own Commercial Pilot licence. At the Escuela para Apoyo de Combate General Lemos they receive military instruction, before entering the Escuela de Aviación del Ejército flying the Cessna T-41D Mescalero. The Escuela also has some dual-command OV-1Ds for advanced training. Helicopter training is on the

elderly Hiller UH-12E.4 light helicopter.

The Instituto Geográfico Militar has single examples of the Beech 65 Queen Air and Cessna 500 Citation for aerial photography.

Above: The Argentine army uses the OV-1D Mohawk for reconnaissance missions from Campo de Mayo. The aircraft retain the US Army grey scheme.

Above left: Eight Fairchild-Hiller UH-12Es are used for primary helicopter training by the Escuela de Aviación de Ejército.

Comando de Aviación del Ejército Argentino

Agrupación Aviación de Ejército 601	Campo de Mayo
Escuadrón de Aviación de Exploración y Reconocimiento 601	
OV-1D	Campo de Mayo
Escuadrón de Aviación de Exploración y Ataque 602	
Agusta A 109A, UH-1H, Bell 205 A-1	Campo de Mayo
Escuadrón de Aviación de Apoyo 603	
Alenia G222, CASA C.212-200, DHC-6-200/300, U-17A, Cessna T 207A, SA-226T Merlin IIIA, SA-226AT Merlin IVA, Sabreliner 75A	Campo de Mayo
(aircraft loaned to various Army Corps – see below)	
Escuela de Aviación del Ejército	
T-41D, OV-1D, Hiller (Soloy) UH-12E4	Campo de Mayo
Grupo de Helicópteros de Asalto 601	
Compañia A	AS 332C Super Puma
Compañia B	UH-1H, Bell 205A, Bell 212
Instituto Geográfico Militar	
Cessna 500, Queen Air B80	Campo de Mayo

Corps-level aviation		
II Army Corps		
Sección Aviación del Ejército 2	Rosario	Merlin III
Sección Aviación del Ejército 11	Rio Gallegos	UH-1H, Bell 205A-1
III Army Corps		
Sección Aviación del Ejército 4	Córdoba	Cessna T 207A, Merlin III
IV Army Corps		
Sección Aviación del Ejército 8	Mendoza	SA 315B Lama
V Army Corps		
Sección Aviación del Ejército 5	Bahia Blanca	Cessna T 207A, Merlin III
VI Infantry Brigade		
Sección Aviación del Ejército 6	Neuquen	SA 315B Lama
IX Infantry Brigade		
Sección Aviación del Ejército 9	Comodoro Rivadavia	UH-1H, Bell 205A-1
XI Mechanised Brigade		
Sección Aviación del Ejército 11	Resistencia	UH-1H, Bell 205A-1

Gendarmeria Nacional

This force of more than 40,000 troops is deployed along the borders. The Departamento de Aviación has its headquarters at Campo de Mayo Army Airfield, where the administrative, logistic and maintenance facilities are based. The helicopter force is assigned on demand to reinforce the activity of the ground forces.

Gendarmeria Nacional

Departamento de Aviación	HQ Campo de Mayo
AS 350B Ecureuil, Cessna U-206, 5 x Hughes 500C/D, L-21B, 5 x PC-6/B, PA-23 Aztec, PA-28-236 Dakota, PA-31T Cheyenne II, 5 x SA 315B Lama	

Prefectura Naval Argentina

The Prefectura Naval has three aviation detachments to facilitate the control of seaborne traffic in the 200-nm (230-mile; 370-km) limit as well as to support SAR and oil pollution control activities. The main base is Estación Aérea San Fernando which houses the two Schweizer 300Cs and two Piper PA-28-181 Warriors used for training by the Centro de Extensión Profesional Aeronáutica and the nucleus of the CASA Aviocar fleet. The other bases are Estación Aérea Comodoro Rivadavia, established in 1990, and Estación Aérea Mar del Plata, activated during 1996.

The Prefectura Naval Argentina uses five CASA 212-300 Aviocars. The first pair was delivered in December 1988, and replaced the service's surviving Short Skyvans, which were sold in late 1993.

Prefectura Naval Argentina

Estación Aérea Comodoro Rivadavia	
det from Estación Aérea San Fernando of C.212 Patrulleros	
Estación Aérea Mar del Plata	SA 330L, AS 365M2
Estación Aérea San Fernando	
CASA C.212-300M, CASA C.212 Patrulleros	
Centro de Extensión Profesional Aeronáutica	
EA San Fernando	PA-28 Warrior II, Schweizer 300C

BOLIVIA

Capital: La Paz
Population: 7.4 million
Land area: 1.099 million km² (424,000 sq miles)
Major cities: Sucre, Santa Cruz, Cochabamba, Oruro

Escuadrón 310 at Base Aérea 'General Walter Arze' El Alto, La Paz operates Bolivia's AT-33ANs and T-33SFs. The squadron maintains two detachments at Cochabamba and Santa Cruz de la Sierra.

Fuerza Aérea Boliviana

The Fuerza Aérea Boliviana (FAB) divides the country into three Regiones Aéreas: Iª Brigada Aérea is located at El Alto, La Paz; IIª Brigada Aérea is at Cochabamba, Colcapiru; and IIIª Brigada Aérea is at El Trompillo, Santa Cruz. Within these regional areas are 11 operational Grupos and the Colegio Militar de Aviación.

Base Aérea 'General Jorge Jordan', El Alto, is the principal air base, close to the seat of government and the major city of La Paz. Here the bulk of the transport system can be found, much of which is operated by the Transporte Aéreo Militar (TAM) which also retains the FAB title of Grupo Aéreo de Transporte 71. Equipped with a mixture of assorted Lockheed Hercules, Fokker F27 Friendships, IAI Aravas, Rockwell Commanders, CASA C.212 Aviocars and other minor types, the unit is tasked with support of the interior and out-based units. The Transporte Aéreo Boliviano (TAB), a pseudo-civilian operation, is also attached to GAdT 71.

To undertake the COIN role, the FAB utilises a number of Canadair-built AT-33As, initially acquired in 1973. The surviving AT-33As were supplemented by additional deliveries, and are assigned to Escuadrón 310 under Grupo Aéreo de Caza 31 at El Alto, with an operating detachment at El Trompillo under Grupo de Caza 32, and at Grupo Aéreo de Caza 33 at Cochabamba. Another five were received in 1977 in an exchange by North West Industries for the last six airworthy P-51D Mustangs, and in 1985 an additional 18 former French T-33ANs were acquired. The FAB is due to have 18 of its AT-33As upgraded by Kelowna, of Canada, with a glass cockpit.

Above: Wearing an attractive eagle's head scheme, this is one of the Fuerza Aérea Boliviana's 22 PC-7s. The aircraft are being armed for COIN duties.

Rotary-wing operations are conducted principally by Grupo Aéreo 51, which is equipped with the Bell UH-1H, Hughes 369 and Helibras HB 315 Lama. Located at the former international airport of El Trompillo, Santa Cruz, they meet army, navy and air force requirements. Also at El Trompillo is the FAB's only surviving airworthy Douglas C-47 Dakota, which is a Basler Turbo BT-67 conversion. This aircraft is tasked primarily in anti-narcotics-related operations.

TRAINING OPERATIONS

At El Trompillo is the College of Military Aviation (COLMILAV) with its variety of Cessna aircraft, many of which have been acquired following confiscation in drug-related prosecutions. Pilot primary training is now carried out at Santa Rosa utilising 15 T-23 Uirapurus donated by Brazil.

With the bulk of the helicopter assets having vacated Cochabamba, Grupo Aéreo Mixto is left with a small search and rescue detachment of SA 315B Lamas, together with the survivors of the 16 Pilatus PC.6B Turbo-Porters procured for COIN operations. They will be supported by the 22 Pilatus PC.7s once those aircraft have been converted to this task.

The FAB has a number of other operating locations, including Base Aérea de Rebore which is still thought to be home to the Grupo de Operaciones Aéreas Especiales. This unit had originally received the Hughes 369M, although whether they have relocated to within the structure of Grupo Aéreo 51 is unclear. This applies equally to Grupo Aéreo de Cobertura No 1 at El Tejar, as its former T-6Gs are thought to have been withdrawn. Conversion of the

Fuerza Aérea Boliviana

Iª Brigada Aérea

Base Aérea 'General Walter Arze' El Alto, La Paz

Grupo Aérea de Caza 31

Escuadrón 310	AT-33AN, T-33SF
Escuadrón 311	B55/B58 Baron, King Air F90/200C, CASA C.212-200, Turbo Commander 680

Grupo Aéreo 71

Escuadrón 711	C-130B/H
Escuadrón 712	Convair 580, F27-400M, Arava 201, L-188A Electra
Escuadrilla Ejecutiva	PA-34 Seneca, Sabreliner 60
Servicio Nacional de Aerofotogrametria	Learjet 25B/25D/35A, PA-23 Aztec

IIª Brigada Aérea

Base Aérea de Colcapiru, Cochabamba

Grupo Aéreo Mixto PC-7, SA 316B/HB 315B
Grupo Aéreo de Caza 33 AT-33AN, T-33SF
(aircraft det from Escuadrón 310)

IIIª Brigada Aérea

Base Aérea El Trompillo, Santa Cruz de la Sierra

Grupo Aéreo de Caza 32 AT-33AN, T-33SF
(aircraft det from Escuadrón 310)
Grupo Aéreo 51 UH-1H, SA 315B/HB 315B
Grupo Aéreo 72

Escuadrón 721	Basler Turbo 67

Colegio Militar de Aviación

Base Aérea El Trompillo, Santa Cruz de la Sierra

Escuadrón Básico	Beech V35/36 Bonanza, Lancair 320, Cessna 152/172K/182R/U206R/U206C/TU206C
Escuadrón Primario	A-122 Uirapuru, A-132 Tangara

PC.7 to COIN operation will probably see a detachment here.

Future equipment for the FAB may include 12 ex-US Navy TA-4J Skyhawks. In mid-1998 it was planned to acquire 18 aircraft (six for spares) and use them for training purposes. They would also make effective strike aircraft, providing a significant asset alongside, or replacing, the AT-33s.

Aviación de las Fuerzas Navales Bolivianas

Lake Titicaca, the world's highest navigable lake, is virtually a land-locked sea (with an area of 8288 km²/3,200 sq miles) and is the reason why Bolivia, a land-locked country, has a navy. Relations with neighbouring Peru are good and funds are slim, so the navy is a modest force. Its sole Cessna 402C patrol/communication aircraft is kept at El Alto.

Aviación de las Fuerzas Navales Bolivianas
El Alto, La Paz   1 x Cessna 402C

Aviación del Ejército Boliviano

Bolivian army aviation maintains a few communication and support aircraft from the FAB facility at El Alto, La Paz, and has one transport, a CASA C.212 (EB-002). For all other operations it relies on FAB assets for support.

Aviación del Ejército Boliviano
El Alto, La Paz   CASA C.212-300M, Cessna 210, King Air C90

Carabineros Bolivianos

For most operations the Bolivian police (Carabineros Bolivianos) relies on FAB aviation assets for support and transport. It does maintain one Cessna 421B, CB-001, for communication duties.

Carabineros Bolivianos
Cessna 421B   1

Above left: The Bolivian navy uses this Cessna 402C.

Below: CASA 212 EB-50 was used by the Bolivian army, but was destroyed in April 1995.

BRAZIL

Capital: Brasília
Population: 153.3 million
Land area: 8.512 million km² (3,286,000 sq miles)
Major cities: São Paulo, Rio de Janeiro, Belo Horizonte, Recife, Porto Alegre, Salvador

Above: Flying over Rio de Janeiro's crowded bay and famous landmark is a pair of FAB F-5Es. The country received 36 new-build F-5Es as the first export customer for the variant, and 24 ex-USAF examples, supported by six new F-5Bs and four F-5Fs.

The fifth largest country in the world, Brazil also has the biggest armed forces in its region. Supported by about 700 aircraft of the Força Aérea Brasileira (FAB), manned by 59,000 personnel, these forces have the job of defending Brazil's frontiers with 10 South American countries, as well as the coastline.

Força Aérea Brasileira

The FAB has an independent status among the Brazilian armed forces, its roles being that of national defence and supporting the opening up of the vast hinterland for economic development. It currently comprises five commands. The Comando Geral do Air (COMGAR), or General Air Command, supervises most of the flying operations of the FAB, reporting to the Ministry of Aeronautics. Its parallel organisations, which also report directly to the Ministry of Aeronautics, comprise the Comando Geral de Apoio (COMGAP) or Support Command; Comando Geral do Pessoal (COMGEP) or Personnel Command; Departamento de Pesquisa e Desenvolvimento (DEPED) or R&D Department; and Departamento de Ensino (DEPENS) or Training Department.

As the main FAB flying element, COMGAR administers several sub-formations in the form of seven Comandos Aéreos Regionais (COMARs) or Regional Air Commands; three Forças Aéreas (FAes) or Air Forces which have specific individual tasks; a Comando Aéreo de Treinamento (CATRE) or Air Training Command; Comando Geral de Apoio (COMGAP), or General Support Command; and the Nucleo do Comando de Defesa Aéroespacial Brasileiro (NuCOMDAER) or HQ Brazilian Aerospace Defence Command. At unit levels, Grupos de Aviação (GAv) or wings usually comprise anything from one to five consecutively-numbered Esquadrões (squadrons), each with varying numbers of aircraft from half a dozen to 12 or more, smaller formations being known as Esquadrilhas (flights).

COMGAR's SUBORDINATE FORCES

The Segunda Força Aérea (IIª FAe), with its HQ in Rio de Janeiro, undertakes maritime patrol, shipborne ASW, SAR and related roles, as well as helicopter tactical support and observation tasks in co-operation with the Brazilian army. Its 1° Grupo de Aviação Embarcada (1° GAE) at Santa Cruz is responsible for operating the FAB's Grumman S-2E (P-16E/H) Tracker fleet for both land- and carrier-based ASW roles. The Trackers rarely fly. Three EMBRAER EMB-111A (P-95A) Bandeirante Patrulhas (or 'Bandeirulhas') also operate with 1° GAE. Others are used by the 1° Esquadrão of the 7° Grupo de Aviação 'Orungan' (1°/7° GAv) of IIª FAe, at Salvador air base in Bahia state, and 2°/7° GAv 'Phoenix' at Florianópolis in the southern Santa Catarina state, and 3°/7° 'Netuno' at Belém in the northern Pará state.

Completing the FAB's maritime assets, within IIª FAe, is an independent fleet support unit, 2ª Esquadrilha de Ligacão e Observacão 'Duelo', operating half a dozen armed EMBRAER EMB-312 Tucano AT-27 turboprop trainers from Santa Cruz. They spent much of their time on anti-drug-smuggling aircraft interdiction patrols.

Maritime patrol will be performed solely by the P-95 A/B Bandeirante Patrulhas upon the imminent retirement of the P-16E/H Tracker fleet.

SAR AND ARMY SUPPORT

Five EMB-110P1Ks, or SC-95Bs, have equipped 2°/10° GAv 'Pelicano' at Campo Grande in Mato Grosso for dedicated search and rescue roles. This small unit is intended to cover Brazil's entire vast inland area and coastal waters. For shorter-range and specialised SAR roles, 2°/10° GAv is also equipped with six Bell UH-1H Iroquois, known locally as 'Sapões' (or 'big frogs'). UH-1Hs are also operated by some of the five squadrons in IIa FAe's 8° Grupo, which is mainly concerned with army support.

At Belém (Para), 1°/8° GAv 'Falcão Pioneiro' is the sole FAB unit to operate the Aérospatiale AS 355M Twin Ecureuil or CH-55 Esquilo in attack and observation roles. At Recife, in Pernambuco state, 2°/8° GAv 'Poti' is equipped with about 10 HB-350B Esquilos (UH-50), and also operates five indigenous Neiva T-25C Universal two-seat basic trainers for liaison and observation. 3°/8° GAv 'Puma' at Campo dos Afonsos uses eight Eurocopter AS 332M Super Pumas (CH-34s) for general and assault transport roles and three Neiva Regentes for army

Força Aérea Brasileira

Esquadrão de Demonstração Aérea 'Esquadrilha da Fumaça'
EMB-312 Tucano · Pirassununga

Grupo de Transporte Especial · Brasília
737-2N3, BAe 125-400/403B, EMB-121E, HB 355M Esquilo, VU-35 Learjet

Comando Geral de Apoio (General Support Command)

Diretoria de Electrônica e Proteção. Ao Vôo
Grupo Especial de Inspeção em Vôo · Santos Dumont AP
BAe 125 Srs 3B/RC, EMB-110/A

Parques de Material Aeronáutico (PAMA) (maintenance units)
PAMA-AF	Campo dos Afonsos
PAMA-BE	Belém
PAMA-GL	Galeão
PAMA-LS	Lagoa Santa
PAMA-RF	Recife
PAMA-SP	Campo de Marte

(PAMAs may use EMB-110s or DHC-5s as spares ferries)

Departamento de Pesquisa e Desenvolvimento
(Department of Research and Development)
Centro Tecnico Aeroespacial (Aerospace Technical Centre)
various · São Jose Dos Campos
Centro de Lançamento de Alcântara *(for Ariane support)*
Cessna 208 · São Luis do Maranhão

Comando Geral do Ar (General Air Command)

COMANDOS AÉREOS REGIONAIS
1º Comando Aéreo Regional · **Belém AB**
1 Esquadrão de Transporte Aéreo
EMB-110P1K, Cessna 208A

2º Comando Aéreo Regional · **Recife**
2 Esquadrão de Transporte Aéreo · Pernambuco
EMB-110K1

3º Comando Aéreo Regional
3 Esquadrão de Transporte Aéreo · Galeão
EMB-110K1/P1A

4º Comando Aéreo Regional
4 Esquadrão de Transporte Aéreo · São Paulo/
EMB-110K1/P1A · Cumbica AFB

5º Comando Aéreo Regional
5 Esquadrão de Transporte Aéreo · Porto Alegre/
EMB-110K1 · Canoas
Santa Maria AB base flight · Cessna 208 · Santa Maria

6º Comando Aéreo Regional
6 Esquadrão de Transporte Aéreo · Brasília AB
EMB-110, EMB-120RT, EMB-121E
Brasília AB base flight · Cessna 208 · Brasília AB

7º Comando Aéreo Regional · **Manaus**
1 Esquadrilha · Boa Vista
EMB-312, Cessna 208A
2 Esquadrilha · Puerto Velho
EMB-312, Cessna 208A
7 Esquadrão de Transporte Aéreo · Manaus
EMB-110

Segunda Força Aérea (2nd Air Force) · Rio de Janeiro

2º Esquadrilha de Ligação & Observação 'Duelo'
EMB-312 Tucano · Santa Cruz
1º Grupo de Aviação Embarcada
1º GAE 'Cardeal' · EMB-111B · Santa Cruz
7º Grupo de Aviação
1º Esq 'Orungan'	EMB-111B	Salvador
2º Esq 'Phoenix'	EMB-111B	Florianópolis
3º Esq 'Netuno'	EMB-111B	Belém

8º Grupo de Aviação
1º Esq 'Falcão Pioneiro'	HB 355FS Esquilo	Belém
2º Esq 'Potí'	N-621 Universal, HB 350B Esquilo	Recife
3º Esq 'Puma'	AS 332M, N-591 Regente	Afonsos
5º Esq 'Pantera'	UH-1H, EMB-810C, N-591 Regente	Santa Maria
7º Esq 'Falcão'	UH-1H, EMB-810C	Manaus

10º Grupo de Aviação
2º Esq 'Pelicano'	UH-1H, EMB-110P1K/P-L	Campo Grande
	UH-1H, EMB-110P1K/P-L	

Terceira Força Aérea (3rd Air Force) · Brasília

1º Grupo de Defesa Aérea · **Anápolis**
1º Esq 'Jaguares' · Anápolis
Mirage IIIE/EBR/BE/DBR, EMB-312, HB350B Esquilo
Anápolis Base Flt · Neiva U-42 Regente · Anápolis
1º Grupo de Aviação de Caça · **Santa Cruz**
1º Esq 'Jambock'	F-5E/B, EMB-312	Santa Cruz
2º Esq 'Pif Paf'	F-5E/B, EMB-312	Santa Cruz

4º Grupo de Aviação · **Fortaleza**
1º Esq 'Pacau' · Fortaleza
EMB-326, HB 350B Esquilo
6º Grupo de Aviação · **Recife**
1º Esq 'Carcara' · EMB-110B, Learjet R-35 · Recife
10º Grupo de Aviação · **Santa Maria**
1º Esq 'Poker'	EMB-326	Santa Maria
3º Esq 'Centauro'	EMB-326	Santa Maria

14º Grupo de Aviação · **Canoas**
1º Esq 'Pampa' · F-5E/F, EMB-312 · Canoas
16º Grupo de Aviação · **Santa Cruz**
1º Esq	AMX, AMX-T	Santa Cruz
2º Esq	AMX, AMX-T	Santa Cruz

Quinta Força Aérea (5th Air Force) · Rio de Janeiro

1º Grupo de Transporte · **Galeão**
1º Esq 'Coral' · C-130E/H Hercules · Galeão
1º Grupo de Transporte de Tropas · **Afonsos**
1º Esq 'Gordo'	C-130E Hercules	Afonsos
2º Esq 'Cascavel'	DHC-5A Buffalo	Afonsos

2º Grupo de Transporte · **Galeão**
1º Esq 'Condor'	BAe 748 Srs 200	Galeão
2º Esq 'Corsário'	KC-137	Galeão

9º Grupo de Aviação · **Manaus**
1º Esq 'Arara' · DHC-5A Buffalo · Manaus
15º Grupo de Aviação · **Campo Grande**
1º Esq 'Onça' · EMB-110P1K · Campo Grande

Comando Aéreo de Treinamento

HQ Natal
Natal Base Flt · U-42 Regente, EMB-110P1K, HB 350B (on det from 2º/8º GAv) · Natal
1º/5º GAv 'Rumba' · EMB-312 Tucano · Natal
2º/5º GAv 'Joker' · EMB-326 Xavante · Natal
11º Grupo de Aviação · **Santos**
1º Esq 'Gavião' · HB 350B Esquilo · Santos

Departamento de Ensino (Department of Training)

Academia da Força Aérea · **Pirassununga**
EMB-810D, EMB-110, HB 350B
1º Esquadrão de Instrução Aérea · EMB-312
2º Esquadrão de Instrução Aérea 'Apollo' · N-621 Universal
Clube Vôo a Vela · Libelle 201B, TPE KW 1b2, EMB-201R, Scheicher, ASW 20, Let L-13, IPE-2b
Clube de Ultraleves · Microleve MXL/MXL-II
Escola do Especialistas de Aeronáutico
(technical training plus single EMB-110) · Guaratingueta

Above: Basic training is undertaken on the indigenous T-27 Tucano. This pair flies with the 1º Esquadrão de Instrução Aérea of the Academia da Força Aérea at Pirassununga.

Below: The Brazilian F-103E (Mirage IIIEDR) fleet operates as a single squadron, 1º/1ºGDA at Anápolis.

remain in service until about 2005.

Three squadrons of Northrop F-5 fighter-bombers are now operated within IIIa FAe, two multi-role squadrons of the FAB's historic 1º Grupo de Aviação de Caça (Fighter Air Wing) at Santa Cruz, and 1º/14º GAv 'Pampa' at Canoas/Porto Alegre, which operates solely in air defence roles with aircraft armed with first-generation AIM-9B Sidewinders. According to the Brazilian Commander of the Armed Forces, General Staff General Benedito Leonel, in late 1995 a decision had been made to upgrade 45 F-5Es (plus six later) and three F-5Fs (plus six later) with new fire-control radar, avionics and related structural components to extend their useful lives well into the next century. Brazil also has an FX programme to find a new fighter aircraft, which is currently examining several types.

GROUND ATTACK AND RECCE

Dedicated long-range ground attack and interdiction are the roles of the Alenia/EMBRAER AMX (or A-1). Brazil's first two production A-1s were delivered in October 1989 to a new unit, 1º Esquadrão of 16º Grupo de Aviação, at Santa Cruz. FAB orders for 65 single-seat A-1s plus 14 two-seat AMX-T TA-1 (A-1B) combat and training versions are to equip up to five squadrons. The aircraft is intended to take over some of the ground-attack roles now performed by AT-26 Xavante armed trainers.

IIIª FAe ground-attack and reconnaissance forces also include three squadrons equipped with EMB-326GB or AT-26 (Ataque Treinamento) Xavante armed jet trainers. 1º/10º GAv 'Poker' and 3º/10º GAv 'Centauro' operate within 10º Grupo from Santa Maria in Rio Grande do Sul. Tasked with a tactical reconnaissance role, 1º/10º GAv is equipped with a dozen or so RT-26 (Reconhecimento) Xavante versions, carrying a Vinten camera pod under the port inboard pylon. This role, plus operational conversion and tactical training, is also undertaken in northeastern Brazil by IIIª FAe's third Xavante unit, 1º/4º GAv 'Pacau' at Fortazela, in Ceara province. A few RT-26s are also operated by 1º/4º, as well as one or two UH-50 Esquilo helicopters for SAR, alongside the base flight's Neiva T-25 communications lightplanes.

As IIIª FAe's final unit, the activities of 1º/6º GAv 'Carcara' at the FAB's most easterly air base – Recife – are officially classified, although nominally involve photographic survey, ground-mapping and reconnaissance. For these roles, the unit has been equipped with six EMBRAER EMB-110B (R-95) Bandeirantes and three R-35A Learjets.

TRANSPORT ASSETS

Transport support for the FAB's combat forces is provided by the Quinta Força Aérea (Vª FAe) from its HQ in Rio de Janeiro, through seven squadrons in five air wings. Heavy logistic support is provided by

observation and forward air control roles. About four are also operated in similar roles by 5º/8º GAv 'Pantera' from Santa Maria, near the southern border, in conjunction with eight Bell UH-1Hs and four EMBRAER EMB-810 Senecas (U-7). UH-1Hs and U-7s are similarly operated by 7º/8º GAv 'Falcão', in the Amazon region at Manaus.

AIR DEFENCE

Brazil's air combat units are grouped within the Terceira Força Aérea (IIIª FAe), with HQ in Brasília, although also integrated with the independent Comando Geral de Apoio (COMGAP), or General Support Command. From its HQ, this command manages the national air defence system, which is further integrated through COMGAR's NuCOMDAER HQ in Brasília.

For interception roles 1º Grupo de Defesa Aérea (1º GDA) operates with a single squadron (1º/1º GDA 'Jaguares') of Dassault Mirage IIIEs within IIIa FAe, based at Anápolis in Goias province, for protection of the capital, Brasília. By late 1995, a dozen F-103E and four F-103D Mirage IIIs remained in FAB service. They are scheduled to

Left: The FAB uses the Hercules for transport (five C-130Es and five Hs), air-to-air refuelling (two KC-130Hs) and search and rescue (two SC-130Es).

Below: The main helicopter in service with the FAB is the UH-1H. Numbers of this helicopter are rising as ex-US Army examples are acquired.

1° Esquadrão of the 1° Grupo de Transporte de Tropas (1°/1° GTT 'Cascavel' – 1 Troop Transport Wing) at Campo dos Afonsos. This unit operates four Lockheed C-130E Hercules on mainly cargo movements. Troop transport and paratrooping missions are usually undertaken from the same base by five de Havilland Canada DHC-5A Buffalos (C-115) operating with 2°/1° GTT 'Cascavel', while up to 10 fly with 1°/9° GAv 'Arara' from Manaus. Five C-130Hs and two SC-130E search and rescue versions are now operated by 1°/1° Grupo de Transporte (1°/1° GT 'Coral') from Galeão. 1°/2° GT 'Condor' at Galeão is unique in operating 12 Hawker Siddeley (BAe) 748 Series 2/2A (C-91).

Light transport support in the Mato Grosso area within Vª Força Aérea is provided from Campo Grande by the 1°/15° GAv 'Onça' with C-95Bs.

Long-range operations are also undertaken by the four ex-VARIG airline Boeing 707-320Cs (KC-137s) with 2° Esquadrão of 2° Grupo de Transporte (2°/2° GT 'Corsario') from Galeão, although their main activity is to act as tankers for the combat force.

VIP TRANSPORT

Grupo de Transporte Especial (GTE), the FAB's VIP and government transport unit, is an independent unit and reports directly to the Aeronautical Ministry in Brasília, where it is based. It includes the Presidential Flight, equipped with two Boeing 737-200s, eight HS 125 Series 3B/RC and nine VU-35A Learjets. GTE's aircraft inventory is rounded off by two VH-55 Twin-Esquilo light helicopters. Six EMB-121 Xingus are flown by 6° ETA.

FAB's REGIONAL AIR COMMANDS

In addition to its three operational air forces, the FAB has a superimposed Regional Air Command administrative structure, comprising seven Comandos Aéreos Regionais (COMARs), covering all 27 states of Brazil. These are also aircraft-operating units, each regional HQ having its own similarly numbered light transport and tactical support squadron (Esquadrão de Transporte Aéreo – ETA), which are the FAB's main users of the C-95 Bandeirante in its various versions.

NON-COMGAR AIR UNITS

The Comando Geral de Apoio (COMGAP), or General Support Command, controls the Grupo Especial de Inspecção em Voo (GEIV), or Special Flight Inspection Wing, responsible for checking and calibrating the FAB's ground-based navigation and approach aids. It operates four specially-equipped EC-95/B Bandeirantes, and two EU-93 BAe 125s from its base at Rio's downtown Santos Dumont airport. COMGAP also provides depot-level maintenance for FAB aircraft and helicopters in six regional Parques de Material Aeronáutico or Aeronautical Equipment Parks (PAMAs).

An important FAB unit operated by the Departamento de Pesquisa e Desenvolvimento (DEPED), or R&D Department, is the Centro Tecnico Aéroespacial (Aerospace Technical Centre) located at São José dos Campos, São Paulo. The CTA undertakes service trials and more basic research with a variety of aircraft which includes examples of the XC-95 Bandeirante, XT-27 Tucano, YT-27 Tucano, YTA-1 AMX trainer, XC-97 Brasília, AT-26 Xavante, XU-93 BAe 125 radar testbed, U-7 Seneca and U-42 Regente. An associated DEPED unit is Centro de Lançamento de Alcantara (CLA), which is responsible for telemetry equipment used in association with Brazil's participation in the Ariane space programme, and the rockets and satellites launched from French Guiana, which uses a single Cessna 208 (C-98).

One other independent unit is the Esquadrão de Demonstração Aérea (EDA), also known as the 'Esquadrilha da Fumaça' ('Smoke Squadron'), which is Brazil's national aerobatic team. It has nine bright red-and-white Tucanos.

FAB TRAINING ORGANISATION

All FAB pilots undergo a four-year academic and flying training course at the Academia da Força Aérea (AFA), or Air Force College, which is located at São Paulo's sprawling Pirassununga Air Base; they begin with an initial 16 hours of grading instruction in the Neiva T-25 Universal. No further formal flying training is undertaken during the first two years of the course, but cadets are encouraged to participate in the activities of the AFA's microlight and gliding clubs (Clube de Ultraleves/CU and Clube a Vôo à Vela/CVV). Sixty-five hours of basic training is then flown during the third year on the Universal, operated by 2° Esquadrão de Instrução Aérea (EIA) 'Apollo'. The fourth year is for advanced instruction, using the EMB-312 Tucano.

FINAL TRAINING STAGES

Successful students go to COMGAR's Comando Aéreo de Treinamento (CATRE), which has its HQ in Natal, on Brazil's northeastern tip. Some pilots are streamed at this stage, those selected for helicopter training being posted to 1°/11° GAv 'Gavião' at Santos, near São Paulo, to convert to the Aérospatiale/Helibras HB 350B (UH-50) Esquilo. All other personnel continue to the next stage of a year's tactical and weapons training on armed AT-27 Tucanos of the resident former 5° Grupo at Natal, 1°/5° 'Rumba' and 2°/5° 'Joker' GAv. During the AT-27 course with 1°/5°, students are progressively assessed for their suitability as potential fast-jet combat pilots, some being judged more suitable for transport flying. Fast-jet students then move across to 2°/5° GAv to complete the third year on the Universal, operated by 2° Esquadrão de Instrução Aérea (EIA) 'Apollo'. The fourth year is for advanced instruction, using the EMB-312 Tucano.

FUTURE DEVELOPMENTS

The Xavante is now considered obsolete by the FAB, which is planning its replacement with the ALX Super Tucano. Developed from the stretched EMB-312H, it is fitted with five underwing weapons pylons and associated nav/attack systems, head-up and multi-function displays, cockpit armour and night-vision systems. About 100 ALXs in both single-seat A-29 and two-seat TA-29 armed trainer versions are to be acquired, with the single-seaters to be used as the armed interception platform for the SIVAM project.

The Sistema de Vigilancia da Amazonia (SIVAM) project will comprise integrated ground and airborne radars, mounted on EMBRAER ERJ-145SAs, plus other sensors and communication systems. First deliveries are expected in 2000. It is intended to monitor the vast and remote areas of the Amazon for drug trafficking, illegal mining and logging, plus unauthorised destruction of the rain forests, as well as providing early warning, air defence and air traffic control facilities.

Aviação do Exército Brasileiro

The Brazilian army relied on the air force for its air support until 1986, when it started to form its own air wing. Orders were placed for 16 Helibras HB 350L1 Esquilos (HA-1) for light attack and observation duties and 36 Eurocopter AS 565AA Panthers (HM-1) for use as general transport, and deliveries were underway by 1989. In 1992 20 Eurocopter AS 550A2 Fennecs were ordered to supplement the Esquilos.

The Brazilian army utilises examples of both locally assembled and French-built Esquilo/Fennecs as HA-1s.

The army aimed to provide each of the 14 calvary regiments with its own aviation component, but presently only one is known to be so equipped. The

Aviação do Exército Brasileiro

Headquarters – Santa Maria, Rio Grande del Sul

1ª Brigada de Aviáćo Exército   Taubate, São Paulo
1 Batalhão de Helicópteros   Taubate, São Paulo
AS 550A2 Fennec, HB 350L1 Esquilo
Companhia de Helicópteros de Reconhecimento e Ataque
AS 550A2 Fennec, HB 350L1 Esquilo
Companhia de Helicópteros de Manobra
AS 565AA Panther
Plus:
3 x EMB-110 Bandeirante

air wing gained a light transport capability when it received three C-95 Bandeirantes from the air force, in January 1996.

The Brazilian army would like to acquire the S-70 Blackhawk, but no orders have yet been placed.

South America

Força Aeronaval da Marinha do Brazil

The Escola de Aviação Naval (Naval Aviation School) was founded in August 1916, becoming the Corpo de Aviação (Navy Aviation Corps) in October 1931. The creation of the Forças Aéreas Nacionais (National Air Forces) in January 1941 ended independent naval aviation until 1958. The recent talks concerning the ex-Kuwaiti Skyhawks will transform the navy, especially if current negotiations for the carrier USS *Saratoga* come to fruition. Currently, the Força Aéronaval da Marinha do Brasil operates 55 helicopters, spread among eight units at four bases, and has a sizeable number of helicopter-capable vessels including, still, the carrier *Minas Gerais*.

The Marinha's oldest aviation asset are the Sea Kings of Esquadrão de Helicópteros Anti-Submarino 1 (HS-1). Seven survivors are currently being modified to ASH-3A (SH-3H) standard, while three ex-US Navy SH-3Hs (as ASH-3Bs) arrived in 1996. Normally, three to four aircraft are assigned to the carrier.

The real teeth are the Westland Lynx Mk 21s (SAH-11s) of Esquadrão de Helicópteros de Esclarecimento e Ataque 1 (HA-1). The original Lynx Mk 21s will be upgraded to Mk 21A standard by Westland, while nine Super Lynx Mk 21As are being delivered.

The majority of the Marinha's helicopter force is devoted to transport duties, although some have a secondary armed role. Between them, Esquadrões de

Esquadrão de Helicópteros de Emprego Geral 2 at São Pedro da Aldeia operates the UH-14 (AS 332M Super Puma) in the SAR and utility transport roles.

Brazil's surviving Lynx Mk 21s were converted to Super Lynx standard as Mk 21As, while another nine examples were delivered to this standard.

Helicópteros de Emprego Geral 1, 3, 4 and 5 (HU-1, -3 -4 and -5) operate 17 licence-built Aérospatiale (Eurocopter France) Squirrels and Twin Squirrels, eight (of 11 ordered) single-engined UH-12s (Helibras HB-350/AS 550U2) and nine twin-engined UH-13s

Força Aeronaval da Marinha do Brazil

Esquadrão de Helicópteros Anti-Submarino 1	
ASH-3D, SH-3D/H	São Pedro da Aldeia
Esquadrão de Helicópteros de Esclarecimento e Ataque Anti-Submarino 1	
Lynx HAS.Mk 21A	São Pedro da Aldeia
Esquadrão de Helicópteros de Emprego Geral 1	
HB 355F2 Esquilo II, HB 350BA Esquilo	São Pedro da Aldeia
Esquadrão de Helicópteros de Emprego Geral 2	
AS 332M Cougar	São Pedro da Aldeia
Esquadrão de Helicópteros de Emprego Geral 3	
HB 350BA Esquilo	Manaus
Esquadrão de Helicópteros de Emprego Geral 4	
HB 350BA Esquilo	Ladário
Esquadrão de Helicópteros de Emprego Geral 5	
HB 350BA Esquilo	Rio Grande
Esquadrão de Helicópteros de Instrução 1	
Bell 206B JetRanger III	São Pedro da Aldeia

(Helibras HB-355F/AS 555U2). Both versions are known as Esquilo (Ecureuil/Squirrel).

The Eurocopter AS 332M Cougars (UH-14s) of HU-2 provide heavy-lift capability, in addition to troop transport and SAR duties. The large UH-14s can only be deployed aboard the *Minas Gerais* or the two 'Mattoso Maia'-class assault ships.

Pilot training at São Pedro da Aldeia with Esquadrão de Helicópteros Instrução 1 (HI-1) takes place on Bell 206B Jetranger IIIs (IH-6B). Students undertake a 120- to 140-hour flying course.

In late 1998 Marsh Aviation put forward a proposal to provide around 12 S-2F3T/E-1T TurboTrackers to fulfil its AEW requirement. The aircraft would be conversions of ex-US Navy E-1B Tracers powered by Allied Signal TPE-331-14G/R turboprops. A decision to proceed (or not) will be made in 1999.

CHILE

Capital: Santiago de Chile
Population: 13.4 million
Land area: 752000 km² (290,000 sq miles)
Major cities: Vinã del Mar, Valparaiso, Talcahuano, Concepción

Fuerza Aérea de Chile

The FACH, which numbers approximately 13,000 men and women, operates about 320 fixed- and rotary-winged aircraft, and is organised into three functional commands – Combat, Personnel and Logistics. Combat Command controls the five air brigades and the five wings which operate most of the flying equipment of the FACH. They are deployed between a total of 13 groups (squadrons). There is reasonable AEW radar cover over most of the country but this extends effectively beyond the country's land frontiers only in the extreme north and south, due to the physical barrier of the Andes to the east. The recent acquisition of a Phalcon AEW aircraft (modified from a former LAN-Chile Boeing 707 by IAI and designated Condor in Chilean service) will go some way to remedy this deficiency. This aircraft was used to monitor French nuclear testing in the Pacific during the early part of 1996.

The Brigada Aérea (Air Brigade), which consists of two or more Grupos (squadrons), is the main operational formation of the FACH, the Ala Base (Base Wing) being primarily an administrative and logistic support unit which is generally concentrated at a single base. Each wing includes a liaison flight, equipped with assorted light aircraft and helicopters, plus an anti-aircraft artillery group. There is also an anti-aircraft artillery regiment at La Colina, near Santiago, which serves primarily as an administrative headquarters and training school for the five dispersed A/A artillery groups. Each wing also has an electronic communications group, and the groups attached to the 1st and 4th Wings (31st and 34th) also include an

A-37Bs serve with Grupo 12 from Punta Arenas. A total of 44 was delivered to the air force, starting in 1975 with a batch of 34, followed by a further 10 ex-USAF examples in 1992.

Chile acquired 24 of the upgraded ex-Belgian air force Mirage 5s between March 1995 and April 1996. In Chilean service the aircraft have been further upgraded and are known as Elkans.

The ENAER update of the Mirage 50C, the Pantera C, flies with the 4th squadron of Air Brigade IV from 'Carlos Ibánez' Military Air Base, Punta Arenas.

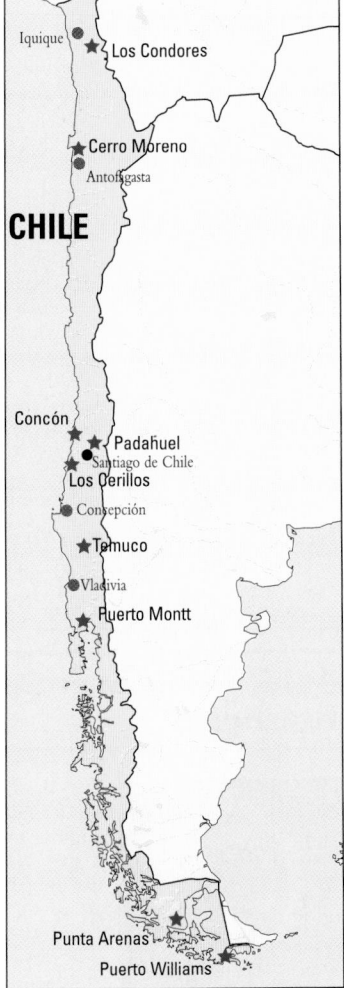

electronic warfare element.

Air Brigade I (HQ Los Condores, Iquique) covers the northern part of the country, from the Peruvian frontier to the Huasco River in southern Atacama

Province, and controls the 4th Wing, at Los Condores. This brigade consists of the 1st (Attack/Training) and the 2nd (Reconnaissance and Photographic) Squadrons. The 1st Squadron, which

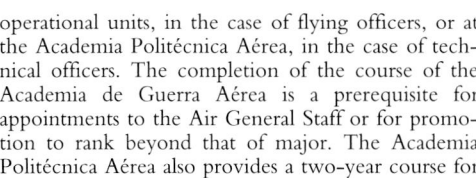

serves as a combined light strike and operational training unit, is equipped with ENAER T-36 trainers and A-36 light strike aircraft. The 2nd Squadron, formerly based at Los Cerrillos, Santiago, is a special unit operating five modified Beech 99As, primarily in Elint/Sigint roles. The 1st Wing also includes the 11th Cerro Moreno Liaison Flight, the 21st A/A Artillery Group, and the 31st Electronic Communications Group.

Air Brigade II (HQ Los Cerrillos, Santiago) covers the region southward from Huasco to the Bío-Bío River, combining the 9th, 10th and 11th Squadrons, and consists of the 2nd Wing, also embracing the two Learjet 35As of the Aerial Photogrammetric Service. The 9th Squadron was reactivated at Los Cerrillos in 1993 and is now the major helicopter element of the FACH, equipped with 16 Bell UH-1Hs, six MBB BO 105s and a single BK 117. The FACH is proposing to replace its UH-1H fleet with S-70A-39 Black Hawks on a one for three basis, receiving a single example in July 1997 which is used in support of Antarctic operations. The 10th (Transport) Squadron, based at Comandante Arturo Merino Benitez International Airport, Pudahuel, Santiago, is the main transport unit of the FACH and is equipped with four Lockheed C-130Bs and two Hs; three Boeing 707s, of which two have been modified to KC-137 standard for inflight refuelling (for which all the first-line combat aircraft of the FACH are now equipped); and a fourth Boeing 707, modified to Phalcon AEW standards by IAI. A single Boeing 737-58N arrived in September 1997 for the Presidential Flight, which previously operated one of the existing Boeing 707s. An ex-USAF C-20B Gulfstream III and a Beech King Air 200 also operate with the flight. The 11th (Training) Squadron, which was primarily a refresher training unit disbanded early in 1998, transferring its ENAER T-35 Pilláns to El Salvador (five), Guatemala (five) and El Bosque. It also controlled both the Piper PA-28 Dakotas of the Air Force Specialists' School and the five Extra 300s of the 'Halcones' aerobatic team. The 2nd Wing also includes the 12th Liaison Flight, the un-numbered and un-named air force paratroop battalion and anti-aircraft artillery regiment, the 22nd A/A Artillery Group, and the 32nd Electronic Communications Group.

Air Brigade III (HQ El Tepual Military Air Base, Puerto Montt) covers the region between the Bío-Bío River and Cerro San Valentín, in southern Aysén Province and consists of the 3rd and 5th Squadrons, also comprising the 5th Wing, at Puerto Montt. The 3rd (Attack) Squadron is a light strike unit, equipped with CASA/ENAER A-36Bs and based at Temuco, and the 5th (Communications) Squadron is a light

The FACH is receiving ENAER-assembled two-seat CASA 101BBs (14 ordered as T-36s) and single-seat CASA 101CCs (23 ordered as A-36) at a slow rate.

Three 40 year-old L-19A Bird Dogs are in use to tow gliders (above). A single C-20B Gulfstream III (ex-86-0200) belongs to the Presidential Flight (left).

transport unit, equipped with four CASA C.212s and based at Puerto Montt. The 5th Wing also comprises the 15th Liaison Flight, the 25th A/A Artillery Group, and the 35th Electronic Communications Group.

Air Brigade IV (HQ 'Carlos Ibáñez' Military Air Base, Punta Arenas) covers the region south from Cerro San Valentín to Cape Horn and is made up of the 4th, 6th and 12th Squadrons, all based at Punta Arenas. It also includes the 3rd Wing. The 4th (Fighter) Squadron operates 15 Dassault Mirage 50CHs, DCHs and FCHs which have been upgraded to Pantera standards, plus five unmodified ex-Belgian Mirage 5s that also will be eventually upgraded. The 6th (Special Operations) Squadron is equipped with four armed DHC-6s, and the 12th (Attack) Squadron operates Cessna A-37Bs in the light strike role. The Fourth Brigade also controls the 19th (Antarctic Exploration) Squadron which is based at 'Teniente Marsh' Military Air Base in the Chilean Antarctic territory, where it operates single examples of the DHC-6, BO 105 and Bell UH-1H. The 3rd Wing also includes the 23rd A/A Artillery Group and the 33rd Electronic Communications Group.

Air Brigade V, which was formed only at the end of 1995, is based at Cerro Moreno, Antofagasta. This is the main fighter unit of the FACH, originally forming part of Air Brigade I, and comprises the 7th and 8th Squadrons and the 1st Wing. The 7th (Fighter) Squadron is equipped with Northrop F-5Es (12) and Fs (three). They have been upgraded to Tigre III standards and are presumably to be joined by the mysterious additional F-5Es believed to have been acquired from Honduras. The 8th (Fighter) Squadron has recently re-equipped with 15 ex-Belgian Mirage/MirSIP 5BAs that have been upgraded to Elkan standards. The 4th Wing also includes the 14th 'Los Condores' Liaison Flight, the 24th A/A Artillery Group, and the 34th Electronic Communications Group.

Logistics Command controls the non-flying supply and maintenance wings, both which also have their headquarters at El Bosque.

Air Force Personnel Command controls the Escuela de Aviación Capitán Avalos (Air Force College), the Specialists' School, the Academia de Guerra Aérea (Air Force Staff College) and the Academia Politécnica Aérea (Air Force Technical School). All are located at El Bosque, Santiago.

The Escuela de Aviación Capitán Avalos is at El Bosque, Santiago, and its flying elements are equipped with ENAER T-35 Pilláns, without any squadron organisation. The school offers a basic three-year course to officer cadets, followed by two years of specialised training before commissioning, in the rank of second lieutenant. This is carried out either in

operational units, in the case of flying officers, or at the Academia Politécnica Aérea, in the case of technical officers. The completion of the course of the Academia de Guerra Aérea is a prerequisite for appointments to the Air General Staff or for promotion to rank beyond that of major. The Academia Politécnica Aérea also provides a two-year course for

Fuerza Aérea de Chile

Combat Command

Anti-Aircraft Artillery Regiment	La Colina

Brigada Aérea I HQ Los Condores, Iquique

Grupo 1	A-36B/T-36 Halcón
Grupo 2	Beech 99A Petrel Alpha/Beta

Ala Base 4 HQ Cerro Moreno, Antofagasta

11th Cerro Moreno Liaison Flight
	DHC-6-100, SA 315B Lama, PA-28-236 Dakota

21st Anti-Aircraft Artillery Group
31st Electronic Communications Group

Brigada Aérea II HQ Los Cerrillos, Santiago

Grupo 9	UH-1H, BO 105CB-4/CBS, BK 117B-1
Grupo 10	KC-137, 707-351C, IAI/Elta Phalcon 1, C-130B/H
	based at Comandante Arturo Merino Benitez Int AP

Air Force Specialists' School
	PA-28-236 Dakota

Esc de Planeadores   L-19A, L-13 Blanik, Discus 6T, Janus C,
                                    Nimbus 3DT/4M, Ventus CM

'Halcones'	Extra 300

Aerial Photogrammetric Service
	Learjet 35A

Presidential Flight   King Air B200, C-20B, 737-58N
                              based at Comandante Arturo Merino Benitez Int AP

Ala Base 2, HQ Los Cerrillos, Santiago

12th Liaison Flight   DHC-6-300, Beech 99A, King Air 100/B200,
                               PA-28-236 Dakota
                               based at Comandante Arturo Merino Benitez Int AP

air force paratroop battalion
anti-aircraft artillery regiment
22nd Anti-Aircraft Artillery Group
32nd Electronic Artillery Group

Brigada Aérea III HQ El Tepual Puerto Montt

Grupo 3	A-36B Halcón	Temuco
Grupo 5	CASA C.212-100	Puerto Montt

Ala Base 5, HQ El Tepual Military Air Base Puerto Montt

15th Liaison Flight   T-35A/B Pillán, BO 105CB-4/CBS
25th Anti-Aircraft Artillery Group
35th Electronic Communications Group

Brigada Aérea IV HQ 'Carlos Ibanez', Punta Arenas

Grupo 4	Pantera C, Mirage 50DC	
Grupo 6	DHC-6-300	
Grupo 12	A-37B	
Grupo 19	UH-1H, DHC-6-300,	'Teniente Marsh'
	BO 105CB-4/CBS	Antarctica

Ala Base 3 HQ 'Carlos Ibanez', Punta Arenas

23rd Anti-Aircraft Artillery Group
33rd Electronic Communications Group

Brigada Aérea V HQ Cerro Moreno, Antofagasta

Grupo 7	F-5E/F Tigre III
Grupo 8	Mirage 5MA/MD Elkan

Ala Base 1 HQ Cerro Moreno, Antofagasta

14th 'Los Condores' Liaison Flight
	DHC-6-300, SA 315B Lama

24th Anti-Aircraft Artillery Group
34th Electronic Communications Group

Logistics Command

Supply Wing	HQ El Bosque
Maintenance Wing	HQ El Bosque

Air Force Personnel Command

Academia de Guerra Aérea	El Bosque
Academia Politécnica Aérea	
PA-28-236 Dakota	El Bosque
Escuela de Aviación Capitán Avalos	
T-35A/B Pillán	El Bosque

NCOs and is equipped with Cessna T-41s and Piper PA-28 Dakotas.

An order for between 15 and 20 T-6A Texan IIs spelt the end for the T-37B/C in Chilean service. The type was retired on 23 December 1998, even though the T-6s will not enter service for some time.

UPGRADE PROGRAMMES

In 1985, with the assistance of Israel Aircraft Industries, Empresa Nacional de Aeronáutica de Chile (ENAER) commenced the modernisation of the Chilean air force's Mirage 50s to a configuration resembling that of the IAI Kfir, the upgraded aircraft being known as the Pantera. A comparable upgrade was subsequently carried out on the air force's Northrop F-5 Tigers, which were then known as Tigres. Most of the Mirage 5s, recently purchased from Belgium, have also already undergone a degree of modernisation and are known as Elkans, which

Under the Tigre III programme the FAC's F-5E/Fs avionics were upgraded. The first Tigre was redelivered to the air force in September 1993.

nevertheless leaves them short of the ENAER/IAI Pantera upgrade.

The existing fleet of F-5s is also being augmented. Originally this was believed to be by the transfer of the 12 examples of this aircraft operated by the Honduran air force and which, after the relative pacification of the Central American region, were deemed by the United States to upset the local balance of air power. Although several F-5s of apparently Honduran origin were delivered to Chile during the early months of 1996, at least six aircraft of this type were still noted to be in Honduran service during the following September. A possibility is that Honduras yielded partially to US pressure and sold half its F-5 inventory to Chile, but the matter remains shrouded in mystery.

Comandancia de Aviacíon de la Armada de Chile

The Chilean navy is traditionally Latin America's best in qualitative terms, even though in size it now lags behind the navies of Argentina, Brazil and Peru. It includes two helicopter-carrying destroyers each equipped to handle and support two AS 332 Super Pumas, two other helicopter-capable destroyers, and four helicopter-capable frigates. Several auxiliary, logistic support and amphibious vessels can also accommodate and/or support helicopters.

Chilean naval aviation accounts for a modest 800 or so people from a total naval manpower of about 24,000, operating approximately 60 aircraft. It is organised into two Naval Air Forces. Naval Air Force 1 based at Concón controls (General Purpose) Squadron VC-1, which is equipped with CASA C.212s, EMBRAER EMB-110CNs (three examples of each, for transport and communications tasks) and UP-3As (two); (General Purpose Helicopter) Squadron HU-1 with MBB BO 105s (eight in service with the navy) which are replacing the surviving Bell 206Bs (six still in service); and (Maritime Reconnaissance) Squadron VP-1 with Lockheed P-3A Orions (four) and EMBRAER EMB-111ANs (six in service). Iquique and Talcahuano air stations each have an EMB-111AN and a BO 105 attached to the Station Flight. Training

Squadron VT-1 is equipped with Pilatus PC-7s (10 aircraft) which can also double in the light strike role, and eight Cessna O-2As from surplus US stocks. Ten arrived in 1997 (with two being used for spares) and are used for coastal surveillance and SAR.

Naval Air Force 2 controls (Attack Helicopter) Squadron HA-1, equipped with seven AS 332 Super Pumas tasked with ASW duties, based at Puenta Arenas. The Puenta Arenas Base Flight has a CASA C.212 attached, while a BO 105 operates at the Puerto Williams Air Station Flight.

Having won the battle with the air force for the control of all military aircraft which operate over the sea, the Chilean navy would like to operate a small number of Sea Harriers or about 20 A-36M Halcóns.

Comandancia de Aviacíon de la Armada de Chile

Naval Air Force 1, Concón, Vina del Mar

HU-1	Bell 206B, BO 105S/LSA-1/CBS-5
VC-1	C.212A-200, EMB-110CN, UP-3A
VP-1	EMB-111AN, P-3A
VT-1	PC-7, O-2A
Iquique Air Station Flight	EMB-111AN, BO 105
Puerto Montt Air Station Flight	EMB-111AN, BO 105
Talcahuano Air Station Flight	EMB-111AN, BO 105

Naval Air Force 2, Punta Arenas

HA-1	AS 332B, AS 532SC
Punta Arenas Base Flight	EMB-111AN, BO 105
Puerto Williams Air Station Flight	EMB-111AN, BO 105

Above: HA-1 operates seven AS 332 Super Pumas in the anti-submarine warfare role from Punta Arenas.

Below: The Chilean army uses 24 MD.530Fs as its primary combat and combat support helicopters.

Aviacíon del Ejército de Chile

The aviation element of the 57,000-strong Chilean army consists mainly of the 1st 'Independencia' Aviation Brigade, which was promoted from regimental to brigade status at the beginning of 1996 and is based at Rancagua, near Santiago. It deploys approximately 90 aircraft, most of which are concentrated within the brigade. Others are deployed on an *ad hoc* basis in accordance with operational requirements, in support of five of the army's seven divisions.

Carabineros de Chile

Since 1927 all law enforcement agencies in Chile have been incorporated into a single force – the Carabineros de Chile – which has a paramilitary organisation and forms a potential reserve for the army. In addition to normal police functions, the Carabineros are also responsible for customs control, and provide the Presidential Guard. The Air Police, which ranks as a separate prefecture within the overall structure of the 32,000-strong Carabineros, operates 20 to 30 fixed- and rotary-winged aircraft. There is no known subordinate group or squadron organisation.

Carabineros de Chile

Prefectura Aérea de los Carabineros de Chile

Bell 206L3 JetRanger	2
Cessna 182Q	4
Cessna U206G Stationair	2
Cessna 210M Centurion II	2
Eurocopter BO 105CBS/LSA-3	8
Piper Navajo	4
Swearingen SA-226TC Metro	4

Aviacíon del Ejército de Chile

1st 'Independencia' Aviation Brigade, Rancagua

Uses the following aircraft:

Aérospatiale SA 315B Lama	16	observation helicopter
Aérospatiale SA 330F Puma	8	transport helicopter
Aérospatiale SA 330L Puma	2	transport helicopter
Airtech CN.235M-100	5	transport
Beech 58 Baron	1	communications
Beech King Air B90	1	communications
Bell UH-1H Iroquois	3	transport helicopter
Bell 206	2	communications
CASA C.212A-100 Aviocar	3	transport
CASA C.212-300 Aviocar	3	transport
Cessna R 172K Hawk XP	16	primary training
Cessna R 182 Skyline RGs		(on order)
Cessna 208/B	8	transport
Cessna 337	3	observation
Cessna Citation II	1	VIP
Enstrom 280FX	15	helicopter training
Eurocopter AS 332B Super Puma	2	transport helicopter
Eurocopter AS 332M1 Super Puma	1	transport helicopter
McDonnell Douglas MD 530F	24	combat helicopter
Piper Navajo	4	communications

Aircraft are deployed to the units below from the above unit as required:

1st Division

Sección de Aviación del Ejército 1	Antofagasta

3rd Division

Sección de Aviación del Ejército 3	Temuco

4th Division

Sección de Aviación del Ejército 4	Aerodromo Las Marias

6th Division

Sección de Aviación del Ejército 6	Arica

7th Division

Sección de Aviación del Ejército 7	Las Bandurrias

COLOMBIA

Capital: Bogotá
Population: 33 million
Land area: 1.139 million km² (440,000 sq miles)
Major cities: Medellín, Cali, Barranquilla

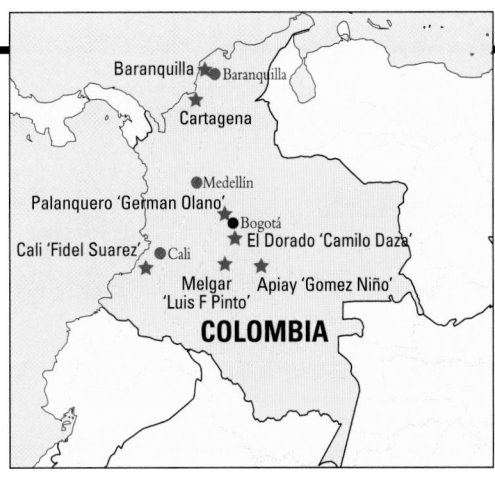

Fuerza Aérea Colombiana

The Fuerza Aérea Colombiana (FAC) currently comprises six main commands: Comando Aéreo de Combate (CACOM), Comando Aéreo de Apoyo Táctico (CAATA), Comando Aéreo de Transporte Militar (CATAM), Escuela Militar de Aviación (EMAVI), Comando Aéreo de Mantenimiento (CAMAN), and Servicio de Aeronavegación a Territorios Nacionales (SATENA). They are principally self-contained, each having a single Grupo and a number of subordinate Escuadrones. Both CACOM and CAATA have a number of elements which have their own separate headquarters and are differentiated by the title CACOM-1, CAATA-2, etc.

The country's air defence is centred around the US-supplied Peace Panorama integrated air-surveillance system that became operational in August 1993. Air assets are provided by CACOM-1 at Base Aérea 'German Olano', Palanquero. Here Grupo 21 has under its control two squadrons of Dassault Mirage and Kfir fighters. Escuadrón de Combate 212 operates 10 Mirage 5COA and two 5COD fighters. They are being upgraded in a joint programme with IAI to a standard similar in avionics to the 12 Kfir C7s and single two-seat TC2 operated by Escuadrón 212's sister unit, Escuadrón 213. Another unit currently parented by Grupo 21 is believed to be Escuadrón de Combate 215, equipped with the Cessna T-37B/C and employed in the advanced flying phase of flight training.

ANTI-TERRORIST/DRUG OPERATIONS

Also located at Palanquero is an Escuadrón Aerotáctico, a unit with a primary anti-terrorist or anti-narcotics tasking. The squadron is equipped with the recently received Basler Turbo 67 AC-47 Dakota and supported by a number of helicopters drawn from the Comando Aéreo de Apoyo Táctico. It is the FAC's intention to set up four of these units to counter known threats at the strategic locations of Palanquero, Barranquilla, Apiay and Cali. Internal problems provoked by the Revolutionary Armed Forces of Colombia (FARC), the leftist National Liberation Army (ELN) and the People's Liberation Army (EPL) have necessitated the establishment of these air units to support both the police and army.

CACOM-2 is to be found at Base Aérea 'Luis F. Gomez Niño', Apiay, in the Los Llanos region; Grupo 31 is the controlling element. Within this command lies the responsibility for COIN operations and advanced pilot training.

Escuadrón 311 has the main COIN task and is equipped with 11 OV-10A Broncos. An additional three ex-USMC aircraft were delivered in 1993, but the type suffers from poor serviceability. As a result, some EMBRAER T-27 Tucanos of Escuadrón de Combate 312 have been fitted with hardpoints to help bolster the COIN commitment. Three IA-58 Pucarás did equip Escuadrón de Operaciones Especiales 314 but have now effectively been withdrawn from use.

CACOM-3 is to be found at the most northerly Colombian mainland base, Base Aérea del Atlántico, Barranquilla. Here, under its Escuadrón de Combate 411, Grupo 41 operates the A-37B. The jet is used in the CAS role, for coastal patrols and the interception of narcotic traffickers. Although 20 aircraft are still on strength, probably just over half could be considered

operational due to limited availability of spares. From these the unit has to maintain a two/three-aircraft detachment on San Andres Island, just off the coast of Nicaragua. They operate from this location under the control of Grupo Aéreo del Caribe.

HELICOPTERS

Tactical helicopter assets are assembled under the Comando Aéreo de Apoyo Táctico, which is divided into two primary operating bases. CAATA-1 at Base Aérea 'Luis F. Pinto', Melgar is the primary helicopter site. Under Grupo 51 the full range of helicopter operations is undertaken, from basic training through to night attack, including NVG training.

Prospective helicopter pilots undergo 75-80 hours of training using the 12 Enstrom F-28Fs at Melgar or on detachment to Velasquez. Pilots then progress to either the Bell UH-1H/Bell 205A with Escuadrón Medianos 512 or the Hughes 369 with Escuadrón 514. Operational taskings take the crews into the various Escuadrones Aerotácticos or to Escuadrón Artillados 515 at Melgar. This is the operational parent unit and is equipped with armed Bell UH-1H/212s and Hughes 369s, both types being NVG-capable. The FAC currently has 24 Bell UH-1Hs. The US funded upgrading of 50 UH-1Hs to 'Super Huey' standard in late 1998 to help fight the drug war, thought to include all those in air force service. In addition to these, there are six surviving Bell 205A-1s and 13 Bell 212s. The Hughes 369s are a mixture of 369D/E/F/HN/HS/Ms, the bulk of which are referred to as Hughes 500/530s. A single OH-13S remains airworthy at Melgar and is displayed on suitable occasions.

CAATA-2 is at Base Aérea 'Rio Negro', located at Medellín's Jose Maria Cordova International Airport. The base is home to Grupo 61 and its subordinate Escuadrón 611 with nine Sikorsky UH-60As and four UH-60Ls. Tasked with anti-

narcotic/-terrorist operations, the helicopters frequently deploy to other locations.

TRANSPORT AND COMMUNICATION

Air transport is split between two commands: CATAM, which has purely military taskings, and SATENA, which is a military-run airline serving remote areas deemed to be unprofitable by airlines in the public sector. CATAM – Comando Aéreo de Transporte Militar – has its headquarters at Bogotá's El Dorado International airport.

The backbone of CATAM's fleet is the C-130 Hercules, supplemented by a pair of CASA C.212 Aviocars, three CASA CN.235s and two EMBRAER EMB-110P Bandeirantes ordered in December 1992, along with a single Cessna Citation 550 and Beech King Air 300. The unit also has a specially modified Rockwell Turbo Commander A695A which it uses for photographic work.

Parented by CATAM is Escuadrón Presidencial, which operates a single Fokker F28 Fellowship acquired in 1971. The former Korean Air Lines Boeing 707-373C, which had been converted by IAI in Israel to a tanker/transport configuration during 1991, and the two Bell 412s normally reside at Base Aérea Madrid.

Other communications aircraft include Cessna and Piper types mainly acquired following confiscations. Most are assigned to the various commands in the liaison role and often adopt an autonomous squadron title within the Grupo. An initial four examples of the indigenous Gavilán 358 utility transport have been ordered, with plans to acquire 12 over three years.

Above: One of the assets in the fight against the narcotic producers is the OV-10A Broncos of Escuadrón de Operaciones Especiales 311 based at Apiay. All the Broncos were acquired from surplus US military stocks, the first arriving in Colombia in 1991. The fleet suffers from a lack of spares.

Right: Like many South American countries, Colombia has a paramilitary airline, SATENA, which provides transport to remote areas in-country using this PC-6B, CASA 212s, F28s and BAe 748s.

Below: The FAC's Tucanos are used in the advanced training and light strike roles with Escuadrón de Combate 312 of Grupo 31 from Apiay.

South America

The FAC has used three Cessna 208 over the years. A number of light aircraft pass through the FAC after confiscation from narcotic traffickers.

The FAC has a mix of C-130Bs and C-130Hs (such as 1005 above) in Escuadrón de Transporte 711 based at BA 'Camilo Daze' El Dorado, Bogotá.

The Servicio de Aeronavigación a Territorios Nacionales (SATENA) is run as a structured airline using military crews. The single Fokker F28, a pair of Boeing 727s impounded from drug smugglers, and a number of CASA C.212 Aviocars operate with military registrations. During November 1996 SATENA received the first two of six Dornier Do 328-120s, the last arriving in June 1998.

Not directly aligned under CATAM, Grupo Aéreo del Sur (GASUR) is located at Base Aérea 'Ernesto Esguerra', Tres Esquinas, and maintains its own transport element close to the Ecuadorian border. Here the unit utilises a single C-47 and the last two remaining DHC-2 Beavers.

TRAINING

Aircrew, officer and trade training comes under the responsibility of Comando Aéreo de Entrenamiento through its three main training institutes. The Instituto Militar Aeronáutico based in Bogotá is responsible for officer ground training, while the Escuela de Suboficiales (ESUFA) based at Base Aérea Madrid handles technical training.

Flying training begins at Cali with the Escuela Militar de Aviación (EMAVI) located within the Base Aérea 'Marco Fidel Suarez' complex. Under Grupo de Vuelos, the school's constituent training squadrons include Escuadrón Primario (formerly Esc 611), flying the Cessna T-41D. It is on these 11 T-41Ds that first-year students undertake 35 hours of primary flying training. From there they move to Escuadrón Básico (formerly Esc 612) and the Beech T-34B for 120 hours of basic flying over the remaining three years of the course. Students are then streamed either to stay

on the Beech T-34B Mentor, to go to Apiay to fly the EMB-312 Tucano, or to Palanquero to fly the Cessna T-37B. Those destined for the rotary-wing world move to Melgar.

Also in use with the school are two and possibly four IAR IS-28B2 gliders. One of the two preserved airworthy PT-17 Stearmans is kept for sport flying, the other being at Palanquero.

MAINTENANCE

The last major command is Comando Aéreo de Mantenimiento (CAMAN). Located 19 km (12 miles) west of Bogotá, Base Aérea Madrid is responsible for the overhaul and repair of most types in the FAC inventory. Operationally, the base is the official home of the two Bell 412s of Escuadrón Presidencial, although they frequently are at the main FAC facility at El Dorado International.

The Escuadrón Presidencial (presidential squadron) uses a Fokker F28 Fellowship 1000 (above) and a 707-373C for long distance flights, while a Bell 206L-3 and a pair of Bell 412s, of three delivered (below), are used for shorter flights.

Ejército de Colombia

The first reports of an army aviation element emerged in 1991, and in the next five years its assets increased to around 20 aircraft. The mixed collection of Piper, Beech and Cessna aircraft appear to be used in the communication and liaison roles and were acquired following confiscation in anti-narcotics operations. Refused UH-60 Blackhawks, the army turned its attention elsewhere and in 1997 it received around 10 Mil Mi-17s, which it operates in conjunction with a number of leased civil-registered Mi-8MTVs. The bulk of the aviation assets are to be found operating from El Dorado International Airport, Bogotá.

Ejército de Colombia

Edificio Centro Administrativo Nacional   **Bogota**
Apiay
  1 x Cessna U206G
Base Aerea 'Luis F. Pinto', Melgar
  Mil Mi-8MTV (civil aircraft leased until Mi-17 pilots are proficient), 10 x Mil Mi-17 'Hip'
El Dorado International Airport
  1 x Beech King Air 200, 1 x Cessna 404, 1 x Convair 580, 4 x Piper PA-34, 1 x Rockwell Commander 695A, 1 x Rockwell Turbo Commander 1000

Fuerza Aérea Colombiana

HQ Bogotá

Comando Aéreo de Combate (CACOM)

Comando Aéreo de Combate 1

Grupo 21	**Base Aéreo 'German Olano', Palanquero**
Escuadrón de Combate 212	Mirage 5COA/COD
Escuadrón de Combate 213	IAI Kfir C-2/C-7/TC-2
Escuadrón Aerotáctico 214	Basler Turbo 67 gunship, Hughes 369D/E/F/HN/HS/M (loaned from Grupo 51), UH-1H (loaned from Gr 51)
Escuadrón Aerotransporte 214	PA-31T Turbo Cheyenne II
Escuadrón de Combate 215	T-37B/C

Comando Aéreo de Combate 2

Grupo 31	**Base Aéreo 'Luis F. Gomez Niño', Apiay**
Escuadrón de Operaciones Especiales 311	OV-10A Bronco
Escuadrón de Combate 312	EMB-312 Tucano
Escuadrón Aerotáctico 313	Hughes 369D/E/F/HN/HS/M (loaned from Grupo 51), UH-1H (loaned from Grupo 51)

Comando Aéreo de Combate 3

Grupo 41	**Base Aéreo del Atlántico, Barranquilla**
Escuadrón de Combate 411	OA/A-37B
Escuadrón Aerotransporte 412	Beech B80 Queen Air

Comando Aéreo de Apoyo Táctico (CAATA)

Comando Aéreo de Apoyo Táctico 1

Grupo 51	**Base Aéreo 'Luis F. Pinto', Melgar**
Escuadrón Medianos 512	UH-1H Iroquois, Bell 205A-1
Escuadrón 513	Hughes 369D/E/F/HN/HS/M, Hughes 500/OH-6A/530FF
Escuadrón Entrenamiento (514)	Enstrom F28F, Bell OH-13S
Escuadrón Aerotáctico 515	Hughes 369D/E/F/HN/HS/M (loaned from Esc 513)
Escuadrón Aerotransporte 515	Cessna 206 Stationair

Comando Aéreo de Apoyo Táctico 2

Grupo 61	**Base Aéreo Rio Negro**
Escuadrón 611	S-70A/UH-60A/L Black Hawk

Comando Aéreo de Transporte Militar (CATAM)

Base Aéreo 'Camilo Daze' El Dorado, Bogotá

Escuadrón de Transporte 711	C-130B/H Hercules, CASA CN.235
Escuadrón de Transporte 712	Arava 201, King Air C90/300, Cessna 550 Citation II, EMB-110P1A Bandeirante
Escuadrón Aerofotográfico	Turbo Commander A695A
Escuadrón Presidencial	Bell 206L-3 Jetranger 707-373C, F28-1000

Comando Aéreo de Mantenimiento (CAMAN)

Base Aéreo 'Justin Marino Cueto', Madrid

Escuadrón Presidencial det	Bell 412

Escuela Militar de Aviación (EMAVI)

Grupo de Vuelos, Base Aéreo 'Marco Fidel Suarez', Cali

EMA	PT-17 (1 at Cali and Palanquero)
Escuadrón Primario (611)	T-41D, IAR IS-28B2
Escuadrón Básico (612)	T-34B Mentor
Escuadrón Avanzado (613)	Cessna 310R
Escuadrón Aeromovil (614)	S-70A (loaned from Gr 61), Hughes 500/OH-6A/530FF (loaned from EscA 515)

Grupo Aéreo del Caribe (GACAR)

Base Aéreo San Andres

det Escuadrón de Combate 411	A-37B/OA-37B

Grupo Aéreo del Sur GASUR

Base Aéreo 'Ernesto Esguerra', Tres Esquinas

Escuadrón de Enlace	C-47 Skytrain, DHC-2 Beaver, Gulfstream 1000

Servicio de Aeronavigación a Territorios Nacionales

Boeing 727-2B7, CASA C.212-200/300 Aviocar,
Dornier Do 328-100, Fokker F28 Fellowship 3000C,
Helio HST-550 Stallion, PC-6/B2-H2 Turbo-Porter

Other types

Cessna 208 Caravan	1	liaison
Gavilán 358	4 (on order)	light transport
Piper PA-31 Navajo	2	communications
Piper PA-32 Cherokee Six	1	communications
Schweizer SA 2-37A	1	surveillance

Aviación Naval Armada de Colombia

Although Colombia has major coastlines bordering both the Pacific Ocean and Caribbean, the Aviación Naval Armada de Colombia is a small force. Until 1984 the navy was dependent solely on the FAC for aviation support. The arrival of four MBB-Bölkow BO 105CB helicopters in late 1983 gave the navy a limited SAR/ASW capability from its frigates and corvettes. A number of fixed-wing twins were added during 1991 to complement the maritime surveillance and communications duties. The Naval Air Arm's headquarters is located in Bogotá alongside all the other military commands. The Grupo Aeronaval del Atlántico (Atlantic Naval Air Group) is home-based at the Rafael Nunoz International Airport, Cartagena.

Policía Nacional de Colombia

The Policía Nacional is responsible for civilian control while working jointly with other organisations in the anti-terrorist and anti-narcotics task. For general activities and limited anti-guerrilla operations, the PNC is a self-sufficient force. Its primary maintenance facility is at Guaymaral in the suburbs of Bogotá, and the force has a number of detached operating locations. Aircrew training is carried out in-house. A number of Ayres Thrush Commander crop-spraying aircraft are used in the Policía Nacional's battle against narcotics. The United States funded the purchase of six UH-60Ls for the PNC in late 1998, and a pair of 'DC-3s' (thought to be Basler Turbo 67s).

Aviación Naval Armada de Colombia

Naval Air Arms HQ Bogotá

Grupo Aeronaval del Atlántico
HQ Rafael Nunoz Int'l Airport, Cartagena
2 x Beech B33 Bonanza, 2 x Eurocopter BO 105CB, 1 x Piper PA-28,
2 x Piper PA-31 Navajo, >3 x Rockwell Commander 680,
1 x King Air 300

Policía Nacional de Colombia

Bogotá det		
Ayres Thrush	3	crop sprayer
Basler Turbo-Dakota	2	transport
Beech 99	1	communications
King Air 200	1	communications
King Air 300	1	communications
Cessna 150	1	communications
Cessna 206 Stationair	4	communications
Cessna 208 Caravan I	2	communications
Cessna 421 Golden Eagle	1	communications
Cessna 441 Conquest	1	communications
DHC DHC-6 Twin Otter	2	communications
Fairchild C-26	2	communications
Piper PA-31 Navajo	1	communications
Piper PA-32 Saratoga	1	communications
Piper PA-36 Pawnee Brave	1	crop sprayer
Cartagena det		
Bell 206L-3 LongRanger	7	communications
Guaymaral det		
Bell 205	3	communications
Bell 212	9	communications
Bell UH-1H Iroquois	2	communications

Above: The Colombian navy has a single Beech King Air 300 among its communications fleet.

The Policía Nacional Colombia use two Basler BT67 Turbo Dakotas (above) for logistics support alongside a single Beech 99 (below) in a very mixed fleet of light aircraft and helicopters.

ECUADOR

Capital: Quito
Population: 10.8 million
Land area: 461000 km² (178,000 sq miles)
Major cities: Cuenca, Guayaquil, Riobamba

The continuing tension with neighbouring Peru over the 1941 loss of a 400-km (250-mile) wide strip of its Amazonian territory has resulted in a number of border skirmishes. They culminated in early 1995 with aerial combat between the opposing forces, leading to at least three confirmed kills for the Mirage F1 and Kfir fighters that spearhead the Fuerza Aérea Ecuatoriana (FAE).

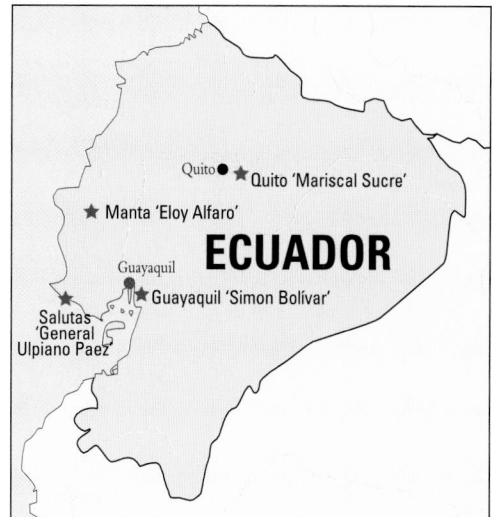

Fuerza Aérea Ecuatoriana

The FAE comprises a balanced force of interceptors, fighter-bombers and COIN aircraft operating from 'hardened' bases and forward operating locations constructed during the 1980s in its remaining part of Amazonia. From its headquarters in Quito, the FAE administers some 4,000 personnel and approximately 120 aircraft. Included within this total are a number of passenger/transport aircraft operated by the paramilitary airline TAME; Ecuador is unusual in having external airline operations as part of the FAE.

Below air staff level, the force is divided into two elements: I Zona Aérea, administering the transport and communications force, and II Zona Aérea with three combat wings and a flying school. Each wing (Ala) has a two-figure designation number beginning with either 1 or 2 depending on the element to which

it is assigned, and occupies a separate base, its responsibilities embracing the management of the complete flying structure. The flying operations are undertaken by an individual Grupo, which has its own three-figure designation conforming with the wing number. Subordinate to this are a number of flying squadrons with designated four-digit numbers commensurate with that of the Grupo.

I ZONA AÉREA

Base Aérea 'Mariscal Sucre', otherwise known as Quito International Airport, is the principal operating location of I Zona Aérea and its component Ala de Transporte 11. Here, under Escuadrón de Transporte 1111, the air force centralises its mixed force of four Lockheed C-130Bs, single examples of the C-130H and L-100-30, along with the TAME-operated Boeing 727 series, Fokker F28, BAe 748 Series 2 and McDonnell Douglas DC-10.

Short-field and remote location support is provided

Above: With Ala de Combate 21 is Escuadrón de Caza 2113, which flies the FAE's Kfir C-2/TC-2s, blooded in combat against the Peruvian air force.

Below: The FAE received A-37Bs under the FMS programme, and others during the 1980s as the type was withdrawn from the USAF.

by the unit's three DHC-6s acquired in the early 1980s which supplement the surviving DHC-5D. One of the three surviving BAe 748s, 001, is operated by the co-located Presidential Flight, which also has on strength a number of Rockwell Sabreliner 40R/60 executive jets.

South America

Fuerza Aérea Ecuatoriana

Headquarters Quito

I Zona Aérea   HQ Base Aérea 'Mariscal Sucre', Quito
Ala de Transporte 11, Quito

Escuadrón de Transporte 1111	BAe 748 Srs 2A, C-130B/H/ L-100-30 Hercules, DHC-5D Buffalo, DHC-6-300 Twin Otter
Escuadrón Presidencial	BAe 748 Srs 2A, Sabreliner 40/40A/60
Transporte Aéreos Militares Ecuatorianos	Boeing 727-17/134/ 230/2T3, F28 Fellowship 4000

II Zona Aérea   HQ Base Aérea 'Simon Bolivar', Guayaquil
Ala de Combate 21, HQ Base Aérea Taura

Escuadrón de Combate 2111	Jaguar International ES/EB
Escuadrón de Caza 2112	Mirage F1JA/JE
Escuadrón de Caza 2113	Kfir C-2/TC-2

Ala de Combate 22, HQ Base Aérea 'Simon Bolivar'

Manta det	SA 316B Alouette III (loaned from other Esc)
Salinas det	SA 316B Alouette III (loaned from other Esc)
Escuadrón de Caza 2211	SA 316B Alouette III, Bell 212
Escuadrón de Rescate 2212	DHC-6-300 Twin Otter, SA 316B Alouette III
Escuadrón de Entrenamiento 2213	Cessna 150L, SA 316B Alouette II

Ala de Combate 23, HQ Base Aérea 'Alfaro', Manta

Escuadrón de Combate 2311	A-37B Dragonfly
Escuadrón de Combate 2313	Strikemaster Mk 89/89A/90

Escuela Superior Militar de Aviación
Base Aérea 'General Ulpiano Paez', Salinas

Escuadrón de Entrenamiento Aéreo	T-34C Turbo-Mentor, T-41D Mescalero

Of the three Bell 212s that the FAE has operated, only this example, 823, remains in service with Escuadrón de Caza 2211, the others having been written off.

Ecuador's Mirage F1JA/JE fleet was ordered in 1977, uses French AAMs for the air defence role and Israeli bombs for ground-attack missions.

II Zona Aérea

II Zona Aérea has three main components which are central to the FAE's defensive and offensive capabilities. The first of these is Ala de Combate 21 and its subordinate Grupo 211, located at the hardened base of Taura, near the main port and largest city, Guayaquil. The first component of Grupo 211 is Escuadrón de Combate 2111, equipped with a dwindling force of SEPECAT Jaguar strike/attack aircraft from a purchase of 10 single-seat fighters and two twin-seat conversion trainers.

The Dassault Mirage F1JA/JEs of Escuadrón de Caza 2112 are the second component of Grupo 211. The unit received 16 single-seat aircraft and a pair of twin-seat trainers between December 1978 and November 1980 armed with MATRA Super 530 and R.550 missiles.

In 1979 the United States relaxed its controls on the sale of the General Electric J79 engine, which renewed interest on the part of the FAE in the Kfir. The initial order in mid-1982 comprised 10 Kfir C2s and two twin-seat Kfir TC2s, plus a number of Rafael Shafrir missiles. They equipped Escuadrón de Caza 2113. It is thought that a third Kfir TC2 was received at some point, while during 1996 an order was placed for four more Kfirs, with an option for another four. It was reported in late 1997 that Ecuador was after between 40 to 50 extra fighters to counter Peruvian purchases and was looking at the F-16 and MiG-29.

The Escuadrón Presidencial of Ala de Transporte 11 uses this single BAe 748 Srs 2A and several Sabreliners for VVIP transport.

Ala de Combate 23 at Base Aérea 'Alfaro' at Manta oversees Grupo 231's Strikemaster-equipped Escuadrón de Combate 2313, and Escuadrón de Combate 2311s A-37s. Twelve new-build A-37Bs were received in 1978, for which attrition has been relatively high. During the mid- to late 1980s the United States released a number of A-37Bs; six, followed by a further three, were delivered to Ecuador.

Surviving Strikemasters

Working alongside the Dragonflies from Manta are the BAe Strikemasters of Escuadrón de Combate 2313. Re-equipment with the Strikemaster occurred in December 1972 with eight Mk 89s, supplemented by an equal number of Mk 89As in 1977. Six Mk 90s were acquired as attrition replacements in the late 1980s, 10 having been lost. Strikemasters serve in a dual capacity of advanced flying training and close air support.

The last element of II Zona Aérea is the rotary-winged assets of Ala de Combate 22 at Base Aérea 'Simon Bolivar', Guayaquil, which also serves as the administrative headquarters of II Zona Aérea. Ala 22 has three assigned squadrons which share a mix of three SAR-configured SA 315 Lamas, five SA 316 Alouette IIIs and one Bell 212, as well as a DHC-6 and four Cessna 150Ls. The SAR-configured helicopters maintain detachments at both Salinas and Manta, while Taura's needs are met from Guayaquil.

The former base of 'General Ulpiano Paez', Salinas also houses the Escuela Superior Militar de Aviación and its attendant Escuadrón de Entrenamiento Aéreo. Equipped with two Cessna T-41Ds for use in the liaison role, the squadron also has 17 Beech T-34Cs which it uses in both the primary and basic flying stages.

Aviación Naval Ecuatoriana

Formed in 1967 to provide a communications link, the Ecuadorian navy has remained a small operating force with around a dozen aircraft of all types administered through three operational and one training squadron. Aviación Naval Ecuatoriana is located at Base Aérea 'Simon Bolivar', Guayaquil, alongside both air force and army assets. It undertakes its own pilot training through the Escuadrilla de Entrenamiento with three Beech T-34C Turbo-Mentors procured in 1980.

Liaison flying is carried out by 1 Escuadrilla de Enlace while the transport tasks fall to 2 Escuadrilla de Transporte. Rotary-wing support is provided by three Bell 206B JetRangers from a batch of five received in 1986, equipping 3 Escuadrilla de Helicópteros.

Servicio Aéreo del Ejército Ecuatoriano

Equipped with 25-30 helicopters and 15-20 fixed-wing aircraft, the army flying branch is divided into three separate operating functions, two of which are centralised on the 19ª Brigada Aérea del Ejército, and the Instituto Geográfico Militar with equal military and government requirements.

The two units under 19ª Brigada are split between two primary operating locations. Grupo Aéreo 43 is to be found alongside navy and air force assets at BA

'Simon Bolivar', Guayaquil, while Grupo Aéreo 45 is located at BA 'Mariscal Sucre', Quito. Both units are believed to also provide other detachments in support of the 11 infantry battalions, 10 independent infantry companies and three artillery regiments.

Servicio Aéreo del Ejército Ecuatoriano

Instituto Geográfico Militar   HQ BA 'Mariscal Sucre'
Cessna 550 Citation II, King Air A100, PC-6B Turbo-Porter, SA 315B Lama

19ª Brigada Aérea del Ejército

Grupo Aéreo 43	**HQ BA 'Simon Bolivar'**
Airtech CN.235M, AS 332B Super Puma, IAI Arava 201, SA 315B Lama, SA 330L Puma, SA 342K/L Gazelle	
Grupo Aéreo 45	**HQ BA 'Mariscal Sucre'**
AS 332B Super Puma, AS 350B Ecureuil, Bell 214B, Cessna 172G, DHC-5D Buffalo, IAI Arava Series 201, King Air 200, SA 315B Lama	
Military Academy	**Quito**
Officers School	**Quito**

The Ecuadorean army received four AS 350B Ecureuils in 1986, but lost one in 1994.

Aviación Naval Ecuatoriana

Base Aérea 'Simon Bolivar', Guayaquil

1 Escuadrilla de Enlace	1 x Cessna 320E Skyknight, 1 x Cessna 500 Citation I
2 Escuadrilla de Transporte	1 x Airtech CN.235M-100, 1 x King Air 200, 1 x King Air 300
3 Escuadrilla de Helicópteros	3 x Bell 206B JetRanger, 2 x Bell 222, 2 x Bell 412 (?)
Escuadrilla de Entrenamiento	3 x T-34C Turbo-Mentor

GUYANA

Capital: Georgetown
Population: 1 million
Land area: 215000 km² (83,000 sq miles)
Major cities: New Amsterdam, Linden

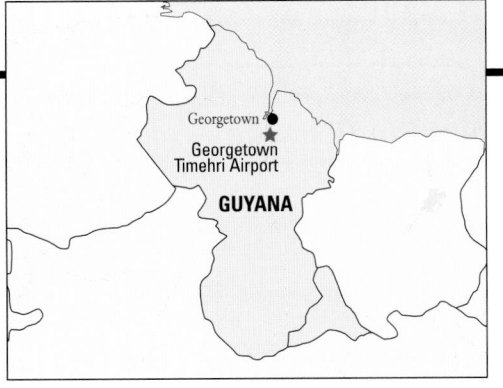

Guyana Defence Force Air Command

Guyana became a 'Co-operative Republic' in 1970. A Defence Force Air Wing was formed in 1968 and renamed Air Command in 1973. With its headquarters at Timehri Airport near the capital of Georgetown, the Air Command operates aircraft with civilian registrations although they

Air Command

HQ Timehri Airport

Bell 206B JetRanger II	2	communications
Bell 412	2	VIP
Pilatus/B-N BN-2A-2 Islander	1	transport
Shorts SC-7 Skyvan 3M	2	transport

carry the inscription 'G.D.F.' on the fuselage sides. Six Britten-Norman BN-2A Islanders were procured over a five-year period for patrol and transport duties, later supplemented by two Short Skyvan Series 3Ms. Rotary-winged assets include two Bell 206B JetRangers and a pair of Bell 212s; the status of three Mil Mi-8s received in 1986 is unclear.

PARAGUAY

Capital: Asuncíon
Population: 4.3 million
Land area: 407000 km² (157,000 sq miles)
Major cities: Concepción, Paraguarí, Pilar, San Pedro, Villarica

Fuerza Aérea Paraguaya

Formed in 1927 as the Fuerzas Aéreas del Ejército Nacional Paraguayo, as part of the army, the Fuerza Aérea Paraguaya became an independent service in 1946. Today FAP Headquarters is at BAM Nhu Guazu, Campo Grande. The flying elements are controlled by the 1ª Brigada Aérea, with five subordinate Grupos, each having a number of Escuadrones.

The combat element of the FAP is Grupo Aerotáctico (GAT) located at Asunción-Silvio Pettirossi International Airport with three

Escuadrones. The 1° Escuadrón de Caza 'Guaraní' is equipped with the EMBRAER EMB-326GB Xavante and is tasked with the COIN/attack role. The Escuadrón of six aircraft is divided into two Escuadrillas, 'Orion' and 'Centauro'. Operations and maintenance are conducted from Asunción although the Escuadrón detaches to both Ciudad del Este and Concepción, where a new facility was opened in December 1996.

The 2° Escuadrón de Caza 'Indios', and its attendant Escuadrillas 'Taurus' and 'Scorpio', were equipped with six Lockheed AT-33As, but the aircraft were grounded in 1996. Twelve former Taiwanese F-5E/Fs were due but have not been noted and, with the Chinese reported to have offered 12 Shenyang J-5s and two JJ-6s in 1998, the Taiwanese deal may have folded.

The 3° Escuadrón de Caza 'Moros' operated the four surviving EMBRAER T-27 Tucanos in the COIN role. Severe funding shortages saw the aircraft grounded in late 1996 while awaiting overhaul of the Martin-Baker seats. 'Gamma' and 'Omega', the two subordinate Escuadrillas, were operating ENAER T-35B Pillán aircraft to maintain flight currency.

Grupo de Transporte Aéreo (GdTA) is the direct descendant of the Transporte Aéreo Militar (TAM) that was formed in 1954 to operate flights to remote areas of the country. The unit today is equipped with two airworthy C-47s and four CASA C.212s.

Grupo Aéreo de Transportes Especiales (GATE) has a number of primary responsibilities. Its major function is the support of the Presidential and government aircraft such as the single Boeing 707-321B, a DHC-6 Twin Otter 200, a civilian-registered Cessna Citation 550 and a Beech King Air 90. It also provides communication and liaison flying for the air force with a collection of Beech, Piper and Cessna types, some of which it took over from Lineas Aéreas de Transporte Nacional (LATN) in 1989.

BAM Nhu Guazu, Campo Grande, with its grass runway, is the home of all instructional activities. With the Grupo Aéreo de Instrucción (GAI), students begin their *ab initio* training of 30-40 hours on the three surviving Aerotec A-122 Uirapurus or T-23s which were acquired from Brazil. Twelve ECH-35A/B Pilláns were received from an original

order for 15 and students undertake another 120 hours of flying training on the Pillán with Escuadrilla 'Antares' before moving to the tactical phase, which is conducted by GAT with 80-90 hours on the T-27 Tucano. The unit also retains two airworthy T-6G Texans which it uses for display flying.

Campo Grande houses all FAP helicopter assets as part of the Grupo Aéreo de Helicópteros (GAH). It currently has three Helibras HB 350B Esquilos and two Bell UH-1Hs received in 1996 as a gift from Taiwan. GAH also provides a home for the Presidential Agusta A 109-II and three ageing Bell UH-1Bs supplied by the US for anti-narcotics operations.

Fuerza Aérea Paraguaya

FAP Headquarters, BAM Nhu Guazu, Campo Grande

1ª Brigada Aerea

Grupo Aerotáctico

1º Escuadrón de Caza 'Guaraní'
Asunción-Silvio Pettirossi International Airport

Escuadrilla 'Centauro'	EMB-326GB Xavante
Escuadrilla 'Orion'	EMB-326GB Xavante
det at Ciudad del Este	EMB-326GB Xavante *(loaned from above)*
det at Concepción	EMB-326GB Xavante *(loaned from above)*

2º Escuadrón de Caza 'Indios', HQ Concepción

Escuadrilla 'Scorpio'	
Escuadrilla 'Taurus'	

(10 F-5Es and two F-5Fs were due during 1997 to equip these units. No evidence that they have arrived from Taiwan)

3º Escuadrón de Caza 'Moros'
Asunción-Silvio Pettirossi International Airport

Escuadrilla 'Gamma'	ENAER ECH-35B Pillán
Escuadrilla 'Omega'	ENAER ECH-35B Pillán

(The units operate the four EMBRAER EMB-312 Tucanos which were grounded awaiting ejection seats to be overhauled. Pilláns are used to keep up flight time)

Grupo Aéreo de Helicópteros

BAM Nhu Guazu, Campo Grande

Escuadrón	UH-1B/H Iroquois, HB 350B Esquilo
Escuadrón Presidencial det	Agusta A 109A

Grupo Aéreo de Instrucción

BAM Nhu Guazu, Campo Grande

Escuadrón Aerotec	A-122 Uirapuru
Escuadrón 'Antares'	ENAER ECH-35A/B Pillán, T-6G Texan

Grupo de Transporte Aéreo

Asunción-Silvio Pettirossi International Airport

Transporte Aéreo Militar	CASA C.212-200 Aviocar, C-47A

Grupo Aéreo de Transporte Especiales

Asunción-Silvio Pettirossi International Airport

Escuadrilla Presidencial	King Air 90, 707-321B, Cessna 550 Citation, DHC-6-200
Escuadrilla	Cessna 185/U206C/210, Cessna 402B, PA-32R Lance

The Pillán serves the Fuerza Aérea Paraguaya in both the ECH-35A and 'B versions with the Escuadrón 'Antares'. The 'A is a primary trainer while the 'B serves in the instrument training role.

The Grupo Aéreo de Helicópteros at BAM Nhu Guazu, Campo Grande, uses the three surviving HB 350B Esquilos of four delivered in 1985. The unit flies a mix of Esquilos and UH-1Hs.

Arma Aérea del Ejército Paraguayo

Following the separation of the Fuerza Aérea Paraguaya from the army in 1946, the army's aerial requirements have in most cases been met by the FAP. The army currently has three liaison aircraft which operate from within the FAP facility at Asunción-Silvio Pettirossi IAP. Maintenance is carried out by FAP technicians.

Arma Aérea del Ejército Paraguayo

Asunción-Silvio Pettirossi International Airport
1 x Beech Baron, 1 x Cessna 206, 1 x Cessna 310R

Aviación de la Armada Nacional Paraguaya

The service was formed as the Servicio de Aeronáutica de la Marina (Naval Air Service) in 1927 and was absorbed into the army air force before the outbreak of World War II. Re-established in the mid-1960s, the naval air service exists to provide aerial patrol of Paraguay's river systems and a limited amount of communication flying in support of its outlying patrol stations. The service has only two principal operating locations: Asunción-Silvo Pettirossi IAP, where the fixed-wing element can be found, and Sajonia Naval Aviation Base, where the rotary-wing element is stationed.

Aviación de la Armada Nacional Paraguaya

SAJONIA NAVAL AVIATION BASE		
Bell OH-13H Sioux	1	helicopter training
Helibras HB 350B Esquilo	2	communications
Hiller UH-12E	1	communications
ASUNCIÓN-SILVO PETTIROSSI INTERNATIONAL AIRPORT		
GAEN (Training Air Naval Group)		
Cessna 150M	2	primary training
GAPROGEN (General Purpose Air Naval Group)		
Cessna U206A	1	communications
Cessna 310	2	communications
Cessna 401B	1	communications

PERU

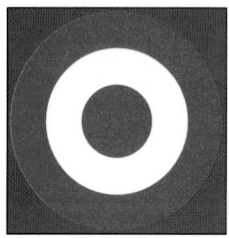

Capital: Lima
Population: 22.3 million
Land area: 1.285 million km² (496,000 sq miles)
Major cities: Callao, Arequipa, Trujillo

The Servicio de Aviación Militar del Ejército was formed in January 1919. The military and naval air arms merged in May 1929 to form the Cuerpo de Aeronáutica del Peru, the forerunner of the current Fuerza Aérea del Peru. Skirmishes between ground forces over a border dispute with Ecuador erupted in 1981 and heightened in early 1995 when the air forces of both countries met in combat, with Peru coming out worse.

Fuerza Aérea del Peru

Administered by the Ministerio de Aeronáutica, Campo de Marte, Lima, the air force is still relatively small. The command structure is sub-divided into a number of air regions, within which are one or more air force groups. Each group is made up of a number of squadrons, for example, Grupo Aéreo 7 located at Piura Air Base has under its command 1,018 personnel assigned to seven squadrons. They are Nos 705 Communications Squadron, 706 Maintenance

Squadron, 707 Support Squadron, 708 Anti-Aircraft Squadron, 709 Security Squadron, 711 Fighter Training Squadron and 712 Fighter Squadron. All groups are similarly structured although few have more than one flying squadron.

All units within the FAP are charged with three primary taskings: to support the civilian community, to maintain the defence of the country, and support

(map of Peru with the following locations marked:)
Iquito 'Secada Vignetta' • Iquito
Talara/El Pata
Piurra 'Capitán Concha'
Chicolayo 'Quinones Gonzales'
Trujillo
PERU
Lima
Lima-Callao 'Jorge Chavez'
La Joya 'Mariano Melgar'
Cuzco
Pisco 'Renan Elias Oliviera'
Arequipa
Arequipa 'Rodriguez Ballon'

Peru uses a mix of C-130As, Ds and L-100-20s (such as 397) within Grupo Aéreo de Transporte No 8 at Base Aérea 'Jorge Chavez', Lima-Callao. A pair of C-130As is being converted to tankers.

the anti-narcotics and anti-terrorism units. In the latter tasking, support is provided by other outside agencies including the US Drug Enforcement Agency (DEA) and US Customs.

COMBAT ASSETS

By mid-1997 only 16 Cessna A-37Bs (of 36 delivered) were thought to remain in service, assigned to Grupo 7 at Base Aérea 'Capitán Concha', Piura. A number of other airframes appear to have been withdrawn, resulting in Escuadrón de Caza 712 being temporarily inactivated, leaving Escuadrón 711 to undertake the dual responsibility of training and operations. The A-37B is widely used in the anti-narcotics field.

The air force is said to have received 32 Su-20 'Fitter-Fs' and four Su-22UM3 'Fitter-G' twin-seat trainers at the low cost of $250 million. Deliveries began during 1978 to re-equip Grupo de Caza 12 at Limatambo. They were followed by an additional order in 1980 for 16 aircraft, which appear to have been the more capable Su-22M2K 'Fitter-J' version. It is thought that this second batch of aircraft was assigned to Escuadrón de Caza-Bombardeo 411 'Eagles' at Base Aérea 'Mariano Melgar', La Joya. Due to border tensions, the Su-20s of Escuadrón de Caza-Bombardeo 111 have now relocated on a more or less permanent basis to Base Aérea 'Capitán Montes', Talara (El Pata) on Peru's northern border with Ecuador, where they are joined periodically by their sister unit from La Joya.

Grupo de Caza 6 at Base Aérea 'Capitán José Abelardo Quinones Gonzales', Chiclayo uses the

Peru has operated Su-22M2K 'Fitter-Js' (above) for some time, although they have recently been upgraded with bolt-on refuelling probes. Fighter forces have been bolstered by the arrival of MiG-29s (below).

Below: The most capable air defence aircraft in the Fuerza Aérea del Peru's diverse fleet is the Mirage 2000P. A dozen years after being delivered, all of the aircraft are believed to survive, flying with Escuadrón de Caza-Bombardeo 412.

Fuerza Aérea del Perú

Grupo Aéreo n°3, Base Aérea 'Jorge Chavez', Lima-Callao	
Escuadrón de Helicópteros 332	Mil Mi-8T/8MTV/17, AS 350B Ecureuil, BO 105CBS/LS
Escuadrón de Helicópteros 341	UH-1H Iroquois, Bell 212, Bell 214ST
Grupo Aéreo n°4, Base Aérea 'Mariano Melgar', La Joya	
Escuadrón de Caza-Bombardeo 411	Sukhoi Su-20 'Fitter-F', Su-22M-2K 'Fitter-J', Su-22UM-3 'Fitter-G'
Escuadrón de Caza-Bombardeo 412	Mirage 2000P/DP
Grupo Aéreo n°6, Base Aérea 'Capitán Jose Abelardo Quinones Gonzales' Chiclayo	
Escuadrón de Caza 611	Mirage 5P/DP
Grupo Aéreo n°7, Base Aérea 'Capitán Concha', Piura	
Escuadrón de Caza-Bombardeo 711	A-37B Dragonfly
Grupo Aéreo de Transporte n°8, Base Aérea 'Jorge Chavez', Lima-Callao	
Escuadrón de Transporte 841	Boeing 707-323C, L-100-20
Escuadrón de Transporte 842	An-32 'Cline', An-74 'Coaler-B'
Escuadrón de Transporte 843	Queen Air A80, King Air C90/300, Cessna 185, Cessna 421
Escuadrilla Presidencial	Bell 412HP, 737-528, DC-8-62CF, F28 Fellowship 1000
Grupo Aéreo n°9, Base Aérea 'Renan Elias Olivera', Pisco	
Escuadrón de Bombardeo 921	Canberra B(I).12/B.58/B.62/T.4/T.54
Grupo Aéreo n°11, Base Aérea 'Capitán Montes', Talara (El Pata)	
Escuadrón de Caza-Bombardeo 111	Sukhoi Su-20 'Fitter-F', Su-22M-2K 'Fitter-J', Su-22UM-3 'Fitter-G'
Grupo Aéreo de Transporte n°42, Base Aérea 'Coronel Francisco Secade Vignette', Iquitos	
Transportes Aéreos Nacionales de la Selva	DHC-6-300 Twin Otter, Harbin Y-12 II, PC-6/B2-H2 Turbo-Porter
Grupo de Fuerzas Especiales, Base Aérea 'Rodriguez Ballon', Arequipa	
Escuadrón Aéreo 211	Mil Mi-24 'Hind-D', Mil Mi-8T/-8MTV/-17

Academia del Aire

Grupo Aéreo de Entrenamiento n°51, Base Aérea 'Las Palmas', Lima	
Escuadrón de Instrucción Primaria 511	T-41D Mescalero, Queen Air A80
Escuadrón de Instrucción Básica 512	EMB-312 Tucano
Escuadrón de Instrucción Avanzada 513	MB-339AP
Escuadrón Aéreo Táctico 514	EMB-312 Tucano

Servicio Aerofotográfico Nacional

Base Aérea 'Las Palmas', Lima	
Escuadrón Aerofotográfico 331	Learjet 25B/36A, Falcon 20F

Other types – units not known

Mikoyan MiG-29/UB 'Fulcrum'	18	air defence
Sukhoi Su-25 'Frogfoot'	14	close air support
de Havilland Canada DHC-5D Buffalo	12	(stored)
Lockheed C-130A Hercules	2	(under conv. to 'KC-130A')
Fairchild C-26	4	communications
Cessna 150F	2	communications
+ microlights		

The FAP uses a mix of Mil Mi-8s and Mi-17s, including FAP 627, a Mi-8T. The helicopter is flown by the Peruvian army and navy as well the air force.

The FAP's MB-339As are used by Escuadrón de Entrenamiento 513 of Grupo 51 for advanced training. Sixteen aircraft were delivered: 13 survive.

survivors of the 40 different Mirage 5Ps that have seen service with the FAP over the years. Later deliveries were of the 5P3/5P4 standard with fixed refuelling probes, new RWR, HF radio and laser rangefinder, and earlier aircraft were updated to this standard by Seman Aerospace. Peru's 10 single-seat Mirage 2000Ps and two Mirage 2000DPs acquired in 1986 are assigned to Escuadrón de Caza-Bombardeo 412 'Hawks' as part of Grupo Aéreo 4 at Base Aérea 'Mariano Melgar', La Joya. Grupo 4, although maintaining a constant alert state with two aircraft, often deploys to Madre de Dios, Ucayali, Huallaga, Iquitos and Lima itself.

Peru received 18 1987/89-vintage MiG-29 and MiG-29UBs from Belarus from November 1996 onwards. Problems with spares were overcome in July 1998 when a delegation from Rosvoorouzhenie and MAPO visited Peru and a spares package was arranged. In August 1998 three more were ordered. Exactly where the aircraft are based is unknown.

By late 1996 only 11 Canberras of all versions were with Grupo Aéreo 9 and they may have been retired.

TRANSPORT AND HELICOPTERS

Grupo Aéreo de Transporte 8 at Lima-Callao is divided into three flying squadrons and also parents the Escuadrilla Presidencial with its newly-arrived Boeing 737-300 and two VIP-configured Douglas DC-8-62CFs. The workhorses of Grupo 8 are the Antonov An-32s. Escuadrón de Transporte 841 is responsible for the operation of the Grupo's Lockheed L-100-20 and C-130As, as well as the former Presidential Fokker F28 and the Boeing 707, which has a dual tanker-transport role. The final flying unit of Grupo 8 is Escuadrón de Transporte 843, tasked with transport/liaison.

Grupo Aéreo de Transporte 42, better known as Transportes Aéreos Nacionales de la Selva, or TANS 42 for short, is based at Antiguo Aeropuerto (Laga Morona Cocha), Iquitos. TANS 42 is charged with support of remote areas, maintaining detachments at both Piura and Pucallpa with land-based PC-6s and Y-12s, while the float-equipped PC-6s and DHC-6s operate from either the Instalaciones Club de Caza y Pesca or the River Nanay.

Co-located alongside Grupo 8 at 'Jorge Chavez' IAP/Lima-Callao are Grupo Aéreo 3's early-build Mi-8Ts, plus new Mi-8MTV or Mi-17 versions. Escuadrón de Helicópteros 332 is equipped with European helicopter types, while Escuadrón de Helicópteros 341 has the American.

Near Arequipa is the Grupo de Fuerzas Especiales at Base Aérea 'Rodrigues Ballon', Vitor, with the FAP's 10 or so Mi-25 'Hind-D' helicopter gunships.

FAP TRAINING

The Academia del Aire (Air Force Academy) is located at Las Palmas. Flying training is undertaken by Grupo Aéreo 51. Primary training is conducted by Escuadrón de Entrenamiento on the Cessna T-41D, from where students progress to the EMBRAER T-27 Tucanos of Escuadrón de Entrenamiento 512. Tucanos are also used operationally by Escuadrón 514, whose crews are drawn from the instructor pilots, in the COIN and anti-narcotics role. The final flying element of Grupo 51 is that of Escuadrón de Entrenamiento 513 with its 13 Macchi MB-339APs.

Also located at Las Palmas is the autonomous Servicio Aerofotográfico Nacional and its Escuadrón 331 equipped with a pair of Learjet 25Bs and a Mystère 20F for survey work, and two Learjet 36As for airfield calibration duties.

Servicio Aeronaval de la Marina Peruana

Although the original naval air service had amalgamated with the army in 1929 to form the Cuerpo de Aeronáutica del Peru, it was re-established as an autonomous flying service in 1950. Relocating its headquarters and flying facilities from Ancon to the 'Jorge Chavez' IAP at Lima-Callao, the service was renamed Servicio Aeronaval de la Marina Peruana. Its main operating location is still at Lima-Callao. Shipborne assets consist of five Agusta-Bell 212AS helicopters and six ASH/ASH-3D Sea Kings with both an anti-submarine capability and anti-surface ability. Maritime patrol and SAR is undertaken by five Beech King Air B200CTs and the sole Fokker F27-500 Friendships. The navy also uses the coast guard Friendships on loan for maritime patrol. Of the service's three Mil Mi-8Ts, one is based at Pucallpa in the armed anti-terrorist role.

Training is provided on Beech T-34C Turbo-Mentors and Bell 206B Jetrangers, while transport support is provided by a pair of Antonov An-32s.

Servicio Aeronaval de la Marina Peruana

Headquarters 'Jorge Chavez' Int'l Airport, Lima-Callao	
Grupo Aeronaval No 1, Lima-Callao	
Escuadrón Aeronaval 11	King Air B200T, F27-500 Friendship
Escuadrón Aeronaval 21	AB 212AS, AS-61D/ASH-3H
Grupo Aeronaval No 3, Lima-Callao	
Escuadrón Aeronaval 31	T-34C-1 Turbo-Mentor
Escuadrón Aeronaval 32	An-32 'Cline', Mil Mi-8T
Escuadrón Aeronaval 33	Bell 206B
(EAN 31 is based at San Juan de Marcona; EAN 33's Mi-8 is used in a detachment at Pucallpa)	

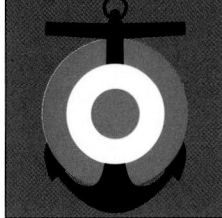

The Servicio Aeronaval de la Marina Peruana bought six T-34C Turbo Mentors in 1978 for basic training with the Escuadrón de Instrucción based at 'Jorge Chavez' International Airport, Lima-Callao, the main operating base of the service.

Guardacosta

The Guardacosta (Peruvian Coast Guard) is associated with the navy but controls its own aircraft and has its own pilots. Two Lake Seawolf amphibians were acquired in 1993 and used for fishery and pollution control, but are not operated today. The current fleet, two Fokker F27-200s based at Callao, and a float-equipped DHC-6-300 based at Iquitos, undertake coastal patrol and search and rescue missions.

Aviación del Ejército Peruano

Separate army aviation in Peru was established in 1971. Army air elements appear to be assigned in small flights in support of regional garrisons.

A purchase of large numbers of Mil Mi-8 'Hip-Cs' for both the air force and the army occurred in the mid-1970s. Tasked primarily with support of its ground forces, these assault helicopters have proved ideal for the needs of the army. This led to a further purchase of the newer Mi-17 derivative, which continues to serve in significant numbers and today comprises the bulk of the army's tactical helicopter force.

A number of Aérospatiale SA 315B Lamas are still used for training, alongside nine Enstrom F28F Falcons received in 1992. Also used are a few Agusta A 109K2s and Bell 412s. The latest acquisition has been three Mil Mi-26 'Halos'. The mainstay of army aviation logistics are three An-32s received in 1994.

The Peruvian army uses the Enstrom F28F for basic helicopter training. This example, seen outside the Enstrom premises at Menominee, Michigan, does not display the army's distinctive markings (left) but rather the FAP's roundel.

Aviación del Ejército Peruano

Headquarters 'Jorges Chavez' Int'l Airport, Lima-Callao
Small flight dets in support of regional garrisons.
Fixed-wing aircraft based at 'Jorge Chavez', where support, maintenance and training facilities are located.

Aérospatiale SA 315B Lama	5	training helicopter
Agusta A 109K2	12	observation helicopter
Antonov An-32B 'Cline'	4	transport
Beech Queen Air A65	1	communications
Bell 412	2	communications
Cessna 172	2	communications
Cessna 185	3	communications
Cessna U206G	2	communications
Enstrom F28F	7	training helicopter
Mil Mi-8/17 'Hip'	40	transport helicopter
Mil Mi-25 'Hind-D'	14	close air support
Mil Mi-26 'Halo'	3	transport helicopter

Policia Nacional Peruana

The police air wing, PNP, was formed in 1983. Helicopter assets include 16 UH-1Hs recently donated by the US to assist in the struggle against both the narcotics trade and terrorist activities. Ten recently acquired Mil Mi-17s are locally deployed in a number of areas, including both Pucallpa and Iquitos. Four MBB-Bolkow BO 105LSA-3s and two BK 117B-1s are used for training and missions around Lima. Fixed-wing assets are predominantly drawn from aircraft confiscated from narcotics traffickers. The exceptions are three An-32s received in 1994 and two surviving Harbin Y-12s.

Policia Nacional Peruana

Escuadrón 200, 'Jorge Chavez' Int'l Airport, Lima

Bell 212	1	paramilitary
Eurocopter BO 105LSA-3	4	training
Eurocopter/Kawasaki BK 117B-1	3	training
Hughes 369D	1	(under restoration)
Mil Mi-17	10	paramilitary
Sikorsky S-76	1	paramilitary

Escuadrón 500, 'Jorge Chavez' Int'l Airport, Lima

Antonov An-32 'Cline'	1	paramilitary
Beech King Air E90	1	paramilitary
Cessna 172	1	paramilitary
Cessna 182N	1	paramilitary
Cessna U206 Stationair	1	paramilitary
Cessna TU206G Stationair	1	paramilitary
Convair VC-131H Samaritan	1	paramilitary
Fairchild C-123K Provider	4	paramilitary
Harbin Y-12	3	paramilitary
Lockheed C-130 Hercules	2	paramilitary
Piper PA-31	2	paramilitary
Piper PA-34-200T Seneca	2	paramilitary
Rockwell Commander 695A	2	paramilitary
Detachment Iquitos		
Harbin Y-12	–	(see above)
Mil Mi-17	–	(see above)
Detachment Pucallpa		
Bell UH-1H Iroquois	16	paramilitary
Pilatus/B-N BN-2A Islander	2	paramilitary

SURINAM

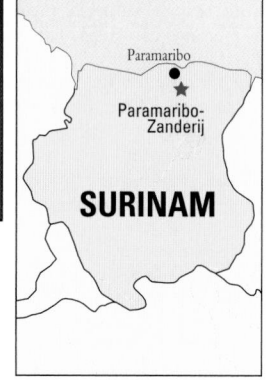

Paramaribo
Paramaribo-Zanderij
SURINAM

Capital: Paramaribo
Population: 400,000
Land area: 163820 km²
(63,235 sq miles)

Surinam Defence Force

Paramaribo-Zanderij

Aérospatiale SA 316B Alouette III	1	SAR
CASA C.212-400	2	coastal patrol
Cessna TU206G Turbo-Stationair	1	border patrol
Cessna 310	1	border patrol
Pilatus/B-N BN-2A Defender	1	coastal patrol
Pilatus PC-7	1	COIN

Surinam achieved self-government in 1954 and full independence from the Netherlands in 1975. In 1982 a small air arm was formed within the Surinam Defence Force, equipped with four PBN BN-2A Defenders. Later, a Cessna 172 and 310 were acquired. All aircraft undertake border patrols and SAR missions from the main base at Paramaribo-Zanderij, and are occasionally detached to both Zorg en Hoop and Moengo. In 1986 anti-government

Of the four BN-2B Defenders delivered to the Surinam Defence Force, one was sold, two were withdrawn and one remains in service.

guerrilla activity prompted the air element to obtain a pair of Aérospatiale SA 316B Alouette IIIs and then two PC-7s for COIN missions; one of the helicopters was later lost in a crash and both of the PC-7s were returned to Switzerland, one being redelivered. A pair of CASA 212-400s was delivered in 1999.

URUGUAY

Capital: Montevideo
Population: 3.1 million
Land area: 187000 km²
(72,000 sq miles)
Major cities: Las Piedras, Salto

Salto
URUGUAY
Durazno Santa Bernadino
BA 'Capitán Curbelo'
Laguna del Sauce
Montevideo

The main offensive weapon of the FAU is the A-37B Dragonfly. Sixteen ex-USAF aircraft were acquired between 1992 and 1993 and most remain in service.

Fuerza Aérea Uruguaya

FAU history dates to March 1913 when the Minister for War and Navy issued a request for military aviators. The Escuela Militar de Aviación was formed on 20 November 1916. Today, the FAU is structured to provide an offensive arm that is tasked with border patrol and with countering the narcotics trade. With

around 3,000 personnel, the FAU is organised into three major commands: Comando Aéreo Táctico (CAT, Tactical Air Command); Comando Aéreo de Entrenamiento (CADE, Air Training Command);

and the Comando Aéreo de Material (CAM, Air Material Command) which is responsible for maintenance, supply, communications and airfields.

The offensive element of the FAU is formed around Brigada Aérea II which is located at Base Aérea No 2 'General Urquiza', Durazno-Santa Bernadina, 120 km (75 miles) north of Montevideo.

Fuerza Aérea Uruguaya

Comando Aéreo Táctico	
BRIGADA AÉREA I	Cessna 206
BA N° 1 'Capitán Boiso Lanza', Montevideo-Carrasco	
Grupo de Aviación 3 (Transporte)	CASA C.212-200 Aviocar,
	EMB-110B1/C Bandeirante,
	F27-100 Friendship,
	C-130B Hercules
Grupo de Aviación 5 (Busqueda y Rescate)	
	Bell UH-1H Iroquois,
	Bell 212, Wessex HC.Mk 2,
	Dauphin
BRIGADA AÉREA II	Cessna 206, U-17A,
	PA-18 Super Cub
BA N° 2 'General Urquiza', Durazno-Santa Bernadina	
Grupo de Aviación 1 (Ataque)	FMA IA-58 Pucará
Grupo de Aviación 2 (Caza)	Cessna A-37B Dragonfly
Centro de Instrucción y Entrenamiento de Vuelo Avanzado	
Grupo de Aviación	PC-7U Turbo-Trainer
Commando Aéreo de Material	
HQ Montevideo-Carrasco	
Brigada de Mantenimiento y Abastecimiento, Montevideo-Carrasco	
	T-34A/B Mentor, Cessna
	182D/210, Commander 680
Dirección General de Aviación Civil	
Instituto de Adiestramiento Aeronáutico, Aerodromo	
'Angel S. Adami', Melilla	Queen Air A80, Cessna 310L
Comando Aéreo de Entrenamiento	
Escuela Militar de Aeronáutica, Pando	
Escuela Básico	T-34A/B Mentor, Queen Air A80
Escuela Primaria	T-41D Mescalero
	Cessna 172/182A
Escuela Técnica de Aeronáutica (Technical Training School)	
Escuela de Comando y Estado-Mayor (Command and Staff College)	

There are two primary operating units assigned to Brigada II: Grupo de Aviación 1 (Ataque) (two Pucarás), and Grupo de Aviación 2 (Caza) (A-37B). Introduction of NVGs and compatible instrumentation has given the A-37B a day/night ability that is especially important in its anti-narcotic operations. In support of Brigada II, the unit has a small liaison flight

Above: Two Pucarás form the equipment of Grupo de Aviacíon 1 (Ataque) at Durazno-Santa Bernadina.

Right: The FAU's PC-7Us support the Pucará and A-37s by providing a light strike capability. All six remain in use, one having been rebuilt during 1995/96 after suffering an accident in April 1994.

comprising three Cessna 185s or U-17As, a single Cessna 206, and a Piper PA-18 Cub.

Designated AT-92, the six Pilatus PC-7U Turbotrainers were received to initially provide advanced fixed-wing training. They appear to have also taken on a more offensive COIN role to assist the FAU's dwindling assets, while still undertaking their primary function within the Centro de Instrucción y Entrenamiento de Vuelo Avanzado (CIEVA), where students undertake a 120-hour, approximately two-year course before receiving their operational posting.

The transport, helicopter and communication elements are collectively assigned to Brigada Aérea I at Base Aérea No 1, 'Capitán Boiso Lanza', Montevideo-Carrasco International Airport. The fixed-wing assets operate under the auspices of Grupo de Aviación No 3 (Transporte) and the rotary-wing with Grupo de Aviación No 5 (Busqueda y Rescate).

The three grouped transport squadrons are still operated autonomously within a flight structure, operating two Lockheed C-130B, three CASA C.212 Aviocars, a single Fokker F27 Mk 100 (used on TAMU flights to neighbouring countries) and three EMB-110C Bandeirantes. The Hercules are used heavily in Antarctic resupply support flights. All FAU transport assets operate with a dual military/civil identification. The Aviocars maintain the resupply routes in and around the country as part of the military airline Transporte Aéreo Militar Uruguayo (TAMU). One Bandeirante is an EMB-110B1 variant and is equipped for photo-mapping.

Grupo de Aviación 5 maintains the rotary-wing assets, comprising two Bell 212s (one used as a Presidential transport), an Aérospatiale Dauphin, three Bell UH-1Hs and six ex-No. 28 Sqn, RAF Westland Wessex HC.Mk 2s. Tactical taskings are assigned to the ex-US Army Bell UH-1Hs and are being supplemented by the Wessexes.

A mixture of communication aircraft can also be found at Carrasco, acquired through confiscation. They are used by BA 1 for communication flying and probably by the Instituto de Adiestramiento Aeronáutico under the Dirección General de Aviación Civil from Aerodromo 'Angel S. Adami', Melilla. This is a civilian flying training school of the Uruguayan Aviation Authority but has instructor pilots assigned from the FAU.

The Air Force Academy is located a few miles to the north of Carrasco at Aeropuerto Militar General Artigas, Pando, where the Escuela Militar de Aeronáutica undertakes flying tuition. Students undertake a four-year course that includes academics. Screening takes place on the T-41D, with the T-34 Mentor being used for the bulk of the 100 hours basic tuition before students move to the advanced flying phase on the Pilatus PC-7U at Durazno.

Ten Cessna 206H Stationairs were delivered on 4 December 1998, to replace Stationairs that had been in service since the 1960s. The aircraft will be used for training, medical evacuation and surveillance.

Aviación Naval Uruguaya

The Uruguayan Naval Air Arm was formed on 7 February 1925. Reorganisation in 1951 saw the service's name change to Aviación Naval.

Currently, the only ASW asset is a single Beech 200T serving with the Escuadrón Antisubmarino y Exploración, while a pair of Piper PA-34-200Ts carries out coastal patrols. The two Senecas also provide advanced twin-engined training. An attack capability is provided by two T-28Ps which re-entered service in mid-1997.

The Escuadrón de Helicópteros operates three Wessex Mk 60s acquired from Bristow Helicopters, five recently acquired ex-No. 2 FTS RAF Wessex HC.Mk 2s and two 'new' Bell 47G-3Bs.

Training of naval pilots is carried out with four

Beech T-34s of three sub-types at the Escuela de Aviación Naval. Three Cessna 182s are used for support and liaison duties. Latest deliveries are two ex-No. 750 Sqn Jetstream T.Mk 2s, which were undertaking crew training at RNAS Culdrose, UK in 1999.

A total of nine T-28 Fennecs is known to have served with the Uruguayan navy, acquired from Argentina in 1979/80. Although out of service by 1995, two were airworthy again by mid-1997.

Aviación Naval Uruguaya

Grupo de Escuadrones, BA 2 'Capitán Curbelo', Laguna del Sauce	
Escuadrón de Entrenamiento Avanzado	
	2 x Sud T-28P Fennec, 3 x Cessna 182H/J/K
Escuadrón Antisubmarino y Exploración	
	1 x King Air 200T, 2 x PA-34-200T, 2 x Jetstream T.Mk 2
Escuadrón de Helicópteros	2 x Bell 47G-3B, 5 x Wessex
	HC.Mk 2, 3 x Wessex Mk 60
Grupo de Escuela 'Mayo Villagram', BA 2 'Capitán Curbelo', Laguna del Sauce	
Escuela de Aviación Naval	1x T-34A Mentor, 1x T-34B Mentor,
	2x T-34C-1 Turbo-Mentor

VENEZUELA

Capital: Caracas
Population: 19.7 million
Land area: 912000 km²
(352,000 sq miles)
Major cities: Maracaibo, Valencia, Maracay

Fuerza Aérea Venezolana

Military aviation was inaugurated in December 1920 with the creation of an Academia Aeronáutica at

Marcy. In 1949 the air service became an autonomous organisation, independent of army control, as the Fuerza Aérea Venezolana. Venezuela's recent past has been marred by two coup attempts in 1992, both centred on the air force. More recently, stability has returned to the region and a number of modernisation plans are beginning to come to fruition. The air force today is one of the most advanced in South America, although its operational capacity is still restrained by severe budgetary considerations.

The FAV's headquarters are at La Carlota, Caracas. With six main operating locations, the FAV is divided into three major commands responsible for 10 wings (Grupos), comprising three fighter, two close support, three transport and two training. Each Grupo is then sub-divided into a number of operating squadrons (Escuadrones), two of which will generally be flying units and the remainder support squadrons.

VF-5A 9124 is one of the 10 aircraft (plus two two-seaters) upgraded by Singapore Technologies Aerospace to have a new HUD, INS, databus and computer, as well as having its airframe refurbished.

COMANDO AÉREO DE COMBATE

Venezuela is unique in South America, being the first and, to date, the only Latin American country to operate the Lockheed Martin F-16A/B Fighting Falcon. Twenty-four aircraft – 18 single-seat F-16As and six twin-seat F-16Bs – were purchased from 1982 to 1984 for use by Grupo Aéreo de Caza No 16. Sixteen are stationed at Base Aérea 'El Libertador', Maracay, the FAV's main air force base with Escuadrón 161 and Escuadrón 162, operated on a pool basis. The FAV was the first South American air force to participate in a Red Flag exercise at Nellis AFB, doing so in 1991. The FAV is looking at the possibility of funding the upgrade of its F-16 fleet to the European Mid-Life Update standard.

FIGHTER FORCE

Operating alongside the Fighting Falcons at 'El Libertador' are the Dassault Mirage 50EVs of Grupo Aéreo de Caza 11. The aircraft are operated by Escuadrón 33 in the air defence role. Under a deal signed in March 1988, the surviving Mirage IIIEVs, 5Vs and 5DVs were returned to Dassault at Bordeaux for modernisation with new avionics, the uprated SNECMA ATAR 9K-50 engine, and canards. The package also included the purchase of a number of new airframes, six 'new' and three former Armée de l'Air machines. Escuadrón 33's sister unit, Escuadrón 34, also operates the Dassault Mirage 50EV/DV although it is tasked with the air-to-ground role.

Grupo de Caza No 12 at Barquisimeto and its attendant Escuadrón, 36, have on strength the Canadair VF-5 and NF-5 Freedom Fighter. In the late 1980s and early 1990s this squadron had been another to suffer from a long period of limited funding, and by May 1990 had been effectively grounded due to fatigue problems on its surviving one VF-5D and 13 VF-5A/R Freedom Fighters. Singapore Aerospace was contracted to upgrade two pattern aircraft and seven others, while six single-seat and one two-seat Dutch NF-5B were acquired and upgraded. At that time a decision on the remaining five aircraft was awaited, but events of October 1992's attempted coup meant that three of the five aircraft which were stored at Barquisimeto were destroyed by strafing Mirage 50s and OV-10 Broncos.

COIN ASSETS

The final unit assigned to the Comando Aéreo de

Delivered under the Peace Delta programme, Venezuela's F-16A/Bs are operated as pooled aircraft from Base Aérea 'El Libertador', Maracay.

Combate is Grupo Aéreo de Operaciones Especiales No 15 at BA 'General en Jefe Rafael Urdaneta', Maracaibo, with Escuadrón 151 operating the OV-10A/E Bronco. Both were utilised in the COIN role with a semi-permanent detachment of three aircraft to Base Aérea 'Mayor Buenaventura Vivas', Santo Domingo, where they work in association with other agencies, including the DEA, and with neighbouring Colombia in the anti-narcotics task. Other detachments are to Puerto Ayncucho in January, Puerto Ordaz in February, Maracay in March and November to work with Grupo de Paracaidistas 'Argua', and Barquisimeto in July. The acquisition of 10 surplus USAF OV-10A Broncos in April 1991 allowed Escuadrón 151 to supplement its surviving OV-10Es. It would appear that most of the OV-10Es have now been retired, leaving the wing to continue to operate the higher-houred but upgraded ex-USAF OV-10As. About 15 Broncos continue to be used.

COMANDO AÉREO LOGÍSTICO

The FAV's transport assets come under the purview of Comando Aéreo Logístico, the bulk of which is also based at 'El Libertador'. Grupo de Transporte No 6 exercises this responsibility through Escuadrón de Transporte No 1, which operates the six surviving Lockheed C-130H Hercules and two Boeing 707-346C transport/tanker aircraft. Escuadrón de Transporte No 2, using the survivors of eight Alenia G222s (six acquired for the FAV and two transferred from the army), has not only a transport tasking but also is responsible for multi-engine conversion. Only a single G222 was airworthy early in 1999, but others are due to be returned to service.

Grupo Aéreo de Operaciones Especiales No 10 controls most FAV helicopters. Escuadrón 101 is assigned a single UH-1B and eight UH-1H Iroquois. Escuadrón 102 is assigned the six surviving Alouette IIIs and eight AS 332B-1 Super Puma medium lift helicopters delivered in 1989. Ten cougar Mk 1s are due to join the unit in 1999. Both squadrons maintain out-based detachments in support of forces in the interior.

The last of Comando Aéreo Logístico's assets are to be found at Base Aérea 'Generalísimo Francisco de Miranda', La Carlota, in downtown Caracas. Grupo Aéreo de Transporte No 4 and its attendant Escuadrón

Fuerza Aérea Venezolana

FAV Headquarters, La Carlota, Caracas

Comando Aéreo de Combate

Grupo Aéreo de Caza No 11 'Diabolos'
Base Aérea 'El Libertador', Palo Negro
Escuadrón 33 'Halcones'   Mirage 50EV/DV
Escuadrón 34 'Caciques'   Mirage 50EV/DV
Grupo Aéreo de Caza No 12
Base Aérea 'Teniente Vicente Landaeta', Barquisimeto
Escuadrón 36   VF-5A/D, NF-5B, Beech 65
Grupo Aéreo de Operaciones Especiales No 15
Base Aérea 'General Urdaneta', Maracaibo
Escuadrón 151'Los Linces'   OV-10A/E Bronco
Grupo Aéreo de Caza No 16 'Dragons'
Base Aérea 'El Libertador', Palo Negro
Escuadrón 161 'Caribes'   F-16A-15/B-15 Fighting Falcon
Escuadrón 162 'Gavilanes' F-16A-15/B-15 Fighting Falcon

Comando Aéreo Logístico

Grupo Aéreo de Transporte No 4, Base Aérea 'Generalísimo Francisco de Miranda', La Carlota, Caracas
Escuadrón 41   Gulfstream II/III, Boeing 737-2N1
Escuadrón 42   Bell 214ST
Grupo Aéreo de Transporte No 5, Base Aérea 'Generalísimo Francisco de Miranda', La Carlota, Caracas
Escuadrón 51   King Air 200/200C
Escuadrón 52   Cessna 500 Citation I, Cessna 550 Citation II, Falcon 20F, Learjet 35A
Grupo Aéreo de Transporte No 6
Base Aérea 'El Libertador', Palo Negro
Escuadrón T1   Boeing 707-346C, C-130H
Escuadrón T2   Alenia G222
Grupo Aéreo de Operaciones Especiales No 10
Base Aérea 'El Libertador', Palo Negro
Escuadrón 101 'Guerreros' UH-1B/H Iroquois
Escuadrón 102 'Piaros'   AS 332B Super Puma, SA 316 Alouette III

Comando Aéreo de Instrucción

Grupo Aéreo de Entrenamiento de Combate No 13
Base Aérea 'Teniente Luis de Valle Garcia', Barcelona
Escuadrón 131 'Los Aviopones'   T-2D Buckeye
Grupo Aéreo de Entrenamiento No 14
Base Aérea 'Mariscal Sucre', Maracay
Escuadrón Primario   VT-34A Mentor, Cessna 182
Escuadrón Secundario   EMB-312 Tucano

41 serve in the VIP role, utilising two military Gulfstream II/III derivatives and a Boeing 737-2N1. The sister unit, Escuadrón 42, has on charge three Bell 214ST helicopters which it also uses in the VIP role. More general communication duties fall to Grupo de Transporte No 5. Its two associated squadrons, Escuadrones 51 and 52, are equipped with several military-registered Beech King Air 200s, Falcon 20DCs, a Learjet 35A and two Cessna Citation 500/550 series executive jets. Two of the Falcon 20DCs were used in the ECM training role based at 'El Libertador' but have been stored since 1991. A Cessna 550 is used for mapping work.

COMANDO AÉREO DE INSTRUCCIÓN

The final major command is Comando Aéreo de Instrucción, which is responsible for all training with Grupo de Entrenamiento No 14, under which is the Escuela de Aviación Militar. The two subordinate units are known as Escuadrón Primario and Escuadrón Secundario. Primary training is undertaken on the Beech VT-34A Mentor, which is due to be

Venezuela has had eight C-130Hs and has lost a quarter of the fleet. The survivors fly with Escuadrón T1 of Grupo Aéreo de Transporte No 6.

Since June 1981 the FAV has operated this Gulfstream II with Grupo Aéreo de Transporte No 4 (Escuadrón 41) in the VIP role. A GIV was ordered but not delivered.

replaced by 24 Aermacchi (SIAI-Marchetti) SF.260Es. Basic training is conducted on the EMBRAER EMB-312 T-27 Tucano. Eighteen T-27 Tucanos

were assigned to the school; the remaining 12 AT-27 versions were originally tasked in the advanced flying training stage, including weapons conversion. They

are on the strength of Grupo Aéreo de Instrucción Táctico No 13 at Base Aérea 'Teniente Luis de Valle Garcia', Barcelona.

Six surviving Rockwell T-2D Buckeyes, of 24 acquired, operate as tactical trainers with Escuadrón 131 of the Grupo de Entrenamiento de Combate at Barcelona. They were fitted with hardpoints and weapons, and in addition to their utilisation in the advanced training stage they also had a tactical application. The Buckeyes are due to be replaced by Aermacchi MB-339FDs as the new FAV lead-in fighter trainer from 2000 onwards. The FAV is also after two-seat AMX-T close-support aircraft, and hopes for an eventual 24 of both types.

Comando de la Aviación Naval

The Venezuelan naval air service formed in the mid-1970s as an organisation in its own right, although it can trace its ancestry back to 1922. Today, it is still a very small organisation tasked primarily with ship-borne and shore-based ASV. Its headquarters are in Caracas.

The main base is at Puerto Cabello, 60 miles (96 km) north of Caracas on the Caribbean coast. Here, Escuadrón Aeronaval de Helicópteros has a shore base for its eight (of nine delivered) Agusta-Bell 212AS anti-submarine helicopters acquired in 1980 and two ex-FAV Bell 212s. When embarked, the squadron is assigned to Venezuela's 'Sucre'-class frigates and uses the Mk 26 or A244/S torpedoes in the ASW role and the Marte anti-ship missile (in its secondary anti-shipping) role.

Shore-based ASW and a maritime patrol capability are provided by three CASA C.212-200ASW aircraft

Delivered during 1998, these three CASA 212-400s are the latest aircraft of the Aviación de la Marina.

of Escuadrón Aeronaval de Patrulla, of four delivered in 1986.

The Grupo de Apoyo of Escuadrón Aeronaval de Transporte is based at Base Aérea 'Generalísimo Francisco de Miranda', La Carlota, operating in the VIP and communications role. The Grupo de Transporte Táctico uses two CASA 212-200s, three Srs 400s and a single Dash-7 from the 'Simon Bolivar' International Airport at Maiquetia.

Comando de la Aviación Naval

HQ Caracas

Escuadrón Aeronaval de Adiestramiento	**Puerto Cabello**
Cessna 210E, Cessna 310/R, Cessna 402B/C, Bell 206B, TH-57A	
Escuadrón Aeronaval de Helicópteros	**Puerto Cabello**
Bell AB 212AS/212	
Escuadrón Aeronaval de Patrulla	**Puerto Cabello**
CASA C.212-200ASW	
Escuadrón Aeronaval de Transporte	
Grupo de Apoyo	
Beech E90, Beech 200, Rockwell 980	**La Carlota**
Grupo de Transporte Táctico	
CASA 212-200/400, DHC-7	**Maiqueti**

Training is conducted by the Escuadrón Aeronaval de Adiestramiento from Puerto Cabello using a variety of single and twin Cessna types. Helicopter training is undertaken on a single Bell 206B and a pair of ex-AMARC TH-57A Sea Rangers.

Fuerzas Armadas de Cooperación

This paramilitary organisation comprises land, sea and air assets controlled by the Ministry of Defence but is responsible for civilian matters including internal security, anti-terrorism, forestry patrols and customs duties. The air assets belong to Destacamento de Apoyo Aéreo (DAA – Air Detachments), Sección de Apoyo Aereo (SAA – Air Section) or the Centro de Adiestramiento Aéreo Guardia Nacional (CAAGN). The Guardia Nacional, as the organisation is more commonly known, has a number of detachments.

Fuerzas Armadas de Cooperación

DAA-1 Santa Barbara de Barinas	1 x B.206B, 1 x Ce 206, 1 x B.412EP, 1 x Pa 34
DAA-2 Santa Barbara del Zulia	2 x A 109A, 1 x B.206B, 1 x Ce 206, 1 x B.412EP
DAA-3 Maracaibo	1 x Be 55, 2 x AS 355F-2, 1 x B.206B, 2 x B.412EP, 1 x M-28
DAA-4 Barquisimeto	1 x Be 90, 1 x B.206L, 1 x Ce 402, 1 x B.412SP, 1 x AS 355F-2
DAA-5 Caracas	1 x B.206B, 1 x Be 90, 1 x A 109A, 2 x IAI 201, 2 x Be 200, 4 x AS 355F-2, 4 x M-28, 2 B.412EP
DAA-6 San Fernando de Apure	1 x IAI 201, 2 x AS 355F-2, 2 x B.412EP
DAA-7 Porlamar, Isla Margarita	4 x B.206B, 1 x Ce 206
DAA-8 Tucupita	2 x Ce 206, 1 x Be 65-80, 1 x B.206B, 1 x IAI 201, 1 x B.412EP
DAA-9 Puerto Ayacucho	1 x B.412SP, 1 x M-28
SAA-11 Santa Barbara de Barinas	1 x B.412EP
CAAGN Porlamar, Isla Margarita	2 x F-28C, 3 x Ce 152, 1 x F280C, 2 x B.206B, 2 x M-26, 1 x Ce 182.

Servicio Aéreo del Ejército Venezolano

The Servicio Aéreo is a small but highly mobile unit of the Venezuelan army. The original air department of the army was formed in 1970. Flight operations take place from three operating locations.

Fixed-wing assets are based at the 'Dr Oscar Machado Zuloaga' Aeropuerto Caracas airport with the Grupo de Transporte del Centro. The main transport is the IAI Arava, while a pair of Beech twins is operated in the VIP transport role. Three Cessna singles undertake the light transport and liaison tasks.

The army's helicopter fleet is operated by the Grupo a Aéreo de Apoyo y Asalto 'General F. Jimenez' from San Felipe. Four of the Agusta A 109s can carry 7.62-mm or HGM 0.50-in machine-guns for offensive support. The AS 61 Sea Kings are based at La Carlota most of the time due to the congestion at San Felipe. The helicopters are dispatched all over Venezuela during the year.

The army is responsible for its own aircrew training, which it carries out under the auspices of Centro de Instrucción at San Felipe. Helicopter training includes 40 hours on two versions of the Bell 206. Fixed-wing pilots undertake 60 hours on the three Cessna singles.

Maintenance is undertaken by the Comando Logístico at La Carlota.

Wearing Guardia Nacional titles, this Cessna 402 of the Fuerzas Armadas de Cooperación was acquired in 1979. GN-7948 is currently based with DAA-4 at Barquisimeto.The organisation uses a number of light aircraft and helicopters from several detachments in roles of civil security and disaster relief.

Comando Aéreo del Ejército

Grupo a Aéreo de Apoyo y Asalto 'General F. Jimenez' San Felipe		
Agusta A 109A/A-2	9	communications
Agusta-Bell 412EP/SP	4/2	transport helicopter
Agusta-Sikorsky AS-61D	4	transport helicopter
Bell 205A-1	3	transport helicopter
Bell UH-1H Iroquois	6	transport helicopter

Grupo de Transporte del Centro, 'Dr Oscar Machado Zuloaga' Aeropuerto Caracas		
Beech 90 King Air	1	VIP
Beech 200 King Air	1	VIP
Cessna TU 206G Stationair	3	communications
Cessna T 207A	1	communications
IAI Arava 201/202	3/2	transport

Centro de Instrucción Aeronáutica, San Felipe		
Bell 206B JetRanger	1	helicopter training
Bell 206L LongRanger	1	helicopter training
Cessna 172L	1	basic training
Cessna 182R	2	basic training

Africa

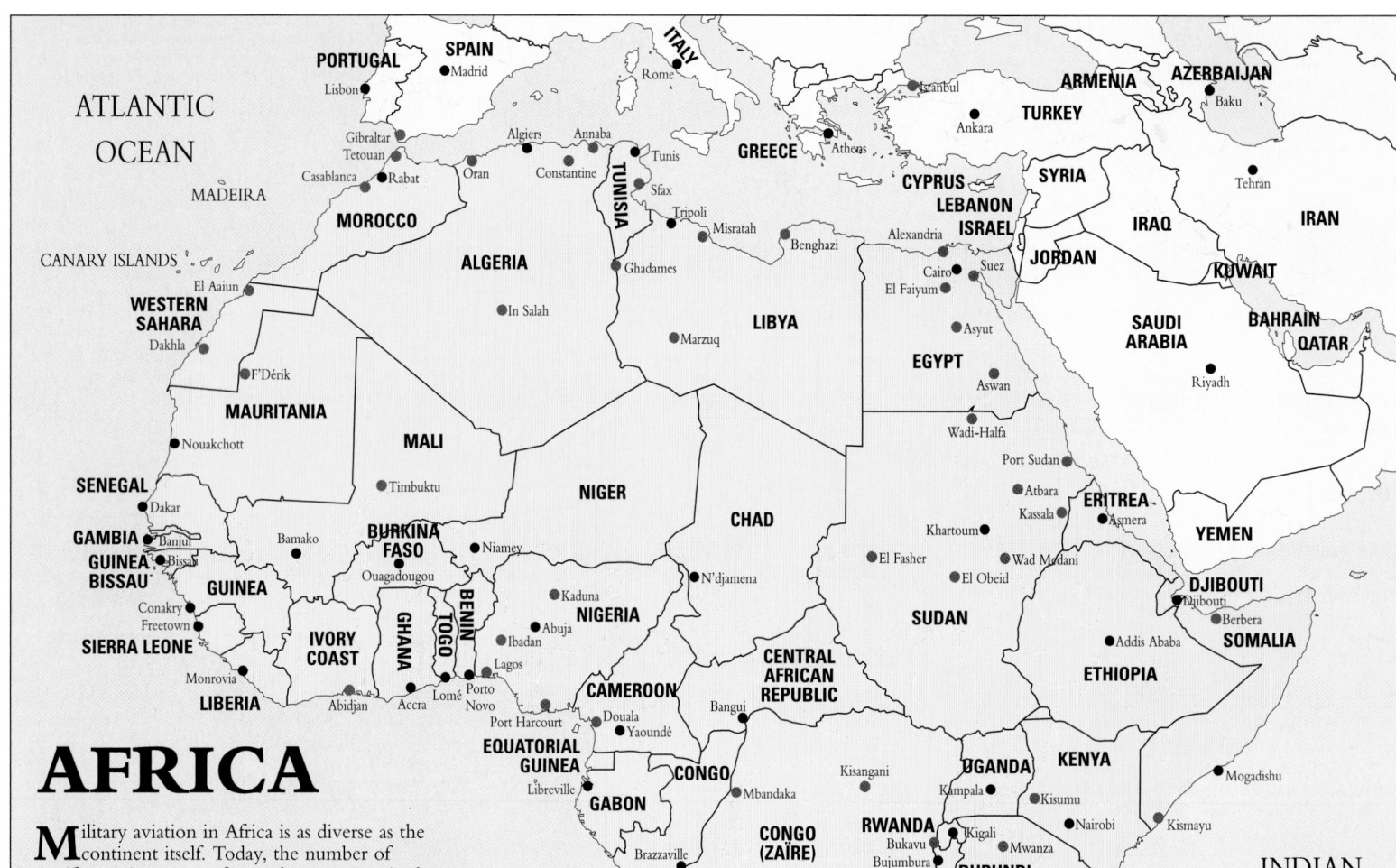

ATLANTIC OCEAN

MADEIRA

CANARY ISLANDS

INDIAN OCEAN

AFRICA

Military aviation in Africa is as diverse as the continent itself. Today, the number of significant air arms are few and most countries have problems maintaining what they have, lacking spares or the skilled manpower to operate the equipment effectively. The end of the Cold War ended the supply of subsidised military aircraft and training from the superpowers, which had largely been supplied to further their own political aims. Budgets for military hardware are small or non-existent, and few aircraft will be purchased new. Geographically, Africa can be divided into several distinct regions.

North Africa consists of Morocco, Algeria, Tunisia, Libya and Egypt. All have air forces that are large when compared to the average for the continent. The area has been the setting for many wars, but with Libyan military adventurism at an end, Egypt's peace accord with Israel still in force and Morocco's war in Western Sahara over, the current problems are on a smaller scale, such as Islamic fundamentalist terrorists operating in Algeria and Egypt.

Sub-Saharan African countries consist of huge tracts of land with small populations. Most are burdened with debts to foreign banks and lack modern infrastructure. The air forces of these nations are small, and usually tasked with providing transport and communications duties, frequently operating a biz-jet for the head of state. Sub-Saharan air forces cannot afford to acquire large numbers of aircraft at one time and thus operate a motley collection of types acquired in small batches. The sophistication and high levels of skill required to operate modern aircraft means that sortie rates are generally low, so aviation would play a small role in any conflict in the region. Ties to France in this region are strong, with the Armée de l'Air having a presence at DA160 Dakar in Senegal.

East Africa has been inflicted with droughts and war over the last decade. Eritrean independence did not end the fighting in the region, as border skirmishes between Eritrea and Ethiopia continue. The air forces of both sides have been involved and Ethiopian losses to ground-based guns have been recorded. Political unrest in Somalia has continued

since that country's civil war started. Djibouti plays host to a French detachment consisting of the Mirage F1Cs of EC 4/33 and Alouette IIIs, Pumas and C.160 Transalls of ETOM 88, as well as maintaining a small air arm of its own.

West Africa is geographically dominated by Nigeria. Grand plans for regional superpower status were dashed by civil unrest and the falling price of oil, its major export. Today, the majority of the country's air force is unserviceable. The civil war in Liberia is nearing an end. Other West African countries have small air arms and few have anything above the capability of armed training aircraft.

At the moment, Central Africa is dominated by the former Zaïre, which was renamed the Republic of Congo at the end of the civil war in 1997. Unfortunately, peace was short-lived and the

country is yet again embroiled in a war which is starting to involve its neighbours. President Laurent Kabila of Congo ordered Rwandan troops out of his country, triggering a revolt by factions in the army ethnically aligned with Rwanda. Ugandan troops deployed to western Congo to attack the Allied Democratic Forces (a rebel army opposing the current Uganda regime), while Angola sent in troops to support the president and to prevent supplies to National Union for the Total Independence of Angola (UNITA) passing through the Congo. Zimbabwe is also supporting the president with troops. The potential for a large-scale regional war is great.

Compared to a decade ago, Southern Africa is a much more stable region. The end of apartheid in the Republic of South Africa and the conclusion of the various 'bush wars' have reduced expenditure by the country on its military. Even so, South Africa remains the dominant military power in that region. Angola continues to be embroiled in a long civil war against the forces of UNITA, in a war that has seen the frequent use of military air power.

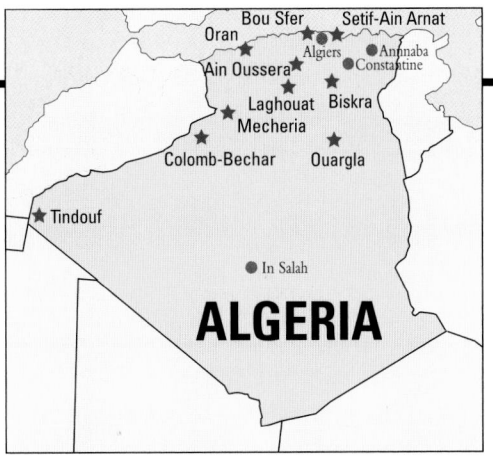

ALGERIA

Capital: El Djezaïr (Algiers)
Population: 25 million
Land area: 2.382 million km² (919,000 sq miles)
Major cities: Constantine, Oran, Annaba

Algerian Air Force (al-Quwwat al-Jawwiya al-Jaza'eriya – QJA)

The QJA was established in 1962 after a long independence war against France. During the French presence, many air bases were built. Egyptian experts advised the Algerians how to use the large French air bases at Mers-el-Kebir (Bou Sfer) and Reganne, and the missile test base at Colomb-Bechar. Egypt gave the QJA its initial equipment: five MiG-15 fighters, 12 Gomhourias (Bü 181 copies) and 12 Yak-11 trainers.

In the summer of 1964, after Algeria's border war with Morocco in 1963, Czechoslovakia and the Soviet Union supplied the Algerian Air Force with one squadron of 12 Il-28s, two squadrons of 20 MiG-17 fighters, and one squadron of 15 MiG-15 fighter-bombers. The overthrow of the Algerian government by a military junta reinforced relations with the Soviets, resulting in direct deliveries of more Il-28s and MiG-17s. In 1965 the first six MiG-21F day fighters were delivered, followed by more military hardware including eight Il-14s, seven An-12s and 18 Mi-4s. Much of the Algerian Air Force inventory was flown by Egyptian and Soviet pilots. Later, the Soviets extended airfields and started to build an air defence missile belt with radars, SAMs and communication equipment. The extended airfields were used as back-up bases for the Egyptian Air Force after the Six-Day War in 1967.

In 1969, 28 refurbished French CM.170 Fouga Magisters were acquired. French SA 330 Pumas and Dutch-built F27 Friendships also found their way to Algeria. During the Yom Kippur War in 1973, the QJA sent one squadron of MiG-17s to the Sinai desert, but they landed on the freshly occupied airfield of El Arish and the pilots were taken prisoner by the Israelis. In 1975 Algeria was involved in the Polisario War with Morocco over the Spanish Sahara. The runway of Tindouf was extended and strengthened, and MiG-21s were based there for air cover.

Algerian transport assets are concentrated at Boufarik. All wear a civil registration to facilitate overseas flights while a majority also have a civil type scheme (as displayed on the Il-76TD below). C-130H 7T-WHS (above) displays the military scheme.

CURRENT SITUATION

The current Algerian Air Force is still based on old Soviet methods and doctrine, and is organised into four sub-divisions – Operations, Technical, Logistics and Schools – on 13 air bases, and has one repair establishment. There are no large specialised commands, but squadrons are specialised by function (interceptor, bombers, etc.), as are some of the bases.

The Air Defence arm became part of the air force in 1986 and includes the SAM brigades under the AD-centre at Cheragas. An Anti-Aircraft-Defence School is located near La Reghaia. The backbone of the Air Defence Force is five squadrons flying the MiG-21bis and -21MF. The MiG-21s are based throughout the country: Tindouf in the southwest (covering the Spanish Sahara), Colomb-Bechar in the west (near the Moroccan border), Bou Sfer (Mers-el-Kebir) in the north (protecting naval facilities) and Laghouat. In 1998 Algeria was offered 36 MiG-29s in exchange for 120 MiG-21s by the Belarussian arms export agency, Beltechexport. The Mach 3-capable MiG-25 interceptors, based at Ain Oussera (Paul-Cazelles), are responsible for the air defence of the capital Algiers and the strategic nuclear site near the air base. It is thought that Algeria intends to produce its own nuclear arsenal. The MiG-25RB reconnaissance aircraft are also located at Ain Oussera. The deep penetration strike Su-24MK were based at Ain Oussera but moved to Laghouat.

Originally most of the Algerian MiG-23s were based at Laghouat with one squadron at Ouargla. From there, permanent detachments were stationed at Tindouf, near the Moroccan border. They are currently having their electronics updated by a Bulgarian firm. One SAM regiment is distributed among various air bases with 20 launchers (with some 48 SA-3, 12 SA-6 and 24 SA-8 missiles). Although three squadrons are equipped with the 'Fitter', it is thought that all the aircraft are grounded owing to a lack of spare parts due to the current situation in Russia. Other Soviet type-equipped squadrons may also have serviceability problems.

For COIN and advanced trainer duties, the QJA operates two squadrons of L-39ZAs, which replaced the CM.170 Magisters. For ground support the Soviets supplied 37 Mi-24 'Hinds', used in two squadrons based at Biskra. A helicopter wing of two squadrons flies 24 Mi-17 and 16 Mi-8MT in the air assault role, while another at Ech Cheliff has two squadrons with Mi-8s and Mi-17s. It is believed that the air force is planning to replace about 30 of its Mi-8/17 fleet and upgrade the rest.

The Algerian Transport Wing is based at Boufarik, near Algiers. The workhorses of this wing are 18 C-130 Hercules, six Il-76s and five An-12s. The Maritime Patrol Squadron is part of the Transport Wing, flying eight F27-400s from Bou Sfer and Boufarik. A VIP communications squadron has three Gulfstream IIIs, two F27-400, three Beech Queen Air B.80s and a pair of Falcon 900s. A liaison squadron operates five Bell 206L and nine AS 350 Ecureuils.

Training takes place at two air bases near Oran: Oran-Tafaroui and Oran-Es Senia. Basic training is given on 20 Zlin 142s followed by primary training on six T-34Cs and 12 Beech 24 Sierra 200s at Tafaroui. Advanced training is given on 33 L-39ZAs

Algeria

al-Quwwat al-Jawwiya al-Jaza'eriya – QJA

Operations Command

Air Wing		
19 Fighter Sqn	MiG-21bis/PFM/UM	Colomb-Bechar
Air Wing		
153 Fighter Sqn	MiG-21PFM/UM	Bou Sfer
Air Wing		
140 Fighter Sqn	MiG-21M/UM	Ouargla
Air Wing		
11 Fighter Sqn	MiG-21PFM/UM	Tindouf
det of 140 Fighter Sqn	MiG-21M/UM	Tindouf
Air Wing		
120 Interceptor Sqn	MiG-25PD/PU	Ain Oussera
510 Reconnaissance Sqn	MiG-25RB	Ain Oussera
Air Wing		
28 Attack Sqn	MiG-23BN/UM	Laghouat
29 Interceptor Sqn	MiG-21PFM/UM	Laghouat
274 Fighter-Bomber Sqn	Su-24MK	Laghouat
Air Wing *('Fitters' believed grounded due to lack of spares)*		
Attack Sqn	Su-20	Mecheria
Attack Sqn	Su-20	Mecheria
Attack Sqn	Su-7BM	Mecheria
Air Wing		
31 Transport Sqn	C-130H/H-30	Boufarik
32 Transport Sqn	An-12BP	Boufarik
510 Transport Sqn	Il-76MD	Boufarik
580 VIP/Comm. Sqn	F27-400M	Boufarik
	Gulfstream III, Falcon 900	
? Liaison Sqn	AS 350, Bell 206L	Boufarik
? Maritime Patrol Sqn	F27-400M	Boufarik/Bou Sfer
Air Wing		
420 Attack Heli Sqn	Mi-24 'Hind-A'	Biskra
440 Attack Heli Sqn	Mi-24 'Hind-D'	Biskra
Air Wing		
36 Assault Sqn	Mi-8MT	El Bouleida
45 Assault Sqn	Mi-17	El Bouleida
Air Wing		
460 Transport Sqn	Mi-8	Ech Cheliff
480 Transport Sqn	Mi-17	Ech Cheliff

Training Command

Air Academy		
67 Elem. Training Sqn	Zlin 142	Oran-Tafaroui
68 Prim. Training Sqn	Beech Sierra 200	Oran-Tafaroui
	T-34C, Queen Air B80, King Air B200	
610 Adv. Training Sqn	L-39ZA	Oran-Tafaroui
620 Adv. Training Sqn	L-39ZA	Oran-Tafaroui
660 Heli Training Sqn	Mi-2, Hughes 269	Setif-Ain Arnat

Current aircraft in QJA service

Type	Role	In Service
Aero L-39ZA	advanced trainer	33
AS 350 Ecureuil	liaison	9
Antonov An-12BP	transport	5
Beech B.80/B.200T	crew trainer	3/2
Beech T-34C	primary trainer	6
Beech 24 Sierra 200	primary trainer	12
Bell 206L	liaison	5
Falcon 900	VIP/communication	2
Fokker F27-400M	maritime patrol/VIP	10
Gulfstream III	VIP/communication	3
Hughes 269	helicopter trainer	6
Ilyushin Il-76TD	transport	6
C-130H/H	transport	6/12
Mil Mi-2	helicopter trainer	28
Mil Mi-8C/MT	assault helicopter	16/16
Mil Mi-17	assault helicopter	24
Mil Mi-24 'Hind-A/D'	anti-tank helicopter	15/19
MiG-21bis/MF	interceptor	45/15
MiG-21UM/US	trainer	15
MiG-23BN	fighter-bomber	18
MiG-23MS/UB	interceptor/trainer	36/10
MiG-25PD/PU	interceptor/trainer	10/3
MiG-25RB	reconnaissance	3
Sukhoi Su-7BM/Su-20	fighter-bomber	10/27
Sukhoi Su-24MK	strike	10
Zlin 142	basic trainer	20

at Oran-Es Senia. Initial helicopter training is given on six Hughes 269As and 28 Mi-2s at Setif-Ain Arnat. Multiple-engined transport crew training is also given on the Beech King Airs.

ANGOLA

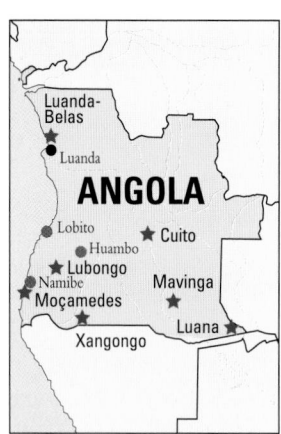

Capital: Luanda
Population:
10 million
Land area:
1.247 million km²
(481,225 sq miles)
Major cities: Lobito,
Namibe

The exact number and sub-variants of Angola's Mil Mi-8/17s is unknown. At least 70 examples and possibly many more have been delivered, seeing action in the fight against UNITA.

Angola operates MiG-23ML 'Flogger-Gs', a sub-type more advanced than the examples supplied by the USSR to former Warsaw Pact countries.

Força Aérea Popular de Angola (Angolan People's Air Force)

Since winning independence from the Portuguese in 1975, Angola has been embroiled in a long and costly civil war. Today the government continues its struggle against the troops of the National Union for the Total Independence of Angola (UNITA) in a war which has seen the use of Cuban and South African soldiers.

The Força Aérea Popular de Angola has been heavily involved in the fighting, and has acquired aircraft from a variety of sources including the Soviet Union, France, Netherlands, Spain and Switzerland. Attrition has been heavy at times – for example, 40 aircraft were lost in the battle for Mavinga in May 1987 – and serviceability of the aircraft is also a problem.

Air defence is provided by MiG-23MLs based at Lubongo but deployed where needed. A report in 1996 suggested the FAPA had ordered a number of MiG-29s, but nothing has been heard of the deal since. Ground attack duties are conducted using MiG-23BNs, Su-22M-4Ks and Su-25s, as well as Mil Mi-24 'Hind D/Es'. PC-9s have also been used in the light strike role.

Força Aérea Popular De Angola

Type	Role	Delivered/ In Service
Aérospatiale SA 342M Gazelle	observation heli.	7/7
SA 316B/IAR-316B Alouette III	communications	35/-
Antonov An-2 'Colt'	utility transport	10/-
Antonov An-12BP 'Cub'	*(poss. wfu)* transport	6/-
Antonov An-26 'Curl'	transport	12/16
Antonov An-32 'Cline'	transport	3/-
CASA C.212-200 Aviocar	transport	8/-
CASA C.212-300M Aviocar	transport	3/-
CASA C.212-300MP Aviocar	maritime patrol	6/-
Eurocopter AS 565AA Panther	combat helicopter	-/5
Eurocopter AS 565UA Panther	utility helicopter	-/8
Fokker F27 Friendship Mk 400MPA	maritime patrol	1/1
MiG-15UTI 'Midget'	advanced training	8/-
MiG-17F 'Fresco'	attack	5/-
MiG-21MF/bis 'Fishbed'	air defence/attack	20/-
MiG-21UM 'Mongol-B'	conversion training	4/-
MiG-23BN 'Flogger-H'	attack	10/-
MiG-23ML 'Flogger-G'	interceptor	10/-
MiG-23UB 'Flogger-C'	conversion training	3/-
Mil Mi-8/17 'Hip'	assault helicopter	25/-
Mil Mi-24 'Hind-D/E'	close air support	25/-
PC-6/B Turbo-Porter	utility transport	4/-
PC-7 Turbo-Trainer	armed trainer	10/23
PC-9	basic training	4/4
BN-2A Islander	utility transport	6/16
Rockwell Turbo Commander 690A	communications	1/1
Su-22M-4K 'Fitter-K'	attack	11/-
Su-22UM-3 'Fitter-G'	conversion training	4/-
Su-25/(UB ?) 'Frogfoot-A(/B ?)'	close air support	-/14

At least 50 per cent of the transport fleet wears civil registrations, including most of the Antonov An-26s and An-32s. The FAPA wishes to purchase some of the ex-RAF Hercules C.MK 1s being replaced by the C-130Js, but delays in the delivery of the Js may have prevented this. Governmental aircraft which fly with civil registrations include single examples of the Gulfstream III (D2-ECB), Rockwell Turbo Commander 690A (D2-EAA), Tupolev Tu-134A (D2-EAA) and a Yak-40 (D2-EAG).

BENIN

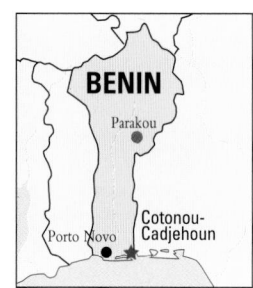

Capital: Porto Novo
Population: 4.7 million
Land area: 112620 km² (43,470 sq miles)
Major city: Parakou

Forces Armées Populaires de Benin (Benin People's Armed Forces)

Known as Dahomey until it achieved independence from France in 1960, Benin undertook a major reorganisation of its armed forces in mid-1990, into land, sea and air elements, plus the Gendarmerie. The aviation element is entirely transport-equipped, initially with seven French-supplied Douglas C-47s, four MH-1521M Broussards and two Agusta-Bell 47Gs. They were supplemented in 1978 by two Fokker F27s (soon transferred to Air Benin) and two Antonov An-26 twin-turboprop transports, plus other Soviet-supplied equipment. A single DHC Twin Otter was transferred to the FAPB in 1989 from Air

Benin. Two Dornier Do 128-2 light twins were delivered from Germany in late 1985 to replace the Douglas C-47s. Most of the older equipment, including the C-47s and Broussards, now appear to have been withdrawn from service.

Forces Armées Populaires de Benin

Type	Role	Delivered/In Service
Antonov An-2	light transport	3/2
Antonov An-26	tactical transport	2/2
de Havilland Canada DHC-6-300 Twin Otter	light transport	1/1
Dornier Do 128-2	light transport	2/2
Eurocopter AS 350B	comm helicopter	3/2

BOTSWANA

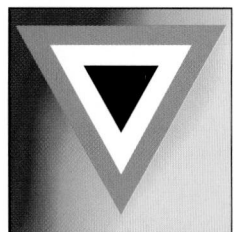

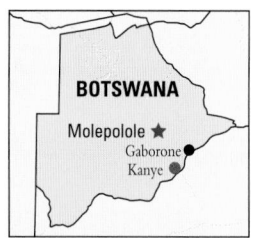

Capital: Gaborone **Population:** 1.3 million
Land area: 575000 km² (221,950 sq miles)
Major cities: Kanye

Botswana Defence Force Air Wing

The Botswana Defence Force Air Wing has recently upgraded its combat efficiency, replacing its Strikemasters with ex-Canadian CF-5 Freedom Fighters, acquired via Bristol Aerospace, which had upgraded the aircraft to serve as CF-18 lead-in trainers before the requirement was cancelled. The first arrived in September 1996.

Transport capabilities have also been rejuvenated with the transfer of a pair of US FMS funded C-130B Hercules from AMARC. Nine Cessna O-2As were reported as having been delivered from the same source in 1993, but have not been reported since.

The Air Wing is believed to have five squadrons (one fighter, two transport, one training and a helicopter squadron).

Botswana Defence Force Air Wing

Sqn	Type	Role	In Service/ Delivered
Z7 Sqn	PC-7 Turbo-Trainer	basic trainer	-/7
	Cessna A152 Aerobat	primary training	1/2
Z10 Sqn	CASA C.212-300 Aviocar	transport	2/2
	C-130B Hercules	transport	2/2
Z12 Sqn	Airtech CN.235M	transport	2/2
	BN-2A-21 Defender	transport	8/12
(One aircraft was w/o on delivery)			
	Gulfstream IV	VIP	1/1
Z21 Sqn	Bell 412/412SP	heli transport	5/5
	AS 350B/BA Ecureuil	utility helicopter	4/5
Z28 Sqn	CF-5A Freedom Fighter	air def./attack	10/10
	CF-5D Freedom Fighter	conv. trainer	3/3
(unit unconfirmed)			

After having taken delivery of five Bell 412s from 1988 for Z21 Squadron, this example, OH-5, was badly damaged and returned for repairs in Canada.

BURKINA FASO

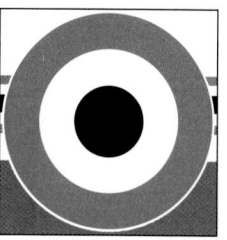

BURKINA FASO

Ougadougou
Koudougou • ★ Ougadougou
★ Bobo Dioulasso

Capital: Ouagadougou **Population:** 9 million
Land area: 274,000 km² (106,000 sq miles)
Major city: Bobo Dioulasso

Force Aérienne de Burkina Faso

Type	Role	Delivered/ In Service
SA 316B Alouette III	utility helicopter	2/1
AS 350B Ecureuil	utility helicopter	1/1
BAe 748 Srs 320	transport	2/2
Cessna F172N	communications	/1
Cessna F337E	communications	2/1
Nord 262C-50P/65	transport	2/2
Mil Mi-8 'Hip'	transport helicopter	/3
SIAI-Marchetti SF.260W	armed trainer	4/4

The old-style Burkino Faso roundel is displayed on this Nord 262 Frégate. A pair of these aircraft continues to provide a light transport capability.

Force Aérienne de Burkina Faso (Burkina Faso Air Force)

Formerly known as Haute-Volta, in the French Union of West Africa, the original Escadrille de Haute-Volta was founded with French aid in 1964 as a transport force, including the transfer of two or three Douglas C-47s and Max Holste MH1521M Broussards. These were later followed by two Alouette III helicopters and a pair of Nord 262 twin-turboprop transports, to form a Transport and a Helicopter Squadron. More recently, the latter unit was reinforced by three former Soviet Mil Mi-8s and two Aérospatiale SA 365Ns, for transport, utility and SAR roles, although the Dauphins are now civil-registered. The FABF is still operating at least one Aérospatiale AS 350B Ecureuil, however.

In mid-1984, Soviet aid to the FABF reportedly included eight MiG-21s and two MiG-21U combat trainers, which were operated in border clashes with Mali in December 1985. These have since been grounded, however, and six ex-Philippine air force SIAI-Marchetti SF.260WP Warrior armed trainers acquired in mid-1986 were bought back by a Belgian dealer in March 1993. Four SF.260Ws bought new in the mid-1980s are believed to remain in Burkina Faso service.

BURUNDI

• Muramvya
• Bujumbura
★ Bujumbura

BURUNDI

Capital: Bujumbura **Population:** 5.4 million
Land area: 27835 km² (10,745 sq miles)
Major city: Muramvya

Armée National de Burundi

Type	Role	Delivered/ In Service
SA 342L Gazelle	utility helicopter	4/2
SA 316B Alouette III	utility helicopter	4/3
Dornier Do 27Q-4	communications	1/1
Reims-Cessna FRA 150L	primary trainer	3/3
SIAI-Marchetti SF.260C/TP/W	armed trainer	10/6

Armée National de Burundi

As a component of the former Belgian-administered UN trust territory of Ruanda-Burundi, Burundi established an army air wing in 1966, following its 1962 independence, with a single Dornier Do 27 liaison aircraft and four Alouette III helicopters. The French air force supplied three Douglas C-47s in 1969, although they were soon transferred to the national airline. Three Reims-Cessna FRA 150Ls followed in the 1970s, but three SIAI-Marchetti SF.260W Warriors delivered from late 1981, together with three SF.260Cs, represented Burundi's first armed aircraft. Four SF.260PT turboprop variants were also ordered, at the same time, although their service status is uncertain. Burundi's helicopter force was augmented in 1982 with two Aérospatiale SA 342L Gazelles, each armed with a single 20-mm GiAT M621 cannon, followed by two more from French aid in 1984.

CAMEROON

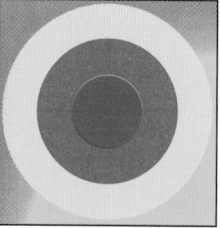

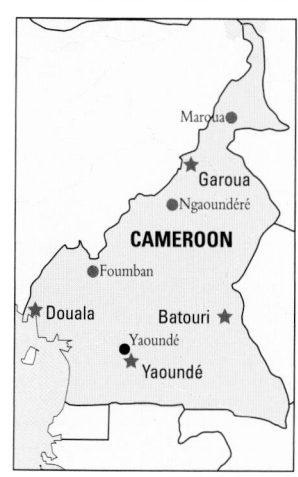

Maroua •
Garoua •
Ngaoundéré •
CAMEROON
Foumban •
★ Douala   Batouri ★
Yaoundé •
★ Yaoundé

Capital: Yaoundé
Population: 11.8 million
Land area: 475500 km² (183,500 sq miles)
Major city: Douala

Impala Mk 1 jet trainers, and four single-seat MB.326K Impala Mk 2 light ground-attack aircraft from South African surplus sources to equip a close air support wing.

Three DHC-4 Caribou piston-engined STOL transports were replaced by a total of four twin-turboprop DHC-5D Buffalos from 1981, following 1977 deliveries of two Lockheed C-130H Hercules. A single DHC-5D and C-130H-30 were subsequently ordered as accident attrition replacements, and three twin-turboprop Do 128s were delivered in 1982 for coastal and maritime patrol. A single civil-registered Boeing 727-2R1, Gulfstream GIII, Eurocopter AS 332L Super Puma and AS 365N Dauphin 2 are also operated for government and VIP transport.

Armée de l'Air Cameroun

Type	Role	Delivered/ In Service
SA 342L Gazelle	utility helicopter	4/4
SA 316B Alouette III	utility helicopter	4/3
SA 318C Alouette II	utility helicopter	/1
Atlas MB.326M Impala Mk 1	armed trainer	2/2
Atlas MB.326K Impala Mk 2	light ground attack	4/4
Bell 206L-3 LongRanger III	communications	2/2
Dassault-Dornier Alpha Jet MS2	ground attack	7/6
Dornier Do 128-6MPA	maritime patrol	3/2
DHC-5D Buffalo	tactical transport	5/4
Fouga CM-170R Magister	basic trainer	*12/6
IAI Arava 201	transport	2/1
Lockheed C-130H/H-30 Hercules	transport	3/3
Piper PA-23 Aztec	communications	2/2

*six in long-term French storage

Left: Cameroon operates both the standard and stretched C-130H Hercules. This aircraft was damaged in France in 1989 and is under rebuild.

Below: The Fouga Magister has served in the Armée de l'Air Cameroun since 1973, examples having been obtained from both France and West Germany. They are used for training purposes.

Armée de l'Air Cameroun (Cameroon Air Force)

Following independence from France in 1960, Cameroon established small defence forces in 1961, with standard French transfers of Douglas C-47s, MH Broussards and Dassault Flamants. Subsequent French procurement included Alouette II/III and ATM-armed Gazelle helicopters, plus half a dozen Fouga Magister and a similar number of Alpha Jet armed jet trainers. In late 1996, they were supplemented by two two-seat Atlas-built MB.326M

CAPE VERDE

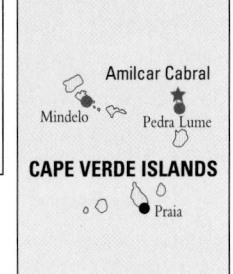

Capital: Praia
Population: 370,000
Land area: 4000 km²
(1,600 sq miles)
Major city: Mindelo

Força Aérea Caboverdaine (Cape Verde Air Force)

Located in the Atlantic some 334 nm (385 miles; 618 km) west-northwest of Senegal, the Cape Verde group of 10 major islands was granted independence from Portugal in 1975, and turned to the Soviet Union to equip its small Popular Revolutionary armed forces and People's Militia. After personnel training with Soviet assistance, three Antonov An-26 twin-turboprop tactical transports were delivered from 1982, to equip an air element of the PRAF. They are based at Amilcar Cabral, and so far appear to be the sole military aircraft in PRAF service.

In the early 1990s they were supplemented by a single civil-registered Dornier Do 228-201 light twin-turboprop utility transport equipped for maritime patrol roles, for operation by the para-military Guardia Costeira (coastguard service). This has since been reinforced by more recent delivery from Brazil of a twin-turboprop EMBRAER EMB-110P1-(K) with maritime surveillance equipment, for similar coastguard operation.

Força Aérea Caboverdaine		
Antonov An-26 'Curl'	3	tactical transport

CENTRAL AFRICAN REPUBLIC

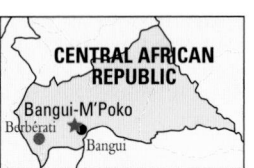

Capital: Bangui
Population: 3 million
Land area: 625000 km²
(241,240 sq miles)
Major city: Berbérati

The active assets of the Escadrille Centrafricaine are limited to a single Rallye 235G Guerrier and this AS 350B Ecureuil, registered as TL-KAZ.

The Central African Republic is ruled by President Ange Patasse, who was democratically elected in 1993. His leadership is weak and the country has collapsed in an economic crisis, induced in part by tribal rifts. The first army mutiny occurred in April 1996 and was followed by a second in May. French forces, which are based in the country, restored order and, after a brief but fierce fight, retook the radio building. The third mutiny started in November 1996

Escadrille Centreafricaine		
Aérospatiale Rallye 235GS Guerrier	1	communications
Britten-Norman BN-2A Islander	2	(u/s)
Cessna 337	1	(u/s)
Eurocopter AS 350B Ecureuil	1	communications

and can be considered as a true military uprising against the unpopular president.

The state of the national air arm, the Escadrille Centreafricaine (Central African Flight), parallels that of the country: complete disarray. Lack of funding has left it with only a Rallye Guerrier and the presidential helicopter (an AS 350 Ecureuil, delivered in 1987) able to fly. A Cessna 337 and two Islanders (one belonging to the Ministry of Water and Forest) are on the inventory but are not in flying condition due to lack of spares. Pilots were trained in France and are proficient, but they suffer from a lack of flying hours.

CHAD

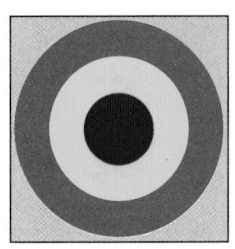

Capital: N'Djamena
Population: 5.7 million
Land area: 1.284 million km² (495,625 sq miles)
Major cities: Moundou

Escadrille Nationale Tchadienne (Chad National Flight)

Mainly French aid in the 1970s expanded the ENT into a relatively comprehensive force with seven ground-attack Douglas AD-4 Skyraiders, three Douglas DC-4/C-54s, a dozen Douglas C-47s, six Reims-Cessna FTB.337 light twins, seven Lockheed C-130s (including two ex-RAAF C-130As), and other supporting types, for operations against Libyan-supported northern rebel guerrillas.

Only a handful of aircraft are now operated, however, following formal ending of hostilities in May 1988. Rebel and Libyan forces were defeated with large-scale French air force support, and Armée de l'Air units still fly on detachments to Chad.

Severe budget limitations have also increased Chad reliance on French support for further defence, and there has been little further procurement since the transfer of two ex-civil Pilatus PC-7 turboprop trainers, now equipped with 20-mm GiAT cannon in underwing pods, from CIPRA in France, in 1985. Three captured Libyan air force SF.260 Warrior armed trainers also entered Chad service in 1987, but one crashed in 1989 and another was sold to the US. In late 1987, the US supplied 24 General Dynamics Stinger shoulder-launched infantry SAMs and seven launchers to Chad, to reinforce its French-established air defence system. Two ex-Dutch air force Alouette III helicopters were acquired in January 1995.

One of a pair of Pilatus PC-6s that has served with the Chad National Flight since 1976 is believed to remain in service, sistership TT-KAB having been sold.

Escadrille Nationale Tchadienne

Type	Role	Delivered/ In Service
Aérospatiale SA 316B Alouette III	utility helicopter	2/2
CASA C.212A-200 Aviocar	transport	2/1
Lockheed C-130A/H/H-30 Hercules	transport	7/3
Pilatus PC-6B2/H2 Turbo Porter	utility	2/1
Pilatus PC-7 Turbo Trainer	armed trainer	2/2
Reims-Cessna F.337F	communications	6/1
SIAI-Marchetti SF.260WL Warrior	armed trainer	3/2

CONGO

Capital: Brazzaville   **Population:** 2.3 million
Land area: 342000 km² (132,000 sq miles)
Major cities: Pointe-Noire

Force Aérienne Congolaise (Congolese Air Force)

From its original French-aided origins in the early 1960s, including supplies of Douglas C-47s, Broussards and Bell 47Gs, followed later by Nord Noratlas tactical transports and Sud Alouette helicopters, the air force of Congo-Brazzaville switched to Soviet equipment in the 1970s. This included five Ilyushin Il-14 and six twin-turboprop Antonov An-24 transports, plus an An-26, while in return for providing bases for Cuban MiG-17 operations over Angola, these fighters, and a few MiG-15UTI combat trainers, were later transferred to the FAC. In 1990,

Force Aérienne Congolaise

Type	Role	Delivered/ In Service
Aérospatiale AS 365C Dauphin	utility helicopter	1/1
Aérospatiale SA 316B Alouette III	utility helicopter	2/2
Aérospatiale SE 3130 Alouette II	utility helicopter	4/2
Antonov An-24V/RV 'Coke'	transport	6/1
Antonov An-26 'Curl'	tactical transport	1/1
Douglas C-47	transport	3/2
MiG-21MF/bis 'Fishbed-J/L'	air defence/GA	16/10*
MiG-21US 'Mongol-B'	combat trainer	2/2*
Mil Mi-8T/17 'Hip'	transport helicopter	10/6
Nord N2501F Noratlas	tactical transport	4/1
*most in storage		

Of the six Antonov An-24s used by the Force Aérienne Congolaise, only one remains in service, while others are stored at Maya-Maya in varying states of disrepair.

the MiG-17s were replaced by 16 USSR-supplied MiG-21MF/bis 'Fishbed-J/Ls', plus a couple of

MiG-21US trainers, together with a Soviet training mission which stayed until late 1991. Several MiG-21s were lost in accidents involving both Soviet and Congolese personnel, and funding limitations resulted in all but two of the remaining dozen or so being

withdrawn from service. They have been in hangar storage at Brazzaville/Maya Maya and Pointe-Noire/Agostino Neto airfields, where four original Mi-8s have also been retired. They were supplemented, however, by six Mi-8/17s acquired from Ukraine in mid-1997, shortly before take-over of the Congo government by Cobra rebels. Airworthiness of the remaining older equipment, including the C-47s and Noratlas, is doubtful.

CONGO (ZAÏRE)

Capital: Kinshasa
Population: 35 million
Land area: 2.345 million km² (905,000 sq miles) **Major cities:** Kananga, Kikwit

(former) Force Aérienne Zaïroise (Zaïrean Air Force)

Originally known as the Belgian Congo until its 1960 independence, the territory of the Congo Basin in central Africa became Zaïre in 1971, until the overthrow of President Mobutu's long-standing administration in 1997. It was then renamed the Republic of Congo, which may result in some confusion with its much smaller northern neighbour, the People's Republic of Congo.

Prior to the recent political events, the FAZ was organised around the two main bases. Kinshasa was home to the training and transport elements while Kamina housed the tactical aircraft. At the time of writing the status of the former FAZ remained unknown. November 1998 deliveries to Zimbabwe of ex-Russian helicopter gunships, fighters and spotter aircraft are believed to be destined for Congo.

(former) Force Aérienne Zaïroise

1 Groupement Aérien	Kinshasa
12e Escadre de Liaison	
122e Escadrille	Aérospatiale SA 330C, Eurocopter AS 332L, Aérospatiale SA 316B, Cessna 310R, Mitsubishi MU-2JE
13e Escadre d'Entrainement	
131ere Escadrille d'Ecolage Elementaire	Reims-Cessna FRA 150M, SIAI-Marchetti SF.260MC
132e Escadrille d'Ecolage Avancé	Aermacchi MB.326GB
19e Escadre d'Appui Logistique	
191ere Escadrille	Douglas C-47, Lockheed C-130H

2 Groupement Aérien	Kamina
21er Escadre de Chasse et d'Assaut	
211ere Escadrille	Dassault Mirage 5M/DM
212e Escadrille	Aermacchi MB.326GB/K
22e Escadre de Transport Tactique	
221ere Escadrille	de Havilland Canada DHC-5D, Reims-Cessna FTB.337FG

Type	Role	Delivered/In Service
Aermacchi MB.326GB	intermediate trainer	19/12
Aermacchi MB.326K	ground attack	6/3
Aérospatiale SA 316 Alouette III	utility helicopter	8/5
Aérospatiale SA 330C Puma	transport helicopter	12/9
Cessna 310R	communications	15/9
Dassault Mirage 5M	air defence/GA	8/5*
Dassault Mirage 5DM	combat trainer	3/1
de Havilland Canada DHC-5D	tactical transport	3/2
Douglas C-47 Dakota	transport	13/8
Eurocopter AS 332L Super Puma	transport helicopter	1/1
Lockheed C-130H Hercules	transport	7/3
Mitsubishi MU-2JE	communications	2/2
Reims-Cessna FTB.337FG	comm/CAS	18/18
Reims-Cessna FRA 150M	primary trainer	16/10
SIAI-Marchetti SF.260MC	basic trainer	12/4

NB: If five Mirage 5s are currently in service, 10 must have been delivered; only eight delivery serials are currently recorded, with five w/o

Below: This AS 332L Super Puma (c/n 2108) was one of the last deliveries before the change of administration.

A pair of Buffalos remains airworthy from the three acquired. The third aircraft is wfu at Kinshasa.

COTE D'IVOIRE

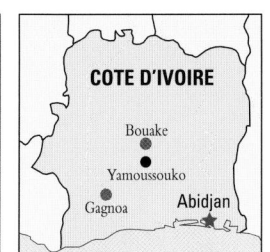

Capital: Yamoussouko   **Population:** 12 million
Land area: 322000 km² (124,000 sq miles)
Major cities: Abidjan, Bouake, Gagnoa

Force Aérienne de la Côte d'Ivoire (Ivory Coast Air Force)

Now officially using only the French version of its name, Ivory Coast achieved independence from France in 1960, but maintains strong links with its former colonial masters through a 1961 bilateral defence agreement. It also has regional agreements with Niger and Benin. French equipment, unit organisation, training and operational techniques have predominated since the formation of the original Air Transport and Liaison Group in 1961, with three Douglas C-47s, followed in 1962-65 by seven MH.1521 Broussard STOL utility aircraft.

With additional equipment, including three Reims-Cessna 337E light twins as Broussard replacements, three Reims-Cessna 150 two-seat lightplanes, and Aérospatiale Alouette II and III, Puma and SA 365 helicopters, the ATLG was organised within Training and Presidential Flights. To them were added a Fighter Flight (Escadrille de Chasse) with deliveries of six Dassault/Dornier Alpha Jet CI light attack/advanced trainers to Bouake from October 1980. Planned orders for six more were cancelled, but an attrition replacement was received in 1983. Only two Alpha Jets are currently thought to be airworthy.

Apart from the Beech F33s and Cessna 421 in a Training and Liaison Flight, current remaining aircraft are operated by the Escadrille Presidentielle.

Force Aérienne de la Côte d'Ivoire

Type	Role	Delivered/In Service
Aérospatiale SA 330H Puma	transport helicopter	3/1
Beech F33C	basic training/liaison	6/4
Cessna 421 Golden Eagle	communications	1/1
Dassault-Dornier Alpha Jet C	advanced trainer	7/5
Eurocopter AS 365C Dauphin 2	transport helicopter	4/2
Fokker 100	VIP transport	1/1
Gulfstream GIII	VIP transport	1/1
Gulfstream GIV	VIP transport	1/1

Below left: Close military ties are maintained with France, as symbolised by this flight of two French EC 13 Mirage F1CTs and Côte d'Ivoire Escadrille de Chasse Alpha Jet.

Below: The Escadrille Presidentielle fleet includes this Gulfstream GIII based at Abidjan airport.

133

DJIBOUTI

Three PZL (Mil) Mi-2s were acquired for the Force Aérienne Djiboutienne, but one was lost on 18 March 1995.

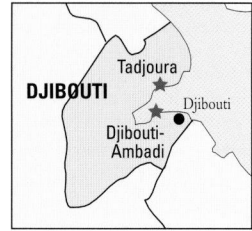

Tadjoura

DJIBOUTI

Djibouti

Djibouti-Ambadi

Capital: Djibouti **Population:** 400,000
Land area: 23000 km² (8,800 sq miles)

Force Aérienne Djiboutienne (Djibouti Air Force)

The former French territory of Afars et Issas gained independence in 1977, but maintains close links with the previous administration, including retention of French forces detachments to undertake air defence, surveillance and logistic support to counter political instability in neighbouring Somalia. Djibouti is one of seven Francophone African nations with which

France has formal defence pacts, as well as military assistance agreements with another 15. Because of major defence economies, however, these are being scaled-down, to limit French military intervention in Africa, except in extreme emergencies.

Hitherto, Djibouti has housed the largest French military contingent in Africa, comprising nearly 4,000 personnel, and permanent deployment of an AA Mirage F1C squadron (EC 4/33 'Vexin' from Reims-Champagne), plus supporting transport and helicopter elements. Reductions of some 40 per cent are now planned for French forces in Africa by 2000, although their current bases in Chad, Gabon, Ivory Coast and Senegal, as well as Djibouti, will be retained for use by military detachments from France for training, exercises and operations when required. Meanwhile, Djibouti's own military aviation element is confined to a handful of communications and liaison aircraft, including a civil-registered Dassault Falcon 50 for government transport.

Force Aérienne Djiboutienne

Type	Role	Delivered/In Service
Aérospatiale SE 313 Alouette II	utility helicopter	1/1
Antonov An-28 'Cash'	light transport	1/1
Bell 206B JetRanger	utility helicopter	1/1
Cessna U206G Stationair	utility	1/1
Cessna 402C	communications	1/1
Eurocopter AS 355F Ecureuil 2	utility helicopter	2/2
PZL-Mil Mi-2 'Hoplite'	transport helicopter	3/2
Mil Mi-8/17 'Hip'	transport helicopter	5/5

An Antonov An-28 provides light transport for the air force. Like all Djibouti military aircraft, it displays a two-letter code derived from its serial (J2-MAT).

EGYPT

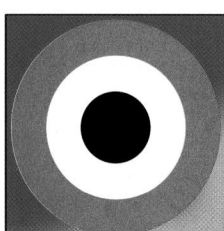

Capital: Cairo
Population: 57 million
Land area: 1 million km² (386,000 sq miles)
Major cities: Alexandria, Giza, Port Said, Suez

Al Quwwat Al Jawwiya Il Misriya (Arab Republic of Egypt Air Force)

From its earlier almost sole dependence on Soviet arms supplies up to the mid-1970s, Egypt has diversified its weapons procurement sources in recent years, while trying to achieve increasing arms self-sufficiency from building up the former multi-nation Arab Organisation for Industrialisation on a national basis. Promised funding from the Gulf states to expand production and buy equipment from the nine AOI factories ended with the 1979 peace treaty signed by Egypt with Israel. However, Egyptian adherence to the US-brokered Camp David peace agreement has ensured long-term continuation of $1.3 billion annually in military aid from Washington, and the provision of advanced US weapons and equipment.

Although still maintaining its basic Soviet organisation and unit structure, exemplified by its discrete Air Defence Command with responsibility for ground-based SAMs, anti-aircraft artillery (AAA), radars, and communications and control systems, the EAF has

EGYPT

El Mansurah
Tanta
Alexandria
Mersa Matruh
Qahira
Port Said
Giza Cairo Abu Sueir
Almaza Suez
Fayid
Beni Suef
El Minya
Hurghada
El Khârga
Aswan

founded its current operational strength on the procurement since 1982 of 196 GD/Lockheed Martin F-16s through five consecutive orders in Egypt's Peace Vector programmes.

The most recent, Peace Vector V, placed on 2 April 1996, was for 21 more GE F110-100B-engined F-16C-40 night precision-strike versions using LM LANTIRN targeting and navigation pods. The aircraft cost $670 million and delivery is scheduled for 1999-2000. The F-16s were intended to replace about 110 MiG-21s, which operate alongside about 50 similar Chinese-supplied Chengdu F-7s, but plans were announced in 1996 for Hindustan Aeronautics to upgrade them and undertake their spares support. Five Grumman E-2C AEW aircraft were delivered in 1990, plus a later attrition replacement, for integration with the EAF fighter force, which also expressed a requirement in 1996 for ex-USAF Boeing KC-135 air-refuelling tankers. Much of the older EAF equipment, notably the Soviet-supplied Tupolev Tu-16 and Ilyushin Il-28 bombers, has been withdrawn from

Above left: No. 35 Sqn continues to operate the Sukhoi Su-20 'Fitter-C' as part of a close air support brigade based at Khatamia. These are periodically returned to eastern Europe for servicing.

Left: The jump from the MiG-21 to the F-4E Phantom proved a difficult task for the Egyptian air force, the Phantom suffering from chronic unserviceability. At one point the fleet was to be sold to Turkey. Today the aircraft are operated by the 222nd Tactical Fighter Brigade at Qahira-West.

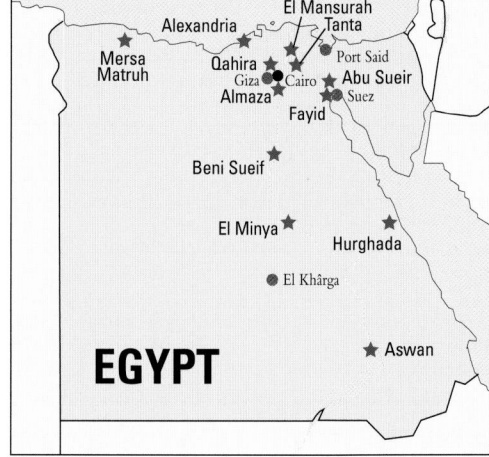

A pair of VIP configured VC-130H Hercules is operated by No. 16 Sqn from Almaza-Qahira-East, resplendent in national colours.

service, and the Chinese-supplied MiG-19-derived Chengdu F-6s are being replaced by F-16s and other new types. Upgrades are also planned for the EAF's Mirage 2000s and MDC Phantoms.

Following the 1990 transfer of about 10 Aero L-39ZO armed trainers from Libya to the EAF, since believed to be withdrawn from use, Egypt ordered 48 upgraded L-59E versions with Progress/Lotarev DV-2 turbofans costing $204 million, delivered in 1993-94. Air Defence Command modernisation has included a $50 million contract with Raytheon in mid-1993 for Egyptian HAWK SAM mobility enhancement kits and upgrades, plus another for $206 million in early 1997 to upgrade eight HAWK fire units to Phase III configuration. Evaluations were being made in 1996-97 of the Raytheon Patriot PAC-2/3 and Russian Antey S-300PMU-1 anti-ballistic-missile systems to replace ADC's ageing SA-2, SA-3, SA-6 and other Soviet-supplied surface-to-air missiles.

Other recent large-scale EAF procurement from the US has included a late 1994 $318 million FMS contract for a further 12 Apache attack helicopters, with a letter of intent for 12 more, as a follow-on to earlier deliveries of 24 AH-64As. Deliveries of the most recent batch started in late 1996, to the EAF's main Apache base at Abu Hammad; items also included more Hellfire ATMs, as well as Target Acquisition and Designation Sights, with Pilot's Night-Vision Sensors (TADS/PNVS), plus the Integrated Helmet and Display Sight System (IHADSS).

All helicopters for Egyptian army support and transport are operated by the EAF, in 15 squadrons. Prior to the Apache deliveries, the EAF's attack helicopter force comprised four squadrons in two brigades with 72 Westland/Aérospatiale SA 342L Gazelles, of which 30 were locally assembled by the AOI's joint-venture Arab-British Helicopter Company (ABHCO). About half the Gazelles were armed with HOT ATMs and the remainder with a 20-mm GiAT cannon for a variety of attack, scout and observation roles.

Three helicopter brigades for army tactical transport were mainly equipped with some 27 Mil Mi-8s in three squadrons from 80 originally delivered, as well as 22 Westland Sea King-derived Commando Mk 1/2 assault helicopters. Six more Commando 2s followed for VIP and ECM use. Fifteen Meridionali-built Boeing CH-47Cs equipping an EAF heavy-lift squadron are now being reinforced by four CH-47Ds ordered in 1997 through a $149 million FMS contract. Plans were also announced in late 1997 to restore and upgrade many of the EAF Commandos and Sea Kings, grounded following Egypt's early 1980s unilateral withdrawal from joint-venture Westland/ABHCO Lynx large-scale production plans, and recent receipt of $320 million compensation by the UK company.

Civil-registered transports operated under EAF auspices for VIP and government flights include an Airbus A340-212, Boeing 707-366C, Dassault Falcon 20E5 and two 20F-5s, two Gulfstream GIIIs and a GIV, plus two GIV-SPs on order.

Backbone of the modern Egyptian air force are the Lockheed Martin(GD) F-16s delivered in successive Peace Vector programmes. These are a TAI-built F-16C Block 40R (foreground) and 40Q.

al-Quwwat al-Jawwiya Il-Misriya

Air Defence Command

102 Fighter Brigade
41 Squadron	Shengyang F-7	Fayid
42 Squadron	Shengyang F-7	Fayid

104 Fighter Brigade
45 Squadron	MiG-21MF/U	El Mansurah
49 Squadron	MiG-21MF/U	El Mansurah

??? FGA Brigade
63 Squadron	MiG-21MF	Mersa Matruh
65 Squadron	MiG-21MF	Mersa Matruh
75 Squadron	Mirage 5SDE	Mersa Matruh

Tactical Fighter Command

??? AEW Brigade
87 Squadron	E-2C	Qahira-West

??? ECM Brigade
?? Squadron	Il-28	Kom Awshim
?? Squadron	Beech 1900 C-1	Almaza-Qahira-East

211 FGA Brigade
20 Squadron	Shengyang F-6/FT-6	Genaclis
21 Squadron	Shengyang F-6/FT-6	Genaclis

222 Tactical Fighter Brigade
76 Squadron	F-4E	Qahira-West
78 Squadron	F-4E	Qahira-West

232 Tactical Fighter Brigade
72 Squadron	F-16A/B Block 15	Inchas
74 Squadron	F-16A/B Block 15	Inchas

236 FGA Brigade
71 Squadron	Mirage 5SDE/SDR/SDD	Tanta
73 Squadron	Mirage 5SDE/SDD	Tanta

242 Tactical Fighter Brigade
68 Squadron	F-16C/D Block 32	Beni Sueif
70 Squadron	F-16C/D Block 32	Beni Sueif

252 Tactical Fighter Brigade
22 Squadron	Shengyang F-7/FT-7	Bir Ket
26 Squadron	Shengyang F-7/FT-7	Bir Ket
82 Squadron	Mirage 2000	Bir Ket

262 Tactical Fighter Brigade
60 Squadron	F-16C/D Block 40	Abu Sueir
64 Squadron	F-16C/D Block 40	Abu Sueir

272 Tactical Fighter Brigade
75 Squadron	F-16C/D Block 40	Genaclis
77 Squadron	F-16C/D Block 40	Genaclis

??? Reconn. Brigade
64 Squadron	MiG-21R/U	Genaclis
69 Squadron	Mirage 5SDR/SDE/SDD	Genaclis

??? CAS Brigade
57 Squadron	Alpha Jet MS.2	Almaza
58 Squadron	Alpha Jet MS.2	Almaza

??? CAS Brigade
?? Squadron	Aero L-29	Khatamia
35 Squadron	Sukhoi Su-20C	Khatamia
56 Squadron	Aero L-29	Khatamia

Helicopters

??? Helicopter Brigade
15 Squadron	SA 342K/L Gazelle	
?? Squadron	SA 342K/L Gazelle	

??? Helicopter Brigade
?? Squadron	SA 342K/L	
?? Squadron	SA 342K/L	

??? Assault Brigade
?? Squadron	Commando Mk 2	
?? Squadron	Commando Mk 1/2E, AS-61, Mi-4	

??? Attack Helicopter Brigade
?? Squadron	AH-64A	Abu Hammad
?? Squadron	AH-64A	Abu Hammad

Transports

??? Transport Brigade
4 Squadron	C-130H	Almaza-Qahira East
16 Squadron	VC/C-130H/H-30, EC-130H	Almaza-Qahira East
?? Squadron	Boeing-707/737, Falcon 20	Almaza-Qahira East
	Gulfstream III/IV, Commando, AS-61, S-70A, King Air	
?? Squadron	SA 342 VIP Gazelle	Almaza-Qahira East

??? Transport Brigade
2 Squadron	DHC-5D	Almaza-Qahira East
?? Squadron	Antonov An-2	Almaza-Qahira East
?? Squadron	Zlin Z-526	Almaza-Qahira East
?? Squadron	PZL-104	Almaza-Qahira East

??? Helicopter Brigade
7 Squadron	Mi-6	Kom Awshim
?? Squadron	CH-47C	Kom Awshim

??? Helicopter Brigade
?? Squadron	Mi-8	Hurghada
?? Squadron	Mi-8	Hurghada

??? Helicopter Brigade
?? Squadron	Mi-8	
?? Squadron	Mi-8	

Air Force Academy

??? Primary Training Brigade
?? Squadron	Gomhouria	Bilbeis
?? Squadron	Gomhouria	Bilbeis

??? Basic Training Brigade
?? Squadron	EMB-312	Bilbeis
?? Squadron	EMB-312	Bilbeis

??? Weapon Training Brigade
?? Squadron	Alpha Jet MS2/MS1	El Minya
?? Squadron	MiG-15UTI	El Minya

??? Flying Training Brigade
?? Squadron	Aero L-29	El Minya
88 Squadron	Aero L-39ZO	El Minya

??? Helicopter Training Brigade
?? Squadron	UH-12E	El Minya
?? Squadron	SA-342L/L	El Minya

Air Navigation School
	DHC-5	Bilbeis

??? OCU
?? Squadron	Shengyang F-6/FT-6	Bilbeis

Flying Training Air Squadron
?? Squadron	Gomhouria	Khatamia

The Polish-built PZL-104 Wilga is used as a general 'hack' in Egyptian service. The exact number in service is not known, but is thought to be about 10.

The Egyptian Navy received 10 SH-2G(E)s which had been upgraded before delivery from SH-2F airframes which had been held in storage at Davis-Monthan AFB, Arizona .

Above: Helwan built 20 Fournier RF-5B Sperbers, some of which assumed military identities and are used for air experience flights.

Above right: Egyptian L-29 Delfins continue to play a role in the training syllabus, based at El Minya. Periodically, the aircraft are sent to the Aero factory for overhaul.

Egyptian Navy Aviation

Having originated with Soviet-supplied Mil Mi-4s in the late 1960s, Egypt's naval helicopter operations underwent a moderate upgrade in the mid-1970s

Republic of Egypt Navy

??? ASW Brigade

?? Squadron	Sea King Mk 47	Borg El-Arab
?? Squadron	SH-2G(E) Seasprite	Borg El-Arab
?? Squadron	SA 342L Gazelle	Borg El-Arab

from receipts of a dozen Westland/ABHCO SA 342Ls armed with Aérospatiale AS12 anti-ship missiles, and five Westland Sea King Mk 47 shore-based ASW helicopters, operating from Alexandria's Borg El-Arab airfield. In 1997, Egyptian Naval Aviation became the first export customer with initial deliveries of the 10 upgraded Kaman SH-2G(E) Super Seasprite ASW ordered from a \$150 million FMS contract. Fitted with new avionics and ASW equipment, including AlliedSignal AQS-18A dipping sonar, the SH-2G(E)s will operate from at least four leased ex-USN frigates, as well as their onshore base. The first arrived at Borg El-Arab on 7 June 1998.

EQUATORIAL GUINEA

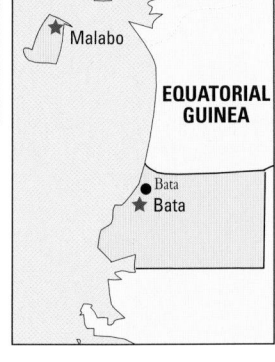

Capital: Malabo
Population: 350,000
Land area: 28000 km²
(11,000 sq miles)

Ala Aérea de Guardia Nacional

Ala Aérea de Guardia Nacional

Antonov An-32 'Cline'	tactical transport	1
Yakovlev Yak-40 'Codling'	transport	1

Located between Cameroon and Gabon, Equatorial Guinea was a Spanish colony until achieving autonomy in 1963, and became an independent republic in 1968. The para-military National Guard was originally formed with Chinese, Cuban and Soviet advisers, although they were later replaced by Moroccan and Spanish military and police personnel. Its only aircraft to date have comprised single Antonov An-32 and Yakovlev Yak-40 transports, operated jointly with the LEASA national airline (Lineas Ecuatoguineanas de Aviación SA) in civil markings.

ERITREA

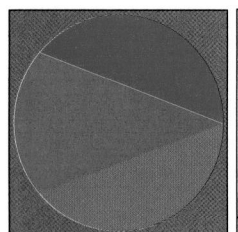

Capital: Asmera
Population: 2.9 million
Land area: 118000 km²
(45,400 sq miles)
Major cities: Mits'iwa

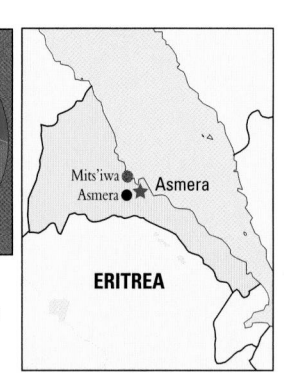

Until the arrival of MiG-29s, the MB.339FD represented the only fixed-wing attack capability of the Eritrean air force. Tensions with Ethiopia have been heightened by both countries acquiring advanced fighters in the form of MiG-29 (Eritrea) and Su-27 (Ethiopia).

Eritrean Air Force

Some military aircraft, including at least six MiG-21s and two Mil Mi-8s, were inherited from Ethiopia when Eritrea gained its independence in May 1993. The fighters were placed in storage and new equipment was ordered for the emergent Eritrean air force, with Italian assistance. The first of eight L-90 light turboprop trainers was delivered from Valmet in Finland in 1994, prior to the take-over of this programme by Aermacchi. Four HAMC Y-12 twin-turboprop light utility transports were also delivered from China by mid-1994, and a \$50 million order placed with Aermacchi for six MB.339FD lead-in fighter trainers, with options on four more. The

MB.339 order included a 12-month training programme in Italy and Eritrea, with deliveries starting in mid-1997. An IAI Astra twin-turbofan light transport was acquired from Israel in mid-1993 for government transport, and other procurement since 1996 has included a few Mil Mi-8 transport helicopters and four Mi-24 attack helicopters from Russian sources.

Border clashes with Ethiopia escalated into an air war on 5 June 1998, with Eritrea losing an MB 339FD the next day. These clashes promoted the purchase of MiG-29s from Moldova in late 1998. Up to 10 examples of the 'Fulcrum' have been acquired, but their exact status remains uncertain.

Eritrean Air Force

Type	Role	Delivered/In Service
Aermacchi MB.339FD	advanced trainer/GA	6/5
Harbin HAMC Y-12 II	light transport	4/4
IAI 1125 Astra SP	light transport	1/1
MiG-29 'Fulcrum'	air defence/GA	10?/?
Mil Mi-8 'Hip'	transport helicopter	5/3?
Mil Mi-24 'Hind'	attack helicopter	4/4
Valmet-Aermacchi L-90TP Redigo	basic trainer	8/8

ETHIOPIA

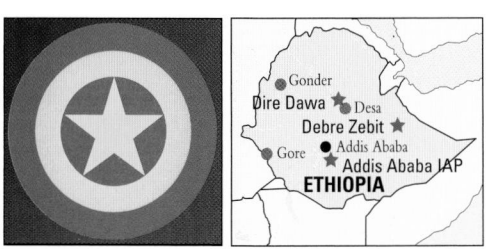

The arrival of four ex-*USAF C-130B Hercules* may mean the end for the Ethiopian air force Antonov An-12s. Many of the fleet have been lost in service.

Capital: Addis Ababa **Population:** 47.9 million
Land area: 905450 km² (349,490 sq miles)
Major cities: Desa, Gonder, Gore

Ye Ityopya Ayer Hayl (Ethiopian Air Force)

Previously equipped over many years from Swedish, British, US and Soviet sources, the Ethiopian armed forces were decimated in the civil war which ended with democratisation of the former Mengistu regime in 1991, and subsequent secession of Eritrea. Between 1987-91, Ethiopia imported nearly $3 billion worth of Soviet arms, but by 1994 annual defence spending had decreased from 13.5 per cent of GNP to only 3.4 per cent. The EthAF was then left with 55 mainly Soviet-supplied combat aircraft and armed helicopters, most no longer airworthy through spares and technical support shortages.

US aid with rebuilding Ethiopia's armed forces, promised in 1996, included personnel training and

FMS deliveries from early 1998 of four surplus USAF Lockheed C-130B transports, for which Lockheed Martin received an $11 million four-year logistic support contract. In late 1997, Israel Aircraft Industries, Ukraine, Elbit and VPK MiG MAPO with Rosvoorouzhenie were competing to upgrade the EthAF's remaining MiG-21s. Four Kazan-built Mil

Mi-8Ts were also bought from long-term storage in Hungary, in late 1997.

In the border clashes with Eritrea during June 1998, Ethiopia lost a pair of MiG-23s and a MiG-21, all to ground fire. In response to increased tensions and Eritrea's acquisition of MiG-29s, Ethiopia purchased between four and 10 Sukhoi Su-27s from the Russian air force. The aircraft were delivered from Krasnodar in 1998. As part of the same deal, Ethiopia is reported to be getting used Mil Mi-8 'Hips' and Mi-24 'Hinds'.

Ye Ityopya Ayer Hayl

Type	Role	Delivered/ In Service
Aero L-39C Albatros	advanced trainer	20/7
Aérospatiale SA 330H Puma	transport helicopter	?/1
Antonov An-12BP 'Cub'	transport	13/5
Antonov An-26 'Curl'	tactical transport	1/1
Antonov An-32 'Cline'	tactical transport	1/1
Lockheed C-130B Hercules	transport	4/4
MiG-21MF 'Fishbed-J'	air defence/GA	?/14
MiG-21US 'Mongol-B'	combat trainer	?/4
MiG-23BN 'Flogger-F'	close air support	?/8
MiG-23UB 'Flogger-C'	combat trainer	?/2
Mil Mi-8T/17 'Hip'	transport helicopter	?/14
Mil Mi-14PL 'Haze'	SAR helicopter	2/2
Mil Mi-24 'Hind'	attack helicopter	/15
SIAI-Marchetti SF.260TP	basic trainer	21/8
SIAI-Marchetti S.208M	communications	?/1
Sukhoi Su-27 'Flanker'	air defence	?/?
Yakovlev Yak-40 'Codling'	transport	1/1

Ethiopian Army Aviation

Bell UH-1H Iroquois	utility helicopter	16/4
Cessna 401	communications	1/1
de Havilland Canada DHC-6-300	light transport	3/2

GABON

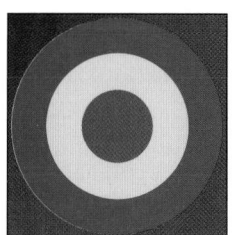

Capital: Libreville **Population:** 1.2 million
Land area: 267665 km² (103,320 sq miles)
Major city: Mayumba

Forces Aériennes Gabonaises (Gabonese Air Forces)

Formed with French assistance and equipment, the three-element FAG continues to rely on military, economic and training assistance from France, which detaches small army and air force units to Gabon. FAG aircraft are mainly deployed at Base Aérienne 01 at Libreville and BA 02 Franceville, the latter housing the Mirages and Fouga Magister of the sole combat unit, Escadron de Chasse 1-02 'Leyou'. Apart from the Presidential Guard and Army Aviation, the remaining FAG assets operate from Libreville, in Light and Heavy Transport Squadrons (the latter with two Hercules, another pair having been sold or scrapped), with detachments at Chibanga.

BA 01 also accommodates a civil-registered Dassault Falcon 900EX (named *Masuka III*), Douglas

Forces Aériennes Gabonaises

Type	Role	Delivered/ In Service
Aérospatiale SA 316B Alouette III	utility helicopter	8/4?
Airtech CN.235M-100	tactical transport	1/1
Dassault Mirage 5G2	air defence/GA	5/3
Dassault Mirage 5DG	combat trainer	4/3
Fouga CM.170R Magister	continuation trainer	1/1
EMBRAER EMB-110P1K Bandeirante	transport	2/1
EMBRAER EMB-111A Bandeirante	maritime patrol	1/1
Gulfstream Aerospace GIII	VIP transport	1/1
Lockheed C-130H/L-100-30 Hercules	transport	4/2

Garde Presidentielle

ATR 42F	transport	1/1
Beech T-34C-1 Turbo-Mentor	basic trainer	4/3
EMBRAER EMB-110P Bandeirante	transport	1/1
Eurocopter AS 332L Super Puma	transport helo	1/1
Fouga CM.170R Magister	advanced trainer/GA	5/4

Aviation Légère des Armées

Aérospatiale SA 330C Puma	transport helo	6/4
Aérospatiale SA 341L1 Gazelle	attack helicopter	5/5
Bell 212	utility helicopter	1/1
Bell 412	utility helicopter	1/1

DC-8-73F and Gulfstream GIII for government use in the French-pattern Groupement de Liaisons Aériennes Ministerielles (GLAM), under FAG auspices. Two Aérospatiale AS 350Bs are operated by the Gendarmerie.

Above: Two *EMB-110P1K Bandeirantes* were delivered to the Gabonese air force in 1980, alongside a single *EMB-111A*.

Gulfstream III TR-KHC was acquired to replace a *GII* that was written off in February 1980. It operates with the GLAM unit from Libreville.

GAMBIA

Since becoming an independent member of the British Commonwealth in 1963, the small enclave

of Gambia within Senegal, with which it was affiliated as Senegambia for some years from 1982, has had no significant defence or security forces, apart from local police, although there is a Ministry of Defence in the capital, Banjul. In August 1996 two surplus Bell AH-1F HueyCobra attack helicopters

were seized by US Customs as they were being loaded into a Lockheed TriStar freighter at Miami Airport, en route for Gambia. They lacked State Department export clearances, although there was no indication as to whether Gambia was their eventual destination.

GHANA

Capital: Accra
Population: 15 million
Land area: 238000 km²
(92,000 sq miles)
Major cities: Kumasi,
Sekondi-Takoradi

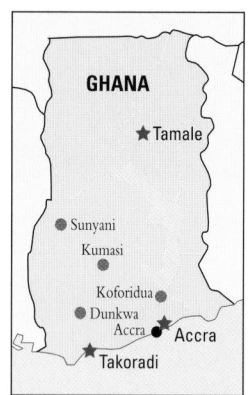

Ghana received the MB.339A in 1988, but has been an operator of Aermacchi's MB.326 since 1978.

Ghana Air Force

Type	Role	Delivered/In Service
Aero L-29 Delfin	intermediate trainer	12/8
Aérospatiale SA 316B Alouette III	utility helicopter	6/4
Aermacchi MB.326F	light GA	9/5*
Aermacchi MB.326KG	light GA	6/4
Aermacchi MB.339A	light GA	4/2
Agusta A 109A	utility helicopter	2/2
Agusta-Bell 412	transport helo	1/1
Bell 212	transport helo	2/2
Fokker F27-400M Troopship	transport	3/3
Fokker F27-600M Friendship	transport	2/2
Fokker F28 Fellowship	transport	1/1
Pilatus-BN BN-2T Turbine Defender	light transport	4/4
Shorts SC-7 Skyvan 3M	transport/patrol	6/4

*in storage

Ghana Air Force

Since its 1959 formation, Ghana's air force has been equipped with a variety of Indian, British, Canadian, Czech, Dutch, French, Italian and US aircraft. In recent years, funding shortages have severely limited procurement, additionally reflected in poor maintenance and serviceability standards. Recent GhAF acquisitions have therefore comprised only two former Ugandan air force Agusta A 109A helicopters,

refurbished in Italy for late 1996 delivery, plus an Agusta-Bell 412, and a couple of ex-South African air force Alouette IIIs.

In 1989, a dozen Aero L-29 Delfins were transferred from the Nigerian air force to the GhAF's Flying School, supplementing earlier deliveries of Aermacchi MB.326s and MB.339s, which now equip a light ground-attack squadron. Four of six grounded Shorts Skyvans were completely overhauled and refurbished by Airwork in the UK in 1990-91, with new avionics and Bendix RDS 81 radar, to extend their transport role in No. 1 Squadron to coastal patrol and maritime surveillance. Five Fokker F27s and all but two of the GhAF's helicopters are operated by No. 2 Squadron, while the Turbine Islanders equip No. 3 Squadron. A Fokker F28 and a Bell 212 serve the Presidential Flight.

Ghana uses five Fokker F27s of two versions. G525 (coded 'F') is a Mk 400M Troopship delivered in 1975 and operated by No. 2 Sqn.

GUINEA-BISSAU

Capital: Bissau   **Population:** 970,000
Land area: 36125 km² (13,945 sq miles)

This Mi-8T 'Hip-C' is one of a pair of these rugged helicopters that has been used by the Força Aérea da Guine-Bissau.

Força Aérea da Guine-Bissau

Type	Role	Delivered/In Service
Aérospatiale SA 313B Alouette II	utility helicopter	1/1
Aérospatiale SA 316B Alouette III	utility helicopter	3/2
Antonov An-24RV 'Coke'	transport	1/1
Dornier Do 27A	communications	2/2
MiG-21MF 'Fishbed-J'	air defence/GA	6/4
MiG-21UM 'Mongol-B'	combat trainer	1/1
Mil Mi-8 'Hip'	transport helo	2/1
Yakovlev Yak-40 'Codling'	transport	1/1

Força Aérea da Guine-Bissau

Having started operations in 1974 after independence from Portugal with former FAP Douglas C-47s,

North American T-6s, two Dornier Do 27s, and one or two Alouette IIIs, the FAGB later received its first combat aircraft from limited Soviet aid. This included five MiG-17s and two MiG-15UTI trainers to equip a single fighter unit at Bissalanca, plus a single Mil Mi-8 transport helicopter, which subsequently crashed. French aid in 1978 included a civil-registered

Reims-Cessna FTB.337 for coastal patrol and a surplus Alouette II helicopter, but a Dassault Falcon 20F presented by Angola for use by Guinea-Bissau's Marxist government was soon sold to the US.

In the late 1980s, the FAGB's MiG-17s were replaced by a similar number of MiG-21MFs and -21UMs, plus an Antonov An-24, a Yakovlev Yak-40, and another Mi-8. The current state of airworthiness of all Guinea-Bissau's Soviet-supplied aircraft is doubtful, however, because of spares shortages and technical support problems.

GUINEA REPUBLIC

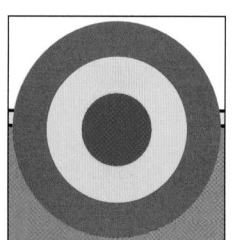

Capital: Conakry
Population: 5.8 million
Land area: 245855 km²
(94,900 sq miles)
Major cities: Kankan,
Labe, Nzérékoré

Force Aérienne de Guinée

As a former French colony, the West African Republic of Guinea formed a small air arm with Soviet assistance following its 1958 independence. Soviet supplies included at least 10 MiG-17F fighters and one or two MiG-15UTI trainers, plus later Antonov An-2 (two), An-14 (four), and An-12

(two), Ilyushin Il-14 (four) and Il-18V (two) transports, and Mil Mi-4 helicopters (four). Among other Eastern Bloc deliveries were three Aero L-29 Delfin jet trainers and six piston-engined Yakovlev Yak-11s, and Romania contributed a licence-built Aérospatiale/ICA-Brasov IAR-316 Alouette III and two IAR-330L Puma transport helicopters.

Additional Soviet aid then requested for Guinea's 800-man air force, in return for access to air base facilities at Conakry for SovAF maritime reconnaissance, resulted in deliveries of eight MiG-21PFMs and a MiG-21U in 1986, to replace the remaining MiG-17s. Occasional military tasks are also undertaken by the transport fleet of Air Guinée, which has included a single An-12, two An-24s, an Il-18V, a Yak-40, a DHC-7 and the government's civil-registered Gulfstream GII. Most of the Soviet-supplied equipment is in a poor state of serviceability.

Force Aérienne de Guinée

Type	Role	Delivered/In Service
Aérospatiale SA 342L Gazelle	utility helicopter	1/1
Antonov An-24RV 'Coke'	transport	1/1
ICA-Brasov IAR-316 Alouette III	utility helicopter	1/1
ICA-Brasov IAR-330L Puma	transport helo	2/1
Eurocopter AS 350B Ecureuil	comm helicopter	1/1
MiG-15UTI 'Midget'	combat trainer	2/1
MiG-21PFM 'Fishbed-F'	air defence	8/?
MiG-21U 'Mongol-A'	combat trainer	1/1

KENYA

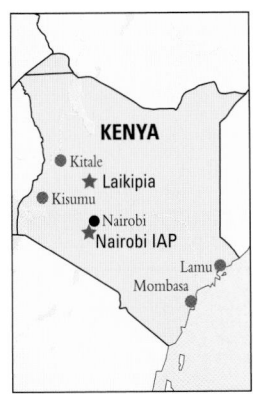

Capital: Nairobi
Population:
24 million
Land area:
583000 km²
(225,000 sq miles)
Major cities: Mombasa

Kenya Air Force

Kenya gained its independence from the United Kingdom in 1963, and sought and received British aid to set up its military. Kenyan Air Force personnel were trained in the UK, its first aircraft being six ex-RAF Chipmunks. Beavers and Caribous were added during the late 1960s, while the first combat aircraft, six Strikemaster Mk 87s were ordered in late 1969. They served until four survivors were sold to Botswana between 1993 and 1994. They were augmented by four Hunter FGA.Mk 80s and two T.Mk 81s, all refurbished ex-RAF aircraft. Five survivors left for the Air force of Zimbabwe in 1980. Soviet support for Somalia and Uganda lead to the purchase of 10 F-5Es Tiger II and a pair of F-5F combat trainers from 1978. Another pair of F-5Fs arrived as attrition replacements in 1982. Today the Tiger II represents the sharp end of the Kenyan Air Force although only six F-5Es and two F-5Fs remain, used for both interception and ground attack.

The development of the Kenyan Air Force was dealt a severe blow on 1 August 1982, when a failed coup attempt was led by junior officers of the air arm.

Six Do 28D-2 Skyservants (redesignated as 128-2s after delivery) were received between July 1977 and January 1978, bolstered by another pair in mid-1980.

The air force was disbanded on 22 August, but limited operations were permitted under the administrative control of the army as 'The '82 Air Force'. It was not until August 1994 the air force was allowed to revert to its old title.

Current transports include DHC-5D Buffalos, purchased in two batches in 1977/78 and 1986/87, and three Dash-8s acquired in 1990, of which at least one is used for VIP transport, alongside a single Fokker 70ER. Lighter tasks are undertaken by seven surviving Do 28D-2 Skyservants. Most of the transport aircraft are based at Eastleigh-Moi. Helicopter transportation is provided by the Puma, with both Aérospatiale and IAR-built examples in service.

Flight training is currently initiated on the Bulldog Mk 103/127s delivered between 1972 and mid-1976. Strikemasters were used as the basic trainers until replaced by 12 Shorts-built Tucano Mk 51s delivered in 1990. Twelve Tucanos had been ordered but one was lost on 22 February 1990, before it had been delivered, and was replaced. Both the *ab initio* and basic phases of training are conducted at Eastleigh-Moi. Pilots progress to the Hawk Mk 52 for the advanced training phase. The Hawks are also used in the light strike role.

The Kenyan Army Flight at Nairobi International Airport uses a number of versions of the McDonnell Douglas (Hughes) 500 series. A pair of 500Ds was delivered in 1979, followed by 15 500MD Scout Defenders the next year. Fifteen 500MD TOW Defenders arrived in 1981, followed by two batches of 500MEs (eight in 1984/85 plus another eight in 1991). They are used for anti-armour warfare, observation and helicopter training.

Right: This MD 500MD TOW Defender is operated by the Kenyan Army Flight in the anti-tank role. Other versions are used for observation and training.

Kenya became the first export recipient of the Hawk in 1980, when it received the first of an order for 12 Mk 52s for advanced training and light strike duties. At least two have been written-off.

Kenya Air Force

Type	Role	Delivered/In Service
Aérospatiale SA 330G Puma	transport helicopter	4/3
British Aerospace Hawk Mk 52	light attack/training	12/10
DHC-5D Buffalo	transport	10/6
DHC-8-103 Dash 8	transport	3/3
Dornier Do 28D-2 Skyservant	utility	8/7
Eurocopter BO 105CBS	communications	?/1
Fokker 70ER	VIP	1/1
Hughes 500D	helicopter training	2/2
Hughes 500MD Scout Defender	observation helicopter	15/12
Hughes 500MD/TOW Defender	anti-tank helicopter	15/12
ICA-Brasov IAR-330L Puma	transport helicopter	16?/9
McDonnell Douglas MD 500ME	observation	16?/
Northrop F-5E Tiger II	air defence/attack	10/6
Northrop F-5F Tiger II	conversion training	4/2
Bulldog Series 103/127	primary training	14/11
Shorts Tucano T.Mk 51	advanced training	13/12
(One w/o before delivery)		

LESOTHO

Lesotho Defence Force Air Wing

Capital: Maseru **Population:** 1.7 million
Land area: 30345 km² (11,715 sq miles)

As the former British-administered enclave of Basutoland, bounded on all sides by South African territory, Lesotho has continued to maintain the totally independent Commonwealth status that it achieved from Britain in 1966. The small air wing – originally a 1978 offshoot of the paramilitary Police Mobile Unit – began operations with two Shorts Skyvan twin-turboprop STOL transports, a leased Cessna A152 Aerobat, two MBB BO 105 helicopters, and a Westland-built Bell 47G converted to Soloy turboshaft power. Two Mil Mi-2 twin-turbine helicopters were donated by Libya in 1983, but were retired from service by 1986.

Deliveries of one Bell and three Agusta-Bell AB 412 helicopters to the air wing of the successively renamed Lesotho Paramilitary Force and Royal Lesotho Defence Force were delayed from 1983 by South Africa, until a 1986 coup resulted in new security agreements with Pretoria signed by

the emergent military government. Exile of the former king then resulted in deletion of the LDF's 'Royal' prefix. In 1989, the Skyvans were replaced by two CASA C.212 light turboprop transports, one of which crashed almost immediately, requiring a third delivered in 1992. A fifth Bell 412 (an EP model) was delivered in May 1998, replacing one written off in January 1998.

Lesotho Defence Force

Type	Role	Delivered/In Service
Bell/Agusta-Bell 412	transport helicopter	5/3
C.212-300 Aviocar	transport	3/2
Cessna 182Q	communications	2/1
Eurocopter-MBB BO 105C/S	utility helicopter	4/2
Soloy-Westland-Bell 47G-3B-1	utility helicopter	1/1

Africa

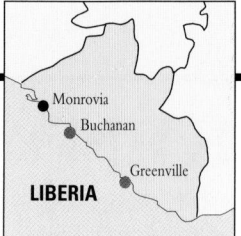

LIBERIA

Liberian Army Air Unit

Prior to its protracted civil war, which began in 1990 and lasted until the intervention of an African peacekeeping force, Liberia had close links with the US and even used the US dollar as legal currency. It founded an army Air Reconnaissance Unit following the 1970 delivery

of three MAP-funded Cessna U-17C light aircraft. A single surviving U-17C, plus a Cessna 150K, 180, 206 and a 337G light twin, transferred to civil status in 1974, were reallocated to the paramilitary Justice Air Wing in 1977.

Other ARU equipment included more Cessna lightplanes, including three 172s, a 206, 207, and two single-turboprop 208s, delivered in the 1980s, together with a Piper Aztec light twin. Major transport reinforcement resulted from 1989 deliveries of two refurbished DHC-4 Caribou and three IAI Arava STOL twins, but most ARU aircraft were destroyed on the ground or crashed during the civil

Capital: Monrovia
Population: 2.6 million
Land area: 111370 km² (42,990 sq miles)
Major cities: Buchanan, Greenville

war. One of the few survivors, then out of the country, was the civil-registered Boeing 707-351B used for government transport.

LIBYA

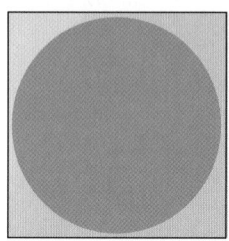

Capital: Hun
Population: 4.5 million
Land area: 1.8 million km² (680,000 sq miles)
Major cities: Al Bayda', Benghazi, Misratah, Tripoli, Tubruq

Libyan Arab Air Force (LARAF) – al-Quwwat al-Jawwiya al-Arabiya al-Libiya (QJAL)

Libya became independent in 1962, after a history of being occupied by Italian, British and French forces. Royal Libyan Air Force (RLAF) personnel were trained by the USAF at Wheelus AFB (from where they used the A-Whatiya bombing range) on T-33As which, along with C-47 Dakotas, were later transferred to the RLAF. A military coup led by Captain Khadaffi resulted in the end of ties with the British and USA, and the air force was renamed the Libyan Arab Air Force (LARAF). The USAF left Wheelus AFB and the British left their base at El Adem, near the Egyptian border. These bases were renamed as Okba ben Nafi and Gamal Abdel Nasser.

After the Yom Kippur War of 1973 the price of oil increased tremendously, and Libya, rich in oil, seized the chance to buy the best military hardware from the West. A bilateral defence agreement was signed with the Soviet Union, allowing Libyan aircrew to attend training courses in the Soviet Union. The USSR also assisted in establishing air defence missile bases and radar stations along the coastline, as well as basing a VVS contingent of Tu-22 'Blinder' bombers at Okba ben Nafi. From 1979 VVS MiG-25R 'Foxbat' reconnaissance aircraft were stationed at Okba ben Nafi to monitor the US 6th Fleet in the Mediterranean. The 12 Soviet Tu-22 'Blinders' were later transferred to the LARAF and a second squadron was formed in 1980 after additional deliveries.

In 1977 during a border dispute, Libya attacked Egyptian positions around Mersa Matruh to which Egypt retaliated by attacking Gamal Abdel Nasser and El Kufra air bases. During the 1980s the LARAF re-equipped, receiving 26 Mirage F1EDs, 60 MiG-25s,

75 Su-22s and 100 MiG-23s. For transport tasks, Il-76s and Alenia G222Ts were delivered.

Libya annexed the northern part of Chad (the Aozou strip) in 1973. Confrontation with Chad during the 1980s resulted in close air support aircraft – SF.260s, Su-22s, Mirage F1s and MiG-23s – being based in the southern part of Libya at El Kufra. In 1987 the LARAF came out worse when it attacked French Jaguars and Mirages supporting Chad government troops. Many aircraft – mostly SF.260WL, L-39ZO and Mi-24 – were lost at Faya Largeau, Ouadi Doum, Aozou and even at El Kufra within Libya (when 22 MiGs, Mirages and Mi-25s were reported destroyed).

The regime of Colonel Khadaffi hosted and promoted terrorist activities by offering training camps and facilities in the desert. This policy, and a unilateral extension of the Economic Exclusion Zone, resulted in direct confrontation with the United States over sea lane navigation rights. On 19 August 1981 two Su-22s from Ghurdabiya were shot down by US Navy VF-41 F-14s. The US launched a massive attack on barracks, air bases (Tripoli and Benina) and SAM installations in 1986 using carrier assets and UK-based F-111s. Further fighting occurred in 1989 when two MiG-23MS from Al Bumbah were downed by F-14s on 4 January.

Libya has always been willing to help its allies by supplying trainers and fighters. For example, after the heavy Beka'a Valley battle with Israel in 1982, Libya donated between 30 and 40 MiG-21s and MiG-23s to compensate for the losses of the Syrian Arab Air Force. Egypt received 10 L-39ZOs in 1990.

CURRENT STRENGTH

The LARAF has always depended strongly on allied countries such as Pakistan, Cuba, Czechoslovakia, Syria, North Korea, Russia, former Yugoslavia, France and Italy. As a result of the US arms embargo, a 1989 French arms embargo and the collapse of the Eastern European regimes, most aircraft are grounded due to a lack of spares.

LARAF organisation mimic the Soviet model of one specific regiment on each base. New air bases – Umm Aitiqah, Martubah, Misurata-Zawiya, Labraq, Ghurdabiya, Al Bumbah, Ghadames and Al Jufra – were built by East Germans, Yugoslavians, Czechoslovakians and Soviets. In the 1980s a major reorganisation took place to move the units to recently opened bases. Before the embargoes, Gamal Abdel Nasser (El Adem) was a major Mirage base, but currently all Mirage 5 aircraft are stored at Okba ben Nafi, while the Mirage F1s today fulfil air defence duties (two squadrons) and ground attack (one squadron) from Okba ben Nafi. Today, the LARAF is

Libya has been a major user of the Sukhoi Su-22M-2 and today has three squadrons with the 'Fitter'.

organised on an air base/squadron structure instead of a regiment/squadron structure – there are no specific three-squadron regiments left, but one air base may host several squadrons for different purposes. For example, Ghurdabiya air base houses one interceptor squadron, one bomber squadron, one fighter-bomber squadron and two counter insurgency squadrons, plus a base flight and a coastal patrol flight. The serviceability of LARAF aircraft is very low. Libya itself does not have enough qualified pilots to fly all its aircraft, and is still dependent on countries like Cuba, Iran, North Korea, Syria and Yemen.

Squadron designations seems to indicate that fighters are flown by 1000-range squadrons, bombers in the 1100 range and the transports in the 1200 range. Possibly, helicopters are numbered in the 1300 range.

Based on the PVO of Russia, the LARAF founded its own Libyan Arab Air Defence Command, controlling all fighter-interceptor units, SAMs and radar installations in the north of the country. Before the US attack on Libya in 1986, there were three regional Air Defence zones: Tripoli, Benghazi and Tobruk. After the attack, Libya's capital was moved from Tripoli to Hun in the desert, and a fourth zone was established. The backbone of the interceptor squadrons are the MiG-23MS, MLD and MF, flown by squadrons from Gamal Abdel Nasser, Al Bumbah, Benina, Misurata-Zawiya and Umm Aitiqah. Four other interceptor squadrons fly the MiG-25 fighters from Ghurdabiya, Al Jufra and Sebha.

Su-22M-2s 'Fitters' equip three squadrons at Okba ben Nafi, Ghadames and Ghurdabiya. Other ground attack aircraft in LARAF service include the MiG-23BN (two squadrons at Labraq, one at Al Bumbah) and the L-39ZO in the light attack and trainer role from Ghurdabiya, Gamal Abdel Nasser, El Khufra (Maaten es Serra) and Sebha and other small bases. One bomber squadron still nominally operates the Tu-22 from Al Jufra. In 1989 Su-24s were delivered to Umm Aitiqah, but the squadron later moved to Ghurdabiya. It is rumoured the Su-24s were to be delivered to Syria but budgetary problems resulted in a sale to Libya instead. Israeli sources stated that Syrian pilots fly the 'Fencers'. Two squadrons are reported to fly the MiG-21 in an interceptor/fighter-bomber role from Gamal Abdel Nasser. Older COIN aircraft include the J-1 Jastrebs in two squadrons from Misurata-Zawiya. The SF.260WL basic trainers are also being used in the counter-insurgency role in the southern part of Libya and Chad. These aircraft once equipped nine other squadrons, mainly based at Faya

Largeau, Aozou, Ouadi Doum and El Khufra, but heavy fighting in Chad resulted in many losses.

TRANSPORT AND TRAINING

Heavy transport tasks are fulfilled by two squadrons of Il-76 'Candids' from Tripoli. Medium transport assets include two squadrons of Alenia G222Ts at Al Jufra, delivered from 1981. Most have been grounded since the Italian arms embargo of 1986, leaving two squadrons of An-26s to operate from Umm Aitiqah. The US delivered 11 C-130s in 1971 and, through the Philippines and Luxembourg, four L-100-30s in 1981. They are based at Benina, but their operational status is very doubtful. For VIP transportation the LARAF can use a Gulfstream II, two JetStars, a Boeing 707 and an S-61 in a single squadron based at Umm Aitiqah.

The LARAF also operates a large number of helicopters for transportation, army support, logistic support to the border posts and army-manned missile and radar installations, as well as fulfilling SAR, communications and liaison roles. In 1976 20 CH-47 Chinook heavy transport helicopters were delivered, 14 being transferred to the Army component in the 1990s. For liaison and transport duties, Libya purchased around 50 Mi-8 'Hips' and 60 Mi-2 'Hoplites'. Some 37 Mi-24 'Hinds' in two squadrons are based at Misurata-Zawiya and Okba ben Nafi, and two patrol flights based at Ghurdabiya and Gamal Abdel Nasser.

The Air Force Academy at Misurata-Zawiya was opened in 1975 and an Air Force Secondary College at Okba ben Nafi in 1978, but the latter moved to Sebha. Training provision was extended by the creation of a network of flying clubs across the country, intended to provide basic flight training for future pilots of the three armed forces. These flying clubs mainly use the SF.260WL basic trainer and L-39ZO advanced trainer. The secondary role of these 'schools' is to provide close air support. A total of 240 SF.260 Warriors was delivered. About 110 aircraft were destined for the two Air Force Schools at Misurata and Sebha, the other 130 for basic training in the paramilitary flying clubs. For armament training and counter-insurgency operations, Libya received more than 110 L-39ZOs from Czechoslovakia.

Libyan Arab Naval Aviation (LARNA)

Libyan Arab Naval Aviation (LARNA) was founded in 1962 with British assistance. The LARNA operates SA 321 Super Frelons and Mi-14 'Hazes' from shore bases near the naval ports at Misurata, Ghurdabiya and Benina. Ten SA 321M and six SA 321GM Super Frelons were delivered in 1971 and 1981, respectively, and can be equipped with Exocet missiles and torpedoes. The helicopters' main tasks are search and rescue and anti-submarine warfare.

Libyan Arab Air Defence Command

Tripoli Air Defence Sector		
1023 Interceptor Sqn	MiG-23/UB	Umm Aitiqah
2 x Interceptor Sqn	Mirage F1ED/BD	Okba ben Nafi
1 x Interceptor Sqn	MiG-23MF/MS/UB	Misurata
SAM-Brigade	SA-5 and SA-3	Okba ben Nafi
Hun Air Defence Sector		
1025 Interceptor Sqn	MiG-25	Al Jufra – Hun
1055 Interceptor Sqn	MiG-25	Ghurdabiya-Sirte
1 x Interceptor Sqn	MiG-25	Al Jufra - Hun
1 x Interceptor Sqn	MiG-25	Sebha
SAM-Brigade	SA-5 and SA-3	Sirte
Benghazi Air Defence Sector		
1040 Interceptor Sqn	MiG-23	Benina
1 x Interceptor Sqn	MiG-23	Benina
SAM-Brigade	SA-5 and SA-3	Benina
Tobruk Air Defence Sector		
1060 Interceptor Sqn	MiG-23/UB	Gamal Abdel Nasser
2 x Interceptor Sqn	MiG-23/UB	Al Bumbah

Libyan Arab Air Force

1032 Fighter Bomber Sqn	Su-22M-2	Okba ben Nafi
1124 Bomber Sqn	Su-24MK	Ghurdabiya-Sirte
1 x Bomber Sqn	Tu-22B	Al Jufra-Hun
1 x Fighter Bomber Sqn	Mirage F1AD/BD	Okba ben Nafi
1 x Fighter Bomber Sqn	Su-22M-2	Ghurdabiya-Sirte
1 x Fighter Bomber Sqn	Su-22M-2	Ghadames
2 x Fighter Bomber Sqn	MiG-21	Gamal Abdel Nasser
1 x Fighter Bomber Sqn	MiG-23BN/UB	Al Bumbah
2 x Fighter Bomber Sqn	MiG-23BN/UB	Labraq
1276 Heavy Transport Sqn	Il-76	Tripoli-Tarrabalus
1 x Heavy Transport Sqn	Il-76	Tripoli-Tarrabalus
1 x Tactical Transport Sqn	C-130H/L-100-20/30	Benina
1 x Tactical Transport Sqn	G222T	Al Jufra-Hun
2 x Medium Transport Sqn	An-26	Umm Aitiqah
1 x VIP-Transport Sqn	Gulfstream II, Jetstar, Boeing 727, AB 212	Tripoli-Tarrabalus
1 x Heavy Trans Heli Sqn	CH-47C	Okba ben Nafi
1 x Armed Helicopter Sqn	Mi-24	Okba ben Nafi
1 x Armed Helicopter Sqn	Mi-24	Misurata
1 x Border Patrol Flight	Mi-24	Gamal Abdel Nasser
1 x Coast Patrol Flight	Mi-24	Ghurdabiya
1 x Medium Trans Heli Sqn	Mi-8	Tripoli-Tarrabalus
2 x Light Helicopter Sqn	Mi-2	Martubah
Base Flights at Gamal Abdel Nasser, Al Bumbah, Benina, Ghurdabiya, Okba ben Nafi, Sebha and Martubah use Mi-8s.		

The following squadrons' aircraft are held in storage:

1030 Interceptor Sqn	Mirage 5DE	Okba ben Nafi
1 x Interceptor Sqn	Mirage 5DE	Okba ben Nafi
2 x Fighter Bomber Sqn	Mirage 5D	Okba ben Nafi
1 x Reconnaissance Sqn	Mirage 5DR	Okba ben Nafi
1 x Operational Conversion Unit	Mirage 5DD	Okba ben Nafi
2 x Counter Insurgency Sqn	L-39ZO	Ghurdabiya-Sirte
2 x Counter Insurgency Sqn	J-1 Jastreb	Misurata

Libyan Arab Army Aviation (LARAA)

Libyan Arab Army Aviation (LARAA) was established in 1970. It operates several types of helicopters in the observation and scouting role, one squadron using 10 Alouette IIIs and five Bell 206 Jet Rangers. Heavy transport of troops and army equipment is undertaken by 12 CH-47C Chinooks, passed down in 1990 from the LARAF, which still operates the Mi-24 'Hinds'. Three army squadrons are equipped with some 35 SA 342 Gazelles for attack helicopter duties. Helicopter tuition is given on eight Bell 47Gs at Misurata-Zawiya.

When Libyan troops were forced to withdraw from Chad, they left behind this Mi-24 'Hind-D'.

Air Force Academy		
3 x Basic Training Sqn	G-2 Galeb	Misurata
2 x Heli Training Sqn	Mi-2	Misurata
1 x Navig Training Sqn	L-410T/UVP	Beni Walid
2 x COIN/Training Sqn	SF.260WL	Misurata
1 x Drone Flight guidance	A 109A Hirundo	Misurata

Paramilitary Flying Clubs
Two training squadrons, one with SF.260WLs and one with L-39ZOs are each based at Labraq, Gamal Abdel Nasser, El Kufra, Sebha, Hattin and Usaman.

Libyan Arab Naval Aviation

1 x SAR/ASW Helo Sqn	SA 321M/GM	Misurata-Zawiya
1 x ASW Helicopter Sqn	Mi-14	Ghurdabiya
1 x SAR/ASW Sqn with Dets at Benina, Umm Aitiqah and Ghurdabiya with Mil Mi-14s		

Libyan Arab Army Aviation

1 x Observation Sqn	Alouette III, AB 206A	Tripoli-Tarrabalus
1 x Heavy Transport Sqn	CH-47C	Sebha
1 x Observation Sqn	Cessna O-1E	
1 x Heli Training Sqn	Bell 47G	Misurata-Zawiya
3 x Anti-Tank Sqn	SA 342 Gazelle	

MADAGASCAR

Capital: Antananarivo
Population: 11 million
Land area: 594000 km² (229,000 sq miles)

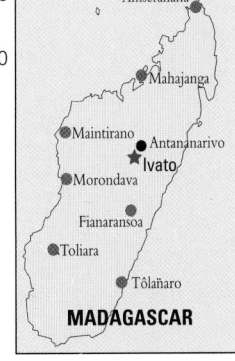

Armée de l'Air Malgache (Malagasy Air Force)

Following its 1958 independence from France, the Malagasy Republic entered a bilateral defence agreement which included the supply of 10 surplus French Douglas C-47/53s, 11 MH.1521 Broussards, six Dassault Flamant light twins and a Bell 47G, to equip an emergent national air force. Later deliveries included Alouette II and III helicopters, three Reims-Cessna F 337 light twins, other light aircraft and a BAe 748.

Soviet supplies started in 1978 with two Yakovlev Yak-40 tri-turbofan light transports, and two Mil Mi-8 transport helicopters. They were accompanied by four MiG-17Fs and eight MiG-21FLs from North Korea as the AAM's first combat equipment, the latter to form a single Fighter Flight (Escadrille de Chasse), alongside the MiG-17s and a MiG-21U in a Fighter School. Douglas C-47s and Antonov An-26 are still listed in some inventory records of the AAM's Escadrille de Transport, and at least one Dakota was still operating in mid-1995, when seven AAM C-47s

Armée de l'Air Malgache

Type	Role	Delivered/ In Service
Antonov An-26 'Curl'	tactical transport	5/3*
BAe 748 Srs 2B	transport	1/1
Cessna 172M	primary trainer	4/4
Cessna 310R	communications	1/1
Douglas C-47	transport	10/3*
MiG-17F 'Fresco-C'	advanced trainer	4/2
MiG-21FL 'Fishbed-D'	air defence/GA	8/
MiG-21UM 'Mongol-B'	combat trainer	1/1
Mil Mi-8 'Hip'	transport helicopter	/5
Pilatus B-N BN-2A Islander	light transport	1/1
Piper PA-23-250 Aztec D	communications	1/1
*believed wfu		

and five An-26s were seen derelict at Ivato airfield. Two civil-registered Yak-40 'Codlings' are operated by the Malgache government, alongside the AAM's Piper Aztec and PNB Islander in its Liaison Flight.

MALAWI

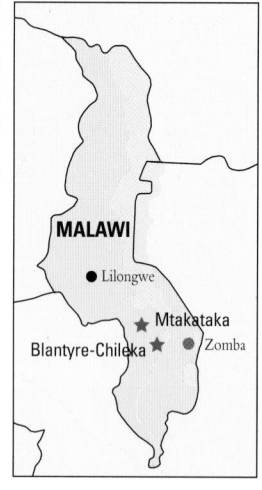

Capital: Lilongwe
Population:
8.6 million
Land area:
94000 km²
(36,000 sq miles)
Major cities:
Blantyre, Zomba

Malawi has operated a pair of BAe 125s as MAAW-J1 for government transportation, registered to the Malawi Army Air Wing. The first aircraft was a Series 700B which was sold as G-BMWW in late 1986, being replaced by this, a Series 800.

Malawi Army Air Wing

Established with German help, including the supply of six Dornier single-engined Do 27s and eight Do 28 light twins in 1976-80, the Blantyre-Chileka-based MAAW in the former UK protectorate of Nyasaland also received an Alouette III, an AS 350 and an AS 355 Ecureuil, plus three SA 330H/L Puma helicopters from France in the same period. A single BAe 125-800 light jet transport was received in 1986 to replace an earlier Srs 700B. Four Dornier

Malawi Army Air Wing		
Type	Role	In Service
Aérospatiale SA 330J Puma	transport helicopter	2
Aérospatiale SA 365N Dauphin 2	transport helicopter	1
BAe 125 Srs 800B	transport	1
Basler-Douglas Turbo 67	transport	2
Dornier Do 228-201/202K	transport	4
Eurocopter AS 355 Twin Ecureuil	utility helicopter	1

Do 228 light turboprop twins delivered between 1986 and 1989 involved part exchange or disposal of older Dornier types, and were followed in 1990 by conversion of two Douglas C-47s to Basler Turbo 67 standard with P&WC PT6A turboprops in the US.

Several government-operated aircraft are also in Malawi service, notably with the Police Air Wing at Mtakataka. This now operates three Britten-Norman BN-2T Turbo Islanders and single BN-2B-26 Islander from three of each originally delivered, plus an AS 350B Ecureuil.

MALI

Capital: Bamako **Population:** 8 million
Land area: 1.24 million km² (479,000 sq miles)
Major cities: Ségou, Sikasso, Timbuktu

Force Aérienne de la République du Mali

As a former French colony, Mali started military aviation operations from 1961 with a surplus AA MH.1521 Broussard, followed by two Douglas C-47s, until Soviet aid began in 1962 with four Antonov An-2 biplane light transports and two Mil Mi-4 helicopters. Five MiG-17F fighters and a MiG-15UTI two-seat jet trainer followed in the mid-

Mali makes use of a pair of Basler Turbo Dakotas, TZ-389 and TZ-390. The original TZ-389 was written off before delivery and replaced by another with the same serial. TZ-390 was converted from a Dakota originally ordered by the USAAF in 1942.

1960s, to equip a combat squadron at Bamako/Senou, initially with Soviet pilots.

Two Ilyushin Il-14 twin piston-engined transports and a Mil Mi-8 helicopter were delivered in 1971, supplemented by two Antonov An-24s and an An-26 in 1976. Delivery was completed in the same year of a dozen MiG-21MFs and two MiG-21UM trainers (the latter apparently since crashed), but the remaining MiG-17s and at least one MiG-15UTI from reported additional deliveries are believed to have been retained in Mali's sole Escadrille de Chasse, or Fighter Flight. Six Aero L-29 jet trainers were also delivered by 1983, to equip an Ecole de Pilotage, or Pilot School, alongside two Yakovlev Yak-18 piston-engined trainers remaining from six originally received. One of Mali's two An-26s crashed in Greece when returning from overhaul in Russia in August 1995.

Force Aérienne de la République du Mali

Type	Role	Delivered/In Service
Aero L-29 Delfin	advanced trainer	6/6
Antonov An-2P 'Colt'	transport	4/2
Antonov An-24 'Coke'	transport	3/2
Antonov An-26 'Curl'	tactical transport	2/1
Basler-Douglas BT-67 Turbo	transport	2/2
Eurocopter AS 350B Ecureuil	utility helicopter	1/1
MiG-17F 'Fresco-C'	advanced trainer	6/5
MiG-21MF 'Fishbed-J'	air defence/GA	12/10
MiG-15UTI 'Midget'	combat trainer	5/1
Mil Mi-8 'Hip'	transport helicopter	1/1
Yakovlev Yak-18 'Max'	primary trainer	6/2

MAURITANIA

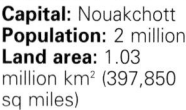

Capital: Nouakchott
Population: 2 million
Land area: 1.03 million km² (397,850 sq miles)

Force Aérienne Islamique de Mauritanie

Initial French supplies of Douglas C-47s and MH.521 Broussards – following Mauritania's independence in 1960 – were later supplemented by other transport and support aircraft, including two Douglas DC-4s, an SE 210 Caravelle, two Shorts Skyvans and six Reims-Cessna 337 tandem twins. The FAIdeM's most

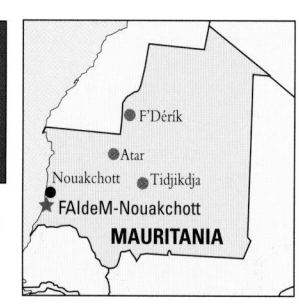

widely-used type and main combat element has been the Britten-Norman BN-2-1A Defender light piston-engined twin, of which five remaining from nine delivered in 1976-78 operate with the Escadrille de Transport and Escadrille de Surveillance.

Two specially-equipped Piper Cheyennes and the remaining Reims-Cessna 337s also equip the latter unit, while two DHC-5 Buffalo STOL turboprop twins were delivered to the Transport Flight in 1977-78. One crashed almost immediately, the other being returned to de Havilland Canada in 1979. After

Polisario rebels shot down one Defender and damaged two more beyond repair in 1978, six IA-58A Pucará ground-attack aircraft were ordered from Argentina, although later cancelled after a Mauritanian military coup.

Recent procurement has been from China, comprising a Xian Y-7-100C version of the Antonov An-26 tactical transport, delivered in October 1997, which crashed in May 1998, killing members of the president's ceremonial army band. A smaller Harbin (HAMC) Y-12 II turboprop-twin arrived in September 1995, but crashed into the Atlantic in 1996 (as did the presidential Douglas DC-9); the Y-12 has since been replaced.

Force Aérienne Islamique de Mauritanie

Type	Role	Delivered/In Service
Britten-Norman BN-2A Defender	armed transport	9/5
Harbin (HAMC) Y-12 II	light transport	2/1
Piper PA-31T Turbo-Cheyenne II	coastal patrol	2/2
Reims-Cessna FTB.337F	patrol/liaison	6/2

The Mauritanian air force operated two Harbin Y-12s but one was destroyed in 1996.

MOROCCO

Capital: Rabat
Population: 25 million
Land area: 711000 km²
(274,000 sq miles)
Major cities: Agadir, Casablanca, Fez, Marrakesh, Oujda, Sidi Infi

Left: The Fighter Pilots School at Meknes uses the majority of 24 Alpha Jet Hs delivered from 1979. This one carries a fuel-tank on its hardpoint. .

Royal Moroccan Air Force – Al-Quwwat al-Jawwiya al-Malakiya Marakishiya

The Royal Moroccan Air Force is totally Western-orientated, and the air forces of Spain, France and Belgium sometimes hold exercises in the country, such as low-level training on Alpha Jets. The organisation of the QJMM is based on French/Spanish patterns, organised by air bases and squadrons. Although the air base numbering is known, little has been determined about the squadron numbering. The Royal Moroccan Air Force is split into three separate commands: Fighter, Transport and Helicopter Command. It is well equipped with Western material.

Above: Based at Sidi Slimane, these Mirage F1EHs are seen on an exchange visit to Orange in France.

Right: Heavy lift is provided by 15 C-130Hs. At least two of the fleet have had SLAR pods fitted.

The majority of the squadrons fulfil transport and liaison duties instead of interception or attack tasks.

The Fighter Aviation Command is responsible for the defence of the country and air support for the Royal Moroccan Army. The fighters are based at two large air bases: Meknes (Air Base 2) and Sidi Slimane (Air Base 5). The Mirage F1s are organised into two squadrons – one fighter (F1CH) and one ground attack (F1EH/-200) – at Sidi Slimane. In the annexed part of Spanish Sahara, El Aioun (Air Base 4) is used by detachments of Mirages and other aircraft. Overseas deployments have included a visit in 1997 by six Mirage F1EHs to Cazaux for joint defence exercises.

The F-5Es and F-5As operate as two squadrons from Meknes. The F-5A squadron also has two RF-5As for reconnaissance duties, but it is believed that most of the F-5As are grounded or stored. In the counter-insurgency role, one squadron operates the CM.170 Magister from Meknes. The Fighter Pilot School, opened in April 1980, is co-located at Meknes, equipped with the Alpha Jet H.

TRANSPORTS

The Transport Command consists of transport squadrons based at Kenitra (Air Base 3). One squadron has operated the C-130H Hercules since 1974, and another was established in 1990 with seven CN.235Ms. The QJMM also has an aerial-refuelling capability, courtesy of two Boeing 707-138/3W6Cs and two KC-130Hs which fly with the Hercules squadron at Kenitra. For electronic warfare two Falcon 20s, two EC-130Hs equipped with Side-Looking Airborne Radar (SLAR) capabilities and one EC-130H Elint aircraft are also Kenitra-based. The VIP squadron flies several jet aircraft (Gulfstreams and Falcons) and propeller aircraft from Rabat-Sale. For training purposes, the Multi Engine Transport School uses five King Air A100s.

Rabat-Sale (Air Base 1), near the capital Rabat, houses the Headquarters of the QJMM and is also the

home of the Helicopter Command, along with the VIP squadron. Heavy helicopter transport is performed by seven CH-47C Chinooks. For medium transport the QJMM purchased 27 SA 330F Pumas from 1978, while light transportation is conducted by two squadrons using 27 AB 205As and five AB 212s. For Army support, 12 HOT anti-tank missile-equipped and 12 20-mm gun-armed SA342L Gazelles fly in two squadrons. The Specialised Helicopter School is also based at Rabat-Sale, where new helicopter pilots receive training on the AB 205, AB 206 and the SA 342L Gazelle.

A single flight at Rabat-Sale of two Do 28D-2s undertakes maritime patrol tasks.

Training Command has a presence at almost every air base. The majority of the training units are located at Marrakech, where initial selection takes place on the AS 202/18 Bravo, after which the cadets learn basic flying techniques on the T-34C-1. After passing this stage, a segregation is made between future training as a fighter pilot, transport pilot or as a helicopter pilot. Fighter pilots get their first jet experience on the CM.170, before passing the course at the Fighter Pilot

Al-Quwwat al-Jawwiya al-Malakiya Marakishiya

Fighter Aviation Command

2 Air Base

2 Fighter Squadron	F-5E/F	Meknes
? Fighter Squadron	F-5A, RF-5A, F-5B	Meknes
? COIN Squadron	CM.170 Magister, OV-10A Bronco	Meknes
? Fighter Squadron	Alpha Jet H	Meknes

5 Air Base

? Fighter Squadron	Mirage F1CH	Sidi Slimane
? Fighter Squadron	Mirage F1EH/-200	Sidi Slimane

Helicopter Command

1 Air Base

? Medium Heli Squadron	SA 330F Puma	Rabat-Sale
? Medium Heli Squadron	SA 330F Puma	Rabat-Sale
? Heavy Heli Squadron	CH-47C Chinook	Rabat-Sale
? Light Heli Squadron	Bell 205A Iroquois	Rabat-Sale
? Light Heli Squadron	Bell 205A Iroquois	Rabat-Sale
? Liaison Heli Squadron	Bell 206B Jet Ranger	Rabat-Sale
? Anti-Tank Heli Squadron	SA 342L-HOT	Rabat-Sale
? Armed Heli Squadron	SA 342L-20mm	Rabat-Sale
Maritime Patrol Flight	Do 28D-2	Rabat-Sale
? VIP Transport Squadron	Gulfstream II/III Falcon 50, King Air 200/300	Rabat-Sale

Transport Aviation Command

3 Air Base

? Transport/Tanker Squadron	C-130H, KC-130H 707-138/3W6C	Kenitra
? Transport Squadron	CN.235M	Kenitra
? ECM Squadron	Falcon 20, EC-130H	Kenitra

Training Command

School Base

Aerobatics Team 'Equipe Voltage'	CAP 10	Marrakech
Aerobatics Team 'Marche Verte'	CAP 230	Marrakech
Initial Selection	AS 202/18 Bravo	Marrakech
1st Stage – Basic Training	T-34C-1	Marrakech
2nd Stage – Jet Training	CM.170R	Marrakech
3rd Stage – Advanced Training	T-34C-1	Marrakech
Fighter Pilot School	Alpha Jet H	Meknes
Multi Engine Transport School	King Air A100	Kenitra
Heli Training School	Bell 205A Iroquois, Bell 206B Jet Ranger, SA 342L Gazelle	Rabat-Sale

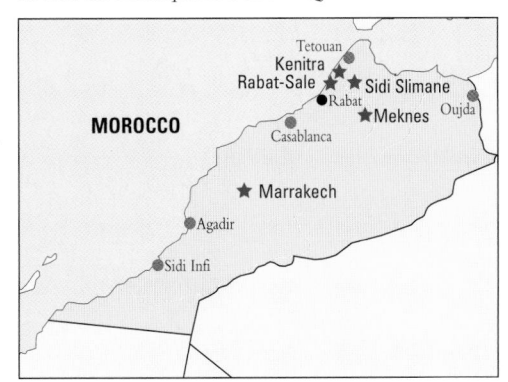

Royal Gendarmerie

The Royal Gendarmerie, or State Police, is an 'umbrella' organisation for the various uniformed, and in some cases non-uniformed, branches of the state security forces. It is organised into a Special Brigade, two Mobile Groups (north and south), an Air Squadron and Coast Guard Support, and operates from various bases in all major cities, strategic locations and harbours. The Air Squadron operates all aircraft on behalf of the various branches. The primary role is to support the government agencies, such as the Ministry of Fisheries or Ministry of Internal Affairs, with sea, land and air operations.

Royal Gendarmerie

Air Squadron, Marrakech 3 x SA 315B Lama, 7 x SA 330H Puma, 6 x SA 342K Gazelle, 2 x S-70B Blackhawk
'Coast Guard', Marrakech 2 x SA 365N Dauphin, 14 x BN-2T
'Special Brigade', Marrakech 2 x Thrush, 2 x Rallye

Africa

School, Meknes, on the Alpha Jet. For transport pilots, a course is held at the Multi Engine Transport School at Kenitra. Future helicopter pilots go to the Specialised Helicopter School at Rabat-Sale.

Proud of its history and abilities, the QJMM has founded two aerobatic display teams for international and local air shows. The 'Equipe Voltage' uses two CAP 10s, while the 'Marche Verte' displays in four CAP 231s.

Flying CAP 231s, the 'Marche Verte' (Green March) aerobatic display team is named after the unarmed occupation by Moroccan civilians of parts of Western Sahara during 1975, leading to a Spanish withdrawal.

MOZAMBIQUE

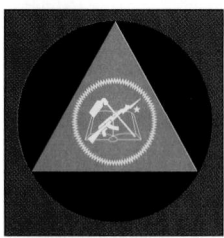

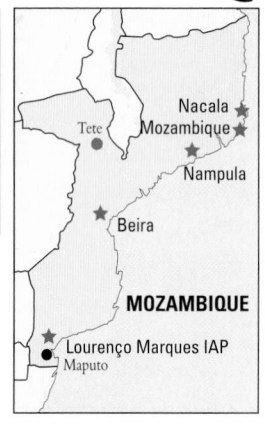

Capital: Maputo
Population: 16 million
Land Area: 785000 km²
(303,000 sq miles)
Major cities: Beira, Nampula

Força Aérea de Moçambique

Initially part of the army, and known until 1990 as the People's Liberation air force, the FAM was equipped with former Portuguese military aircraft, including Douglas C-47 and Nord Noratlas transports, North American T-6 trainers, Dornier Do 27s, and Aérospatiale Alouette III helicopters in 1985, when independence was gained by the Marxist FRELIMO organisation. This was supported by USSR, Cuban and other Communist bloc military aid, including large-scale supplies of trained personnel, plus Soviet arms and equipment, in civil war operations against RENAMO guerrillas.

Following the 1990 ceasefire, and change in government policies to Western-style free-market

economies and democratisation, most of this equipment has fallen into disrepair at the three main air bases of Beira, Nacala, and Nampula, and the FAM is now effectively only a token force. No recent aircraft procurement has been reported, and annual defence budgets have been cut to only about 1.5 per cent of GNP.

Força Aérea de Moçambique

Type	Role	Delivered/In Service
Antonov An-26 'Curl'	tactical transport	c.12/6
MiG-17F 'Fresco-C'	ground attack	23/5
MiG-21MF 'Fishbed-J'	air defence/GA	48/15
MiG-15UTI 'Midget'	combat trainer	3/1
Mil Mi-8 'Hip'	transport helicopter	c.16/6
Mil Mi-24 'Hind'	attack helicopter	16/4
Piper PA-32-300 Cherokee Six	communications	4/3
Zlin 326	primary trainer	7/5

NAMIBIA

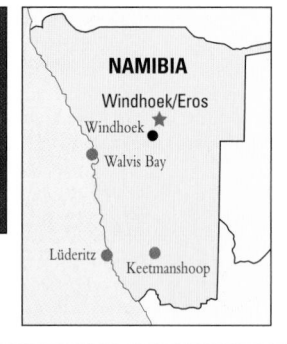

Capital: Windhoek
Population: 1.8 million
Land Area: 824000 km²
(318,000 sq miles)

Air Squadron, Namibia Defence Force

Following independence from South Africa in 1991, Namibia established a small air element of the Defence Force in 1993-94 from deliveries of at least six ex-AMARC Cessna O-2As acquired from the US. They are operated from Windhoek/Eros for border and coastal patrols, and were supplemented in 1994 by two HAL Chetaks and two Cheetah utility helicopters bought from India for Rs1.65 billion ($5 million). In late 1994, Namibia purchased a used civil-registered Learjet 31A to supplement a new Dassault Falcon 900B acquired

in 1992 for presidential and government transport. A civil-registered Reims-Cessna F406 Caravan light twin is also used by the Namibian Sea Fisheries Department for coastal patrol. Interest has been expressed in acquiring the new HAL multi-role ALH.

Namibia Defence Force

Type	Role	Delivered/In Service
Cessna O-2A Super Skymaster	utility	6/6
HAL-Aérospatiale SA 315B Alouette II/Cheetah	utility helicopter	2/2
HAL-Aérospatiale SA 316B Alouette II/Chetak	utility helicopter	2/2

NIGER

Capital: Niamey **Population:** 7.7 million
Land area: 1.186 million km² (458,000 sq miles)

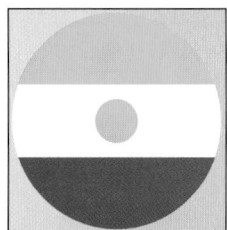

The second C-130H Hercules of the Escadrille Nationale du Niger, 5U-MBH, has been held in store at Brussels Airport in Belgium since the crash of its sistership in 1997. The pair was purchased to replace Noratlases and C-47s in the transport role.

Escadrille Nationale du Niger (Niger National Flight)

Since its 1961 formation with ex-AA Douglas C-47s, Broussards and a Flamant, the ENN has continued transport and support operations with French and German assistance. The latter included the supply of four surplus Noratlas transports, and later deliveries of two Dornier Do 128-2 Skyservants and a single Dornier Do 228-201. The Do 228 is operated alongside a civil-registered Boeing 737-2N9C which replaced an ex-French Douglas C-54B within the ENN for

presidential and government transport. Main ENN transport roles were undertaken by two Lockheed C-130Hs delivered in 1979, although one crashed at Niamey in 1997. An Antonov An-26 was donated by Libya to take its place in June 1997.

Escadrille Nationale du Niger

Type	Role	Delivered/In Service
Antonov An-26 'Curl'	tactical transport	1/1
Dornier Do 28D-2 Skyservant	light transport	2/2
Dornier Do 228-201	transport	1/1
Lockheed C-130H Hercules	transport	2/1

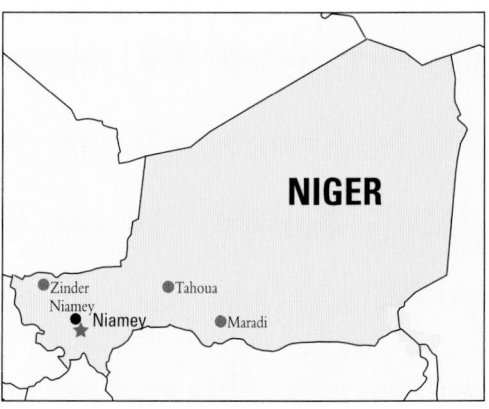

NIGERIA

Capital: Abuja
Population: 108.5 million
Land area: 924000 km²
(357,000 sq miles)
Major cities: Lagos,
Ibadan, Ogbomosho,
Kaduna, Oshogho, Ilorin,
Abeokuta, Port Harcourt

Nigerian Air Force

Nigerian arms procurement has been severely limited by successive military coups and human rights infringements which have resulted in UN arms embargoes and International Monetary Fund (IMF) limitations on acquisitions of military equipment with offensive capabilities. Accompanying economic problems, further complicated by falling oil prices and the necessity of IMF intervention, have also resulted in most of the NAF's aircraft being grounded because of spares and technical support deficiencies. Particularly affected in this respect are the 30 or so MiG-21s which replaced 25 MiG-17Fs at Kano in 1975, and the 18 Jaguars delivered from the UK to Makurdi in 1984. BAe considered repossession of the mainly-grounded NAF Jaguars in 1993, to settle large outstanding payments, but the aircraft remain in Nigeria.

Following initial deliveries of 24 Aero L-39ZA armed jet trainers from Czechoslovakia in 1986-87, negotiations continued in 1991 for a further 27, although IMF restrictions obviated their procurement. Financial problems have similarly prevented fulfilment of a 1993 NAF order for seven Pilatus PC-7 turboprop trainers, despite Swiss government approval. In the same year, NAF plans to buy 36 Slingsby Firefly primary

Nigeria ordered 24 Aero L-39ZAs for delivery between 1986 and 1987, which were received. A further batch of 27 aircraft was ordered in 1990, but had still not been delivered as of 1998, due to the country's continuing financial difficulties.

trainers as BAe Bulldog replacements were cancelled in favour of acquiring 59 Air Beetles developed with Dornier assistance from US Van RF-6A home-built kits. Assembly is planned by the Aeronautical Industrial Engineering Project Ltd at Kaduna, where three prototypes had flown by 1995. Future military aircraft procurement is planned from China.

Civil-registered types operated by the NAF for government transport include a BAe 125 Srs 1000B, Boeing 727-2N6, two Dassault Falcon 900s, a Gulfstream GII, and a Gulfstream GIV.

The sharp edge of the Nigerian Air Force's ground attack fleet should be the Jaguars delivered by British Aerospace from 1984. Unpaid debts have terminated the supply of spares, effectively grounding the fleet.

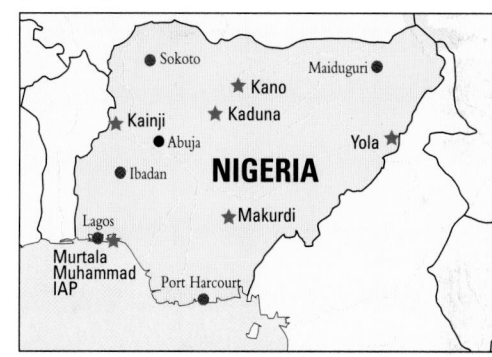

Nigerian Air Force

Type	Role	Delivered/In Service
Aeritalia G222	tactical transport	5/2
Aermacchi MB.339A	intermediate trainer	12/10
Aero L-39ZA Albatros	advanced trainer	24/18
Aérospatiale SA 330H Puma	transport helo	11/2
AIEP T-18 Air Beetle	primary trainer	59/*
Dassault-Dornier Alpha Jet N	advanced trainer	24/20
Dornier Do 28D-1/2 Skyservant	light transport	20/15
Do 128-2/6 Turbo Skyservant	light transport	16/12
Dornier Do 228-100	transport	3/3
Dornier Do 228-212	transport	6/*
Eurocopter AS 332M1 Super Puma	transport helo	12/10
Eurocopter BO 105CB	armed utility helo	30/24
Fokker F27-200MPA Friendship	maritime patrol	2/2
Lockheed C-130H Hercules	transport	6/5
Lockheed C-130H-30 Hercules	transport	3/2
MiG-21MF 'Fishbed-J'	air defence/GA	25/12
MiG-21UM 'Mongol-B'	combat trainer	6/5
Pilatus PC-7 Turbo-Trainer	basic trainer	7/*
Schweizer-Hughes 300C	trainer helo	14/12
SEPECAT Jaguar International SN	ground attack	13/10
SEPECAT Jaguar International BN	combat trainer	5/4
*on order		

Nigerian Naval Air Arm

Supplementing 1986 NAF procurement of two Fokker F27MPA maritime patrol aircraft, the Nigerian navy took delivery of three Westland Lynx Mk 89s for shipborne ASW roles in 1985, to establish No. 101 Naval Air Squadron. The Lynx then deployed singly to NNS *Aradu*, the sole MEKO 360-type frigate in Nigerian service at that time.

Type	Role	Delivered/In Service
Westland Lynx	ASW helicopter	3/2

RWANDA

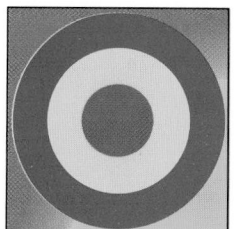

Capital: Kigali   **Population:** 7 million
Land area: 26330 km² (10,165 sq miles)

Force Aérienne Rwandaise

As part of former Ruanda-Burundi territory administered by Belgium, Rwanda became independent in 1962, and Belgian instructors helped to form a national army. An air arm was added in 1972 under army administration; it received its first equipment from France in the 1970s, in the form of seven Aérospatiale Alouette IIIs, but three Aermacchi AM-3C utility lightplanes were returned to their manufacturers soon after entering Rwandan service in 1974 because of funding problems. Spares shortages also resulted in the disposal of all the Alouettes.

Other deliveries included four SA 342L Gazelles, two Britten-Norman Islanders, two Nord 2501 Noratlas, two SOCATA Guerrier armed lightplanes, and two Aérospatiale AS 350B Ecureuils. However, in the tribal war between Hutus and invading Tutsis from Uganda, which began in 1990 and resulted in UN intervention with peacekeeping forces, most aircraft were destroyed on the ground, crashed, or were shot down. A Rwandan government Dassault Falcon 50 was shot down on the approach to its home base of Kanombe Airport in Kigali, on 6 April 1994, killing the presidents of both Rwanda and neighbouring Burundi, plus their senior cabinet ministers, who were onboard. FAR reorganisation and re-equipment has yet to take place.

SENEGAL

Capital: Dakar
Population: 7.3 million
Land area: 197000 km²
(76,000 sq miles)
Major cities: Kaolack,
St Louis, Thies

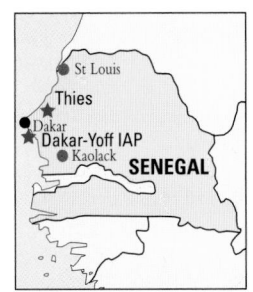

Medium transport duties for the Senegalese air force are carried out by the survivors of six Fokker F27M-400 Troopships delivered. 6W-STA has been in service for over 20 years, arriving in Senegal in November 1977.

Armée de l'Air du Senegal

Following independence from France in 1958, Senegal formed a small military aviation element in 1961 with French assistance, which included the standard 'colonial' provision of surplus Douglas C-47s and MH.1521 Broussards, plus Sud Alouette II and Agusta-Bell 47G helicopters. Close. ties have been maintained with France through training and base facilities agreements. Further French deliveries from the early 1970s of the first AAS jet equipment comprised seven Fouga Magisters, as well as an SA 341H Gazelle and four SA 330F Puma helicopters. One of the latter was shot down during an attempted coup and two more were lost in accidents.

Later expansion of the 1st Senegalese Air Group included six Fokker F27 twin-turboprop transports as C-47 replacements from 1977, when four SOCATA Rallye lightplanes were also acquired. Four armed Rallye 235 Guerrier versions followed in 1984, while government/VIP flights are operated by one Dakar-based civil-registered Boeing 727-2M1 and a de Havilland Canada DHC-6-300MR Twin Otter.

Armée de l'Air du Senegal

Type	Role	Delivered/ In Service
Aérospatiale SA 318C Alouette II	utility helicopter	5/2
Aérospatiale SA 319 Alouette III	utility helicopter	?/?
Aérospatiale SA 330F Puma	transport helo	4/1
Fokker F27-400M Friendship	transport	6/2+
Fouga CM 170 Magister	basic trainer	7/5
SOCATA Rallye/Guerrier R235A	armed trainer	4/2?
SOCATA Rallye 160ST	primary trainer	2/2
SOCATA Rallye 235E	primary trainer	2/2

An ex-French colony, Senegal naturally has predominantly French equipment. Light attack duties are performed by Rallye Guerriers (below) and light helicopter transport by a mix of Alouette IIs and IIIs (left).

SIERRA LEONE

Capital: Freetown
Population: 4 million
Land area: 72000 km² (28,000 sq miles)

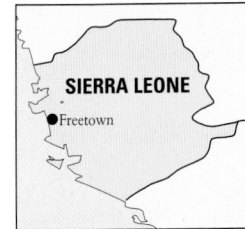

Sierra Leone Military Forces

Sierra Leone achieved independence from the UK in 1961, but not until early 1973 did it form a small air arm, with the delivery of the first two production Saab MFI-15 two-seat trainers. As Scandinavian distributor for Hughes, Saab also supplied two Model 300 (269C) light helicopters for additional training with Ghanaian help, plus another for presidential use, which was replaced in 1976 by an MBB BO 105.

In 1978, the remaining MFI-15 and all three Hughes 300s were sold, and the air arm of the SLM effectively disappeared, with transfer of the BO 105 to the civil register in 1985. Two Aérospatiale SA 355F Ecureuil II helicopters delivered in 1984 were also civil-registered, although operated for the national Defence Ministry. Five mercenary-operated Mil Mi-24V attack helicopters and one or two MiG-17s, used from 1995 to attack

Revolutionary United Front rebels, were reportedly involved in a military coup which seized control of Sierra Leone in May 1997, but seem unlikely to be retained.

SOMALIA

Dayuuradaha Xoogga Dalka Somaliyeed (Somalian Aeronautical Corps)

Formed in 1960 from the British Somali Protectorate and Italian Trusteeship Territory of Somalia, the Somali Republic established an Air Corps that year with former Italian Beech C-45s, Douglas Dakotas, a North American T-6 and other aircraft. From 1963, in exchange for port and base facilities at Berbera,

Zeila and Misimais, the SAC was expanded and re-equipped with Soviet aid, including 54 MiG-15s and -17s, four Ilyushin Il-28s, four Antonov An-24/26 transports, Mil Mi-4/8 helicopters, and SA-3 'Goa' SAMs, followed by the first MiG-21s in 1974. Over 60 MiG-21s were eventually received, plus some 30 Shenyang F/FT-6s (MiG-19s) from China.

In 1977-78, when Somalia's Marxist president Siad Barre tried to seize the Ogaden region from Ethiopia, the US provided military aid in exchange for access to Berbera, after the USSR switched support to Ethiopia. Somalia then received two Aeritalia G222 transports, six SIAI-Marchetti SF.260W Warrior armed trainers and four Agusta-Bell 212 utility helicopters from Italy. Virtually all SAC aircraft were destroyed or abandoned in the Ogaden operations and 1988 civil war (which resulted in President

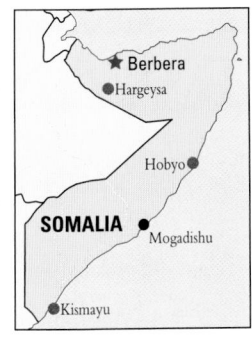

Capital: Mogadishu
Population: 7.5 million
Land area: 630000 km² (243,000 sq miles)

Barre's 1991 ousting), and in continued feuding between rebel warlords – against which ineffective US-led UN intervention ended in 1994.

SOUTH AFRICA

Capital: Pretoria/Cape Town
Population: 35 million
Land area: 1.185 million km² (457,000 sq miles)
Major cities: Johannesburg, Durban

The sole fast-jet squadron in the SAAF today is No. 2 Sqn, which flies the Cheetah C and the Cheetah D two-seat conversion trainer version. A combination of advanced avionics and missiles makes the aircraft more than a match for the current opponents that it is likely to encounter.

Suid-Afrikaanse Lugmag

HISTORY

Strategically positioned at the tip of southern Africa and blessed with large mineral deposits, South Africa is emerging as a respected member of the international community after a long period of conflict and isolation resulting from its apartheid policies.

South African military aviation commenced in 1914 when five South African pilots were attached to the (British) Royal Flying Corps in the United Kingdom. They were soon recalled to South Africa to fly Farman F27s against the German-occupied territory of South West Africa, which was successfully taken by South Africa in July 1915.

The South African Air Force itself was formed on 1 February 1920 and consisted mainly of aircraft which had been presented by the British as an Imperial Gift after World War I. During World War II, the country narrowly voted against neutrality and fought on the side of the Allies, establishing training schools on South African territory under the Joint Air Training Scheme, as well as having personnel fly in the RAF and forming its own combat squadrons. Post-war, the SAAF was scaled down, but did acquire 137 Spitfire Mk IXs for home defence and in 1948 Vampire FB.Mk 5s, its first jets. With the outbreak of the Korean War in 1950, South African pilots flew combat missions with the USAF's 18th Fighter-Bomber Wing in F-51D Mustangs, and later the SAAF's No. 2 Sqn used F-86F Sabres loaned from the USAF. After the war the SAAF continued to rely on Vampires before acquiring its own Sabres from the Canadair production line. In 1957 eight Avro Shackleton MR.Mk 3s were procured to patrol the sea lanes, serving until 1984 when the last examples were retired.

During the 1960s the SAAF was strengthened as opposition to the system of apartheid increased internally and externally, and the country was increasingly drawn into the wars of its neighbours. In the early 1960s the UK provided English Electric Canberra bombers and Hawker Siddeley Buccaneer S.Mk 50 strike aircraft, as well as Westland Wasp anti-submarine shipborne helicopters. The imposition of sanctions on South Africa by the United Nations from 1963, which were adhered to by the majority of the world's nations, made it difficult to re-equip the air force, but it was still able to acquire Mirage IIIs, Super Frelons and Alouette IIIs from France, while Italy

provided Aermacchi MB.326Ks, which were later produced locally as the Atlas Impala.

Bush wars in South West Africa against the South West Africa People's Organisation (SWAPO) infiltration commenced in the mid-1960s, and operations in Rhodesia in support of the white minority rulers were also undertaken. The imposition of sanctions encouraged the country to provide its own weapons, and the foundations of a indigenous aerospace industry were laid. Even so, France was willing to supply Mirage F1AZs and F1CZs during the mid-1970s. The chaos created by the withdrawal of Portugal from its African empire during this period left a power vacuum, allowing SWAPO to establish bases in the south of Angola, drawing South Africa into confrontation with Angolan government forces. The period of armed confrontation continued until the political settlement at the end of the 1980s which resulted in the withdrawal of South African forces from what had become Namibia (South West Africa) – which also gained its independence – and a withdrawal of South African forces from Angola.

The past 10 years have seen a great change in the political and military structures of South Africa. During 1990 severe cuts in the strength of the South African Air Force were announced. The end of the policy of apartheid and the beginnings of free elections for all – the first taking place on 10 May 1994 – opened the way for the lifting of the sanctions. The start of free elections ended the existence of the so-called independent homelands of Transkei, Bophutatswana, Venda and Ciskei, and the assets of their air wings were transferred to the SAAF. The ending of the period of confrontation with its northern neighbours also finished the need for the large armed forces that had put a strain on the economy of the country, and the SAAF was at the forefront of a draw-down of the military. Money saved by the cutbacks was required to fund a programme aimed at improving the living standards for the majority black population.

Even after the large reduction of types and squadrons over the last decade, South Africa is still a

regional superpower and the South African Air Force is probably the most powerful and best trained in the region, keeping a balance between combat roles and essential support capabilities.

AIR DEFENCE

A small retirement ceremony held at Hoedspruit on 25 November 1997 marked the retirement of the Mirage F1AZ from SAAF service, when No. 1 Sqn relinquished the type after 22 years. The Mirage F1AZ was used for both air defence and attack, and with its passing the air force was left with one type of fast jet. South African air defence is entrusted to the Atlas Cheetah Cs based at Louis Trichardt with No. 2 Squadron, appropriately nicknamed 'The Flying Cheetahs'. A total of 38 aircraft was produced, the first aircraft arriving at Louis Trichardt in January 1993 and the last in June 1995. Although the aircraft is basically a variation on the Mirage III theme, the use of canards has increased manoeuvrability. The advanced radar, coupled with air-to-air missiles – the Kentron V3C Darter with a 5-km (3-mile) range and compatibility with a helmet-mounted sight – makes the Cheetah C more than a match for any aircraft likely to be encountered in the region. Pilot conversion to the type is undertaken within the squadron on its two-seat Cheetah Ds.

TRANSPORT ASSETS

A shortfall in transport capability resulted in seven Transall C.160Zs re-entering service early in 1995, after being retired in January 1993. This reprieve was short-lived and the type was again withdrawn, being put up for disposal in September 1997. The Hercules of No. 28 Sqn were left to soldier on from Waterkloof as the sole heavylift transport. The original seven C-130Bs were delivered from 1963 and all remain in service today. To replace the Transalls, a pair of ex-USAF C-130Bs was delivered in May and October

The C-130Bs that arrived in 1963 have remained in service, supplemented recently by ex-USAF examples and ex-US Navy C-130Fs. They all fly with No. 28 Sqn from Waterkloof.

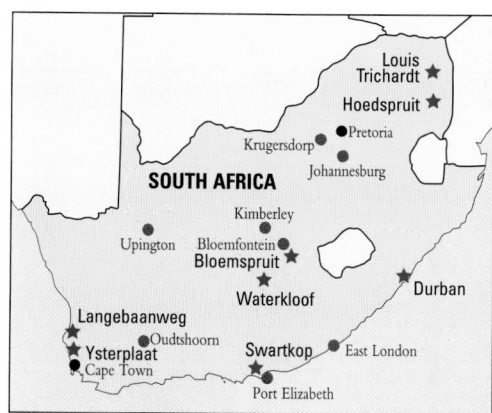

SOUTH AFRICA

Louis Trichardt ★
Hoedspruit ★
● Pretoria
Krugersdorp ●
Johannesburg ●
Kimberley ●
Upington ●   Bloemfontein ●
Bloemspruit ★
Waterkloof ★
Durban
Langebaanweg ★
● Oudtshoorn
Ysterplaat ★   Swartkop ★   East London
● Cape Town
Port Elizabeth

Africa

The Impala Mk II is a veteran of the many bush wars in which South Africa has been involved. The type is due to be replaced in the near future.

Suid-Afrikaanse Lugmag

2 Sqn	Cheetah C/D	Louis Trichardt
8 Sqn	Impala I/II	Bloemspruit
15 Sqn	Oryx, BK 117A	Durban
16 Sqn	Rooivalk AH-2A (due in 1999)	Bloemspruit
17 Sqn	Oryx, SA 316B/SE 3160 Alouette III, AS 365N Dauphin 2	Swartkop
19 Sqn	Oryx, SA 316B/SE 3160 Alouette III	Louis Trichardt
21 Sqn	BAe 125 Srs 400B/403B, Cessna 551, Falcon 50, Falcon 900	Waterkloof
22 Sqn	Oryx, SA 316B/SE 3160 Alouette III	Ysterplaat
28 Sqn	C-130B/F	Waterkloof
35 Sqn	C-47TP, King Air 200C	Cape Town
41 Sqn	Cessna 208, King Air 200C, King Air 300, PC-12	Waterkloof
42 Sqn	Cessna 185A/D/E, PC-6B	Swartkop
44 Sqn	C-47TP, CASA C.212	Swartkop
60 Sqn	Boeing 707-320B	Waterkloof

Training Command		
85 AFS	Impala I/II	Hoedspruit
86 MEFS	C-47TP, CN.235M-100	Bloemspruit
87 HFS	SA 316B/SE 3160 Alouette III	Bloemspruit
CFS	PC-7 Mk II Astra	Langebaanweg

1996. Three ex-US Navy C-130Fs – two arriving in December 1996 and the third in November 1997 – were acquired from the AMARC facility in Arizona. An avionics update programme for the fleet is being undertaken by Marshalls of Cambridge in the UK and the SAAF is keen to acquire more of the type.

At least 34 Dakotas have been remanufactured as C-47TP Turbo Dakotas, the modification involving a fuselage stretch as well as the engine change. The type is used not only for transport duties with No. 44 Sqn at Swartkop but also for maritime patrol with No. 35 Sqn from Cape Town airport. Some examples are also used by the 86 Multi-Engined Flying School (86 MEFS) for multi-engined training at Bloemspruit. Some criticisms have been levelled at the conversion, including the standard of airworthiness and the type's suitability for parachute operations. Early in 1998, the SAAF begun to dispose of some of its C-47TPs, 12 aircraft being offered for sale; Dodson Aviation of Ottawa, Kansas has acquired at least three examples. Since March 1998 No. 44 Sqn has also operated four CASA C.212s transferred from 86 MEFS. They had been acquired from the independent homelands' air wings. A single Airtech CN.235M-100 impressed from the Bophutatswana Defence Force continues to serve in 86 MEFS.

Light transport tasks are undertaken by a variety of types. The survivors of the Cessna 185s that were delivered from 1962 and the Cessna 208 Caravan Is which were delivered in some secrecy from 1988 constitute the majority of the light aircraft in service. The Cessna 185s serve with No. 42 Sqn from Swartkop, flying alongside an ex-Bophutatswanan Pilatus PC-6B. The Caravans, which wear a very civil-looking scheme, were acquired through a local Cessna dealer and flew with No. 41 Sqn from Waterkloof for six years before adopting military serials. They operate alongside three Beech King Air 200Cs, also acquired on the civil market, which had previously operated in the fisheries protection role with No. 35 Sqn; a King Air 300 previ-

ously with the Ciskei Defence Force; and a single Pilatus PC-12 acquired from Pilatus in exchange for the three ex-Bophutatswanan PC-7 Turbo-Trainers which had seen service with the CFS before being stored.

VIP flights are undertaken by No. 21 Sqn from Swartkop. The unit uses a mix of business jets, all of which carry civil registrations. The aircraft comprise two Citations (one a Cessna 550 Citation II, the other a 551 Citation IISP) from the Venda Defence Force, a pair of Falcon 50s and a single 900, plus five BAe 125 Srs 400B/403B Mercurius.

No. 60 Sqn at Waterkloof operates five Boeing 707-320Bs, four as tankers/Elint-gatherers and one as a general freighter. Equipped with three-point hose-and-drogue gear, the tankers are used to increase the endurance of the Cheetahs. Four of the aircraft were acquired through Israel, and became operational in 1986 but were not revealed until 1991. Details of the intelligence mission of the 707s have not been released.

HELICOPTERS

Helicopters continue to play a major part in the modern SAAF, which had used large numbers in the past, being highly suited for the kind of bush wars that South Africa has experienced over the years. Recent cuts and retirements have seen the air force lose the SA 321L Super Frelon heavylift helicopter and the Puma, which served in large numbers with the SAAF.

The Aérospatiale Puma was finally retired in 1995, when No. 15 Sqn, the last operator, replaced them with the Oryx. A total of 50 Atlas TP-1 Oryx was built, looking very similar to the Eurocopter Super Puma/Cougar. They operate with Nos 15 (based at Durban), 17 (Swartkop), 19 (Louis Trichardt) and 22 Squadrons (at Ysterplaat) in the tactical transport role, but undertake other duties as well. For example, No. 22 Sqn has the responsibility for providing helicopters for navy ships, and is also tasked with fire-fighting and SAR missions.

The Alouette III has been in service since 1966 and is beginning to show its age. Having been used in both utility transport and gunship roles, the type continues to serve with Nos 17, 19 and 22 Squadrons alongside the Oryx, as well as with 87 Helicopter Flying School at Bloemspruit for helicopter pilot training duties.

No. 15 Sqn uses nine BK 117A-3s acquired from

the tribal homelands (two from Venda, two from Transkei, two from Bophutatswana and three from Ciskei). A single AS 365N Dauphin 2 from Bophutatswana was operated by No. 17 Sqn, but was offered for sale in 1997.

The most important helicopter for the SAAF's future is the Denel CSH-2 Rooivalk AH-2A attack helicopter, of which two production batches, totalling 20 helicopters, have been ordered. A pair of prototypes has been flying for some time and the type is being actively promoted abroad. The SAAF has earmarked No. 16 Sqn at Bloemspruit to be the first Rooivalk operator.

TRAINING

The Pilatus PC-7 Mk II Astra replaced the Harvard in service as the air force's *ab initio* trainer. Of the 60 aircraft ordered in 1993, two have been written off and a third was damaged in May 1997. Undertaken by the Central Flying School (CFS) at Langebaanweg, the first course on the new type began in 1996. From the CFS, SAAF pilots go to either 86 Multi-Engined Flying School at Bloemspruit for conversion to transport aircraft, or to 85 Combat Flying School at Hoedspruit for advanced training on the Impala Mk I and Mk II. The Impala Mk I is on the way out, the Astra having assumed many of its roles. The Impala Mk II is an armed version and also serves with No. 8 Squadron at Bloemspruit in the light attack role. About 50 remain, of which approximately half are stored. Helicopter training is undertaken by 87 Helicopter Training School, using Alouette IIIs.

TESTING AND PRESERVATION

Located at Bredasdorp is the Test Flight and Development Centre, which tests various new equipment, weapons and aircraft. The unit has on strength examples of various types in service with the SAAF, as well as a pair of BN-2 Islanders transferred from the Ciskei Defence Force Air Wing.

The SAAF Museum is a fully fledged unit in the air force with headquarters at Swartkop. The collection aims to gather one type of each aircraft type operated by the force, and maintains several in flying condition. The recent round of extensive cutbacks has released types such as the C.160Z, Super Frelon, Buccaneer S.Mk 50 and Mirage F1AZ for preservation.

FUTURE PROCUREMENT

On 18 November 1998, the deputy President, Thabo Mbeki, announced the preferred suppliers for six defence programmes representing potential orders worth R29.8 billion ($5.2 billion). All programmes included significant offset arrangements. The deal included the selection of the Saab/BAe JAS 39 Gripen for the Light Fighter Aircraft requirement, the Cheetah Cs replacement. A total of 28 is to be procured between 2002 and 2015. The type was selected in preference to the Mirage 2000-5 and MiG-29.

The winner of the competition to replace the Impala Mk II is the Hawk LIFT (Lead-In Fighter Trainer), which was chosen in preference to the MB 339FD, YAK/AEM-130 and MiG-AT. Final assembly of the Hawks, of which 24 are to be ordered, will be undertaken by Denel from components manufactured by BAe.

For the light utility helicopter requirement – the Alouette III replacement – Agusta will receive an order for 40 A 109s. The South African company General Aviation will assemble the helicopters while Denel will add South African systems. Eurocopter had offered 60 EC 635s.

Four GKN Westland Super Lynx 300 maritime helicopters are to be ordered. Fitted with a full 'glass' cockpit and Rolls-Royce/AlliedSignal CTS800 engines (30 per cent more powerful than the usual Gem 42s), the type was selected in place of the competing Eurocopter AS 532SC Cougar.

The South African Air Force is due to form its first squadron of Denel CSH-2 Rooivalk attack helicopters in 1999. The first production example was handed over to the SAAF on 17 November 1998.

SUDAN

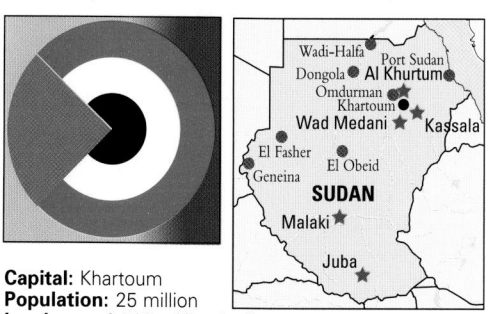

Capital: Khartoum
Population: 25 million
Land area: 2.506 million km²
(967,000 sq miles) **Major cities:** Omdurman, Port Sudan

Silakh al Jawwiya As'Sudaniya (Sudanese Air Force)

Widely-varying Sudanese political alignments and military equipment receipts began with Soviet- and Chinese-supplied MiG-21PFMs, Antonov An-12s, An-24s, Mil Mi-4s and Mi-8s, plus 18 Shenyang F-5 (MiG-17F) fighters and 10 FT-5 combat trainers, replacing original UK-sourced Douglas C-47s, and Hunting Provosts, Jet Provosts and Pembrokes after a 1969 revolutionary left-wing coup. US and Saudi aid then took over, following Soviet involvement in an attempted 1971 presidential coup. Saudi funding ended, however, after Sudan's support of Egypt in the 1979 peace agreement with Israel.

US arms worth $52 million followed 1981 Sudan border clashes with Libya, until Libyan support of a

1985 military coup in Sudan and its new Islamic government's campaign against the mainly Christian Sudan People's Liberation Army southern rebel group. The 1982 US supply of 10 Northrop F-5Es and two F-5Fs was then halted at two each F-5E/Fs, as were orders for six BAC 167 Strikemasters, after three were delivered in 1983. Deliveries were completed, however, of four DHC-5D Buffalo and six Lockheed C-130H transports from 1977, 12 Agusta-Bell 212s from 1982, and 15 IAR-AS 330 Pumas from Romania in 1984.

More than a dozen MiG-23s and pilots were also supplied from Libya from 1987, and others later from Iraq, plus seven Iran-financed Chengdu F-7Bs from China in 1996, supplementing earlier deliveries of 15, and Shenyang F-6s via Iran. Two Shaanxi Y-8D turboprop transports, also from China, and a few Mil Mi-24s arrived from Kyrgyzstan in 1995. Many SAF aircraft have been lost in prolonged and continuing attacks on SPLA rebels, while few others are currently serviceable. Previously unrecorded Sudanese receipts of the Antonov An-32 turboprop transport were revealed from the February 1998 crash of one, which

killed 13 of 57 occupants, including Sudan's first vice-president.

The air force currently uses the survivors of 22 Chengdu F-7Bs delivered. It is possible that some of the MiG-21PFMs they replaced remain in service.

Four of the six Sudanese air force C-130H Hercules are reported as having been shot down.

Silakh al Jawwiya As'Sudaniya		
Type	**Role**	**Delivered/ In Service**
Agusta-Bell AB 212	utility helicopter	12/9
Antonov An-24 'Coke'	transport	6/2
Chengdu F-7B	air defence/GA	22/10
de Havilland DHC-5D Buffalo	tactical transport	4/2
de Havilland DHC-6-300 Twin Otter	light transport	1/1
Eurocopter BO 105CB	armed helicopter	12/8
ICA-Brasov IAR 330L Puma	transport helicopter	15/10
Lockheed C-130H Hercules	transport	6/2
Shaanxi Y-8D	transport	2/2
MiG-23BN 'Flogger-H'	ground attack	15/6
Mil Mi-8 'Hip'	transport helicopter	10/6
Mil Mi-24 'Hind'	attack helicopter	4/4
Shenyang F-5/FT-5	GA/combat trainer	28/16
Shenyang F-6/FT-6	air defence/ combat trainer	21/10

SWAZILAND

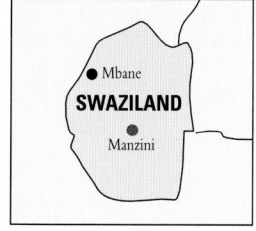

Capital: Mbane   **Population:** 800,000
Land area: 17365 km² (6,705 sq miles)
Major cities: Manzini

Two IAI Arava 102s are used by the Umbutfo Swaziland Defence Force, including 3D-DAC.

Umbutfo Swaziland Defence Force		
Type	**Role**	**Delivered/In Service**
IAI 202 Arava	transport	3/2

Umbutfo Swaziland Defence Force

Although an independent member of the British Commonwealth since 1968, Swaziland had no military forces for over 10 years, relying on some 800 civil police for maintaining public order. In 1979, however, two IAI Arava twin-turboprop light transports, with provision for underwing weapons for light ground-attack roles, were acquired from Israel to supplement new ground defence forces, and based at Matsapa airport. One soon crashed while being demonstrated by IAI in Malawi, but was replaced by a third example in 1980. A seven-seat Cessna U206G utility lightplane has also been reportedly delivered for Swazi government use.

TANZANIA

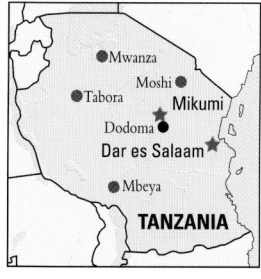

Jeshi La Wananchi La Tanzania (Tanzanian People's Defence Force Air Wing)

Tanzania has favoured its army over the naval and air arms, reflecting the number of border disputes which have affected the region.

Capital: Dodoma
Population: 26 million

The majority of the air force's aircraft are from China, with which Tanzania has a close relationship, while maintaining its political independence. Exactly how many of the surviving FT-2s, F-5s, F-6s and F-7As are serviceable is open to speculation. Most of the combat aircraft are based at Dar es Salaam-Mikumi.

Five Buffalos and a pair of Harbin Y-12s, delivered

Land area: 939760 km² (362,750 sq miles)
Major cities: Dar es Salaam, Mbeya, Tabora

in late 1994, provide the transport capability of the air force. The sole King Air A100 was noted at Nairobi-Wilson airport in Kenya in July 1997 awaiting disposal. Helicopter support is provided by four Agusta-Bell AB 205Bs.

The government of Tanzania has two aircraft which use civil registrations, a Fokker F28 Fellowship 3000 (5H-CCM) and a Fokker 50 (5H-TGF).

TANZANIAN POLICE AIR WING

The Tanzanian Police Air Wing has a total of four helicopters and an aircraft for general paramilitary duties.

Jeshi La Wananchi La Tanzania		
Type	**Role**	**In Service/ Delivered**
Agusta-Bell 205B	utility helicopter	4/-
Beech King Air A100	communications	1/1
DHC-5D Buffalo	transport	5/6
Harbin Y-12 (II)	transport	2/2
Shenyang FT-2	conversion training	-/2
Shenyang F-5	attack	8/12
Shenyang F-6	air defence/attack	10/16
Xian F-7A	air defence/attack	11/16

Tanzania Police Air Wing		
Type	**Role**	**In Service**
AB 206L LongRanger	paramilitary	2
Bell 47G-3B2	paramilitary	1
Cessna U206 Stationair	communications	1
Eurocopter BO 105CBS	paramilitary	1

TOGO

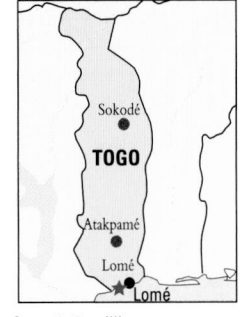

Capital: Lomé **Population:** 3.5 million
Land area: 56785 km² (21,920 sq miles)
Major cities: Atakpamé, Sokodé

Force Aérienne Togolaise
(Togolese Air Force)

Togo is one of three military operators of the Epsilon. Delivered in 1986, the aircraft are used as armed lead-in trainers for the Alpha Jets.

After initial French equipment and organisation following its 1960 independence, Togo replaced its original Douglas C-47s with two DHC-5D Buffalo STOL transports in 1976, as the basis of an Escadrille de Transport. In the same year, a combat element, or Escadrille de Chasse, was formed with five ex-Luftwaffe Fouga Magister armed jet trainers and the first three of seven EMB-326GBs from Brazil. Togo's armed jet trainer fleet was further expanded by five Dassault-Dornier Alpha Jets from France in 1981, and by three piston-engined Aérospatiale TB-30 Epsilons for the Ecole Militaire in 1986. Single examples of each were later delivered as attrition replacements, although

the Magisters were returned to France in 1985.

An Escadrille de Liaison operates one or two Beech King Airs, Cessna/Reims-Cessna 337s, and Aérospatiale Lama helicopters, while the presidential Boeing 707, flown in military markings, is supplemented by a civil-registered Fokker F28 Fellowship 1000 and a Eurocopter AS 332L Super Puma for government transport.

Force Aérienne Togolaise

Type	Role	Delivered/ In Service
Aérospatiale SA 315B Lama	utility helicopter	3/3
Beech King Air 200	light transport	/2
Boeing 707-312B	transport	1/1
Cessna/Reims-Cessna 337E	communications	3/2
Dassault-Dornier Alpha Jet E	advanced trainer/GA	5/4
de Havilland DHC-5D Buffalo	tactical transport	2/1
EMBRAER EMB-326GB Xavante	advanced trainer/GA	7/4
SOCATA TB-30 Epsilon	armed basic trainer	4/3

TUNISIA

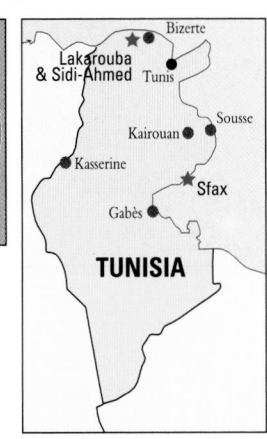

Capital: Tunis
Population: 8.2 million
Land area: 164000 km² (63,000 sq miles)
Major cities: Bizerte, Kairouan, Sfax, Sousse

Al-Quwwat al-Jawwiya al-Jamahiriyah At'Tunisia
(Tunisian Republic Air Force)

Although currently the most politically stable North African state, Tunisia is devoting increasing sums from its limited defence budget to internal security in an attempt to counter potential domestic disturbances from Islamic militants. Continued reliance is also being placed on US and French military aid, the

former recently including up to seven surplus USAF Lockheed C-130Bs from 1997, to supplement 1985 US deliveries of two C-130Hs and 24 Northrop F-5E/Fs.

Deliveries also started in late 1995 of 12 Aero Vodochody L-59 uprated armed jet trainers with Western avionics, as part of a $70 million Czech procurement package, which further included three Let L.410UVP twin-turboprop transports received in September 1994. The L-59s operate alongside four remaining Aermacchi MB.326s in advanced training roles, while further combat training and operational capabilities derive from four two-seat MB.326LTs and eight single-seat MB.326KTs delivered in 1977-78. Basic training is undertaken on the remaining 14 of nine SIAI-Marchetti SF.260Cs and 12 armed SF.260WT Warriors received between 1974 and 1978.

The surviving Tunisian F-5 Tigers perform the air defence and ground attack role with No. 15 Squadron based at Sidi-Ahmed.

Al-Quwwat al-Jawwiya al-Jamahiriyah At'Tunisia

Known current principal TRAF units include the following:

Unit	Equipment	Base
11 Squadron	Aermacchi MB.326KT/LT	
15 Squadron	Northrop F-5E/F	Sidi-Ahmed
21 Squadron	Let L.410UVP, Lockheed C-130	
31 Squadron	Agusta-Bell/Bell 205A, Bell UH-1	
32 Squadron	Aérospatiale Alouette II/III, Eurocopter Ecureuil	
Flying School	Aero L-59F, Aermacchi MB.326B, SF.260	

Type	Role	Delivered/ In Service
Agusta-Bell AB 205A	utility helicopter	18/15
Aero Vodochody L-59F	advanced trainer/GA	12/12
Aermacchi MB.326KT/LT	advanced trainer/GA	12/11
Aermacchi MB.326B	advanced trainer	8/4
SA 3130 Alouette II	utility helicopter	8/4
SA 316B Alouette III	utility helicopter	8/5
AS 350B Ecureuil	utility helicopter	6/6
Bell 205A-1	utility helicopter	12/10
Bell UH-1H Iroquois	utility helicopter	4/4
Let L.410UVP-E20G	transport	3/3
C-130B Hercules	transport	5/5
C-130H Hercules	transport	2/2
Northrop F-5E Tiger II	air defence/GA	20/10
Northrop F-5F Tiger II	combat trainer	4/2
SIAI-Marchetti S.208A	communications	4/4
SF.260CT/WT Warrior	armed basic trainer	21/14

UGANDA

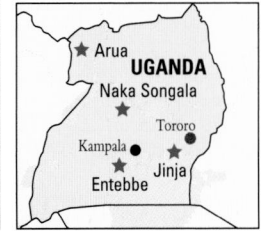

Capital: Kampala **Population:** 17 million
Land area: 237000 km² (91,000 sq miles)
Major cities: Jinja

Ugandan Air Force

The original Ugandan Army Air Force was formed in 1964 as a spin-off from the Police Air Wing, with initial aid from Israel, including supplies of 12 surplus IAI Bedek-built Fouga Magisters, six

Douglas C-47s, a single Nord N.2502D Noratlas and 16 Piper Cubs. Eight ex-Luftwaffe Piaggio P.149Ds were also acquired. Soviet links resulted in first combat aircraft deliveries from 1966 of single squadrons of MiG-17Fs and MiG-21MFs, plus Mil Mi-8 helicopters and Czech Aero L-29 Delfin jet trainers. The switch to Soviet supplies, and repayment deficiencies to Israel, resulted in withdrawal of the Israeli training mission in 1972 and return of the remaining Magisters and C-47s. Four MiG-17s and seven MiG-21s were also destroyed on the ground at Entebbe by Israeli forces in 1976 by Israeli commandos rescuing hijacked El Al passengers.

Training and equipment aid followed from North Korea and several Arab countries, including transfers of small numbers of SIAI-Marchetti SF.260 and Aero L-39 trainers, plus Agusta-Bell 206 and other helicopters, from Libya. Additional helicopters were procured from Italy and the US, and in 1977 six FFA AS 202 primary trainers were acquired. Successive rebel guerrilla campaigns and military coups since the 1971 overthrow of President Obote by General Idi Amin have resulted in most UAF aircraft being destroyed, scrapped or becoming barely serviceable

through spares shortages. Today the air arm suffers from an acute spares shortages, the L-39s being very rarely flown. This state also applies to the civil-registered National Police Air Wing (NPAW) inventory, which included a DHC-2 Turbo-Beaver, DHC-4 Caribou, DHC-6 Twin Otter, two Cessna 180s, a Piper Aztec, and Bell 205, 206 and 212 helicopters. A civil-registered Gulfstream G.III is still flown for government transport, and orders for more Agusta-Bell 205s were reported in late 1996.

Ugandan Air Force

Type	Role	Delivered/ In Service
Aero L-39ZA	armed jet trainer	4/3
Agusta-Bell AB 412SP	utility helicopter	6/5
Agusta-Bell 206 JetRanger	utility helicopter	/3
Bell 206 JetRanger	utility helicopter	/2
Bell 412	utility helicopter	3/2
FFA AS 202-18A-1 Bravo	primary trainer	6/1
Mil Mi-8 'Hip'	transport helicopter	7/5
Saab MFI-17 Supporter	primary trainer	/1
SIAI-Marchetti SF.260W Warrior	armed basic trainer	3/3

ZAMBIA

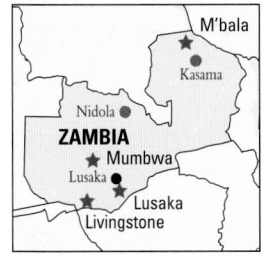

Capital: Lusaka   **Population:** 7.8 million
Land area: 752762 km² (290,510 sq miles)
Major cities: Kasama, Livingstone, Nidola

Zambian Air Force & Air Defence Command

After initial equipment from RAF aid in 1964 with Douglas C-47s and Hunting Pembrokes, plus DHC Chipmunks, Beavers and Caribou from Canada, further ZAF expansion followed with Italian training help from supplies of SF.260 and MB.326 trainers, and Agusta-Bell 205 helicopters. Similar aid also came from Yugoslavia, with planned 1971 supply of 15 Galeb jet trainers and 18 single-seat Jastreb light

A total of 20 SIAI SF.260MZs was ordered by the Zambian Air Force, but only nine are thought to have been delivered. In service they are used for basic training and light attack.

ground attack aircraft, as well as two Douglas DC-6Bs, although not all jet deliveries were believed to have been completed.

Despite unification of the Zambian National Defence Forces under central command in 1976, receipt of an $80 million USSR arms package from 1981 comprising MiG-21s, An-26s, Mi-8 helicopters, SA-3 SAMs, AAA, radar and other ground equipment resulted in the formation of a Soviet-style semi-autonomous Air Defence Command. After the acquisition of more Agusta-Bell helicopters, DHC-5D Buffalo and Dornier Do 28 transports, and 20 Saab-MFI 17 armed light trainers, procurement of Western equipment virtually ceased. In 1977-78 China presented Zambia with 12 each of Shenyang F-6 (MiG-19) fighters and Nanchang BT-6 piston-trainers, followed by three Harbin Y-12 transports in 1996.

Zambian Air Force

Type	Role	Delivered/In Service
Aermacchi MB.326GB	advanced trainer	23/15
Agusta-Bell AB 47G-4A	utility helicopter	16/12
Agusta-Bell AB 205A	utility helicopter	13/12
Agusta-Bell AB 206	utility helicopter	4/3
Agusta-Bell AB 212	utility helicopter	2/2
Antonov An-26 'Curl'	tactical transport	4/4
de Havilland DHC-5D Buffalo	tactical transport	7/3
Dornier Do 28D-2 Skyservant	light transport	10/5
Harbin Y-12 II	light transport	3/3
MiG-21MF 'Fishbed-J'	air defence/GA	16/13
MiG-21US 'Mongol-B'	combat trainer	4/3
Mil Mi-8 'Hip'	transport helicopter	8/3
Nanchang BT-6	primary trainer	12/10
Saab MFI-17 Supporter	primary trainer	17/15
Shenyang F-6/ FT-6	air defence/GA	12/9
SIAI-Marchetti SF.260MZ Warrior	armed basic trainer	9/8
SOKO G-2A Galeb	advanced trainer	15/2
SOKO J-1E Jastreb	ground attack	18/6
Yakovlev Yak-40 'Codling'	transport	3/2

Funding limitations have also limited further procurement, but 13 ZAF MiG-21s were reportedly being upgraded to Lancer standard by IAI and Aerostar in Romania in 1997-98, when at least five MFI-17s were also being overhauled in Harare, Zimbabwe. Two civil-registered Beech King Air C90s supplement a tri-turbofan ZAF Yak-40 for government transport.

ZIMBABWE

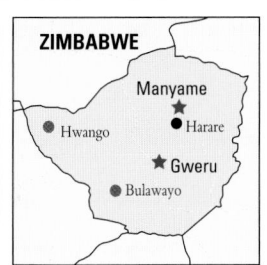

Capital: Harare   **Population:** 9.3 million
Land area: 390000 km² (150,000 sq miles)
Major cities: Bulawayo, Gweru, Mutare

Air Force of Zimbabwe

The Unilateral Declaration of Independence in 1965, a long guerrilla campaign which started in 1972 and UN sanctions helped shape the air arm of Zimbabwe. Starting as the Royal Rhodesian Air Force (Rhodesian Air Force post-1969), it became the Zimbabwe-Rhodesia Air Force in 1979 and finally, in 1980, the Air Force of Zimbabwe.

Today's Air Force of Zimbabwe is based at two main bases and consists of eight squadrons. Some early aircraft types – such as the Aermacchi AL.60C-5 Trojan, Agusta-Bell AB 205A Cheetah, Canberra and C-47 Dakota, inherited from the state's previous incarnations – have disappeared, while others, including the Alouette IIIs, Genets, Hunters and Lynxes, continue to provide efficient service.

In 1962/63 12 Hunter FGA.Mk 9s were supplied from RAF stocks to form the equipment of No. 1 Sqn and were used extensively during the war for independence. They were augmented in 1981 by the delivery of five Hunters (three FGA.Mk 9s and a pair of T.Mk 81s) which had previously served in the Kenyan Air Force. More ex-RAF FGA.Mk 9s were added in 1984 (five) and 1987 (four). From these aircraft, No. 1 Sqn flies six single-seaters and an ex-Kenyan T.Mk 81, 1084 (ex-802 and XL604). No replacement is in sight for the Hunter, which is judged to be an excellent ground attack aircraft.

Operating from Thornhill/Gweru alongside No. 1

Sqn are the Hawk Mk 60/60As of No. 2 Sqn. Tasked with fast-jet training and light ground attack duties, eight Mk 60s were ordered in 1981, the first arriving in-country in July 1982. On the night of 25 July 1982 a sabotage attack on Thornhill Air Base damaged four Hawks, as well as nine Hunters and a single FTB-337G. One Hawk was a write-off, another was repaired on site and the other two had to be returned to British Aerospace for rebuild. A follow-on order for five Mk 60As was completed by September 1992. The Jet Flying Training School within No. 2 Sqn prepares pilots for the Hunters or Chengdu F-7s operated by No. 5 Sqn. The first supersonic interceptor operated by the Air force of Zimbabwe, a dozen F-7 Airguards in two sub-types (F-7IIN with two hard-points under the wing and F-7II with only a single hardpoint) were delivered in 1986, accompanied by two Chengdu FT-5s. The FT-5s were withdrawn in 1991 and replaced by a pair of Guizhou FT-7BZs conversion trainers. The Airguards are used for point defence and interception.

No. 4 Sqn flies the Reims-Cessna FTB-337G Lynx as a light strike, observation and light transport aircraft. Eighteen aircraft were delivered in 1976 to the Rhodesian Air Force, with three others following in 1977. Around 12 aircraft are serviceable at one time. A pair of O-2As was delivered in 1993 as part of the FMS programme. The O-2As are operated by No. 4 Sqn on behalf of the National Parks Board and undertake anti-poaching patrols.

Transport duties are undertaken by No. 3 Sqn from Manyame/Harare, the renamed New Sarum, using five BN-2A Islanders (from six delivered) and 13 CASA 212 Aviocars (with another example having been lost). One of the Aviocars is in use as a VIP aircraft in a smart white scheme, while the majority are used for the tactical transport role. Two ex-AMARC C-130B Hercules were offered to the air force in 1996, but may not have been accepted. Helicopter transport is provided by the SA 316 Alouette IIIs and AS 532 Cougars of No. 7 Sqn and the Agusta-Bell AB 412SPs of No. 8 Sqn, both based at Manyame/Harare. Alouette IIIs were acquired from a variety of sources, including ex-Portuguese air force and Romanian IAR-built examples. South African examples are also believed to have served with the air force during periods of crisis. Alouette IIIs are used as both troop transport (G-car) and gunships (K-car). Two Cougars were reported in service during 1997, one

Zimbabwe's main combat aircraft are the Hunter FGA.Mk 9s of No. 1 Sqn and the Chengdu F-7s of No. 5 Sqn. Both squadrons are based at Thornhill/Gweru.

having been delivered during April 1995 and the second in September 1996. They are used for the VIP role, along with two AB 412s. The other AB 412s are used for assault, transport and battlefield support roles.

No. 6 Sqn is the basic training squadron, flying the SF.260 Genet in 'M, 'TP (turboprop conversion of the 'W) and 'W form. Six new SF.260Fs were ordered in June 1997.

The Air Force of Zimbabwe is keen to preserve its heritage and maintains an airworthy Hunting Provost T.Mk 1/52 hybrid for displays in its Historic Flight.

In November 1998 it was reported that a $54 million arms shipment of helicopter gunships, fighters and spotter aircraft had arrived in Zimbabwe to assist the President of the Democratic Republic of Congo. Early in 1999 it was unclear which types were delivered and who would actually operate the aircraft.

Air Force of Zimbabwe

1 Sqn	Hunter FGA.Mk 9/T.Mk 81	Thornhill/Gweru
2 Sqn	Hawk Mk 60/60A	Thornhill/Gweru
3 Sqn	CASA C.212-200	Manyame/Harare
	BN-2A Islander	
4 Sqn	Cessna FTB 337G	Thornhill/Gweru
5 Sqn	Chengdu F-7 II/IIN	Thornhill/Gweru
	Guizhou FT-7BZ	
6 Sqn	SF.260M/TP/W	Thornhill/Gweru
7 Sqn	SA 316 Alouette III	Manyame/Harare
	AS 532	
8 Sqn	Agusta-Bell 412 SP	Manyame/Harare
National Parks Board (looked after by 4 Sqn)		
	O-2A	Thornhill/Gweru
Historic Flight	Provost T.Mk 1/52	Thornhill/Gweru

Jayapura

**PAPUA
NEW GUINEA**

Bahjarmasin
Ujung Pandang

INDONESIA

SOLOMON ISLANDS

**PACIFIC
OCEAN**

Mataram

Kupang

Port Moresby

TIMOR SEA

Darwin

CORAL SEA

VANUATU

FIJI

Wyndham

Derby

AUSTRALIA

Cairns

Townsville

NEW CALEDONIA

Port Headland

NORTHERN
TERRITORY

Rockhampton

Carnarvon
Exmouth

WESTERN
AUSTRALIA

Alice Springs

QUEENSLAND

Geraldton

Brisbane

Kalgoorlie-Boulder

SOUTH
AUSTRALIA

Nimbin

Perth
Fremantle

Whyalla

**NEW
SOUTH
WALES**

Newcastle
Sydney

Albany

Elizabeth
Adelaide

Canberra

Wollongong

TASMAN SEA

VICTORIA
Ballarat
Geelong   Melbourne

**AUSTRALIAN
CAPITAL
TERRITOTRY**

Auckland

AUSTRALASIA
& PACIFIC ISLANDS

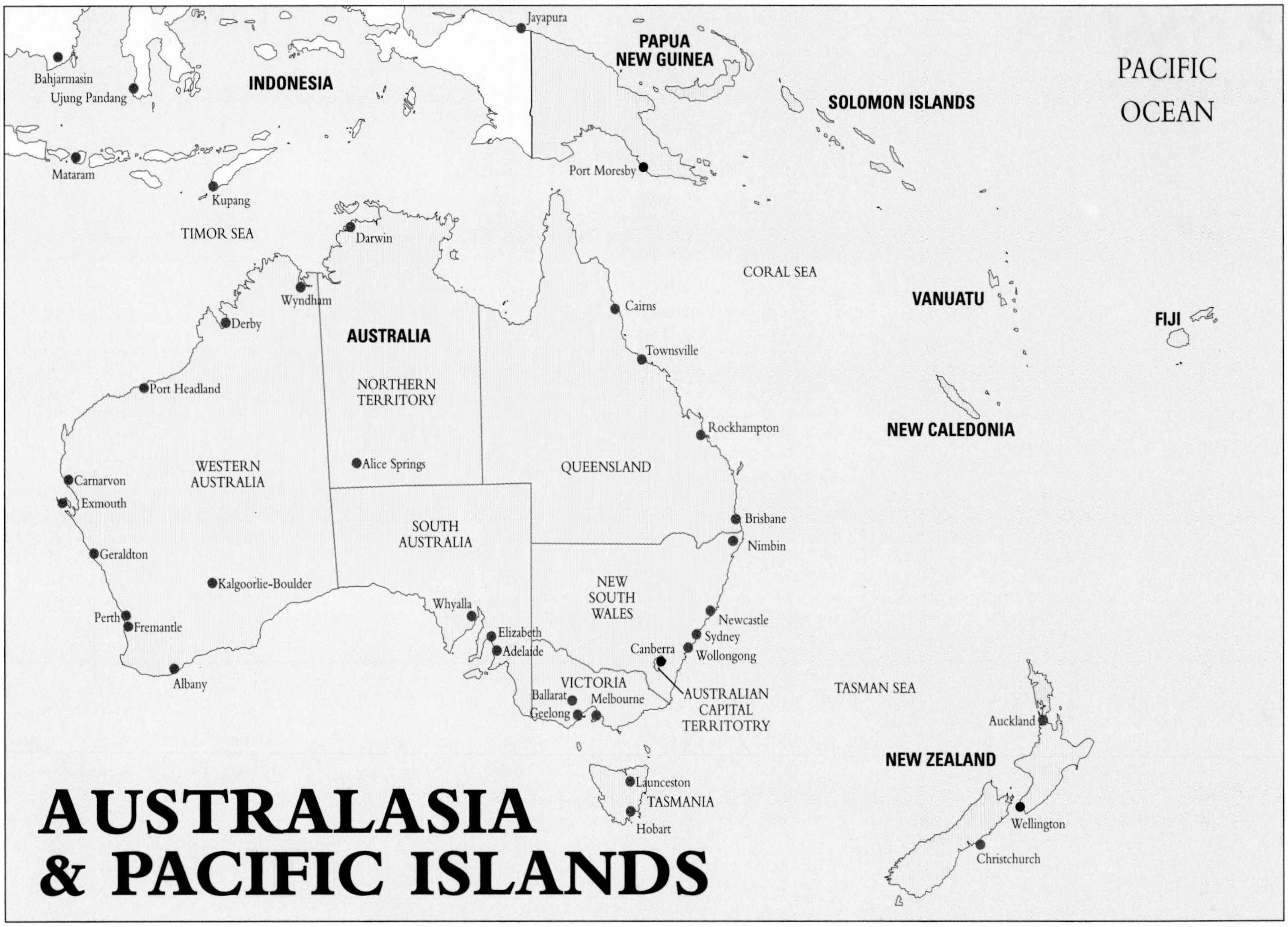

Launceston
TASMANIA
Hobart

NEW ZEALAND

Wellington

Christchurch

Australasia comprises the islands of Australia, Fiji, New Zealand, Tonga and the half of New Guinea that includes Papua New Guinea. None of the armed forces in the area is excessively large, as the lack of a credible potential enemy has allowed successive politicians to keep the military lean.

Left: Symbolising the close links between the two nations, a RAAF No. 77 Sqn F/A-18A takes on fuel while RNZAF No. 2 Sqn A-4Ks await their turn.

Below: In late 1998 New Zealand announced that it was to acquire the 28 F-16A/Bs originally ordered for Pakistan. They will be delivered from 2002 to replace the upgraded A-4K Skyhawks.

Australasia is dominated by the largest land mass, Australia, with New Zealand playing a parallel but smaller role in the defence of the region. The years after World War II generated a strategy of defence through regional agreements and taking action when the countries' vital interests were deemed threatened. The policy of forward defence placed Australian troops in the combat zones of Vietnam to help stem the perceived encroachment of Communist troops southward. Post-Vietnam, a policy of military self-reliance emerged. As an island nation, Australia considers the defence and control of the surrounding waters to be of extreme importance.

With Australia and the United States, New Zealand is part of the ANZUS defence agreement. The disagreement between New Zealand and the USA about the use of New Zealand's ports by ships that are either nuclear-powered or carrying nuclear weapons has resulted in the agreement being reduced to a pair of bilateral pacts. The dissolution of the Soviet Union, which had such a big effect in North America and Europe, resulted in little change in the defensive postures of Australia and New Zealand. The contraction of the US presence and influence in the area has resulted in these two countries undertaking regional duties which in the past might have been the purview of the superpowers. The current ceasefire following the unrest in Papua New Guinea has been policed by Australians and New Zealanders – the RAAF and RNZAF both having helicopters based there.

The mid-1990s saw the Fiji Air Wing dispose of its helicopters after re-establishing defence ties with Australia and New Zealand, while the Kingdom of Tonga acquired a Beech 18, its first aircraft since the Victa Airtourer was withdrawn from service.

AUSTRALIA

Capital: Canberra
Population: 18.5 million
Land area: 7.682 million
km² (2.966 million sq miles)
Major cities: Sydney,
Melbourne, Brisbane,
Adelaide, Darwin

The Australian Defence Force prides itself on self-reliance (although allies such as the USA are acknowledged as crucial in a major conflict), and because Australia is an island, control of the maritime approaches is a priority. The land battle scenario has not been neglected, with emphasis on the country's north, and considerable effort has been dedicated to mobility and flexibility, enabling a rapid response to any incursion.

Australia is a member of the ANZUS Alliance (with New Zealand and the USA), and also a participant in the Five Power Defence Arrangements (with the UK, New Zealand, Malaysia and Singapore). In addition to national commitments, Australia has been a staunch supporter of the United Nations, and all three armed services have contributed to UN operations throughout the world.

Royal Australian Air Force (RAAF)

Formed on 31 March 1921, the Australian Air Force ('Royal' from 31 August that year) became the second independent air arm in the world, predated only by Britain's RAF. The RAAF's history includes fighting not only in World War II, but also in Korea (1950-1953), and Vietnam (1964-1972). Since then, it has supported the UN in such places as Kashmir, the Sinai, Cambodia, and (most recently) the Persian Gulf.

The 1990s have seen re-equipment and restructuring designed to retain a technological edge over any potential adversary, also reflecting the direction of the perceived threat. A string of bases has been built along the northern edge of the country, but, of these, only Tindal is home to a flying unit. The others (Learmonth and Curtin in Western Australia, and Scherger in Queensland) are 'bare bases' that would host south-based squadrons in times of tension.

The aircraft-operating units of the RAAF come under the jurisdiction of two major Commands, Air Command at Glenbrook in the Blue Mountains west of Sydney, and Training Command at Point Cook (RAAF Williams) in Victoria.

Air Command controls four aircraft-operating groups, each with their own headquarters, and often co-located with the assets under their control. Subordinate to these groups are a number of wings that oversee the operation of the flying squadrons.

The RAAF is undertaking an upgrade package for its F/A-18s, to counter aircraft like the Sukhoi Su-27. ASRAAM and AMRAAM missiles will replace Sidewinder and Sparrow.

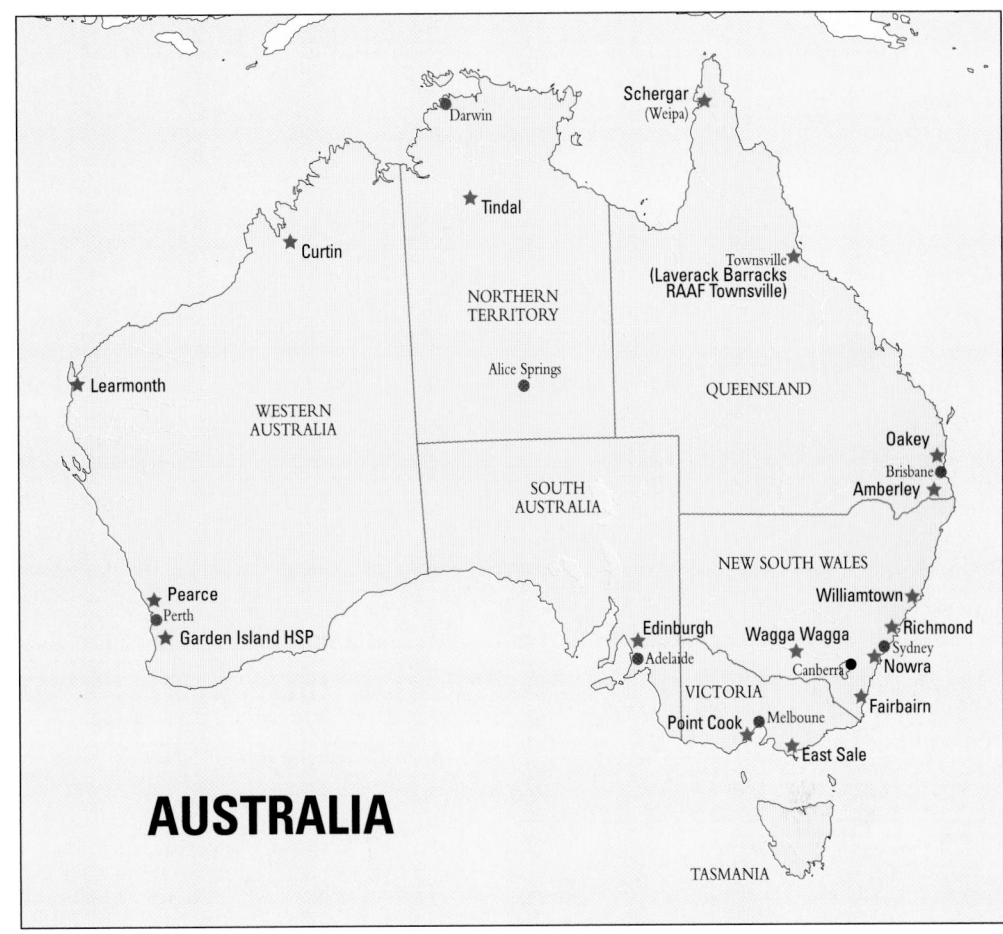

AUSTRALIA

Training Command, as its name suggests, is responsible for the training organisation of the RAAF, and controls two flying schools.

STRIKE AND RECONNAISSANCE GROUP

The wartime task of the Strike and Reconnaissance Group would be to fly interdiction sorties within Australia's maritime approaches, and provide precision strike capability with guided weapons. No. 82 Wing, comprising Nos 1 and 6 Sqns, operates the General Dynamics F-111C/G and RF-111C.

Weapons at the disposal of the Group include the AGM-84D Harpoon, Mk 36/41 Destructor mines, and GBU-10/24 and GBU-12 LGBs. All F-111Cs are Pave Tack capable, and provided with AIM-9M Sidewinders for self-protection (at present the F-111Gs lack these features). This capability will be enhanced with the introduction of the AGM-142 RAPTOR in 1999, and the completion of a digital

Displaying the new all-over grey scheme, this No. 6 Sqn F-111C is seen flying at low level over southern Queensland with a warload of 12 Mk 82 bombs.

avionics upgrade (AUP) for the F-111C/RF-111C. The AUP is well underway, with the 'digital' aircraft currently being operated alongside the similar F-111Gs by No. 6 Sqn (the F-111 OCU).

The four reconnaissance-tasked RF-111Cs are normally operated by No. 1 Sqn, but the AUP has seen some fleet-swapping, and at least two have operated (as bombers) with No. 6 Sqn.

Australia is the last operator of the very capable 'Aardvark'. They fly with Nos 1 (below) and 6 Squadrons of No. 82 Wing, based at Amberley.

Above: One training and two operational squadrons operate 18 P-3Cs and three TAP-3 crew trainers. The seabird on the tail is the badge of No. 11 Sqn.

Left: The RAAF uses five Boeing 707s. Four Series 338Cs have tanker pods on the wings, while a single Series 368C is used solely as a transport.

TACTICAL FIGHTER GROUP

Primarily equipped with the F/A-18A Hornet, the Tactical Fighter Group controls Air Defence and Ground Attack assets within the RAAF, through the subordinate No. 81 Wing at Williamtown. This Wing oversees Williamtown's three Hornet units, and also No. 75 Sqn at Tindal in the Northern Territory. Each squadron specialises in a particular role, but maintains proficiency in another. Nos 3 and 75 Squadrons have air defence as their primary role, whereas No. 77 Sqn is optimised for close air support. This latter squadron also operates a trio of PC-9/As in the forward air control role, a capability inherited from the Macchi-equipped No. 76 Sqn in 1997.

The Hornet is currently fitted with the APG-65 radar and armed (in the air defence role) with the AIM-7M/AIM-9M, studies have recently been undertaken definingan upgrade package that will enable the aircraft to operate in an environment where the opposing force may be flying aircraft such as the Su-27. The upgrade will take place as two programmes, the first focusing on navigation, communication and IFF systems, while the second will concentrate on the radar and flight software. The AN/APG-73 radar has been selected to replace the APG-65 from 2001, and the AIM-132 will be the new short-range missile.

In addition to the Hornet squadrons, No. 81 Wing provides fleet co-operation, close air support and introductory fighter courses with the last airworthy examples of the CAC-built Macchi MB.326H. Based on opposite coasts (close to the major fleet bases), No. 76 Sqn at Williamtown, NSW and No. 79 Sqn at Pearce, WA eagerly await delivery of the BAe Hawk Mk 127 from 2000 to replace the Macchi. No. 79 Sqn took over the duties of No. 25 Sqn in 1998, allowing the latter to revert to its Reserve status.

AIR LIFT GROUP

Consisting of Nos 84 and 86 Wings at Richmond, the ALG provides the ADF with a multi-level transport capability and is unique in being the only Group to control two flying Wings. No. 84 Wing can call

upon the HS.748s of No. 32 Sqn for light transport tasks when they are not operating for the School of Air Navigation, and also controls No. 33 Sqn's five Boeing 707-320Cs and No. 34 Sqn's VIP Falcon 900s. Four of the Boeing 707s provide the RAAF's air-refuelling capability, albeit probe-and-drogue only.

With the exception of No. 37 Sqn, No. 86 Wing is the only tactical transport wing, and is currently equipped with the C-130 Hercules and DHC-4A Caribou.

Of the Caribou units, No. 35 Sqn is based at Townsville to support the Army's 3rd Brigade (part of the Operational Deployment Force), and No. 38 Sqn provides training and transport from its Amberley home.

Richmond is the home of the Hercules fleet, with No. 36 Sqn flying the camouflaged C-130H in the tactical role, and sister unit No. 37 Sqn providing strategic airlift with the C-130E.

The C-130J-30 is about to begin replacing the elderly E-model Hercules (further options are held on the C-130J to replace the newer C-130H); the HS748 is to be upgraded to allow continued effectiveness in the navigator training role; 707 replacement is being accelerated due to increased maintenance costs; and the Falcon 900s are now approaching the end of their lease agreement (although a five-year extension option exists).

Practically all of the Group's aircraft are due to be either replaced or upgraded. Replacement of the Caribou is the most pressing, and a selection from the shortlisted CASA CN.235 or C.295 and Lockheed Martin/Alenia C-27J is expected during 1999.

MARITIME PATROL GROUP

Responsible for safeguarding Australia's sea lanes and fishing zones, the MPG would also provide much intelligence in times of conflict and, apart from the traditional ASW role, has an important over-the-horizon targeting capability and anti-shipping strike potential with the AGM-84D Harpoon.

Solely equipped with the P-3 Orion, Edinburgh-based No. 92 Wing's flying units are Nos 10, 11 and 292 Sqns, with the latter being the P-3 training unit. Although they wear squadron markings, all of the Orions are pooled, with aircraft allocated to crews as required.

Royal Australian Air Force

Headquarters Air Command, Glenbrook NSW

Direct Reporting Unit

ARDU	F/A-18A/B, PC-9/A *(trials & testing)* *(plus various types as required)*	Edinburgh

Strike & Reconnaissance Group, Amberley Qld

82 Wing, Amberley Qld

1 Sqn	F-111C, RF-111C* *(strike/recce)*	Amberley
6 Sqn	F-111C (AUP), F-111G *(strike/training)*	Amberley

* will also operate AUP aircraft on completion of programme

Tactical Fighter Group, Williamtown NSW

81 Wing, Williamtown NSW

3 Sqn	F/A-18A, F/A-18B *(air defence)*	Williamtown
75 Sqn	F/A-18A, F/A-18B *(air defence)*	Tindal
76 Sqn	MB.326* *(fleet co-operation, close air support, introductory fighter training)*	Williamtown
77 Sqn	F/A-18A, F/A-18B, PC-9/A *(ground attack, FAC)*	Williamtown
79 Sqn	MB.326H* *(fleet co-op, close air support, introductory fighter training)*	Pearce
2 OCU	F/A-18A, F/A-18B *(Hornet conversion)*	Williamtown

* to be replaced by Hawk Mk 127 commencing 2000

41 Wing, Williamtown NSW (Radar Units)

1 RSU	*(Radar Surveillance Unit, Jindalee OTHRN)*	Alice Springs
2 CRU	*(Control & Reporting Unit)*	Darwin *
3 CRU	*(Control & Reporting Unit, includes Software Development Unit)*	Williamtown
114 MCRU	*(Mobile Control & Reporting Unit)*	Tindal **

* to relocate to Tindal by 2001 ** to relocate to Darwin by 2001

Air Lift Group, Richmond NSW

84 Wing, Richmond NSW

32 Sqn	HS.748, B200 King Air *(navigator training, transport support)*	East Sale
33 Sqn	Boeing 707 *(strategic air transport special purpose {VIP} transport, AAR)*	Richmond
34 Sqn	Falcon 900 *(special purpose transport)*	Fairbairn

86 Wing, Richmond NSW

35 Sqn	DHC-4A *(light tactical transport)*	Townsville (Darwin Det)
36 Sqn	C-130H *(tactical airlift)*	Richmond
37 Sqn	C-130E* *(strategic airlift)*	Richmond
38 Sqn	DHC-4A *(light tactical transport, training & conversion)*	Amberley (Pearce Det)

* to be replaced by C-130J commencing 1999

Maritime Patrol Group

92 Wing, Edinburgh SA (Butterworth Det)

10 Sqn	P-3C*, TAP-3 *(as required) (maritime patrol)*	Edinburgh
11 Sqn	P-3C*, TAP-3 *(as required) (maritime patrol)*	Edinburgh
292 Sqn	TAP-3, P-3C *(as required) (training)* **	Edinburgh

* to operate AP-3C upgrade aircraft from 1999
** operates 10 and 11 Sqn aircraft as required

Training Command, HQ Williams (Laverton) Vic

ADFBFTS	CT-4B, CAP-10* *(basic flight training)*	Tamworth
CFS	PC-9/A *(standards, instructor training, 'Roulettes')*	East Sale
2 FTS	PC-9/A *(advanced training)*	Pearce
SAN	B200 King Air, HS.748** *(navigator training)*	East Sale
RSTT	various *(technical training)*	Wagga Wagga

*operated by BAe Flight Training Australia on behalf of the ADF
** operated by 32 Sqn on behalf of SAN as required

The RAAF acquired 67 Pilatus PC-9/As to partially replace the fatigue-plagued MB.326Hs. They entered service in 1987.

Three transport TAP-3s (converted ex-USN P-3Bs) were delivered in 1997/98 to relieve P-3Cs of the rigours of circuit training. They also provide an in-house transport capability for regular deployments in which MPG aircraft operate as far away as Butterworth in Malaysia, or Learmonth in the north-west of Western Australia.

To be designated AP-3C on completion, the P-3Cs will be upgraded by E-Systems (all but the prototype to be completed in Australia), maintaining operational viability until at least 2015.

AIRCRAFT RESEARCH & DEVELOPMENT UNIT (ARDU)

This unit does not report to a particular group, but directly to HQ Air Command. Based at Edinburgh, ARDU provides for the trials and testing of aircraft and equipment. Most RAAF – and indeed ADF – types have been operated (either on charge or loan), and at the time of writing included the F/A-18A and PC-9/A. The ARDU operates four C-47B Dakotas. The aircraft have seen 50 years of service with the RAAF (and displayed nose art claiming that they are *50 Years Young*) and were due for retirement in December 1998. However, no successor has been found, and they continue in service into 1999.

TRAINING COMMAND

Having no operational squadrons, RAAF Training Command is responsible for the operation of the flying training units within the Air Force. At present, prospective RAAF and RAN pilots undergo 15 hours of flight screening with BAe Australia's Flying Training Academy at Tamworth, flying civilian CT-4B and CAP 10s. Upon successful completion, they then pass to No. 2 FTS at Pearce to fly the PC-9/A. From January 1999, however, BAe Australia has also provided basic flying training for these services, the students then progressing to No. 2 FTS for the advanced flying training phase.

Responsible for instructor training and the maintenance of flying standards within the RAAF, the PC-9/A-equipped Central Flying School at East Sale also has the (more public) function of operating the 'Roulettes' aerobatic team.

Also at East Sale, the RAAF School of Air Navigation (SAN) does not operate aircraft in its own right, but utilises HS.748s and King Airs from No. 32 Sqn as required (two leased civilian King Airs were delivered to the SAN in 1997, but passed to No. 32 Sqn in 1998).

THE FUTURE

Almost all of the RAAF's major assets are currently subject to either an upgrade or replacement programme and the next decade will see several types added to the order of battle.

Surveillance of the maritime approaches is most important and although the troubled Jindalee Over-The-Horizon Radar (JORN) is under development, one major shortfall is a lack of AEW&C capability. This is being addressed, with some degree of urgency, as Project Air 5077 (commonly known as Project Wedgetail). The Boeing 737-700 (Northrop-Grumman MESA radar), Lockheed Martin C-130J (APS-145), and Airbus A310 (Elta) are shortlisted, with the winner to be announced in late 1999.

Royal Australian Navy – Fleet Air Arm

The Fleet Air Arm of the RAN was officially formed on 28 August 1948 to operate Fireflies and Sea Furies from the aircraft-carrier HMAS *Sydney*, and after only three years found itself fighting in Korea. HMAS *Melbourne* replaced *Sydney* in 1956, introducing the jet age with the Sea Venom and Gannet, but it was not until the introduction of the Skyhawk and Tracker in 1967 that the FAA became a modern force. *Melbourne's* retirement was announced in 1982, and, although Britain offered HMS *Invincible* (an offer later withdrawn in the post-Falklands era), the fixed-wing element was withdrawn in 1984.

Today's Fleet Air Arm is almost an all-helicopter force and is responsible for operations from RAN ships. Playing a major role in the Navy's contribution to the maritime defence of the country, it would fly anti-submarine, over-the-horizon targeting and, with the introduction of the Penguin-armed SeaSprite, anti-shipping sorties.

Based at RANAS Nowra (HMAS *Albatross*) on the south coast of New South Wales, the Fleet Air Arm currently operates five types, consolidated in three flying squadrons.

HC723 operates the Aérospatiale Ecureuil (Squirrel), CAC-built Bell 206 Kiowa, and the HS.748. A multi-role unit, its duties include SAR, medevac, fleet support, communications, electronic warfare training and the supply of aircraft to ships' flights. The unit is due to merge into HS817 by 2000.

The Kiowas operated from the hydrographic survey ship HMAS *Moresby*, but, since it was decommissioned at the end of 1997, their future must be limited.

The Squirrels have been upgraded to AS 350BA standard, and regularly detach aboard 'Adelaide'-class frigates, often embarked alongside the Seahawk. This mix is standard for the Persian Gulf deployments carried out during and since the Gulf War in support of UN sanctions.

The only two fixed-wing aircraft in the Navy, the HS.748s are operated by an Electronic Warfare Flight within HC723, and provide EW training primarily for the fleet (although they often operate on behalf of the other services) and are able to confer a limited transport capability.

The 'teeth' of the service at present belong to the S-70B-2 Seahawks of HS816. More comprehensively equipped than their US Navy brethren, the Seahawks provide an important anti-submarine capability. At present, only 12 of the 16 purchased are in service (the balance held as attrition replacements), but all will be active by 2000. A detachment at Garden Island (HMAS *Stirling*) in Western Australia supports the Fleet Base West FFGs.

A Mid-Life Upgrade programme is in hand to give the Seahawk FLIR, ESM and enhanced countermeasures, aimed at ensuring that the aircraft remain effec-

The Sea King Mk 50s have been replaced in the anti-submarine warfare role by the S-70Bs, and have been relegated to the utility role with HS817.

tive in the anti-submarine warfare (ASW) and anti-surface surveillance and targeting (ASST) roles.

The final squadron of the Fleet Air Arm is HS817, with the Westland Sea King. The six RAN aircraft (Mk 50A), plus one ex-RN example (Mk 50B), have recently completed a Life Of Type Extension that has seen, among other modifications, the dunking sonar removed and the aircraft configured for the utility role. The Sea Kings retain the MEL 5955 radar, coupled to a new signal processor, allowing the aircraft to continue to perform ASST missions. In their new-found utility role, they equip a detachment aboard HMAS *Success*, and also deploy aboard HMAS *Tobruk* and the ex-USN amphibious transports HMAS *Kanimbla* and *Manoora*.

The new 'Anzac' frigates are optimised for anti-surface warfare and a new type of helicopter was deemed necessary (though interim Seahawk opera-

RAN Fleet Air Arm

HC723	AS 350BA, 206B-1 Kiowa, HS.748 Nowra (SAR, medevac, EW training, fleet support, communications, ships flights)
HS816	S-70B-2 Nowra (ASW, ASTT) (Garden Island Det)
HS817	Sea King Mk 50A/B (fleet utility) Nowra

Notes:
1. HC723 is due to be absorbed by HS817 by 2000
2. 805 Sqn will be raised prior to 2000 to operate the SH-2G(A) in the anti-surface warfare role.

Helicopter-capable ships of the RAN

'Adelaide' class (USN 'Oliver Hazard Perry')
FFG 01 HMAS *Adelaide*, FFG 02 HMAS *Canberra*, FFG 03 HMAS *Sydney*, FFG 04 HMAS *Darwin*, FFG 05 HMAS *Melbourne*, FFG 06 HMAS *Newcastle*

'Anzac' class (Meko 200)
FFH 150 HMAS *Anzac*, FFH 151 HMAS *Arunta*, FFH 152 HMAS *Warramunga**, FFH 153 HMAS *Stuart**, FFH 154 HMAS *Parramatta**, FFH 155 HMAS *Ballarat**, FFH 156 HMAS *Toowoomba**, FFH 157 HMAS *Perth**
* not yet completed

'Kanimbla' class (USN 'Newport')
LPA 51 HMAS *Kanimbla*, LPA 52 HMAS *Manoora*

'Tobruk' (Imp. *Sir Bedivere*)
LSH 50 HMAS *Tobruk*

'Success' (FNS *Durance*)
AOR 304 HMAS *Success*

'Westralia' (Start 32)
AO 195 HMAS *Westralia* (platform)

Of the 16 Sikorsky S-70B-2s purchased for the RAN, only 12 are operational with HS816 at any one time, the other four being held as attrition spares.

The Kiowas of HC723 comprise a mixture of survivors of the original naval order, such as N17-006/'896', and examples on loan from the Army.

tions are currently being carried out). Following evaluation of the Westland Lynx and Kaman SeaSprite, 11 of the latter type were ordered in 1997.

Delivery of the SH-2G(A)s (remanufactured SH-2Fs) will commence in 2001, and they will be extensively equipped, including Telephonics APS-143 ISAR radar, Elisra integrated electronic warfare suite (with RWR, chaff/flare dispenser, missile warning receiver and FLIR). Perhaps most importantly, the SeaSprite will be armed with the Konsgberg Penguin Mk 2 Mod 7 anti-ship missile, the first such weapon to serve with the Fleet Air Arm. To operate the new type, No. 805 Sqn will stand up prior to 2000.

Although not operated by the FAA itself, brief mention must also be made of the Skyhawks of No. 2 Sqn, RNZAF, and the civilian Learjets that are Nowra-based to provide fleet support training and target towing. Finally, the RAN contracts a civilian company to provide a Fokker F27-500 and Dash 8 (delivered in 1998) for Laser Aerial Depth Sounder (LADS) hydrographic survey work.

THE FUTURE

Almost the entire inventory has been (or shortly will be) upgraded, and although there are no new aircraft types in the pipeline after the delivery of the last SeaSprite, the RAN can be proud of its small aviation arm.

Australian Army Aviation Corps (AAAC)

Army Aviation officially commenced on 1 December 1960 with the formation of 16 Army Light Aircraft Squadron (ALAS), operating the Bell Sioux and Cessna 180. The Corps itself did not come into being until 1 July 1968, and in the meantime Army Aviation had become immersed in the Vietnam War, when No. 161 Independent Recce Flight (later 161 Sqn) was deployed in 1965.

The 1st Divisional Aviation Regiment (later 1 Aviation Regiment) formed from 16 ALAS in 1967, and the Sioux and Cessna 180 were replaced by the Turboporter, Kiowa and Nomad over the next decade.

The battlefield helicopter role transferred from the RAAF in 1989, forcing a rapid expansion to embrace the Blackhawk, Iroquois and (later) Chinook. As a result, 5 Aviation Regiment was formed at Townsville, and apart from the (arguably) premature retirement of the Turboporter in 1992, and withdrawal of the Nomad in 1994, this is the situation today.

Oakey-based 1 Aviation Regiment controls two reconnaissance squadrons of the CAC-built Kiowa, No. 161 (Recce) Sqn at Darwin and No. 162 (Recce) Sqn at Townsville. At Oakey, No. 171 (Operational Support) Sqn operates the battlefield support-tasked UH-1H, and No. 173 (Surveillance) Sqn provides general support with leased civilian Twin Otters and B.200 King Airs.

The newer 5 Aviation Regiment 'numbers' its squadrons alphabetically. A and B Sqns perform troop lift in support of the airmobile 3 Brigade with the S-70A-9 Blackhawk, and C Sqn operates the Chinook for medium lift/airmobile support, and a troop of Bushranger (gunship) Iroquois. The Chinooks are four of the 11 ex-RAAF CH-47Cs converted to CH-47D and redelivered in 1995.

Under the Army's Training Command, the Australian Defence Force Helicopter School at Fairbairn (AS 350BA), and School of Army Aviation at Oakey (Kiowa, Iroquois, Blackhawk), fulfil the AAC's training needs. The ADFHS provides rotary-wing training for all three services, Army students then passing to the SAA for further training.

THE FUTURE

The Kiowa is no longer suited to the reconnaissance role and, with the Iroquois, is due for replacement. Project Air 87 seeks to replace both with a single type. Termed an 'armed reconnaissance' (reconnaissance/aerial fire support) helicopter, candidates include the Bell AH-1Z, Boeing AH-64D, Eurocopter Tigre, Denel Rooivalk and Agusta 129. Selection is expected in 2000. However, in the immediate future, an order for an additional two CH-47Ds has been placed for delivery in 2000.

It would be fair to say that the influx of complex equipment in the late 1980s has presented the AAAC with some difficulties. These problems have now been largely overcome, and Army Aviation is an indispensable tool at the disposal of the ADF.

The CA-32 Kiowa has been in service with the AAAC since 1971 and is deemed due for replacement – the new helicopter type will also replace the Iroquois.

Australian Army Aviation

1 Division

1 Aviation Regiment (Oakey Qld)

Regt. HQ	Bell 206B-1*	Oakey
161 Sqn	Bell 206B-1 *(reconnaissance)*	RAAF Darwin
162 Sqn	Bell 206B-1 *(reconnaissance)*	Townsville/ Laverack Barracks
171 Sqn	UH-1H *(battlefield support, aerial fire support)*	Oakey
173 Sqn	B.200 King Air, DHC-6 Twin Otter *(surveillance)*	Oakey

* RHQ/1 Avn Regt operates one Kiowa, maintained by 171 Sqn

5 Aviation Regiment (RAAF Townsville Qld)

A Sqn	S-70A-9 *(troop lift, airmobile support, training, special forces support)*	Townsville
B Sqn	S-70A-9 *(troop lift, airmobile support)*	Townsville
C Sqn	CH-47D *(medium lift, airmobile support)*	Townsville

Army Training Command

ADFHS	AS 350BA *(basic rotary wing training)* Fairbairn
School of Army Aviation *(operational training)*	Oakey
	Bell 206B-1, S-70A-9, UH-1H
	(also B.200 King Air & DHC-6 Twin Otter as required)

Note: The Army Squadron roles noted here describe the actual tasks undertaken, see text for official Army description

The Army has 35 S-70A-9 Blackhawks, which serve with A and B Squadrons of the 5th Aviation Regiment and the School of Army Aviation.

The helicopter training needs of the Australian defence forces are met by the Army's fleet of 18 AS 350BA Ecureuils.

FIJI

Capital: Suva
Population: 700,000
Land area: 18330 km²
(7,075 sq miles)

Air Wing, Republic of Fiji Military Forces

Founded in 1987, the air arm of the Republic of Fiji was equipped with two helicopters during its relatively brief existence. Fiji consists of 110 inhabited islands upon which there are only 21 airfields, so helicopters were seen as a more useful acquisition than

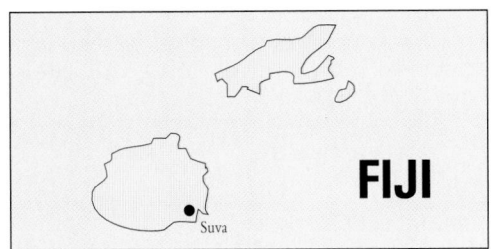
FIJI

fixed-wing aircraft. The Australian and New Zealand governments suspended defence co-operation (mainly relating to army training) following the bloodless

military-led coups of 1987. Fiji then turned to France, which was happy to supply an Aérospatiale AS 355F-2 Twin Squirrel in early 1989, along with 53 army trucks. The aircraft (with the civil registration DQ-FGH) was used in the medevac, VIP and surveillance roles. In May 1991, a Eurocopter (Aérospatiale) SA 365N2 Dauphin 2 was delivered and was operated in the same roles with the registration DQ-FGD. Both aircraft were based at Suva Nausori airport. Military ties with Australia and New Zealand were re-established in 1992, but no further development of the air arm occurred and, by March 1996, both helicopters had been withdrawn and sold, one possibly in a damaged condition.

NEW ZEALAND

Capital: Wellington
Population: 3.9 million
Land area: 265000 km²
(102,000 sq miles)
Major cities: Auckland, Christchurch, Dunedin

With a population of only 3.9 million people, but one of the largest Exclusive Economic Zones in the world and a tradition of involvement in international affairs, New Zealand maintains a small but professional armed forces for the defence of its security interests and those of its South Pacific neighbours. At the core of this is the Royal New Zealand Air Force (RNZAF), which maintains strike, patrol, transport and helicopter elements in support of the New Zealand Defence Force (NZDF) as a whole and New Zealand's wider foreign policy objectives.

Royal New Zealand Air Force

Active military aviation in New Zealand dates back to 1919 when the government took delivery of 29 'Imperial gift' aircraft under the title New Zealand Air Service. The New Zealand Permanent Air Force was established in 1923 under Army control and was equipped with a variety of mostly obsolete biplanes. In 1934 the service was renamed the Royal New Zealand Air Force, and was made an independent service in 1937. At the time of the outbreak of war in September 1939, New Zealand pilots were training in the UK on the first modern combat aircraft in the RNZAF, the Vickers Wellington. These were immediately donated to the RAF together with crews as No. 75 (NZ) Squadron. New Zealand-manned squadrons served in Europe and Africa and RNZAF squadrons in the Pacific and on home defence duties until the end of hostilities. After the war, RNZAF flying units or aircrew served in Japan (1946-48), Malaya (1949-60), Cyprus (1952-55), Malaysia/Borneo (1964-66), Vietnam (aircrew only 1967-71), Sinai (1982-86), Iran (1988-91), Kuwait and Iraq (1990-91), Somalia (1993), Bougainville (1997-98) and the Persian Gulf (1998). Disaster relief work has been undertaken on frequent occasions in the Pacific Islands.

New Zealand is a member of the Five-Power Defence Alliance and ANZUS (Australia-New Zealand-United States) Arrangements, although this is now effectively a pair of bilateral treaties between Australia and New Zealand and Australia and the US because of US displeasure at the Nuclear-Free policy adopted by New Zealand in 1985.

Conversion training to the 14 A-4K Skyhawks is undertaken using the five surviving TA-4Ks. This example is a No. 2 Squadron TA-4K.

The RNZAF's No. 40 Sqn has operated five C-130Hs since 1965 as part of its mixed fleet. Options for eight Js were taken in 1996, later revised to six.

Since 1987, NZ's defence policy has been organised around the notions of self-reliance, co-operation with Australia and defence of South Pacific nations, while acknowledging that in any forseeable circumstance combat air operations would be undertaken as part of a coalition with Australia and/or other partners. New Zealand is also responsible for the defence of Samoa, the Cook Islands, Niue and the New Zealand Pacific Territories (Tokelau).

The RNZAF is organised into three operational elements.

AIR ATTACK FORCE

Consisting of the 17 A-4K/TA-4K Skyhawks of No. 75 Squadron with the 17 Aermacchi MB. 339CBs of No. 14 Squadron potentially available for light attack. The Skyhawks are optimised for close air support, interdiction and (limited) anti-shipping and were upgraded between 1986-90 under Project Kahu (Hawk) with APG-66(NZ) radar, HOTAS flight controls and the capability to use AIM-9L Sidewinder AAMs, AGM-65B Maverick AGMs and GBU series LGBs.

LONG RANGE MARITIME PATROL FORCE

Six P-3K Orions of No. 5 Squadron are used for surveillance and reconnaissance of New Zealand's EEZ and the South Pacific Islands. Subsidiary roles of the P-3s include ASW, SAR and medevac of patients

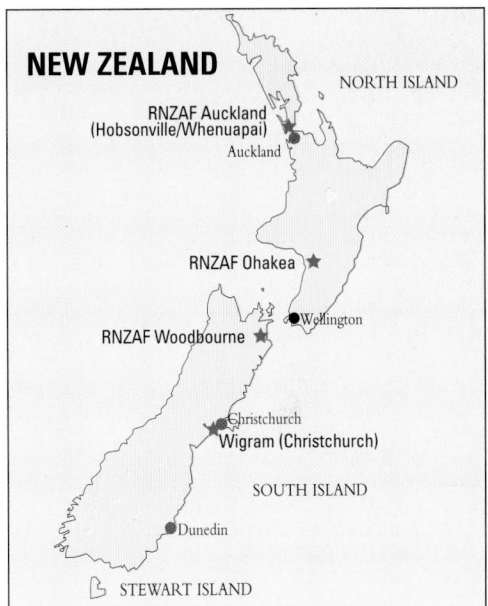
NEW ZEALAND
NORTH ISLAND
RNZAF Auckland (Hobsonville/Whenuapai)
Auckland
RNZAF Ohakea
Wellington
RNZAF Woodbourne
Christchurch
Wigram (Christchurch)
SOUTH ISLAND
Dunedin
STEWART ISLAND

needing specialist care from Pacific Islands. P-3s act as navigation escort for A-4 deployments to Australia, the Pacific and Southeast Asia.

AIR TRANSPORT FORCE

For the long-distance movement of NZDF elements, tactical transport support for operations in NZ and the Pacific, support of embassies and peacekeeping operations abroad and VIP transport, the RNZAF utilises a mixed fleet of five C-130Hs and two Boeing 727s flown by No. 40 Squadron. The air transport force includes the Bell UH-1Hs of No. 3 Squadron, which also act in the SAR role and in the disaster relief role both in New Zealand and the Pacific. The transport force supports New Zealand's scientific programme in Antarctica with the annual 'Ice Cube' deployments of Bell UH-1s and Lockheed C-130s.

Aircrew training is concentrated at Ohakea with the 17 CT/4B Airtrainers of the Pilot Training Squadron and the 17 MB.339 'Macchis' of No. 14

Above: The wide variety of operating locations demanded of the UH-1Hs of No. 3 Sqn includes the Southern Alps range.

Left: Advanced training and light strike roles are undertaken by the Aermacchi MB.339CB, which replaced the BAC Strikemaster Mk 88 in No. 14 Sqn.

Squadron for jet conversion. The CT-4s and their instructor pilots form the 'Red Checkers' aerobatic team. Pacific Aerospace Corporation is in the process of buying back the CT-4Bs and replacing them with 13 CT-4E versions. Multi-engine training is now undertaken by No. 42 Squadron at Whenuapai with the three Beech 200 King Airs leased from Pacific Aerospace, while helicopter training (for both Air Force and Navy crews) is done by No. 3 Squadron on the Bell 47 and UH-1.

The effects of an economic downturn and the ending of the Cold War led to a reduction in the number of types operated by the RNZAF in the early 1990s. The Cessna 421C, Victa Airtourer, Fokker F27 and Andover have all gone without having been replaced by new types. The Scheduled Air Transport Service (SATS) between the main bases and Wellington is now fulfilled using C-130s and the 727s.

Currently, the RNZAF is concentrated around the North Island bases of Whenuapai/Hobsonville (Base Auckland) and Ohakea. The Treasury would like Defence land assets reduced further and is pushing for the sale of Base Auckland and the relocation of its units to Ohakea. The airspace there is already becoming very crowded and there have been suggestions that the Pilot Training Squadron might move to Woodbourne, which is currently without any based flying units. No. 2 Squadron is based at RANAS Nowra in New South Wales, Australia, where its primary function is mutual training with units of the Royal Australian Navy. This arrangement has recently been extended for another five years. Other roles undertaken by No. 2 Squadron include pilot conversion and trials.

Upgrading of the fleet of six P-3K Orions is proceeding under projects Kestrel and Sirius. The former involves the replacement of outer wings and horizontal stabilisers by Hawker Pacific Pty at Richmond, Australia and the latter is the upgrading of the aircraft's avionics fit to current standards. Under project Delphi, the C-130H aircraft will be equipped with defensive countermeasures and cockpit armour. Previous proposals to acquire tanker packages for the C-130s appear to have fallen by the wayside. For the first time, the use of outside contractors and leasing has made a significant impact on RNZAF operations, with the new CT-4E Airtrainers and Beech King Airs both being leased from and maintained by Pacific Aerospace Corporation at Hamilton.

There have only ever been two helicopter types used exclusively by the RNZAF, the Bell UH-1D/H Iroquois and the Bell 47G Sioux. Both continue to operate from Hobsonville, Auckland, with the survivors of the latter in the training role. The UH-1s are undergoing communications upgrades and minor airframe modifications to assist 'hot-and-high' performance. This had been useful during Operation Belisi in Bougainville where three aircraft (initially assisted by RNZN Wasps) had been involved in transporting personnel and equipment of the TMG (Truce

Monitoring Group), maintaining the truce in the civil war negotiated at Burnham, New Zealand in 1997 and the permanent ceasefire (Lincoln Agreement) signed in January 1998. New Zealand Defence Force personnel returned home after the 30 April 1998 ceasefire, having been replaced by an Australian-led Peace Monitoring Group (PMG).

The 1997 Defence Review re-emphasised the current roles of the RNZAF and called for an increase in the capital equipment budget for the air force. The document upheld the need for the air attack force, regarding a force of 18 combat aircraft as the 'critical mass' in time of crisis or conflict to allow for 10 aircraft deployed, six for training, and two for maintenance. This is without allowing for losses. A December 1998 announcment proclaimed the lease-purchase of the 13 F-16A-15 OCUs and 15 F-16B-15 OCUs of the embargoed Pakistani order currently stored at AMARC Davis-Monthan AFB, Arizona, which are due for delivery in 2002. Funds for a Kahu II upgrade, including a new anti-ship missile, have been allocated to the F-16 although there are no firm plans as to the nature of an F-16 upgrade. Initial plans to acquire up to eight C-130Js to replace all the RNZAF's transport fleet have been revised with an option held with Lockheed-Martin (set to expire in 2000) for six new-generation Hercules and replacement of the B727s by 737-300s or similar early in the next century.

New Zealand can be said to get good value for its defence purchases. The average time since service entry of air force aircraft types is 23 years, the oldest (Sioux and Hercules) being 34 years old in 1999. The 1997 Defence Review has, to a small degree, reversed the trend of cuts to the RNZAF budget, although a replacement en masse of the ageing types in service – in the same way that 1965-70 and 1976-77 saw large-scale re-equipment of units – seems unlikely, with various upgrade programmes supplemented by limited aircraft purchases being the policy for the forseeable future.

The close co-operation between New Zealand and Australia sees P-3Ks of No. 5 Sqn frequently working with Australian warships. Here NZ4203 flies over HMAS Adelaide (pennant no. 01).

Royal New Zealand Air Force

New Zealand Defence Force Headquarters, Wellington
RNZAF Air Staff Defence HQ

AIR COMMAND, RNZAF AUCKLAND

RNZAF Auckland (Hobsonville/Whenuapai)

Operations Wing

No. 3 Sqn	UH-1H, Bell 47G-3B2, SH-2F
No. 5 Sqn	P-3K
No. 40 Sqn	727-100C, C-130H
No. 42 Sqn	King Air B 200

Notes: No. 3 Sqn based at Whenuapai operates SH-2F on behalf of RNZN and maintains a SAR detachment of UH-1Hs at Wigram (Christchurch). Nos 5, 40 and 42 Sqns at Whenuapai

Support Units
Aviation Medicine Unit
Parachute Training and Support Unit
RNZAF Air Command HQ
Staff College

RNZAF Ohakea

Flying Wing

No. 2 Sqn	A-4K, TA-4K
No. 14 Sqn	MB.339CB
No. 75 Sqn	A-4K, TA-4K

Note: No. 2 Sqn flies from RANAS Nowra, Australia, on various duties including Skyhawk conversion, trials and fleet support for RAN

Central Flying School

Pilot Training Squadron	CT-4E Airtrainer
Historic Flight	Avro 626 (u/s), DH 82A Tiger Moth, Harvard Mk III

RNZAF Woodbourne

Ground Training Wing
(Recruit, Technical, Administrative and Command Training)
No. 1 Repair Depot

The CT-4B Airtrainer, colloquially referred to as 'The Plastic Rat', is in the process of being replaced by the upgraded CT-4E. This aircraft, painted in the recently introduced yellow/black scheme, wears the nose-band of the 'Red Checkers' display team.

Royal New Zealand Navy

The Royal New Zealand Navy maintains a small fleet of SH-2F Seasprites for service aboard its fleet and for training. The Westland Wasps, in service since 1966, were retired in 1998, and the F-model Seasprites will be replaced by four new-build SH-2G Super Seasprites (with a fifth on option) in 2000. The aircraft are crewed by RNZN crews, but are operated as the Naval Support Flight of No. 3 Squadron, RNZAF at Whenuapai. The G models will be equipped with AGM-65 Mavericks and a more advanced avionics fit, but not the 'glass' cockpits of the Australian SH-2Gs.

The Royal New Zealand Navy has recently exchanged its Wasps for SH-2F Seasprites.

PAPUA NEW GUINEA

Capital: Port Moresby
Population: 3.7 million
Land area: 462840 km² (178,655 sq miles)
Major city: Rabaul

Papua New Guinea Defence Force

AIR TRANSPORT SQUADRON, PNGDF

From a peak strength in the late 1980s of six C-47s, six or seven GAF Nomads, including radar-equipped Searchmasters, and three IAI Arava 201s, the air component of the PNGDF has declined markedly, due in large part to the drain on resources caused by the long-runing seccesionist rebellion on the island of Bougainville. This conflict saw the first use of helicopters by the PNGDF, with first four and then a further two ex-RAAF UH-1Hs being donated by Australia in 1989/90. Mostly flown by Australian pilots, the aircraft were involved in some controversial actions involving attacks on villages and the deaths of captured Bougainville Revolutionary Army (BRA) members. Only two are now believed to be in service, the remainder stored due to lack of funds. An ill-advised attempt by then-Prime Minister Sir Julius Chan to tip the military balance in Bougainville by hiring the services of British mercenary firm Sandline International failed when arms shipments destined for the government forces on Bougainville were intercepted. These included two Mi-24 'Hinds' and two Mi-17 'Hips' impounded by Australia when the An-124 that was carrying them was forced to land at RAAF Tindal by No. 77 Squadron F/A-18 Hornets

in March 1997. Another pair of Mi-24s had arrived at Port Moresby in early February, for delivery to the army special forces based at Wewak. These were used briefly before Sandline was ejected following the revelation of mercenary involvement, a virtual army rebellion against the Prime Minister and violent street demonstrations, culminating in Chan's resignation.

Following the election of a new government in June 1997, talks were conducted at Burnham army camp, and later Lincoln University in New Zealand, culminating in the Lincoln Agreement which has brought about the cessation of hostilities on Bougainville. Military aviation there has since been restricted to truce monitoring flights by RNZAF/RNZN and RAAF helicopters, all painted orange to distinguish them from armed aircraft used in the civil war (and known as 'Orange Roughies').

The Air Transport Squadron formerly had an 'A' Flight (Aravas) and a 'B' Flight (Nomads), but now consists of an indeterminate number of operational Nomads and CN.235s. All aircraft may be grounded due to lack of funds. The C-47Bs were sold on to the Australian civil market in late 1992. A single King Air 350 (P2-PNG) is operated by the Government Flying Unit.

Papua New Guinea Defence Force

Base: Jackson International Airport, Port Moresby
Headquarters: Boroko

IPTN/CASA CN.235M-100	2 new delivered 1992. Replaced C-47s and Aravas.
Bell UH-1H	6 ex-RAAF
GAF N22 Nomad	6 delivered, 2 believed w/o
Beech King Air 350	1 (P2-PNG) VIP and charter use
Mil Mi-17	4? reportedly delivered 1990
Mil Mi-24	2 briefly used by Sandline, 2? others impounded in Australia
CASA 212-200 Aviocar	reportedly used by Sandline

PAPUA NEW GUINEA

Port Moresby

The Papua New Guinea Defence Force has used six Hueys since 1989. This UH-1H Huey, P2-0403 (c/n 9510), is ex-A2-510 of the RAAF.

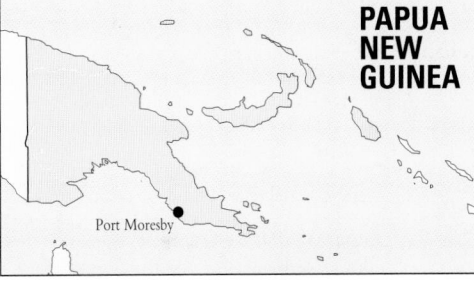

At least two of the N22 Nomads have been written off in accidents. A total of four more could be operational, but this is doubtful at the present time.

 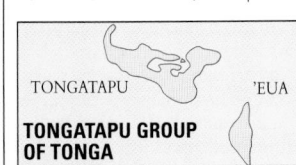

Capital: Nuku'alofa
Population: 100,000
Land area: 747 km² (288 sq miles)

TONGATAPU   'EUA
TONGATAPU GROUP OF TONGA

TONGA

Tongan Defence Services – Air Wing

The Kingdom of Tonga maintains a small defence service for the maintenance of public order, coastal

and fisheries patrol, and civil defence. In addition to the Land Force and Maritime Force, the Tonga Defence Force has an Air Wing consisting of a single aircraft, making it the smallest active air arm in the world. The Air Wing's first aircraft, a Victa Airtourer four-seat lightplane, was used for liaison, observation and training from 1986. It was reported in storage by 1996. Despite 1987 reports of the purchase of a number of IAI Aravas, new equipment did not arrive until 1996 when the TDS Air Wing took delivery of a 1960 (possibly 1961 or 1962) Beech G.18S.

The aircraft was acquired for SAR, patrol and surveillance of the nation's important fisheries resource, roles usually undertaken by RNZAF P-3s. The Beech, formerly N9644R with an Alaskan owner, was purchased for $NZ90,000 and was outfitted in Seattle with $NZ250,000 worth of avionics including Bendix King radar and Garmin GPS. It arrived in Tonga on 4 May 1996, the birthday of defence minister Crown Prince Tupouto'a, whose

interest in 'classical historical items' led to the choice of the Beech 18. The Air Wing hopes to eventually have three aircraft. Maintenance is undertaken by DH Beaver operator Pacific Island Seaplanes.

1 x Beech G-18S	AW-01	ex-N9644R c/n BA-483

Tonga's Air Wing only operates one aircraft – this Beech 18 – although there are plans to acquire at least two more aircraft.

Europe

Europe can be divided into members of the North Atlantic Treaty Organisation (NATO) and of the former Warsaw Pact, former Soviet Republics, and those that can not be classified, such as neutral countries. For over 40 years the majority of the countries in Europe had been preparing for an all-out East-West confrontation, so the end of this prospect following the break-up of the Soviet Union has caused significant changes in the air forces of the countries of Europe. NATO countries (Belgium, Denmark, France, Germany, Greece, Iceland, Italy, Luxembourg, the Netherlands, Norway, Portugal, Spain, Turkey and the United Kingdom) have reduced military expenditure. Many armed forces have undergone drastic restructuring, and new systems – such the French Rafale, multinational Eurofighter EF2000 Typhoon, Franco-German Tiger attack helicopter and the NH 90 multi-role helicopter – have had longer gestation periods than anticipated as the threat has receded. European manufacturers such as France's Aérospatiale and Dassault, Germany's DASA, Italy's Alenia, Spain's CASA, the United Kingdom's British Aerospace, and multinationals such as the Franco-German Eurocopter and Airbus consortia are major producers of aerospace materials. Many of the European NATO countries are also members of the European Union, an organisation which has sought to bring sovereign countries closer together; the establishment of a single currency is a primary aim.

The former Warsaw Pact countries are re-establishing a national identity after living in the shadow of Soviet domination since the end of World War II. Free elections and self-determination have not brought instant solutions to years of stagnation, and many of the former Warsaw Pact countries are experiencing economic and social problems. Czechoslovakia decided to revert to being the Czech and Slovak Republics, in a peaceful dissolution. Like their former adversaries in NATO, the air forces of Bulgaria, the Czech Republic, Hungary, Poland, Romania and the Slovak Republic have undergone considerable change in the last 10 years. The dependence on Russian aircraft has gone and many of the countries are considering acquiring Western defence technology. Perhaps the event most unthinkable a decade ago is that Poland, Hungary and the Czech Republic are due to join NATO during 1999.

The break-up of the Soviet Union had a profound effect on Europe, signalling the end of the Cold War. As the political control that held the union together disintegrated, the Russian Republics began a campaign of independence. Armenia, Azerbaijan, Belarus, Georgia, Moldova and Ukraine all declared independence from Moscow, led by the Baltic states of Estonia, Latvia and Lithuania. Most inherited what remained of the assets of the Soviet forces based on their territories.

Russia's vast size means it is a political and military concern not only within Europe, but for most of the nations of the world. Geographically, the majority of the country is in Asia, but it has had its greatest effect on the countries of Europe. The end of Communist control in Russia and the advent of reform has been a period of immense turmoil for the Russian military. Military budgets were slashed, advanced programmes were cancelled or reduced in scope, personnel have gone unpaid for long periods and the general operational ability of the forces has declined drastically. Military restructuring has been undertaken, and many units have disbanded.

Switzerland and Sweden armed themselves to maintain national sovereignty without joining either side during the Cold War, while Austria and Finland maintained small forces constrained by treaties signed soon after World War II. Yugoslavia disintegrated in the wars which started in 1991 into the Republics of Bosnia-Herzegovina, Croatia, Macedonia, Serbia and Slovenia. This process is still continuing as various ethnic groups aspire to independence. Albania has recently emerged from years of isolation after pursuing an independent Communist ideology.

The Mediterranean islands of Malta and Cyprus have small armed forces, and the potential for conflict between Cyprus's divided population is ever-present.

ALBANIA

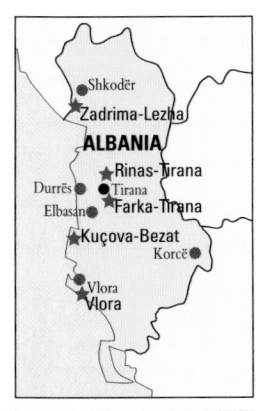

Capital: Tirana
Population: 3.3 million
Land area: 28750 km²
(11,100 sq miles)
Major cities: Durrës, Vlorë, Korcë, Shkodër, Elbasan

Forcat Ashtarake Ajore Shgipetare
(Albanian Air Force)

The Albanian Air Force (Forcat Ashtarake Ajore Shgipetare) was created at Tirana Aerodrome on 24 April 1951. The new air force's first aircraft were 11 Yak-9Ps and a single Yak-9V, which were transferred by the Soviets that same year. The pilots were Albanians who had been trained on Yak-3s by the Yugoslavian air force. Also during 1951, four Polikarpov Po-2s were supplied by the Soviets for crop-spraying. The first pilot training course in Albania was made possible by the delivery of four ex-Soviet, early type Yak-18 basic trainers and four Yak-11 advanced trainers. To give student pilots practice in flying nosewheel-equipped aircraft such as the MiG-15bis/UTI and Il-14 transport, six tricycle-undercarriage Yak-18As were delivered in 1953. The jet age for the Albanian air force began on 31 January 1955, with the delivery of 24 MiG-15bis and four MiG-15UTIs. In 1956 some Chinese-built F-2s, four Czech- and four Chinese-built MiG-15UTIs/FT-2s were acquired. The Yaks and MiGs were by this time based at Albania's oldest airfield, Tirana Aerodrome, which had been built by the Italians in 1939.

The first twin jet bomber, an Ilyushin Il-28, arrived in 1957 from the Soviet Union. In service it was used as a bomber and as a target-tug with the anti-aircraft units. The aircraft was exchanged with the Chinese in

Above: Basic training is undertaken on Nanchang CJ-6s at Vlora. Powered by the Ivchenko AI-14R, the type has been in service for over 35 years.

Above: Europe's (and probably the world's) last operational military Ilyushin Il-14s are flown, along with the last Harbin H-5, by 4020 Regiment.

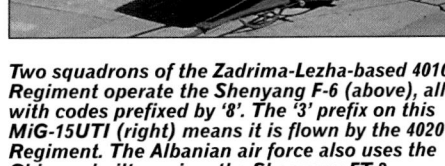

Two squadrons of the Zadrima-Lezha-based 4010 Regiment operate the Shenyang F-6 (above), all with codes prefixed by '8'. The '3' prefix on this MiG-15UTI (right) means it is flown by the 4020 Regiment. The Albanian air force also uses the Chinese-built version, the Shenyang FT-2.

1971 for an H-5, which is still used as a target-tug. The first transport type, the Ilyushin Il-14, arrived the same year as the Il-28, also from the Soviets. Three similar aircraft were delivered later, after being overhauled in China: two from the German Democratic Republic (East Germany) and one from Czechoslovakia. All are still operational today with squadrons at Rinas.

Rotary-wing operations also commenced in 1957, with the delivery of three Mil Mi-1s and four Mi-4As from the Soviets. These helicopters were used in transport, search and rescue and air observation duties. During the winter, especially, they were used to supply food and other goods to remote farms in mountainous areas and to transport sick people to hospitals.

Twelve more Yak-18As were delivered in 1959, to facilitate the training of the additional pilots who were required in expectation of the delivery of MiG-17/19s three years later. The first limited all-weather capability was acquired with the delivery of 12 ex-Soviet MiG-19PMs in October 1959. They were in service until 1965 and were exchanged for Chinese-built F-6s (MiG-19S). It is most likely that the Chinese were after the aircraft's radar system. Conversion from the MiG-15UTI to MiG-19 proved to be very difficult, so in 1961 eight FT-5s (MiG-17UTI) were bought from the Chinese. One year later the Albanians also bought a squadron of 12 F-5s, 11 of which are still operational at Kuçova.

With the foundation of an Air Academy at Vlora on 11 May 1962, 20 Nanchang Type 61 (CJ-6) trainers were acquired. Flight training for the cadets entailed one year's flying on the CJ-6 and two years on the MiG-15UTI/bis before an airman got his wings.

Thirteen Chinese Y-5s (Antonov An-2 copies) were delivered in 1963/64 and have been used as crop sprayers, as well as utility transports. Late in the 1960s, and in 1971, more F-6s from China were delivered, bringing the total to at least 70 aircraft. Except for the 12 MiG-19PMs, the Soviets had also delivered a few MiG-19Fs. As Albania broke totally with the Soviet Union in the 1960s, the country turned to the Chinese Communists for military hardware.

In 1967 the helicopter capacity was boosted by the delivery of 30 Harbin Z-5s (24 Mil Mi-4A and six Mi-4S sub-types), which are still in service. Their pilots have expressed a desire to replace the type with the more modern Bell 412 or Aérospatiale Puma.

The most 'modern' aircraft in the inventory are the 10 (from 12 delivered) F-7s – Chinese-built MiG-12F-13s – delivered in November 1970. Their pilots are the only ones in the air force to

Forcat Ashtarake Ajore Shgipetare

4004 Regiment/Air Academy		
1 Sqn	Nanchang CJ-6	Vlora
2 Sqn	MiG-15bis, Shenyang F-2/FT-2 CS 103	Kuçova-Bezat
4010 Regiment		
1 Sqn	Chengdu F-7A Shenyang FT-5	Zadrima-Lezha
2 Sqn	Shenyang F-6	Zadrima-Lezha
3 Sqn	Shenyang F-6	Zadrima-Lezha
4020 Regiment		
1 Sqn	Shenyang F-6 Shenyang FT-2, Il-14	Rinas-Tirana
2 Sqn	Shenyang F-6 Shenyang FT-2, Il-14, Harbin H-5	Rinas-Tirana
4030 Regiment		
1 Sqn	Shenyang F-6 Shenyang FT-2	Kuçova-Bezat
2 Sqn	Shenyang F-5/FT-5, Shenyang FT-2	Kuçova-Bezat
4040 Regiment		
1 Sqn	Mil Mi-4S, Alouette III	Farka-Tirana
2 Sqn	Harbin Z-5	Farka-Tirana
3 Sqn	Harbin Z-5	Farka-Tirana
4050 Regiment		
1 Sqn	Yunshuli Y-5/Antonov An-2 Nanchang CJ-6, AS 350 Ecureuil, Bell 222UT, Mil Mi-8, Alouette III	Tirana

wear plastic helmets, all others still using World War II-style leather helmets.

At the beginning of the 1990s the Albanian government acquired a single Bell 222 and three Ecureuils, which are used for VIP transport, search and rescue, and police operations. Recent acquisitions are some Alouette IIIs and Mi-8s, based at Tirana Aerodrome, used for the same tasks.

Because of the current economic situation and an almost total breakdown of law and order, with a resultant low tax income for the government, the purchase of new and modern aircraft in the future is unlikely.

Helicopter transport is provided by the Mi-4S/ Harbin Z-5s of the Farka-Tirana based 4040 Regiment. The regiment also uses Alouette IIIs which are thought to have come from Switzerland.

ARMENIA

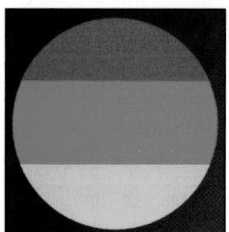

Capital: Yerevan   **Population:** 3.3 million
Land area: 30000 km² (11,580 sq miles)
Major cities: Gymuri, Kirovakan

Republic of Armenia Air Component

Located between the eastern borders of Georgia and Turkey, Armenia has experienced major economic problems which have limited its military expansion plans, following armed clashes with Azerbaijan in 1993-94 over the disputed border areas of Nagorno Karabakh. Earlier, in December 1992, its Conventional Forces in Europe air strength was listed as only three combat aircraft and 13 armed Mil Mi-8/17 and Mi-24 helicopters, although more than half the latter types were subsequently claimed shot-down by Azeri ground fire.

Since then, however, arms worth around $1 billion have been airlifted by Russia to Armenia without charge. Apart from eight 'Scud-B' SSM launchers and 24 missiles, these mainly comprised defensive and infantry weapons, including 26 SA-4 'Ganef' SAM launchers and 349 missiles, 40 SA-8A 'Gecko' SAMs, and 40 SA-18 'Grouse' infantry SAM launchers, plus 200 missiles. No aircraft were reported in these arms transfers, although Armenia is allowed up to 100 combat types and 50 attack helicopters under CFE Treaty limits.

Current Armenian military aviation equipment is believed to be limited to about a dozen Mil Mi-8/17 and Mi-24 helicopters, plus a few fixed-wing aircraft, possibly including three Sukhoi Su-25 'Frogfoots' and one or two Antonov An-2s, for army support.

AUSTRIA

Capital: Vienna
Population: 7.8 million
Land area: 83855 km² (32,370 sq miles)
Major cities: Graz, Innsbruck, Linz, Salzburg

Österreichische Luftstreitkräfte

The 1955 State Treaty with the four World War II Allies replaced the occupational forces with the Army of the 2nd Austrian Republic, and fixed Austria's position during the Cold War period as neutral. The government still follows this policy, which helps to explain why Austria is Saab's most important foreign customer, with 119 airframes ordered.

The Air Arm (later Fliegerdivision) was founded on 18 October 1955 as an integral part of the Bundesheer (Federal Army), to which it still belongs. First equipment was four Yak-11s, four Yak-18s and one Bell 47G-2, all donated by the departing occupation troops of the Allied powers. More capable aircraft were sought following the crushed uprising in neighbouring Hungary in 1956, and eight de Havilland DH.115 Vampires were delivered from March 1957, giving the first jet experience to the young force. During the next years, the so called 'butterfly-collection' was procured, consisting of 10 LT-6G Texans, 29 Cessna L-19s, six DHC-2/L-20 Beavers and 18 Fouga Magister, plus 16 Alouette II and 10 Westland S-55 helicopters. More aircraft were added

in 1960 and 1967 – 30 used Saab J 29F and 40 new Saab 105Ös, respectively – but no political decision was reached to purchase a supersonic interceptor.

In the 1970s the variety of piston types was replaced with 13 Pilatus PC-6 Turbo-Porters and 16 PC-7 Turbo-Trainers, two Short Skyvans and 26 AB 204B, 13 AB 206, 24 AB 212, 12 OH-58, 28 Alouette III and two Sikorsky S-65 helicopters. In 1986 the new coalition government introduced a true interceptor after ordering 24 used Saab J 35Ö Drakens (upgraded D models with F-model canopies), and the Fliegerdivision finally made the transition to the supersonic era.

Serious border violations by JRV jets during the fight for Slovenian independence in 1991 prompted the purchase of AIM-9P-3s for the Drakens, from 1993 onwards. Following the retirement of the Danish Drakens, RWRs and chaff/flare dispensers were added from Danish stocks and – assisted by VALMET of Finland – integrated into the rear fintip and the afterburner section of the Austrian aircraft. For fatigue-prevention purposes, as well as in response to the reduced threat, weaponry availability, ageing sensors and cameras, the Saab 105s ceased carrying all underwing stores in 1996, and today they are used only for jet-conversion training.

Responding to the changes in Europe, a major restructuring of Austria's army began in 1992. The air arm lost 20 per cent of its personnel and 25 per cent of the administration. All the structures at wing level (except in the 2nd Regiment) were axed and all the A/A units – previously under regional corps command – were fully integrated (with the exception of the Mistrals of the mechanised brigades). Currently, the Kommando Luftstreitkräfte consists of the command which reports to the air department of the Austrian MoD, the air communication battalion, three active flying regiments (Fliegerregimenter), three air defence regiments (each with one battery of Mistral SAMs and three or four of AAA), the air surveillance command (operating the fixed

Above: A total of 24 Saab J 35Ö Drakens was delivered from May 1988, and operated without air-to-air missiles until Sidewinders arrived in 1993.

Below: The OH-58B Kiowa can be armed with a Minigun for infantry support missions. All serve with 3. Hubschrauberstaffel of Fliegerregiment 3.

Goldhaube radar surveillance system and several mobile radar-sets, along with the underground combat centre), the air reconnaissance command (waiting for the purchase of UAVs), the Flying School (Fliegerschule), the maintenance yards and the logistical centre of aviation technology.

A total of eight bases and airfields is operated. The only military-dedicated ones are Zeltweg (fighters) in the upper Styrian mountains, Langenlebarn

Glossary
Düsenstaffel	*Jet Squadron*
Flächenstaffel	*Winged Squadron*
Fliegerdivision	*Air Division*
Fliegerregiment	*Aviation Regiment*
Fliegerschule	*Flying School*
Hubschrauberstaffel	*Helicopter Squadron*
Luftabteilung	*Air Department*
Überwachungsgeschwader	*Surveillance Wing*
Übungsstaffel	*Training Squadron*

Above: The Saab 105OEs are used by the Austrian air force for jet conversion training, having given up their more war-like duties in 1996. This pair belongs to the 3 (Düsen-) Staffel of Fliegerregiment 3.

Below: Pilot training is undertaken on the PC-7s of the Übungsstaffel. All 16 acquired in 1983 remain in service, based at Zeltweg.

(helicopters and light fixed-wing aircraft) north of the capital Vienna and its satellite at Wiener Neustadt, and the helicopter bases at Schwaz/Tirol and Aigen/Ennstal. The others – Linz-Hörsching (jet trainers and helicopters), Graz-Thalerhof (fighters) and Klagenfurt (helicopters) – are dual-use airfields.

Austria has mostly alpine terrain, so its mountain flying capabilities are well respected and the alpine training camps for helicopter crews are regularly joined by foreign army pilots. Additionally, the army helicopters are used for civil purposes such as SAR, environmental, archeological and survey work. Public opinion and media sentiment have preferred rotary-winged units to fixed-wing ones because of their potential for 'humanitarian' operations. The helicopter units have highlighted the need for a new multi-role helicopter, but replacements have yet to be ordered.

With 168 airframes and a peacetime cadre of approximately 3,700, the Fliegerdivision is one of the smallest air arms in Europe. It is strictly defence-oriented, having no offensive strike ordnance for its aircraft, and the only armed helicopters are the Minigun-equipped OH-58s. Despite Austria's long-term UN contributions and rising participation in international missions abroad, a significant fixed-wing transport capability is totally missing. The Drakens were planned to have been replaced by a follow-on type from 1996, so the F-16, F/A-18, Mirage 2000-5, JAS 39 Gripen and MiG-29SE are under evaluation. A public discussion about the necessity of having a new fighter has since started, but Austria – although it is a Partnership for Peace member – has still not decided its future security arrangements, and the political will or ability to find the funding required seems not to be available.

Österreichische Luftstreitkräfte

Luftabteilung (within Ministry of Defence)	Vienna
Command of Fliegerdivision	Tulln-Langenlebarn

Fliegerregiment 1	**Tulln-Langenlebarn**
1. Hubschrauberstaffel	Agusta-Bell AB 212
2. Hubschrauberstaffel	Agusta-Bell AB 206A Jet Ranger
3. Hubschrauberstaffel	Bell OH-58B Kiowa
4. (Flächen-) Staffel 'Elise'	Pilatus PC-6/B2H2 Turbo Porter, Short 3M Skyvan
(occasional detachments for parachute/special forces support at Wiener Neustadt-West)	

Fliegerregiment 2		**Zeltweg**
Überwachungsgeschwader		**Zeltweg**
1. Staffel	Saab J 35Ö Draken	Zeltweg
2. Staffel	Saab J 35Ö Draken	Graz-Thalerhof
(Saab 105Ös are borrowed by both squadrons from FlR 3 for training and communication)		
Hubschraubergeschwader		**Aigen in Ennstal**
1. Hubschrauberstaffel	SA 316B Alouette III	
2. Hubschrauberstaffel	SA 316B Alouette III	
(regular detachments of one Alouette to Klagenfurt and Schwaz in Tirol)		

Fliegerregiment 3	**Linz-Hörsching**
1. Hubschrauberstaffel	Agusta-Bell AB 212
2. Hubschrauberstaffel	Agusta-Bell AB 204B
3. (Düsen-) Staffel	Saab 105Ö
(former JaBo Geschwader)	

Fliegerschule	**Zeltweg**
Übungsstaffel	Pilatus PC-7 Turbo Trainer

AZERBAIJAN

Capital: Baku
Population: 7 million
Land area: 87000 km² (33,580 sq miles)
Major cities: Gyandzha, Mingechaur, Nakhickevan, Sumgait

Azeri Air Forces & Air Defence Forces

Flanked by Georgia and Iran on the Caspian Sea, the mainly Muslim ex-Soviet state of Azerbaijan is much larger than neighbouring Armenia, with which an uneasy ceasefire has been maintained since May 1994, following hostilities in the disputed Nagorno Karabakh border region. Although potentially the world's third largest oil supplier, Azerbaijan's economic development is still restricted by international investment limits pending settlement of its border problems.

Military aviation development has been similarly limited, since the 1991-92 Soviet withdrawal of most of the locally-based 124 combat aircraft and 24 armed helicopters. Only five unserviceable MiG-25RB 'Foxbat-B' fighter/reconnaissance and 10 Sukhoi Su-24 'Fencer' strike aircraft, plus nine armed Mil

Mi-8 'Hip' and Mi-24 'Hind' helicopters, were left behind. All were later claimed to have been made airworthy, although they are in poor condition, as are other Azeri aircraft, after experiencing maintenance and spares problems. Designed to fulfil mainly strategic roles, the technically complex MiG-25s and Su-24s are also unsuited to Azeri requirements for mainly tactical and counter-insurgency roles, and additionally must operate on limited military budgets and with unskilled manpower.

In border clashes with Armenian forces, Azerbaijan captured six SA-2 'Guideline' SAM launch units on 20 March 1992, and sought surplus ex-Soviet combat aircraft and pilots from elsewhere in the CIS. Its air force establishment then reportedly increased to about 50 aircraft, including a few MiG-21 'Fishbeds', plus about four Mi-24 attack helicopters and more Mi-8s. By late 1992, however, at least a dozen Azeri helicopters – mostly Mi-8s, but including two Mi-24s –

AZERBAIJAN

plus two Sukhoi Su-25 'Frogfoots' and a MiG-21 – were shot down by ground-fire over Nagorno Karabakh and Armenia.

By August 1995, transfers to Azerbaijan of former Soviet CFE-limited equipment were planned to coincide with Conventional Forces in Europe Treaty ceilings of 100 combat aircraft and 50 armed helicopters, in addition to notified procurement of 10 unspecified combat aircraft and other arms from Ukraine in 1993-95. Only the latter appear to have been delivered, although, in early 1997, Armenia accused Azerbaijan of covertly acquiring 12 MiG-21s, four Sukhoi Su-15 'Flagon' interceptors and eight Su-24 and Su-25 attack aircraft, plus KAB-500L laser-guided bombs, and Kh-25ML 'Kegler' and Kh-29L 'Kedge' anti-radar missiles from CIS sources.

Azeri Air Forces & Air Defence Forces

Type	Role	In Service
Aero L-29 Delfin	basic trainer	?
Aero L-39C Albatros	advanced trainer	?
Antonov An-2 'Colt'	utility transport	?
Antonov An-26 'Curl'	tactical transport	?
MiG-21M 'Fishbed'	interceptor	18
MiG-25RB/PD 'Foxbat'	recce/air defence	5
Mil Mi-8/17 'Hip'	utility helicopter	10
Mi-24D 'Hind-D'	attack helicopter	4
Sukhoi Su-15 'Flagon'	air defence	4
Sukhoi Su-24M 'Fencer'	ground attack	12
Sukhoi Su-25/UB 'Frogfoot'	ground attack/trainer	6

BELARUS

Capital: Minsk
Population: 10.2 million
Land area: 208000 km² (80,290 sq miles)
Major cities: Homel, Hrodno, Mahilyov, Vitebsk

Voyenno-Vozdushnyye Sily (Military Air Forces)

Following the dissolution of the Soviet Union, Byelorussia (White Russia) was declared an independent state, the Republic of Belarus, on 26 July 1991. When still part of the Soviet Union, Byelorussia constituted the Byelorussian Military District (BoMD), one of the most heavily armed USSR military districts, with headquarters in Minsk.

The aviation assets based in the BoMD comprised units of the 26th Air Army (Frontal Aviation) headquartered in Minsk, the 24th and 46th Strategic Air Armies (headquartered in Vinnitsa, Ukraine, and Smolensk, Russia, respectively), the 2nd and Moscow Air Defence Armies (PVO, headquartered in Minsk and Moscow, respectively), and Military Transport Aviation (VTA). The army aviation assets were subordinated to two tank armies and one all-arms army.

The Byelorussian Military District was deactivated in May 1992. All units stationed on Byelorussian territory which were not subordinated to the Soviet strategic forces were placed under the command of the Ministry of Defence of the Republic of Belarus. Since the formation of the Air Force of the Republic of Belarus, the assets of the former Frontal Aviation Army and Air Defence Army have been integrated into a single Air Force command. The subordinated regiments have since been designated Aviatsionnaya Baza (Air Base, sometimes shortened to Aviabaza), whereby all previously existing separate supporting units on the base have been united with the flying unit under a single command. During 1997, the generic term Aviatsionnaya Baza was further defined with a prefix to denote the role of the unit based.

In 1991, no fewer than four regiments of strategic bombers and strategic reconnaissance aircraft (equipped with Tu-16 'Badgers', Tu-22 'Blinders' and Tu-22M3 'Backfires'), as well as four tactical bomber

Above: Despite displaying 'Il-18' on the nose, this is the sole Belarus air force Il-22M command post.

Below: Four An-12s are operated by the 50 TAB from Minsk/Machulishche. 'Yellow 10' is believed to have once been an An-12BK-IS. It retains a large radome, two probes on the nose, flare dispensers and various other non-standard antennas.

regiments (equipped with Su-24 'Fencer-B/Cs' and Su-24M 'Fencer-Ds') were based on Belarussian territory. The air force of Belarus has retained one bomber regiment, equipped with Su-24Ms, and all remaining Russian 'Badgers', 'Blinders', 'Backfires' and 'Fencers' departed Belarus for the Russian Federation during 1994.

The Treaty on Conventional Armed Forces in Europe (CFE) limits Belarus to 294 fixed-wing combat aircraft and 80 helicopters; through destruction and sale of airframes, numbers in the inventory dropped below this limit during 1997. In early 1998, the inventory of Belarus numbered 250 combat aircraft and 64 attack helicopters. The close military co-operation between the armed forces of the Republic of Belarus and the Russian Federation is reflected, among others ways, by the continued use of the Soviet red star on the combat aircraft and helicopters of Belarus; the country has not adopted an indigenous national insignia for its military aircraft.

For air defence, Belarus depends on the MiG-29 'Fulcrum' and the Su-27 'Flanker', which equip two regiments. A number of MiG-29s were surplus to requirements and possibly as many as 18 of them were sold to Peru. Belarus was particularly well-endowed with ground-attack aircraft. The large number of Su-25 'Frogfoots' of the three independent ground-attack regiments based in Belarus in Soviet times have since been concentrated at a single Aviabaza, and two regiments have been disbanded. A number of surplus Su-25s (18) are thought to have been delivered to Peru. Tactical reconnaissance missions are performed by Su-24MR 'Fencer-Es', which are co-located with Su-24M 'Fencer-Ds' at 116 Bomber Reconnaissance Aviation Base (Ross).

Transport duties are undertaken by a single transport regiment, which operates Il-76 'Candids', An-12 'Cubs' and An-26 'Curls', as well as an An-24 'Coke', Tu-134 (Tu-135) 'Crusty' and Il-22M 'Coot-B' (Bizon). A number of the Il-76s of the Belarus Air Force are operated by the co-located air transport company Trans Avia Export.

The army aviation assets of the Byelorussian Military District comprised three full assault and attack regiments (one attached to each of the three armies), several liaison and transport squadrons, as well as a squadron operating the dedicated Mi-8SMV 'Hip-J' and Mi-8PPA 'Hip-K' electronic warfare helicopters. Since 1991, a number of smaller units have been disbanded and a redistribution of airframes has taken place, and all rotary-winged assets are operated by the air force. Three independent helicopter regiments have been restructured into (transport) combat helicopter bases; in addition, an independent liaison squadron and an independent combat command and control squadron remain operational.

At present, Belarus possesses no air academy, and *ab initio* training of candidate pilots for the air force of Belarus takes place under contract in the Russian Federation.

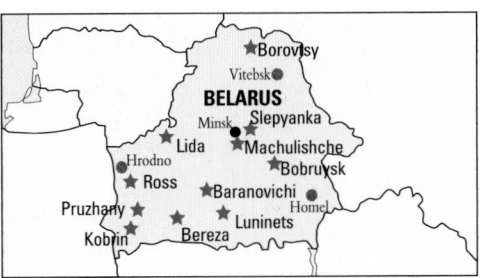

The Belarus air force has two fighter regiments. The 927 IAB is equipped solely with MiG-29s (above). This aircraft is carrying a pair of R-60s (AA-8) and an R-27R (AA-10) under each wing. The 61 IAB operates MiG-23/25/29s and Su-27s (below). Red '62' is an Su-27UB two-seat operational conversion trainer.

The Belarus 'Hind' fleet includes examples of the Mi-24P 'Hind-F', armed with the GSh-30-2 cannon.

Air Force of the Republic of Belarus

Air Force of the Republic of Belarus	HQ: Minsk
50 TAB	**Machulishche**
An-12 (4), An-24 (1), An-26 (8), Il-22 (1), Tu-134 (Tu-135) (1), Il-76 (18)	
61 IAB	**Baranovichi**
MiG-23MLD/-UB (44), MiG-25 (1), MiG-29 (26), Su-27 (23)	
116 BRAB	**Ross**
Su-24M/Su-24MR (36)	
206 ShAB	**Lida**
Su-25 (80)	
927 IAB	**Bereza**
MiG-29 (40)	
13 OVE BU	**Bobruysk**
Mi-6 (6), Mi-8 (7)	
65 TBVB	**Kobrin**
Mi-8 (55), Mi-26 (14)	
181 BVB	**Pruzhany**
Mi-8 (18), Mi-24 (22)	
248 OVE	**Minsk-Slepyanka**
Mi-6/Mi-22 (3), Mi-8/9 (23)	
276 BVB	**Borovtsy (aka Polotsk)**
Mi-8 (27), Mi-24 (23), Mi-24K (4), Mi-24R (4)	
1169 BRAT	**Luninets**
Mi-6 (24), Mi-8 (15), Mi-24 (16), Mi-24K (5), Mi-24R (2)	

Glossary	
BRAB	Bomber Reconnaissance Aviation Base
BRAT	Aircraft Reserve Base
BVB	Combat Helicopter Base
IAB	Fighter Aviation Base
OVE BU	Independent Combat Command & Control Helicopter Squadron
OVE	Independent Helicopter Squadron
ShAB	Ground Attack Aviation Base
TAB	Transport Aviation Base
TBVB	Transport Combat Helicopter Base

BELGIUM

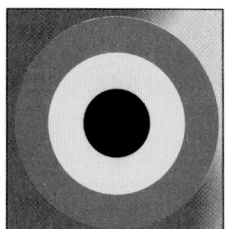

Capital: Brussels
Population: 9.9 million
Land area: 30520 km²
(11,780 sq miles)
Major cities: Antwerp,
Bruges, Charleroi, Ghent,
Liège

Belgische Luchtmacht/ Force Aérienne Belge (Belgium Air Force)

The establishment of a military flying school on 23 March 1910 marks the start of the Belgium air force, although the first aircraft did not arrive until 5 May. The force fought valiantly in World Wars I and II, and Belgium became a founding member of NATO.

The planned restructuring of the armed forces of Belgium was announced early in 1993 by the then-Minister of Defence, Leo Delcroix. The Belgische Luchtmacht/Force Aérienne Belge (Belgium Air Force) faced a reduction of aircraft and personnel. The Mission Statement of the Belgium air force is "The defence of the airspace and the support of operations on land and sea", and the BAF, in co-operation with allied forces, has to contribute a credible and effective means to the peacekeeping and the prevention or control of crisis situations in Europe or elsewhere in the world. Furthermore, the BAF assists people in distress by bringing rescue helicopters into action on land and sea and participates in humanitarian operations at different locations through out the world, offering help to people in need.

The F-16, the only remaining fighter aircraft after the withdrawal of the Mirage 5, is a good example of the effect of this policy. Of the 180 originally purchased, 90 remain in active service. The six operational squadrons (Smaldeel or Escadrille) each operate 12 F-16s. Eighteen are used as an operational reserve fleet. All other surviving Belgian F-16s are stored at Weelde, pending their sale. The F-16 in Belgium is based at two air bases: Florennes (2 Wing Tac) and Kleine Brogel (10 Wing Tac). 2, 23 and 31 Smaldeel are part of the Main Defence Forces, while 1, 349 and 350 Smaldeel are part of NATO's Reaction Forces. 1 Smaldeel is part of the Immediate

Reaction Forces (IRF) and tasked with strike and reconnaissance roles, while both 349 and 350 Smaldeel form part of the Rapid Reaction Forces (RRF) and have an air defence task. 2 Smaldeel also has an air defence task, while 23 and 31 Smaldeel have a conventional and a nuclear fighter-bomber role.

For the reconnaissance task the BAF used Orpheus pods loaned from the Koninklijke Luchtmacht (Royal Netherlands Air Force) until the delivery of eight Per Udsen Modular Reconnaissance Pods (MRP), which arrived from January 1998. The Mid-Life Update (MLU) programme adopted by the BAF involves all 90 F-16s. First deliveries have been made and the first Smaldeel should become operational in 1999. Dassault's Carapace passive ECM system is another standard modification for BAF F-16s and, together with the acquisition of 25 ALQ-131 pods, gives the type good threat protection. AIM-120B AMRAAM missiles have been ordered for delivery in 1999, and for the air-to-ground role, AGM-65 Maverick missiles have already been delivered. Since 18 October 1996 four BAF F-16s have operated in support of Operation Deliberate Guard within the Belgian-Dutch-Luxembourg DATF (Deployable Air Task Force) from Villafranca, in Italy.

All flying training activities of the BAF are conducted at Beauvechain, with 1 Wing. Primary training is done by 5 Smaldeel with a total of 34 examples of the SIAI-Marchetti SF 260. Advanced flying training is done by 7 Smaldeel and initial operational training by 11 Smaldeel, both operating the Alpha Jet. The Alpha Jet was bought to replace the Lockheed

Right: The need for a heavy-lift capability prompted the FAB to invest in a pair of used A310s.

Below: This C-130H of 20 Sm is seen dropping World Food Programme packages while wearing the colours of the European Union. Training for humanitarian and operations other than war plays an increasing role in European military aviation.

Six squadrons and an OCU operate the F-16 in the FAB. Half of the original purchase remains in service, and is being upgraded in the Mid-Life Update programme. These aircraft from 1 Sm and 2 Sm (nearest aircraft) represent 2 Wing Tac at Florennes.

T-33 and the Fouga Magister from 1978. The VEC is tasked with the evaluation of trainees and the training of Alpha Jet instructors. All remaining Fouga Magisters (11) are concentrated within 33 Smaldeel and are mainly used by staff officers to gain flight hours.

15 Wing Tpt Aé at Melsbroek (Brussels) operates a mix of transport aircraft. For VIP and personnel transport, 21 Smaldeel of 15 Wing Tpt Aé operates two Falcon 20Es (Mystère 20Es) which were acquired in 1973, a single Dassault Falcon 900B long-range executive jet, five Swearingen Merlin IIIAs (from six delivered in 1976) and a couple of A310s. The A310s were bought from Belgium's national airline, SABENA, to replace the two Boeing 727-29Cs which were acquired in 1976/77. Both Airbuses had been delivered to 21 Smaldeel by January 1998. Three Hawker Siddeley HS748 Series 288 turboprop aircraft were delivered during 1976/77. These machines operate in the personnel and cargo transportation role, for which their interiors can be very quickly converted.

A force of 11 C-130Hs based at Melsbroek is part of 20 Smaldeel. A total of 12 was delivered, but one

Belgische Luchtmacht/ Force Aérienne Belge

Tactical Air Force		
2 Wing Tac		**Florennes**
1 Smaldeel	(R)F-16A, F-16B	
2 Smaldeel	F-16A/B	
350 Smaldeel	F-16A/B	
10 Wing Tac		**Kleine Brogel**
23 Smaldeel	F-16A/B	
31 Smaldeel	F-16A/B	
349 Smaldeel	F-16A/B	
OCU(S)	F-16A/B	
15 Wing Tpt Aè		**Melsbroek**
20 Smaldeel	C-130H	
21 Smaldeel	A310-222, BAe 748-288, Falcon 20E, Falcon 900B, Swearingen Merlin IIIA	
40 Smaldeel	Sea King Mk 48 (based at Koksijde)	
Training and Support Command		
1 Wing		**Beauvechain**
5 Smaldeel	SF 260D/M	
7 Smaldeel	Alpha Jet E	
11 Smaldeel	Alpha Jet E	
VEC	Alpha Jet E	
33 Smaldeel	CM 170	

Basic training is conducted in Belgium on the SF.260D/M before pilots move on to the Alpha Jet E. All training takes places at Beauvechain.

The Belgian army's Alouettes are a mixture of SA 318Cs and SA 313Bs. These examples are SA 318C Alouette II Astazous in service with 16 BnHLn.

was lost in 1996 at Eindhoven air base in the Netherlands, and an additional second-hand C-130 will be bought to replace it. SABENA Technics initiated a programme for the C-130Hs to extend their operational lives by an additional 15 years. Completed by May 1994, it involved dismantling and replacing all outer wing sections, using kits supplied by Lockheed Martin, and will keep the BAF's C-130Hs operational until at least 2010. In 1992 SABENA Technics was awarded a contract to undertake an avionics upgrade programme for these aircraft. Each aircraft will be retrofitted with a Honeywell EFIS cockpit and flight management system, INS and GPS navigation systems, colour weather radar and an associated chaff/flare self-defence system. The Belgian C-130Hs can also be equipped with the ALQ-131 for self-protection.

The five Sea King Mk 48s have been in service since 1976. An additional sixth example (second-hand) will be added to this SAR fleet, which is based at Koksijde. A modernisation programme for the Sea King Mk 48 which started in 1995 included new search radars, navigation systems and a FLIR camera on the port side of the helicopter. These helicopters will remain in service at least until 2005.

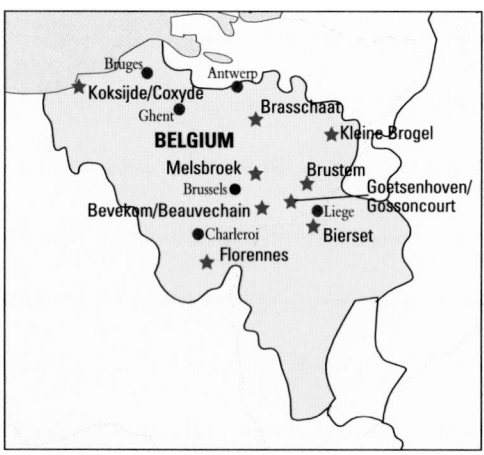

Groepering Licht Vliegwezen/Le Groupement d'Aviation Légère (Belgian Army Aviation Group)

Light Aviation, which upon its foundation in 1946 was part of the FAB, was put under Army command on 1 April 1954. At that time it consisted of the 15th Escadrille at Brasschaat near Antwerp and the 16th Escadrille at Bützweilerhof (Germany). In 1956 the 17th Escadrille was founded at Werl and the 18th Escadrille at Merzbrück (both in Germany). To save funds, the units based in Germany were withdrawn to Liège-Bierset from 1993. The Groepering Licht Vliegwezen (Army Aviation Group) has existed in its current form since 19 May 1993.

The main mission of the Army Aviation Group is to execute anti-tank actions as a mobile combat element of the Army Operational Command. It also has to be constantly ready to be put into action under an international framework for crisis management, as was the case in Somalia and Rwanda. In peacetime, the Army Aviation Group uses 15 Alouette II light liaison helicopters, four Britten Norman BN-2A Islander aircraft and 38 Agusta A 109BA combat helicopters. In wartime this fleet would be augmented by the aircraft and helicopters of the Vliegwezen (Army Aviation Group School).

The 16th Battalion's main mission is liaison. For this purpose it operates Alouette IIs and the Britten Norman BN-2A Islanders. The 17th and 18th Battalions, whose main mission is destroying enemy tanks, operate the A 109BA in six different versions. A reconnaissance version is equipped with the HELIOS observation system, and can also be equipped with two MAG machine-guns. The transport version has a minimum of equipment to allow maximum transportation of personnel. The anti-tank version carries the HELITOW observation and firing system and can be armed with eight TOW 2A missiles. Two versions were in their final phase of development early in 1998, one for medical evacuation and one which will be provided with rocket pods.

Above: The BN-2A Islander has been in service with the Belgian army since 1976. This example flies with the SLV at Brasschaat.

Groepering Licht Vliegwezen/Le Groupement d'Aviation Légère

Liège-Bierset	
16 BnHLn	Alouette II, BN-2A
17 BnHATk	A 109HO, A 109HA
18 BnHATk	A 109HO, A 109HA
Brasschaat	
SLV	Alouette II, A 109HO, BN-2A

Glossary	
BnHLn	Battalion Helicopters Liaison
BnHATk	Battalion Helicopters Anti-Tank
SLV	School van het Lichte Vliegwezen

Belgische Marine/Marine Belge (Belgian Navy)

Since 1971 the Belgische Marine/Marine Belge (Belgian Navy) has operated three Alouette III helicopters. The small helicopters are ideal for maintaining contact with ships at sea. When operating from the logistical supply vessels, BNS *Zinnia* and BNS *Godetia*, they provide the escorted ships with food and spare parts. They also undertake liaison flights for shore-based commands and, in the case of accidents at sea, they make a rapid evacuation possible. The heliflight is based with 40 Smaldeel of the BAF, but undertakes independent training and maintenance.

Belgische Marine/ Marine Belge

Koksijde	
Heli-flight	Alouette III

Belgian Rijkswacht/ Gendarmerie (Police)

The Rijkswacht/Gendarmerie (Police) has its own independent luchtsteundetachement (Air Support Unit). A few years ago it was transferred from Brasschaat to Melsbroek air base, which it shares with 15 Wing Tpt Aé. For many years it operated three SA 330Hs, which were delivered in 1973, and a couple of SA 318C Alouette Astazous. One Puma was sold to Aérospatiale and one was written off in an accident, and the last one was withdrawn by mid-1998. To cover all the police tasks this unit has to fulfil, the fleet has been expanded over the last few years with a Britten Norman BN-2T Islander, an MDH 900 and a couple of Cessna 182s.

Rijkswacht/Gendarmerie

Luchtsteundetachement (Air Support Unit)	
Brasschaat	Alouette II
Brussels-Melsbroek	SA 318C Alouette Astazou, BN-2T, Cessna 182Q/R, MDH-900

Below: The MDH-900 Explorer has replaced the SA 330H Puma in the Belgian Gendarmerie.

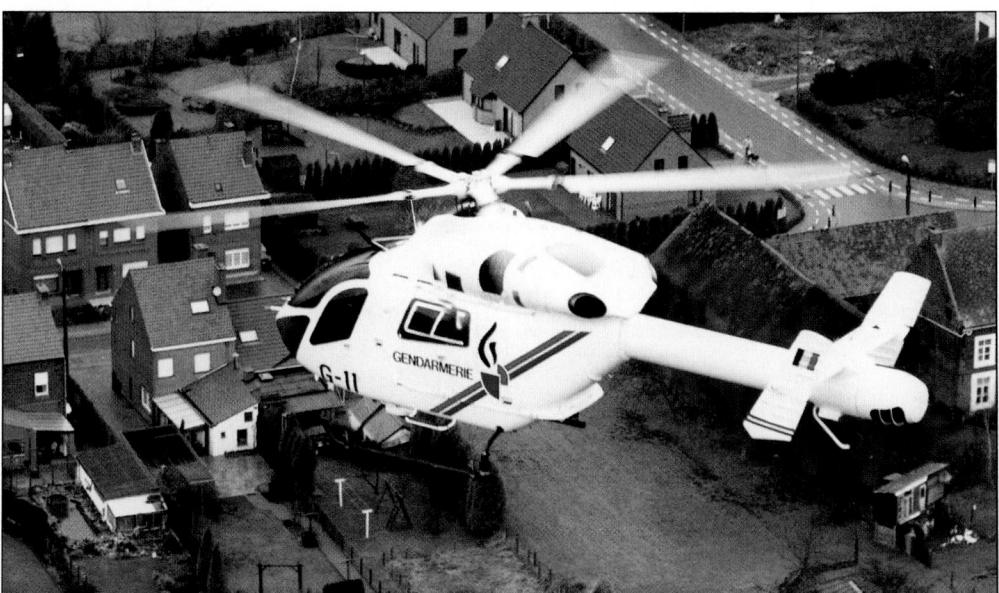

BOSNIA-HERZEGOVINA

Capital: Sarajevo **Population:** 4.5 million
Land area: 51130 km² (19,735 sq miles)
Major cities: Banja Luka, Zenica

The signing of the Dayton Peace Accords in November 1995 brought to an end the three-year Bosnian war. This treaty and the associated Organisation for Security and Co-operation in Europe (OSCE) arms control regime placed strict limits on the armed forces of Bosnia's two mini-states or entities.

US military aid to Bosnia-Herzegovina was to have included 15 Hueys, delivered via Bückeburg, Germany. This UH-1H, 98, is ex-67-17198.

The Federation

The wartime alliance of the Muslim and Croatian parts of Bosnian was formalised in February 1994 when the Washington accords established the Federation. The predominately Muslim Army of Bosnia and Herzegovina (ABiH) and the Croatian Defence Council (HVO) were supposed to have merged into a joint Federation armed forces, but to a large extent this has not happened. The joint units that have been established have only been of a symbolic nature.

The ABiH made extensive use of some 10 aircraft and half a dozen helicopters during the war to supply the beleaguered Bosnian enclaves of Srebrenica, Zepa, Gorazde and Bihac. These aircraft were not part of a separate air force but assigned to various army corps of the ABiH. As part of the US 'Train and Equip' programme, the Federation was to receive some 15 ex-US Army UH-1H/V Hueys, but disputes between the HVO and ABiH has dogged the forma-

ABiH

Visoko/Tuzla
    4 x Mil Mi-8/17,
    15 x UH-1V/H *(temporarily in Germany for crew training)*
Sarajevo Airport
    1 x Cessna Citation
Field Locations
    4 x UTVA-66/75, 1 x Zlin 526, 1 x Piper PA-18 Super Club
Based outside Bosnia
    1 x CASA 212, 1 or 2 Lockheed Martin C-130

tion of the new air unit. The HVO operated a number of helicopters during the war but they have since been transferred to the Croatian air force.

The Republika Srpska

As the Yugoslav National Army (JNA) withdrew from Bosnia in the spring of 1992, local Serb forces were provided with a small force of close air support aircraft and helicopters to defend the newly established Serb mini-state.

Most of the air arm of the Republika Srpska armed forces (VRS) is based around the northwestern city of Banja Luka. Like the Federation forces, the activities of the VRS air force is closely monitored and controlled by NATO forces.

VRS Air and Air Defence Force

92nd Air Brigade   **Banja Luka International Airport**
238th Fighter Squadron
    5 x UTVA G-4M Super Galeb,
    10 x Soko J-21 Jastreb
    9 x Soko J-22 Orao, ? x Soko G-2 Galeb
92nd Multi-role Squadron
    UTVA-66/75, Cessna 172,
    Piper PA-18 Super Cub
111th Helicopter Regiment   **Zaluzani**
    16 x Soko SA-341H/342L Gazelle, 18 x Mil Mi-8

BULGARIA

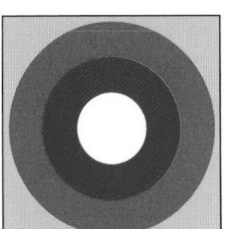

Capital: Sofia
Population: 9 million
Land area: 110910 km²
(42,810 sq miles)
Major cities: Burgas, Plovdiv, Ruse, Varna

Above: Three air bases house four MiG-21bis squadrons. This example is based at Graph Ignatievo.

Below: The pride of the Bulgarian air force are the MiG-29s based at Ravnetz. The 1/5 Eskadrila's complement includes MiG-29UB two-seaters.

Bulgarski Voennovazdushni Sili (BVVS)

In mid-1940s, immediately after the end of World War II, the BVVS was rapidly re-equipped with Soviet-made combat aircraft – initially piston-engined fighters, bombers and attack aircraft, and then in the early 1950s the first generation of jet fighters (Yak-17/23 and MiG-15s) as well as with Il-28 jet bombers. Until the 1990s the air arm was modelled after the Soviet style and it received the most modern front-line aircraft available for export, including the MiG-23MF/ML/MLD/BN, MiG-25RBT, Su-22M4, Su-25K and MiG-29. The sharp economic crisis and disappearance of the low-cost Soviet support after the break-up of the Communist system in Eastern Europe resulted in drastic falls in serviceability, and stagnation of the modernisation process.

POST-COLD WAR CHANGES

The first steps of the reorganisation from a Soviet-style air arm into a modern and flexible Western-modelled air force began in 1992 under the leadership of the forward-thinking C-in-C Lieutenant General Mikho Mikhov, who was appointed Chief of the General Staff of the Bulgarian Armed Forces in June 1997. These steps comprised renaming the air arm,

which had been known since 1961 as the PVOiVVS (Air Defence and Air Force) to BVVS (Bulgarian Air Force), introduction of new national insignia, flag, unit badges and flying suits. The roles of the 20,000-strong BVVS were also changed and today they are as follows: in peacetime – air traffic control and air policing, SAR, medevac/casevac and emergency liaison; in wartime – air defence of the national industrial and urban assets, air reconnaissance, close air support, battlefield co-operation with the land forces and navy, and special operations support.

The next important step of the reform was the introduction of a new air base structure close to that adopted by the USAF and the Turkish AF, which replaced the old Soviet-style regimental organisation in September 1994.

AIR DEFENCE ASSETS

In September 1996 the next step of the reform followed with the disbandment of the existing air defence divisions and the establishment of the sole CPVO (Corpus PVO – Air Defence Corps), head-quartered in Sofia. It incorporated all BVVS air defence assets – fighter air bases as well as SAM and early warning radar brigades.

The fighter fleet of the CPVO is spread over five air bases, housing a total of six squadrons – three

equipped with the MiG-21bis/UM (approximately 50/12 aircraft), two with MiG-23MF/ML/MLD/UB (30 MF/ML/MLD and six two-seaters) and one with MiG-29 (17/4). In mid-1997 the Bulgarian government declined the MiG MAPO offer of delivery of one more 'Fulcrum' squadron of 14 multi-role MiG-29SMs plus a spare parts package and weapons, worth $450 million.

The MiG-21bis upgrade has been discussed since

Above: Bezmer houses the two Eskadrilas of the 22 Shturmova Aviobasa, which operate Bulgaria's Sukhoi Su-25Ks (seen here) and two-seat -25UBKs.

Below: The Su-22M4s are concentrated in a single squadron, the 2/26 Eskadrila, based at Dobritch. Serviceability of the type in BVVS service is low.

Bulgarski Voennovazdushni Sili

Bulgarski Voennovazdushni Sili	HQ, Sofia
16th Transportna Aviobasa (Transport Air Base)	
	Vrazhdebna
1/16 Eskadrila	An-24RV, An-26, An-30, L-410UVP/ UVP-E, Yak-40, Tu-134A/B3

Corpus Protivovazdushna Otbarana (Air Defence Corps)	
An-2	**HQ Sofia**

1st Iztrebitelna Aviobasa (Fighter Air Base)	**Dobroslavtzy**
1/1 IAE 'Nebesni Ritzari'	MiG-23MF/ML/MLD/UB, L-29
2nd Iztrebitelna Aviobasa	**Gabrovnitza**
1/2 Eskadrila	MiG-23ML/UB, L-29
3rd Iztrebitelna Aviobasa	**Graph Ignatievo**
1/3 Eskadrila	MiG-21bis/UM
2/3 Eskadrila	MiG-21bis/UM
5th Iztrebitelna Aviobasa	**Ravnetz**
1/5 Eskadrila	MiG-29/UB, L-39ZA
6th Iztrebitelna Aviobasa	**Baltchik**
1/6 Eskadrila	MiG-21bis/UM, L-29

Corpus Takticheska Aviatzia (Tactical Aviation Corps)	
An-2	**HQ Plovdiv**

21st Iztrebitelna-Bombardirovachna Aviobasa	
(Fighter-Bomber Air Base)	**Uzundzhovo**
1/21 Eskadrila	MiG-21bis/UM
22nd Shturmova Aviobasa (Attack Air Base)	**Bezmer**
1/22 Eskadrila	Su-25K/UBK
2/22 Eskadrila	Su-25K/UBK
23rd Vertoletna Aviobasa Boyni Vertolety	
(Helicopter Base of Combat Helicopters)	**Stara Zagora**
1/23 Eskadrila	Mi-24D/V
2/22 Eskadrila	Mi-24D/V
24th Vertoletna Aviobasa (Helicopter Air Base)	
1/24 Eskadrila	Mi-17	**Krumovo**
2/24 Eskadrila	Mi-8
25th Iztrebitelna-Bombardirovachna Aviobasa	
	Tcheshnegirovo
1/25 Eskadrila	MiG-23BN/UB, L-39ZA
2/25 Eskadrila	MiG-23BN/UB, L-39ZA
26th Razuznavatelna Aviobasa (Recce Air Base)	
1/26 Eskadrila	MiG-21MF/UM	**Dobitch**
2/26 Eskadrila	Su-22M-4/UM-3

Visshe Voennovazdushno Utchilishte 'Georgy Benkovsky'	
(Higher Air Force School)	HQ Metropolya Dolna
Zveno za Motivatzionni Poleti (Air Experience Flight)	
	L-29
11th Utchebna Aviobasa (Training Air Base)	
1/11 Eskadrila	L-29	**Shtraklevo**
2/11 Eskadrila	L-29
12th Utchebna Aviobasa	**Kamenetz**
1/12 Eskadrila	L-39ZA
2/12 Eskadrila	L-39ZA

1993 when the Mikoyan OKB presented its first offer to modernise part of the 72-strong 'Fishbed-N' fleet to the MiG-21-93 standard. Talks on the 'Fishbed' upgrade, which was to be done locally by the Bulgarian overhaul and maintenance company TEREM-VRZ 'Georgy Benkovsky', ended unsuccessfully in mid-1997, simultaneously with the refusal of the MiG-29SM offer. In 1998 the BVVS announced that either IAI or VPK MAPO will upgrade a limited number of MiG-21bis.

STRIKE AND RECCE ASSETS

The Plovdiv-headquartered CTA (Corpus Takticheska Aviatzia – Tactical Aviation Corps), known until 1996 as 10th Composite Aviation Corps, incorporated the BVVS strike/reconnaissance assets grouped in four air bases with total of seven fixed-wing squadrons – two equipped with the MiG-23BN/UB (20-30 aircraft), two with Su-25K/UBK (35/4), and one Su-22M4/UM3 (18/3), MiG-21bis/UM (18/4) and MiG-21MF/UB (13/4), as well as an air base housing two squadrons of Mi-24D/V attack helicopters (37/6 helicopters). A transport helicopter air base of three squadrons – Mi-8 (seven) and Mi-17 (19) – is also under the CTA control. Fourteen PZL Mi-2s were offered for sale in early 1999, along with four of the Mi-8s. The BVVS hopes to acquire six light helicopters with the money it will get from the sale.

In 1995-1996 the Sukhoi OKB offered to sell or lease on favourable terms, including barter payment, a squadron of 10-12 Su-25TK attack aircraft for replacement of the BVVS MiG-23BN fleet, and also to upgrade the existing BVVS Su-25Ks. The Bulgarian answers to both of the proposals were negative. In September 1997 the Bulgarian MoD attempted unsuccessfully to sell up to 14 of the BVVS's 18 Su-22M4s to an unspecified Third World customer because of the type's extremely high maintenance costs, but the deal never materialised due to the low price which the customer offered. After withdrawal from use in late 1995 of the last three MiG-21Rs of the 26th RAB, a limited number of MiG-21MFs were modified to use the MiG-21R's underfuselage reconnaissance pod, thus receiving the MiG-21MFR designation.

Mil Mi-17s constitute the bulk of Bulgaria's helicopter airlift assets. This example has six hardpoints for external stores.

TRANSPORT AND TRAINING ASSETS

The BVVS transport aviation is grouped in the 16th TAB at Sofia Airport-Vrazdebna and comprises a squadron of five An-26s, two An-24RVs, one An-30, seven L-410UVP/UVP-Es, one Yak-40 and two Tu-134A/B-3s. The An-30 is Bulgaria's Open Skies aircraft; in 1994-95 it underwent an extensive equipment upgrade including replacement of its original camera suite with Western-made Leika and Vinten cameras featuring computerised control as well as a new GPS/INS-based navigation system.

Ground and aircrew training is provided by the Dolna Metroplya-based Higher Air Force School – the VVVU 'Georgy Benkovsky', which controls two air bases. Mainstay of the trainer fleet is the Aero L-39ZA (35 aircraft), delivered between 1986 and 1991. The last of the veteran L-29s (about 50-60 surviving aircraft, from 90-plus delivered in the 1960s and 1970s), which still serve in rapidly dwindling numbers, are expected to remain in use until the early 2000s. In 1995-96 the BVVS issued a requirement for up to 10 prop trainers for initial flying training. There were expectations that six Yak-18Ts would be supplied as a write-off of the trade debt owed to Bulgaria by Russia, but the deal did not materialise.

THE LATEST CUTS

In September 1997 the new BVVS C-in-C Major General Popov announced the first wave of cuts imposed for budgetary reasons. They comprised closing down four air bases until September 1999 and 10 per cent reduction of the 20,000-strong BVVS manpower. The bases destined for closure were 2nd IAB Gabrovnitza, 6th IAB Baltchik, 21st IBAB Uzundzhovo and 11th UAB Shtraclevo. Up to 80

Vrazhdebna houses the 1/16 Eskadrila which operates the Bulgarian transport fleet. Five Antonov An-26s are in service alongside three similar An-24s.

front-line aircraft were placed in storage due to lack of funds for maintenance.

THE FUTURE

The near-term development of the BVVS will depend on Bulgaria's timescale for joining NATO, a move which would be possible by at least 2003-05. The BVVS started participation in Partnership for Peace exercises in 1996.

Primary short-term effort will be placed in improving the current 50 per cent serviceability rate of the fleet and increasing pilot flying hours. To fulfil these two tasks, the air arm has requested $180 million until 2000. The replacement of the 100-strong MiG-21 and 50-strong MiG-23 fleets is expected to start around 2003-05. The requirement list for the future multi-role fighter type will emphasise NATO compatibility first; the same is true for the fixed- and rotary-wing transport fleets.

Used for anti-submarine warfare and search and rescue, nine Mi-14PLs serve with the Bulgarian Navy, based at Tchaika. A single Mi-14BT is used for transport roles.

Aviatzia na Bulgaskia Voennomorski Flot (Bulgarian Naval Aviation)

The modern Bulgarian naval aviation was born in October 1959 when the OPLEV-VMF (Otdelna Protivolodachna Eskadrila na VMF – Independent ASW Squadron of the Navy) was formed, equipped initially with four Mi-4 general transport helicopters. In 1962 OPLEV was rebased at Tchaika Air Station at Varna and three years later it received the first of its six Mi-4M shore-based ASW helicopters.

The turbine-powered Mi-14PL 'Haze-A' replaced the Mi-4M in 1979. A total of six was delivered in this first batch, followed by two Mi-14BT mine countermeasures helicopters in 1983, one of which survives today, in use as a general transport only. In 1994 the OPLEV took delivery of a single Ka-25Tz early warning helicopter, used actively until 1991. At present, the machine is held in storage due to technical support problems.

The second 'Haze-A' batch comprising four second-hand Mi-14PLs was delivered in 1990 as part of the last Soviet military aid package. In September 1990 the squadron was renamed OMEV (Otdelna Morska Eskadrila Vertoleti – Independent Naval Helicopter Squadron) due to the shift in its role

following the end of the Cold War. Due to funding problems, the serviceability of OMEV's Mi-14s has been poor since the early 1990s, but the type is expected to be kept in service until at least 2003-05.

Aviatzia na Bulgaskia Voennomorski Flot (Bulgarian Naval Aviation)

Bulgarska Voennomorska Aviatzia Otdelna Morska Eskadrila Vertoleti (Independent Naval Helicopter Squadron)
Mi-14PL/BT   **Tchaika**

CROATIA

Capital: Zagreb
Population: 4.7 million
Land area: 56540 km² (21,825 sq miles)
Major cities: Osijel, Rijeka, Split, Zadar

Hrvatsko Ratno Zrakoplovstvo i Protu Zracna Obrana

After the collapse of Yugoslavia, Croatia had to establish its own defence force. Subsequently, the Hrvatsko Ratno Zrakoplovstvo i Protu Zracna Obrana (HRZ i PZO, or Croatian War Aviation and Anti-Aircraft Defence) was created in 1992. Currently, the HRZ i PZO is reorganising its structure to a more Western standard and it is expected that the new air force will be renamed to Hrvatske Zracne Snage (HZS, or Croatian Air Force).

During the first years of the civil war, the HRZ i PZO used mainly light civil aircraft types (such as An-2, UTVA-75s and various Piper variants) to attack

This Antonov An-2 is believed to be one of two that have been converted to this configuration for an unspecified electronic role. The aircraft may be Elint platforms or possibly a 'mini-AWACS'.

targets. These aircraft were insufficient to establish an effective air force. An unknown number of Mi-8 helicopters were bought on the civil market in the former Soviet Union. First operated in civil marks, they are all now painted in overall white or camouflaged colours. They operate from two bases and are used for transporting both people and cargo.

Croatia operates nine Mi-24 'Hind' attack helicopters that are believed to be ex-Ukraine Air Force, although some have had their weapons removed. Also believed to be from Ukraine are two dozen MiG-21 fighters, which saw actual combat missions over the Krajina region during 1995. All MiGs are divided between two bases, Zagreb-Pleso and Pula, but can operate from various deployment bases all over the country. During March 1998 four aircraft were deployed to Split for a two-week exercise.

Ten Bell 206B-3s, with a total value of approximately US$15 million, were purchased from Bell Helicopter Textron at the end of 1996; the first delivery took place in January 1997 to the helicopter training squadron at Zemunik. This base, close to the town of Zadar, is the main training base of the HRZ i PZO and also houses the total UTVA-75 and PC-9 fleet. All UTVA-75 light training aircraft were obtained from the various aeroclubs in Croatia and are used for initial flying training. Pilatus delivered 20 new-build PC-9 trainers. After initial flying training, students proceed to their course in the PC-9.

Above: Croatia has a total of 20 MiG-21bis split between two fighter squadrons. No. 22 Eskadrila, to which this example belongs, is based at Pula.

Above: Displaying prominent mosquito nose art between the cockpits is a Mi-24V of the 29 Eskadrila. Croatia acquired a mix of Mi-24Ds and Vs in 1993.

Hrvatsko Ratno Zrakoplovstvo i Protu Zracna Obrana

Zracna baza Zagreb-Pleso	
Eskadrila Transportnih Aviona	An-2 (4), An-32 (2), CL-601 (2)
29 Eskadrila Borbenih Helikoptera	Mi-24D/V (9)
21 Eskadrila Lovacka	MiG-21bis (13), MiG-21UM (2)
Zracna baza Zagreb-Lucko	
Eskadrila Transportnih Helikoptera	Mi-8(T and S) (6), Mi-8MTV-1 (10)
Zracna baza Pula	
22 Eskadrila Lovacka	MiG-21bis (7), MiG-21UM (2)
Eskadrila Transportnih Helikoptera	MD500 (4)
Zracna baza Zadar-Zemunik	
Eskadrila Trenaznih Zrakoplovo	PC-9/M (20), UTVA-75 (10)
Eskadrila Trenaznih Helikoptera?	Bell 206B-3 (9)
Zracna baza Divulje	
Eskadrila Transportnih Helikoptera	Mi-8 (2), Mi-8MTV-1 (4)

CYPRUS

Capital: Nicosia
Population: 0.7 million
Land area: 9250 km²
(3,570 sq miles)
Major cities: Famagusta, Limassol

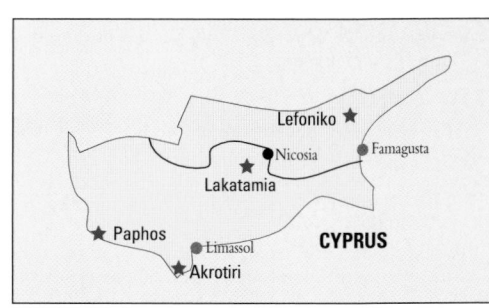

Cyprus National Guard/ Kibris Türk Emniyet Kuvetleri

The Cyprus National Guard Air Wing was formed on 16 August 1960, but was disbanded after the Turkish occupation in 1974. Since then, Cyprus has been divided into a Turkish part in the north and a Greek-Cypriot part in the south. A United Nations peacekeeping force (UNFICYP) was established to prevent a recurrence of fighting between the Greek-Cypriot and Turkish communities. Two Hughes 500 helicopters are used by the UNFICYP, and they could be the two that are often referred to as the H500s of the Greek-Cypriot National Guard. The UK's Army Air Corps also has some helicopters assigned to UNFICYP. The Gazelles belong to 16 Flight, which operates from Dhekelia. The other British unit on Cyprus is RAF No. 84 Sqn, flying Wessex helicopters out of Akrotiri.

The Greek-Cypriot National Guard was reformed in 1982, operating a single BN-2B-21 Islander. During 1988, the force was augmented by four HOT-equipped SA 342L-1 Gazelles, with a possible delivery of two more during 1990. Two PC-9s were acquired in 1989 and are used for training and perhaps light attack. Both are based at the military airfield of Paphos. The most recent acquisitions are two PZL-Swidnik Kanias, a further development of the Mi-2 'Hoplite', delivered by the end of 1990.

The Greek-Cypriot Police operate a small force of a single BN-2T Islander and two Bell 412EP helicopters. The police operate from Lakatamia.

The 35,000-strong Turkish Cypriot Security Force (Kibris Türk Emniyet Kuvetleri) in the northern part of Cyprus is always on a 100 per cent alert status. Two Turkish Mi-17 helicopters should be on strength, as are some UH-1Ds.

Cyprus National Guard

Bell 206L-3 (2), SA 342L-1 (6), Kania (2)	Lakatamia
BN-2B-21 (1), PC-9 (2)	Paphos

Cyprus Police

Bell 412EP (2), BN-2T (1)	Lakatamia

Kibris Türk Emniyet Kuvetleri

Mi-17 (2), UH-1D	Lefkoniko (Geçitkale)

CZECH REPUBLIC

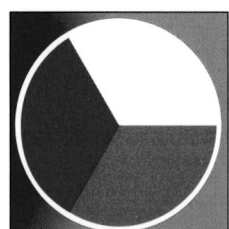

Capital: Prague
Population: 10.3 million
Land area: 78864 km²
(30,433 sq miles)
Major cities: Brno, Olomouc, Ostrava, Pilsen

Vzdusné Síly ACR – Air Force of the Army of the Czech Republic

The Ceskoslovenske Letectvo was re-established in spring 1945 around Czech pilots returning with their aircraft after service within the RAF and in regiments on the Soviet side. The Czechs began to build a strong air force consisting of British aircraft and products from intact German production lines, a considerable amount of which had been left on Czechoslovak territory. The Avia-built S.199 (Bf 109G) and the S.92 (Me 262) were examples that emerged from these factories in the early post-war period. Several long-established indigenous aviation companies, such as Aero-Vodochody, Zlin and LET-Kunovice, were on their feet quickly and provided the young force with basic trainers and light transports.

The locally produced LET L-410 Turbolet is used for survey, VIP and light transport work in four sub-types. This is an L-410T transport of 61 DLT.

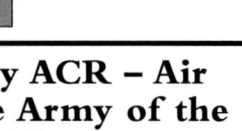

Above: With the retirement of the MiG-23ML, the sole fighter in service is the MiG-21MF. This example is carrying four AA-2C 'Atoll' AAMs.

Below: Mi-24V 'Hind-E' 0837 belongs to the 33 ZVrL at Prerov. As well as 19 'Hind-Es', the Czech air force uses 16 Mi-24D 'Hind-Ds'.

Between 1945 and 1948 Czechoslovakia enjoyed its independence and began to build some of the German types and to develop new aircraft. After the Communist take-over of 1948, the general direction changed totally. Under the Gottwald government, the Letectvo became organised and equipped in the Soviet style, and all the ex-RAF pilots were sent to re-education camps as reactionary personnel. The Spitfires (S-89) and Mosquitoes (B-36) were scrapped just because they were British. During the 1950s, Yak and MiG fighter series succeeded the German types, culminating in Avia's licence-production of the MiG-15, MiG-19S and MiG-21F-13. First permanent jet aircraft bases were Pardubice, Namest nad Oslovou, Hradec-Kralove and Caslav.

Prior to 1968, the Ceskoslovenske Letectvo was divided into two Air Armies. The 7th Air Army, headquartered at Stara-Boleslav, was independent of Army command and reported directly to the Ministry of Defence. It was split into the 3rd Fighter Division at Brno and the 7th at Zatec, and controlled all the air-defence fighters/interceptors. The 10th Air Army carried out tactical support under the direction of the Army's front commander, headquartered at Hradec-Kralove. This was divided into the 1st Fighter Division at Bechyne and the 34th Fighter-Bomber Division at Pardubice. Following the Soviet pattern, the 10th Air Army had its own frontal fighter protection for its strike elements.

The country had joined the Warsaw Pact in May

Czech success with the Aero L-39 has resulted in an order for the L-159 single-seat development. This L-39ZA is one of 37 L-39s in service.

The LSZc2 of the Letecky Zkusebni Odbor (LZO) looks after the Czech air force small transport fleet. The unit uses three of the four Antonov An-26s.

1955 but almost no standing Soviet forces were permanently based in Czechoslovakia. This changed radically from August 1968 onwards, when Soviet and other WarPac troops crushed the so-called Prague Spring, in which the Czech people tried to gain a more liberal state. The Soviet Central Group of Forces was established in Czechoslovakia with headquarters at Milovice near Prague, which would have controlled the Czech Air Armies in the event of a large assault towards the West. The 7th (Air Defence) Air Army would have fallen directly under Soviet PVO control and the 10th Air Army would have been attached to the frontal commanders at Milovice. Despite Poland becoming a trusted ally of the Soviets in the following years, new types were also delivered to the Czechs, finally replacing older MiGs and Su-7s. Thus, Mi-24s (1978), MiG-23BNs (1978), MiG-23MF/MLs (1979), Su-25s (1984), Su-22s (1985/86) and MiG-29s (1987/88) were introduced. Additional air bases included Bechyne, Zatec, Prerov, Ceske Budejovice, Dobrany-Line and Mosnov. Before 1989 the Letectvo was the strongest WarPac air force, with over 600 aircraft on strength and pilots flying between 120 and 150 annual hours.

After the bloodless Velvet Revolution of 1989, during which in November of that year the Communist government leaders stepped down, all Communist guidelines were abandoned. The future of the 'CSFR-Letectvo' for the next years was dictated by severe restructuring and base closure programmes by the new government, dictated by the dramatically changed economic situation. First, the two air armies were disbanded and a combined Air Corps was created. Programmes were stopped before completion and even more radical ones came into focus. By 1993 Bechyne – by then the most modern airfield – had been closed, as was Mosnov and Prostejov. During the changes, an even deeper, and unexpected, cut had to be managed.

When the two national parliaments of the new states – Czech Republic and Slovakia – dissolved their union on 1 January 1993, the assets of the air force of

the former Czechoslovakia were divided 2:1 in favour of the Czechs, with the exception of the 18 MiG-29s which were shared equally, since the Slovaks received no 'Floggers'. Since separation, the Ceske Letectva a Protivzdusna Obrana has undergone a further major reduction of numbers, types and bases. Shortly after separation, the MiG-23BNs and the MiG-23MFs were phased out. The 4th Air Defence Corps at Stara-Boleslav and the 3rd Tactical Air Force Corps were created in 1994. Ceske-Budejovice and Zatec were closed, and due to cost-cutting measures even the sharp edge of the air force, the nine MiG-29s, ceased operations at the end of June 1994 and were later exchanged for PZL Sokol helicopters. 1994 and 1995 saw a major decrease in pilot flying hours, down to 40-50 per year. The Czechoslovakian aerobatic team, 'Biely Albatrosy', had relocated to Kosice in Slovakia, so there is no aerobatic team in the Czech Republic.

The axeing of the wing/regimental level created the Zakladna (air base) orientated organisational structure and on 1 November 1997 the two Corps were united under the Headquarters of the Air Force, which changed its name to Vzdusné Síly ACR. Caslav remains the sole fighter base, all fighter-bomber and reconnaissance assets being concentrated at Namest nad Oslavou, all the air force helicopters at Prerov, the transports at Prague-Kbely, and all training at Pardubice. Every year, classes of around 25 students start their four-year training at Hradec-Kralove and then move to Pardubice. Being a NATO candidate and Partnership for Peace member since 1996, language training is seen as vital and every year students are sent to the USA; general staff members also undergo staff college classes abroad. After several fatal accidents, a government resolution directed extra funds for flying hours, which resulted in approximately 70 to 80 hours in the active units (except advanced training) being flown by current regular squadron pilots.

As the MiG-23MLs became time-expired during 1998, the Letectvo is limited to MiG-21MFs as its main type. All plans for extensive modernisation of the MiG-21s by Elbit have been abandoned, but they will be modified for operations within Western airspace. Building on the very successful L-39 programme developed to meet former WarPac countries' advanced training needs, 72 Aero L-159 single-seat light-attack aircraft are on order to replace most of the combat fixed-wing fleet in the future. The first five should be delivered to Caslav in 1999. The administration is now considering the purchase of 24 to 36 Western fighters. According to the government, the new fighters should become operational in 2002 or 2003 as part of an overall plan for the armed forces. Despite the unavailability of funds until later in 1998, contenders tested include the JAS 39, later-block F-16Cs and the F/A-18. Many higher-level personnel consider C3 assets and new L-band radars more important if the Czech Republic is to join NATO in 1999.

The PZL Swidnik W-3 Sokol is used by 62. vrlt at Praha-Kbely. Of a total of 11 examples operated, four are dedicated to the search and rescue role.

Resplendent in the national colours, this Mil Mi-17 is used for VIP transport with the Praha-Kbely based 62. vrlt/6 ZDL.

Air Force of the Army of the Czech Republic

Velitelství vzdusnych síly ACR	Stara-Boleslav

(Headquarter of the Air Force of the Army of the Czech-Republic)

4. Základna Tactickeho Letectva (ZTL), Caslav

41. Stihaci Letka (SL) 'Tigri'	MiG-21MF/UM, L-39ZA
(planned after MiG-23 retired)	
42. Stihaci Letka (SL) 'Pegasus'	MiG-21MF/UM, L-39ZA
43. Stihaci Letka (SL) 'Wolves'	MiG-21UM, L-39ZA
Base flight	Mi-17

32. Základna Tactickeho Letectva (ZTL), Namest nad Oslavou

321. tacticka a pruzkumná letka (tpzlt) 'Biskajsky'	Su-22M-4K/UM-3K, L-29
322. tacticka Letka (tlt)	Su-25K/UBK, L-39ZA
Base flight	Mi-17

34. Základna Skolniho Letectva (ZSL), Pardubice

341. vycvikova letka (vlt)	Aero L-39C/MS, L-29
342. vlt	Zlin 142CAF, Mi-2
(based at Hradec-Kralove in the summer)	
343. pruzkumná dopravni letka (pzdlt)	An-26Z1-M, An-30FG, L-410FG / -UVP, Mi-17

33. Základna Vtrulnikoveho Letectva (ZVrL), Prerov

331. vtrulnikova letka (vlt)	Mi-24D/V/DU
detachment at Line (Army Airfield)	Mi-24D/V
332. Dopravnich a specialnich vrtulnikova letka (ltDsVrt)	Mi-8PPA, Mi-9, Mi-17, Mi-17Z-II, Mi-2

6. Základna Dopravniho Letectva (ZDL), Praha-Kbely

61. Dopravna letka (DLT)	An-26, L-410M/T/UVP, Tu-154B2, Tu-134A
62. vrlt	Mi-8S/T, Mi-17, W-3A, Mi-2

Letecky Zkusebni Odbor (LZO) Praha-Kbely

LZSc1 (based at Line)	MiG-21MF/-UM, L-29, L-39C
LSZc2	An-24, An-26, L-410UVPE, L-610

41. protiletadlová raketová brigáda (plrb) Slany-Drnov

	SAM-3/6/8/10

42. plrb Brno

	SA-3/6/8/10

with SAM regiments at Ostrava (43.plrp), Rozmitál (44.plrp), Kromerice (45 plrp), Strakonice (46 plrp) and radiolocation brigades at Chomutov (41.rtb), Ceske Budejovice (42 smrtb) and Brno

Glossary

Armady Ceske Republiky (ACR)	Army of the Czech Republic
dopravna letka (dlt)	Transport Squadron
Letecky Zkusebni Odbor	Aviation Test Department
letka Dopravních a specialních vrtulnikova (ltDsVrt)	Transport and Special Helicopter Squadron
protiletadlová raketová brigáda	Anti-aircraft missile brigade
pruzkumná dopravni letka (pzdlt)	Recce Transport Squadron
stihaci letka	Fighter Squadron
tacticka a pruzkumná letka (tpzlt)	Tactical and Recce Squadron
tacticka letka (tlt)	Tactical Squadron
Velitelství vzdusnych sil	Headquarters of the Air Force
vrtulniková letka (vrlt)	Helicopter Squadron
vycvikova letka (vlt)	Training Squadron
Základna dopravniho letectva	Transport Air Base
Základna skolniho letectva (ZSL)	Training Air Base
Základna taktického letectva (ZTL)	Tactical Air Base
Základna vtrulnikoveho letectva (ZVrL)	Helicopter Air Base

Czech Republic Police – Aviation

Organised on paramilitary lines, the Czech Republic Police – Aviation branch has its headquarters in Prague. Duties involve aiding the local authorities and disaster relief.

Czech Republic Police – Aviation

Letecka sluzba Policie Ceské Republiky	Praha-Ruzyne

(Czech Republic Police – Aviation)

Bell 412HP	B-4362, B-4363, B-4369, B-4370
Eurocopter BO105CBS-4	B-5265, B-5278, B-5292
Mil Mi-8P	B-8733, B-8938

DENMARK

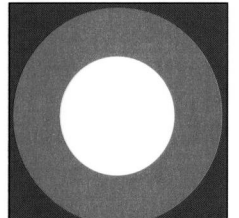

Capital: Copenhagen
Population: 5.2 million
Land area: 43075 km² (16,625 sq miles)
Major cities: Ålborg, Århus, Odense

The Flyvevåbnet (Air Force)

The roots of Danish military flying can be traced back to the very early days of aviation, with the Marinen (Navy) and Hæren (Army) having built up their own flying components since December 1911 and April 1912, respectively. On 1 October 1950 the Flyvevåbnet (Air Force) was officially established as an independent force by fusion of the former Marinens Flyvevæsen (Naval Air Service) and Hærens Flyvertropper (Army Aviation Troops).

COMMAND STRUCTURE

The aim of the Danish armed forces is specified by an act passed by the Folketinget (Parliament) in 1993 as follows: to prevent conflicts and war, to maintain Danish sovereignty and ensure the continuous existence and integrity of the country, and to promote peaceful development in the world with respect for human rights. The act identifies the following main mission areas for the armed forces: conflict prevention, crisis management, defence, peacekeeping, peacemaking and humanitarian missions in the context of NATO, UN and OSCE.

The Flyvevåbnet is under direct command of the tri-service (Army, Navy and Air Force) Chief of Defence, currently General Christian Vit, and is headquartered at the Forsvarskommando (Defence Command) in Vedbæk and subordinate to the Forsvarsministeriet (Ministry of Defence). The Flyvevåbnet has a peacetime strength of 5,700 military personnel, rising to 18,600 in the event of mobilisation.

Organisationally, the Flyvevåbnet consists of the Flyvertaktisk Kommando (Tactical Air Command Denmark, commonly known as TACDEN) with headquarters at Karup and the Flyvemateriel Kommando (Air Material Command) with headquarters at Værløse. TACDEN is responsible for operational control of all manned flying squadrons, SAM missile squadrons, radar surveillance sites and supporting elements, while the Flyvemateriel Kommando is responsible for technical support and maintenance of all aviation-related equipment.

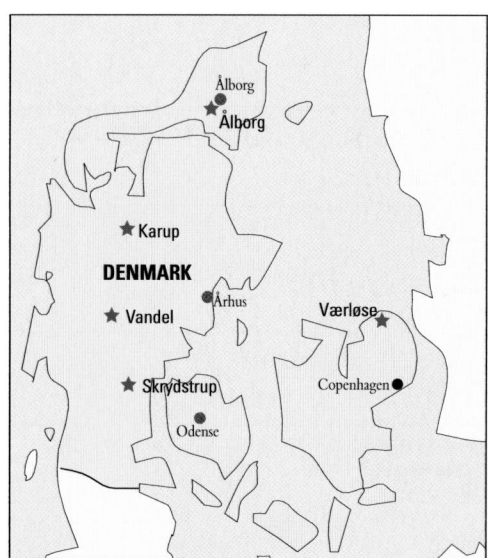

Above: Centrepiece of the Danish air force are the F-16A/Bs operated in four squadrons. These two Block 15 OCU F-16As belong to Esk 726. F-16Bs (right) operate alongside the F-16As.

Within the NATO command structure, the Flyvevåbnet is assigned to Allied Forces Central Europe (AFCENT) and its subordinate command is Allied Forces Baltic Approaches (BALTAP), headquartered at Karup. BALTAP is responsible for all military operations in the area covering Denmark, Schleswig-Holstein north of the river Elbe, parts of Skagerrak, Kattegat, the Danish straits and the Baltic Sea. NATO air operations in the BALTAP area are controlled by the Interim Combined Air Operations Centre (ICAOC) at Finderup in Jutland. The commander of ICAOC is dual-hatted, also being the national commander of TACDEN.

THE COMBAT ELEMENT

The Flyvevåbnet has the General Dynamics (now Lockheed Martin) F-16 Fighting Falcon as its fighter backbone, with all four of its front-line squadrons equipped with this versatile combat aircraft. Denmark signed the initial contract for 58 Fighting Falcons in June 1975 (46 F-16As, 12 F-16Bs) from the SABCA assembly line for 1980-85 delivery. A follow-on order was signed in June 1984 for delivery during 1988/89 of eight F-16As and four F-16Bs, from the Fokker assembly line. When USAF-surplus F-16A/Bs became available, the Flyvevåbnet purchased two batches of three (June 1994) and four aircraft (March 1997).

All four of the Flyvevåbnet F-16 squadrons are tasked with and fully qualified for both air defence and ground/sea attack. One squadron is due to disband, leaving one at Aalborg and two at Skrydstrup.

On delivery, the Flyvevåbnet F-16s were of Block 1, 5, 10 and 15 standard, but during 1981-82 all Block 1 and 5 aircraft were upgraded to Block 10 standard in the Pacer Loft programme carried out by the workshop at Ålborg. Denmark participates in the five-nation NATO F-16 Mid-Life Update (MLU) programme and will upgrade a total of 61 airframes (48 F-16As and 13 F-16Bs) to MLU standard. The Flyvevåbnet MLU Trial Verification Installation airframe, F-16B serial ET204, made history on 12 December 1997 when it completed the first-ever launch of an AIM-120 AMRAAM missile by a European F-16.

The F-16's fine short-field performance has allowed the Flyvevåbnet to make use of small civil airfields for dispersed operations, with eight civilian fields having been designated as Minimum Facility Bases (MFB).

RECONNAISSANCE F-16s

With the closure of Karup by the end of 1993, and the retirement of the Saab 35 Draken from Eskadrille 729 in 1993, the Danish armed forces were left without a national airborne reconnaissance capability. Consequently, the Flyvevåbnet set about developing a reconnaissance pod for the F-16, and on 1 January 1994 Eskadrille 726 at Ålborg took over the reconnaissance role, in addition to its normal multi-role assignment of combined air defence and ground/

sea attack. The squadron employs about one-third of its operational capability in the role of photo-reconnaissance and maintains two reconnaissance pod-equipped F-16s on constant alert.

GROUND-BASED AIR DEFENCE

In 1962 all SAM batteries were transferred from Army to Air Force control, forming the Luftværnsgruppe (Air Defence Group) within the Flyvevåbnet command structure. Headquartered at Skalstrup, south of Copenhagen, it operates eight squadrons of MIM-23B Improved HAWKs (IHAWK), each squadron consisting of six launchers with three missiles each as well as target acquisition radar and system control units. Presently, the IHAWK system is being upgraded to DEHAWK standard (Danish Enhanced HAWK), a purely national programme that will be completed in 2001.

Luftværnsgruppen also is responsible for the Nærluftforsvaret (close air defence) and is tasked with pinpoint air defence of the Flyvevåbnet air bases. In 1996 the anti-aircraft guns were replaced by hand-launched Stinger SAM units, giving the Nærluftforsvaret a much more effective weapon system.

Radar surveillance of Danish airspace is performed by the Kontrol og Varslingsgruppe (Control and Warning Group) which reports to TACDEN. It is integrated with NATO E-3A AWACS radar coverage and is linked to the NATO Air Defence Ground Environment (NADGE).

TRANSPORT AND ARCTIC COMMITMENTS

Three Lockheed C-130H Hercules were delivered to Eskadrille 721 at Værløse in 1975. The C-130s are flown intensively, as three airframes are hardly enough to fulfil the Flyvevåbnet heavylift requirements. Recently, a Danish defence review committee recommended that two additional C-130s should be acquired to ease pressure on the fleet.

In 1982 Eskadrille 721 received three Gulfstream Aerospace Gulfstream III Special Mission Aircraft (SMA) as a result of increased sea surveillance commitments following the 1977 expansion of Denmark's economic zone to 200 nm (370 km; 230 miles). The GIII SMA is a heavily modified Gulfstream III version tailor-made for Eskadrille 721's Arctic operations, for which the unit has one GIII permanently deployed to Luftgruppe Vest (Air Group West) at Søndre Strømfjord in Greenland on a rotational basis. Although the main duties of the GIII lie in the North Atlantic, the aircraft has retained its

(map of Denmark showing Ålborg, Karup, Vandel, Skrydstrup, Århus, Værløse, Odense, Copenhagen)

Flyvevåbnet

Ålborg	
Esk 723	F-16A/B
Esk 726	F-16A/B
Karup	
Flyveskolen	Saab T-17
Skrydstrup	
Esk 727	F-16A/B
Esk 730	F-16A/B
Værløse	
Esk 721	C-130H, Challenger 604, Gulfstream III
Esk 722	Sikorsky S-61A-1/5

Above: Heavy transport is undertaken by the three C-130Hs of Esk 721, which joined the RDAF in 1975.

Below: The Sikorsky S-61A fleet adopted the overall dark grey scheme during the 1990s after wearing silver with Dayglo patches. It is planned to place all helicopter units under joint command.

Søværnets Flyvetjeneste (Navy Flying Service)

Danish Naval Aviation was formally established by the creation of Marinens Flyvevæsen (Naval Air Service) on 14 December 1911. It went through a pre-World War II evolutionary process similar to that of Danish Army Aviation, eventually leading to the termination of naval flying by the 1950 formation of the Flyvevåbnet.

A tragic shipwreck resulted in the need for new naval inspection vessels, the 'Hvidbjørnen'-class, which could deploy helicopters. A total of eight Alouette IIIs was used between 1962 and 1982 with, first, the Alouette Flight of Eskadrille 722 at Værløse, which finally achieved full squadron status on 29 April 1977 when it became the Søværnets Flyvetjeneste (Navy Flying Service), reporting directly to Danish Navy headquarters as an independent unit.

A contract was signed for the purchase of eight Westland Lynx Mk 80s for delivery in 1980/81. Compared with the Alouette III, the Lynx Mk 80 represented a new generation of technology, permitting the Søværnets Flyvetjeneste to fulfil its assigned tasks much more effectively, more rapidly and more securely than had previously been possible. Two Lynx Mk 87s originally built for the Comando Aviación Naval Argentina (Argentine Navy), upgraded to Mk 90 standard, were delivered in 1987/88 to cover losses. The operational capability of the Danish Lynx fleet has been kept continually up to date through various modification programmes. Another major mid-life update has recently been announced, which will effectively extend the operational lifetime of the Lynx until 2015. The mid-life update is expected to be completed by 2004 and the upgrade will convert the Danish Lynx fleet to Mk 90B standard, equivalent to Westland Super Lynx or Royal Navy Lynx HMA.Mk 8 standard.

Søværnets Flyvetjeneste

Værløse	
Lynx Sqn	Lynx HAS.Mk 80A
(with the planned closure of Værløse, the Lynx will go to Karup)	

The only helicopter type currently used by the Danish navy, the Lynx is operated for EEZ and surface search tasks. They carry no weapons.

Esk 721 operates two surviving (of three acquired) Gulfstream IIIs (above). The latest type to join the squadron is the Challenger 604 (below), arriving in 1998. It is configured for VIP transportation.

executive jet capability and is frequently used for VIP transport of the Royal Family and cabinet ministers.

After losing one aircraft, the Flyvevåbnet decided to replace the entire GIII fleet rather than carry out an expensive mid-life update on the type. In December 1997 a contract was signed for the purchase of one specially equipped Challenger 604 for delivery in 1999, with an option on two more. In addition, a VIP-configured Challenger 604 has been leased with a view to eventually buying it, and was delivered to Eskadrille 721 in 1998. Eskadrille 721 is due to move to Aalborg as Værløse is due to close.

SEARCH AND RESCUE

Eskadrille 722 at Værløse air base outside Copenhagen (but due to move to Karup) is responsible for SAR in mainland Denmark. Secondary missions include medical evacuation, transport of personnel, environmental protection and general support to military as well as civil authorities. Eskadrille 722 flies the Sikorsky S-61A Sea King, with eight S-61A1s delivered in 1965 and an attrition S-61A5 acquired in 1971. Eskadrille 722 continuously maintains SAR readiness at Ålborg, Skrydstrup and Værløse, each with one S-61 on 15-minute standby during the day and 60-minute at night. During hard

Only basic flight training is undertaken in Denmark, using the Saab T-17 Supporters of the Flyveskolen. The type also undertakes liaison duties.

winters, the Værløse SAR readiness is relocated to Rønne on the island of Bornholm, reducing flying time to the many fishing areas in the Baltic. The S-61 is scheduled to remain in service until 2003 and the Flyvevåbnet is presently evaluating possible replacements, giving high priority to acquiring combat SAR capability.

TRAINING SYLLABUS

The Flyveskolen (Flying School) is tasked with basic flying training and screening of pilot candidates for the Flyvevåbnet as well as for the flying components of the Danish Army and Navy. Aircraft are drawn from a pool of 28 T-17 Supporters, which are the survivors of 32 T-17s originally purchased in 1975-77. In addition to basic flying training, T-17s are used for local liaison and communication with Eskadrille 721 at Værløse, usually having three aircraft on strength, and Station Flights at the major Flyvevåbnet air bases of Ålborg, Karup and Skrydstrup, each with a single T-17.

Following basic flying training, pilot candidates go through two years of officer education at Værløse, before dispersing to various flying schools in the United States for continued flying training. Fighter pilots undergo 260 flying hours on Cessna T-37 Tweets and Northrop T-38 Talons during a 13-month course with the EuroNATO Joint Jet Pilot Training (ENJJPT) programme at Sheppard AFB, Texas.

Candidates destined for transport aircraft are posted to the US Navy for six months of basic flying training on Beech T-34C Turbo Mentors (92 flying hours) at NAS Whiting Field, Florida, followed by five months of advanced flying training on Beech T-44A Pegasus (87 flying hours) at NAS Corpus Christi, Texas. After that they undergo either 35 hours of C-130 conversion training during a two-month stay with the USAF at Little Rock AFB, Arkansas, or one month of Gulfstream III simulator training with Flight Safety International at Savannah, Georgia.

Flyvevåbnet helicopter pilots start with the same six-month US Navy basic flying training course as transport pilots, but then go through six months of US Navy helicopter training at NAS North Island, California, receiving 116 flying hours on the Bell TH-57B Sea Ranger and 42 hours on the Sikorsky SH-3D Sea King. Hæren (Army) helicopter pilots follow the 11-month training syllabus of the US Army Aviation Center at Fort Rucker, Alabama, while Søværnet (Navy) helicopter pilots are trained at various US Navy flying schools.

Hærens Flyvetjeneste (Army Flying Service)

Danish army aviation dates to 1 April 1912 when a Hæren (Army) officer completed flying training at a private flying school in Copenhagen. It was formally created as an independent unit on 1 February 1923 as the Hærens Flyverkorps (Army Air Corps), later changing its name to Hærens Flyvertropper (Army Aviation Troops). All flying activities within the framework of the Danish Army were closed down by the formation of the Flyvevåbnet (Air Force) in 1950, but on 1 April 1958 Danish army aviation was revived by the creation of the Artilleri-flyvebatteri (Artillery Flying Battery) at Vandel air base in central Jutland, gaining its present designation – Hærens Flyvetjeneste (Army Flying Service) – on 1 July 1971.

Today, the Hærens Flyvetjeneste is tasked with such roles as anti-armour, reconnaissance, light transport and medical evacuation, but a more civil role of ferrying police observers on traffic surveillance is also undertaken.

During 1971 12 Hughes 500M Cayuse helicopters were acquired, followed by another three in July 1974, two having been lost in crashes. As the H.500M originally was acquired as an observation helicopter, the Danish Parliament never granted any funding for purchase of weapons, and to this day the H.500M remains unarmed.

In the 1980s, the Danish Army expressed a wish for an airborne anti-tank capability, leading to a contract for the purchase of 12 AS 350L1 Ecureuil (AS 550C2 Fennec) helicopters for 1990/91 delivery. As Danish Army field units already operated the TOW (Tube-launched, Optically-sighted, Wire-guided) missile system for anti-tank defence, one of the conditions for selection was that the new helicopter had to be compatible with the TOW system.

Organisationally, Hærens Flyvetjeneste is divided into two companies, the Panserværnshelikopter-kopagni (PVHKmp – anti-tank helicopter company) equipped with 12 Fennecs and four H.500Ms, and the observationshelikopterkompagni (OBSHELKm – observation helicopter company) flying the remaining nine H.500Ms. Both companies are presently based at Vandel in peacetime (but due to move to Karup), and in the event of hostilities Hærens Flyvetjeneste would split its entire fleet into several Delinger (platoons) and disperse to various parts of the country. As small independent components, they would operate under direct command of local Army Field Brigades.

All Danish Army AS 550C2s are operated by the PVHKmp, based at Vandel during peacetime. With the planned closure of Vandel, the army helicopters will move to Karup air base.

Hærens Flyvetjeneste

Vandel
OBSHELKm	Hughes 500M
PVHKmp	AS.550C-2, Hughes 500M

ESTONIA

Capital: Tallinn **Population:** 1.6 million
Land area: 45100 km² (17,413 sq miles)
Major cities: Kohtla-Jarve, Narva, Parnu, Tartu

More than 50 years of Soviet occupation of Estonia ended with the August 1991 collapse of the Soviet Union, after which Estonia finally regained national independence. Estonia had been heavily militarised by the Soviets, with some 132,000 Soviet troops (the equivalent of almost 10 per cent of Estonia's population) based at more than 500 military installations of various types and sizes (covering almost 2 per cent of the nation's territory). Estonia is the smallest of the three Baltic states, and steps towards forming a military aviation structure have progressed rather slowly. Important defence and security assistance has been supplied by Western nations and a small Air Force has been created; aviation equipment is also operated by the pseudo-military Border Guard.

Below: The Estonian air force operates two ex-Russian PZL Mi-2Us which display the pre-occupation triangle national insignia.

Below right: All Estonian border guard aircraft display Piirivalve Lennusalk titles and civil registrations. This Let L-410UVP(T) is seen at the Tallinn base and is ex-Luftwaffe 5301 and East German air force 313.

Eesti Õhuvägi (Estonian Air Force)

The Eesti Õhuvägi (Estonian Air Force) was formed on 13 April 1994 by the creation of a separate air force staff within the national defence forces. Initially, the air force was only tasked with ground-based air surveillance and air defence, using old Soviet radar equipment and anti-aircraft artillery.

On 15 May 1997 the Estonian Air Force moved into the ex-Soviet Su-24 base at Ämari, south of Tallinn, with a small fleet of two An-2 biplanes and three Mi-2 helicopters (of which only two are airworthy, as one is used for spares). All were ex-Soviet DOSAAF equipment, and the Mi-2s had previously been operated by the Estonian Border Guard. In May 1998 the air force acquired a PZL-104 Wilga 35 lightplane and a pair of Blanik gliders from a local flying club, and during 1999 it is planned to purchase a pair of Yak-52 trainers.

Presently, consideration is being given to fusing the Estonian Air Force and the Border Guard Aviation Group into a single military air force structure. Future plans include the operation of three squadrons equipped for fighter, rescue and transport roles. However, so far the acquisition of jet fighters is considered unrealistic due to the cost of purchase and operation.

Eesti Õhuvägi (Estonian Air Force)

Piievalve Lennu Eskadril (Air Operations Squadron)
Antonov An-2 (2), LET L-13 Blanik (2),
PZL Swidnik (Mil) Mi-2U (2), PZL-104 Wilga 35 (1)

Bright colours and a prominent 'SAR' marking leave little doubt as to the use to which this Mil Mi-8T of the Border Guards is put.

Eesti Piirivalve (Estonian Border Guard)

The Eesti Piirivalve (Estonian Border Guard) was formed in 1990, mainly tasked with national border patrol as well as SAR, and reports directly to the Ministry of Internal Affairs. In February 1993 Germany donated to Estonia two ex-Luftwaffe Let 410UVPs, which were organised within the Riiklik Lennusalk (National Aviation Group) based at Tallinn International Airport, operating in support of the Border Guard. In October 1994 helicopter operations were initiated with the delivery of three ex-DOSAAF Mi-2s, followed in November 1995 by four Mi-8s (three ex-Luftwaffe and one ex-East German civil). In April 1997 the Riiklik Lennusalk was renamed Piirivalve Lennusalk (Border Guard Aviation Group), still based at Tallinn. By that time the Mi-2 helicopters had been passed to the Estonian Air Force, and during 1997/98 two of the Mi-8s were updated with new radar, IR sensor and hoist to make them more effective in the SAR role. The remaining two Mi-8s are stored at Pärnu airfield in southern Estonia awaiting funds for their update.

Eesti Piirivalve (Estonian Border Guard)

Piirivalve Lennusalk (Border Guards Aviation Group)
Piievalveamet (Flight Operations)	Tallinn
Let L-410UVP Turbolet (2), Mil Mi-8S/T/TB (1/2/2)	

FINLAND

Capital: Helsinki
Population: 5 million
Land area: 337030 km²
(130,095 sq miles)
Major cities: Tampere,
Turku

Suomen Ilmavoimat
(Finnish Air Force)

Finland, the sixth largest European nation, has been independent since December 1917. Finnish military aviation started with the donation of a Morane Saulnier Type D in March 1918 by the Swedish Count Kreivi von Rosen. Its first major action was the Winter War against an opportunistic Soviet invasion on 30 November 1939; a ceasefire was declared after the Soviet Union had taken the territory it had demanded. The German invasion of the Soviet Union led to the Continuation War with the USSR from 22 June 1941 until 4 September 1944, overwhelming numbers leading to a Soviet victory. Finland was restricted to 60 combat aircraft by the 1947 Paris Treaty, and the 1948 Treaty of Friendship, Co-operation and Mutual Assistance gave the Soviet Union undue influence over its national affairs. Today, Finland, like all its neighbours, is faced with a much-changed political situation, military uncertainty in the former Soviet Union, and very tight budgets.

CURRENT COMMAND STRUCTURE

With the recent introduction of the Hornet, the peacetime structure of the Ilmavoimat was changed. Three Air Commands have replaced the previous Wings. Each Air Command comprises one Hävittäjälentolaivue (HäLLv, squadron) and a radar network. Each squadron is made up of four flights, which report directly to it. The north of Finland falls under control of the Lapland Air Command, with the HQ at Rovaniemi; the southeast under Karelian Air Command, HQ at Kuopio-Rissala; and the southwest under Satakunta Air Command, HQ at Tampere-Pirkkala.

The Ilmavoimat HQ is based at Tikkakoski-Jyväskylä, along with its associated Air Support Squadron. The Ilmasotakoulu (Air Force Academy) is based at Kauhava, while the Koelentue (Test Flight) is based at the Ilmavoimat Flight Test Centre, Halli.

MODERN FIGHTER FORCE

Finland is currently replacing its MiG-21s and Drakens with 64 Boeing (formerly McDonnell Douglas) F-18 Hornets (57 single-seat F-18Cs and seven F-18D trainers). The first Finnish Hornet made its maiden flight from the McDonnell Douglas plant at St Louis on 21 April 1995. All seven F-18Ds have now been delivered to the air force. In October 1995 the Finnish aerospace firm Valmet (now renamed Finavitec) started to assemble its first F-18C, which was delivered to the Ilmavoimat in June 1996. The

Top: The F-18C (fore) is due to be Finland's main fighter type well into the 21st century. Equipped with the AN/APG-73 radar, the aircraft is a quantum leap over the MiG-21s and Drakens it replaces.

Above: Armed with Falcon missiles, this Saab 35FS Draken serves with HavLLv 11 at Rovaniemi.

final F/A-18C is expected to be delivered in August 2000. Finland's aircraft are designated F-18, not F/A-18, due to their exclusive air defence tasking.

Since 1972, HäLLv 21 at Tampere-Pirkkala has been a Saab Draken operator. After 25 years the Draken was withdrawn from service, when HäLLv 21 became the first operational Ilmavoimat Hornet unit. The first three US-built F-18Ds were flown to Pirkkala in November 1995, while the first Valmet-built F-18C arrived in July 1996, starting the full conversion to the F-18. The first operational training flight completely converted to the Hornet in the summer of 1997. A second flight operated both the

Above: Finnish Learjet 35As undertake transport, target-towing, photo-survey and EW training roles.

Below: Hawk Mk 51s are used by every wing of the air force. Some have adopted an overall grey scheme.

Suomen Ilmavoimat

Lapin Lennosto (Lapland Air Command)	**Rovaniemi**
Hävittäjälentolaivue 11 (11th Fighter Squadron)	
1st Flight	Saab 35S/FS
2nd Flight	Saab 35S/FS/CS
3rd Flight	Hawk Mk 51/51A
4th Flight	Arrow II, PA-31, Redigo, Vinka
Satakunnan Lennosto (Satakunta Air Command)	**Pirkkala**
Hävittäjälentolaivue 21 (21st Fighter Squadron)	
1st Flight	F-18C/D
2nd Flight	F-18C/D
3rd Flight	Hawk Mk 51/51A
4th Flight	Arrow II & IV, PA-31, Redigo, Vinka
Karjalan Lennosto (Karelian Air Command)	**Rissala**
Hävittäjälentolaivue 31 (31st Fighter Squadron)	
1st Flight	F-18C/D
2nd Flight	F-18C/D
3rd Flight	Hawk Mk 51/51A
4th Flight	Arrow IV, PA-31, Redigo, Vinka
Tukilentolaivue (Air Support Command)	
	Jyväskylä-Luonetjärvi
Tukilentolaivue (Air Support Squadron)	
1st Flight	F27 Elint/Comint, Learjet 35A/S
2nd Flight	Hawk Mk 51/51A
3rd Flight	F27
4th Flight	Arrow II, PA-31, Redigo, Vinka
Ilmasotakoulu (Air Academy)	**Kauhava**
Koulutuslentolaivue (Flight Training Squadron)	
4 x Flights	Vinka, Hawk Mk 51/51A, Arrow II, PA-31, Redigo
Ilmavoimien Koelentokeskus (Air Force Flight Test Center)	
	Halli
F-18C/D, Saab 35FS/S, Hawk Mk 51, Redigo, Vinka, Arrow IV	

Three PA-28RT-201 Cherokee Arrow IVs and four PA-28R-180R Arrow IIs continue to undertake light transport duties in the liaison flights.

Only two Valmet L-70 Vinkas have been lost by the air force since 1980, from 30 acquired. They serve in the basic training and liaison roles.

Redigos are used by the liaison flights of each air command alongside various Piper designs. The air force was the launch customer for the design.

Hornet and the Saab 35FS/CS Draken until September 1997, after which it, too, switched completely to the Hornet. The final Draken mission of HäLLv21 was performed on 30 November 1997 by two Saab 35FSs and one Saab 35CS. The redundant Drakens were transferred to HäLLv 11, at Rovaniemi, which will operate the type for a few more years. In late 1997 HäLLv 21 was declared fully operational on the F-18.

On 7 March 1998 Finland's MiG-21bis fighters roared through Finland's airspace on an operational mission for the last time. Until that final day, HäLLv 31 managed to keep about 14 MiG-21bis 'Fishbed-N' fighters operational, of the 27 originally received in 1980 and 1981. Due to the low number of operational MiG-21s, the transition to the F-18 was welcomed at HäLLv 31. The first Hornets arrived at Kuopio-Rissala in September 1996, making HäLLv 31 the second Ilmavoimat F-18 operator. Conversion from MiG-21 to F-18 takes about one month of computer-based training, followed by just four flights in the F-18D and then the first F-18 solo. The conversion has proved to be quite straightforward. The Ilmavoimat foresaw the problems with the Russian cockpit technology at an early stage and refitted them with Western instrument systems, giving the 'Fishbed' cockpit a similar look to that of the Hawk. In the autumn of 1997 one flight of the squadron was declared operational, and in the summer of 1998 the second flight followed.

Rovaniemi-based HäLLv 11 of the Lapland Air Command is the last Finnish operator of the Saab Draken. In 1972 the first Drakens entered Ilmavoimat service, these being six former Flygvapnet (Swedish air force) Saab J 35Bs (known in Finland as the Saab 35BS). They were leased, pending the delivery of a dozen Valmet-built Saab 35Ss (new-build aircraft, with the Saab designation Saab 35XS), between April 1974 and July 1975. Then, the six leased Saab 35Bs were purchased outright, together with another six Saab 35FSs (ex-Swedish J 35Fs) and three two-seat Saab 35CS trainers (ex-Flygvapnet Sk 35Cs). In 1984 another order was made for 18 (refurbished) Saab 35FSs and two Saab 35CSs. From the 48 Drakens delivered about 23 machines survive today. Three Draken variants currently used by HäLLv 11 are the Valmet-built Saab 35S, former Swedish Air Force Saab 35FS and the Saab 35CS. The last Draken overhaul was completed by Finavitec in December 1998. HäLLv 11 is destined to become the third and last Hornet operator, in the next millennium. According to the F-18C delivery schedule, the Lapland squadron will receive the first Hornets in the autumn of 1998. It is expected that the Draken will be flown until 2001, although the FAF has plans to fly it until 2005.

The third flight of each Air Command is a training flight equipped with the BAe Hawk. In 1980 the first four BAe-built Hawk Mk 51s arrived in Finland, while Valmet assembled the remaining 46. Between 1993 and 1994 seven additional Hawk Mk 51As were delivered as a 'top-up' batch. Although the Hawk is officially registered as a trainer (and not counted among the 60 fighter aircraft of the Paris Treaty), it is of course a great asset as a small fighter.

Each Air Command's fourth flight is the liaison flight, normally comprising one Piper PA-28 Arrow, one Piper PA-31 Chieftain, two Valmet L-90TP Redigos and one or two Valmet L-70 Vinkas. The oldest liaison aircraft are four surviving PA-28R-180R Piper Arrow IIs of the six delivered since 1974. In 1980 four PA-28RT-201 Piper Arrow IVs were added to the fleet. From 1983 six Piper PA-31-350 Chieftains further boosted capacity. The newest aircraft to enter service are 10 Valmet L-90TP Redigos which were taken on strength from 1991 onwards.

Ilmavoimat cadets start their four-year course at the Ilmasotakoulu (Air Force Academy) at Kauhava. Basic flight operations take place with the Koulutuslentolaivue (Training Squadron). This unit is equipped with the Valmet L-70 Vinka, of which the Ilmavoimat has received 30 since 1980. Students fly the Vinka for 11 months for a total of 45 flights. During the following three years both Vinka and Hawk are flown at Kauhava. After 60 hours on the Vinka and 100 hours on the Hawk the cadet graduates to pilot status. To develop operational readiness the new pilot is transferred to one of the three Air Commands, to fly Hawks for approximately 12 months and 150 hours. In the second year, 120 flight hours on the Hornet are required. After this, the pilot has reached the final phase of full operational readiness in which he is certificated to fly the F-18 and the Hawk.

The Tukilentolaivue (Air Support Squadron), established in early 1997, has Hawks fitted with reconnaissance pods taken from MiG-21bis/Ts and three Fokker F27s. One, a F27-100 delivered in 1982, had been based in a special hangar at Jyväskylä since its modification to serve as an Elint/Comint platform in the late 1980s. The other two Fokkers – another F27-100 delivered in 1980 and an F27-400M delivered in 1984 – are tasked with regular transport. Replacement of the F27s is being considered, with interest shown in purchasing three civil ATR72s from Finnair. The barter of Finnish-built Sisu amphibious vehicles for the Indonesian-built IPTN CN.235s is also a possibility.

Three Learjets delivered in 1982 perform a multitude of tasks including passenger transport, maritime patrol, target-towing, photo-reconnaissance, aerial mapping, passive electronic warfare training, calibration, AWACS duty, fall-out surveillance/air sampling, NBC reconnaissance, SAR and Elint.

Maavoimat (Army Aviation)

In January 1997 the Ilmavoimat's Helikopterilentue (Helicopter Flight) at Utti was disbanded and its two Hughes 500D, six Mi-8T and two Mi-8P 'Hip' helicopters were handed over to the Maavoimat, the Finnish Army, which also took over the base at Utti. One Hughes 500D was destroyed in a landing accident during 1998, so it is planned to acquire two used examples during 1999.

The Maavoimat is at the heart of current military aviation expansion plans in Finland. In March 1997 the government announced its intention to establish three rapid-reaction brigades. Preparation had already started in 1996 under an army training initiative. Experience in Bosnia as part of the IFOR/SFOR operation and the prospect of future operations and exercise under NATO's PfP (Partnership for Peace) initiative underlined the need to make Finland's armed forces more flexible and less devoted to all-out war scenarios.

To accomplish this, the Army hopes to acquire between 30 and 45 transport helicopters and 10 to 15 combat helicopters.

Maavoimat (Army Aviation)	
Utin Jääkärirykmentti/Helikopterilentue	Utti
H-500D, Mi-8T/P	

Rajavartiolaitos (Frontier Guard)

The paramilitary Rajavartiolaitos operates helicopters and aircraft to protect Finland's lengthy borders, in conjunction with a large force of ground vehicles and naval vessels. The flying branch of the Rajavartiolaitos, the Air Patrol Squadron, operates from three airfields – Helsinki, Turku and Rovaniemi – to cover southern, western and northern Finland, respectively. Currently, the Air Patrol Squadron is equipped with four AB 206s, four AB 412s, three AS 332L1 Super Pumas and two Dornier Do 228s.

One of the four AB 206 helicopters is an ex-Ilmavoimat AB 206A, while the remainder are AB 206Bs. In addition to border patrol, the AB 206s are also used for training duties. These airframes are due to be replaced soon by three new helicopters, most likely newly purchased AB 206s. The AB 412 fleet is divided among three standard AB 412SPs and one AB 412EP, equipped with a Bendix 1500. Both AB 412s and Super Pumas are fully equipped for SAR duties with searchlights and rescue winch installation. The two Dornier Do 228s are equipped with a SLAR and other sensors for maritime patrol duties. All Finnish Frontier Guard pilots have served with the Ilmavoimat before joining the Frontier Guard. Its aircraft have been purchased from the Finnish Defence budget and are civil registered to avoid document problems with emergency diversions to neighbouring countries, which can occur due to rapidly-changing Finnish weather conditions. In time of crisis, the Frontier Guard would come under command of the Finnish Navy.

Rajavartiolaitos (Frontier Guard)	
Air Patrol Squadron detachments at Helsinki, Turku and Rovaniemi	
AB 206A/B, AB 412SP/EP, AS 332L1, Dornier Do 228	

Above: Both Dornier Do 228-213s of the Finnish Frontier Guard are equipped with maritime sensors.

Below: Finnish army Mil Mi-8s served from 1973 until 1997 in the air force. This example is an Mi-8T.

FRANCE

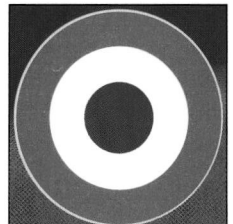

Capital: Paris
Population: 56.6 million
Land area: 543965 km² (209,970 sq miles)
Major cities: Bordeaux, Lyon, Marseilles, Nantes, Nice, Strasbourg, Toulouse

Armée de l'Air (French Air Force)

France's Armée de l'Air (AA) is organised on three levels: the high command, the major commands and the air bases. The high command is subordinate to the chief of the air staff, or Chef d'Etat-Major de l'AA (CEMAA), Général Jean Rannou. Under his authority, five organic commands are in charge of the Air Force operational readiness: Air Combat Command, or Commandement de la Force Aérienne de Combat (CFAC); Air Mobility Command, or Commandement de la Force Aérienne de Projection (CFAP); Air Surveillance, Information and Communication Systems Command, or Commandement Air des Systèmes de Surveillance, d'Information et de Communication (CASSIC); Air Force Education and Training Command, or Commandement des Ecoles de l'AA (CEAA); and Air Force Ground Security Command, or Commandement des Fusiliers Commandos de l'Air (CFCA). Two other commands are listed as operational commands: Strategic Air Command, or Commandement des Forces Aériennes Stratégiques (CFAS); and Air Defence and Air Operations Command, or Commandement de la Défense Aérienne et des Opérations Aériennes (CDAOA). The CFAS controls all Air Force nuclear assets, while the CDAOA is tasked with defining the missions and the deployment of the Air Force.

The reorganisation of French armed forces introduced in 1997 will considerably change the appearance of the AA. From 83,000 personnel today, total strength is due to drop to 71,000 by 2002, with civilians and 'military air technicians' replacing the current 26,000 conscripts fulfilling operational or

The main air defence aircraft of the French air force is the Mirage 2000C. A total of 124 has been received. This example flies with EC 1/12.

Armée de l'Air

Etat-Major de l'Armée de l'Air (EMAA)

Direct Reporting Units

Centre d'Expérimentations Aériennes Militaires 00.330 Mirage 2000B/C/D/N, Mirage F1C/CR/CT/B, Alpha Jet E, DHC-6-200/300, MS 760 Paris, TBM 700	BA118 Mont-de-Marsan
Escadron de Convoyage 00.070 DHC-6-300	BA279 Chateaudun

Air Force Regions

Région Parisienne
No base flights are currently known.

Région Aérienne Atlantique

Base Aérienne 120 Falcon 20C	BA120 Cazaux
Base Aérienne 273 – CVAA 55/273 Jodel D140E/R	BA273 Romorantin
Base Aérienne 721/SAVV Jodel D140E/R	BA721 Rochefort
Base Aérienne 722/SAVV Jodel D140E/R	EETAA722 Saintes

Région Aérienne Méditerranée

Base Aérienne 126/SAVV Jodel D140E/R, MS 760 Paris	BA126 Solenzara
Base Aérienne 200/SAVV Jodel D140E/R	BA200 Apt
Base Aérienne 278/SAVV Jodel D140E/R	BA278 Amberieu
Base Aérienne 749/SAVV Jodel D140E/R	EPA749 Grenoble

Région Aérienne Nord-Est
No base flights are currently known.

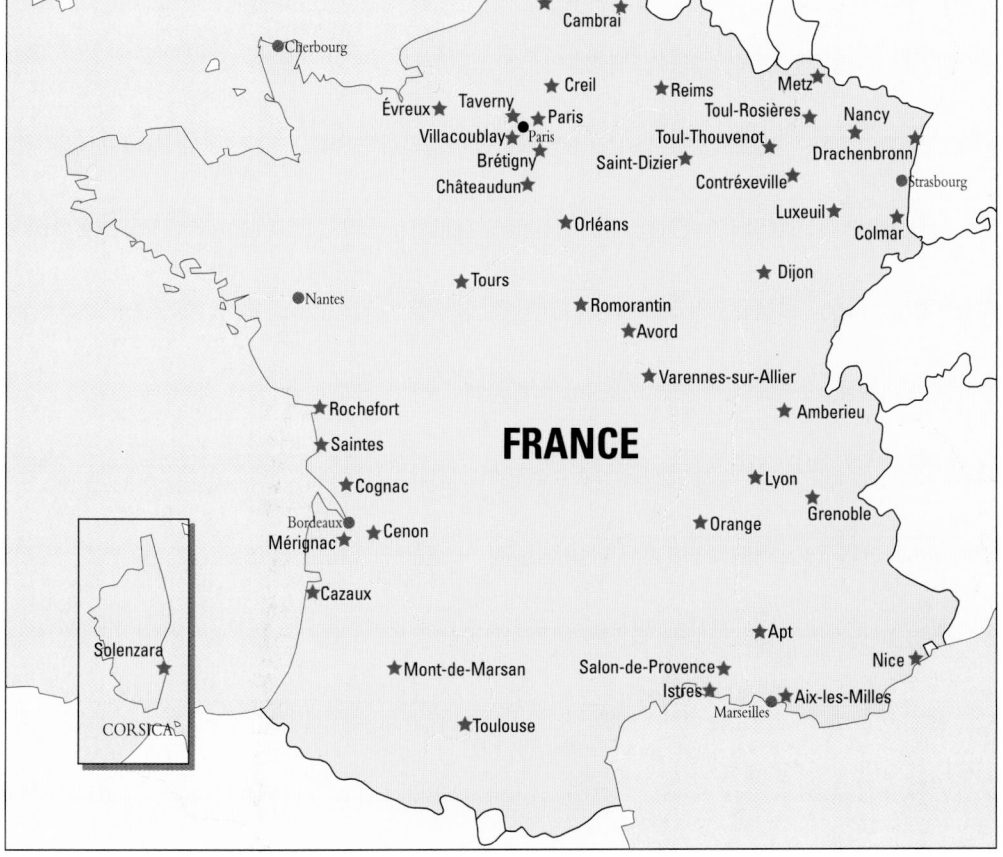

Calais			
Doullens			
Cambrai			
Cherbourg			
Creil	Reims	Metz	
Évreux	Taverny	Toul-Rosières	Nancy
Villacoublay	Paris	Toul-Thouvenot	
Brétigny	Saint-Dizier	Drachenbronn	Strasbourg
Châteaudun	Contréxeville		
Orléans	Luxeuil	Colmar	
Tours	Dijon		
Nantes	Romorantin		
	Avord		
	Varennes-sur-Allier		
Rochefort	Amberieu		
Saintes	**FRANCE**		
Cognac	Lyon		
Bordeaux	Cenon	Orange	Grenoble
Mérignac			
Cazaux	Apt	Nice	
Solenzara	Mont-de-Marsan	Salon-de-Provence	
CORSICA	Toulouse	Istres	Aix-les-Milles
		Marseilles	

Three squadrons of Mirage 2000Ds based at Nancy are primarily tasked with conventional strike missions. They can be supplemented by the three squadrons of 2000Ns which are primarily tasked with 'pre-strategic' strike missions. The 2000D is optimised for attack using precision-guided weapons such as the BGL-1000 laser-guided bomb and AS30L laser-guided missile. Autonomous designation for both is provided by the PDLCT pod, carried on a pylon under an intake. In the near future Mirage 2000Ds will field the APACHE stand-off munitions dispenser.

Europe

Above: Mirage 2000B conversion trainers retain an operational capacity. This example displays the famous stork emblem of EC 1/5 on the tail.

Above right: The Jaguar A's survival in service is partly due to its suitability for overseas deployment. Wearing a temporary desert scheme, this EC 1/7 aircraft carries a practice bomb dispenser under the centreline hardpoint.

Above: A true multi-role aircraft, the Mirage F1CT serves with EC 30 based at Colmar. This variant has been in service since 1992.

Commandement de la Force Aérienne de Combat (CFAC)	
Escadron de Chasse 01.002 'Cigogne' Mirage 2000C	BA102 Dijon
Escadron de Chasse 02.002 'Côte d'Or' Mirage 2000B/C	BA102 Dijon
Escadron de Chasse 01.003 'Navarre' Mirage 2000D	BA133 Nancy
Escadron de Chasse 02.003 'Champagne' Mirage 2000D	BA133 Nancy
Escadron de Chasse 03.003 'Ardennes' Mirage 2000D	BA133 Nancy
Escadron de Chasse 01.005 'Vendée' Mirage 2000B/C	BA115 Orange
Escadron de Chasse 02.005 'Ile de France' Mirage 2000C	BA 115 Orange
Escadron de Chasse 01.007 'Provence' Jaguar A	BA113 St Dizier
Escadron de Chasse 02.007 'Argonne' Jaguar A/E, Alpha Jet E	BA113 St Dizier
Escadron de Chasse 03.007 'Languedoc' Jaguar A	BA113 St Dizier
Escadron de Chasse 01.012 'Cambrésis' Mirage 2000B/C	BA103 Cambrai
Escadron de Chasse 02.012 'Picardie' Mirage 2000B/C	BA103 Cambrai
Escadron de Chasse 01.030 'Alsace' Mirage F1CT	BA132 Colmar
Escadron de Chasse 02.030 'Normandie-Niemen' Mirage F1CT	BA132 Colmar
Escadron de Chasse 03.033 'Lorraine' Mirage F1B/F1C	BA122 Reims
Escadron de Reconnaissance 01.033 'Belfort' Mirage F1CR	BA122 Reims
Escadron de Reconnaissance 02.033 'Savoie' Mirage F1CR	BA122 Reims
Escadron Electronique Tactique 01.054 'Dunkerque' C.160G Gabriel (EET 01.054 is 'on loan' to CFAC from CFAP)	BA128 Metz
10 air defence SAM squadrons using Aspic and Crotale NG	

Commandement de la Défense Aérienne et des Opérations Aériennes (CDAOA)   HQ Taverny
Responsible for air defence, using the assets of CFAC

support activities. At the same time, the number of air bases will drop from 42 in 1998 to 32 in 2002, although the number of bases hosting combat units will remain unchanged. Since the dismantling of the wing (escadre) organisation, all the squadrons (escadrons) have been reinforced (each combat unit now boasts 20 aircraft) and have gained a certain autonomy.

AIR COMBAT COMMAND (CFAC)

The CFAC is primarily tasked with the defence of French national airspace. Within the framework of particular military actions (international actions, overseas deployment, etc.), it conducts the air battle for air superiority and offensive actions on land and sea. The CFAC is also tasked with electronic and photographic intelligence gathering.

The CFAC operates a total of 315 combat aircraft assigned to 15 front-line squadrons and two reconnaissance squadrons. In order to fulfil its electromagnetic intelligence role, the CFAC also includes the 54th Air Intelligence Squadron with two C.160G Transall Gabriels at its disposal.

All the CFAC units now have an inflight-refuelling capability which enables them to meet France's world-wide commitments. The pre-positioning of combat aircraft units, mainly in Africa and in the French overseas departments and territories, considerably increases the operating radius of these assets.

Following the withdrawal of the last Mirage III and Mirage 5F in the early 1990s, the combat aircraft fleet now consists of Jaguar, Mirage F1 and Mirage 2000. From 2002, the retirement of the last Jaguars and Mirage F1s in their different versions will greatly increase the fleet's homogenisation. At that point, the AA will operate a fleet of some 370 combat aircraft (including 60 Mirage 2000Ns from the strategic air command, CFAS), a figure similar to that of today.

Initially scheduled for 1996, the arrival of the Rafale will not take place until 2005 (20 years after the demonstrator's first flight). Deliveries will be spread over nearly 15 years.

Because of very restrictive budgets, the CFAC is currently encountering severe difficulties in buying spare parts and maintaining its fleet. For this reason, the equivalent of nearly 20 Mirage 2000s are currently kept on a non-operational status.

MIRAGE 2000C

Air defence squadrons use the Mirage 2000C-RDI (Cambrai-based 1/12 and 2/12, and Orange-based 1/5 and 2/5), the first Mirage 2000 to enter operational service with the AA, in 1984. This situation is currently changing quickly with the introduction of the 'new' Mirage 2000-5, Dijon-based 1/2 and 2/2 being the first squadrons to be re-equipped, having previously flown the 2000C-RDM.

Although a very capable aircraft, the Mirage 2000-5 was not unanimously welcomed within the AA. When the idea was raised of modifying some 2000C-RDIs to 2000-5 standard, some observers feared that it would further delay the arrival of the Rafale, which turned out to be the case. The number may change in the future, but the AA is scheduled to receive a total of 37 Mirage 2000-5Fs (F for France),

all retrofitted from 2000C-RDI. The first aircraft (serial number 38) was delivered in December 1997. Eleven machines were due to be delivered in 1998, 22 in 1999 and the remaining three in 2000. Thirty aircraft will be given to squadrons and seven aircraft will be kept in reserve.

The squadrons to be re-equipped will be 1/2 'Cigogne' and 2/2 'Côte d'Or', and the first squadron is expected to reach operational status by January 2000. Thirty-seven aircraft are not enough to equip two squadrons, so further 2000-5 orders could be considered in the near future.

The utilisation of the first operational aircraft at Dijon will be shared between French and Taiwanese pilots trained in France. In order to fill the gap before the delivery of the complete 2000-5 batch, Alpha Jets will be lent to 1/2 and 2/2 so that a sufficient number of machines will be available for the Dijon-based pilots to fly their annual 180 hours.

To optimise the aircraft, the Mirage 2000-5s are being retrofitted from the most recent Mirage 2000C versions, i.e., those fitted with the RDI radar. Later, the older RDM radars still in service will be replaced with RDI radars.

The main asset of the Mirage 2000-5 is its RDY radar, which turns the Mirage into the first operational multi-target European combat aircraft. Combined with the MATRA/BAe Dynamics MICA, the Mirage 2000-5 is able to follow and simultaneously engage four different aerial targets. The MICA is a fire-and-forget missile with a range from a few hundreds metres to more than 60 km (37 miles), from the dogfight to the medium-range interception. A light missile (110 kg/242 lb), four MICAs can be carried under the belly of the Mirage 2000. Compared with the previous configuration – two Super 530Ds and two Magic 2s under the wings – this configuration leaves enough space for two 1700- or 2000-litre (374- or 440-Imp gal) underwing fuel tanks. When added to the centre-line 1300-litre (286-Imp gal) fuel tank, air defence endurance is doubled, from 90 minutes to more than three hours, without inflight refuelling.

Initially planned for 1997, the first deliveries of the MICA missiles have been rescheduled for 1999. Simultaneously, procurement by the AA has been slashed to 1,000 missiles from 2,300.

For self-protection, the aircraft are fitted with a Thomson Serval RWR, a Dassault Electronique Sabre jammer and MATRA/BAe Spirale flare dispenser. The AA aircraft can also carry an additional MATRA Eclair flare dispenser in the parachute cone under the vertical tail.

The AA claims that the Mirage 2000-5 is not a transition aircraft, and will form the framework of air defence units well beyond 2010. The first Rafales the AA will receive in 2005 will be two seat-strike aircraft.

MIRAGE 2000D

The AA is expected to receive a total of 86 Mirage 2000Ds by 2002. Three squadrons (1/3 'Navarre', 2/3 'Champagne' and 3/3 'Ardennes') are equipped with 20 machines each, the remaining aircraft being used to make up for the losses. 1/3 'Navarre' was the first squadron to receive the state-of-the-art strike aircraft in 1993. Just a few weeks after receiving its

first six aircraft, 1/3 was launching them from Cervia during Opération Crécerelle over former Yugoslavia.

The presence of a weapon system operator (WSO) in the backseat, and a very complete avionics suite, means that the Mirage 2000D has the capability to fulfil any strike missions previously undertaken by the Mirage F1CT or the Jaguar, with an increased precision and a true all-weather capability. The crews spend an average of 20 per cent of their 180 yearly flying hours on night missions training, twice as much as other combat units.

Of note is the Mach limit of the Mirage 2000D, compared with that of the 2000C from which it was extrapolated. An all-weather strike aircraft designed for conventional low-altitude penetration, the 2000D has a top speed of Mach 1.4. In order to reduce its radar cross-section, the Mirage 2000D has received a thin gold layer on its frontal canopy.

Like every modern strike aircraft, all mission preparation can be done on a computer and the data downloaded into the aircraft's computer before take-off. The onboard computer can keep the equivalent of 1 million km² (386,000 sq miles) of terrain in its memory for nap-of-the-earth flying without using the Antilope 5 radar in its terrain-following mode.

Like the Mirage 2000-5, which lacks the MICA, the Mirage 2000D lacks, for the time being, a conventional cruise missile which would increase stand-off strike capabilities. Its main armament remains the AS30L laser-guided air-to-ground missile, fired with the PDLCT (laser designation and thermal camera) pod, and 250-, 400- and 1000-kg laser-guided bombs.

One hundred anti-runway Apache cruise missiles were ordered in 1997 and are due to enter service in 2000. The Scalp EG (EG for Emploi Général – general purpose) unitary warhead cruise missiles will not be available before 2003.

JAGUAR

Workhorse of the AA for the past 25 years, the Jaguar has been supplemented by more modern aircraft but will continue well beyond 2000. Sixty aircraft are still operational (and 20 others are stored in flying condition) and around 50 should still be available between 2002-2005.

In order to regain some potential and power, the Adour jet engines are being overhauled; for economic reasons, replacing them was not possible. The Jaguar has participated in all conflicts in which France has been involved over the past 25 years, and has logged a vast quantity of flying hours at all altitudes. Still, this tough aircraft seems to be irreplaceable, especially on the African continent where its ruggedness is much appreciated by pilots and technical crews.

Some efforts have been made in the past years to improve its capabilities, especially the avionics. The Jaguars have been fitted with numeric navigation aids, GPS receivers and large flare dispensers. All are now wired to use the ATLIS target designation system (first used in 1987) for laser-guided armament (AS30L and LGBs). The Jaguar cockpit accommodates a video screen on the right-hand side and a joystick, for directing the camera, on the left. The ATLIS-equipped Jaguars are commonly used to designate and illuminate targets for Mirage F1CRs and F1CTs.

Ground attack with laser-guided armament is the primary mission of Jaguar-equipped 3/7 'Languedoc', whereas 1/7 'Provence' is mainly tasked with anti-radar missions using the AS37 Martel.

The secondary mission of the Jaguar is reconnais-

Above: Seen refuelling a Mirage 2000D, the Istres-based C-135FRs gained CFM56 engines from 1985.

Below: Five active (and five stored) Mirage IVPs remain to provide a long-range strategic reconnaissance capability.

sance, for which it is equipped with the OMERA 40 camera. The Jaguar can also be fitted with the RP-36 reconnaissance pod, which is a 1700-litre (374-Imp gal) external fuel tank modified to carry a full batch of cameras. The RP-36 pod has been used since 1984.

EC 3/7 'Languedoc' from St Dizier should be the first squadron to trade its Jaguars for two-seat Rafales (for low-level penetration) in 2005, when a first squadron with 20 aircraft will be formed.

The final retirement of the Jaguar and the loss of the AS37/Jaguar combination should also mean an end to all AA anti-radar missions. It is understood that France will rely on the expertise of its allies (UK and Germany) in this specific field, although there is some mention of the possible use of HARM or ALARM missiles with the Rafale.

MIRAGE F1

In 1998, the AA boasted a force of 115 Mirage F1CRs and CTs. This figure should drop to 85 in 2002-2005 (40 years after the F1's first flight, in December 1966), before the type retires around 2020. The last F1Cs fly with the Djibouti-based 4/33 'Vexin', which runs a small detachment of fewer than 10 aircraft, and the Reims-based 3/33 'Lorraine', the operational conversion unit. 3/33 flies 15 F1Cs along with 10 Mirage F1B two-seaters, and is also tasked with the fighter training of the Mirage F1CR reconnaissance pilots.

Like the Jaguar, the reconnaissance version of the Mirage F1C, the F1CR, has been present in every AA action from Africa to the Gulf War and Kurdistan. It equips 1/33 'Belfort' and 2/33 'Savoie', each of which has 20 examples. The F1CR can carry a very large range of equipment. The aircraft is permanently fitted with a camera, either the panoramic OMERA 40 or the OMERA 33. A Super Cyclope infra-red sensor replaces the right-hand 30-mm gun. The data obtained from the Super Cyclope can be relayed in flight to a SARA tactical ground processing unit. On the centreline hardpoint, the F1CR can be fitted with either a Raphaël SLAR or an ASTAC Elint pod, the latter being used to monitor, pinpoint and catalogue radars emissions.

In the near future, the F1CR will receive an electro-optical (video) reconnaissance capability, replacing the current film-loaded cameras.

Colmar-based 1/30 'Alsace' and 2/30 'Normandie-Niemen' fly the Mirage F1CT tactical fighter-bomber. The F1CT (Chasse et Tactique – Fighter and Tactical) was conceived in order to exploit the poten-

tial of Mirage F1C-200 fighters from EC 5 when they began to be replaced with the new Mirage 2000 RDI. F1Cs were thus turned into ground attack aircraft by adding new capabilities, and leaving untouched the existing air defence potential. The aircraft are equipped with a new attack and navigation system, similar to the one fitted to the F1CR, allowing pinpoint navigation and delivery of precision armament. The F1CT was supposed to fill the gap between the retirement of the Mirage 3 and Mirage 5F and the arrival of the Rafale, initially expected in 1996. With Rafale deliveries being postponed to 2005, at the earliest, the 55 Mirage F1CTs produced will have to soldier on at least 10 years more than expected.

In the near future, the installation of a camera in a small equipment bay aft of the forward landing gear should allow the F1CT to fulfil some reconnaissance missions, apart from its fighter-bomber capabilities. There is little doubt that the F1CT will be the 'jack of all trades' of the AA in the early stage of the 21st century. Especially in the overseas theatres, the Mirage F1CR is overtasked and the Mirage 2000 too sophisticated.

In order to fulfil its electronic intelligence role, Air Combat Command also includes the 54th Air Intelligence Wing equipped with two C.160G Transall Gabriels (the aircraft are flown by CFAP crew). Specialised in collecting electromagnetic and photographic intelligence, the Gabriels are due to be modernised. They carry a crew of 14, including nine system operators. Its highly classified equipment includes two Elint ASTAC systems, similar to the one carried on the Mirage F1CR, one Comint EPICEA system and two panoramic OMERA 51 cameras in a rear lateral bay to photograph the identified sites.

STRATEGIC AIR COMMAND (CFAS)

All AA nuclear assets are now controlled by the CFAS (Commandement des Forces Aériennes Stratégiques). Since the retirement of the last Mirage IVP in the nuclear role in 1996, and the disbandment of 2/91 'Bretagne', nuclear missions are undertaken by the three Mirage 2000N squadrons: Luxeuil-based 1/4 'Dauphiné' and 2/4 'La Fayette', and the Istres-based 3/4 'Limousin'. The 75 Mirage 2000Ns ordered have been delivered to the AA. Primary armament is the medium-range air-to-ground nuclear missile ASMP which entered in service in May 1986 with the Mirage IVP. An improved and 'hardened' version with a greater range, the ASMP-A, is expected to enter service around 2007.

Secondary mission is all-weather conventional strike and deep penetration. The Mirage 2000Ns were used in this role above the former Yugoslavia in 1994. Launched from the Italian base of Cervia, they were the first war missions undertaken by CFAS forces since their creation. However, five Mirage IVPs belonging to strategic reconnaissance squadron 1/91 'Gascogne' are still used for strategic reconnaissance, with the CT 52 photo pod carried in place of the former AN52 free-fall atomic bomb. These aircraft are regularly used to supplement the Mirage F1CR on long-range reconnaissance missions (Somalia, Bosnia, Rwanda, Chad and Zaïre, for example).

Historically, the tanker fleet has always belonged to the CFAS, although the aircraft are now commonly used by Air Combat Command's machines. All the tankers are now regrouped into a single air refuelling squadron, ERV 93 'Bretagne', which runs a fleet of 11 C-135FRs and five KC-135s. All the KC-135s will receive new engines (CFM-56s). The idea of buying a small batch of second-hand Airbus A310s and turning them into tankers has been abandoned for the time being, on economical grounds.

The CFAS includes the GA (Groupe Aérien) 59 'Bigorre' and its four C.160H Transall Astartés, which are used for relaying orders from the highest authorities to the French SSBN. The very low frequencies involved in these communications oblige the Transall to unwind special antennas while orbiting at high altitude. One antenna is 8500 m (27,887 ft) long and the other one 1800 m (5,905 ft). Like the Gabriel, the Astarté are due to be modernised in the near future.

AIR MOBILITY COMMAND (CFAP)

Responsible for the projection and deployment of men and equipment, and the logistic support of forces, the CFAP undertakes nearly one-third of the total AA flying hours. The CFAP is also tasked with inflight-refuelling missions, using its tactical refuelling aircraft (C.160NG) either for buddy-buddy refuelling or for refuelling CFAC aircraft on operations. The total strength of the command is 4,500 personnel, including 1,460 crews, 155 aircraft of 15 different types and 112 helicopters of six different types.

The core of the command is the 85-aircraft tactical transport fleet, which comprises 71 C.160 Transalls and 14 C-130H Hercules.

RENOVATED TRANSALL

The CFAP is currently involved in an upgrade programme for the ageing C.160 Transall. Sixty-six examples of this very capable tactical aircraft are being upgraded with a new avionics suite and a complete self-protection kit. This effort was triggered by participation in operations in former Yugoslavia, and particularly the Sarajevo air bridge, which highlighted a dire need for self-protection for the aircraft.

The first C.160R (R for Rénové) was delivered in 1994 and the programme will be completed in 1999 with the delivery of the 66th and last aircraft. The aircraft are fitted with a Sherloc RWR, a Spirit chaff and flare dispenser and an Incoming Missile Warning System. The cockpit is also being renovated and made compatible with the use of NVGs. A head-up display is fitted on the left-hand seat. The new Transall also benefits from the computerised mission preparation system which, until now, has exclusively been used by combat types.

The CFAP has recently reorganised, giving more autonomy to each squadron. Each tactical transport squadron (1/61, 2/61 and 3/61 at Orléans; 1/64 and 2/64 at Evreux) takes turns on different 'reaction positions' dubbed PR1, PR2 and PR3. PR1 means the squadron must be able to project its strength overseas in less than 24 hours. If the need arises, the unit must be able to deploy at very short notice. Introduced for the first time early in 1997, this organisation was quickly put into action during Opération Pélican when Westerners had to be evacuated from Congo in June 1997.

If there is no alert, the squadron devotes all its activity to operational training. The PR1 position lasts six weeks, as does PR2. In PR2, the squadron can follow normal activity but it must be able to reach PR1 status in less than 48 hours. After 12 weeks of PR1 and PR2 positions, the squadron reaches PR3 and holds it for six months. The unit is then tasked with normal activity, like providing aircraft and crew for regular overseas deployment such as to Bangui, Libreville, N'djamena and Djibouti.

Some transport aircraft crews also receive specific training for special operations. A 'special ops division' was created in September 1993 within the Toulouse-based CIET (Centre d'Instruction des Equipages de Transport) training centre. Four highly specialised crews are currently available and CFAP keeps a C.160R in Toulouse and two C-130s in Orléans on constant readiness for special operations. According to some sources, these C-130s were received by the CFAP only to fulfil the need for special ops. These missions include commando infiltration and exfiltration, mainly at night.

The transport fleet also incorporates other types of aircraft to meet specific requirements (liaison flights, VIPs, light transport, etc.) They include two Falcon 900s, four Falcon 50s, six Falcon 20s, 11 TBM 700s, 19 Nord 262s, six DHC-6 Twin Otters and eight CASA 235s. Before 2001, CFAP will receive another seven CASA 235s, reaching a total fleet of 15 aircraft. In the future, however, CFAF would need more aircraft of this type to rationalise the fleet and replace the Evreux-based DHC-6 and the last 19 Nord 262s at Villacoublay.

The long-distance transport capacity is provided by 3/60 'Esterel' operating from Paris Charles de Gaulle airport. Since the retirement of its old DC-8-55, the squadron uses two CFM56-powered DC-8-72s and two Airbus A310s. Two other Airbus aircraft may join the fleet in 1999.

HELICOPTERS

The CFAP also boasts a force of nearly 110 helicopters. The most numerous type is now the twin-engined Ecureuil/Fennec, 43 of which are in service. These aircraft are mostly tasked with MASA (Mesures Actives de Sûreté Aérienne) – air defence against terrorist threats – missions. The helicopters, armed with a 20-mm gun or snipers, patrol vulnerable sites (such as the Kourou space centre in French Guyana) or important events (VIP visits or public events.)

For liaison missions, the AA still operates 24 Alouette IIIs and six Ecureuils. Three VIP Super Pumas are flown by Villacoublay-based 3/67 'Parisis', while another Super Puma and three Pumas are used with Aix-based 5/67 'Alpilles'. Of note are the three AS 532 Cougars employed by Evreux-based GAM 56 'Vaucluse' for special missions.

Above: Limited procurement of the CASA 235s fills a niche role for smaller loads, freeing up Transalls and Hercules for other tasks. CFAP CASA 235s serve with ETL 1/62 based at Creil.

Above far left: CFAP's workhorse is the C.160 Transall, in service since 1967 and reinstated into production in 1982. The fleet is being upgraded.

Above left: Light transport duties have been performed by the SOCATA TBM 700 since 1991.

Below: The French air force is a major helicopter user. Since 1990 the AS 555AN Fennec has been used for base security patrols. This example displays the codes for EH 5/67 from Aix-les-Milles.

Commandement de la Force Aérienne de Projection (CFAP)	
Escadron d'Hélicoptères 01.067 'Pyrénées' SA 330Ba Puma, AS 555AN Fennec	BA120 Cazaux
Escadron d'Hélicoptères 02.067 'Valmy' SA 319 Alouette III, AS 555AN Fennec	BA128 Metz
Escadron d'Hélicoptères 03.067 'Parisis' SA 330Ba Puma, SA 319 Alouette III, AS 355F1 Ecureuil, AS 555AN Fennec	BA107 Villacoublay
Escadron d'Hélicoptères 04.067 'Durance' AS 555AN Fennec	BA200 Apt
Escadron d'Hélicoptères 05.067 'Alpilles' SA 330Ba Puma, AS 555AN Fennec, AS 332 Super Puma	BA114 Aix
Escadron d'Hélicoptères 06.067 'Solenzara' SA 330Ba Puma	BA126 Solenzara
Escadron de Transport 03.060 'Esterel' A310-304, DC-8-72CF	BA110 Creil
Escadron de Transport 01.061 'Touraine' C.160F/R	BA123 Orléans
Escadron de Transport 02.061 'Franche-Comté' C-130H/H-30	BA123 Orléans
Escadron de Transport 03.061 'Poitou' C.160F/R	BA123 Orléans
Escadron de Transport Léger 01.062 'Vercors' CN.235-100, AS 555AN Fennec, DHC-6-200/300	BA110 Creil
Escadron de Transport 01.064 'Béarn' C.160NG	BA105 Evreux
Escadron de Transport 02.064 'Anjou' C.160NG	BA105 Evreux
Escadron de Transport et d'Entrainement 00.041 'Verdun' TBM 700, Nord N 262AEN/D *(aircraft loaned from ETEC 65)*	BA128 Metz
Escadron de Transport et d'Entrainement 00.043 'Médoc' AS 555AN Fennec, Nord N 262AEN/D, TBM 700 *(Helicopters loaned from EH 67, '262s from ETEC 65)*	BA106 Bordeaux-Mérignac
Escadron de Transport et d'Entrainement 00.044 'Mistral' Nord N 262AEN/D, TBM 700 *('262s loaned from ETEC 65)*	BA114 Aix
Escadron de Transport, d'Entrainement et de Calibration 00.065 'Gael' Falcon 20C/E/F, Falcon 50, Falcon 900, Nord N 262AEN/D, TBM 700	BA107 Villacoublay
Escadron Electronique 00.051 'Aubrac' DC-8-53 Sarigue	BA105 Evreux
Groupe Aérien Mixte 00.056 'Vaucluse' AS 532UL Cougar, C.160F/R (loaned from ET 61), DHC-6-200/300	BA105 Evreux

Aviation Légère de l'Armée de Terre

Direct-reporting units		
Escadrille de COMALAT		
	SA 341F Gazelle	Les Mureaux
Escadrille de EMAT	SA 341F Gazelle	Les Mureaux
Ecole d'Application de l'ALAT		**Le Luc**
EHLA 3	SA 342M Gazelle	Le Luc
	SA 341F Gazelle	
EHM 6	SA 330B/Ba Puma, AS 555UN Fennec	Le Luc
Ecole de Spécialisation de l'ALAT		**Dax**
EHLE 2	SA 341F Gazelle	Dax
EL FFA	SA 316/SE 3130 Alouette III	Baden-Oos
GAM/ETAMAT-Versailles		Les Mureaux
	SA 341F Gazellle, SA 316/SE 3130 Alouette III	
1 GHL		
Escadrille d'Hélicoptères 'Ile de France'		Les Mureaux
	SA 341F Gazelle	
3 GHL	Cessna F 406 Caravan II, TBM 700	Rennes
5 GHL		
1 EHL	SA 316/SE 3130 Alouette III	Lyon
det	SA 316/SE 3130 Alouette III	Ajaccio
det	SA 316/SE 3130 Alouette III	Le Luc
6 GHL	SA 316/SE 3130 Alouette III	Metz
1 GSALAT/ERGM	PC-6/B2-H4	Montauban
3e Brigade Aéromobile		
6 RHCM		
1 EHAP	SA 341F Gazelle	Compiègne
2 EHL	SA 316B/SE 3130 Alouette III	Lille
3 EHA	SA 342M Gazelle	Compiègne
4 EHA	SA 342M Gazelle	Compiègne
5 EHA	SA 342M Gazelle	Compiègne
6 EHM	SA 330B/Ba Puma	Compiègne
7 EHM	SA 330B/Ba Puma	Compiègne
7 RHC		
1 EHAP	SA 341F Gazelle	
3 EHA	SA 342M Gazelle	
4 EHA	SA 342M Gazelle	
5 EHA	SA 342M Gazelle	
6 EHM	SA 330B/Ba Puma	
4e Division Aeromobile		
1 RHC		
1 EHAP	SA 341F Gazelle	Phalsbourg
2 EHAP	SA 341F Gazelle	Phalsbourg
3 EHA	SA 342M Gazelle	Phalsbourg
4 EHA	SA 342M Gazelle	Phalsbourg
5 EHA	SA 342M Gazelle	Phalsbourg
6 EHM	SA 330B/Ba Puma	Phalsbourg
EHR	SA 341F/342L Gazelle	Phalsbourg
3 RHC		
1 EHAP	SA 341F Gazelle	Etain
3 EHA	SA 342M Gazelle	Etain
4 EHA	SA 342M Gazelle	Etain
5 EHA	SA 342M Gazelle	Etain
6 EHM	SA 330B/Ba Puma	Etain
EHR	SA 341F/342L Gazelle	Etain
4 RHCM		
1 EHC	SA 341F Gazelle	Nancy-Essey
2 EHM	SA 330B/Ba Puma	Phalsbourg
3 EHM	AS 532UL Cougar	Phalsbourg
4 EHM	SA 330B/Ba Puma	Pau
5 EHM	AS 532UL Cougar	Phalsbourg
EOS	SA 330B/Ba Puma	Pau
????	AS 532UL Cougar HORIZON	Phalsbourg
5 RHC		
1 EHLR	SA 342L Gazelle	Pau
2 EHAP	SA 341F Gazelle	Pau
3 EHA	SA 342M Gazelle	Pau
4 EHA	SA 342M Gazelle	Pau
5 EHA	SA 342M Gazelle	Pau
6 EHM	SA 330B/Ba Puma	Pau
Overseas units		
Central African Republic		
det ALAT CAR	SA 330B/Ba Puma	Bangui, CAR
Chad		
det ALAT Tchad	SA 330B/Ba Puma	N'Djamena
Djibouti		
det ALAT 188	SA 330B/Ba Puma	Djibouti
Croatia		
det ALAT	AS 532UL Cougar, SA 341 Gazelle	Ploce

A pair of Cessna F 406 Caravan IIs is used by 3 GHL based at Rennes. They undertake target-towing and light transport duties.

which were delivered from 1968. Since these helicopters have been constantly upgraded, they are expected to remain in service for another 10 or 15 years, by when the average Puma will be 35 years old. All ALAT Pumas have received composite rotor blades, NADIR Mk 1 navigation and mission management computers, and NVG-compatible cockpits. All Pumas are also to receive new-generation tactical radio and chaff dispensers. With NH 90 (dubbed TTH in the army version) deliveries delayed, the ALAT is also considering a third major overhaul

Above: As part of France's commitment to the Stabilisation Force, the ALAT has a detachment based in Croatia which reports to the Multi-National Division South East. AS 532ULs are used to provide air mobility to the troops in the field.

Below: Tested in the Gulf War, where it was carried by Pumas (such as this example), the HORIZON is deployed operationally on Cougars. The 4 RHCM is scheduled to establish an intelligence squadron to operate the system.

(including the re-engining with the Cougar's more powerful Makila turbines) to some of the Pumas, which would extend their service life. A number of Puma have been mothballed and could replace machines reaching the end of their useful lives, providing a transition until the arrival of the TTH.

The Pumas are also supplemented by 24 Cougar transport helicopters, which will last until 2030.

COUGAR HORIZON

The HORIZON (Hélicoptère d'Observation Radar et d'Investigation sur ZONe) heliborne radar system was initially known as Orchidée. It was supposed to be used at army corps level and linked to the utilisation of the HADES 'pre-strategic' nuclear weapon, for targeting purposes. The HADES programme was cancelled but the Orchidée, now called HORIZON, gained a renewed interest in the intelligence and electronic battlefield surveillance role.

The first prototype, fitted to a Puma, was used effectively during the Gulf War against Iraqi ground troops. The first operational system fitted to a Cougar was delivered in 1996. The second one came in 1998, and two more will follow. Before joining a future airmobile intelligence regiment, the HORIZON-equipped aircraft are based at Phalsbourg with the 4th RHCM.

Within the future intelligence squadron, HORIZON will be used in co-operation with attack helicopter squadrons. These squadrons will be tasked with either protecting the HORIZON system, or attacking the targets spotted with the radar. It is also said that the HORIZON could be teamed with reconnaissance Gazelles fitted with the IR and daylight sensors of the CL-289 drone. Two Gazelles initially received the front section (with the sensors) of the jet-powered drone for training purposes. This installation finally gave way to a new reconnaissance system for low-threat areas. This system was deployed around Mostar, Bosnia, with the UN IFOR.

TIGRE AND NH 90

The first deliveries of the NH 90 in its tactical cargo version (TTH) have been pushed back to 2011. No

more than 133 machines should enter service with the ALAT. Tigre combat helicopters seem to be more within reach for the ALAT. France still plans to buy 215 Tigres, 115 escort machines (HAP – Hélicoptère d'Appui Protection) with a turreted gun and a roof-mounted sight, and 100 anti-tank versions (HAC – Hélicoptère Anti-Char) with a mast-mounted sight and the TriGAT anti-tank system. The first 25 Tigres, all escort machines fitted with air-air missiles and a 30-mm turreted gun, will be delivered from 2001, giving an initial operational capability in 2002. On the other hand, delivery of the first anti-tank HAC has slipped until 2011, a decade later than planned.

The existence of two different versions of the Tigre is a result of French doctrine. Rather than having a single multi-mission machine and differently-tasked crews, the ALAT prefers to have its crews highly competent in a specific mission – anti-tank or escort and air-air combat – thus enabling them to get the most from their machines.

A new Franco-German Tigre flying school will be set up at Le Luc, in southern France. The school, which will train about 100 German pilots and as many French pilots each year, is expected to start its operation in 2001. It will use 25 machines and be commanded alternatively by a French and a German officer. The ALAT is closely studying the British experience of transferring its basic training to civilian schools; however, such a move seems unlikely in France.

The ALAT also uses some fixed-wing machines, such as Cessna 406 Caravans and TBM 700s. Five Montauban-based Pilatus PC-6 Turbo-Porters are used for parachuting and light transport. The ALAT may expand the fixed-wing fleet in the future, in order to compensate for the disbanding of the light helicopter groups in the liaison mission.

Within the army structure, but not part of ALAT, are a number of units that use microlights for road-traffic spotting duties. The most common type operated is the HM.1000 Balerit.

Outside the ALAT structure the army uses microlights in the following organisations:-		
EAT	HM.1000 Balerit	Tours
516 RCR	HM.1000 Balerit	Toul
601 RCR	HM.1000 Balerit	Arras
602 RCR	HM.1000 Balerit	Fontainebleu
2 RCS	HM.1000 Balerit	Versailles-Satory
6 RCS	HM.1000 Balerit	Nîmes
7 RCS	HM.1000 Balerit	Besançon
9 RCS	HM.1000 Balerit	Nantes
10 RCS	HM.1000 Balerit	Chalons-sur-Marne
14 RCS	HM.1000 Balerit	Toulouse
27 RCS	HM.1000 Balerit	Grenoble
3 REI	SMAM Petrel	Kourou, French Guyana
STAT	Chereau J-300 Srs 2, HM.1000 Balerit, SMAM Petrel	Toulouse
EAT	Escadrille de l'Armée de Terre (Army Flight)	
RCR	Régiment de Circulation Routiére (Transport Regiment)	
RCS	Régiment de Commandement et de Soutien (Command and Supply Regiment)	
REI	Régiment Etranger d'Infanterie (Overseas Infantry Regiment)	
STAT	Section Technique de l'Armée de Terre (Army Technical Section)	

Europe

Formations Aériennes de la Gendarmerie

Fitted with snow skis for operations in mountainous terrain, this Gendarmerie SA 319B Alouette III Astazou is seen flying through the Alps.

The Gendarmerie is a military force specialising in police missions, mainly in rural areas. The Formations Aériennes de la Gendarmerie (FAG) was created in 1954 with a single Bell 47G. Today, the FAG boast a total force of 42 helicopters: 12 Alouette IIIs and 30 single-engined Ecureuils. Three fixed-wing Cessna 206s were flown until 1995. The type was not replaced when retired from service.

The headquarters are based at Villacoublay air base, which also hosts the training flight (Groupement d'instruction et de Sécurité des Vols) and the Paris area squadron ('Section Ile de France'). Although basic training of the Gendarmerie pilots is performed by the ALAT at Dax and Le Luc, the Villacoublay training flight gives pilots a four-week specialisation programme. Pilots are taught the basic skills required by the missions specific to the Gendarmerie:
– surveillance and criminal investigation: this mission has been increasingly important over the last few years. FLIR is used on these missions.

– search and rescue, mainly in mountainous areas, provided for civilians in difficulty. One third of the total missions are accounted for by the SAR role.

Training sessions for mountain flight are organised several times a year at Briançon (Alps), intended to give pilots a basic knowledge of mountain operations and to weed out those who will receive a comprehensive training.

The FAG is organised into 26 autonomous units covering the whole of French territory, including five in overseas departments and territories. Each unit is linked to a Gendarmerie district and remains at the disposal of the ground forces. The FAG fly around 15,000 hours per year.

The Gendarmerie is seeking to replace its ageing fleet of Alouette IIIs. The future helicopter will have to display excellent mountain operation capability, and, as of early 1999, no choice had been made.

Centre d'Essais en Vol

The CEV (Centre d'Essais en Vol) belongs to the DGA, a government organisation that deals with military programmes for the French forces. A recent reorganisation brought together the 40 or so state-owned test centres, including the CEV, into a newly created DCE (Direction des Centres d'Expertises et d'Essais).

Today the CEV flies nearly 60 aircraft, both fixed-wing and helicopter types. However, due to a decreasing budget and a reduced workload, the fleet will soon be reduced by one-third. The 12 Mirage IIIs still flying at Brétigny and Istres, the two CEV active bases, will be replaced in the near future with a reduced number of Mirage F1s. Three of the five CASA 212s will also be sold. Another budget-saving measure will be the closure of the Brétigny flight test centre. Flying activity will be relocated at Cazaux and Istres air force bases.

In order to maintain a good level of activity, the CEV is now marketing its expertise and ground measurements installations to foreign air forces. A radar cross-section measurements programme took place in early 1998 with RAF Tornado F.Mk 3s, Luftwaffe MiG-29s and KLu F-16s.

The EPNER (Ecole du Personnel Navigant d'Essai et de Réception) also reports to the CEV.

The CEV fleet includes 14 Falcon 20s, some of which are modified as testbeds. No. 262/'CA' is used to evaluate avionics and has an RWR and a modified nose.

Since its creation in 1946, the EPNER has trained 1,700 test pilots and engineers, including 364 from 24 different nations. Candidates for the EPNER must have excellent professional records: 80 per cent of the pilots are fighter pilots, whereas engineers come from the aeronautics industry. Both are taught flight-testing of fixed-wing aircraft and helicopters. A regular course at the EPNER takes 11 months and includes lectures and practical experimentation. The students fly between 100 and 125 hours on 25 different aircraft types (or 15 different helicopters). The course ends with a two-week stay

The test centres use a number of French-built light aircraft for liaison and training. They include 13 Robin HR.100s (above) and 16 Wassmer CE.43 Guépards (below).

in a foreign test centre and, as a final test, six evaluation flights with a previously unknown type.

GEORGIA

Capital: Tbilisi   **Population:** 5.4 million
Land area: 69700 km² (26,905 sq miles)
Major cities: Batumi, Kutaisi, Rustavi, Sukhumi

Republic of Georgia Air Forces & Air Defence Forces

Having operated about 55 interceptors, 190 tactical aircraft and 48 armed helicopters from more than a dozen air bases in Georgia, all Soviet aviation units were withdrawn by mid-1992. Georgia was then left with only four fixed-wing aircraft and a few helicopters, including four Mi-24 gunships operating from Telavi. It also took over some new Su-25 attack aircraft from 18 completed locally at the Tbilisi factory for its national forces, as well as 12 Su-25TMs upgraded with new night/all-weather avionics for the former Soviet air forces (VVS), before further production was suspended. In March 1993, Russian Defence Minister General Pavel Grachev claimed in Moscow to have 'reliable intelligence information' that the Georgian air force was operating seven ex-Tbilisi Su-25s finished in Russian camouflage and markings.

Some of these were used in several Georgian air attacks in the Sukhumi area after the breakaway Muslim minority in the Abkhazia Black Sea region declared unilateral sovereignty in July 1992. Several were shot down in continuing operations up to mid-1993, but, in early 1997, the Georgian Defence Ministry announced the planned procurement over a seven-year period from resumed production at the Tbilisi factory of 50 Su-25TM upgraded attack aircraft.

Under current CFE arms control agreements, Georgia is allowed to operate 100 fixed-wing combat aircraft and 50 armed helicopters, but, having signed a security pact with Moscow and received Russian military assistance, currently has insufficient military and economic resources to reach this establishment. Although possessing excellent air bases and associated infrastructure, which is currently being reinforced by a $15.7 million 1997 Northrop Grumman contract for a new nationwide air traffic control ground radar and communication system, Georgia is still in need of a wide range of combat and support types.

Although little has recently been heard of the Abkhazian separatist air wing formed in October 1992 from the seizure in northern Georgia of several ex-Soviet military aircraft, including two Sukhoi Su-25s, two Aero L-39C armed jet trainers, Mil Mi-8s, plus a few supporting types and infantry SAMs, Tbilisi claimed in November 1996 that two Sukhoi Su-27 'Flanker' advanced air superiority fighters had been added to its inventory.

Several of its aircraft, including two Mi-8s and a two-seat Yakovlev Yak-52 basic trainer, operating in reconnaissance and assault roles, were shot down by Georgian forces in the Sukhumi, Shroma and Thvarchcheli areas in mid-1993. Surviving Abkhazian aircraft are currently thought to include a couple of Su-25s, a similar number of L-39s, one or two Mi-8s and Antonov An-2s, and four Yak-52 trainers.

Republic of Georgia Air Forces & Air Defence Forces

Current Georgian aircraft are thought to include the following:

Type	Role	In Service
Antonov An-2 'Colt'	light transport	?
Mil Mi-8 'Hip'	utility helicopter	?
Mi-24 'Hind'	attack helicopter	?
Sukhoi Su-25 'Frogfoot'	attack aircraft	8
Tupolev Tu-134 'Crusty'	transport	1
Tupolev Tu-154 'Careless'	transport	1
Yakovlev Yak-52	basic trainer	6

GERMANY

Capital: Berlin
Population: 78.5 million
Land area: 356840 km²
(137,740 sq miles)
Major cities: Bonn,
Dresden, Frankfurt,
Hamburg, Leipzig, Munich

Luftwaffe

The first air vehicles used by the German armed forces were lighter-than-air machines in 1911. From the outbreak of World War I in 1914, the procurement of military aircraft climbed steeply to reach a peak of almost 20,000 aircraft in 1917. Led by colourful pilots such as von Richthofen and Immelmann, and equipped with famous fighter aircraft such as the Fokker Eindecker with its synchronised machineguns, the army's air corps had a great impact on the conflict. After that war, Germany was forbidden to have any form of military aircraft, a situation which lasted for more than 15 years until Hitler became leader of Germany in 1934, when he quickly ordered the establishment of the Luftwaffe. At that time, hundreds of pilots were already being trained secretly in Russia. Two years later, during the Spanish civil war, Luftwaffe pilots gained invaluable experience that

took them a step ahead of other European pilots during the first years of World War II. Throughout the war the Luftwaffe fought in many theatres and fielded the first rocket- and jet-propelled fighter and bomber aircraft, revolutionising air combat. At the end of the war, the Luftwaffe was broken and Germany was forbidden to have armed forces for the next 10 years.

The modern Luftwaffe was established in 1956. An enterprising programme was set in motion to produce an air arm consisting of 20 wings with more than 1,300 fighters and transport aircraft. Within a few years the majority of the (R)F-84s, F-86s and N2501s – to name a few types – were delivered. The Germans caught up rapidly with the NATO allies. In the 1960s the F-104 Starfighter was purchased in large quantities and in the 1970s the (R)F-4 Phantom came on line to replace reconnaissance and fighter Starfighters. The end of the Starfighter era started in the early 1980s when the Panavia Tornado Interdiction Strike fighter-bomber entered service.

After the reunification of East and West Germany in 1990 the Luftwaffe entered a tough financial climate brought about by the expensive reunification process and economic malaise. Simultaneously, the world entered a new security environment with the break-up of the Soviet Union and end of the Cold War. Consequently, the air arm had to initiate a drastic restructuring process, resulting in disposals of numerous ageing aircraft and disbandment of associated units. Most of the former East German air force aircraft were retired, but its MiG-29s and some transport aircraft are still operational, while a few Su-22s remained in service as test and evaluation aircraft.

The Tornado ECR (Elektronische Kampfführung und Aufklärung – Electronic Combat and Reconnaissance) version usually carries a pair of AGM-88 HARM anti-radiation missiles on the under fuselage stations. This example is seen serving with the Piacenza-based Einsatzgeschwader 1.

FIGHTERS

The main body of the air defence force still consists of a large F-4F Phantom fleet. The severe programme delays in the EF2000, the long-awaited successor to the McDonnell Douglas F-4F Phantom and MiG-29 'Fulcrum', forced the Luftwaffe to execute an operational life extension programme for a large portion of the Phantom fleet. This programme was completed at the end of 1997. Called the Improved Combat Efficiency (ICE) programme, it involved 110 Phantoms and was conducted by DASA at its Manching facility. The modifications include the new APG-65 radar, the ability to fire the AIM-120 AMRAAM and a new navigation suite that includes GPS. The 37 operational Phantoms not part of the programme received only the new navigation suite. They fly with JG 72 'Westfalen' and the training squadron at Holloman AFB, N.M. The Luftwaffe intends to keep the Phantom flying until 2012, nearly 40 years after McDonnell delivered the first example.

Under the umbrella of Jagdgeschwader 73 'Steinhoff', a squadron of 16 F-4Fs operates alongside a squadron of 23 MiG-29 'Fulcrums' at Laage air base north of Berlin. The Phantom squadron was originally based at Pferdsfeld but relocated to Laage in 1997. The 'Fulcrum' squadron is among the busiest units in NATO, being officially NATO-assigned and

The fast combat jets of the Luftwaffe comprise the Tornado, a small number of MiG-29s and the F-4F Phantom. In service since 1973, the Phantom equips a total of seven squadrons in four wings, a training establishment and the test organisation (WTD 61). Luftwaffe aircraft are assigned to a wing and display badges to that effect. On the Phantom the Jagdgeschwader badge is located on the intake side, as displayed by the JG 74 example (right). Although assigned to the wing, only the pilots of 732 Staffel operate the F-4F of JG 73 (below) as 731 Staffel is the sole MiG-29 unit. Both the Phantom and the MiG-29 will be replaced by the Eurofighter.

responsible for the air defence of eastern Germany. It also provides realistic adversary training for NATO fighter pilots. The Luftwaffe MiGs have been adapted to meet NATO regulations, modifications consisting of identification and navigation systems as well as engine adjustments to increase life expectancy. Six to eight MiG-29s have received a GPS system, and provision for two 1150-litre (253-Imp gal) underwing fuel tanks increases the range to 3000 km (1,865 miles), enabling them to cross the Atlantic.

EUROFIGHTER

The Luftwaffe has ordered 140 Eurofighter EF2000s to replace the Phantom and MiG-29, plus 40 for the air-to-ground role to replace the oldest Tornados from 2012. It is expected that the first four to six aircraft will be temporarily stationed at the German government's test facility at Manching, where the Luftwaffe will evaluate the new fighter and where the first operational Luftwaffe pilots and maintenance personnel will receive initial training. Then they will go to Laage, where the training syllabus is produced. Laage is already adapting its infrastructure to the new fighter. The Phantom and 'Fulcrum' will be simultaneously replaced, a process that will take at

least two years since the EF2000 build-rate is only 15 aircraft a year, production being limited by the high unit cost of the aircraft. JG 74 will be the second wing to convert, followed by JG 71 and JG 72. The latter will receive the ICE Phantoms of JG 73 when the Eurofighter enters service.

TORNADO ATTACK

In all, the Luftwaffe has 276 Tornados (150 are assigned to NATO) spread over interdictor/strike (IDS) Tornado geschwaders, a dedicated reconnaissance wing and a defence-suppression wing. In order to meet the demands of the first 25 years of the 21st century, the Luftwaffe initiated a Tornado mid-life modernisation programme for a large part of the fleet. The programme has been split into two parts. The Kampfwertanpassung (KWA, combat efficiency enhancement) incorporates a Litton GPS/laser INS navigation system and a revised display concept including colour displays; the rolling map has been replaced by a digital map, two additional board computers supplement the existing computers, the software language has been changed from Spirit 3 to ADA, and a new Mil Std 1760 databus has been added. Rafael Litening laser designator pods provide

precision attack (primarily) for the reaction force squadrons. The Luftwaffe hopes to acquire a sufficient stock for the whole fleet in the future.

Additional modernisation is conducted under the Kampfwerterhaltung (KWE, combat efficiency upgrade programme) that concerns the electronic warfare suite. The radar warning receiver and electronic defence systems (chaff/flare, jammer) are enhanced so that they can cope with the newest Western and Eastern air defence systems. The Tornado Self-Protection Jammer (TSPJ) entered service in 1998 as successor to the Cerberus III. For the future, the Luftwaffe has planned a missile warning system and towed radar decoys (TRD) carried in modified BOZ 101 chaff/flare pods.

The ground-mapping radar has been enhanced to give better results on medium-level flight operations. Along with the need for better electronic warfare capability, a requirement has been determined for six Elint pods. The upgrade programme, to be finished in 2002, does not apply to the whole fleet; for obvious reasons, the reaction force units are the priority.

NEW TORNADO WEAPONS

The Luftwaffe has acquired Texas Instruments (Lockheed Martin) Paveway III guidance kits for Mk 84/BLU-109 2,000-lb penetrating bombs and GBU-22 (Mk 82). Another precision weapon is the DASA/Bofors KEPD 350 Taurus tactical cruise missile, to enter service in the first decade of the 21st century. The Luftwaffe will use Taurus – which has similarities to the Apache/Storm Shadow – as a dispenser system against soft targets, and as a penetrating point target weapon against hardened objects.

The self-defence armament today comprises AIM-9L Sidewinders but it is likely the Luftwaffe will get the Bodensee Gerätetechnik IRIS-T as its Sidewinder successor. In combination with a helmet mounted sight (HMS), the self-defence capability will be greatly enhanced.

SPECIAL MISSION TORNADOS

In the Tornado ECR (electronic combat, reconnaissance) the Luftwaffe has the most capable lethal defence suppression aircraft in NATO, and probably in the world. Thirty-five ECRs were delivered in the first half of the 1990s to JBG 32 at Lechfeld and JBG 38 at Jever to form a combined IDS/ECR wings but, to improve efficiency, the Luftwaffe has since concentrated all ECRs at Lechfeld.

The heart of the ECR is the Texas Instruments (TI) Emitter Locating System (ELS), a sensitive system that detects, analyses and targets enemy radar-guided air defence systems. The ELS feeds targeting data directly into the processor of the TI AGM-88 HARM, which has been upgraded to Block 6 standard in co-operation with the USA, involving adding a GPS

The Eurofighter is the future of the Luftwaffe's fast-jet combat fleet. Germany's prototypes are DA1 and DA5 (seen left). DA5 is fitted with the Eurojet EJ200 engines and ECR 90 radar.

receiver and an inertial measurement unit that memorises the position of the target, significantly increasing kill probability and reducing fratricide.

The dedicated reconnaissance wing is equipped with the Tornado IDS. Shortly after the RF-4E reconnaissance Phantoms of Aufklärungsgeschwaders (AG) 51 and 52 were phased out in 1993, the Luftwaffe adopted 40 Marineflieger Tornados and equipped them with 24 MBB reconnaissance pods borrowed from the Marineflieger and Italian air force, and re-established AG 51 at Schleswig-Jagel. The unit retained the tiger badge and gained the name and traditions of the disbanded AG 52 'Immelmann'.

The MBB reconnaissance pods were an interim solution until a new pod developed by DASA was fielded; the first of 37 new pods and two ground stations entered service in 1998 at Schleswig-Jagel. Initially the new pod accommodates two Zeiss conventional wet film cameras for low- and medium-level reconnaissance and an infra-red line scanner (IRLS), taken from the Tornado ECR, for use between 200 and 2,000 ft (60 and 610 m). Early in the 21st century the new pods will accommodate electro-optical sensors and a datalink, allowing the crew to send essential information in near-real-time to a ground station. For stand-off reconnaissance from medium altitudes, the Luftwaffe is aiming at a specially modified version called the TELE Lens Pod (long-range oblique photography). One of the two conventional cameras has been replaced by a camera with a long focal distance that can be swivelled to several positions on the horizontal axis.

A radar reconnaissance system will be available from 2000. Ten so-called Radar Aufklärungs Behälter (RABE) pods with Synthetic Aperture Radar/Moving Target Indicator (SAR/MTI) will give the Luftwaffe an all-weather reconnaissance capability.

The establishment in August 1995 of Einsatzgeschwader 1 (EG 1) at San Damiano air base (Piacenza, Italy) with ECR and dedicated reconnaissance Tornados, as part of Operation Deny Flight, was

Resplendent in a two-tone pale grey scheme, this AKG 51 recce-configured Tornado IDS carries the Cerberus ECM pod, AIM-9Ls and the DASA/Alenia reconnaissance pod.

a major milestone for the Luftwaffe. It was the first time since its post-World War II reinstitution that its combat aircraft have operated outside the NATO sphere. EG 1 was preceded by a Bundestag (parliament) constitutional adjustment.

The Bundeswehr (German armed forces) is required to be able to defend Germany, to provide humanitarian assistance, to support NATO and Western European Union (WEU) operations for collective defence or crisis management, and to participate in international military actions in response to a crisis, or pre-emptively. A Luftwaffe crisis detachment is always called einsatzgeschwader (literally, 'application wing').

German participation in the allied efforts was welcomed by NATO, not only because yet another major European service participated in the combined air efforts, but also because the ECR and reconnaissance Tornados were a badly needed asset. EG 1 hardly had time to get accustomed to the operation and flying conditions before Deliberate Force commenced. The Luftwaffe withdrew the ECRs on 22 November 1996 after flying 920 missions for the RRF and 2,042 for IFOR.

Both AG 51 and JBG 32 belong to the German Krisenreaktionskräfte (KRK, Rapid Reaction Forces). Other Luftwaffe KRK wings are JGs 71 and 74, JBGs 31 and 34, the transport wings and one helicopter squadron (for SAR, reconnaissance, and support duties), and Air Defence Missile Wings 1 and 3.

SAMs

Ground-based air defence is also the responsibility of the Luftwaffe. Six Flugabwehrraketengeschwaders (SAM wings) equipped with a mix of Raytheon Patriots and Hawks are positioned throughout the country. Euromissile Roland short-range point defence SAMs are also available at high-value objects such as air bases. Two of the SAM wings (1 and 3) belong to the Rapid Reaction Forces.

Germany is pursuing a capable ballistic missile protection capability. The Luftwaffe hopes to upgrade the Patriots to PAC-3, allowing engagement of theatre ballistic missiles with a range of up to 1000 km (620 miles). Thirty-six firing units are on strength.

Above left: Germany is alone among European air arms in having a dedicated SEAD (suppression of enemy air defences) aircraft. The Luftwaffe Tornado ECRs of JBG 32 have been much in demand for recent NATO/United Nations missions.

Above: The majority of German training is undertaken in the southern states of the United States. This IDS Tornado belongs to the Ausbildungsstaffel Tornado of the Taktische Ausbildungskommando Luftwaffe USA, based at Holloman AFB.

A more efficient theatre missile defence (TMD) capability is being pursued with the Medium Extended Air Defence System (MEADS), due to replace the Raytheon Hawk after 2005. MEADS is a tri-national programme involving Germany, Italy and the USA.

TRAINING CHANGES

Throughout the late 1990s, fighter crew training was extensively restructured and shifted almost completely to the USA because the often adverse central European weather made advanced weapon and basic training difficult. The Taktische Ausbildungskommando USA at Holloman Air Force Base, New Mexico, is becoming the Luftwaffe's training linchpin; no fewer than 48 Tornados and 24 F-4Fs will be based at the New Mexican air base by the end of 1999. The first eight arrived in April 1996 to start the Fighter Weapon Instructor Course formerly held at Jever, in northwestern Germany. Basic Tornado training has subsequently begun.

The Tornado community benefits more from the operational conditions in the USA than does the Phantom community. The Tornado was designed for low-level tactical flying, but Germany offers no such opportunities for noise complaints by residents have led to a minimum authorised altitude almost everywhere of 1,000 ft (304 m). The optimal altitude for the Tornado is 200 ft (60 m) – at night. In the US, the Tornados are cleared to fly as low as 100 ft (30 m) and to perform hot (live) weapon delivery.

The 20th Fighter Squadron, known in Germany as Ausbildungsstaffel Phantom, is responsible for German Phantom crews, and Ausbildungsstaffel

Below: Formed at Büchel in 1958 with F-84F Thunderstreaks, and re-equipping with F-104G Starfighters in 1964, JBG 33 gained its Tornado IDSs in May 1985. This aircraft displays the three-tone camouflage scheme adopted for operations over central Europe.

Left: Designed for European operations, the C.160D Transall has served the Luftwaffe since 1968. Its replacement will be designed for longer-ranged operations, reflecting the increased role the Luftwaffe has in peacekeeping operations. This C.160D is from LTG 61, and is seen over the Royal Bavarian Neuschwanstein Schloss.

Above and below: LTG 63 based at Hohn has one squadron of C.160Ds (631 Staffel) operated alongside one squadron of UH-1Ds.

Above: Deutsche Luftwaffen Ausbildungsstaffel at Tucson, Arizona, undertakes the primary training for future Luftwaffe pilots on civil Beech F33 Bonanzas. Arizona offers ideal weather for all-year training in uncrowded airspace.

Tornado is responsible for Tornado crews. Luftwaffe Tornados are the only foreign aircraft based in the US that are allowed to retain their national insignia.

Texan II

The Luftwaffe conducts almost all its training in the USA. The candidate pilots first go to Goodyear, Arizona for initial flight training on the Beech Bonanza. After graduation, they continue at the Euro NATO Joint Jet Pilot Training at Sheppard AFB, Texas, starting on the Cessna T-37 before advancing to the Northrop T-38 Talon. The best become fighter pilots, going on to Holloman AFB for type conversion.

The Luftwaffe purchased 40 T-37s in 1965. The USAF is scheduled to replace its T-37s with Raytheon T-6 Texan II JPATS at Sheppard AFB

between 2006 and 2008 and, as it is likely the Luftwaffe will stay at Sheppard, acquiring its own T-6s is a logical move.

Weapon System Officers commence flight training at NAS Pensacola on US Navy training aircraft before heading for Holloman, where they team up with a Tornado or Phantom pilot.

Tactical training is also 'exported'. Every operational crew member undertakes low-level training at Goose Bay, Canada and ACMI training at Decimomannu, Italy, each year for a two/three-week period. The unit at Goose Bay is called Taktische Ausbildungskommando Canada, but is also referred to as German Air Force Training in Canada (GAFTIC).

The Tornados and Phantoms (and C.160Ds) are based there from April until the end of September. MiG-29s are expected to follow in 1999.

Luftwaffe training on the Italian island of Sardinia is called Taktische Ausbildungskommando Italien. For the Phantom and 'Fulcrum', it involves air combat manoeuvring at the ACMI range, while the Tornados perform bombing practice at the Capo di Frasca range.

The immediate consequence of the expanding Holloman training centre is the decrease in size of the wings in Germany in terms of personnel and aircraft. One of the two Jever-based JBG 38 squadrons will be disbanded by the end of 1999.

As well as operating alongside the Transall units, the UH-1D serves in the SAR (left) and VIP transport (above) roles with HTS 2 of the FBS. Plans to standardise on the Eurocopter Cougar to replace the VIP Hueys were shelved in 1997.

Luftwaffe

Luftwaffenkommando Süd	HQ Meßstetten

1 Luftwaffendivision *(controls JBG 32 and JBG 34)* **Karlsruhe**

Jagdbombergeschwader 32		**Lechfeld**
321 Staffel	Tornado ECR/IDS	
322 Staffel	Tornado ECR/IDS	
Jagdbombergeschwader 34 'Allgäu'		**Memmingen**
341 Staffel	Tornado IDS	
342 Staffel	Tornado IDS	

2 Luftwaffendivision *(controls JBG 33 and JG 74)* **Birkenfeld**

Jagdbombergeschwader 33		**Büchel**
331 Staffel	Tornado IDS	
332 Staffel	Tornado IDS	
Jagdgeschwader 74 'Mölders'		**Neuberg**
741 Staffel	F-4F	
742 Staffel	F-4F	

Luftwaffenkommando Nord	HQ Kalkar

3 Luftwaffendivision *(controls AG 51, JG 72 and JG 73)* **Berlin-Gatow**

Aufklärungsgeschwader 51 'Immelmann'		**Schleswig-Jagel**
511 Staffel	Tornado IDS	
512 Staffel	Tornado IDS	
Jagdgeschwader 72 'Westfalen'		**Hopsten**
721 Staffel	F-4F	
722 Staffel	F-4F	
Jagdgeschwader 73 'Steinhoff'		**Laage**
731 Staffel	MiG-29 'Fulcrum-A'/MiG-29UB 'Fulcrum-B'	
732 Staffel	F-4F	

4 Luftwaffendivision *(controls JBG 31, JBG 38 and JG 71)* **Aurich**

Jagdbombergeschwader 31 'Boelcke'		**Nörvenich**
311 Staffel	Tornado IDS	
312 Staffel	Tornado IDS	
Jagdbombergeschwader 38 'Friesland'		**Jever**
381 Staffel	Tornado IDS	
382 Staffel	Tornado IDS	
Jagdgeschwader 71 'Richthofen'		**Wittmundhaven**
711 Staffel	F-4F	
712 Staffel	F-4F	

Lufttransportkommando Münster	

Lufttransportgeschwader 61		**Landsberg**
611 Staffel	C.160D	
612 Staffel	UH-1D	
Lufttransportgeschwader 62		
621 Staffel	C.160D	Wunsdorf
622 Staffel	C.160D	Wunsdorf
623 Staffel	UH-1D	Wunsdorf
Lufttransportgruppe Diepholz	UH-1D	Diepholz
Lufttransportgeschwader 63		**Hohn**
631 Staffel	C.160D	
632 Staffel	UH-1D	
Flugbereitschaftsstaffel		
LTS 1	A310-304, 707-320C CL-601 Challenger Tu-154M 'Careless'	Köln-Bonn
HTS 2	UH-1D	Nörvenich
3 Staffel	AS 532U2 Cougar L-410UVP(S) Turbolet	Berlin/Tegel

Luftwaffenamt Köln	

80th Flying Training Wing		**Sheppard AFB, USA**
89th Flying Training Squadron	T-37B	
90th Flying Training Squadron	T-38A	
Taktische Ausbildungskommando Luftwaffe USA		**Holloman AFB, USA**
Ausbildungsstaffel	F-4F	
Ausbildungsstaffel Tornado	Tornado IDS	
Taktische Ausbildungskommando Luftwaffe Canada		**CFB Goose Bay**
various types		
2. Deutsche Luftwaffen Ausbildungsstaffel NAS Pensacola, Fla. T-34C, T-1A, T-2C, T-39N		
3. Deutsche Luftwaffen Ausbildungsstaffel Tucson, Arizona Beech F33 Bonanza		

Above: Seven Canadair CL-601 Challengers were acquired between 1986 and 1987, initially to replace Jetstars and Hansa Jets. Another pair of Challengers was scheduled to replace the three VFW-614s retired in 1998, but the order was cancelled.

Below: Of the 12 Let 410s inherited from the former East German air force, the Luftwaffe continues to use four with the FBS.

AIR TRANSPORT

As is the case with the fighter force, the air transport fleet is getting old and needs modernisation and replacement. Additionally, the location of the training units and crisis operations puts an extra pressure on the available transport assets. The Luftwaffe maintains three tactical transport geschwaders, each operating both C.160Ds and UH-1Ds, both of which types are more than 25 years old. The wing at Wunstorf will relocate to the former NVA MiG-21 base at Holzdorf in the former East Germany, a process that advances slowly because of lack of funding. The UH-1Ds of this wing are currently spread over Wunstorf, Diepholz and Holzdorf.

The Luftwaffe has 85 C.160D Transalls. The fleet is currently being modernised with a new cockpit design that includes a GPS/laser gyro navigation system, in a programme due to be completed by early 2000. Twenty-four aircraft are being modified for special operations. Twelve already had received chaff/flare dispensers, a radar jammer and a radar warning system for duties in the Balkans. In 1999 the 24 will receive a full defensive aid sub-systems (DASS) package that also includes a missile warning system. Despite these modernisations, the Luftwaffe wants to replace the Transall with the Future Large Aircraft when it becomes available. The requirement calls for 75 aircraft to enter service from 2008. As the FLA will be a much bigger aircraft than the Transall, a full purchase will significantly enhance the transport capacity of the Luftwaffe.

HELICOPTERS

The Huey, too, is being modernised because its successor, the NH Industries NH 90, has been delayed. The Luftwaffe has a total of 114 Hueys in the inventory. Fifteen are at facilities such as technical schools and will never fly again; the 99 airworthy examples have all undergone a service life extension programme that included new composite rotor blades. Dornier Luftfahrt is responsible for further upgrades to 24 Hueys called the Nachttief-flugfähigkeit/Flugsicherheitmassname (NTH/FSM), which give the Huey a night vision capability for low-level night operations. The airframe will be remodelled, and better avionics, a new VOR/ILS, a spotlight (IR/white light) and probably an HF radio will be added. A GPS receiver is currently under evaluation. The last Huey is scheduled for retirement in 2008, and the NH 90 is planned to enter service in 2003. The Luftwaffe wants 85 as the Light Transport Helicopter and as a search and rescue asset. The first eight Luftwaffe NH 90s will be specially equipped for combat SAR, with the UH-1D squadron at Diepholz already under going training for this mission with specially prepared Hueys. Another eight are destined for VIP transportation, while 15 will get provisions for aeromedical equipment, helmet-mounted displays and electronic warfare equipment.

AERIAL REFUELLING

The Luftwaffe plans to have a small aerial tanker force by early in the 21st century when two Airbus A310s (already flying for the FLB as passenger transports) are to be converted as tankers. The A310s will be first modified to convertible freighter (CF) configuration and later to tankers by Airbus Industries. Three other A310s are on strength, one in VIP configuration and the rest passenger-only. The new tankers will receive only the hose-drogue refuelling system, which is incompatible with the Phantom's boom-receptacle system. Plans to give the Phantom a refuelling probe have been abandoned because the cost of converting two Luftwaffe Boeing 707s as tankers would have been as high as buying new aircraft. Instead, the FLB will retire the two Boeings at the end of 1999.

Apart from the transport aircraft mentioned above, the FLB has seven Canadair CL-601 Challengers, and three UH-1Ds for VIP duties at the FLB home base of Köln-Bonn. Of 16 basic Hueys based at Nörvenich and used as passenger transports, three VIP variants are painted white but will receive the standard green colours in the near future. The Challengers will move to Berlin-Tegel in 1999/2000 in anticipation of the government's move from Bonn to Berlin during the upcoming years. Tegel is already home base for four Let 410s and three Eurocopter AS 532U2 Cougars.

LUFTWAFFE ORDER OF BATTLE

The Luftwaffe has organised its wings (geschwader) slightly differently than other air forces. A geschwader usually consists of two staffeln (squadrons), which itself comprises only pilots, technicians and support people; the aircraft belong to the wing. Pilots usually wear the wing patch in addition to the staffel patch, but the aircraft only carry the wing patch.

Glossary	
Aufklärungsgeschwader (AKG)	Reconnaissance Wing
Flugberreitschaft	Special Air Missions
Flying Training Squadron (FTS)	USAF unit
Jagdbombergeschwader (JBG)	Fighter-Bomber Wing
Jagdgeschwader (JG)	Fighter Wing
Lufttransportgeschwader (LTG)	Air Transport Wing
Staffel	Squadron
Wehrtechnische Dienststelle (WTD)	Military Technology Aviation Establishment

Left: MFG 2 is the German Navy's surviving Tornado IDS wing, adopting the badge of MFG 1 on the disbandment of that unit on 1 January 1994 . MFG 2 replaced its Starfighters from September 1986 and will continue to operate the Tornado for the foreseeable future.

Below: Operating alongside the standard maritime patrol Atlantics are the five surviving (from six converted) Peace Peek Elint platforms (top). The conversion displays several external differences including additional radomes under the fuselage and modified wing pods.

Marineflieger

The first aircraft for German naval aviation were acquired in 1910 and the Marine-Fliegerabteilung was established in 1913. At the end of World War I some 1,500 aircraft and 16 airships were on strength. The Abteilung was deactivated in 1920 after the Versailles treaty. In 1932 training for a new Marineflieger began on waterplanes and flying-boats, but in 1939 Hermann Göring determined that all German air assets should come under Luftwaffe command and be flown by Luftwaffe pilots. The modern Marineflieger was established in 1956 with Sea Hawks, Fouga Magisters, Fairey Gannets, Sycamore helicopters and the Grumman Albatross amphibious plane.

The Flottille der Marineflieger (Naval Air Service) of the German Bundesmarine today has its headquarters at Kiel-Holtenau on the Baltic Sea coast and controls three subordinate wings. In the early 1990s the Marineflieger was reorganised and drawn down. It lost one Tornado wing and associated air base (Schleswig-Jagel) to the Luftwaffe, which re-established AG 51 at the base. The surviving wing is Marinefliegergeschwader 2 at Tarp/Eggebeck with 52 Tornado IDS aircraft. MFG 2 is a multi-purpose wing, its three Staffeln (squadrons) being tasked with anti-shipping operations and reconnaissance. As a fighter-bomber, the Tornado conducts offensive air operations with the Kormoran stand-off anti-ship missiles and dumb bombs (Mk 82, 83). In the escort attack role it is equipped with the AGM-88 HARM to suppress enemy surface-based radar-guided air

defence systems. Another important mission is reconnaissance, for which the Tornado carries an MBB pod containing a panoramic camera for low-level operations, a steerable camera with a narrow field-of-view for long-range reconnaissance, and an IRLS for low-level operations. MFG 2 is equipped with a mobile photo-processing unit allowing rapid off-base photo processing and interpretation. Finally, Marineflieger Tornados have the capability to refuel other drogue-equipped aircraft with the buddy pod, as do their Luftwaffe colleagues.

MFG 3 is based at Nordholz and has three types of aircraft. In 1966 it received the first of 18 Breguet Br.1150 Atlantic maritime patrol aircraft, six of which were extensively modified for Sigint collection, primarily for operations over the Baltic Sea. They detect electro-magnetic signals in order to compile electronic orders of battle and have some distinct external modifications that clearly identify their mission. The majority of the Atlantics act as long-range reconnaissance aircraft for aerial patrol operations, shadowing threat aircraft and intercepting target

data transmissions. They also autonomously conduct anti-submarine warfare, as well as co-operate with friendly submarines.

The second type is also employed for ASW purposes. The Lynx Mk 88 has been operated since 1981. Two examples with 18 flight and technical support personnel form the air assets on the Type

MFG 3's 2 Staffel operates the Lynx Mk 88. The type is currently the subject of a life-extension programme to bring the type to Mk 88A standard.

Marineflieger (Naval Air Arm)

Marinefliegergeschwader 2	**Eggebek**
1 Staffel	Tornado IDS
2 Staffel	Tornado IDS
Marinefliegergeschwader 3 'Graf Zeppelin'	**Nordholz**
1 Staffel	Atlantic/Atlantic KWS
2 Staffel	Lynx Mk 88/88A
3 Staffel	Dornier 228-212/1(LM)
Marinefliegergeschwader 5	**Kiel/Holtenau**
1 Staffel	Sea King Mk 41

Originally intended for SAR only, the Sea King Mk 41s have been progressively modified to equip them also for the anti-surface vessel and transport roles.

Four Dornier Do 228-212s serve with the navy. This example is used for light transport duties from Nordholz.

F 122 and 123 frigates. Other Lynx missions include SAR, boarding team operations utilising the fast rope technique, aerial reconnaissance, targeting and transport.

Severe delays to deliveries of the new helicopter, the NH 90, mean that the Marineflieger has had to keep the Lynx in service longer than anticipated, and even had to order an additional seven Mk 88As in 1996 for the new 'Brandenburg'-class frigates. The Lynx fleet will total 24 in 1999 (two have been lost since 1981), by when the 17 older helicopters will have undergone a life extension programme to bring them to Mk 88A standard. This programme includes changing the tail rotor, engine and fuselage strengthening, replacing the metal rotor-blades with composite blades, and engine modifications. The German Lynxes are receiving the Marconi Multi Sensor Turret FLIR and GPS coupled to the navigation system. The German navy expects to fly the Lynx beyond 2010.

The third MFG 3 type is the Dornier Do 228, two variants of which are used for completely different missions. One pair is in use for liaison and transport, and the other is specially equipped for pollution control. They have Forward-Looking and Sideways-Looking Airborne Radar (FLAR, SLAR), Laser Fluor Sensor, IR/UV sensor, video camera and datalink.

The Do 228 played a key role in the response to the 1996 floods in eastern Germany.

Twenty-two Sea Kings have been in service since 1975 with MFG 5 at Kiel. For years, SAR has been their primary mission with detachments stationed at Westerland, Borkum, Parow in the former DDR and, occasionally, in Helgoland. The Sea Kings are now also active in tactical air transport roles for the Navy, Air Force and Army during peace support operations. They were utilised in this role for the first time during the Gulf War.

In addition to the ship-based Lynx, the Marineflieger has a number of Tornados, Atlantics and Sea Kings available for the Krisenreaktionkräfte (rapid reaction forces).

THE FUTURE

After 2008, 38 NH Industries NH 90s will replace all Marineflieger helicopters. The type will be equipped with a FLIR, ESM, dipping sonar, sonobuoys, lightweight torpedoes, and 360° search and surveillance radar. The radar has an IFF with interrogation mode.

Being more than 30 years old, the Breguet Atlantic needs a successor. The Italian AF has a similar requirement, so the countries have jointly examined possibilities and have decided to opt for an existing aircraft that will be modified for maritime missions. In the meantime, the Atlantic is undergoing a modernisation which will keep it in service until 2010 by enabling the airframe to fly another 12,000 hours. The patrol aircraft will also receive a FLIR, ESM systems, and better navigation and communications equipment.

The Tornados, too, are being modified. The German Navy has adopted part of the Luftwaffe's upgrade package which will allow the use of laser-guided bombs.

Wehrtechnische Dienststelle 61

Based at Manching/Ingolstadt near Munich is the technical unit of the German armed forces. WTD 61 is responsible for testing and evaluating new aircraft and equipment. It directly reports to the Ministry of Defence and has most types currently in operational service with the Marineflieger, Heeresflieger and Luftwaffe. These include the CH-53, UH-1D, Tornado, F-4F and Do 228. The unit will also be the first to receive operational Eurofighters, and is closely involved in the ongoing EF2000 development programme being undertaken by DASA at the same base.

The single Dornier Do 228-201 evaluated by the Luftwaffe is still in use with WTD 61 for light transport duties.

Wehrtechnische Dienststelle 61   Manching/Ingolstadt
F-4F, Tornado ECR/IDS, UH-1D, C.160D, Dornier 228-201, Mil Mi-24D/P 'Hind'

Heeresflieger

German army aviation, called Heeresflieger, was established in 1957, just two years after the Bundesrepublik Deutschland became a NATO member. The first aircraft were the Dornier Do 27, Vertol H-21 and Sikorsky H-34. Most of the current types entered service in the 1960s and 1970s. The mission of today's Heeresflieger is to provide mobility for the army and to undertake independent air combat operations in support of the army. It has three components: an air transport regiment, an air assault regiment and a large flying school.

HEERESFLIEGERBRIGADE 3

Heeresfliegerbrigade 3 (HFB 3) at Mendig owns almost all the transport helicopters and reports directly to the Heeresführungskommando (HFuKdo), the second highest command level of the German Army, and is part of the main defence forces. It plans, prepares and takes care of the employment of air transport forces of the army; its assets perform liaison, guard, air reconnaissance, medevac, logistics, airborne landing and regular personnel transport operations.

The inventory of HFB 3 consists of 96 Sikorsky CH-53Gs, 104 Bell UH-1Ds and 69 BO 105Ms. They are assigned to five regiments at Hohenlockstedt (Itzehoe) in the far north of Germany, Rheine in the northwest, Laupheim and Niederstetten in the south, and Mendig in central Germany. The CH-53G and the UH-1D are old helicopters and both are subjects of life extension modification programmes. The CH-53G should linger until 2030, but the successor to the UH-1D – the NH Industries NH 90 – is scheduled to enter service in 2008. The Heeresflieger has a requirement for 150 NH 90s in the Tactical Transport Helicopter (TTH) model.

Operations in the former Yugoslavia revealed that the CH-53Gs were not properly equipped for peace support and peace enforcement operations. The most pressing need was for a self-protection system against missiles, so 20 CH-53Gs are being modified to CH-53GS (S for special) standard with six laser warning receivers, eight radar warning antennas and four missile warning sensors. The sensors, placed all over the fuselage, are integrated into one computer. An Advanced Digital Dispensing System automatically decoys incoming missiles with chaff/flares from dispensers placed on both sides of the tailboom and in the forward part of the gear bays. Both pilots have a threat warning display in front of them, while the flight engineer, seated between the pilots and slightly behind them, provides assistance by looking for missiles and monitoring the displays.

Another part of the modifications concerns the range, which has proved to be too short. The

The primary combat helicopter of the German army is the BO 105P Panzerabwehrhubschrauber 1 (PAH-1). A total of 212 BO 105Ps was delivered from 1979.

CH-53G has a range of 360 km (224 miles) when loaded with 36 armed soldiers or 5500 kg (12,125 lb) cargo. Two large external 4920-litre (1,082-Imp gal) fuel tanks are mounted on outriggers attached to the wheel bays; they extend flying time by five hours, giving the large helicopters a range of 1800 km (1,118 miles). The third main modification concerns night low-level flying capabilities in the form of NVG-compatible cockpit and external lighting. The helicopters also receive an emergency power supply for ground radio operations, for which two batteries are installed under the cockpit entrance.

The first two CH-53GSs had only the EW suite. One was sent to Rajlovac for SFOR operations in March 1997, the second was used for crew training.

The whole upgrade programme should be completed in 2001.

HFB 3 itself is not a Krisenreaktionskräfte (KRK, rapid reaction forces) brigade, although elements are. From 1 January 2000, a substantial number of all three helicopter types must be able to deploy within 15 days. Fifteen of the 32 CH-53s, including one equipped for medevac, should be capable of deploying within three days. Thirty-eight UH-1Ds also have a KRK mission and would support the Luftechanisierte Brigade 1 (air mobile brigade) should that unit be employed for KRK purposes. Fifteen Hueys must be ready within three days, and also have a responsibility towards the air mobile forces of the multinational division; eight have a three-day reaction time. The number of available Hueys will be reduced to 76 in 2000. Only 10 Bölkow BO 105Ms have a limited KRK task.

Since 1990 HFB 3 has been involved in various out-of-area operations. Apart from three fire-fighting operations in Greece, it participated in Operation Provide Comfort in northern Iraq and Turkey in 1990 with nine UH-1Ds and 24 CH-53Gs. Less known is the operation in Iraq in support of UNSCOM between 1991 and 1996 (three CH-53Gs) and UNOSOM in Somalia in 1993-94 (four Hueys). In 1995 and 1996 up to 17 UH-1Ds and 15 CH-53Gs flew numerous missions in Croatia in support of NATO's Implementation Forces and were assigned to the German Contingent IFOR (GECONIFOR). When IFOR was replaced by SFOR (Stabilisation Forces) only seven CH-53s remained in Bosnia (GECONSFOR). HFB 3 also has an important civilian mission in the event of disaster in Germany, e.g., nine Hueys and 14 CH-53Gs flew 1,250 missions in 1996 when enormous floods struck eastern Germany.

AIR ASSAULT

The operational air assault assets are assigned to Luftmechanisierte Brigade 1 (air mechanised brigade), which is headquartered at Fritzlar and part of IV Corps. The brigade is part of the KRK and is primarly equipped with BO 105P (Panzerabwehrhubschrauber 1, PAH-1), with some BO 105Ms and UH-1Ds. The armament of the BO 105P consists of six HOT anti-tank missiles. The LuMechBrig also has two Huey squadrons at Fassberg for logistic purposes. In 2000 their number will be decreased to 38, from 52.

The Eurocopter Tiger will be a much more powerful machine than the BO 105 and UH-1 it replaces, and the first of a total of 212 for the Heeresflieger will come on line in 2001. The first

batch consists of 80 Unterstützungshubschrauber Tigers (UHT). Although developed as an anti-tank helicopter, the German Tiger will also be used in combat support reconnaissance and escort roles. Its weapons will be the HOT and TriGAT anti-tank missiles, two FN Herstal 12.7-mm machine-gun pods and two pods with 22 68-mm unguided rockets. Avionics include a FLIR, helmet-mounted sight, mast-mounted sensors, a laser rangefinder and TV camera. The first UHTs will be delivered in autumn 2001. The first seven go to Le Luc in France for training, work-ups and evaluation, and delivery to operational units in Germany will begin in 2003.

SCHOOL

The Heeresfliegerwaffenschule (HFlgWaS, army aviation weapon school) at Bückeburg is responsible for helicopter training for the Heeresflieger, Marineflieger, Luftwaffe, and foreign air arms. The Alouette II has been the primary trainer for decades but will be replaced in 1999 by 15 Eurocopter EC 135s, which will feature an NVG-compatible cockpit. The reduced number of helicopters for basic training is countered by the use of frequent and improved simulation assets. All helicopter types in operational service with the Heeresflieger are also available for advanced training. In total, the school has about 100 helicopters (42 Alouette IIs, 11 CH-53Gs, nine BO 105Ms, 15 BO 105Ps and 25 UH-1Ds). Bückeburg has auxiliary air fields at Sulingen, Düdinghausen, Leierberg and Loccum.

Also based at Bückeburg and part of HflgWaS is the Heeresfliegerversuchsstaffel 910 (HFVS, army aviation test squadron). This squadron is responsible for test and evaluation of new equipment; it has a number of BO 105Ps permanently assigned and may borrow other types as necessary from other units.

Operations in support of the UN and NATO have highlighted the need for self-protection measures for the army's heavy helicopter fleet. This CH-53GS is seen firing anti-heat-seeking missile flares.

Above: Seen following a canal across the German countryside is a pair of BO 105Ps (PAH-1s).
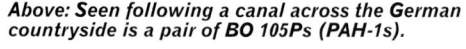

Below: Since 1967 the standard transport helicopter has been the Dornier-assembled UH-1D. Of the 204 delivered to the army, just over half remain.

Heeresflieger (Army Aviation)

Heeresfliegerbrigade 3		Mendig
Heeresfliegerregiment 6		Hohenlockstedt
1 Staffel	UH-1D	
2 Staffel	UH-1D	
Heeresfliegerregiment 15		Rheine-Bentlage
1 Staffel	CH-53G	
2 Staffel	BO 105M (VBH)	
Heeresfliegerregiment 25		Laupheim
1 Staffel	CH-53G	
2 Staffel	BO 105M (VBH)	
Heeresfliegerregiment 30		Niederstetten
1 Staffel	UH-1D	
2 Staffel	UH-1D	
Heeresfliegerregiment 35		Mendig
1 Staffel	CH-53G	
2 Staffel	BO 105M (VBH)	
Luftmechanisierte Brigade		**Fritzlar**
Heeresfliegerregiment 10		Fassberg
1 Staffel	UH-1D	
2 Staffel	UH-1D	
Heeresfliegerregiment 16		Celle
1 Staffel	BO 105P (PAH-1)	
2 Staffel	BO 105P (PAH-1)	
3 Staffel	BO 105P (PAH-1)	
Heeresfliegerregiment 36		Fritzlar
1 Staffel	BO 105P (PAH-1)	
2 Staffel	BO 105P (PAH-1)	
3 Staffel	BO 105P (PAH-1)	
Heeresflieger Verbindungs-und Aufklärungsstaffel 400		
	BO 105M (VBH)	Cottbus
II Korps		**Ulm**
Heeresfliegerregiment 26		Roth
1 Staffel	BO 105P (PAH-1)	
2 Staffel	BO 105P (PAH-1)	
3 Staffel	BO 105P (PAH-1)	
Heeresamt		
Heeresfliegerwaffenschule		Bückeburg
	BO 105P (PAH-1), BO 105M (VBH), UH-1D, CH-53G, SE 3130 Alouette II	
Heeresflieger Versuchsstaffel 910		Bückeburg
	BO 105P (PAH-1)	
plus Eurocopter EC 135 (15) *(on order) (Training)*		

Glossary

Heeresfliegerregiment	Army Aviation Regiment
Heeresfliegerstaffel	Army Aviation Squadron
Heeresfliegerversuchsstaffel	Army Aviation Experimental Squadron
Heeresfliegerwaffenschule	Army Aviation Weapons School
Heeresfliegerstaffel	Army Aviation Squadron

GREECE

Capital: Athens
Population: 10.3 million
Land area: 131985 km²
(50,945 sq miles)
Major cities: Patras,
Pireaus, Thessalonika

Elliniki Polimiki Aeroporia

The first Greek military aircraft was a Farman biplane which arrived at Larisa in 1912. More than 60 aircraft types had been operated prior to the formation of the autonomous Hellenic Air Force in 1931. After World War II the Elliniki Polimiki Aeroporia (EPA) received former RAF and USAAF aircraft, such as Harvards, Spitfires, Helldivers, Tiger Moths, Baltimore and Wellington bombers, and transport types including Dakotas, Oxfords, Austers and Ansons. The first jet aircraft (RT/T-33As) arrived in 1951 and soon replaced all the piston-engined fighters. Greece joined NATO in 1952. From the 1950s to the late 1960s the EPA received large numbers of jet fighters via MAP or via Western NATO countries: 200 F-84G Thunderjets, 104 F-86E(M) Sabres, 149 F-84F Thunderstreaks, 25 RF-84F Thunderflashes, 50 F-86D Sabres, 25 F-102 Delta Daggers, in excess of 150 F-104

Right: 343 Mira continues to operate first-generation Northrop Freedom Fighters, the last Greek squadron to do so. This is an RF-5A.

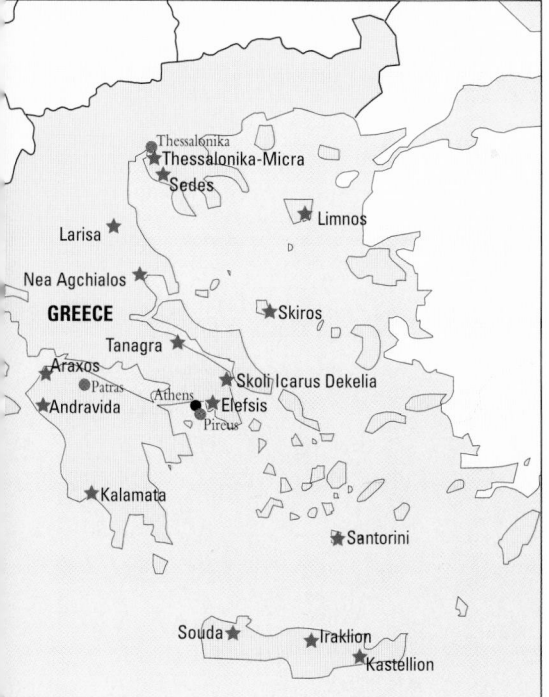

Starfighters and more than 150 F-5 Freedom Fighters.

The current Elliniki Polimiki Aeroporia (EPA, Hellenic Air Force) comprises three Commands. The Taktikis Aeroporiki Dynamis or Tactical Air Force commands seven Pteriga Machis (PM, Combat Wings) and 126 Sminarchia Machis (SM, Combat Group). The Diikissi Aeroporiki Ipostirixis or Air Support Command commands 112 PM. The Diikissi Aeroporikis Ekpedefsis or Air Training Command controls 120 Pteriga Ekpedefsis Aeros (PEA, Air Training Wing) and the Ethniki Aeroporia Academia or Air Academy.

FIGHTERS

From 1989, 40 Block 30 F-16C/Ds were delivered to 330 and 346 Mira Pantos Kerou (MPK, All Weather Fighter Squadron) at Nea Agchialos. The first of 40 Block 50 F-16C/Ds arrived in July 1997 to equip 347 MPK at Nea Agchialos. In mid-1997 346 MPK was relocated to Larisa to make room for the second Block 50 squadron. 341 MPK was receiving Block 50 F-16 aircraft, expecting the last to be delivered in early 1999. When 341 MPK becomes operational it is expected to be tasked with SEAD, armed with AGM-88B HARM and AIM-120B AMRAAM missiles, while 347 MPK will be fitted with LANTIRN pods. Meanwhile, the Block 30 F-16s are undergoing an HAI Falcon-Up upgrade programme.

Delivered in Block 30 and 50 standards, 80 F-16C/DG are due to be delivered to the Greek air force. This is the first Block 50 aircraft delivered, equipped with a visual identification searchlight.

The other main interceptor is the Mirage 2000EG/BG, of which 40 were received from 1988, operated by 331 and 332 MPK at Tanagra.

From 1974, 92 F-4E Phantoms were received and are flown by three squadrons: 337 MPK at Larisa, and 338 and 339 MPK at Andravida, the latter operating a batch of 28 ex-US ANG F-4Es delivered in 1991. To prolong Greek F-4 operations to 2015, DASA is updating 39 F-4Es to a standard similar to F-4F-ICE, including the APG-65 radar and AIM-120 capacity, while HAI has a structural life extension programme for 70 Phantoms. The F-16, Mirage 2000 and F-4E units provide air defence for the whole of Greece, and also operate from several forward operating locations at Limnos, Santorini, Simnos, Skiros and Kastelli.

Less capable fighters are the 30 remaining Mirage F1CGs of the 40 delivered in 1975, operated by 342 MPK at Tanagra and 334 MPK at Iraklion. The Northrop F-5 is the oldest Greek fighter, entering EPA service in 1965. With the deactivation of 349 Mira Anachetisis Imeras (Daylight Interception Squadron) in September 1997, all remaining F-5A/Bs, NF-5A/Bs and RF-5As are now operated by 343 MAI based at Thessalonika-Macedonia.

New-built Greek air force RF-4Es delivered in 1978 were later supplemented by ex-German air force examples (such as this pair of aircraft). The 'Rhino' is used by 348 MTA based at Larisa.

Elliniki Polimiki Aeroporia

110 PM	**Larisa**
330 MPK 'Keraunos'	F-16C/D
337 MPK 'Fantasma'	F-4E
348 MTA 'Matia'	RF-4E
ATA Flight	C-47 (Det 355/1STM)
111 PM	**Nea Agchialos**
341 MPK 'Assos'	F-16C/D
346 MPK 'Iason'	F-16C/D
347 MPK 'Perseos'	F-16C/D
Det 358 MED	AB 205A
112 PM	**Elefsis**
353 MNAS 'Albatross'	P-3B, HU-16B(EW)
355 MTM 'Ifaistos'	CL-215, Do 28D
356 MTM 'Iraklis'	C-130B/H, YS-11A
358 MED 'Faethon'	AB 205, AB 206, AB 212
113 PM	**Thessaloniki-Macedonia**
343 MAI 'Asteri'	(N)F-5A/B, RF-5A
355/1 STM	C-47 Thessaloniki-Sedes
114 PM	**Tanagra**
331 MPK 'Aegeas'	Mirage 2000EG/BG
332 MPK 'Geraki'	Mirage 2000EG/BG
342 MPK 'Sparta'	Mirage F1CG
366 SEE	T-33A
115 PM	**Souda**
340 MV 'Alepou'	T/A-7H
345 MV 'Lailaps'	T/A-7H
222 MEE –	T-33A
Det 355 MTM	Do 28D
Det 358 MED	AB 205A
116 PM	**Araxos**
335 MV 'Tigreis'	A-7E, TA-7H
336 MV 'Olympus'	A-7E, TA-7H
Det 358 MED	AB 205A
117 PM	**Andravida**
338 MPK 'Aris'	F-4E
339 MPK 'Ajax'	F-4E
126 SM	**Iraklion**
334 MPK 'Thalos'	Mirage F1CG
360 MEA -	T-41D, Grob 103 **Dekelia**
359 MAEDY	Ag-Cat, Dromader, Bell 47
120 PEA	**Kalamata**
361 MVE 'Mystras'	T-37B/C
362 MPE 'Nestor'	T-2E
363 MEE 'Danaos'	T-2E
Det 358 MED	AB 205A
130 SM	**Limnos**
FOL 337 MPK	F-4E
134 SM	**Santorini**
FOL 342 MPK	Mirage F1CG
FOL 349 MAI	F-5A
137 SM	**Simnos Airport**
FOL 114 PM	Mirage 2000EG
Det 358 MED	AB 205A
135 SM	**Skiros**
FOL 114 PM	Mirage 2000EG
Det 358 MED	AB 205A **Chios Airport**
Det 358 MED	AB 205A **Rhodos Airport**

Below right: The only European country to use the Buckeye, Greece employs its T-2Es for advanced training, wearing the same green and brown camouflage scheme applied to the Corsair IIs. The aircraft are pooled between 362 MPE and 363 MEE.

Below: The T-33A Shooting Star is expected to celebrate 50 years' service with the Greek air force. Retirement will see the end of military service for the type in Europe. This line-up of 222 MEE 'Stars' is headed by a red example received from the Luftwaffe in May 1976.

Greece has a wide variety of combat aircraft in service. Delivered between 1988 and 1991, Mirage 2000EGs (top) equip squadrons with 114 Pterix, primarily charged with the defence of Athens and armed with Super 530D and Magic 2 AAMs. An older Dassault design, the Mirage F1CG (above), was delivered in 1974 when tension with Turkey was high and the United States would not supply combat aircraft. F1CG 118 serves with 334 MPAK at Iraklion on the island of Crete. For fighter-bomber missions the air force has a large fleet of Corsair IIs, operated since 1975. The original A-7Hs have been supplemented by ex-US Navy A-7Es (left).

FIGHTER-BOMBERS

The EPA has five FBA units, of which four operate the A-7 Corsair. About 45 TA/A-7Hs of the original 65 Corsairs delivered during 1975-77 are split between 340 and 345 MV at Souda. At Araxos, 335 and 336 MV operate 70 ex-US Navy A-7Es and TA-7Hs (modified TA-7C), delivered during 1993/94 to replace the TF/F-104G Starfighter.

RECONNAISSANCE

Until 1991, 348 Mira Taktikis Anagnoriseos (MTA, Tactical Reconnaissance Squadron) still operated a handful of RF-84Fs, a type that entered service in 1956. The Thunderflashes flew alongside five operational RF-4Es of the original eight delivered in 1978. From 1992 the five RF-4Es were finally relieved, following a delivery of 29 ex-German Air Force RF-4Es.

DIIKISSI AEROPORIKI IPOSTIRIXIS

The Dakota is the oldest aircraft in service with the EPA and a handful of the more than 80 delivered are still going strong as target-towing aircraft in 355/1 Sminos Taktikon Metaforon or Tactical Transport Flight at Sedes, and another is detached to 110 PM for HQ liaison. The main transport squadron is 356 Mira Taktikon Metaforon, Tactical Transport Squadron at Elefsis, which has operated 10 C-130Hs since 1975, augmented by the delivery in 1992 of five former US ANG C-130Bs. For personnel transport two YS-11A

are flown, survivors of the six delivered in 1981. The sole Gulfstream I was struck off charge in 1995 after three decades of service. Since 1974 355 MTM has operated a dozen surviving Canadair CL-215s, augmented by four former Yugoslavian CL-215s in 1995. Only four of the 15 ex-German Air Force Do 28Ds delivered in 1985 are still operated by 335 MTM, one detached with 115 PM. 358 Mira Evrenas Diasosis, Search and Rescue Squadron at Elefsis uses about 20 AB 205As for SAR duty, for which the unit has several detachments all over Greece. For VIP transport, four Bell 212s and a couple of AB 206As remain from six delivered in 1971, flown from Elefsis.

After the deactivation of 370 Mira Epichirisiakis Ekpedefsis (MEE, Operational Training Squadron) at Larisa in the summer of 1996, 222 MEE at Souda became the main T-33A operator. From 1951 onwards more than 150 T-33As were delivered, of which the final 18 are now used for target-towing in 222 MEE and a few for training in 366 SEE at Tanagra. The T-33 is expected to remain in EPA service until 2000.

359 Mira Exipiretisis Dimosion Ypiresion (MAEDY, Civil Service Assistance Squadron) at Dekelia is tasked with agricultural duties. For crop-spraying, 10 ageing Bell 47Gs, delivered in 1968, will be used into the next millennium, due to the low operating costs and large stock of spare parts. Half of the 23 Grumman G164 Ag-Cats received in 1974 struggle on, together with 22 surviving PZL M-18A Dromaders of 30 delivered in 1983.

The civil service assistance squadron based at Dekelia is tasked with providing crop-sprayers for the agricultural sector. Three types are operated by the unit (359 MAEDY), the two fixed-wing aircraft being the PZL M-18A Dromader (above) and the Grumman G-164 Ag-Cat (below). During the spraying season the aircraft are deployed all over Greece. Agricultural flying is a high-risk business and both types have suffered from high attrition.

DIIKISSI AEROPORIKIS EKPEDEFSIS

Future EPA pilots have their aptitude tests and basic flight training with 360 MAE Mira Archikis Ekpedefsis (Primary Training Squadron) at Dekelia, which is equipped with five Grob 103 gliders delivered in 1983 and 20 T-41Ds delivered in 1969. Primary jet training since 1964 has been performed with 35 T-37B/Cs of 361 Mira Vasikis Ekpedefsis (Basic Training Squadron) at Kalamata, while 36 North American (Rockwell) T-2E Buckeyes remain of the 40 delivered in 1976, used for advanced training by 362 Mira Prochorimenis Ekpedefsis (Advanced Training Squadron) and 363 MEE at Kalamata.

FUTURE PROCUREMENTS

The EPA needs 60 new fighter aircraft to replace the F-5 and Mirage F1s and to augment or replace the F-4E fleet. In 1998 the EPA evaluated both the Block 50+ F-16 and F-15H, and the Mirage 2000-5 is also a possible contender, but decided to defer the decision. An order for between 60 and 80 Eurofighter Typhoons was placed in 1999, for delivery commencing in 2005. Greece's is the first Typhoon export order.

The HAF requirement for four airborne AEW&C aircraft, fought between the Northrop Grumman E-2C Hawkeye, Lockheed Martin C-130J AEW and EMBRAER ERJ-145 Regional Jet fitted with the Ericsson Erieye S-band phased array radar, was won by the latter. The first is due to be delivered between 2003 and 2004. In addition, up to six transport aircraft are required, with the Hercules most favoured.

The training fleet is to be updated with 45 T-6A Texan IIs ordered for delivery between July 1999 and 2002, replacing the T-37Cs in service. The choice of a new advanced trainer to replace the T-2E Buckeye has been narrowed down to the Aermacchi MB 339FD, Yak/AEM-130, Aero Vodochody L-59F or the British Aerospace Hawk LIFT.

Fire-fighting assets are due to receive a boost with the delivery of 10 CL-415GRs from 1999 onwards.

Below: 356 MTM operates the Greek fleet of Hercules, consisting of C-130Bs and Hs. C-130H 749 has been in service since April 1977 and displays the standard Greek tactical camouflage with a light undersurface.

Elliniko Polimiko Naftikon (Hellenic Naval Aviation)

The current Elliniko Polimiko Naftikon or Hellenic Naval Aviation was established in March 1976, but its roots go back to 1910. 353 Mira Naftikis Aeroporikis Sinergasias (MNAS, Naval Air Co-operation Squadron) officially stopped operating the Grumman HU-16B Albatross in 1996, although one Electronic Albatross ECM variant was noted as being operational late in 1997. The first of six P-3Bs was officially handed over to the EPN in May 1996, following an earlier delivery of four Lockheed P-3A Orions for spare parts. The 'Naval Air Co-operation' in the squadron's name dictates that system operators are EPN personnel, while the pilots are from the EPA.

The Dioikisi Elikopteron Naftikon (DEN, Naval Helicopter Command) operates its helicopters from the land base at Kotroni as well as from Navy vessels. The Alouette Sminos (Flight) operates the two SA 319B Alouettes remaining of the four delivered in 1975. In 1995 the SA 319Bs were relieved of the ASW task and are now used only for training. During 1979/84 the DEN was expanded with a dozen Agusta-Bell AB 212s, forming the AB 212 Sminos. Today, eight AB 212ASWs are tasked with anti-submarine warfare and two AB 212EWs are operated for electronic warfare.

In 1995 the first S-70B-6 Aegean Hawk LAMPS (Light Airborne Multi-Purpose System) arrived at the

Helicopter training and shipborne SAR is undertaken by the pair of surviving SA 319B Alouette IIIs operated by Alouette Sminos.

Above: The coast guard's airborne assets are limited to three light aircraft, but expansion is planned. Cessna 172 RG Cutlass AC-1 was the first aircraft operated by the organisation and is still in service.

Above: The latest helicopters to join the Greek navy are the Aegean Hawks which operate from the MEKO 200HN frigates Hydra, Spetsai, Psara and Salamis.

S-70 Sminos at Kotroni. The Aegean Hawk is a hybrid of the SH-60B Sea Hawk and SH-60F Ocean Hawk, equipped with a Bendix AN/ASQ-18(V)3 active dipping sonar for ASW, and can fire the NFT Penguin Mk 2 Mod 7 anti-ship missile. In September 1997 the sixth Aegean Hawk was received. The DEN will soon receive two additional S-70B-6s which were ordered in October 1997.

Since 1996 the Skoli Elikopteron Naftikou or Naval Helicopter School has trained all DEN pilots, using helicopters loaned from the operational Sminos. The Alouette IIIs and AB 212s can operate from one MEKO-class frigate, three LCVs and one naval cadet training ship, while the S-70B-6s can only be embarked on the new MEKO 200 frigates.

Elliniko Polimiko Naftikon

353 MNAS (EPA)	P-3B, HU-16B(EW)	Elefsis
Dioikisi Elikopteron Naftikon		**Kotroni**
Alouette Sminos	Alouette III	Kotroni
AB 212 Sminos	AB 212ASW, AB 212EW	Kotroni
S-70 Sminos	S-70B-6	Kotroni
SEN	all types of helicopter	Kotroni

SEN is Skoli Elikopteron Naftikou (Naval Helicopter School)

The AB 212ASWs usually embark on the navy's five ex-Dutch navy 'Elli'-class frigates and three ex-US Navy 'Epirus'-class frigates. All navy helicopters are land-based at Kotroni.

Liminiki Astonomia (Coast Guard)

Since 1982 the Liminiki Astonomia (Coast Guard), based at Dekelia, has operated two Cessna 172RG Cutlasses. In 1988 they were augmented by two SOCATA TB-20 Trinidads. The Liminiki Astonomia wants to expand its fleet in the near future with three to four unspecified aircraft.

Liminiki Astonomia

Liminiki Astonomia	Cessna 172RG, SOCATA TB-20	Dekelia

Above: 2 TEAS at Megara has control over the army's heavylift capability, provided by the Chinook. This example is a CH-47DG.

Above left: Greece was the first European customer for the AH-64A Apache. With Turkey hoping to expand its combat helicopter force, Greece is likely to order more Apaches in the future.

Elliniki Aeroporia Stratou

In 1956 the Elliniki Aeroporia Stratou (EAS, Hellenic Army Aviation) was formed with a small fixed-wing flying unit operating U-17s and a Beaver. Later equipment included Bell 47Gs, OH-13As and three AB 204B helicopters.

The EAS comprises three Tagmas (Battalions) and the Scholi Aeroporia Stratou (Army Aviation School). 1 Tagma Elliniki Aeroporia Stratou (TEAS) is based at Stefanoviklio, near Larisa and is sub-divided into three Lokos (Companies). 2 TEAS is based at Megara, near Athens and is the largest with five Lokos, while 3 TEAS is the smallest and is located at Alexandropouli in northern Greece, comprising three Lokos. The Scholi Aeroporia Stratou (SAS) is based at Stefanoviklio and commands three Lokos.

From the original 50 ex-USAF Cessna U-17A/Bs built in the early 1960s, fewer than 20 remain in use with the three TEAS for reconnaissance and artillery spotting. As the two Aero Commander 680FLs of Stefanoviklio were retired during 1996/97, the only other fixed-wing aircraft is a Beech C-12 based at Megara for VIP transport, as is one AB 212.

Since 1975 the Huey has been the workhorse of the EAS. About 140 UH-1Hs and AB 205As have been delivered for troop transport and are flown by all three Tagmas. In 1992-93 another 10 ex-US Army UH-1Hs were added to the fleet. In the early 1980s 16 AB 206B were delivered for the observation role, of which only a small number remain operational. To supply Army units in the eastern Aegean and for the paradropping role, five Meridionali-build CH-47C

Chinooks were delivered during 1981-83. In 1988 another five CH-47Cs were received from the EPA. Recently, all have been updated to CH-47DG standard by Boeing, with more powerful engines, enlarged fuel capacity, composite rotor-blades and new avionics, including NVG compatibility. All Chinooks are operated by the third Lokos of 2 TEAS. Midway through 1998 seven new CH-47Ds were added to the fleet.

The Skoli Aeroporias Stratou (SAS, Army Aviation School) at Stefanoviklio uses Nardi-Hughes 300Cs for basic training, of which only half of the original 30 delivered in 1969 remain in use. After the NH-300Cs, pilots are trained in the SAS on the UH-1H, AB 205A, AB 206A or U-17.

Until 1995 the army's airborne firepower was no more intimidating than cabin-mounted M60 machine-guns in the Hueys, and M24 machine-guns in the side door and M41 machine-guns in the rear door of the Chinooks. In mid-1995 the first AH-64A Apache was delivered to the EAS, with the 20th and last arriving in December 1995. All Greek AH-64As are under the command of 1 Tagma Epidolkon Elikopteron (TEEP, Attack Helicopter Regiment) at Stefanovikio. The EAS has expressed a need for another 10 Apaches.

Elliniki Aeroporia Stratou

1 TEAS		**Stefanoviklio**
1 Lokos	UH-1H, AB 205A	
2 Lokos	UH-1H, AB 205A	
3 Lokos	AB 206B, U-17A/B	
1 TEEP		**Stefanoviklio**
1 Lokos	AH-64A	
2 TEAS		**Megara**
1 Lokos	UH-1H, AB 205A	
2 Lokos	UH-1H, AB 205A	
3 Lokos	CH-47D/DG	
4 Lokos	AB 206B, U-17A/B	
5 Lokos	AB 212, C-12C	
3 TEAS		**Alexandroupoli**
1 Lokos	UH-1H, AB 205B	
2 Lokos	UH-1H, AB 205B	
3 Lokos	AB 206B, U-17A/B	
SAS		**Stefanoviklio**
1 Lokos	NH-300C	
2 Lokos	UH-1H, AB 206B	
3 Lokos	U-17A/B	

The sharp edge of the Hungarian air force comprises 22 single-seat MiG-29s and six two-seaters in two squadrons of the 59th Tactical Fighter Regiment. With the retirement of the Su-22M, MiG-23 and MiG-21MF, the Hungarian air force has been reduced to only four jet combat squadrons.

HUNGARY

Capital: Budapest
Population: 10.4 million
Land area: 93030 km² (35,910 sq miles)
Major cities: Debrecen, Györ, Miskolc, Pecs, Szeged

Magyar Légierö (Hungarian Air Force) – Air Units of the Hungarian Defence Forces

Hungary fought World War II on the side of the Axis powers until the end of the war in Europe. In June 1947 the Allied Control Commission allowed Hungary to have an air force, the then-new Hungarian Air Force (Magyar Légierö) first operating

a mixture of German and Soviet equipment. In the early 1950s the Soviet Union became the sole supplier, and the first MiG-15s and MiG-15UTIs arrived in March 1951 followed by Mil Mi-4s in 1956. Hungary joined the Warsaw Pact in May 1955 and the Légierö was then the strongest air force in the Balkans.

During the 13-day uprising of Hungarians against the Communists in October 1956, there was little air activity; some Mil-Mi 4s were operated by the troops of the counter-revolution and wore a new national insignia. After the suppression by Soviet troops, Hungarian units were demobilised. A new air force emerged, under tight Soviet control and guidance, without local training of pilot officers, with reduced numbers and fewer personnel. The MiG-15bis was delivered in 1958, prompting the return of older versions, and following MiG-17PFs came the first supersonic type, the MiG-19, in 1959. Delivered from 1967, the MiG-21 remained the backbone of the force for almost 30 years, with all three generations of 'Fishbeds' being flown by the Légierö. At the peak of its use, the type gave Hungary a kind of local air superiority, due to the long-term lack of advanced interceptors in neighbouring Austria. Violations of Austrian airspace were common and at one occasion a MiG-21 pursuing a refugee aircraft crashed over the border after an engine failure.

Hungary, as a reliable Warsaw Pact member, enjoyed a stream of newer Soviet types like the Mi-8TB in 1971, the Mi-24D and the MiG-23MF from the early 1980s onwards. Following a doctrine to build up front-line ground-attack power, the Su-22M3 was added by the end of 1983, making Hungary the sole operator of this sub-type in eastern Europe.

The Hungarians were a prime mover in the 1989 political changes towards democratisation and liberalisation. The resulting economic difficulties led to drastically decreased military funds, forcing the HDF

Above: Hungarian 'Hinds' are operated by two squadrons of the Combat Helicopter Regiment based at Szentkirályszabadja. This is an Mi-24D.

Below: Former Luftwaffe and East German air force Aero L-39ZOs were transferred to the Hungarian air force and serve with the third squadron of the tactical fighter regiment.

Above: Wearing a Red Cross badge on the cabin door, this Mi-8T belongs to the Mixed Transport Aircraft Brigade based at Szolnok.

into its most radical changes since 1956. Each pilot had flown over 100 annual hours at the beginning of the 1980s, but this descended to an average of 50 hours in 1995. The end of Cold War secrecy revealed a high attrition rate, showing the loss of 49 pilots in 29 accidents from 1970 to 1995. Currently, a programme is running under which some pilots fly 80-90 hours, and others fly only enough to retain their flying capability while undergoing English and theoretical training.

Main bases are Pápa, Kecskemét, Taszár, Szentkirályszabadja and Szolnok. The old civil airfield of Budaörs near the capital was used only by a small liaison unit, unlike Tököl, from where Soviet MiG-29s flew (withdrawn in 1991) and where the Danubia Aircraft company is located today. Débrecen air base, a former Soviet Su-24 base, was not taken into active service.

During the years of the Croatian War across Hungary's southern border, a number of provocative airspace violations made by Serbian jets were countered by Hungarian pilots. From 1993/94 onwards, all operational combat aircraft received Western-compatible IFF and transponder units made by Allied-Signal.

Through a Russian debt reduction scheme for Hungary, 28 MiG-29s were delivered to Kecskemét in three batches from mid-1993 onwards, but they led to a new demand for even more funds and were not how the air force wished to spend its money. Due to the MiGs' high operating costs, and the need to move the Taszár-based regiment when the base became NATO's IFOR/SFOR staging area from December 1995, further changes were unavoidable. In 1994, 20 Mi-24D/Ps and 20 L-39ZOs were delivered as a gift by the German government, of which 19 L-39s are in use in Kecskemét. Using these and Yak-52s as basic trainers, for the first time since 1956 in-country pilot-training was established in 1993/94 – only to be halted in 1997 due to the cost, after only one course. Today there are plans to exchange the ex-German L-39ZOs

The 2nd Fighter Squadron of the 47th Tactical Fighter Regiment undertakes the conversion training role for the MiG-21 units, operating twin-stick MiG-21UMs.

for L-59s or later L-159s from the manufacturer, Aero-Vodochody. In 1997 the name of the force reverted from 'Magyar Honvédség Repülő Csapatai' to 'Magyar Légierö', and the MiG-23MFs and Su-22M3s were withdrawn. Those which are still serviceable remain in storage, although some of the aircraft could return to service on short notice. Only the Su-22s are considered to be airworthy, since before their sudden retirement they underwent a mid-life overhaul. The cuts in combat aircraft types and numbers are illustrated by the 47th Tactical Fighter Wing at Papa, which previously had flown one squadron of 'Floggers' and one of 'Fishbeds' but now operates 10 MiG-21UMs in the ground-attack and 12 MiG-21bis in the air-defence roles.

The present situation will last until the air force's requirement for new fighters can be funded by the politicians. Until the Boeing F/A-18, Lockheed Martin F-16 or Saab/BAe JAS 39 Gripen enters service, the 'Puma' and 'Döngó' (Wasp) squadrons with their MiG-29s will have to protect Hungary's airspace into the new millennium, and well past the nation's entry into NATO, expected to take place in April 1999.

Magyar Légierö

HQ Magyar Honvédség Vezérkar	Budapest
GS Légierö Vezérkar	Veszprém
59th Harcászati Repülö Ezred 'Szentgyörgy Deszö'	**Kecskemét**
1st Vadászrepülö Század (VS) 'Puma'	Kecskemét
	MiG-29A/UB
2nd VS 'Döngó'	MiG-29A/UB Kecskemét
3rd Század	L-39ZO Kecskemét
47th Harcászati Repülö Ezred 'Papa'	**Pápa**
1st VS 'Sámán'	MiG-21bis Pápa
2nd VS 'Griff'	MiG-21UM Pápa
87th Harci Helikopter Ezred (HHE) 'Bakony'	**Szentkirályszabadja**
1st Harci Helikopter Század (HHS) 'Kerecsen'	Szentkirályszabadja
	MiG-24D/V
2nd HHS 'Fönix'	Mi-24D/V Szentkirályszabadja
3rd Szállitó Helikopter Század (SHS) 'Borz'	Szentkirályszabadja
	Mi-8TB
4th Szállitó Helikopter Század Veyges	Szentkirályszabadja
	Mi-8TB, Mi-9, Mi-17/PP
89th Vegyes Szállitórepülö Ezred 'Szolnok'	**Szolnok**
1st Szállitórepülö Század (SRS) 'Teve'	Szolnok
	An-26
2nd SHS 'Sárkány'	Mi-8S/T Szolnok
3rd SHS	Mi-8S/T Szolnok
4th Futárrepülö Helikopter Század (FHS)	Szolnok
	Mi-2
5th FHS	Mi-2 Szolnok
93rd Vegyes Szállitórepülö Osztaly 'Vitéz Hári Lászlo'	**Tököl**
1st SRS	An-26, L-410UVP Tököl
2nd Futárrepülö Raj	Zlin 43 Budaörs

Direct reporting units:

Bázisrepülötér 'Kapos'	Taszár
11. Vegyes Légvédelmi Raketa Ezred	Budapest
12. Vegyes Légvédelmi Raketa Ezred	Györ
54. Légtérellenörzö Ezred 'Veszprém'	Veszprém
1. Logisztikai Ezred	Veszprém
HQ Repülö Akádemia	Szolnok
(under control of the military faculty of the University of Budapest – former military academy)	
1 st Kikepzo Század (KS)	Szolnok
	Aerostar Yak-52 (stored)

Glossary

Bázisrepülötér	Airfield Command
Futárrepülö Helicopter Század (FHS)	Liaison Helicopter Squadron
Futárrepülö Raj	Liaison Flight
GS Légierö Vezérkar	General Staff of the Air Force
Harcászati Repülö Ezred	Tactical (Fighter) Regiment
Harci Helikopter Ezred (HHE)	Combat Helicopter Regiment
Harci Helikopter Század (HHS)	Combat Helicopter Squadron
Kikepzo Század	Basic Training Squadron
Légtérellenörzö Ezred	Radiolocation Regiment
Logisztikai Ezred	Logistic Regiment
Szállitó Helikopter Szazad (SHS)	Transport Helicopter Squadron
Szállitó Helikopter Szazad Vegyes	Mixed Transport Helicopter Squadron
Szállitórepülö Század (SRS)	Transport Aircraft Squadron
Vadászrepülö Szazad (VS)	Fighter Squadron
Vegyes Szállitórepülö Ezred	Mixed Transport Aircraft Brigade
Vegyes Szállitórepülö Oztaly	Mixed (Transport) Aircraft Unit
Vegyes Légvédelmi Raketa Ezred	Mixed SAM Regiment

Squadron Names

'Borz' = *Badger*; 'Döngó' = *Wasp*; 'Fönix' = *Phoenix*; 'Griff' = *Griffon*; 'Kerecsen' = *Falcon*; 'Sámán' = *Shaman*; 'Sárkány' = *Dragon*; 'Teve' = *Dromader (camel)*

ICELAND

Capital: Reykjavík   **Population:** 300,000
Land area: 102820 km² (39,690 sq miles)
Major cities: Akureyri, Hafnarfjördhur, Kópavogur

Landhelgisgæslan (Icelandic Coast Guard)

A small country with fewer than 300,000 inhabitants, located halfway between North America and Europe, Iceland has a strategic position on the route between the New and the Old World. It does not have any armed forces; the only similar body is the Icelandic Coast Guard. This law enforcement organisation is tasked with both fisheries protection and general search and rescue. In the past the organisation has operated a PBY-6A Catalina, which was replaced by a C-54 Skymaster. After rejecting the HU-16C Albatross, the ICG acquired a Fokker F27-200 (TF-SYR) in 1972 and used it until 1980.

The ICG now operates four aircraft, all from its home base at Reykjavík Airport. It has one Fokker F27-200 Friendship (TF-SYN), which features many of the modifications included in the maritime patrol version of the F27, such as the maritime radar mounted under the forward fuselage, and underwing tanks to extend endurance. The aircraft regularly makes long reconnaissance patrols around the 200-mile (320-km) economic zone.

The ICG is equipped with three helicopters. An Aérospatiale AS 350B Ecureuil (TF-GRO) is also used for road patrol duties during the summer with the police, and entered service in 1986. An Aérospatiale SA 365N Dauphin (TF-SIF) has been used for rescue and patrol work since 1985. The remaining helicopter, an AS 332L2 Super Puma, arrived in Iceland in June 1995. The larger helicopters have been instrumental in saving a number of seamen in distress and have also been called upon as general ambulance and distress-relieving aircraft when accidents and other calamities have occurred in the country.

Iceland Coast Guard

ICG	Fokker F27-200,	Reykjavík Airport
	Aérospatiale AS 350B Ecureuil,	
	SA 365N Dauphin, AS 332L2 Super Puma	

IRELAND

Capital: Dublin
Population: 3.6 million
Land area: 68895 km² (26,595 sq miles)
Major cities: Cork, Galway, Limerick, Sligo, Waterford

Aer Chor na h-Eireann (Irish Air Corps)

The Irish Air Corps can trace its origins back to the delivery of the first military aircraft – a Martinsyde Type A Mk II – to the then-Irish Free State on 16 June 1922 (Ireland did not become a fully-fledged republic until 1949). A military air arm was established in 1922 and this officially became the Army Air Corps on 31 October 1924. It was not until the 1970s that 'Army' was dropped from the official title of the service, and not until 1994 that a new blue dress uniform and insignia was adopted, replacing the previous army green. Ties with the Army have always been strong and support of Army operations has been a key Air Corps task. While notionally charged with defending the state from outside aggression, the Air Corps is not equipped to do so. Since the early 1970s Irish security concerns have been driven largely by the threat of cross-border paramilitary terrorism. The signing of the Belfast Agreement peace accord on 10 April 1998 has brought an end to overt paramilitary violence in the six counties of Northern Ireland. This in turn has concentrated attention on the future roles and requirements of the Irish Air Corps, which had already been under review during the mid-/late 1990s.

The Air Corps is commanded by its own GOC, a brigadier-general, and Air Corps HQ is co-located with the overall Defence Forces Headquarters. The primary operational base is Casement Aerodrome, Baldonnel, with a secondary airfield at Gormanstown. Personnel currently numbers 1,072, with 24 fixed-wing aircraft and 15 helicopters in the inventory. The

Arguably Ireland's most important military aircraft are the two CN.235s operated by the Maritime Squadron. They undertake coastal patrol work.

IAC has two primary flying wings, supported by a training wing, engineering wing, administrative wing and an air support signals company. All are based at Casement Aerodrome. Only the Army Co-operation Squadron is located elsewhere, at Gormanstown.

The progressive withdrawal from service during 1998 of the CM.170 Super Magister has left the IAC without its only jet trainer and primary light strike aircraft. Six CM.170-1s were delivered in 1975/76. In the light-attack role the Magisters could carry two 7.62-mm machine-guns in the nose and underwing 68-mm rocket pods – though the latter were rarely seen. The Super Magisters were backed up in the training and light attack roles by 10 SIAI-Marchetti SF.260WE Warriors delivered in 1977, and an 11th aircraft acquired in 1979. The Marchettis are charged with primary training and weapons training and can be armed with 7.62-mm gun pods or 68-mm rocket pods. They can also be equipped with a Vinten VICON 70 camera pod.

A total of eight SE 316 Alouette IIIs was acquired between 1963 and 1974 and they undertake Army support and inland/short-range SAR duties. SAR, especially maritime SAR, is an important role for the IAC. Its primary SAR helicopter is the SA 365N Dauphin II, six of which were acquired in 1986. The Dauphins are equipped to a high standard with

Both of Ireland's helicopter types are seen here, the SA 365N and SE 316. The types are employed on SAR duties, the Alouettes also supporting the army.

Irish Air Corps

**Air Corps HQ, Park House, Dublin
Air Corps Group HQ, Casement Aerodrome, Baldonnel**

No. 1 Support Wing, Casement Aerodrome

Light Strike Squadron   CM.170 Super Magister
By August 1998 only two of the six CM.170s remained in service. The last of these is scheduled to be withdrawn, when its airframe life expires, in September 1999.
Transport and Training Squadron   Gulfstream IV (1)
    Beech 200 Super King Air (1)
Provides support for Ministerial Air Transport Service (MATS)
Maritime Squadron   Airtech CN.235MPA (2)

No. 3 Support Wing, Casement Aerodrome

Army Support Squadron   SE.316B Alouette III (8)
One Alouette (202) is on long-term rebuild. One aircraft is routinely split-based between Finner Camp, Donegal, and Monaghan Barracks, for border duties.
Naval Support Squadron   SA 365N Dauphin II (2)
Aircraft (244 and 245) can operate from naval service fisheries protection vessel L.E. Eithne (P.31)
Search and Rescue Squadron   SA 365N Dauphin II (3)
One Dauphin is detached to Finner Camp, Donegal, on SAR alert.

Training Wing, Casement Aerodrome

Basic Flying Training Squadron   SF-260WE Warrior (7)
Army Co-operation Squadron, Gormanstown, Co. Meath
    FR.172 Reims Rocket (6)
Note: Figures in brackets refer to actual operational aircraft.

Garda Air Support Unit (GASU), Casement Airdrome

AS 355N Twin Squirrel (1), PBN Defender 4000 (1)
Flown and maintained by Air Corps personnel (Nos 3 and 1 Support Wing, respectively), with police observer crews.

Bendix RDR-1500 search radar, EFIS cockpit and a fully-coupled autopilot. Helicopter training is undertaken by two SA 342L Gazelles.

In 1994 the IAC greatly extended its maritime patrol capability through the acquisition of two CN.235-100MP patrol aircraft. These aircraft carry an AN/APS-504(V)5 search radar, a FLIR Systems FLIR 200HP, datalink and Agiflite stabilised camera system. The can also airdrop SAR supplies. The dedicated maritime patrol CN.235s replaced a single interim CN.235M transport, which had in turn taken over the patrol role from Beech Super King Airs.

The IAC's transport arm has a primary VIP tasking and operates a single Beech 200 Super King Air (delivered in 1980, the last of three once in service), BAe 125-700B (delivered in 1980, replacing an earlier aircraft) and a Gulfstream IV (delivered in 1991, replacing a Gulfstream III).

The Army Co-operation Squadron flies Reims-Cessna FR.172H Rockets, and single FR.172K on surveillance, survey/reconnaissance, ground escort, target-towing, parachuting and training tasks.

ITALY

Capital: Rome
Population: 57.7 million
Land area: 301245 km²
(116,280 sq miles)
Major cities: Bologna,
Catania, Florence, Genoa,
Milan, Naples, Palermo,
Turin, Venice

Aeronautica Militare Italiana (AMI) (Italian Air Force)

In 1998 the Aeronautica Militare Italiana celebrated its 75th anniversary. Its forebear, the Regia Aeronautica, was officially created on 28 March 1923. The first commander was General Piccio, a former World War I ace. The 'golden age' of the Regia Aeronautica was undoubtedly the 10 years after its constitution, during which many records were broken by Italian aviators with locally-built aircraft. One of the most famous records was the transcontinental flight of 14 SIAI Marchetti S.55As; commanded by the ace Italo Balbo, they arrived in Rio de Janeiro after flying 10000 km (6,210 miles) in 61 flying hours.

World War II saw the Regia Aeronautica flying and fighting over the North African and Russian theatres, most of the time with obsolete war machines. The only high-performance aircraft, such as the MC.205 and the Fiat G55, arrived too late and were built in limited numbers. After the war, the new Aeronautica Militare Italiana was re-equipped with Allied aircraft such as British Spitfires and American P-51 Mustangs, P-38 Lightnings and P-47 Thunderbolts.

The jet age was heralded with the introduction into service of the DH.100 Vampire, and in 1951 the first Italian-built jet, the Fiat G80, made it maiden flight. In 1952 the AMI received its first F-84G Thunderjet, followed in the later years by the F-86E and F-86K Sabre, F-84F Thunderstreak and RF-84F Thunderflash. The transport fleet was improved with the acquisition of the C-119 Boxcar, and in 1952 the AMI received the T-33A for pilot training. The Fiat G91R, a single-seat light fighter which had won a NATO contest for a lightweight strike fighter, was introduced into service in the 1960s. The two-seat version, the G91T, was assigned to the Advanced

Italy's world famous aerobatic team, the 'Frecce Tricolori' ('Three-coloured Arrows'), has displayed the Aermacchi MB.339PAN at air shows since 1982.

Flying Schools. Pilots completed training on the G91T, having started on the Italian-built Aermacchi MB.326 that had been used by the Initial Training Schools from 1962 and had replaced the T-33A. From 1963 the fighter force was improved with the introduction of the F-104G, substituted by the Fiat-built F-104S from 1970.

During the 1970s the Italian AF acquired the C-130H and the G222 for the transport role, and the Br 1150 Atlantic for anti-submarine warfare.

The beginning of the 1980s saw the arrival of the Aermacchi MB.339A, replacing the aged MB.326, and the Panavia Tornado IDS for ground attack. Some years later it was followed by the new Italian fighter-bomber AMX, developed in conjunction with the Brazilian EMBRAER company.

THE AMI TODAY

The AMI is geographically divided into three Regioni Aeree, which report directly to the Chief of the Air Force at the Aeronautica Militare Headquarters in Rome. Main roles include the air defence of Italian airspace, support of NATO operations, provision of SAR operation, and support of the civilian population, working with other government organisations and military corps.

The basic unit of the AMI is the Stormo (wing), which is composed of one Gruppo (air group) or more. The AMI has also three Brigate Aeree (air brigades), composed of two Stormi or more than two Gruppi.

For the interceptor role and for air defence, the AMI's prime unit – the 1ª Brigata Aerea Missile Interceptor Group – is equipped with the very old Nike Hercules missiles. The 1st Air Brigade is composed of two Stormi, the 16º Stormo at Padova and the 17º Stormo at Treviso, which have six Gruppi assigned, with many detachments in the North East of the Country. For the defence of the various main airfields, the AMI also has the SPADA system, comprising a group of missile batteries specialising in destroying low-level enemy intruders.

The air defence aircraft force of Italy consists of two operational Gruppi equipped with 24 Tornado ADVs, on lease from the Royal Air Force, delivered in 1996/97. The Tornados are assigned to the 21º Gruppo of the 53º Stormo and the 12º Gruppo of 36º Stormo, and are equipped with AIM-9L and

Top: AMXs supplement the Tornado IDS in the attack role in the AMI. This example, exhibiting a shark-mouth marking under the nose, is operated by the 13º Gruppo of the 32º Stormo.

Above: Six squadrons still use the ageing Starfighter. This F-104S ASA-M of the 23º Gruppo/5º Stormo displays the current all-over grey scheme.

Top: Primary all-weather attack platform for the AMI is the Tornado IDS, in service since 1981. '6-18' belongs to the 6º Stormo, based at Ghedi.

SkyFlash missiles. Also for air defence, the AMI has five all-weather interceptor groups equipped with the F-104S ASA Starfighter. These Gruppi are: the 22º of 51º Stormo at Istrana, 23º of 5º Stormo at Cervia, 9º of 4º Stormo at Grosseto, 10º of 9º Stormo at Grazzanise and 18º of 37º Stormo at Trapani. Forty-nine Starfighters are being upgraded to the ASA-M standard, with new avionics and various other electronic modifications.

The Starfighter in the air-to-air role is equipped with AIM-9L and with the radar-guided Aspide missile (an Italian-built variant of the Sparrow). The F-104S ASA-M will remain in service after 2000, and will be replaced by the new Eurofighter Typhoon. The AMI has ordered 121 of these aircraft, including some two-seaters for the operational conversion role. For conversion training to the Starfighter the AMI uses the TF-104G, which is assigned only to the 20º Gruppo of 4º Stormo at Grosseto.

Sixteen aircraft will be upgraded to TF-104G ASA-M standard during the next few years, in order to keep some two-seat Starfighters active until the arrival of the Eurofighter Typhoon.

The ground-attack role is assigned to the Tornado IDS, 100 of which were bought by the AMI from the Panavia consortium. The Tornados are assigned to three fighter-bomber groups: the 102º and the 154º are assigned to the 6º Stormo based at Ghedi, and the 156º Gruppo of 36º Stormo based at Gioia del Colle. Having the capability to carry 9000 kg (19,840 lb) of a wide range of offensive bombs, including laser-guided bombs, the Tornado can be equipped with a sensor pod under the fuselage for the reconnaissance mission. The aircraft of the 156º Gruppo are also assigned the anti-ship mission, with the Kormoran missile. Another Tornado Gruppo, the 155º of the 50º Stormo based at Piacenza, is equipped with the

Aeronautica Militare

1ª Brigata Aerea

16° Stormo	Nike Hercules	Padova
17° Stormo	Nike Hercules	Treviso
2° Stormo		
14° Gruppo	AMX	Rivolto
602ª Sq. Coll.	SIAI S.208M	Rivolto
3° Stormo		
132° Gruppo	AMX	Villafranca
603ª Sq. Coll.	SIAI S.208M	Villafranca
4° Stormo		
9° Gruppo	F-104S ASA-M	Grosseto
20° Gruppo	TF-104G/F-104S	Grosseto
604ª Sq. Coll.	SIAI S.208M, AB 212 MB.339	Grosseto
5° Stormo		
23° Gruppo	F-104S ASA-M	Cervia
605ª Sq. Coll.	SIAI S.208M	Cervia
6° Stormo		
154° Gruppo	Tornado IDS	Ghedi
102° Gruppo	Tornado IDS	Ghedi
606ª Sq Coll.	SIAI S.208M	Ghedi
9° Stormo		
10° Gruppo	F-104S ASA-M	Grazzanise
609ª Sq Coll.	SIAI S.208M	Grazzanise

9ª Brigata Aerea

14° Stormo		
8° Gruppo	G222RM, B 707T/T	Practica di Mare
71° Gruppo	G222VS, PD 808GE	Practica di Mare
15° Stormo		
85° Gruppo SAR	HH-3F, AB 212	Practica di Mare
81° C.A.E SAR		Practica di Mare
615ª Sq Coll.	AB 212E	Practica di Mare
15° Stormo maintains the following SAR dets:		
83° Centro SAR	HH-3F	Rimini
82° Centro SAR	HH-3F	Trapani
RSV (Test wing)		
311° Gruppo	various	Practica di Mare
30° Stormo		
86° Gruppo	Br 1150 Atlantic	Cagliari
318° Stormo		
93° Gruppo	Falcon 50, SH-3D TS	Ciampino
306° Gruppo	DC-9-32, Gulfstream III	Ciampino
32° Stormo		
13° Gruppo	AMX	Amendola
101° Gruppo	AMX-T	Amendola
632ª Sq. Coll.	SIAI S.208M	Amendola
36° Stormo		
12° Gruppo	Tornado ADV	Gioia del Colle
156° Gruppo	Tornado IDS	Gioia del Colle
636ª Sq. Coll.	SIAI S.208M, MB.339, P 180	Gioia del Colle
37° Stormo		
18° Gruppo	F-104S ASA-M	Trapani
637ª Sq. Coll.	SIAI S.208M	Trapani

46ª Brigata Aerea

2° Gruppo	G222	Pisa
98° Gruppo	G222	Pisa
50° Gruppo	C-130H	Pisa
41° Stormo		
88° Gruppo	Br 1150 Atlantic	Sigonella
50° Stormo		
155° Gruppo	Tornado ECR	Piacenza
655ª Sq. Coll.	SIAI S.208M	Piacenza
51° Stormo		
22° Gruppo	F-104S ASA-M	Istrana
103° Gruppo	AMX	Istrana
651ª Sq. Coll.	AB 212E, MB.339, SIAI S.208M	Istrana
53° Stormo		
21° Gruppo	Tornado ADV	Cameri
653ª Sq. Coll.	SIAI S.208M, P 180	Cameri
	AB 212E	Linate
61° Stormo		
212° Gruppo	MB.339A/C	Lecce
213° Gruppo	MB.339A/C	Lecce
70° Stormo		
207° Gruppo	SF 260AM	Latina
674ª Sq. Coll.	MB.339A	Latina
72° Stormo		
208° Gruppo	NH 500E	Frosinone
672ª Sq. Coll.		Frosinone
303° Gruppo Autonomo		
	P 166DL3/APH	Guidonia
313° Gruppo Addestramento Acrobatico		
'Frecce Tricolori'	MB.339PAN	Rivolto
Centro Volo a Vela		
423ª Squadriglia	Twin Astir, L 13 Blanik, A 21 Calif, Ventus 2B, Nimbus 4, SIAI S.208M	Guidonia

latest version of the Tornado, the IT ECR. The first was officially received on 7 April 1997, and by the end of 1998 the group had six. This version is dedicated to the SEAD role, and the AMI will modify 15 Tornado IDSs to this configuration, equipping them with new dedicated electronic installations with the ability to operate the AGM-88 HARM missile.

The AMI also acquired 110 Italian-built AMX fighter-bombers, assigning them to four fighter-bomber groups: the 13° of the 32° Stormo, the 14° of the 2° Stormo, the 103° of the 51° Stormo and the 132° of the 3° Stormo. The AMX is used for ground attack, and can carry around 3800 kg (8,380 lb) of bombs and rockets, include laser-guided bombs. For

the reconnaissance mission the AMX can use the Orpheus pod. The AMI also bought 26 AMX-Ts, the two-seater version, for the operational conversion role. This variant maintains the operational attack capability of the single-seat version. The two-seaters are mainly assigned to the 101° Gruppo of the 32° Stormo but one or two aircraft are also assigned to each AMX group.

The basic training role is undertaken by the Aermacchi MB.339s assigned to the 212° and 213° Gruppi of the 61° Stormo Volo Basico Iniziale Aviogetti at Lecce. The MB.339 can also be used for the close air support and anti-helicopter missions; the MB.339-equipped Italian aerobatic team 'Frecce

Tricolori' has also been assigned a secondary mission of close air support, and periodically the pilots are trained in this discipline. In 1997 the AMI received the first of 15 new Aermacchi MB.339CDs. This new version has upgraded avionics, including a new mission computer, HUD, MFDs and HOTAS. The aircraft are also equipped with a new navigation system and provision for an air-to-air refuelling probe. For initial training the AMI uses the SIAI-Marchetti SF 260AM, assigned to the 70° Stormo Volo Basico Avanzato Elica based at Latina. Each year, the AMI sends some students to

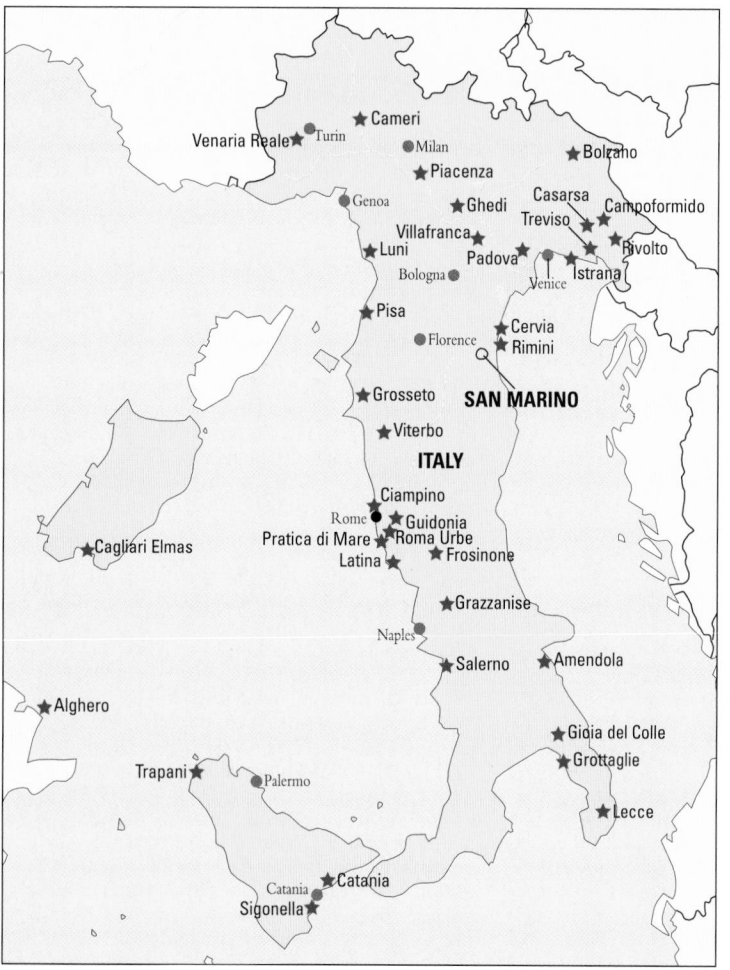

Right: The 8° Gruppo/ 14° Stormo operates four Boeing 707 tanker/ transports. They have been in service since 1992, and were previously operated by Air Portugal.

Right: Two Gulfstream IIIs are operated by the 306° Gruppo Transporto Speciali for VIP transportation. MM62025 entered service in January 1987.

Right: The Italian Air Force is the last user of the Piaggio PD 808. This example is used by the 14° Stormo for electronic warfare training, the grey scheme having replaced the green and blue/grey camouflage. Acquired new, the aircraft have seen more than 30 years of service.

Right: Sharing the transport role with the C-130H, the G222 equips two groups based at Pisa and Practica di Mare. Of the 51 G222s of all variants acquired, 44 remain. This example, a G222TCM, serves with the RSV at Practica di Mare.

Since 1972 the air force has operated 18 Atlantics for the navy; it provides the flight crew while the navy provides the system operators.

train at the Euro NATO Joint Jet Pilot Training School (ENJJPT) at Sheppard AFB in the USA.

The transport fleet is comprised of a C-130H group and two Alenia G222 groups, assigned to the 46ª Brigata Aerea at Pisa. The Hercules and the G222 are also used in the fire-fighting roles, equipped with the MAFFS system. After the tragic loss of a G222 over Bosnia from a Serbian missile, the G222s and the C-130s were fitted with a new self-protection suite. In 1999 the AMI will receive the first of 18 C-130Js. Some of the G222s are modified for radio calibration and electronic warfare tasks and assigned to the 14º Stormo, part of the 9ª Brigata Aerea based at Pratica di Mare near Rome, comprising the 15º Stormo SAR and the Reparto Sperimentale Volo, the test wing of Italian Air Force. The 14º Stormo also has the B 707T/T fleet, used for long-range transport and air-refuelling with underwing pods, and some PD 808s used for ECM and EW. They are the only variant of this type still in use. The RSV (Test Wing) has one or more aircraft of each type in service for test and various experimental projects.

The SAR fleet comprises 35 HH-3Fs and 32 AB 212Es, the Pelicans being assigned to the 15º Stormo. Fifteen helicopters are to Bravo standard, with new avionics and equipment for the combat SAR mission. This version of the HH-3F is camouflaged and the crew can use NVGs. The VIP fleet is concentrated within the Ciampino-based 31º Stormo, which has three flying groups equipped with two DC-9-32s (to be replaced by two Airbus A 319CJs from November 1999), Falcon 50 and two Gulfstream IIIs. Two Falcon 900EXs will replace the Gulfstreams from September 1999. Two Sikorsky SH-3D/TS VIP versions of the Sea King are also in service, and they are often used to transport the Pope within Italy. The AMI also acquired six Piaggio P 180s, which are assigned to the commands of the three Regioni Aeree for transport and liaison tasks. It is planned to buy 18 more P 180s in the future. Each operational unit also has a small Squadriglia Collegamenti, equipped with the SIAI-Marchetti S.208M. A number of these Squadriglie also have a dual role, with a SAR detachment assigned with some AB 212s.

Five S.208Ms have been specially modified as glider tugs and come under the control of the Centro di Volo a Vela at Guidonia. This unit has also two Let L 13 Blaniks, two Caproni A 21 Califs, nine Grob Twin Astirs, two Shemp Hirths, one Ventus and one Nimbus 4, which are mainly used for training but are also entered into glider competitions. For aerial photographic duties and light transport, six Piaggio P 166DL-3/APHs equipped with five cameras are based at Guidonia and assigned to the 303º Gruppo Autonomo.

Frosinone Air Base is the home base of the 70º Stormo, the unit tasked with rotary-wing training, equipped with the Nardi Hughes NH 500E. The helicopters are also used for liaison and provide escort for the HH-3F during combat SAR missions. The AMI purchased 50 NH 500Es in 1990, and some helicopters are detached to operational units for liaison tasks.

Used for maritime patrol, anti-submarine missions and maritime SAR, 18 Br 1150 Atlantics are assigned to the 86º Gruppo of the 30º Stormo based at Cagliari and the 88º Gruppo of the 41º Stormo based at Sigonella. The Atlantics have been upgraded recently with new radar and avionics equipment, and the aircraft are operated jointly with the Marina Militare, the Italian Navy.

The AMI also has some units based in Sardinia in support of the Poligono Sperimentale (Experimental Test Range) of Salto di Quirra/Perdasdefogu and the Reparto e Standardizzazione Tiro Aereo e Air (Weapons Training Installation) (RSST/AWTI) at Decimomannu, which has a number of AB 212s assigned.

Marina Militare Italiana

The Marina Militare has a fleet of helicopters and a fixed-wing component equipped with VSTOL Harriers. The aircraft of the Marina Militare are assigned to two different organisations: the first is the Aviazione per la Marina, and includes the Br 1150 Atlantics operated jointly with the AMI, while the second is the Componente Volo Marina Militare or Aviazione Navale. The Aviazione Navale is under the control of the Comando di Squadra Navale (CINCNAV), except for SAR and fishery patrols when it is under the control of the Alto Comando Periferico, which is responsible for a specific local area.

The Aviazione Navale has three main Maristaeli (naval air base), where the helicopters and Harriers are based. Maristaeli Luni (near la Spezia) is home base for two flying groups. GRUPELICOT 1 is equipped with the AB 212ASW, an anti-submarine version also used for light transport and liaison duties. The other flying group is GRUPELICOT 5 equipped with the Agusta/Sikorsky SH-3D/H. Main task of the Sea Kings is ASW, but they are also heavily tasked with transport and for operations with the Naval Intruder Force, the COMSUBIN.

Maristaeli Grottaglie (near Taranto) is the home base of the Harrier fleet. The Marina Militare bought 18 AV-8B Harrier II Plus, including two two-seaters, and the main mission is to ensure the air defence of the fleet, flying from the Marina Militare carrier Garibaldi. The secondary role of the Harrier is close air support. Grottaglie also houses GRUPELICOT 4 equipped with the AB 212ASW, used also to grant support to the operational group 'San Marco' marines battalion and COMSUBIN intruder battalion.

Maristaeli Catania is the third main naval air base and is home to GRUPELICOT 2's AB 212ASW, and GRUPELICOT 3 with SH 3D/Hs.

Capitanerie di Porto (Coast Guard)

The Capitanerie di Porto (the Coast Guard) has a fleet of Piaggio P 166DL-3s and AB 412s used mainly for maritime patrol and SAR, assigned to various detachments along the coast.

Above: After receiving the government's approval to operate fixed-wing aircraft, the Marina Militare Italiana received its first Harrier II, a TAV-8B, in August 1991. They serve with the GRUPAER.

Below: Three squadrons operate the Agusta-Bell AB 212ASW, which first entered service in 1968. About 50 remain from the 68 ordered. They are due to be replaced by the NH Industries NH 90 from 2004.

All of the Air Groups detach helicopters regularly to the carrier Garibaldi. Some AB 212ASWs are specially modified to launch the Teseo anti-ship missile and some of the Sea Kings can launch the Marte anti-ship missile. For the anti-ship and anti-submarine missions, all the helicopters can carry torpedoes.

The helicopter fleet is also heavily used for fire-fighting during the summer season, in support of both military or civilian authorities.

Aviazione Navale of Marina Militare

Maristaeli Luni	
GRUPELICOT 1	AB 212ASW
GRUPELICOT 5	SH-3D/H
Maristaeli Catania	
GRUPELICOT 2	AB 212ASW
GRUPELICOT 3	SH-3D/H
Maristaeli Grottaglie	
GRUPELICOT 4	AB 212ASW
GRUPAER	AV-8B/TAV-8B Harrier Plus
Cagliari Elmas Air Base	
30° Stormo	Br 1150 Atlantic
	(operated jointly with AMI)
Sigonella Air Base	
41° Stormo	Br 1150 Atlantic
	(operated jointly with AMI)

Nineteen Agusta-built SH-3Ds and 10 SH-3Hs are in service with the MMI. This SH-3D of GRUPELICOT 3 wears markings celebrating 30 years of Italian Sea King operations.

Capitanerie di Porto/ Guardia Costiera

1° Nucleo Aereo	AB 412, P 166DL-3	Sarzana
2° Nucleo Aereo	P 166DL-3	Pescara
3° Nucleo Aereo	P 166DL-3	Catania Fontanarossa
1 Sezione Elicotteri		Sarzana

Aviazione dell'Esercito (AvES)

The Italian Army has a large helicopter fleet. The only fixed-wing aircraft are the Dornier Do 228s based at Viterbo, used for transport and liaison duties. All the Cessna O-1 Bird Dogs and L 1019s have been retired from service. The most important helicopter in service is the Agusta Bell AB 205, used for light transport and liaison, medevac and fire support missions for the ground troops, for which it employs a Minigun pod and rockets. In the scout and observation role, the Italian Army uses the AB 206, which is also used for liaison, and the Agusta A 109EOA. Also in service is the AB 412, with most concentrated at Viterbo air base. For the anti-tank role the Italian Army has the new Agusta A 129 Mangusta, a modern gunship assault helicopter that can be equipped with Hughes TOW anti-tank missiles. All of these new helicopters are based in northeast Italy, with a few Mangustas assigned to the Viterbo Training School.

Heavy transport duties are undertaken by the CH-47C Chinook, based at Viterbo. The Chinooks

Anti-tank and armed escort tasks have been undertaken by Agusta A 129 Mangustas since 1990.

are also used by the Pisa-based paratroopers of the Unit Folgore. Most of the important units are based in the north of Italy, as it is vital to defend this strategic industrial area. To accomplish this successfully, many units are trained to fly in mountainous areas and work closely with the mountain troops, the Alpini.

Viterbo air base is the main training base of the Italian Army, home to the Aviazione dell'Esercito School, which uses all the Army helicopter types.

Agusta-built Bell-designed helicopters, including this AB 412, feature prominently in the Italian Army. The majority are AB 205s.

Aviazione dell'Esercito

Centro Aviazione Esercito		
Raggruppamento Mezzi aerei various		Viterbo
1° Reggimento 'Antares'		
11° Gr Sq 'Ercole'	CH-47C, AB 412	Viterbo
51° Gr Sq 'Leone'	CH-47C, AB 412	Viterbo
Squadrone ACTL	Do 228	Viterbo
Squadrone EM 'Italair'	AB 205	Beirut Libano
2° Reggimento 'Sirio'		
2° Gr Sq 'Gru'	AB 206	Lamezia Terme
20° Gr Sq 'Andromeda'	AB 212, AB 206	Salerno
3° Reggimento 'Aldebaran'		
53° Gr Sq 'Cassiopea'	AB 205, AB 206	Bresso
4° Reggimento 'Altair'		
34° Gr Sq 'Toro'	AB 205, AB 206	Venaria Reale
54° Gr Sq 'Cefeo'	AB 205, AB 206	Bolzano
5° Reggimento 'Rigel'		
25° Gr Sq 'Cigno'	AB 205, AB 206	Campoformido
7° Reggimento 'Vega'		
48° Gr Sq 'Pavone'	A 129, A 109EOA, AB 205	Belluno
49° Gr Sq 'Capricorno'	A 129, A 109EOA, AB 205	Casarsa
Direct-reporting Units		
21° Gr Sq 'Orsa Maggiore'	AB 205, AB 206	Cagliari
26° Gr Sq 'Giove'	AB 205, AB 206, AB 412	Pisa
28° Gr Sq 'Tucano'	AB 206C, A 109EOA	Roma Urbe
30° Gr Sq 'Pegaso'	AB 206, A 109EOA	Catania
39° Gr Sq 'Drago'	AB 412	Alghero
55° Gr Sq 'SOATTC'	AB 206	Padova
419° Squadrone 'Perseo'	AB 206	Salerno
427° Reparto 'Mercurio'	AB 206	Firenze
441° Squadrone 'Fenice'	AB 206	Venaria Reale
551° Reparto 'Dragone'	AB 206	Padova

Police forces

For the four Italian police forces – Carabinieri, Polizia di Stato, Guardia di Finanza and the Corpo Forestale dello Stato – the helicopter is the most frequently used asset. The Carabinieri has the main Centro Elicotteri based at Practica di Mare and maintains various other detachments (Elinuclei) around Italy. The service uses the AB 206, the A 109 and the AB 412 for many tasks, including transport, support of the civilian population and, of course, for patrol and law enforcement duties.

The Polizia also has its main central unit, the Centro Addestramento e Standardizzazione Volo (CASV), based at Practica di Mare, and has various other detachments in Italy. The Polizia uses the AB 206, the A 109 and the AB 412 for various duties including patrol and surveillance of the roads. For transport, liaison and research, the Polizia also has some Partenavia Observer aircraft.

The Guardia di Finanza's main unit, the Centro Aviazione, is based at Practica di Mare with various other detachments in Italy. It has a fixed-wing fleet consisting of 12 Piaggio P 166DL-3s specially equipped for the maritime patrol and control role, using radar and navigation systems. Helicopters in use are the NH 500D, A 109 and AB 412, which are also used for border and maritime control and law enforcement. Also assigned to the Guardia di Finanza are three ATR 42-400MPs, specially modified for long-range maritime patrol.

Corpo Forestale dello Stato

Centro Operativo Aeromobili	H 369HS, NH 500D, AB 412, CL-215	Roma Urbe

Helicopter summer deployment bases are Caprera, Cecina, Frosinone, Sabaudia, Vieste, Pescara, Maratea, Lamezia Terme, Verona. Aircraft summer deployment bases are Ciampino, Genova, Alghero, Cagliari, Lamezia Terme.

The Corpo Forestale dello Stato has its main unit, the Centro Operativo Aeromobili del Corpo Forestale, based at the Roma Urbe airport. The Corpo Forestale uses the NH 500C/D and the AB 412SP mainly for fire-fighting missions and for anti-poacher patrols, helicopters being assigned to various detachments in Italy.

Carabinieri

Centro Elicotteri	Practica di Mare
1° Nucleo Elicotteri	Volpiano
2° Nucleo Elicotteri	Bergamo Orio al Serio
3° Nucleo Elicotteri	Bolzano
4° Nucleo Elicotteri	Pisa San Giusto
5° Nucleo Elicotteri	Falconara
6° Nucleo Elicotteri	Bari Palese
7° Nucleo Elicotteri	Pontecagnamo
8° Nucleo Elicotteri	Vibo Valentia
9° Nucleo Elicotteri	Palermo Boccadifalco
10° Nucleo Elicotteri	Olbia
11° Nucleo Elicotteri	Cagliari Elmas
12° Nucleo Elicotteri	Catania Fontanarossa
13° Nucleo Elicotteri	Forlì
14° Nucleo Elicotteri	Treviso Sant'Angelo
15° Nucleo Elicotteri	Albenga

Polizia di Stato

Centro Addestramento e Standardizzazione Volo	Practica di Mare
1° Reparto Volo	Practica di Mare
2° Reparto Volo	Milano Malpensa
3° Reparto Volo	Bologna Borgo Panigale
4° Reparto Volo	Palermo Boccadifalco
5° Reparto Volo	Reggio Calabria
6° Reparto Volo	Napoli Capodichino
7° Reparto Volo	Abbasanta
8° Reparto Volo	Firenze Peretola
9° Reparto Volo	Bari Palese
10° Reparto Volo	Venezia Tessera
11° Reparto Volo	Pescara

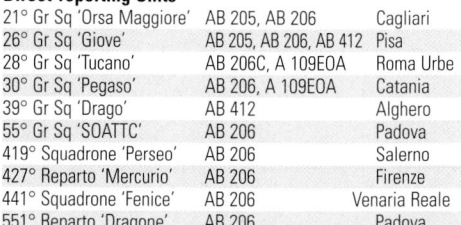

The most common helicopter in Guardia di Finanza service is the Nardi-built NH 500MC, of which about 35 serve alongside NH 500M and NH 500MD versions.

Guardia di Finanza

Centro Aviazione	Practica di Mare
1° Gruppo Aereo Addestramento Avanzato	P 166
2° Gruppo Aereo Esplorazione Aeromarittima	ATR 42
3° Gruppo Aereo Funzione Tecnica	
4° Gruppo Aereo Funzione Logistica	
Sezione Aerea di Manovra Gruppo Aeronavale Napoli	
AB 412HP	Napoli Capodichino
Sezione Aerea di Manovra Gruppo Aeronavale Taranto	
AB 412HP	Grottaglie
Sezione Aerea Bari	A 109, NH 500	Bari Palese
Sezione Aerea Bolzano	NH 500	Bolzano
Sezione Aerea Cagliari	A 109, NH 500	Cagliari Elmas
Sezione Aerea Catania	NH 500	Catania Fontanarossa
Sezione Aerea Como	NH 500	Venegono Inferiore
Sezione Aerea Genova	NH 500	Genova Sestri
Sezione Aerea Palermo	NH 500	Palermo Boccadifalco
Sezione Aerea Pescara	NH 500	Pescara
Sezione Aerea Pisa	NH 500	Pisa San Giusto
Sezione Aerea Roma	NH 500	Practica di Mare
Sezione Aerea Rimini	NH 500	Rimini
Sezione Aerea Venezia	NH 500	Venezia Tessera

Both of the air force Antonov An-2s are seen parked next to each other at Lielvarde. In the background is the sole air force Let 410UVP-T.

The majority of the air force consists of helicopters, including three PZL-built Mil Mi-2s. The type is primarily used for communications and SAR roles.

LATVIA

Capital: Riga **Population:** 2.7 million
Land area: 63700 km² (24,590 sq miles)
Major cities: Daugavpils, Jelgava, Jurmala, Liepaja

Latvia declared its independence from the Soviet Union on 4 May 1990. However, it was not until after the January 1991 unrest and the subsequent failed coup attempt against Mikhail Gorbachev, which led to the August 1991 break-up of the Soviet Union, that Latvia finally achieved its desired independence. The nation quickly started to rebuild a military aviation structure, but, with very little financial assets, Latvian commanders initially had to make do with equipment left by the Russians. Newer equipment, as well as defence and security assistance, was later received from Western nations, and today military aviation in Latvia is divided between the Air Force and the National Guard.

LATVIJAS GAISASPEKI

The Latvijas Gaisaspeki (Latvian Air Force) was formed on 24 February 1992 at Riga-Spilve airport, but in August 1994 the air force moved into the ex-Soviet MiG-27 base at Lielvarde east of Riga. At present, a traditional air force is too expensive for Latvia to operate, and instead the air force is tasked with a mix of civil and military duties, including aerial surveillance, light transport, liaison, SAR, paratroop drop and pilot training.

Today the Latvian Air Force operates a small fleet of light aircraft from Lielvarde, comprising two An-2 biplanes (of a batch of 12 ex-Aeroflot An-2s transferred from Latvian Airlines to the air force in 1992, but 10 have been sold overseas to raise much-needed funds), one Let 410UVP-T (from two ex-Luftwaffe examples donated by Germany in February 1993 – the other one crashed at Lielvarde on 7 June 1995) and three Mi-2 helicopters (of eight ex-Aeroflot Mi-2s received from Latvian Airlines in 1992, two have been sold, one reduced to spares and two withdrawn from use). Future fast jet operations have not been ruled out, but so far plans to acquire former Soviet L-39s or Saab 105s have not come to fruition.

Aviacijas Un Pretgaisa Aizsardzibas Speki (Latvian Air Force)

2nd Helikopteru Eskadrila		Lielvarde
Mil Mi-8T	6	transport helicopter
PZL Swidnik (Mil) Mi-2R/S/U	3	communications
3rd Transporta Eskadrila		**Lielvarde**
Antonov An-2	2	transport
Let L410UVP-T Turbolet	1	transport

Glossary	
Helikopteru Eskadrila	Helicopter Squadron
Transporta Eskadrila	Transport Squadron

LATVIJAS ZEMESSARDZE

The Latvijas Zemessardze (Latvian National Guard) was formed as a self-defence organisation on 23 August 1991, immediately after the nation was reborn following Soviet occupation. Today the National Guard is Latvia's largest military force, with 1,500 full-time and 15,000 part-time voluntary personnel organised into five brigades, mainly tasked with national border patrol and emergency relief during natural or man-made disasters. To support its tasks the Latvian National Guard has operated an aviation component since 1993, equipped with light aircraft and gliders, most of which had been left behind by Soviet DOSAAF units. The total fleet consists of 12 An-2s and five PZL-104 Wilga 35s (a sixth Wilga crashed during 1996) as well as 39 gliders of various types (mostly Czech-built Blaniks), deployed at bases in Cesis, Daugavpils, Limbazi, Rezekne and Riga-Spilve. A pair of ex-Aeroflot Mi-2 helicopters was also acquired, but have never been taken on charge and remain stored at Riga.

Latvijas Zemessardze (Latvian National Guard)

Type	Number	Role
Antonov An-2	10	communications
Jantar ST-3	–	gliding
LAK-12	–	gliding
Let L13	20	gliding
PZL-104	5	communications

Operated from Cesis, Daugavpils, Rezekne, Limbazi and Riga-Spilve.

LITHUANIA

Capital: Vilnius
Population: 3.8 million
Land area: 65200 km² (25,165 sq miles)
Major cities: Kaunas, Klaipeda

The former Soviet Socialist Republic of Lithuania was the first Baltic State to declare its independence from the Soviet Union, on 11 March 1990, and was formerly recognised as an independent state on 19 August 1991. The Republic of Lithuania borders the Baltic Sea, Latvia, Belarus, Poland and the Russian enclave of Kaliningrad. Together with the other two Baltic states of Latvia and Estonia, Lithuania is developing joint defence positions and structures. The Krasto Apsaugos Ministerija (Ministry of National Defence) signed bilateral co-operation agreements with the defence ministries of eight NATO countries: Denmark, France, Germany, Norway, the Netherlands, Turkey, the United Kingdom and the USA. Agreements have also been signed with the Czech Republic, Estonia, Latvia and Poland and in January 1994 the Republic of Lithuania officially applied for NATO membership.

The Karinès Oro Pajègos (Lithuanian Air Force)

The sole jet-powered aircraft of the Lithuanian air force are the four Aero L-39Cs of No. 11 Squadron. Equipped with hardpoints, they provide a light strike capability.

Light transport duties are undertaken by a pair of Zokniai-based No. 12 Squadron Let L410UVPs, augmenting the fleet of Antonov An-2s.

was established on 1 March 1993. An initial form of military aviation (Karo Aviacija) was created by the Ministry of National Defence which had acquired 25 An-2 transport aircraft in January 1992. Four L-39C Albatros aircraft were bought from Kyrgyzstan and delivered in February 1993. In addition to these aircraft, two L410 Turbolets were handed over by the German Luftwaffe and five Mi-2s were donated from Poland. Three An-26s, one An-24 and three Mi-8s were also acquired. One Mi-8 was written off in an accident and the An-24 was donated to a museum in Kaunas-Aleksotas. Twenty-two An-2s remain in possession of the KOP today, but only a handful are still operational. Future plans are to have two An-2s at each air base.

Future procurement of new military hardware depends on the availability of funds. Poland offered some MiG-21s a while ago. There is a need for two squadrons (24 aircraft), but currently funds exist for only 12 aircraft. Sweden has also offered Saab 105s.

The Savanoriskoji Krasto Apsaugos Tarnyba (SKAT or Voluntary National Defence Guard) operates a mix of single-engined aircraft and gliders from two former Soviet DOSAAF airfields. The SKAT works closely with the Karinès Oro Pajègos, but falls under direct command of the Ministry of National Defence. Some of the duties of SKAT pilots include police-missions, border and coastal patrol and the interception of low-flying smugglers' aircraft.

Karinès Oro Pajègos

Krasto Apsaugos Ministerija	Vilnius
(Ministry of National Defence)	

Karinès Oro Pajègos (Military Air Forces) HQ Kaunas	

Pirmoji Aviacijos Baze (1st Air Base)	Zokniai
11 Squadron	4 x L-39C
12 Squadron	2 x L410UVP, 10 x An-2
13 Squadron	1 x Mi-8T, 1 x Mi-8MTV
Antroji Aviacijos Baze (2nd Air Base)	**Pajuostis**
21 Squadron	no aircraft assigned
22 Squadron	3 x An-26B, 12 x An-2
23 Squadron	5 x Mi-2
Trecioji Aviacijos Baze (3rd Air Base)	**Kazlu Ruda**
no units assigned yet	

Savanoriskoji Krasto Apsaugos Tarnyba (Voluntary National Defence Guard)

Vilnaus Dariaus ir Gireno Eskadrile	Kyviskes
Yak-18T/52/55, An-2, PZL-104, PA-38	
Silutes Aviacijos Eskadrile	Silute
Yak-52, An-2, PZL-104, L13, LAK-12, Jantard Standard 3	

FORMER YUGOSLAV REPUBLIC OF MACEDONIA (FYROM)

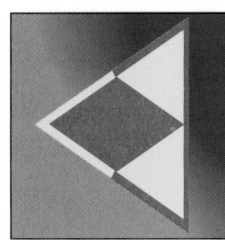

Capital: Skopje   **Population:** 2.1 million
Land area: 25715 km² (9,925 sq miles)
Major cities: Bitolj, Kumanovo, Prilep

After being claimed several times by neighbouring Bulgaria and Serbia, this former Ottoman territory was finally declared independent in January 1992. It is known as the Former Yugoslav Republic of Macedonia. The name 'Macedonia' is the same as that of a northern Greek province, leading to the compromise of the name FYROM.

The current air arms of FYROM comprise the Air Force of the Macedonian Army and the Border Police Aviation, known as Milicija. The Air Force of the Macedonian Army is still in an organisational state and is receiving help and advice from Turkey and the USA. In July 1994 the Air Force of the Macedonian Army was established, and took delivery of its first helicopters in the same month. For transport and utility tasks six Mil Mi-17 'Hip-Hs' have been delivered from an unknown source. FYROM 'Hips' wear the colourful national flag – a red sun with beams on a yellow background – on the side of the fuselage. For staff and VIP transport the Air Force operates one Learjet 25B and one Beech King Air 200. All aircraft and helicopters are stationed at the airport of the capital, Skopje. The Special Border Police Unit operates a few helicopters for border surveillance and standard policing tasks. The Milicija fleet, comprised only of helicopters, has its main base at Skopski-Petrovec near Skopje. Currently it uses two Agusta-Bell AB 206 Jetrangers, two AB

The air force of Macedonia has no offensive assets at the present, but uses several Mil Mi-17 'Hip-Hs' for troop transport.

FYROM Air Arms

Air Force of the Macedonian Army	**Skopski-Petrovec**
Mi-17 'Hip-H', Learjet 25B, Beech King Air 200	
Milicija	**Skopski-Petrovec**
Agusta-Bell AB 206, AB 212, Mi-8	

212s and two or three ex-Yugoslavian Mil Mi-8 'Hips'.

MALTA

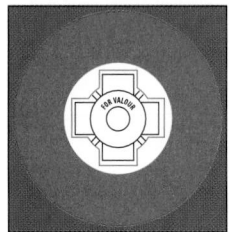

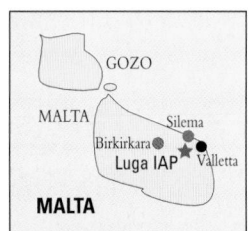

Capital: Valletta   **Population:** 400,000
Land area: 316 km² (122 sq miles)
Major cities: Sliema, Birkirkara

Armed Forces of Malta, the Air Squadron

Malta, a small group of islands consisting of the (main) islands of Malta, Gozo and Comino, is situated in the Mediterranean, between Italy and Libya. Both coun-

tries offered assistance to guarantee Maltese neutrality, although that of Libya was cancelled during 1980 due to quarrels about territorial waters. Subsequently, all Libyan forces were withdrawn, leaving only three Alouette IIIs in storage.

Plans for creating the Air Squadron were formulated in 1970; the service was then known as the Helicopter Flight, and became reality when West Germany donated four Bell 47 helicopters in 1971. Today, the Armed Forces of Malta comprise three regiments, of which the second has the Air Squadron (formerly the Helicopter Flight) attached to it. The Air Squadron has on current strength four Bell 47s, five Alouette IIIs (two of which were purchased from the Netherlands at a cost of $295,000 during 1996, and the other three being the examples left by Libya), two Hughes 369HMs, five Cessna O-1Es and one BN-2B-26 Islander. Some aircraft, however, have been withdrawn from active service and are used as spare parts.

Because of the limited range of the previous aircraft, a Britten-Norman BN-2B-26 was bought from the civil market. The twin-engined aircraft was delivered in December 1995 and subsequently

modified with a larger nose to house a search radar. At the end of 1997 a second Islander was bought, but it will not be delivered until Pilatus has modified the aircraft from a BN-2T to a BN-2B-26.

The Air Squadron does not have a real military role, and, as Malta has no defence ministry, the squadron reports directly to the prime minister. The main tasks of the Air Squadron are SAR operations, coastal patrol and medical evacuation flights. To augment the small squadron, Italy permanently bases two AB 212s of 9ª Brigata Aerea at Luqa. The AB 212s are flown by Italian pilots with Maltese observers, providing a SAR flight with a wide radius of operations. The Italian helicopters operate alongside those of the Air Squadron at the southern corner of the international airport at Luqa.

Armed Forces of Malta

2nd Regiment	
Air Squadron	Luqa
Bell 47 (2), SA 3160 (2), SA 316B (2),	
H 369HM (1), O-1E (3), BN-2B (2)	

MOLDOVA

Capital: Chisinau
Population: 4.4 million
Land area: 33700 km²
(13,010 sq miles)
Major cities: Rabnita, Tiraspol

Upon its independence in 1991, the former Soviet republic of Moldova inherited 34 MiG-29s from the Soviet navy's 86 IAP, 119 Diviziya, Black Sea Fleet. Moldova hoped initially to exchange the aircraft for armed and transport helicopters; some MiG-29s were temporary loaned to Yemen, from where the majority returned. In June 1997 the United States acquired 21 Moldovan MiG-29s as part of its Co-operative Threat Reduction Program. The aircraft consisted of six 'Fulcrum-As', 16 'Fulcrum-Cs' and a single MiG-29UB 'Fulcrum-B', in addition to 500 AA-11 'Archer' air-to-air missiles and a large quantity of spares. Moldova sold the aircraft because of its current

poor financial state; it is alleged that the US acquired them to stop them falling into the hands of so-called 'rogue states'.

About 10 other examples of the MiG-29 were sold to Eritrea in late 1998. Exactly what state these aircraft were in is open to speculation – they may not have been airworthy.

The Moldovan air force is structured along old Soviet lines with separate air force and air defence

Air Moldova operates a single Tupolev Tu 134A-3 on behalf of the government of Moldova.

organisations. Current Moldovan assets consist of eight Mil Mi-8s and about a dozen Antonov An-2s, all operating from Kishinev. As part of the Conventional Forces in Europe Treaty, Moldova is allowed up to 50 armed helicopters and 50 armed aircraft.

NETHERLANDS

Above: Tactical reconnaissance tasks are presently undertaken by No. 306 Squadron, using F-16s configured to operated the Orpheus pod as F-16A(R)s. The squadron is due to disband and the reconnaissance task is to be shared among the surviving squadrons.

Below: No. 323 Squadron at Leeuwarden AB has been operational on the F-16 since April 1982. This F-16B does not display the squadron's 'Diana the Huntress' badge but does have a partial blue/white tail strip containing red, a representation of the Frisian flag.

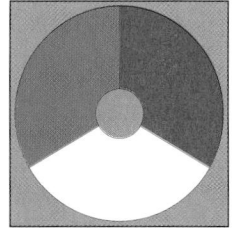

Capital: Amsterdam
Population: 15.1 million
Land area: 41160 km² (15,890 sq miles)
Major cities: Eindhoven, Groningen, The Hague, Haarlem, Nijmegan, Rotterdam, Tilburg, Utrecht

Koninklijke Luchtmacht (Royal Netherlands Air Force)

The Koninklijke Luchtmacht (KLu, Royal Netherlands Air Force) was born in 1913 as the military Luchtvaartafdeling (LVA, Aviation Department) of the Royal Netherlands Army. Its primary mission was the defence of the national airspace. Between World War I and II, Anthony Fokker, who had returned to Holland at the end of World War I, became the main supplier of its combat aircraft, including bombers. The foundation of the modern KLu was laid in World War II in the United Kingdom. Under the RAF's flag, Nos 320 and 322 Sqn were activated with British and American combat aircraft. Shortly after the war ended the Koninklijke got its present name. The Netherlands has been a NATO member from the beginning of the alliance's existence. National defence policy determines that the Dutch armed forces must be able to carry out four peacekeeping/crisis management operations simultaneously at battalion level. Depending on the intensity of the crisis, the air force must be able to carry out two operations simultaneously. Three operational F-16 squadrons and two SAM squadrons are assigned to NATO's Reaction Forces. Presently, the KLu has a strength of about 13,000 personnel and a budget of DFl. 2.6 billion.

FIGHTER FORCE

The KLu has six operational F-16A/B squadrons and one training squadron based at three Main Operating Bases. The operational squadrons are swing-role tasked, the pilots being trained for both the air-to-ground as well as the air defence mission. No. 306 Sqn based at Volkel is the exception, as it has a reconnaissance rather than air-to-ground tasking, operating F-16A(R)s. The early 1999 Headline Memorandum published by the Dutch Ministry of Defence outlined the disbandment of one F-16 squadron (down to six in total) and the retirement of 18 aircraft, with the axe falling on No. 306 Sqn. In the future the reconnaissance role will be undertaken by all squadrons when a new reconnaissance pod enters service early in the next century, enhancing the swing-role nature of the F-16 force. It will leave 90 F-16s (from 108) assigned to NATO.

No. 323 Sqn at Leeuwarden AB is also the tactical training, evaluation and standardisation squadron (TACTESS). In this capacity, it is responsible for training weapon instructors, operational testing of new systems, procedures and the like, and is the air warfare nucleus of the F-16 community. Nos 311 and 312 Sqn both have a nuclear strike mission. The training squadron (No. 313 Sqn) is responsible for training new F-16 pilots under European conditions.

By 2001 all aircraft are to have undergone an F-16 Mid-life Update (MLU), and will emerge as F-16AM/BMs, as they are designated by Lockheed Martin. A total of 138 of the original 213 F-16A/Bs (Block 10 and 15) delivered in the late 1970s and throughout the 1980s will receive the update package. The remainder will be withdrawn from service and put up for sale, storage or exhibition. More than 30 have been lost in accidents. The MLU programme is being conducted by the air forces of Belgium, the Netherlands, Norway, Denmark and the USA. It incorporates the APG-66 radar upgrade to APG-66V2, Advanced Identification Friend or Foe system, the Improved Data Modem, GPS, the Modular Mission Computer, Digital Terrain System, provisions for inlet hardpoints, EW management system, colour displays, a wide-angle HUD, night-vision-compatible cockpit and AIM-120 AMRAAMs. Basically, the MLU F-16 is a Block 40/50 hybrid.

In addition to the standard, five-country, MLU package, the KLu has purchased several other systems, including 10 Lockheed Martin Sharpshooter targeting pods and 60 GEC-Marconi Atlantic FLIR navigation pods, GBU-24 LGBs and AGM-65G Mavericks. Simultaneously with the MLU upgrades, the KLu will standardise the Pratt & Whitney F100-PW-200 engines to the more reliable and powerful PW-220.

Entering service early in the next century will be a new reconnaissance pod for low- to medium-altitude

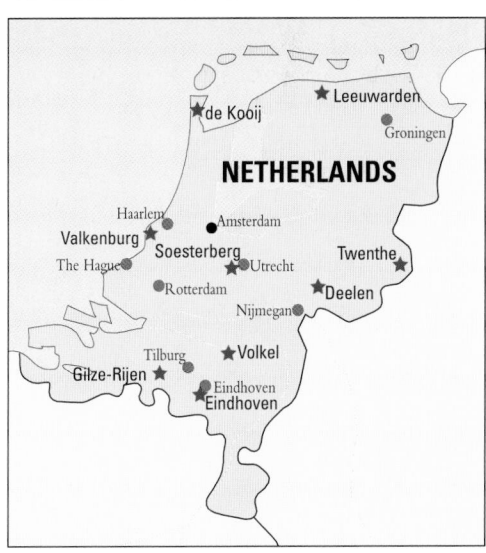

operations, a replacement for the ageing Delft Orpheus and MARS pods in service with No. 306 Sqn. The new pod will have electro-optical sensors, a datalink for near real-time recce, and an IRLS. MARS (Medium-Altitude Reconnaissance System) is comprised of two KS-87 cameras placed in a Per Udsen pod. Four were acquired in an urgent programme for operations at medium altitude over Bosnia.

An ambitious KLu requirement is for an anti-radiation missile (delayed due to budget cuts), helmet-mounted cueing system, an AIM-9 Sidewinder successor, microwave landing system, night-vision goggles and an Elint capacity. All should be in service early in the 21st century.

Europe

The Tactical Helicopter Group operates AS 532U2 Cougar Mk 2s (above) and CH-47Ds (above right) in the medium and heavy transport roles, while the 'punch' is provided by the fleet of NAH-64Ds (left).

In service since 1975, the BO 105CB is due to be replaced in the liaison and observation role in the near future.

The KLu plans to replace the F-16 from 2010, and strongly favours the US Joint Strike Fighter. Together with Norway and Denmark, the MoD has signed a memorandum of understanding with the USA to be a multi-national associated partner in the JSF programme. The initial requirement called for 140 aircraft, but the actual deliveries will probably be less. The KLu expects to retire the last F-16s in 2020/25.

TACTICAL HELICOPTER GROUP

The Tactical Helicopter Group RNLAF (THG/KLu) was formed in 1994 with a primary mission of supporting the newly established Dutch Army's 11th Air Mobile Brigade. Heavy and medium transport helicopters are responsible for transport of troops and materiel, casevac, support of special forces and tactical operations. Attack helicopters provide combat support and light helicopters are used for liaison, observation and transportation of members of the Royal Family. Apart from working with the 11th AMB, the THG also undertakes national and international general support tasks for the air force, navy and civilian authorities. The majority of the pilots are from the air force, but a small number of army pilots started flight training in 1997 on an experimental basis.

The transport helicopters are based at Soesterberg. Seven CH-47D Chinooks were procured from the Canadian Forces and extensively rebuilt by Boeing to Dutch requirements. Boeing delivered six new Chinooks in 1998 to No. 298 Sqn. The Dutch Chinooks differ from standard CH-47Ds in several aspects: they are the first with a fully-integrated 'glass' cockpit, have more powerful engines and a nose-mounted weather radar. For operations in winter conditions, skis can be attached to prevent the helicopter from sinking into the snow.

No. 300 Sqn has 17 Eurocopter AS 532U2 Cougar Mk 2s, delivered in 1996 and 1997. Like the Chinook, the Cougar has a fully night-vision-compatible glass cockpit and provisions for skis and a rescue winch. The first seven delivered have flotation gear for maritime operations. Amphibious operations with the Cougar are planned to take place from the

HMrS *Rotterdam*, an amphibious transport ship delivered to the Royal Dutch Navy in mid-1998. The KLu has also envisaged a combat SAR capability for No. 300 Sqn in the future.

The attack and observation/reconnaissance helicopters are based at Gilze Rijen. The first Dutch AH-64D Apache was delivered in May 1998, and the last of 30 ordered are expected in the Netherlands in April 2002. Pending delivery, the Dutch AF leased 12 AH-64As from the US Army for the symbolic fee of US$12; they fly with No. 301 Sqn at Gilze Rijen. This enabled the KLu to quickly establish an attack helicopter force. The 12 A models arrived in 1996 and are to be returned to the US Army when sufficient D models have arrived. The Dutch AH-64D is dubbed NAH-64D and does not have the fire control radar, which the Dutch AF hopes to procure at a later stage, nor does it have radar-guided Hellfire missiles, only laser-guided. For liaison and observation duties, the THG uses the MBB BO 105s. The little helicopters have seen service in the former Yugoslavia in support of IFOR and SFOR. Six Alouette IIIs remain in service at Soesterberg providing, among other tasks, transportation for members of the Royal Family. It is planned to retire the BO 105s and remaining Alouette IIIs and replace them with a single type. SAR is provided by three Agusta-Bell AB 412SPs, based at Leeuwarden AB with No. 303 Sqn. One AB 412 is always detached to the isle of Vlieland, located north of Leeuwarden, when the shooting range (Vliehors) at this island is active. Other helicopters are planned to complement the AB 412 for SAR, possibly other examples of the same design.

AIR TRANSPORT

In recent years the KLu has greatly increased its air transport capacity. From the mid-1960s until 1996, only 12 F27 Troopships/Friendships formed the airlift fleet. In anticipation of the dramatically changed world situation, the KLu purchased two Lockheed Martin C-130H-30s, four Fokker 60s, two Fokker 50s, two KDC-10s and a single Grumman C-20 Gulfstream IV. All Dutch transport aircraft fly with No. 334 Sqn at Eindhoven Air Base.

The strategic transport/tanker KDC-10s are second-hand converted DC-10CFs (Convertible Freighters). They have accommodation for up to 334 passengers or a maximum cargo load of 65 tons, or a combination of both. Main missions are transportation of troops and cargo, support for F-16 operations and aerial refuelling. Additional tasks are humanitarian relief operations.

The KDC-10s are the first tankers with a remotely controlled boom operating system. A unique stereo-scopic video system enables the boom operator to place the boom in the receptacle of the receiving aircraft. He is located in a RARO (Remote Air Refuelling Operator) station placed just aft of the cockpit, so there is no KC-10-style aft-facing compartment with a large window in the back of the fuselage. The RARO system is based on five video cameras – two on the wingtip and three under the fuselage just in front of the boom. The two mounted on the wingtips give the RARO operator a near-180° field of view behind the aircraft. Three-dimensional vision is achieved via two of three fuselage-mounted systems, which display their images on a single screen, and the boom operator wears special 3-D glasses. Almost all receptacle-equipped NATO aircraft active in the European theatre can be refuelled, including the C-130 and E-3. The KDC-10s are not equipped for hose-drogue refuelling.

The Fokker 60s and C-130s are the tactical transports. The C-130H-30s have provisions for a radar warning system, chaff/flare dispensers and ALQ-131 ECM pods. The Fokker 60 is used for regular cargo/passenger transport, medical evacuation, and paradropping, and has capacity for up to 45 passengers, or 7325 kg (16,150 lb) of cargo, including F-16 engines. They have a missile approach warning system and chaff/flare dispensers for self-defence. The KLu is the only user of this enlarged Fokker 50, which itself was born from the F27. The two Fokker 50s are configured for passenger/VIP transport. The sole Grumman Gulfstream IV is exclusively used for VIP transport.

BOSNIAN OPS

The KLu was among the first NATO members to fly over the former Yugoslavia enforcing the 'No-Fly Zone' of Operation Deny Flight. The first F-16s for reconnaissance and fighter operations arrived in 1993 at Villafranca air base, Italy. They were involved in the first combat in NATO's history when, in 1994, the raid against Udbina air base in Bosnia was conducted. Dutch F-16s also took part in Operation Deliberate Force late in the summer of 1995. Since 1996 they have operated with the Belgian Air Force under the Deployable Air Task Force. The DATF concept was born from the wish to save costs when both countries deploy for the same operations or exercises.

Left: A pair of DC-10CFs was acquired from Martinair Holland in the mid-1990s and converted to tanker/transports. They serve with No. 334 Sqn.

Below: Probably the last in a very long line of Fokker products to enter service with the KLu are the four Fokker 60UTAs of No. 334 Squadron.

Koninklijke Luchtmacht (Royal Netherlands Air Force)

Volkel AB

306 Sqn°	F-16A/B	reconnaissance/air defence
311 Sqn	F-16A/B	strike/air defence
312 Sqn	F-16A/B	strike/air defence
640 Sqn	Hawk, 40L70	air base defence
900 Sqn	F-16B	Testgroep KLu

Twenthe AB

313 Sqn	F-16A/B	training
315 Sqn*	F-16A/B	strike/air defence
620 Sqn	Hawk, 40L70	air base defence

Leeuwarden AB

303 Sqn	AB 412SP	SAR
322 Sqn*	F-16A/B	strike/air defence
323 Sqn	F-16A/B	tactical training, evaluation, standardisation
630 Sqn	Hawk, 40L70	air base defence

Eindhoven AB

334 Sqn	KDC-10, G-IV, Fokker 50/60, C-130H-30	transport

Hato AB, Curaçao, Netherlands Antilles

336 Sqn°	F27M-200MPA	maritime surveillance

Soesterberg AB

Tactische Helikopter Groep KLu

298 Sqn	CH-47D	air transport
300 Sqn	AS 532U2	air transport

Gilze Rijen AB

299 Sqn	BO 105CB4	reconnaissance/observation
301 Sqn	AH-64A*	attack
302 Sqn	AH-64D*	attack
670 Sqn	Hawk, 40L70	air base defence

Woensdrecht AB

131 EMVO Sqn	PC-7	basic flight training

De Peel AB

Groep Geleide Wapens

801 Sqn*	Patriot/Hawk/Stinger	air defence
802 Sqn*	Patriot/Hawk/Stinger	air defence
803 Sqn*	Patriot/Hawk/Stinger	air defence
804 Sqn*	Patriot/Hawk/Stinger	air defence

* assigned to NATO's reaction forces
° due to disband

Fokker F27s – and later the C-130s and Fokker 60s – have flown many air transport missions into Croatia and Bosnia. The Alouette IIIs and BO 105s have been used as liaison, reconnaissance and observation assets in support of NATO forces. The Dutch MoD policy calls for the F-16 presence as long as there are NATO forces in theatre.

TRAINING

The Elementaire Vliegopleiding (EMVO, Elementary Flight Training) at Woensdrecht provides basic training for air force, navy and army pilots with 13 Pilatus PC-7s. Pilots earmarked for the F-16 advance to the Euro NATO Joint Jet Pilot Training (ENJJPT) at Sheppard AFB to fly the T-37 and T-38, before heading to the Arizona Air National Guard at Tucson for initial F-16 training. The KLu is also present at Goose Bay, Canada. From April until October, F-16 squadrons deploy there for annual low flying training.

Helicopter pilots first take an advanced course on the PC-7 at Woensdrecht. Chinook and Apache pilots receive basic and type training at Fort Rucker, Alabama, but BO 105 and AB 412 pilots are trained in the Netherlands. Cougar pilots train with Helicopter Service, a civilian Norwegian company. Apart from the AB 412 pilots, all go to the bureau TACTESS (Tactical Training, Evaluation and Simulation Standardisation) which borrows helicopters from the operational squadrons.

GROEP GELEIDE WAPENS

The Groep Geleide Wapens (Group Guided Weapons) at De Peel is responsible for ground-based air defence. The KLu has four such squadrons, each employing the Raytheon Patriot, Hawk and Stinger missiles in the TRIAD (Triple Air Defence) concept, covering high-altitude, medium-altitude and close-in ranges commanded and controlled by one integrated unit. The KLu is to get the Patriot PAC-3 missile, which will give it a greater theatre missile defence capability. All four squadrons are part of NATO's main defence forces, and at any time two (they rotate) are also part of the rapid-reaction forces. During the Gulf War, Dutch Patriot units were deployed at Diyarbakir, Turkey and in Israel to protect the city of Jerusalem. In mid-1999 a trilateral Extended Air Defence Task Force (EADTF) of the Netherlands, Germany and the United States is due to be established.

Hawk and Stinger missiles plus 40L70 anti-aircraft artillery are based at the three F-16 air bases for air base protection.

CARIBBEAN

No. 336 Sqn at Hato, Curaçao, is the only Dutch unit permanently based in the Dutch Antilles. Equipped with two F27M Maritimes since 1981, the squadron performs a variety of overwater missions, such as SAR, surveillance, medevac, counter-drug operations, transport roles and fishery inspection. No. 336 Sqn also trains Dutch Navy P-3 Orion observers. It was announced in the early 1998 defence review that No. 336 Sqn would disband and the F27Ms sold.

The 1989 delivery of 13 Pilatus PC-7s to the EMVO marked the rebirth of basic training in the Netherlands.

Marineluchtvaartdienst

The Netherlands has long been dependent on sea trade (Rotterdam is the world's biggest harbour) and a strong navy is seen as necessary to protect the country's vital maritime interests. The naval air arm is called Marineluchtvaartdienst (MLD, Naval Air Service) and received its first aircraft in 1914. As with the air force, the foundations of the modern MLD were laid by the RAF in World War II. No. 320 Sqn was assigned to RAF Coastal Command and flew various patrol and bomber aircraft throughout the war. After the war, the MLD played a major role in the former Dutch colony now known as Indonesia until the last part, Irian Jaya, came under Indonesian control in 1962. Indonesia itself had fought for independence in 1949. Until 1968 the MLD also flew from the HMrS *Karel Doorman*, an aircraft-carrier that served until very recently with the Argentine navy.

The current MLD has two aircraft types flying from two naval air stations. The Group Maritime Patrol (MARPAT) has 13 Lockheed Martin P-3C II½ based at NAS Valkenburg, near the city of The Hague. The operational tasks are performed by No. 320 Sqn and the training tasks by No. 321 Sqn. Missions include ASW, ASuW, fishery inspection, surveillance for the coast guard, counter-drug operations and SAR. The early 1999 Headline Memorandum announced that the MLD is to lose three of its P-3Cs, which are to be sold. This may be a prelude to the closure of NAS Valkenburg. An option under study is a joint Dutch-German maritime patrol group based in Germany, about which a decision will be made in 1999.

One P-3C is always based at Keflavik, Iceland, and another two at Hato, Curaçao. The latter are employed for counter-drugs operations in co-operation with US forces and agencies and (for the time being) the KLu F-27Ms of No. 336 Sqn, and regular maritime operations. All four aircraft report to the Antilles and Aruba coast guard. MLD P-3Cs also started flying reconnaissance missions over Kosovo, as part of Operation Eagle Eye, in mid-January 1999.

The standard armament of the Orion is limited to torpedoes and sea mines but, with a few small cockpit modifications, RNlN Harpoon missiles (normally based on ships) can be fired from wing pylons if their boosters are removed.

The MLD Orions will fly until at least 2015 and seven are poised to undergo a modernisation programme called CUP (Capability Upkeep Programme). CUP, focused on the modernisation of the sensor, weapon and commando systems, would include a new FLIR, new ESM equipment, new imaging radar, and new acoustic signal computer. The cockpit is not upgraded, in order to save money.

NAVAL HELICOPTERS

The MLD has 22 SH-14D Lynx helicopters based at De Kooy near the naval port of Den Helder. The helicopters entered service in the second half of the 1970s and are tasked with ASW, ASuW, SAR and special marines operations. The Lynx weapons are torpedoes and, if necessary, a single MAG 7.62-mm machine-gun for boarding operations.

No. 860 Sqn is responsible for operational tasking onboard the numerous RNlN ships. When detached to a ship, the main mission is ASW. Training and land-based SAR is provided by No. 7 Sqn. The RNlN has 16 frigates, two supply ships capable of helicopter operations, and an amphibious ship (HMrS *Rotterdam*, a 12,800-ton ship with accommodation for four utility landing craft and four to six helicopters). The Lynx is more than 20 years old and, from 2003, 20 NH 90s are to enter service as replacement. The MLD will be the first operational user of the helicopter, if it can meet the requirements.

Naval pilots for the Lynx helicopter go to Hato, Curaçao, Netherlands Antilles after graduating from the EMVO, Woensdrecht. There, a civilian company operates two Aérospatiale AS 355F1 light helicopters for basic helicopter training. No. 7 Sqn at De Kooy takes care of Lynx conversion training.

Orion pilots follow a different route. After graduation from the EMVO they go to NAS Valkenburg to fly the Beech 200 Super King Air before commencing training on the P-3 at the same air base.

No. 303 Sqn operates three AB 412SPs in the SAR role. They have been in service since 1994.

Of the 13 P-3Cs in service with the MLD's MARPAT, three are due to be offered for sale.

Marineluchtvaartdienst (MLD)

7 Sqn, De Kooy	SH-14D	training, SAR
860 Sqn, De Kooy	SH-14D	shipborne operations
Groep Maritieme Patrouillevliegtuigen (MARPAT)		
320 Sqn, Valkenburg	P-3C	maritime operations
321 Sqn, Valkenburg	P-3C	training

NORWAY

Capital: Oslo
Population: 4.3 million
Land area: 323895 km²
(125,025 sq miles)
Major cities: Bergen,
Stavanger, Trondheim

Luftforsvaret (Air Force)

The Norwegian Air Force, Luftforsvaret, was officially formed on 10 November 1944 by the amalgamation of the Hærens Flyvåpen (Army Air Force) and Marinens Flyvevæsen (Naval Air Service), both of which had existed since 1915. Four Luftforsvaret squadrons were originally formed as Norwegian squadrons within the Royal Air Force command structure early in World War II to fight Nazi Germany. Post-war, Norway was a founding member of NATO.

COMMAND STRUCTURE

Luftforsvaret is responsible for all Norwegian military aviation and the Kystvakt (Coast Guard). Headquarters are part of Forsvarets Overkommando (Defence High Command) located at Huseby in Oslo, alongside Hæren (Army), Marinen (Navy) and Heimevernet (Home Guard). All are subordinates to the Forsvarsdepartementet (Ministry of Defence).

Units based north of latitude 65° north (Andøya, Bardufoss, Banak and Bodø) report to Forsvarskommando Nord-Norge (Defence Command Northern Norway), while units based south of 65° north (Gardermoen, Rygge, Sola, Værnes and Ørland) report to Forsvarskommando Sør-Norge (Defence Command Southern Norway). Each geographical command has the ability to operate as an independent air force, fully equipped with fighter squadrons, transport and rescue units. All technical support and maintenance of aviation equipment is the responsibility of Luftforsvarets Forsyningskommando (Air Force Supply Command), a separate command with headquarters at Kjeller.

Within NATO's structure, Luftforsvaret is assigned to Allied Forces North Europe (AFNORTH), headquartered at Stavanger, Norway, which is subordinate to Allied Forces Northwestern Europe (AFNORTH-WEST) in High Wycombe, England. If the command of national forces is transferred to Allied authorities, all military forces in Norway will be placed under the command of a Norwegian commander within the NATO chain of command.

PRIMARY COMBAT AIRCRAFT

Luftforsvaret operates the General Dynamics (now Lockheed Martin) F-16 Fighting Falcon as its primary combat aircraft. The first Luftforsvaret F-16 – F-16B serial 301 – was delivered from the Fokker assembly line at Schiphol in Holland on 15 January 1980. During the following four years the initial Norwegian contract for 72 aircraft, comprising 60 F-16As and 12 F-16Bs, was completed. Two more attrition replacement F-16Bs were delivered in 1989.

When factory-fresh Falcons began flowing into Rygge, 332 Skvadron already had five pilots and a number of ground crew with F-16 experience, trained by the USAF at Hill AFB, Utah, and having participated in the MOT&E programme (Multi-national Operational Test & Evaluation) during 1979. From May 1980 onwards, 332 Skvadron became responsible for F-16 transition training (in addition to its operational roles), and all future Luftforsvaret F-16 pilots went through a four-/five-month Transition Training Programme with 332 Skvadron at Rygge.

The F-16 had originally been acquired to replace Norwegian Lockheed F-104 Starfighters. During

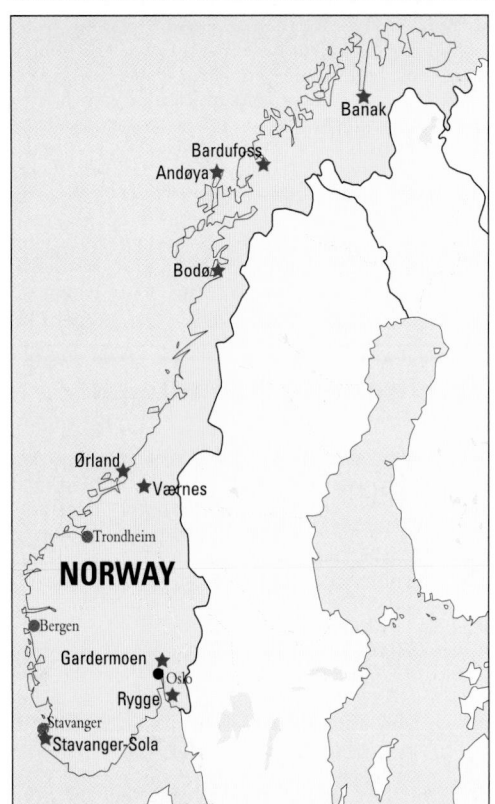

1981-82, 331 and 334 Skvadron at Bodø went into transition status while their pilots converted to the new fighter at Rygge. By the summer of 1982, 331 Skvadron was declared combat ready, followed by 334 Skvadron a year later. Next came the turn of Northrop F-5 Freedom Fighters to give way to the Falcon, and, during the summer of 1985, 338 Skvadron at Ørland also received the F-16, bringing the number of Luftforsvaret F-16 squadrons to four. 331, 332, 334 and 338 Skvadron are all fully qualified for both air-to-air and air-to-surface missions, utilising the F-16's multi-role capability to the full.

Norwegian Falcons were the first to be equipped with the extended finroot fairing, housing a braking parachute for use on hard-packed ice-covered runways. This also allows the aircraft to use some 20 small civil airstrips throughout the country, for dispersed operation.

Luftforsvaret F-16s were constructed to Block 1, 5, 10 and 15 standard, but all Block 1 and 5 aircraft have been upgraded to Block 10 standard. Norway is a participant in the five-nation NATO F-16 Mid-Life Update (MLU) programme and will upgrade a total of 56 airframes (45 F-16As and 11 F-16Bs) to MLU standard.

SECONDARY JET FIGHTER

336 Skvadron at Rygge continues to operate the F-5s updated under the Service Life Extension Programme (SLEP – 1984/85) and the Programme for Avionics and Weapon System (PAWS – 1993/95) updated F-5s which are now approaching an impressive 30 years of service. Currently the main mission of

The Lynx Mk 86s operate on behalf of the Norwegian coast guard from Bardufoss. None has been lost in nearly 20 years of service.

Above: F-16s form the bulk of the Luftforsvaret's combat aircraft. A blue lightning flash below the cockpit and a fin-stripe reveals this F-16A to be from 331 Skvadron.

Below: Freedom Fighters continue to be operated by 336 Skvadron. Of the 14 F-5Bs purchased, eight continue to serve, while seven (of 78) As are in use.

Kongelige Norske Luftforsvaret (Royal Norwegian Air Force)

Forsvarskommando Nord-Norge		
330 Skvadron - A Flight	Sea King Mk 43	Bodø
330 Skvadron - B Flight	Sea King Mk 43	Banak
331 Skvadron	F-16A/B	Bodø
333 Skvadron	P-3C U.III/P-3N Orion	Andøya
334 Skvadron	F-16A/B	Bodø
717 Skvadron	Falcon 20ECM-5	Bodø
719 Skvadron	DHC-6-100/200	Bodø

Forsvarskommando Sør-Norge		
330 Skvadron - C Flight	Sea King Mk 43	Ørland
330 Skvadron - D Flight	Sea King Mk 43	Sola
332 Skvadron	F-16A/B	Rygge
335 Skvadron	C-130H Hercules	Oslo/ Gardermoen
336 Skvadron	F-5A/B Freedom Fighter	Rygge
337 Skvadron	Lynx Mk 86	Bardufoss
338 Skvadron	F-16A/B	Ørland
339 Skvadron	Bell 412SP	Bardufoss
720 Skvadron	Bell 412SP	Rygge
Flygeskolan	MFI-15 Safari MFI-17 Supporter	Vaernes
MOT&E	F-16B	Leeuwarden

Glossary	
Flygeskolan	*Flying School*
Skvadron	*Squadron*

Right: Bodø's 719 Skvadron uses two DHC-6-100s and a single -200. The type has served on light transport tasks since 1967.

Below right: Flight screening is undertaken on the Saab MFI-15s of the Flygeskolen.

336 Skvadron is air defence and electronic tactical support. The F-5 is overdue for replacement and the Norwegian Ministry of Defence has announced that it plans to purchase 48 new fighters with deliveries starting in 2003, to replace the F-5 and cover F-16 attrition. Luftforsvaret has shortlisted the F-16C Block 50N and Eurofighter 2000 Typhoon, eliminating the F/A-18C and Rafale, and a decision is expected during 1999.

MARITIME OPERATIONS

333 Skvadron at Andøya has a primary mission of maritime surveillance and anti-submarine warfare, plus a secondary mission of search and rescue over northern waters. The Skvadron has recently become available to NATO's Supreme Allied Commander in the Atlantic area (SACLANT) and the European theatre (SACEUR) as Immediate Reaction Force and Rapid Reaction Force. Four new P-3C Update IIIs were delivered in 1989, replacing seven P-3Bs. Two P-3Bs were modified for sea surveillance, environmental control and passenger transport as P-3Ns, and are operated by 333 Skvadron on behalf of the Kystvakt (Coast Guard).

330 Skvadron, with its fleet of Westland Sea King Mk 43 helicopters, is tasked with SAR. Ten were delivered in 1972, but two have crashed, and were replaced by single examples delivered in 1978 and 1989. In 1996 the fleet was expanded to 12 Sea Kings with the delivery of two additional Mk 43Bs. Headquartered at Bodø, 330 Skvadron maintains SAR readiness at bases throughout the country, with two operational helicopters assigned to each of four flights as follows: A Flight at Bodø, B Flight at Banak, C Flight at Ørland and D Flight at Sola. Operational tasking comes from the Rescue Co-ordination Centres at Bodø and Sola.

In 1981 Luftforsvaret acquired six Westland Lynx Mk 86s financed from the civilian budget for Kystvakt (Coast Guard) duties. They were delivered to 337 Skvadron based at Bardufoss but operationally they fly from the stern platform of three Kystvakt ships of the 'Nordkap'-class.

TRANSPORT AND SUPPORT

Six Lockheed C-130H Hercules transports, delivered in 1969, provide transport support with 335

Four flights of Sea King Mk 43s provide search and rescue cover around the Norwegian coast. After the delivery of a single Mk 43B in early 1992, all Mk 43/43A survivors were due to be upgraded to the new standard. This involved the addition of a Bendix nose radar, MEL Sea Searcher radar, FLIR and advanced avionics.

Skvadron at Gardermoen. Six C-130J-30 Hercules II, required for delivery around 2000, may replace the C-130H fleet. Light transport and communication is the task of the three de Havilland Canada DHC-6 Twin Otters operated by 719 Skvadron at Bodø.

Two Dassault Falcon 20ECMs operate in the ECM role, in conjunction with air defence fighters and ground-based radar installations, and calibrate military air navigational aids. A third Jet Falcon was acquired in 1978 for VIP flying. All three are operated by 717 Skvadron at Rygge.

Eighteen Bell 412SP Arapahos were delivered in 1987-89, plus one attrition replacement. They are primarily used by 339 Skvadron at Bardufoss (for army support) and 720 Skvadron at Rygge (for SAR), but 719 Skvadron at Bodø also has a few examples.

PILOT SCREENING

All pilot candidates must pass a 20-hour screening with Flygeskolen (primary flying school) at Værnes, using the 20 Saab/MFI-15 Safaris. As Flygeskolen's requirement is for only 16 aircraft, four Safaris are hired out for civilian use as an operational reserve.

Screening is part of an 18-month officer training course in Norway, after which the trainees go to various flying schools in the United States.

KYSTVAKT

Kystvakt (the Coast Guard) was established in April 1977 for fishery protection, environmental control and SAR in the economic zone around mainland Norway and around the islands of Svalbard and Jan Mayen in the North Atlantic. It is defined as a non-military force and as such can only carry out civilian duties, although organisationally it reports to Forsvarets Overkommando (Defence High Command) as part of the Norwegian Sjøforsvaret (Naval Defence) comprising Marinen (Navy), Kystartilleriet (Coastal Artillery) and Kystvakt.

Divided into Kystvaktskvadron Nord and Kystvaktskvadron Sør (Coast Guard Squadron North and South), with headquarters at Sortland and Bergen, Kystvakt operates a large number of ships and even has military aircraft and helicopters at its disposal, including the two P-3N Orions of 333 Skvadron and six Lynx helicopters of 337 Skvadron.

POLAND

Capital: Warsaw
Population: 38.2 million
Land area: 312685 km²
(120,695 sq miles)
Major cities: Krakow,
Lodz, Poznan, Wroclaw

The most numerous fighter in the Polish arsenal is the MiG-21; this example, a MiG-21bis 'Fishbed-L', belongs to 9. PLM of 2. KOP. Poland has operated this version since 1980.

Poland has a long aviation tradition, having its own aviation industries that manufacture helicopters and fixed-wing aircraft for the armed forces.

Since the demise of the Warsaw Pact, Poland has actively pursued NATO membership. After years of Poland being a Partnership for Peace member, NATO decided in 1996 that the country could become a full member in 1999. It is the biggest of the three former Warsaw Pact countries to join, with correspondingly large armed forces. In the previous nine years the Polish armed forces have invested much in areas such as interoperability, standardisation, adaptation of military infrastructure, military equipment and the education and training of personnel. Still, in the years ahead, billions more dollars must be invested while force restructuring continues. This will lead to a leaner structure, new Western fighters, new transport aircraft and new helicopters for the air force and the air arms of the navy and army. The personnel strength will be reduced from 56,000 in 1997 to 38,000 in 2004.

Wojska Lotnicze I Obrony Powietrznej (Air and Air Defence Forces of Poland)

The Polish armed forces were formed immediately after Poland gained independence in 1918. In November that year, the Aeronautical Section of the Technical Department of the Military Ministry was established. At that time, aircraft of this section were already involved in combat in defence of the new borders, in what was called the Polish-Ukrainian War. Between World Wars I and II, the air arm was further enlarged, supported by a booming aviation industry. When Nazi Germany invaded Poland late in the

summer of 1939, the outnumbered and outperformed Polish aircraft managed to shoot down 130 German aircraft in 17 days of fighting. Many pilots fled to France and the UK in order to continue fighting, and at the end of the war 15 Polish squadrons were flying under the RAF flag. Throughout the war the Poles were credited with 764 enemy aircraft and 130 V-1 flying bomb kills.

Meanwhile, the German forces were forced back from Poland by Russian forces. A 17-regiment strong air force mainly equipped with Russian-made aircraft was in place by early 1945. The first jet aircraft were introduced in 1951 (Yak-23 and MiG-15) followed two years later by Il-28 light bombers and Mi-1 helicopters. When democracy finally superseded Communist rule in 1990, Poland had the largest air force in the Warsaw Pact after the Soviet Union. The oldest aircraft, notably Polish-built MiG-15 and MiG-17s, have since been withdrawn from use. In 1990 the air force became the Air and Air Defence Forces, and had been preparing for full NATO membership since the early 1990s.

The basic tasks of the modern WLOP are detection of and fighting the enemy's air power; protection of important administration and economic centres, troops, naval bases and other key assets; destroying land and sea targets and fighting seaborne landing operations; reconnaissance; and air transportation. The WLOP also supervises air traffic over Poland and conducts SAR operations. The WLOP has three basic services: aviation, air defence missiles and radiotechnical units (AADF). Aviation is divided between two air defence corps (Korpus Obrony Powietrznej, KOP) with headquarters at Bydgoszcz (2nd KOP) and Wroclaw (3rd KOP). All tactical units report to these two commands. Directly reporting to the WLOP headquarters are the 36. SPLT at Warszawa Okecie equipped with VIP transport aircraft, and the 45. EL at Modlin. The latter is the test and evaluation unit of the WLOP. The air force academy has all the fixed- and rotary-wing training units under its command.

FIGHTERS

Fighter aviation concerns the operational fighter regiments equipped with some 200 MiG-21s, 36 MiG-23s and 22 MiG-29s. A regiment usually has between 30 and 40 jets divided between three squadrons. The 'Floggers' are scheduled for retirement in 1999; the fleet of MiG-21s, the most numerous fighter, is gradually being reduced as the aircraft reach the end of their lifetimes. The most modern fighter is still the MiG-29 'Fulcrum'. The first 12 were procured in 1989 directly from the factory and operate with 1. PLM. This regiment received another 10 'Fulcrums' from the Czech Republic in late 1995 in exchange for 11 W-3 Sokol helicopters.

Fighter-bomber aviation has a fleet of about 100 Su-22 'Fitters'. Since the disbandment of 6. PLMB at Pila and the closure of this base in 1997, the 'Fitters' have been assigned to the remaining three wings. 7. PLBR at Powidz also employs 'Fitters' as reconnaissance assets with the KKR-1 recce pod adopted from the retired Su-20s in 1997. Subsequently, the Su-22 became the only aerial reconnaissance asset when the MiG-21R was retired from the reconnaissance mission during the summer of 1997. Each tactical fighter regiment has a few TS-11 Iskras and An-2s available for liaison and training.

The Polish Air Force is working towards a fighter force of about 100 Western-style fighter aircraft, to be

procured in the first decade of the 21st century, and the current inventory of Sukhoi Su-22 fighter-bombers. Of the tactical jet fleet, only the 'Fitters' and 'Fulcrums' have been equipped with Western-style navigation aids and identification equipment.

TRANSPORT AND TRAINING

Transport aviation is comprised of a regular air transport regiment and a special-purpose regiment equipped with the An-2, An-26, An-28, Yak-40, Tu-154, Mi-2, Mi-8 and W-3 Sokol. Replacement is being sought for the An-2s and An-26s. The Yak-40s are used for liaison, VIP transport and target towing. The latter task is the responsibility of 17. EL based at Poznan-Lawica. Both air defence corps headquarters have a squadron equipped with the Mi-2 and An-2 for liaison/transport duties.

All navy, army, air force and ministry of defence pilots are trained at the Air Force Academy at Deblin. Pilots train on locally-built TS-11 Iskras and PZL-130 Orlik turboprops. 58. LPSz used to be equipped with the Mielec I-22 Iryda but structural problems forced the factory to take back all the aircraft for modification to I-22M96 configuration. However, the fate of the Iryda remains unsure because of the lack of funds.

The air defence missile units are equipped with a variety of Soviet-made SAM and AAA systems.

The PZL-130TC-1 Turbo Orlik (Spotted Eaglet) is used by 60. LPSz for basic training. The Polish air force has a requirement for 48 of the type.

Above: Polish MiG-23MFs carry the AA-7 'Apex' AAM. All 36 aircraft are operated by 23. PLM.

Below: Adorned with a tiger tail-stripe, this MiG-29 belongs to Minsk-Mazowiecki-based 1. PLM.

Wojska Lotnicze I Obrony Powietrznej

Direct-reporting units		
36. SPLT	Tu-154M, Yak-40, Mi-8P/8S, Bell 412	Warszawa-Okecie
45. EL	PZL-130TM/130TC2, MiG-21PFM/21R/21US	Modlin
2. Korpus Obrony Powietrznej		
1. PLM	MiG-29/29UB	Minsk-Mazowiecki
3. PLM	MiG-21PFM/21R/21US	Poznan-Kresiny
8. PLMB	Su-22M4/22UM3	Miroslawiec
9. PLM	MiG-21bis/21UM	Zegrze Pomorskie
28. PLM	MiG-23MF/23UB	Slupsk
40. PLMB	Su-22M4/22UM3	Swidwin
41. PLM	MiG-21MF/21UM	Malbork
2. EL	Mi-2, An-2	Bydgoszcz
17. EL	Yak-40	Poznan-Lawica
3. Korpus Obrony Powietrznej		
7. PLBR	Su-22M4/22UM3	Powidz
10. PLM	MiG-21MF/21UM	Lask
11. PLM	MiG-21MF/21UM	Wroclaw
13. PLT	An-26, PZL An-2, PZL An-28, PZL Mi-2	Krakow
Wyzsza Szkola Oficerska Sil Powietrznych (Air Force Academy)		**Deblin**
47. SzPS	PZL W-3, PZL Mi-2	Nowe Miasto
58. LPSz	TS-11	Deblin
60. LPSz	PZL-130TB/130TC-1	Radom
23. LESz	An-2	Deblin
61. LPSz	TS-11	Biala Podlaska
(combat units also use various types for liaison and training)		

The indigenous PZL Mielec TS-11 Iskra (Spark) lost out to the Aero L-39 in the competition to provide a new advanced trainer for WarPac countries, but was adopted in that role by Poland. This aircraft displays the Crest of Warsaw squadron badge of 1. PLM, the MiG-29 unit.

Glossary		
EL	Eskadra Lotnicza	*Air Squadron*
DLMW	Dywizjon Lotniczy Marynarki Wojennej	
		Navy Aviation Squadron
LESz	Lotnicza Eskadra Szkolny	
		Air School Flight
LPSz	Lotniczy Pulk Szkolno	*Air School Regiment*
LPSzB	Lotniczy Pulk Szkolno-Bojowy	
		Air School/Combat Regiment
PL	Pulk Lotniczy	*Aviation Regiment*
PLBR	Pulk Lotniczy Bombowo-Rozpoznawczego	
		Bomber Reconnaissance Regiment
PLM	Pulk Lotnictwa Mysliwskiego	
		Fighter Aviation Regiment
PLMB	Pulk Lotnictwa Mysliwsko Bombowego	
		Fighter-Bomber Aviation Regiment
PLT	Pulk Lotnictwa Transportowego	
		Transport Aviation Regiment
PSB	Pulk Smiglowcow Bojowych	
		Combat Helicopter Regiment
PSzZL	Pulk Szwolezerow Ziemi Leczyckiej	
		Cavalry Land of Leczyca Regiment
PUL	Pulk Ulanow Lubelskich	*Lublin's Ulan Regiment*
SPLT	Specjalny Pulk Lotnictwa Transportowego	
		Special Transport Aviation Regiment
SzPS	Szkolny Pulk Smiglowcow	
		Helicopter School Regiment
SSzwZL	Szwadron Szwolezerow Ziemi Leczyckiej	
		Cavalry Land of Leczyca Squadron

Long-range air defence is provided by SA-5s (Antey S-200), medium-range by the SA-2s (S-75M Wolchow), and short-range by the SA-3s (S-125M Newa) and SA-7s (Strzala 2-M). AAA systems include 37-mm and 57-mm systems.

The radiotechnical units provide continuous control of airspace over Poland, using radars, control and command and signal intelligence equipment.

The reorganisation of the air force that began in the early 1990s is far-reaching, but it will take many years before Poland reaches the status of an equal NATO member. In the modernisation process the national aerospace industry will provide a substantial portion of the helicopter fleet and will licence-build new Western helicopter and fighter aircraft. The new fighter is a national priority, and the budget for this project would not come from the MoD but from a special government fund.

Right: Sukhoi Su-22M4s have been operated since 1985 and serve with two fighter-bomber and a bomber-reconnaissance regiment. About 15 Su-22UM3s (such as this 6. PLMB example) provide pilot training for just under 100 Su-22M4s.

Below: Principal airborne ASW platforms of the Polish navy are 2. DLMW's 10 Mi-14PLs.

Lotnictwo Marynarki Wojennej (Polish Naval Aviation)

Poland has a long coastline on the Baltic Sea, a sea that is of utmost economic and strategic importance for many countries in the region. Consequently, the Polish navy and its air arm play a crucial role in the country's defence and its economic interests. Polish Naval Aviation consists of three squadrons reporting to the Brygada Lotnictwa Marynarki Wojennej (Naval Aviation Brigade). The 1. Wywizjon Lotniczy Marynarki Wojennej at Gdynia Babie Doly near Gdansk has three squadrons, of which two are equipped with the MiG-21bis/UM in the strike role. The wing inherited the MiG-21s from the Polish AF in the early 1990s to replace ageing licence-built MiG-17s, and now has 28 on strength. The third squadron operates the PZL W-3/RM SAR helicopters and the PZL An-28RM Bryza. The An-28RM (Ratowhictwa Morskiego, Maritime Reconnaissance) is licence-built in Poland and is equipped with improved avionics, a radio search system, GPS, Doppler and weather radar, and rescue equipment such as dinghy, stretchers, and illumination devices for SAR in darkness.

Darlowek is the home base of 2. DLMW equipped with the ASW/SAR Mi-14PL/PSs and SAR An-2s. 3. DLMW flies the TS-11(R) Iskra, PZL An-2 and An-28 at Siemirowice in the reconnaissance and maritime patrol duties.

Lotnictwo Marynarki Wojennej

Brygada Lotnictwa Marynarki Wojennej		
1. DLMW		**Gdynia Babie Doly**
Eskadra A/B	MiG-21bis/21UM	
Eskadra C	PZL W-3/W-3RM, PZL An-28	
2. DLMW	Mi-14PL/14PS, Mi-2RM	Darlowo
3. DLMW		**Siemirowice**
1. Eskadra	TS-11/TS-11R	
2. Eskadra	PZL An-2, PZL An-28	

The Polish army is in the process of acquiring a large number of Sokol derivatives to replace its current fleet. The armed fleet will consist of PZL Swidnik W-3W Huzars, replacing the 'Hinds'.

Lotnictwo MSWiA (Ministry of Interior and Administration Aviation)

The Ministry of Interior and Administration Aviation bases 103. PL at Warsaw-Bemowo for several police and other law enforcement duties. This regiment is equipped with the Mi-8/17, Mi-2, PZL W-3, Bell 206 and PZL M-20.

Lotnictwo Wojsk Ladowych (Army Aviation)

The mission of army aviation units is to support land forces and carry out transport tasks. Most helicopters were transferred from the Air Force to the Army in 1995. Spearpoint of the post-Communist army is the 25th Air Cavalry Division, formed in 1994. The 25th ADC is slowly building up to a force of some 115 helicopters by early in the 21st century, when it should attain full combat readiness. The 25th ADC should be able to deploy anywhere in the country within several hours. It has reconnaissance (W-3 Sokol), combat (W-3W Huzar) and transport (Mi-8/17 'Hip', Mi-2 'Hoplite') helicopters.

Air assault troops are to operate in the enemy's rear areas. There are two regiments at Inowroclaw and Pruszcz Gdanski equipped with Mi-24D/V 'Hinds' and Mi-2s. Germany has donated 16 'Hinds' in recent years. All 'Hinds' and 'Hoplites' are earmarked for retirement by 2005. A drastic modernisation programme calls for up to 96 W-3 Huzars, more than 100 W-3 Sokols, a handful of W-3RR Elint/ECM helicopters and up to 50 WS-4 light helicopters. The Polish Army also has a requirement for attack helicopters

Lotnictwo Wojsk Ladowych

49. PSB	Mi-24D, PZL Mi-2	Pruszcz Gdanski
56. PSB	Mi-24V, PZL Mi-2	Inowroclaw
25. Dywizja Kawalerii Powietrznej		
1. PSzZL	Mi-8/17, PZL W-3	Leczyca
7. PUL	PZL W-3/3W/3WA	Tomaszów Mazowiecki

PORTUGAL

Capital: Lisbon
Population: 10.6 million
Land area: 91630 km²
(35,370 sq miles)
Major cities: Faro, Oporto

Força Aérea Portuguesa (Portuguese Air Force)

The Força Aérea Portuguesa (FAP) was formed on 27 May 1952 by the merger of the Aeronáutica Militar and the Aviação Naval, both of which came into existence in 1917. Portugal was a founding member of NATO in 1949. Today the Força Aérea Portuguesa has a strength of some 230 aircraft/helicopters and 10,000 personnel. It is structured into three main commands: COFA – Comando Operacional de la FA (AF Operational Command); CLAFA – Comando Logístico y Administrativa de la FA (AF Logistic and Administrative Command); and CPESFA – Comando de Personal de la FA (AF Personnel Command). Several of its squadrons are assigned to the NATO commands of CINCSOUTH, SACLANT and SACEUR. The Portuguese Air Force also provides technicians and flying personnel for the NATO Early Warning Force's Boeing E-3 Sentries.

The FAP is organised into numbered Bases Aéreas (BA – Air Bases). Each BA has three Grupos (Groups). Operativo (Operational) is responsible for the flying squadrons, airfield services, air traffic control, radar, communications, meteorology and firefighting/ambulance services. Material is concerned with maintenance, weapons, electronics and supply, while Apoyo (Support) is in charge of the base infrastructure. Finally, under the direct command of the base CO there is a services squadron.

AIR DEFENCE

The air defence system was upgraded in 1994 by the SICCAP (Sistema de Mando y Control Aéreo de Portugal – Portuguese Air Command and Control System), which includes centralising command and control operations in a single Operations Centre. The commissioning of Esquadra 201 'Falcóes' (Falcons) took place at BA5 Monte Real (Grupo Operativo 51) in January 1994. The 'Falcóes' operates the FAP's 17 F-16A Block 15 OCUs (Operational Capability Upgrade) and three F-16B Block 15 OCUs, the first fighter type in service since the retirement of the last F-86Fs in the 1980s. Operations commenced on 19

July 1994 and by the end of 1995 the 20 aircraft were in service. Their primary mission is all-weather interception armed with AIM-9L/P Sidewinders, AIM-7F Sparrows and the M61 gun, with detachments to the Azores and Madeira. Two hardened shelters for two QRA aircraft have been built at Monte Real.

CORSAIR

BA5 Monte Real received the FAP's first A-7 Corsairs in 1981. A total of 50 Corsair strike aircraft, comprising 44 A-7P single-seaters and six TA-7P two-seaters, was received to equip Esquadras 302 and 304. They are refurbished A-7As equipped with the A-7E nav/attack system and the Pratt & Whitney TF30 turbofan; another 20 A-7As were acquired as spares sources. Today, both squadrons share the 35 surviving A-7Ps and five TA-7Ps. Their tasked missions cover tactical air support for maritime operations (TASMO), air interdiction (AI) and offensive/defensive air support (OAS/DAS). For defensive purposes, they have AN/ALQ-101 jamming pods besides their RWR and chaff/flare dispensers. The FAP has secured 16 ex-AMARC F-16As (plus five for spares use only) and four F-16Bs to replace a squadron of Corsairs under the Peace Atlantis II programme. The new F-16s will receive the Falcon-Up structural improvements, the F100-PW-220E engine and Mid-Life Update (MLU) avionics, and are due in service from 2001/2002.

Primarily used in the maritime strike role, the Portuguese A-7Ps are armed with the AGM-65A Maverick and AGM-84A Harpoon.

The F-16A-15 OCUs of Esquada 201 are armed with AIM-9L Sidewinders for the air defence role. The nearest example also carries an AGM-65.

TRANSPORT AND MR BASE

BA6 at Montijo of Grupo Operativo 61 is the home of five squadrons. Esquadra 501 is equipped with the six C-130Hs delivered during the late 1970s, two of which were recently stretched to C-130H-30 standard by the air force rework facility, Oficinas Gerais de Material Aeronáutico (OGMA). Two additional H-30s have also been purchased. The Hercules perform SAR and tactical and general airlift duties, including a regular shuttle service to the Azores and Madeira, and recently they have acted as forest-fire bombers carrying the MAFFS kit. Esquadra 502 is tasked with tactical and general airlift roles with eight Aviocars and is also the operational transition unit for future Aviocar crews, who must have a given number of hours in the Spanish transport before being posted to the Hercules squadron. Two CASA 212s form a semi-autonomous flight within Esquadra 502 which undertakes electronic warfare tasks, both for training purposes and front-line operations. Esquadra 504 has one Falcon 20 and two Falcon 50s for VIP transport.

The maritime patrol mission including anti-surface unit warfare (ASUW) and ASW roles are the tasks of Esquadra 601, equipped with six Lockheed P-3P

Força Aérea Portuguesa

Comando Operacional de la Força Aérea

Grupo Operativo 12	BA1 Sintra
Esquadra 401	C.212B Aviocar
Esquadra 505	Cessna FTB.337G
Esquadra 802	RF-10, ASK.21, Chipmunk T.Mk 20, L13
Grupo Operativo 41	**BA4 Lajes**
Esquadra 503	C.212A Aviocar
Esquadra 752	SA 330 Puma
Grupo Operativo 51	**BA5 Monte Real**
Esquadra 201	F-16A/B Block 15OCU Falcon
Esquadra 302	A-7P, TA-7P Corsair
Esquadra 304	A-7P, TA-7P Corsair
Grupo Operativo 61	**BA6 Montijo**
Esquadra 501	C-130H/H-30 Hercules
Esquadra 502	C.212A Aviocar
Esquadra 504	Falcon 20C, Falcon 50
Esquadra 601	P-3P Orion
Esquadra 751	SA 330 Puma
Grupo Operativo 111	**BA11 Beja**
Esquadra 101	Epsilon, Cessna 337
Esquadra 103	Alpha Jet
Esquadra 301	Alpha Jet
Esquadra 552	Alouette III

Orions (ex-RAAF P-3Bs). The last unit at Montijo is Esquadra 751, with five Pumas for SAR and tactical transport roles.

GRUPO OPERATIVO 111

BA11 at Beja, home of Grupo Operativo 111, is in the Alentejo region of southern Portugal. Beja is the FAP's youngest air base, having been commissioned in 1964. Today it is home to elements of the FAP after having been used by the Luftwaffe for weapons training until 30 June 1993. Towards the end of 1993 Esquadra 103 and 301 started to re-equip with Alpha Jets, sharing between them 45 aircraft (50 having been delivered, with five being used as spares sources). Esquadra 103 has the task of advanced training, and also provides the aircraft and pilots for the FAP's flight demonstration team 'Asas de Portugal', which for many years used the T-37Cs of Esquadra 102. Today the team operates the Alpha Jet. A basic training syllabus is provided by Esquadra 101, equipped with 16 Aérospatiale TB.30 Epsilons. Portugal also makes use of the NATO Undergraduate Pilot Training scheme operated in the United States.

Esquadra 301 'Jaguares' continues in its traditional tasks of battlefield air interdiction (BAI) and offensive/defensive air support (OAS/DAS). The Alpha Jet has better weapons capability than its previous mount (the G91R Gina) and has the ability to carry electronic warfare sets, six of which have been acquired for the squadron's exclusive use. Esquadra 301 has been a member of the NATO Tiger Squadron Association since the mid-1980s.

The fourth unit at Beja is Esquadra 552, equipped with 24 Alouette IIIs used for helicopter pilot training, troop transport and SAR. Its Alouettes are the last in service of a total of 144 delivered since 1963. There is a FAP requirement for 30-40 medium-lift helicopters replacements, either Blackhawks or Super Pumas/Cougars, for supporting army operations.

The FAP has two reserve air bases, Aeródromo de

Portugal's Epsilon basic trainers were assembled locally by OGMA. All serve with Esquadra 101.

Above: The workhorse of the helicopter fleet is the Aérospatiale SA 330 Puma. Originally delivered in 1969 as SA 330Cs, the majority were upgraded to SA 330L standard. Of the 13 received, 10 remain in service.

Below: Portugal operates three standard and three stretched C-130Hs flown by 501 Esquadra 'Bisontes' from BA6 Montijo. This example is one of the C-130H-30s. Portugal has acquired one stretched Hercules and modified the other two from C-130Hs.

Manobra (AM1) at Ovar, and AM2 at San Jacinto. BA2 Ota has no aircraft assigned, being home to the CFMTA (AF Military & Technical Training Centre).

SINTRA AND LAJES

Grupo Operativo 1 at BA1, near Sintra, houses the Air Force Academy, the FAP's Staff College and three flying squadrons. Esquadra 401 flies six CASA C.212 Aviocars – two for photo-survey roles with mapping cameras, one earth resources aircraft, one navigational trainer and two in the SAR/maritime patrol role. Esquadra 505, with around 12 Cessna-Reims 337G

Miliroles, is tasked with light transport, liaison, visual and photographic reconnaissance. The third squadron is Esquadra 802 with five ASK.21 sailplanes and three powered Aerostructure RF-10s, and provides initial air experience for the Academy's cadets.

Based at BA4 Lajes (Grupo Operativo 41), on Terceira Island of the Azores archipelago, is Esquadra 503 with eight C.212 Aviocar transports and Esquadra 752 with about five SA 330 Pumas for SAR and logistic/troop transport. The A-7 Corsairs of Esquadras 302 and 304 and the F-16s of Esquadra 201 frequently have detachments on the base.

Aviação Naval

Nearly 50 years after it was merged with the Aeronáutica Militar, the delivery of five Westland Super Lynx Mk 95s signalled the reactivation of the Aviação Naval. The first example, one of two ex-Royal Navy aircraft in the order, first flew on 27 March 1992. The Marinha's aircrews were trained in initial/basic helicopter operations by Esquadra 111 and then Esquadra 552, later going to RNAS Portland to undertake the full Lynx course with the FAA. The Lynxes are operated from three MEKO 200-class frigates, *Vasco da Gama*, *Alvares Cabral* and *Corte Real*.

Each 'Vasco da Gama'-class guided-missile frigate can deploy a pair of Lynx Mk 95s. With three frigates in the class, plus training commitments, the five helicopters are kept busy.

Aviação Naval		
Esquadrilha de Helicopteros de Marhina		
	Lynx Mk 95	Montijo

Europe

ROMANIA

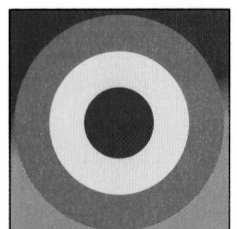

Capital: Bucharest
Population: 23.2 million
Land area: 237500 km²
(91,675 sq miles)
Major cities: Braila,
Brasov, Cluj Napoca,
Constanta, Craiova, Galati,
Iasi, Ploesti, Timisoara

Aviatez Militaire Romane (Fortele Aeriene Romane)

Romania took part in World War II as a strong German ally, but following the end of the war the country fell within the Soviet zone of influence. From 1967 to 1989 the dictator Nicolae Ceasescu carried out a policy of independence from the Soviet Union, and as a result the country became a half-hearted Warsaw Pact Treaty Organisation member. The efforts towards self-sufficiency resulted in licence-production of Aérospatiale helicopters like the SA 316 Alouette III and the SA 330 Puma. A strike and inter-diction fighter – the IAR-93 – was developed and produced in co-operation with Yugoslavia. A new-generation indigenous jet trainer and light attack aircraft – the IAR-99 Soim – entered service in late 1980s.

THE POST-CEASESCU ERA

The Romanian AF underwent a process of large-scale restructuring in the early and mid-1990s. The service's budget was increased in 1990-1991, allowing more flying hours for the front-line pilots; however, the combination of rapidly-increased flying activity of predominantly inexperienced pilots, with the low reliability of the considerably aged fleet, resulted in a very poor safety record. From July 1991 until the end of 1994 at least 44 aircraft were lost in various accidents, including two MiG-29s. In 1994 alone, the Romanian AF lost eight fixed-wing aircraft – one MiG-29, three MiG-23s, four MiG-21s – and two helicopters. In 1997 the Romanian CinC General Yon Stan was killed in an L-39ZA crash. For aircraft attrition replacement two MiG-23UBs and two MiG-29s were acquired from Russia in 1995-96.

Front-line assets are grouped in two air force and air defence corps, headquartered in Bucharest and Cluj, respectively, which control a total of 430 combat aircraft and a number of support types. The 1990 Conventional Forces in Europe (CFE) Treaty forced Romania to withdraw 70 combat aircraft from use in order to reach the allowed level of 430 units. The CFE allows expansion of the combat helicopter fleet from 13 to 120 units.

Spearheading the Romanian fighter force is the MiG-29, which serves with two squadrons of the 57th Regiment de Vinatoure.

AIR DEFENCE ASSETS

The mainstay of the Romanian air defence fleet is the MiG-29. The Mikhail Kogalniceanu-based squadron of the 57th Fighter Regiment received the first of its 12 MiG-29s in late 1989. Forty-eight MiG-23MF/UBs were originally delivered in the late 1970s and now equip two squadrons of the 57th Fighter Regiment and one squadron of the 93rd Fighter Regiment at Timisoara. The attrition of the MiG-23 has been high in recent years and at least 10 aircraft are believed to have been lost in accidents since the type entered service.

The most numerous air defence fighter is the MiG-21MF/UM, about 150 of which are still in use with eight squadrons. The Romanian 'Fishbed' fleet suffered from heavy attrition in the early 1990s. In November 1993 Romania accepted the offer by the Israeli company Elbit Defence Systems to upgrade 100 single-seat MiG-21MFs and 10 twin-seat MiG-21UMs under a $325 million contract. The upgraded fighter, which received the name Lancer, underwent extensive systems and avionics modernisation in Romania by a combined Aerostar/Elbit team. Twenty-five of the single-seaters will be equipped for both interception and ground-attack roles, while the reminder will be optimised for close support operations. Press reports indicate that the Romanian AF procured the MATRA Magic 2 short-range AAMs to equip the Lancer fleet. The service is also interested in procuring new-generation radar-guided BVR air-to-air missiles.

The first production Lancer was delivered in October 1996 and initial operational capability was declared on 8 May 1997 for the first squadron of the Bacau-based fighter regiment. By early 1998, 40 upgraded MiG-21s had been delivered and had accu-mulated more than 1,000 flights. In May 1997 a new variant of the IAR-99 Soim jet trainer was demon-strated at Le Bourget, developed specially for lead-in-fighter training of Lancer pilots, featuring a high degree of avionics commonality.

The main strike aircraft, equipping five squadrons, is the SOKO/CNIAR IAR-93A/B, a SEPECAT

Current upgrades to the MiG-21 fleet to bring them up to the Lancer standard provide the type with a new Elta radar, HOTAS, multifunctional displays and other advanced avionics.

Jaguar look-alike developed jointly with Yugoslavia in 1970s. The first aircraft was delivered to the Craiova-based 67th Fighter Bomber Regiment in 1982. The initial IAR-93A variant featured non-afterburning Viper Mk 632 turbojets and its produc-tion totalled 35 units. The improved IAR-93B fitted with the afterburning Viper Mk 633 entered service later; 165 units of this version were initially ordered, but the UN weapons embargo imposed on former Yugoslav states meant the end of IAR-93B produc-tion in Romania due to lack of specific parts and components produced in Yugoslavia.

The indigenous IAR-99 Soim jet trainer and light attack aircraft entered service in the late 1980s, ini-tially as an advanced and weaponry trainer. In late 1989 a number of the 50 produced units were assigned to the IAR-93-equipped regiments where the IAR-99s provide cheaper flight hours.

Romania has been a major user of IAR-built Pumas. This example operates from Baza de Transport Aerian Otopeni, with 19 FMT.

Aviatez Militaire Romane

Aviatez Militaire Romane (Fortele Aeriene Romane)		
		HQ Bucharest
19 Flotila Militara de Transport		**Bucharest/Otopeni**
?? Squadron	An-24RT/TV, An-26, An-30, C-130B	
52 Squadron	Boeing 707, Tu-154, SA 365N	
	Mi-8PS, Mi-17-1VA, IAR-330, IAR-316B	
57th Regiment de Vinatoure (Air Base 01981)		
1/57 Squadron	MiG-29/29UB	**Mikhail**
1/57 Squadron	MiG-29/29UB	**Kogalniceanu**
2/57 Squadron	MiG-23MF/23UB	
3/57 Squadron	MiG-23MF/23UB	
58 Flotila de Lupta Elicopter		**Sibiu**
?? Squadron	IAR-316/IAR-330	
59 Flotila de Lupta Elicopter		**Tuzla**
1/59 Squadron	IAR-316/IAR-330	
2/59 Squadron	IAR-316/IAR-330	
67 Regiment de Vinatoare		**Craiova**
1/67 Squadron	IAR-93A/93B, IAR-99	
2/67 Squadron	IAR-93A/93B, IAR-99	
3/67 Squadron	MiG-21MF/21UM	
86 Regiment de Vinatoare		**Borcea-Fetesti**
1/86 Squadron	Lancer I, MiG-21UM	
2/86 Squadron	MiG-21PFM/21US	
3/86 Squadron	Harbin H-5R	
93 Fighter Regiment (Air Base 01981)		**Timisoara**
1/93 Squadron	MiG-23MF/23UB	
2/93 Squadron	MiG-21MF/21UM	
?? Regiment de Vinatoare		**Cluj-Luni**
?? Squadron	MiG-21MF/21UM	
?? Squadron	MiG-21MF/21UM	
?? Fighter Bomber Regiment		**Ianca**
?? Squadron	IAR-93B	
?? Squadron	IAR-93B	
?? Squadron	IAR-93B	
Institute for Aviation/SMOA		**Boboc**
?? Squadron	L-39ZA	
?? Squadron	L-39ZA	
?? Squadron	L-29	
?? Squadron	L-29	
?? Squadron	Iak-52	
?? Squadron	IAR-823	
SMP	An-2	Buzau
SMAS		**Bacau**
?? Squadron	MiG-21, Lancer	
?? Squadron	MiG-21, Lancer	
SMP (Scola Militara de Parasutism)		Parachute School
SMOA (Scola Militara de Ofiteril de Aviatia)		Officers Aviation School
SMAS (Scola Militara de Aviatia Supacsonic)		Supersonic Aviation School

Romania has three Antonov An-30s which are used for survey duties, including flights as part of the Open Skies Treaty, to which Romania is a signatory.

The first Eastern European country to acquire the C-130, Romania hopes to gain more of the type to replace Soviet types in service with 19 FMT.

Reconnaissance assets of the Romanian AF comprise one MiG-21R-equipped squadron (up to 15 aircraft) assigned to the 93rd Fighter Regiment (Air Base 01981) at Timisoara. Romania also has the survivors of 12 obsolete Harbin H-5Rs (the Chinese copy of the Il-28R) in one squadron at Borcea. There were reports that the L-39ZAs of the Boboc-based Institute for Aviation are used for reconnaissance, probably equipped with an underwing camera pod.

TRANSPORT AND TRAINING ASSETS

The main air transport unit of the Romanian AF is the Bucharest/Otopeni-based 19 FMT, equipped with 10 An-24s, six An-26s, three An-30s and four C-130Bs. The VIP transport unit is the 52nd Squadron, equipped with two Boeing 707s, one Tu-154, one Dolphin, IAR-330s, IAR-316s, Mi-8s and, from 1996, one Mi-17-1VA.

The mainstay of the rotary-winged fleet is the licence-built Puma variant (the IAR-330), some 90 of which were ordered. A team from Elbit/IAR-Brasow currently is upgrading 24 IAR-330Ls with the Elbit's SOCAT system. The programme covers the integrated development for avionics, sensors and armament for an attack helicopter, thus considerably increasing the combat capabilities of the Puma. Some 95 examples of the licence-built Alouette III variant (the IAR-316) are still in service. A small number of Mi-2s and Mi-8s, about six of each, are also used for transport tasks.

Air and ground crew training is provided by the Boboc-based Institute for Aviation 'Aurel Vilacu', which operates a fleet of some 32 L-39ZAs, 35 L-29s, 40 IAR-823s and 12 Yak-52s, and other support types.

THE FUTURE

Romania was the first former Warsaw Pact member to sign NATO's Partnership for Peace programme, allowing closer co-operation between former Communist countries and NATO.

In 1996 the Romanian AF received its first two of the four C-130B Hercules transports, given by the US government under FMS auspices from AMARC storage. The restoration of the aircraft to flyable condition, the crew training and the spare parts amounted to US$2 million. The Westernisation of the service will be expanded considerably in the early 2000s by when the country is expected to have achieved full NATO membership. Although no orders have yet been placed, a decision in principle to buy a 'small number' of F-16s as MiG-21 replacements was announced by the Ministry of Defence in October 1996. Romanian interest appears to be in new, rather than ex-USAF, F-16A/Bs. In 1997 the MoD announced plans for the acquisition of an airborne early warning system and additional C-130 transports.

The next ambitious programme that is awaiting funding to begin is the AH-1RO Dracula, a derivative of the Bell AH-1W Super Cobra. Ninety-six attack helicopters will be built under licence by IAR Brasov. Progress of the programme was halted in 1997 when Romania cut defence spending to meet International Monetary Fund targets, but in early 1998 Romania invited tenders from partners willing to finance the programme. Eleven banks and financial institutions were invited to offer loans of up to US$1.5 billion; if successful, the work on the Dracula could begin in 1999.

Romanian Naval Air Arm

The Romanian Navy operates a small fleet of helicopters for anti-submarine warfare, cargo and personnel transportation, special operations support and SAR, grouped in an independent squadron reporting directly to the Navy headquarters. The main ASW type is the Mil Mi-14PL 'Haze-A' acquired in the early 1980s, and six are believed to be still in service. Six IAR-316 Alouette IIIs are used for support tasks. During NATO's Strong Resolve exercise in mid-March 1998, an IAR-316 of the air force's 59 Flotila de Lupta Elicopter and one IAR-330 from Bucharest/Otopeni were embarked on the destroyer *Maracesti* and used for a wide variety of missions.

Flotila de Marina Elicopter

Mi-14PL, IAR-316B	Tuzla

Air force helicopters are sometimes deployed on naval vessels. This IAR-316B Alouette III is seen aboard the destroyer Maracesti *during the exercise* Strong Resolve.

RUSSIAN FEDERATION

Capital: Moscow
Population: 148.1 million
Land area: 17.078 million km² (6,592,110 sq miles)
Major cities: Chelyabinsk, Kazan, Nizhny Novgorod, Novosibirsk, Omsk, Perm, Samara, St Petersburg, Ufa, Yekaterinburg

The Russian Air Force has inherited the lineage and honours of the Soviet air forces – which can be traced back to 1917 – and their massive assets. Historically, the Soviet air arms were subordinated to each of the six branches of the Armed Forces: Air Forces (VVS), Air Defence Forces (PVO), Strategic Rocket Forces (RSVN), Ground Forces (SV), Naval Forces (VMF) and Airborne Forces (VDV). The largest air arm has been the Air Force, which until recently comprised four commands (Long-range Aviation, Frontal Aviation, Military Transport Aviation, and a reserve and training command). The Air Defence Forces were until recently made up of a fighter component (IA-PVO), as well as surface-to-air missile, anti-aircraft artillery and radar forces. The Ground Forces operate a massive Army Aviation component, whereas the air assets of the Strategic Rocket Forces are limited to command and control and liaison detachments with a small number of helicopters. Finally, Naval Aviation operates a wide variety of shipborne and land-based fixed-wing aircraft and helicopters.

DISSOLUTION AND REDEPLOYMENT

The Union of Soviet Socialist Republics (USSR) was formally dissolved in December 1991 and, by a decree of President Boris Yeltsin dated 7 May 1992, the Russian Federation established its own armed forces. The dissolution of the Soviet Union and other dramatic political changes which took place during 1989-1991 – including the collapse of Communism in Eastern Europe, the reunification of Germany, and the dissolution of the Warsaw Pact – necessitated a massive regrouping of the Russian armed forces.

From 1990, in a massive withdrawal on an unprecedented scale, all Soviet and Russian armed forces were withdrawn from East Germany, Poland, Czechoslovakia and Hungary. Initially, units were also relocated to Belarussian and Ukrainian territory but, following those countries' independence, all units were withdrawn to the Russian Federation, which had claimed control of all armed forces remaining on

Above: Armed with a pair of R-27R medium-range semi-active radar-guided AAMs under the wings and two wingtip-mounted R-73 medium-range IR-guided AAMs, this Sukhoi Su-27 is relatively lightly loaded.

Right: Unlike the Su-27, which was designed from the outset as a pure air superiority fighter, the MiG-29 was designed for the tactical fighter role. Bort number '02' is a 9-13 MiG-29 'Fulcrum-C'.

the territory of the ex-Warsaw Pact countries. The headquarters of the powerful 16th Air Army (Western Group of Forces) was relocated from Wunsdorf, south of Berlin, to Kubinka, west of Moscow, but its subordinate units were widely dispersed over the vast territory of the Russian Federation. The headquarters of the 4th Air Army (Northern Group of Forces) relocated from Legnica to Rostov-na-Donu, in the North Caucasus military district. Between 1991 and 1994, a considerable number of units were disbanded, with older types of aircraft being withdrawn from service, while other, more modern aircraft were used to re-equip units already based in the Russian Federation, or to re-equip units that were withdrawn to Russia and remained operational. In general, units were often withdrawn to air bases which previously housed a training unit which was either disbanded, or was converted to a combat role as part of the reorganisation of the Air Force.

During the same period, the Russian armed forces also relocated from the Baltic Military District (which had been renamed Northwestern Group of Forces), but they retain a military presence in the strategically important Russian enclave of Kaliningrad. The withdrawal from the Baltic States also included elements of Naval Aviation of the Baltic Fleet, Fighter Aviation of the Air Defence Forces, Long-range Aviation and Military Transport Aviation.

However, on the territories of the newly independent states of Belarus, Ukraine and Moldova, as well as the new republics in the Caucasus and Central Asia, the situation was different. A number of new republics claimed and eventually retained the elements of the Soviet armed forces (including strategic

Air Force (VVS)

GK VVS		Moscow
4 TsBP I PeLS		**Lipetsk**
160 IIAP	Su-25 (21), MiG-29 (28)	Borisoglebsk
968 IISAP	MiG-25 (4), MiG-29 (18),	Lipetsk
	Su-24 (13), Su-25 (11), Su-27 (14)	
976 IBAP	Su-24 (38)	Totskoye
269 OVP	Mi-8 (53)	Malino
8 ADON		**Chkalovskiy**
	An-12/-24/-26/-72, Il-18/-20/-22,	
	Tu-134/-154, Il-62/-76/-86	

(squadrons of 8 ADON are 353 APON and 354 APON – both based at Chkalovskiy – but which uses which is unknown)

The Russian Air Force uses small numbers of Tupolev Tu-154As for staff transportation.

elements) based on their territory (such as in Ukraine), whereas in other cases units were hastily withdrawn to Russian in a speedy operation (such as in Azerbaijan in 1992).

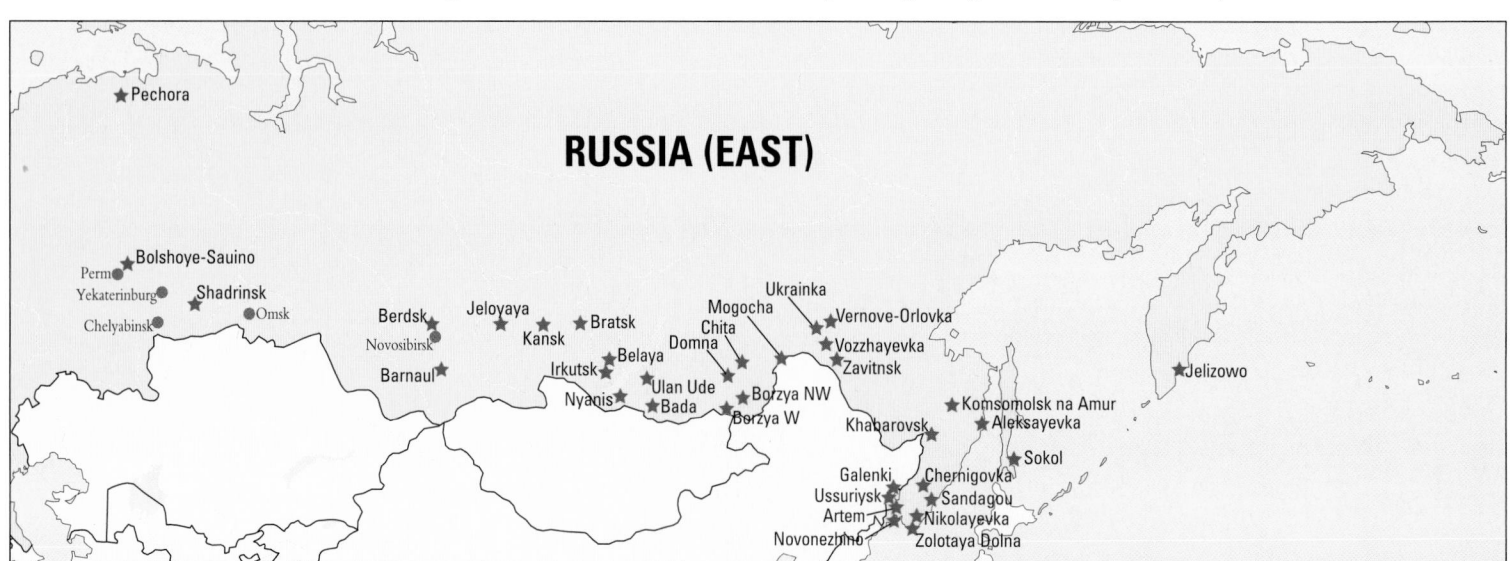

RUSSIA (EAST)

Pechora
Perm
Bolshoye-Sauino
Yekaterinburg
Shadrinsk
Chelyabinsk
Omsk
Berdsk
Novosibirsk
Barnaul
Jeloyaya
Kansk
Bratsk
Irkutsk
Belaya
Nyanis
Ulan Ude
Bada
Borzya NW
Borzya W
Mogocha
Chita
Domna
Ukrainka
Vernove-Orlovka
Vozzhayevka
Zavitnsk
Jelizowo
Komsomolsk na Amur
Aleksayevka
Khabarovsk
Sokol
Galenki
Chernigovka
Ussuriysk
Sandagou
Artem
Nikolayevka
Novonezhino
Zolotaya Dolna

REFORM AND REORGANISATION

Following the collapse of the Soviet Union, the Russian government realised that it could no longer afford the massive armed forces and military doctrine of the Soviet and Cold War eras. As a consequence, sweeping military reforms have been introduced, leaving no branch of the armed forces unaffected. Between 1 January 1991 and 31 December 1996, the number of aviation regiments in the Russian Air Force was reduced by approximately 55 per cent, the number of combat aircraft by nearly 50 per cent, and personnel strength by over 60 per cent, when compared with the former Soviet Air Force.

The Treaty on Conventional Armed Forces in Europe (CFE) limits the Russian Federation to operating 3,450 combat aircraft and 890 attack helicopters (as of 1 January 1998, the Russian inventory numbered 2,868 and 805, respectively). However, the Treaty covers the ATTU (Atlantic to the Urals) area only, and Naval Aviation is exempt from the terms of the Treaty; strategic aircraft (Tu-95 and Tu-160) are subject to the terms of the SALT and START treaties.

By decree of President Boris Yeltsin of 16 June 1997, the Air Force (VVS) and Fighter Aviation of the Air Defence Forces (IA-PVO) were to be fully integrated into a single Air Force command by 1 January 1999. This reorganisation has resulted in the formation of three air force and air defence armies (OA VVS I PVO, headquartered in St Petersburg, Rostov-na-Donu and Khabarovsk), and three air force and air defence Corps (OK VVS I PVO, headquartered in Chita, Novosibirsk and Yekaterinburg). The Moscow MD retains its special status, and on the basis of the 16th Air Army and the Moscow PVO Air Army, the Moscow Air Force and Air Defence District (MO VVS I PVO) have been formed. The new combined air armies and corps come under the operational control of the commanders of the military districts during peacetime, and of the commander of the fronts at time of war. The new-style Air Force is commanded by Colonel General Anatoliy Mikhaylovich Kornukov, who was appointed on 20 January 1998 and succeeded Colonel General Pyotr Deynekin, who had reached the retirement age of 60. The merger has resulted in a reduction of the personnel strength and combat potential of the Russian Air Force by 45 and 40 per cent, respectively.

The new operations centre of the joint command, located at Zarya (Balashikha District, southeast of Moscow), became operational on 1 March 1998. Pending the completion of the planned integration, VVS and IA-PVO assets will be treated separately in the following account.

Furthermore, Long-Range Aviation (DA) ceases to exist as a separate Air Force command, and has been transformed into two air armies. Likewise, Military Transport Aviation (VTA) is being transformed from an air force command to an air army with 10 subordinate regiments

STRATEGIC DIRECTIONS AND MILITARY DISTRICTS

Prior to 1991, three strategic directions existed (Western, Southern and Far Eastern GTVD, headquartered in Kiev, Tashkent and Irkutsk, respectively), which were divided into five regional theatres. Two Military Strategic Commands replace the former Western and Eastern Strategic Directions (GTVD). Subordinated to these are six military districts, with the status of territorial commands, covering the entire territory of the Russian Federation. The Leningrad MD (LenVO) becomes the Northern ('Northwestern') MD, the Moscow MD (MoMD) becomes Central MD, the Volga and Ural military districts (PriVO and UVO) are merged into a single Southeastern MD, the Transbaikal and Siberian MDs (ZabVO and SibVO) are merged into a single Eastern MD, the North Caucasus MD (SKVO) becomes Southwestern MD and, finally, the Far Eastern MD (DVO) retains its name.

Europe

Air Force (VVS)

The history of the Russian Air Force can be traced to the Imperial Russian Air Force of World War I. In 1991, the Air Force (VVS) comprised 20 operational and tactical organisations, 38 divisions and 211 aviation regiments, and had an inventory of more than 10,000 aircraft. Until recently, the Air Force comprised four commands: Dal'naya Aviatsiya (DA, Long-range or Strategic Aviation), Frontovaya Aviatsiya (FA, Frontal or Tactical Aviation), Voenno-Transportnaya Aviatsiya (VTA, Military Transport Aviation) and Komandovanie Rezerva I Podgotovki Kadrov (KR I PK, Reserve and Personnel Training Command); the latter two have been disbanded as air force commands, and the subordinated elements were placed under the direct control of air force headquarters and the commanders of the military districts.

Frontal Aviation has traditionally been the largest component of the Air Force, and its assets were formed into tactical air armies, which during the Soviet era were attached to each of the 16 military districts and four groups of forces. Its subordinated echelons are the corps (korpus), division (diviziya), regiment (polk) and squadron (eskadril'ya).

Under Soviet and Russian military doctrine, the assets of the Air Force are traditionally divided into the following categories: fighter, bomber, fighter-bomber, ground-attack ('assault'), reconnaissance, transport, and special purpose and support aircraft.

FIGHTER AVIATION

Prior to the merger of VVS and IA-PVO, the fighter aviation assets of Frontal Aviation (FIA, Frontovaya Istrebitel'naya Aviatsiya) had a dual function – both air superiority and attack on ground targets – and these missions largely fell on the MiG-23 and its successor, the MiG-29, which made up the bulk of the fighter regiments' aircraft. During 1991-94, a number of fighter regiments converted from the MiG-23 to the MiG-29, large numbers of

which became available when units were withdrawn from Germany, Hungary and elsewhere. In contrast to IA-PVO, which was solely tasked with area defence, the Air Force operated a smaller number of Su-27s, which among other tasks served as escorts for strike missions by the Su-24.

Pending the production of a fifth-generation fighter, more than 150 MiG-29s (Izdelye 9-13) will be upgraded to a standard comparable to Izdelye 9-17 ('MiG-29SMT'). The upgrade will comprise a new radar, changes in cockpit displays and layout,

Above: Diminishing in numbers in its fighter role, the MiG-25 remains an important asset for reconnaissance duties. This example is a MiG-25RB.

Left: The principal tactical bomber in the Russian Air Force is the Sukhoi Su-24. This Su-24M 'Fencer-D' has the sub-variant's retractable refuelling probe and recontoured nose. About 330 fighter-bomber 'Fencers' remain in service.

Above: Mainstay of the shturmovoy (ground-attack) regiments is the Sukhoi Su-25. About 250 of the type remain in service, and are due to be supplemented by the Su-25TM.

Below: Large numbers of older aircraft types are held in storage or used for test purposes. This Aero L-29 is employed by the air force's test base at Akhtubinsk, modified for pilotless drone operations.

increased internal fuel capacity, addition of an inflight-refuelling system, and increased combat load (to include R-77s and precision-guided munitions). The first batch of upgraded aircraft is scheduled for delivery before the end of 1999.

The development of the MIG MAPO 1-42 multi-functional tactical fighter (*mnogofunktsionalyi frontovi istrebitel*) has been delayed for budgetary reasons. Sukhoi is concentrating on the development of the Su-37 with thrust-vectoring nozzles, and has since fielded the S-37 ('S-32') Berkut, which made its first flight on 25 September 1997.

BOMBER AVIATION

Bomber aviation is divided into strategic (intercontinental – Tu-95 and Tu-160), long-range (continental – Tu-22 and Tu-22M) and frontal (Su-24) bomber

Above: Wearing Aeroflot colours, this Il-78 undertakes trial hook-ups with a test MiG-29.

Below: Fitted with an air sampling pod on their fuselage sides, small numbers of Antonov An-24RKRs undertake the radiation reconnaissance role.

Frontal Aviation (FA)

1 VA, Khabarovsk

u/i ORAP	Su-24MR	Vozzhayevka
u/i OSAP	An-12, An-24, An-26	Khabarovsk
u/i OVE REB	Mi-8PPA/8SMV	
u/i OShAP	Su-25	Chernigovka
18 GvOShAP	Su-25	Galenki
u/i IAP	Su-27	Vernoye-Orlovka
u/i BAD		
207 BAP	Su-24	Komsomolsk
u/i BAP	Su-24	u/i
u/i BAP	Su-24	u/i

4 VA, Rostov-na-Donu

11 ORAP	Su-24MR (36)	Marinovka
535 OSAP	An-12, An-24, An-26, Mi-8 (12), Tu-134 (1)	Rostov-na-Donu
286 OVE REB	Mi-8/8PPA/8SMV (29)	Zernograd
16 GvIAD, Millerovo		
19 GvIAP	MiG-29 (40)	Millerovo
31 GvIAP	MiG-29 (40)	Zernograd
960 IAP	MiG-29 (27)	Primorsko-Aktharsk
1 ShAD, Krasnodar		
16 ShAP	Su-25 (33)	Taganrog
368 ShAP	Su-25 (34)	Budennovsk
461 ShAP	Su-25 (34)	Krasnodar
10 BAD, Yeysk		
296 BAP	Su-24 (31)	Marinovka
559 BAP	Su-24M (30)	Morozovsk
959 BAP	Su-24 (32)	Yeysk

16 VA, Kubinka

47 ORAP	MiG-25 (46), Su-24MR (13)	Shatalovo
226 OSAP	Mi-8 (32), An-12, An-24, An-26, Tu-134	Kubinka

285 OVE REB	Mi-8 (20)	Saransk
237 GTsPAT*	Su-25 (1), Su-27 (21), MiG-29 (31)	Kubinka
899 OShAP	Su-25 (53)	Buturlinovka
9 IAD, Shaykovka		
14 GvIAP	MiG-29 (38)	Zherdevka
73 GvIAP	MiG-29 (39)	Shaykovka
343 IAP	MiG-29 (49)	Sennoy
105 BAD, Voronezh		
20 GvBAP	Su-24 (30)	Kamenka
164 GvBAP	Su-24 (30)	Shatalovo
455 BAP	Su-24 (30)	Voronezh
1 BAP	Su-24 (40)	Lebyazhye

23 VA, Chita

125 ORAP	Su-24MR	Domna
u/i OSAP	An-12, An-24, An-26	Chita
u/i OVE REB	Mi-8PPA/SMV	
u/i OShAP	Su-25	
120 IAP	MiG-29	Domna
u/i BAD		
u/i BAP	Su-24M	Nyangi
u/i BAP	Su-24	Bada
u/i BAP	Su-24	Borzya NW
76VA, St Petersburg		
98 ORAP	MiG-25 (26), Su-24 (20)	Monchegorsk
202 OSAE	An-12, An-26, Mi-8 (9)	Levashovo
227 OVE REB	Mi-8 (19)	Apatity
239 IAD, Besovets (aka Petrozavodsk)		
28 GvIAP	MiG-29 (62)	Andreapol
159 GvIAP	Su-27 (36)	Besovets
871 IAP	MiG-29 (38)	Smolensk
149 BAD, Smuravyevo		
67 BAP	Su-24 (37)	Siverskiy
722 BAP	Su-24 (41)	Smuravyevo

PriVO, Samara

144 OSAE	Mi-8 (4)	Samara

aviation. Strategic and long-range missions are typically flown by units of the Long-range Aviation command, whereas the Su-24 is the principal asset of the frontal (tactical) air armies.

ATTACK AVIATION

Within the new organisational structure of the Air Force, attack aviation (*udarnaya aviatsiya*) elements, comprising bomber (*bombarirovochnoy*) and ground-attack (*shturmovoy*) regiments will represent one-third of total Air Force and Air Defence Force strength.

Between 1991 and 1994, the MiG-27 and Su-17, which had formed the mainstay of the attack force of the Tactical Frontal Aviation armies, were phased out of service, and large numbers of these fighter-bombers have since been stored at a number of aircraft reserve bases. Those Su-17- and MiG-27-equipped units that were withdrawn from the ex-Warsaw Pact countries and subsequently remained operational in Russia largely converted to the Su-24M, whereby their designation was changed from APIB (fighter-bomber regiment) to BAP (bomber regiment).

Ukraine claimed six regiments equipped with Su-24s which were based on its territory in 1991. Belarus retained one regiment equipped with Su-24Ms, out of four bomber regiments (equipped with Su-24s and Su-24Ms) based on its territory; the remaining three had relocated to Russia by 1994.

Three regiments of Su-25s stationed on Belarussian territory were transferred to the air force of Belarus, and one regiment of Su-25s was claimed by Ukraine. The Su-25 was operated by independent ground attack regiments in the air armies of the Western Group of Forces and peripheral Military Districts, but in 1993 a thus far unique ground-attack division (ShAD) was formed in the North Caucasus MD. It was the first of its kind since numerous ground attack divisions operated the original Shturmovik, the Il-2, during and after the Great Patriotic War, highlighting the great importance attached to the 'Frogfoot' following extensive combat experience with this potent attack aircraft during the war in Afghanistan. Both the Su-25 and Su-24 were widely used during the conflict in Chechnya.

The Su-25 will reportedly play a major role in the rapid-deployment groups that will be formed in each of the six military districts. In addition to attack and assault helicopters, these units will consist of four Su-25TMs (Su-39s) and 12 Su-25/25BMs. During 1990-1991, 20 Su-25Ts have been produced in Tbilisi (Georgia) for the Russian Air Force, of which a number have been delivered and are undergoing testing at Akhtubinsk and Lipetsk.

After the collapse of the Soviet Union, production of the Su-25T was transferred to the Sukhoi production facility in Ulan-Ude (Russia), which had hitherto built Su-25UBs and Su-25UTGs. The first Su-25T produced in Ulan-Ude first flew in 1995, and has since been redesignated Su-25TM; the factory will also upgrade the 20 Su-25Ts built in Georgia to Su-25TM standard. Delivery of the first series-produced Su-25TMs and the upgrade of the Su-25Ts was scheduled for the end of 1998.

The Su-24M is currently the only tactical aircraft capable of inflight refuelling, and will continue to be the mainstay of the Russian tactical bomber force until replaced by the Su-34, which made its first flight in 1990 as the 'Su-27IB'. The Su-34 was scheduled to enter service in 1998 and to have replaced the Su-24 by 2005.

RECONNAISSANCE AVIATION

In the Russian Air Force, reconnaissance aircraft were traditionally divided into strategic (Tu-22R and derivatives), operational (Il-20M, MiG-25RB variants and Su-24MR) and tactical (Su-17M3R and Su-17M4R). Since the mid-1980s, operational and

tactical reconnaissance missions have been performed by independent reconnaissance regiments and squadrons (ODRAP and ODRAE), which were directly subordinated to the headquarters of the respective tactical air armies. These units existed in several configurations, either as single-type units equipped with two or three squadrons of Su-17M3Rs/-M4Rs, Su-24MRs or several variants of the MiG-25R and MiG-25RB, or a combination of the three. The Su-17 has been withdrawn from service, and currently – with a very limited quantity of MiG-25s remaining operational – the primary type tasked with tactical reconnaissance is the Su-24MR. A dedicated reconnaissance variant of the Su-34 is under development.

The Il-20M is operated as an Elint (electronic intelligence; RTR – *radiotekhnicheskaya razvedka*) platform by detachments directly subordinated to headquarters of the air armies. The highly specialised M-55 Geofizika (M-17M), which first flew in August 1988, is reportedly in service with a detachment at Smolensk. The An-30 is operated by an independent long-range reconnaissance squadron (ODRAE) with six aircraft, based at Yelovaya (Krasnoyarsk).

TRANSPORT AVIATION

Transport aircraft are categorised as heavy (An-22 and An-124), medium (An-12 and Il-76) and light (An-26), plus multi-purpose helicopters (Mi-6, Mi-8 and Mi-26). The transport mission is flown by units subordinated to the headquarters of the Military Districts, and VTA. In addition to the heavy and medium transport aircraft of the VTA, medium and short-haul transport and liaison aircraft are traditionally operated in the Air Force (and other branches of the Russian Armed Forces) by independent composite aviation regiments and squadrons, which are directly subordinated to the headquarters of the air armies and other Air Force commands. These units typically operate a mix of An-12s, An-24s and An-26s, and a variety of support aircraft and helicopters.

SPECIAL PURPOSE AND SUPPORT TYPES

In addition to the above-mentioned missions, the Soviet and Russian Air Force has traditionally operated a wide variety of specialised support aircraft, often with a bewildering array of impromptu modifications, which have given rise to a plethora of NATO ASCC designations, while many other modifications went either unnoticed or have not been 'honoured' with such a designation. A special role is played by 8 ADON (Special Purpose Aviation Division), based at Chkalovskiy and directly subordinated to Air Force headquarters. Its two subordinated regiments operate a large variety of the aircraft mentioned below, including the unique Il-76 Skalpel-MT and Il-76MDK cosmonaut trainer ('vomit Candid').

AIRBORNE COMMAND POSTS

A variety of airborne command posts and communications relay platforms, comprising both fixed-wing aircraft and helicopters, is in service, such as the An-26RT, Il-22/Il-22M, Il-76VKP (Il-76SK, Il-82), Il-86VKP, Tu-135 (Tu-134AK), Mi-8VzPU, Mi-9, Mi-6A VKP and Mi-22 (Mi-6AYa).

VIP TRANSPORT AND LIAISON

For VIP transport and liaison, the Air Force operates the following types: An-2, An-24, An-26PS, An-32, An-72, Il-62, Yak-40, L-410, Mi-8S and Mi-8TPS, Tu-134 and Tu-154.

ELECTRONIC WARFARE

Electronic warfare missions are performed by EW variants of the An-12 (An-12PP, An-12PPS, An-12BK-PPS), as well as the Su-24MP, Tu-22P/PD Mi-8PPA and Mi-8SMV. The Tu-22MP, based on the Tu-22M3, has apparently not progressed beyond

The two long-range bombers in service are the Tupolev Tu-95MS (above left) and the Tu-160 (above). Two versions of the Tu-95MS are in service, able to carry six (Tu-95MS6) and 16 (Tu-95MS16) AS-15A 'Kent' cruise missiles. The Tu-160 can carry six of the same missiles.

Long-range Aviation (DA)

(to be reorganised into VA VGK SN)

GK DA, Moscow		
3317 AB	An-12/-24/-26	Ostafyevo
43 TsBP I PeLS, Ryazan (aka Dyagilevo)		
652 UAVP	Tu-134UB	Tambov
	(may not be operational)	
?	Tu-95, Tu-22M3	Ryazan
22 GvTBAD, Engels		
203 GvAPSZ	Il-78M (20)	Engels
121 GvTBAP	Tu-160 (6)	Engels
326 TBAD, Soltsy		
52 GvTBAP	Tu-22M2/-M3	Shaykovka
		(aka Anisovo-Gorodische)
840 TBAP	Tu-22M3	Soltsy
u/i TBAD, Belaya (aka Irkutsk)		
u/i TBAP	Tu-22M3	Belaya
u/i TBAP	Tu-22M3	Belaya
u/i TBAD, Ukrainka (aka Seryshevo)		
u/i TBAP	Tu-95MS	Ukrainka
u/i TBAP	Tu-95MS	Ukrainka
u/i TBAD, Ussuriysk (aka Vozdvizhenka)		
u/i TBAP	Tu-22M3	Ussuriysk
u/i TBAP	Tu-22M3	Zavitinsk

prototype stage. Since the MiG-25BM, which was produced in small numbers, was withdrawn from service in the early 1990s, no dedicated 'Wild Weasel' aircraft is operational. A number of other aircraft (notably the Su-24 and Su-25) can be armed with anti-radiation missiles, and the projected replacement for the MiG-25BM is the Su-30M.

AERIAL REFUELLING

The Il-78 has replaced the 3MS-2 as the principal aerial tanker. In addition, Su-24s equipped with the UPAZ pod are also capable of refuelling other aircraft in flight, notably other Su-24s.

SAR AND CASUALTY EVACUATION

At airfields throughout the Russian Federation, small detachments of Mi-8s are based for SAR duties.

For evacuation of casualties, the Air Force operates a number of specialised fixed-wing aircraft and helicopters, such as the An-26M Spasatel, a unique Il-76MD Skalpel-MT, the Mi-6, Mi-8 and Mi-26.

LONG-RANGE AVIATION (DA)

The history of Long-range (Strategic) Aviation as a separate Air Force command goes back to 1946. During the Soviet era, the strategic assets of the Air Force were heavily concentrated on Belarussian and

Approximately 150 Tu-22Ms are in service with the Russian Air Force and Navy. About two-thirds of the aircraft are based in 'European' Russia.

221

Along with the Antonov An-22, the An-124 Ruslan undertakes the heavy-lift transport role. Few of the aircraft actually wear military colours.

Ukrainian territory as part of the Western and Southwestern Directions. The units were subordinated to the 24th and 46th Strategic Air Armies, headquartered at Vinnitsa and Smolensk, respectively, or under direct control of DA headquarters in Moscow.

By the end of 1994, the four regiments of strategic bombers and strategic reconnaissance aircraft (equipped with Tu-16s, Tu-22s and Tu-22M3s) based in Belarus had all been relocated to Russia, but DA assets based on Ukrainian territory were claimed by the Republic of Ukraine. They comprised three heavy bomber regiments with Tu-16s, Tu-22s and Tu-22M3s, one independent long-range reconnaissance regiment with Tu-22s, one heavy bomber regiment with Tu-160s and one heavy bomber regiment equipped with the Tu-95MS. Discussions between Ukraine and Russia about the disposition of the 19 Tu-160s formerly operated by 184 GvTBAP at Priluki and 23 Tu-95MSs of 1006 TBAP based at Uzin had been going on since 1991. Eventually, Ukraine decided to use money provided by the USA to scrap these aircraft. In addition, 409 APSZ, one of only two air-refuelling regiments equipped with the Il-78, was retained by Ukraine.

'BLINDER' AND 'BEAR'

The bomber versions of the Tu-22 'Blinder' only served with units of 46 VA, and were retired in 1994, when a large number of 'Blinders' were withdrawn from Belarus to Engels. Currently, of 92 Tu-22s stored at 6213 BLAT (Engels), 63 are held in reserve and the remainder are due to be scrapped.

Strategic reconnaissance missions are performed by the reconnaissance variants of the Tu-22. The Tu-22R entered Air Force service – as the first operational version of the 'Blinder' – in 1962, and is reportedly still operational in small numbers, pending the development of a successor. A small number of a reconnaissance variant of the Tu-22M2 (Tu-22M2R) are in service, as are a limited number of the electronic warfare variants of the Tu-22.

The venerable 'Bear' has been in service since 1956, and the currently operational version, the Tu-95MS, entered service with 79 TBAD at Semipalatinsk in 1984. The last Tu-95MS was delivered from the Kuybyshev plant in 1992. Tu-95MS6s are armed with six Kh-55 air-launched cruise missiles (ALCMs). About 60 aircraft remain in service with 182 GvTBAP (which relocated from Mozdok to Engels) and the heavy bomber division based at Ukrainka; the Tu-95K-22s of the latter unit were replaced by the 40 Tu-95MSs of 79 TBAD (1223 and 1226 TBAP) which were transferred from Semipalatinsk (Kazakhstan) to Russia in 1993.

'BACKFIRE' AND 'BLACKJACK'

The Tu-22M3, which became operational with 185 GvTBAP at Poltava in 1981, is numerically the most important bomber in the DA inventory, and serves with six regiments. Because of delays in the development of the Sukhoi T-60, the intended replacement of the Tu-22M3, it has been decided to embark on a major upgrade of the 'Backfire'. The Tu-22M2s and Tu-22M3s of both the Air Force and Naval Aviation will be upgraded to Tu-245 standard, with a new radar, new missile systems and an automatic terrain-following capability.

The prestigious Tu-160 became operational in May 1987, with 184 GvTBAP at Priluki (Ukraine), but the 19 aircraft on strength with the regiment were claimed by the Republic of Ukraine. The Russian Air Force has six aircraft on strength, based at Engels; six other airframes are located at Zhukhovskii in various states of disrepair.

MILITARY TRANSPORT AVIATION (VTA)

The history of the Military Transport Aviation command (VTA) goes back to the Great Patriotic War. In 1946, VTA was subordinated to the Airborne Troops Command as Desantnogo-Transportnogo Aviatsiya VDV, but in 1955 it was transferred to the Air Force. During the 1950s and 1960s, in addition to helicopters, the An-8 and An-12 were the mainstay of VTA. In 1969 the command gained a heavy-lift capacity when the first two An-22s entered service with a squadron of 229 VTAP at Ivanovo, on the basis of which 81 VTAP was formed in 1970 at Ivanovo. In November 1972, a second regiment with An-22s

Since 1959 the Antonov An-12 has been the medium transport of the Soviet and later Russian Air Force. Supplemented by the Il-76, approximately 200 remain in the transport fleet, the majority being An-12BPs such as Bort '16'.

(556 VTAP) was activated at Seshcha, and a third (8 VTAP) was formed in 1975 at Migalovo (Tver). In 1974, the Il-76 became operational, and during 1987 the An-124 entered service with 556 VTAP at Seshcha.

In 1991, at the time of the dissolution of the Soviet Union, the Military Transport Aviation command operated more than 20 transport and training regiments, of which only eight were based on Russian territory. Of these, no fewer than six regiments were claimed by Ukraine, and one was transferred to Belarus. One regiment relocated to Russia from Azerbaijan, and three were withdrawn from the Baltic republics. VTA assets currently comprise 25 An-124s, 45 An-22s, and a large number of Il-76s.

Instead of a VTA command, an air army with 10 regiments will be formed, equipped with the An-124, An-22 and Il-76MD 'Candid'. Two divisions, headquartered at Ulan-Ude and Orenburg, are scheduled for disbandment, as is the regiment based at Ulyanovsk. Using aircraft and personnel of the disbanded units, commercial joint-stock transport companies will be set up, initially at Ulyanovsk and Kursk. About one-quarter of the VTA air bases will be made available for operations of the commercial companies. The An-12 is to be phased out from Air Force service in favour of the new An-70.

ELEMENTARY FLIGHT TRAINING

ROSTO (formerly DOSAAF) is a paramilitary organisation with a large number of aeroclubs for elementary flying training, where cadets can get a taste of flying in various light aircraft and jets (Yak-52, Yak-55, L-29s, L-39s) and helicopters before entering one of the air academies.

AIR ACADEMIES

Two air academies for pilots (at Balashov and Volgograd, with branches at Barnaul and Yeysk, respectively) and one for navigators (at Chelyabinsk) operate training regiments at a number of locations, notably Barnaul, Bataysk, Chelyabinsk, Kotel'nikovo,

Military Transport Aviation (VTA)

(to be transformed into VA VGK VTA)

GK VTA		Moscow
610 TsBP I PLS, Ivanovo- (Severnyy)		
517 IVTAP	Il-76	Ivanovo- (Severnyy)
u/i IVTAE	An-12	Ivanovo- (Severnyy)
708 OVTAP	Il-76	Taganrog NW
3 GvTAD, Krechevitsy (aka Novgorod)		
103 VTAP	Il-76	Smolensk
110 VTAP	Il-76	Krechevitsy (aka Novgorod)
196 GTAP	Il-76	Migalovo (aka Tver)
334 VTAP	Il-76	Pskov
6 GvTAD, Orenburg		
128 VTAP *	Il-76	Orenburg
u/i VTAP *	Il-76	Shadrinsk
12 VTAD, Migalovo (aka Tver)		
8 VTAP	An-22	Migalovo (aka Tver)
81VTAP	An-22	Ivanovo (Severnyy)
566 VTAP	An-124	Seshcha
235 VTAP *	An-124, Il-76	Ulyanovsk
u/i VTAD, Ulan-Ude *		
u/i VTAP *	An-12, Il-76	Nikolayevka
u/i VTAP *	An-12	Ulan-Ude
u/i VTAP *	An-12	Zavitinsk
(* to be disbanded)		

Air Academies

GK VVS		
KVVAU, Krasnodar		
797 UAVP	MiG-29 (20), Su-25 (5), Su-27 (17), L-39 (33)	Kushchevskaya
802 UAVP	Su-17 (14), Su-25 (3), L-39 (50)	Krasnodar
FKVVAU, Yeysk		
u/i UAVP	MiG-29 (15), L-39 (21), Mi-8 (1)	Yeysk
ChVVAUSh, Chelyabinsk		
u/i UAVP	Tu-134UBSh	Chelyabinsk

PriVO		
BVVAUL, Balashov		
u/i UAVP	An-24/-26	Balashov
u/i UAVP	L410	Petrovsk
u/i UAVP	L410	Rtischevo
127 UAVP	L-39	Ryazhsk
644 UAVP	L-39	Michurinsk
FBVVAUL, Barnaul		
u/i UAVP	L-39	Barnaul
u/i UAVP	L-39	u/i
KVVAUL, Volgograd		
704 UAVP	L-39	Kotelnikovo
706 UAVP	L-39	Volgograd
801 UAVP	L-39	Bataysk

Above: Potential pilots get their first flights in the Yak-52s of ROSTO. This example still has the initials of ROSTO's predecessor, DOSAAF, on its tail.

Below: Advanced training is undertaken on the Aero L-39C. About 1,250 of the type remain on strength.

Above: This An-12BK-PPS of the Gosudarstvenny Lyotno Ispitatelny Tsentr (GLIT – State Flight Test Centre) based at Akhtubinsk is used to test electronic warfare systems.

Right: The new Russian Air Force is much leaner than its predecessor, and relies on assets such as its fleet of Beriev/Ilyushin A-50 AWACS platforms to act as 'force-multipliers'.

Kushchevskaya, Krasnodar, Volgograd and Yeysk. The flying schools of Orenburg and Tambov (for training of pilots for A-VMF and DA, respectively) have been closed, and since 1995 the academy at Balashov has been responsible for the training of all pilots for heavy aircraft of all the Russian air arms, with subordinated training units based at Balashov, Petrovsk, Rtischevo, Ryazhsk and Michurinsk. It is planned that the PVO Air Academy at Armavir will also be fully integrated in the new combined Air Force and Air Defence Force structure.

The academy at Krasnodar (with a branch at Yeysk) performs a special role because, here, pilots and technical personnel from foreign countries that operate Russian-built military aircraft receive their training.

The Zhukhovskii Air Force Engineering Academy for scientific research is located in Moscow. Engineers are trained at Voronezh, Borisoglebsk, Irkutsk, Tambov; non-commissioned technical personnel are trained at Achinsk, Krasnodar and Perm. Moreover, there are special boarding schools for young cadets at Yeysk, Barnaul, Akhtubinsk, Taganrog and Orenburg.

The above-mentioned schools continue to operate large numbers of L-39 trainer aircraft, as well as L-410s, An-24s and An-26s. Navigators train on the Tu-134UBSh at Chelyabinsk. Two survival training centres for aircrews are located near Khabarovsk and in Dzhubga, on the Black Sea coast near Krasnodar.

The school for test pilots (SchLI), which is named after A. V. Fedotov, is attached to the Gromov Flight Research Institute (LII im. M.M. Gromova) at Zhukhovskii, and operates the MiG-21UM, MiG-23UB, MiG-25PU and MiG-29UB.

Instructors for tactical aircraft are trained at the Combat Training and Flight Crew Conversion Centre (4 TsBP i PeLS) at Lipetsk, with subordinated training units for fighter and bomber instructors at Borisoglebsk and Totskoye, respectively. The research-instructor regiment based at Lipetsk has four squadrons (equipped with Su-27s, MiG-29s, Su-24s and Su-25s, including a small number of MiG-25BMs and Su-24MPs) and is responsible for conversion training, development of operational tactics and various research projects. Instructors for military transport aircraft are trained at Ivanovo, and those for Long-Range Aviation at Shaykovka.

In addition to the aircraft operated by the flight training centres, the operational units of the Russian Air Force use a variety of two-seat versions of combat aircraft for conversion and continuation training. They comprise a dwindling number of MiG-23UBs and MiG-25PUs/-RUs, and small numbers of Su-27UBs, MiG-29UBs and Su-25UBs. In heavy bomber regiments, frequent use is made of the Tu-134UBL.

Instead of ordering the Czech-built Aero L-59/159 as a successor for the 2,094 L-39s delivered to the Soviet Union between 1973 and 1989, Russia is evaluating the indigenous MiG-AT and the Yak-130 advanced trainers. The fleet of L-39s is to be refurbished in Russia, following problems with spares.

AIRCRAFT RESERVE BASES

Considerable numbers of Tu-22s, MiG-23s, MiG-25s, MiG-27s and Su-17s and helicopters are held at several reserve bases (BRS, BRSV, BRV) throughout the Russian Federation.

MAINTENANCE FACILITIES

In 1991, the Soviet Air Force possessed a total of 58 repair facilities (ARZ). Currently, the Air Force operates a network of 33 maintenance facilities located throughout the Russian Federation. It is commanded by Major General Dmitry Alexandrovich Morozov and has a staff of 1,200 officers and a work force of about 30,000 employees.

TEST CENTRES

In addition to the Gromov Flight Research Institute at Zhukhovskii (LII im. M.M. Gromova, 'Ramenskoye'), the prime test centre for military aircraft is the State Flight-Test Centre named after V. P. Chkalov, located at Akhtubinsk, with subordinated bases at Chkalovskiy, Volskye and Nalchik. The test centre for weapons is located at Nizhny Tagil.

IA-PVO

In 1948, the Air Defence Forces (PVO) became a separate branch of the Soviet armed forces. Its fighter aviation assets, Istrebitel'naya Aviatsiya (IA-PVO), have traditionally been subordinated to nine Independent Air Defence Armies, covering Air Defence Zones throughout the Soviet Union, but by 1 January 1999 they were fully integrated with the Air Force in combined Air Force and Air Defence Corps.

Between May 1992 and May 1994, 11 IA-PVO regiments were disbanded and five were relocated. During this period, the Su-15 and MiG-25PD/PDS were withdrawn from service, and several regiments converted from the MiG-23 and Su-15 to the Su-27, or from the MiG-25 to the MiG-31.

In early 1998, 28 regiments (including four training regiments with L-39s and MiG-23s and an airborne early warning regiment with A-50s) were declared as remaining operational. Of these, the front-line fighter units comprise nine regiments with Su-27s, 12 regiments with MiG-31s, and one with MiG-23s, subordinated to PVO headquarters and the three Independent Air Defence Armies and three Independent Air Defence Corps. The regiment based at Kursk is the last remaining front-line PVO unit operating the MiG-23, which is rapidly being phased out of PVO – and VVS – service. MiG-23s are held in reserve at several air bases. The current fourth-generation interceptors, the Su-27 and the MiG-31, entered service in 1986 and 1982, respectively. The Su-30 (Su-30PU) two-seat long-range interceptor and the MiG-31M are undergoing operational testing in regiments subordinated to the IA-PVO training and combat conversion centre at Savostleyka.

Strategic Rocket Forces (RSVN)

The strategic rocket forces, which traditionally rank first in importance among the five branches of the Russian armed forces, operate a number of detachments, which usually consist of two aerial command posts (Mi-8 VzPU or Mi-9) and six to eight transport helicopters (Mi-8s).

Air Defence Forces Fighter Aviation (IA-PVO)

GK IA-PVO, Moscow

116 UtsBP	MiG-23 (31), MiG-29 (16), Mi-8 (6)	Astrakhan
148 TsBP I PeLS		**Savostleyka (aka Murom)**
54 GvIAP	Su-27 (40)	Savostleyka
786 IAP	MiG-25 (7), MiG-31 (31), Su-24 (4)	Pravdinsk
144 AP*	A-50 (16)	Pechora-Kamenka
*with detachments of 2-3 A-50s at Klin and a location in the Far East		

AVVAUL, Armavir

163 OAE	Mi-8 (9), An-24, An-26	Armavir
218 UAVP	MiG-23 (103)	Salsk
627 UAVP	L-39 (81)	Tikhoretsk
713 UAVP	L-39 (96)	Armavir
761 UAVP	MiG-23 (84)	Maykop

MO PVO, Moscow

106 OTAE	Mi-8 (6), An-12, An-26	Stupino
790 IAP	MiG-31 (28), MiG-25 (5)	Khotilovo
611 IAP	Su-27 (27)	Dorokhovo (aka Bezhetsk)
153 IAP	MiG-31 (24)	Morzhansk
472 IAP	MiG-23 (68)	Kursk

5 OK PVO, Yekaterinburg-Koltsovo

764 IAP	MiG-31 (24), MiG-25 (3)	Bolshoye Savino (aka Perm)
u/i OSAE		Yekaterinburg-Koltsovo

6 OA PVO, St Petersburg

174 GvIAP	MiG-31 (28), MiG-25 (1)	Monchegorsk
132 OTAE	Mi-8 (9), Mi-26 (3)	Vaskovo
177 IAP	Su-27 (29)	Lodeynoye-Pole
180 GvIAP	MiG-31 (25)	Gromovo
458 IAP	MiG-25 (6), MiG-31 (24)	Kotlas
470 IAP	Su-27 (28)	Afrikanda
518 IAP	MiG-31 (27), MiG-25 (1)	Talagi
941 IAP	Su-27 (29), Mi-8 (3)	Kilp-Yavr

11 OA PVO, Khabarovsk

u/i IAP	MiG-31	Jelizowo (aka Petropavlovsk-Kamchatki)
u/i IAP	Su-27	Komsomolsk-na-Amure
u/i IAP	MiG-31	Sandagou
u/i IAP	MiG-31	Dolinsk-Sokol
47 GvIAP	Su-27	Zolotaya Dolna (aka Unashi, Nakhodka)

12 OK PVO, Rostov-na-Donu

83 IAP	MiG-31 (26), MiG-25 (5)	Rostov-na-Donu
209 IAP	Su-27 (29)	Astrakhan
562 IAP	Su-27 (29)	Krymskaya

u/i OK PVO, Novosibirsk

64 IAP	MiG-31	Omsk-Severnyy
u/i IAP	MiG-31	Kansk South
350 IAP	MiG-31	Bratsk

Europe

Naval Aviation (A-VMF)

The history of Naval Aviation can be traced to before World War I. In 1938, Naval Aviation (Aviatsiya Voenno Morskogo Flota, A-VMF) was formed as a command of the Soviet Navy and, since then, the four Soviet fleets (Northern, Baltic, Black Sea and Pacific Ocean) have each operated their own aviation component. Typical missions performed by Naval Aviation are anti-ship missile attack, anti-submarine warfare (ASW), SAR and maritime reconnaissance.

Prior to 1991, the Naval Aviation element of the fleets was generally composed of a division (MRAD) with two subordinated regiments (MRAP) of missile-carrying bombers (Tu-16K-26 or Tu-22M2/M3), a long-range ASW regiment (OPLAP DD) of Tu-142 (Northern and Pacific Fleets only), a fixed-wing ASW regiment (OPLAP) of Be-12 and Il-38, a helicopter ASW regiment (OPLVP) with Ka-25, Ka-27 and Mi-14, a reconnaissance regiment (ORAP) of Su-17M3 and/or Su-14MR, an attack regiment (OMShAP) of Su-17M3 or Su-24, and transport and support units, equipped with Mi-8, An-12, An-24, An-26, Il-20, Il-22 and Tu-134 (Tu-135). In addition, the VTOL Yak-38 and Yak-38M were operated by shipborne ground-attack regiments (KBAP) deployed aboard three carriers.

Since then, the assets of the Baltic and Black Sea Fleets (the latter divided between Russia and Ukraine) have been considerably reduced, but those of the Northern and Pacific Fleets have remained relatively intact. Following the transfer of the training and test centres at Nikolaev (Kulbakino), Kirovskoye and Saki to Ukrainian control, Ostrov serves as the main training base for Naval Aviation crews.

BLACK SEA FLEET (ChF)

As a result of the division of Black Sea Fleet Naval Aviation assets between Russia and Ukraine, Russia received 19 Tu-22M3s, as well as 10 Be-12s. A transport unit, equipped with An-12s and An-26s, remains based at Kacha in support of the Baltic Fleet. Furthermore, Russia retained 872 OPLVP, a long-

time resident of Kacha, and the Su-17M3s of 43 OMShAP at Gvardeyskoye, which has since been reduced to squadron strength. The 10 Su-25UTGs were divided in equal numbers between Ukraine and Russia. The training facility at Saki (now under Ukrainian control) is periodically leased by Russia for the use of its simulated carrier jump deck and arresting gear by the Su-33s, Su-25UBPs and Su-25UTGs of 279 KIAP.

BALTIC FLEET (BF)

In January 1995, all Naval Aviation and Air Force assets based in the newly created KOR (Kaliningrad Special District) were combined into a single military structure under the command of the C-in-C of the Baltic Fleet, including the former IA-PVO fighter regiment, equipped with the Su-27. As part of the reorganisation, the Be-12s of 49 OPLAE were resubordinated to the transport regiment at Khabrovo.

MISSILE-CARRYING AVIATION

In the early 1960s, Naval Aviation converted from torpedo-carrying aircraft to missile-carrying platforms, and in March 1961 all navy units operating the Tu-16KS were renamed 'morskie raketonosnye'. Following the withdrawal from service of the Tu-16

Principal combat aircraft of the carrier **Kuznetsov** are the Sukhoi **Su-33s** of the 279 Shipborne Fighter-Aviation Regiment. Eighteen of the type are deployed during a cruise, with four **Su-25UTGs**, 15 **Ka-27s** and a pair of AEW **Ka-31RLDs**. The cancellation of the other three carriers in the class has limited the procurement of carrier aircraft.

in the early 1990s, the Tu-22M2 and Tu-22M3, armed with the medium-range Kh-22 air-to-surface missile (ASM), are the principal missile-carrying platforms in naval aviation service. They are currently operated by two divisions of two regiments each, subordinated to the Northern and Pacific Fleets.

ANTI-SUBMARINE WARFARE

The ASW mission (protivolodochnaya aviatsiya) is traditionally divided into short-range (using the Be-12PL, in service since 1965), medium-range (using the Il-38, in service since 1967) and long-range (using the Tu-142, in service since 1970). The Be-12PL and Il-38/Il-38M serve in ASW regiments attached to the Northern and Pacific Fleets, and the Be-12PL is operated in small numbers by units of the Baltic and Black Sea Fleets.

The long-range ASW regiment based at Kipelovo operates two squadrons with about 40 anti-submarine

Glossary

Abbr.	Full (Russian)	Translation
APIB	Aviatsion'nyi Polk Istrebitelei-Bombirovchikov	
	fighter-bomber aviation regiment	
A	Armiya	Army
AB	Aviatsionnaya Baza	Aviation Base
A-VMF	Aviatsiya-Voenno-Morskogo Flota	Naval Aviation
ACK	Aviatsionno-Sportivnyy Klub	Flying Club
ADON	aviatsionnaya diviziya osobogo naznacheniya	
	aviation division for special purposes	
AK	Armeyskiy Korpus	Army Corps
APON	Aviatsionnyy Polk Osobogo Naznacheniya	
	aviation regiment for special purposes	
APSZ	Aviatsionnyy Polk Samoletov-Zapravshchikov	
	air refuelling aviation regiment	
ARZ	Aviatsionnyi Remontnnyy Zavod	Aviation Repair Factory
AVVAUL	Armavirskoe VVAUL	Armavir HMAS for Pilots
BAD	Bombardirovchnaya Aviatsionnaya Diviziya	
	bomber aviation division	
BAP	Bombardirovchnyy Aviatsionnyy Polk	
	bomber aviation regiment	
BF	Baltiyskiy Flot	Baltic Fleet
BLAT	Baza Likvidatsii Aviatsionnoy Tekniki	
	Aircraft Destruction Base	
BRS	Baza Reserva Samoletov	Aircraft Reserve Base
BRSV	Baza Reserva Samoletov i Vooruzheniya	
	Aircraft and Armament Reserve Base	
BRV	Baza Reserva Vertoletov	
	Helicopter Reserve Base	
BVVAUL	Balashovkoye VVAUL	
	Balashov HMAS for Pilots	
ChVVAUSh	Chelyabinskoe Vysshee Voennoe Aviatsionnoe Uchilishche Shturmanov	
	Chelyabinsk HMAS for navigators	
DA	Dal'naya Aviatsiya	Long-range Aviation
DVO	Dal'nevostochniy Voennyy Okrug	
	Far East Military District	
FKVVAU	Filial Krasnodarskogo VVAUL	
	Branch of Krasnodar HMAS for Pilots	
FPS	Federal'naya Pogranichnaya Sluzhba	

Abbr.	Full (Russian)	Translation
	Federal Border Guards Service	
GK SV	Glavnoe Komandovanie Sukhoputnhykh Voysk	
	High Command Ground Forces	
GK Voysk PVO	Glavnoe Komandovanie Voysk Protivovozdushnoy Oborony	
	High Command Air Defence Forces	
GK VVS	Glavnoe Komandovanie Voenno-Vozdushnykh Sil	
	High Command Air Force	
GRVZ	Gruppa Rossiyskykh Voysk v Zakavkaz'e	
	Group of Russian Forces in Transcaucasus	
Gv	Gvardeyskaya	Guards
GvIAD	Gvardeyskaya Istrebitel'naya Aviatsionnaya Diviziya	
	Guards fighter aviation division	
GvIAP	Gvradeyskyy Istrebitel'nyy Aviatsionnyy Polk	
	Guards fighter aviation regiment	
GvTBAP	Gvardeyskyy Tyazhelyy Bombardirovochnyy Aviatsionnyy Polk	
	Guards heavy bomber aviation regiment	
HMAS	Higher Military Aviation School	
IAD	Istrebitel'naya Aviatsionnaya Diviziya	
	fighter aviation division	
IAO	Istrebitel'nyy Aviatsionnyy Polk	
	fighter aviation regiment	
IAP PVO	Istrebitel'nyy Aviatsionyy Polk PVO	
	Air Defence Forces fighter aviation regiment	
IBAP	Instruktorskiy Bombardirovochnyy Aviatsionnyy Polk	
	bomber-instructor aviation regiment	
IIAP	Instruktorskiy Istrebitel'nyy Aviatsionnyy Polk	
	fighter-instructor aviation regiment	
IISAP	Issledovatel'skogo- Instruktorskiy Smeshannyy Aviatsionnyy Polk	
	instructor-research composite aviation regiment	
IIVP	Instruktorsko-Issledovatel'skiy Vertoletnyy Polk	
	instructor-research helicopter regiment	
IVTAP	Instruktorskiy Voenno-Transportnyy Aviatsionnyy Polk	
	instructor military transport aviation regiment	
KIAP	Korabel'nyy Istrebitel'nyy Aviatsionnyy Polk	
	shipborne fighter aviation regiment	
KSHAP	Korabel'nyy Shturmovoy Aviatsionnyy Polk	
	shipborne ground-attack aviation regiment	
KOR	Kaliningradskiy Osobyi Rayon	
	Kaliningrad Special District	

Abbr.	Full (Russian)	Translation
KVVAU	Krasnodarskoe Vysshee Voennoye Aviatsionnoe Uchilishche	
	Krasnodar Higher Military Aviation School	
KVVAUL	Kachinskoe VVAUL	Kacha HMAS for Pilots
LenVO	Leningradskiy Voennyy Okrug	
	Leningrad Military District	
MO VVS i PVO	Moskovskiy Okrug VVS i PVO	
	Moscow Air Force and Air Defence District	
MO PVO	Moskovskii Okrug Protivovozdushnoy Oborony	
	Moscow Air Defence District	
MO RF	Ministerstvo Oborony Rossiyskoy Federatsii	
	Ministry of Defence of the Russian Federation	
MRAD	Morskaya Raketonosnaya Aviatsionnaya Diviziya	
	Naval missile-carrying aviation division	
MRAP	Morskaya Raketonosnyy Aviatsionnyy Polk	
	Naval missile-carrying aviation regiment	
MVO	Moskovskiy Voennyy Okrug	
	Moscow Military District	
OA PVO	Otdel'naya Armiya Protivovozdushnoy Oborony	
	Independent Air Defence Army	
OA VVS i PVO	Otdel'naya Armiya VVS i PVO	
	independent Air Force and Air Defence Army	
OAE	Otdel'naya Aviatsionnaya Eskadril'ya	
	independent aviation squadron	
OBVP	Otdel'nyy Boevoy Vertoletnyy Polk	
	independent combat helicopter aviation regiment	
ODRAP	Otdel'nyi Dal'nyy Razvedivatel'nyy Aviatsionnyy Polk	
	independent long-range reconnaissance regiment	
OGRV-vPRRM	Operativnaya Gruppa Rossiyskykh Voysk v 'Dniester Regn Moldova'	
	Operations Group of Russian Forces in the Dniester Region of the Republic of Moldova	
OK PVO	Otdel'nyy Korpus Protivovozdushnoy Oborony	
	Independent Air Defence Corps	
OK VVS i PVO	Otdel'nyy Korpus VVS i PVO	
	Independent Air Force and Air Defence Corps	
OMSHAE	Otdel'naya Morskaya Shturmovaya Aviatsionnaya Eskadril'ya	
	independent naval ground attack aviation squadron	
OMSHAP	Otdel'nyy Morskoy Shturmovoy Aviatsionnyy Polk	
	independent naval ground attack aviation regiment	

The Russian navy uses the Kamov Ka-27PL 'Helix-A' in the anti-submarine warfare role. In operational service since 1982, the type is scheduled to be replaced by the Kamov Ka-40, a development of the basic design. About 85 Ka-27PLs are in service.

Tu-142Ms and Tu-142MZs, and a third squadron with the Tu-142MR. In addition, a number of specialised shipborne and shore-based ASW helicopters (Mi-14PL, Ka-25PL and Ka-27PL) are operated by composite ASW helicopter regiments and squadrons in all four fleets.

MARITIME STRIKE/GROUND ATTACK

In Naval Aviation, maritime strike and ground-attack missions are flown by Su-17M3 and Su-24 of the independent ground attack and bomber regiments and squadrons directly subordinated to the headquarters of the four fleets. The Su-32FN will replace the Naval Aviation Su-17 and Su-24 in the long-range maritime strike role.

MARITIME RECONNAISSANCE

The maritime reconnaissance mission (*morskaya razvedka*) is carried out by the Su-24MR. Reportedly a small number of Tu-22M2Rs are in Naval Aviation service. The Tu-95RT, which was operated for a long time as a long-range maritime reconnaissance aircraft and target designator for missiles, was withdrawn from service in the early 1990s. The status of the development of a variant of the Tu-142 with a similar mission (Tu-142MRT) is uncertain. A few Elint Il-20s are in Naval Aviation service, as well as the Il-18 SIP (Il-20RT) radiotelemetry platform.

SPECIAL PURPOSE (SUPPORT)

About 10 Tu-142MRs are used for VLF communications relay between nuclear submarines and command centres with the two long-range ASW regiments at Kipelovo (SF) and Alekseyevka (TOF). A number of specialised variants of fixed-wing aircraft

and helicopters (An-12PS, Be-12PS, Mi-14PS, Ka-25PS, Ka-27PS) are used for SAR operations, and the Mi-14BT is employed as a minesweeper. The Ka-29 combat transport helicopter is in service in small numbers. A few transport units subordinated to the Navy and the four fleet headquarters operate An-12, An-24, An-26 (An-26PS), Tu-134, Mi-8 and Mi-6s in the transport and liaison role.

CARRIER AVIATION

Following the withdrawal from service of the Yak-38M and the decommissioning of the aircraft-carriers *Minsk*, *Kiev* and *Novorossiysk* in the early 1990s, carrier aviation (Korabel'naya Aviatsiya) is concentrated on the TABKR *Admiral Flota Sovetskogo Soyuza Kuznetsov* (formerly *Tbilisi*). The unit

deployed aboard the carrier is 279 KIAP, which is equipped with 24 Su-33s (Su-27K), and a dozen Su-25UTGs and Su-25UBPs which serve as shipborne two-seat trainers for the Su-33 crews.

The MiG-29K, of which two prototypes were extensively tested on the *Kuznetsov*, did not progress beyond the prototype stage.

The commander of Naval Aviation, Colonel General Vladimir Deyneka, has expressed the need for a two-seat version of the Su-27K, and an upgrade of the Su-25UTG, as well as a new-generation maritime patrol aircraft, based on the Tu-204 airliner, to replace the Be-12, Il-38 and Tu-142. Furthermore, the Ka-27 is in need of replacement by a new multi-purpose helicopter, and the A-40 amphibian is under study as a replacement for the ageing SAR Be-12s.

Naval Aviation (A-VMF)

GK AVMF, Moscow

u/i OTAE	An-12, An-24, An-26, An-72	Ostafyevo
240 GvOSAP	Su-24 (18), Tu-22M3 (3), Il-38 (3), Tu-142 (4), Be-12 (2), Il-20	Ostrov

SF (Northern Fleet), Severomorsk

279 KIAP	Su-33 (24), Su-25UBP/25UTG (7/4)	Malyavr (Severomorsk-3)
u/i OPLAP DD	Tu-142M/142MZ/142MR (39)	Kipelovo
u/i MOSAP	An-12*/-24/-26 (incl. 2 An-12PS)	Pechenga
u/i OPLVP	Ka-25/-27, Mi-14	Murmansk NE
u/i OPLAP	Il-38, Be-12	Severomorsk
5 MRAD, Olenya		
574 MRAP	Tu-22M3 (18)	Lakhta (aka Arkhangelsk-Kholm)
924 MRAP	Tu-22M3 (20)	Olenya

ChF (Black Sea Fleet) HQ West at Sevastopol
HQ East at Novorossiysk

43 OMShAE	Su-17M3 (9)	Gvardeyskoye
872 OPLVP	Ka-25/-27 (30), Mi-14	Kacha
u/i OPLAE	Be-12 (10)	Kacha
u/i OSAPOROSAE	An-12, An-26, Mi-8	Kacha

BF (Baltic Fleet), Kaliningrad

u/i OSAP	An-26 (10), An-24 (3), An-12 (5), Be-12 (10)	Khabrovo
4 OMShAP	Su-24M (42)	Chernyakhovsk
846 ORAE	Su-24 (20), Su-24MR (12)	Chkalovsk (aka Proveren)
689 IAP	Su-27 (28)	Nivenskoye (aka Jushniy)
u/i OPLVP	Ka-25, Ka-27PL (30), Ka-29	Donskoye (aka Brusteryort)

TOF (Pacific Fleet), Vladivostok

u/i ORAE	Su-24 (recce & attack)	Artem
u/i OSAP	An-12, An-24, An-26, Tu-134	Artem
u/i OPLAP DD	Tu-142	Alekseyevka
u/i OPLAP	Il-38	Nikolayevka
u/i OPLVP	Ka-25/-27/-29, Mi-14	Novo-Nezhino
u/i OPLAP	Be-12 (10), Ka-25/-27/-29, Mi-14	Jelizowo
u/i MRAD, Artem		
u/i MRAP	Tu-22M2	Artem
u/i MRAD, Alekseyevka		
u/i MRAP	Tu-22M3	Alekseyevka
u/i MRAP	Tu-22M3	Alekseyevka

OPLAE	Otdel'naya Protivolodochnaya Aviatsionnaya Eskadril'ya	
	independent anti-submarine aviation squadron	
OPLAP DD	Otdel'nyy Protivolodochnyy Aviatsionyy Polk Dalnego Deystviya	
	independent long-range anti-submarine aviation regiment	
OPLVP	Otdel'nyy Protivolodochnyy Vertoletnyy Polk	
	independent anti-submarine helicopter regiment	
ORAE	Otdel'naya Razvedyatel'naya Aviatsionnnaya Eskadril'ya	
	independent reconnaissance aviation squadron	
ORAP	Otdel'nyi Razvedyatel'nyy Aviatsionnnyi Polk	
	independent reconnaissance aviation regiment	
OSAE	Otdel'naya Smeshannaya Aviatsionnaya Eskadril'ya	
	Independent mixed aviation squadron	
OSAP	Otdel'nyy Smeshannyy Aviatsionnyy Polk	
	independent mixed aviation regiment	
OSHAP	Otdel'nyy Shturmovoy Aviatsionnyy Polk	
	independent ground attack regiment	
OTAE	Otdel'naya Transportnaya Aviatsionnaya Eskadril'ya	
	independent transport aviation squadron	
OTAP	Otdel'nyy Transportnyy Aviatsionnyy Polk	
	independent transport aviation regiment	
OTBVP	Otdel'nyy Transportno-Boyevoy Vertoletnyy Polk	
	independent combat transport helicopter regiment	
OUAE	Otdel'naya Uchebnaya Aviatsionnaya Eskadril'ya	
	independent training aviation squadron	
OVE REB	Ootdel'naya Vertoletnaya Eskadril'ya Radioelektronnoy Bor'by	
	independent EW helicopter squadron	
OVE	Otdel'naya Vertoletnaya Eskadril'ya	
	independent helicopter squadron	
OVP BU	Otdel'nyy Vertoletnyy Polk Boevogo Upravleniya	
	independent combat command and control helicopter regiment	
OVP	Otdel'nyy Vertoletnyy Polk	
	independent helicopter regiment	
OVTAP	Otdel'nyy Voenno- Transportnyy Aviatsionnyy Polk	
	independent military transport aviation regiment	
PLSAP	Protivolodochnyy Semshannyy Aviatsionnyy Polk	
	mixed anti-submarine aviation regiment	
PriVO	Privolzhkiy Voennyy Okrug	*Volga Military District*
REB	Radioelektronnaya Bor'ba	*Electronic Warfare*
ROSTO	Rossiyskaya Oboronnaya Sportivno-Tekhnicheskaya	

	Organizatsiya	
RVSN	Raketnye Voyska Strategicheskogo Naznacheniya	
	Strategic Rocket Forces	
RVVDU	Ryazanskoe Vysshee Vozdushno-Desantnoe Uchilishche Ryazan	
	Higher School for Airborne Troops	
SchLI	Schkol Letchikov-Ispytateley	*Test Pilots School*
SF	Severnyy Flot	*Northern Fleet*
SibVO	Sibirskiy Voennyy Okrug	
	Siberian Military District	
SHAD	Shturmovoya Aviatsionnaya Diviziya	
	ground attack aviation division	
SHAP	Shturmovoy Aviatsionnyy Polk	
	ground attack aviation regiment	
SKAD	Smeshannaya Korabel'naya Aviatsionnaya Diviziya	
	mixed carrierborne aviation division	
SKVO	Severo-Kavkazkiy Voennyy Okrug	
	North Caucasus Military District	
SV	Sukhoputnye Voyska	*Ground Forces*
SVVAUL	Syzranskoe VVAUL	*Syzran HMAS for pilots*
TABKR	Tyazhelyy Avianesushchiy Kreyser	
	heavy aircraft-carrying cruiser	
TBAD	Tyazhelaya Bombadirovochnaya Aviatsionnaya Diviziya	
	heavy bomber aviation division	
TBAP	Tyazhelyy Bombarovochnyy Aviatsionnyy Polk	
	heavy bomber aviation regiment	
TOF	Tikhookeanskiy Flot	*Pacific Fleet*
TsBP i PeLS VTA	Tsentr Boevogo Primenniya i Pereuchivaniya Letnogo Sostava VTA	
	Combat Training and Conversion Centre for Flight Personnel of Military Transport Aviation	
TsBP i PeLS	Tsentr Boevogo Primeneniya i Pereuchivaniya Letnogo Sostava	
	Combat Training and Conversion Centre	
TsBP i PeLS AA	Tsentr Boevogo Primeneniya i Pereuchivaniya Letnogo Sostava Armeyskoy Aviatsii	
	Combat Training and Conversion Centre for Flight Personnel of Army Aviation	
TsBP i PLS	Tsentr Boevogo Primeneniya i Podgotovki Letnogo Sostava	
	Combat Training Centre for Flight Personnel	
TsF	Chernomorskiy Flot	*Black Sea Fleet*

TsPAT	Tsentr Pokaza Aviatsionnoy Tekhniki	
TVVAUL	Tambovskoye VVAUL	
	Tambov HMAS for Pilots	
UAP	Uchebnyy Aviatsionnyy Polk	
	training aviation regiment	
UATs (PLS)	Uchebnyy Aviatsionnyy Tsentr (Podgotovki Letnogo Sostava)	
UBAP	Uchebr'nyi Bombardirovch'nyi Aviatsion'nyi Polk	
	Bomber Aviation Training Regiment	
UfVVAUL	Ufa VVAUL	*Ufa HMAS for Pilots*
UrVO	Ural'skiy Voennyy Okrug	*Ural Military District*
USiBV BF	Upravlenie Sukhoputnykh i Beregovykh Voysk Baltiyskogo Flota	
	Baltic Fleet Ground- & Shore-based Troops Directorate	
UVP	Uchebnyy Vertoletnyy Polk	
	training helicopter regiment	
VA VVS i PVO	Vozdushnaya Armiya VVS i PVO	
	Air Force and Air Defence Army	
VA VGK DA	Vozdushnaya Armiya Verkhovnogo Glavnogo Komandovaniya Dal'naya Aviatsiya	
	Long-range Air Army of the Supreme Command	
VA VGK VTA	Vozdushnaya Armiya Verkhovnogo Glavnogo Komandovaniya Voenno-Transportnaya Aviatsiya	
	Transport Air Army of the Supreme Command	
VDD	Vozdushno-Desantnaya Diviziya	*airborne division*
VDV	Vozdushno-Desantnye Voyska	*Airborne Troops*
VMF	Voenno-Morskoy Flot	*Navy*
VO	Voennyy Okrug	*Military District (MD)*
VS RF	Vooruzhennykh Sil RF	
	Armed Forces of the Russian Federation	
VTA	Voenno-Transportnaya Aviatsiya	
	Military Transport Aviation	
VTAD	Voenno-Transportnaya Aviatsionnaya Diviziya	
	military transport aviation division	
VTAP	Voenno-Transportnyy Aviatsionnyy Polk	
	military transport aviation regiment	
VVAUL	Vysshee Voennoe Aviatsionnoe Uchilishche Letchikov	
	HMAS for Pilots	
VV MVD RF	Vnutrennie Voyska Ministerstva Vnutrennikh Del RF	
	Interior Forces of the Ministry of the Interior of the Russian Federation	

Armeyskaya Aviatsiya (Army Aviation)

The first helicopter squadron, equipped with the Mi-1, was activated in the early 1950s at Serpukhov (MoMD). Soon after, the first regiments (equipped with 60 Mi-1s and Mi-4s each) were formed at Torzhok and Kaunas, and the Soviet rotary-winged force has continued to grow ever since. The first attack regiments equipped with Mi-24As were formed at Chernigovka (DVO), Brody (PrikVO), Parchim and Stendal (Group of Soviet Forces in Germany) in the early 1970s.

The Mi-24 and Mi-8 saw widespread service during the war in Afghanistan and, just prior to the collapse of the Soviet Union, Army Aviation strength reached its zenith. Its assets were particularly heavily concentrated in the former German Democratic Republic, where 10 combat helicopter regiments confronted NATO forces.

The Mi-24V and Mi-24P, and the Mi-8T and Mi-8MT, still form the backbone of the combat and combat transport helicopter regiments. In addition, specialised variants of the Mi-24, Mi-24K and Mi-24R are operated as reconnaissance/fire-control and NBC reconnaissance helicopters.

Currently, Army Aviation units are based in all military districts of the Russian Federation (with the exception of the Ural MD). In addition, a few regiments and squadrons serve as detachments at

Above left: The Mi-24P 'Hind-F' replaced the single 12.7-mm with a twin-barrelled 30-mm cannon. About 700 'Hinds' of all types serve with Army Aviation.

Above right: Operational since 1983, the Mil Mi-26 can carry up to 80 combat-equipped troops, or 20000 kg (44,090 lb) of freight.

Moldova, Kaliningrad and Georgia under direct control of the Ministry of Defence.

Army Aviation, which is commanded by Colonel General Vitaliy Pavlov, currently operates two flight training centres for helicopter pilots (at Ufa and Syzran), and one for flight engineers (at Kirov). During 1999, the three above-mentioned academies will be integrated into a single flight training centre for helicopter pilots, flight engineers and ground personnel, located at Syzran.

A NEW GENERATION OF HELICOPTER

It is widely admitted that by 2000 a large part of the inventory of the Army Aviation helicopters will have exceeded their service lives, or will be in need of major overhaul. However, the introduction of new types has been considerably delayed. Both the Mi-28 and Ka-50, intended as replacements for the Mi-24, made their first flights in 1982; upgraded versions with night-vision capability of both types (Ka-50N and Mi-28N) are currently undergoing flight testing, as is the two-seat Kamov Ka-52 Alligator, which first flew in 1997. The latest modifications of the Mi-8 are Mi-8MTV-5 (Mi-17MD) and Mi-8AMTSh; new multi-purpose helicopters proposed as replacements for the Mi-8 are the Ka-60 and Mi-38.

Used for communication jamming and electronic warfare, the Mil Mi-8PPA 'Hip-K' has six cross-dipole antennas on the sides and a row of six heat exchangers under the fuselage.

Airborne Troops (VDV)

The Airborne Forces (VDV – Vozdushno-Desantnye Voyska) traditionally relied heavily on the Il-76s of VTA for transport, but large paradroppings are a sign of the past and, under the current military doctrine, airborne troops are inserted into the battle zone with helicopters (Mi-8s). A number of detachments (OTRYAD) are subordinated to VDV headquarters and each of the airborne divisions. These small units generally comprise one Mi-8 aerial command post (either an Mi-8VzPU or an Mi-9) and a dozen or so An-2s.

Sukhoputnykh Voysk – Armeyskaya Aviatsiya (Ground Forces – Army Aviation)

GU SV		Kaluga
344 TsBP I PeLS AA		**Torzhok**
361 IIVP	Mi-8 (16), Mi-24 (25)	Sokol (aka Vologda)
696 IIVP	Ka-50 (4), Mi-6 (3), Mi-8 (15), Mi-24 (16), Mi-24K/24R (2/3), Mi-26 (15)	Torzhok
113 OSAE	Mi-8 (8), Mi-24R (2)	Kaluga
2881 BRV	Mi-2 (203), Mi-8 (28), Mi-24 (100), Mi-24K/24R (25/14)	Totskoye
USiBV BF		
288 OVP	Mi-8, Mi-24	Nivenskoye (*aka Jushniy)
LeVO		**St Petersburg**
332 OTBVP	Mi-6 (12), Mi-8 (26)	Pribylovo (aka Kluchevoye)
6 A		**Petrozavodsk**
485 OVP BU	Mi-8 (30), Mi-24 (23), Mi-24K/24R (6/6)	Alakurtti
30 AK		**Vyborg**
172 OBVP	Mi-8 (21), Mi-24 (34), Mi-24R (2)	Kasimovo
MoVO		**Moscow**
239 GvOBP	Mi-6 (13), Mi-8 (37)	Yefremov
41 OVE	Mi-8 (8), Mi-24 (12)	Klokovo (Tula)
22 A		**Nizhny Novgorod**
225 OVP	Mi-8 (17), Mi-24 (33)	Protasovo
439 OVP BU	Mi-8 (27), Mi-24 (24), Mi-24K/24R (7/7)	Kostroma
20 AK		**Voronezh**
178 OBVP	Mi-8 (14), Mi-24 (40)	Kursk
490 OVP BU	Mi-8 (25), Mi-24 (16), Mi-24K/-R (6/6)	Klokovo
1 A		**Smolensk**
440 OVP BU	Mi-8 (42), Mi-24 (23), Mi-24K/24R (6/6)	Vyazma
336 OBVP	Mi-8 (18), Mi-24 (36)	Kaluga
OGRV v PRRM		**Tiraspol**
36 OVE	Mi-8 (5), Mi-24K (2)	Tiraspol

SKVO		Krasnodar
326 OSAE		**Bataysk**
325 OTBVP	Mi-6 (20), Mi-8 (30), Mi-26 (10)	Yegorlykskaya
67 AK		**Krasnodar**
55 OVP	Mi-8 (20), Mi-24 (39), Mi-24R (2)	Korenovsk
58 A		**Vladikavkaz**
487 OVP	Mi-8 (17), Mi-24 (27), Mi-24K/24R (4/6)	Budennovsk
GRVZ		**Tbilisi**
311 OVE	Mi-8 (5), Mi-24 (5)	Vaziani
PriVO, Samara (ex Kuybyshev)		
793 OTBVP	Mi-8 (37), Mi-26 (24)	Kinel Cherkassy
367 OTBVP	Mi-6 (9), Mi-8 (22)	Serdobsk
SVVAUL		**Syzran**
131 UVP	Mi-8 (116)	Sokol
626 UVP	Mi-8 (4), Mi-24 (91)	Pugachev
484 UVP	Mi-8 (5), Mi-24 (86), Mi-24R (15)	Syzran
UfVVAUL		**Ufa**
330 UVP	Mi-8 (98)	Ufa
851 UVP	Mi-8 (76)	Bezenchuk
2 A		**Chernorechnye**
437 OVP	Mi-8 (14), Mi-24 (36)	Ozinki
SibVO		

The Army Corps headquartered at Kemerovo (28 AK) has no subordinated Army Aviation units. An unidentified helicopter unit, directly subordinated to the Siberian Military District, is based at Berdsk.

ZabVO		

Two Army Corps (55 AK and 57 AK) are based in the Transbaikal Military District, headquartered at Borzya and Ulan-Ude, respectively. The subordinated helicopter units include a regiment based at Mogocha.

DVO		

Army Aviation assets based in the Far Eastern Military District are subordinated to two armies (5 A and 35 A, headquartered at Ussuryisk and Belogorsk, respectively), and to 68 AK, with headquarters at Yuzhno-Sakhalinsk.

Vozdushno-Desantnye Voyska

VDV		
283 OVE	Mi-8 (15)	Podolsk
RVVDU		**Ryazan**
58 OVTAE	An-2 (10), Mi-8	Ryazan
7 VDD		**Novorossiysk**
185 OVTAE	An-2, Mi-8	Krymskaya
76 VDD		**Pskov**
242 OVTAE	An-2, Mi-8	Pskov
98 VDD		**Ivanovo**
243 OVTAE	u/i	Ivanovo
104 VDD		**Ulyanovsk**
116 OVTAE	u/i	Belyy Klyuch
106 VDD		**Tula**
110 OVTAE	u/i	Tula

Aviatsiya FPS (Border Guards Aviation)

The aviation element of the paramilitary Federal Border Guards Service (FPS) is tasked with patrolling the borders of the Russian Federation and the Kaliningrad exclusive economic zone. For this purpose, the service operates a fleet of general aviation aircraft, multi-purpose aircraft and helicopters, notably An-2s, An-26BRLs, Yak-18Ts, Mi-8s, Mi-24s and Mi-26s. The SM 92P Finist armed light utility aircraft has entered service. In 1996, several dozen Yak-18Ts were received which had been given a new lease of life at the Smolensk aircraft factory.

In addition to the units stationed in the various Border Guards districts, including the Pacific Ocean, Northwest and Central Asian Frontier Districts, FPS operates the Kaliningrad Group of Border Guards and the Group of Russian Frontier Guards in the Republic of Tajikistan, and its forces are also deployed on the border between Georgia and Turkey.

The Mi-8s of the Frontier Guards can be recognised by a white horizontal band on the rear cargo doors. The helicopters of the Interior Forces of the Ministry of the Interior of the Russian Federation (VV MVD RF) are distinguished by a white vertical band around the tailboom.

SLOVAK REPUBLIC

Capital: Bratislava
Population: 5.2 million
Land area: 49035 km² (18,927 sq miles)
Major cities: Banska Bystrica, Kosice, Nitra, Presov, Zilina

Above: Liberally adorned with the national colours, this MiG-29 is one of the original batch shared between the Czech and Slovak Republics.

Left: 313. Stihaci Letka at Sliac continues to operate the MiG-21MF. Conversion training is undertaken on the MiG-21UM and US (seen here).

Letectva a Protivzdusnej Obrany Slovenskeho (Slovak Air and Air-Defence Force)

The history of Slovak military aviation must be divided into periods before 1945 and after 1993.

When Germany occupied the remainder of former Czechoslovakia in 1938 and the Clero-Fascist Slovak Republic became a belligerent ally to Nazi Germany, an independent Slovak Air Arm (Slovenskych Vzdusnych Zbraní) was created. Trying to exploit the lack of preparedness of the newly formed Slovak forces, the (also) right-wing Hungarians tried to occupy the Carpathian east of the country from March 1939 onwards. In the dogfights between Slovak and Hungarian fighters which occurred until mid-April 1939, 18 Slovak aircraft were destroyed. Under pressure from Germany, a border agreement was reached soon afterwards. Two years later, the Slovaks, equipped with German products, joined Hungary as part of the German assault on the Soviet Union. By the end of World War II, the Slovak Air Force had made use of 175 German airframes against US bomber formations over Slovakia and several dozen Bf 109F/Gs in the 13th Fighter Flight operated alongside the Germans on the Southern Russian front. Defecting pilots from these units, together with Slovak Communists, formed the backbone of the La-5-equipped Czecho-Slovak Mixed Air Division on the Russian side. After the national uprising in 1944 was suppressed by the Germans, all the units on the German side were disbanded.

On the Allied side, about 450 exiled airmen of Slovak origin fought within the RAF, most of them in Wellingtons and Liberators of No. 311 (Bomber) Squadron and Spitfires of Nos 310 and 312 (Fighter) Squadrons. After VE-Day, the four Allied units, together with the crews from the Soviet side, returned to a liberated and reunited Czechoslovakia. It took 44 years before a Slovak national insignia appeared again on a military aircraft.

A Communist state prevailed in Czechoslovakia until 1989, when in November of that year the Communist government leaders stepped down and all Communist guidelines were abandoned. When the two national parliaments of the new states – Czech and Slovak Republics – dissolved their union on 1 January 1993, the assets of the air force of the former Czechoslovakia were divided. This marks the birth of the new, independent Slovak Air and Air-Defence Force.

Since having a number of different aircraft types was not considered desirable, the agreed rule of a 2:1 split on all military assets was broken for aircraft assets. The Slovaks received no MiG-23s but gained half of the MiG-29s. In the case of the An-12s, only two airframes were in the inventory so each country gained one. Only one unit of the former Czechoslovakian forces had been stationed on Slovak territory, in Kosice, so the new state had to construct new bases and hangars before it transported tons of equipment to them, and had to accommodate all the aircrews and their families. A formerly rarely-used airfield within the huge bomb-

ing-range of Malacky, near the Austrian border, became home to the Su-22 fighter-bomber and reconnaissance aircraft and MiG-21s. The former aviation academy at Kosice now houses all the training squadrons and is also home to the display team 'Biely Albatrosy'. As the large LOT (Letecke Opravovne Trencin) overhaul facility is a licensed Sukhoi rework plant, the Su-25s were based there. All the combat helicopters were concentrated at Presov, north of Kosice in the east of the country. The transport and special-purpose aircraft are stationed at Piestany in the west. Sliac is located right in the centre of Slovakia and was therefore selected to house all the fighters/interceptors for the defence of Slovakian airspace.

After solving the problems of aircraft and personnel relocation, pilot training and combat readiness received more attention in the new air force. An original annual flight time of 50 hours was increased to the current 80-90 hours. Compared to the former (larger) Czechoslovakian air force, there are fewer accidents reported and there is more daily activity on the flight lines.

The 10 MiG-29s inherited from the separation are beginning to age, so early in 1994 the Slovak Air Force obtained six new MiG-29s, including one UB, as part-payment in a debt reduction scheme from Moscow. These aircraft were joined later by eight additional examples, giving Slovakia a total of 24. They are all based at Sliac, which often hosts multinational exercises that are undertaken because of Slovakia's involvement as a Partnership for Peace nation since 1996. In the same year, the Su-25s left Trencin and were relocated to the fighter-bomber base of Malacky, while the MiG-21s were transferred from there to Sliac. Also in that period the air force was reorganised on a Zakladna (air base) structure.

Russia has stepped up efforts to sell a range of aviation and aerospace equipment to Slovakia. The country is the only nation in Central Europe obviously not planning to switch from its former Soviet-made hardware to Western designs. According to Slovak sources, there are plans to acquire additional MiG-29s and Yak-130s and to replace the older Mi-24Ds with Mi-35s. Although live-firing trial demonstrations of the Ka-50 in Malacky in 1996 have not yet led to a purchase, the command of the Air Force wants to renegotiate the CFE limits (115 fixed-wing combat aircraft and 25 combat helicopters) to prepare for the acquisition of approximately 25-30 modern battlefield helicopters by reducing its number of fixed-winged aircraft.

Glossary

Letectva a Protivzdu-snej Obrany Slovenskeho	Slovak Air and Air-Defence Force
Letecka Zakladna (LZ)	Air Force Base
Stihaci Letecke Kridlo (SLK)	Fighter Wing
Zmiesany Dopravny Kridlo (ZmDLK)	Mixed Transport Wing
Stihaci Bombardovacie Letecky Kridlo (SBLK)	Fighter-Bomber Wing
Vtrulnikovy Kridlo (VRK)	Helicopter Wing
Vycvikove Stredisko Letectvo	Training Unit
Statni Letecky Utvar	Governmental Flying Service
Policia	Police

Letectva a Protivzdusnej Obrany Slovenskeho

Velitelstvo 3.zLaPVO Slovenskeho	Zvolen
31. Letecka Zakladna (LZ)	**Sliac**
31. Stihaci Letecke Kridlo (SLK)	
311. Stihaci Letka (SL)	MiG-29A/29UB
312. Stihaci Letka (SL)	MiG-29A/29UB
313. Stihaci Letka (SL)	MiG-21MF/21US/21UM, L-39ZA
314. Letka	Mi-2, Mi-17, L-410T
32. LZ	**Piestany**
32. Zmiesany Dopravny Kridlo (ZmDLK)	
321. Letka	An-24B, An-26, L410FG/MA/T/UVP, Tu-154B2, Yak-40
322. Letka	Mi-8P/8PPA/8T, Mi-17Z-2, Mi-2
33. LZ	**Malacky-Kuchyna**
33. Stihaci Bombardovacie Letecky Kridlo (SBLK)	
331. Letka	Su-22M-4/22UM-3K, L-29
332. Letka	Su-25K/25UBK, L-39C
333. Letka	L-29
34. LZ	**Presov**
34. Vtrulnikovy Kridlo (VRK)	
341. Letka	Mi-24V
342. Letka	Mi-24D/24DU
343. Letka	Mi-8T, Mi-17
344. Letka	Mi-2
Vycvikove Stredisko Letectvo	**Kosice**
5th Letecky Skolsky Pluk	
51. Letka	Aero L-39C/39MS/39V
52. Letka	Aero L-29
53. Letka	L410T/410UVP
Statni Letecky Utvar, Bratislava/M.R.Stefanik Airport	
Statni Letecky Utvar	Mi-8S (B-8231 and B-8427), Tu-154M (OM-BYO), Yak-40 (OM-BYE, -BYL)
Policia	PZL Mi-2 (B-2048, -2405, -2406, -2744, -2950), Mi-8P (B-8231, -8427, -8532), MD-902 Explorer (on order)

Eight Mil Mi-24Ds, 10 Mi-24Vs and a single Mi-24DU operate from Presov as two squadrons of 34. VRK. This Mi-24V is seen with a torpedo-like fuel tank on the stub wing.

SLOVENIA

PC-9s arrived in Slovenia in 1995. The three aircraft are the examples returned to the manufacturer by the US Army.

Wearing Stabilisation Force titles, this Bell 412EP is named **Novo Mesto**. *The majority of Slovenian aircraft have individual names.*

Capital: Ljubljana
Land area: 20250 km² (7,815 sq miles)
Population: 2 million
Major cities: Celje, Kranj, Maribor, Velenje

15 Brigada Vojaskega Letalstva

When Slovenia declared its independence on 25 June 1991, the government of Yugoslavia planned an attack on the new republic to prevent its breakaway. The war, which lasted for 10 days, resulted in a victory for Slovenia. Military aviation in Slovenia was established as soon as the war was over. On 9 June 1992, the Air Force Unit of the Slovenian Army (formerly know as Territorial Defence Force) was renamed 15 Brigada Vojaskega Letalstva, becoming the only brigade within the Slovenian Army operating aircraft and helicopters. 15 Brigada is divided into two squadrons – one operating fixed-wing aircraft and the other helicopters – and operates from two bases at Ljubljana-Brnik and Cerklje.

Two Zlin 143L and eight Zlin 242L light trainers (four- and two-seaters, respectively) are used for pilot training and are based close to the Croatian border at Cerklje Air Base. Also at Cerklje are the three Bell 206B-3s, which are used to train helicopter pilots.

A single Let 410 is in use as a transport, but can also be used for aerial photography. As Slovenia has no fighter aircraft on strength, fixed-wing pilots maintain their flying hours on the PC-9. At the moment three PC-9s are on strength, with a further nine on order which should be delivered by mid-1999. Two PC-6 Turbo-Porters should also be delivered imminently. The main tasks for these aircraft will be paradropping and fire-fighting duties. All Bell 412s used by the Air Force Unit are now in service with the Brigada, supplemented by a further five examples. Three of these helicopters were offered for SFOR-duties from 1 October 1997, as was the Let 410, in an attempt to help gain NATO membership.

Slovenia has an outstanding requirement for an interceptor, the F-16 being the favoured type. An attack helicopter is also high on the agenda, but, like the fighter plans, a decision is not expected within the near future.

Slovensko Vojska

15 Brigada Vojaskega Letalstva	
Letalska Eskadrilja	
L410UVP-E (1), PC-6 (2), PC-9 (3)	Ljubljana-Brnik
Zlin 143 (3), Zlin 242L (8)	Cerklje
Helikopterska Eskadrilja	
Bell 412EP (5), Bell 412SP (1), Bell 412HP (2),	Ljubljana-Brnik
Bell 206B-3 (3)	Cerklje

SPAIN

Capital: Madrid
Population: 39 million
Land area: 504880 km² (194,885 sq miles)
Major cities: Barcelona, Seville, Valencia

Ejército del Aire (Spanish Air Force)

Spanish military aviation stems from the Servicio Militar de Aerostación (Military Aerostation Service), an army lighter-than-air unit established in December 1896. In 1911 the Servicio Militar de Aeronáutica (Military Aviation Service) was formed, and in March an Aviation School was inaugurated at Cuatro Vientos, near Madrid.

An uprising by part of the army in July 1936 marked the start of the Spanish Civil War. The Aeronáutica Militar title was used by the Spanish Nationalist aviation units which merged with those of the Navy, who had supported the uprising. Loyal government aviation units became the new Fuerzas Aéreas de la República Española (FARE). The Nationalists received help from Italy (including the contingent known as Aviación Legionaria) and Germany (which sent the Legion Condor), while the Republicans received the majority of their aircraft and equipment from Russian sources. The Civil War ended on 29 March 1939 with victory by the Nationalist forces, and on 9 November 1939 a new air arm was formed as an independent service, the Ejército del Aire.

Since then the Spanish Air Force has passed through periods of neutrality (during World War II) and isolation (between 1945 to 1953). The signing of

The Mirage F1 fleet is finished in air superiority grey, as typified by this F1EE of Ala 14.

the defence agreement with the United States saw massive infusions of F-86 Sabres, T-33s and T-6s. Consolidation during the 1970s and 1980s resulted in Spain becoming a member of NATO in 1982. Recently, Spain became fully integrated into the alliance's military structure and today the Ejército del Aire has a strength in excess of 500 aircraft.

COMBAT JETS

From 1986 to 1990 Spain received a total of 72 F/A-18 Hornets, comprising 60 single-seat F/A-18As and 12 two-seat F/A-18Bs, known as C.15s and CE.15s in Spanish service. The surviving 69 are shared between four front-line squadrons and one operational training unit. An upgrade programme has bought the whole fleet to the same standard, with all aircraft now able to fire AIM-120 AMRAAMs. A further upgrade in 1999-2002 will add new weapons capability and radar improvements.

In December 1996 the EdA received the first of a batch of 30 more Hornets. These are ex-US Navy F/A-18As, but are being upgraded to the same standard as the rest of the Spanish fleet. The last of the batch is expected to be delivered in 1999. One squadron has become operational with these 'new' aircraft.

The F/A-18 performs both air-to-air (60 per cent of hours flown) and air-to-surface (40 per cent) missions. 121 Escuadrón is dedicated to tactical support for maritime operations with Harpoon missiles and laser-guided bombs.

The Ejército del Aire (EdA) was one of the early customers for Dassault's second-generation Mirage F1. The type currently equips two squadrons at Los Llanos. The air force has received 10 ex-Qatari F1EDAs and two F1DDAs to supplement the surviving aircraft. Recently, in place of the standard Thomson-CSF systems, six F1EEs received new Indra AN/ALR-300 RWR sets designed and manufactured in Spain, with performance similar to, and in some

Spain's primary multi-role fighters are the five squadrons of F/A-18A/Bs. This aircraft of 211 Escuadrón shows evidence of its past career with the US Navy, retaining an aggressor scheme with the unit designation on the fuselage overpainted.

instances better, than the ALR-67 system which equips the F/A-18s. The whole Spanish Mirage F1 fleet is to be fitted with this equipment.

For air-to-ground operations the Mirage carries dumb bombs from 275 lb to 1,984 lb (125 kg to 900 kg), GBU-10 and -16 LGBs and Mk 20 Rockeye and BME-300 cluster weapons.

The oldest serving fast-jet design is the McDonnell Douglas RF-4C Phantom II. With the benefit of a major upgrade, the Phantom still soldiers on in the reconnaissance role with 123 Escuadrón, and is slated to stay in service until 2010.

MARITIME PATROL AND TRANSPORT

Currently, 221 Escuadrón of Grupo 22, Ala 21 has on strength five ex-RNAF P-3B 'Super Bravo' Orions and two P-3As. Recently, the P-3Bs have received AGM-65G Maverick missiles for anti-surface unit warfare (ASUW) operations.

The Ejército del Aire has seven transport squadrons in three wings, and the real workhorse is the Lockheed C-130 Hercules. The 12 Hercules are shared by

The national aerobatic team is the 'Patrulla Aguila' formed in 1985, and is co-located with the Air Academy at San Javier. The team, instructors at the academy, fly the E.25 (CASA 101 Aviojet).

RF-4Cs are operated by 123 Escuadrón from Torrejón. The 'Rhino' has served with the air force since 1978, when four were delivered, followed by eight ex-123rd TRW, KY ANG aircraft in 1989.

The CASA-assembled SF-5Bs continue to be used in the lead-in fighter training role. Of the 34 two-seaters built, just over 20 survive in service with Ala 23 based at Talavera.

Ejército del Aire

Mando Aéreo del Centro (MACEN)

Brigada Aérea Número 1
Ala 12

Grupo 12	U.9 (Do 27/C.127)	Torrejón
121 Escuadrón	C.15C/CE.15 (F-18A+/B+)	Torrejón
122 Escuadrón	C.15C/CE.15 (F-18A+/B+)	Torrejón
123 Escuadrón	CR.12 (RF-4C)	Torrejón
Grupo 43	UD.13T (CL215T)	Torrejón
Grupo 45		
451 Escuadrón	T.17 (B.707), T.18 (Falcon 900)	Torrejón
452 Escuadrón	T.11 (Falcon 20), T.16 (Falcon 50)	Torrejón
CLAEX		
Grupo 44	C.14B (Mirage F1EE), C.15 (F-18A+) E.25 (C.101 Aviojet), E.26 (T-35 Tamíz), XT.12/T.12B (C.212 Aviocar)	Torrejón
Centro de Inteligencia Aérea		
408 Escuadrón	TM.12D (C.212 Aviocar) TM.11 (Falcon 20), TM.17 (B.707)	Torrejón
Ala 35		
351 Escuadrón	T.19A/B (CN.235)	Getafe
352 Escuadrón	T.19A/B (CN.235)	Getafe
Ala 37		
371 Escuadrón	T.12B/C (C.212 Aviocar)	Villanubla
372 Escuadrón	T.12B (C.212 Aviocar)	Villanubla
373 Escuadrón	T.12B (C.212 Aviocar)	Villanubla
Ala 48		
402 Escuadrón	HT.21 (SA 332B Super Puma) HT.21A (AS 532 Cougar)	Cuatro Vientos
403 Escuadrón	TR.12A (C.212 Aviocar), TR.20 (Citation V), U.9 (Do 27/C.127)	Cuatro Vientos
803 Escuadrón	D.3A/B (C.212 Aviocar) HD.21 (SA 332B Super Puma)	Cuatro Vientos
Grupo 74		
744 Escuadrón	E.25 (C.101 Aviojet)	Matacán
745 Escuadrón	T.12B (C.212 Aviocar)	Matacán
Grupo 42		
421 Escuadrón	E.20 (B55 Baron)	Getafe
422 Escuadrón	E.24A (F33A Bonanza)	Getafe

Mando Aéreo del Estrecho (MAEST)

Ala 21
Grupo 21

211 Escuadrón	C.15A (F/A-18A)	Morón
Grupo 22	U.9 (Do 27/C127)	Morón
221 Escuadrón	P.3 (P-3A/B Orion)	Morón
Ala 23	U.9 (Do 27/C.127)	Talavera
231 Escuadrón	AE.9 (F-5B)	Talavera
232 Escuadrón	AE.9 (F-5B)	Talavera
Ala 78	U.9 (Do 27/C.127)	Armilla
781 Escuadrón	HE.24 (Sikorsky S-76C)	Armilla
782 Escuadrón	HE.20 (Hughes H 269C)	Armilla

Mando Aéreo de Levante (MALEV)

Brigada Aérea Número 2
Ala 31
Grupo 15

151 Escuadrón	C.15C (F-18A+)	Zaragoza
152 Escuadrón	C.15C (F-18A+)	Zaragoza
153 Escuadrón	CE.15/C.15C (F-18B+/A+)	Zaragoza
Grupo 31	U.9 (Do 27)	Zaragoza
311 Escuadrón	T.10/TL.10 (C-130H/C-130H-30)	Zaragoza
312 Escuadrón	T.10/TK.10 (C/KC-130H)	Zaragoza
801 Escuadrón	HD.19 (SA 330 Puma) D.3A/B (C.212 Aviocar)	Son San Joan

Long-range SAR coverage of the Atlantic is provided by the Gando-based F27MPAs of 803 Esc.

Academia General del Aire

Escuela de Vuelo Elemental	San Javier
E.26 (T-35 Tamíz)	
Escuela de Vuelo Básico	San Javier
E.25 (C.101 Aviojet)	
'Patrulla Aguila' E.25 (C.101 Aviojet)	
Escuela de Navegación	
TE.12B (C.212 Aviocar), U.9 (C.127/Do 27)	
Escuela Militar de Paracaidismo	
721 Escuadrón T.12B (C.212 Aviocar)	Alcantarilla
Ala 14 U.9 (Do 27/C.127)	Los Llanos
141 Escuadrón C.14A/B (Mirage F1C/CE/EE)	Los Llanos
142 Escuadrón C.14A/B (Mirage F1C/CE/B/BE/EDA/DDA)	Los Llanos

Mando Aéreo de Canarias

Ala 46

461 Escuadrón	T.12B/C (CASA C.212)	Gando
462 Escuadrón	C.15A/B (F/A-18A/B)	Gando
802 Escuadrón	HD.21 (SA 332B Super Puma)	Gando
803 Escuadrón	D.2 (Fokker F27MPA)	Gando

Escuadrones 311 and 312, the first specialising in cargo and the latter in tanker/cargo operations.

Ala 35, based at Getafe, south of Madrid, started to re-equip with the CASA CN.235 in 1988, and deliveries were completed in 1994. Both of the unit's squadrons, 351 and 352, share 20 CN.235-100Ms, two of which are in VIP configuration; the remaining 18 serve as tactical transports (delivered in lizard scheme and being repainted in the omnipresent grey). Ala 35 will start to receive the first of 18 new C.295s in about five years and, in turn, will pass its CN.235s to Ala 37, which operates 24 C.212 Aviocars, including four in VIP configuration.

461 Escuadrón, part of Ala Mixta 46, based at Gando air base, Canary Islands, is equipped with 11 Aviocars, of which one is in VIP configuration. Its duties are varied and include performing shuttle military flights between the several islands which have military garrisons, as well as the usual tactical flights, and SAR missions in support of 802 Escuadrón.

SAR

At present, there are three Ejército del Aire SAR-dedicated squadrons, which operate Puma and Super Puma helicopters, and Aviocar and Fokker F27MPA fixed-wing aircraft. They provide SAR cover for the assigned Flight Information Region (FIR) zones in mainland Spain, the Mediterranean and the Atlantic Ocean; they also have humanitarian taskings as well as more warlike combat SAR duties. 801 Escuadrón is based at Son San Juan air base on the island of Mallorca, and is equipped with five SA 330 Pumas, four SAR-configured C.212-200 Aviocars, and one C.212-200, without radar, in cargo configuration. 802 Escuadrón based at Gando, Canary Islands has four Aérospatiale AS 332B Super Pumas and three Fokker F27MPAs, the latter used for long-range operations. 803 Escuadrón, at Cuatro Vientos, has four Super Pumas, two radar-equipped Aviocars and one in cargo configuration.

402 Escuadrón at Cuatro Vientos is currently equipped with two Super Pumas and four Cougars, tasked with VIP and government transportation roles, as well as SAR and combat SAR missions. This unit, together with 803 and 403 Escuadrones, all operate Super Pumas. 403 Escuadrón is the Ejército del Aire's dedicated photo-mapping and aerial survey squadron, being equipped with six Aviocars fitted with De Wilde cameras and three Cessna Citation Vs.

FIRE-FIGHTERS

Grupo 43 is probably the most popular Ejército del Aire unit, with its yellow-and-red Canadair CL-215Ts that fight forest fires. This unit is funded by the Ministry of Agriculture and Fishing. A total of 30 of the unique Canadair water-bombers has been acquired, seven of which have been lost. Fifteen CL-215s have been converted to CL-215Ts, which entails replacing the P&W R2800 CA3 reciprocating engines with PW123F turboprops.

SPAIN
- La Coruna
- Pontevedra
- Santander
- Bilbao
- Agoncillo
- Villanubla
- Gerona
- Zaragoza
- Barcelona
- Matacán
- Madrid
- Colmenar
- Torrejón
- Cuatro Vientos
- Getafe
- Toledo
- Talavera
- Bétera
- Manises
- MENORCA
- MALLORCA
- Mahon
- Son San Juan
- Almagro
- Valencia
- Palma
- Los Llanos
- IBIZA
- Ibiza
- Cordoba
- Alicante
- El Copero
- Seville
- Alcantarilla
- Morón
- San Javier
- Granada
- Armilla
- Rota
- Cadiz
- Malaga

CANARY ISLANDS
- TENERIFE
- Los Rodeos
- LAS PALMAS
- Gando
- GRAN CANARIA

Europe

Above: Spain has a single squadron of Orions, 221 Escuadrón. This is one of the unit's P-3Bs.

Above: Tactical transport is undertaken by the CASA CN.235s of Ala 35.

Below: Both locally-assembled CASA 127s and Dornier Do 27s continue to undertake liaison tasks.

ELECTRONIC FALCONS

Grupo 45, based at Torrejón, is equipped with three Falcon 20s, two fitted with advanced electronic systems to check navigation aids, and one as a VIP transport; one Falcon 50 as VIP; two Falcon 900s also as VIPs; and three Boeing 707s, one in VIP configuration and the other two in tanker-transport configuration. Grupo 45 has two flying squadrons: 451 Escuadrón operates the Boeing 707s and Falcon 900s, and 452 Escuadrón is in charge of the Falcon 20 and 50 operations.

Besides its three Falcon 20s, the unit had another two which, in 1994, were transferred to 408 Escuadrón. This squadron, which is the Ejército del Aire's dedicated electronic warfare (EW) unit, was activated at Getafe as 408 Escuadrilla (Flight) during the mid-1980s. Its initial equipment consisted of two specially modified Aviocars, which were used for training fighter pilots and controllers in ECM avoidance tactics, but were also used as Elint platforms. Undoubtedly, the star of 408 Escuadrón is the highly-modified Boeing 707, which has been refitted in Israel with advanced equipment for Elint/Sigint/Comint roles.

TRAINING

The Ejército del Aire academy, the Academia General del Aire, is located at San Javier air base in the province of Murcia, southeast Spain. Students begin the Elementary Flying Syllabus on the T-35 Tamíz trainer, followed by the Basic Flying Syllabus with the C.101 Aviojet in the fourth year. Those pupils unable to attain the necessary flying qualifications at this stage transfer to navigation training. The fifth year is spent in the Specialisation Schools – Ala 23 for fast jets, Grupo 74 for multi-engine conversion and Ala 78 for helicopters.

The Academia is divided into three groups: Grupo de Apoyo for infrastructure tasks; Grupo de Material for maintenance tasks; and Grupo de Estudios for academic purposes. The latter has a Department of Flight and Navigation Techniques which is in charge of the three flight schools – primary, basic and navigation.

The Escuela de Vuelo Elemental (primary) has an aircraft inventory of 36 T-35 Tamíz (40 ENAER Pilláns assembled in Spain by CASA). The pupils undertake a total of 48 flight hours, split in two phases, Primary Selective and Primary Advanced, plus 12 hours in the classroom.

The following year, they pass to Escuela de Vuelo Básica (basic) which has 40 C.101 Aviojets on strength (the Ejército del Aire received 88 of these jets). The pupils perform a total of 111 hours, split into four phases consisting of the transition stage, formation flying and IFR training, plus the fourth stage which comprises 50 classroom hours spent studying the aircraft and its systems, and 34 simulator hours.

Ala 23, based at Talavera air base, near Badajoz, is the fast-jet training school, operating the supersonic F-5B conversion trainer. By the end of 1996, all 22 twin-stickers had received an upgrade by CASA.

Ala 78 is in charge of basic/advanced helicopter training both for the air force and army. 782 Escuadrón is equipped with 14 Hughes H 269Cs (TH-55s) and, with this squadron, future helicopter pilots amass 50 flight hours. They then go to 781 Escuadrón, which is equipped with eight Sikorsky S-76Cs.

The last training unit is Grupo 42, which has two flying squadrons – 421 Escuadrón equipped with five Beechcraft B55 Barons, and 422 Escuadrón with 25 F33A Bonanzas. They are used as refresher trainers for transport pilots on staff jobs and in the communications and liaison roles. Some 30 Dornier Do 27/ CASA 127 liaison aircraft remain in service, being used at most air bases for liaison and hack duties.

The air force will undergo further modernisation in the new millennium with the introduction of the EF2000 Typhoon and updates to the sensors of the RF-4 and F/A-18 fleets. The Hornet will undergo a major mid-life upgrade between 1999 and 2002, receiving new weapons capability and some of the elements of the later APG-73 radar system. Also slated for an upgrade is the C-130H fleet, which will receive new avionics, a 'glass' cockpit and advanced ECM systems.

Flotilla de Aeronaves (Fleet Air Arm, Spanish Navy)

The modern Naval Air Force (Flotilla de Aeronaves) was formed in November 1954, but the original force was formed in 1917 and lasted until the start of the civil war. It is the aerial component of the Spanish Navy Fleet Air Arm (Arma Aérea de la Armada) which, with seven squadrons (escuadrillas), is based at Rota Naval Air Station, together with the carrier *Príncipe de Asturias* (R-11) and the five Spanish-built FFG frigates, forming Battle Group Alpha. The *Príncipe de Asturias* – laid down on 8 October 1979, launched on 22 May 1982 and commissioned on 30 May 1988 – is fitted with a 12° ski-jump for Harrier operations. The composition of the Air Group (Unidad Aérea Embarcada – UNAEMB) changes according to operational needs, but a typical UNAEMB composition could be five AB 212s, six SH-3Hs, two SH-3H(AEW)s and eight Harriers.

CARRIER ASSETS

Fleet defence is handled by 9ª Escuadrilla's Harrier II/II+s (locally called Matador IIs). Twelve AV-8Bs were delivered, of which nine remain in service today alongside eight AV-8B Plus. The squadron is in the process of having its AV-8Bs converted to the Plus

standard. The Harriers usually carry four AIM-9Ms (and from 1999 the AIM-120B AMRRAM), but in more offensive roles the type can carry laser-guided GBU-10/16 bombs and the highly effective GAU-12 25-mm gun.

The 5ª Escuadrilla has operated the SH-3D Sea King since being officially commissioned on 29 June 1966. Deliveries of SH-3Ds comprised three received during 1966, three in 1967, two in 1972, four in 1974 and six in 1981. Six have been lost in accidents, and three underwent conversion to AEW standard with the Thorn-EMI Searchwater radar. The ASW fleet was upgraded to SH-3H standards starting in the late 1980s. The main mission of 5ª Escuadrilla SH-3Hs is antisubmarine warfare and SAR, but they can also function as troop transports with the ASW gear removed.

By mid-1998 the 3ª Escuadrilla had 10 AB 212s on strength. The tasking assignments are varied, the most important being support for the Marines, logistic transport, SAR, light attack, electronic warfare and casualty evacuation.

SEAHAWKS AND CITATIONS

All six SH-60Bs that arrived at Rota on 5 December 1988 remain in service with 10ª Escuadrilla. The squadron's primary roles are antisubmarine and anti-surface warfare, plus SAR, medevac evacuation and logistics transport. The helicopters form a fully integrated weapons systems with the

'Santa María'-class FFG frigates known as LAMPS III. The squadron is due to receive six more SH-60Bs in late 1999/early 2000 to equip the new F100 frigates.

With three Cessna Citation IIs, 4ª Escuadrilla's missions are varied, including VIP transport, fleet logistic support, medical evacuations, photographic/IR reconnaissance, ship's radar calibrations/targets, plus the two weekly surveillance sorties to cover the Atlantic and Mediterranean approaches to the Gibraltar Straits.

TRAINING

All future pilots are trained in the United States with the US Navy, following a 10-hour orientation syllabus on 6ª Escuadrilla's Hughes 500. Activated on 11 April 1972, 6ª Escuadrilla received 14 Hughes (McDonnell Douglas) 500ASWs, of which 10 remain in service today. They are used for training, ship's radar calibration, liaison, reconnaissance, light transport and FAC with the Ferranti 306 laser designators, having abandoned the ASW task.

Future Harrier pilots undertake the complete fighter/strike course, which includes the primary syllabus with 75 hours on the T-34C, 110 hours in the T-2C (including carrier qualifications) and 110 hours in the TA-4J. Future helicopters pilots fly 110-130 hours in the T-34C, followed by 137 hours in the Bell TH-57B/Cs. Pilots for the SH-3s or SH-60s squadrons receive operational training with the US Navy's replacement air group squadrons.

Fitted with a hoist for its secondary SAR role, this 5ª Esc SH-3H displays the low-visibility markings in vogue with the Flotilla de Aeronaves.

The Príncipe de Asturias's strike assets comprise the Matador IIs of 9ª Escuadrilla. This example is an EAV-8B, due to be upgraded to Harrier II Plus standard.

Arma Aérea de la Armada Española

3ª Escuadrilla	AB 212ASW	Rota AB
4ª Escuadrilla	Cessna 550 Citation II	Rota AB
5ª Escuadrilla	SH-3H Sea King, SH-3 AEW Sea King	Rota AB
6ª Escuadrilla	Hughes 500	Rota AB
9ª Escuadrilla	EAV-8B/B+, TAV-8B	Rota AB
10ª Escuadrilla	S-70B Seahawk	Rota AB

Above left: The most numerous type in FAMET service is the UH-1H Iroquois, which equips the medium transport companies.

Above: Principal anti-armour helicopters are the BO 105ATHs operated by BHELA-1 from Almagro.

Above: The Guardia Civil's helicopters are distributed among five units located around the country. Though only operating helicopters (such as this BK 117A3) at present, the Guardia aspires to operate fixed-wing transports in the near future, probably CASA 212s.

Fuerzas Aeromoviles del Ejército de Tierra (FAMET – Spanish Army Aviation)

While Army flying commenced in 1958, it was not until July 1965 that the Aviación Ligera del Ejército de Tierra – ALET (Army Light Aviation Group) was formed to establish helicopter units and develop their doctrines and tactics. The ALET became the FAMET on 20 March 1973 and today, headquartered at Colmenar Viejo near Madrid, has a personnel strength of 2,277 and 170 helicopters. As part of the Army's Rapid Reaction Force, FAMET consists of one attack, four field (Maniobra) and one heavy battalion (Batallón), while Colmenar houses three support units, training and maintenance facilities.

MBB BO 105

The BO 105 is used in the anti-tank role armed with HOT missiles as the BO 105ATH (28 being received), as ground support helicopters armed with a belly-mounted Rheinmetall RH202 20-mm cannon as the BO 105GSH (18 helicopters) and as an unarmed light observation platform, the BO 105LOH (14 delivered). A total of 11 ex-German Army BO 105s and an ex-MBB BO 105P demonstrator was also acquired.

In service they operate with the FAMET's only attack unit, Batallón de Helicópteros de Ataque I (BHELA I - Attack Helicopter Battalion I) deployed at Almagro base. Current strength is 28 BO 105ATHs/LOHs and five BO 105GSHs. The battalion has three anti-tank companies each equipped with six BO 105ATHs and an LOH, while six BO 105LOHs are used by the Headquarters Company and four BO 105GSHs by an Armed Reconnaissance Section. BHELA-I will eventually re-equip with 30 purpose-built attack helicopters, either the AH-64D Longbow Apache or the Eurocopter Tiger.

BO 105GSHs are also used by the reconnaissance companies of the Batallón de Helicópteros de Maniobras (BHELMA – Field Helicopter Battalions).

TRANSPORT HELICOPTERS

Battlefield mobility is provided by UH-1Hs, AB 212s, AS 532 Super Puma/Cougars and Chinooks. Apart from its main roles of cargo and personnel transportation, the UH-1H has been employed for parachuting, medical evacuation and as a gunship. About 50 examples remain in use in two transport companies of BHELMA III, while BHELMA II and IV each have one company that operates the Huey. The second company of BHELMA VI is equipped with six AB 212s.

Recently, the second company of BHELMA II converted to the AS 532UL Cougar, 18 of which are due to be delivered. BHELMA II is scheduled to get nine while the rest will re-equip one company of BHELMA III. Fifteen earlier AS 532UC Super Pumas are operated by BHELMA IV. The battalion maintains three detachments at the moment, two in Africa at Ceuta and Melilla and one at Mostar, Bosnia, in support of the UN Stabilisation Force (SFOR).

FAMET's Chinooks are operated by two heavy transport companies reporting to Batallón de Helicópteros de Transporte V (BHELTRA V – Transport Helicopter Battalion). The battalion

currently operates five BV-414s, nine CH-47Ds and two CH-47Cs, but the CH-47Cs and BV-414s are in the process of being upgraded to D standard.

OTHER UNITS

FAMET's three other helicopter units are based at Colmenar. Batallón de Transporte (BATRANS) is a liaison unit which operates four UH-1Hs. The Centro de Enseñanza de las FAMET (CEFAMET) is the FAMET's training centre. After initial screening is done at the CEFAMET, prospective pilots undergo a course at the Spanish air force helicopter school, Ala 78, at Armilla, Granada on the Hughes TH-55. If successful, they then take CEFAMET's specialised course for future army fliers. CEFAMET operates eight OH-58As, five BO 105LOHs and eight UH-1Hs.

Colmenar is also home to the Servicio de Helicópteros (SHEL) with its three Super Pumas, one BO 105LOH and one UH-1H. It loans these aircraft to the centralised maintenance facility on the base (the Unidad de Mantenimiento y Apoyo), which is in charge of third-level maintenance of the whole fleet, and to the FAMET's HQ (Jefatura FAMET – JEFAMET) for command and control duties and VIP transportation.

Fuerzas Aeromoviles del Ejército de Tierra (FAMET)

Jefatura FAMET (JEFAMET)	HQ Colmenar Viejo
loans helicopters from SHEL	
Batallon de Helicópteros de Ataque I	**Almagro/Ciudad-Real**
HQ Compañia	BO 105LOH
1ª Compañia Helicópteros de Ataque	BO 105ATH/LOH
2ª Compañia Helicópteros de Ataque	BO 105ATH/LOH
3ª Compañia Helicópteros de Ataque	BO 105ATH/LOH
Sección Reconocimiento	BO 105GSH
Batallón de Helicópteros de Maniobra II	**Betera**
HQ Compañia	UH-1H, BO 105LOH
Seccion Reconocimiento	BO 105GSH
Compañia Transporte Medio	AS 532UL Cougar
Compañia Transporte Medio	UH-1H
Batallón de Helicópteros de Maniobra III	**Agoncillo/Logrono**
Compañia Transporte Medio	UH-1H
Compañia Transporte Medio	UH-1H
Sección Reconocimiento	BO 105GSH, BO 105LOH
Batallón de Helicópteros de Maniobra IV	**El Copero**
HQ Compañia	AS 532UC Super Puma
Compañia Transporte Medio	AS 532UC Super Puma
Compañia Transporte Medio	AS 532UC Super Puma
det Ceuta	AS 532UC Super Puma
det Melilla	AS 532UC Super Puma
det SFOR, Mostar	AS 532UC Super Puma
Batallón de Helicópteros de Transporte V	**Colmenar Viejo**
Compañia Transporte Pesado	CH-47C/D, Bv-414
Compañia Transporte Pesado	CH-47C/D, Bv-414
Batallón de Helicópteros de Maniobra VI	**Los Rodeos IAP**
Compañia Transporte Medio	UH-1H
Compañia Transporte Medio	AB 212
Batallón de Transporte	Colmenar Viejo
	UH-1H
Centro de Enseñanza de las FAMET	Colmenar Viejo
	OH-58A, BO-105LOH,
	UH-1H
Servicio de Helicopteros	Colmenar
	AS 532UC, BO 105LOH,
	UH-1H
Unidad de Mantenimiento y Apoyo	Colmenar
	borrows helicopters from SHEL

Guardia Civil (Civil Guard)

The Guardia Civil is a militarised police force, like the French Gendarmerie, under order of the Spanish Home Ministry. Among its several departments is the Servicio de Vuelo (Flying Service), which parents the Agrupación de Helicópteros (Helicopter Group) which has on strength 17 BO 105C/CB/CB4/CBS4 and eight BK 117A3s. The Agrupación de Helicópteros received its first helicopters early in 1973 and, since then, has continued to grow slowly but steadily. The HQ of the Helicopter Group is at Torrejón air base, with helicopter units as follows: Unidad de Helicópteros 1 (UHEL-1) at Los Rodeos, Tenerife Island; UHEL-21 at El Copero, Seville; UHEL-31 at Manises, Valencia; UHEL-41 at Monflorite, Huesca; and UHEL-51 at Agoncillo, Logroño. The HQ unit covers the Madrid area for operational roles. Each unit has one or two helicopters on detachment.

Sixty pilots are assigned to the Agrupación de Helicópteros, who fly about 7,000 hours per year in public order and humanitarian roles. To become a pilot, he or she must be a graduate of the Guardia Civil Academy, taking the basic course with the Ejército de Aire's Ala 78 (the helicopter school) at Armilla air base, Granada. After that, they are posted to a training unit at Torrejón for the BO 105 and BK 117 courses.

If the budget allows, the Servicio de Vuelo will buy equipment for at least two fixed-wing units. The department is also in charge of providing security at Spanish civilian airports and navigation aids sites.

Servicio de Vuelo

Agrupación de Helicópteros	**HQ Torrejón**
UHEL-11	Los Rodeos, Tenerife Island
UHEL-21	El Copero, Seville
UHEL-31	Manises, Valencia
UHEL-41	Monflorite, Huesca
UHEL-51	Agoncillo, Logroño
(UHEL – Unidad de Helicópteros – Helicopter Unit)	

SWEDEN

Capital: Stockholm
Population: 8.7 million
Land area: 449790 km²
(173,620 sq miles)
Major cities: Göteburg,
Malmö

Flygvapnet (Air Force)

The Swedish Air Force, Flygvapnet, was established as an independent force on 1 July 1926 by the amalgamation of Marinens Flygväsen (Naval Flying Service) and Arméns Flygkompani (Army Flying Company), which had received their first aircraft in 1911 and 1912, respectively. By 1936, when the Flottilj (Wing) designation was introduced, only five Flottiljer existed, including a flying school. However, during World War II the force was expanded considerably to ensure the country's neutrality, and in 1946 Flygvapnet organisation peaked with 21 active Flottiljer. Today only eight Flottiljer remain, comprising six active flying wings and two ground schools.

COMMAND STRUCTURE

Swedish security policy is based on military non-alignment and an adequate national defence, aimed at the country staying neutral in the event of war. In peacetime, Flygvapnet is manned by 6,600 military personnel (with 70,000 available in the event of wartime), at present under the command of Generallöjtnant Kent Harrskog. Flygvapnet headquarters are part of the joint army, navy and air force Högkvarteret (Headquarters) in Stockholm, which is the central body of command for the Swedish armed forces and reports directly to the government in the Riksdag (Swedish Parliament).

Operationally, Flygvapnet is divided into three regional air commands, comprising Norra Flygkommando (Northern Air Command) including F4 Östersund and F21 Luleå, Mellarsta Flygkommando (Middle Air Command) with F16 Uppsala and Södra Flygkommando (Southern Air Command) with F7 Såtenäs, F10 Ängelholm and F17 Ronneby. Each air command has one underground Stridledningsoch Luftbevakningscentral (fighter control and airspace surveillance centre).

The basic operational unit in Flygvapnet organisation is the Flottilj (wing) made up of two or three Jakt (fighter), Attack (strike) or Spanings (reconnaissance) Divisioner (squadrons). Each Divisionen consists of some 14-16 airframes and maintains an operational

strength of eight combat-ready aircraft, the remainder usually undergoing maintenance. Also assigned to the Flottiljer are a number of smaller units called Gruppar (groups), each one having a Sambandsflyggrupp (liaison group) for local communication and liaison flying as well as a Helikoptergrupp (helicopter group) providing regional search and rescue. In addition, some Flottiljer have groups specialised for Typinflygning (type conversion), Radarspaning (airborne early warning), Signalspaning (electronic intelligence), or transport.

VIGGEN FIGHTER BACKBONE

For many years the Saab 37 Viggen constituted the backbone of Swedish air defence, and today five of Flygvapnet's six front-line Flottiljer continue to fly this powerful combat aircraft.

The first batch of 180 Viggens was delivered to Flygvapnet from 1971-79 (comprising 108 AJ 37, 27 SF 37, 28 SH 37 and 17 Sk 37 variants). Afterwards, production switched to the JA 37 version, of which Flygvapnet received 149 between 1979 and 1990. The newer JA 37 is a second-generation Viggen optimised for air defence.

During 1993-97 some of the first-generation Viggens were given a new lease of life through the AJS 37 upgrade programme, comprising installation of a new mission computer and digital databus to provide the aircraft with integrated attack, fighter and reconnaissance capabilities. The programme eventually resulted in conversion of 48 AJS 37s (former AJ 37 with additional radar surveillance capability), 25 AJSH 37s (former SH 37 with additional ground attack capability) and 25 AJSF 37s (former SF 37 with additional fighter capability). Today, only 1 Divisionen of F10 and 1 Divisionen of F21 continue to fly the AJS/AJSF/AJSH 37 Viggens.

Top: Latest in a long range of Saab fighter to join the Swedish air force is the JAS 39 Gripen. This pair of 'G'-coded (F7) aircraft is armed with Rb 99s (AIM-120 AMRAAM) and Rb 74s (AIM-9L) missiles.

Above: Like nearly all modern two-seat conversion trainers, the JAS 39B is fully combat-capable. Originally the air force wanted 25 JAS 39Bs, but budget restraints limited the number to 14.

The second-generation JA 37 is operated by F4, F16, F17 and F21, each with two Divisioner. The type has been kept continually up-to-date, with the most recent upgrade programme being a mid-life update referred to as the JA 37 Mod D package, which includes new ANP 37 weapon interface and stores management computer enabling the use of Rb 99 (AIM-120 AMRAAM) 'fire-and-forget' missiles, updated PS46A radar, integration of a new tactical radio system, and new data transfer unit for mission planning and tactical data recording. The first Mod D upgraded JA 37 is expected to enter service in 1998.

DRAKEN BOWS OUT

The Draken remained in Flygvapnet front-line service with its last operater, 2 Divisionen of F10 at Ängelholm, which flew the J 35J version in the air defence role, until 12 December 1998. It is planned that F10 will convert to the new JAS 39 Gripen during 1999/2000.

NEW-GENERATION GRIPEN FORCE

The Saab JAS 39 Gripen is intended to replace both the Draken and Viggen in Flygvapnet service. It is a fourth-generation lightweight multi-role combat aircraft developed by Saab Military Aircraft and marketed jointly by Saab and British Aerospace. To date, Flygvapnet has a total of 204 Gripens on order, comprising 176 single-seaters and 28 fully combat-capable two-seaters. Flygvapnet has already announced that the third production batch will be to improved JAS 39C and JAS 39D standard, with the planned improvements also being retrofitted to earlier aircraft and thereby gradually converting the JAS 39A/B to JAS 39C/D as well.

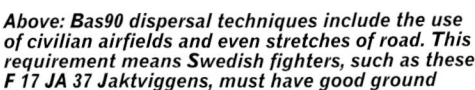

Above: Bas90 dispersal techniques include the use of civilian airfields and even stretches of road. This requirement means Swedish fighters, such as these F 17 JA 37 Jaktviggens, must have good ground handling.

Right: Since its entry into service in 1972, the Viggen has been the most important combat type in service, being used as a fighter, attack and reconnaissance platform.

The first Gripens were taken on charge by the TU JAS 39 (Taktisk Utprovning JAS 39; operational test and evaluation JAS 39) in 1994, tasked with developing mission tactics and operational guidelines as well as preparing a detailed programme for type conversion. Initially, TU JAS 39 was based at Malmslätt, but from 1998 TU JAS 39 transferred to F7 Såtenäs. 2 Divisionen of F7 at Såtenäs was declared combat-ready on the JAS 39 in November 1997, with 1 Divisionen expected to follow during 1998. F7 will provide centralised type conversion for all future Swedish Gripen pilots (as well as for the pilots of other countries that acquire Gripens).

When the JAS 39B entered Flygvapnet service in 1999 it is planned that Gripen conversion training will be split 70/30 between single- and two-seaters. Next to receive the JAS 39 Gripen will be two Divisioner of F10 Ängelholm in 1999/2000, followed by F16 with two Divisioner during 2000/01. Between 2001 and 2006 another six Divisioner will convert to the Gripen, giving a total of 12 JAS 39 Divisioner in Flygvapnet service.

DISPERSED OPERATIONS

Sweden relies on a concept of protection by dispersion called Bas90 (Base 90), continually developed since the 1960s. Experience from the Gulf War proved the viability of the Bas90 concept, as Iraqi aircraft were destroyed one by one inside their HASs by modern precision-bombing techniques. In the event of war, each Flottilj would disperse its aircraft into small groups operating from strips of highway scattered all over the Swedish countryside.

Their are only six main peacetime air bases, but 24 reserve war bases are available for dispersed operations. In addition to the Bas90 war bases, some 50 other sites are available, including civilian light aircraft runways.

ELECTRONIC WARFARE SYSTEMS

Electronic and signal intelligence (Elint/Sigint) is the responsibility of the Signal/radarspaningsflyg-grupp (radio/radar reconnaissance group) of F16M based at Malmslätt which, until recently, used two specially-equipped Tp 85 Sud Aviation Caravelle IIIs. These aircraft were retired in November 1998 and January 1999, the last aircraft going to La Caravelle Club who hope to maintain it in airworthy condition. They were replaced by a pair of S102B Korpen (Elint/Sigint-configured Gulfstream IV) with the individual names of *Hugin* and *Munin*.

The retirement of the Saab J 32E Lansen in the target-towing and electronic warfare role with the F16M's MFD (Målflygdivision – target flying squadron) at Malmslätt left no dedicated electronic warfare training unit. Plans are being implemented to convert 10 two-seat Viggens to EW training plat-forms as Sk 37Es, and they are due to enter service with F4 at Östersund during 1999. The Flygvapnet is also considering converting some Sk 60s (Saab 105) for target-towing and ECM training.

Fulfilling the airborne early warning and control role of the Signal/radarspaningsflyggrupp are six S 100B Argus (Saab 340B airliners configured for AEW&C) taken on charge during 1997/98. Flygvapnet plans call for the Signal/radarspaningsflyg-grupp to relocate its fleet from Malmslätt to Uppsala, forming a new unit there in 1999 with the transport-configured Tp 100A and Tp 102A currently based at Bromma. The S 100B AEW&C platforms are operating as an integrated part of Flygvapnet's StriL 90 system (Stridsledning och Luftbevakning; fighter control and airspace surveillance), which also includes a network of ground-based PS15, PS65, PS66, PS860 and PS870 radars, LOMOS visual observation posts (Luft och Markobservationssystem; air and ground observation system) and StriC command and control centres (Stridledningscentral; fighter control centre).

TRANSPORT, SUPPORT AND SAR

Heavylift transport is carried out by the Transportflyggrupp (transport group) of F7 at Såtenäs using eight Tp 84s (Lockheed C-130 Hercules). Two C-130Es (latter updated to C-130H standard) and six C-130Hs have been purchased.

Another Transportflyggrupp is assigned to F16 Uppsala, but is based at Bromma airport in Stockholm and due to move to Uppsala in 1999. Mostly tasked with VIP flying for military and civilian authorities, the unit is equipped with one Tp 100A (Saab 340B) delivered in 1989 and one Tp 102A (Gulfstream IV) delivered in 1992. Four-seat Sk 60Es on loan from the Flying School are also used for transporting passengers. One Tp 88C (Fairchild SA.227AC Metro III) delivered in 1987 was sold as SE-IVP in 1998.

Each of the operational Flottiljer is allocated one Sambandsflyggrupp (liaison and communication group) equipped with varying numbers of Sk 60s (Saab 105) and Sk 61s (Scottish Aviation Bulldog) on loan from the Flying School. The Sambandsflyg-gruppar of F7, F17 and F21 also have a Tp 101 (Beech 200 King Air) on charge. A single Cessna 550 Citation II (Tp 103) was leased from late 1998 for an unspecified time for evaluation as a potential replacement for the Tp 101s.

For regional search and rescue, the Flygvapnet operated a fleet of seven Hkp 3Bs (Agusta-Bell 204B) received in 1962 and 12 Hkp 10s (Aérospatiale AS 332M1 Super Puma). With the creation of the Försvarsmaktens Helikopterflottilj (Swedish Armed Forces Helicopter Wing) on 1 January 1998, control of the helicopters was devolved to the new organisation. From its headquarters at Malmslätt the Försvarsmak-tens Helikopterflottilj is responsible for operational command and control of Flygvapnet, Armén and Marinen helicopter resources, as well as for co-ordinating joint maintenance and acquisition of helicopters and training of personnel. Before 1 January

Europe

Above: Named the Korpen (Raven) in Swedish service, the S 102B Gulfstream GIV-SPs replaced the secretive Elint/Sigint Tp 85 Caravelles from 1998.

Above right: For over 30 years the Hercules has been operated by the Swedish air force. The eight aircraft are all based at Såtenäs in the Transportflygdivision.

Flygvapnet

FlygKommando Syd (FKS)	
F7 Skaraborgs Flygflottilj	**Såtenäs**
1.JAS-division	JAS 39A, JAS 39B
2.JAS-division	JAS 39A, JAS 39B
Transportflygdivision	Tp 84
Sambandsflyggrupp	Sk 60, Sk 61, Tp 101
F10 Skånska Flygflottiljen	**Ängelholm-Barkåkra**
1.Attackdivision/Spaningsdivision	AJS 37, SH 37, AJSF 37
2.Jaktflygdivision	JAS 39A
Grundläggande Flygutbildning	Sk 60, Sk 61
F17 Blekinge Flygflottilj	**Ronneby-Kallinge**
1.Jaktflygdivision	JA 37
2.Jaktflygdivision	JA 37
Sambandsflyggrupp	Sk 60, Sk 61, Tp 101, Tp 103
FlygKommando Mitt (FKM)	
F16 Upplands Flygflottilj	**Uppsala**
2.Jaktflygdivision	JA 37
3.Jaktflygdivision	JA 37
GrundläggandeTaktisk Utbildning	Sk 60
Sambandsflyggrupp	Sk 60, Sk 61
F16M	**Malmslätt**
Målflygdivision Flygenhet	Tp 100, S 100B, Tp 102, S 102B
FlygKommando Norr (FKN)	
F4 Jämtlands Flygflottilj	**Östersund-Frösön**
1.Jaktflygdivision	JA 37
Typinflygningsdivision	JA 37, Sk 37
Sambandsflyggrupp	Sk 60, Sk 61
F21 Norrbottens Flygflottilj	**Luleå-Kallax**
1.Spaningsflygdivision	AJS 37, SH 37, AJSF 37
2.Jaktflygdivision	JA 37
3.Jaktflygdivision	JA 37
Sambandsflyggrupp	Sk 60, Sk 61, Tp 101
Försvarets Materielverk (FMV)	**Malmslätt (Linköping-Malmen)**
Försökscentralen (FC)	J 32B, J 35F, AJ 37, JA 37, Sk 37, SH 37, Sk 60, Sk 61, Tp 86
Flygvapnets Bomb- och Skjutskola	Vidsel
	J 32D, Hkp 3, Hkp 6C

each of the operational Flottiljer had one Helikoptergrupp (helicopter group) assigned, equipped with a pair of SAR helicopters. Hkp 3Bs were based with the Helikoptergrupp of F10 and F16 while Hkp 10s equipped the F4, F7, F17 and F21 Helikoptergruppar. In addition, F17 maintained a SAR detachment at Visby on the island of Gotland to cover the Baltic area, and therefore the F17 Helikoptergrupp had a larger complement than the normal two helicopters. The obsolete Hkp 3B is planned for retirement during 1999. Tasking of SAR missions is done by the combined civil/military Air Rescue Co-ordination Centre located at Arlanda. The new organisational structure is not planned to cause any changes in the distribution of Flygvapnet's six Helikoptergruppar.

FAST-JET TRAINING

The Krigsflygskolan (Central Flying School) relocated to F10 Ängelholm from F5 Ljungbyhed in 1997. Today Flygvapnet pilot candidates go through an all-jet primary training syllabus with Krigsflygskolan, using a fleet of Sk 60 (Saab 105) twin-engined

Sweden's airborne early warning and control aircraft, the S 100B Argus, is a successful mating of the Saab 340B and Ericsson PS-890 Erieye side-looking airborne reconnaissance radar.

jet trainers. Originally 150 Sk 60 were delivered to Flygvapnet in 1966/68, and during the years only 10 have been written off in crashes. The type has a secondary light ground attack capability and 5 Lättattackdivision (light attack squadron) of F16 at Uppsala is tasked with this role, using aircraft from the Krigsflygskolan Sk 60 pool. The Sk 60 fleet has recently gone through major upgrade programmes to make the type serviceable until 2015. During 1988/91, structural overhauls (including rewinging) were carried out, and during 1995/98 the old RM9B jet engines were replaced by new RM15 (Williams/Rolls-Royce FJ44) turbofans which give better performance and are less noisy. However, only 105 Sk 60A/B/Cs will be modified.

Krigsflygskolan also operates a large fleet of Sk 61 Bulldogs which, until 1987, were used for early phases of the basic flying training course. Today, all Flygvapnet pilot candidates start flying jets on the Sk 60, so the Sk 61s only train civil pilot candidates of the Trafikflygarhögskolan (commercial pilot school) located at Ljungbyhed.

TEST AND DEVELOPMENT

Testing, supply, maintenance and administration of all Swedish defence equipment (for the air force, army and navy) is the responsibility of FMV (Försvarets Materielverk; defence materiel administration), which is an authority directly subordinated to the Swedish government. FMV has approximately 2,700 employees, who at present are involved in the following main projects for Flygvapnet: System JAS 39, S 100B, S 102B, Sk 60 re-engining, Bas90, StriC and TARAS. Test and evaluation is undertaken by a section of FMV called FMV:Prov (FMV:Test), with airborne activities taking place at its two primary aviation test centres: ProvFC at Malmslätt and ProvRFN at Vidsel.

FC (Försökscentralen; test centre) was originally formed in 1933 and had gone through many organisational changes until its incorporation within FMV:Prov in 1974. FC is tasked with test and evaluation of aviation-related equipment, including development of new aircraft systems, integration of new weaponry and sensor systems, command, control and information systems, mission planning and analysis

equipment, simulators and verification of tactical system functions. FC operates a large fleet of different aircraft and helicopters, including most types presently in Swedish military service. Some 25/30 aircraft and helicopters are permanently assigned to FC at Malmslätt, but in addition extra equipment is loaned from Flygvapnet units when needed.

RFN (Robotförsöksplats Norrland; missile test centre Northern Sweden) was formed in the late 1950s as a missile test-firing range north of Vidsel close to the Arctic Circle, and since 1974 has been part of the FMV:Prov organisation. Today, RFN consists of the Vidsel air base with a 2300-m (7,545-ft) runway, and the nearby 1650-km² (637-sq mile) test range (the largest land-based test range in Western Europe) equipped with sophisticated observation and measuring equipment. From Vidsel, RFN operates a few civil-registered Saab J 32 Lansens for target towing over the range as well as a couple of helicopters for transport and surveillance. In addition, operational Flygvapnet fighter units often deploy to Vidsel to practise missile firing on the range or cold weather training during winter.

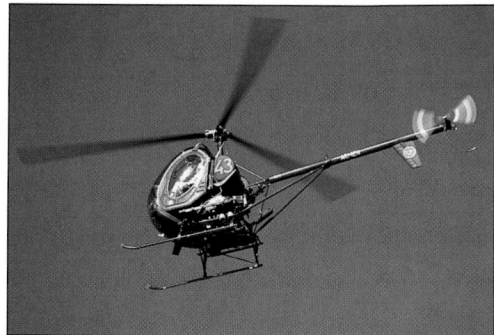

Helicopter training for all branches of the Swedish armed forces is undertaken on the Hkp 5Bs based at Malmslätt as part of Division Malmen.

Marinflyget
(Navy Flying Service)

The Svenska Marinen (Swedish Navy) began operating aircraft in 1911 and on 1 January 1916 the Marinens Flygväsen (Naval Flying Service) became a separate organisation of the Marinen. In July 1926 Flygvapnet (the Air Force) was created as an independent force by fusing the aviation components of the Navy and Army. On 1 June 1957 naval flying activities were revived by creation of the present Marinflyget, initially equipped with a fleet of Hkp 1 (Boeing Vertol 44A) and Hkp 2 (Sud Aviation Alouette II) helicopters.

The largest rotor-winged craft in Marinflyget use is the Hkp 4 (Kawasaki/Vertol 107). A total of 14 Hkp 4B/C/Ds remains in service, all more or less updated to the same technical standard (despite the different sub-type designations).

The Hkp 6B (Agusta-Bell 206B Jet Ranger) is used for training, liaison, reconnaissance and limited ASW. Originally, 10 were purchased in 1970; one was subsequently lost, but was replaced during 1991.

Trials with a leased Tp 87 (Cessna 404 Titan) for patrol duties in 1983 led to the acquisition of a radar-equipped SH 89 (CASA 212M-200 Aviocar) delivered in 1985, together with two more for the Kystvakt (Coast Guard) which were allocated civil registrations. The naval SH 89 is based with 13 Hkp div at Kallinge where it is part of a special Ubåtsjaktflygplangrupp (submarine hunting fixed-wing group). Since 1990 the Marinflyget has leased five Tp 54s (Piper PA-31 Chieftain) from a private company in Göteborg (two were returned to their owner in 1995) for liaison, light transport and visual maritime surveillance. The aircraft are shared between the two Helikopterdivisioner, based at Säve and Kallinge.

The Marinflyget constitutes an effective force of 24 helicopters and four fixed-wing aircraft, operated by some 200 officers, 120 conscripts and 30 civilians.

Commanded by the Marinflygledningen headquartered within Högkvarteret in Stockholm, the organisation consisted, until January 1998, of three Helikopterdivisioner (helicopter squadrons): 11 Hkp div at Berga near Stockholm, 12 Hkp div at Säve near Göteborg and 13 Hkp div at Kallinge near Ronneby.

The formation of the Försvarsmaktens Helikopterflottilj (Armed Forces Helicopter Wing) on 1 January 1998 resulted in all the Marinflyget helicopters and the CASA 212 being put under the control of Division Syd of the new wing. 11 Helikopterdivision at Berga was disbanded before the creation of the new organisation and its Hkp 4B/Cs and Hkp 6Bs were redistributed.

Above left: In service with the navy since 1963 for anti-submarine-warfare, the Hkp 4s were used during 1998 for SAR at air force bases.

Left: AF1 operates the Agusta-Bell AB 412HP (as the Hkp 11) in the 2nd Helicopter Transport Company for medical flights and transportation.

Arméflyget
(Army Flying Service)

Aviation activities within the Svenska Armén (Swedish Army) began in the late 1890s using manned balloons for observation, followed in 1912 by the first light aircraft being acquired for the army's Flygavdelning (flying section). On 1 January 1916 Arméns Flygkompani (Army Flying Company) was formed as a separate organisation within the army, only to be disbanded 10 years later by the July 1926 creation of the Flygvapnet (air force), which operated all Swedish military flying components. However, on 23 April 1954 flying activities were resumed by the Swedish army, giving birth to the Arméflyget organisation extant until the creation of the Försvarsmaktens Helikopterflottilj (Swedish Armed Forces Helicopter Wing).

The Arméflyget force (before the establishment of the new helicopter wing) totalled 86 helicopters of five different types operated by some 300 officers and 150 conscripts, rising to 1,300 personnel upon mobilisation. Arméflyget peacetime organisation consisted of two Arméflygbataljoner (army flying battalions): AF1 at Boden near Luleå (radio callsign ZÄTA) and AF2 at Malmslätt (radio callsigns KALLE and WILHELM).

Each Arméflygbataljon is divided into Pansarvärnshelikopterkompanier (anti-tank helicopter companies), Transporthelikopterkompanier (transport helicopter companies) or Flygskolar (flying schools). In wartime, Arméflyget would split into a number of smaller Plutoner (platoons) and deploy into the field, operating in support of army land forces. Arméflyget is tasked with such roles as anti-armour, transport of personnel and equipment, reconnaissance and casevac.

The Hkp 3C (Agusta-Bell 204B Iroquois) is the oldest helicopter in Arméflyget use and is due to be withdrawn by 2001. Replacement will be Agusta-Bell

Försvarsmaktens Helikopterflottilj (Swedish Armed Forces – Helicopter Wing)

Created on the 1 January 1998, the Försvarsmaktens Helikopterflottilj (Swedish Armed Forces Helicopter Wing) controls the helicopters of the air force, navy and army. This has allowed greater flexibility in their use and saving in areas such as maintenance and training.

Organised along geographical lines, the wing provides SAR cover for the air force's air bases, stationing two Hkp 10s at each base. The grounding of the type in mid-1999 while engine problems were cured meant that (naval) Hkp 4s could be drafted in as replacements.

The navy's contribution also includes the fixed-wing SH 89 based at Ronneby, for administration purposes.

If required, the Division Ost will use the assets of Division Malmen based at Malmslätt.

Equipped with the Saab-Emerson HeliTow System, the BO 105CD is the Swedish anti-tank helicopter.

412s (already in use as Hkp 11), Eurocopter AS 532 Cougars or Sikorsky UH-60 Black Hawks.

Twenty-six Hkp 5Bs (Hughes/Schweizer 300C) are based with AF2 at Malmslätt, and in peacetime they are primarily used for basic helicopter training, with pilot candidates flying 85 hours on the Hkp 5 during a four-month GHUA course (Grundläggande Helikopter UtbildningArmén; basic helicopter training army). In wartime, the Hkp 5 fleet would set up seven Fördelningshelikopterplutoner (distribution helicopter platoons) for FAC and scout duty.

Mainly tasked with liaison and reconnaissance are 19 Hkp 6As (Agusta-Bell 206A Jet Ranger). All are based with AF1 at Boden. The type is also used for tactical helicopter training, and pilot candidates coming directly from GHUA receive 120 flying hours on the Hkp 6 during a six-month GTUA course (Grundläggande Taktisk UtbildningArmén; basic tactical training army).

During 1987/88 Arméflyget acquired 20 Hkp 9As (MBB BO 105CB) armed with Rb 55H (BGM-71C ITOW and BGM-71D TOW2) anti-tank missiles. The Hkp 9 fleet is distributed between two Pansarvärnshelikopterkompanier (anti-tank helicopter companies) allocated two AF1 and AF2, but tactically they operate in Plutoner (platoons) of five helicopters each.

The final helicopter to join the Arméflyget was the Hkp 11 (Agusta-Bell 412HP), originally acquired to undertake military as well as civil medevac duty in the sparsely populated and difficult-to-access regions of northern Sweden. In 1993 three Hkp 11s were leased from Agusta for one year, awaiting the 1994 delivery of the Arméflyget's own five Hkp 11s. All are based with AF1 at Boden, and a medevac detachment is maintained at the hospital in Lycksele.

The creation of the Swedish Armed Forces Helicopter Wing in January 1998 placed all the Arméflyget helicopter under its operational control.

Försvarsmaktens Helikopterflottilj

HQ Malmslätt

Division Syd		HQ: Ronneby-Kallinge
12 Helikopterdivision	Hkp 4B/C, Hkp 6B	Goteborg-Säve
13 Helikopterdivision	Hkp 4D, Hkp 6B, SH 89	Ronneby-Kallinge
Flygräddningsgrupp F17	Hkp 10	Ronneby-Kallinge
Flygräddningsgrupp F7	Hkp 10	Sätenäs
Flygräddningsgrupp F10	Hkp 10	Ängelholm-Barkåkra

Division Malmen		HQ: Malmslätt
AF2 Östgöta Arméflygbataljon		**Malmslätt**
Flyskola 1	Hkp 5B	Malmslätt
Flyskola 2	Hkp 3C, Hkp 5B	Malmslätt
3. Pansarvärnshelikopterkompagni		Malmslätt
	Hkp 9A	

Division Ost		HQ: Berga
11 Helikopterdivision	Hkp 4, Hkp 6	Berga
Flygräddningsgrupp F16	Hkp 10	Uppsala

Division Norr		HQ: Boden
Flygräddningsgrupp F4	Hkp 10	Östersund - Frösön
Flygräddningsgrupp F21	Hkp 10	Luleå-Kallax
AF1 Norrbottens Arméflygbataljon		Boden
1. Transporthelikopterkompani		Boden
	Hkp 3C, Hkp 6A	
2. Transporthelikopterkompani		Boden
	Hkp 3C, Hkp 6A, Hkp 11	
3. Pansarvärnshelikopterkompagni	Hkp 9A	Boden

SWITZERLAND

Capital: Berne
Population: 6.8 million
Land area: 41285 km² (15,935 sq miles)
Major cities: Basel, Geneva, Lausanne, Zürich

Schweizer Flieger und Fliegerabwehrtruppen (Swiss Air Force)

In July 1914 the Fliegertruppe was established, to be renamed as the Schweizerische Flugwaffe in October 1936. The name Schweizer Flieger und Fliegerabwehrtruppen, or Swiss Air Force, was adopted in the 1950s. After World War II the Flugwaffe comprised more than 500 operational aircraft, with the most important fighters being the Messerschmitt Bf 109 and Morane-Saulnier MS.406, plus American aircraft like the P-51D Mustang and T-6 Harvard. In 1946 the first jets arrived in Switzerland when a small number of de Havilland Vampires were evaluated, followed three years later

Above: Winner of the Neue Jagdflugzeug competition in 1988, the Swiss Hornets are committed to the air defence task and are thus designated as F-18C/Ds (instead of F/A-18). They are fitted with the F404-GE-402 engines and AN/APG-73 radar and are capable of firing AIM-120 AAMs. This F-18D is the first example to be assembled by SF at Emmen, Switzerland.

by 25 Vampires T.Mk 55s. In December 1990 the last Vampires, which were used in the target-towing role, were retired. During 1953-55 the Swiss AF received 150 Venom Mk 1s and 100 Mk 4s, which were later modified to FB.Mk 50 standard. In 1984 the last Venom FB.Mk 50s and FB.Mk 54s were withdrawn from service. From 1958 the Swiss Air Force received the Hawker Hunter, comprising 100 F.Mk 58s, 52 F.Mk 58As and eight T.Mk 68s which were flown until December 1994.

STRUCTURE

The Schweizer Flieger und Fliegerabwehrtruppen is the flying branch of the Swiss Army. During peacetime the Swiss Air Force is under command of the Uberwachungsgeschwader or Surveillance Wing with its headquarters at Dubendorf. This command is divided in five Brigades: Flugwaffenbrigade 31 for aircraft and equipment, Flugplatzbrigade 32 for airfield maintenance and airfield construction, Fliegerabwehrbrigade 33 for air defence systems, Informatikbrigade 34 for radar-intelligence and non-flying units, and Flieger und Flugabwehrbrigade 35 for maintenance and civilian personnel. Flugwaffenbrigade 31 comprises the active fighter Staffeln (squadrons) based at two air bases. The sub-division Fliegerregiment 4 commands support and training Staffeln on about 15 other air bases. Second-line units, especially the helicopter Staffeln, operate from

detachments on bases spread over Switzerland.

Most personnel of the Swiss Air Force are reservists (75 per cent) who become active during large bi-annual peacetime exercises. During crises the order of battle of the Swiss Air Force would change dramatically, with many sleeping squadrons activated and operated from several wartime bases spread over the country. In addition to the peacetime bases of Dübendorf and Payerne, fighters would then operate from reserve bases Alpnach, Ambri, Meiringen, Mollis, Sion, Stans/Buochs, Raron and Turtmann. Helicop-ters would deploy to bases such as Alpnach, Dübendorf, Interlaken, Raron and Ulrichen.

However, the Swiss wartime order of battle is not known precisely. The sleeping Staffeln regularly deploy to their wartime bases for a few weeks each year during the peacetime exercises, manned by Militia (reserve) personnel. Several wartime bases are hidden deep in mountain passes and use motorways as runway. Most have tunnels, called Kavernen, cut out of the mountains, able to shelter up to a whole squadron. Some Kavernen comprise an H-shaped tunnel system with one exit, and have a number of aircraft hanging on steel cables from the roof to optimise the use of space.

FIGHTERS

In January 1997 a new fighter aircraft was delivered to the Swiss Air Force. A total of 26 F-18C single-seat and eight F-18D dual-seat Hornets will enter Swiss

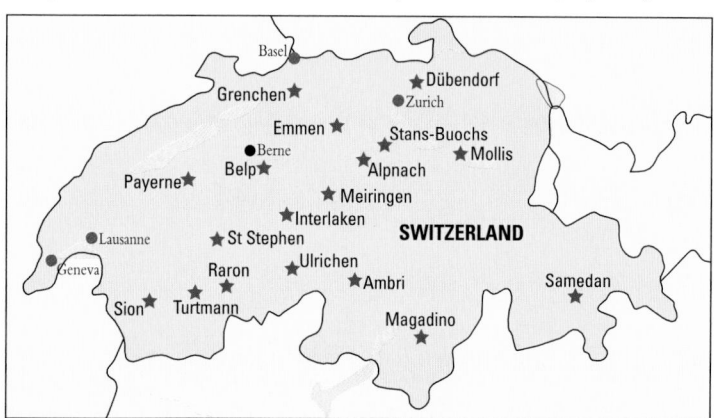

Above: Initially delivered in the air defence role armed with AIM-9Ps, the retirement of the Hunter saw some F-5Es being re-assigned to the ground attack role.

Below: The upgraded Mirage IIIS features a fixed canard and strake on each side of the nose, new ejector seats and improved avionics.

Schweizer Flieger Und Fliegerabwehrtruppen

Flugwaffenbrigade 31		
Uberwachungsgeschwader		
FlSt 1	F-5E	Dübendorf
FlSt 10	Mirage IIIRS	Dübendorf
FlSt 11	F-5E	Dübendorf
FlSt 16	Mirage IIIS	Payerne
FlSt 17	F-18C/D	Payerne
FlSt 18	F-5E	Payerne
14 IFlSt	PC-7	Dübendorf
Gruppe fur Rustungsdienste	all types	Dübendorf
Fliegerregiment 4		
Pilotenrekrutenschule	PC-7	Magadino
Pilotenschule 1	PC-7, Hawk	Sion
Pilotenschule 2	PC-7, Hawk	Emmen
Mirage Instr St	Mirage IIIDS	Payerne
Tiger Instr St	F-5F	Dübendorf
LFlSt 1	Alouette III	several bases
LFlSt 2	Alouette III	several bases
LFlSt 3	Alouette III	several bases
LFlSt 4	Alouette III	several bases
LTSt 5	Alouette III/Super Puma	Interlaken
LTSt 6	Alouette III/Super Puma	Alpnach
LTSt 7	Pilatus PC-6	Sarnen
LTSt 8	Alouette III/Super Puma	Ulrichen
ZFlSt 12	PC-9	Sion, Samedan
VIP Flight	Learjet 35A, Falcon 50	Dübendorf

Above: Swiss Vampire T. Mk 55s were replaced by the Hawk Mk 66 from 1990. Future Swiss Air Force pilots complete 115 hours on the type, after having undertaken 100 hours on the PC-7 (right).

service; the first two aircraft were assembled by McDonnell Douglas (Boeing) in St Louis, while the remainder will be assembled by the Schweizer Unternehmung fur Flugzeuge und Systeme (SF) in Emmen. On 31 October 1996 the first Swiss-built Hornet made its first flight. The SF delivery schedule is one Hornet per month with deliveries continuing until 1999. Swiss Hornets will have to stay in service for several decades, the main spars are made of a stronger titanium alloy to extend the aircraft's lifespan from 3,000 to 5,000 flight hours. As the Swiss Air Force will operate the F/A-18 mainly in the air defence role, so SF developed a low-drag pylon instead of the standard heavy SUU-63 pylons, which generate high drag. The first Hornets replaced the Mirage IIIS of FlSt 17 at Payerne, while 16 FlSt will convert to the Hornet during 1998-99.

With the delivery of the last Hornet, the Swiss Air Force will comprise 80 combat aircraft, a large reduction compared to the 300 in 1994. The Swiss Air Force Staff has already indicated a need for a similar batch to replace the ground-attack and reconnaissance Hunters retired in 1994.

To replace the Hunter fleet the first F-5E/F Tiger IIs were delivered in 1978. The first batch of 19 aircraft (13 F-5E and six F-5F) was delivered by a USAF C-5 Galaxy. The remaining 43 aircraft were locally assembled by SF at Emmen. A third batch of 32 F-5Es and six F-5Fs was procured in 1979 and, except for one F-5E, they were also built by FFA at Emmen. In total, the Swiss Air Force received 110

F-5 Tiger II fighters which now serve in three active units in the air defence role. FlSt 18 at Payerne was the first to receive the F-5, in 1978, followed one year later by 11 FlSt at Dübendorf. In 1982 the FlSt 1 at Payerne received the F-5. The 'Patrouille de Suisse' display team at Dübendorf started to fly the F-5E in place of the Hunter in 1995.

Wartime F-5 Tiger operators would be FlSt 6 to be based at Sion, FlSt 8 and FlSt 13 at Meiringen, FlSt 3 at Ambri, FlSt 4 at Raron and FlSt 19 at Mollis. The active F-5 units would also deploy to wartime bases: FlSt 1 would move to Turtmann, FlSt 11 to Alpnach and FlSt 18 F-5E to Payerne.

MIRAGE III

During 1966-69 the Swiss Air Force received 57 Mirage IIIs (36 Mirage IIIS fighters, 18 Mirage IIIRS for reconnaissance, two Mirage IIIBS trainers and one Mirage IIIC testbed). The Swiss Mirages were modified with a strengthened airframe, better brakes, hinging nosecone, adjustable nosewheel and a tailhook. After one Mirage IIIBS trainer was built by the Swiss from spares, another two were built in 1983. From 1983 the remaining Mirage IIIs were updated and received canards. After the update the Mirage IIIBS was designated Mirage IIIDS. As the Mirage IIIS is now being replaced by the F-18, 16 FlSt at Payerne will operate all Mirage IIIS fighters until it converts to the Hornet in 1998-99. In 1967 18 Mirage IIIRS arrived at FlSt 10 at Dübendorf for photo-reconnaissance. FlSt 10 has wartime deployments at Payerne, Sion and Stans/Buochs, while FlSt 16 will deploy to Stans/Buochs. The Gruppe fur Rustungsdienste or Equipment Group at Duubendorf operates several aircraft types.

TRAINING

From 1981 40 Pilatus PC-7 Turbo Trainers replaced the P-3 in the basic trainer role. The PC-7 is used by the Pilotenrekrutenschule at Magadino and by Pilotenschule 1 at Sion. The Instrument-fliegerstaffel 14 at Dubendorf loans PC-7s to develop the students' IFR skills. From 1991 19 BAe Hawk Mk 66s replaced the old Vampires in the Pilotenschule 2 at Emmen. The Hawk would be used in wartime as a fighter armed with AIM-9 Sidewinders and a belly-mounted 30-mm canon. Fighter conver-

sion takes place at the Mirage Instruction Staffel at Payerne and the Tiger Instruction Staffel at Dübendorf.

TRANSPORT

A dozen Pilatus PC-6 transport aircraft were purchased in 1965, augmented a decade later by PC-6-B2 Turbo Porters, and are all operated by Militia pilots in the sleeping Leichtfliegerstaffel 7 at Sarnen for paradropping. To support the Swiss Army, 84 Alouette IIIs were delivered from the early 1960s, of which about 70 remain, operated in the utility and SAR role by Leichtfliegerstaffeln 1, 2, 3 and 4 without a fixed base. Four Alouette IIIs and four AS 332M1 Super Pumas, of which 15 were delivered during 1987-93, form the inventory of the three Lufttransportstaffeln: LTSt 5 is at Interlaken, LTSt 6 at Alpnach and LTSt 8 at Ulrichen. The Swiss Air Force ordered another 12 AS 532UL Cougars in December 1998 to be delivered from mid-1999. Ten will be manufactured by SASC at Emmen. From 1989 13 PC-9s have been used by Zielfliegerstaffel 12 from Sion and Samedan as target-towing aircraft as a replacement for the C-3605 Schlepp. The secondary task of ZFlSt 12 is ECM training using the VISTA 5 pod. The VIP Flight at Dübendorf is equipped with one Learjet 35A and one Falcon 50. With its participation in 'shooting' deployments abroad, the Swiss Air Force has discovered a need for a small number of medium-range transport aircraft such as the C-130J or the US/Italian built C-27J.

Light tactical transport is undertaken by the Pilatus PC-6s of LTSt 7 based at Sarnen (above). The type's STOL performance is especially useful in the Alps. The Swiss Air Force VIP flight consists of a single Falcon 50 (left) and a Learjet 35A (below). The Falcon replaced a second Learjet in 1996.

TURKEY

Capital: Ankara
Population: 58.7 million
Land area: 779450 km²
(300,870 sq miles)
Major cities: Adana, Bursa,
Caziantep, Ismir, Istanbul

Türk Hava Kuvvetleri
(Turkish Air Force)

The Türk Hava Kuvvetleri (TuAF, Turkish Air Force) is one of Europe's oldest air forces – its roots go back as far as 1911 – and has operated 183 different aircraft types. Originally under command of the Army, the TuAF became independent on 31 January 1944, operating a mix of fighter types. Turkey joined NATO in 1952, becoming the most easterly NATO member. The TuAF entered the jet era with the T-33A in 1951, followed by large numbers of F-84G Thunderjets, F-86E Sabres, F-84F Thunderstreaks, F-100 Super Sabres, F-102 Delta Daggers and F-104 Starfighters. Not least because of the shared borders with the former Soviet Union, Syria and Iraq, Turkey's Air Force has always been substantial.

The TuAF has been divided in two Air Force Commands: 1nci Taktik Hava Kuvveti Komutanliği (1 THKK) or 1st Tactical Air Force with headquarters at Eskişehir, and 2 THKK with headquarters at Diyarbakir. 1 THKK controls the Turkish territory west of the 35th Meridian, roughly the virtual north-south line over Ankara, while 2 THKK controls eastern Turkey. Each THKK is divided in several Ana Jet Üs (AJÜ, Main Jet Base), each operating several Filos or squadrons. The headquarters at Eskişehir commands 1, 3, 6 and 9 AJÜ, while the headquarters at Diyarbakir commands 5, 7 and 8 AJÜ.

Above: This pair of Block 30 F-16Cs was delivered as part of Peace Onyx 1, the original F-16 programme. Seen armed with wingtip AIM-9P Sidewinders, the early F-16s are operated by 4 AJÜ.

Below: Displaying a small squadron badge depicting a leopard above the fin flash of its tail, this Block 40 F-16C belongs to 181 Filo 'Pars', part of 8 AJÜ based at Diyarbakir.

In addition to the main air bases, about 20 secondary bases are regularly used by the TuAF.

F-16 FIGHTING FALCON

In 1984 the modernisation of the TuAF was started with the Peace Onyx I programme, under which Turkish Aerospace Industries licence-produced 160 Block 30/40 F-16s (132 F-16Cs and 28 F-16Ds) to replace the huge F-104 Starfighter fleet. The first F-16 recipients were Akinci-based 141 and 142 Filos at 4 Ana Jet Üs, in 1988. 6 AJÜ at Bandirma followed in 1991. Since 1994 co-located 161 Filo has operated the LANTIRN system, while sister unit 162 Filo will be LANTIRN-equipped by 1999. The third F-16 wing became 9 AJÜ at Balikesir, commanding 191 and 192 Filo, receiving Block 40 F-16s during 1993-94. After the final Turkish F-104 flight by 182 Filo from Diyarbakir in September 1994, the conversion of the 8 AJÜ to the F-16 followed during 1994-95.

Peace Onyx II comprised the delivery of 80 TAI-built Block 50 F-16s (68 F-16Cs and 12 F-16Ds) which were used to replace the F-5 fleet of 5 AJÜ at Merzifon. In 1997 152 Filo received the first Turkish Block 50 F-16s, and sister unit 151 Filo was expected to be fully converted during 1998. The approximately 40 remaining Block 50 F-16s are thought to be used for the reactivation of 184 Filo at Diyarbakir and an F-16 training flight at Konya. With the delivery of the last Block 50 F-16 in late 1999, the TuAF will have received 240 F-16s.

F-4 PHANTOM

Six combat fighter squadrons operate the Phantom. The TuAF has received about 197 F-4Es and 40 RF-4Es in eight batches from 1973 to 1994. 1 AJÜ at Eskişehir operates the F-4E in 111 and 112 Filo in the FBA, AWI tasks, respectively, while 113 Filo is tasked with photo-reconnaissance with the RF-4E. To

Above: 7 AJÜ operates the F-4E Phantom II from Erhaç in both three-tone camouflage and overall grey.

Left: Resplendent in shark-mouth and eye markings, and armed with AIM-9Ps for the all-weather interception role, this F-4E is operated by 172 Filo.

Türk Hava Kuvvetleri

1nci Taktik Hava Kuvveti Komutanliği, HQ Eskişehir		
1THKK Irtibat Kita	CN.235, UH-1H	
1 AJÜ		**Eskişehir**
111 Filo 'Panter'	F-4E	
112 Filo 'Seytan'	F-4E	
113 Filo 'Isik'	RF-4E	
4 AJÜ		**Akinci**
141 Filo 'Kurt'	F-16C/D	
Öncel Filo	F-16C/D	Öncel
Irtibat Kita	UH-1H	
6 AJÜ		**Bandirma**
161 Filo 'Yarasa'	F-16C/D	(LANTIRN equipped)
162 Filo 'Zipkin'	F-16C/D	
Irtibat Kita	UH-1H	
9 AJÜ		**Balikesir**
191 Filo 'Kobra'	F-16C/D	
192 Filo 'Kaplan'	F-16C/D	
Irtibat Kita	UH-1H	
2nci Taktik Hava Kuvveti Komutanliği, HQ Diyarbakir		
2THKK Irtibat Kita	CN.235, UH-1H	
5 AJÜ		**Merzifon**
151 Filo 'Tunç'	F-16C/D	
152 Filo 'Akinci'	F-16C/D	
Irtibat Kita	UH-1H	
7 AJÜ		**Erhaç**
171 Filo 'Korsar'	F-4E	
172 Filo 'Sahin'	F-4E	
173 Filo 'Safak'	RF-4E	
Irtibat Kita	UH-1H	
8 AJÜ		**Diyarbakir**
181 Filo 'Pars'	F-16C/D	
182 Filo 'Atmaca'	F-16C/D	

Hava Eğitim Komutanliği, HQ Adnan Menderes		
HEK Irtibat Kita	CN.235, UH-1H	
Hava Harp Okulu		**Istanbul-Yesilky**
HvHO	T-41D	
Irtibat Kita	UH-1H	
2 AJÜ		**Izmir-Çigli**
121 Filo 'Ari'	T-38A	
122 Filo 'Akrep'	T-37B/C	
123 Filo 'Palaz'	SF.260D	Izmir-Kakli
124 Filo -	T-38A	
3 AJÜ		**Konya**
131 Filo 'Ejder'	F-4E	
132 Filo 'Hançer'	F-4E/F-5A	
133 Filo 'Pençe'	(N)F-5A/B	
134 Filo 'Türk Yildizlari'	NF-5A/B	
Irtibat Kita	UH-1H	
Hava Ulastirma Komutanliği, HQ Erkilet		
12 AU		**Kayseri-Erkilet**
221 Filo 'Esen'	C-160D	
222 Filo 'Alev'	C-130B/E	
12 UG		**Ankara-Etimesgut**
223 Filo 'Gezgin'	CN.235	
224 Özel Filo	Ce550, Ce650, G1159C, CN.235, UH-1H	
Tanker Filo	KC-135R	Incirlik

Right: Turkey operates a total of 40 SF 260Ds as primary trainers, which fly from Izmir-Kakli with 123 Filo 'Palaz' (Gosling), part of 2 AJÜ. The aircraft were co-produced and assembled locally.

Above: 133 Filo uses the F-5A Freedom Fighter as a lead-in trainer for fast-jet pilots. Aircraft have come from a variety of sources, including the Netherlands, Norway, Taiwan and the United States.

Right: Two squadrons of 2 AJÜ fly the T-38A Talon. This example belongs to 121 Filo.

Below: For years the Turkish Hercules fleet retained a natural silver finish, but has recently opted for a USAF-style camouflage, as worn on this C-130B.

augment the few surviving RF-4Es of the first batch of 1973, the TuAF received 32 operational former German Air Force RF-4Es. A similar operation is 7 AJÜ at Erhaç, where 171 Filo is tasked with FBA, 172 Filo has AWI-tasked F-4Es and 173 Filo is equipped with the RF-4E. The TuAF ordered the modernisation of 54 Phantoms by IAI in Israel and 1 HIBM at Eskişehir. The upgraded F-4E's first flight was undertaken in February 1999, with the first delivery following in February 2000.

TRAINING

The Hava Eğitim Komutanligi (HEK, Turkish Air Training Command) comprises eight squadrons and four air bases, with the HQ at Ankara. The Hava Harp Okulu (HvHO, Turkish Air Force Academy) at Istanbul-Yesilky uses gliders and Cessna T-41Ds for aptitude tests. After HvHO the student is transferred to 123 Filo at Izmir-Kakli, which has operated SIAI SF.260Ds since 1991. The next squadron will be Izmir-Çigli-based 122 Filo, which has operated the T-37B/C since 1964. The TuAF is pleased with the T-37 and hopes to acquire more when the JPATS enters service in the US early next century. 121 Filo is equipped with the T-38A Talon which augmented the T-33A in 1979. The T-33A was withdrawn from service in 1993 and in that year additional ex-USAF T-38As were received, bringing the total to 70. The

Above: Alongside various executive jets, 224 Özel Filo operates a fleet of TAI-built CASA CN.235s in a smart overall-white scheme.

124 Standardizasyon Egitim Filo at Izmir-Çigli loans aircraft for evaluation and research and development tasks. After Izmir, 131 Filo at Konya, with a large number of F-4Es, runs tactical training for the Phantom pilot. F-4 weapons and tactics is the task of 132 Filo, borrowing F-4Es from 131 Filo. Future F-16 pilots transfer to the Öncel Filo at Akinci. The F-5 was withdrawn from service at Merzifon in 1997, and today all F-5s are based at Konya, where 133 Filo operates a large mix of RF-5As, F-5A/Bs and NF-5A/Bs for combat training, while 132 Filo loans F-5s for tactical weapon training and instructor training. Also based at Konya is 134 Filo 'Türk Yildizlari' ('Turkish Stars') demonstration team operating NF-5A/Bs. The TuAF is modernising 48 NF-5A/Bs in an IAI/Elbit/Singapore Technologies Aerospace programme to become F-16 lead-in trainers at the 1 HIBM at Eskişehir. The F-5s will re-enter service from 2000 onwards.

TRANSPORT

The Hava Nakil Kuvveti Komutanliği (HNKK, Turkish Transport Command) comprises four squadrons. Since 1971, 221 Filo at Erkilet has operated 19 ex-German Air Force C.160Ds. 222 Filo has been equipped with seven C-130Es since 1964. In 1991 six ex-USAF C-130Bs were added to the Hercules fleet. To replace the Dakota from 1991, the TAI-built CN.235M entered service with 223 Filo at Etimesgut. VIP CN.235Ms belong to 224 Özel (Special) Filo which also operates two Citation IIs, two Citation VIIs and one Gulfstream IV for VIP transport. UH-1Hs are also operated by 224 Özel Filo for special operations missions. 7 Elektronik Filo at Etimesgut operated three intelligence-gathering ECM-47 Dakotas until the Dakota was withdrawn from service in 1998. The first TuAF KC-135R was

Above: Since 1972 the T-41D has been used for screening future pilots. Thirty-two are still used.

Most Turkish air force bases are equipped with a couple of UH-1Hs for local SAR and general transport duties. In service since 1969, around 45 helicopters of this type are employed.

officially handed over in October 1997; they belong to the Tanker Filo temporarily based at Incirlik, which will eventually operate seven Stratotankers. For liaison and SAR most bases are left with a UH-1H flight for liaison, as the T-33As were withdrawn from service in 1997.

FUTURE PROCUREMENT

To replace the UH-1H, 20 TAI-built combat SAR AS 532 Cougars will be delivered from 1999. Another 10 will serve in the transport role. Turkey has a requirement for four AEW aircraft. The Boeing 707-based Elta Phalcon and Northrop Grumman E-2C have been evaluated, other competitors being the Boeing E-3 AWACS and Lockheed Martin C-130J. As a reaction to the Hellenic Air Force F-15E request, the TuAF in turn requested the delivery of 20 to 40 F-15Es in early 1998.

Türk Kara Kuvvetleri

The flying branch of the Turkish Army was founded as a flying branch of the Artillery School at Polati in 1948. The first aircraft were 15 ex-US Army L-4J Cubs followed by 149 L-18B Super Cubs. Later, Cessna O-1E Bird Dogs and Dornier Do 27s were received, with the first helicopter (the AB 204) arriving in 1966. Currently, the flying branch of the TKK comprises four Hava Alayi (Air Regiments) and the Türk Kara Havacilik Okulu (TKHO, Turkish Army Aviation School).

TÜRK KARA HAVACILIK OKULU

The TKHO at Ankara-Güvercinlik operates all aircraft types in the TKK, about 180 helicopters and fixed-wing aircraft in total. Beside Army pilots, the pilots of the Navy, Jandarma and Polis are trained at Güvercinlik. Fixed-wing basic training is performed with 30 Bellanca 7GCBC Citabrias and 25 Cessna T-41D Mescaleros, while rotary-wing pilots train on 25 Hughes H 269Cs and 10 Robinson R-22Bs. The latter helicopters were bought in 1992 to replace the TH-13T trainer, which was struck off charge in 1991. In addition, 20 AB 206 trainers were delivered in 1996.

Apart from the Türk Kara Kuvvetleri titles, no external markings are carried on this S-70A-28 Black Hawk, making the helicopter look very much like a standard US Army example.

The Jetrangers are also used for observation and liaison training to augment three Bell OH-58B Kiowas.

Other helicopter types include a dozen surviving AB 204Bs, but the main types are the UH-1H and AB 205A, of which more than 25 are operated by the TKHO. In 1993 the Hueys were augmented by four S-70A Black Hawks, of which the TKK has ordered 95 to eventually replace all Hueys. In 1996 the Black Hawk fleet was joined by 20 Eurocopter AS 532UL Cougars, of which a few serve in the TKHO. Another 10 Cougars will be assembled by TAI during 1999-2002.

For fixed-wing observation training the TKHO uses 25 Cessna U-17B Skywagons. For VIP transport three Cessna 421B Golden Eagles and five Beech King Air 200s, two AB 212s and a single AB 206L are operated. Four Beech T-42A Cochises are used for VIP transport, twin-engine and instrument training. The Dornier Do 28D Skyservants were withdrawn from service in 1995. Initially, the TKK had hoped to replace them with CN.235Ms but this plan was abandoned in 1996.

TAARUZ HELIKOPTERI TABURU

Since 1990 the TKK has operated a Taaruz Helikopteri Taburu (Attack Helicopter Battalion) from Ankara-Güvercinlik. Ten ex-US Marine Corps AH-1W Super Cobras were augmented by 30 ex-US Army AH-1P/Ss, including four TAH-1P trainers, during 1993-95. The Cobras regularly deploy to eastern Turkey to combat PKK terrorists. Recently, the TKK has upgraded the AH-1P/S fleet with a night targeting system, RWR, GPS kit and a three-barrelled 20-mm gun. Though Turkey has a desire for more attack helicopters, the US was unwilling to provide them until recently.

HAVA ALAYI

The four Hava Alayi are geographically spread over Turkey: the first Hava Alayi at Samandra in the west, the second at Malatya in the southeast, the third at Erzincan in the northeast, and the Ege Hava Alayi in the southwest and the Turkish Islands in the Aegean. Each Hava Alayi comprises two or three Hava Taburu (Air Battalions) of 10 to 30 helicopters and aircraft. OH-58Bs, AB 206Rs and U-17Bs are used for observation, while UH-1Hs and AB 205As take care of troop transport. In the third Hava Alayi, the Hueys have already been replaced by the Black Hawks, while the Ege Hava Alayi is re-equipping with the AS 532UL Cougar.

FUTURE PROCUREMENT

The TKK requires a new attack helicopter, candidates being the Kamov/IAI Ka-50 'Hokum-A', Boeing AH-64D Longbow Apache, Bell Textron King Cobra and Eurocopter Tiger. One condition for selection is that TAI must co-produce 50 to 145 of the chosen helicopter.

Twenty heavy transport helicopters are also needed, and the TKK has shown interest in the Boeing-Vertol CH-47D and the Sikorsky CH-53E Super Stallion, with the Mil Mi-26 'Halo' being viewed as an outside contender. If selected, 10 CH-53Es, updated to the Israeli Yasur 2000 standard by IAI, would be required for use by the special forces.

Above: The most advanced combat helicopters in the Turkish army fleet are the AH-1W SuperCobras of the Helicopter Attack Battalion.

Above: The most numerous fixed-wing aircraft in army service is the Cessna U-17B Skywagon.

Three types of Huey are used by the Turkish army, but the most numerous is the UH-1H. This example is finished in an unusual tactical camouflage scheme.

Türk Kara Kuvvetleri

Türk Kara Havacilik Okulu	Güvercinlik
Basic Training Battalion	T-41D, Citabria, R-22B, H-269C, AB 206R/A
Observation and Liaison Training Battalion	U-17B, OH-58B, AB 204B, UH-1H, AB 205A, S-70A
Transport and Miscellaneous Tasks Battalion	T-42A, Beech 200, Cessna 421B, AB 206L, AB 212
Taaruz Helikopteri Taburu	AH-1W, AH-1P	Güvercinlik
1 Hava Alayi	**Samandra**
2 or 3 Hava Taburu	OH-58B, AB 206R, U-17B, UH-1H/AB 205A
2 Hava Alayi	**Malatya**
2 or 3 Hava Taburu	OH-58B, AB 206R, U-17B, UH-1H/AB 205A
3 Hava Alayi	**Erzincan**
2 or 3 Hava Taburu	OH-58B, AB 206R, U-17B, S-70A, UH-1H/AB 205A
Ege Hava Alayi	**Izmir**
2 or 3 Hava Taburu	OH-58B, AB 206R, U-17B, UH-1H/AB 205A, AS 532UL

Three Agusta AB 204ASs (left) are used in the utility role, having been replaced in the shipborne anti-submarine role by the AB 212ASW (above).

Polis (Police)

The Türk Polis has its own helicopter branch, using four types of helicopter for general police tasks. Four AS 330L Pumas are used for VIP transport, as are two S-70A Black Hawks. Four more S-70As are employed in the utility role. Since 1993, training of the Polis pilots has been performed at Eskişehir-Anadolu airfield with two Alouette II helicopters; previously, the TKHO at Ankara-Güvercinlik undertook flight training for the Polis. The Alouette IIs at Anadolu are part of a batch of 18 ex-German army SA 318Cs and were originally delivered in 1984 to the TKK, which passed them to the Polis. The Polis also possesses two Hughes H 300 training helicopters. The main base is at Ankara-Golbasi with other hubs at Diyarbakir and Istanbul-Yesilky. A detachment of one or two helicopters is maintained at every major city.

Polis

Mainbase HQ Ankara-Glbasi	AS 330L, SA 318C, S-70A, H 300
Eskişehir-Anadolu	SA 318C
Istanbul-Yesilky	AS 330L, SA 318C, S-70A
Diyarbakir	AS 330L, SA 318C, S-70A

Türk Donama Havaciligi

The roots of the Türk Donama Havaciligi (TDH) go back to the Ottoman navy in 1914 with the arrival of the first aircraft, a Curtiss MF. In 1925 the TDH was incorporated in the Turkish air force and during the 1930s the fleet comprised one Junkers A-20W, two Rohrbach RO111A, one Rodras, 20 Savoia S-16bBis/MS and eight Savoia S-59s. Later, the fleet was supplemented by six Supermarine Southampton Mk IIs and six Walrus Mk IIs. In 1971 an independent TDH was re-established.

MARITIME PATROL AIRCRAFT

The 301 Deniz Hava Filosu (301 Naval Air Squadron) operated the S-2A/E Tracker from 1971 until 1993 when, after a succession of fatal crashes, the TDH ceased Tracker operations. The S-2Es were kept in flying condition for potential buyers until

1997, when 15 S-2Es were trucked on huge trailers to TAI at Akinci for modification as fire-fighters. This plan was cancelled in 1998. To stay current, 301 Filo pilots flew SF-260Ds of 123 Filo at Izmir-Kakli until the TDH received its own trainers in 1995, when seven SOCATA TB-20 Trinidads were delivered to the new Eğitim Filo. The Trinidads are now flown by the TDH, pending the delivery of nine CN.235MPAs Maritime Patrol Aircraft. The CN.235MPA was chosen because the P-3 Orion proved to be too expensive and the CN.235MPA can be built by TAI.

HELICOPTERS

In 1972-73 three AB 204AS helicopters were delivered to 351 Filo Deniz Helikopter Filosu, followed by a dozen AB 212ASWs in 1977-78. The first six AB 212ASWs were fitted with the SMA APS-705 search radar, while the remaining six were equipped with the Ferranti Sea Spray Mk 3 radar. The AB 212s took over the ASW task from the

AB 204ASs, which are now used for training, transport and SAR duties. In 1987-88 three AB 212EWs arrived at 351 Filo for electronic warfare duties. The AB 212s can be based on frigates of the MEKO and 'Knox' classes. During 1994 Turkey wished to purchase 14 ex-US Navy Kaman SH-2F Seasprites stored at AMARC. This deal apparently has been cancelled, for eight Sikorsky S-70B-28 Seahawks have been ordered, to be delivered from 1999.

Türk Donama Havaciligi

301 Deniz Hava Filosu	NAS Topel
no aircraft (CN.235MPA to be delivered)	
351 Deniz Helikopter Filosu	NAS Topel
AB 204AS, AB 212ASW, AB 212EW	
Eğitim Filo	NAS Topel
TB 20 Trinidad	

Türk Sahil Gvenlik

Since 1993 the Türk Sahil Gvenlik (TSG, Turkish Coast Guard) has operated maritime surveillance aircraft to track environmental polluters and smugglers. The current fleet comprises three AB 206Bs and one Maule MX 7. Soon the TSG fleet will be augmented by four Agusta-Bell 412 helicopters and three TAI-built CN.235MPAs.

Türk Sahil Gvenlik

Türk Sahil Gvenlik	NAS Topel
AB 206B, Maule MX 7	

Türk Jandarma Teskilati

In peacetime the Türk Jandarma Teskilati (Turkish Paramilitary Police) is under command of the Ministry of the Interior at Ankara. In wartime the Jandarma comes under direct command of the Turkish General Staff for national military operations. The Jandarma helicopter fleet has also been involved in the anti-PKK campaigns in eastern Turkey. Main bases are Ankara-Güvercinlik and Diyarbakir, plus a helicopter base at Van.

The flying branch was established in 1968 with the arrival of four AB 206Bs, followed a few years later by AB 204Bs, AB 205s, AB 212s, Piper Pa-32s, Cessna F182Ps and one Rockwell 690A Turbo Commander. In 1989 six S-70As were delivered to augment the ageing helicopter fleet, followed by an order for

another 30. The latest addition to the fleet was made in 1995 with 19 Mi-17V 'Hip-Hs', of which two are equipped as ambulances, while another three can be armed. All are new production models fitted with equipment meeting Western standards, but, apart from the two ambulance Mi-17s which are flown regularly, the fleet has a low operational record.

Türk Jandarma Teskilati

Jandarma Hava Grup Komutanlikari	Ankara-Güvercinlik
AB 204B, AB 205A, AB 206A, AB 212, Mi-17V, S-70A, Rockwell 690A Turbo Commander	
Jandarma Hava Grup Komutanlikari	Diyarbakir
AB 204B, AB 205A, AB 206A, AB 212, S-70A, Mi-17V	
Helikopter Filo	Van
AB 204B, AB 205A, AB 206A, AB 212, S-70A, Mi-17V	

UKRAINE

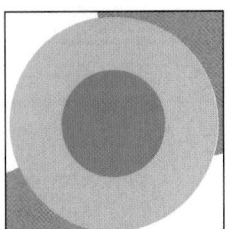

Capital: Kiev
Population: 51.8 million
Land area: 603700 km²
(233,030 sq miles)
Major cities:
Dnipropetrovsk, Donetsk,
Odessa, Kharkiv

Ukrainian Armed Forces (Zbroyni Sily Ukrainy)

Following the dissolution of the Soviet Union, Ukraine was declared an independent republic on 24 August 1991. The Ukrainian Ministry of Defence was formed in 1992 on the basis of the Kiev Military District (which had been disbanded in November 1991). The armed forces of Ukraine (Zbroyni Syly Ukrainy) comprise four branches: Land Forces (Sukhoputni Viys'ka, SV), Air Force (Viys'kovo-Povitryani Syly, VPS), Air Defence Forces (Syly Protypovitryanoyi Oborony, Syly PPO) and Navy (Voienno-Mors'kykh Syl, VMS.)

Ukraine claimed the assets of the former Soviet Armed Forces based on its territory, comprising three Frontal Aviation air armies (5, 14 and 17 VA), plus units subordinated to two Strategic Air Armies (24 and 46 VA), two Air Defence Armies (2 and 8 VA PVO), two Military Transport divisions, army aviation units subordinated to the armies of the three Military Districts, and part of the naval aviation assets of the Black Sea Fleet.

The Treaty on Conventional Armed Forces in Europe (CFE) limits Ukraine to operating 1,090 combat aircraft and 330 attack helicopters. In January 1998, the Ukrainian inventory numbered 966 combat aircraft and 290 attack helicopters.

Fixed-wing aircraft and helicopters are operated by all four branches, but it should be noted that the fighter assets of the Air Defence Forces have been

Above: Spearheading the bomber fleet are the 50 Tupolev Tu-22Ms in service with three units. This example, an ex-Russian navy Tu-22M3, belongs to 185 TBAP based at Poltava.

Below: Displaying the Ulan-Ude factory bear and Ukrainian naval flag on the nose, this Sukhoi Su-25UB combat-capable trainer is thought to be operated by 299 OSHAP at Saky.

Approximately 60 Sukhoi Su-27s are operated by the Ukrainian air force's 831 IAP from Myrhorod. The aircraft can carry the R-27 (AA-10 'Alamo') and the R-73 (AA-11 'Archer') AAMs.

transferred to the air force. Henceforth, the Air Defence Forces comprise missile and radar forces only; a single support squadron is subordinated to PPO headquarters at Kyiv (Kiev).

UKRAINIAN AIR FORCE

The Ukrainian Air Force (Viys'kovo-Povitrayani Syly) was created on 17 March 1992 and its command structure was formed on the basis of the 24th Air Army, headquartered in Vinnytsya (Vinnitsa). Since then, the assets of the Air Force and Fighter Aviation of the Air Defence Forces have been combined into a single command. A number of units have been designated Aviatsionnaya Baza (Air Base), whereby all previously existing separate supporting units on the base have been united with the flying unit under a single command. Several of these have since been given a new regimental number. Commander-in-Chief of the Ukrainian Air Force is Colonel General Volodymyr Antonyets.

FIGHTER, BOMBER AND ASSAULT

Vynyschuval'na Aviasiia (Fighter Aviation) until recently used the MiG-25PD and Su-15, but they have been withdrawn from service. The air defence mission is performed by the MiG-23, MiG-29 and the Su-27, which equip a total of nine fighter regiments.

The VPS's Bombarduval'na Aviatsiia (bomber aviation) operates a heavy bomber division with two heavy bomber regiments equipped with Tu-22M3s, and five tactical bomber regiments equipped with the Su-24 and Su-24M.

Protracted (and ultimately fruitless) discussions between Ukraine and the Russian Federation about the disposition of the 19 Tu-160s formerly operated by 184 Guards Heavy Bomber Regiment (GvTBAP) at Priluki and 23 Tu-95MSs of 1006 TBAP based at Uzin started in 1991. Scrapping of these aircraft commenced in November 1998 under a contract issued by the US government. In a surprise move, it was announced in March 1999 that three Tu-160s had been sold to a US company as satellite launchers.

Ukraine operates two independent ground-attack regiments, termed Shturmuval'na Aviatsiia (assault aviation) equipped with the Su-25.

RECONNAISSANCE AVIATION

Tactical and operational reconnaissance missions are flown by the Su-17M4R and Su-24MR of Rozviduval'na Aviatsiia (Reconnaissance Aviation). They equip two independent reconnaissance regiments (ORAP). One of these regiments also operates a small number of the Su-24MP dedicated ECM-variant. Long-range (strategic) reconnaissance missions are flown by a single independent long-range reconnaissance squadron (ODAE), which continues to operate several reconnaissance and electronic warfare (EW) variants of the Tu-22, and also several An-30s which were formerly operated by 86 ODRAE at Chernivtsi (Chernovtsy).

TRANSPORT AVIATION

In 1991, Ukraine inherited six military transport regiments with about 175 Il-76 Candids from the Soviet Voenno-Transportnaya Aviatsiia (Military Transport Aviation). Of these, three regiments remain, which are subordinated to a single transport division. Also subordinated to this division are the Il-78 tankers of 409 air refuelling regiment (APSZ), a

Advanced and weapons training is undertaken on the Aero L-39C, of which about 450 are in service. Like many Ukrainian aircraft, this example shows signs of having the Soviet Red Star on its tail painted over.

Viys'kovo-Povitryani Syly

Air Force (VPO)		Vinnytsya (Vinnitsa)
15 AvBP SN		**Boryspil (Borispol)**
1 OTAP	An-12, An-24 An-26, Il-76 Mi-6, Mi-8, Tu-134	Boryspil
10 OAE	An-30, An-26	Boryspil
456 GvOSAP	An-12, An-24, An-26, Il-22, Mi-22, Mi-8	Vinnytsya (Vinnitsa)
11 OAE	?	Vinnytsya (Vinnitsa)
13 TBAD		**Poltava**
18 ODRAE	An-30, Tu-22R/RD/RDK/UD/PD/KD	Nizhyn (Nezhin)
184 TBAP	Tu-22M3, Tu-134UBL	Pryluky (Priluki)
185 TBAP	Tu-22M3	Poltava
33 TsBP		**Mykolayiv (Nikolaev)**
6 AvB	Su-24, Tu-22M2	Mykolayiv (Nikolaev)
316 OPLAE	Be-12	Mykolayiv (Nikolaev)
8 AvB	L-39	Baherove (Bagerovo)
7 VTAD		**Melitopol'**
16 AvB	Il-76MD	Kryvyi Rih (Krivoy Rog)
25 GvVTAP	Il-76MD	Melitopol'
338 VTAP	Il-76MD	Zaporizhzhia (Zaporozhye)
409 APSZ	Il-78	Uzyn (Uzin)
5 AK		**Odesa (Odessa)**
2 OSAP	An-12, An-26, Mi-6/VKP Mi-8	Odesa (Odessa)
6 IAP	MiG-23MLD	Chervono glinske (Chervono glinskoye)
149 AvB	L-39	Kupians'k (Kupyansk)
208 OVE REB	Mi-8/PPA/SMV,	Buyalyk (Buyalik)
299 OSHAP	Su-25	Saky (Saki)
511 ORAP	Su-17, Su-24MR	Buyalyk (Buyalik)
32 BAD		**Starokostiantyniv (Starokonstantinov)**
7 BAP	Su-24M	Starokostiantyniv (Starokonstantinov)
44 BAP	Su-24	Kanatovo
138 IAD		**Myrhorod (Mirgorod)**
62 BAP	MiG-29, Su-27	Belbek
161 IAP	MiG-29	Lymansk'e (Limanskoye)
642 IAP	MiG-23, MiG-29	Martynovskaya
831 IAP	Su-27	Myrhorod (Mirgorod)
14 AK		**L'viv (L'vov)**
48 GvORAP	Su-17, Su-24MR/MP	Kolomyia (Kolomiya)
202 AvB	L-39	Uman'
209 OVE REB	Mi-8PPA/-SMV (29)	Luts'k
243 OSAP	An-12, An-24, An-26, Mi-8	L'viv (L'vov)
452 OSHAP	Su-25	Chortkiv (Chortkov)
6 IAD		**Ivano-Frankivs'k (Ivano-Frankovsk)**
8 IAP	MiG-29	Vasyl'kiv (Vasilkov)
9 IAP	MiG-23MLD	Ozerne (Ozernoye)
85IAP	MiG-29	Starokostiantyniv (Starokonstantinov)
114 IAP	MiG-29	Ivano-Frankivs'k (Ivano-Frankovsk)
289 BAD		**Luts'k**
69 BAP	Su-24	Cherlyani (aka Gorodok/Horodok)
806 BAP	Su-24	Luts'k
947 BAP	Su-24M	Dubno
Air Academy		**Kharkiv (Kharkov)**
31 UATs	Mi-2	Sumy (Sumi)
32 UATs	Mi-2	Bohodukhiv (Bogodukhov)
72 AvB	Mi-8, Mi-24	Konotop
201 AvB	L-39, Mi-8	Chernihiv (Chernigov)
203 AvB	L-39	Chuhuyiv (Chuguyev)
204 AvB	L-39	Okhtyrka (Akthyrka)

Air Defence Forces (SPPO)		Kyiv (Kiev)
223 OAE	An-24, An-26, Mi-8	Zhulyany

Main combat helicopters of the Ukrainian army are the 250 Mil Mi-24s in D/V/P ('Hind-D/E/F') versions. Bort 'Yellow 77' is an Mi-24P.

Army airlift is provided by 300 Mil Mi-8T/TVs. A limited number are used by the air force in the utility role.

number of which are often operated on cargo flights, with their inflight-refuelling equipment removed.

Three independent regiments (OSAP and OTAP) and two independent squadrons (OAE) operate a variety of short- and medium-range transport aircraft, aerial command posts and helicopters, and are directly subordinated to Air Force headquarters and the headquarters of the two air corps (L'viv and Odesa). The 'Blakitna Stezha' squadron (10 OAE) was formed after Ukraine decided in 1992 to participate in the Open Skies programme.

NAVAL AVIATION

At the time of the dissolution of the Soviet Union, Ukraine received three Be-12s, which were based at Mikolayiv (Kulbakino), the long-time base of the 33rd Combat Training Center of Soviet Naval Aviation (TsBP A-VMF). As a result of the division of Black Sea Fleet Naval Aviation assets between Ukraine and the Russian Federation, Ukraine received 18 Tu-22M2s and 20 Tu-22M3s, and a further 11 Be-12s . Of these, the Be-12s have been transferred to Air Force control at Mikolayiv; likewise, the Tu-22M2s and Tu-22M3s have also been transferred to the Air Force, and are now based at Mikolayiv and Poltava, respectively.

Ukraine retained 555 OPLVP (Independent Antisubmarine Helicopter Regiment), a long-time resident of Ochakiv. The 10 Naval Aviation Su-25UTGs were divided in equal numbers between Ukraine and Russia.

Kirovs'ke (Kirovskoye) air base is directly subordinated to the Ukrainian Ministry of Defence, and home to the International Fighter Pilots Academy (IFPA), which is officially designated 3rd Fighter Squadron. At this base a wide variety of naval and Air Force aircraft types are based (including An-12, An-72/74, Il-38, Ka-27, L-39, Mi-6, Mi-8, Mi-14, Mi-24, MiG-29, Su-25UTG, Su-27, Tu-142MZ and Yak-38).

ARMY AVIATION

In 1991, at least nine helicopter regiments and seven helicopter squadrons were based in the Carpathian, Kiev and Odessa military districts, subordinated to five armies. Following the restructuring of

Army Aviation, the helicopter units have been organised into seven consecutively numbered Army Aviation Brigades (BrAA), which are subordinated to the newly created Northern, Western and Southern Operational Commands and their respective armies.

TRAINING

Candidate pilots for both fixed-wing aircraft and helicopters of all branches of the Ukrainian Armed Forces receive their training at the Kharkov Pilot Institute, which in 1995 integrated the three former flying schools based in Kharkov (Chuguyev), Chernigov and Lughansk (the latter for navigators). Aviation engineers are trained at the Kiev Air Force Institute (KIVVS), which combines the two former schools in Kiev and Vasilkov. In addition, the 33rd Combat Training Centre at Nikolayev was transferred to the air force and operates a number of Su-24Ms, Tu-22M2s and Be-12s formerly belonging to Naval Aviation.

UKRAINIAN NATIONAL GUARD

The Ukrainian National Guard (Natsional noi Gvardii Ukrainy, NGU) was formed in October 1995. Its aviation assets comprise an independent helicopter brigade, which is made up of a regiment with two transport squadrons with Mi-6s and two combat-transport squadrons equipped with Mi-8s, and an independent helicopter squadron equipped with Mi-24s and Mi-8s.

TSOU

The former DOSAAF of Soviet times has been renamed TSOU (Tovaristvo Spriyanniya Oboroni Ukrayini, Society for the Support of the Defence of Ukraine), and comprises more than 30 aeroclubs, equipped with over 600 An-2s, Mi-2s, PZL-104s, Yak-52s and Yak-55s.

Above: About 30 Antonov An-26 medium transports fly with the Ukrainian air force.

Above right: Much of the Ukrainian Il-76 fleet operates in a quasi-civil scheme comprising the national colours, as displayed by this Il-76MD.

Armiys'Ka Aviatsiia

Army Aviation (AA)		Kyiv (Kiev)
ZOK		**L'viv (L'vov)**
7 BrAA	Mi-2, Mi-8, Mi-24, Mi-26	Kaliniv (Kalinov)
13 AK		**Rivne (Rovno)**
5 BrAA	Mi-8, Mi-24	Zhovtneve (Zhovtnevoye)
38 AK		**Ivano-Frankivs'k (Ivano-Frankovsk)**
3 BrAA	Mi-8, Mi-24	Brody (Brodi)
YuOK		**Odesa (Odessa)**
2 BrAA	Mi-6, Mi-8	Chornobaivka (Chernobayevka)
6 AK		**Dnipropetrovs'k (Dnepropetrovsk)**
6 BrAA	Mi-8, Mi-24	Raukhovka
SOK		**Chernigov (Chernihiv)**
8 AK		**Zhitomir (Zhytomyr)**
1 BrAA	Mi-8, Mi-24	Radjans'ke (Radyanskoye*) *aka Berdichev
4 BrAA	Mi-8, Mi-24	Vapniarka (Vapnyarka)

Aviatsiia Voienno-Morsk'kykh Syl

Naval Aviation (AVMS)		Kyiv (Kiev)
u/i OSAE	An-12, An-26, Mi-8	Saky (Saki)
555 OPLVP	Ka-25, Ka-27, Ka-29, Mi-14	Ochakiv (Ochakov)

Natsional noi vardii Ukrainy

National Guard (NGU)		
ovb NGU		
51 OVP	Mi-6, Mi-8	Oleksandriia (Aleksandriya East)
u/i OVE	Mi-8, Mi-24	Bila Tserkva (Belaya Tserkov')

Glossary:

AK	Army Corps
APSZ	Air Refuelling Aviation Regiment
AvB	Aviation Base
AvBP SN	Aviation Brigade for Special Purposes
AvBR	Aviation Base for Reduction
AvK	Aviation Corps
BAD	Bomber Aviation Division
BAP	Bomber Aviation Regiment
BrAA	Army Aviation Brigade
Gv	Guards
IAD	Fighter Aviation Division
IAP	Fighter Aviation Regiment
KhIL	Kharkov Institute for Pilots
MO	Ministry of Defence
NGU	Ukrainian National Guard
OAE	Independent Aviation Squadron
ODRAE	Independent Long Range Reconnaissance Aviation Sqn
OPLAE	Independent Anti-Submarine Aviation Squadron
OPLVP	Independent Anti-Submarine Helicopter Regiment
ORAP	Independent Reconnaissance Aviation Regiment
OSAE	Independent Composite Aviation Squadron
OSAP	Independent Composite Aviation Regiment
OShAP	Independent Attack Aviation Regiment
OVB	Independent Helicopter Brigade
OVE	Independent Helicopter Squadron
REB	Electronic Warfare
SOK	Northern Operational Command
SPPO	Air Defence Forces
TBAD	Heavy Bomber Aviation Division
TBAP	Heavy Bomber Aviation Regiment
TsBP	Combat Training Centre
TSOU	Society for the Support of the Defence of Ukraine
UATs	Aviation Training Centre
VTAD	Military Transport Aviation Division
VTAP	Military Transport Aviation Regiment
VMS	Naval Forces
VPS	Air Force
YuOK	Southern Operational Command
ZOK	Western Operational Command

UNITED KINGDOM

Capital: London
Population: 55.6 million
Land area: 244755 km² (94,475 sq miles)
Major cities: Belfast, Birmingham, Bradford, Bristol, Cardiff, Edinburgh, Glasgow, Leeds, Liverpool, Manchester, Sheffield

Royal Air Force

The amalgamation of the Royal Flying Corps and Royal Naval Air Service to form the Royal Air Force on 1 April 1918 gave the United Kingdom an air arm with a personnel strength in excess of 291,000 men, equipped with 22,000 aircraft. The cessation of hostilities later that year led to a rapid reduction in size, and in under 18 months only 25 squadrons and 31,500 officers and men remained. The majority of these were stationed overseas as an effective means of policing the remoter parts of the British Empire, and

air power was regularly used against local uprisings in the Middle East.

It was not until the recognition of the German threat in the mid-1930s that the lack of funds and development allocated to both aircraft and engine technology was rectified. The RAF Expansion Scheme enabled a doubling in the number of front-line squadrons within five years and a reorganisation which saw the inception of Fighter, Bomber, Coastal and Training Commands in 1936. Advances in the development of airframes and engines accompanied the introduction into service of such famous types as the Blenheim (faster than the 'modern' fighters then in service), Hurricane, Spitfire, Wellington and Sunderland, all of which were to have a profound influence on the conflict which was soon to follow.

With the outbreak of hostilities in 1939, the RAF was pitched into battle with an enemy which had swept all before it and had a fighter force three times its size. Nevertheless, the use of radar and the fact the RAF was fighting over its own territory enabled Fighter Command between July and October 1940 (the Battle of Britain) to prevent Germany achieving the condition Hitler had stipulated for the German invasion of the United Kingdom – i.e., the destruction of the RAF.

The RAF played a vital role in the Battle of the Atlantic against the U-boat menace, employing American-built B-24 Liberators supplied under Lend-Lease in the very-long-range patrol role. In North

Designed to intercept Soviet long-range bombers over the northern approaches rather than for close-in air combat, the Tornado F.Mk 3 has had to adapt after the end of the Cold War. This aircraft is from Leuchars-based No. 111 Squadron.

Africa, the RAF and Army developed close co-operation tactics which continued to play an important role throughout the Desert Campaign and the subsequent invasion of Italy. The defence of Malta – which, in the middle of the Mediterranean, provided a base for Beaufort torpedo strike bombers to attack Rommel's supply lines – was a major factor in the North African campaign.

One of the primary roles of the RAF during World War II was strategic bombing. With the introduction of four-engined heavy bombers, Bomber Command was able to take the war into the very heart of Germany on a daily basis. Tonnage dropped during the war reached a peak of 525000 tonnes (517,000 tons) of bombs in 1944.

With virtual air supremacy gained, the tactical squadrons with Typhoons and Tempests were able to strike enemy communications and transport centres by day. In the Far East, the RAF suffered from a lack of effective combat types and it was not until later in the war that it was able to go on the offensive against the Japanese. In 1944 the RAF had formed its first jet fighter squadron, No. 616 Squadron operating the Gloster Meteor.

By the end of the war the RAF possessed 55,000 aircraft and over one million personnel. The price had been heavy, the RAF losing some 70,253 killed in action and a further 22,924 wounded. As in the aftermath of World War I, the service contracted rapidly. It was the start of the Cold War, the Berlin Blockade relief effort in 1948 and the Korean War which brought a halt to the decline.

By 1950 Fighter Command had converted fully to jet aircraft in the day-fighter role, followed by the night-fighter force in 1952. Bomber Command was undergoing dramatic re-equipment, with the first jet-powered bomber, the Canberra, having entered service in 1951. Faster and able to fly higher than contemporary fighters, the Canberra entered widespread service as both a bomber and photo-reconnaissance type, not only serving with the home squadrons but in Germany, the Near and Far East.

In RAF service the Tornado undertakes the long-range interdiction and defence suppression (GR.Mk 1 and 4), maritime strike (GR.Mk 1B), reconnaissance (GR.Mk 1A), air defence and long-range interception (F.Mk 3) roles. ZA490 is a No. 12 Squadron GR.Mk 1B, complete with a pair of underfuselage Sea Eagle anti-ship missiles.

The intensification of the Cold War saw the introduction in 1955 of the first V-bomber, the Valiant. It was followed by the Vulcan (1956) and Victor (1957), which together provided the UK with a strategic nuclear deterrent. The construction of a Ballistic Missile Early Warning Station provided a four-minute warning of attack, sufficient time for the bombers on Quick Reaction Alert (QRA) to become airborne. The V-force was supplemented between 1958 and 1963 by Thor Intermediate Range Ballistic Missiles equipping 20 Bomber Command Squadrons.

The cancellation of the American Skybolt missile led to the UK government to replace the Victor- and Vulcan-launched Blue Steel missiles with the submarine-launched Polaris missile system. The V-force continued with its strategic nuclear deterrent role until 1969, when that was transferred to the Royal Navy, and the bombers adopted a tactical role and were assigned to NATO.

During the 1950s, the Meteors and Vampires of Fighter Command were replaced by the Sabre, Swift and classic Hawker Hunter in the day interceptor role and by the delta-winged Javelin at night. The infamous 1957 Duncan Sandys Defence White Paper dictated that the English Electric P1 Lightning would be the last manned fighter to be ordered for the RAF, as missile systems would be able to replace fighter aircraft. Defence of V-bomber and Thor missile sites was provided by Bloodhound SAMs, but the 1957 paper did much damage to the RAF and the national aircraft industry. In 1960 the RAF's first supersonic fighter, the Lightning, entered service. Throughout the postwar period Transport Command maintained a large fleet to support operations throughout the world, equipped with English designs including the Beverley, Britannia, Comet, Hastings and Valetta. In the mid-1960s a large re-equipment programme began, with the Buccaneer, Phantom, Harrier, Hercules, Nimrod and VC10 all entering service. Joint projects with France also led to the Jaguar, Puma and Gazelle.

As the United Kingdom reduced its overseas bases and the number of its squadrons, with first the withdrawal of forces from east of Suez (Far East Air Force disbanded in 1971) and then from the Middle East (during 1978), the RAF was focused on the need to stem a possible Soviet invasion of the European mainland. Massive cuts in the RAF's transport and support units in the mid-1970s reduced the ability of the nation to deploy its forces overseas.

The 1982 invasion of the Falkland Islands by Argentina stretched the RAF to the limit in some respects. Strategic airlift had to be undertaken by the Hercules fleet or chartered aircraft (including the Belfasts sold by the RAF in the 1976 defence cuts), while the ageing fleet of Victor tankers used up a large proportion of their remaining fatigue life. Vulcan bombers undertook the (then) longest bombing missions in history. As British forces regained the

The Chinook is the backbone of the RAF's Support Helicopter Force. This aircraft is one of two on the strength of No. 78 Squadron, which supports the Falklands garrison from its Mount Pleasant base.

Strike Command

HQ RAF High Wycombe

No. 1 Group, RAF High Wycombe		
No. 1 Sqn	Harrier GR.Mk 7, T.Mk 10	RAF Wittering
No. 2 Sqn	Tornado GR.Mk 1A/1T	RAF Marham
No. 3 Sqn	Harrier GR.Mk 7, T.Mk 10	RAF Laarbruch
No. 4 Sqn	Harrier GR.Mk 7, T.Mk 10	RAF Laarbruch
No. 6 Sqn	Jaguar GR.Mk 1A, T.Mk 2A	RAF Coltishall
No. 7 Sqn	Chinook HC.Mk 2/2A Gazelle HT.Mk 3	RAF Odiham
No. 9 Sqn	Tornado GR.Mk 1/1T/4	RAF Brüggen
No. 12 Sqn	Tornado GR.Mk 1B/1T	RAF Lossiemouth
No. 13 Sqn	Tornado GR.Mk 1A/1T/4A	RAF Marham
No. 14 Sqn	Tornado GR.Mk 1/1T	RAF Brüggen
No. 15(R)Sqn	Tornado GR.Mk 1/1T	RAF Lossiemouth
No. 16(R)Sqn	Jaguar GR.Mk 1A, T.Mk 2A	RAF Lossiemouth
No. 18 Sqn	Chinook HC.Mk 2	RAF Odiham
No. 20(R) Sqn	Harrier GR.Mk 7, T.Mk 10	RAF Wittering
No. 27 Sqn	Chinook HC.Mk 2	RAF Odiham
No. 31 Sqn	Tornado GR.Mk 1/1T	RAF Brüggen
No. 33 Sqn	Puma HC.Mk 1	RAF Benson
No. 39(1 PRU) Sqn	Canberra PR.Mk 7, PR.Mk 9, T.Mk 4	RAF Marham
No. 41 Sqn	Jaguar GR.Mk 1A, T.Mk 2A	RAF Coltishall
No. 54 Sqn	Jaguar GR.Mk 1A, T.Mk 2A/T.Mk 4	RAF Coltishall
No. 72 Sqn	Puma HC.Mk 1, Wessex HC.Mk 2	RAF Aldergrove
No. 230 Sqn	Puma HC.Mk 1	RAF Aldergrove
No. 617 Sqn	Tornado GR.Mk 1B/1T	RAF Lossiemouth
No. 11/18 Group, RAF Bentley Priory & RAF Northwood		
No. 5 Sqn	Tornado F.Mk 3/3T	RAF Coningsby
No. 8 Sqn	Sentry AEW.Mk 1	RAF Waddington
No. 11 Sqn	Tornado F.Mk 3/3T	RAF Leeming
No. 22 Sqn	Sea King HAR.Mk 3/3A	
A Flt		RAF Chivenor
B Flt		RAF Wattisham
C Flt		RAF Valley
No. 23 Sqn	Sentry AEW.Mk 1	RAF Waddington
No. 25 Sqn	Tornado F.Mk 3/3T	RAF Leeming
No. 43 Sqn	Tornado F.Mk 3/3T	RAF Leuchars
No. 51 Sqn	Nimrod R.Mk 1	RAF Waddington
No. 56(R) Sqn	Tornado F.Mk 3/3T	RAF Coningsby
No. 100 Sqn	Hawk T.Mk 1/1A	RAF Leeming

No. 111 Sqn	Tornado F.Mk 3/3T	RAF Leuchars
No. 202 Sqn	Sea King HAR.Mk 3	
A Flt		RAF Boulmer
D Flt		RAF Lossiemouth
E Flt		RAF Leconfield
No.203(R) Sqn	Sea King HAR.Mk 3/3A	RAF St.Mawgan
Battle of Britain Memorial Flight		RAF Coningsby
	Spitfire Mk IIa/Vb/IX/XIX/PR.Mk XIX	
	Hurricane Mk IIc, Lancaster B.Mk I	
	Chipmunk T.Mk 10, Dakota C.Mk 4	
Kinloss MR Wing		**RAF Kinloss**
No. 42(R) Sqn	Nimrod MR.Mk 2	RAF Kinloss
No. 120 Sqn	Nimrod MR.Mk 2	RAF Kinloss
No. 201 Sqn	Nimrod MR.Mk 2	RAF Kinloss
No. 206 Sqn	Nimrod MR.Mk 2	RAF Kinloss
38 Group, RAF High Wycombe		
No. 10 Sqn	VC10 C.Mk 1K	RAF Brize Norton
No. 32 (The Royal) Sqn	BAe 125 CC.Mk 3, BAe 146 CC.Mk 2, Twin Squirrel, S.76C	RAF Northolt
No. 101 Sqn	VC10 K.Mk 2/3/4	RAF Brize Norton
No. 216 Sqn	Tristar K.Mk 1, KC.Mk 1, C.Mk 2/2A	RAF Brize Norton
Northolt SF	Islander CC.Mk 2/2A	RAF Northolt
Lyneham Transport Wing (LTW)		**RAF Lyneham**
No. 24 Sqn	Hercules C.Mk 1/3	RAF Lyneham
No. 30 Sqn	Hercules C.Mk 1/3	RAF Lyneham
No. 47 Sqn	Hercules C.Mk 1/3	RAF Lyneham
No. 57(R) Sqn	Hercules C.Mk 1/3	RAF Lyneham
No. 70 Sqn	Hercules C.Mk 1/3	RAF Lyneham
Direct-reporting units		
No. 78 Sqn	Chinook HC.Mk 2 Sea King HAR.Mk 3	Mount Pleasant, Falk.
No. 84 Sqn	Wessex HC.Mk 2	RAF Akrotiri, Cyprus
No. 1310 Flt	Chinook HC.Mk 2	Split, Croatia
No. 1312 Flt	Hercules C.Mk 1, VC10 K.Mk 2/3/4	Mount Pleasant, Falk.
No. 1435 Flt	Tornado F.Mk 3	Mount Pleasant, Falk.
School of Aviation Medicine		**Farnborough**
	Hawk T.Mk 1	Boscombe Down
Air Warfare Centre		**RAF Waddington**
Tornado F.3 Operational Evaluation Unit		RAF Coningsby
	Tornado F.Mk 3/3T	
Strike/Attack Operational Evaluation Unit		Boscombe Down
	Harrier GR.Mk 7, Jaguar GR.Mk 1A/3, T.Mk 4	
	Tornado GR.Mk 1/1A/1B	
E-3D OEU	E-3D	RAF Waddington
	(loaned from Nos 8 & 23 Sqn)	

islands, the need for the RAF to have a strategic airlift capability was restated.

There followed a further re-equipment programme and a variety of new types entered service, including TriStar tanker/transports and VC10 tankers. Most notably, the Tornado GR.Mk 1 and F.Mk 3 replaced a number of existing types in the air defence and interdiction roles. The ability of the RAF to operate outside the European mainland was demonstrated during the Gulf War in 1991, when RAF Tornado GR.Mk 1s were among the first aircraft to strike Iraqi targets.

In common with the majority of Western air arms, the end of the Cold War led to a dramatic down-

sizing of the RAF. Current personnel strength totals some 56,000, a reduction of 36,000 from 1990. A reserve force consisting of former serving members would add 46,000 personnel in time of war, while the Royal Auxiliary Air Force, amalgamated with the RAF Volunteer Reserve in April 1997, would give an additional 1,500 personnel. The fall in manpower has followed a reduction in funds allocated to the service, the RAF in 1998 receiving an annual budget of £4 billion, representing 18 per cent of the UK's overall defence budget.

Celebrating its 80th anniversary in 1998, the RAF enters the 21st century adapting itself for its new roles with a series of new aircraft and upgrades.

STRIKE COMMAND

Formed on 30 April 1968, Strike Command embraced a number of other commands which had contracted with the UK's withdrawal from its overseas commitments. In April 1975 it became fully integrated into NATO, a commitment which continues to this day. Strike Command controls all the RAF's front-line strike/attack and air defence assets, as well as supplying transports, tankers, maritime patrol and support helicopters.

In time of conflict the Air Officer Commander-in-Chief of RAF Strike Command would become CINCUKAIR and would be responsible to the Supreme Commander Europe (SACEUR) for the defence of the UK and NATO's northwestern flank, and to Supreme Allied Commander Atlantic (SACLANT) for the defence of the eastern Atlantic. With an established strength of 35,000 military and 7,000 civilian personnel, Strike Command forms the

The Tornado fleet has been heavily involved in overseas deployments. This No. 617 Sqn GR.Mk 1B waits on a Kuwaiti airfield for missions over Iraq.

No. 9 Sqn is an overland attack unit with a defence suppression speciality, employing the BAe ALARM missile. It is one of three Tornado squadrons at RAF Brüggen which are shortly to return to the UK, to be split between the Tornado bases of Lossiemouth and Marham.

largest single command within the RAF. Headquartered at RAF High Wycombe, the Command has a number of groups and direct reporting units, the largest of which is the Air Warfare Centre (AWC).

GROUPS

Reporting to Strike Command are three Groups which operate the majority of front-line squadrons. No. 1 Group at RAF High Wycombe contains the RAF strike, reconnaissance and helicopter support force, including the Germany-based units inherited after the demise of RAF Germany. Established in 1996, No. 11/18 Group is responsible for air defence, maritime patrol, electronic warfare and SAR operations and, while operated as a single Group, each element has maintained its own individual headquarters, No. 11 at Bentley Priory and No. 18 at Northwood as part of the Royal Navy Fleet HQ. Finally, No. 38 Group at RAF High Wycombe controls tanker and transport assets.

OFFENSIVE AIR SUPPORT

The RAF's three squadrons of front-line Jaguars are based at RAF Coltishall with pooled aircraft. During the Gulf War the type distinguished itself in the daylight ground-attack role, and a series of rolling upgrades to avionics and engines is expected to keep the Jaguar in service until 2008, when it will be replaced by the Eurofighter. An Urgent Operational Requirement (UOR) for operations over Bosnia led to 11 aircraft (nine single-seaters as GR.Mk 1Bs and a pair of two-seaters as T.Mk 2Bs) being modified to carry the Thermal Imaging and Laser Designation (TIALD) system. The rest of the fleet is being brought up to this standard as GR.Mk 3s, following state-of-the-art cockpit upgrades including multi-function LCD, an integrated mission-planner, wide-angle HUD and a helmet-mounted sight. The two-seat Jaguars on strength will also receive some upgrades but will remain in the training role, having been redesignated T.Mk 4s. Pilot conversion training is undertaken by No. 16(R) Squadron at RAF Lossiemouth, but this will move to Coltishall in 1999 to centralise Jaguar operations at a single base.

TORNADO FORCE

The 142-strong Tornado GR.Mk 1/4 fleet carries the burden of long-range interdiction, maritime strike, defence suppression and reconnaissance, although it

lost its nuclear strike role on 31 March 1998 with the withdrawal of the WE177 free-fall munition. The Tornado force was in the thick of the action during Operation Granby, six being lost during the conflict as they struck Iraqi airfields and other targets with the JP233 airfield denial weapon, 1,000-lb bombs and laser-guided bombs (LGBs). Shortly after the end of the conflict, the Laarbruch Wing of four Tornado Squadrons was withdrawn, with one squadron transferring to Marham while the other three squadrons disbanded. No. 17 Sqn has since disbanded, leaving three squadrons in Germany at RAF Brüggen. Nos 9 and 31 Squadrons are responsible for defence suppression with the BAe ALARM, while No. 14 Squadrons is equipped with TIALD pods for the autonomous delivery of laser-guided bombs. In addition to their specialised tasks, all four squadrons undertake the conventional interdiction role. The squadrons based at RAF Brüggen, Germany, are scheduled to be withdrawn to RAF Marham (Nos 9 and 31) and Lossiemouth (No. 14) prior to the scheduled closure of the base in 2002.

The withdrawal of the Laarbruch Wing from Germany provided the opportunity to replace the Buccaneer S.Mk 2 in the maritime strike role with the Tornado. Using the Sea Eagle AShM-capable Tornado GR.Mk 1B, the two based squadrons (Nos 12 and 617 'Dambusters') are assigned to SACLANT in the maritime strike mission, although this represents only 40 per cent of their tasking, the remainder comprising conventional overland attack.

Work commenced in 1996 on the Tornado mid-life update programme, under which the type is being redesignated as the Tornado GR.Mk 4 in RAF service. Upgrades include new cockpit displays, digital map database and integration of night-vision goggle capability with the FLIR equipment. The first of the converted aircraft was delivered to DERA Boscombe Down in October 1997, and the first two units to receive the more potent aircraft are No. 13 Squadron at Marham with the reconnaissance GR.Mk 4A and No. 9 Squadron at Brüggen, Germany.

Based at RAF Cottesmore, the Tri-national Tornado Training Establishment (TTTE) provided conversion training on the Tornado IDS version for the RAF, German and Italian air forces. Established in 1980 as the first Tornado unit, the TTTE disbanded in March 1999, when the RAF element merged with No. 15(R) Squadron at Lossiemouth, giving the service a centralised training syllabus. No. 15 Squadron already undertakes the weapons conversion role. Like

all shadow squadrons in the RAF, during times of war the unit would assume a more aggressive role and be manned by the unit's instructors.

HARRIER

The RAF was the first air force to deploy an operational VTOL aircraft, the Hawker Siddeley (later BAe) Harrier. The first generation Harrier, the GR.Mk 1/T.Mk 2, was upgraded as the GR.Mk 3/T.Mk 4 before being replaced by the second-generation GR.Mk 5 from the late 1980s. Night attack capabilities were introduced in the GR.Mk 7, to which standard all surviving GR.Mk 5/5As were upgraded. As well as 94 single-seat second-generation Harriers, the RAF also acquired the trainer version, the T.Mk 10.

The TTTE will be replaced at Cottesmore by Harrier GR.Mk 7-equipped Nos 3 and 4 Squadrons currently based at Laarbruch, Germany, which are scheduled to return to the UK by December 1999. They will be joined at the same time by the third front-line Harrier unit, No. 1(F) Squadron, currently at nearby Wittering. The Harrier OCU, No. 20(R) Squadron, will remain at Wittering with its mixed fleet of Harrier GR.Mk 7 and two-seat T.Mk 10s.

Closer integration and basing of the RAF and FAA Harrier and Sea Harrier fleet was called for in the Strategic Defence Review, as part of the Joint Force 2000 concept. Both types are to be replaced by the Future Carrier-Borne Aircraft, for which the Joint Strike Fighter (JSF) is seen as the logical contender.

AIR DEFENCE

When the RAF looked for a replacement for its ageing air defence fleet of Lightnings and Phantoms, primary consideration was given to the threat of Soviet bombers launching nuclear and conventional missiles from considerable range. The affordable answer was thought to be a version of the Tornado IDS optimised for the air defence mission; this became the Tornado ADV, whose long range and loiter capability was considered ideal for the role. Today the type is in service as the F.Mk 3, but with the demise of the Soviet threat the RAF has five front-line squadrons of an aircraft less than ideally suited to the kind of conflict the RAF expects to fight in future.

The Tornado F.Mk 3 fleet is operated from three air bases, each of which has two squadrons based. A 15-minute alert facility is maintained by the Leuchars-based squadrons, but the once-commonplace interception of Soviet aircraft is now an extremely rare

Including the operational conversion unit, the RAF has four Harrier GR.Mk 7 squadrons. These are No. 1 Sqn (above left), No. 3 Sqn (below left), No. 4 Sqn (above right), and No. 20(R) Sqn (below right). The night attack-capable GR.Mk 7 replaced the GR.Mk 5 in service from 1987. Thirteen second-generation two-seat conversion trainers were delivered from 1995, of which the majority serve with No. 20(R) Sqn.

Three UK air bases house Tornado F.Mk 3s, each with two squadrons. Nos 111 (above) and 43 (below) are based at Leuchars, where both squadrons previously operated the Phantom.

The Leeming Tornado F.Mk 3 squadrons are Nos 11 (above) and 25 (below) Squadrons. A total of 152 F.Mk 3s was delivered to the RAF, the type having entered service in 1986.

Above: No. 5 Sqn, previously a Lightning F.Mk 6 operator, has flown the Tornado from Coningsby since December 1987.

Below: The operational conversion unit for the ADV fleet is No. 56(R) Sqn. This aircraft is a twin-sticker from the last production batch for the RAF.

Above: The RAF Sentries are named after the characters in the 'Seven Dwarfs'. This example displays No. 23 Squadron's colours.

Below: The RAF's target facilities squadron is No. 100 Sqn, which operates the Hawk T.Mk 1/1A.

event. The UK Air Defence Region (ADR), which stretches from the waters off Norway to the South West approaches, is co-ordinated by the Air Defence Operations Centre at HQ Strike Command, with RAF Buchan controlling the northern sector and RAF Neatishead the southern.

The most northerly air defence base is RAF Leuchars in Scotland, home to Nos 43 and 111 Squadrons. RAF Leeming in North Yorkshire is home to Nos 11 and 25 Squadrons; RAF Coningsby in Lincolnshire hosts Nos 5 and 56(R) Squadrons. No. 56(R) squadron is the operational conversion unit for the type. A forward operating base at RAF Stornoway in the Hebrides, which would have been used to defend the UK from an attack from the north, was closed in 1998.

The air defence fleet is committed to supporting the NATO Immediate Reaction Force and, although no specific squadrons are assigned, aircraft would be supplied as stipulated by NATO in time of crisis. It was mooted that the Tornado F.Mk 3 should be replaced by an interim type pending delivery of the first Eurofighter, but this has been dismissed in favour of upgrading the current fleet. Under a Capability Sustainment Programme (CSP), the 100 Tornado F.Mk 3s in service will receive an AMRAAM and ASRAAM capability, the existing SkyFlash missile becoming increasingly obsolete. Other modifications include radar and engine upgrades, improved defensive aids, an NVG-compatible cockpit and the ability to receive the Joint Tactical Information Distribution System (JTIDS), which currently can only be accepted by a limited number of aircraft. The programme should be completed in 2001, although a shortage of the more capable missiles and the JTIDS equipment will mean only two squadrons will have fully upgraded aircraft and, accordingly, will be assigned to the allied rapid reaction force.

Based at the former Vulcan base of RAF Waddington are the RAF's seven Sentry AEW.Mk 1s (E-3Ds). Since 1992 the Sentries have operated as the UK component of the NATO Airborne Early Warning Force (AEWF). The two squadrons (Nos 8 and 23) are tasked to provide airborne early warning and control services to national and NATO air defence authorities, as well as support to offensive air operations.

The BAe Hawk T.Mk 1/1A forms the backbone of RAF flying training, and is also assigned to a single Strike Command squadron. Operating alongside the Leeming Tornado F.Mk 3 squadrons, No. 100 Squadron uses its 19 aircraft to provide dissimilar air combat training (DACT) and target towing for air defence units, in addition to training the crews of support aircraft in defensive tactics.

EUROFIGHTER

Designed and manufactured by a consortium of BAe, DASA of Germany, Alenia of Italy and CASA of Spain, the Eurofighter will form the linchpin of the RAF's future fighter capability. A total of 232 aircraft will be ordered for the RAF. Initially replacing the Tornado F.Mk 3 when it enters service in 2005, it will subsequently replace the Jaguar, also, although the

Eurofighter is unlikely to be based at Coltishall as it does not have hardened facilities.

AIR WARFARE CENTRE

Formerly the Central Tactics and Trials Organisation, the AWC is concentrated at impressive new facilities at RAF Waddington in Lincolnshire. Through the Tactics and Electronic Warfare Centre (T&EW), its role is to provide mission support at both the operational and tactical levels to commanders and to front-line squadrons. This includes the development of air power doctrine and the provision of air warfare training for all who might be involved in operations, from aircrews to commanders. The AWC also provides EW support to all three services and executes trials and evaluations of new and existing equipment. It administers a number of Operational Evaluation Units (OEU) established for specific new aircraft or systems, currently consisting of the Tornado F.Mk 3 OEU at Coningsby with four aircraft assigned, and the E-3D OEU at Waddington. The latter draws aircraft as required from the two Waddington-based Sentry squadrons (Nos 8 and 23) for use in operational evaluation and development tasks. It operates closely with the similarly titled E-3D OEU DERA which undertakes trials and development work on the type at Waddington.

The largest AWC flying unit is the Strike Attack Operational Evaluation Unit (SAOEU) which oper-

Europe

Four squadrons operate the Jaguar GR.Mk 1As. In the reconnaissance role is No. 41 Sqn (above), while No. 54 Sqn (left) is one of two squadrons tasked with offensive air support. The type has been deployed overseas many times since the end of the Cold War in support of NATO and UN actions, usually gaining overwing hardpoints for AIM-9s (below) or ECM.

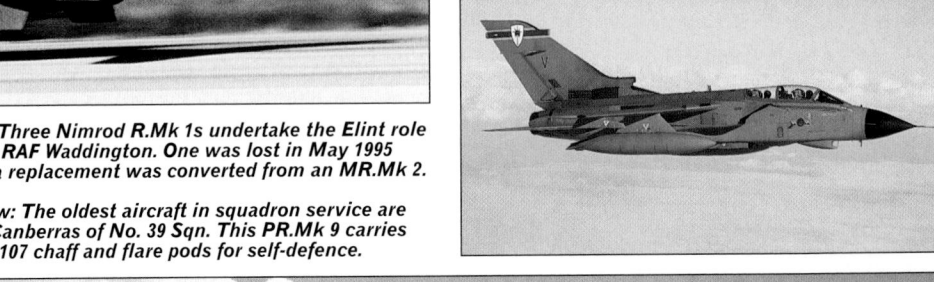

Only two squadrons of Tornado GR.Mk 1A/4As are tasked with reconnaissance, and they also have a secondary attack role. Since entering service in 1989 both Nos II(AC) (above) and 13 (below) have been constantly in demand thanks to their filmless IR reconnaissance suite. The RAF has 24 GR.Mk 1As and 4As.

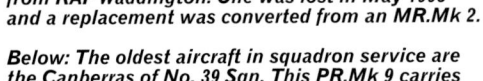

Left: Three Nimrod R.Mk 1s undertake the Elint role from RAF Waddington. One was lost in May 1995 and a replacement was converted from an MR.Mk 2.

Below: The oldest aircraft in squadron service are the Canberras of No. 39 Sqn. This PR.Mk 9 carries BOZ-107 chaff and flare pods for self-defence.

ates a mixed fleet of Harriers, Jaguars and Tornados from Boscombe Down. The unit is responsible for the development of tactical doctrine for each type and the practical application of new aircraft systems and weapons, deploying overseas on occasion to use test facilities. The AWC is also responsible for the Operations Support Branch which, following an administrative reorganisation, was established in April 1997 to co-ordinate air traffic control, fighter control, intelligence, RAF Regiment and flight operations.

RECONNAISSANCE

The RAF operates four types in the reconnaissance role. The centre for reconnaissance operations is RAF Marham in Norfolk, home to two Tornado GR.Mk 1A squadrons (Nos II(AC) and 13 Squadrons) and No. 39 (No. 1 Photo Reconnaissance Unit) Squadron operating the ageing Canberra PR.Mk 9. The specialised reconnaissance version of the Tornado, the GR.Mk 1A/4A, give the two Marham squadrons a world-leading reconnaissance platform. With a revolutionary IR/video system, the aircraft excelled during the Gulf War when used to hunt 'Scud' missile launchers in the Iraqi desert. Still possessing a secondary (20 per cent) attack role, both units were the only RAF nuclear strike squadrons until the 1998 withdrawal of the last RAF nuclear munitions.

Operating alongside the latest in reconnaissance technology are the oldest aircraft still in active service in the RAF. Since entering service in 1959, the Canberra PR.Mk 9 has provided a primarily 'wet-film' capability and, although not assigned to NATO, the unit is committed to the Joint Rapid Deployment

Force. Pairs of Canberra T.Mk 4s and PR.Mk 7s are used for training and type-conversion purposes, rotated through periods of storage to preserve airframe hours. In recent years the value of the Canberra has become apparent, having been used to detect mass graves in Bosnia and the location of refugees in Rwanda. System upgrades currently under development will extend Canberra service until the new century, when they will be replaced by the Airborne Stand-Off Radar battlefield surveillance system (ASTOR). In March 1999 the RAF chose Raytheon E-Systems to fulfil the contract, using the Bombardier (Canadair) Global Express airframe.

A single Jaguar squadron – No. 41 Squadron at RAF Coltishall – is tasked with the reconnaissance mission. The unit's Jaguar GR.Mk 1s are to be equipped with the Jaguar Replacement Reconnaissance Pod to replace the current wet-film equipment, giving them the ability to record directly onto video tape. They will operate alongside the two Tornado squadrons and the RAF will then have the most advanced reconnaissance equipment currently available.

The most secretive aircraft operated by the RAF are the three Nimrod R.Mk 1s flown by No. 51 Squadron from RAF Waddington. Specific details of the missions flown have not been disclosed, but it is assumed that they undertake a stand-off

electronic surveillance role, monitoring radio and other broadcasts and electronic emissions. The aircraft are due to receive JTIDS.

TRANSPORT AND TANKER ASSETS

Entering service in April 1967, 66 C-130K Hercules C.Mk 1s were delivered to the RAF. Thirty had their fuselages stretched to become Hercules C.Mk 3s. Most of the survivors now serve with the Lyneham Transport Wing (LTW), which consists of five squadrons operating a total of 26 Hercules C.Mk 1s and 29 C.Mk 3 variants. By the early 1990s, the sterling service provided over the years by the RAF's Hercules squadrons had begun to take its toll on the aircraft as corrosion and reliability problems manifested themselves. In a contract signed in November 1993, the UK ordered 25 new C-130Js, which are due to enter service with the Lyneham Transport Wing from late 1999. The order is split between 10 standard-length C-130Js and 15 stretched C-130J-30s, to be known in RAF service as Hercules C.Mk 5 and C.Mk 4, respectively. The former will be delivered to No. 24 Squadron and the latter will join No. 70 Squadron, flying alongside older models. As both the special forces Chinook HC.Mk 3s and the Merlin HC.Mk 3 are due to be delivered with the ability to carry air-to-air refuelling probes, it is highly likely

Above: The RAF is the last operator of the VC10, using the type as a tanker and transport. 'P' is one of five VC10 K.Mk 4 three-point tankers of No. 101 Sqn. All K.Mk 4s initially served with British Airways.

Below: Stalwart of the RAF's transport assets for over 30 years, the Hercules fleet continues to operate as a pool from RAF Lyneham. This Hercules C.Mk 3 displays the latest overall-grey scheme of the type.

Above: A single TriStar C.Mk 2A operated by No. 216 Sqn undertakes strategic airlift missions.

No. 32 (The Royal) Squadron operates a mixed fleet of VIP-configured aircraft. The unit uses six BAe 125 CC.Mk 3s (above) for short-range transportation. Since the squadron retired its Wessex HCC.Mk 4s, a lease has been issued to Air Hanson to maintain a Sikorsky S-76C+ (below) for the Royal Family's use.

that some of the new Hercules will be delivered with the ability to carry tanker equipment.

Future plans initially provided for the replacement of the remainder of the older Hercules fleet with more C-130Js (the MoD has an option on five), and the European Future Large Aircraft (FLA). However, uncertainty over the FLA's development and the urgent requirement for an aircraft capable of supporting the UK armed services' role of rapid reaction has resulted in the Short Term Strategic Airlift (STSA) tender. The requirement is for four (probably leased) C-17 aircraft, or equivalent, and the RAF is also considering the An-124 Ruslan and the Il-76, but the requirement seems to be written around the Boeing C-17A Globemaster III or its (as yet unbuilt) civil version, the MD-17. A new squadron (most likely No. 53 Squadron, the old Belfast C.Mk 1 unit) will probably be formed to operate the new aircraft at Brize Norton.

The large hardstandings at this base in Oxfordshire, which was once home to deployed USAF Strategic Air Command aircraft, are ideally suited for large aircraft operations and, accordingly, today it hosts the TriStars and VC10s of the RAF combined medium transport and tanker fleet. The largest aircraft in the RAF inventory, the Lockheed TriStar is operated by No. 216 Squadron, which has in service four different versions totalling nine aircraft. All but three of these (two C.Mk 2s and a single C.Mk 2A) are adapted for air refuelling in addition to the carriage of personnel and cargo.

The other two Brize Norton-based squadrons are committed to the air refuelling mission. Using 12 VC10 C.Mk 1Ks, No. 10 Squadron has a dual tanker/transport role, while No. 101 Squadron is dedicated solely to tanking, operating a mix of 14 aircraft of various marks (five K.Mk 2s, four K.Mk 3s and five K.Mk 4s). The RAF air refuelling fleet has been heavily committed to various overseas detachments in recent years, operating aircraft in the Middle East, Italy, Turkey and the Falkland Islands, in addition to its regular RAF and NATO missions. With an ageing and increasingly expensive-to-operate VC10 fleet, consideration will need to be given soon to choosing a replacement, as the RAF's overseas involvement in NATO and UN operations (highlighting its need for a tanker/transport type) are unlikely to diminish in the short term.

The RAF unit with perhaps the highest profile is No. 32 (The Royal) Squadron, based at RAF Northolt. This West London base provides a convenient and secure airfield for members of the Royal Family and UK government. For this role, the squadron has a fleet of six BAe 125 CC.Mk 3 executive jets and a trio of larger BAe 146 CC.Mk 2s. One of the squadron's most publicised duties in recent years was carrying the body of Diana, Princess of Wales back to the UK from Paris in 1997. No. 32 (The Royal) Squadron is the only RAF squadron to operate both fixed- and rotary-winged aircraft, using two (leased) AS 355F1 Twin Squirrels primarily for the transport of VIPs. When the squadron retired its pair of Wessex HCC.Mk 4s in 1998, a 10-year lease was issued to Air Hanson for the provision of a single Sikorsky S-76C+ for the use by the Royal household.

Above: RAF Odiham is the home of the Chinook force, housing Nos 7, 18 and 27 Squadrons. Type conversion is handled by a flight within No. 27 Sqn (illustrated).

Right: The support Helicopter Force has a standard two-tone green wrap-round scheme, as displayed on these Puma HC.Mk 1s of No. 33 Sqn.

Below: Even with the imminent arrival of the Merlin HC.Mk 3, the Wessex HC.Mk 2 will probably remain in service well into the next century. No. 72 Sqn has operated the Wessex since August 1964. This example carries the name Henley-on-Thames.

Above: Since RAF St Mawgan lost its Nimrods, the fleet has been concentrated at RAF Kinloss. In service since 1969, the aircraft are due to be remanufactured for further service as MRA.Mk 4s.

Left: Since the retirement of the Wessex in the role, the sole search and rescue helicopters are the Sea King HAR.Mk 3/3As. This HAR.Mk 3 is operated by one of the flights of No. 202 Squadron.

The RAF's two Islanders (ZH536 being the CC.Mk 2) are operated by the Northolt Station Flight. They may be tasked with mapping or surveillance duties.

As part of the continuing search for cost savings, the government is exploring ways of increasing the involvement of private enterprise in the squadron. For long-range international journeys, the VC10s of No. 10 Squadron at Brize Norton are equipped with a VIP interior. Northolt also houses a Station Flight which operates a pair of secretive Islanders (a CC.Mk 2 and a C.Mk 2A), ostensibly in the light communications role, one of which is currently based at Waddington.

SUPPORT HELICOPTER FORCE

Unlike the majority of Western air arms, the RAF – not the Army Air Corps – supplies medium-lift helicopter support to the ground forces. Under the Strategic Defence Review of July 1998, the establishment of a tri-service Joint Helicopter Command will bring together all battlefield helicopters under a single command. Reorganisation of the Support Helicopter Force (SHF) in 1997 rationalised assets, centralising the operation of the 32 Chinook HC.Mk 2s at a single base, RAF Odiham in Hampshire. Here, the special forces-tasked No. 7 Squadron (which also operates a single Gazelle HT.Mk 3 for hack duties) was joined by No. 18 Squadron in April 1997. Following its withdrawal from RAF Laarbruch in Germany, the unit lost its flight of Puma HC.Mk 1s,

which was transferred to No. 72 Squadron.

No. 27 Squadron is the third Chinook unit and returned to operational status in early 1998. Prior to that it had reserve status, being responsible for the operational conversion of both Puma and Chinook crews. It maintains a training role for Chinook aircrew although new simulators will help keep the actual use of squadron aircraft to a minimum.

The first of six new-build Chinook HC.Mk 2As was delivered for acceptance trials at Boscombe Down in December 1997 and delivery is expected to be completed by 2001. A total of eight special-forces Chinook HC.Mk 3s with an inflight-refuelling capability are part of the order.

Based at RAF Aldergrove are Nos 72 and 230 Squadrons which form the SHF Northern Ireland contingent. No. 72 Squadron is the last UK-based operator of the venerable Wessex HC.Mk 2, 16 of which serve alongside a small number of ex-No. 18 Squadron Puma HC.Mk 1s. No. 230 Squadron uses only Pumas. Both provide an air mobility service to the army based in the province, a task considered of paramount importance for supporting remote garrisons and inserting patrols in rural areas.

On the mainland, No. 33 Squadron operates 13 Puma HC.Mk 1s from RAF Benson. The unit transferred to the Oxfordshire base in June 1997 from RAF Odiham, following the disbanding of No. 60 Squadron and retirement of its Wessex HC.Mk 2s the previous month. The unit undertakes operational conversion training on the Puma, having inherited the role along with a small number of helicopters from No. 27(R) Squadron.

The Support Helicopter Force is reaping the benefits of a standardisation of mission equipment, common weapons, ECM and defensive aids. This precedent will continue with the introduction of the first of 22 EHI (Westland) Merlin HC.Mk 3s ordered in 1995. Capable of carrying 30 troops and their equipment, the new helicopters will be equipped with rear-loading ramps and an optional refuelling probe for use with the service's 'new Hercules'. The Merlin will initially equip a reformed No. 28 Squadron at RAF Benson, a unit which until June 1997 operated Wessex helicopters in the former British colony of Hong Kong. The unit will act as the operational conversion unit for the type, which is due to initially replace the Wessex and then the Puma in RAF service. The first was rolled out on 25 November 1998.

MARITIME PATROL

A descendant of the former Coastal Command, No. 18 Group commands the RAF assets responsible for the maritime role, including anti-submarine warfare, SAR and the monitoring of the seas around the UK, in particular the oil and gas rigs which pepper the North Sea. While the Group no longer operates the maritime strike-designated Tornado squadrons it has, in the Nimrod MR.Mk 2, what is widely regarded as the best anti-submarine platform in the world. The aircraft also regularly undertake long-range SAR missions and act as support aircraft for overseas deployments of fighters.

Following a tender process, the UK government elected to upgrade the RAF Nimrod fleet instead of acquiring a new type. The £26 million Nimrod 2000 project will involve 21 aircraft being modified by Cobham plc at its facilities in Bournemouth. The first three airframes were delivered in February 1997, the fourth following in November 1998. Only 20 per cent of the original airframe will remain after the upgrade, which will involve the installation of the Rolls-Royce BR710 turbofan engines, a new Searchwater 2000MR main surveillance radar and modifications to the fuel tanks, wings and undercarriage. Due to re-enter service from 2001, the rebuilt aircraft will be designated Nimrod MRA.Mk 4 (MRA for Maritime Reconnaissance Attack) to signify the type's offensive role.

The RAF's SAR fleet consists of two squadrons of Westland Sea King HAR.Mk 3/3As. No. 202 Squadron has operated the Sea King since August 1978. After No. 22 Squadron had withdrawn the Wessex HC. Mk 2 from SAR duties it received the new-build Sea King HAR.Mk 3As from May 1997, capable of carrying 17 passengers or six stretchers in the casevac role. Each unit has three flights located around the UK coastline at strategic points to enable SAR coverage to be maintained in concert with the Fleet Air Arm and Coastguard. Although primarily tasked with military SAR, the vast majority of missions involve civilian rescues. All operations in the UK Search & Rescue Region (UKSRR) are the responsibility of the Aeronautical Rescue Co-ordination Centre at RAF Kinloss. Established in December 1997, the UKSRR covers some 3 million sq miles (7.8 million km²) and calls daily upon the 15-minute response alert Sea King crews to assist in rescues (which are often a considerable distance into the Atlantic) by day or night.

OVERSEAS OPERATIONS

Strike Command has additional commitments in Cyprus, the Falkland Islands, Gibraltar and Ascension Island.

Following the successful Falklands campaign of 1982, the government elected to build an airfield on the British dependency to allow fast-jet operations for the defence of the islands and rapid reinforcement

Four Tornado F.Mk 3s are detached to Mount Pleasant to provide air defence of the Falklands with No. 1435 Flight.

from the UK in time of crisis. Today Mount Pleasant Airfield is the permanent home of No. 78 Squadron, which operates a single Chinook HC.Mk 2 in the troop support role (with one in-use reserve) and a pair of Sea King HAR.Mk 3s in the SAR role, in addition to assisting with civilian rescues and other emergencies. Permanent detachments also use the base, including No. 1435 Flight operating four Tornado F.Mk 3s drawn from the UK-based squadrons, No. 1312 Flight with a single Hercules C.Mk 1 from the LTW and a single VC10 K.Mk 4 from No. 101 Squadron. The garrison is resupplied by weekly TriStar flights from the UK which call at the Ascension Islands en route to refuel.

The RAF places a great deal of importance on its Tornado strike force receiving the intensive low-level training required to effectively penetrate enemy defences in times of war. Flying restrictions in the UK and on mainland Europe have forced the RAF to look further afield for a viable alternative. The answer was to maintain a detachment at Goose Bay in Canada, which is within easy access of large areas of sparsely populated terrain ideal for fast-jet tactical training. Each front-line Tornado squadron rotates personnel to the detachment and makes use of the aircraft deployed in Canada during the duration of the nine-monthly training periods.

The sole RAF flying unit in Cyprus, No. 84 Squadron, operates five Wessex HC.Mk 2s in a number of diverse roles. The light blue fuselage bands on the helicopters are an indication of their role, which is assisting United Nations forces monitoring the division between the Greek and Turkish communities on the Mediterranean island. A SAR standby facility is maintained every day of the year and in the summer months the squadron adopts a fire-fighting role using underslung buckets. The unit is based at RAF Akrotiri, which has provided a vital secure staging post for aircraft and troops deploying to the Middle East area.

The UK government's seat on the United Nations Security Council has ensured UK involvement in recent years in many multinational peacekeeping efforts across the world. Administered by the Joint Rapid Deployment Force (JRDF) at Northwood, this component of the RAF was actively involved in peacekeeping missions in many parts of the world in

The retirement of the Gazelle HT.Mk 3 and the Wessex HC.Mk 2s of No. 2 FTS at RAF Shawbury heralded the introduction of tri-service helicopter training. The advanced phase at the joint-service Defence Helicopter Flying School syllabus is undertaken on the Griffin HT.Mk 1s, which operate as No. 60(R) Squadron.

early 1998. No. 1310 Flight operates a detachment of Chinook HC.Mk 2s on Stabilisation Force (SFOR) duties in Bosnia under Operation Lodestar. Tornado GR.Mk 1/1As are based at Al Kharj, Saudi Arabia and at Incirlik, Turkey as part of Operations Jural and Warden, respectively, enforcing the 'No-Fly Zones' over Iraq. At Gioia del Colle, Italy, are Jaguars assigned to Operation Deliberate Guard, with TriStars at Ancona and Sentries at Aviano. The Italian Jaguar and Tristar detachments have since come to an end, but such operations are likely to remain commonplace in the coming years.

With such contingencies in mind, the UK and French governments agreed in 1995 to establish the Franco-British European Air Group as a steering group to sponsor annual exercises to train the air arms of both countries in joint operations. With an emphasis on humanitarian relief, hostage rescue and peacekeeping missions, its success encouraged Italy and Germany to join a retitled European Air Group in 1998, and there is a strong likelihood of Spain joining in the foreseeable future.

One of the most famous units in the RAF is the 'Red Arrows' aerobatic display team. The addition of the belly-mounted ADEN Mk 4 cannon and a pair of AIM-9L Sidewinders on this Hawk T.Mk 1A illustrates the team's air defence war role.

FLYING TRAINING

The end of the Cold War had a significant effect on the organisation of RAF flying training. The reduction in front-line squadrons led to a 15 per cent reduction in the number of aircrew required, and two Flying Training Reviews and a Defence Costs Study during the early and mid-1990 saw funding fall by one-third, a series of base closures, expansion of contracted civilian services and of joint services training. As a consequence, the flying training community has adapted to the forced changes quicker and better than any element of the UK air arms. Personnel and Training Command (headquartered at RAF Innsworth in Gloucestershire) is responsible for the whole of the RAF flight training.

The primary sources of future aircrew are the

Personnel and Training Command

HQ RAF Innsworth

Training Group Defence Agency

Joint Elementary Flying Training School *(civilian contract)*		
	Firefly 160/200/260	RAF Barkston Heath
No. 1 FTS	Tucano T.Mk 1	RAF Linton-on-Ouse
Defence Helicopter Flying School, RAF Shawbury		
	Squirrel HT.Mk 1	
No. 60(R) Sqn	Griffin HT.Mk 1	RAF Shawbury
SARTU	Griffin HT.Mk 1	RAF Valley
3 FTS, RAF Cranwell		
CFS Bulldog Sqn	Bulldog T.Mk 1	RAF Cranwell
No. 45(R) Sqn	Jetstream T.Mk 1	RAF Cranwell
No. 55(R) Sqn	Dominie T.Mk 1	RAF Cranwell
4 FTS/Advanced Training & Tactics Unit, RAF Valley		
No. 19(R) Sqn	Hawk T.Mk1/1A	RAF Valley
No. 74(R) Sqn	Hawk T.Mk1/1A	RAF Valley
No. 208(R) Sqn	Hawk T.Mk1/1A	RAF Valley
CFS*	Tucano T.Mk 1	RAF Linton-on-Ouse
	Hawk T.Mk 1/1A	RAF Valley
*aircraft drawn from 1 FTS at Linton-on-Ouse and No.19(R) Sqn at Valley		

University Air Squadrons/Air Experience Flights

University of Birmingham AS/8 AEF		RAF Cosford
	Bulldog T.Mk 1	
Bristol UAS	Bulldog T.Mk 1	RAF Colerne
Cambridge UAS/5 AEF		Cambridge Airport
	Bulldog T.Mk 1	
East Lowlands UAS		RAF Leuchars
	Bulldog T.Mk 1	
East Midlands UAS/7 AEF		RAF Newton
	Bulldog T.Mk 1	
Universities of Glasgow & Strathclyde AS		Glasgow Airport
	Bulldog T.Mk 1	
Liverpool UAS/10 AEF		RAF Woodvale
	Bulldog T.Mk 1	
University of London AS/6 AEF		RAF Benson
	Bulldog T.Mk 1	
Manchester & Salford Universities AS		RAF Woodvale
	Bulldog T.Mk 1	
Northumbria Universities AS/11 AEF		RAF Leeming
	Bulldog T.Mk 1	
Oxford UAS	Bulldog T.Mk 1	RAF Benson

Southampton UAS/2 AEF		Boscombe Down
	Bulldog T.Mk 1	
University of Wales AS		RAF St Athans
	Bulldog T.Mk 1	
Yorkshire Universities AS/9 AEF		RAF Church Fenton
	Bulldog T.Mk 1	

Volunteer Gliding Schools

No. 611 VGS	Viking T.Mk 1	RAF Watton
No. 612 VGS	Vigilant T.Mk 1	RAF Abingdon
No. 613 VGS	Vigilant T.Mk 1	RAF Halton
No. 614 VGS	Viking T.Mk 1	RAF Wethersfield
No. 615 VGS	Viking T.Mk 1	RAF Kenley
No. 616 VGS	Vigilant T.Mk 1	RAF Henlow
No. 617 VGS	Viking T.Mk 1	RAF Manston
No. 621 VGS	Viking T.Mk 1	RAF Hullavington
No. 622 VGS	Viking T.Mk 1	RAF Upavon
	Vigilant T.Mk 1	
No. 624 VGS	Vigilant T.Mk 1	RAF Chivenor
No. 625 VGS	Viking T.Mk 1	RAF Hullavington
No. 626 VGS	Viking T.Mk 1	Predannack
No. 631 VGS	Viking T.Mk 1	RAF Sealand
	Vigilant T.Mk 1	
No. 632 VGS	Vigilant T.Mk 1	RAF Ternhill
No. 633 VGS	Vigilant T.Mk 1	RAF Cosford
No. 634 VGS	Viking T.Mk 1	RAF St Athan
No. 635 VGS	Vigilant T.Mk 1	Samlesbury
No. 636 VGS	Viking T.Mk 1	Aberporth
No. 637 VGS	Vigilant T.Mk 1	RAF Little Rissington
No. 642 VGS	Vigilant T.Mk 1	RAF Linton-on-Ouse
No. 645 VGS	Viking T.Mk 1	RAF Syerston
No. 661 VGS	Viking T.Mk 1	RAF Kirknewton
No. 662 VGS	Viking T.Mk 1	Arbroath
No. 663 VGS	Vigilant T.Mk 1	RAF Kinloss
No. 664 VGS	Vigilant T.Mk 1	Belfast City/Newtonards

Air Cadet Central Gliding School/644 VGS		RAF Syerston
	Kestrel T.Mk 1, Vigilant T.Mk 1,	
	Valiant T.Mk 1, Viking T.Mk 1	

Display Teams

'Red Arrows'	Hawk T.Mk 1/1A	RAF Cranwell

Ground Training

1 School of Technical Training	RAF Cosford
CTTS/4 School of Technical Training	RAF St Athan
Airframe Technology Flight	RAF Cranwell
Servicing Instruction Flight	RAF Cranwell

Above: Of the 130 Tucano T.Mk 1s that the RAF received, just over a third are in store at any one time at RAF Shawbury. These belong to the CFS.

Above: Due to be replaced by the Grob 115 Tutor, the Bulldog T.Mk 1 serves with the UAS/AEFs and the CFS. This aircraft belongs to the Southampton UAS/2 AEF based at Boscombe Down.

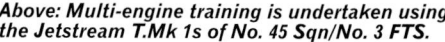

Above: Multi-engine training is undertaken using the Jetstream T.Mk 1s of No. 45 Sqn/No. 3 FTS.

Below: The oldest aircraft on the strength of Personnel and Training Command are the Dominie T.Mk 1s of No. 55(R) Sqn/No. 3 FTS.

University Air Squadrons (UAS) based at various airfields nationwide. The system of free RAF flying training for university students continues to prove an effective tool for securing graduate entrants to the service who will already have accumulated 90 flying hours on the Bulldog T.Mk 1 prior to commencing their formal training. Non-graduate entrants complete a 24-week course at the tri-service Joint Elementary Flying Training School (JEFTS) at RAF Barkston Heath in Lincolnshire. A civilian contractor provides Slingsby T-67M Firefly aircraft, maintenance and the majority of flying instructors to teach the RAF training syllabus. One consequence of the reorganisation in the training structure has been an earlier selection of pilots for specific service roles. Accordingly, at the conclusion of the JEFTS course trainee pilots join their UAS counterparts for streaming into either fast-jet, multi-engine or rotary-wing training.

All basic fast-jet training is initially undertaken on the turboprop-powered Tucano T.Mk 1 at RAF Linton-on-Ouse, near York. Pilots then progress to RAF Valley in North Wales for training on the Hawk T.Mk 1/1A with No. 4 Flying Training School (FTS), also known as the Advanced Training and Tactics Unit. No. 4 FTS has three reserve squadrons which fulfil different roles in the unit. Pilots initially join No. 208(R) Squadron for conversion to the Hawk and basic training on the aircraft. Tactics and weapons training, formerly the preserve of Tactical Weapons Units (TWUs) at RAF Chivenor and RAF Brawdy, is undertaken by No. 74(R) Squadron ('Tigers') after which successful pilots earn their wings and progress to Operational Conversion Units (OCU). A third Hawk squadron at Valley, No. 19(R) Squadron, operates to train instructors on behalf of the Central Flying School (CFS) and as an OCU for pilots destined for No. 100 Squadron.

The ageing Hawk fleet suffers limited availability and aircraft are rotated through storage for maximum usage. A limited fatigue life extension programme and new centre and rear fuselage sections will enable 80 Hawks to continue in service until 2010. There is every possibility that it will then be replaced by an upgraded version of the same basic design.

Reporting to the Commandant of the RAF College Cranwell, No. 3 FTS incorporates three different units with three different training missions. Under the guise of No. 45(R) Squadron, the Multi-Engine Training Squadron (METS) has 11 Jetstream T.Mk 1s in which student pilots destined for the service's fleet of transport, tanker and anti-submarine aircraft are tutored. Progressing from either the Bulldog or Firefly, multi-engine students earn their wings at the conclusion of a 20-week course prior to joining an OCU for their final training on the designated service aircraft.

Also part of No. 3 FTS is the Air Navigation School at RAF Cranwell which trains RAF navigators, and under the banner of No. 55(R) Squadron flies 11 Dominie T.Mk 1s. The aircraft have recently undergone upgrades which will keep them capable in their role for a number of years. Navigators destined for the fast-jet squadrons undergo a nine-week course on the Dominie prior to completing training in the Hawk at RAF Valley, being streamed half way through the 29-week course into either air defence or strike/attack classes. Multi-engine student navigators remain on the Dominie throughout, undergoing 23 weeks of tuition.

The final No. 3 FTS unit has no shadow squadron designation. Known simply as the Central Flying School Bulldog Squadron, it uses 19 of this primary trainer type in the instructor training role and in the basic training of helicopter student navigators.

A move to civilian contractorisation for helicopter training brought the establishment of the Defence Helicopter Training School (DHFS) at RAF Shawbury in April 1997. A fleet of Squirrel HT.Mk 1s and Griffin HT.Mk 1s replaced the Gazelle HT.Mk 2s and Wessex HC.Mk 2s of No. 2 Flying Training School for single- and multi-engine training, respectively. While students from each of the services receive training of varying duration on the single-engined Squirrel HT.Mk 1s, the DHFS's nine Griffins HT.Mk 1s of No. 60(R) Squadron/DHFS are used solely to train RAF aircrew on multi-engine operations. Two Griffins are detached to the Search & Rescue Training Unit (SARTU) at RAF Valley close to the Snowdonia mountain range. Each helicopter is equipped with a winch installation and flotation bags and each student is taught the demanding art of SAR in both mountain and maritime environments. Once qualified, aircrew complete OCU training within the front-line squadrons as there are no longer separate designated reserve squadrons fulfilling this role.

TECHNICAL TRAINING AND STORAGE

The closure of No. 1 School of Technical Training (SoTT) at RAF Halton in 1995 led to No. 2 SoTT at RAF Cosford in Shropshire becoming the RAF's largest ground training station. Some 2,000 students at any one time are undergoing airframe, avionics, electrical, propulsion or weapons training on airframes withdrawn from front-line service.

Deep maintenance and modification of the majority of aircraft types in the RAF inventory is undertaken at its largest air base, RAF St Athan in South Wales, which is commanded by an air commodore. Aircraft maintenance activity is undertaken within the Engineering Division, which is led by a group captain. This Division contains a fully integrated service/civilian workforce of some 3,400 personnel.

Initial experience of flying for many future aircrew is provided by the Volunteer Gliding Schools. They are equipped with Vigilant T.Mk 1s (above, from No. 612 VGS, Abingdon) or Viking T.Mk 1s (below, from No. 626 VGS, Predannack).

The sole RAF four-engined long-range bomber still in service, Lancaster B.Mk 1 PA474 (in No. 9 Sqn colours) is the flagship of the Battle of Britain Memorial Flight.

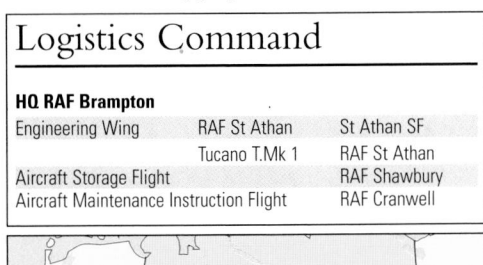

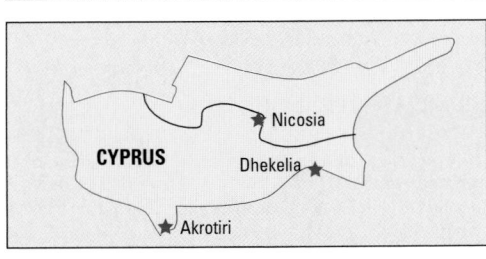

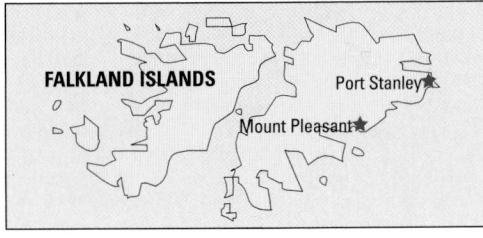

The Eurofighter Typhoon will be the RAF's main combat aircraft well into the 21st century. This example is the British Aerospace-assembled DA2 prototype, seen undergoing dry refuelling trials with a VC10 K.Mk 3 on loan from No. 101 Squadron.

Work is also undertaken on the Fleet Air Arm Sea Harrier and Jetstream aircraft. Two Tucano T.Mk 1s are operated by the Station Flight to ferry the based pilots when delivering or collecting aircraft. The base also hosts the Civilian Technical Training School/ No. 4 SoTT, providing training of personnel in a variety of ground trades, and also provides storage for a variety of types pending a return to service, sale or scrapping. A similar function is undertaken at Shawbury. The Aircraft Maintenance Instruction Flight at Cranwell uses a number of redundant airframes for training purposes.

Logistics Command

HQ RAF Brampton

Engineering Wing	RAF St Athan	St Athan SF
	Tucano T.Mk 1	RAF St Athan
Aircraft Storage Flight		RAF Shawbury
Aircraft Maintenance Instruction Flight		RAF Cranwell

GERMANY

Brüggen

CYPRUS

Nicosia

Dhekelia

Akrotiri

FALKLAND ISLANDS

Port Stanley

Mount Pleasant

UNITED KINGDOM

Benbecula
Kinloss
Lossiemouth
Leuchars
Glasgow
Edinburgh
Glasgow-Abbotsinch
Prestwick
Boulmer
Ballykelly
West Freugh
Sydenham
Aldergrove
Belfast
Catterick
Leeming
Topcliffe
Dishforth
Linton-on-Ouse
Church Fenton
York
Leconfield
Bradford
Leeds
Woodvale
Manchester
Sheffield
Mona
Liverpool
Sturgate
Coningsby
Valley
Scampton
Cranwell
Waddington
Barkston Heath
Ternhill
Newton
Marham
Coltishall
Llanbedr
Chetwynd
Cottesmore
Honington
Shawbury
Cosford
Birmingham
Wittering
Wyton
Oakington
Wattisham
Aberporth
Hereford
Bedford
Cambridge
Bawdsey
Brawdy
Brize Norton
Abingdon
Lyneham
Fairford
Halton
Cardiff
Benson
Northolt
London
St Athan
Bristol
Wroughton
Bristol-Filton
Manston
Middle Wallop
Farnborough
Chivenor
Netheravon
Odiham
Yeovilton
Boscombe Down
Fleetlands
Hurn
Lee-on-Solent
St Mawgan
Exeter
Portland
Roborough
Culdrose
Predannack

253

The only Fleet Air Arm fixed-wing aircraft to operate from Royal Navy aircraft-carriers is the Sea Harrier FA.Mk 2. The three squadrons that operate the type are No. 800 Sqn (above – from April 1995), No. 801 Sqn (since October 1994) and the training squadron, No. 899 (left – from January 1994). No. 899 Sqn operates the five two-seat Harrier T.Mk 8 (below) conversion trainers operated by the navy.

for which a large training centre is being established.

What was the busiest heliport in Europe, RNAS Portland in Dorset, closed in March 1999 following the transfer to Yeovilton of the FAA's fleet of Lynxes earlier in 1999.

Fleet Air Arm

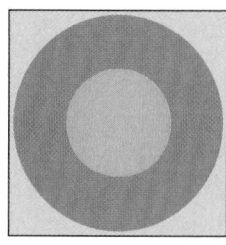

British naval aviation was born in 1914 when the Naval Wing of the Royal Flying Corps (RFC) became the Royal Naval Air Service. In 1918 the Service merged with the RFC into a single service, the Royal Air Force. There followed a long struggle until the Naval Air Branch, later known as the Fleet Air Arm (FAA), was returned eventually to Admiralty control in 1939. World War II gave a new impetus to naval flying which gradually changed naval tactics from a ship versus ship conflict to aircraft versus ships, with devastating effect.

By the end of World War II the strength of the FAA comprised 59 aircraft-carriers, 3,700 aircraft and 56 air stations located all over the world. In the Korean War, FAA Sea Furies held their own against jet-powered opposition. The service's own jet aircraft entered service on new large strike carriers at the end of that conflict, accompanied by an expansion of helicopter flying. The FAA was the first service in the world, in 1943, to order helicopters for operational use.

The new faster and heavier aircraft led to revolutionary advances in carrier operating techniques, such as the angled flight deck, steam catapult and mirror landing sight. Following the ill-fated Suez operation in 1956, the FAA entered another era with the introduction of more powerful all-weather aircraft such as the Sea Vixen, Scimitar and Buccaneer. They were followed in 1968 by the McDonnell F-4 Phantom, which provided a true multi-role capability for the first time.

The operation of large carriers came to an end with the withdrawal of HMS *Ark Royal* and in 1980 the first of three light aircraft-carriers was commissioned. They were equipped with another British innovation, the ski jump, enabling the new BAe Sea Harriers to carry a much greater load when taking off with forward thrust.

The new system was successfully put to the test in 1982 in the Falklands War, when British forces recovered the South Atlantic Islands from Argentina in a bitter campaign. In 1991, Sea King helicopters supported coalition forces in the liberation of Kuwait while Lynxes were responsible for sinking a number of Iraqi naval vessels. Immediately after the Gulf crisis, naval commando Sea Kings were deployed to assist in the Kurdistan relief operation where humanitarian aid was provided to thousands of refugees. Peacekeeping in the former Yugoslavia has seen heavy involvement from Royal Marines Sea King HC.Mk 4s and reconnaissance missions by Sea Harrier FA.Mk 2s.

NAVAL AIR COMMAND

Naval Air Command provides the Commander-in-Chief Fleet with a multi-role aviation combat capability, able to operate autonomously, at short notice, world-wide. Commanded and administered by the Flag Officer Naval Aviation (FONA) at RNAS Yeovilton, support is provided by the chief of Fleet Support through Director General Aircraft (Navy)'s Aircraft Support Executive (Navy) and the tri-service Defence Helicopter Support Authority.

ROYAL NAVAL AIR STATIONS

The FAA's support of the maritime fleet is achieved from the three principal air bases at Yeovilton, Culdrose and Portland, designated Typed Air Stations. Each is a functional entity combining all aspects of naval aviation, including operations, aircrew and engineering training, first and second line support, logistic support, warfare skills and infrastructure. Another air station is located at Prestwick in Scotland, hosting the largest Sea King squadron in the FAA and the second busiest SAR operation in the UK.

The largest military helicopter base in Europe is located at RNAS Culdrose in Cornwall with five rotary- and two fixed-wing squadrons. The base also hosts the RN School of Flight Deck Operations and is the operating base for the Hawk T.Mk 1/1As of the Fleet Requirements Air Direction Unit (FRADU), which provide airborne target, electronic warfare and air direction facilities to warships. The base has been designated as the future host of the FAA Merlin fleet

AIRCRAFT-CARRIERS

The Royal Navy enters the 21st century with three 20000-tonne 'Invincible'-class aircraft-carriers (HMS *Invincible*, *Illustrious* and *Ark Royal*), home ported in Portsmouth on the south coast of England. Two carriers are operational at any one time, the third in storage. Standard complement of a Carrier Air Group (CAG) is eight Sea Harrier FA.Mk 2s for air defence, strike and reconnaissance, three Sea King AEW.Mk 2 or 2As from No. 849 Sqn for airborne early warning, and nine Sea King HAS.Mk 6s for anti-submarine warfare. During the Falklands conflict RAF Harrier GR.Mk 3s operated from carrier decks, a concept that was resurrected in 1998 when tension with Iraq saw a British carrier deploy to the region with Harrier GR.Mk 7s from the RAF's No. 1(F) Sqn.

The latest 'flat top' is HMS *Ocean*, a Commando Helicopter Support Carrier designed to support amphibious warfare operations. Able to carry 12 Commando HC.Mk 4 transport and six attack helicopters (for example Army WAH-64D Apaches), the vessel provides the capability to mount a significant, large-scale helicopter assault in support of an amphibious landing. Elsewhere in the fleet, each of the 35 destroyers and frigates in the Royal Navy carries a ship's flight equipped with either a Lynx (No. 815 Sqn) or Sea King (No. 810 Sqn) while at sea.

SEA HARRIER FORCE

Combat-proven during the Falklands War, the latest version of the Sea Harrier, the FA.Mk 2, is equipped with a look-down/shoot-down radar tied to the Advanced Medium Range Air-to-Air Missile (AMRAAM), providing the FAA with a small and effective air defence fighter for its light carrier fleet. One squadron is assigned to each of the two carriers in service at any one time, each flying from Yeovilton when not at sea. Secondary roles of anti-ship, offensive counter-air and close air support and reconnaissance are flown, although in time of conflict it is probable that RAF Harrier GR.Mk 7s would fly from the carriers to provide more firepower in the air-ground environment. The navy's two operational Sea Harrier squadrons are due to combine with the RAF's three operational Harrier GR.Mk 7 squadrons

Above: No. 702 Sqn moved from RNAS Portland to RNAS Yeovilton during January 1999. The unit uses a few early model Lynxes for conversion training duties, adorned with the squadron's Wild Cat badge.

Below: The Sea King replacement is the EHI Merlin HMA.Mk 1. Initially, the type was to replace the Lynx as well, but this has been abandoned. The Intensive Flight Trials Unit, No. 700M Squadron, stood up at Culdrose in 1998, while the operational training squadron will be No. 824 Squadron.

Above: The original Lynx entered naval service in December 1977, while the latest HMA.Mk 8 joined in 1994. It is intended to convert 38 Lynxes to this configuration.

Below: Based at RNAS Prestwick, No. 819 Sqn has the role of supporting the Clyde-based submarines, SAR and provision of flights for frigates and RFAs. The squadron uses the Sea King HAS.Mk 6.

into a new integrated Joint Force 2000 unit by about 2005. The unit will operate the Future Carrier-Borne Aircraft (FBCA), which is likely to be a version of the American STOVL Joint Strike Fighter (JSF). Plans call for the current force level of 26 aircraft to be retained until 2012. Training on this complex weapons system is the responsibility of No. 899 Sqn, which maintains 10 single-seat and five twin-seat (Harrier T.Mk 8) aircraft at Yeovilton.

ASW AND SAR

The UK's role within NATO of protecting the eastern Atlantic meant great emphasis during the Cold War on anti-submarine warfare (ASW). Backbone of the ASW fleet is the rugged Westland Sea King which, over the years, has undergone a series of rolling upgrades that has so far culminated in the Sea King HAS.Mk 6. This provides the fleet with an autonomous ASW platform equipped with an integrated passive sonar system using sonar buoys, an active dipping sonar and electronic support measures. Weapons consist of Sting Ray light torpedoes and depth charges. Two Culdrose-based squadrons (Nos 814 and 820) are assigned to the two 'Invincible'-class carriers in service. If, in time of crisis, the third carrier were to be brought back into operational use, then No. 810 Squadron at Culdrose would form the ASW element of its CAG. Its peacetime role is aircrew and engineer training, which concludes with an embarkation on the aviation training ship, RFA *Argus*. The squadron also supports two Frigate Flights.

Responsibility for SAR in the UK is split between the FAA, RAF and the Coast Guard, which uses civilian contracted helicopters. Located in both a popular holiday destination for tourists in summer and in a dangerous area for shipping in winter, it is no surprise that RNAS Culdrose is home to the busiest SAR operation in England. The only dedicated SAR unit in the FAA, No. 771 Sqn operates a single Sea King HAS.Mk 5 for training purposes in addition to its five HAS.Mk 5Us for real missions. These latter aircraft will be among the last Sea Kings to be retired from service, current plans calling for service to continue

until 2010. A single Sea King HAS.Mk 5U is also operated by No. 819 Sqn at RNAS Prestwick to assist the unit's secondary role of SAR. Its main duty is ASW, however, being based near the important access to and from the Royal Navy's nuclear submarine base at Faslane.

The replacement of the ASW Sea King fleet will be the EHI Merlin, initial derivatives of which will be designated HMA.Mk 1. This large, three-engined helicopter is due to enter front-line service in 2001 and is considered to have a deck landing agility equal to the much smaller Lynx. An initial order of 44 aircraft is unlikely to be increased. The Merlin will provide a significant improvement in operational capability, reliability and ease of maintenance. It is equipped with an extensive sensor package including a 360° radar, passive sonar processor, sonobuoy dispensing carousel and low-frequency active dipping sonar. A fully integrated tactical and secure communications system enables the Merlin to link into the surface ships' active information systems, providing a potent addition to the fleet's capabilities.

The first example of this highly capable helicopter will be delivered following acceptance trials at Boscombe Down. It will be assigned to No. 700 Merlin IFTU, which formed at RNAS Culdrose in December 1998.

AEW

The requirement for airborne early warning was underscored during the Falklands conflict, but the size of the RN's carriers and their ski ramps made a fixed-wing AEW platform impractical. Accordingly, the Sea King was modified to fulfil the role, having a

The Sea King HC.Mk 4s are tasked with the deployment of troops and their associated equipment by air, usually working closely with the Royal Marines. This examples belongs to No. 846 Squadron, based at RNAS Yeovilton.

Searchwater radar mounted on the starboard side of the fuselage. The AEW variants are operated exclusively by No. 849 Sqn at Culdrose, which operates a flight of three helicopters on each of the two carriers in service at any one time, providing tactical control to the Sea Harriers and other shore- and carrier-based aircraft. The AEW fleet is due to be updated as AEW.Mk 7s, while a replacement, as envisaged in the the Future Organic AEW (FOAEW) programme, is due to be in service by 2012. The FOAEW is likely to be a version of the EHI Merlin or Bell/Boeing V-22.

LYNX

Two Yeovilton-based squadrons operate various marks of the maritime Westland Lynx. The first, No. 702 Sqn, is responsible for training aircrew and maintainers, while No. 815 Sqn is the sole front-line unit undertaking anti-surface, anti-submarine, SAR and communications duties while aboard destroyers and

Europe

Fleet Air Arm

Naval Air Command, RNAS Yeovilton	
RNAS Yeovilton (HMS *Heron*)	
702 Sqn	Lynx HAS.Mk 3/3S
800 Sqn	Sea Harrier FA.Mk 2
801 Sqn	Sea Harrier FA.Mk 2
815 Sqn	Lynx HAS.Mk 3/3S/3S(ICE)/3SGM, HMA.Mk 8
815 Operational Evaluation Unit	
	Lynx HMA.Mk 8
845 Sqn	Sea King HC.Mk 4
846 Sqn	Sea King HC.Mk 4
847 Sqn (3 Commando Brigade Air Squadron)	
	Gazelle AH.Mk 1, Lynx AH.Mk 7
848 Sqn	Sea King HC.Mk 4
899 Sqn	Sea Harrier FA.Mk 2, T.Mk 8
Heron Flight	Jetstream T.Mk 3
Engineer Training School	various ground instruction airframes
RN Historic Flt	Swordfish Mk II, Chipmunk T.Mk 10, Sea Fury FB.Mk 11 (under restoration) Firefly AS.Mk 5
RNAS Culdrose (HMS *Seahawk*)	
700(M) Sqn	Merlin HMA.Mk 1
750 Sqn	Jetstream T.Mk 2
771 Sqn	Sea King HAR.Mk 5/HAS.Mk 5U
810 Sqn	Sea King HAS.Mk 6
814 Sqn	Sea King HAS.Mk 6
820 Sqn	Sea King HAS.Mk 6
849 Sqn	Sea King HAS.Mk 5U/AEW.Mk 2/2A
Fleet Requirements & Directions Unit (FRADU)	
	Hawk T.Mk 1/1A
School of Flight Deck Operations	
	various ground instruction airframes
Engineer Training School	various ground instruction airframes
RNAS Prestwick (HMS *Gannet*)	
819 Sqn	Sea King HAS.Mk 5U/6
Naval Aircraft Repair Organisation	**Fleetlands**
	Gazelle AH.Mk 1 Fleetlands
Air Engineering & Survival School (HMS *Sultan*)	
Civilian Contract Flying	
Flag Officer Sea Training	Culdrose
	AS 365N2 Dauphin Roborough
Contract Units	
Flight Refuelling Aviation	Falcon 20C/DC/ Bournemouth E/ECM
Joint Elementary Flying Training School	
	Firefly Barkston Heath
Naval Flying Grading Flight	Roborough
	Grob G115 D-2 Heron
Defence Helicopter Flying School	
705 Sqn	Squirrel HT.Mk 1 Shawbury

The Royal Marine's organic air squadron, No. 847 Sqn, operates virtually army-standard Lynx AH.Mk 7s (above) and Gazelle AH.Mk 1s from RNAS Yeovilton when not embarked or in the field.

frigates. Each individual flight comprises a pilot, observer and seven maintainers. Offensive weapons carried consist of the Sea Skua anti-ship missile, Sting Ray torpedoes and depth charges.

The latest version of the Lynx, the HMA.Mk 8, equipped with a central tactical system and passive identification device, continues to progressively enter service. The July 1998 Strategic Defence Review limited Merlin HMA.Mk 1 procurement – which was due ultimately to replace the Lynx in service with the navy – to 44, and increased the number of conversions of Lynx HMA.Mk 8s by 10 to 38. Two Lynxes modified for cold weather conditions (Lynx HAS. Mk 3S(ICE)) are carried by the ice patrol ship HMS

Above: Using the same scheme adopted by the RAF Hawk fleet, the Fleet Requirements and Direction Unit's Hawks are used to simulate attacking aircraft and cruise missiles for ship's gunners and radar operators. The unit is run by Flight Refuelling Ltd.

Below: The FAA has retired its fleet of Sea Heron C.Mk 20s, but the Heron name lives on with its adoption of the Grob 115D-2 for the Navy Flying Grading Flight operated by Shorts. The Herons replaced Chipmunk T.Mk 10s in 1994.

Endurance, which supports British interests in the South Atlantic.

COMMANDO ASSAULT

While Culdrose is home to the ASW and SAR mission Sea Kings, Yeovilton also operates three squadrons of the medium-lift assault version of the helicopter, the HC.Mk 4 or Commando. In addition to their primary role of troop carrying, Nos 845, 846 and 848 Sqns regularly carry underslung loads and 105-mm guns from ship to shore. With a full NVG capability, the aircraft offer effective support to amphibious operations and regularly deploy to Norway for training in Arctic conditions.

The replacement for the Sea King HC.Mk 4 is the Future Amphibious Support Helicopter (FASH) for which contenders are likely to be the Bell-Boeing V-22 Osprey, EHI Merlin and the Sikorsky S-92. An in-service date of 2008 is expected.

Further support for the Royal Marines is provided by No. 847 Sqn (3 Commando Air Squadron) which provides communications and anti-tank support during amphibious operations with its complement of Gazelle AH.Mk 1 and Lynx AH.Mk 7 helicopters.

FLYING TRAINING

Initial flying grading is undertaken by the Britannia Royal Naval College at Dartmouth, using contractor-operated Grob 115 Herons. Students progress to the JEFTS for a 24-week basic flying training course. Rotary-wing students progress to the DHFS for an 11-week course on Squirrel HT.Mk 1s before receiving operational conversion training to the multi-engined front-line types. Pilots destined for the Sea

The reduced need for multi-engine training following the 1970s defence papers allowed No. 750 Sqn to replace its Sea Princes with ex-RAF Jetstreams modified to T.Mk 2 standard, the first entering service in 1978. Of the 16 converted, two were sold to the Uruguayan navy in 1998, two are stored and one other was written off.

Above: Flag Office Training has access to two AS 365Ns provided by Bond Helicopters. Both are painted in a bright red scheme and have rescue winch attachments.

Harrier squadrons train alongside RAF students on Tucanos and Hawks before transferring to No. 899 Sqn for operational conversion.

SUPPORT UNITS

The Naval Aircraft Repair Organisation at Fleetlands near Gosport undertakes the maintenance of the FAA's large helicopter fleet. Both here, and at Almondbank, helicopters are stored, many of which could be returned to service quickly if the need arose, while Sea Harriers are both maintained and stored at RAF St Athan. Using a number of retired airframes for training purposes, the Air Engineering and Survival School at HMS *Sultan*, Gosport and the two Engineer Training Schools at Culdrose and Yeovilton provide training for ground trades. The School of Flight Deck Operations, also at Culdrose, trains students destined for the busy decks of the aircraft-carriers.

The Fleet Requirements and Direction Unit (FRADU) provides aerial targets for the training of ground-based fighter direction operators with a fleet of 12 Hawk T.Mk 1/1As. Operated by civilian personnel, the unit is supplemented by a fleet of 18 civilian-registered Dassault Falcon 20 aircraft which operate in the electronic warfare training role.

THE FUTURE

The three existing carriers are scheduled to be replaced between 2012 and 2015, by two 30,000-40,000-tonne vessels (CVFs) capable of carrying 50 aircraft each. About 100-120 Future Carrier-Borne Aircraft (FCBA) are to be acquired to operate from the new CVFs. HMS *Ocean* will be supplemented by two new LPDs, *Albion* and *Bulwark*, which will replace the ageing assault ships *Intrepid* and *Fearless*, only one of which is kept operational at present.

The UK government considers the next-generation US Joint Strike Fighter (JSF) to be the ideal replacement for both the FAA's Sea Harrier and RAF Harrier GR.Mk 7 fleet. A competition between two consortia will see a contract awarded in 2001 for the manufacture of a STOVL aircraft which will replace a variety of current types in the USAF, USN and USMC inventories. In FAA service the JSF will provide the very latest in aircraft and missile technology and will enable the service to meet the likely calls by the UK government to project power overseas.

Army Air Corps

The current Army Air Corps (AAC) was formed on 1 September 1957 and today, through its involvement in operations in Northern Ireland, the Gulf War and former Yugoslavia, it arguably has more operational experience than any Western army air arm.

The very earliest military flying in the United Kingdom was conducted by the army's Royal Engineers which formed an air battalion in 1911, having experimented with balloons for a number of years. It was absorbed into the Royal Flying Corps in 1912. World War I saw the emergence of air power as an integral part of the support of army operations. The creation of the RAF at the end of the conflict led to all flying in support of the British Army becoming one of its tasks, for which it established Army Co-operation Squadrons. World War II and the emergence of the airborne assault mission saw the AAC, and subsequently the Glider Pilot Regiment, operate Horsa and Hamilcar gliders, most famously in Operation Overlord, the D-Day invasion in 1944. During this period the RAF continued to operate 12 Air Observation Post Squadrons in the artillery observation role.

Following a lengthy struggle to become independent, the AAC finally broke away in 1957 and almost immediately received its first helicopter, the Saro Skeeter. Since then, the AAC has concentrated predominantly on rotary-wing operations, although small numbers of fixed-wing aircraft – the Beaver AL.Mk 1 and more recently the Islander AL.Mk 1 – have been retained for observation and liaison duties. During the Cold War the anti-tank role come to the fore, but the necessary change in emphasis when the Cold War ended dictated a more mobile and flexible force. Fortunately, this has come at a time when extensive re-equipment was already necessary, enabling the AAC to enter the 21st century as a well-equipped, modern and potent force.

ROLES AND ORGANISATION

Constituting some 1.5 per cent of the personnel of the British Army (3 per cent if the Royal Electrical & Mechanical Engineers/REME component is included), the AAC has five defined roles: armed action, observation and reconnaissance, direction of fire and forward air control, assistance in command and control, and the (limited) movement of men and materiel. Unlike most Western army air arms, the AAC does not possess a medium- or heavy-lift helicopter capability, for which role the RAF operates the Support Helicopter Force of Puma and Chinooks.

The Headquarters Director Army Aviation at Middle Wallop in Hampshire commands a fleet of helicopters greater in size than those of both the RAF and FAA combined. It has only one squadron reporting directly to it – 667 (Development & Trials) Sqn – which, as its title suggests, undertakes test work with its pairs of Gazelle AH.Mk 1s and Lynx AH.Mk 7s. The front-line helicopter squadrons are formed into five regular and one volunteer AAC regiment which, in turn, report to the aviation branch of Land Command (LandCom) at Netheravon in Wiltshire.

REGIMENTS

Based at the former RAF airfield at Gütersloh, 1 Regiment AAC is assigned to the 1st (UK) Armoured Division at Herford, Germany. The UK maintains three armoured Brigades and a divisional headquarters in Germany as part of British Forces Germany, a title adopted following the demise of the British Army of the Rhine and RAF Germany. The regiment has three component AAC squadrons, each equipped with six TOW-fitted Lynx AH.Mk 7s and six Gazelle AH.Mk 1s. Fulfilling an identical role in the UK itself is the Dishforth, North Yorkshire-based 9 Regiment AAC. Assigned to the 3rd UK Division at Bulford, Wiltshire, the regiment returned to the UK from Germany in the early 1990s.

The former RAF Phantom base at Wattisham in Suffolk is home to 3 and 4 Regiments which are both assigned to the 24th Airmobile Brigade, a flexible, high-speed anti-tank reserve force based at Catterick, North Yorkshire. The Brigade forms part of the Multi-National Division which includes elements from the German, Belgian and Dutch armies. Each of the assigned AAC Regiments operates two squadrons with a mixed complement of six Gazelle AH.Mk 1s and six TOW-armed Lynx AH.Mk 7s. The remaining squadron in each regiment operates 11 Lynx AH.Mk 9s in the light battlefield mission. Medium lift is provided to the Brigade by the RAF, which is tasked to supply 18 Pumas and Chinooks.

The British Army continues to maintain a heavy commitment in Northern Ireland, where 5 Regiment at Aldergrove operates the only front-line fixed-wing

For well over 20 years the main helicopters in Army service have been products of the 1967 Anglo-French helicopter agreement. A total of 212 Gazelle AH.Mk 1s (above) and 113 Lynx AH.Mk 1s were procured for the AAC, entering service in 1973 and 1977 respectively. The Lynxes were upgraded as AH.Mk 7s from 1987 onwards (top).

aircraft in the AAC. 1 Flt flies four Islander AL.Mk 1s on surveillance and general support duties in the province alongside two helicopter squadrons. While both 655 and 665 Squadrons have six Lynx AH.Mk 7s and nine Gazelle AH.Mk 1s assigned, each operates a much larger complement, borrowing helicopters as required from other regiments. The nature of the threat to the Army in Northern Ireland sees a greater emphasis on troop-carrying consequent to the need to support and resupply army garrisons by air. The Gazelles are also used extensively in the observation role.

Overseas, the AAC has been heavily committed to operations in the former Yugoslavia since 1992. Originally commanded by the United Nations Protection Force and subsequently NATO's Implementation Force (IFOR), in 1996 NATO established a Stabilisation Force (SFOR) which included as part of its multi-national make-up a single AAC squadron in rotation. British ground troops have also been supported by RAF Chinooks and FAA Sea King HC.Mk 4s.

VOLUNTEER FORCES

The Territorial Army is the reserve component of the British Army. Its aviation element is controlled by 7 Regiment (Volunteers) at Netheravon which operates two squadrons equipped with six Gazelle AH.Mk 1 each from the Wiltshire airfield. Two other territorial units are located at the RAF bases at Leuchars (3 Flt) and Shawbury (6 Flt) and, in common with their Netheravon counterparts, are tasked to support the TA and Field Army. All aircrew are former regular AAC personnel. In conjunction with contracted maintenance, this is regarded as an economic way of fulfilling a necessary role while freeing the regular field army units to continue with their operational

1 Flight's four Islander AL.MK 1s (above) are equipped for the surveillance mission. The type has served with the army since 1989. They operate from Aldergrove alongside a Lynx AH.Mk 7 squadron and a Gazelle AH.Mk 1 squadron (No. 665, below) providing support for troops in Northern Ireland.

training. Each flight comes under the administrative control of 7 Regiment but under the operational command of the local Army headquarters. For example, 3 Flight is commanded by a full-time major under the operational control of Headquarters Army Scotland, with a complement of 12 ex-regular pilots, 12 ex-regular aircrew and two regular airtroopers.

Army Air Corps

Headquarters Director Army Aviation		Middle Wallop
667 (Development and Trials) Sqn		Middle Wallop
	Gazelle AH.Mk 1, Lynx AH.Mk 7	
1 Regiment		
651 Sqn	Gazelle AH.Mk 1, Lynx AH.Mk 7	Gütersloh
652 Sqn	Gazelle AH.Mk 1, Lynx AH.Mk 7	Gütersloh
661 Sqn	Gazelle AH.Mk 1, Lynx AH.Mk 7	Gütersloh
2 (Training) Regiment/School of Army Aviation		
Defence Helicopter Flying Squadron/660 Sqn		
	Squirrel HT.Mk 1	Shawbury
670 Sqn	Squirrel HT.Mk 2	Middle Wallop
671 Sqn	Gazelle AH.Mk 1,	Middle Wallop
	Lynx AH.Mk 7	
Joint Elementary Flying Training Squadron *(civilian contract)*		
	Firefly 160	RAF Newton
	Firefly 260	RAF Barkston Heath
Advanced Fixed Wing Flight		Middle Wallop
	Islander AL.Mk 1	
3 Regiment		
653 Sqn	Lynx AH.Mk 9	Wattisham
662 Sqn	Gazelle AH.Mk 1, Lynx AH.Mk 7	Wattisham
663 Sqn	Gazelle AH.Mk 1, Lynx AH.Mk 7	Wattisham
4 Regiment		
654 Sqn	Gazelle AH.Mk 1, Lynx AH.Mk 7	Wattisham
659 Sqn	Lynx AH.Mk 9	Wattisham
669 Sqn	Gazelle AH.Mk 1, Lynx AH.Mk 7	Wattisham
5 Regiment		
655 Sqn	Lynx AH.Mk 7	Aldergrove
665 Sqn	Gazelle AH.Mk 1	Aldergrove
1 Flight	Islander AL.Mk 1	Aldergrove
7 Regiment (Volunteers)		
658 Sqn (V)	Gazelle AH.Mk 1	Netheravon
666 Sqn (V)	Gazelle AH.Mk 1	Netheravon
3 Flight (V)	Gazelle AH.Mk 1	RAF Leuchars
6 Flight (V)	Gazelle AH.Mk 1	RAF Shawbury
9 Regiment		
656 Sqn	Gazelle AH.Mk 1, Lynx AH.Mk 7	Dishforth
657 Sqn	Gazelle AH.Mk 1, Lynx AH.Mk 7	Dishforth
664 Sqn	Gazelle AH.Mk 1, Lynx AH.Mk 7	Dishforth
Independent Units		
7 Flight	Bell 212	Brunei
8 Flight	Agusta A 109A, Gazelle AH.Mk 1	Hereford
12 Flight	Gazelle AH.Mk 1	RAF Brüggen
16 Flight	Gazelle AH.Mk 1	Dhekalia, Cyprus
25 Flight	Gazelle AH.Mk 1	Belize
BATU	Gazelle AH.Mk 1	Suffield, Canada
Museum of Army Flying Historical Flight		Middle Wallop
	various	

Below: Two Wattisham squadrons use the Lynx AH.Mk 9 in the light battlefield support mission; this example is attached to the 4 Regiment's 659 Sqn. Some 25 AH.Mk 9s have been delivered.

TRAINING

The training of aircrew is the responsibility of 2 (Training) Regiment at Middle Wallop. With the retirement of the DHC Chipmunk from AAC service, flying training is now undertaken in the tri-service Joint Elementary Flying Training School. Initial flying grading is undertaken at RAF Newton while basic training is concentrated at RAF Barkston Heath. Students progress to the Defence Helicopter Flying Squadron (DHFS) for a 17-week course of training on the single-engined Squirrel HT.Mk 1. While basic training is undertaken alongside RAF and FAA aircrew, army student pilots undertake an army advanced syllabus prior to transferring to 670 Sqn at Middle Wallop. Here, using the Squirrel HT.Mk 2, students receive tactical operations training in a 30-week course. Upon obtaining their wings, students progress to operational conversion on either the Gazelle or Lynx at 671 Sqn, also based at the Hampshire airfield. Training for ground trades is provided at the School of Army Aviation at Middle Wallop. REME technicians are trained at the School of Aeronautical Engineering at Arborfield in Berkshire, which uses a number of retired helicopters for this purpose.

INDEPENDENT UNITS

A number of small flights undertake unique missions on behalf of the AAC at various locations around the world. Supporting British Forces exercising in the Sultanate of Brunei is 7 Flight, which operates a type unique within the AAC. Its three twin-engined Bell 212s, dry-leased from Bristow Helicopters, are similar to the RAF's Griffin HT.Mk 1s and are considered to be ideal for operating over the dense jungle of the Asian country. Despite the RAF's withdrawal of a permanent presence in Belize, the AAC maintains 25 Flight there. Consideration is currently being given to acquiring additional Bell 212s, as single-engined Gazelles are not ideal when operating over decidedly inhospitable terrain; recently, two Lynx AH.Mk 7s have been deployed to help overcome this problem.

The British forces in Cyprus are supported by three Gazelles assigned to 16 Flight at Dhekalia, and the various headquarters around Rheindahlen in Germany have the services of 12 Flight at RAF Brüggen. The largest permanent overseas detachment is in Canada, where five Gazelles operate as part of the

Above: Middle Wallop is the home of the Army Air Corps and the base of the operational conversion unit, No. 671 Sqn, which uses both Gazelles and Lynx AH.Mk 7s. The AH.Mk 9 differs from the AH.Mk 7 by having wheels instead of skids, advanced BERP rotors and a reversed-direction tail rotor.

British Army Training Unit (BATU). The large live firing ranges on the Alberta plains enable six battle-group-sized exercises to be conducted each year on a scale simply not possible in Europe. To save the cost of shipping helicopters to Canada for each exercise, three of the assigned Gazelles are used by the aviation element of the deploying force. The remaining two Gazelles are used by range safety staff and are accordingly painted in high-visibility markings. They are often called upon to act in their secondary role of casualty evacuation, the nearest hospitals being some distance from the range areas. In 1997 Lynx AH.Mk 7s were deployed for the first time to BATU and active consideration is now being given to basing some examples of the type in Canada.

Four Agusta A 109As and a pair of Gazelle AH.Mk 1s are used by 8 Flight at Hereford. Two of the A 109As were captured from Argentine forces during the 1982 Falklands War and returned to the UK, while a second pair was purchased new from the manufacturer. 8 Flight is due to merge with 657 Sqn.

DAWN OF THE APACHE

The delivery of the first of 67 WAH-64Ds in May 2000 will see the beginning of marked changes in the AAC's organisational structure, prompted primarily by the dramatic increase in capability the Longbow Apache will bring. The helicopter will be powered by Rolls-Royce RTM.322 engines, common to RAF and FAA Merlins. Anti-armour missiles will consist of an advanced development of the Rockwell AGM-114F Hellfire. The initial eight Apaches delivered will form a Fielding Squadron for the training of the first instructors on the type, following which the helicopters will transfer to the operational conversion role. Single Apaches will also join the DERA trials fleet at Boscombe Down and 667 (D&T) Squadron for ongoing development work.

A new Air Manoeuvre Brigade will be established to which 3 and 4 Attack Regiments at Wattisham and 9 Attack Regiment at Dishforth will report. Each will have two squadrons of eight WAH-64Ds alongside a squadron operating a similar number of modified Lynx AH.Mk 7s. These Lynxes are scheduled to lose their TOW equipment and fulfil a new role as light utility helicopters. In turn, the existing Lynx AH.Mk 9s will transfer to 1 Regiment in Germany, whose three squadrons will each operate eight of the type. With the reduction in the number of Gazelle and Lynx helicopters operated, the AAC's troop-carrying capacity will fall by some 50 per cent. Elsewhere, 5 Regiment is scheduled to expand on paper, adopting the aircraft it currently operates on loan.

Tentative plans to operate a Skyship 600 in a surveillance role have been abandoned following extensive trials, and the dirigible has been returned to the manufacturer.

Prospective army pilots pass from the Slingsby T.67M-260 (below left) of the Hunting Aviation Ltd-run Joint Elementary Flying Training School to the Squirrel HT.Mk 1s of the Defence Helicopter Flying School at RAF Shawbury, operated by FBS Ltd. From there they move to Middle Wallop and fly the DHFS's Squirrel HT.Mk 2s (below right) of No. 670 Squadron.

Defence Evaluation and Research Agency

The largest military research agency in Europe, the Defence Evaluation and Research Agency (DERA) was formed in 1995 with the amalgamation of a number of different agencies. A semi-independent organisation, it controls the core of the UK government's Ministry of Defence non-nuclear scientific and technical assets.

DERA's flying activities are controlled by the Aircraft Test & Evaluation Sector which centres its flying activities at RAF Boscombe Down in Wiltshire. This airfield, long the home of test and evaluation flying in the UK, now also conducts research and development flying following the end of test flying at Bedford and Farnborough. At Boscombe Down a trials fleet of some 40 aircraft of diverse types are operated by three platform departments – Combat

Flanked by a Jaguar T.Mk 2 in the distinctive 'raspberry ripple' colour scheme, this BAe Dunsfold Harrier T.Mk 10 wears large calibration marks on its tail and nose.

Aircraft, Rotary Wing and Patrol & Support. Test flying is undertaken by serving military test pilots in test squadrons within each department. The fleet is supplemented at any one time by some 20 additional aircraft on loan from the UK armed forces for contract work.

Test pilots from throughout the world are trained at the Empire Test Pilots School at Boscombe Down, which can trace its roots to 1914. On average, 18 test pilots and engineers graduate from each 40-week course, having flown the School's unique fleet of diverse aircraft. Service aircraft are loaned as and when required and training is also undertaken abroad at overseas test facilities.

Lodger units at Boscombe Down include the Meteorological Research Flight (MRF) and School of Aviation Medicine (SAM). The former operates a

Defence Evaluation and Research Agency

DERA Headquarters	Farnborough

Aircraft Test & Evaluation Sector, Boscombe Down
Combat Aircraft Department (Fast Jet Test Sqn)
Rotary Wing Aircraft Department (Rotary Wing Test Sqn)
Patrol & Support Aircraft Department (Heavy Wing Test Sqn)
Empire Test Pilots School
(above squadrons use Andover, BAC 1-11, Basset, Gazelle, Harrier, Harvard, Hawk, Hunter, Jaguar, Lynx, Navajo, Sea King, Tornado and Tucano, plus aircraft loaned from the services)
Ranges
Air & Sea Capabilities Sector

DERA Llanbedr	Canberra B.Mk 2, Hawk T.Mk 1, Meteor D.Mk 16
DERA West Freugh	Jetstream T.Mk 2
Meteorological Office	**Bracknell**

Meteorological Research Flight, Boscombe Down
Hercules W.Mk 2

single Hercules W.Mk 2, a unique research platform for the Meteorological Office at Bracknell; the School of Aviation Medicine, part of Strike Command, operates two Hawk T.Mk 1s in support of Farnborough-based research projects.

FEDERAL REPUBLIC OF YUGOSLAVIA

Capital: Belgrade
Population: 10.5 million
Land area: 102170 km² (39,435 sq miles)
Major cities: Kragujevac, Nis, Novisad, Subotica

Ratno Vazduhoplovstvo i Protiv Vazdusna Odbrana (Air Force and Air Defence Force)

The remains of the old Communist Yugoslavia were reinvented as the Federal Republic of Yugoslavia (FRY) after Slovenia, Bosnia, Croatia and Macedonia broke away from Belgrade in 1991-92. The new Federal Air and Air Defence Force (RV i PVO) had to undergo considerable reorganisation as it withdrew to bases in Serbia and Montenegro. At the same time, elements of the JRV, particularly transport helicopters, continued to support the Bosnian Serb and Krajina Serb air forces in their continuing wars with Bosnian and Croat forces through late 1995.

In the aftermath of the Dayton Peace Accords, the RV i PVO has had to face reductions in front-line airframe numbers imposed by the Organisation for Security and Co-operation in Europe (OSCE) arms control treaty. Under this treaty the RV i PVO is limited to 155 combat aircraft and 53 attack helicopters. This has not been altogether unwelcome because it has allowed the RV i PVO to dispose of some of its older

RV i PVO

Direct-reporting Units		
det 353rd IAE 'Hawks'	MiG-21R/UM	Batajnica
det 353rd IAE 'Hawks'	IJ-22, NJ-22	Ladjevci
677th TRA Eskadrila	An-2TD, An-26, Do 28D	Nis
890 MHE 'Pegasus'	Mi-8, SA 341/GAMA	Batajnica
comms flight	UTVA-75	Beograd-Sucrin
SUKL (Federal Flying unit)	Yak-40	Beograd-Sucrin
VOC (test centre)	various	Batajnica

83rd Lovacki Avijackijski Puk		
123rd LAE 'Lions'	MiG-21bis/UM	Pristina
124th LAE 'Thunders'	MiG-21bis	Pristina
comms flight	SA 341, UTVA-75	Pristina

98th Lovacki Bombardersli Avijacijski Puk		
241st LBAE 'Tigers'	J-22/NJ-22 Orao	Ladjevci
252nd LBAE	J-22/NJ-22 Orao	Batajnica
comms flight	UTVA-75	Ladjevci

119th Helikopterska Brigada		
712th POHE 'Scorpions'	SA 341 GAMA	Nis
714th POHE 'Shadows'	SA 341 GAMA	Ladjevci
787th TRE	Mi-8	Nis

172nd Avijacijska Brigada		
229th LBAE 'Swords'	G-4 Super Galeb	Podgorica
239th LBAE 'Vampires'	G-4 Super Galeb	Podgorica
242nd LBAE 'Eagles'	G-4 Super Galeb	Podgorica
251st LBAE 'Puma'	G-2A Galeb	Podgorica
333rd NAE	UTVA-75	Kovin
897th MHE 'Hornets'	Mi-8, SA 341	Podgorica
(provides Gazelles for Hornets display team)		
comms flight	UTVA-75	Podgorica
'Letece Zvezde' (Flying Stars team)	G-4 Super Galeb	Podgorica
VIP flight	Mi-8	Podgotica

204th Lovacki Avijacijski Puk		
127th LAE 'Knights'	MiG-29/UB	Batajnica
126th LAE 'Delta'	MiG-21R/UM	Batajnica
comms flight	UTVA-75	Batajnica

airframes to museums, exports and the breakers yard. Some helicopters (including Mi-24s) have also been transferred to the paramilitary police for use in the crisis-torn region of Kosovo, for internal security duties against Albanian guerrilla groups. The mainstream RV i PVO has concentrated on close air support and air defence missions. Helicopter operations have been devolved to the ground forces and navy, which retired the three Mi-14 'Haze-As', two Ka-27 'Helix-As' and six Ka-25 'Hormones' in 1998. In early 1998 it emerged that Yugoslavia was in the process of negotiating a deal with the Russian Federation to supply a large consignment of weapons, including MiG-29 fighters and a new generation surface-to-air missiles.

RV i PVO Antonov An-26s are believed to be based at Nis as part of a composite transport squadron. Currently, the air force operates about 25.

The aerobatic team of the RV i PVO is the 'Letece Zvezde' (Flying Stars) which operates six SOKO G-4A Super Galebs for its display routine.

Middle East

The Middle East is strategically placed between Africa, Asia and Europe, and has a long and varied history which has been dominated in the last 60 years by warfare and the oil industry. Today, the Middle East is the largest arms market in the world.

The emergence of the state of Israel and the subsequent wars it has fought (1948, 1956, 1967 and 1973) have left a heavy legacy in the region. Israel is one of the best-armed countries in the world, and few air forces have gained as much combat experience. The lack of a comprehensive peace treaty at the 'conclusion' of each conflict did not allow any draw-down of forces or relaxation of the tension in the area. This changed with the Egyptian-Israeli Peace Treaty of 1979 and the Israeli-Jordanian treaty of 1994, allowing the start of peaceful discussions on mutual matters of interest. The plight of the Palestinians has been a subject high on the list of issues in the Middle East for over half a century. The Israeli-occupied West Bank and Gaza

Strip are subject to the Israeli-Palestinian Interim Agreement; signed in September 1993, this treaty provided a transitional period of Palestinian interim self-government in the areas. Negotiations for a permanent settlement started in May 1996.

Since the end of the second Gulf War in 1991, Iraq has been monitored by the United Nations Special Commission (UNSCOM), which is charged with the destruction of Iraq's remaining weapons of mass destruction and its facilities to produce and operate them. Saddam Hussein's brinkmanship has resulted in American and British air force aircraft being used in combat many times since the end of that war, most visibly in Operation Desert Fox (16-20 December 1998). Continued enforcement of the northern and southern 'No-Fly Zones' will keep American and British forces based in the region for some time to come.

Iran continues to be a pariah state in the West, but the election of a more moderate cleric leader

(Mohammad Khatami) in August 1997 may yet see a new direction taken in its foreign policy. Syria is also isolated, due to its non-acceptance of the peace process between Israel and the Arab states, and its support of terrorist groups.

The end of the civil war in the Lebanon has allowed the country to return to some kind of normality. Israeli attacks on suspected Hizbollah bases in the country in 1996 – which resulted in the deaths of over 100 people – led to a strained ceasefire.

Saudi Arabia, with just over one-quarter of the world's oil reserves, maintains a large and modern air force. The Saudi air force and the Gulf states rely heavily on foreign nations to maintain and operate their aircraft. Bahrain, Kuwait, Qatar and the United Arab Emirates all reinforced their military forces after the 1990/1991 Gulf War.

Civil war in Yemen resulted in a victory for the north in July 1995, but unrest is rife in the country.

BAHRAIN

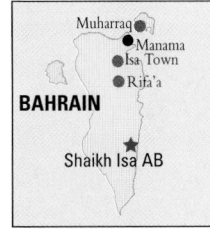

Capital: Manama   **Population:** 500,000
Land area: 661 km² (255 sq miles)
Major cities: Isa Town, Muharraq, Rifa'a

Bahrain Amiri Air Force

The Emirate of Bahrain, which gained its independence from Britain in 1971, has amassed wealth from its oil exports. The Bahrain Amiri Air Force can trace its existence back to a small State Police air arm established in 1965 and equipped with Westland Scouts, but the modern air force was organised during 1976 when three BO 105Cs were delivered. Today, the air force operates from the massive Shaikh Isa Air Base

Armed with a pair of wingtip AIM-9Ls, this is the first F-16D received by the Bahrain Amiri Air Force. All serve with the 1st Fighter Sqn at Shaikh Isa AB.

facility on the main island of Sintrah, constructed with American assistance from 1987.

The sharp edge of the service are the fighters of the 1st Fighter Wing, consisting of two squadrons operating the F-5E/F Tiger II and the F-16C/D Block 40D/E. Operated by the 1st Fighter Squadron, eight F-16Cs and four F-16Ds were ordered in 1987 and delivered under the Peace Crown programme during 1990. In service the type carries the AIM-9L Sidewinder and the AIM-7F Sparrow, and also uses the Sharpshooter targeting pod. No F-16s have been lost. In the aftermath of the Gulf War, the air force considered acquiring more F-16s, including the embargoed Pakistani examples, the ex-US Navy F-16Ns and 20 ex-USAF F-16A/Bs; in 1998 Bahrain ordered 10 new production F-16C/Ds.

Bahrain's Tiger IIs belong to the 6th Fighter Squadron. After being refused the F-20A Tigershark, eight F-5Es and two F-5Fs were purchased, the first arriving in December 1985, and were based at Manama Airport until Shaikh Isa AB opened. They were armed with AIM-9P-3s, but may have taken on

Bahrain Amiri Air Force

1st Fighter Wing	Shaikh Isa AB
1st Fighter Sqn	F-16C/D
6th Fighter Sqn	F-5E/F
Combat Helicopter Sqn	AH-1E, TAH-1P
Amiri Royal Flight	Gulfstream II(TT)/III, 727-2M7, UH-60L.

Bahrain Public Security Flying Wing	
	Bell 412, BO 105C
Bahrain Amiri Navy	
	BO 105

a ground attack role since the F-16s were delivered.

A combat helicopter squadron is believed to have operated eight AH-1Es (also quoted as P models) and six TAH-1Ps since 1994.

The Amiri Royal Flight operates a civil-registered Gulfstream II(TT), a GIII and a Boeing 727-2M7 as VIP transports for the Bahrain government. A single UH-60L equipped with ESSS pylons is also used for VIP transport. Twelve AB 212s which fly in full air force colours are used on transport and civilian policing tasks, but are thought to be used by the Public Security Flying Wing, alongside three Bell 412s and a pair of BO 105s. A pair of BO 105CBS-4s is used by the Navy's air element for SAR.

IRAN

Capital: Tehran
Population: 58 million
Land area: 1.648 million km² (636,130 sq miles)
Major cities: Isfahan, Meshed, Tabriz, Shiraz

Islamic Republic of Iran Air Force

The Imperial Iranian Air Force was the most powerful force in the Middle East until the fall of the Shah. The roots of Iranian air power were created in 1924 when aircraft were used to put down warlords in the south of the country. Training and equipment was acquired from Britain. Iran (Persia) was occupied during World War II by the allies, and the air arm was reformed soon after when the country joined the Central Treaty Organisation (CENTRO), receiving F-47 Thunderbolts from the US.

The air arm became an independent air force in August 1955 under the Aviation Department of the Ministry of War, and gained F-86F Sabres, its first combat jets. The strategic position of Iran did not go unnoticed by the USA, which began to supply large numbers of aircraft to the country from the early 1960s. The oil in the south of the country provided the money to fuel the expanding ambitions of the

Top right: Speculation has been rife over the last 20 years about the status of Iran's Tomcats. Operational aircraft carry a much reduced warload of AIM-54s and AIM-7s.

Below: Sukhoi Su-24MK 'Fencer-Ds' give the IRIAF a long-range strike capability. The second Gulf War provided the air force with 24 examples which escaped from Iraq.

Islamic Republic of Iran Air Force

Western Area Command	
TAB 1, Tehran-Mehrabad	
11 TFS	MiG-29A/29UB 'Fulcrum'
12 TFS (?) *	F-5A/B Freedom Fighter
83 TFS (?)	F-14A Tomcat
? TFS	Su-24MK 'Fencer-D'
11 TS	C-130H Hercules
12 TS	C-130H Hercules
13 TS (?)	Boeing 707-3J9C
14 TS (?)	Boeing 747F-131
TAB 2, Tabriz	
21 TFS	F-5E/F Tiger II
22 TFS	F-5E/F Tiger II
23 TFS (?)	MiG-29A/29UB 'Fulcrum'
Base Flight	AB 212, Bell 214
TAB 3, Hamadan-Shahroki	
31 TFS	F-4D/E Phantom
32 TFS	F-4D/E Phantom
33 TFS	F-4E Phantom
Base Flight	AB 212
TAB 4, Dezful-Vahdati	
41 TFS	F-5E/F Tiger II
42 TFS	F-5E/F Tiger II
43 TFS *	F-5E Tiger II
TAB 5, Umidiyeh	
51 TFS	F-7M Airguard
52 TFS	F-7M Airguard
53 TFS (?) *	F-7M Airguard
TAB 7, Shiraz	
71 TFS	F-5E/F Tiger II
72 TFS (?)	Su-24MK 'Fencer-D'
71 TS	C-130H Hercules
72 TS	C-130H Hercules
73 TS (?)	Il-76MD 'Candid'
83 TFS det	F-14A Tomcat
83 TFS *	F-14A Tomcat

? HTS (?)	CH-47C, AB 212
? ASW Sqn	P-3F
TAB 8, Khatami-Isfahan	
81 TFS *	F-14A Tomcat
82 TFS	F-14A Tomcat
Base Flight	AB 212
TAB 12, Tehran-Dashan Tappeh	
? VIP TS	various types
? SOS	PC-6B Turbo-Porter

Southern Area Command	
TAB 6, Bushehr	
61 TFS	F-4D/E Phantom
62 TFS	F-4D/E Phantom
82 TFS	F-14A Tomcat
TAB 9, Bandar Abbas	
91 TFS	F-4E Phantom
92 TFS	F-4D/E Phantom
? SAR Flight	AB 212 ?
? Army Sqn	Rockwell Commander 681B, AB 212
TAB 10, Chah Bahar	
101 TFS	F-4D/E Phantom

Eastern Area Command	
TAB 13, Zahedan	
? TFS (Pasdaran)	Shenyang J-6

Flying Training School	
TAB 11, Tehran-Ghale Morghi	
? TTS	MFI-17 Mushshak
? TTS	Beech F33A/C Bonanza
? TTS	EMB-312 Tucano
? TTS	PC-7 Turbo-Trainer
? TTS	PC-7 Turbo-Trainer
TAB 7, Shiraz	
? TTS	T-33A Shooting Star

* aircraft of this squadron currently believed to be in storage

Middle East

The IRIAF continues to use large numbers of Tiger IIs (including the F-5F conversion trainer). The type was originally destined to be replaced by the F-16 but political events prevented this purchase.

Above: Iranian F-4E Phantoms are used in the ground attack role, and can carry AGM-65 Mavericks, Standards, BL755 cluster bombs and indigenous missiles and bombs. It is also believed that the type has gained an anti-shipping capability.

Right: Iran receive six Boeing 707-3JCs fitted with air refuelling booms. Wingtip-mounted hose-and-drogue systems were added 'in-country'. This aircraft is seen leading an F-5F (with a fixed probe), an RF-4E and an F-14A.

Shah. Northrop F-5 Freedom Fighters were supplied from 1965 onwards and F-4 Phantom IIs followed from 1968. By the early 1970s Iran was militarily the most powerful country in the region. The willingness of the US to supply its top-of-the-range military hardware, such as the F-14A Tomcat and AIM-54 Phoenix AAM, demonstrated the close relations between the two countries.

The 1979 Islamic revolution changed the situation in the country overnight and dramatically altered the balance of power in the region. No longer was Iran loyal to the US, but openly hostile. American assistance ceased and the air force suffered as pro-Shah officers were removed, resulting in very low serviceability levels. The air force was renamed the Islamic Republic of Iran Air Force (IRIAF). Iraq saw its chance to gain large tracts of Iranian territory and launched an attack on 22 September 1980, starting the first Gulf War, which was to last for eight years and turned into a war of attrition. Computer codes were broken for the vast spares inventory purchased by the Shah during the early 1980s, allowing the IRIAF to

fend off repeated attacks by the Iraqi air force.

Defections to Saudi Arabia resulted after purges were undertaken among the pilots of the air force. Covert supplies of American arms (later to become public knowledge during the 'Iran-gate' affair) allowed the air force to increase its sortie rate.

The country was officially neutral during the second Gulf War but benefited when a proportion of the Iraqi air force sought shelter in the country. These aircraft were seized and some were impressed into service. The types involved included Falcon 50s, Mirage F1EQs, Il-76MDs, MiG-29s, Mil Mi-8/17 and Mi-Mi-24s, Sukhoi Su-22M-2s, Su-24MKs, Su-25Ks and the sole Il-76 'Adnan' AEW platform.

CURRENT SITUATION

The exact number of aircraft in service with the IRIAF today is difficult to judge, as is its order of battle. Organisation is by regional command, with each airfield within that command having units assigned to it under a numbered Tactical Air Base (TAB).

The primary air defence aircraft of the IRIAF are the F-14A Tomcats, MiG-29s and the F-7Ms. Cut off from (official) spares suppliers for nearly 20 years, it is

a testament to the ingenuity of the service that any Tomcats are still flying at all. In 1985 a formation of 25 Tomcats overflew Tehran. Iran's ability to reverse-engineer should ensure that the type is still available to undertake limited long-range interception missions at low sortie rates. Photographs of Tomcats armed with modified MIM-23 Hawk surface-to-air missiles in the

Islamic Republic of Iran Army Aviation

The army aviation branch still has a large number of its pre-revolutionary helicopters, based at five bases throughout the country. Exactly what levels of serviceability can be achieved is open to question. Developments towards self-sufficiency have resulted in the Shabaviz (Owl) 2-75 helicopter, which looks (externally) very much like the UH-1H, and the Shabaviz 2061 (a reverse-engineered Bell 206). As all the components (except engines) were Iranian-made, it is likely that Iran has been able to keep a large proportion of its Agusta-Bell and Bell helicopters airworthy. Embargoed Iraqi Mil Mi-8/17 and Mi-24s have also been pressed into service.

Current equipment of the Islamic Republic of Iran Air Force

Type	Role	In service
Beech F33A/C	trainer	26
AB 206 JetRanger	liaison	2
Agusta-Bell AS 61A	VIP transport	2
Agusta-Bell AB 212	support	10
Bell 214B/C	support	30
Boeing 707-3J9C	tanker/transport	8
Boeing 747F-131	transport	8
CH-47C Chinook	support	2
Chengdu F-7M	interceptor	30
EMB-312 Tucano	trainer	10
F-14A Tomcat	interceptor	60
Guizhou FT-7	trainer	5
Fokker F27 400M/600M	transport	3/2
Il-76MD 'Candid'	transport	15
R/C-130H Hercules	recon/transport	1/25
Lockheed T-33A	trainer	7
P-3F Orion	ASW	5
F-4D/E Phantom	ground attack	18/46
RF-4E Phantom	reconnaissance	8
MiG-29A/UB 'Fulcrum'	interceptor	35/6
Northrop F-5A/B/E/F	ground attack	10/25/57/18
Northrop RF-5E	reconnaissance	5
Pilatus PC-6B	utility transport	15
Pilatus PC-7	basic trainer	45
Rockwell 681	communications	3
Shenyang J-6	ground attack	12
Su-24MK 'Fencer-D'	penetration strike	30

★ Tabriz
★ Dahan-Tappeh
Mehrabad ★ ● Tehran
★ Ghale Morghi
★ Hamadan-Shahroli   **IRAN**
★ Khatami-Isfahan
★ Dezful-Vahdati
Zahedan ★
★ Shiraz
★ Bushehr
★ Bandar Abbas

Islamic Republic of Iran Naval Aviation

It is believed that the naval air arm is manned by air force personnel and works closely with the IRIAF, which operates the long-range maritime patrol P-3F Orions from Shiraz, and also some C-130Hs in this role. Short-range ASW is undertaken by the squadron of Agusta-built Sea Kings based at Bandar Abbas.

The Agusta-Bell AB 205A and AB 212ASs are the only helicopters that can be deployed onboard the two 'Babr'-class destroyers, but spend most of their time supporting the defence of oil installations. It is believed that one example may have been converted for early warning and control duties. The surviving RH-53Ds are used for logistical transport duties only.

Islamic Republic of Iran Naval Aviation (IRINA)

TAB 6		**Bushehr (heliport)**
? ASW Sqn	RH-53D, AB 212AS	
TAB 7		**Shiraz Int'l**
? Transport Sqn	Falcon 20, F27-400M/600 Shrike Commander	
TAB 9		**Bandar Abbas**
1 ASW Sqn	ASH-3D Sea King	
? Liaison Sqn	AB 205A, AB 206	

Type	Role	In service
AB 205A	liaison	5
AB 206 JetRanger	liaison	12
AB 212AS	ASW	6
Agusta ASH-3D	ASW	14
Falcon 20E	transport	2
Fokker F27-400M	transport	1
Fokker F27-600	transport	1
Shrike Commander	liaison	3
RH-53D Sea Stallion	ASW	5

Islamic Republic of Iran Army Aviation (IRIAA)

1 x Support Sqn	AB 205	Bakhtaran
2 x Liaison Sqn	AB 206	Bakhtaran
1 x Utility Sqn	Cessna O-2A	Bakhtaran
3 x Attack Sqn	AH-1J	Bakhtaran
3 x Support Sqn	Bell 214A Isfahan	Bakhtaran
1 x Support Sqn	AB 205	Isfahan
2 x Liaison Sqn	AB 206	Isfahan
3 x Transport Sqn	CH-47C	Isfahan
4 x Support Sqn	Bell 214A Isfahan	Isfahan
3 x Attack Sqn	AH-1J	Isfahan
1 x Support Sqn	AB 205	Kerman
1 x Liaison Sqn	AB 206	Kerman
2 x Transport Sqn	Mi-8, Mi-17	Kerman
1 x Support Sqn	Bell 214A Isfahan	Kerman
1 x Attack Sqn	AH-1J	Kerman
1 x Attack Sqn	AB 206, Mi-24, AH-1J	Masjed Suleyman
1 x Helicopter Train Sqn	AB 205, AB 206	Tehran-Ghale Morghi
1 x Helicopter Train Sqn	Hughes 300C	Tehran-Ghale Morghi
1 x Fixed Wing Train Sqn	Falcon 50	Tehran-Ghale Morghi

Type	Role	In service
AB 205	support	52
AB 206A/B1 Jet Ranger	liaison	103
AH-1J Super Cobra	attack helicopter	120
Bell 214	support	30
Bell 214A Isfahan	support	167
CH-47C Chinook	medium transport	51
Cessna 185 Skywagon	liaison	10
Cessna 310	liaison	10
Falcon 50	VIP transport	2
Fokker F27 Mk 200	medium transport	2
Hughes 300C	training	5
Mil Mi-8/17 'Hip'	support (ex-Iraq)	25
Mil Mi-24 'Hind'	attack (ex-Iraq)	10
PC-6B Turbo Porter	light transport	10
Shrike Commander	light transport	4

air-to-air role have surfaced, and the air force still has stocks of US-supplied missiles. MiG-29s were acquired from 1990 and have been topped up with impounded Iraqi examples, while the Chinese- built Chengdu F-7Ms have served since 1987.

Long-range strike missions can be undertaken by the Sukhoi Su-24MK 'Fencer-Ds' based at Shiraz and Mehrabad. The aircraft are able to carry the Upaz-A buddy refuelling system to extend the range of the

others. Freedom Fighters/Tiger IIs are also used in the strike role, alongside the surviving Phantoms.

In 1992, after the second Gulf War, Iran ordered 12 Tupolev Tu-22M 'Backfires', 24 MiG-31 'Foxhound-As', 48 more MiG-29/UB 'Fulcrum-A/Bs', 24 MiG-27 'Flogger-Ds' and a pair of Beriev A-50 'Mainstays' but there is no evidence of any deliveries.

Experience in maintaining its fleet of aircraft without external help has given Iran a large industrial base

from which to design and build aircraft. The first of these was reported to be the Parastou (Dove), an aircraft that looked like the Beech Bonanza. It is believed to be in service with the air force. In February 1998 it was further reported that an indigenous fighter, the Azarakhsk, had entered series production. An advanced trainer, the Tondar, has also been noted, but the status of these programmes is unclear.

IRAQ

Capital: Baghdad
Population: 18.9 million
Land area: 438445 km²
(169,240 sq miles)
Major cities: Basra, Kirkuk, Mosul

Al Quwwat Al Jawwiya Al Iraqiya (Iraqi Air Force & Air Defence Forces)

Iraq had the largest armed forces in the Persian Gulf region prior to the 1990-91 Gulf War, including over 800 combat aircraft, plus a complex and integrated network of Soviet-supplied SA-2, -3, -6, -7, -8, -9, -11, -13 and -14 SAMs, and Euromissile Roland air defence missiles, but the country now has been substantially weakened by its rout in the war. While still fielding a 400,000-strong army in some 30 divisions, its air forces are now estimated to have only 300 or so combat aircraft on strength, from a total inventory of about 500, with poor serviceability and mission availability because of maintenance and spares problems.

These problems have worsened due to the continuing UN arms embargo and economic sanctions against Iraq, limiting its military procurement to covert supplies from Chinese, Libyan (via Jordan),

North Korean and some CIS sources. UN sanctions have also prevented access to maintenance and overhaul facilities formerly utilised by Iraq in the CIS and Eastern Europe, where some IrAF fighters and helicopters have been impounded because of current embargoes.

From the remaining Iraqi MiG-21s (including some Chinese-supplied F-7 versions), MiG-23s, MiG-25s, MiG-29s, Mirage F1s and Sukhoi Su-22s, the US estimated in 1992 that only about 150 were airworthy. They were said to comprise some 15 MiG-29s, 30 Mirage F1s and 50 MiG-23s, as well as 20 Su-25 and 30 Su-20/-22 ground-attack aircraft. Deliveries of another 70 MiG-29s, ordered to follow initial IrAF contracts for 26, were then embargoed by the UN. A similar situation faced Dassault, with some 20 Mirage F1EQs, including at least three two-seat versions, completed but undelivered to Iraq by 1990. Between 1981-1990, Iraq acquired 94 single-seat Mirage F1EQs, including 38 AM39 Exocet-armed F1EQ5/EQ6-200s also equipped for air refuelling, and 14 two-seat F1BQ operational trainers.

More than 100 Iraqi aircraft were claimed destroyed on the ground in the 1990-91 Gulf War, about 40 were lost in air combat, and 115 fled to Iran, from where few returned. Apart from 24 Mirage F1EQs, four each MiG-23UB/UMs, MiG-23BNs, MiG-23MLs, MiG-29s and Su-20s, plus 40 Su-22Ms, all 24 IrAF Su-24s, and seven Su-25s, a further 32 Iraqi civil-registered aircraft, including 15 Ilyushin Il-76s, also sought temporary refuge in Iran.

Further limitations in Iraq's military recovery have come from continuing UN inspection team surveys of its weapons of mass destruction facilities, despite prolonged deception and obstructive tactics. From original Soviet supplies of 819 'Scud' tactical ballistic missiles to Iraq, of which over 500 were launched against Iran by 1988, plus a further 88 against Israel and Saudi Arabia in the 1990-91 war, the UN team has tracked down and destroyed all but an estimated 25 'Scud-B' and longer-range (600 km/373 miles) Al Hussein versions believed to be in secret storage.

Iraq's longer-range air strike element at one time comprised about 10 Tupolev Tu-22 'Blinder' supersonic bombers and six older ASM-armed Tu-16 'Badgers', supplemented from 1987 by four Xianbuilt B-6D 'Badgers' from China carrying C-601 anti-ship missiles. Most were destroyed on the ground by coalition forces.

Iraq's army aviation element was relatively inactive, but only about 140 of its 450 or so original helicopters are in current service. Some were used to attack Kurdish and Shiite rebels immediately following the

The start of the second Gulf War stranded Iraqi aircraft at European overhaul facilities, such as this MiG-21bis in Yugoslavia. It is unlikely that these aircraft will be returned to Iraq in the near future.

1991 ceasefire, and before the UN imposition of 'No-Fly Zones' under coalition protection in north and south Iraq. Iraqi armed forces' organisation and tactics are believed to continue on their original Soviet lines, but few details are currently available of air force or army aviation unit deployments. All aircraft totals are necessarily speculative.

Iraq has been a major purchaser of the Mirage F1, with 16 F1EQ, 16 F1EQ-2, 28 F1EQ-4-200, 20 F1EQ-5-200 and 18 F1EQ-6-600 single-seaters delivered and 15 F1BQ two-seaters. 4010 is a Mirage F1EQ.

Al Quwwat Al Jawwiya Al Iraqiya

Type	Role	In Service
Aero L-29 Delfin	basic trainer	c20
Aero L-39ZO Albatros	armed jet trainer	c50
SA 321GV Super Frelon	transport helicopter	c8
Aérospatiale SA 330F Puma	transport helicopter	c15
Agusta-Sikorsky AS-61TS	transport helicopter	c3
Antonov An-12 'Cub'	transport/tanker	c6
Antonov An-24 'Coke'	medium transport	c5
Antonov An-26 'Curl'	tactical transport	c2
Bell 214ST Super Transport	utility helicopter	c25
Chengdu F-7A	air defence	c40
Mirage F1EQ/EQ2/EQ4-200/ EQ5-200/EQ6-200	multi-role fighter	30-35
Dassault Mirage F1BQ	combat trainer	c10
EMBRAER EMB-312 Tucano	basic trainer/GA/recce	c70
FFA AS.202/18A2 Bravo	primary trainer/GA	c20+
Hughes 300C	light training helicopter	c18
Hughes/MDC H.500D	utility helicopter	c20
Hughes/MDC 530F	utility helicopter	c18
Ilyushin Il-76T/M/MD 'Candid-A/B'	transport/tanker	c10
Learjet 35A	light transport	c6
MBB BK 117B-1	utility helicopter	c18
MiG-21PFM/MF 'Fishbed-F/J'	air defence	c60
MiG-21U 'Mongol'	combat trainer	c10
MiG-23MF/BK 'Flogger'	multi-role fighter	c50
MiG-23UB 'Flogger-C'	combat trainer	c10
MiG-25PD 'Foxbat-A'	air defence	c5
MiG-29 'Fulcrum'	air defence	c8
MiG-29UB 'Fulcrum-B'	combat trainer	c2
Mil Mi-6 'Hook'	heavy-lift helicopter	c5
Mil Mi-8/-17 'Hip'	transport/assault helicop.	c80*
Sukhoi Su-22M3/M4 'Fitter-H/J'	strike-interceptor	c30
Sukhoi Su-22U 'Fitter-E'	combat trainer	c6
Sukhoi Su-25K 'Frogfoot-A'	ground attack	c18
Sukhoi Su-25UBK 'Frogfoot-B'	combat trainer	c2
Zlin Z.326 Trener Master	primary trainer	c12

includes some army-operated versions

Army Aviation

SA 315C Alouette III	armed utility helicopter	c21
Aérospatiale SA 342K/L Gazelle	light attack helicopter	c30
MBB BO 105C	armed utility helicopter	c40
Mil Mi-24/25 'Hind'	attack helicopter	c20

ISRAEL

Capital: Jerusalem
Population: 4.8 million
Land area: 20770 km²
(8,015 sq miles)
Major cities: Haifa, Ran,
Ramat, Tel Aviv

Tsvah Haganah le Israel/ Heyl Ha'Avir – (Israeli Defence Force/Air Force)

Few of the world's air arms can have gained as much combat experience since 1945 as the Israeli air force. Often battling for the very survival of the state it is tasked with defending, the IDF/AF has built up an enviable reputation for professionalism. The IDF/AF is larger and better equipped than the air arms of many bigger, and more populous, nations. This imposes a heavy financial burden on Israel's 4 million people, even with massive US aid.

Despite its large size, the IDF/AF is outnumbered by neighbouring air arms, while the edge it once enjoyed in quality has been eroded by deliveries of the latest advanced aircraft types to its potential enemies.

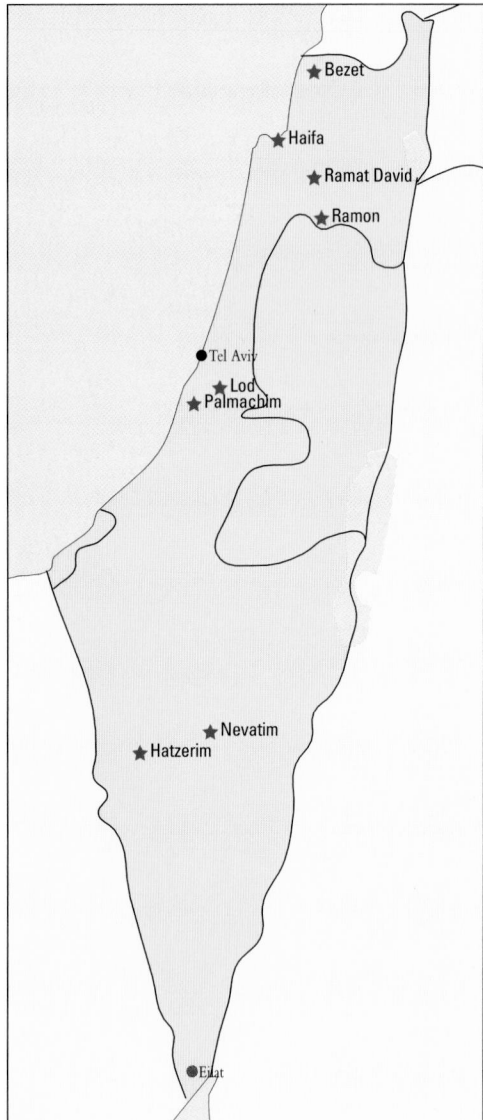

The F-16 remains the backbone of the Israeli air force. A total of 280 has been delivered. This aircraft is one of the early Block 5 F-16As that were received from July 1980.

Israel is small and compact, and the entire state can be reached in minutes by incoming enemy aircraft. From the beginning, it has been impossible to guarantee that all incoming air raids would be intercepted, and the IDF/AF has always had a vital secondary role of ground attack, allowing it to help stem any incursions by enemy ground forces and to retaliate against hostile air attacks.

Emigration of persecuted European Jews began before World War II to what was then Palestine, administered by Britain under a League of Nations mandate. After the war the new United Nations decided to partition Palestine into separate Jewish and Arab states. When Britain withdrew the Jews began taking over Arab areas, and the new State of Israel (claiming the whole area of the former mandate) was born on 14 May 1948. Arab states intervened, but were decisively defeated in battle.

Even before Britain withdrew from Palestine, Israel had formed an air arm. In the face of a UN arms embargo, Israel acquired aircraft clandestinely. Following the 1948 war, however, Israel was able to purchase military aircraft normally, initially turning to Britain, and then to France, with whom it began a long-term relationship under which it received Dassault Ouragans and Mystère IVAs.

Israel continued to be the target of guerrilla raids and cross-border attacks. The accession of the nationalist Nasser in Egypt prompted Israel to become involved (secretly) in the Anglo-French invasion of the Suez Canal zone. Israel attacked Egypt to provide the Anglo-French alliance with a pretext for intervention (separating the combatants). The allies occupied the canal zone to 'safeguard it', establishing a ceasefire line which allowed Israeli units to advance, before US pressure forced a withdrawal.

Following the 1956 war, Israel continued to receive combat aircraft from France, including Vautour fighter-bombers, Super Mystères and Mirage IIICJs, as well as Super Frelon helicopters, Magister trainers and Noratlas transports. French-supplied types formed the core of the IDF/AF into the 1960s.

On 5 June 1967 Israel mounted what it claimed was a pre-emptive attack on Egypt, Jordan, Syria and Iraq. Although it virtually destroyed the Egyptian and

Above: The F-4E Phantom has served since 1969, and about 55 of the fleet have been upgraded to Kurnass 2000 standard with new avionics. 586 of 201 Tayeset 'The One Squadron' displays Kurnass 2000 titles in Hebrew on its flank.

Below: Israeli RF-4Es differ from other RF-4Es in that they can carry AIM-9s on underwing pylons.

Jordanian air forces on the ground on the first day of the war, Israel did not have it all its own way, losing many aircraft to SAMs and Arab fighters. Israel invaded and seized large areas of Jordan (the West Bank and East Jerusalem), as well as the Sinai, the Gaza Strip and the Golan Heights. France imposed an embargo immediately before the war, cancelling the delivery of 50 Mirage 5Js. This forced Israel to turn to the USA for aircraft in the short term, and to establish an indigenous aircraft industry.

The first front-line product of Israel's indigenous industry was the Nesher, an unlicensed copy of the Mirage 5. The Nesher was developed into the Kfir, with new avionics, a US-supplied J79 engine, and canard foreplanes. The Kfir entered service in 1975, most successfully in the fighter-bomber role.

The first US combat aircraft delivered included A-4H and A-4E Skyhawks, plus F-4E and RF-4E Phantoms. Bell 205s began replacing French-supplied S-58s, and Sikorsky S-65s began to augment the IDF/AF's Super Frelons. From 1971, the IDF/AF began operating the Lockheed C-130 Hercules in the tactical transport role. A state of undeclared war existed between Israel and Egypt between the end of the Six Day War in June 1967 and 8 August 1970, when a ceasefire was signed. The so-called War of Attrition was marked by commando raids, artillery duels and a series of air attacks and dogfights, in which the IDF/AF generally gained the upper hand.

In 1973, Egypt and Syria launched simultaneous attacks on Israel on two fronts, aiming to recover territories lost in 1967. Striking on Yom Kippur (6 June 1973, the Jewish Day of Atonement and the 10th day of Ramadan), the Arab offensive was blunted by the IDF/AF, and, after a massive resupply effort by the USA, Israel eventually turned the tide. The Yom Kippur War marked the first Israeli use of RPVs in combat, and the IDF/AF has placed heavy emphasis on the use of drones for reconnaissance ever since. Losses of aircraft were massive on both sides, especially to SAMs. Although Arab forces suffered heavier losses (made good by a massive resupply effort by the USSR), they acquitted themselves well, and the war prompted both sides to look to diplomatic solutions to their problems. This led eventually to the lasting peace between Egypt and Israel.

Israel received 25 F-15As and F-15Bs after the war, and immediately began using them in support of operations in the Lebanon. The end of the war was also marked by the delivery of AH-1 Cobra attack helicopters and a number of other aircraft types. The historic formal peace agreement with Egypt opened

Below: The most capable interceptor in service is the F-15, based at Tel Nof. Israeli Eagles have seen considerable combat over the years and many have made kills. This 133 Tayeset 'The Twin Tailed Squadron' F-15A displays four.

Tsvah Haganah le Israel
Heyl Ha'Avir

Headquarters Tel Aviv

Air Defence Corps, HQ Kfar Sirkin

controls SAMs (Hawk, Patriot) and various radar sites.

Northern Command, HQ Ramat David

Canaf I

109 Tayeset	F-16C/D Block 30	Ramat David
110 Tayeset	F-16C/D Block 30	Ramat David
117 Tayeset	F-16C/D Block 30	Ramat David
190 Tayeset	AH-64A Apache	Ramat David

Bacha 21

123 Tayeset Det	AB 212	Haifa
124 Tayeset Det	AB 212, UH-60	Haifa
Technical Maintenance School		Haifa

Shachar 7

505 Tayeset Det	microlights	Megiddo
505 Tayeset Det	microlights	Ein Shemer

Shachar 11

160 Tayeset Det	AH-1F	Biraneet
161 Tayeset Det	AH-1F	Biraneet

Shachar 33

123 Tayeset Det	AB 212	Bezet
124 Tayeset Det	AB 212, UH-60	Bezet

Central Command, HQ Hatzor

Canaf 4

101 Tayeset	F-16D/C Block 40	Hatzor
105 Tayeset	F-16D/C Block 40	Hatzor
144 Tayeset	F-16A/B Block 10	Hatzor
149 Tayeset*	Kfir C7	Hatzor

Bacha 8

69 Tayeset	F-15I Thunder	Tel Nof
106 Tayeset	F-15C/D	Tel Nof
114 Tayeset	CH-53D Yasur 2000	Tel Nof
118 Tayeset	CH-53D Yasur 2000	Tel Nof
119 Tayeset	F-4E-2000, RF-4E	Tel Nof
133 Tayeset	F-15A/B	Tel Nof
148 Tayeset	F-15A/B/D	Tel Nof
201 Tayeset	F-4E-2000, RF-4E	Tel Nof
505 Squadron	microlights	Tel Nof
601 Tayeset	test unit with various types	Tel Nof

Bacha 22

YAA	maintenance unit	Tel Nof
216 Tayeset	airframe maintenance	Tel Nof
226 Tayeset	engine maintenance	Tel Nof

Southern Command, HQ Tel Nof

Bacha 6

102 Tayeset	A-4H	Hatzerim
123 Tayeset	AB 212, UH-60	Hatzerim
142 Tayeset*	F-4E-2000, RF-4E	Hatzerim
155 Tayeset	Samson UAV, Delilah UAV	Hatzerim

Canaf 28

104 Tayeset	F-16A/B	Nevatim
115 Tayeset	F-16A/B	Nevatim
116 Tayeset	F-16C/D	Nevatim
132 Tayeset*	Kfir C7	Nevatim
141 Tayeset*	Kfir C2	Nevatim
251 Tayeset*	Kfir C2	Nevatim

Canaf 25

113 Tayeset	AH-64A Apache	Ramon
140 Tayeset	F-16A/B Block 15	Ramon
146 Tayeset	Samson UAV, Delilah UAV	Ramon
147 Tayeset	F-16A/B Block 15	Ramon
253 Tayeset	F-16A/B Block 5/10	Ramon

Bacha 10 **

137 Tayeset*	A-4H/N	Ovda
143 Tayeset*	Kfir C2	Ovda
145 Tayeset*	A-4H/N	Ovda
202 Tayeset*	A-4H/N	Ovda

Bacha 30

124 Tayeset	UH-60A, AB-212	Palmachim
127 Tayeset	AH-64A Apache	Palmachim
160 Tayeset	Bell AH-1F/S	Palmachim
161 Tayeset	Bell AH-1F/S	Palmachim
193 Tayeset	Aérospatiale HH-65A, AS 565	Palmachim
200 Tayeset	Scout, Hunter, Searcher & Silver Arrow UAVs	Palmachim

Transport Command, HQ Lod

Bacha 27

103 Tayeset	Lockheed C-130E/H, KC-130H	Lod
120 Tayeset	Douglas C-47, RC-47	Lod
122 Tayeset	B-707, KC-707, IAI Westwind	Lod
126 Tayeset	IAI Arava 101/202 ECM	Lod
131 Tayeset	Lockheed C-130E/H	Lod
134 Tayeset	EC-707, RC-707	Lod
195 Tayeset	IAI 1124N Seascan	Lod

Canaf 15

100 Tayeset	SOCATA TB-20 Trinidad	Sde Dov
125 Tayeset	Bell 206L, AB 206B	Sde Dov
128 Tayeset	Beech U-21, RU-21A	Sde Dov
129 Tayeset	SOCATA TB-20 Trinidad	Sde Dov
135 Tayeset	Dornier Do 28D-1	Sde Dov
191 Tayeset	RC-12D/K, Beech King Air 200	Sde Dov

Training Command, HQ Hatzerim

12 Bist		
Primary School	Piper Super Cub	Hatzerim
Secondary School	IAI Tzukit	Hatzerim
130 Tayeset	Bell 206	Hatzerim
162 Tayeset	Hughes 500MD	Hatzerim
247 Tayeset	Do 28B	Hatzerim
252 Bihiys Latisah-Mata Kadet Krav	TA-4H	Hatzerim

* Reserve Tayeset (Squadron)

** Ovda is believed to house the IDF/AF's stored IAI Kfirs. With its location deep in the Negev, Ovda is also believed to be the home for various top-secret units, perhaps including aircraft types not generally acknowledged to be in IDF/AF service.

Israel's preoccupation with security means that squadron designations, locations and equipment are a state secret, while squadron locations change periodically. Thus, any order of battle must inevitably include an element of speculation and informed guesswork.

Above: Israel uses a mix of four RC-12Ds and two RC-12Ks with 191 Tayeset. 980 is one of the former.

Below: 120 Tayeset 'The Dakota Squadron', based at Lod, continues to use the C-47 Pe're (Savage) on a variety of tasks, including electronic reconnaissance.

1974-76 civil war. Syrian intervention stopped the civil war, but did nothing to halt PLO activity. Israel therefore invaded southern Lebanon in 1978, and has operated in the country ever since, using air power to attack SAM sites, PLO training camps, and sometimes even refugee camps. With Syria in the north of Lebanon and Israel in the south, conflict between the two was inevitable. The IDF/AF mounted major strikes against Syrian SAM sites and GCI stations in the Beka'a Valley in June 1982, followed by intensive air-to-air combat.

In June 1981, IDF/AF F-16s and F-15As attacked the Iraqi Osirak nuclear reactor. The raid provoked American ire, however, and delivery of the last 22 F-16As was delayed in retaliation. On 1 October 1985, the IDF/AF carried out a long-range attack against the PLO Headquarters in Tunis. This flagrant act did not delay Israel's order for 51 F-16Cs and 24 F-16Ds, which was actually increased by 60 in 1987, following the cancellation of the indigenous IAI Lavi.

Israel's attitude to its neighbours has been dismissed as paranoiac, but should be seen in the light of continuing Arab hostility. Although Israel remained aloof from the Gulf War of 1991 it still found itself the target of Iraqi 'Scud' missiles. As a reward for its 'forbearance', Israel received 50 USAF-surplus F-16As and F-16Bs, and 13 F-15As and F-15Bs. Since the Gulf War, Israel has taken delivery of Sikorsky S-70 assault helicopters, and has continued to receive AH-64 Apache attack helicopters. Most recently, the air force has started to replace F-4E-2000s in the interdiction role, having received the first batch of 25 F-15I Thunders from the USA. They are broadly equivalent to the USAF's F-15Es, and mark a dramatic improvement in air-to-ground capability.

Today the IDF/AF operates from eight main bases, although transport and liaison aircraft operate from military enclaves at civil airfields (Sde Dov and Lod), and a large number of other airfields are used as forward operating bases or to support small helicopter detachments or deployments. Palmachim has only

---

the door to further arms deliveries from the USA, and the IDF/AF entered a new era in 1980, with the delivery of the first of 67 F-16As and eight F-16Bs, together with 31 F-15Cs and F-15Ds. The Camp David accord also saw Israel losing the Sinai, which had been a useful buffer zone and a vital expanse of virtually unpopulated airspace for realistic flying training. Israel also lost three major air bases, necessitating the construction of new airfields at Ramon and Ovda.

While peace with Egypt brought a welcome respite

from PLO attacks in the south, Palestinian aspirations remained unfulfilled, and the PLO continued operations. When the PLO hijacked an Air France Airbus en route from Tel Aviv to Paris in June 1976, they forced it to fly to Entebbe. Rather than Israel give in to demands for the release of prisoners, the IDF mounted a successful hostage rescue mission, flying special forces teams in aboard four C-130s.

The PLO also attacked Israel from the Lebanon, where it had sided with Muslim forces during the

---

Basic training is undertaken from Hatzerim on the IAI Tzukit (Merlin), a modernised and rebuilt Fouga Magister. About 40 remain on the active roster.

From 1995 22 TB-20 Trinidads were delivered to replace Do 28s and Beech 65s in the light communications role.

Mainstay of the Israeli transport fleet are its C-130 Hercules. Three KC-130Hs were delivered alongside the transport, but may have been deconverted.

Above: 114 Tayeset 'The Super Frelon Squadron' based at Tel Nof operates the CH-53D Yasur 2000 upgrade alongside 118 Tayeset. The IAF received 42 Yasur 2000 upgrades.

helicopters and UAVs, but the remaining airfields are fully equipped for fast-jet operations, with underground hangars, hardened shelters, and redundant/reserve runways. The IDF/AF has increasingly moved away from single-role bases, and several airfields accommodate a mix of fighters, fighter-bombers and attack helicopters (and UAVs, too, at Ramon). Only Nevatim and Hatzor remain essentially single-type bases (both with F-16s).

If one believed official Israeli claims, one might assume that the IDF/AF had never lost a fighter in air-to-air combat, and that it has fought only against half-trained imbeciles. In fact, the IDF/AF has often been pitted against quality opponents, and on rare

occasions has even been bettered. This underlines its great achievement. There can be no doubt that the IDF/AF today is arguably the world's best trained and most capable air arm.

Above left: Israeli has three squadrons of AH-64A Apaches. This Apache is carrying four AGM-114s (complete with US Army titles) and long-range tanks.

JORDAN

Capital: Amman
Population: 3.2 million
Land area: 96000 km² (37,055 sq miles)
Major cities: Irbid, Zarka

Al Quwwat Al Jawwiya Al Malakiya Al Urduniya (Royal Jordanian Air Force)

Jordan's armed forces trace their origins back to the largely British-officered Arab Legion, which formed as the army of the emerging state of Transjordan during the British Mandate. The Royal Jordanian Air Force began life in 1948 as the Arab Legion Air Force, using British-supplied aircraft and organised along RAF lines. Initially equipped with a de Havilland Dragon Rapide, the force gained Tiger Moths, Percival Proctors, Austers, Doves and Chipmunks. Two Vampire T.Mk 11s were added in 1955, together with nine Vampire FB.Mk 9s donated by the RAF.

Links with Britain were loosened considerably in the wake of the 1956 Anglo-French Suez operation when Jordan became more closely aligned with its Arab neighbours, although relations with Iraq cooled following the murder of the Iraqi Royal Family and with Syria following attempts to assassinate King Hussein. The USA funded a Jordanian Hawker Hunter purchase, and later oversaw the transfer of F-104A Starfighters from Taiwan under a Mutual Defense Assistance Program in 1967.

Having been the only nation to hold territory (and even make gains) against Israel in 1948, Jordan was a priority target when Israel launched the Six Day War against its Arab neighbours in 1967. The war began with pre-emptive and unprovoked airstrikes, and

these virtually destroyed the RJAF on the ground. Despite this, the Jordanian Hunter pilots were sent to Iraq to fly Iraqi Hunters for the remainder of the war. In so doing, they scored four confirmed victories (using rigorous kill confirmation procedures) and two probables (for only one air-to-air loss (the pilot ejecting safely). This was a creditable achievement against the much-vaunted Israeli air force.

Losses in the 1967 war were made up by transfers of Hunters from Iraq and Saudi Arabia, and by new purchases. Large numbers of F-5Es entered service from 1972, and they remain numerically the most important type in the inventory. US refusals to supply

Above: Jordan has a total of 44 F-5Es and 13 F-5Fs shared between three squadrons. This pair of No. 9 Squadron F-5Es carries AIM-9P Sidewinders.

Below: Mirage F1EJs have served with the RJAF's No. 1 Squadron since 1982, and are currently being supplemented by ex-No. 25 Squadron F1CJs.

F-16s or A-4s (almost certainly the result of pressure from Israel) prompted the RJAF to turn to Dassault for an eventual total of 36 Mirage F1s, delivered in 1981-82.

Still a monarchy, ruled by the same family that was installed in power in 1922, Jordan is today arguably the most stable nation in the Middle East, although it has not been immune to the instability which has gripped the region since 1945. Generally pro-Western and politically moderate, Jordan remained neutral during the Gulf War of 1991, and maintains closer links with Israel than do most of its Arab neighbours.

With no oil deposits, the Jordanian economy relies

Above: The latest combat aircraft in service are the 12 F-16As and four F-16Bs delivered during 1998. They are optimised Air Defence Fighter variants.

Below: The CASA C. 101CC Aviojet fleet has been concentrated in No. 11 Squadron after No. 2 Sqn became the new F-16 unit. Twelve are in service.

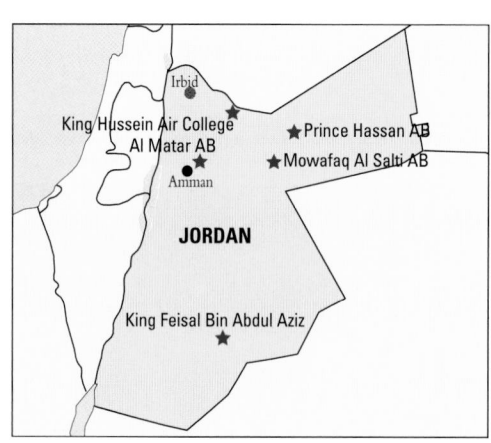

Al Quwwat Al Jawwiya Al Malakiya Al Urudniya

Operations Command

Al Matar Air Base, Amman/Marka
No. 3 Sqn	C-130H, CASA C.212-100, PBN BN-2A Islander
No. 7 Sqn	AS 332M-1 Super Puma
No. 8 Sqn	UH-1H Iroquois
No. 10 Sqn	Bell AH-1F
No. 12 Sqn	Bell AH-1F

King Feisal Bin Abdul Aziz, Al Jafr
No. 9 Sqn	F-5E/F Tiger II

Mowafaq Al Salti Air Base, El Azraq
No. 1 Sqn	Mirage F1CJ/EJ, F1BJ
No. 2 Sqn	F-16A/B

Prince Hassan Air Base, H5
No. 6 Sqn	F-5E/F Tiger II
No. 17 Sqn	F-5E/F Tiger II

Another UH-1H Iroquois squadron has been formed using the latest helicopters delivered. Its base is unknown.

Training Command

King Hussein Air College, Al Mafraq
No. 4 Sqn	Bulldog Mk 125/125A
No. 5 Sqn	MD 500D
No. 11 Sqn	CASA C-101CC

No. 4 Squadron is responsible for screening and primary training, No. 11 conducts basic flying training, weapons and tactical training. No. 5 conducts rotary-wing training.

Direct-reporting Units

King Abdullah/Al Matar Air Base
Royal Flight,	L-1011 TriStar 500, G.1159A Gulfstream III, Dove 7, S-70A-11

Al Matar Air Base, Amman/Marka
Royal Jordanian 'Falcons' Aerobatic Team	Extra 300
Historic Flight	Alouette III, Hunter F.Mk 6A*/T.Mk 7/F.Mk 58* Vampire FB.Mk 6/T.Mk 55

* One F.Mk 58 and the F.Mk 6A remain in the UK at present.

Twenty-four AH-1F Cobras were delivered to Jordan in 1985, but at least one has been written off since. The helicopters are armed with BGM-71 TOW missiles, and are the RJAF's primary anti-tank asset. All are based at Al Matar Air Base.

chiefly on phosphates and agriculture, and the country is poorer than most of its neighbours. It has traditionally relied on Saudi (and sometimes Iraqi) funding for many of its defence purchases, though this has dwindled in recent years. As a result, some re-equipment programmes have failed to come to fruition, and the country's armed forces have become the operators of increasingly obsolescent equipment, and have failed to keep up with either allies or potential enemies. The failure of allies to fulfil their funding obligations led to the cancellation of Jordanian orders for both the

Mirage 2000 and the Panavia Tornado. While it lacks the most modern combat aircraft, the RJAF is exceptionally well trained and keeps abreast of the latest tactical developments, making it a tougher opponent than many better-equipped air forces.

Despite the quality and professionalism of its armed forces, Jordan lost the West Bank and East Jerusalem to Israel in 1967, areas which have remained under Israeli occupation ever since. The West Bank included Jordan's best agricultural land, and had been a homeland for dispossessed Palestinians when the

state of Israel was established. Jordan signed a formal peace treaty with Israel in October 1994, opening the way to a resumption of unlimited US military aid.

The first products of the new MDAP included 18 surplus Bell UH-1H helicopters, delivered in December 1994, followed by another 18 in 1996, and a C-130H in March 1997. Most significantly, the resumption of US aid has finally allowed the delivery of 12 surplus ex-USAF F-16As and four F-16Bs to replace the elderly Mirage F1CJs in the air defence role. Jordan contributed $80 million to the cost of pre-delivery modifications and upgrades, but the remaining $140 million was provided by the US DoD under FMS (Foreign Military Sales) provisions. Deliveries of F-16s to No. 25 Squadron began in December 1997, but the aircraft now serve with No. 2 Squadron.

The Kuwaiti flag and toned-down squadron badge visible on the tail of this Hornet are the only features identifying this aircraft as one of the 32 Kuwaiti F/A-18Cs. No roundels are worn. This example operates with No. 9 Squadron.

sold to Brazil. No. 9 Squadron has an air defence role and No. 25 Squadron is tasked with air-to-ground missions. Options for a further 32 aircraft were cancelled in 1992, but the air force would still like to acquire at least a dozen more.

Despite reports that the Mirage F1CK-2 fleet was withdrawn from use and offered for sale, Nos 18 and 61 Squadrons are nominally equipped with the type. They were refurbished by Dassault following the Gulf War and are believed to be flown – very infrequently – by high-ranking officers.

Advanced training and close support roles are undertaken by the Hawk Mk 64s operated by No. 12 Squadron. Ten of the 12 delivered survived the war, but are currently grounded following serviceability problems. Kuwait has reported it would like to acquire more Hawks, possibly 100/200 series aircraft, for lead-in fighter instruction for its Hornets.

Basic training is undertaken on 16 Tucano Mk 52s. The aircraft were ordered pre-war but only delivered in 1995. Anti-tank duties are undertaken by the SA 342L Gazelle fleet of No. 33 Squadron, some of which have the ability to fire HOT anti-tank missiles. Kuwait had 16 AH-64D Apaches on order, to be delivered without the Longbow radar, but this order was subsequently cancelled.

KUWAIT

Capital: Kuwait
Population: 2.6 million
Land area: 24280 km² (9,370 sq miles)

Kuwait Air Force

Kuwait's original air component was formed in the 1950s as an extension of the Security Department of the Kuwaiti government. The Kuwait Air Force was set up in 1960 and its first aircraft (Jet Provost T.Mk 51s) arrived in 1961. Personnel were trained by the RAF. Its first combat aircraft – Hunter FGA.Mk 57s – arrived in 1964-65. A 1969 order for Lightning F.Mk 53/T.Mk 55s and Strikemaster Mk 83s was

eclipsed by the 1973 defence expansion programme, which oversaw the acquisition of Mirage F1CK/BKs and A-4KU/TA-4KU Skyhawks for the service.

The Iraqi invasion of Kuwait on 2 August 1990 sparked off the second Gulf War, resulting in an American-led coalition formed to oust the invading Iraqi forces. Kuwait was liberated on 25 February 1991, allowing that country's authorities to set about rebuilding the air force.

The cornerstone of the new air force was the September 1988 order for 32 F/A-18C and eight F/A-18D Hornets, 200 AIM-7F Sparrows and 120 AIM-9L Sidewinders air-to-air missiles, 344 AGM-65G Mavericks and 40 AGM-84D Harpoon. USA support for Kuwait, which had been lukewarm before the Gulf War, was reaffirmed, resulting in the US Army Corps of Engineering helping to rebuild the air bases at Ali al Salerm and Ahmed al Jaber in preparation for the Hornets' arrival. Delivered from January 1992, the aircraft were operated from Kuwait International Airport while the bases were completed. Final deliveries were completed in August 1993, by which time the aircraft had begun their operational service as part of Operation Southern Watch.

Today, Hornets are operated by the two squadrons which formerly operated the A-4KU Skyhawks, since

Kuwait Air Force

9 Sqn	F/A-18C/D	Ahamad Al Jaber AB
12 Sqn	Hawk Mk 64	Ali Salim Sabah AB
18 Sqn	Mirage F1CK	Ali Salim Sabah AB
19 Sqn	Tucano Mk 52	Ali Salim Sabah AB
25 Sqn	F/A-18C/D	Ahamad Al Jaber AB
32 Sqn	SA 330F Puma	Ali Salim Sabah AB
33 Sqn	SA 342L Gazelle	Ali Salim Sabah AB
41 Sqn	L100-30, DC-9-32CF	Ali Salim Sabah AB
61 Sqn	Mirage F1CK/BK	Ali Salim Sabah AB
62 Sqn	AS 532SC Cougar	Ali Salim Sabah AB

Shorts supplied Kuwait with its Tucano Mk 52s during 1991, but they were stored in the UK until 1995 while the air force was reformed.

Kuwaiti Hawks were delivered from 1985 and serve with No. 12 Squadron at Ali Salim Sabah Air Base. Kuwait may acquire more in the future.

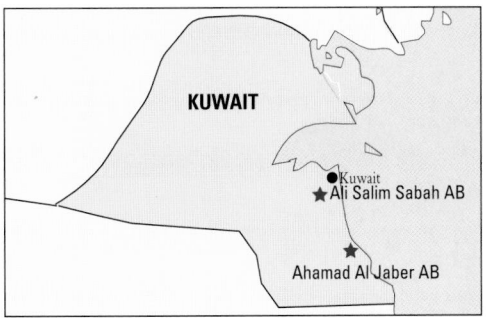

KUWAIT

Kuwait
★ Ali Salim Sabah AB

★ Ahamad Al Jaber AB

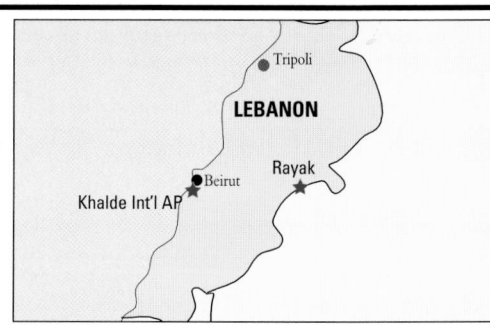

Helicopter airlift is provided by one squadron of eight SA 330H Pumas. Three Exocet-armed AS 532SC Cougars are used by No. 62 Squadron for anti-shipping and SAR missions.

Fixed-wing transport assets are limited to No. 41 Squadron's three L-100-30 Hercules and a single DC-9-32CF. An MD-83 acquired to replace the second DC-9 flies in civil marking for the government of Kuwait.

Six L-100 Hercules have served with the Kuwait Air Force. Two L-100-20s were acquired, and four L-100-30s in 1982, one being lost during the Gulf War.

LEBANON

Capital: Beirut
Population: 3.2 million
Land area: 10400 km²
(4,015 sq miles)
Major cities: Tripoli

Al Quwwat Al Jawwiya Al Lubnaniya
(Lebanese Air Force)

Lebanon's national armed forces are being reorganised with US assistance following the 1990 end of the 15-year civil war between Christian and Muslim factions, in which the minuscule Lebanese Air Force

Al Quwwat Al Jawwiya Al Lubnaniya

Type	Role	Delivered/In Service
SA 316B Alouette III	utility helicopter	14/6
Aérospatiale SA 342L Gazelle	light attack helicopter	4/3
Aérospatiale SA 330L Puma	transport helicopter	12/9
Agusta-Bell AB 212	utility helicopter	12/7
Bell UH-1H Iroquois	utility helicopter	16/16
Fouga CM.170R Magister	basic jet-trainer	9/3
Hawker Hunter FGA.70/70A	air defence/GA	10/5
Hawker Hunter T.66C	combat trainer	3/1
Scottish Aviation Bulldog 126	primary trainer	6/5
Sud SE 3130 Alouette II	utility helicopter	4/2

took little part, although losing many of its assets. By late 1997, the forces had acquired $140 million worth of arms, equipment and training, including 16 ex-US Army Bell UH-1Hs, 800 M113 APCs and 3,000 other vehicles, delivered from mid-1995. Another 16 UH-1Hs were due to follow in 1998.

With only $400 million available in annual defence spending, further equipment procurement will be

severely limited. However, 12 Romanian-built SA 330 Pumas and French-built Gazelles which survived the war, but were grounded in 1992 through lack of spares, are being overhauled and upgraded, together with three remaining Hunter F.70s. Nine Lebanese Dassault Mirage IIIELs and one IIIBL two-seat combat trainer, which spent many years in local storage, are believed to have been acquired by Pakistan. A single Dassault Falcon 20F light jet transport is air force-operated for the Lebanese government, although civil-registered.

OMAN

Capital: Muscat
Population: 2 million
Land area: 271950 km²
(104,970 sq miles)
Major cities: Nizwa, Salalah

Al Quwwat Al Jawwiya Al Malakiya As Omaniya
(Royal Air Force of Oman)

Initially formed with UK aircraft, personnel and assistance in 1959, the original Sultan of Oman's Air Force was organised mainly on RAF lines. Its mid-1990 change of name to the Royal Air Force of Oman and appointment of an Omani commander-in-

Two AS 202/18A4 Bravos were delivered in 1988 to augment an earlier pair in service from 1976. They are used by No. 1 Squadron for primary training duties.

chief reflected its increasing independence and autonomy. Expansion and re-equipment plans have been limited by budget problems resulting from the fall in world oil prices, although in 1993-94 the RAFO was able to replace its venerable Hunters with four two-seat BAe Hawk Mk 103 lead-in fighter trainers and 12 single-seat Westinghouse APG-66H radar-equipped Hawk Mk 203 light ground-attack/interceptors. After evaluations of new combat aircraft, the RAFO elected instead in September 1997 to extend the service lives of its 17 remaining BAe Jaguar ground-attack fighters to around 2008 from a £40 million digital avionics upgrade contract with the UK MoD.

In 1977-78 Oman's 10 original single-seat Jaguar OS versions and two OB trainers were delivered, followed by a similar mid-1980 follow-up order, augmented by an ex-RAF GR.Mk 1 and a T.Mk 2 trainer in 1986. All have been upgraded to GR.Mk 1A/T.Mk 2A standards by the installation of Ferranti FIN1064 inertial nav/attack systems. They are now being fitted by a DERA and UK industrial team with a multi-function cockpit display, HUD, HOTAS and associated new systems, with provision for

GEC-Marconi's TIALD targeting pod and ASRAAM, to Jaguar '97 or GR.Mk 1B/T.Mk 2B standards by 2000.

Based at Thumrait, the Omani Jaguar fleet is divided between Nos 8 and 20 Squadrons. They were delivered from 1977 and, as well as performing the strike mission, are tasked with air defence. Initially, MATRA 550s were carried, before later switching to AIM-9P-4s.

Al Quwwat Al Jawwiya Al Malakiya As Omaniya

Squadron	Aircraft	Base
No. 1 Sqn	Strikemaster, Mushshak, Falke Bravo	Masirah
No. 2 Sqn	Skyvan	Seeb
No. 3 Sqn	AB 205A, AB 212, Bell 212, 214	Salalah
No. 4 Sqn	BAe One-Eleven 485GD	Seeb
No. 5 Sqn	Skyvan	Salalah
No. 6 Sqn	Hawk Mk 103/203	Masirah
No. 8 Sqn	Jaguar OS/OB	Thumrait
No. 10 Sqn	Rapier FSB1, Javelin SAM	
No. 12 Sqn	Rapier FSB1, Javelin SAM	
No. 14 Sqn	AB 205A, AB 206A	Salalah
No. 16 Sqn	C-130H	Seeb
No. 20 Sqn	Jaguar OS/OB	Thumrait
Royal Flight	Boeing 747SP, Eurocopter AS 332C/L1, Gulfstream GIV	Seeb

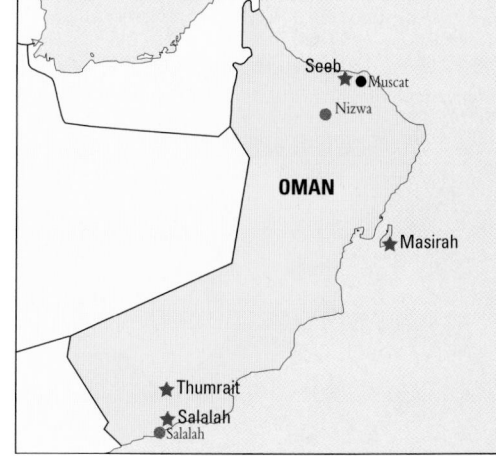

Al Quwwat Al Jawwiya Al Malakiya As Omaniya (Royal Air Force of Oman)

Type	Role	Delivered/In service
Agusta-Bell AB 205A-1	utility helicopter	31/18
Agusta-Bell AB 212	utility helicopter	1/1
Bell 206B JetRanger	utility helicopter	3/3
Bell 212	utility helicopter	2/2
Bell 214B	utility helicopter	6/5
BAe Hawk Mk 103	lead-in fighter-trainer	4/4
BAe Hawk Mk 203	air defence/GA	12/12
One-Eleven 485GD	transport	3/3
Strikemaster Mk 82/82A	armed jet-trainer	24/12
FFA AS 202	primary trainer	4/4
Lockheed C-130H Hercules	tactical transport	3/3
PAC MFI-17 Mushshak	primary trainer	7/7
Scheibe SF.25C Falke	primary trainer	2/2
Jaguar OS/GR.Mk 1	ground attack	21/14
Jaguar OB/T.Mk 2	combat trainer	5/3
Shorts SC.7 Skyvan 3M	light transport/SAR	16/14

Above: The purchase of APG-66H-equipped Hawk Mk 203s greatly enhanced the air defence capability of the RAFO. They replaced Hunter F.Mk 7A/Bs with No. 6 Squadron from late 1994.

Left: Operating alongside the Hawk Mk 203s are four Mk 103 lead-in fighter trainers. They are equipped with FLIR and laser rangefinders.

aircraft. These include the Royal Flight with a Boeing 747SP-27, Gulfstream GIV, three AS 332C/L Super Pumas, a Dassault Falcon 20E and four FFA AS.202/18A-4 Bravo trainers. The Royal Oman Police Wing, formed in 1970, operates two civil-registered Airtech CN.235M-100, a Dornier Do 228-100 and a Pilatus Britten-Norman PBN-2T Turbo Islander, all twin-turboprop transports, a Pilatus PC-6B Turbo Porter, three Bell 205A-1s and six SAR-equipped Bell 214ST helicopters.

PALESTINE

Air element of the Palestine Authority

Following the Middle East peace agreements with Israel, Palestine began the formation of an air element for its public security forces in 1996-97. As yet mainly helicopter-equipped, the air element operates a Gaza-based Agusta-Bell AB 212 with an Egyptian civil registration, supplemented by two Mil Mi-8s and two Mi-17s from CIS sources. A civil-registered Lockheed L-1329 JetStar is also used for Palestine government transport roles.

QATAR

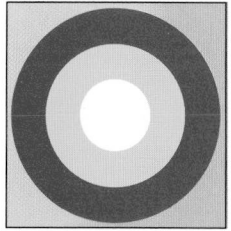

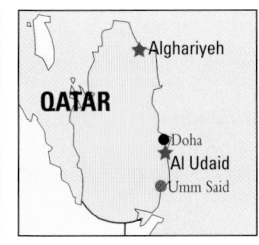

Capital: Doha **Population:** 0.4 million
Land area: 11435 km² (4,415 sq miles)
Major cities: Umm Said, Zakrit

Qatar Emiri Air Force

Having abstained from the Union of Arab Emirates after independence from Britain in 1971, Qatar began forming its own defence forces with UK aid. In 1974, the mainly helicopter-equipped Air Wing of the Public Security Forces achieved air force status, following receipt of more British rotary-wing aircraft and ex-RAF Hunters as the QEAF's first fixed-wing and combat types. Modernisation and reorganisation

Carrying a single AM39 Exocet anti-ship missile, this Commando Mk 3 has the Sea King-type sponsons and a radar, reflecting its main role of sea surveillance.

plans then resulted in 1979 orders for six Dassault/Dornier Alpha Jets and 14 Mirage F1 multi-role fighters delivered during 1980-84.

Based at Doha International Airport, with several outlying desert strips available for helicopter operations, QEAF units experienced severe congestion and evident vulnerability in the 1990-91 Gulf War, resulting in new infrastructure contracts worth over $200 million in France. They involved construction of a dedicated military air base and HQ, with hardened aircraft shelters, air defence radars and Roland missile batteries, southwest of the capital at Al Udaid.

A 1987 military co-operation agreement with France also led to $1.6 billion contracts in 1994 for nine single-seat Dassault Mirage 2000-5EDA multi-role combat aircraft and three two-seat 2000-5DDA combat trainers, with associated MATRA/BAeDynamcis Magic 2 and MICA AAMs. Deliveries started in December 1997, and Qatar's remaining 11 Mirage F1EDAs and two two-seat F1DDAs were returned to Dassault in part-exchange, and subsequently sold on to Spain.

Orders from a £500 million late 1996 letter of agreement for a UK arms package including 18 BAe Hawk 100s and 15 Shorts Starburst SAM systems have been delayed by budget reductions accompanying falling oil prices. Upgrades have nevertheless been completed of Qatar's Westland Commando helicopter force, which includes at least two Mk 3s equipped to

Armed with Euromissile HOT anti-tank missiles, the Qatari Gazelles serve with No. 6 Close Support Squadron. They support motorised and tank battalions of the Qatari army.

launch Aérospatiale AM39 Exocet anti-ship missiles for ASV roles, supplementing normal ASW equipment and armament.

For government and VIP transport, the Emir Wing operates single civil-registered examples of the Airbus A340-211, Boeing 707-336C and -3P1C, and Boeing 727-2P1, plus two Dassault Falcon 900s.

The Alpha Jet Cs of No. 11 Close Support Squadron provide weapons training for pilots who have undertaken pilot courses overseas, but in times of war would be used for attack missions.

Qatar Emiri Air Force

No. 1 Fighter Wing		Al Udaid
No. 7 Air Superiority Sqn		Mirage 2000-5EDA/DDA
No. 11 Close Support Sqn		Alpha Jet
No. 2 Rotary Wing		**Al Udaid**
No. 6 Close Support Sqn		SA 342G/L Gazelle/HOT
No. 8 Anti-Surface Vessel Sqn		Commando Mk 3
No. 9 Multi-Role Sqn		Commando Mk 2A/C

Type	Role	Delivered/In Service
Aérospatiale SA 342G Gazelle	utility helicopter	2/2
Aérospatiale SA 342L Gazelle	armed helicopter	14/12
Dassault Mirage 2000-5EDA	multi-role fighter	9/9
Dassault Mirage 2000-5DDA	combat trainer	3/3
Dassault/Dornier Alpha Jet C	lead-in fighter trainer	6/6
Westland Commando Mk 2A	assault helicopter	3/3
Westland Commando Mk 2C	transport helicopter	1/1
Westland Commando Mk 3	ASW/ASV helicopter	8/8

Middle East

SAUDI ARABIA

Capital: Riyadh
Population: 10.5 million
Land area: 2.4 million km² (926,745 sq miles)
Major cities: Jeddah, Mecca

Al Quwwat Al Jawwiya Al Malakiya As Sa'udiya (Royal Saudi Air Force)

Over the past 15 or so years, major procurement programmes from France, the UK and the US have resulted in deliveries of 98 MDC F-15C/Ds, 72 F-15Ss, 96 Panavia Tornado IDSs, 24 Tornado ADVs, 50 Pilatus PC-9s, 50 BAe Hawk Mk 65/65As, their associated weapons, and many supporting aircraft and helicopter. They are being supplemented by current RSAF orders for 44 Agusta-Bell 412EP SAR helicopters costing $340 million. Discussions were also being held with Boeing for procurement of up to 25 more F-15S aircraft. In February 1998, RSAF selection of the Block 50 F-16C/D was confirmed by Lockheed Martin, which was then finalising negotiations for up to 90 examples with the Saudi government as Northrop F-5 replacements, then put on hold. Unfilled requirements in the Al Yamamah II procurement programme with the UK include one for at least 20 BAe Hawk 100 lead-in fighter trainers, for which funding is still being sought. Studies are also being made of maritime patrol aircraft for surveillance of the Gulf area, in co-operation with Saudi Arabia's GCC partners.

Under the Armed Forces Chief of Staff, the RSAF is one of four equal status components, also comprising Naval Forces, Land Forces and Defence Command. This last includes an Air Defence Command element to control ground and moored

Above: F-5E Tiger IIs provided the Saudi air force with the bulk of its combat fleet during the late 1970s and early 1980s, having replaced Lightning F.Mk 53s in 1975. Five squadrons still use the type.

Below: The first of 72 F-15S Eagles for Saudi Arabia, this aircraft is seen fitted with conformal fuel tanks. The type is only slightly downrated from the F-15E.

balloon-carried radars, AAA, surface-to-air missiles and communications networks, plus AEW coverage from the RSAF's Boeing E-3A AWACS fleet, supported by KE-3A tankers from the 1982 US Peace Shield and Peace Sentinel contracts.

Major orders also placed by Saudi Arabia for airfield defence systems included a $675.7 million mid-1992 contract with Oerlikon-Buehrle to supply over 100 GDF-002/005 twin 35-mm cannon units with associated Skyguard fire control radars, plus options for Oerlikon's ADATS SAM system. On 23 December 1992, via the US Army Missile Command, the Saudi government supplemented a $513 million 1990 order for eight Raytheon Patriot air defence fire units and 300 PAC-2 standard missiles with a follow-up contract for a further 13 launch units and 761 upgraded PAC 2 missiles costing $1.03 billion. Patriot deliveries were completed to Saudi Arabia between 1994-95.

These elements are further integrated with Air Operations and Air Intelligence Commands, and linked with neighbouring Gulf Co-operation Council defence forces. With limited indigenous personnel resources, however, the Saudi armed forces and the RSAF, in particular, continue to be largely dependent on expatriate technicians to man their support and infrastructure services.

The RSAF is believed to be the operating agency for about 40 CSS-2 (Dong Feng-3A) IRBMs acquired from China in a surprise purchase in late 1987, and deployed in north and south desert silos at Al Jaffer and As Sulayyil. With a 2800-km (1,740-mile) range and a conventional HE warhead of up to 2000 kg (4,409 lb), these IRBMs have sub-strategic strike and deterrent roles against neighbouring Arab states and Israel, although they were not used in retaliation against Iraqi 'Scud' SSM attacks against Saudi cities in the 1990-91 Gulf War.

A major RSAF transport element, No. 1 Sqn also operates as the Royal Flight, supplementing several civil-registered transports led by two Boeing 747SP-68/3G1s (alternating as 'Air Force One'), and also including two Boeing 707-368Cs, a Boeing 737-268, two Gulfstream Aerospace GIVs, two Lockheed L-1011 TriStar 500s and two McDonnell Douglas MD-11, for VIP and government flights. They operate alongside several other fixed- and rotary-winged aircraft with special joint civil/military registrations in No. 1 Sqn, comprising four Airtech CN.235Ms, four BAe 125-800Bs, a Beech King Air 200, a Boeing 707-138B, two Gulfstream GIIIs, two Learjet 35As, two VC-130H Hercules, two Lockheed L-1329 JetStar 8s, and an Agusta-Bell AB 212 and three

The Hercules serves in several versions with the Saudis, but the majority – such as 1614 – are C-130Hs based at Prince Sultan Air Base.

Saudi Arabia was the only export customer for the Tornado. Some 24 ADVs (above) and 96 IDSs (below), including 12 reconnaissance-configured aircraft, were acquired from 1986 in two batches.

Agusta-Sikorsky AS 61 helicopters.

In related transport activities from Riyadh, the Saudi Armed Forces Medical Services also operate a number of civil-registered aircraft, including two Gulfstream GIIs, a GIII, a Learjet 35A, three Lockheed C-130H, a C-130H-30 and three L100-30 Hercules, plus five Aérospatiale AS 365N Dauphins, one Agusta-Bell 212, and 16 specially-equipped S-70-1L Black Hawks in air ambulance, casualty evacuation and SAR roles. Some 18 Kawasaki-Vertol KV-107 helicopters are also operated by the RSAF on behalf of the Security Ministry on SAR and fire-fighting roles.

As part of the Al Yamamah arms contract, Saudi Arabia received 30 Hawk Mk 65s. These were followed by 20 combat-capable Mk 65As (above) as part of the Al Yamamah II deal.

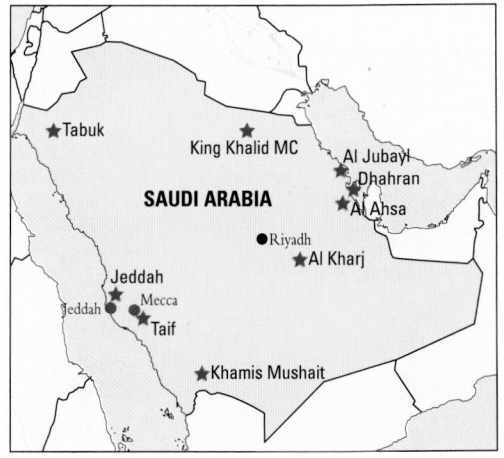

Al Quwwat Al Jawwiya Al Malakiya As Sa'udiya

King Abdullah Aziz AB, Dhahran
No. 7 Sqn	Tornado IDS
No. 13 Sqn	F-15C/D
No. 21 Sqn	Hawk Mk 65
No. 29 Sqn	Tornado ADV
No. 34 Sqn	F-15C/D
No. 35 Sqn	Jetstream 31M
No. 37 Sqn	Hawk Mk 65
No. 66 Sqn	Tornado IDS/(R)
No. 83 Sqn	Tornado IDS

King Faisal Air Academy, Al Kharj
No. 1 Sqn (Royal Flight)	707-138B, VC-130H, L-100-30, CN.235M-10
No. 8 Sqn	Cessna FR 172G/H/M
No. 9 Sqn	PC-9
No. 11 Sqn	Hawk Mk 65
No. 22 Sqn	PC-9
No. 79 Sqn	Hawk Mk 65A
No. 88 Sqn	Hawk Mk 65/65A

King Fahd AB, Taif
No. 3 Sqn	F-5B/E/F
No. 5 Sqn	F-15C/D
No. 10 Sqn	F-5B/E/F
No. 12 Sqn	AB 205A, AB 206A
No. 14 Sqn	AB 206A, AB 212, KV-107-II

King Khalid Military City
VIP Helicopter	UH-60VH-1

King Khalid AB, Khamis Mushait
No. 6 Sqn	F-15C/D
No. 15 Sqn	F-5B/E/F
No. 55 Sqn	F-15S

Prince Abdullah AB, Jeddah
No. 4 Sqn	C-130, KV-107, AB 212

Prince Sultan AB, Al Kharj/Riyadh
No. 16 Sqn	C-130E/H
No. 18 Sqn	E-3A, KE-3A
No. 32 Sqn	KC-130H
No. 42 Sqn	F-15C/D

Tabuk
No. 2 Sqn	F-5E/F
No. 17 Sqn	RF-5E, F-5F

Two Jetstream 31Ms were acquired, fitted out as flying classrooms for Tornado WSOs. One was lost in October 1989 and the second serves with No. 35 Squadron at King Abdullah Aziz Air Base.

Four CASA CN.235-10s have been operated by No. 1 Squadron since early 1987. Two are configured as VIP transports while the other two are convertible passenger/freight versions.

Royal Saudi Navy Force

RSNF helicopter operations started in 1980 with an order for 25 Aérospatiale AS 565SA/SC Panthers, mainly to equip four new F2000 frigates in a $3.45 billion French arms package. The 19 remaining armed AS 565SAs are equipped with Agrion 15 radar to launch AS 15T anti-ship missiles and ASW weapons, while the six unarmed AS 565MAs have OMERA ORB 32 radar for SAR and utility roles. When disembarked, they operate from Al Jubayl Naval Airport, alongside King Abdul Aziz Naval Base. They were reinforced from a follow-on French contract placed in mid-1998 for a dozen shore-based Eurocopter AS 532UC/SC Cougars. Six AS 532UCs are equipped for armed transport and utility roles, while the SC Cougars can carry AM39 Exocet anti-ship missiles as well as homing torpedoes and other weapons for anti-surface and anti-submarine warfare roles.

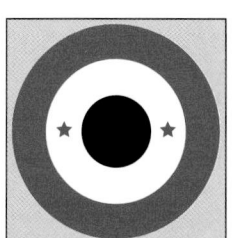

The Royal Saudi Land Forces Bell 406CS Combat Scout armed reconnaissance helicopters are based on the OH-58D airframe, but fitted with a lower standard of equipment. They arrived in Saudi Arabia in 1990. Five of the helicopters were delivered with the ability to fire TOW anti-tank missiles, as carried here.

Royal Saudi Land Forces Army Aviation Command

The Royal Saudi Land Forces Army Aviation Command (RSLFAAC) is one of two currently small subsidiary combat helicopter-equipped aviation branches of the Saudi armed forces. It received 15 Bell 406CS Combat Scouts and a dozen Sikorsky UH-60A-1 utility helicopters plus a single VIP version from 1990, and has a requirement for at least 50 more Black Hawks or similar types. Further RSLFAAC combat reinforcement started in April 1993 with deliveries of the 12 Boeing AH-64A Apache attack helicopters. They were ordered in late 1990 from an estimated $300 million contract as part of US emergency aid programmes to the GCC states, and were accompanied by six spare GE T700-701 engines, 155 Rockwell Hellfire ATMs, spare TADS/PNV, and integrated helmet-mounted display sighting systems.

A $1.8 billion mid-1992 US military support package for Saudi Arabia included $606 million in hangar and infrastructure funding, plus 362 more Hellfire missiles and 3,500 Hydra-70 rockets, for the RSLFAAC Apaches. Contracts for the Saudi Apache package were finalised with McDonnell Douglas Helicopters through the US Army's Aviation and Troop Command in the spring of 1993 for delivery in 1995 from Production Lot 11. Two AS 365N Dauphin helicopters are operated mainly in medical evacuation roles.

Royal Saudi Land Forces Army Aviation Command

Type	Role	Delivered/In Service
Bell 406CS Combat Scout	light attack helicopter	15/13
Eurocopter AS 365N Dauphin 2	utility helicopter	2/2
MDH AH-64A Apache	attack helicopter	12/12
UH-60A-1 Desert Hawk	transport helicopter	12/12

SYRIA

Capital: Damascus
Population: 12.2 million
Land area: 185680 km² (71,675 sq miles)
Major cities: Aleppo, Homs

Al Quwwat Al Jawwiya Al Arabiya As'Souriya (Syrian Arab Air Force)

Syrian support for extremist political organisations, and its absence from Middle East peace negotiations, has gained it international pariah status alongside such countries as Iran, Libya and North Korea, with corresponding restrictions on arms procurement and strategic supplies. Being almost entirely Soviet-equipped, its armed forces are suffering from a lack of spares and technical support, and many of its 500 or more combat aircraft are being cannibalised to keep a small proportion airworthy.

Russia is still owed some $11 billion by Syria for previous arms supplies up to 1988-89, but has recently been discussing the possible release of defensive equipment, including more MiG-29s, and Antey S-300PMU (SA-12 'Gladiator'/'Giant') surface-to-air and anti-ballistic missile systems to supplement older SA-6s and SA-8s. Agreement in principle was also reported in early 1998 with India for repair, overhaul and upgrade of some Syrian MiGs and Mil helicopters. This would be on a barter basis, because of Syrian funding problems resulting from falling oil prices, economic down-turns and lack of financial support from other Middle Eastern countries, apart from fellow pariah Iran.

Syrian military airlift commitments are undertaken entirely by civil-registered aircraft in Syrianair colours, comprising one Antonov An-24 'Coke' and five An-26 'Curl' twin-turboprop cargo/passenger transports, plus four Ilyushin Il-76M 'Candid' four-turbofan freighters. Two civil-registered twin-turbofan Dassault Falcon 20F, two Tupolev Tu-134B-3 'Crusty' and six tri-turbofan Yak-40 'Codling' passenger transports are also flown by the air force in Syrianair colours, together with a single Piper PA-31-310 Navajo light piston-engined twin.

A small air arm equipped solely with helicopters is operated by the Syrian navy on anti-submarine warfare, SAR and general support roles. Its current equipment is reported to be about 20 Kamov Ka-25Bsh 'Hormone-A' helicopters, with a shipboard capability, plus about five Mil Mi-14PL 'Haze-A' amphibious ASW helicopters, although few are believed to be fully serviceable at the present time.

Type	Role	In Service
Aero L-29 Delfin	basic trainer	40
Aero L-39ZA Albatros	armed advanced trainer	40
Aero L-39ZO Albatros	advanced trainer	50
Aérospatiale SA 342L Gazelle	light attack helicopter	50
MBB/SIAT 223 Flamingo	primary trainer	45
MiG-15UTI 'Midget'	combat trainer	10
MiG-17F 'Fresco'	combat trainer	30
MiG-21PF/MF/bis 'Fishbed'	air defence	220
MiG-21U/UM 'Mongol-B'	combat trainer	20
MiG-23MF/ML/MS 'Flogger'	air defence	80
MiG-23BN/UB 'Flogger-F'	ground attack	60
MiG-23UB 'Flogger-C'	combat trainer	6
MiG-25RB 'Foxbat-B'	reconnaissance	8
MiG-25PD 'Foxbat-E'	air defence	30
MiG-25PU 'Foxbat-C'	combat trainer	2
MiG-29 'Fulcrum-A'	air defence/com trainer	42
MiG-29UB 'Fulcrum-B'	combat trainer	6
Mil Mi-6 'Hook'	heavy-lift helicopter	10
Mil Mi-8/17 'Hip'	transport helicopter	100
Mil Mi-24D/V 'Hind-D/E'	attack helicopter	30
PAC MFI-17 Mushshak	primary trainer	6
PZL/Mil Mi-2 'Hoplite'	transport helicopter	20
Sukhoi Su-20 'Fitter-F'	ground attack	20
Sukhoi Su-22M2/3/4 'Fitter-H/K'	ground attack	43
Sukhoi Su-22U/UM 'Fitter-G'	combat trainer	5
Sukhoi Su-24MK 'Fencer-D'	ground attack	20
Sukhoi Su-27 'Flanker'	air defence	14
Sukhoi Su-27UB 'Flanker-C'	combat trainer	2

Al Quwwat Al Jawwiya Al Arabiya As'Souriya

Northern Air Defence Zone

Air Defence Division
? Air Regiment	MiG-23	Abu-ad-Duhor
	MiG-21	Hamah
? Air Regiment	MiG-25PD/25PU	T4
? Air Regiment	MiG-21	Deir az Zawr
	MiG-21	Tabqa
	MiG-21	Jirah

Southern Air Defence Zone

Air Defence Division
? Air Regiment	MiG-21	Halhul
? Air Regiment	MiG-21	Al-Quasyr
	Su-27/27UB (?)	Al-Qusayr (?)
? Air Regiment	MiG-29/29UB	Seikal

Tactical Air Command

Air Division
? Air Regiment	MiG-23ML/23MF	Nazariah
? Air Regiment	MiG-23ML/23MF	Blaj
	Su-22M/22U/22UM	Blaj
? Air Regiment	MiG-23ML/23MF	Dmeir
	MiG-25R	Dmeir
	Su-20, Su-22M/U/UM	Dmeir

Air Division
? Air Regiment	MiG-23BN/23UB	Sarat
	Su-20, Su-22M/U/UM	Sarat
? Air Regiment	Su-24MK	T4
	Su-20, Su-22M/U/UM	T4

Air Division
? Air Brigade	Mil Mi-8	
? Air Brigade	Mil Mi-8	
? Air Brigade	Mil Mi-24D/V	Sueda
? Air Brigade	SA 342L Gazelle	Mezze
	Mil Mi-8	Mezze
? Air Brigade	Mil Mi-8	Aleppo
? Air Regiment	Tu-134*, B-727*	Damascus Int'l
	An-24B*, An-26/26B*	Mezze
	Il-76M/T*	Damascus-Int'l
	Falcon 20*, Yak-40*	Mezze

Training Command

Air Force Academy
? Basic Train Regiment	MB-223, Aero L-39	
? Basic Train Regiment	MB-223, Mil Mi-8	
? Basic Train Regiment	Mil Mi-8, PZL Mi-2	Quaar-as-Sitt
? Prim. Train Regiment	Aero L-29	Aleppo
? Weap. Train Regiment	Aero L-39	Aleppo
? Weap. Train Regiment	Aero L-39	

Naval Aviation

? Naval Regiment	Ka-25BSh	Latakia
	Mi-14PL	Latakia

* operated in civil colours by Syrianair

Syrian Naval Aviation

Ka-25BSh 'Hormone-A'	shipborne ASW	20
Mil Mi-14PL 'Haze-A'	ASW helicopter	5

UNITED ARAB EMIRATES

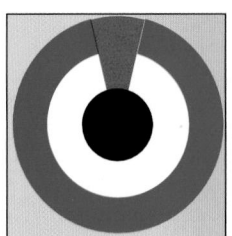

Capital: Abu Dhabi
Population: 1.6 million
Land area: 75150 km² (29,010 sq miles)
Major cities: Dubai, Sharjah

UAE Air Forces

Of the seven UAE Gulf states – comprising Abu Dhabi, Dubai, Ajman, Fujairah, Ras al Khaimah, Sharjah and Umm al Qaiwain – only the first two have sizeable armed forces and air components. While formally merged in 1976, with integrated military planning, the UAE defence forces maintain mainly discrete procurement, organisational and non-operational training policies. Contributing some 80 per cent of the union's defence budget and personnel, Abu Dhabi is the dominant partner. The main national air components, comprising the Abu Dhabi Air Force (ADAF) and Dubai Air Wing (DAW), respectively operate as the UAE's Western and Central Air Commands, with local HQ and individual operational organisations. Some UAE units are also deployed to Sharjah within Central Command, among the remaining emirates.

Apart from a nominal light attack element, the DAW is mainly helicopter-equipped, and only Abu Dhabi has so far operated dedicated air combat equipment. Both main UAE air forces also operate their own training and support elements. As in some other Gulf states, their manning includes a proportion of expatriate personnel, from the UK as well as other Arab countries and Pakistan.

A major ADAF modernisation programme, including new air bases at Al Hamra and Tarif, started in December 1997 with a $3 billion order for 30 Dassault Mirage 2000-9s and a $77 million contract to upgrade 33 Mirage 2000s with Thomson-CSF RDY radar and new digital avionics to similar standards. New weapons are expected to include more GEC-Marconi Al Hakim or new MATRA/BAeD Black Shahine versions of the Storm Shadow/SCALP air-launched cruise missiles, with a range of 350 km (217 miles).

After competitive evaluations, the ADAF selected Lockheed Martin's new Block 60 F-16C/D development in May 1998, over the shortlisted Eurofighter Typhoon and Rafale. Eighty F-16 Block 60s, costing some $7 billion with AMRAAMs, smart weapons and support equipment, will replace Abu Dhabi's remaining Mirage 5s and provide a quantum increase in its combat strength in 2002-04.

The ADAF also needs four to six maritime patrol aircraft to supplement the Abu Dhabi navy's small helicopter-equipped air wing, which acquired combat capability from 1995 orders for Eurocopter's conversion of five of its AS 532F Cougars for ASW/ASV roles, with radar, dipping sonar, and AM39 Exocet AShM launch capability, plus procurement of seven Eurocopter AS 565SA Panthers armed with Aérospatiale AS15TT anti-ship missiles.

Civil-registered aircraft of the Abu Dhabi Royal Flight include two Airbus A300-620s, and single BAe 146-100, Boeing 707-3L6B, Boeing 737-2P6, Boeing 747SP-Z5, and Dassault Falcon 900. A similar VIP Flight in Dubai operates a civil-registered Boeing

Above: A single Short Skyvan 3M was acquired by the Sharjah Amiri Guard Air Wing in 1986, being transferred to the Dubai Air Wing in 1995, with which it still serves.

Below: The Mirage 2000 is Abu Dhabi's principal fighter. The force is used for air defence, ground attack and reconnaissance.

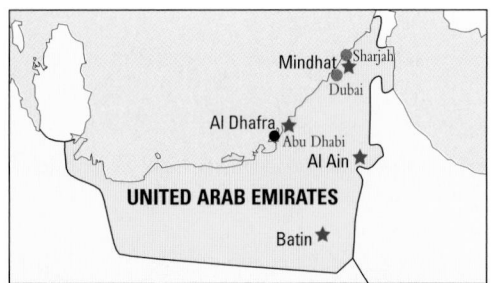

UNITED ARAB EMIRATES

The United Arab Emirates' 1980s decision to standardise on the Hawk resulted in orders from both Dubai (in 1981) and Abu Dhabi (in 1983). 1001 (above) was the first Mk 63 delivered to Abu Dhabi.

Below: Dubai also uses the Aermacchi MB-339A for advanced training, based at the Flying Training Academy at Mindhat.

Right: The United Arab Emirates ordered 20 Apaches in 1991, with the first handed over in October 1993. An additional 10 were ordered but subsequently cancelled.

707-3L6C, Boeing 747SP-31, Dassault Falcon 900, Gulfstream GIV, plus a Bell 212 and a Sikorsky S-76 helicopter. The paramilitary Dubai Police Air Wing (DPAW) is exclusively helicopter-equipped, with three Agusta A 109K2s, four Agusta-Bell/Bell 212s, two Agusta-Bell 412EPs, a Bell 206B, and four Eurocopter BO 105CBS.

Among other UAE military and paramilitary air assets in the Central Air Command, apart from ADAF Mirage 5 and 2000 units on detachment, Sharjah's Amiri Guard Air Wing (AGAW) has a VIP Boeing

Dubai Air Wing

UAEAF Central Air Command		
Mindhat		
III Shaheen Sqn	Hawk	
Transport Sqn	BN-2T, C-130, PC-6B, Skyvan, Shorts 330	
Helicopter Sqn	AB 205, AB 206, AB 212, AB 214, AB 412, IAR 330	
Flying Training Academy SF.260, MB-326, MB-339		

Type	Role	Delivered/In Service
Aermacchi MB-326KD	advanced trainer/GA	6/3
Aermacchi MB-326LD	basic trainer	2/2
Aermacchi MB-339A	advanced trainer	7/4
AS 350B Ecureuil	utility helicopter	1/1
AS 365N1 Dauphin 2	transport helicopter	1/1
Agusta-Bell AB 205A-1	utility helicopter	6/6
Agusta-Bell AB412 Gryphon	utility helicopter	3/3
BAe Hawk T.Mk 61	advanced trainer/GA	9/6
Bell 206B JetRanger	utility helicopter	6/5
Bell 206L-1 LongRanger	utility helicopter	2/2
Bell 214B	utility helicopter	5/5
IAR-330L Puma	transport helicopter	10/10
L-100-30 Hercules	tactical transport	1/1
C-130H-30 Hercules	tactical transport	1/1
BO 105CBS	utility helicopter	3/3
BN.2T Turbo Islander	light transport	5/1
PC-6B2 Turbo Porter	light transport	2/2
Short SC.7 Skyvan 3M	light transport	1/1
Shorts 330UTT	light transport	1/1
SIAI-Marchetti SF.260TP	primary trainer	6/5

737-2W8, two Beech King Air 350s and three Bell 206B JetRangers. Ras al Khaimah operates a civil-

Abu Dhabi Air Force

UAEAF Western Air Command		
Al Dhafra		
II Shaheen Sqn	Mirage 5RAD*, Mirage 2000RAD	
Apache Sqn	AH-64A	
Al Ain		
Khalifa Bin Zayed Air College	Grob G 115TA	
Batin		
Puma Sqn/Royal Flt	SA 330, AS 532	
Spray Unit	Piper PA-36	
Transport Sqn	C.212, CN.235, C-130	
Maqatra		
Al Ghezelle Sqn	SA 342L	
Flying Training School	Hawk 63/102, PC-7	
Sharjah		
I Shaheen Sqn	Mirage 5AD/DAD/EAD*, Mirage 2000EAD	

Type	Role	Delivered/In Service
SA 330C/F Puma	transport helicopter	10/8
SA 342L Gazelle	armed helicopter	11/10
Airtech CN.235M	tactical transport	7/7
BAe Hawk T.Mk 63A/B/C	advanced trainer	20/19
BAe Hawk Mk 102	lead-in fighter trainer/GA	18/18
CASA C.212-200	tactical transport	4/4
Mirage 5AD/5EAD	ground attack	26/21*
Mirage 5RAD	tactical recce	3/3*
Mirage 5DAD	combat trainer	2/1*
Mirage 2000EAD	air defence	22/20
Mirage 2000RAD	tactical recce	8/8
Mirage 2000DAD	combat trainer	6/5
Mirage 2000-9	multi-role	30**
AS 532UC/UL Super Puma	transport helicopter	8/7
Grob G 115TA	primary trainer	12/12
C-130H Hercules	tactical transport	6/3
AH-64A Apache	attack helicopter	30/30
BO 105CBS	utility helicopter	4/4
Pilatus PC-7 Turbo Trainer	basic trainer	24/23
*most in reserve	**on order	

Abu Dhabi navy

ASW Squadron, Al Dhafra		
AS 532F Cougar	ASW helicopter	5/5
AS 565SA Panther	ASV helicopter	7/7

registered Cessna 500 Citation I, and Umm al Qaiwain Royal Flight has a VIP Bell 222UT.

YEMEN

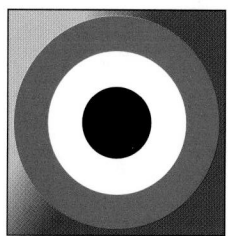

Capital: San'a
Population: 12 million
Land area: 477530 km² (184,325 sq miles)
Major cities: Al Mukalla, Ash Shaykh Uthman

Republic of Yemen Air Force

Unification on 22 May 1990 of the Islamic Yemen Arab Republic in the north and the Marxist People's Democratic Republic of Yemen in the south included amalgamation of their armed forces. The substantially larger PDRY air forces were entirely Soviet-equipped and -organised, with MiG-21 and MiG-23BM fighter-bombers, Su-22 'Fitter' ground-attack aircraft, and even a few Mach 2.7 MiG-25R 'Foxbat' tactical-reconnaissance fighters, mostly with Soviet crews. The former IYAR operated fewer and slightly less-sophisticated weapons and equipment from Western and Chinese, as well as Soviet, sources, the most advanced of which comprised a few Chengdu F-7

(MiG-21) Airguards, plus some Sukhoi Su-22s and Northrop F-5E/Fs, the latter transferred from Saudi Arabia in the late 1970s.

With the collapse of Communism in Eastern Europe and the USSR, most of the advisers and military technicians who assisted Yemeni military

Republic of Yemen Air Force

Type	Role	In Service
Agusta-Bell AB 206B JetRanger	utility helicopter	5
Agusta-Bell AB 212	utility helicopter	5
Antonov An-12B 'Cub '	transport	1
Antonov An-24V 'Coke'	transport	6
Antonov An-26 'Curl'	transport	7
Chengdu F-7M Airguard	air defence/GA	4
Lockheed C-130H Hercules	transport	2
MiG-15UTI 'Midget'	advanced trainer	4
MiG-21MF/bis 'Fishbed-J/N'	air defence/GA	40
MiG-21UM 'Mongol'	combat trainer	12
MiG-23ML 'Flogger-G'	air defence	20
MiG-23UB 'Flogger-C'	combat trainer	5
Mil Mi-8 'Hip'	transport helicopter	30
Mil Mi-24 'Hind'	attack helicopter	15
Northrop F-5E Tiger II	air defence/GA	8
Northrop F-5B	combat trainer	3
Sukhoi Su-22M2 'Fitter-D'	ground attack	15
Sukhoi Su-22UM3 'Fitter-E'	combat trainer	3
Yakovlev Yak-11 'Moose'	basic trainer	20

aircraft operation have been withdrawn, while further supplies of CIS arms and spares have also been greatly reduced. Previous Yemeni military links with China, Iran and Iraq appear to have been maintained, however, with the PRC providing an alternative source of spares, equipment and military advisers.

Long-standing tribal hostilities between the former North and South Yemen republics resulted in renewed civil war in 1994, and brief transfer of a squadron of MiG-29s to the northern elements. Several examples were lost in combat and the remainder were then withdrawn to the CIS.

Many other RYAF aircraft are no longer serviceable because of spares and technical support problems, as well as shortages of trained indigenous air and ground crews. A single civil-registered VIP Boeing 727-2N8 in Yemenia colours is used for government transport.

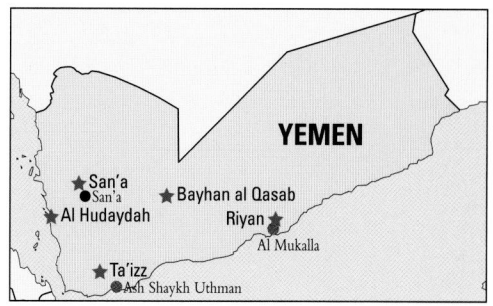

Indian Sub-continent

The area known as the Indian sub-continent contains the nations of India, Pakistan and Bangladesh, and the kingdoms of Bhutan and Nepal. In the wider arena of the Indian Ocean are the island nations of the Comores, Maldives, Seychelles and Mauritius.

The present military situation in the area reflects, to a large extent, decisions made over 50 years ago when independence for India was first announced. The war that followed ended without settling recognised borders and this has remained a large area of friction between India and Pakistan ever since, especially over the disputed Kashmir region.

The explosion of five nuclear devices by India in May 1998 and the 'counter' tests by Pakistan soon after heralded the start of a new chapter in the region's history. Coupled with the development of long-range ballistic missiles, the tests cast further doubt over the already shaky stability of the region.

Even without a nuclear capability, India had long

been a regional superpower. It has the world's fourth-largest air force and is the only country in the region to possess an aircraft-carrier. External threats to India's security come not only from Pakistan, with which India has fought three wars since independence, but also from China. Border skirmishes are common. The country's mix of different religious, ethnic and political groups also poses internal security problems.

Pakistan's air force is suffering for the nuclear weapons programme, especially after the passing of the Pressler amendment in the USA, which effectively ended F-16 sales to the country and resulted in the embargo of 28 F-16A/Bs paid for by the country. Pakistan has forged closer links with China to maintain a credible force, but joint development of combat and training aircraft has not been as straightforward as originally planned. Having tried to maintain a qualitative edge against the numerically superior Indian forces, Pakistan is in the

unenviable position of seeing it being eroded as India plans large-scale upgrades and purchases of new aircraft.

Bangladesh maintains a small air force for its own security, using Chinese combat aircraft. Bhutan and Nepal, sandwiched between India and China, have forces that support the internal operation of their small armies, and have no offensive capabilities.

The island state of Sri Lanka continues to be embroiled in the long-running civil war between government forces and the Liberation Tigers of the Tamil Eelam – the 'Tamil Tigers'. Air power has become an integral part of this struggle, and the Sri Lankan Air Force has grown during the conflict from being a transport and training organisation to being an operator of fast jets.

The need for military forces on the island states of the Comores, Maldives, Seychelles and Mauritius is minimal, and this is reflected in their allocation of resources to military aviation.

The most potent aircraft in service are the F-7M Airguards of the 'Supersonics' and 'Thundercats' squadrons (Nos 5 and 35). F-7Ms are pooled for maintenance and allocated to a squadron when required for operational duties; thus, this aircraft wears the badge of No. 5 Sqn and logo of No. 35.

BANGLADESH

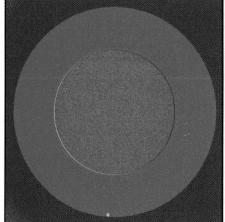

Capital: Dhaka   **Population:** 109.3 million
Land area: 144000 km² (55,585 sq miles)
Major cities: Chittagong, Khulna

Bangladesh Biman Bahini (Bangladesh Air Force)

Bangladesh, which achieved fiery independence from Pakistan in December 1971, has settled down as Southern Asia's second most populous country, with its Bengali aspirations clearly making it distinctly different from being a mere eastern wing of the Islamic State of Pakistan. Geographically flanked by the Indian provinces of West Bengal, Assam, Meghalaya and Tripura, Bangladesh is periodically lashed by cyclones originating in the Bay of Bengal or ravaged by floods during the monsoons when the mighty rivers that flow through it into the bay swell to tidal proportions. Some years ago, the bulk of the fighter aircraft based at Chittagong were destroyed during such a cyclonic storm.

The Bangladesh Biman Bahini (BBB) has been built up and equipped for air defence of the republic and close air support of ground and naval forces, but also for limited aid to the civil administration. There are some 170 aircraft in the inventory, mostly of Chinese origin, with some Russian-supplied types and a miscellany of others. Finance – or more precisely, the lack of it – has remained a major hurdle to expansion. Presently, the BBB operates from three air bases: Bashar AB at Tezgaon, Dhaka, Matiur AB at Jessore and Zahurul Haque AB at Chittagong.

COMBAT ASSETS

Initially, the BBB received a few light transport aircraft and helicopters gifted from India. After repatriation of ex-PAF Bengali personnel, the new air arm geared up very quickly and achieved combat

Below: All three of the Bangladesh Air Force An-32s serve with the 'Unicorns' – No. 3 Sqn. Each carries Bengali titles on the port side and English on the starboard.

status with an initial batch of MiG-21MFs and MiG-21Us received from the Soviet Union. The historical links with Pakistan were soon obvious and the BBB thereafter received a 'gift' of F-6s phased out by the PAF, later augmented by more F-6s and A-5s from China. They were followed a decade later by Chengdu F-7Ms, which today constitute the back-bone of the BBB's air defence force, equipping two squadrons at Dhaka and Chittagong. It is possible that the earlier MiG-21MFs/UMs are held in storage. A single squadron operates the A-5 from Dhaka in the ground attack role. Initially, two squadrons operated the type, but No. 8 Sqn based at Chittagong has since disbanded.

In late December 1997, the Bangladesh Prime Minister announced that the BBB would shortly acquire MiG-29 air superiority fighters from Russia and more F-7Ms from China, as part of its military modernisation programme.

TRAINING AND TRANSPORTS

Flying training is concentrated at Jessore – home of the Air Force Academy – with pilots proceeding from Nanchang CJ-6s (PT-6 in BBB service) to Cessna T-37Bs. (The T-37Bs had replaced Fouga Magisters by late 1997.) Eight Aero L-39ZA Albatros jet trainers were acquired in 1996 and are based at Chittagong for both advanced training and light attack tasks, alongside the few remaining Shenyang FT-6s still with the squadron. Rotary training is provided by a pair of Bell 206L LongRanger IIs from Jessore.

The BBB has only acquired a handful of transport aircraft, being limited to two Antonov An-32s delivered in 1990, with a further aircraft delivered in 1995. The scheduled arrival of up to four ex-USAF Hercules will give the air force a significant boost in its transport capabilities.

The helicopter fleet includes nine Mi-17 'Hip-Hs' (which replaced Mi-8s) at Chittagong and Dhaka for transport and VIP duties. About a dozen Bell 212s based at Chittagong and Dhaka undertake SAR missions as well as transport. Bangladesh Bell 212s have undertaken United Nations missions in Kuwait. No. 9 Sqn maintains a detachment of Bell 212s at Chittagong to augment the based examples of No. 1 Squadron.

The Bangladesh Army, with a tiny air element, has four Cessna 152s and two Cessna 337Fs.

Above: Armament training is undertaken by No. 25 Sqn – the 'Trendsetters' – on Aero L-39ZAs fitted with four underwing hardpoints, and the few FT-6s still used in this role. The Albatros has been operated by the Bangladesh Air Force since 1995.

Below: Ground attack duties are undertaken by the survivors of the Nanchang A-5Cs delivered in the late 1980s, which serve with No. 21 Squadron.

Bangladesh Biman Bahini

Bashar Air Base, (Tezgaon) Dhaka	
No. 5 Sqn	Chengdu F-7M Airguard, Guizhou FT-7
No. 9 Sqn	Bell 212
No. 21 Sqn	Nanchang A-5C, Shenyang FT-6
No. 31 Sqn	Mil Mi-17
No. 35 Sqn	Chengdu F-7M Airguard, Guizhou FT-7
Matiur Rahman Air Base, Jessore	
No. 3 Sqn	Antonov An-32
Air Force Academy – Flying Training Wing	
No. 11 Sqn	Nanchang PT-6
No. 15 Sqn	Cessna T-37B
No. 18 Sqn	Bell 206L-1 LongRanger II
Zahurul Haque Air Base, Chittagong	
No. 1 Sqn	Mil Mi-17, Bell 212
No. 25 Sqn	Aero L-39ZA, Shenyang FT-6

Above: The Bangladesh air force received 12 T-37Bs in 1995, replacing the Magisters by 1997. The aircraft, of No. 15 Sqn, still carry the blue and white training scheme of the USAF.

Above: Primary training is undertaken on the Nanchang PT-6s of No. 11 Sqn. Half of the fleet are painted in the olive green/brown colours in which they arrived from China.

BHUTAN

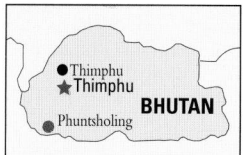

Capital: Thimphu
Population: 1.5 million
Land area: 46620 km² (17,995 sq miles)
Major cities: Phuntsholing

Bhutan Air Arm

Dornier Do 228 (1), Mil Mi-8 (2)	Thimphu

Bhutan Air Arm

Situated in the eastern Himalayas between India and China, the tiny Indian protectorate monarchy of Bhutan has established a small air arm in support of its 5,000-strong national army, which is primarily concerned with internal security. It has recently taken delivery of a single Dornier Do 228 twin-turboprop light utility transport (which is most likely to be a HAL-assembled version), and two Mil Mi-8 transport helicopters, which operate from Bhutan's main airport at Thimphu.

Above: The Bell 212 fleet is divided between Nos 1 and 9 Squadrons. This example – fitted with flotation gear for overwater SAR – is from No. 1 Sqn.

COMORES

Capital: Moroni
Population: 600,000
Land area: 1860 km²
(718 sq miles)

Comores Military Aviation Command

Three of the four Comoro Islands, off Madagascar, declared independence in July 1975 as the state of Comoro, although the fourth, Mayotte, voted to retain links with France as an Overseas Department. A Cessna 402B light twin acquired in 1976 for government use was subsequently replaced by a Eurocopter AS 350B Ecureuil helicopter, delivered in mid-1987 and based at Moroni. Three SIAI-

Marchetti SF.260W Warrior armed light trainers were delivered in January 1977, from Belgium, to which they were resold in 1988.

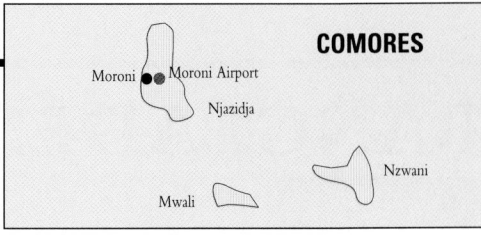

Type	Role	Delivered/In Service
Eurocopter AS 350B Ecureuil	comm helicopter	1/1

INDIA

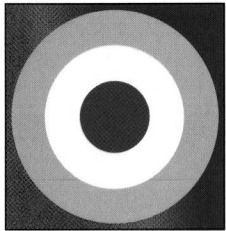

Capital: New Delhi
Population: 844 million
Land area: 3.17 million km²
(1,222,395 sq miles)
Major cities: Agra, Allahabad, Amritsar, Bangalore, Calcutta, Hyderabad, Kanpur, Lucknow, Madras, Mumbai (Bombay), Patna, Pune

Bharatiya Vayu Sena (Indian Air Force)

In the late 1990s the Indian Air Force finds itself in the situation where, on paper, it remains a formidable air arm with over 1,400 aircraft in its inventory (750 being combat types). Lack of timely decision-taking in the recent past is likely to result in major force level reductions over the next few years, while modernisation plans are piling up but not being realised.

Almost exactly 10 years ago, the IAF order of battle included top-of-the-line third-generation combat aircraft of Soviet, French and British origin; the 10 squadrons of MiG-29s, Mirage 2000s and Jaguars constituted a more modern inventory than any other Asian or Third World air arm. The backbone of the IAF combat force was an equivalent of 20 squadrons equipped with MiG-21 variants (-bis, -M, -FL) which gave the Indian Air Force a quality-plus-quantity advantage over its potential adversaries. Another 10 squadrons flew ageing types such as the Hunter, Ajeet (improved licence-built Folland Gnat) and Canberra, giving a total of some 40 combat squadrons based mostly in the north, west and east of India.

RE-EQUIPMENT PLANS

In the late 1980s the Indian Air Force embarked upon several projects to consolidate its combat strength and introduce force multipliers. These included the selection of an Advanced Jet Trainer (AJT) for third-stage lead-in fighter training, an airborne early warning (AEW) system to enhance air defence capabilities (especially against threats from the north and northeast), inflight-refuelling tankers to enhance endurance and range of its combat aircraft, and procurement of smart weaponry for precision attack against ground targets. As for the 'high-tech' component of its combat arm, the IAF was to decide upon either the Mirage 2000 or MiG-29 for larger induction or in-country licence-production, after experience gained with the two squadrons of each type already in service. In the event, another squadron of MiG-29s was raised in 1990 and it was surmised that the IAF would procure large numbers of the 'Fulcrum' to re-equip another half a dozen squadrons still flying obsolescent MiG-21FLs/Ms, with Hindustan Aeronautics Limited licence-producing this type at Nasik even as production of the swing-wing MiG-27ML was tapering off. The two squadrons of Mirage 2000s were 'topped up' to make good attrition, and in 1993 Hindustan Aeronautics

Top: Three squadrons fly the 9-12B MiG-29B 'Fulcrum-A', called the Baaz in Indian service. KB742 belongs to the 'Archers' – No. 47 Sqn based at Pune. The first Indian Air Force MiG-29 arrived in 1986.

Above: No. 24 Sqn ('Hawks'), based at Pune, is the first to gain the Sukhoi Su-30MK. India hopes to eventually build the type, but the 40 now on order will be delivered from the Irkutsk factory.

Limited (HAL) received orders to manufacture an additional batch of Jaguars for similar purposes.

In 1994-95, the Indian Air Force began evaluation of the Sukhoi Su-27, with test pilots and senior personnel visiting Russia on several occasions and reportedly filing very favourable reports. The Sukhoi Bureau undertook to develop the basic Su-27 long-range interceptor into the Su-30MKI multi-role version to meet the IAF's long-term needs, in place of the Mirage 2000Hs, which were judged to be too expensive to procure and maintain. Experience with the MiG-29B had been somewhat poor, particularly the high premature withdrawal rates of its RD-33 engines. Although MIG MAPO was working on a follow-on variant, known variously as the MiG-29M, the MiG-33 and subsequently the MiG-29SMT, the IAF was apparently uninterested and actively pursued the Su-30 project. The government of India formalised an agreement in November 1996 for the supply of 40 Su-30MKIs. Ten more were ordered in December 1998.

This was to be a fairly extensive, and complex, programme consisting of seven phases, the first aircraft

(Su-30Ks) being essentially the two-seat Su-27PUs built for the Russian Air Force but never delivered. They were to be supplemented by Su-30MKs, incorporating canards, while the final Su-30MKIs – with higher-thrust engines, TVC, new radar, modern avionics and provision for precision air-to-ground weaponry – were to be received in 2001, the earlier aircraft being sent back to Russia for modifications.

The first eight Su-30Ks were inducted into the IAF at Pune air base in June 1997. Delivery of the next batch of Su-30MKs has reportedly been delayed, for several reasons, including protracted time to make a decision about the suite of new French-, Israeli- and Indian-built avionics, extended integration times, and apparent economic difficulties hitting Russia's aviation industry. Therefore, it is possible that the IAF will receive another batch of 10 Mirage 2000Hs to augment its combat inventory in the short term while HAL has recently completed manufacture of another 15 single-seat Jaguars (with more two-seaters due) to supplement the 100-odd remaining Jaguars operating with five squadrons since 1979. The Su-30MKI, once fully developed, will equip two IAF squadrons, and

Above: Displaying a large squadron badge under the cockpit, this 'Battle Axes' Mirage 2000H Vajra of No. 7 Sqn operates from Gwalior alongside No. 1 Sqn, as part of Central Command.

Above right: A total of 162 MiG-27ML Bahadurs was produced for the Indian Air Force by HAL – this is the last example completed, which serves with No. 18 Sqn at Hindan.

Right: India is currently the largest operator of the Jaguar (Shamsher), the winner of the 1978 Deep Strike Penetration Aircraft programme. Jaguar ISs use overwing rails to carry self-defence R550 Magic AAMs.

Below: A Flight, No. 6 Sqn ('Dragons') uses the Jaguar IM tasked with maritime attack. The aircraft are fitted with the Agave radar and carry BAe Sea Eagle anti-ship missiles.

the 50 aircraft on order will possibly be supplemented, in the long term, by another 100-plus to be licence-built by HAL at its Nasik facilities.

MiG-23 AND MiG-27

The variable-sweep MiG-23 has had a relatively short service life with the Indian Air Force. In the early 1980s, the IAF received 40 MiG-23MF air defence fighters and 80 MiG-23BN close air support aircraft to equip a total of six squadrons. At the same time, HAL geared up to produce 165 MiG-27ML 'optimised' ground attack fighters, the last of which was delivered in 1994 to re-equip squadrons which had flown the HAL Ajeet light fighter and, subsequently, to supplant MiG-21FLs and MiG-21Ms. One MiG-23MF squadron has been re-equipped with the MiG-29 and the other relegated for air defence training. One MiG-23BN squadron has re-equipped with the HAL-built MiG-27ML. The IAF is keen to upgrade the weapons/avionics fit on its MiG-27MLs but this must first await the successful upgrade of the earlier MiG-21bis fleet, now considered as an absolute priority by the service.

'FISHBED' LEGACY

The MiG-21 has had a long (35 years plus) and chequered career in the Indian Air Force since the first squadron was equipped with this Mach 2 fighter in 1963. The metamorphosis of the MiG-21 – from the limited endurance, lightly-armed day-interceptor -21F version to the basic -21FL version (built under licence by HAL), through the M series (also built by HAL) to the definitive -21bis variant (220 built by HAL) – has made this the most important combat aircraft type to serve with the IAF. At its peak, some 20 squadrons were equipped with MiG-21 variants,

Indian Sub-continent

Acquired as a stopgap until the arrival of the Mirage 2000 and the MiG-29, 40 MiG-23MF 'Flogger-Bs' were used by Nos 223 ('Tridents') and 224 ('Warlords') Squadrons. Today the survivors continue in service with the later squadron from Jamnagar.

The MiG-21 equips more squadrons than any other combat type in the air force, and will remain in service for many years to come. This MiG-21bis, late of No. 15 Squadron, is seen at Jodhpur air base for maintenance.

Most of India's two-seat MiG-21UMs serve with the MOFTU at Tezpur. This unit provides advanced fighter training, also employing the first-generation MiG-21FL single-seater. This aircraft has an unusual camouflaged spine originally fitted to another aircraft.

the total number received by the IAF (both direct supplies from the Soviet Union and built under licence at Nasik for airframes, Koraput for engines and Hyderabad for avionics) being nearly 1,000 over the period 1963 to 1985.

The early model MiG-21FL (Type 77) is still flown by two squadrons and also equips the MiG Operational Flying Training Unit (MOFTU) at Tezpur in the Assam Valley. The responsibility for Stage III flying training has been shouldered by the MOFTU since the Hunter OFTU (HOFTU) was disbanded at Kalaikunda (near Calcutta) in 1994, but this is not considered a satisfactory arrangement by the IAF which has been pressing for a dedicated lead-in fighter trainer (or Advanced Jet Trainer) since 1986. The air force has shortlisted the Hawk and Alpha Jet for the role.

The MiG-21M (Type 96) continues with four squadrons in the ground-attack role but is considered underpowered and is a candidate for early phase-out. The multi-role MiG-21bis (Type 75), which still

equips 10 squadrons, is the subject of a major upgrade programme but one which has been experiencing continued delays owing to numerous factors, both Indian and Russian in origin. Since timely replacement of the MiG-21 has been considered a priority by IAF planners, and the genesis of the indigenous Light Combat Aircraft (LCA) lies in this vital requirement, it is necessary to examine both the MiG-21bis upgrade and LCA development programmes simultaneously.

NEW AND UPGRADED AIRCRAFT

The IAF has had to plan for the long-term identification of a replacement for the MiG-21s, not an easy task considering the quantity involved and costs thereof. The expanded aeronautics industrial base,

increasingly capable R&D organisations and massive home market were the factors that gave birth to the Light Combat Aircraft (LCA) programme. From the mid-1980s, the indigenous LCA has been considered as the type which would eventually supplant the massive force of MiG-21s in IAF service. The original concept of the LCA had been a more modest one. An 'improved' version of the Gnat (or Ajeet) would have met the initial requirement for a cost-effective front-line fighter, essentially for close air support with adequate self-defence capability.

A reassessment of both the Air Force's requirement and the indigenous design/development capability was attempted in the early 1980s. As the original LCA timetable had slipped by a decade, entry into service was projected from the mid-1990s. Accordingly, a

Bharatiya Vayu Sena

Air Headquarters

Air Research Centre & Analysis Wing		**Palam**
	An-32, Astra SPX, Boeing 707-337C, Gulfstream III, Il-76	
'Pegasus' Sqn		**Palam**
	Boeing 737-2A81, HAL 748 Srs 2, Mi-8S	

Central Air Command, HQ Allahabad

1 Sqn	Mirage 2000H/TH	Gwalior
7 Sqn	Mirage 2000H/TH	Gwalior
11 Sqn	Antonov An-32	Gwalior
12 Sqn	Antonov An-32	Agra
16 Sqn	Jaguar IS/IB	Gorakhpur
27 Sqn	Jaguar IS/IB	Gorakhpur
35 Sqn	MiG-21M/21MF, Canberra T.Mk 54, PR.Mk 57	Bakshi-Ka-Talab
44 Sqn	Ilyushin Il-76MD	Agra
102 Sqn	MiG-25R/25RU	Bakshi-Ka-Talab
106 Sqn	Canberra PR.Mk 57, HAL 748 Srs 2	Agra
105HU	Mil Mi-8	Gorakhpur
1 TTU	Canberra B(I).Mk 12/B(I).Mk58	Pune

Eastern Air Command, HQ Shillong

8 Sqn	MiG-21FL	Bagdogra
20 Sqn	MiG-27ML	Kalaikunda
22 Sqn	MiG-27ML	Hashimara
30 Sqn	MiG-21FL	Tezpur
33 Sqn*	Antonov An-32	Gauhati
43 Sqn	Antonov An-32	Jorhat
49 Sqn	Antonov An-32	Jorhat
59 Sqn	HAL 228-201	Gauhati
222 Sqn	MiG-27ML	Hashimara
122 HU	Mil Mi-8	Port Blair
* may have disbanded		

Maintenance Command, HQ Nagpur

HQ Flight	HAL 748	Nagpur
1 Base Repair Depot		Kanpur
3 Base Repair Depot	Mil Mi-17	Chandigarh
11 Base Repair Depot		Nasik

Southern Air Command, HQ Tiruvanathapuram

109 HU	Mil Mi-8	

South Western Air Command, HQ Gandhinager

4 Sqn	MiG-21bis	Jaisalmer
51 Sqn	MiG-21M/MF, MiG-27ML	Jamnagar
224 Sqn	MiG-23MF/23UB	Jamnagar
2 TTU	Canberra B(I).Mk12/B(I).Mk 58	Agra
2nd Wing		**Pune**
6 Sqn	Jaguar IM/IS/IT	Pune
24 Sqn	Su-30MK/30MKI	Pune
28 Sqn	MiG-29A/UB	Pune
32nd Wing		**Jodhpur**
10 Sqn	MiG-27M	Jodhpur
29 Sqn	MiG-27M/27UB	Jodhpur
32 Sqn	MiG-21bis	Jodhpur
107 HU	Mil Mi-8	Jodhpur
116 HU	Chetak	Jodhpur

Training Command, HQ Bangalore

Air Force Academy	TS-11, HJT-16 Kiran II Cheetah, Chetak	Hakimpet
FTS	HJT-16 Kiran I/IA/II	Bidar
FTS	HJT-16 Kiran I/IA	Dundigal
EFTS	HPT-32 Deepak	Allahabad
Flying Instructors School	HPT-32 Deepak, HJT-16 Kiran I	Tambaram
MiG Operational Flying Training Unit	MiG-21FL/21UM	Tezpur
Navigation & Signals School	HAL 748 Srs 2	Begumpet
Paratroop Training School	Antonov An-32	Agra
Tactics & Air Combat Development Est.	various	Jamnagar
Test Pilots School	various	Bangalore Airport
Transport Training Wing		**Yelahanka**
	Antonov An-32, HAL 748 Srs 2, Do 228	
112 HU	Mil Mi-8	Yelahanka
Aircraft & Systems Testing Establishment	HAL 748 Srs 2, Jaguar IS	**Bangalore Airport**

Western Air Command, HQ Palam

2 Sqn	MiG-27ML	Hindan
3 Sqn	MiG-21bis	Pathankot
5 Sqn	Jaguar IS/IB	Ambala
9 Sqn	MiG-27ML	Hindan
14 Sqn	Jaguar IS/IB	Ambala
15 Sqn	MiG-21bis	Chandigarh
17 Sqn	MiG-21M/21MF	Bathinda
18 Sqn	MiG-27ML	Hindan
21 Sqn	MiG-21bis	Chandigarh
23 Sqn	MiG-21bis	Ambala
25 Sqn	Ilyushin Il-76MD	Changigarh
31 Sqn	MiG-23BN	Halwara
41 Sqn	HAL 228-201, HAL 748	Palam
47 Sqn	MiG-29A/29UB	Adampur
48 Sqn	Antonov An-32	Changigarh
101 Sqn	MiG-21M/21MF	Adampur
108 Sqn	MiG-21M/21MF	
220 Sqn	MiG-23BN	Adampur
221 Sqn	MiG-23BN	Halwara
223 Sqn	MiG-29/29UB	Adampur
104 HU	Mil Mi-24/35	Pathankot
114 HU	Cheetah, Chetak	Siachen Glacier
125 HU	Mil Mi-25	Pathankot
126 HU	Mil Mi-26	Pathankot

Units with unknown assignments

26 Sqn	MiG-21bis
36 Sqn	MiG-21bis
37 Sqn	MiG-21bis
45 Sqn	MiG-21bis
52 Sqn	MiG-21bis
226 Sqn	MiG-23MF/23UB
110 HU	Mil Mi-8
111 HU	Mil Mi-8
117 HU	Mil Mi-8
118 HU	Mil Mi-8
119 HU	Mil Mi-8
120 HU	Mil Mi-8
121 HU	Mil Mi-8
130 HU	Mil Mi-8
151 HU	Mil Mi-8
152 HU	Mil Mi-8

Indian Air Force Museum Historic Flight

DH.82 Tiger Moth, DH.115 Vampire, HAL Ajeet, HAL HT-2, Harvard, Spitfire LF.Mk VIIIC

Light transportation tasks are undertaken by the IAF's fleet of HAL-assembled Dornier Do 228s. The Transport Training Wing uses the type for multi-engine training.

significant upgrade was made to the ASR: the IAF now wanted BVR-capability, software-controlled systems, advanced weaponry and so on, all of which were incorporated into the Air Staff Requirement that crystallised in October 1985. An immense amount of effort and funding requirements lay ahead and, inevitably, there were multiple slippages, over-runs in time and cost estimates, so the first LCA prototype was rescheduled to fly in late 1996, with first squadron deliveries not until 2005. The first flight is now planned for mid-1999 and IOC will not be until 2008, if all goes well. US sanctions imposed after India's nuclear tests in May 1998 will also directly affect the LCA's development, as the powerplant (GE F404) and flight controls (Lockheed Martin) are of US origin.

The Indian Air Force has had to work out some urgent interim solutions but its options have remained severely limited. Keeping the MiG-21bis in service for at least another decade became not only a necessity but imperative if the IAF was not to lose some 50 per cent of its combat strength in the period 2000-2005.

Air Headquarters at New Delhi examined a number of alternatives for the MiG-21's upgrade, including an attractive one from Israel for the MiG-21-2000 development, but in late 1993 an evaluation team visited Russia to examine the MIG MAPO concept, designated the MiG-21-93. This combined sophisticated Western avionics with the most advanced Russian airborne radar and air-launched missiles plus an essentially unchanged MiG-21bis airframe and R-25 powerplant.

MIG MAPO had a simple objective: to convert the second-generation front-line aircraft into a fourth-generation tactical aircraft. The heart of the MiG-21-93's new weapons control system would be the Phazotron Kopyo radar, developed from the Zhuk radar installed in the MiG-29M fighter. Development of this new multi-function, multi-target pulse-Doppler X-band radar, weighing 165 kg (363 lb) and having a 500-mm (20-in) flat slotted-array antenna, had some French connections. Thomson-CSF reportedly supplied the data and signal processors, while the

Winner of the Heavy Transport Aircraft requirement to replace the Antonov An-12BK, a total of 24 Il-76MDs is operated by the air force as the Gajraj (King Elephant) in two squadrons.

Russians provided the antenna, transmitter, receiver and primary power distribution, the complete system being named the Kopyo.

The MiG-21-93 would be capable of intercepting airborne targets between altitudes of 30 and 20000 m (98 and 65,616 ft), flying at 1600-2300 km/h (995-1,430 mph) and carrying some of the most advanced air-to-air missiles in the Russian arsenal. The main punch of the MiG-21-93 would be provided by an array of advanced and hitherto 'under wraps' weapons, giving it BVR capability, provided by two Vympel R-27RI semi-active radar missiles or four Vympel RVV-AE active-radar missiles or four R-73E close combat air-to-air missiles. Alternately, six R-60 AAMs could be carried.

The Kopyo's effective air-to-ground/sea modes would make it possible for the MiG-21-93 to carry an impressive array of precision, stand-off weapons, including one X-31P and two X-25MP high-speed anti-radar missiles; one X-31A and two X-35 anti-ship missiles: two KAB-500 KR television-guided bombs, 14 x FAB 100-500 kg conventional bombs, or 10 S5M, S5K and S24 rocket pods.

The Indian Air Force identified several Western avionics systems to be incorporated, including a ring laser-gyro inertial navigation system with satellite correction, and a lightweight radar warning receiver (RWR) – probably the Dassault EWS-A – which would also control the chaff-and-flare dispensers located along the wingroots. Indian Government Furnished Equipment (GFE) to be included in the MiG-21 upgrade would comprise two radio stations, master radio receiver, radio altimeter and an IFF, most of these systems being already incorporated in the HAL-built Jaguar and MiG-27ML.

The Indian government signed an intention to proceed (ITP) with Mikoyan in March 1994 and a joint MIG MAPO/IAF/HAL team reportedly began work in earnest from mid-July 1994 under a tight cost-and-time schedule: the entire programme was to be completed in 24 months, at an estimated cost of Rs1200 crore (US$400 million), the retro-mod of

The Antonov An-32 Sutlej replaced the C-119 Packet, C-47 Dakota and DHC-4 Caribou in service. This example is from the Transport Training Wing.

Of the 72 HAL-assembled BAe 748s delivered to the Indian Air Force, 20 are dedicated military freighters with a large cargo door on the port side.

each MiG-21bis being estimated at Rs10 crore (US$3 million). The first two MiG-21-93 'prototype' aircraft were to be built and the test flight programme completed in Russia, after which the technology would be transferred to India for HAL to begin retro-modification of the IAF's MiG-21bis fleet.

However, according to recent reports from Moscow, MIG MAPO and the Sokol plant have had some differences about responsibility for the Indian programme. They have been compounded by alleged financial mismanagement, resulting not only in a two-year delay to the development programme but in continued uncertainty about the transfer of technology which would enable HAL to begin retro-modification of the 120-plus MiG-21bis in India. The first upgraded aircraft, designated as the MiG-21I, first flew on 6 October 1998, under the control of the ANPK-MAPO bureau.

The other combat aircraft in the IAF's inventory are two reconnaissance squadrons with the MiG-25RB and Canberra PR.Mk 57, the latter soldiering on after over 40 years in service. There are another score or so Canberras of various marks in the target training and ECM roles, and various life-extension schemes have been examined to keep this perennial type in service for more years.

SAMs

The total personnel strength of the IAF is some 110,000 and a large proportion of these man the many Signal Units (radar station complexes under the extensive Air Defence Ground Environment System) and nearly 60 SAM squadrons.

In the early 1980s, the Indian Air Force massively expanded its surface-to-air missile inventory, supplanting the dozen-odd squadrons of SA-2s with twice as many squadrons equipped with the SA-3 Pechora missile. They form the core of the Base Air Defence System and are also deployed to protect a number of vital areas and economic installations in northern, western and northeastern India. For air defence against fast, low-flying threats, the IAF has a number of mobile squadrons with the SA-8 Osa-AK-M quick-reaction low-level air defence missile. Finally, for close range, 'last-ditch', air defence, there

The Chetak (HAL-built Alouette III) serves with all branches of the Indian armed forces. In the IAF they serve as light utility helicopters with the Helicopter Units, and as trainers. The majority of the air force's helicopters were transferred to the army.

Above: No. 112 Helicopter Unit 'The Thoroughbreds' is part of the Transport Training Wing, teaching pilots to fly the Mil Mi-8 'Hip', the main transport helicopter type.

Right: The Mil Mi-8Ss of the Air Force Headquarters Communications Squadron based at Palam wear a smart blue and white scheme.

are numerous detachments with the shoulder-fired SA-16 Igla missile.

TRANSPORTS

The Indian Air Force has a relatively large transport and helicopter fleet, primarily engaged in support of Indian Army and paramilitary forces manning the mountain frontiers with Pakistan and China, and in the rugged jungle terrain bordering Myanmar (formerly Burma) in the east. The IAF has a true continental responsibility, with over 3000 km (1,865 miles) of Indian territory to cover north to south and an equal distance west to east, with all manner of geographic terrain in between, ranging from the world's highest mountain ranges to the extensive deserts of Rajasthan, and southward to where the great Indian peninsula divides the Arabian Sea from the Bay of Bengal. There are also offshore oil assets to support. The island territories of the Laccadives in the west and Andaman & Nicobar Islands in the east are additional responsibilities.

There are nearly 200 transport aircraft in the inventory, including two squadrons with the Ilyushin Il-76 heavy-lift strategic transport, five squadrons with the Antonov An-32 medium-lift tactical transport, two squadrons with the HAL-Dornier 228 light transport aircraft and one with the HAL-built BAe (Avro) 748. There are also several Communication Flights with the 748, attached to Air Command Headquarters, while the Air Headquarters Communication Squadron at Palam (Delhi) operates a number of VIP-configured Boeing 737-200s, BAe (Avro) 748s, Mi-17s and Mi-8s.

Some 110 Antonov An-32s were contracted for in the early 1980s and, at their peak, served with seven squadrons, two of which have since been number-plated and their aircraft held in reserve. The large airlift capacity provided by the IAF was demonstrated to good effect during the Indian Peace Keeping Force's 33-month operational deployment in Sri Lanka (1987-90), the long-range deployment of a Parachute Battalion Group in the distant Maldive

Islands, plus various missions to Central Asia, the Middle East and within the sub-continent itself.

The Ilyushin Il-76 has boosted the heavy-lift capacity and will remain the IAF's mainstay for several decades, and the possibility of re-engining the type with Western turbofan engines has been examined. The Il-76 would probably also be the platform for an airborne early warning version which is reportedly under development, with Israeli involvement, although the Soviet Union offered the A-50 version in the late 1980s. The Il-78 version is being evaluated by the IAF as an air-to-air refueller, but the IAF has been overly conservative in exploiting this inbuilt potential of its Jaguars and Mirage 2000s. The first inflight-refuelling trials were carried out between IAF Jaguars and a RAF VC10 tanker 'on loan' as recently as 1996. It is primarily the Su-30, with its already very long range characteristics, that will be given the benefit of such operations.

Twenty-five HAL-built Dornier Do 228s have, so far, been delivered to the IAF and equip two logistic air support squadrons, allocated to Western and Eastern Air Commands. In 1996, a number of Dornier Do 228s were transferred to the Transport Training Wing at Yellahanka (near Bangalore), supplanting the BAe (Avro) 748s and complementing the An-32s at the station, this being a far more cost-effective method for multi-engine conversion training.

HELICOPTERS

The IAF's large rotorcraft force comprises some 15 squadrons (known as Helicopter Units) with the ubiquitous Mil Mi-8 and Mi-17 medium-lift helicopters, serving in all theatres and being the virtual lifeline for personnel and logistic supplies for the Indian Army and, in difficult mountain terrain, the civil administration. About 150 Mi-8/-17s are in the IAF's inventory. Some 50 remaining HAL-built Chetak (Alouette III) and Cheetah (Lama) helicopters equip three squadrons and a number of FAC and case-vac flights. The bulk of the IAF's Chetaks/Cheetahs (approximately 170) were transferred to the Indian Army under a major decision taken in 1986.

Two squadrons are equipped with the armed Chetak (with the AS-11 wire-guided missile) but it is the Mi-25 and Mi-35 that give the IAF serious anti-armour capability, two squadrons flying the types in

Left: The HAL HJT-16 Kiran Mk II is used for basic armament training from Hakimpet. The type is due to be withdrawn from service within 10 years.

Below: India bought the TS-11 Iskra from Poland to fulfil an advanced/weapons training requirement. It shares this role with the Kiran Mk II at the Air Force Academy.

close co-operation with the ground forces. The Indian Army, however, has been fairly insistent on having its own dedicated anti-tank helicopter units, for which various options have been evaluated. Indian Army Aviation will be the first to receive the HAL-developed Advanced Light Helicopter (ALH), in 1999.

Finally, there is the world's largest helicopter, the Mil Mi-26, which equips a Helicopter Flight for transportation of outsize equipment, including AFVs, in very difficult mountainous jungle areas.

TRAINING

The IAF's Training Command has had to struggle with *ad hoc* flying training schemes for over a decade and urgently awaits the introduction of a suitable Advanced Jet Trainer for lead-in fighter (or Stage 3) training. At present, after some 40 hours of primary training on HAL HPT-32 piston-engined trainers, student pilots are given 90 hours on the HAL HJT-16 Kiran Mk I basic jet trainer and then another 90 hours on either the Kiran Mk II or Iskra jet trainer for applied and basic armament training. Until 1995, future fighter pilots were sent to the HOFTU to fly Hunters (60 hours) or to the MOFTU for MiG-21s, but following the phasing-out of most of the Hunters from service (a few are reported as still being active at Kaliakunda) they have had to contend with the unforgiving MiG-21UM and FL before being posted to various combat squadrons where they go into type-conversion trainers.

Various committees set up by Air Headquarters since 1986 have warned that this gap in advanced jet training is primarily responsible not only for unacceptable rates of flying accidents or incidents but has also contributed to a lower quality of flying training. Ideally, the trainee pilots would move on to a basic jet trainer (after grading on a piston or light turboprop trainer) and then to an advanced jet trainer. While HAL has started development of the HJT-36 tandem-seat basic jet trainer to replace the Kiran from about 2005 – a mock-up was displayed at Aero India '98 – the BAe Hawk Series 100 has consistently been considered the ideal aircraft for Stage 3 advanced jet training.

Border Security Force (Air Wing)

This vast paramilitary force formed a small air wing in the 1980s, with six HAL (BAe) 748s transferred from Indian Airlines, plus some Douglas DC-3s/C-47s phased out by the IAF and private airlines. A few Beech King Airs were added for senior staff transportation and in the last few years they have become the only airworthy types, the C-47s and HAL (BAe) 748s having been grounded. The BSF's Air Wing has been examining various possibilities and is likely to receive six 50-seaters once a final decision is taken on the aircraft type selected for the overall Indian requirement. A joint evaluation team headed by HAL has examined a number of options and the ATR-42/-72 is likely to be selected for co-production in India.

Indian Naval Aviation

The Navy's aviation arm is well equipped and highly trained even if, in size, it remains relatively modest compared with the IAF. Owing to financial constraints, the Navy's total budget has remained static over the past decade. After final decommissioning of the light fleet carrier INS *Vikrant* (ex-HMS *Hercules*) some years ago, the navy has just one aircraft-carrier in service, the INS *Viraat* (ex-HMS *Hermes*), on which are embarked Sea Harriers and Sea Kings. The shore bases are INAS Hansa (Goa) for the VTOL fighters and INAS Garuda (Cochin) for the ASW helicopters. There are plans to construct three light carriers (30,000 tonnes) in Indian shipyards. In 1998 India announced that would be acquiring the Russian carrier *Admiral Gorshkov*, completed but not put into service. The vessel is nominally free, but India must pay for a major refit and for the equipment, thought to include Sukhoi Su-33s, Kamov Ka-28s and Ka-31s. The Ka-28 is already in service although more would be required.

The Indian Navy has recently contracted for two additional Harrier trainers, to replace peacetime losses, and these ex-RAF Harrier T.Mk 4s were refurbished in the UK for delivery to India in 1999. The navy originally obtained 28 Sea Harrier FRS.Mk 51/T.Mk 60s which have now been in service for 15 years, supplanting the Hawker Sea Hawks that were original equipment for INAS 300 'White Tigers'. The other carrierborne type, the Breguet Alizé of INAS 310 'Cobras', was phased out in the late 1980s and has been replaced by HAL-built Dornier 228s with maritime surveillance radar and other sensors, which are likely to be armed with short-range air-to-surface missiles. The Dornier 228s are shore-based and the

The Indian Navy operates eight Tupolev Tu-142M 'Bear-F' long-range maritime patrol and anti-submarine aircraft over the Indian Ocean. All aircraft fly with INAS 312 from Arrakonam.

navy could increase the present inventory (15 aircraft) in the coming decade.

There are some 40 Westland Sea Kings in service, Mks 42, 42A, 42B and 42C, the latter for 'commando'-type assault operations, the rest for ASW tasks. Some Sea Kings, and all the Sea Harriers, are configured to carry the BAe Sea Eagle AShM.

Russian-built Kamov Ka-25 and Ka-28 ASW helicopters are both shore-based and embarked upon the Navy's 'Kashin'-class GU destroyers. Indian-built frigates carry HAL Chetak helicopters while the bigger 'Delhi'-class GW destroyers embark two Sea Kings. Three Ka-31T AEW helicopters are reportedly on order for 'picket' duties for the fleet at sea.

The shore-based, long-range MR/ASW component comprises one squadron each of Tupolev Tu-142Ms and Ilyushin Il-38s, the latter now in need of urgent replacement. Several options have been examined, including ex-Aéronavale Atlantics or ex-USN Orions, but no decision has been taken.

The navy conducts its own flying training, starting with the piston-engined HPT-32 at Wellington Island (pilots are also sent to the IAF's Basic Flying Training School) and proceeding to the HJT-16 Kiran at Dabolim (Goa). Both Kiran Mks I and II are in service but the navy has long awaited a decision on the IAF's Advanced Jet Trainer (AJT) so as to order some aircraft for itself.

There are some remaining PBN Islander/Defenders in service, used for fleet support duties and utility tasks. From HAL have been received over 40 Chetaks (Alouette IIIs), which will be replaced by the Advanced Light Helicopter (ALH) within the next few years.

Above: The Sea Harrier FRS.Mk 51, which replaced the Hawker Sea Hawk in Indian Navy service, is shore-based at Goa-Dablomin when not operating from the carrier INS Viraat.

Above: INAS 310 at Goa-Dablomin uses the HAL-assembled Dornier 228-101 for its coastal patrols.

Below: A total of four versions of Sea Kings is in service. IN529 is an ASW- and ASuW-capable Mk 42B.

Aviation Research Centre (ARC)

Now in its 35th year of existence, this largely unknown aviation organisation is directly controlled by the Cabinet Secretariat and the Prime Minister's Office at New Delhi. Established in the wake of India's disastrous frontier clash with China in late 1962, the Aviation Research Centre initially received some Curtiss C-46 Commandos and Helio STOL utility aircraft from the United States for support of

special forces in the northern frontiers bordering Tibet and China. Over the years, the ARC has assumed additional responsibilities including very-long-range transportation, Elint/Sigint functions and border reconnaissance, all flown by aircraft with civil registration – although aircrew are largely retired (or seconded) Air Force personnel.

The ARC currently operates a motley mix of aircraft types, including two ex-Air India Boeing 707s, three Gulfstream IIIs, two Learjet 29s, five Beech King Airs, two Ilyushin Il-76s, six Antonov An-32s and a dozen helicopters including Mi-17s and Chetaks.

Coast Guard Air Wing

Working in close co-ordination with the Navy since its establishment in 1976, the Indian Coast Guard organisation has expanded greatly in size and assumed major responsibility for protection of the EEZ around the Indian peninsula and offshore islands in the Southern Arabian Sea and Bay of Bengal. The Coast Guard Air Wing is primarily equipped with the HAL Dornier 228MPA and HAL Chetak, operating from three main air bases and a number of air enclaves along India's coasts.

The first dedicated air station was established at the former Portuguese enclave of Daman (north of Bombay), from where Dornier 228s patrol the long western coastline from the Rann of Kutch southward to the Malabar coast. Patrolling the eastern seas are

Dornier 228s based at Meenambakkam (Madras International Airport), and a third Dornier 228 squadron, recently established, is at Port Blair, in the Andaman Islands.

Air enclaves exist at Dablomin (Goa), Wellington Island (Cochin), Vishakhapatnam and Dum Dum (Calcutta), and support not only transiting Dornier 228s but the many Coast Guard Chetaks.

The Coast Guard's inventory includes 18 Dornier 228s (with another seven on order) plus two dozen

HAL Chetaks. The first two HAL ALHs to be delivered to a service customer are earmarked for the Coast Guard, in late 1999.

The Coast Guard Air Wing uses the HAL-built Dornier Do 228-101 for maritime patrol of the EEZ, and for search and rescue.

Indian Sub-continent

Indian Army Aviation

Army aviators have flown light fixed-wing AOP aircraft and helicopters since the first AOP flight came into existence in August 1947. They were part of the Indian Air Force which was responsible for command and support of the AOP squadrons until 1986 when the independent Army Aviation establishment came into being. At this point some 170 Chetaks (Alouette III) and Cheetahs (Lama) were transferred to the Army but, under a special directive, the IAF was to remain responsible for maintenance support for 10 years.

The Indian Army has reorganised the AOP squadrons and flights into 10 R&O (Reconnaissance and Observation) Squadrons, with several independent flights, mostly equipped with HAL Cheetahs, and fewer HAL Chetaks, the latter for communication and staff transportation. To make good attrition, the army has ordered some 40 additional Cheetahs over the years and now awaits delivery of the first of 100 ALHs ordered from HAL.

Despite sustained efforts to get the Mi-25/Mi-35

Around half of the Indian Army squadrons are equipped with the HAL Cheetah, which is in use for reconnaissance and observation tasks in the R & O squadrons and flights. Z3212 (above) serves with No. 665 Sqn from Nasik.

dedicated anti-tank helicopters transferred, the Army has yet to suceed, although the Air Force operates them in close co-operation with the land forces. The Army has long evaluated armed helicopters of various types and the selection is reportedly now between the Mi-28 and Ka-50.

Indian Army
(known information only)

3 R & O Flight	Cheetah	
4 R & O Flight		
6 R & O Flight	Cheetah	
7 R & O Flight		
8 R & O Flight		
10 R & O Flight		
11 R & O Flight	Chetak	
16 R & O Flight		
17 R & O Flight	Chetak	
18 R & O Flight	Cheetah	
21 R & O (I) Flight		
22 R & O (I) Flight	Cheetah	
23 R & O Flight	Cheetah	
25 R & O (I) Flight		
30 R & O (I) Flight		
31 R & O Flight	Cheetah	
32 R&O Flight		
660 R & O Sqn		
665 Sqn	Cheetah, Chetak	Nasik
R & O = Reconnaissance and Observation		

The Cheetah is a licence-built version of the Aérospatiale SA 315 Lama, a type noted for its outstanding high-altitude performance.

The Indian Army uses the HAL Chetak, a licensed copy of the Sud Aviation Alouette III, for utility tasks such as staff transportation. About 130 are used.

MALDIVES

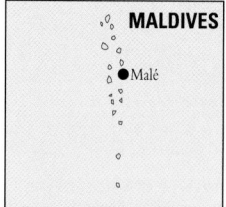

Capital: Malé
Population: 200,000
Land area: 298 km²
(115 sq miles)

National Security Service

The 1,200 Indian Ocean coral islands of the Maldives, which were under British protection until 1965, have no specific military or paramilitary aviation organisations. A civil-registered Dornier Do 228-212 of Air Maldives has been used for island and ocean surveillance when not carrying passengers.

NEPAL

Capital: Kathmandu
Population: 19 million
Land area: 141415 km²
(54,585 sq miles)
Major cities: Bhatgaon, Patan

Royal Nepal Army Aviation

This Himalayan Kingdom has no air force but its army includes an aviation element, mostly for communication, light logistic support and casualty evacuation – its helicopters are often engaged in dramatic high mountain rescues of stricken climbers.

MAURITIUS

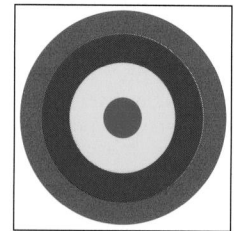

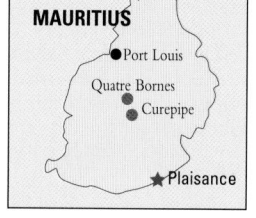

Capital: Port Louis
Land area: 1865 km² (720 sq miles)
Major cities: Curepipe, Quatre Bornes
Population: 1.1 million

Mauritius Coast Guard

Type	Role	Delivered/In Service
Dornier-HAL Do 228-101	coastal patrol	2/2
BN-2T Maritime Defender	coastal patrol	1/1

The 11th Brigade, at Kathmandu's Tribhuvan airport, is administratively responsible for the maintenance and operation of the handful of helicopters. There are two Aérospatiale SA 330C/G Puma medium-lift helicopters and three HAL Chetaks (Alouette III) plus a single AS 350B2. Three fixed-wing aircraft (Short Skyvan 3M-400) operated previously have since been withdrawn from use.

The Royal Flight, also maintained and operated by Nepalese Army personnel, has one BAe 748 Srs 275 transport aircraft, two Aérospatiale AS 332L/4 Super

Mauritius Coast Guard

Located some 430 nm (495 miles; 795 km) east of Madagascar in the Indian Ocean, Mauritius became an independent member of the British Commonwealth in 1968. With no defence forces other than the armed police, Mauritius ordered two radar-equipped Dornier/HAL Do 228-101 in March 1990 to form the nucleus of an airborne maritime surveillance element from July 1991, for a national Coast Guard force. This was reinforced a year later with the delivery of a twin-turboprop Pilatus/BN BN-2T Maritime Defender, similarly-equipped for coastal patrol and surveillance.

Royal Nepalese Army

Air Battalion		
Aérospatiale SA 330C Puma	1	transport helicopter
Aérospatiale SA 330G Puma	1	transport helicopter
HAL Chetak	2	communications
Royal VIP Flight		
Bell 206L-3 LongRanger III	1	(civil reg) VIP
Bell 206L-4 LongRanger IV	1	(civil reg) VIP
British Aerospace 748 Srs 275	1	VIP
Eurocopter AS 332L Super Puma	1	(civil reg) VIP
Eurocopter AS 332L1 Super Puma	1	(civil reg) VIP

Pumas (in civil registrations) fitted out in VIP configuration and two Bell 206L3/L4 Long Rangers, also in civil markings.

Nepalese aircrew receive flying training at Indian Air Force establishments and, in the past, maintenance support has been provided by contracted personnel (mostly civilians) from India.

It is believed that two Pumas are operated by the Air Battalion of the Nepalese Army. One may have been written off in August 1995.

PAKISTAN

Capital: Islamabad
Population: 112.1 million
Land area: 803940 km² (310,320 sq miles)
Major cities: Hyderabad, Karachi, Lahore, Lyalpur, Multan, Peshawar, Rawalpindi

Pakistan Fiza'ya (Pakistan Air Force)

Doctrine of the Pakistan Air Force (PAF) is essentially to train, equip and prepare itself for confrontation with its main adversary, the Indian Air Force (IAF). As a recent Air Chief has stated, the PAF has always been far behind the IAF in numerical strength but had striven to maintain a qualitative edge, particularly after getting F-16s in the early 1980s. "That qualitative edge has now eroded ... after the IAF acquired the Mirage 2000, Jaguar, MiG-29 and Su-30." In January 1998, the new PAF Air Chief raised concern over the widening technological lead in combat aircraft by India. After the US embargo on further supplies of F-16s, the PAF has tried very hard to acquire Mirage 2000-5s, Su-27s or JAS 39 Gripens, but has been thwarted by prices and politics.

Thus, apart from the remaining F-16A/Bs in its inventory, the Pakistan Air Force today comprises a mixture of second-hand Mirage IIIs and 5s procured from a variety of sources, plus Chinese-built F-7 copies of the MiG-21, in its struggle to maintain credible force levels. The United States Pressler Amendment, strictly enforced since 1990 when Pakistan's nuclear weapon programme was first suspected, has stringently curbed the PAF's plans which were to have a modern multi-purpose combat force, built around the F-16 Fighting Falcon, supplemented by a larger inventory of Chinese-supplied F-7s and A-5s. The latter types, in the longer term, were to be replaced by the Sino-Pak joint development 'Super 7' or what has since evolved into the FC-1 fighter.

Re-equipment plans for the Pakistan Air Force had been succinctly spelt out in early 1988 by the then-Minister of State for Defence in Islamabad, who had hoped for more F-16 Fighting Falcons and introduction of the Xian F-7 fighter to augment the front-line units and replace the weary F-6s. In the longer term, a joint aircraft development programme with the US and Chinese aeronautical industries was to be promoted, initially known as the Sabre 2, based on the F-7 but with extensive redesign, US engines and contemporary avionics and weapon systems. However, by the early 1990s, the Sabre 2 was finally abandoned, and replaced by the Sino-Pak FC-1 programme. Particularly troublesome during this time has been the performance of the Chinese A-5 attack aircraft, compounding the PAF's already difficult re-equipment prospects.

Above: The Pakistani F-16 purchase was curtailed after Pakistan refused to abandon its nuclear weapons programme. A total of 28 F-16As and 12 F-16Bs was delivered, but a second batch, already paid for, was embargoed by the USA and placed in storage at AMARC, Arizona.

Right: Pakistan is the last major operator of the Mirage III/5 family, having acquired many second-hand examples over the last few years to supplement survivors. This example is a Mirage IIIDP acquired directly from Dassault.

From 1980 onwards, the PAF was clear that it required five F-16 squadrons as the core of its fighting force. The Soviet invasion of Afghanistan made possible the resumption of supply of top-of-the-line US aircraft. When the US lifted the arms embargo that had prevailed since 1965, it had offered first the A-7 Corsair II (110 to equip five squadrons including an OCU) and later a similar number of F-5Es. The first offer, made in 1976, was tied to Pakistan accepting American controls on its nuclear weapons programme but the then-Prime Minister refused to pay the political price, and the deal fell through.

With the 1978 Communist coup in Kabul, Pakistan once again became important to the US and the Carter administration began considering selling fighter aircraft. The US wanted to release the F-5E, but the PAF was adamant that the offer had come a decade too late and insisted on the F-16, in line with its thinking that it should possess a smaller number of the best aircraft, rather than go for larger numbers of less expensive but inferior ones.

The F-16 sale would never have come about but for the Reagan administration's decision, taken belatedly in 1981-82, to massively intervene in Afghanistan. The first batch of 40 aircraft included 12 F-16B trainers, showing that Pakistan had every intention, even then, of somehow building up to its postulated level of five squadrons. Whereas US reluctance to upset the military balance in the subcontinent by supplying its most advanced fighters had been the previous inhibition, affordability now became the main issue.

Forty aircraft were ordered, to re-equip two squadrons. In 1988, the US cleared another 11 aircraft (Block 15) and, in anticipation of their arrival, Pakistan re-equipped yet another squadron. Although this meant that all three squadrons were under strength, it permitted a saving of time in getting the

The Shenyang FT-5 is near the end of its service life with the air force, being operated as a lead-in trainer with No. 1 Fighter Conversion Unit.

third squadron into service. In 1988 the US agreed to sell another 60 F-16A/Bs, with the upgraded engines, at a unit cost of $23 million over the period 1992-95, but the Pressler Amendment then effectively put a stop to all further supplies.

Introduction of the F-16 from 1982 solved one PAF equipment problem, but the combination of limited numbers which the country could afford and the large Shenyang F-6 inventory required the purchase of another type. In 1983 the PAF began to receive 52 Shenyang A-5Cs (Q-5 III in Chinese parlance) to re-equip, in turn, three F-6 squadrons. Options were placed for another 90 aircraft for delivery in 1986-87. The PAF has reportedly been dissatisfied with this aircraft, forcing its cancellation, while a higher-than-expected accident rate had led to the elimination of one squadron. Still, despite reservations, the incredibly low price of $1 million per A-5 must have been an inducement. The first A-5 rebuilt in Pakistan rolled out from Kamra in April 1988, after some four years. The A-5 facility at Kamra will have an eventual capacity for rebuilding 10 A-5s annually. In early 1991 reports circulated of fresh negotiations between the PAF and CATIC for the purchase of up to 100 uprated A-5Ms following completion of this aircraft's development programme. Main features were the A-5M's AMX-type all-weather nav/attack system and digital avionics installation integrated through Alenia as prime contractor, plus improved WP6A turbojets and 12 external stores stations. However, the A-5 matter has remained ambiguous.

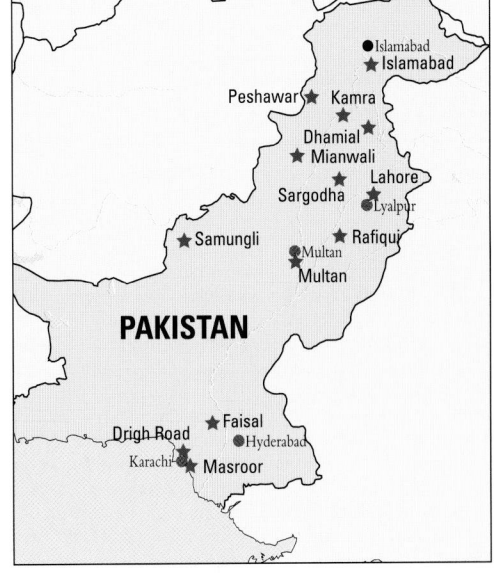

The Hercules fleet includes seven C-130Bs (including 58739, above), four C-130Es and an L-100. All are operated by No. 6 Sqn from Chaklala.

The PAC Mushshak is a version of the Saab MFI-17 assembled in Pakistan. They are used by both the air force and army for training and general communications duties.

The air force shares about 30 Alouette IIIs with the army. In service since 1967, they are used by the air force as general utility helicopters and for SAR. This examples flies with No. 84 Sqn.

Procurement by the Pakistan Air Force of the Xian F-7, a Chinese copy of the MiG-21F fighter, was a logical step and continued the familiar pattern established in the mid-1960s: relatively large numbers of less expensive fighters would give the PAF some strength in depth. The Chinese, in any case, were vigorously marketing products of their aeronautical industry and the F-7M was regarded as an attractive alternative to the more sophisticated but far more costly combat aircraft from the West.

The version for Pakistan was designated F-7P, christened by the service as the Skybolt and embodying some 20 PAF-specified changes including provision for four PL-SB or AIM-9 Sidewinder AAMs, and retaining the two wingroot-mounted Type 30-1 cannon. Most Western-origin systems of the F-7M were incorporated, although some equipment, including the IFF, was installed in Pakistan. The new avionics plus the improved performance makes the F-7P an effective fighter in an environment where the adversary's air inventory still consists largely of MiG-21s. The F-7P has been a very important addition to the PAF; nowhere near an F-16 in capability, but in so far as three F-7Ps can be purchased for the price of one F-16, the PAF was more than willing to go for over 100 F-7s, to equip four squadrons plus an OCU.

The PAF has had a sustained, and pragmatic, interest in acquiring as many second-hand Mirage III/5s as it could identify, with which to build up a parallel, reserve combat force. The first success was with ex-Royal Australian Air Force Mirage IIIOs when, in early 1990 in a deal worth A$36 million, the PAF finalised orders for 50 ex-RAAF Mirage IIIOs. The package included 45 additional Atar 09C engines, a flight simulator and spare parts which would "help establish a Mirage overhaul plant at Karachi", according to the then-PAF Chief of Air Staff. The first batch of 32 Mirage IIIOs was shipped from Wyalla Port in Australia and arrived at Karachi in late November 1990, the balance being shipped to Pakistan by the end of that year.

The RAAF Mirage IIIOs had earlier been phased out, mothballed and stored at the Woomera Rocket Range for several years. Although the Australians maintained that the aircraft were at the end of their fatigue lives and would require extensive overhaul

and structural work to make them airworthy, the PAF was quite confident that these Mirages could be virtually rebuilt and kept in service for another decade. In fact, the PAF had pulled off a strategic coup that would stand it in good stead through the 1990s. The Mirage IIIOs will predictably be assigned the air defence role, protecting air bases and other strategic targets, releasing the F-16s for offensive strike tasks.

In mid-1996, the PAF contracted for another 40 Mirage III/5s via the French Defence Ministry's maintenance division under a deal known as Blue Flash 6. Sagem and Sogerma were responsible for upgrading of avionics, and nav-attack systems. Another dozen-odd Mirage IIIBLs had been bought from Lebanon some years earlier while PAF teams scouted for more Mirages being phased out in Spain, Belgium and Zaïre. In early 1998, Belgium reportedly agreed to sell 24 Mirage 5s to Pakistan. These aircraft had been in varying storage conditions for some time, so required refurbishing and upgrading, but were still very cost-effective.

There is an enigma regarding Arab fighters being made available to the PAF during an 'emergency'. The PAF has for a long time provided commanders, pilots and technicians to various countries, particularly Abu Dhabi – a tiny state which acquired 31 Mirage 5s in the 1970s. Upon introduction of new Mirage 2000s, many of the remaining UAE Mirage 5s were reportedly transferred to the PAF in the early 1990s.

According to reliable reports, the Pakistan Air Force order of battle currently comprises three (under-strength) squadrons of F-16A/Bs, five squadrons of Mirage IIIs/5s, five squadrons of Xian (Chengdu) F-7P/Ms, three squadrons of A-5s and two squadrons with the F-6, a total of 18 combat squadrons with some 385 aircraft. An ESM squadron is equipped with the Dassault Falcon 20G, and two composite units fly a mix of FT-5s, F-7Ps and F-6s for fighter conversion and combat leader training.

The PAF has a modest transport fleet, with one squadron of C-130B/Es and L-100s and a VIP transport squadron flying F27s, Falcon 20s and Boeing 707s. A liaison/communication squadron operates Beech Barons and Aero Commanders. The bulk of the rotorcraft are operated by the Army, apart from some Alouette IIIs for liaison and casevac. Flying training is carried out on PAC Mushshaks (licence-built MFI-17s) and SF-25C Falkes, moving on to Cessna T-37B/Cs and recently acquired Karakoram K-8s.

Pakistan Fiza'ya

Air Force Academy		Risalpur
PFTS	Mushshak, SF-25C	Risalpur
BFTS	T-37B/C, K-8	Risalpur
35 Wing		**Chaklala**
6 Sqn	C-130B/E, L-100, HAMC Y-12-II	Chaklala
12 Sqn	F27-200, Falcon 20E, 707-320, Citation V	Chaklala
41 Sqn	Baron, Seneca, Cessna 172	Chaklala

Central Air Command		
34 Wing		**Rafiqui/Shorkot Rd**
5 Sqn	Mirage IIIEP/RP/RDP	Rafiqui/Shorkot Rd
18 Sqn	F-7P	Kamra
20 Sqn	F-7P	Rafiqui/Shorkot Rd
83 Sqn	Alouette III	Rafiqui/Shorkot Rd
38 Wing		**Sargodha**
9 Sqn	F-16A/B	Sargodha
11 Sqn	F-16A/B	Sargodha
24 Sqn	Falcon 20F (ECM equipped)	Sargodha
82 Sqn	Alouette III	Sargodha

Northern Air Command		
33 Wing		**Kamra**
14 Sqn	F-16A/B	Kamra
15 Sqn	Shenyang F-6, FT-6	Peshawar
36 Wing		**Peshawar**
16 Sqn	Nanchang A-5C	Peshawar
26 Sqn	Nanchang A-5C	Peshawar
81 Sqn	Alouette III	Peshawar
37 Wing		**Miawali**
1 Sqn	Shenyang FT-5	Mianwali
19 Sqn	Chengdu F-7P, FT-7P	Masroor
25 Sqn	Chengdu F-7P, FT-7P	Mianwali

Southern Air Command		
31 Wing		**Quetta/Samungli**
17 Sqn	Chengdu F-7P, FT-7P	Rafiqui/Shorkot Rd
23 Sqn	Shenyang F-6, FT-6	Quetta/Samungli
85 Sqn	Alouette III	Quetta/Samungli
32 Wing		**Masroor**
2 Sqn	R/T-33A, Chengdu F-7P	Masroor
7 Sqn	Nanchang A-5C	Masroor
8 Sqn	Mirage 5PA3	Masroor
22 Sqn	Mirage 5PA1, Mirage IIIDP	Masroor
84 Sqn	Alouette III	Masroor

Pakistan Army Aviation Corps

The Army's organic air element consists of some 15 squadrons, flying a large variety of rotorcraft and a few fixed-wing types. The ubiquitous PAC Mushshak is used for primary flying training and there are some O-1Es for AOP tasks. Two squadrons are equipped with Bell AH-1S Cobra anti-tank helicopters, with some Bell 206B Jet Rangers for continuation training/communication. Earlier Russian supplies include Mi-8s and Mi-17s (numbers have been augmented by defecting Afghan aircraft). The Bell UH-1H Huey is used for utility tasks and two squadrons have the SA 330J Puma. There are considerable numbers of SA 315 Lamas for logistic support of the ground forces in the high Karakorams. While the Army's helicopter units are scattered throughout the country in support of corps and divisions in the field, the main base is at Multan and headquarters is at Dhamial (Rawalpindi)

Pakistan Army Aviation Corps

Sqn	Aircraft	Base
2 Sqn	UH-1H, O-1	Lahore
3 Sqn	Mushshak, O-1	Multan
4 Sqn	Mil Mi-8, Mil Mi-17	Dhamial
5 Sqn	Alouette III	Dhamial
6 Sqn	UH-1H, Mi-17	Dhamial
7 Sqn	Mushshak, O-1	Faisal
8 Sqn	SA 315B Lama, Alouette III	Dhamial
9 Sqn	Mushshak, O-1, Alouette III	Peshawar
13 Sqn	Mushshak, O-1	Dhamial
21 Sqn	SA 330J, UH-1H	Multan
24 Sqn	SA 330J	Multan
25 Sqn	SA 330J	Dhamial
31 Sqn	AH-1S, Bell 206B	Multan
32 Sqn	AH-1S, Bell 206B	Multan
Aviation School	Mushshak, Schweizer 269C, Alouette III, OH-13S, Bell 206B	Rahwali
VIP Flight	Cessna 421C, Commander 840, SA 330J	Dhamial

Pakistan Naval Aviation

The naval air arm has recently received a boost after delivery of the three long-embargoed Lockheed P-3C Orions (equipped with Harpoon missiles). Other MR/ASW squadrons are equipped with refurbished Breguet Atlantics, Fokker F27Ms, and BN-2T Maritime Defenders.

One squadron operates the Sea King HAS.Mk 45, with Exocet air-to-surface missiles, and there are a number of Westland Lynx HAS.Mk 3s and Alouette IIIs, the former embarked on a number of navy frigates. The main naval air base is at Karachi (Drigh Road).

Pakistan Navy

Sqn	Aircraft	Base
27 Sqn	F27-200/400M	Drigh Road
29 Sqn	P-3C, Atlantic	Drigh Road
93 Sqn	PBN BN-2T Maritime Defender	Drigh Road
111 Sqn	Sea King HAS.Mk 45/45B	Drigh Road
333 Sqn	Lynx HAS.Mk 3, Alouette III	Drigh Road

SEYCHELLES

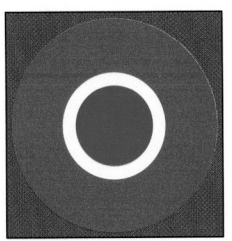

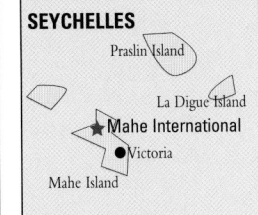

Capital: Victoria
Population: 100,000
Land area: 404 km² (156 sq miles)

Seychelles Air Force

Following independence from the UK in 1976, an air arm of the Seychelles Defence Force was formed in 1980 with two SOCATA Rallye lightplanes donated by Libya for training and liaison. India also presented two HAL-built Chetak (Alouette III) helicopters in 1982, followed by a third attrition replacement in 1988. Unfortunately another Chetak crashed in 1992. An air force-operated used Swearingen Merlin IIIB twin-turboprop light transport, bought in 1983 for government use, was replaced in 1989 by a new Cessna Citation V, which was sold in 1994.

A Reims-Cessna 406 Caravan II light turboprop

Seychelles Air Force

Type	Role	Delivered/In Service
Britten-Norman BN-2B Defender	light transport	1/1
Cessna A.150M	primary trainer	1/1
Reims-Cessna 406 Caravan II	utility transport	1/1
SOCATA Rallye 235	communications	2/1
HAL-Aérospatiale SA 316B Alouette III/Chetak	utility helicopter	3/1

twin recently acquired for air force SAR roles was accompanied by a second example to supplement a Britten-Norman BN-2A Islander operated by the Seychelles Coast Guard, alongside a Cessna 152.

SRI LANKA

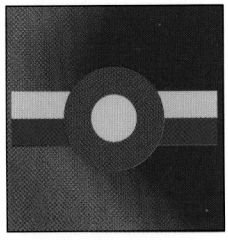

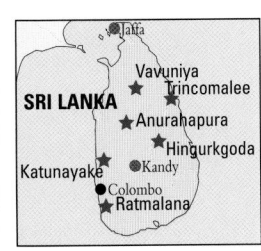

Capital: Colombo
Population: 17 million
Land area: 65610 km² (25,325 sq miles)
Major cities: Dehiwela-Mount Lavinia, Jaffna, Kandy, Kotte, Moratuwa

Sri Lanka Air Force (SLAF)

Until the mid-1980s, the air arm of the idyllic island of Sri Lanka, with its handful of light transport and training aircraft, was employed in support of the growing tourist industry, flying holiday-makers from the capital Colombo to the beautiful beaches of northeastern Sri Lanka. Now, transport aircraft of the much expanded SLAF are engaged in flying in troops and supplies to embattled zones and flying out casualties, while its helicopter gunships lend close air support even as the land forces fight it out against fanatical LTTE (Liberation Tigers of Tamil Eelam) cadres.

The Sri Lanka Air Force acquired its first combat aircraft (two Shenyang FT-5 operational fighter-trainers) from China, which were soon supplemented by the more modern and effective Chengdu FT-7 trainer and F-7BS fighters (MiG-21s). Since then, the SLAF has expanded with aircraft acquisitions of a bewildering variety from China, Russia, Israel, Italy and Argentina. Today, it operates some 20 different aircraft types and is actively engaged in combat operations throughout the island. The SLAF has recently had very high aircraft losses, some to hostile small arms fire but increasingly to a variety of accidents, some attributed to lack of experience, others to subterfuge or even sabotage.

The SIAI SF.260TPs serve with No. 1 Flying Training Wing in both the training and light strike role, alongside examples of the SF.260WB.

Since early 1997, SLAF Pucarás and Kfirs have seen considerable action in support of the Army and Navy, fighting for control of key areas and towns in the Jaffna peninsula and also around the major harbour of Trincomalee in eastern Sri Lanka. The ex–FAA IA-58A Pucarás were to equip an attack squadron at Vavuniya (No. 7 Sqn) but this never occurred. Two were written off in 1995 and 1997. Of the survivors, one is used for spares, with the airworthy example belonging to the No. 1 Flying Training Wing.

The Israelis have provided more than hardware, also training SLAF aircrew and providing initial maintenance support. The IAI Kfir C2s/TC2s equip one squadron at Katunayake (near Colombo), where No. 5 Squadron operates the F-7BS and FT-7s. A pair of Shenyang FT-5s used by this squadron was retired by mid-1997.

Minneriya-Hinngurakgoda is home to the Mi-24D gunships and Mi-17 assault squadrons. The Mi-24s, reportedly delivered from Ukraine, have seen strenuous action in the heavy fighting that continues and three have been lost. Russian personnel are said to be assisting the Sri Lankans in flying and maintaining the increasing numbers of helicopters and transport aircraft from that source. Bell 212s and 412s provide mobility for the army, as well as undertaking combat SAR missions.

The Heavy Transport Squadron at Ratmalana has a motley mix of Antonov An-32s, BAe 748s and Shaanxi Y-8s. Heavy losses have resulted in the air force hiring civilian contractors who have operated

In service from November 1995, the Mil Mi-24V 'Hind-Es' have been very active supporting troops fighting the 'Tamil Tigers'. Of the 13 acquired, three have been claimed by the rebels as destroyed and three have been returned to the supplier, leaving the air force with seven still in service.

An-24RVs (Kazakhstan Airlines) or An-32s. Three demodified ex-RAF Hercules C.Mk 1Ks are to be received. The co-located No. 8 Squadron uses four surviving Harbin Y-12-IIs and a Cessna 150.

Flying training is carried out using SF.260WB/TPs, which are also employed for light attack (fitted with locally-assembled bombs) and reconnaissance. There are also Cessna 150s used for primary training. Maritime patrol is undertaken by a single Beech 200T, converted locally for the task. An order for four second-hand examples was cancelled in 1997, but several Cessna 337F Skymasters are believed to operate alongside the King Air. A pair of Dauphin 2s used by the squadron was retired in 1997.

Sri Lanka Air Force

Headquarters: Colombo

Anuradhapura

No. 1 Flying Training Wing	SF.260WB/TP (10), Cessna 150 (5), Pucará (1)
No. 11 UAV Sqn	IAI Super Scout

Trincomalee (China Bay)

No. 3 Maritime Sqn	Beech 200T (1), Cessna 337F (3)

Katunayake

No. 4 Helicopter Sqn	Bell 412SP (4)
No. 5 Jet Sqn	Chengdu F-7BS (4), Guizhou FT-7B (1)
No. 10 Fighter Sqn	IAI Kfir C2 (6), IAI Kfir TC2 (2)
Aircraft Engineering Wing	–

Minneriya-Hinngurkgoda

No. 7 Helicopter Sqn	Bell 206B JetRanger (6), Bell 212 (10)
No. 9 Attack Helicopter Sqn	Mi-24V (7)

Rathmalana

No. 2 Heavy Transport Sqn	BAe 748 (2), Shaanxi Y-8D (2), An-32B (1), An-24RV (2)*, Cessna 421C
No. 8 Light Transport Sqn	Harbin Y-12-II (4), Cessna 150 (1)
Aircraft Preservation and Storage Unit	

Vavuniya

No. 6 Helicopter Sqn	Mi-17 (11)

* contractor's aircraft

Nine Harbin Y-12-IIs were delivered from 1987, of which two have been lost and three grounded for spares. The survivors operate from Ratmalana with No. 8 Light Transport Squadron.

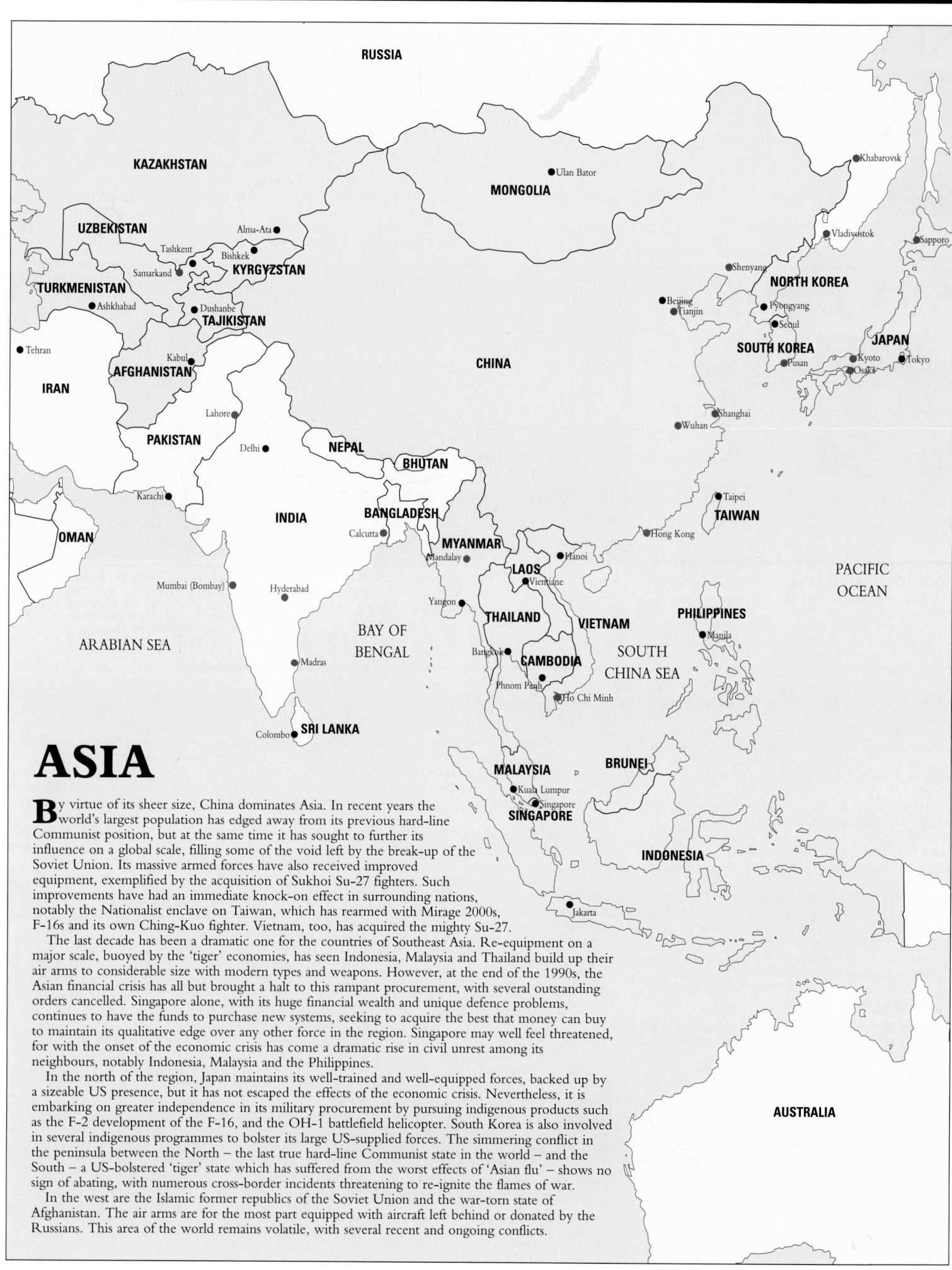

ASIA

By virtue of its sheer size, China dominates Asia. In recent years the world's largest population has edged away from its previous hard-line Communist position, but at the same time it has sought to further its influence on a global scale, filling some of the void left by the break-up of the Soviet Union. Its massive armed forces have also received improved equipment, exemplified by the acquisition of Sukhoi Su-27 fighters. Such improvements have had an immediate knock-on effect in surrounding nations, notably the Nationalist enclave on Taiwan, which has rearmed with Mirage 2000s, F-16s and its own Ching-Kuo fighter. Vietnam, too, has acquired the mighty Su-27.

The last decade has been a dramatic one for the countries of Southeast Asia. Re-equipment on a major scale, buoyed by the 'tiger' economies, has seen Indonesia, Malaysia and Thailand build up their air arms to considerable size with modern types and weapons. However, at the end of the 1990s, the Asian financial crisis has all but brought a halt to this rampant procurement, with several outstanding orders cancelled. Singapore alone, with its huge financial wealth and unique defence problems, continues to have the funds to purchase new systems, seeking to acquire the best that money can buy to maintain its qualitative edge over any other force in the region. Singapore may well feel threatened, for with the onset of the economic crisis has come a dramatic rise in civil unrest among its neighbours, notably Indonesia, Malaysia and the Philippines.

In the north of the region, Japan maintains its well-trained and well-equipped forces, backed up by a sizeable US presence, but it has not escaped the effects of the economic crisis. Nevertheless, it is embarking on greater independence in its military procurement by pursuing indigenous products such as the F-2 development of the F-16, and the OH-1 battlefield helicopter. South Korea is also involved in several indigenous programmes to bolster its large US-supplied forces. The simmering conflict in the peninsula between the North – the last true hard-line Communist state in the world – and the South – a US-bolstered 'tiger' state which has suffered from the worst effects of 'Asian flu' – shows no sign of abating, with numerous cross-border incidents threatening to re-ignite the flames of war.

In the west are the Islamic former republics of the Soviet Union and the war-torn state of Afghanistan. The air arms are for the most part equipped with aircraft left behind or donated by the Russians. This area of the world remains volatile, with several recent and ongoing conflicts.

AFGHANISTAN

Capital: Kabul
Population: 16.5 million
Land area: 652225 km² (251,773 sq miles)
Major cities: Herat, Kandahar, Mazar-i-Sharif

De Afghan Hauai Ouvah

Current principal DAHO equipment is thought to include the following:

Type	Role	In service
MiG-21MF/bis 'Fishbed-J/L'	air defence/GA	50
MiG-21U/UM 'Mongol-B'	combat trainer	15
MiG-23BN 'Flogger-F'	ground attack	25
MiG-23UB 'Flogger-C'	combat trainer	5
Sukhoi Su-7BM 'Fitter-A'	ground attack	10
Sukhoi Su-7U 'Moujik'	combat trainer	?
Sukhoi Su-20/22M-3/4 'Fitter-C/K'	ground attack	50
Sukhoi Su-22U 'Fitter-E'	combat trainer	?
Antonov An-2 'Colt'	utility transport	12
Antonov An-12BP 'Cub'	transport	10
Antonov An-26 'Curl'	tactical transport	15
Antonov An-30 'Clank'	survey transport	1
Antonov An-32 'Cline'	tactical transport	6
Ilyushin Il-18D 'Coot'	transport	1
Aero L-39C Albatros	advanced trainer	10
Yakovlev Yak-18A 'Max'	basic trainer	12
Mil Mi-8/17 'Hip'	transport helicopter	45
Mil Mi-24 'Hind'	attack helicopter	30

Seen during earthquake relief operations in Afghanistan in February 1998, this An-12 wears the markings of the Northern Alliance air force which opposed the Taliban government.

De Afghan Hauai Ouvah – DAHO (Afghan Army Air Force)

Military aviation assets in Afghanistan have been fragmented in continuous civil wars by various internal rebel factions since the withdrawal of Soviet forces in 1989, and overthrow by the Mujahideen of the Communist-backed Najibullah government in March 1992. Most Afghan air force aircraft and associated assets are now operated by the Pakistani-backed Islamic fundamentalist Taliban government, following its August 1995 capture of Herat air base and overthrow in September 1996 of President Rabbani in Kabul, despite his Russian arms support. Other aircraft, seized or defected from Afghan airfields, or supplied from neighbouring CIS countries, were also reported in use from time to time by the anti-Taliban alliance in northern areas of Afghanistan (Northern Alliance) held by rebel warlords and their forces, until their almost complete expulsion by August 1998.

Further complication resulted from splits in the former anti-government alliance, known until early 1998 as the Dostum Golboddin Militia. In northwest Afghanistan, along the Turkmenistan border, Jowzjan Province was held by the forces of ex-Communist Uzbek warlord Abdul Rashid Dostum, which began operating a few Sukhoi Su-22s, three MiG-21s, three Aero L-39s and a few Mil Mi-24s from Mazar-e-Sharif air base against Taliban targets, including Bagram air base, in 1994-95. Several MiG-21s and Su-22s were reportedly shot down, although replaced by others from Uzbekistan, but seven were flown by defectors to Kabul in early 1994, while others were believed withdrawn to Tajikistan. Mazar-e-Sharif was recaptured by the Taliban in July 1998.

In northeastern Afghanistan, adjoining Tajikistan, Takhar Province and the Panjshir Valley were held by the forces of Ahmed Shah Massoud, former military leader to ex-President Rabbani, until his deposition by the Taliban. Massoud reportedly received at least 12 MiG-21s, via Tajikistan, for operation from Taloqan air base, in Takhar, from where operations were reported against targets near Kabul in mid-1997, until its recapture by Taliban forces in August 1998. As leader of yet another Afghan rebel faction, ex-President Rabbani was said to have been offered up to 20 ex-Iraqi air force MiG-21s still held by Iran since the 1991 Gulf War.

BRUNEI

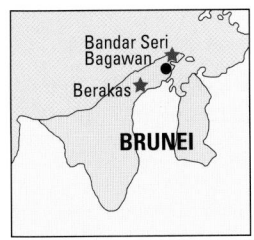

Capital: Bandar Seri Begawan **Population:** 300,000
Land area: 5765 km² (2,225 sq miles)

Tentara Udara Diraja Brunei (Royal Brunei Air Force)

Tentara Udara Diraja Brunei (Royal Brunei Air Force)

1 Sqn	Bell 212, 214ST	Berakas Camp
2 Sqn	BO 105	Brunei Airport
3 Sqn	SF.260, Bell 206	Brunei Airport
4 Sqn	CN.235M, CN.235MPA	Bandar Seri Begawan

Enjoying the world's highest per capita income thanks to massive oil resources, Brunei has plenty of money to spend on defence, should it wish. In fact, although Brunei's air arm has been operating helicopters since 1965, expansion has only gathered pace since the country assumed full responsibility (from the UK) for external as well as internal affairs, in 1984.

Having previously operated Bell 205A Iroquois, the air wing later standardised on twin-engined Bell 212s, and now flies 12 on oil rig patrol and army support duties. Six MBB BO 105CB armed helicopters were acquired in 1981. A single Bell 214ST is used for SAR, while two Bell 206B JetRangers are trainers. The latest helicopters to join the fleet are four UH-60Ls.

Fixed-wing training is conducted on four SIAI-Marchetti SF.260W Warriors, which are reportedly

Brunei's small air force consists primarily of rotary-wing assets. The most numerous is the Bell 212, used for tactical transport.

due to be replaced by four Pilatus PC-7s. In late 1989, an intention to order 16 BAe Hawk 100/200s was announced, but the deal was still under negotiation in late 1998. Three CN-235MPAs were delivered for maritime patrol alongside a single CN-235M for transport duties, operating from the new base at Bandar Seri Begawan. An additional three CN-235Ms were ordered in 1998.

CAMBODIA

Capital: Phnom Penh
Population: 8.7 million
Land area: 181000 km² (69,865 sq miles)
Major cities: Battambang, Kampot, Kratie

Force Aérienne Royale Cambodge (Royal Cambodian Air Force)

With the final diminution of the threat from Khmer Rouge guerrillas, reactivation of its national armed forces in 1993, and failure of an attempted military coup in mid-1997, Cambodia is now expected to enter a period of stability and consolidation. Some $1.6 billion in Soviet military arms and assistance received from 1980-90 included the supply of MiG-21s, various transport aircraft, Pechora export versions of the SA-3 'Goa' medium-range mobile SAM, and much other aviation-related equipment. Six Tecnam P-92 Echo high-wing two-seat micro-lights were also delivered from Italy from September 1994 for FARC training and observation, followed in 1995 by two Harbin Y-12 II light turboprop transports from China.

An $80 million contract was signed in February 1996 with Israel's IAI to upgrade eight of 15 Cambodian MiG-21bis/UM with new digital avion-

Asia

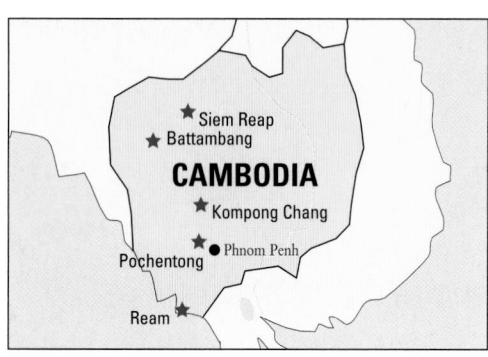

Force Aérienne Royale Cambodge		
Type	Role	Delivered/In Service
Aero L-39ZA Albatros	advanced trainer	6/5
Antonov An-24RV 'Coke'	transport	3/3
Antonov An-26RV 'Curl'	tactical transport	3/1
Beech 200 King Air	light transport	1/1
Cessna 401/421	light transport	1/1
Dassault Falcon 20	light transport	1/1
Eurocopter AS 350B Ecureuil	utility helicopter	1/1
Eurocopter AS 365 Dauphin 2	utility helicopter	1/1
Fokker F28 Fellowship	transport	1/1
Harbin Y-12 II	utility transport	2/2
MiG-21bis 'Fishbed-N'	air defence	19/15
MiG-21UM 'Mongol-B'	combat trainer	3/1
Mil Mi-8S/T 'Hip-C'	transport helicopter	8/4
Mil Mi-17 'Hip-H'	transport helicopter	9/7
Mil Mi-24 'Hind'	attack helicopter	3/0
Mil Mi-26 'Halo'	heavy-lift helicopter	2/2
Pilatus B/N BN.2A Islander	utility transport	3/2
SOCATA TB.10	light transport	/1
Tecnam P92 Echo	primary trainer	6/5
Tupolev Tu-134A3 'Crusty'	transport	2/2

Seen in the IAI Lahav workshop in Israel, these MiG-21s of the Cambodian air force are undergoing upgrade to MiG-21-2000 standard.

1997 delivery. Other 1996 procurement included two Mil Mi-17 and two Mi-26 heavy-lift helicopters from Russia, followed by two more Mi-8s bought from the Czech Republic for government transport in March 1997.

ics to MiG-21-2000 standard, plus the supply of six refurbished used Aero L-39ZA armed jet-trainers for

CHINA

Capital: Beijing
Population: 1,089 million
Land area: 9.6 million km² (3,704,440 sq miles)
Major cities: Canton, Shanghai, Shenyang, Tianjin, Wuhan

AFPLA – Air Forces of China's People's Liberation Army (PLAAF)

China's three major air arms – the People's Liberation Army Air Force (PLAAF), the People's Liberation Army Navy-Air Force (PLAN-AF) and the Army Aviation Corp (PLA-AAC) – are all in the midst of long-overdue modernisation programmes. China's air forces are still largely comprised of obsolete platforms and hobbled by insufficient training and logistical support, plus a doctrine that prevents the full realisation of their potential. Its military aircraft design and production sector lags behind that of the West, resulting in continued reliance on foreign technology, which China is quickly assimilating. It remains to be seen whether China can build an air arm capable of sustained long-range all-weather joint-force operations. In the near future China will rely on more capable missile forces for theatre power-projection. As long as China's political leadership remains committed to the goal of building the military forces necessary to subdue democratic Taiwan and consolidate control over the South China Sea, there will be sustained investment in air force modernisation. The demonstration of US air power during the 1990-91

Originally entrusted to Harbin, production and subsequent development of the H-6 (Tu-16) has been undertaken by Xian. Around 100 serve with the PLAAF in both conventional and nuclear bombing roles. Among the more unusual tasks assigned to the H-6 is the bombing of ice jams which form annually on the Yellow River, threatening surrounding areas with severe flooding.

Gulf War forced PLA leaders to recognise the utility of modern air power and the need for China to reform its air forces for modern warfare.

HISTORY

The People's Liberation Army Air Force was established on 11 November 1949, soon after the Communist Party's victory over the Nationalists, who fled to Taiwan. It began with less than 160 aircraft and an established subordination to the Army. War on the Korean peninsula provided the impetus for a rapid expansion – up to 3,000 aircraft by early 1954. Chinese pilots joined Russian pilots flying for North Korea against US and allied air forces. In the late 1950s the PLAAF skirmished with Taiwan's US-supplied air force, usually MiG-17s against F-86s and F-84s. During the Vietnam War the PLAAF shot down several US Firebee reconnaissance drones and at least one US Navy A-6 Intruder that entered Chinese airspace. The PLAAF was used only lightly

during the brief 1979 border war with Vietnam. Its next (and most recent use) did not entail firing in anger, but comprised a series of exercises from 1994 to 1996 designed to intimidate Taiwan into not contemplating formal independence from China.

PLAAF ORGANISATION

The PLAAF is China's largest air arm, with combat, transport and training aircraft units, plus anti-aircraft and logistic support organisations. The PLAAF structure reflects its subordination to the Army and, mirroring that service, it is divided into seven Air Force Districts which coincide with the seven military regions (Beijing, Chengdu, Guangzhou, Jian, Lanshou, Nanjing, and Shenyang). Air Force Districts are further composed of Divisions, which contain two to three Regiments (each normally contains about 70 to 124 fighters or bombers), broken down into Squadrons which are responsible for three or four Air Sections, with three

Drawing heavily on the MiG-21 design, the Shenyang J-8 is a much larger twin-engined aircraft optimised for high-altitude interception work. The initial J-8 and J-8I (left) versions have been supplemented by the J-8II (above) with side intakes and larger nose radar.

to four aircraft each. Military Regions normally control a theatre of operations while Regiments are responsible for a specific geographic area. For example, the Nanjing Military Region would control all PLA service operations against Taiwan.

Since the late 1950s the PLAAF has controlled anti-aircraft artillery and surface-to-air missile units, usually dedicated to city or airfield defence. The 15th Airborne Army was created in 1961, and contains three Brigades, two combat and one training. These airborne units are intended for use beyond China's borders or for internal action, as during the 1989 Tiananmen Square incident.

Logistic support and maintenance is subordinate to PLAAF Headquarters. The PLAAF has long laboured under poor logistic and maintenance conditions, partially due to a dispersed manufacturing sector in which aircraft and engine manufacturers are co-located, resulting in an inability to exchange parts between similar airframes and engines. Aircraft manufacturers include Chengdu, Shenyang and Xian, plus helicopter makers Harbin and Changhe. China has been unable to develop advanced jet engines and is only beginning to learn advanced, fourth-generation aircraft manufacturing techniques. A proliferation of new foreign systems in the 1990s is likely to only compound logistic and maintenance support problems.

Perhaps the most important organisation modernisation challenge facing the PLAAF is whether it can develop doctrine, roles and missions necessary for modern air warfare. It has yet to create an integrated air defence system that melds fighters, missiles and C4I elements into an efficient network. While clearly a goal, the PLAAF has not yet acquired AEW, tanker, and long-range interdiction aircraft armed with modern missiles with which it can fashion an all-

Two 'old soldiers' – the Harbin H-5 (Il-28, above) and Shenyang J-6 (MiG-19S, right) – still form the backbone of the fighter and bomber forces, and their replacement remains the main procurement problem for the PLAAF. Both aircraft types also undertake reconnaissance missions (as the HZ-5 and JZ-5), while the H-5 also serves in modified form as a missile carrier. Sizeable numbers of JJ-6 trainers are in use, this version being an indigenous development of the basic single-seat design.

weather strike capability. Training is just beginning to move beyond old Soviet-style positive ground control tactics to stress air combat manoeuvring on instrumented ranges, aggressor units and advanced air combat simulators. For the future, China has an interest in building missile defences; it is not clear whether the PLAAF will perform this mission.

PLAAF MODERNISATION

Before the revolution, China relied on successive waves of imported aircraft technology, a tradition continued by the Communists. By the mid- to late 1950s, China had received productions rights for several Soviet designs: MiG-15 (J-2), MiG-17 (J-5), MiG-19 (J-6), MiG-21 (J-7), Il-28 (H-5) and Tu-16 (H-6). Most transports and helicopters also relied on Soviet designs, like the An-2 (Y-5), An-12 (Y-8),

An-24 (Y-7) and Mil-4 (Z-5). The end of Soviet assistance in 1960 was a severe blow to the PLAAF. Despite the aircrafts' obsolescence, China was not able to produce new designs to replace its 1950s-era Soviet aircraft, and they accounted for most of the PLAAF's combat inventory until the mid-1990s. The 1980s saw a reintroduction of US and European technology, but it was cut off with the embargoes that followed the Tiananmen Square massacre. Nevertheless, China was able to produce extended Soviet-based designs like the Q-5 (based on the J-6) and the J-8 (based largely on the J-7). The 1990s have brought a reintroduction of Russian aircraft and sub-systems, plus extensive help from Israel and some European technology.

Influenced largely by the success of US air forces

CHINA

[Map of China with the following locations marked:]

Hailar · Yalu · Yichun · Qiqihar · Harbin · Changchun · Mudanjiang · Fuxin · Chifeng · Shenyang · Zhangjiakou · Jinzhou · Xingcheng · Bautou · Beijing · Qinhangdao · Datong · Tangshan · Luda · Yinchuan · Tianjin · Canzhou · Dongying · Yantai · Shijiazhuang · Qingdao · Yining · Kuqa · Urumqi · Hami · Xingrenbu · Jinan · Zhucheng · Wensu · Uxxatal · Shanshan · Xining · Lanzhou · Taiyuan · Jining · Kashi · Gonghe · Liangyungang · Korla · Golmud · Wugon · Xian · Jiyuan · Nanjing · Danyang · Yushu · Chenggu · Zhengzhou · Xinyang · Wuhu · Wuxi · Shanghai · Chengdu · Anquing · Hangzhou · Xigaze · Lhasa · Chongqing · Wuhan · Tunxi · Ningbo · Xichang · Suiyang · Nanchang · Wenzhou · Guiyang · Changsha · Lianchiang · Kunming · Leiyang · Fuzhou · Guilin · Zhangzhou · Huian · Xiamen · Mengzi · Nanning · Guangzhou · Anhai · Simao · Shantou · Ningming · Huiyang · Zhanjiang · Hong Kong · Haikou · Lingshui · Sanya

Using the J-6 (MiG-19) as a basis, Nanchang developed the Q-5 for low-level strike/attack missions. The Q-5A is the nuclear version, with a belly recess to house the special weapon. The first drop was undertaken in January 1972.

during the Gulf War, plus the Western embargoes, in 1992 China purchased its first batch of 26 Russian Sukhoi Su-27SK fighters, followed by 24 more in 1996, and then made an agreement to co-produce 200 more that same year. They will be produced in Shenyang under the designation J-11. The first two apparently flew in late 1998 and the reported goal is to increase production to 15 a year by 2002 – an ambitious goal that may require extensive Russian help. The Su-27 is China's first modern fourth-generation fighter, with long-range, helmet-sighted R-73 short-range AAMs and R-27 semi-active medium-range AAMs. Successive batches have been upgraded with ECM pods, and the co-produced variants may include upgraded radar that can handle the R-77 active-guided medium-range AAM and the Kh-31P anti-radiation missile that will be co-produced in China under the designation KR-1.

Russian Phazotron radar and R-27 missiles have been incorporated into the Shenyang J-8IIM which may only be offered for export. The Russian RD-33 engine is likely to form the core of the Chengdu FC-1 fighter designed to compete in the F-5/MiG-21 replacement market. Co-funded by Pakistan, the FC-1's future is not secure, as it remains uncertain whether the type will be purchased by the PLAAF. Chengdu is also marketing the J-7MG, the latest in a long line of J-7s that incorporates a new cranked wing, upgraded radar, defensive electronics, and a helmet-sighted version of the PL-9 short-range AAM, which is a copy of the Israeli Python-3.

For the future, Shenyang apparently has the lead for a new fifth-generation stealthy fighter known in the West as the XXJ. Projections by the US Office of Naval Intelligence show the XXJ bearing a resemblance to the US F-15. This project could evolve much differently should China succeed in gaining access to Russia's MiG 1.42 fifth-generation project, but recent reports suggest Russia is not yet willing to let China into this programme.

The most important current domestic fighter project for the PLAAF is the Chengdu J-10, China's first indigenous fourth-generation fighter; reports suggest the country may buy up 300. After a lengthy development it reportedly flew for the first time in April 1998. Due to Israeli assistance the J-10 is expected to bear a strong resemblance to the cancelled Lavi. Israel and Russia are competing to provide the radar, electronic and missile sub-systems for the J-10, though new Chinese radar and missiles could be used. The aircraft is thought to be powered by the Russian AL-31 engine, but there are reports that a twin-RD-33-engined J-10 could eventually go the PLA Navy Air Force.

This air arm, rather than the PLAAF, may be the first customer for the Xian JH-7 attack fighter that has been in development since the mid-1970s and was revealed with much fanfare at the 1988 Zhuhai air show. Resembling a beefed-up twin-seat Jaguar, JH-7 promises a respectable 900-mile (1450-km) radius, an all-weather attack radar and a low-level targeting pod to employ laser-guided bombs. Xian may also be planning an electronic warfare version of the JH-7. While Xian says it has mastered the WS-9 engines based on Rolls-Royce Spey Mk 202s, recent reports suggest the manufacturer is trying to buy more Speys to complete the first batch of JH-7s.

The JH-7 programme has survived despite the PLAAF's clear preference for the Sukhoi Su-30 attack fighter. At the beginning of 1999 reports were circulating of closure on a deal to purchase 20 to 50 of these fighters, comparable in capability to the US F-15E.

Left: The J-7E is a day-fighter development of the J-7II with a cranked wing, marketed for export as the F-7MG. This aircraft flies with the 'August 1' aerobatic display team, which until 1997 was equipped with the Chengdu FT-5.

Below: Manufactured over the years by Shenyang, Xian and Chengdu, the J-7 is the Chinese-built version of the MiG-21. The Chinese have developed their own versions, including the J-7III, an all-weather version similar to the MiG-21bis, which first flew on 26 April 1984. Guizhou builds the two-seater.

Full exploitation of these new fighter and attack aircraft will depend on the PLAAF being able to absorb new support aircraft. The most important is a joint Israeli-Russian project to build an AWACS system based on the Russian A-50 with an Israeli IAI Phalcon phased-array radar. China has committed to buying the prototype and may purchase three to seven more. This AWACS may be competing with a GEC-Marconi Argus radar-equipped Il-76 which has been the subject of long discussion. In 1996 China purchased a reported six to eight Racal Searchwater AEW radars to be fitted on the Y-8 transport. The first prototype may have flown in mid-1998.

The PLAAF has long sought an aerial-refuelling capability and has converted a small number of H-6 bombers for this purpose. Russia is heavily marketing the Il-78M tanker, which has appeared at both Zhuhai shows. As India has done, it remains possible that China will purchase the Il-78M after it begins to receive its Su-30s. For ECM/Elint missions China has converted two Tu-154 transports, with electronic systems that may have been purchased or derived from Israeli systems. This does not represent an extensive capability. At Zhuhai '98 China revealed a new series of EW/Elint pods for fighter or attack aircraft.

The PLAAF has had a long-standing requirement to replace its obsolete Xian H-6 medium bombers. There is occasional mention of Chinese interest in purchasing a small number of Russian Tu-22M-3 supersonic medium bombers, but no order has materialised. It is also possible that Xian may be developing a new design that has not yet been revealed.

The PLAAF controls most of China's long-range surface-to-air missiles, like the S-300PMU system purchased from Russia in the early 1990s. These missiles are mainly stationed around Beijing, though they were featured in the 1996 exercises near Taiwan. At the 1998 Farnborough air show, China revealed its new FT-2000 long-range SAM, which may be based on the S-300, clandestinely acquired US Patriot SAM, and Chinese technology. It will initially be guided by passive anti-radiation systems, although an active-guided version may be completed after 2000. Sources interviewed at the Zhuhai '98 show indicated that this missile eventually will have an anti-tactical ballistic missile (ATBM) capability. For the future, this raises the question of whether the PLAAF will inherit anti-missile missions.

PLA Army Aviation Corps

In 1988 the Army took control of most of the PLAAF's helicopters to form the PLA Army Aviation Corp (PLA-AAC), and by 1996 they were organised into five helicopter Brigades with selected Group Army formations (Beijing Shi, Shanyang, Tianjin, Guangzhou and Shanghai Shi). Rather late, the PLA is actively developing doctrine and tactics for helicopter assault and ground-support operations; heliborne assaults were part of the March 1996 combined-arms exercises near Taiwan. A large fleet of Harbin Z-5s (Mi-4) was supplemented in the late 1960s by Soviet Mi-8 medium-lift and a small number of Mi-6 heavy-lift helicopters stolen from rail shipments to Vietnam. In the 1980s China purchased 24 Sikorsky S-70 medium helicopters which now suffer from a lack of spare parts. During this period China secured co-production rights for three French helicopters: Harbin produces several versions of the Dauphin as the Z-9; Changhe makes the Super Frelon (Z-8) and in 1996 it was revealed to be producing a version of the Ecureuil (Z-11). Harbin's latest Z-9G is a dedicated attack helicopter equipped with advanced sights, anti-tank missiles and new TY-90 anti-aircraft missile. Russia has also sold about 24 Mi-17 medium transport helicopters and may be a source for a future PLA advanced attack helicopter. Russia is known to have marketed versions of the Mi-24 and Kamov Ka-50 modified with French avionics. There is also a good possibility that China may be developing its own advanced attack helicopter, as seen by recent reports of Chinese interest in the South African Rooivalk and the Italian Agusta A 129.

Above: These Harbin Z-9s are part of the Hong Kong garrison, established in the former UK colony after it was handed back to the Chinese in 1997. The Z-9 is a licence-built version of the Eurocopter AS 365. The Z-9Z is equipped with wire-guided missiles.

Below: First flying on 16 December 1994, the Changhe Z-11 is an Ecureuil derivative which entered service in 1997. It is used for a variety of utility roles, including training.

Current equipment PLAAF and PLA-AAC

Type	Western designation	Role	In service	No. of regiments
Z-11	AS 355	utility	–	–
Z-9	SA 365	medium transport	60	2
Z-8	SA 321	heavy transport	75	3
Z-6	Mi-4 modified	transport	100	4
Z-5	Mi-4	transport	250	10
Mi-17	Mi-17	assault	24	1
Mi-8	Mi-8	transport	30	1
Mi-6	Mi-6	transport	>10	
S-70C-II	S-70C-II	transport	24	1
AS 332M	AS 332	VIP transport	6	#
Bell 214	Bell 214	VIP transport	4	#
# = one VIP squadron				

PLA Navy-Air Force

Since at least 1953 the PLA Navy-Air Force has performed fleet defence and coastal patrol duties. It is organised in all provinces along the coast, following the Navy's division into the Northern, Eastern and Southern Fleets. It is also the principal air arm to be used in the South China Sea, especially from the base on Woody Island in the Paracel chain, which has a 7,000-ft (2134-m) runway.

The PLAN-AF mainly uses aircraft already in service with the PLAAF. Fighter units use the J-6, J-7 and J-8II, while attack and bomber units use the Q-5, H-5 and H-6, the latter specially fitted with radar and C-601 anti-ship missiles. In recent years the PLAN-AF has acquired 24 J-8D fighters equipped with refuelling probes to be used in conjunction with H-6 tankers, extending the J-8's radius to over 600 miles (965 km). The initial batch of JH-7 attack fight-

ers has gone to the PLAN-AF, conferring a much greater interdiction capability when armed with C-802 anti-ship missiles or the supersonic Kh-31/KR-1 missile. These new fighters will also soon benefit from Y-8 transports now being outfitted with British Racal Searchwater AEW systems.

The PLAN-AF took to sea first with the Z-8 helicopter. It has since acquired a small number of Z-9s for frigates and destroyers, some of which may have radar and some anti-submarine capability. The Z-9 is also likely to be the first helicopter to carry the new C-701 TV-guided anti-ship missile. In 1997 China purchased at least eight Russian Ka-27 ASW helicopters, which will be used from China's soon-to-be acquired 'Sovereminyi'-class missile destroyers and the new 'Luhu'-class destroyer. PLAN-AF helicopters will use a new 10,000-ton training ship similar to the British *Argus*. The PLAN dearly wants an aircraft-carrier but it is not clear that the government is ready to pay for it. Meanwhile, the PLAN is studying conventional and V-STOL carrier options. The former could be built around a J-10 or a Sukhoi-centred carrier wing. The PLA-AF lacks a dedicated long-range patrol aircraft, but has expressed an interest in the Russian Beriev Be-200 turbofan-powered amphibian.

Below: The Harbin PS-5 serves in small numbers on maritime patrol duties. It is equipped with MAD, radar and sonar for the ASW mission.

The latest type to enter PLANAF service is the Xian JH-7, a powerful two-seat fighter-bomber. The large nose radar acquires targets for the sea-skimming C-802 anti-ship missile.

The PLANAF's main shipborne helicopter is the Changhe Z-8, a licence-built version of the SA 321J Frelon. It is equipped with nose search radar.

Current equipment PLANAF

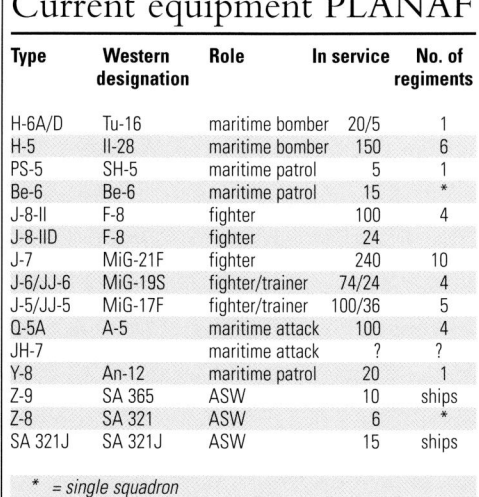

Type	Western designation	Role	In service	No. of regiments
H-6A/D	Tu-16	maritime bomber	20/5	1
H-5	Il-28	maritime bomber	150	6
PS-5	SH-5	maritime patrol	5	1
Be-6	Be-6	maritime patrol	15	*
J-8-II	F-8	fighter	100	4
J-8-IID	F-8	fighter	24	
J-7	MiG-21F	fighter	240	10
J-6/JJ-6	MiG-19S	fighter/trainer	74/24	4
J-5/JJ-5	MiG-17F	fighter/trainer	100/36	5
Q-5A	A-5	maritime attack	100	4
JH-7		maritime attack	?	?
Y-8	An-12	maritime patrol	20	1
Z-9	SA 365	ASW	10	ships
Z-8	SA 321	ASW	6	*
SA 321J	SA 321J	ASW	15	ships

* = single squadron
ships = helicopters based on navy vessels

INDONESIA

Capital: Jakarta
Population: 179.4 million
Land area: 1.9 million km² (740,905 sq miles)
Major cities: Bandung, Banjarmasin, Madiun, Padang, Palembang, Surabaja, Ujung Pandang

Indonesia's main fighter base is at Madiun-Ishwahyudi on the island of Java, from where single squadrons of F-5Es (illustrated) and F-16s operate. The F-5 user is SkU.14.

An Indonesian oddity is the Boeing 737-2X9 Surveiller, which carries antennas for the Motorola SLAMMR radar above the rear fuselage. This provides a maritime reconnaissance capability.

The A-4 Skyhawk has been central to the TNI-AU's plans for many years, and the fleet is being swelled by further purchases. SkU.11 is based at Hasanuddin on the island of Sulawesi. The A-4 has seen action on internal missions against insurgent and separatist forces, notably in East Timor.

Indonesia is the world's fifth most-populous nation, spread across 13,000 islands in a chain 3,000 miles (4828 km) long. Despite this, the Indonesian air force is one of the region's smallest, with only three combat jet squadrons. It was not always thus, since the air force was the recipient of over 170 Soviet fighters and transports during the period following independence, when President Sukarno ruled. Soviet influence was eradicated when he was removed in 1965 following an abortive Communist coup. The unpopular President Suharto was ousted in 1998 following civil unrest, fuelled by chronic shortages caused by the Asian recession.

All the air force squadrons were concentrated into two operational commands on 1 April 1985: Komando Operasi Angkatan Udara (KOOPSAU) I in Jakarta, for the western part of Indonesia, and KOOPSAU II in Ujung Pandang, for the eastern part of Indonesia.

FIGHTER FORCE

Over the past years, the air force has repeatedly expressed its need for additional fighter aircraft in order to protect the archipelago, comparing its small force with those of Singapore and Malaysia.

The current major upgrade programme involves the eight F-5Es and four F-5Fs of SkU.14. On 8 June 1995 a \$40 million contract was signed with SABCA of Belgium. The programme is called MACAN (Indonesian for 'tiger'), which stands for Modernisation of Avionics Capabilities for Armament and Navigation. It includes a radar upgrade, Litton INS, GEC Marconi Avionics Sky Guardian RWR and HUD/WAC, HOTAS controls and a Mil Std 1553B databus. The first two aircraft, F-5E TS-0501 and F-5F TL-0516, arrived at Gosselies on 31 May 1995 for a period of 18 months. The remainder of the TNI-AU's F-5s will be upgraded in Indonesia with SABCA support, all to act as F-16 lead-in trainers.

With the delivery on 11 December 1989 of the first of eight F-16As and four F-16Bs, ordered in August 1986, Indonesia started preparing its forces for the 21st century. The aircraft replaced the OV-10F in SkU.3, the premier fighter squadron. The Broncos were used to reactivate SkU.1, a former bomber unit.

Indonesia's premier fighter is the F-16A/B, 20 of which are in service. The aircraft wear this unique three-tone blue/grey scheme. SkU.3 undertakes both air-to-air and attack missions.

In November 1995, the air force expressed a total requirement for 64 F-16s to equip four squadrons, and showed interest in the (28) Pakistani air force F-16s stored at the AMARC facility in the US. Following an unsolicited US offer, the TNI-AU announced the purchase of eight F-16As and one F-16B on 26 April 1996, for a total cost of \$110 million. All the aircraft will be updated to Indonesian standard and include a drag-chute housing, ILS, software update and AGM-65 Maverick capability. The TNI-AU will now have a total of 16 F-16As and four F-16Bs, as F-16B TS-1604 had crashed by 1994. The order for 12 Sukhoi Su-30Ks has probably spelled the end of F-16 procurement for Indonesia, although the Sukhoi deal has stalled through lack of funds, perhaps permanently.

ATTACK FORCE

A total of 32 former Israeli DF/AF A-4Es and TA-4Hs were delivered between 1980 and 1982, forming SkU.11 and SkU.12 in the tactical fighter

Tentara Nasional Indonesia – Angkatan Udara

KOOPSAU I

Skwadron Udara 2	CN.235M-100, F27-400M Friendship	Halim-Perdanakusuma
Skwadron Udara 6	Sikorsky S-58T, NBO 105CB/CB4	Atang Senjaya
Skwadron Udara 8	NAS 330L Puma	Atang Senjaya
Skwadron Udara 12	Hawk Mk 109/209	Pekanbaru
Skwadron Udara 17	F27-400M Friendship, L-100-30/C-130H-30 Hercules, 707-3MC1, F28 Fellowship 3000/R, NAS 332L1 Super Puma	Halim-Perdanakusuma
Skwadron Udara 31	C-130H-30 Hercules	Halim-Perdanakusuma

KOOPSAU II

Skwadron Udara 1	Hawk Mk 109/209, OV-10F Bronco	Abdulrachman Saleh
Skwadron Udara 3	F-16A/B-15 OCU	Ishwahyudi
Skwadron Udara 4	NC.212-100/200 Aviocar, Cessna 401A/402	Abdulrachman Saleh
Skwadron Udara 5	737-2X9 Surveiller	Hasanuddin
Skwadron Udara 8	SA 330J	Atang Senjaya
Skwadron Udara 11	TA-4H/J, A-4E Skyhawk	Hasanuddin
Skwadron Udara 14	F-5E/F Tiger II	Ishwahyudi
Skwadron Udara 15	Hawk T.Mk 53	Abdulrachman Saleh

(aircraft borrowed from SkaPen 103)

Skwadron Udara 32	K/C-130B/H Hercules	Ishwahyudi

Wing Pendidikan 1/Sekolah Penerbang

Skwadron Pendidikan 101	FFA AS 202/18A-3 Bravo, T-41D Mescalero	Adisumarmo
Skwadron Pendidikan 102	T-34C-1 Turbo Mentor	Adisumarmo
Skwadron Pendidikan 103	Hawk T.Mk 53	Ishwahyudi
Akademi ABRI Bagian Udara	Cessna 172	Adisucipto
Satuan Udara Pertanian	PC-6/B Turbo-Porter	Halim-Perdanakusuma

MD 500MD	12	utility
Sukhoi Su-30K	12	(on order)*
Transall C.160NG	3	transport
Transall C.160P	3	transport
SME MD3-160	20	(on order
IPTN (Bell) 412	1	(on order)
+ gliders		

* order postponed indefinitely or possibly even cancelled

Glossary

Akademi Angkatan Bersenjata Republik Indonesia Bagian Udara (Armed Forces Academy)
KOOPSAU   Komando Operasi Angkatan Udara (Air Force Operational Command)
Skuadron Pendidikan   (Training Squadron)
Skuadron Udara   (Air Squadron)
Satuan Udara Pertanian

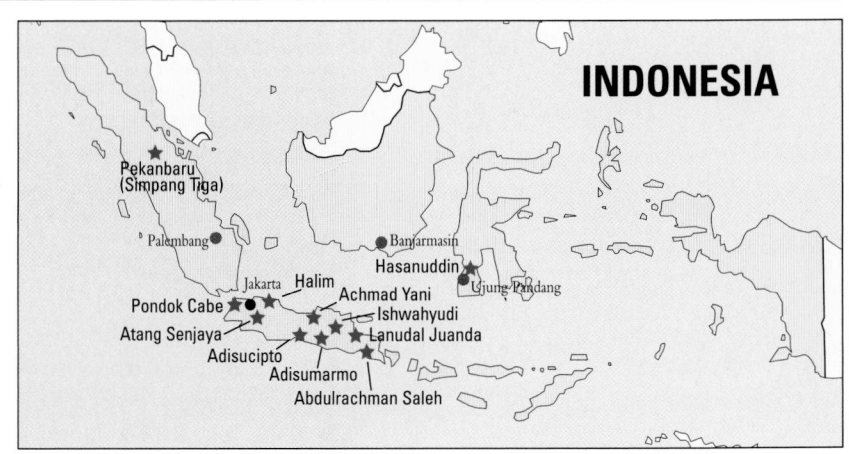

INDONESIA

Pekanbaru (Simpang Tiga)

Palembang

Banjarmasin

Hasanuddin

Ujung Pandang

Jakarta Halim

Pondok Cabe

Achmad Yani

Ishwahyudi

Atang Senjaya

Lanudal Juanda

Adisucipto

Adisumarmo

Abdulrachman Saleh

Indonesia was an early customer for the Hawk, buying 20 Mk 53s in several batches. Experience with these in the advanced training role led to the adoption of the Hawk Mk 109/209 for the attack role, a total of 24 being ordered with a stated requirement for another 72.

role. After 14 years of intense (combat) use, the A-4s of SkU.12 were grounded in late 1995, and the best aircraft from approximately 27 survivors are now concentrated in SkU.11. The OV-10Fs, delivered in 1976/77 and now part of SkU.1, were withdrawn from service during 1996. The 12 remaining examples are veterans of numerous operations in East Timor and Irian Jaya, but were also used as IPTN chase planes.

In June 1993, the TNI-AU ordered eight Hawk Mk 109s and 16 Hawk Mk 209s for £500 million. The total requirement for the next 25 years is 96 armed Hawks in eight squadrons. In order to accommodate the new Hawks, the air base at Pekanbaru, home of SkU.12, was upgraded from 1993 and received a new hangar, aircraft shelters, workshops, simulator and housing for the RAF instructor pilots. The first five TNI-AU pilots were sent to England in late 1995, and two returned to Pekanbaru on 17 May 1996, together with the first three Mk 109s.

The second delivery was planned for 27 May 1996, and included the first three Hawk Mk 109s for SkU.1. The last of these 24 Hawks was delivered by January 1997. In FY 1996/97, $676 million had been reserved for a second batch of Hawks, and in June 1996 an option for 16 Hawk Mk 209s was exercised by the TNI-AU. SkU.1 is intended to relocate to Pontianak, West Kalimantan, by 1999, thus extending the TNI-AU's combat reach to the widely-disputed Spratley Islands.

TRANSPORT FLEET

On the transport front, the six remaining F27-400Ms of SkU.2 have been supplemented by IPTN-produced CN.235M-100s. The first two paratroop variants arrived on 12 January 1993, followed by two LAPES (low-altitude parachute extraction system) aircraft in 1993 and two medevac aircraft in 1994. The VIP unit SkU.17 has also added a number of aircraft to its fleet. Two IPTN NAS 332L1s (VVIP models) were delivered on 22 February 1993, followed by two ex-Garuda F28-3000Rs during December 1994. Finally, the two ex-Merpati L-100-30s were passed on to SkU.17 by August 1995, after overhaul at Bandung.

SkU.4 has passed on six of its 10 NC.212s to other branches of the armed forces. The police received the two old NC.212M-100s on 1 July 1995, while PENERBAD (the army) and DISNERBAL (the navy) were each to receive two NC.212M-200s.

AIRBORNE RADAR MODERNISATION

The 1990s will witness a steady upgrading and expansion. In 1989 a $117 million contract was signed with Boeing to upgrade the three 737-2X9s in use with SkU.5. The modifications included an update of the Motorola SLAMMR (side-looking airborne modular multi-mission radar), a new nose radar, infrared detection system, GPS, IFF and improved data processing and displays. Aircraft AI-7301 was modified by Boeing and delivered by October 1993. The

Primary training is accomplished using the FFA AS 202A3 Bravo. This lightplane was designed by SIAI-Marchetti but built in Switzerland. Forty were ordered for Indonesia, serving with SkaDik.101. Pupils then progress to the T-34C Turbo Mentor.

remaining two (AI-7302 and AI-7303) were modified by IPTN at the Air Force Maintenance Depot 010 at Bandung. AI-7302 was completed by April 1994, and AI-7303 by late 1994.

PILOT TRAINING PROGRAMME

Would-be officers enter the Air Force Academy for a period of three years, during which they receive basic military training and undertake studies. They receive their first flying lessons on gliders and in the AS 202. Basic technical knowledge is acquired with the Ground School, SkaDik. 104. Those who pass start their Elementary Training (Latih Mula, LM) with SkaDik. 101 at Adisumarmo, Surakarta. Here they undertake 40 hours' flying in the AS 202 Bravo (60 hours for pilots destined for the Naval Aviation and Police Services).

Dinas Penerbangan Angkatan Laut (Naval Aviation Service)

Three types are operated by 800 Skuadron Udara (RON 800) for maritime patrol. The naval air arm received 12 GAF Nomad Searchmaster B and six Searchmaster L twin-turboprops from 1975-79 to form a maritime patrol squadron. Based at the naval headquarters, Surabaya, the Nomads are also detached to Tanjung Pinang and Manado, and are being supplemented by the ex-Australian army aircraft grounded in 1995. Six NC.212-MPAs ordered as part of the On Top II programme joined the squadron in mid-1996. At least one of the Aviocars is fitted for Elint duties. Six longer-range CN.235-MPAs, part of a $151 million joint TNI-AU and TNI-AL order revealed in June 1996, are in the process of delivery to RON 800.

For ASW/ASV operations, 10 Westland Wasp HAS.Mk 1 helicopters were acquired from the Netherlands and refurbished by Westland. They fly from three former Royal Navy 'Tribal'-class frigates. Six anti-submarine helicopters are required to be operated from four new naval vessels. The Westland Super Lynx is the preferred choice, but has not been ordered yet. They would replace the Wasps with RON 400, four of which are thought to remain airworthy, and will operate next to IPTN NB-412S helicopters delivered in 1989/90.

Following receipt of four NAS 332B Super Puma transport helicopters from IPTN, the naval air arm embarked on an ambitious programme to acquire 22 radar- and Exocet missile-equipped NAS 332Fs from the same source. The French machines are capable of flying from the navy's four LSTs. Six IPTN NBO 105CBs are used for SAR and liaison. Another four NBO105Ss are on order.

Providing transport to the navy, RON 600 replaced its last two C-47s with two ex-UAE DHC-5 Buffalos. A pair of former TNI-AU NC.212M-200 transports was also delivered in early 1996.

The Akademi TNI-AL (AAL) at Morokrembangan (Bumi Moro), Surabaya, houses the Sekolah Penerbang TNI-AL (SENERBANG), or Pilot School. Students first enter Ground School, and fixed-wing pilots then continue with RON 200.

Students then return to Adisucipto for 110 hours of Basic Training (Latih Dasar, LD), with SkaDik.102 on the T-34C Turbo Mentor. Finally, they move to the large air base of Iswahyudi, Madiun, where they join SkaDik.103 for 90 hours of Advanced Training (Latih Lanjut, LL) on the BAe Hawk Mk 53.

Those students who successfully complete the course become pilot officers and are divided into three groups: fighter, transport and helicopter pilots. Fighter pilots undergo conversion at their assigned squadron (but SkU.11 is the dedicated A-4 conversion squadron). Transport pilots generally begin with SkU.4, and helicopter pilots all go to SkU.7 for their transition and training on the Bell 47G and H500.

The TNI-AU has strong ties with civil aviation in Indonesia. Before students graduate from the academy, they all have the chance to take the Commercial Pilot Licence Course at Curug. Many Garuda and Merpati airline pilots are from the TNI-AU, and they also act as instructors with several civil pilot schools in Java.

Dinas Penerbangan Angkatan Laut (Naval Aviation Service)

Skwadron Udara (RON) 200/Laith	**Lanudal Juanda**
F33A Bonanza, PA-34 Seneca, PA-38-112 Tomahawk Rockwell Lark Commander 100, TB 9 Tampico	
Skwadron Udara (RON) 400/Heli	**Lanudal Juanda**
IPTN (Bell) 412, IPTN (Eurocopter) NBO 105CB, NAS 332B/F Super Puma, SA 313 Alouette II, Wasp HAS.Mk 1	
Skwadron Udara (RON) 600/Angkut	**Lanudal Juanda**
DHC-5 Buffalo, NC.212-100/200 Aviocar,	
Skwadron Udara (RON) 800/Patroli Maritim	**Lanudal Juanda**
Airtech CN.235M, Nomad N22/24, NC.212-200 Elint/MPA	
IPTN (Eurocopter) NBO 105S	4 (on order)

Here they receive 80 hours of elementary training on the DC-100 and F-33A, and 40 hours of basic training on the PA-38. Advanced training is type related, and takes place with one of the other fixed-wing squadrons. There, they start as co-pilots, and after a minimum of 700 hours qualify as captains. Helicopter pilots receive all their training with RON 400.

The TNI-AL would like to order a modern seagoing helicopter such as the Lynx, but in the current financial climate will have to soldier on with the few remaining Wasps (above).

Skuad.2 operates a mixed bag of fixed-wing types in support of the army. This NC.212 is being used for parachute training.

Dinas Penerbangan Angkatan Darat (Army Aviation Service)

Although large, the army operated few helicopters until the mid-1970s, but has since more vigorously embraced the airmobile doctrine. Over the past 10 years, PENERBAD has primarily focused on expanding its helicopter fleet. On 8 September 1988, four IPTN NB-412Ss were delivered to the attack helicopter squadron, Skuad.1, supplemented in 1995 by three IPTN NB-412HP (High Performance) models. Also delivered to Skuad.1 were six IPTN NBO 105CB-4s, between 1990 and 1994. PENERBAD plans to purchase 20 Bell 205s on the US civil market and convert them for military use, starting in 1996. Eight Mil Mi-171V 'Hips' are also on order.

Three (of five) ex-UAE air force DHC-5 Buffaloes

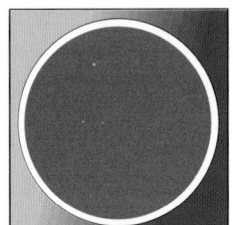

The locally-built NBO 105 serves the army as both a light attack helicopter and as a trainer.

delivered to IPTN in September 1995 were to be transferred to the general support squadron, Skuad.2, in March 1996, replacing DC-3s withdrawn by October 1995. Early in 1996, PENERBAD's Skuad.2 also received two NC.212M-200s from the TNI-AU.

Pilot training starts with 90 hours of combined primary and basic training on the H300C. Those students who are to become fixed-wing pilots go to the civil Juanda Flying School (JFS) in Surabaya. After their course there, they return to SEBANG for 70 hours of advanced training on the CN.212, BN-2A Grand- or Turbo Commander. Helicopter pilots receive 70 hours of advanced training on the Bell 206, NBO 105 or Bell 205. After graduation, they all begin as co-pilots. They will need 450 flying hours (including 350 hours type-rated) to qualify for the captain-pilot course. This course includes another 60 hours of flying training.

Dinas Penerbangan Angkatan Darat

Skwadron Udara Angkatan Darat (Skuad) 1/Heli Serbu
(Attack Helicopter Squadron), Achmad Yani
Bell 205A-1, NBO 105C/CB, IPTN/Bell 412HP/S, SA 316B
Alouette III
Skwadron Udara Angkatan Darat (Skuad) 2/Bantuan Umum
(General Support Squadron), Pondok Cabe
DHC-5D Buffalo, NC.212-200 Aviocar, BN-2A Islander,
Commander 680FL, Turbo Commander
Sekolah Penerbang (Flying School), Achmad Yani
Bell 205A-1, Bell 206 JetRanger, NBO 105C/CB, Schweizer-
Hughes 300C

Tentara Nasional Indonesia - Angkatan Kepolisian (Indonesian National Defence – Police Force)

The service's main tasks are SAR, medevac and air detection. Aircraft have been used since at least the mid-1950s. Types that have been used include an ex-Shell Grumman G-73 Mallard, two Rockwell Aero Commander 500A/560, Beech Super 18, Piper PA-31T and Cessna 180 and 185. Based at Pondok Cabe, the mainstay of today's fleet are IPTN-built BO 105s, supplemented by a pair of Aviocars that arrived from the TNI-AU on 1 July 1995, becoming operational on 13 October. The service has expressed interest in NB-412s, although production had to be halted due to quality problems.

Tentara Nasional Indonesia - Angkatan Kepolisian

Subdir Polisi Udara (Police Air Squadron)
Bell 206B JetRanger (1), Cessna U206 (1), IPTN (Eurocopter)
NBO 105CB (21), IPTN (CASA) NC.212M-100 Aviocar (2)

JAPAN

Capital: Tokyo
Population: 123.7 million
Land area: 369700 km² (142,705 sq miles)
Major cities: Kobe, Kita-Kyushu, Kyoto, Nagoya, Osaka, Sapporo, Yokohama

Koku-Jieitai (Japan Air Self-Defence Force, JASDF)

Koku-Jieitai was established on 1 July 1954 as one of the three defence forces, having the responsibility for air defence. In Boei-keikaku no Taiko (National Defence Programme Outline), the JASDF must have 10 intercept-fighter (FI) squadrons and three support fighter (FS) squadrons.

In FY75, 10 FI squadrons consisted of six with the F-4EJ and four with the F-104J/DJ. All three FS squadrons were equipped with the F-86F, replaced by increasing numbers of F-104Js and F-4EJs since the late 1960s. Because of the age of the F-86F, the JDA opted to develop an original support fighter at the end of the 1960s. It was decided to save development and procurement costs by developing both an advanced trainer and a support fighter from one basic airframe. Thus was born the Mitsubishi T-2 and F-1. Two prototypes of the F-1 were made by modifying T-2s, called T-2 Tokubetu Siyo-ki (Special Spec. T-2), which made a first flight on 3 June 1975. After series tests, 77 production F-1s were built, forming three squadrons between FY77 and FY80.

On 28 December 1977, the McDonnell Douglas F-15 was named as the new interceptor fighter to replace, initially, the F-104, and then the F-4EJ. The single-seat F-15J is almost identical to the F-15C used by the US Air Force, and likewise the dual-seat F-15DJ to the F-15D. Due to the withholding of part of the American electronic warfare suite from Japanese F-15J/DJs, the aircraft lack the TEWS and use instead the Japanese system consisting of the

J/APR-4 RWR, AN/ALE-45 chaff/flare dispenser, J/ALQ-8 ECM and J/APQ-1 rear warning system; it is known as JTEWS. The procurement of F-15J/DJs started in FY78 and continued until FY96. In total, 213 F-15J/DJs were purchased and equipped mainly FI squadrons. The first two F-15Js and also the first 14 F-15DJs were made by McDonnell Douglas, while the rest were manufactured by Mitsubishi Heavy Industries under licence.

The JASDF has 15 sub-divisions, including commands, groups and a hospital, all under the direct control of the Chief of Staff, Air Self-Defence Force. Most aircraft are assigned to four major commands, Koku Sotai (Air Defence Command), Koku Sien Shudan (Air Support Command), Koku Kyoiku Shudan (Air Training Command) and Koku Kaihatu Jikken Shudan (Air Development and Test Command).

Koku Sotai (Air Defence Command, ADC), with headquarters located at Fuchu, Tokyo, is the front-line defence force. ADC divides Japan into four regions, each with fighter wings, AC&W wings and air defence missile groups. Three out of four regional air defence forces are called Homentai. Hokubu Koku Homentai (Northern Air Defence Force) covers northeast Japan, Chubu Koku Homentai (Central Air Defence Force) covers central Japan, and Seibu Koku Homentai (Western Air Defence Force) covers western Japan. The other is the Nansei Koku Konsei-dan (Southwestern Composite Air Division), covering Okinawa and the west, and only this division has a single FI squadron. Every Homentai has two fighter wings, each of which has two fighter squadrons.

Currently, 10 FI squadrons consist of eight with F-15s and two with F-4EJ Kais. Two of the eight F-15 squadrons (Dai 201 and Dai 203 Hiko-tai –

Japan's Eagle force consists of eight front-line squadrons spaced strategically around the nation. Deliveries totalled 163 F-15J single-seaters and 50 F-15DJ two-seaters. This special-scheme example is from 304 Hiko-tai.

Above: A pair of F-1s from 8 Hiko-tai escorts a 601 Hiko-tai E-2C. Both types are based at Misawa AB in the north.

Left: Various RF-4E variants are flown by 501 Hiko-tai for reconnaissance.

Below: The replacement for the Mitsubishi F-1 is the F-2, based on the F-16. This is the fourth prototype, an XF-2B.

JAPAN

Hiko-tai meaning squadron) are located at Chitose Air Base under the control of Hokubu Koku Homentai, two squadrons each at Komatsu AB (Dai 303 and Dai 306 Hiko-tai) and Hyakuri AB (Dai 204 and Dai 305 Hiko-tai) both under the control of Chubu Koku Homentai, and the final two F-15 squadrons are located at Tsuiki AB (Dai 304 Hiko-tai) and Nyutabaru AB (Dai 202 Hiko-tai) under the control of Seibu Koku Homentai. Dai 202 Hiko-tai based at Nyutabaru AB also has the role of operational conversion for pilots as well as its routine FI squadron's roles.

Two F-4EJ Kai squadrons are located at Nyutabaru AB (Dai 301 Hiko-tai) and Naha AB (Dai 302 Hiko-tai), the latter squadron under the control of Nansei Koku Konsei-dan. Dai 301 Hiko-tai is also the conversion unit for the F-4EJ. Both F-15 and F-4EJ Kai conversion squadrons are based at Nyutabaru AB, which consequently is referred to as the 'Mother Base of JASDF's Fighter Pilot'.

The word 'Kai' in F-4EJ Kai means modification (with upgrading). JDA began researching a means to upgrade the F-4EJ fleet in February 1982. The programme includes a change of radar, from the AN/APQ-120 to the AN/APG-66J; a new central computer – the indigenous J/AYK-1 digital unit; new INS (from AN/ASN-63 analog INS to J/ASN-4 digital INS); and radar warning receiver changed from J/APR-2 to J/APR-6A. All these changes increase the F-4EJ's attack capability and add the ability to carry the ASM-1/-2 anti-ship missile. This multi-mission capability is needed to fill the gap between the F-1's retirement and the entry into service of the new support fighter (FS-X). In total, about 90 F-4EJs were modified to F-4EJ Kai and are now operated by three squadrons. Each F-4EJ Kai unit also operates a

A few T-33As are retained by the JASDF, flying as 'hacks' with the Sotai Sireibu Hiko-tai (air defence headquarters flight) at Iruma.

few original F-4EJs for training support missions such as target towing.

The first F-4EJ Kai squadron was Dai 306 Hiko-tai at Komatsu AB. Because of the decreasing number of operational F-1s, these aircraft were transferred to Dai 8 Hiko-tai at Misawa AB, and Dai 306 Hiko-tai changed its aircraft for F-15s in FY 1996. Thus, the current three FS squadrons consist of two F-1 squadrons (Dai 3 Hiko-tai at Misawa AB and Dai 6 Hiko-tai at Tsuiki AB) and one F-4EJ Kai (Dai 8 Hiko-tai). Each FI and FS squadron has a small number of T-4 trainers as liaison aircraft, except the F-1 squadrons, which use T-2s.

Replacement of the F-1 was planned as the FS-X programme and a decision on a modified version of the F-16 was made on 21 October 1987. The changes from the original F-16 include installation of the GE F110-GE-129 engine, a new wing with an increased area (35.0 m²/115 sq ft) made of a one-piece composite material with radar-absorbent material at the

The Kawasaki T-4 has completely replaced the Fuji T-1 in advanced training and fighter squadron 'hack' duties. These are from 1 Koku-dan.

leading edge, application of advanced material and structural technologies to the stretched fuselage and tailplanes, advanced avionics (such as active phased-array radar leading to a changed nose shape, mission computer, inertial reference system and integrated electronic warfare system), a strengthened windshield, and the addition of a drag chute. Fitting two canard wings to the underside of the front fuselage was also considered but later rejected. The FS-X was officially named the F-2.

The first prototype F-2 (XF-2) flew for the first time on 7 October 1995. Four flying XF-2s were constructed for the series flight tests along with two static test airframes. The first two XF-2s are single-seat XF-2As, while Nos 3 and 4 are dual-seat XF-2Bs. The four XF-2s are now undergoing flight tests with the Hiko Kaihatsu Jikkenn-dan (Air Development and Test Wing) at Gifu AB. First production F-2s will be delivered to an operational squadron (Dai 3 Hiko-tai at Misawa AB is planned) in FY 1999 and will

The JASDF was the first customer for the Boeing E-767 – essentially the system from the E-3 AWACS installed in a 767 airframe. The operational squadron formed in 1999 with its four aircraft, based at Hamamatsu.

achieve full operational capability in FY 2000. Three F-2 squadrons will be formed as FS squadrons by FY 2004, with the JASDF also planning to use F-2s as an advanced trainer with the Hiko Kyodotai (Tactical Fighter Training Group). Therefore, total procurement of F-2s will probably reach around 130 aircraft.

The JASDF is now planning to upgrade its F-15Js, for which prototype development works started in FY 1997. This upgrade programme includes changing the radar and central computer to the latest American equipment. Japanese-developed expendable ECM and IRST equipment are planned for the aircraft. A prototype will fly in FY 2001 and about 100 F-15Js are planned to be upgraded in total.

Peacetime air defence is provided by the JASDF using fighter aircraft (including support fighter units) on alert. Standing by to scramble at each fighter base are two pairs of aircraft, making four pairs of aircraft in each Homentai and two pairs in the Southwestern Composite Air Division. Only one pair in each region is on a five-minute alert status, and if it were to be scrambled, another pair would then change its alert status to five minutes. These alert fighters carry two live IR-guided AAMs (AIM-9L or AAM-3) and Vulcan munitions. The JASDF also deploys 24 Kosha-tai (Air Defence Missile Squadron) in every district in Japan. All Kosha-tai have completed re-equipping their missile batteries from the Nike-J system to the Patriot.

Units reporting directly to the Koku Sotai include Sireibu Hiko-tai (ADC Headquarters Squadron) at Iruma AB, Teisatsu Koku-tai (Tactical Reconnaissance Group) at Hyakuri AB, Hiko Kyodo-tai (Tactical Fighter Training Group) at Nyutabaru AB, and Keikai Koku-tai (Airborne Early Warning Group) at Misawa AB.

Headquarters Squadron operates T-33As, T-4s and Gulfstream U-4s for transportation of command staff

Similar to the BAe C-29A operated by the US FAA, the U-125 is a navigation aids checking aircraft based on the BAe 125-800. Three serve with the Hiko Tenken-tai at Iruma, alongside three YS-11FCs configured for the same mission.

For VVIP transport (principally of the Japanese Royal Family), 701 Hiko-tai operates two Boeing 747-47Cs from Chitose. They are fitted with additional communications equipment.

and liaison. This squadron also has the Denshi-sen Sien-tai (ECM Support Unit) under its control, equipped with an EC-1 and YS-11E/AE for ECM training of ground-base radar sites and two YS-11ELs for the Elint mission.

Teisatsu Koku-tai is the only tactical reconnaissance unit in the JASDF and has one squadron (Dai 501 Hiko-tai) which is equipped with RF-4s. JDA purchased 14 RF-4Es in FY 1973 and received them in FY 1975 and FY 1976, bought directly from USA. Upgraded to RF-4E Kai to increase mission effectiveness, the main modifications are a new radar, AN/APQ-172 digital image process radar replacing the AN/APQ-99, and RWR changing from J/APR-2 to J/APR-5. Because of budget restrictions, not all RF-4E Kais currently have the AN/APQ-172, and modification of radars is ongoing.

Teisatsu Koku-tai also increased its reconnaissance aircraft numbers by adding RF-4EJs converted from F-4EJs for the tactical reconnaissance role. Not all RF-4EJs have internal reconnaissance equipment but they are capable of carrying three types of sensor pods which were developed specially for it: (1) Tactical Reconnaissance (TAC) pod, (2) Long-Range Oblique Photography (LOROP) pod, and (3) Tactical Electronic Reconnaissance (TACER) pod. Seventeen F-4EJs are planned to be converted to RF-4EJ standard and all are assigned to Dai 501 Hiko-tai alongside the RF-4E/RF-4E Kais.

Keikai Koku-tai at Misawa AB currently operates 13 E-2Cs. The aircraft gained operational capability in FY 1983 and were gradually upgraded by, for example, changing the surveillance radar. Now, all E-2Cs of the JASDF have the AN/APS-138 radar

JASDF

Koku Sotai (Air Defence Command)	**HQ Fuchu**
Hokubu Koku Homentai **(Northern Air Defence Force)**	**HQ Misawa**
Dai 2 Koku-dan (2nd Air Wing)	**Chitose AB**
Dai 201 Hiko-tai (201st Squadron)	F-15J/DJ, T-4
Dai 203 Hiko-tai (203rd Squadron)	F-15J/DJ, T-4
Dai 3 Koku-dan (3rd Air Wing)	**Misawa AB**
Dai 3 Hiko-tai (3rd Squadron)	F-1, T-2
Dai 8 Hiko-tai (8th Squadron)	F-4EJ Kai, T-4
Hokubu Sien Hiko-han (Northern Air Support Flight)	T-4
Chubu Koku Homentai **(Central Air Defence Force)**	**HQ Iruma**
Dai 6 Koku-dan (6th Air Wing)	**Komatsu AB**
Dai 303 Hiko-tai (303rd Squadron)	F-15J/DJ, T-4
Dai 306 Hiko-tai (306th Squadron)	F-15J/DJ, T-4
Dai 7 Koku-dan (7th Air Wing)	**Hyakuri AB)**
Dai 204 Hiko-tai (204th Squadron)	F-15J/DJ, T-4
Dai 305 Hiko-tai (305th Squadron)	F-15J/DJ, T-4
Seibu Koku Homentai **(Western Air Defence Force)**	**HQ Kasuga**
Seiku Sien Hiko-tai (WADF HQ Support Flight)	Kasuga, T-4
Dai 5 Koku-dan (5th Air Wing)	**Nyutabaru AB**
Dai 202 Hiko-tai (202nd Squadron)	F-15J/DJ, T-4
Dai 301 Hiko-tai (301st Squadron)	F-4EJ Kai, F-4EJ, T-4
Dai 8 Koku-dan (8th Air Wing)	**Tsuiki AB**
Dai 304 Hiko-tai (304th Squadron)	F-15J/DJ, T-4
Dai 6 Hiko-tai (6th Squadron)	F-1, T-2
Nansei Koku Konsei-dan **(Southwestern Composite Air Division)**	**HQ Naha**
Dai 83 Koku-gun (83rd Air Group)	**Naha AB**
Dai 302 Hiko-tai (302nd Squadron)	F-4EJ Kai, F-4EJ, T-4
Nansei Sien Hiko-han (SW Support Flight)	T-4, B-65
Chokkatu-butai (Direct Reporting Unit)	
Sotai Sireibu Hiko-tai (ADC HQ Flight Group)	Iruma AB, T-4, T-33A, U-4
Densi-sen Sien-tai (ECM Support Unit)	Iruma AB, YS-11E/AE/EL, EC-1
Teisatu Koku-tai **(Tactical Reconnaissance Group)**	**Hyakuri AB**
Dai 501 Hiko-tai (501st Squadron)	RF-4E, RF-4E Kai, RF-4EJ, T-4
Hiko Kyodo-tai (Tactical Fighter Training Group)	Nyutabaru AB, F-15DJ, T-4
Keikai Koku-tai **(Airborne Early Warning Group)**	**Misawa AB**
Dai 601 Hiko-tai (601st Squadron)	E-2C
Koku Sien Shudan (Air Support Command)	**HQ Fuchu**
Koku Kyunan-dan (Air Rescue Wing)	HQ Iruma
Kyunan Kyoiku-tai (Air Rescue Training Squadron)	Komaki AB, UH-60J, U-125A
Chitose Kyunan-tai (ARS Chitose)	Chitose AB, UH-60J, U-125A
Matsushima Kyunan-tai (ARS Matsushima)	Matsusima AB, KV-107, MU-2S
Hyakuri Kyunan-tai (ARS Hyakuri)	Hyakuri AB, UH-60J, MU-2S
Hamamatsu Kyunan-tai (ARS Hamamatsu)	Hamamatsu AB, KV-107, MU-2S
Akita Kyunan-tai (ARS Akita)	Akita Airport, KV-107, MU-2S
Komatsu Kyunan-tai (ARS Komatsu)	Komatsu AB, UH-60J, U-125A
Niigata Kyunan-tai (ARS Niigata)	Niigata Airport, KV-107, MU-2S
Ashiya Kyunan-tai (ARS Ashiya)	Ashiya AB, KV-107, MU-2S
Nyutabaru Kyunan-tai (ARS Nyutabaru)	Nyutabaru AB, KV-107, MU-2S
Naha Kyunan-tai (ARS Naha)	Naha AB, KV-107, MU-2S
Misawa Herikoputa Kuyu-tai (Misawa Helicopter Airlift Squadron)	Misawa AB, CH-47J
Iruma Herikoputa Kuyu-tai (Iruma Helicopter Airlift Squadron)	Iruma AB, CH-47J
Kasuga Herikoputa Kuyu-tai (Kasuga Helicopter Airlift Squadron)	Kasuga, CH-47J
Naha Herikoputa Kuyu-tai (Naha Helicopter Airlift Squadron)	Naha AB, CH-47J
Dai 1 Yuso Koku-tai **(1st Tactical Airlift Group)**	**Komaki AB**
Dai 401 Hiko-tai (401st Squadron)	C-130H
Dai 2 Yuso Koku-tai **(2nd Tactical Airlift Group)**	**Iruma AB**
Dai 402 Hiko-tai (402nd Squadron)	C-1, YS-11P/C, U-4
Dai 3 Yuso Koku-tai **(3rd Tactical Airlift Group)**	**Miho AB**
Dai 403 Hiko-tai (403rd Squadron)	C-1, YS-11P/PC/NT
Dai 41 Kyoiku Hiko-tai (41st Training Squadron)	T-400
Hiko Tenken-tai (Flight Check Squadron)	Iruma AB, YS-11FC, U-125
Tokubetu Koku Yuso-tai **(Special Air Transport Group)**	**Chitose AB**
Dai 701 Hiko-tai (701st Squadron)	Boeing 747-47C
Koku Kyoiku Shudan (Air Training Command)	**HQ Hamamatsu**
Dai 1 Koku-dan (1st Air Wing)	**Hamamatsu AB**
Dai 31 Kyoiku Hiko-tai (31st Air Training Sqn)	T-4
Dai 32 Kyoiku Hiko-tai (32nd Air Training Sqn)	T-4
Dai 4 Koku-dan (4th Air Wing)	**Matsushima AB**
Dai 21 Hiko-tai (21st Squadron)	T-2
Dai 22 Hiko-tai (22nd Squadron)	T-2
Dai 11 Hiko-tai (11th Squadron)	T-4 ('Blue Impulse')
Dai 11 Hiko Kyoiku-dan (11th Flying Training Wing)	Shizuhama AB, T-3
Dai 12 Hiko Kyoiku-dan (12th Flying Training Wing)	Hofu-kita AB, T-3
Dai 13 Hiko Kyoiku-dan (13th Flying Training Wing)	Ashiya AB, T-1A/B, T-4
Koku Kaihatsu Jikken Shudan **(Air Development and Test Command)**	**HQ Iruma**
Hiko Kaihatsu Jikken-dan (Air Development and Test Wing)	Gifu AB, T-1, T-2, T-3, T-4, C-1, F-4EJ, F-4EJ Kai, F-15J/DJ, FS-T2 Mod, T-2 CCV, XF-2A/B, E-767

(only E-767 Test Squadron is based at Hamamatsu AB)

The JASDF has three squadrons assigned to tactical transport, one equipped with the C-130H and two with the Kawasaki C-1 (illustrated). The latter are based at Iruma and Miho.

After completion of primary training in the Fuji T-3 and basic training on the Kawasaki T-4, pilots destined for transport operations undertake multi-engine training on the Raytheon/Beech T-400.

The Iruma-based Densi-sen Sien-tai (ECM support unit) operates a variety of YS-11 'specials' on electronic warfare tasks. This is a YS-11E used as an ECM 'aggressor' for ground radar training.

with enhanced high speed processor (EHSP), designated AN/APS-138 Plus.

The JDA decided to introduce an AWACS aircraft in FY 1990 and ordered four Boeing E-767s in FY 1993 and FY 1994 (two aircraft each year). The first two E-767s arrived at Hamamatsu AB on 25 March 1998 and started a one-year practical use test with the Hiko Kaihatu Jikken-dan. It is planned that an operational squadron of E-767s will be formed at Hamamatsu AB in March 1999 and that the headquarters of Keikai Koku-tai will move there at the same time, while the E-2Cs will remain at Misawa.

Hiko Kyodo-tai, stationed at Nyutabaru AB, operates seven F-15DJs and two T-4s for tactical combat training against every fighter squadron.

Koku Sien Shudan (Air Support Command) operates transport and SAR aircraft. Search and rescue units, which all have both fixed-wing aircraft and helicopter(s), are deployed to 12 bases. Main equipment is the Mitsubishi MU-2S and KV-107, but they are beginning to be replaced by the Raytheon U-125A and Sikorsky UH-60J. Koku Kyunan-dan (Air Rescue Wing) is also responsible for air transport to such sites as radar stations, using CH-47Js of the 4

Herikoputa Kuyu-tai (Helicopter Airlift Squadron).

Three Yuso Koku-tai (Tactical Airlift Groups) are stationed at Komaki AB, Iruma AB and Misawa AB. Dai 1 Yuso Koku-tai operates C-130Hs and the other two use Kawasaki C-1s and YS-11s. Dai 2 Yuso Koku-tai at Iruma AB is also receiving Gulfstream U-4s for staff and light cargo transportation. Dai 3 Yuso Koku-tai has the responsibility for training pilots of transport aircraft on its T-400s (Diamond 1). VIP transport is a role of Tokubetsu Koku Yuso-tai (Special Air Transport Group) at Chitose AB, operating two Boeing 747-400s. Hiko Tenken-tai (Flight Check Squadron) is also under the Koku Shien Shudan which is stationed at Iruma AB and operates three YS-11FCs and three Raytheon U-125s.

All crews (not only pilots but also, for example, maintenance staff) are trained in the Koku Kyoiku Shudan (Air Training Command). Current pilot training is conducted in four steps. Elementary training is conducted on piston-engined Fuji T-3s, flying about 70 hours of elementary flight training. After this, student pilots are divided for the fighter pilot course and others (transport and helicopter). On the fighter pilot course, students move to primary jet

training on the Fuji T-1, flying about 80 hours. After completing this section, they proceed to medium jet training on the Kawasaki T-4 with about 115 hours training, but some students will be trained in the United States on the T-38A, flying 80 hours. Finally, they will be gathered together for advanced training on the Mitsubishi T-2 for 140 hours.

The JASDF is reviewing its training syllabus to simplify and increase efficiency. The T-4 will be more widely used, and F-15DJs and F-2Bs will be used in advanced training roles including fighter conversion training, after the retirement of the T-2. Fuji T-3s are approaching the end of their lives and the selection of a new basic trainer – a turboprop aircraft – will soon be required.

Koku Kaihatsu Jikken Shudan (Air Development and Test Command) is the only research and development organisation in the JASDF. The flight unit, named Hiko Kaihatsu Jikken-dan (Air Development and Test Wing), at Gifu AB is equipped with most types of aircraft operated by the JASDF. This unit conducts not only trials before an aircraft enters service but also development tests of new equipment and armament.

Kaijo Jieitai (Japan Maritime Self-Defence Force, JMSDF)

The Kaijo Jieitai (JMSDF) has the responsibility of defending Japan from maritime threats, and the mission of its aircraft force is mainly anti-submarine warfare. Therefore, most of its aviation assets are anti-submarine aircraft and helicopters, and there are no fighter or attack aircraft.

The JMSDF includes the Jiei Kantai (Self-Defence Fleet), Koku Kyoiku Shudan (Air Training Command), Renshu Kantai (Training Squadron), several Chiho-tai (Regional District) and other units. The main part of the aircraft force is Koku Shudan (Fleet Air Force) which is under the control of Jiei Kantai. Koku Kyoiku Shudan are independent organisations and each Chiho-tai has one Koku-tai (squadron) or Hiko-han (flight) in its organisation.

Koku Shudan, whose headquarters are located at Atsugi, consists of seven Koku-gun (Fleet Air Wing), three Koku-tai (squadron), Koku Kansei-tai (Air Control Service Group) and Koku Sisetsu-tai (Air Construction Engineers Group). Four of the seven Koku-gun are fixed-wing aircraft wings, with P-3Cs. Two are helicopter forces and the remaining Koku-gun is a composite force. Dai 1 Koku-gun (Kanoya AB), Dai 2 Koku-gun (Hachinohe AB), Dai 4 Koku-gun (Atsugi AB) and Dai 5 Koku-gun (Naha AB) are P-3C wings and all Koku-gun have two Koku-tai (Air Patrol Squadron, VP) each. Some Koku-tai also operate Beech LC-90 King Airs for liaison duty. Each Koku-gun also has a Kichi Koku-tai (Air Base Flight) which operates small numbers of UH-60Js or S-61As tasked mainly with SAR. The coverage of Dai 4 Koku-gun includes Io-jima (which is located about 650 nm/1200 km south of Tokyo), for which a special unit operating S-61As, named Io-jima Koku Kichi-tai, is under Dai 4 Koku-gun's control.

Dai 21 Koku-gun (Tateyama AB) and Dai 22 Koku-gun (Omura AB) are helicopter wings. Dai 21 Koku-gun currently has three Koku-tai (Air Patrol Squadron Helicopter, HS) which comprise two

SH-60J squadrons and an HSS-2B/S-61A squadron; the latter will disband within a few years, after which each ASW-helicopter-equipped Koku-gun will have two HSs of SH-60Js each. The HSs are deployable to Herikoputa Goei-kan (Helicopter Destroyer, DDH), with Dai 121 Koku-tai for the DDHs of Dai 1 Goeitai-gun (1st Escort Flotilla), Dai 124 Koku-tai for Dai 2 Goeitai-gun (4th Escort Flotilla), Dai 122 Koku-tai for Dai 2 Goeitai-gunn (2nd Escort Flotilla) and Dai 123 Koku-tai for Dai 3 Goeitai-gun (3rd Escort Flotilla).

Dai 31 Koku-gun at Iwakuni AB is a composite wing which consists of one P-3C squadron, one US-1 squadron and one training support/Elint squadron. US-1As, which equip Dai 71 Koku-tai, are large search and rescue amphibian flying-boats and have SAR mission coverage all around Japan. For rapid response in northeastern Japan, one US-1A from Dai 71 Koku-tai is deployed to Atsugi AB on a permanent detachment. Dai 81 Koku-tai is a training support squadron which operates six Learjet U-36As. The U-36A simulates high-speed anti-ship missiles by

Right: Operating from its base at Iwakuni and a detachment at Atsugi, 71 Koku-tai is the sole operator of the mighty ShinMaywa US-1A amphibian, used to provide long-range SAR.

Below: The P-3C equips nine patrol squadrons, a test unit and a training squadron. Special-mission Orions also serve with 81 Koku-tai.

using HWQ-2-3T missile seeker simulators which are mounted on the top of the right wingtip tank and used in the target facilities role. The left wingtip tank contains a video camera. U-36As also can carry J/XLQ-2 ECM pod and RM-30/ARS-1P the target-towing equipment. Dai 81 Koku-tai operates one EP-3 which is internally equipped with various Elint equipment. Dai 81 Koku-tai will receive one UP-3D in FY 1998, which will have various ECM packages for electronic warfare training support. A second UP-3D will also be delivered in the near future.

Chokkatsu Butai (Direct Reporting Units) to the Koku Shudan are Dai 61 Koku-tai, Dai 111 Koku-tai, and Dai 51 Koku-tai. Dai 61 Koku-tai (VR-61) based at Atsugi AB is a transport squadron and operates YS-11s and LC-90s. YS-11s are mainly used for

Built by Mitsubishi, the SH-60J Seahawk has virtually replaced the Mitsubishi/Sikorsky HSS-2 (S-61) as the JMSDF's ASW helicopter. The helicopters regularly deploy aboard destroyers.

Six Learjet 36s were modified for JMSDF use by ShinMaywa under the designation U-36A. They are used for target facilities work, including mimicking sea-skimming missiles, by Dai 81 Koku-tai.

The JMSDF undertakes the training of all its pilots and crew. Initial flying training is performed on the Fuji T-5. During this course students are divided into pilots and other aircrew for separate training.

scheduled flights and non-regular staff transport, and the LC-90 as a liaison aircraft. Dai 111 Koku-tai (HM-111) at Iwakuni AB is a mine countermeasures helicopter squadron which operates 11 MH-53Es; it was established in February 1974 using KV-107s and started to convert to the MH-53E in FY 1990. Its mission equipment is the MK-103 Mod 2 towed system, MK-104 Mod 3 acoustic system, MK-105

Mod 2 magnetic system and MK-106 which combines the MK-104 and MK-105.

Dai 51 Koku-tai (VX-51) at Atsugi AB is a research and development squadron using P-3Cs, SH-60Js and HSS-2Bs. Its missions is to evaluate new aircraft and to conduct development tests for new airborne equipment. Each of its P-3Cs, SH-60Js and HSS-2Bs are slightly different in internal system and equipment fits.

VX-51 is currently heavily involved in the R&D for the SH-60J Kai programme, the main purpose of which is improving the ASW capabilities and providing more mission effectiveness in the modern electronic warfare environment. Enlarging the aircraft's cabin by stretching and increasing its height is also being considered. R&D work started in FY 1997 and the first prototype is due to fly in July 2000.

A unique aircraft of VX-51 is the UP-3C, specially developed for the test and evaluation of new airborne equipment and new aircraft systems. It entered service in FY 1996. UP-3s and EP-3s are derivatives of the P-3C, not conversion aircraft. These types were originally developed by the JMSDF and have been manufactured as special aircraft from the outset by Kawasaki Heavy Industries.

Four helicopter squadrons are assigned to the Chiho-tai (Regional District). Three are equipped with ASW helicopters (converting from HSS-2Bs to SH-60Js) and operate from land bases, and the other (under control of Yokosuka Chiho-tai) is a support unit for Antarctic observation activities onboard the Antarctic observation ship *Sirane*.

Kyoiku Koku Shudan (Air Training Command) trains pilots and mission crews. Pilot training courses start with basic flight training with Dai 201 Kyoiku Koku-tai (Training Squadron) on T-5 turboprop side-by-side trainers (about 110 hours), and the students will be streamed as pilots or aircrew (such as TACCO) late in the course. Student pilots move to Dai 202 Kyoiku Koku-tai with TC-90 King Airs for multi-engine and instrument flying training for about 101 hours of flying. Finally, they convert to P-3Cs with Dai 203 Kyoiku Koku-tai of Shimofusa Kyoiku Koku-gun. The Shimofusa Kyoiku Koku-gun trains the mission crews of the P-3C, using first YS-11T-As and then the P-3C. Helicopter crew training is conducted by Dai 211 Kyoiku Koku-tai at Kanoya AB, after basic flight training in Dai 201 Kyoiku Koku-tai. Flying hours are about 230 hours normally but only about 150 hours for students who have finished basic flight training on the T-5. All student helicopter pilots must receive about 89 hours of instrument flying training with Dai 202 Kyoiku Koku-tai for before converting to operational helicopters.

JMSDF

Koku Shudan (Fleet Air Force)	HQ, Atsugi
Dai 1 Koku-gun (1st Fleet Air Wing)	**Kanoya AB**
Dai 1 Koku-tai (VP-1)	P-3C, LC-90
Dai 7 Koku-tai (VP-7)	P-3C
Kanoya Koku Kichi-tai (Kanoya Air Base Flight)	UH-60J
Dai 2 Koku-gun (2nd Fleet Air Wing)	**Hachinohe AB**
Dai 2 Koku-tai (VP-2)	P-3C
Dai 4 Koku-tai (VP-4)	P-3C
Hachinohe Koku Kichi-tai (Hachinohe Air Base Flight)	UH-60J
Dai 4 Koku-gun (4th Fleet Air Wing)	**Atsugi AB**
Dai 3 Koku-tai (VP-3)	P-3C
Dai 6 Koku-tai (VP-6)	P-3C
Atsugi Koku Kichi-tai (Atsugi Air Base Flight)	UH-60J
Io-jima Koku Kichu-tai (Io-jima Air Base Flight)	S-61A
Minami-Torishima Koku Haken-tai (Detachment Minami-Torishima)	S-61A
Dai 5 Koku-gun (5th Fleet Air Wing)	**Naha AB**
Dai 5 Koku-tai (VP-5)	P-3C
Dai 9 Koku-tai (VP-9)	P-3C
Dai 21 Koku-gun (21st Fleet Air Wing)	**Tateyama AB**
Dai 101 Koku-tai (HS-101)	HSS-2B, S-61A
Dai 121 Koku-tai (HS-121)	SH-60J
Dai 123 Koku-tai (HS-124)	SH-60J
Dai 22 Koku-gun (22nd Fleet Air Wing)	**Omura AB**
Dai 122 Koku-tai (HS-122)	SH-60J
Dai 124 Koku-tai (HS-123)	SH-60J
Dai 31 Koku-gun (31st Fleet Air Wing)	**Iwakuni AB**
Dai 8 Koku-tai (VP-8)	P-3C
Dai 71 Koku-tai (71st Squadron)	US-1A
Atsugi Haken-tai (Detachment Atsugi)	US-1A Atsugi AB
Dai 81 Koku-tai (81st Squadron)	EP-3, U-36, LC-90
Chokkatsu Butai (Direct Reporting Unit)	

Dai 51 Koku-tai (VX-51)	P-3C, UP-3C, HSS-2B, SH-60J	Atsugi AB
Dai 61 Koku-tai (VR-61)	YS-11M/M-A, LC-90	Atsugi AB
Dai 111 Koku-tai (HM-111)	MH-53E	Iwakuni AB
Yokosuka Chiho-tai (Yokosuka Regional District)		**HQ, Yokosuka**
Shirase Hiko-han (Shirase Flight)	S-61A, OH-6D	Tateyama AB
Kure Chiho-tai (Kure Regional District)		**HQ, Kure**
Komatsujima Koku-tai (Komatsujima Squadron)	HSS-2B/SH-60J	Komatsujima AB
Sasebo Chiho-tai (Sasebo Regional District)		**HQ, Sasebo**
Omura Koku-tai (Omura Squadron)	HSS-2B/SH-60J	Omur AB
Ominato Chiho-tai (Ominato Regional District)		**HQ, Ominato**
Omitato Koku-tai (Ominato Squadron)	HSS-2B/SH-60J	Ominato AB
Kyoiku Koku Shudan (Air Training Command)		**HQ, Shimofusa**
Shimofusa Kyoiku Koku-gun (Shimofusa Air Training Group)		Shimofusa AB
Dai 205 Kyoiku Koku-tai (VT-205)	YS-11T-A	
Dai 203 Kyoiku Koku-tai (VT-203)	P-3C	
Shimofusa Koku Kichu-tai (Shimofusa Air Base Flight)	UH-60J	
Tokushima Kyoiku Koku-gun (Tokushima Air Training Group)		**Tokushima AB**
Dai 202 Kyoiku Koku-tai (VT-202)	TC-90, UC-90	
Dai 202 Shien Seibi-tai (202nd Support Maintenance Squadron)	UH-60J	
Tokushima Koku Kichi-tai (Tokushima Air Base Flight)	S-61A	
Ozuki Kyoiku Koku-gun (Ozuki Air Training Group)		**Ozuki AB**
Dai 201 Kyoiku Koku-tai (VT-205)	T-5	
Ozuki Koku Kichi-tai (Ozuki Air Base Flight)	S-61A	
Dai 211 Kyoiku Koku-tai (HT-201)	HSS-2B, SH-60J, OH-6D	Kanoya AB

Rikujo-Jieitai (Japan Ground Self-Defence Force, JGSDF)

Organisation of aircraft units of the Rikujo-Jieitai (JGSDF) consists of units which are directly assigned to five Homentai (District Army) and two units which report directly to Rikujo Bakuryo Kanbu (Ground Staff Office).

Each Homentai has several Shidan (Divisions) and one squadron of UH-1s and OH-6Ds under its direct control. The number assigned to these squadrons is reflected in the number of assigned Shidan, thus Dai 2 Hiko-tai (2nd Squadron) is assigned to Dai 2 Shidan (2nd Division). Also, each Homentai operates a Homen Koku-tai (District Army Air Group) which consist of Honbu Zuki-tai (Headquarters Squadron), Herikoputa-tai (Helicopter Squadron) and Tai-Sensha Herikoputa-tai (Anti-Tank Helicopter Squadron). The Seibu-Homentai (Western Army) has control of Dai 1 Konseidan (1st Composite Group) operating

from Naha AAB on Okinawa to cover Okinawa and Nansei Shoto (Southwestern Islands). The Seibu-Homen Herikopta-tai (Western Army Helicopter Squadron) is the only Army Helicopter Squadron which operates a small number of CH-47J Chinooks.

Utility helicopters which are currently operated by the JGSDF are the UH-1H and UH-1J. The UH-1H is the same helicopter which Bell produced except that those of the JGSDF have been manufactured by Fuji Heavy Industries (FHI) under licence since 1964 (having starting with the UH-1B). The UH-1J is an indigenous version developed from the UH-1H by

FHI with a change of engine to the Allison T53-703, the same engine used in the AH-1S. This increased power raised the payload and enhanced the helicopter's manoeuvrability, and a vibration reduction unit known as LIVE increased its speed by 20-30 kt (37-55 km/h). UH-1Js also have an IR counter-measures system and night vision goggle-compatible cockpit. The first example was delivered to the JGSDF on 3 September 1993 and the planned purchase was 78 UH-1Js until FY 1998, to replace about two-thirds of the UH-1Hs.

The LR-1 which each headquarters squadron oper-

The JGSDF's only fixed-wing types are the Mitsubishi LR-1 (left) and Raytheon/Beech LR-2 (right), assigned at Air Group Headquarters level. LR-1 is the designation for the army's version of the MU-2, used for liaison and reconnaissance, employing an oblique camera for the latter. In 1999 the first LR-2 (a version of the King Air 350) was taken on charge to replace the LR-1. Note the prominent underbelly fairing.

The future for the JMSDF lies with the Kawasaki OH-1 battlefield helicopter. It is intended to replace the OH-6 in the observation role, while also adding light attack and air-to-air roles.

Large numbers of OH-6D helicopters are in JGSDF service, flying general observation missions, scouting patrols for the AH-1 fleet and performing training tasks.

The JGSDF's aerial muscle at present comes from the TOW-equipped Bell AH-1S. A replacement is sought, for which the AH-64D is favourite.

Japan

ates is the only fixed-wing aircraft in the JGSDF. Its main mission is liaison but it is also able to carry an oblique reconnaissance camera in its rear fuselage for the aerial reconnaissance role. The LR-1 is a modified Mitsubishi MU-2 turboprop business aircraft, and because of its age the LR-2 has been selected as a replacement. The LR-2 is a modified version of the Raytheon Beech 350 King Air and the first was delivered in the latter half of FY 1998.

Five Tai-Sensha Herikoputa-tai (Anti-Tank Helicopter Squadron, ATH) were formed from FY 1986 to FY 1996. Each ATH has two squadrons with eight AH-1Ss and two OH-6Ds (one ATH operating 16 AH-1Ss and four OH-6D in total). The JGSDF started the development of a new observation helicopter for future ATHs and other units, called the OH-1 and developed by Kawasaki Heavy Industries (KHI). The procurement of production models started in FY 1997. The JGSDF is also planning the replacement of the AH-1S under an AH-X programme, which is expected to be in service from FY 2006. Selection of AH-X will not be completed until FY 2001.

The CH-47J is the only heavy-lift helicopter in the JGSDF and equips mainly Dai 1 Herikoputa-dan (1st Helicopter Brigade) at Kisarazu AAB. It is manufactured by KHI under licence and is almost identical to the Boeing CH-47D. For the longer range required by Dai 1 Konsei-dan at Naha AAB, KHI produced a version of the CH-47J called CH-47JA, which has almost double the internal fuel capacity in enlarged fuselage side tanks. They are also equipped with weather radar and FLIR in the nose, GPS and an NVG-compatible cockpit.

Dai 1 Herikoputa-dan consists of two squadrons each equipped with 16 CH-47Js and two OH-6Ds. In 1997 two CH-47Js from Dai 1 Herikoputa-dan began a northern deployment at Okadama AAB between April and November, after an evaluation made the previous year. The purpose of the deployment is to establish a rapid mission readiness in northern Japan. This brigade also oversees a Tokubetsu Yuso Koku-tai (Special Air Transport Squadron) for VIP transportation using three AS 332L Super Pumas.

Pilot training is conducted by the Koku Gakko (Aviation School) at Akeno AAB and Kita-Utsunomiya AAB. Both schools have the same training course, but officer cadets receive training at Akeno AAB and non-commissioned officer training is conducted at Kita-Utsunomiya AAB. Pilot training starts on the OH-6D with about 200 hours of flying, after which students are then qualified as operational pilots. Conversion training to UH-1H/J and AH-1S is also done in both schools with about 40 hours flight training. Since FY 1996, a new course for training operational UH-1H/J pilots has been underway, consisting of about 65 hours flying on the OH-6D at Kasumigaura School and 135 hours flying at Akeno AAB. Kasumigaura School also trains all mechanical staff for helicopters.

Koku Gakko has Kyoiku Sien Hiko-tai (School Support Squadron) under its control. This unit supports training activities by using a development training syllabus and undertaking the research and development of aircraft and aviation equipment for the JGSDF. In FY 1997, it started test flights of the UH-60JA – a new utility helicopter based on the Black Hawk manufactured by MHI – which will begin to replace the UH-1H in the near future.

A new observation helicopter, the OH-1, has been undergoing series flight testing since FY 1997. Four prototype XOH-1s are used in tests which are due to

end in FY 1999. For development and operational testing, a new organisation, Hiko Kaihatsu Jikken-tai (Air Development and Test Squadron), was formed under the auspices of the Aviation School at Akeno AAB; it currently operates four XOH-1s and two UH-1Js for test support.

The organisation of the Dai 12 Shidan (12th Division) and Dai 12 Hiko-tai (12th Squadron) will be changed to form a new Koku Kido Ryodan (Air Mobility Brigade) from FY 2000. Headquarters of this brigade will be at Somagahara and it will employ eight

AH-Xs, eight CH-47JAs, eight UH-60JAs and five OH-1s. As currently planned, this brigade will have four CH-47JAs, four UH-60JAs and four OH-6Ds at its conception. Modernisation and expansion of the number of helicopters will come later.

JGSDF

Hokubu Homen-tai (Northern Army)		HQ Sapporo
Dai 2 Shidan (2nd Division)		**HQ Asahikawa**
Dai 2 Hiko-tai (2nd Sqn)	OH-6D, UH-1H/J	Asahikawa AAB
Dai 5 Shidan (5th Division)		**HQ Obihiro**
Dai 5 Hiko-tai (5th Sqn)	OH-6D, UH-1H/J	Obihiro AAB
Dai 7 Shidan (7th Division)		**HQ Higashi-Chitose**
Dai 7 Hiko-tai (7th Sqn)	OH-6D, UH-1H/J	Okadama AAB
Dai 11 Shuidan (11th Division)		**HQ Makomanai**
Dai 11 Hiko-tai (11th Sqn)	OH-6D, UH-1H/J	Okadama AAB

Hokubu Homen Koku-tai (Northern Army Air Group)		HQ Okadama
Hokubu Homen Koku-tai Honbu Zuki-tai (Northern Army Air Group HQ Sqn)	LR-1	Okadama AAB
Hokubu Homen Herikoputa-tai (Northern Army Heli Sqn)	OH-6D, UH-1H/J	Okadama AAB
Dai 1 Tai-Sensha Herikoputa-tai (1st Anti-Tank Helicopter Squadron)		**Obihiro AAB**
Dai 1 Hiko-tai (1st Sqn)	AH-1S, OH-6D	
Dai 2 Hiko-tai (2nd Sqn)	AH-1S, OH-6D	

Tohoku Homen-tai (Northeastern Army)		HQ Sendai
Dai 6 Shidan (6th Div.)		**HQ Jinmachi**
Dai 6 Hiko-tai (6th Sqn)	OH-6D	Jinmachi AAB
Dai 9 Shidan (9th Div.)		**HQ Aomori**
Dai 9 Hiko-tai (9th Sqn)	OH-6D	Hachinohe AAB

Tohoku Homen Koku-tai (Northeastern Army Air Group)		HQ Kasuminome
Tohoku Homen Koku-tai Honbu Zuki-tai (Northeastern Army Air Group HQ Sqn)	LR-1	Kasuminome AAB
Tohoku Homen Herikoputa-tai (Northeastern Army Heli Sqn)	OH-6D, UH-1H/J	Kasuminome AAB
Dai 2 Tai-Sensha Herikoputa-tai (2nd Anti-Tank Helicopter Squadron)		**Hachinohe AAB**
Dai 1 Hiko-tai (1st Sqn)	AH-1S, OH-6D	
Dai 2 Hiko-tai (2nd Sqn)	AH-1S, OH-6D	

Tobu Homen-tai (Eastern Army)		HQ Asaka
Dai 1 Shidan (1st Div.)		HQ Nerima
Dai 1 Hiko-tai (1st Sqn)	OH-6D	Tachikawa AAB)
Dai 12 Shidan (12th Div.)		HQ Somagahara
Dai 12 Hiko-tai (12th Sqn)	OH-6D	Kita-Utsunomiya AAB

Tobu Homen Koku-tai (Eastern Army Air Group)		HQ Tachikawa
Tobu Homen Koku-tai Honbu Zuki-tai (Eastern Army Air Group HQ Sqn)	LR-1	Tachikawa AAB
Tobu Homen Herioputa-tai (Eastern Army HQ Sqn)	OH-6D, UH-1H/J	Tachikawa AAB
Dai 4 Tai-Sensha Herikoputa-tai (4th Anti-Tank Helicopter Squadron)		**Kisarazu AAB**
Dai 1 Hiko-tai (1st Sqn)	AH-1S, OH-6D	
Dai 2 Hiko-tai (2nd Sqn)	AH-1S, OH-6D	

Chubu Homen-tai (Central Army)		HQ Itami
Dai 3 Shidan (3rd Div.)		**HQ Senzou**
Dai 3 Hiko-tai (3rd Sqn)	OH-6D	Yao AAB
Dai 10 Shidan (10th Div.)		**HQ Moriyama**
Dai 10 Hiko-tai (10th Sqn)	OH-6D	Akeno AAB
Dai 13 Shidan (13th Div.)		**HQ Kaidaichi**
Dai 13 Hiko-tai (13th Sqn)	OH-6D	Hofu AAB

Chubu Homen Koku-tai (Central Army Air Group)		HQ Yao
Chubu Homen Koku-tai Honbu Zuki-tai (Central Army Air Group HQ Sqn)	LR-1	Yao AAB
Chubu Homen Herioputa-tai (Central Army Heli Sqn)	OH-6D, UH-1H/J	Yao AAB
Dai 5 Tai-Sensha Herioputa-tai (5th Anti-Tank Helicopter Squadron)		**Akeno AAB**
Dai 1 Hiko-tai (1st Sqn)	AH-1S, OH-6D	
Dai 2 Hiko-tai (2nd Sqn)	AH-1S, OH-6D	

Seibu Homen-tai (Western Army)		HQ Kengun
Dai 4 Shidan (4th Div.)		**HQ Fukuoka**
Dai 4 Hiko-tai (4th Sqn)	OH-6D	Metabaru AAB
Dai 8 Shidan (8th Div.)		**HQ Kita-Kumamoto**
Dai 8 Hiko-tai (8th Sqn)	OH-6D	Yakayubara AAB

Seibu Homen Koku-tai (Western Army Air Group)		HQ Takayubara
Seibu Homen Koku-tai Honbu Zuki-tai (Western Army Air Group HQ Sqn)	LR-1	Takayubara AAB
Seibu Homen Herikoputa-tai (Western Army Helicopter Squadron)	OH-6D, UH-1H/J, CH-47J	Metabaru AAB and Takayubara AAB
Dai 3 Tai-Sensha Herikoputa-tai (3rd Anti-Tank Helicopter Squadron)		**Metabaru AAB**
Dai 1 Hiko-tai (1st Sqn)	AH-1S, OH-6D	
Dai 2 Hiko-tai (2nd Sqn)	AH-1S, OH-6D	
Dai 1 Konsei-dan (1st Composite Group)		**Naha AAB**
Dai 101 Hiko-tai (101st Sqn)	UH-1H, CH-47JA, LR-1	
Dai 1 Herikoputa-dan (1st Helicopter Brigade)		**Kisarazu AAB**
Dan-honbu oyobi Honbu Kanri Chu-tai (Headquarters and Management Sqn)	LR-1	
Dai 1 Herikoputa-tai (1st Heli Sqn)	CH-47J, OH-6D	
Dai 2 Herikoputa-tai (2nd Heli Sqn)	CH-47J, OH-6D	
Tokubetu Yuso Hiko-tai (Special Air Transport Squadron)	AS 332L	
Koku Gakko (Aviation School)		**HQ Akeno**
Honko (HQ School)	OH-6D, UH-1H/J, AH-1S	Akeno AAB
Kasumigaura Bunko (Kasumigaura School)	OH-6D, UH-1H/J, AH-1S, CH-47J	Kasumigaura AAB
Utsunomiya Bunko (Utsunomiya School)	OH-6D, UH-1H, LR-1	Kita-Utunomiya AAB
Kyoiku Sien Hiko-tai (School Support Sqn)	UH-1J, OH-6D, AH-1S, CH-47JA, UH-60JA	Akeno AAB
Hiko Kaihatu Jikken-tai (Air Development and Test Sqn)	XOH-1, UH-1J	Akeno AAB

299

KAZAKHSTAN

Capital: Alma-Ata
Population: 16.7 million
Land area: 2.72 million km² (1,048,880 sq miles)
Major cities: Chimkent, Karaganda, Pavlodar, Semipalatinsk

The former Soviet republic of Kazakhstan became an independent nation, the Republic of Kazakhstan (Qazaqstan Respublikasy), on 16 December 1991. During Soviet times, the Kazakh Soviet Socialist Republic was part of the Central Asian Military District. On 8 May 1992, all former Soviet armed forces located on its territory (excluding strategic forces) came under the control of the President of Kazakhstan, Nursultan Nazarbayev.

The Russian Federation continues to operate the Baikonur Cosmodrome, and in return for the use of this facility – and the relocation of the 40 Tu-95MSs of 79 TBAD (heavy bomber division) from Semipalatinsk to Russia in 1993 – Russia supplied Kazakhstan with MiG-29s (21), Su-25s (14) and Su-27s (38, including eight Su-27UBs) during 1995-1997. A further four Su-27s were supplied on 25 January 1999, in part exchange for the continued Russian use of the Baikonur space complex. Russia is due to provide Kazakhstan with a further 12 Su-27s and S-300 anti-aircraft missile systems, which will be used to defend the capital. A large number of MiG-21s, MiG-23s and MiG-25PD/-PUs in storage at Taldy Kurgan, Lugovoy and Semipalatinsk, respectively, have been offered for sale.

AIR DEFENCE FORCE

Kazakhstan is the only former Soviet republic to operate the MiG-31 long-range interceptor, a legacy of the Soviet era, when 356 IAP PVO (air defence fighter regiment) at Semipalatinsk was subordinated to the Tashkent Air Defence Army.

AIR FORCE

The Air Force of the Republic of Kazakhstan, which has been commanded since 1996 by Colonel Mukhametzhan Ibraev (a former Su-24 pilot and regimental commander), comprises fighter, fighter-bomber, reconnaissance and transport aviation units.

A fighter regiment (715 IAP) based at Lugovoy operates a squadron of MiG-29s and a squadron of MiG-23MLDs, which during Soviet times served with the Foreign Pilots Conversion Centre at Lugovoy. The 21 MiG-29s delivered by Russia could bring the fighter regiment at Lugovoy to full MiG-29 strength. The 38 Su-27s supplied by Russia are believed to have been used to form a new regiment, but it is not known at which base.

Kazakhstan operates an attack division with two regiments of MiG-27M fighter-bombers and one regiment of Su-24 bombers. The Su-25s may have been used to form a ground attack squadron.

During Soviet times, 39 ORAP (independent reconnaissance regiment) flew reconnaissance missions with a squadron of MiG-25s and a squadron of Su-24MRs. Since then, the Su-24MRs have been relocated to Nikolayevka, and the status of the reconnaissance unit remains unclear. In addition, a number of An-30s are operated by an independent reconnaissance squadron (ODRAE) at Almaty-Burunday.

Transport duties are performed by a single independent mixed aviation regiment (OSAP) based at Almaty, equipped with An-12s, An-24s, An-26s and a single Il-18 and Tu-134 (Tu-135). The Il-22 aerial command post formerly operated by this unit was converted into a VIP aircraft, but was destroyed in a ground accident in 1995.

The rotary-winged element of the Kazakhstan Air

Air Defence Force (IA-PVO)

356 IAP	MiG-31 (32)	Semipalatinsk

Air Force (VVS)

u/i OSAP	An-12 (12), An-24 (1), An-26 (12), Il-18 (1), Tu-134 (1)	Almaty* (*Alma-Ata)
u/i ORAE	An-24 (1), An-26 (2), An-30 (6)	Almaty-Burunday** (**Alma-Ata NW)
39 ORAP	MiG-25RB (13), MiG-25RU (2)	Balkhash* (*status uncertain)
157 OTBVP	Mi-8MT (40), Mi-8PPA (4), Mi-26 (24)	Stanchi Kambyl (aka Dzhambul)
486 OVP	Mi-6 (6), Mi-8 (14), Mi-24D/V (42)	Ucharal
715 IAP	MiG-23MLD (12), MiG-23UB (4), MiG-29 (12), MiG-29UB (4)	Lugovoy
11 ADIB	?	Taldy Kurgan
129 APIB	MiG-23UB (10), MiG-27M (32)	Taldy Kurgan
134 APIB	MiG-23UB (3), MiG-27M (9)	Zhangiztobe
149 BAP ?	Su-24 (30), Su-24MR (12)* (*ex-39 ORAP)	Nikolayevka (aka Iliysk S or Kapchagay S)

Force comprises a combat transport regiment (OTBVP) equipped with Mi-8s and Mi-26s, and a combat helicopter regiment (OVP) operating Mi-6s, Mi-8s and Mi-24s. The latter unit (486 OVP) relocated in 1991 from Altes Lager (Jüterbog) in the former German Democratic Republic to Kazakhstan.

KAZAKHSTAN — Nikolayevka, Semipalatinsk, Zhangiztobe, Balkhash, Ucharal, Taldy Kurgan, Almaty, Alma-Ata, Stanchi Kambyl, Lugovoy

---

NORTH KOREA

Capital: Pyongyang
Population: 21.8 million
Land area: 122310 km² (47,210 sq miles)
Major cities: Chonctsin, Haeju, Hungnam, Kaesong, Wonsan

Korean People's Army Air Force (KPAAF)

Despite a nominal strength of more than 500 first-line aircraft, with almost as many supporting types, the KPAAF inventory mainly comprises obsolescent equipment. Since further supplies of Russian arms and aircraft effectively ceased in late 1992, closely followed in January 1993 by the withdrawal of special trading terms by China, its other main supplier, standards of serviceability have reportedly dropped to the point where only a small percentage of KPAAF aircraft are airworthy at any one time.

These problems are compounded by North Korea's (Democratic Peoples' Republic of Korea) weakened economy and political instability, and diversion of funding to weapons of mass destruction, including nuclear warhead development and chemical and biological stockpiles. In parallel programmes, ballistic missiles derived from original 'Scud' SSM technology are being developed as delivery systems, ranging from the 'Scud Mod B/C', through the No-dong 1 and

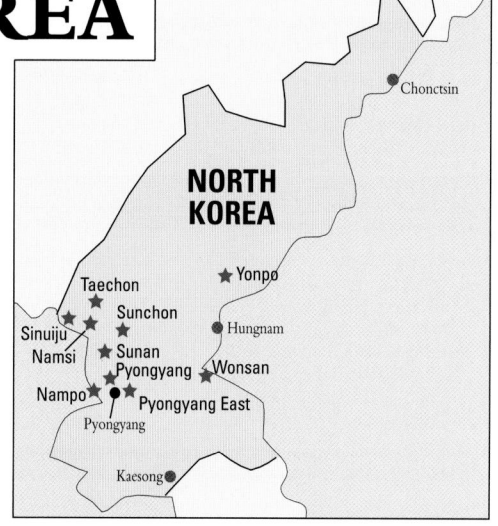

NORTH KOREA — Chonctsin, Yonpo, Taechon, Sunchon, Sinuiju, Namsi, Sunan, Hungnam, Pyongyang, Wonsan, Nampo, Pyongyang East, Pyongyang, Kaesong

two-stage Taepo-dong 2 IRBM, with operational ranges from 200-6000 km (125-3,728 miles). In April 1996, North Korea declared it would no longer observe the terms of the 1953 Korean War armistice, and has since become increasingly hostile and unpredictable in its relations with South Korea.

Korean People's Army Air Force

Estimated totals of current equipment comprise:

Type	Role	In Service
Aero L-39C Albatros	advanced trainer	12
Antonov An-24 'Coke'	transport	10
Harbin H-5 'Beagle'	tactical bomber	40
Hughes 300C	trainer helicopter	10
Ilyushin Il-18D 'Coot'	transport	4
Ilyushin Il-62M 'Classic'	transport	1
Ilyushin Il-76T 'Candid'	transport	2
McDonnell Douglas MD 500D/E	close support helicopter	15/50
MiG-15UTI/FT-2 'Midget'	combat trainer	30
MiG-21PF/PFM 'Fishbed-F'	air defence/attack	120
MiG-21UM 'Mongol-B'	combat trainer	25
MiG-23ML 'Flogger-G'	air defence	40
MiG-23UB 'Flogger-C'	combat trainer	10
MiG-29 'Fulcrum-A'	air defence	25
MiG-29UB 'Fulcrum-B'	combat trainer	5
Mil Mi-2/Hyokshin-2 'Hoplite'	utility helicopter	n/k
Mil Mi-8/17 'Hip'	utility helicopter	20
Mil Mi-14PL 'Haze'	ASW helicopter	10
Mi-24D/V 'Hind-D/E'	attack helicopter	20
Nanchang NAMC CJ-5/6	basic trainer	100
Nanchang Q-5 1A 'Fantan'	ground attack	40
Shenyang F-5	ground attack	120
Shenyang FT-5	advanced trainer	20
Shenyang F-6	air defence/attack	100
Shenyang FT-6	combat trainer	10
Sukhoi Su-7BMK 'Fitter-B'	attack aircraft	30
Sukhoi Su-25K 'Frogfoot'	attack aircraft	18
Sukhoi Su-25UBK 'Frogfoot-B'	combat trainer	2
Shijiazhuang (SAP) Y-5 'Colt'	utility transport	200
Tupolev Tu-154B 'Careless'	transport	3

Four Ilyushin Il-18Ds are used by the North Korean air force and this example, P-835 registered to Air Koryo, is believed to be one of them.

SOUTH KOREA

Capital: Seoul
Population: 43.3 million
Land area: 98445 km² (38,000 sq miles)
Major cities: Inchon, Kwangju, Pusan, Taegu, Taejon

In the final days of World War II the Soviet Union occupied a large portion of the Korean peninsula. The country was divided arbitrarily along the 38th Parallel – Communist control to the north and US control to the south – in order to accept the Japanese surrender. Months of reunification talks between the two sides broke down irrevocably, leading to the establishment of the modern-day Republic of Korea in 1948. All-out war from 1950-53 failed to resolve the situation other than to redraw the border along the ceasefire line reached by November 1953. That border still holds, although there has never been an official ending to hostilities. Cross-border incidents remain rife, with numerous attempts by the North to infiltrate the South. It is the stated aim of both sides to reunite the nation.

Numerically, the F-5 remains the backbone of the RoKAF, the inventory including some elderly F-5As such as this 122nd Sqn example. Despite the arrival of two batches of RF-4Cs, a handful of camera-equipped RF-5As are also in use.

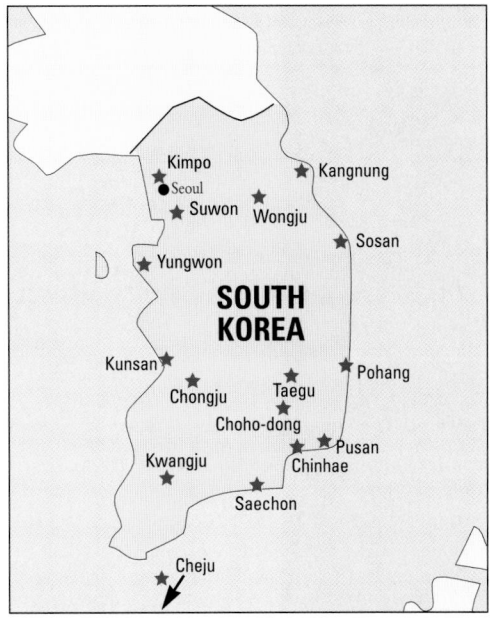

Hankook Kong Goon – Republic of Korea Air Force (RoKAF)

Immediately after the Korean War the United States maintained its sizeable forces in the country, and still stations the 7th Air Force (F-16s and OA/A-10s) at Osan and Kunsan. The US also oversaw the establishment and equipment of the Republic of Korea Air Force (RoKAF) with US types, notably the F-86 and, later, the Northrop F-5A and McDonnell Douglas F-4. The RoKAF and 7th Air Force remain closely integrated through the Combined Forces Command.

South Korea's burgeoning industrial and techno-logical base in recent years has allowed it to play a greater part in the provision of its own forces. Between 1968 and 1974 233 Northrop F-5E/Fs were received, the last 68 of which were assembled by Korean Air. The same company has also assembled large numbers of MD500 helicopters for the Army and, later, the UH-60P Black Hawk. Licence-assembly became a major issue in the Korean Fighter Program competition of the late 1980s. The RoKAF had previously acquired 40 F-16C/D Block 32s from the manufacturer from 1986 (Peace Bridge programme), and this type was pitted against the F/A-18 for a licence-assembly deal. The F/A-18 was chosen initially, in late 1989, but this decision was overturned in 1991, the F-16C/D Block 52 being chosen. The first dozen aircraft came from Fort Worth, followed by 36 kits for assembly. The final 72 aircraft of the 120-unit order were wholly built in Korea, by a consortium led by Samsung Aerospace.

Spurred by this success, and buoyed by a fast-growing 'tiger' economy, Korea began the development of indigenous products. The tangible results are the Daewoo KTX-1 Woong-Bee turboprop basic trainer and the Samsung KTX-2, an F404-powered supersonic advanced jet trainer/light attacker devel-

oped with the aid of Lockheed Martin. The first prototypes of the KTX-1 are flying, and an order for 85 has been placed, with the first due to enter service in March 2000. An armed FAC version, designated KOX, is under development for a slated service entry for the first of 20 aircraft in 2003. The future of the KTX-2 is less certain, although development continues. Hyundai has teamed with DASA to offer the AT-2000 Mako in competition, although development of the KTX-2 has been given the go-ahead, with an expected Lot 1 production batch of 94 aircraft and a follow-on requirement for 100 more. Until the new jet trainer can enter service (2006 at the earliest), the RoKAF has leased 30 Northrop T-38 Talons to fill its advanced training needs.

South Korea was one of the nations gravely afflicted by the 'Asian flu' economic crisis of the late 1990s, throwing into confusion the procurement plan. Nevertheless, the RoKAF is expected in the

The RoKAF is the last user of the F-4D. This cannon-equipped aircraft serves with the 110th TFS.

Ex-USAF RF-4Cs are the RoKAF's main tactical reconnaissance assets, flying with the 131st TRS.

This Cessna O-2A is one of the FAC aircraft assigned to the 237th TCS. The FAC fleet is expected to be replaced by the Daewoo KOX.

Hankook Kong Goon – Republic of Korea Air Force (RoKAF)

Fighter Command

11th TFW		
110th TFS	F-4D	Taegu
151st TFS	F-4D	Taegu
17th TFW		
132nd TFS	F-4E	Chongju
152nd TFS	F-4E	Chongju
153rd TFS	F-4E	Chongju
19th TFW		
161st TFS	F-16C/D	Yungwon
162nd TFS	F-16C/D	Yungwon
155th TFS	F-16C/D	Yungwon
159th TFS	F-16C/D	Yungwon
20th TFW		
120th TFS	F-16C/D	Sosan
123rd TFS	F-16C/D	Sosan

Strike Command

1st TFW		
105th TFS	F-5E	Kwangju
122nd TFS	F-5E	Kwangju
115th OCU	F-5B	Kwangju
5th TFW		
201st TFS	F-5E	Cheju (?)
202nd TFS	F-5E	Cheju (?)
8th TCW		
216th TCS	Hawk T.Mk 67	Wongju
238th TCS	A-37B	Wongju
239th TCS (?)	A-37B, OA-37B	Wongju (?)

10th TFW		
101st Sqn	F-5E	Suwon
102nd FS	F-5A	Suwon
103rd TFS	F-5E	Suwon
111th TFS	F-5E	Suwon
12th TCW		
236th TCS	O-1 A	Kangnun
237th TCS	O-2 A, OV-10D	Kangnun
16th TTW		
205th TTS	F-5E	Saechon
206th TTS	F-5A/B	Saechon
39th TRG		
125th TRS (132nd TRS?)	RF-5A	Suwon
131st TRS	RF-4C	Suwon

Transport Command

? TAW		
255th TAS	C-130H-30	Pusan-Kim Hae
256th TAS	CN.235M	Pusan-Kim Hae
258th TAS	C-130H	Pusan-Kim Hae
259th SOS	S-70C	Pusan-Kim Hae
? MAW		
257th MAS (?)	C-130H-30	Seoul-Kimpo
? VIP-Sqn	Boeing 737-300, BAe HS-748, Aero Commander	Seoul-Kimpo
233rd SAR-Sqn	Bell 212 Bell 214, UH-1H, UH-60P	Pusan
235th Sqn	HH-47D	

Training Command

? FTW		
208th Sqn	CAP 10B	Chongju
212th FTS	T-41B	Chongju
? FTS	T-33A	Chongju
236th FTS	T-37C	Chongju
? FTS	T-38B	Chongju

Winner of the 120-aircraft Korean Fighter Programme, the F-16C/D Block 52 is built locally by Samsung. The aircraft joined the 40 Block 32 aircraft supplied in the late 1980s to form the RoKAF's premier fighter units, the 19th and 20th Tactical Fighter Wings.

next few years to make a decision over a fifth-generation fighter, with various contenders (Rafale, Typhoon, Su-35, F-22 and modernised versions of F-15, F-16 and F-18) vying for what could be a lucrative contract. By necessity, the scale of this purchase is linked with the nation's economic achievements in the near-term, as is the acquisition of an AEW aircraft to increase the effectiveness of the air defence network.

ORGANISATION AND ASSETS

Naturally, the RoKAF is organised along USAF lines, with four role-specific commands, consisting of wings (one wing per base). Each wing has a number of subordinate squadrons and in most front-line cases operates a single type. Air defence is the main task of Fighter Command, which has four wings. The F-16 is the primary fighter, and equips six squadrons in two wings at Yungwon and Sosan. More squadrons are expected to form while deliveries continue. The

other two wings both fly the Phantom: the 17th TFW at Chongju has the F-4E (95 received) while the 11th TFW at Taegu soldiers on with the ancient F-4D (66 received). It was this latter version which the F-16 was originally procured to replace.

Despite the main air defence role of Fighter Command, its aircraft are also tasked with attack missions. The F-16s can carry LANTIRN pods, and AGM-88 HARM, AGM-130 and AGM-142 missiles are available. By the same token, the attack-orientated aircraft of Strike Command can also perform air defence duties. Strike Command has seven wings, three of which fly the F-5E as their front-line equipment. The 1st TFW also has a squadron of F-5Bs which acts as the type OCU. A few surviving F-5As also serve with the 10th TFW's 102nd Fighter Squadron. Another wing of F-5s, the 16th Tactical Training Wing, performs advanced/weapons training with the F-5A/B/E.

Strike Command's other three wings are tasked with reconnaissance and tactical control/training. RF-5As and RF-4Cs (15 delivered) operate in single squadrons under the 39th Tactical Reconnaissance Group at Suwon. The squadron has also been reported as having an electronic role, using 'EF-5As'. The tactical control wings fly a mix of FAC types

(O-1, O-2, OV-10 – 12th TCW) and light attack/training aircraft (OA/A-37B, Hawk T.Mk 67 – 8th TCW). The first of 20 Hawks entered service in June 1993, replacing the Lockheed T-33 in the advanced/weapons training role. An outstanding requirement for more, if filled, would replace the Cessna OA/A-37s, the survivors of 27 handed over by the US after having escaped from South Vietnam. The Hawk Mk 67 is something of a hybrid, having the standard trainer airframe of the Mk 60 but the longer nose of the attack-roled Mk 100. In the case of the Mk 67, the nose houses additional avionics rather than attack sensors.

Transport Command has two wings, one at Pusan equipped with the C-130H, CN.235 and S-70C Black Hawk, and another at Seoul consisting of a tactical unit with C-130Hs and a VIP squadron equipped with a Boeing 737, BAe 748s and five Aero Commanders. A new tactical transport is required, for which the C-130J appears the most likely contender. Helicopter units are primarily concerned with SAR, and operate a variety of types. A recent arrival is the Boeing HH-47D, a SAR-configured Chinook tailored for combat SAR work.

Training Command provides a syllabus which begins with primary training/flight screening on Cessna T-41s and basic training on T-33s and T-37s, the latter being the survivors of 30 acquired from Brazil. The advanced stage is handled by the T-38 Talon and Hawk, before pilots are posted to the F-5 OCU. In the future the KTX-1 will replace the T-37, and the KTX-2 take over from the T-38.

Recognising the importance of cross-border surveillance, the RoKAF has an outstanding order (signed in July 1996) for eight special-mission versions of the Raytheon Hawker 800XP under the Peace Pioneer programme. Four are to be designated 800RA and have a SAR radar fit, while the other four are 800SIG aircraft (codename Paekdu), with an extensive onboard signals intelligence gathering suite. They are expected to be delivered during 1999.

Above: This is the fifth prototype of the Daewoo KTX-1, representative of the production trainer. Deliveries to the RoKAF are due to start in 2000.

Left: Long-range combat SAR is the primary role of the RoKAF's winch-equipped HH-47D force. The 235th Squadron operates the aircraft.

Republic of Korea Army

Although the US Army's 2nd Infantry Division maintains a sizeable aviation element in-country, the RoK Army maintains an organic aviation capability.

The ubiquitous Huey serves as a medium transport, with 25 UH-1Bs received second-hand in 1977 and 20 UH-1Hs received new in 1985. To update its air-

For airborne assault the UH-60P is used alongside Huey versions. More UH-60Ps are being assembled by Korean Air to continue the replacement of the Bell UH-1.

mobility capability the Army ordered 100 UH-60P Black Hawks, which differ from the US Army's UH-60L in minor details. After one was delivered from Sikorsky, Korean Air assembled 19 more from parts supplied from the manufacturer. At least 80 more are to be totally assembled by Korean Air at Kim Hae.

Three AS 332 Super Pumas are used for staff and VIP transport. Heavy-lift transport is provided by 18 CH-47D Chinooks. A requirement existed for a further 12 of the type.

Following receipt of 34 Hughes 500 light helicopters from the US, licensed production of the type was undertaken by Korean Air, and some 200 were eventually delivered. A quarter of these are 500MD Defender versions armed with BGM-71A TOW anti-tank missiles, while the rest are scout versions, which can be lightly armed. Another armed helicopter type is the Eurocopter (MBB) BO 105CBS, used for scout work with a variety of gun and rocket pod options.

AH-1F HueyCobras form the main attack element of the Army's aviation branch. The first 42, delivered

Republic of Korea Army Aviation (RoKAA)

1 x Transport Sqn	14 x CH-47D	Taegu
1 x Assault Sqn	25 x UH-60P	Choho-dong
4 x Assault Sqn	25 x UH-60P	
3 x Attack Sqn	25 x AH-1F/J	
1 x Anti-tank Sqn	22 x Hughes 500MD-TOW	
2 x Anti-tank Sqn	23 x Hughes 500MD-TOW	
1 x Transport Sqn	26 x UH-1B	Kunsan
1 x Transport Sqn	26 x UH-1B	Osan
2 x Transport Sqn	26 x UH-1H	
1 x Transport Sqn	27 x UH-1H	
1 x Liaison Sqn	10 x DHC-2	Taegu
1 x AOP Sqn	5 x O-1A	
6 x AOP Sqn	26 x Hughes 500MD	

Future procurement is expected to include an order for a combat helicopter to replace the AH-1 Cobra currently in service. The AH-64 Apache was ordered in 1992 when an FMS deal for 37 AH-64As was announced, but the deal fell through. Revived interest in 1996 for an initial 18 helicopters and an eventual 38 to 48 saw the Apache as the front-runner, but interest was also expressed in the Eurocopter Tiger. The attraction of licence-production has also led to the Kamov Ka-50 being considered for the new battlefield helicopter contract.

Korea's army aviation is principally concerned with blunting an armoured thrust from the North, and has a sizeable anti-tank force of AH-1s and MD 500s (above right). Scout versions of the MD 500 and BO 105 (above left) augment the main attack force.

from 1988, were equipped with the C-Nite optical target acquisition system for night-time engagements. A second batch of 20 was later acquired. The AH-1F is operated alongside the survivors of eight AH-1J SeaCobras that arrived in 1978.

Republic of Korea Navy

A land-based ASW squadron received 28 Grumman S-2E Trackers from the US Navy in 1976-81. These aircraft were replaced in service by eight Lockheed Martin P-3C Orion Update IIIs. The first aircraft were delivered in April 1995 and the final example was in service by the end of the year, after initial training for the Korean crews was undertaken by the US Navy's VP-30 at NAS Jacksonville, Fla. The aircraft can carry the AGM-84A Harpoon anti-shipping missile.

Main ASW helicopter since 1989 has been the Westland Lynx Mk 99. Original procurement comprised of 12 Mk 99s, armed with Sea Skua anti-

The P-3 Orion is an important asset due to the large number of territorial water violations by North Korean submarines and boats, including some aimed at infiltrating agents.

ship missiles. A second batch of 13 was ordered in 1997, the deal also covering the upgrading of the survivors to Super Lynx standard with the fitting of new mission systems avionics. Lynxes operate aboard the 'Sumner'- and 'Gearing'-class frigates.

Other helicopters used by the Navy include SA 319B Alouette IIIs for general utility tasks, which serve either with 621 or 623 Sqn. These helicopters have in the past operated with the Marines and may continue to provide airlift for this organisation. The navy has operated about 25 torpedo-armed Hughes 500 Defenders, but these are thought to have been replaced by the Lynx. Naval Aviation Unit 1056 operated two Bell 206B JetRangers for unspecified tasks, but they have not been noted recently.

In 1998 the Navy began to take delivery of five Reims-Cessna F 406 Caravan II light transports which are to be used in the target-towing and general utility roles.

Republic of Korea Naval Aviation (RoKNA)

6 Air Wing		
613 ASW-Sqn	P-3C U.III (8)	Pohang
? Air Wing		
62? ASW-Sqn	SA 319B Alouette III (9)	Chinhae
627 ASW-Sqn	Lynx Mk 99 (17)	Chinhae

The RoKNA's rotary-wing force consists of the Lynx Mk 99 (above) and the SA 319B Alouette III (below). The MD 500s are now believed to be out of use. The Alouettes perform utility transport tasks, including support of the Marines, while the Lynxes are the main seagoing type. For the attack role they carry Sea Skua missiles.

Paramilitary forces

South Korea has three government agencies operating rotary-wing equipment. The Coast Guard has AS 365

Dauphins for SAR work, and may still use MD 500s. The sizeable Korean National Police air unit is known to have five Bell 206L LongRangers, two Bell 212s, a Bell 412SP and seven Hughes 369s. A single Kamov Ka-26 is believed to be operated by the maritime

police. Ka-32Ts and AS 350s are employed by the Forest Department, fitted with belly-mounted fire-fighting equipment.

The Coast Guard operates helicopters in close co-operation with the Navy. This is an AS 365F.

Among the Korean police force's equipment is the Bell 206L, complete with Heli-Tele surveillance kit.

The Forest Department operates the Ka-32T on fire-fighting duties.

KYRGYZSTAN

Capital: Bishkek
Population: 4.4 million
Land area: 198500 km² (76,620 sq miles)
Major cities: Dzhalal-Abad, Osh, Prezhevalsk, Tokmak

The country inherited an aviation training school from the Soviet Union, which used the Aero L-39C (67), Mil Mi-8 (28), Mi-17 (two) and Mi-24/25/35 (17). Kyrgyzstan intended to maintain the school and train foreign pilots in an attempt to earn foreign currency, but has not been able to maintain the service. A large number of MiG-21/21UMs are stored in the country and are unlikely to fly for the new air force.

The air defence force has no aircraft, relying on SA-2, SA-3 and SA-7 surface-to-air missiles for defence of its national airspace.

Kyrgyzstan participates in the NATO Partnership for Peace programme and is a member of the Euro-Atlantic Partnership Council, and is also a signatory to

Formerly the Kirghiz Soviet Socialist Republic, Kyrgyzstan (the Kyrgyz Republic) gained its independence on 31 August 1991 after the 'Silk Revolution'.

a confidence-building measure agreement with Russia, aimed at preventing future conflict in the region.

LAOS

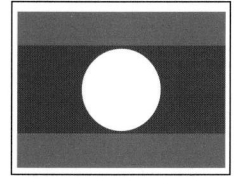

Capital: Vientiane
Population: 4.2 million
Land area: 236725 km²
(91,375 sq miles)
Major cities: Luang Prabang, Savannakhet, Pakse

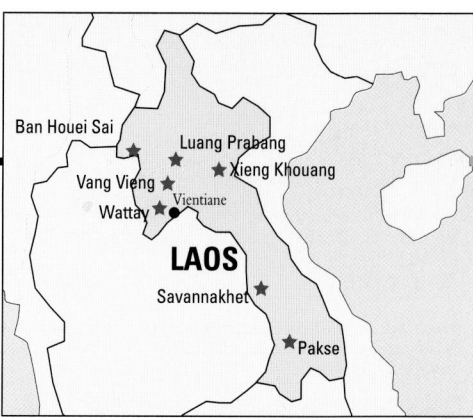

Lao People's Liberation Army Air Force

Type	Role	Delivered/In Service
Antonov An-2 'Colt'	transport	c12/10
Antonov An-24V/RV 'Coke'	transport	7/6
Antonov An-26 'Curl'	transport	3/3
Douglas C-47	transport	10/6
Harbin Y-12 II	transport	7/7
MiG-21PFM/MF 'Fishbed-F/J'	combat aircraft	25/19
MiG-21U 'Mongol'	combat trainer	c6/3
Mil Mi-6 'Hook'	heavy-lift helicopter	/1
Mil Mi-8S 'Hip'	utility helicopter	10/9
Xian Y-7-100C	transport	2/2
Yakovlev Yak-40 'Codling'	transport	3/2
Yakovlev Yak-18A 'Max'	basic trainer	8/6

Lao People's Liberation Army Air Force (LPLAAF)

The former kingdom of Laos was proclaimed a people's republic in November 1975, following the Communist take-over in the wake of the US withdrawal from the region, and its armed forces have replaced most of their former American equipment with Soviet/Russian and Chinese assistance. Main LPLAAF strength now comprises two squadrons of MiG-21PFMs in a fighter regiment at Viangchan-Wattay, which have undergone major overhauls with soft loans from China worth some $16 million. Further maintenance, upgrades and technical support

are being negotiated with India.

Military co-operation agreements with Russia in 1997 resulted in LPLAAF procurement of 12 Mil Mi-17 transport helicopters for delivery by mid-1999, to follow earlier deliveries of a dozen or so Mi-8s. Other Russian arms procurement has included S-125 Neva (SA-3 'Goa'), 9K32M Strela 2D/M (SA-7B 'Grail'), and other SAM systems. Most LPLAAF

Many of the LPLAAF transport aircraft, including this Antonov An-26, operate in the colour scheme of Laos Aviation from Wattay-Vientiane.

transports operate in dual civil/military roles, while seven Harbin HAMC Y-12 II and two Xian Y-7-100C turboprop twins received from China are flown in Lao Aviation colours.

MALAYSIA

Capital: Kuala Lumpur
Population: 17.9 million
Land area: 332965 km²
(128,525 sq miles)
Major cities: George Town, Ipoh, Kota Kinabalu, Kuching

Tentara Udara Diraja Malaysia

Markas Besar Operasi Udara

6 Sku	Hawk Mk 108/208	Kuantan
9 Sku	Hawk Mk 108/208	Labuan
12 Sku	F-5E, F-5F, RF-5E	Butterworth
15 Sku	Hawk Mk 108/208, MB-339AM	Butterworth
16 Sku	Beech B200T King Air	Subang
17 Sku	MiG-29N/NUB	Kuantan
18 Sku	F/A-18D	Butterworth
19 Sku	MiG-29N/NUB	Kuantan

Markas Bantuan Udara

1 Sku	DHC-4A Caribou, CN.235	Kuching
2 Sku	Fokker F28, Falcon 900, Cessna 402B, Challenger SE, S-70A, AS-61N-1	Kuala Lumpur/Subang
3 Sku	S-61, Alouette III	Butterworth
5 Sku	S-61	Labuan
7 Sku	S-61, Alouette III	Kuching
10 Sku	S-61, Alouette III, A 109C	Kuala Lumpur/Subang
14 Sku	C-130H	Kuala Lumpur/Subang
20 Sku	C-130H-30	Kuala Lumpur/Subang

Training Division

1 FTC	PC-7, MD3-160	Alor Setar
2 FTC	S-61, Alouette III	Alor Setar
3 FTC	PC-7	Kuantan

Kuala Lumpur/Subang is also known as Simpang

Tentara Udara Diraja Malaysia – TUDM (Royal Malaysian Air Force – RMAF)

Malaya became an independent member of the British Commonwealth on 31 August 1957 during a prolonged campaign against Communist guerrillas. In September 1963, a new federation, named Malaysia, which included Malaya, Sabah, Sarawak, Brunei and Singapore was formed (the last two subsequently became independent states). The Royal Malayan Air Force was formed on 1 June 1958 to help support the army in the anti-Communist campaign and was

equipped initially with examples of the Scottish Aviation Pioneer and Twin Pioneer, with DHC Chipmunks and Hunting-Percival Provosts for training duties. Renamed on 16 September 1963, to reflect the new federation, the RMAF remained a transport fleet, adding Handley Page Heralds and DHC-4A Caribou, and did not receive its first combat types until 1969, when 16 CA-27 Avon-Sabres were passed on from the RAAF. Today the air force has a modern fleet of combat and support aircraft. Virtually all are based on the Malayan peninsula, where the main centres of population are located, but a few aircraft and bases are maintained on Borneo. Defence agreements with other Southeast Asian states, Australia and the UK further ensure the nation's sovereignty.

FIGHTING FORCE

The withdrawal of the permanent RAAF presence at Butterworth forced the RMAF to look into alternatives. A defence package with the UK resulted in an

The Pilatus PC-7 is flown by two units of the Training Division on basic instruction duties, bridging the gap between primary training on the locally-built MD3-160 lightplane and the MB-339/Hawk 108.

15 Squadron at Butterworth is the main advanced training unit, flying both Aermacchi MB-339s (illustrated) and BAe Hawk Mk 108s. The latter have an emergency war role, and can carry wingtip Sidewinders.

order for the Tornado IDS, which was abandoned on cost grounds. However, Malaysia decided instead to purchase 10 BAe Hawk 100s (for advanced training) and 18 Hawk 200s (for the strike role), which were delivered between 1994 and 1995. The Mk 208s were fitted with a fixed refuelling probe. Equipping three squadrons which operate both versions, the type replaced the A-4PTM Skyhawk in service.

For interception duties the RMAF ordered the MiG-29N, which was delivered from 1995 and today operates with two squadrons from Kuantan. A total of 16 MiG-29Ns and two MiG-29NUB conversion trainers was delivered, the two-seaters serving with the operational conversion unit, No. 17 Squadron. Able to operate R-73 'Archer' and R-27 'Alamo' missiles, the type was recently upgraded to with the ability to fire the R-77 'Adder', and received radar upgrades and an air-to-air refuelling probe. An order for a further 18 examples was announced in 1996, but these will not now be delivered.

The arrival in March 1997 of four F/A-18D Hornets greatly enhanced the capabilities of the RMAF. The aircraft, powered by the F404-GE-402 enhanced performance turbofans and equipped with the APG-73 radar, were joined by a second batch of four aircraft in June 1997. The Hornets are tasked with interdiction, night attack and strike, and although the RMAF would like to acquire a second batch the current economic conditions make this unlikely in the near future.

A single squadron continues to use the F-5 Tiger II in both F-5E/F fighter/conversion trainer (armed with AIM-9J Sidewinder missiles) and RF-5E reconnaissance versions, the MiG-29 having replaced the type in its air defence duties. It is expected the Hawk or Hornet will acquire the type's reconnaissance role in the future.

SUPPORT FORCE

Maritime patrol is undertaken by No. 16 Squadron from Subang operating four Beech 200T King Airs, which replaced three C-130H-MPs from 1994. Two C-130H-MPs were converted to tankers as C-130Ts. The Hercules is also the main heavy transport of the RMAF, the primary operator being No. 20 Squadron at Subang with seven C-130H-30s, a single C-130H-MP as well as the C-130Ts. Five C-130Hs are located at Labuan with No. 15 Squadron to support the forces in Borneo. The other transport type in service is the DHC-4A Caribou, which operates with No. 1 Squadron. About a dozen Caribous remain in service, the type having been operated since 1966. The Caribou were upgraded during 1996 by Airod in Malaysia, after an order for 32 IPTN-built CN.235s placed in December 1993 was downgraded to 18, then just six by mid-1994. The CN.235s arrived in early 1999.

About 40 Sikorsky S-61A-4 Nuri helicopters were received from 1968-78, with most still in service, flying alongside approximately 15 smaller Aérospatiale Alouette IIIs in four squadrons. These are split between bases in Peninsular and East Malaysia. A single Agusta A 109C from two operated flies with No. 10 Squadron. Two S-70A Black Hawks have been delivered for use by the Prime Minister's office

The elderly Sikorsky S-61A-4 Nuri is in desperate need of replacement, yet no funds have been forthcoming. Nevertheless, four squadrons continue to operate the veteran, spread out around the nation.

Malaysia's fighter forces have received a dramatic boost with a two-pronged procurement campaign. Eight F/A-18Ds (above) have been acquired for 18 Squadron at Butterworth, while 17 and 19 Squadrons have received probe-equipped MiG-29Ns (and NUB trainers). The MiGs are expected to receive the RVV-AE missile in the near future.

but are operated by No. 2 Squadron - the type may be ordered to replace the S-61A Nuri. No. 2 Sqn is the staff and VIP transport squadron and, apart from the S-70As, operates a Dassault Falcon 900, a Fokker F28 Fellowship and a Challenger Special Edition, which is due to be replaced by a Global Express during December 1998.

Flight training for the RMAF commences on the MD3-160 AeroTiga, of which 20 are in the process of being delivered. They serve with the 1 FTC (Flying Training Centre). Trainees proceed to the PC-7s (44 received 1983-84) of 1 FTC or 3 FTC, while students destined for helicopter operations go to

2 FTC. Advanced training is undertaken on the MB-339A (13 received including an attrition replacement) or Hawk Mk 108 at No. 15 Squadron based at Butterworth, before proceeding to the fast jets. Cessna 402Bs are used to provide twin training and have further expanded their role to include photo-mapping, VIP transport and medical evacuation.

Future procurement for the RMAF has been seriously curtailed by the current Asian economic crisis. The air force is only maintaining its current high standards with difficulty, and has had to postpone additional purchases and upgrade programmes for the near future.

Although the fighter F-5s have been reroled as advanced trainers, the RF-5E TigerEyes continue in a front-line role for the time being.

Hawk 108s have been acquired for the advanced/weapons training role. The single-seat Hawk 208 is in service in the attack role, replacing the A-4PTM.

Royal Malaysian Navy (RMN)

The RMN received six former Royal Navy Westland Wasp HAS.Mk 1 helicopters in 1989, and formed its first flying unit, No. 499 Squadron, at Lumut. The service is currently after a replacement for the Wasp and has narrowed the choice down to the SH-2G Seasprite or the Super Lynx. A total of six helicopters is expected to initially be ordered.

Royal Malaysian Navy		
499 Sku	Wasp HAS.Mk 1	Lumut

Malaysian Army Air Corps

The Army air corps was inaugurated on 13 March 1997 using former air force Alouette IIIs as No. 881 Squadron. The organisation is tasked with airlift, tactical support and reconnaissance for Malaysia's Rapid Deployment Force, and hopes to be able to acquire up to 300 new attack and transport helicopters over the next 15 years. In the short term, it is planned to equip two further squadrons (Nos 882 and 883) with Alouette IIIs to form an aviation regiment with the existing No. 881 Sqn.

Malaysian Army Air Corps		
881 Sku	Alouette III	Keluang

MONGOLIA

Capital: Ulan Bator
Population: 2.1 million
Land area: 1.56 million km² (604,090 sq miles)
Major cities: Choybalsan, Hovd, Ulyasutay

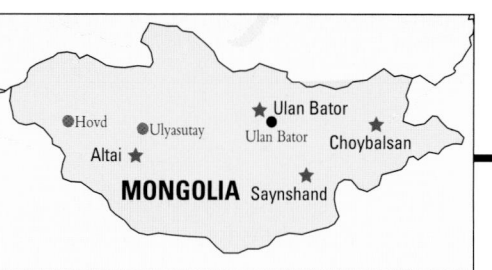

Air Force of the Mongolian People's Republic

Defence HQ, Ulan Bator

Antonov An-2 'Colt'	7	communications
Antonov An-24 'Coke'	6	transport
Antonov An-26 'Curl'	1	transport
Mil Mi-8 'Hip'	12	transport helicopter

Air Force of the Mongolian People's Republic

Landlocked between Russia and China, Mongolia has a population of 2 million, primarily involved in animal herding. Having defeated China's attempts to reassert sovereignty over the territory in 1921, Mongolia became the world's second Marxist-Leninist state.

The arrival of a Junkers F13 at Ulan Bator, the capital, on 25 May 1925 is seen as the founding of Mongolian civil and military aviation. Military aviation in Mongolia was dominated by the Soviets during the 1920s and 1930s, aircraft even wearing red stars until the 'Zoyombo' national insignia started to appear in 1935. Soviet forces based combat aircraft in the country. Flying Polikarpov R-1s, R-5s and U-2s, Mongolians fought indigenous anti-Communist rebels in 1932 and Japanese Manchurian troops in 1936.

The air force was officially named the Mongolian People's Army Air Corps in May 1937, later becoming the Air Force of the Mongolian People's Republic. Towards the end of World War II the air force had a few combat types (I-15s, I-16s, Yak-9s, Il-2s), but by 1953 they had been retired. Mongolian Airlines (or MIAT), established in 1956, operated the air force's non-combat types.

Organised along Soviet lines, the air force was also responsible for SA-2 'Guideline' SAMs, which were noted in the 1966 National Day parade. By the early 1990s they protected the important communications hub of Choibalsan in eastern Mongolia. The first jet combat aircraft, MiG-17 'Fresco-Cs', joined the air force in 1970, and were operated alongside An-2s, Il-14s, Ka-26s, Li-2s, Mi-2s, Mi-4s, Mi-8, PZL-104s, Yak-11s and Yak-18s. MiG-21PFMs and -21USs were delivered in the mid-1970s.

Today, the air force is virtually grounded. The fall of the Soviet regime in Moscow and the debt repayments have resulted in the virtual cutting-off of fuel and spares. The squadron of 12 MiG-21PFMs and two -21USs has been mothballed and a 12-helicopter Mi-24 squadron disbanded. Large numbers of An-2s are stored at Ulan Bator airport. Approximately a dozen Mi-8s and various transport aircraft (An-24, An-26, An-30) are operational. 1994 saw the delivery of five Harbin Y-12 IIs for Mongolian Airlines, plus a replacement aircraft for one written off.

MYANMAR (BURMA)

Capital: Yangon (Rangoon)
Population: 39.3 million
Land area: 678030 km² (261,720 sq miles)
Major cities: Bassein, Henzada, Mandalay, Myingyan, Moulmein, Pegu

Tamdaw Lay (Myanmar Air Force)

Type	Role	Delivered/ In Service
Aérospatiale SA 316B Alouette III	utility helicopter	/8
Bell UH-1H Iroquois	utility helicopter	12/10
Cessna 180	liaison	/6
Cessna 550 Citation II	light transport	1/1
Chengdu F-7M Airguard	air defence	26/24
Fokker/Fairchild F27/FH.227B/E	transport	5/4
Guizhou (GAIC) FT-7	combat trainer	6/6
Mil Mi-17-1V 'Hip H'	transport helicopter	12/12
Nanchang A-5C 'Fantan'	ground attack	24/24
Pilatus PC-6/B2-H2 Turbo-Porter	light transport	6/5
Pilatus PC-7 Turbo Trainer	basic trainer	17/12
Pilatus PC-9	advanced trainer	4/4
PZL-Swidnik Mi-2US/URN Kania	utility helicopter	22/15
PZL Swidnik W-3 Sokol	utility helicopter	15/15
Shaanxi Y-8D	heavy transport	4/4
SOKO G-4 Super Galeb	armed jet trainer	6/2

*The training assets of the Myanmar air force are concentrated at Meiktila air base and include the survivors of 17 **PC**-7s supplied from 1979 (below) and two **G**-4 Super Galebs (above).*

Tamdaw Lay (Myanmar Air Force)

On becoming a Socialist Republic in 1974 and adopting a broadly Marxist constitution, Burma abandoned its former main US arms procurement policies in favour of trade and military aid involvement with the Communist bloc. Renamed Myanmar in 1989, it subsequently became a military dictatorship, but was accepted as an ASEAN member in 1997. Twelve SIAI-Marchetti SF.260 basic trainers were purchased from Italy in 1975-76, followed by 20 SOKO G-4 Super Galeb armed jet-trainers from Yugoslavia in 1990. Main combat procurement has been from China, which delivered 26 Chengdu F-7M air defence fighters, 24 Nanchang A-5M attack aircraft, plus associated weapons, ground radars, and support equipment in the mid-1990s from a $1.2 billion contract.

In 1979-1986 Myanmar acquired from Switzerland six Pilatus Porter/Turbo-Porter STOL utility aircraft, plus 17 Pilatus PC-7 and four PC-9 turboprop train-

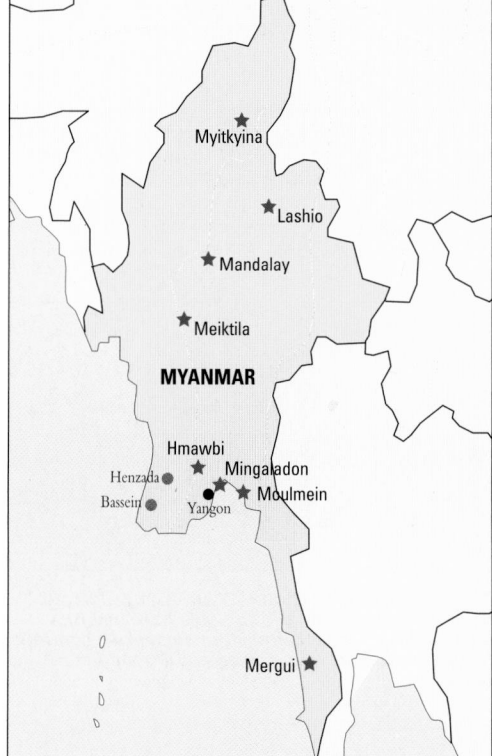

ers. At least 22 PZL-Swidnik Mil Mi-2 and 15 W-3 Sokol helicopters followed from Poland between 1991 and 1997, and 12 Mi-17-1Vs from Russia in 1995-96. An order was placed in 1997 for a single CASA C.212 light twin-turboprop transport. Myanmar is expected to become an early export customer for China's NAMC K-8 Karakoram jet trainer, to replace its Super Galebs which are mostly grounded for lack of spares and technical support. Negotiations have also been reported for the acquisition of MiG-29s from Russia, and F-7 upgrades discussed with Elbit in Israel. Main air bases are at Hmawbi, north of Yangon (Rangoon); Meiktila, near Mandalay; Myitkyina; and Moulmein.

Myanmar's transport helicopter force centres around the Mi-17, of which a dozen are in service. They are augmented by PZL-built Mi-2s and W-3s.

PHILIPPINES

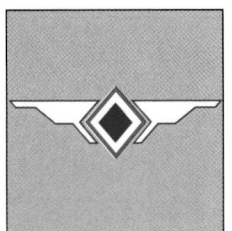

Capital: Manila
Population: 62.9 million
Land area: 300000 km² (115,800 sq miles)
Major cities: Cebu, Davao, Iloilo, Quezon City

Hukbong Himpapawid Ng Pilipinas (Philippine Air Force)

During the 1990s the Philippines has suffered from political unrest, guerrilla activity, large external debts and natural disasters such as hurricanes and volcanic eruptions. Hampered first by the corrupt mismanagement of the Marcos regime, and then by chronic instability following his enforced departure in early 1986, the Philippine Air Force has been involved in fighting the New People's Army and the Moro National Liberation Front by providing helicopter support to the army. Well over 100 UH-1H Iroquois have been received from the US since the 1970s, and some of the survivors are armed with machine-guns. Licence-built BO 105s, armed Sikorsky H-76s, MD 500MDs and MD 520MG Defenders have also seen service against the guerrillas. SAR duties are handled by a variety of helicopters, including UH-1s, Sikorsky S-76Cs and BO 105s. Up to 12 new helicopters are sought as a new SAR type, which is also likely to be selected as the Navy's new ASuW helicopter. The air force would like a heavylift helicopter, too, if the budget will allow.

For many years the Philippines depended on the USAF, located at Clark AFB, to provide its air defence. On 12 June 1991, Mount Pinatubo erupted after 600 years of dormancy, covering the air base in a thick blanket of ash and hastening a withdrawal from the base that had already been instigated by the demand for more 'rent' by the government in Manila.

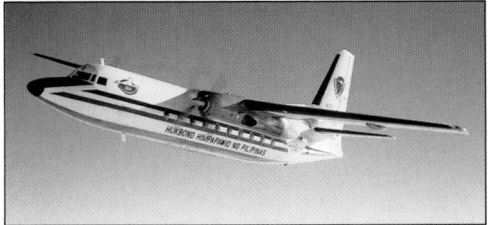

Above: The Fokker F27 provides useful inter-island transport, operating with the 221st Airlift Squadron at Villamor. Among the complement is a VIP-configured aircraft.

Below: Sangley Point is the main COIN base, home to the armed MD 520MGs of the 18th TASS. This unit operates in concert with AUH-76s and OV-10As.

This left the country defended by a token force of Northrop F-5A Freedom Fighters. Nineteen F-5As and three F-5Bs were delivered from 1965, while more recent deliveries have included one from Taiwan in 1996, two from Jordan and three from South Korea. All were grounded in January 1997, but they may be overhauled to act as a lead-in fighter for a new type.

The search for a replacement for the air force's ageing Freedom Fighters has been drawn out. Requests for proposals for 18 single-seat and six two-seat advanced combat aircraft were expected for the Mirage 2000-5, Kfir 2000, F-16C/D, F/A-18C/D, MiG-29 and JAS 39 Gripen. In 1997, the preference of the Philippine armed forces chief of staff, General Arnulfo Acedera, was for the Hornet armed with the AIM-120. Economic realities may preclude such a deal. Recently, the Kuwaiti A-4KU Skyhawks were under consideration, but their sale to Brazil has closed this option. The embargoed Pakistani F-16A/B Block 15s have also been mentioned, but these were snapped up by New Zealand. A parallel requirement under a 15-year defence modernisation budget is for 18 lead-in fighter trainers, for which surplus F-5E/F Tigers are the prime candidate, possibly from Saudi Arabia, South Korea or Taiwan. The Canadian CF-5A/Ds, the disposal of which is being managed by Bristol Aerospace, are also in the running, possibly as part of a barter deal.

Ex-USAF OV-10A Broncos have performed the ground attack mission for the air force since 1991. It is believed that at least 22 aircraft have been delivered, having replaced SF 260MP Warriors and AT-28Ds. Some of the Warriors were sold, but the survivors were upgraded to SF 260TP turboprop standard.

Pilot training with the 100th Training Wing at Fernando air base starts on the T-41D, of which 20 were received, before moving on to the SF.260TP. Advanced training is completed on the SIAI S.211 jet trainers ordered in 1988 for local assembly. The S.211 fleet has suffered from high attrition and is grounded regularly.

The Lockheed Hercules in three sub-types (two C-130Bs, three C-130Hs and an L-100-20), the

Responsibility for the defence of the Philippines rests with the F-5A/Bs of the 6th TFS at Basa, the last F-8 Crusaders having been retired some time ago.

surviving Philippine-assembled BN-2 Islanders (22 received, 1978-82) and GAF Nomad turboprops operate with the 220th Heavy Airlift Wing. The 250th Airlift Wing operates the Fokker Friendships (eight Mk 200s and a VIP Mk 200MPA) in one squadron, while another squadron is tasked with VIP helicopter services using the service's only S-70A Black Hawks and Pumas among its charges. Longer-range VIP transport is the role of the 702nd Presidential Airlift Squadron, which has a Friendship and a Fellowship on charge.

Hukbong Himpapawid Ng Pilipinas

5th Fighter Wing	**Basa**
6th Tactical Fighter Sqn	F-5A/B Freedom Fighter
105th Combat Crew Training Sqn	S.211
15th Strike Wing	**Sangley Point**
16th Attack Squadron	OV-10A Bronco
18th Tactical Air Support Sqn	MD 520MG
20th Air Commando Sqn	AUH-76 Eagle
100th Training Wing	**Fernando**
101st Pilot Training Sqn	T-41D Mescalero
102nd Pilot Training Sqn	SIAI-Marchetti SF.260MP/TP
103rd Pilot Training Sqn	S.211
205th Helicopter Wing	**Villamor**
210th Helicopter Training Sqn	Bell 205A-1
211th Helicopter Sqn	UH-1H, Bell 205A-1
505th Air Rescue Sqn	Bell 214, BO 105C, Sikorsky S-76, UH-1H
220th Heavy Airlift Wing	**Mactan**
222nd Heavy Airlift Sqn	C-130B/H/L-100-20 Hercules
223rd Tactical Airlift Sqn	BN-2A Defender, N22B Nomad Missionmaster/Searchmaster L
250th Airlift Wing	**Villamor**
221st Airlift Squadron	F27-200/200MPA Friendship
252nd Helicopter Sqn	Bell 212, BO 105C, SA 330L Puma, S-76, S-70A-5 Black Hawk
702nd Presidential Airlift Sqn	F27-200, F28 Fellowship 3000
901st Weather Sqn	Cessna T210G

Philippine Police Service

The Philippine Police Service has an aviation unit with a pair of BO 105s and a pair of locally assembled BN-2A Islanders.

Assets

Eurocopter BO 105C	2	paramilitary
P/B-N BN-2A Islander	2	paramilitary

Philippine Naval Aviation

A naval aviation unit uses four BN-2A Islanders. Like others in military use, they were locally assembled. Also in use are four MBB BO 105C helicopters. All are based at Sangley Point. As part of the major upgrade of the indigenous Philippine armed forces since the withdrawal of the US forces, the Philippine Navy has a requirement for up to nine anti-surface warfare helicopters. Contenders include the Bell 412, Eurocopter AS 565 Panther, BO 105 and Westland Super Lynx. The final decision may be affected by the choice of the air force's new SAR helicopter.

Assets

Eurocopter BO 105C	4	SAR helicopter
P/B-N BN-2A-21 Islander	4	transport

SINGAPORE

Capital: Singapore City
Population: 3 million
Land area: 616 km²
(238 sq miles)

Singapore's fighter assets are being bolstered by 42 new F-16C/D Block 50s, although some are retained in the US for training. The F-16Ds have the enlarged spine previously seen on some Israeli two-seaters. It houses extra avionics, perhaps for the SEAD mission.

Republic of Singapore Air Force

Situated at the end of the Malayan peninsula, Singapore holds a commanding position over the trade routes of Southeast Asia. Built up from little more than coastal jungle by the United Kingdom since Sir Stamford Raffles acquired the island from the Sultan of Johore in 1819, Singapore was a bastion of British power in the region, apart from a short and bloody spell under Japanese occupation from 1942 to 1945. In the post-war years Singapore was a British Crown Colony and the airfields at Changi, Sembawang, Seletar and Tengah were highly important as safe bases from which British forces could operate against Communist guerrillas during the Malayan Emergency. From 1959 the island was a self-governing state within the Malayan Federation, and in 1965 chose full independence. The British continued to defend the island state until Singaporean forces could take over in 1971, whereupon the RAF Lightnings of No. 74 Squadron returned to the UK.

Singapore stands at the crossroads of the Far East, and forms an important economic bridge between East and West. Political and economic unrest, exacerbated by the late 1990s 'Asian flu' Far East recession, surround Singapore on all sides, while its neighbours acquire ever more capable military equipment. The nation has seen phenomenal economic growth since independence and is keen to protect its position by maintaining the best-equipped forces in the region. Annual defence expenditure amounts to a huge 6 per cent of GDP. Equipment is cannily sourced from various nations, with the United States the main supplier.

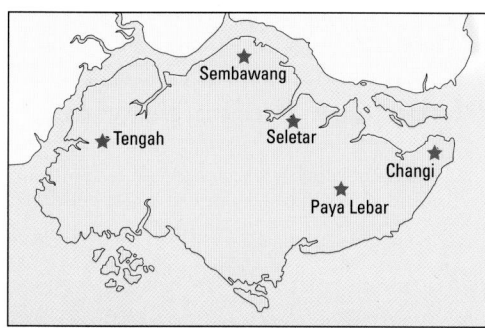

JET EQUIPMENT

Initially equipped with ex-RAF equipment such as the Hunter fighter and the Bloodhound SAM, the Republic of Singapore Air Force then embarked on a major transformation which saw the arrival of the A-4 Skyhawk (from 1975) and Northrop F-5E/F (from 1979). A total of 160 A-4S Skyhawks was acquired over the years, and they have been extensively modified, culminating in the current A-4SU variant, re-engined with the General Electric F404 turbofan. Three squadrons still fly the type within the RepSAF, based at Tengah, although only one (145 Sqn) is fully operational in the attack role.

In the late 1980s the RepSAF embarked on another major upgrade of its capabilities with the arrival of four E-2C Hawkeyes (1987) and eight F-16A/Bs (1989). Based at Tengah, these aircraft provided the principal air defence for the island, augmented by F-5s, until new F-16C/Ds were received in 1998. A batch of 22 F-16Cs and 20 F-16Ds was purchased, all to the latest Block 50 standard. Most, if not all, of the two-seaters feature the 'fat-back' spine previously seen on some Israeli F-16Ds. Singaporean aircraft have been noted carrying Sharpshooter targeting pods, a downgraded version of LANTIRN. A dozen of the aircraft serve with the 428th FS (USAF) at Cannon AFB, New Mexico, on training duties.

During 1998 the first squadron of F-5S/T aircraft achieved IOC. The upgrade programme is performed by Singapore Technologies Aerospace at the F-5's base of Paya Lebar, and involves fitting a new cockpit with two colour MFDs, wide-angle HUD and numerous other improvements. Fighter versions also receive a new radar in the form of the FIAR Grifo-F, at the expense of one of the cannon. The eight Tigereye reconnaissance aircraft also went through the programme, emerging as the RF-5S. They retain the camera nose in place of radar. The upgrade programme has been undertaken in co-operation with the Israeli company Elbit, and unconfirmed reports suggest that the F-5 and F-16 are now armed with the Rafael Python 4 missile and compatible with the Elbit DASH helmet-mounted sight.

Singapore is one of the nations being heavily courted by fighter manufacturers such as Dassault (Rafale), Eurofighter (Typhoon), Sukhoi (Su-35) and the Americans. A major purchase of state-of-the-art

fighters is expected in the near future to maintain Singapore's considerable technological edge over its neighbours.

SUPPORT AIRCRAFT

The Hercules forms the backbone of the small transport fleet, based at Paya Lebar. Two of the Hercules are configured as tankers to support the F-5 and A-4 fleet. With the arrival of large numbers of F-16s, a boom-equipped tanker is a necessity, and four Boeing KC-135Rs are on order. Two of these will be retained at McConnell AFB for crew training.

For maritime reconnaissance and transport duties the RepSAF has eight Fokker 50 Enforcer 2s, armed with AGM-84 Harpoon anti-ship missiles. The Fokker 50s are the only aircraft to be based at Changi, the international airport.

Republic of Singapore AF

111 Squadron 'Jaeger'	E-2C	Tengah
140 Squadron 'Osprey'	F-16A/B, F-16C/D	Tengah
142 Squadron 'Gryphon'	A-4SU, TA-4SU	Tengah
		to move to Cazaux, France
143 Squadron 'Phoenix'	disbanded (ex-A-4SU/TA-4SU)	Tengah
143 Squadron Aerobatics Team 'Black Knights'	A-4SU, TA-4SU	Tengah
145 Squadron 'Hornet'	A-4SU, TA-4SU	Tengah
Air Logistic Squadron (ALS) 'Woodpecker'	none	Tengah
Airfield Maintenance Squadron (AMS)	none	Tengah
120 Squadron 'Kestrel'	UH-1H	Sembawang
123 Squadron 'Red Hawk'	AS 550A2	Sembawang
124 Squadron 'Flaming Arrow'	AS 550C2	Sembawang
125 Squadron 'Puma'	AS 532UL	Sembawang
126 Squadron 'Cougar'	AS 532UL	Sembawang
Air Logistic Squadron (ALS)	none	Sembawang
Airfield Maintenance Squadron (AMS)	none	Sembawang
122 Squadron 'Condor'	C-130H, KC-130B/H	Paya Lebar
141 Squadron 'Merlin'	RF-5S	Paya Lebar
144 Squadron 'Blackite'	F-5S/T	Paya Lebar
149 Squadron 'Shikra'	F-5S/T	Paya Lebar
Air Logistic Squadron (ALS)	none	Paya Lebar
Airfield Maintenance Squadron (AMS)	none	Paya Lebar
121 Squadron 'Brahminy Kite'	Fokker 50 Enforcer Mk 2	Changi
Air Logistic Squadron (ALS)	none	Changi
130 Squadron 'Eagle'	S.211	Pearce (Austr.)
131 Squadron 'Harrier'	S.211	Pearce (Austr.)
150 Squadron 'Falcon'	disbanded (ex-SF-260MS/W)	Seletar

Air Defence Brigade (ADB)

160 Squadron 'Double A'	Oerlikon 35-mm	Seletar
163 Squadron 'Hawk'	I-Hawk	Lin Chu Kang
165 Squadron 'Rapier'	Rapier	Lin Chu Kang
170 Squadron	none (ex-Bloodhound)	Seletar
201 Squadron	ADRU ITT Radar	Bukit Gombak

This formation of Singapore's principal combat aircraft highlights the nation's commitment to maintaining a credible defence. Singapore's tiny airspace means that early warning of attack is vital. The E-2 Hawkeye has greatly extended the effective range of the nation's air defences. Both the A-4 and F-5 have undergone major modification programmes, while the F-16s and their US-trained pilots remain among the region's most respected fighters.

The adoption of the F404 engine in the upgraded A-4SU makes the Singaporean Skyhawks a potent tool. This example is from 145 Squadron.

Singapore Technologies Aerospace modified eight F-5s to RF-5E status. These have been further upgraded to RF-5S configuration.

Below: The bulk of Singapore's training effort is handled overseas. RAAF Pearce in Western Australia plays host to two squadrons of S.211 basic trainers which fly alongside the RAAF's 2 FTS, equipped with the Pilatus PC-9/A.

HELICOPTERS

Eight Alouette IIIs formed the initial rotary-wing complement of the RepSAF, followed swiftly by large numbers of Bell UH-1Bs and UH-1Hs. In the late 1980s the AS 332 Super Puma was purchased as the principal tactical transport type, along with AS 550s for light duties and training. The UH-1H remains in service with one squadron.

In 1999 Singapore announced its much-awaited decision concerning a battlefield helicopter, a new class of aircraft for the force. Not surprisingly, the AH-64D Longbow Apache was chosen, and the first of eight is expected to arrive in 2002. An option exists for a further 12, which is likely to be converted to a firm order soon.

TRAINING

Singapore measures just 42 km by 25 km and much of that space is heavily populated. With Malaysia across a channel to the north and the nearest Indonesian island easily visible from the south coast, airspace is at a premium. To remove as much of the burden from that space as possible, much of the training effort is located overseas. Basic training is under-taken at RAAF Pearce in Australia using 30 SIAI-Marchetti (Agusta) S.211s. Advanced jet training is now performed using the TA/A-4SU Skyhawk, with 10 aircraft based in France at Cazaux since 1998. A further 10 are to be transferred. Singaporean pilots train in the United States on the F-16, while advanced rotary-wing training is carried out at Oakey Army Air Base, Queensland, using 12 AS 332s.

TAIWAN

Capital: Taipei
Population: 19.7 million
Land area: 35990 km² (13,890 sq miles)
Major cities: Kao-Hsiung, Tainan, Taichung

Taiwan's F-16s are to Block 20 standard, the most advanced of the F-16A/B versions. Two wings fly the type, having replaced the F-5E.

Ta Chun Kuo Kung Chuang – Republic of China Air Force

The RoCAF is organised along USAF lines, with a Combat Air Command looking after manned aircraft, an Air Defence Artillery Command and a Logistics Command. The air force has eight front-line wings based at eight major hardened air bases, with a training unit at a ninth similar airfield. Five of the wings are designated as Tactical Fighter Wings, one as a Composite Tactical Fighter Wing, one as a TFW/Tactical Training and Development Centre, and one as a Troop Carrier and Anti-Submarine Combined Wing. A Tactical Control Wing administers the Hughes Skynet air defence ground environment, into which the Hawkeyes are now integrated.

Although planned production totals have been cut by approximately 50 per cent, the AIDC Ching-Kuo IDF (Indigenous Defensive Fighter) continues to be regarded as being of vital importance. Under present plans the aircraft will equip two fighter wings (the 3rd TFW at Chin Chuan Kang and the 1st at Tainan). The 3rd TFW has completed its re-equipment with the aircraft, with No. 28 Squadron being declared operational on 22 November 1995.

The next unit to transition to the IDF was the 1st TFW at Tainan. When the wing completes its re-equipment, production of the Ching-Kuo will almost certainly end (at 130 aircraft, including 28 two-seaters).

Despite its difficulties, the Ching-Kuo represents a major step for the Taiwanese aviation industry. The type equips two tactical fighter wings.

Planned IDF production was cut back when Taiwan gained a one-off opportunity to conclude a $5.8 billion deal to procure 150 F100-PW-220-engined F-16s. The Taiwanese F-16s comprise 130 single-seat F-16As and 30 two-seat F-16Bs, with a number of AIM-7 missiles and Raytheon AN/ALQ-183 ECM pods. Although nominally F-16As, the Taiwanese aircraft are more capable than their designation might suggest, and are in some respects the most advanced Fighting Falcons ever exported. Their weakness lies in the armament with which they were supplied, and repeated attempts to procure AIM-120 AMRAAMs have so far been unsuccessful.

By the end of 1997 the first F-16s had virtually replaced the F-5Es hitherto serving with the three squadrons of the 4th TFW at Chia Yi (Nos 21, 22 and 23 Squadrons) – initial deliveries were made to the base in April 1997. The next batch of new aircraft will re-equip the 8th TFW at Hualien. This unit currently

Tactical reconnaissance needs are met by the RF-5E 'TigerGazer'. AIDC-assembled F-5E fighters remain in use in some strength.

includes one squadron equipped with a mix of F-5Es, F-5Bs and F-5Fs (No. 16 Squadron, the F-5 OCU) and two squadrons equipped (since 1994) with 40 T-38s leased from the USAF to cover a shortage of F-5Es, pending the delivery of the F-16s.

Shortly after signing the F-16 contract, Taiwan also signed a $3.8 billion deal for the supply of Mirage 2000 fighters on 17 November 1992. The contract specified the delivery of 48 Mirage 2000-5Ei single-

Arguably Taiwan's most advanced aircraft are the Mirage 2000-5Eis of the 2nd TFW. The RDY radar and MICA missiles give multi-target capability.

seaters and 12 Mirage 2000-5Di two-seat trainers, and included the supply of 400 Magic 2 and (more importantly) 960 MICA missiles. Taiwan actually received its first MICAs before the Armée de l'Air did. The aircraft themselves were of the advanced Mirage 2000-5 version, with RDY radar, a new onboard CPU and an integrated ECM suite.

In service, the Mirage 2000 is replacing the fighter-tasked F-104s of the 2nd TFW at Hsinchu. Ten Mirage 2000s had been delivered by mid-1997, and by the end of that year all three squadrons were operational (to some degree) on the type.

The almost simultaneous introduction of three front-line fighter types has not been easy, and has imposed some training and logistics challenges. It has also allowed the air force to deploy the aircraft to suit their optimum role, so the Mirage 2000s are primarily responsible for high-altitude long-range air defence, the AIDC Ching-Kuos for low- and medium-level air defence, and the F-16s for medium-level air defence and other tactical fighter missions. All three fighter types are equipped with modern BVR missiles, and all are superior to the PLA Air Force's current fighters, including its first generation 'Flankers'. All three types also have a real air-to-ground capability, and the air force has an effective anti-ship capability using indigenously developed weapons and perhaps the AGM-84 Harpoon.

By the end of 1997, single squadrons of Mirage 2000s and F-16s were in full service, and it is expected that deliveries will be completed during 1999. This will leave all five of the air force's front-line fighter/composite wings equipped with modern fourth-generation fighters, with F-16s at Chia Yi and Hualien, Mirage 2000s at Hsinchu, and IDFs at Chin Chuan Kang and Tainan. F-5Es and RF-5Es will be based at Tao Yuan and Taitung.

Taiwan finally began retiring its ageing RF-104G 'Stargazer' from late 1997 (a process to be completed in 1998), and is replacing these aircraft with seven surplus F-5Es converted to RF-5E 'TigerGazer' configuration by Singapore Aerospace. The new RF-5Es may only be an interim replacement for the RF-104s, and some sources suggest that a number of the F-16s may be delivered with a reconnaissance capability.

Re-equipment of the RoCAF proceeded apace

Taiwan's search and rescue capability received a major boost with the arrival of the S-70C 'Bluehawk'. These are equipped with radar and FLIR, and can carry external tanks.

during the 1990s. Those F-5Es not being replaced by AIDC Ching-Kuos, Lockheed Martin F-16s and Mirage 2000s (about 90 of the 284 built by AIDC) will be upgraded by AIDC. The type will also continue to operate in the adversary training role. The prototype 'Tiger 4' is due to fly for the first time by the end of 1998. Units continuing to operate the F-5E will be the 5th TFW at Tao Yuan (with Nos 17, 26 and 27 Squadrons) and the 7th TFW/TT&DC (Tactical Training and Development Centre) at Taitung. Other F-5E-equipped units are already converting to other aircraft types: the 1st TFW to the Ching-Kuo, and the 4th and 8th TFWs to the F-16. Plans for a more ambitious upgrade, with a new engine, appear to have been dropped.

TRANSPORT, EW AND SAR ASSETS

The RoCAF finally withdrew the last of its C-119s in December 1997. Only one C-119 squadron (No. 103) remained, with No. 101 having converted to the C-130 in 1986, and No. 102 in 1993.

The 6th Troop Carrier and Anti-Submarine Combined Wing also incorporates a single Electronic Warfare and Airborne Early Warfare Squadron (this unit has also been reported in some sources as No. 78 Squadron, but the EW&AEW title is still believed to be accurate). It is equipped with a single EW/Elint-configured Lockheed C-130H (delivered 'green' and equipped by AIDC) and four Grumman E-2T Hawkeyes. It was said that the arrival of the E-2T raised the 'warning of hostile air attack' time available to Taiwan from five to 25 minutes.

The RoCAF did operate at least eight Elint/EW training/calibration-configured C-47s, and two Elint/Sigint C-54s. The last two C-47s in service have now been replaced by two Beech 1900Cs in the airfield/navaid calibration role.

The RoCAF's remaining transport aircraft are based at Sungshan, close to the island's capital of Taipei. The VIP Squadron of the Sungshan Air Base Command is currently undergoing major changes: its long-serving Presidential Boeing 720 and VIP C-47s have been retired and partially replaced by three Fokker 50s. Only one (of two) Boeing 727s remain in use, primarily used for resupplying the garrison on the island of Quemoy. A long-range Presidential aircraft is urgently required, with a new-generation Boeing 737 or 757 being the most favoured option.

Taiwan has always placed heavy emphasis on SAR capabilities, currently using Sikorsky S-70C-1A Blue Hawks in the role. The Blue Hawk is basically a UH-60A, equipped with a rescue winch, similar to the USAF's HH-60As but without folding stabilators or an inflight-refuelling capability. Two of the 14 Blue Hawks delivered are maintained on detachment at Sungshan, where they operate in the SAR role while maintaining a VIP transport capability.

TRAINING PROGRAMME

The Taiwanese training requirement is significant, since the air force has to at least maintain its present strength of 1,000 pilots, while retention (in the face of a booming civil sector) is not easy.

The surviving Pazmany PL-1s of the Flying Cadet School were sold to a US dealer in 1982, and have been replaced by a number of microlight aircraft to

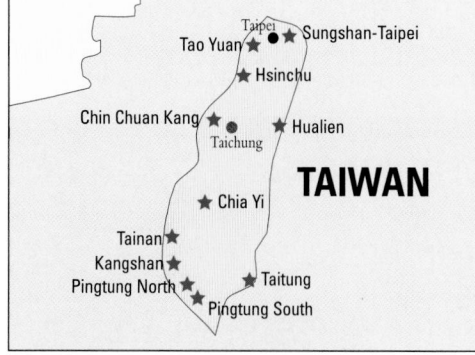

TAIWAN

Taipei
Tao Yuan ● ★ Sungshan-Taipei
★ Hsinchu
Chin Chuan Kang ★ ● ★ Hualien
Taichung
★ Chia Yi
Tainan ★
Kangshan ★ ●
Pingtung North ★ ★ Taitung
Pingtung South

The VIP Squadron at Sungshan continues to operate a single Boeing 727, primarily on the regular shuttle serving the strategic outpost on Quemoy island. A modern replacement is desperately sought.

Ta Chun Kuo Kung Chuang – Republic of China Air Force HQ

1st TFW (443 TFW)		
1st Fighter Group		
1 TFS	IDF Ching-Kuo	Tainan
3 TFS	IDF Ching-Kuo	Tainan
9 TFS	IDF Ching-Kuo	Tainan
71 TFS	AT-3B Tzu Chung	Tainan
72 TFS	ACH-1B Chung-Shing	Tainan
2nd TFW (499 TFW)		
11th Fighter Group		
41 TFS	Mirage 2000-5Ei/5Di	Hsinchu
42 TFS	Mirage 2000-5Ei/5Di	Hsinchu
48 TFS	Mirage 2000-5Ei/5Di	Hsinchu
3rd TFW (427 TFW)		
3rd Fighter Group		
7 TFS	IDF Ching-Kuo	Chin Chuan Kang
8 TFS	IDF Ching-Kuo	Chin Chuan Kang
28 TFS	IDF Ching-Kuo	Chin Chuan Kang
4th TFW (455 TFW)		
4th Fighter Group		
21 TFS	F-16A/B Block 20	Chia Yi
22 TFS	F-16A/B Block 20	Chia Yi
23 TFS	F-16A/B Block 20	Chia Yi
SAR Sqn 'Seagull'	S-70C-2 Blackhawk	Chia Yi
5th TCW (401 TCW)		
5th Fighter Group		
17 TFS	F-5E	Tao Yuan
26 TFS	F-5E	Tao Yuan
27 TFS	F-5E	Tao Yuan
6th TC&ASCW (439 TC&ASCW)		
10th Transport Group		
101 TCS	C-130H	Pingtung North
102 TCS	C-130H	Pingtung North
103 TCS	C-130H	Pingtung North
EW & AEW Sqn	E-2T, C-130H(EW)	Pingtung North
7th TFW/TT&DC (737 TFW)		
7th Fighter Group		
44 TFS	F-5E/F	Taitung
45 TFS	F-5E/F	Taitung
46 TFS	F-5E/F (F-16A/B)	Taitung
8th TFW (828 TFW)		
8th Fighter Group		
4th TRS	RF-5E TigerGazer	Hualien
14 TFS	none	Hualien
15 TFS	none	Hualien
16 TFS	F-5B/E/F	Hualien
Sungshan Air Base Command		
VIP Squadron	Boeing 727-109	Sungshan-Taipei
	Beech 1900C-1	Sungshan-Taipei
	Fokker F-50	Sungshan-Taipei
SAR det 'Seagull'	S-70C-2	Sungshan-Taipei
RoCAF Academy		
35 Squadron	AT-3 Tzu-Chung	Kangshan
Basic Training Group		
?? Squadron	T-34C	Kangshan
?? Squadron	T-34C	Kangshan
Fighter Training Group		
?? Squadron	AT-3 Tzu-Chung	Kangshan
?? Squadron	AT-3 Tzu-Chung	Kangshan
Transport Training Group		
loaned from VIP Sqn	Beech 1900C	Kangshan
Air Defence		
?? Battery (6 launchers)	Patriot SAM	Wan-Li (Taipei NE)
?? Battery (6 launchers)	Patriot SAM	Nan-Kang (Taipei S)
?? Battery (6 launchers)	Patriot SAM	Lin Kow (Taipei SW)
?? Battery (6 launchers)	Patriot SAM	Taichung
?? Battery (6 launchers)	Patriot SAM	Kaoshiung

The AT-3 Tzu-Chung provides advanced training at the Air Force Academy, where all aircraft wear the colours of the 'Thunder Tigers' aerobatic team.

All training assets are located at Kangshan. Two squadrons fly the Beech T-34C Turbo Mentor on basic training tasks.

maintain that portion of the flying training syllabus. Under this, aircrew trainees transfer to basic flying training only after elementary training and screening on the new microlights. From 1984 the indigenous T-CH-1 trainer began to be replaced by 40 Beech T-34C Turbo Mentors. Reversing the trend away from indigenous aircraft types was the replacement of the T-33 by the AIDC AT-3 from 1985.

Pilots not destined for front-line fast jets transition from the T-34C or AT-3 to the Transport Training Group. This unit finally replaced its last five ageing DC-3/C-47s in January 1996, using the Raytheon Beech 1900Cs which had replaced the DC-3/C-47 in the light transport role. The surviving Beech 1900s

(11 of 12 delivered) all belong to the VIP unit but one is flown each day to the Air Academy for multi-engine training, returning to Sunshan each evening.

FUTURE PROCUREMENT

Taiwan currently requires an aircraft that has a rear loading ramp (for paradrops and low-level cargo dropping), and a degree of rough-field/STOL capability. Front-runners to meet the requirement are the CN.235-300 and the AE2100-engined C-27J, with the ATR 52C being viewed as an outside contender. A follow-on fighter is also sought, with Rafale, Eurofighter or F-22 seen as the most likely contenders.

Republic of China Army Aviation

The army on Taiwan has a helicopter force of some 100 Bell UH-1Hs (most of which were licence-built by AIDC) which are used for air mobility. Observation and light attack duties are undertaken by the OH-58D Kiowa Warrior, of which 26 were delivered from 1993. Another 13 examples were ordered in September 1997. The main strike asset of the army is its AH-1W SuperCobras, 42 having been in the initial order, followed by another 21 helicopters in May 1997. Helicopter training is undertaken on 14 Hughes TH-55A Osages. They are in the process of being replaced by 30 Bell TH-67A Creeks, which are also due to replace UH-1Hs in the advanced training role. The army also flies some Cessna O-1 Bird Dog lightplanes in the communication role and three Boeing 234MLR Chinooks in the heavy-lift role.

Republic of China Naval Aviation

Despite the conversion of Taiwan's S-2 Trackers to turboprop power during 1991, the aircraft has proved largely (and increasingly) inadequate, and a replacement is urgently sought. For the time being, though, the 1,645-shp (2205-kW) Garrett TPE331-1-5AW-powered S-2T TurboTracker remains in front-line use. Only 22 conversions are believed to have been undertaken as the programme became mired in a procurement scandal. The S-2Ts served with the air force's 6th TC&ASCW at Pingtung, until transferred to the navy in 1998. Taiwan has made repeated requests for the supply of P-3C Orions but these have just as habitually been rebuffed.

The navy operates a dozen Hughes 500D light helicopters in the ASW role from 'Gearing'-class destroyers, and received 12 Sikorsky S-70C(M)-1 Thunderhawk ASW helicopters to operate from 'Perry'-class frigates. Another 11 S-70C(M)-2s were ordered in early 1998 as part of an FMS contract.

Republic of China Naval Aviation

6th ASW Group		
33 ASS	S-2T Tracker	Pingtung South
34 ASS	S-2T Tracker	Pingtung South
8thTFW (828th TFW)		
ASW Heli Group		
501 Squadron	MD-500	Hualien
701 Squadron	S-70C(M)-1	Hualien

Illustrated here are the three aircraft types operated by the navy: S-2T Tracker (above), S-70C(M)-1 Thunderhawk (below left) and MD-500 (below).

Republic of China Army Aviation

1st Attack Heli Wing		
1 Attack Heli Sqn	AH-1W	Longtang-Tao Yuan
2 Attack Heli Sqn	AH-1W	Longtang-Tao Yuan
1 Heli Recce Sqn	OH-58D	Longtang-Tao Yuan
10 Assault Heli Sqn	UH-1H	Longtang-Tao Yuan
11 Assault Heli Sqn	UH-1H	Longtang-Tao Yuan
2nd Attack Heli Wing		
3 Attack Heli Squadron	AH-1W	Kuejien-Tainan
4 Attack Heli Squadron	AH-1W	Kuejien-Tainan
2 Heli Recce Squadron	OH-58D	Kuejien-Tainan
20 Assault Heli Sqn	UH-1H	Kuejien-Tainan
21 Assault Heli Sqn	UH-1H	Kuejien-Tainan
3 Heli Recce Squadron	OH-58D	Kuejien-Tainan
? Medium Transport Sqn	Bv234MLR	Kuejien-Tainan
Training		
? Heli Training Sqn	TH-55A, TH-67A	Kuejien-Tainan
Communication		
? Communication Squadron	O-1 Bird Dog	Kuejien-Tainan

The OH-58D is the army's main scout helicopter, working closely with the AH-1W attack force.

TAJIKISTAN

Capital: Dushanbe
Population: 5.2 million
Land area: 143100 km² (55,235 sq miles)
Major cities: Khudzhand, Kulyab, Kurgan-Tyube, Ura-Tyube

The Tajik Soviet Socialist Republic gained independence from the Soviet Union on 9 September 1991, following the August *putsch*, and became the Republic of Tajikistan. The country was soon being torn apart by armed conflict and regionalism. Civil war soon followed, with the forces of Emomali Rahmanov (originally Speaker of the Parliament and later President of the Republic) on one side and a range of opposition parties on the other.

The arrival of a Russian-dominated peacekeeping force (consisting mainly of the 201st Motorised Rifle Division) in early 1993 and military support from Uzbekistan dispersed most of the opposition militia over the border into Afghanistan or into the remoter parts of Tajikistan. The United Nations Mission of Observers in Tajikistan (UNMOT) was established in December 1994. Fierce fighting broke out on the Tajik-Afghan border in April 1995, during which Russian bombers were in action. Again, in December 1995, Russian fighter aircraft bombed opposition bases in Tavil-dara. Reports of the Russian 23rd Composite Regiment using Su-25s, Mi-8s, Mi-24s, An-24s and An-26s operating from Dushanbe were received in late 1996, and the Russian Border Guards units in the country are known to have their own Mi-8s.

A UN-brokered accord between the government and the Islamic opposition groups was signed in Moscow on 27 June 1997. Unfortunately, the secular opposition is not part of the peace treaty, and armed groups hostile to the government abound. The Russian presence in the country is to be scaled down

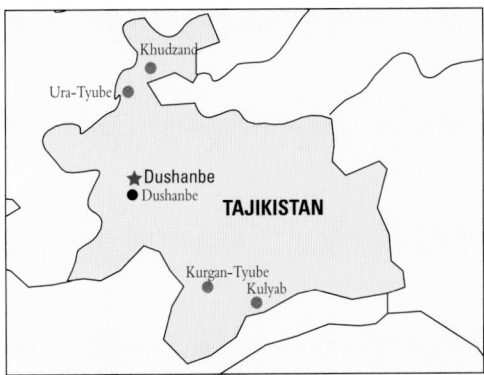

and there are plans to reduce the Tajik armed forces by 10 per cent. Tajikistan's own air force is small, including only about 10 Mi-8MTBs delivered in 1993, and five Mi-24s, all based at Dushanbe. It is planned to acquire Sukhoi Su-25s from Belarus to form an attack squadron. The army provides air defence using SA-2 and SA-3 missiles.

THAILAND

Capital: Bangkok
Population: 54.6 million
Land area: 514000 km²
(198,405 sq miles)
Major cities: Chiang Mai, Thonburi, Ubon Ratchathani

Thailand's most capable aircraft is the F-16A/B (above), which flies with two squadrons. A recent order for F/A-18s was cancelled due to lack of funds, although some money is being spent on upgrading the F-5Es (right). This example is from 904 Squadron, formerly the Royal Flight.

Kongtap Agard Thai/Royal Thai Air Force

The roots of the Kongtap Agard Thai (RTAF, or Royal Thai Air Force) go back to 1911. In 1917 the Royal Siamese Flying Regiment took part in World War I, flying against the Germans. After the invasion by the Japanese in World War II, the Siamese air force became part of the Japanese air force, operating Martin B-10s, Nakajima Ki-21s, Ki-27s, Ki-30s and Ki-43s. After World War II, the Royal Thai Air Force was modernised with Spitfires, F8F Bearcats and AT-6A Texans. From 1957 jets arrived in the shape of 30 F-84G Thunderjets, 47 F-86F and 17 F-86L Sabres. As the American involvement in Southeast Asia increased during the late 1960s, 88 (A)T-28D Trojans arrived, followed by 16 A-37B Dragonflies in the mid-1970s.

The RTAF comprises four geographically-divided Air Divisions and a Flying Training School. The first Air Division is in the Bangkok area, the second in the eastern part of Thailand, the third in the central-northern provinces and the fourth in the stretched-out southern provinces.

FIGHTER SQUADRONS

Since July 1989, 103 Sqn at Korat has been tasked with air defence for the whole of Thailand, equipped with 14 F-16As and four F-16Bs. From September 1995, 403 Sqn at Takhli received a similar number of F-16s. Delivered in 1966, the surviving RF/F-5As and F-5Bs are flown by 231 Sqn at Udon Thani beside the more capable F-5E. From 1980, the RTAF received 38 F-5Es and six F-5Fs, which have been operated by 711 Sqn at Surat Thani (with SRT codes) since 1994, in the fighter-bomber attack/all-weather interceptor and aggressor roles. A third F-5 unit is the Ubon Ratchathani-based 211 Sqn, which retired 16 A-37B Dragonflies in favour of the F-5E/F in 1994. The F-5-equipped Royal Flight was brought to squadron status as 904 Sqn during 1996 wearing VM codes; F-5 pilot Crown Prince Vatsilalongkon Machidon is a frequent flyer.

In September 1993 the first of 36 Aero L-39ZA Albatros was delivered. Three L-39 squadrons are tasked with ground-attack and jet training. 101 Sqn at Korat became the first Albatros user in 1993, a year later followed by sister squadron 102 Sqn. In 1994 401 Sqn at Takhli withdrew the FT-600 Fantrainer in favour of the L-39ZA.

The GAF Nomad is in widespread use for tactical transport work with the army. The aircraft originally had a counter-insurgency role.

At Chiang Mai 411 Sqn operates 19 OV-10C Broncos of the 32 delivered during 1971-73, in the counter-insurgency role. Other COIN aircraft followed during 1973-74, being 33 AU-23A Peacemakers now operated by 202 Sqn at Kokkathium and 531 Sqn at Prachuap Khiri Khan. These squadrons also operate the last remaining air force O-1 Bird Dogs.

TRANSPORT

Since 1980, 601 Sqn has operated the Hercules fleet which now comprises six C-130Hs and six C-130H-30s. Three C-123K Providers were flown by 602 Sqn for rainmaking missions until 1995, when the final three C-47s of 603 Sqn were transferred to 602 Sqn. The Dakotas are to be converted to Basler Turbo-67 standard to continue the rainmaking task. 603 Sqn operates six BAe 748s (of 1964 vintage) besides six Alenia G222s delivered from May 1995 for transport missions. 604 Sqn, nicknamed the 'Thai Flying Club', is tasked with liaison and refresher training, operating about five SF.260MTs, a few Cessna 150Hs and about 20 T-41Ds. 605 Sqn and 461 Sqn at Phitsanulok operate 22 GAF N-22B Nomads delivered in 1982-84. Initially the Nomads were tasked with counter-insurgency to replace AC-47s, but are now used for regular transport. 605 Sqn also performs special missions such as photographic survey using three Merlin IVAs delivered in 1979. Three Learjet 35As delivered in 1982 are tasked with target towing and ECM missions, while three IAI Aravas (delivered from 1988) perform Elint missions. The Royal Flight at Don Muang operates one Boeing 737-3Z6 for the king of Thailand, delivered in 1984. Other VIP transports include an Airbus A310 delivered in 1991, two Bell 412STs and two AS 332L-4s delivered in 1996.

The RTAF has two helicopter squadrons for transport and combat SAR, both based at Kokkathium AB. 201 Sqn flies 14 20-year-old Sikorsky S-58Ts in the combat SAR role, while 203 Sqn flies 30 UH-1Hs. The latter squadron has six SAR detach-

The RTAF has a number of veteran types still performing useful work. The OV-10A (above) flies COIN missions, while the S-58T (below) is used for combat SAR. The S-58s are turbine-engined conversions of CH-34s.

ments at Chiang Mai, Ubon Ratchathani, Udon Thani, Don Muang and Kamphaeng Saen.

TRAINING

Since 1973, 1 Flying Training Squadron of the Flying Training School at Kamphaeng Saen has been equipped with 23 CT-4A/Bs Airtrainers. For advanced training, 2 FTS operates 21 PC-9s delivered in 1991. Jet training is undertaken with seven T-37B/Cs of 1961 vintage. The T-33A was withdrawn from service in 1995, so fighter lead-in training is performed by the L-39 squadrons. Helicopter training is performed on AB 206B-3s and UH-1Hs.

THE FUTURE

The Asian financial crisis of 1998 forced the air force to cancel its order for eight MDC F/A-18C/Ds in April 1998. The funds are being diverted to Elbit, which is due to upgrade 36 F-5E/Fs. At the same time, the government has opened discussions with the United States over the possible purchase of used F-16s, a much cheaper alternative than the new F/A-18s. A replacement is needed for the SF.260MT and interest has been shown in 48 Grob G 115TAs. The air force has identified a requirement for an additional 10 L-39ZA/ARTs and six C-130H-30s.

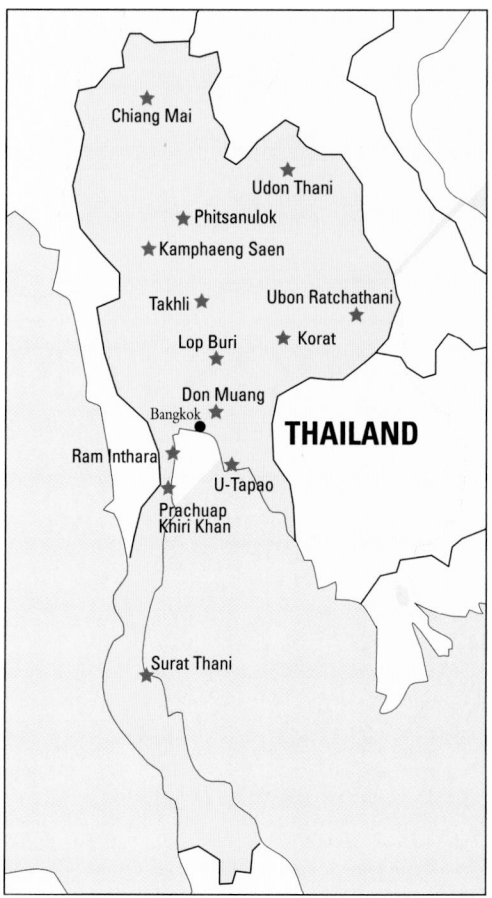

Transport for the RTAF is provided by a variety of types, including the Alenia G222.

605 Sqn employs several types for specialised tasks. The Merlin IVAs on strength fly photo-survey missions.

Royal Thai Air Force

1st Air Division

Wing 2	Kokkathium AB (Lop Buri)
201 Sqn	S-58T
202 Sqn	AU-23A
203 Sqn	UH-1H
Wing 6	Don Muang
601 Sqn	C-130H/C-130H-30
602 Sqn	C-47 (Basler Turbo-67)
603 Sqn	BAe 748, G222
604 Sqn	T-41D, Ce 150, SF.260MT
605 Sqn	Merlin IVA, Learjet 35A, IAI 201, N-22B
904 Sqn	F-5E
Royal Flt	Boeing 737, Airbus A310, Bell 412, AS332L-4

2nd Air Division

Wing 1	Korat AB
101 Sqn	L-39ZA
102 Sqn	L-39ZA
103 Sqn	F-16A/B
Wing 21	Ubon Ratchathani
211 Sqn	F-5E/F
Wing 23	Udon Thani
231 Sqn	(R)F-5A/B, F-5E/F

3rd Air Division

Wing 4	Takhli
401 Sqn	L-39ZA
403 Sqn	F-16A/B
Wing 41	Chiang Mai
411 Sqn	OV-10C
Wing 46	Phitsanulok
461 Sqn	GAF N-22B

4th Air Division

Wing 53	Prachuap Khiri Khan
531 Sqn	AU-23A, O-1
Wing 71	Surat Thani
711 Sqn	F-5E/F

Flying Training School (FTS), Kamphaeng Saen

1 FT Chicken Sqn	CT-4A/B
1 FT Twinny Sqn	T-37B/C
2 FT Mustang Sqn	PC-9
Helicopter Flight	Bell 206B-3, UH-1H

Used for flight experience and primary training, the RTAF operates the survivors of 14 Hoffman H 36 Dimona motor-gliders. This example flies with Tango Squadron, the Historic Flight.

Kongbin Tha Han Lur/Royal Thai Navy Air Division

The roots of the Royal Thai Navy Air Division (RTNAD) extend to June 1938 when a Japanese-built Watanabe aircraft was operated. After World War II the RTN switched to American-built aircraft, such as the SB2C Helldiver. Today the RTNAD comprises three wings, each with specific tasks and operating a diversity of aircraft. No. 1 Wing at U-Tapao has the combat role, No. 2 Wing at Songkhla has second-line duties, and No. 3 Wing at U-Tapao is dedicated to carrier operations.

COMBAT ROLES

In September 1997, ex-Spanish Navy Harriers arrived in Thailand on the new Spanish-built aircraft-carrier HTMS *Chakri Naruebet*, from which the seven AV-8Ss and two TAV-8Ss of No. 301 Sqn will operate. However, due to the economic depression, no more spare parts can be purchased until mid-1999, resulting in a growing number of unserviceable Harriers. The HTMS *Chakri Naruebet* will also embark No. 302 Sqn operating six S-70B-7 Sea Hawks delivered in 1997 for ASW and SAR duty.

At U-Tapao is No. 1 Wing with four squadrons. No. 101 Sqn remains operational with four S-2F Trackers delivered in 1952, using the aircraft only for flight training and surveillance. From 1991 No. 101 Sqn has also operated three Dornier Do 228-212s in the maritime patrol and reconnaissance role. Three Fokker F27 Mk 200 Maritime Enforcers have performed the ASW task from 1984. In 1995 the ASW and maritime patrol assets were boosted by the delivery of two P-3Ts and a UP-3T utility/trainer, while two P-3As were delivered for spare parts in 1993-94. Light attack and short-range support has been the task of No. 103 Sqn since the early 1980s, operating 11 Summit Sentry T-337SPs, four Cessna U-17B Skywagons and five Cessna O-1G Bird Dogs. Tasked with land-based jet attack, No. 104 Sqn is equipped with 18 ex-US Navy Corsairs (14 A-7Es and four TA-7Es), of which the first were received late in 1995.

SECOND-LINE ASSETS

Based at Songkhla in southern Thailand, near the border with Malaysia, No. 2 Wing is tasked with sec-

RTN attack assets consist of the land-based A-7Es of 104 Squadron (below) and the ex-Spanish Navy carrier-capable AV-8S of 301 Squadron (above). The Harriers suffer from a lack of spares.

The S-2F Tracker clings to service in the RTN, although it is expected to be fully replaced by the Do 228. Both types serve in 101 Squadron, the Trackers having a principal training function.

Royal Thai Navy Air Division

Wing 1	U-Tapao
101 Sqn	S-2F, Do 228-212
102 Sqn	P-3T, F27-200ME
103 Sqn	T-337SP, U-17B, O-1G
104 Sqn	A-7E, TA-7C
Wing 2	Sonkhla
201 Sqn	C-47, CL-215, F27-400M, N-24A, Do 228-200
202 Sqn	Bell 212, UH-1H
203 Sqn	Bell 214ST, S-76B
Wing 3	U-Tapao
301 Sqn	T/AV-8S
302 Sqn	S-70B

ond-line duties and operates both fixed- and rotary-winged aircraft. No. 201 Sqn has been equipped with one C-47 transport and one Elint/Sigint C-47 since 1973. For SAR and fire-fighting, a pair of CL-215-IIIs has been used since 1978, together with two F27-400Ms for transport and SAR. Delivered during 1982-84, five GAF N-24L Searchmasters are used by No. 201 Sqn for coastal patrol and SAR. Rotary-winged support is provided by No. 202 Sqn at Sonkhla with UH-1Hs delivered in 1975, augmented three years later by eight Bell 212s. Sister unit No. 203 Sqn operates four Bell 214STs delivered in 1987 and five S-76s received during 1996-97.

FUTURE PROCUREMENTS

In late 1997 the US government offered the RTNAD 10 SH-2Fs from AMARC. The helicopters are free, but Thailand would have to cover the cost of upgrading the helicopters to SH-2G standard and integrating them with the S-70 force; the high cost of this process makes the deal doubtful. The RTNAD is pleased with the Do 228 and is negotiating for three Do 228 MPAs.

The RTN operates two P-3Ts (together with a UP-3T trainer) for long-endurance maritime patrols. These are ex-US Navy P-3Bs and serve with 102 Squadron.

The Bell 212 augments the UH-1 in the tactical transport role. A few undertake staff transport tasks.

Kongbin Tha Han Bo/Royal Thai Army Air Division

In 1952 the Royal Thai Army Aviation Division (RTAAD) was separated from the RTAF, initially operating 22 L-4 Piper Cubs and five Stinson L-5s. Today, the RTAAD operates in excess of 150 aircraft and helicopters from two Air Regiments based at Lop Buri and detachments spread over Thailand.

AVIATION BATTALION

With more than 200 Hueys delivered since the early 1980s (about half still in service), the type is the backbone of the RTAAD, but from 1986 they have been augmented by 56 Bell 212s. In 1972 four CH-47A Chinooks arrived, replaced in 1989 by five new CH-47Ds and augmented in 1991 by three

Two Beech 1900C-1s provide VIP transport to the Army, flying alongside Jetstream 41s and King Airs.

Royal Thai Border Police Aviation

Activated in 1954 with assistance from the US Central Intelligence Agency (CIA), Royal Thai Border Police Aviation currently operates a variety of aircraft at 22 detachments. Home base for fixed-wing types is Don Muang, while the helicopters are based at Ram Inthara. During the 1970s, 18 UH-1Hs and 26 Bell

upgraded CH-47Cs. A dedicated attack flight was added in 1990, comprised of only four AH-1F Cobras. For observation tasks 10 Bell 206B JetRangers are operated, while most of the 60 Cessna O-1E Bird Dogs and 10 U-17Bs delivered from the mid-1950s still perform artillery spotting and utility tasks. For reconnaissance tasks the RTAAD received eight U-27As (FMS-sourced Cessna 208 Caravans) in 1986, wired to carry an electro-optical reconnaissance pod.

TRAINING

Since 1973 the Army Aviation Centre has used the Hughes TH-55 and H-300 as basic trainers, and the UH-1H and Bell 212 for advanced training. Since 1956 the O-1E has operated as the fixed-wing trainer, augmented by Cessna T-41Ds and U-17Bs from 1973. Twenty-five Maule MX-7-25s were added to the training fleet in 1992.

FIXED-WING

Two Short 330-UTTs arrived in 1984, augmented two years later by two CASA C.212-300s. Staff and VIP transport is performed by two Beech 200 King Airs, two Beech 1900C-1s and two Jetstream 41s.

Royal Thai Border Police Aviation

Don Muang
PC-6B, Shorts 330UTT, Fokker 50, CN.235-200
Ram Inthara
UH-1H, Bell 205, Bell 212, Bell 412, Bell 206B, Bell 206L, Hiller UH-12E, SA 365

205s were delivered. In the 1980s these were augmented by 15 Bell 212s, two Bell 412s, two Bell 206Ls, 12 armed Bell 206Bs and 10 Hiller UH-12E trainers. In 1990 two SA 365 Dauphins were delivered. Four PC-6Bs remain from the 1973 delivery, augmented by two Shorts 330UTTs in 1984-85. One Fokker 50 was added in 1992, followed four years later by a single CN.235-200. Another pair of CN.235s are required.

The Thai Border Police maintain a sizeable air wing to patrol the nation's long borders with Burma, Cambodia and Laos. Two Shorts 330s are in use.

Royal Thai Army Air Division

Army Aviation Centre	Lop Buri
Hughes H-300C, Bell 206, UH-1H, Bell 212, Maule M-7-235, O-1G, T-41D, U-17B	
Army Aviation Battalion	**Lop Buri**
UH-1H, Bell 212, AH-1F, Bell 206, CH-47C/D, O-1A, T-41D, U-17B, Cessna U-27A, Shorts 330UTT, CASA C.212-300, BAe Jetstream 41, Beech 200, Beech 1900C-1	

Large numbers of O-1s continue to serve the Army, both in an operational observation role and as a trainer.

KASET/Royal Thai Agricultural Air Division

The KASET is under the control of the Ministry of the Interior and is tasked with crop-spraying and fire-fighting. Home base is at Takhli but the aircraft are widely detached over Thailand. Pre-1990 aircraft comprise one Sikorsky S-58T, one Hughes H-300, two MDH-500s, four Fletcher FU-24-954s, seven PC-6s and one BN-2 Islander. A series of Cessna aircraft is operated, including two 180s, one 310Q, three U206s and two 208s. During the 1990s the KASET received 10 AS 350BA Ecureuils and 11 CASA-Nurtanio C.212 Aviocars (four -100s, one -200 and six -300s).

KASET/Royal Thai Agricultural Air Division

Takhli
S-58T, H-300, MDH-500, AS 350BA Ecureuil, Fletcher FU-24-954, PC-6, BN-2 Islander, Ce180, CeU206, Ce208, Ce310Q, CN212-100/200/300.

TURKMENISTAN

Capital: Ashkhabad
Population: 3.6 million
Land area: 488100 km² (188,405 sq miles)
Major cities: Chardzhou, Krasnovodsk, Mary, Nebit-Dag

Turkmenistan, blessed with the world's third largest natural gas reserves, gained its independence from the USSR on 27 October 1991. By 1997 defence accounted for 11 per cent of government spending (3 per cent gross domestic product) and the country maintains armed forces estimated to comprise 16,000-18,000 people and about 170 combat aircraft. Turkmenistan participates in the NATO Partnership for Peace programme and is a member of the Euro-Atlantic Partnership Council, from which it is likely to get technical assistance and training.

Turkmenistan's air force retains the organisation inherited from the time it was the Soviet's 73rd Air Army. The air force is organised with a separate air

force and air defence force.

Currently, the air force has one composite attack regiment (667 shap - attack aviation regiment) with 22 MiG-29s and a pair of two-seat MiG-29Us, and 65 Sukhoi Su-17Ms and Su-17UM3s. An independent composite aviation squadron (osae) is based at the capital's airport, and uses a single Antonov An-24, 10 Mil Mi-8s and 10 Mi-24s. A third unit is thought to be a military composite combat training regiment and the serviceability of its three Sukhoi Su-7s, three MiG-21s and eight Yakovlev Yak-28 is believed to be poor. The unit also has three Antonov An-12s, which may not be transport versions.

At Kizyl-Arvat, 56 brs (base for reserve aircraft)

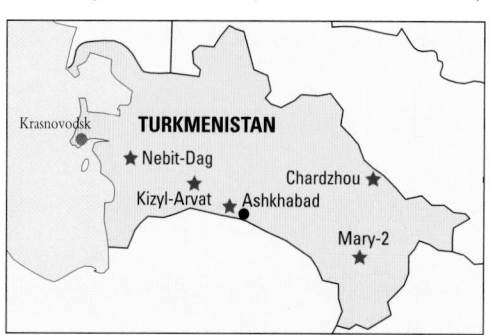

Turkmenistan air force

V-VS		
67 shap	MiG-29/29UB, Su-17M/17UM3	Mary-2
47 osae	An-24, Mi-8, Mi-24	Ashkhabad
31 vshas	Su-7BKP, MiG-21, L-39, Yak-28, An-12	Chardzhou
56 brs	(storage unit)	Kizyl-Arvat
P-VO		
55 iap-PVO	MiG-23M/23UB	Nebit-Dag
107 iap-PVO	MiG-23M/23UB, MiG-25/25PU	Akepe

stores the majority of the aircraft inherited by the country from the Soviet Union, including 172 MiG-23Ms and 45 Su-25s.

The PVO (air defence force) has two regiments based at Nebit-Dag and Akepe operating 48 MiG-23Ms, 10 MiG-23Us and (with the 107 iap-PVO) 24 MiG-25/PUs. The force also uses about 50 anti-aircraft SA-2, SA-3 and SA-5 missiles.

Government transport is undertaken by a special section of the Akhal Aircompany using Mil Mi-8 EZ-25911, BAe 125-1000B EZ-B021 and Boeing 757-23A EZ-A010 based at Ashkhabad airport.

UZBEKISTAN

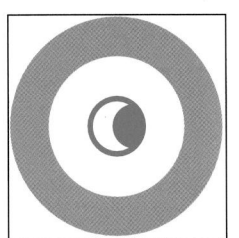

Capital: Tashkent
Population: 20.3 million
Land area: 447400 km²
(172,695 sq miles)
Major cities: Samarkand, Namangan, Andizhan, Bukhara

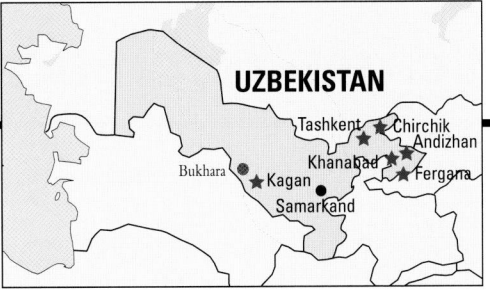

Uzbekistan – formerly the Uzbek Soviet Socialist Republic – is a newly independent Central Asia nation in the midst of profound political and economic change. The Uzbekistan Air Force is built around elements that the Soviets left after the country gained independence following the dissolution of the USSR. The regiments have been renumbered and may be reorganised into the Avia-Baze (Air Base) structure in the near future. The acting Commander of the Air Force – and simultaneously Deputy Minister of Defence – is Colonel Farkhadzan Khashimov.

The Air Force Academy operates approximately a dozen L-39 Albatros jet trainers, from an unknown location. The Air Force consists of a VIP flight based at Tashkent and a transport regiment with the Antonov An-12 and An-26, one combat helicopter regiment with Mi-8 and Mi-24 combat helicopters, and a transport helicopter regiment with a considerable number of Mi-6 and Mi-8 transport helicopters. Two fighter regiments are responsible for the air defence of Uzbekistan and operate the MiG-29 and the Su-27. Ground attack and close air support assets are divided over two regiments, one operating with a mixture of the Su-17 'Fitter' and the Su-25 'Frogfoot'. The second regiment operates the Su-24 'Fencer' and also has a reconnaissance task, for which it uses the Su-24MR.

Uzbekistan Air Force

? OSAE	1 x An-24, 1 x Tu-134	Tashkent
59 APIB	24 x Su-17M3, 6 x Su-17UM-3, 20 x Su-25	Chirchik
60 BAP	23 x Su-24, 1 x Su-24MR	Khanabad
61 IAP	33 x MiG-29, 6 x MiG-29UB	Kakaydy
62 IAP	25 x Su-27P, 6 x Su-27UB	Andizhan
65 OVP	27 x Mi-6, 48 x Mi-8	Kagan
66 OVP	15 x Mi-8, 45 x Mi-24, 1 x Mi-26	Verkhnekomsomolsk
? OSAP	? x An-12, ? x An-26	Fergana

VIETNAM

Capital: Hanoi
Population: 66.2 million
Land area: 329565 km² (127,210 sq miles)
Major cities: Haiphong, Ho Chi Minh

Vietnamese People's Air Force

The Vietnamese People's Air Force (Khong Quan Nhan Dan Viet Nam, VPAF) was founded on 24 January 1959; in May 1959, a transport regiment and a training regiment were formed. The air force entered the jet age in February 1964, when fighter regiment 921 Sao Do (Red Star) was activated at Mengtz, in southern China. The unit was equipped with Chinese-supplied MiG-17s and MiG-15UTIs.

Between 1965 and 1973, North Vietnamese interceptors battled with USAF and US Navy strike forces and their escorts on numerous occasions. Despite the overwhelming American numerical superiority, North Vietnamese MiG-21s periodically gained an advantage over their adversaries, notably during late 1967, early 1968 and mid-1972.

Vietnamese People's Air Force

370 'Hai Van' Su Doan		**HQ Da Nang**
929 Trung Doan	MiG-21bis/UM	Da Nang
935 'Dong Nai' Trung Doan	MiG-21bis/UM	Bien Hoa
937 'Hau Giang' Trung Doan	Su-22M4/UM3	Phan Rang
371 'Thang Long' Su Doan		**HQ Noi Bai**
921 'Sao Do' Trung Doan	MiG-21bis/UM	Noi Bai
927 'Lam Son' Trung Doan	MiG-21bis/UM	Kep
931 Trung Doan	MiG-21bis/UM	Yen Bai
372 'Le Loi' Su Doan		**HQ Tho Xuan**
923 'Yen The' Trung Doan	Su-22M/UM	Tho Xuan
925 Trung Doan	?	?
933 Trung Doan	MiG-21bis/UM	Kien An
Air Force Academy		
910 'Julius Fucik' Trung Doan	Aero L-39ZA	Nha Trang
920 Trung Doan	MiG-21bis/UM	Phu Cat
Transport Brigade		
916 'Ba Vi' Trung Doan	Mi-6, Mi-8/17, Mi-24	Hoa Lac
917 'Dong Thap' Trung Doan	Mi-8/17, An-26	Tan Son Nhat
918 'Hong Ha' Trung Doan	An-2, An-26	Gia Lam
others		
954 Trung Doan	Ka-25, Ka-28, Ka-32	Kien An, Da Nang
??	Su-27SK/UBK	Phan Rang

On 30 April 1975, the North Vietnamese People Army concluded its offensive against South Vietnam with the dramatic capture of Saigon; on 2 July 1976, the Democratic Republic of Vietnam and the Republic of South Vietnam were officially reunified to form the Socialist Republic of Vietnam. Large numbers of abandoned South Vietnamese A-37s, F-5s and UH-1s were incorporated, and were used with success during the 1979 invasion of Cambodia.

On 15 May 1977, the VPAF was given separate status as Quan Chung Khong Quan (Air Force Command). In November 1978, a 25-year Treaty of Friendship, Co-operation and Mutual Assistance between Vietnam and the Soviet Union was signed, assuring economic and military aid. The MiG-21bis, which was supplied in substantial numbers between 1979 and 1984, is numerically still the most important aircraft in the Vietnamese order of battle. The recent acquisition of the Su-27 'Flanker' represents a long-overdue upgrade of the Vietnamese air defence capability. Five Su-27SKs and one Su-27UBK were delivered during 1995/96; Vietnam has ordered a total of 12 'Flankers', which will also include the Su-30.

The Kamov Ka-28, a downgraded export version of the Ka-27PL 'Helix-A', is operated alongside Ka-25s and Ka-32s in the 954 Trung Doan.

Conversion training for the MiG-21bis is undertaken on the MiG-21UMs of the 920 Trung Doan.

Vietnam's transport capability rests squarely on the shoulders of the An-26, which equips two squadrons at Gia Lam and Tan Son Nhut.

The Su-22 'Fitter' equips two attack regiments. Reportedly, approximately 80 'Fitters' are in service: 32 Su-22Ms and Su-22Us were delivered in 1980 (including a limited number of the Su-22MR tactical reconnaissance variant), and the remainder is made up of Su-22M4s and Su-22UM3Ks delivered during 1988-90. The VPAF possesses no heavy-lift capacity in the form of the An-12 or Il-76, and the An-26 medium transport still forms the backbone of the Vietnamese military transport fleet.

Additional transport capability is provided by the Mi-6 and Mi-8. The Mi-6 entered service in 1966; the armed assault versions of the Mi-8 were delivered after 1975 and were used during Vietnamese operations in Kampuchea between 1979 and 1989, alongside 30 Mi-24A gunships delivered in 1979 and 1982. Also in Vietnamese service are the Kamov Ka-25PL ASW helicopter and its successor, the Kamov Ka-28.

Flying training is conducted at the Air Academy at Nha Trang on the Aero L-39C, 24 of which were delivered during 1980-81. Advanced jet training and type conversion takes place at Phu Cat on the MiG-21UM and MiG-21bis.

Vietnam's Aero L-39Cs are operated by the 910 'Julius Fucik' Trung Doan of the Air Force Academy in the flight training role. Fucik was a Czech Marxist.

Aircraft/country cross-reference list

Aircraft are listed by manufacturer. Those that are subjects of major licence-production programmes (and marketed separately as such) are listed under the local producer (e.g. IAR or Helibras) rather than the principal manufacturer (e.g. Eurocopter). Those marked with an asterisk are on firm order.

AERMACCHI *(see also Atlas, EMBRAER)*
MB-326: Argentina, Australia, Congo (Zaïre), Ghana, Tunisia, United Arab Emirates, Zambia
MB-339: Eritrea, Ghana, Italy, Malaysia, New Zealand, Nigeria, Peru, United Arab Emirates

AERO
L-29 Delfin: Azerbaijan, Bulgaria, Cuba, Czech Republic, Egypt, Ghana, Iraq, Mali, Romania, Russia, Slovakia, Syria
L-39/59/159 Albatros: Afghanistan, Algeria, Azerbaijan, Bangladesh, Bulgaria, Cambodia, Cuba, Czech Republic, Egypt, Ethiopia, Georgia (Abkhazia), Hungary, Iraq, North Korea, Kyrgyzstan, Libya, Lithuania, Nigeria, Romania, Russia, Slovakia, Syria, Thailand, Tunisia, Turkmenistan, Uganda, Ukraine, Uzbekistan, Vietnam

AÉROSPATIALE *(see also Changhe, Eurocopter, HAL, IAI, IAR and SOCATA)*
CM 170 Magister: Belgium, Cameroon, Gabon, Ireland, Lebanon, Morocco, El Salvador, Senegal
N 262 Frégate: Burkina Faso, France
SA 313/318/SE 3130 Alouette II: Belgium, Cameroon, Congo, Djibouti, Dominican Republic, France, Germany, Guinea-Bissau, Indonesia, Lebanon, Senegal, Tunisia, Turkey
SA 315 Lama: Argentina, Chile, Ecuador, Morocco, Pakistan, Peru, Togo
SA 316/319 Alouette III: Albania, Angola, Argentina, Austria, Belgium, Bolivia, Burkina Faso, Burundi, Cameroon, Chad, Congo, Congo (Zaïre), Ecuador, France, Gabon, Ghana, Greece, Guinea-Bissau, Indonesia, Iraq, Ireland, South Korea, Lebanon, Libya, Malaysia, Malta, Mexico, Myanmar, Pakistan, Portugal, Senegal, South Africa, Surinam, Switzerland, Tunisia, Venezuela, Zimbabwe
SA 321 Super Frelon: China, France, Iraq, Libya
SA 330 Puma: Argentina, Chile, Congo (Zaïre), Côte d'Ivoire, Ecuador, Ethiopia, France, Gabon, Iraq, Kenya, Kuwait, Lebanon, Malawi, Morocco, Nepal, Nigeria, Pakistan, Philippines, Portugal, Senegal, Spain, Turkey, United Arab Emirates, United Kingdom
SA 341/342 Gazelle: Angola, Bosnia (Serb), Burundi, Cameroon, Cyprus, Ecuador, Egypt, France, Gabon, Guinea-Bissau, Iraq, Ireland, Kuwait, Lebanon, Libya, Morocco, Qatar, Syria, United Arab Emirates, United Kingdom, Yugoslavia
TB 30B Epsilon: France, Portugal, Togo

AEROSTAR SA (YAKOVLEV)
Iak-52: Hungary, Romania, Russia, Ukraine

AÉROSTRUCTURE (FOURNIER)
RF-10: Portugal

AEROTEC
A-122 Uirapuru: Bolivia, Paraguay
A-123 Tangara: Bolivia

AGUSTA
A 109 Hirundo: Argentina, Belgium, Ghana, Italy, Libya, Malaysia, Paraguay, Peru, South Africa*, United Arab Emirates, United Kingdom, Venezuela
A 129 Mangusta: Italy

AGUSTA-BELL
AB 204: Austria, Sweden, Turkey
AB 205: Greece, Iran, Italy, Oman, Saudi Arabia, Tanzania, Tunisia, Turkey, United Arab Emirates, Zambia
AB 206: Austria, Finland, Greece, Iran, Israel, Italy, Libya, Macedonia, Saudi Arabia, Tanzania, Turkey, Uganda, Yemen, Zambia
AB 212: Austria, Bahrain, Greece, Iran, Israel, Italy, Lebanon, Libya, Macedonia, Oman, Palestine, Peru, Saudi Arabia, Spain, Sudan, Turkey, United Arab Emirates, Venezuela, Yemen, Zambia
AB 412: Finland, Ghana, Italy, Lesotho, Netherlands, Saudi Arabia*, Sweden, Turkey*, Uganda, United Arab Emirates, Venezuela, Zimbabwe

AGUSTA (SIAI-MARCHETTI)
S.211: Philippines, Singapore

AGUSTA-SIKORSKY
AS-61: Egypt, Iran, Iraq, Italy, Malaysia, Peru, Saudi Arabia, Venezuela

AIDC
A-CH-1: Taiwan
AT-3 Tsu Chiang: Taiwan
Ching-Kuo: Taiwan

AIEP
Air Beetle: Nigeria

AIRBUS INDUSTRIE
A310: Belgium, Canada, France, Germany, Thailand
A319: Italy*
A340: Qatar

AIRTECH (CASA/IPTN)
CN.235: Botswana, Brunei, Chile, Colombia, Ecuador, France, Gabon, Indonesia, Ireland, South Korea, Malaysia, Morocco, Oman, Panama, Papua New Guinea, Saudi Arabia, South Africa, Spain, Thailand, Turkey, United Arab Emirates

ALENIA (Aeritalia/Fiat)
G222: Argentina, Italy, Libya, Nigeria, Thailand, Venezuela

ALENIA (Aeritalia/Lockheed)
F-104ASA: Italy

AMX INTERNATIONAL
AMX: Brazil, Italy

ANTONOV *(see also PZL, Shaanxi, Shijiazhuang, Xian)*
An-2 'Colt': Afghanistan, Albania, Angola, Armenia, Azerbaijan, Benin, Bulgaria, Croatia, Cuba, Egypt, Estonia, Georgia, Georgia (Abkhazia), Laos, Latvia, Lithuania, Mali, Moldova, Mongolia, Nicaragua, Poland, Romania, Russia, Ukraine, US Army, Vietnam, Yugoslavia
An-12 'Cub': Afghanistan, Algeria, Angola, Belarus, Ethiopia, Iraq, Kazakhstan, Russia, Turkmenistan, Ukraine, Uzbekistan, Yemen
An-22 'Cock': Russia
An-24 'Coke': Belarus, Bulgaria, Cambodia, Congo, Cuba, Czech Republic, Guinea-Bissau, Guinea Republic, Iraq, Kazakhstan, North Korea, Laos, Mali, Mongolia, Romania, Russia, Slovakia, Sri Lanka, Sudan, Syria, Turkmenistan, Ukraine, Uzbekistan, Yemen
An-26 'Curl': Afghanistan, Angola, Azerbaijan, Belarus, Benin, Bulgaria, Cambodia, Cape Verde, China, Congo, Cuba, Czech Republic, Ethiopia, Hungary, Iraq, Kazakhstan, Laos, Libya, Lithuania, Madagascar, Mali, Mongolia, Mozambique, Nicaragua, Niger, Poland, Romania, Russia, Slovakia, Syria, Ukraine, Uzbekistan, Vietnam, Yemen, Yugoslavia, Zambia
An-30 'Clank': Afghanistan, Bulgaria, China, Czech Republic, Kazakhstan, Romania, Russia, Ukraine
An-32 'Cline': Afghanistan, Angola, Bangladesh, Croatia, Cuba, Equatorial Guinea, Ethiopia, India, Peru, Russia, Sri Lanka

An-72/74 'Coaler'): Peru, Russia, Ukraine
An-124 Ruslan ('Condor'): Russia

ASTA (GAF)
N22/24 Nomad: Indonesia, Papua New Guinea, Philippines, Thailand

ATLAS/DENEL
Cheetah: South Africa
Impala: Cameroon, South Africa
Oryx: South Africa
Rooivalk: South Africa*

AVIOANE
IAR-99 Soim: Romania

AVIONS DE TRANSPORT RÉGIONAL
ATR 42/52/72: Gabon, Italy
AVRO *(see British Aerospace)*

AYRES
Turbo-Thrush: Colombia

BAC *(see British Aerospace)*

BASLER
Turbo-67: Bolivia, Colombia, Guatemala, Malawi, Mali, El Salvador, Thailand

BEAGLE
Basset: United Kingdom

BEECH *(see Raytheon)*
Model 18: Papua New Guinea
Model 23/24 Musketeer: Algeria
Model 33/35/36 Bonanza: Bolivia, Colombia, Côte d'Ivoire, Germany, Indonesia, Iran, Mexico, Spain
Model 45 Mentor/T-34: Algeria, Argentina, Colombia, Dominican Republic, Ecuador, Gabon, Indonesia, Morocco, Peru, Uruguay, US Army, USMC, US Navy, Venezuela
Model 50 Twin Bonanza: Haiti, Mexico
Model 55/58 Baron/T-42: Bolivia, Chile, Haiti, Honduras, Mexico, Paraguay, Spain, Turkey, Venezuela
Model 65/80 Queen Air: Algeria, Argentina, Colombia, Dominican Republic, Israel, Japan, Peru, Uruguay, Venezuela
Model 90 King Air/T-44: Bolivia, Canada, Chile, Colombia, Haiti, Japan, Mexico, Paraguay, Peru, US Navy, Venezuela
Model 99: Chile, Colombia
Model 100 King Air: Chile, Ecuador, Jamaica, Morocco, Tanzania, US Army
Model 200 King Air/C-12: Algeria, Argentina, Australia, Bolivia, Cambodia, Chile, Colombia, Ecuador, Egypt, France, Germany, Guatemala, India, Ireland, Israel, Macedonia, Malaysia, Morocco, New Zealand, Peru, Saudi Arabia, South Africa, Sri Lanka, Sweden, Thailand, Togo, Turkey, Uruguay, USAF, US Army, USMC, US Navy, Venezuela
Model 300/350 King Air/LR-2: Colombia, Ecuador, Japan, Morocco, Papua New Guinea, Peru, South Africa
Model 400 Beechjet/T-1: Japan, USAF, US Navy
Model 1900/C-12J: Egypt, Taiwan, Thailand, USAF

BELL *(see also Agusta-Bell, Fuji)*
Model 47 Sioux: Colombia, Dominican Republic, Greece, Lesotho, Libya, Malta, New Zealand, Pakistan, Paraguay, Tanzania, Uruguay, Zambia
Model 204/UH-1 Iroquois: Honduras, South Korea, Paraguay, El Salvador
Model 205/UH-1 Iroquois: Argentina, Australia, Bolivia, Bosnia-Herzegovina, Brazil, Chile, Colombia, Cyprus, Dominican Republic, Ethiopia, Germany, Greece, Guatemala, Honduras, Indonesia, Jamaica, Jordan, South Korea, Lebanon, Mexico, Morocco, Myanmar, New Zealand, Oman, Panama, Papua New Guinea, Pakistan, Paraguay, Peru, Philippines, El Salvador, Singapore, Spain, Taiwan, Thailand, Tunisia, Turkey, Uruguay, US Army, Venezuela
Model 206/TH-57/OH-58/TH-67 JetRanger:

Algeria, Australia, Austria, Bangladesh, Brazil, Brunei, Cameroon, Canada, Chile, Colombia, Croatia, Cyprus, Djibouti, Ecuador, Guatemala, Guyana, Indonesia, Israel, Jamaica, Mexico, Morocco, Nepal, Oman, Pakistan, Peru, Slovenia, Spain, Sri Lanka, Sweden, Taiwan, Thailand, Turkey, Uganda, United Arab Emirates, US Army, US Navy, Venezuela
Model 209/AH-1 Cobra: Bahrain, Iran, Israel, Japan, Jordan, South Korea, Pakistan, Taiwan, Thailand, Turkey, US Army, USMC, US Navy
Model 212/UH-1 Twin Huey: Argentina, Bangladesh, Brunei, Colombia, Ecuador, Gabon, Ghana, Guatemala, Jamaica, South Korea, Mexico, Oman, Panama, Peru, Philippines, Sri Lanka, Thailand, United Kingdom, United States, Uruguay, USAF, USMC, US Navy, Venezuela
Model 214: Brunei, Cambodia, Ecuador, Guatemala, Iran, Iraq, South Korea, Oman, Peru, Philippines, Thailand, United Arab Emirates, Venezuela
Model 222: Albania, Ecuador, Jamaica
Model 406/OH-58D: Saudi Arabia, Taiwan, US Army
Model 412 Griffon: Bahrain, Botswana, Canada, Colombia, Cyprus, Czech Republic, Ecuador, Gabon, Guatemala, Guyana, Honduras, Indonesia, Lesotho, Norway, Peru, Poland, Slovenia, Sri Lanka, Thailand, Turkey*, Uganda, United Kingdom, Venezuela

BELL/BOEING
V-22 Osprey: USAF*, USMC*

BELLANCA
Citabria/Decathlon: Guatemala, Turkey

BERIEV
Be-6 'Madge': China
Be-12 Tchaika 'Mail': Russia, Ukraine

BOEING *(see also Kawasaki, Mitsubishi)*
AH-64 Apache: Egypt, Greece, Israel, Netherlands, Saudi Arabia, Singapore*, United Arab Emirates, US Army
B-52 Stratofortress: USAF
C-17 Globemaster III: USAF
CH-47 Chinook: Argentina, Australia, Egypt, Greece, Iran, Italy, South Korea, Libya, Morocco, Netherlands, Spain, Taiwan, Thailand, United Kingdom, US Army
E-3 Sentry: France, NATO, Saudi Arabia, United Kingdom, USAF
E-4: USAF
E-6 Mercury: US Navy
F-15 Eagle: Israel, Japan, Saudi Arabia, USAF
F/A-18 Hornet: Australia, Canada, Finland, Kuwait, Malaysia, Spain, Switzerland, USMC, US Navy
Model 107/CH-46 Sea Knight: Canada, USMC, US Navy
Model 707/C-137/C-18: Argentina, Australia, Brazil, Chile, Colombia, Egypt, Germany, India, Indonesia, Iran, Israel, Italy, Morocco, NATO, Pakistan, Paraguay, Peru, Qatar, Romania, Saudi Arabia, South Africa, Spain, Togo, USAF, US Navy, Venezuela
Model 717/C-135 Stratotanker: France, Singapore*, Turkey, USAF
Model 727/C-22: Bahrain, Colombia, Ecuador, Libya, Mexico, New Zealand, Panama, Qatar, Taiwan, USAF, Yemen
Model 737/T-43/C-40: Brazil, Chile, Egypt, India, Indonesia, South Korea, Mexico, Peru, Saudi Arabia, Thailand, USAF, US Navy*, Venezuela
Model 747/VC-25: Iran, Japan, Oman, Saudi Arabia, USAF
Model 757/C-32: Argentina, Mexico, Turkmenistan, USAF
Model 767/E-767: Japan, US Army

BOEING (MCDONNELL)/BAE
AV-8A/C/S Harrier: Thailand
AV-8B Harrier II/II Plus: Italy, Spain, United Kingdom, USMC, US Navy
T-45A/C Goshawk: US Navy
BOEING/GRUMMAN
E-8 J-STARS: USAF
BOEING/SIKORSKY
RAH-66 Comanche: US Army*
BOMBARDIER
Global Express: Malaysia*
BREGUET (DASSAULT AVIATION)
Br.1150 Alizé: France
BRITISH AEROSPACE *(including Avro, BAC, Hawker, Hawker Siddeley, etc.)*
125/Dominie: Brazil, Japan, Malawi, Saudi Arabia, South Africa, Turkmenistan, United Kingdom
146: United Kingdom
748/Andover: Australia, Belgium, Brazil, Burkina Faso, Ecuador, India, South Korea, Madagascar, Nepal, Sri Lanka, Thailand, United Kingdom
Bulldog: Jordan, Kenya, Lebanon, Sweden, United Kingdom
Canberra: Argentina, India, Peru, United Kingdom
Harrier: India, United Kingdom
Hawk: Australia*, Canada*, Finland, Indonesia, Kenya, Kuwait, South Korea, Malaysia, Oman, Saudi Arabia, South Africa*, Switzerland, United Arab Emirates, United Kingdom, Zimbabwe
Hunter: Lebanon, United Kingdom, Zimbabwe
Jetstream: Saudi Arabia, Thailand, United Kingdom, Uruguay
Nimrod: United Kingdom
Sea Harrier: India, United Kingdom
Strikemaster: Ecuador, Oman
Trident: China
One-Eleven: Oman, United Kingdom
VC10: United Kingdom
BRITISH AEROSPACE/MCDD *(see Boeing/BAe)*

CANADAIR
Challenger: Canada, China, Croatia, Denmark, Germany, Malaysia
CL-41 Tutor: Canada
CL-215/215T: Greece, Italy, Spain, Thailand
CL-415: Greece*
CASA *(see also ENAER)*
C.101 Aviojet: Honduras, Jordan, Spain
C.212 Aviocar: Angola, Argentina, Bolivia, Bosnia-Herzegovina, Botswana, Chad, Chile, Colombia, France, Indonesia, Jordan, Lesotho, Mexico, Myanmar*, Panama, Papua New Guinea, Paraguay, Portugal, South Africa, Spain, Surinam, Sweden, Thailand, United Arab Emirates, Uruguay, USAF, Venezuela, Zimbabwe
CESSNA *(see also Reims/Summit)*
Model 150/152: Bangladesh, Bolivia, Botswana, Colombia, Ecuador, Haiti, Mexico, Paraguay, Peru, Seychelles, Sri Lanka, Thailand
Model 172/T-41: Argentina, Bolivia, Bosnia (Serb), Burkina Faso, Chile, Colombia, Dominican Republic, Ecuador, Greece, Guatemala, Honduras, Indonesia, Ireland, South Korea, Madagascar, Nicaragua, Pakistan, Peru, Philippines, El Salvador, Thailand, Trinidad & Tobago, Turkey, Uruguay, Venezuela
Model 180/185 Skywagon/U-17: Argentina, Greece, Honduras, Iran, Myanmar, Nicaragua, Paraguay, Peru, South Africa, Thailand, Turkey, Uruguay
Model 182 Skylane: Argentina, Belgium, Bolivia, Chile, Lesotho, Peru, Uruguay, US Army, Venezuela
Model 206 Super Skywagon: Argentina, Bolivia, Chile, Colombia, Costa Rica, Djibouti, Indonesia, Mexico, Paraguay, Peru, Surinam, Tanzania,

Thailand, Uruguay, Venezuela
Model 207 Skywagon: Argentina, Dominican Republic, Venezuela
Model 208 Caravan I/U-27: Brazil, Chile, Colombia, South Africa, Thailand
Model 210 Centurion: Bolivia, Chile, Dominican Republic, Guatemala, Jamaica, Paraguay, Philippines, El Salvador, Uruguay
Model 305/0-1: Chile, South Korea, Libya, Malta, Pakistan, Taiwan, Thailand
Model 310/U-3: Colombia, Congo (Zaïre), Iran, Madagascar, Paraguay, Surinam, Thailand, Trinidad & Tobago, Uruguay, Venezuela
Model 318/A-37 Dragonfly: Chile, Colombia, Dominican Republic, Ecuador, Guatemala, Honduras, South Korea, Peru, El Salvador, Uruguay
Model 318/T-37 Tweet: Bangladesh, Colombia, Germany, Greece, South Korea, Pakistan, Thailand, Turkey, USAF
Model 320 Skyknight: Ecuador
Model 337 Super Skymaster/O-2: Bangladesh, Belize, Burkina Faso, Central African Republic, Chile, Costa Rica, Dominican Republic, Jamaica, South Korea, Mexico, Namibia, Nicaragua, El Salvador, Sri Lanka, Togo, US Army, Venezuela, Zimbabwe
Model 400 series: Bahamas, Barbados, Bolivia, Cambodia, Colombia, Côte d'Ivoire, Djibouti, Ethiopia, Haiti, Indonesia, Malaysia, Mexico, Nicaragua, Pakistan, Paraguay, Peru, Sri Lanka, Trinidad & Tobago, Turkey, Venezuela
Model 500/600 series Citation/T-47: Argentina, Bosnia-Herzegovina, Colombia, Ecuador, Myanmar, Pakistan, Paraguay, South Africa, Spain, Sweden, Turkey, US Army, Venezuela
CHANGHE
Z-8: China
Z-11: China
CHENGDU
J-7/F-7: Albania, Bangladesh, China, Egypt, Iran, Iraq, Myanmar, Pakistan, Sri Lanka, Sudan, Tanzania, Yemen, Zimbabwe
JJ-5/FT-5: China, North Korea, Pakistan, Sudan
CHRISTEN
Pitts S-2: Venezuela
CNAMC/PAC
K-8 Karakorum: Pakistan
CONVAIR
240/440: Peru
580: Bolivia
CURTISS
C-46 Commando: Haiti

DAEWOO
KTX-1: South Korea*
DASSAULT
Atlantic/Atlantique: France, Germany, Italy, Pakistan
Etendard: France
Falcon/Mystère 10: France
Falcon/Mystère 20/HU-25: Belgium, Cambodia, Egypt, France, Iran, Lebanon, Morocco, Norway, Oman, Pakistan, Peru, Portugal, Spain, Syria, USCG, Venezuela
Falcon/Mystère 50: France, Iran, Italy, Morocco, Portugal, South Africa, Spain, Switzerland
Falcon 900: Algeria, Australia, Belgium, France, Gabon, Italy*, Malaysia, Qatar, South Africa, Spain
Mirage III: Argentina, Brazil, France, Pakistan, Switzerland
Mirage IV: France
Mirage 5: Argentina, Chile, Colombia, Congo (Zaïre), Egypt, Gabon, Libya, Pakistan, Peru, United Arab Emirates

Mirage 50: Chile, Venezuela
Mirage 2000: France, Greece, India, Peru, Qatar, Taiwan, United Arab Emirates
Mirage F1: Ecuador, France, Greece, Iraq, Jordan, Kuwait, Libya, Morocco, Spain
Rafale: France*
Super Etendard: Argentina, France
Super Mystère: Honduras
DASSAULT-DORNIER
Alpha Jet: Belgium, Cameroon, Côte d'Ivoire, Egypt, France, Morocco, Nigeria, Portugal, Qatar, Togo
DE HAVILLAND
Dove: Jordan
DE HAVILLAND CANADA (BOMBARDIER)
DHC-1 Chipmunk: Portugal
DHC-2 Beaver: Colombia, Haiti, South Korea, USAF, US Navy
DHC-3 Otter: US Navy
DHC-4 Caribou: Australia, Costa Rica, Malaysia
DHC-5 Buffalo: Brazil, Canada, Cameroon, Congo (Zaïre), Ecuador, Egypt, Indonesia, Kenya, Peru, Sudan, Tanzania, Togo, Zambia
DHC-6 Twin Otter/UV-18: Argentina, Australia, Benin, Canada, Chile, Colombia, Ecuador, Ethiopia, France, Haiti, Norway, Paraguay, Peru, Sudan, USAF, US Army
DHC-7 Dash 7/RC-7/OE-5: US Army, Venezuela
DHC-8 Dash 8/CC-142/E-9: Canada, Kenya, USAF
DORNIER
Do 27: Burundi, Guinea-Bissau, Spain
Do 28/128: Benin, Cameroon, Greece, Israel, Kenya, Morocco, Niger, Nigeria, Yugoslavia, Zambia
Do 228: Bhutan, Cape Verde, Finland, Germany, India, Italy, Malawi, Maldives, Mauritius, Niger, Nigeria, Oman, Thailand
Do 328: Colombia
DOUGLAS *(see also Basler)*
DC-3/C-47: Colombia, Congo, Congo (Zaïre), Dominican Republic, Greece, Guatemala, Haiti, Honduras, Israel, Laos, Madagascar, Mexico, Paraguay, El Salvador, South Africa, Thailand
DC-6/C-118: Mexico
DC-8: France, Gabon, Peru

EH INDUSTRIES
EH101 Merlin: Canada*, Italy*, United Kingdom
EMBRAER
EMB-110/111 Bandeirante: Brazil, Cape Verde, Chile, Colombia, Gabon, Uruguay
EMB-120 Brasilia: Brazil
EMB-121 Xingu: Brazil, France
EMB-312 Tucano: Argentina, Brazil, Colombia, Egypt, France, Honduras, Iran, Iraq, Peru, Venezuela
EMB-326 Xavante: Argentina, Brazil, Paraguay, Togo
ERJ-145: Brazil*, Greece*
EMBRAER (NEIVA)
T-25 Universal: Brazil
ENAER
A-36/T-36 Halcón: Chile, Honduras
Pantera 50CN: Chile
T-35 Pillán: Chile, Guatemala, Panama, Paraguay, El Salvador, Spain
ENSTROM
F-28: Colombia, Peru
Model 280FX: Chile
EUROCOPTER
EC 135: Germany*
Tigre/Tiger: France*, Germany*
EUROCOPTER (AÉROSPATIALE)
(see also Changhe, Helibras, IPTN)
AS 332/532 Super Puma/Cougar: Argentina, Brazil, Chile, China, Congo (Zaïre), Ecuador, Finland, France, Gabon, Germany, Iceland, Japan,

Jordan, Kuwait, Mexico, Nepal, Netherlands, Nigeria, Oman, Saudi Arabia, Singapore, Spain, Sweden, Switzerland, Thailand, Turkey, United Arab Emirates, Venezuela, Zimbabwe
AS 350/355/550/555 Ecureuil/Squirrel/Fennec: Albania, Algeria, Argentina, Australia, Benin, Botswana, Brazil, Burkina Faso, Cambodia, Central African Republic, Comores, Denmark, Djibouti, Ecuador, France, Gabon, Guinea Republic, Iceland, Ireland, Jamaica, South Korea, Malawi, Mali, Mexico, Peru, Singapore, Thailand, Tunisia, United Arab Emirates, United Kingdom, Venezuela
AS 365/565 Dauphin/Dolphin/HH-65: Angola, Argentina, Brazil, Cambodia, Congo, Côte d'Ivoire, Dominican Republic, France, Iceland, Ireland, Israel, South Korea, Malawi, Morocco, Romania, Saudi Arabia, South Africa, Thailand, United Arab Emirates, United Kingdom, Uruguay, USCG
EUROCOPTER (MBB) *(see also IPTN)*
BO 105: Bahrain, Brunei, Chile, Colombia, Czech Republic, Germany, Iraq, Kenya, South Korea, Lesotho, Mexico, Netherlands, Nigeria, Peru, Philippines, Spain, Sudan, Sweden, Tanzania, Trinidad & Tobago, United Arab Emirates
EUROCOPTER (MBB)/KAWASAKI
BK 117: Chile, Iraq, Peru, South Africa, Spain
EUROFIGHTER
EF2000 Typhoon: Germany*, Greece*, Italy*, Spain*, United Kingdom*
EXTRA
300: Chile, Jordan

FAIRCHILD
A-10A Thunderbolt II: USAF
AU-23 Peacemaker: Thailand
C-123 Provider: Peru
FAIRCHILD/FOKKER
F-27J: Mexico
FH-227: Mexico, Myanmar
FAIRCHILD (HILLER)
FH-1100: Costa Rica
FAIRCHILD (SWEARINGEN)
Merlin IIIA: Argentina, Belgium, El Salvador
Merlin IV/Metro/C-26: Argentina, Chile, Colombia, Peru, Thailand, USAF, US Army, US Navy*, Venezuela
FAMA/FMA/DINFIA
IA-46 Ranquel: Argentina
IA-50 Guaraní II: Argentina
IA-58 Pucará: Argentina, Sri Lanka, Uruguay
IA-63 Pampa: Argentina
FFA
AS 202 Bravo: Indonesia, Iraq, Morocco, Oman, Uganda
FLETCHER
FU-24: Thailand
FOKKER
50/60: Netherlands, Singapore, Taiwan, Thailand
70/100: Côte d'Ivoire, Kenya
F27 Friendship/C-31: Algeria, Angola, Argentina, Bolivia, Finland, Ghana, Guatemala, Iceland, India, Indonesia, Iran, Myanmar, Netherlands, Nigeria, Pakistan, Peru, Philippines, Senegal, Spain, Thailand, Uruguay, US Army
F28 Fellowship: Argentina, Cambodia, Colombia, Ecuador, Ghana, Indonesia, Malaysia, Peru, Philippines
FUJI
T-1: Japan
T-3: Japan
T-5: Japan
FUJI-BELL
UH-1: Japan

GAVILAN
358: Colombia
GENERAL DYNAMICS
F-111: Australia
GLOSTER
Meteor: United Kingdom
GROB
G 109 Vigilant: United Kingdom
G 115/Heron: United Arab Emirates, United Kingdom*
GRUMMAN (see also Northrop Grumman)
G-164 Ag-Cat: Greece
HU-16 Albatross: Greece
OV-1/RV-1 Mohawk: Argentina
S-2 Tracker: Argentina, Taiwan, Thailand
GRUMMAN AMERICAN
AA-5A: Costa Rica
GUIZHOU (GAIC)
JJ-7/FT-7: Bangladesh, China, Egypt, Iran, Myanmar, Pakistan, Sri Lanka, Zimbabwe
GULFSTREAM AEROSPACE
Commander 1000: Colombia, Guatemala
Gulfstream I/C-4: USCG, Venezuela
Gulfstream II: Bahrain, Libya, Morocco, Panama, Saudi Arabia, US Army, USCG, Venezuela
Gulfstream III/C-20: Algeria, Bahrain, Chile, Côte d'Ivoire, Denmark, Egypt, Gabon, India, Italy, Jordan, Morocco, Saudi Arabia, USAF, US Army, US Navy, Venezuela
Gulfstream IV/U-4/C-20: Botswana, Côte d'Ivoire, Egypt, India, Ireland, Japan, Netherlands, Oman, Saudi Arabia, Sweden, Turkey, USAF, US Army, USMC, US Navy
Gulfstream V/C-37: USAF

HARBIN
H-5: Albania, China, North Korea, Romania
SH-5: China
Y-11: China
Y-12: Cambodia, Eritrea, Laos, Mauritania, Mongolia, Pakistan, Peru, Sri Lanka, Tanzania, Zambia
Z-5: Albania, China
Z-6: China
Z-9 Haitun: China
HAWKER/HAWKER SIDDELEY (see British Aerospace)
HELIBRAS/EUROCOPTER
HB 315 Gavião: Bolivia
HB 350/355 Esquilo: Brazil, Paraguay
HELIO
AU-24A/Stallion: Colombia
HELIOPOLIS
Gomhouria: Egypt
HILLER
H-23 Raven/UH-12: Argentina, Egypt, Paraguay, Thailand
HINDUSTAN (HAL)
ALH: India*
Cheetah: India, Namibia
Chetak: India, Namibia, Nepal, Seychelles
HJT-16 Kiran I/II: India
HPT-32 Deepak: India
HOFFMAN
H 36 Dimona: Thailand
HUGHES (see McDonnell Douglas, Nardi, Schweizer)

IAI
1123/1124/1125 Westwind/Astra/C-38: Eritrea, Honduras, India, Israel, USAF
Arava: Bolivia, Cameroon, Colombia, Ecuador, Guatemala, Honduras, Israel, Mexico, Swaziland, Thailand, Venezuela
Dagger/Finger: Argentina
Kfir: Colombia, Ecuador, Israel, Sri Lanka
Tzukit: Israel
IAI/ELTA
Phalcon: Chile
IAR (see also Avioane)
IAR-823: Romania
IAR (AÉROSPATIALE/ EUROCOPTER)

IAR-330L Puma: Guinea Republic, Kenya, Romania, Sudan, United Arab Emirates
IAR (SUD/AÉROSPATIALE)
IAR-316B Alouette III: Angola, Guinea Republic, Romania
ILYUSHIN (see also Harbin)
A-50 'Mainstay': China*, Russia
Il-14 'Crate': Albania, China
Il-18/20/22 'Coot': Afghanistan, Belarus, Kazakhstan, North Korea, Russia, Ukraine
Il-28 'Beagle': Egypt
Il-38 'May': India, Russia, Ukraine
Il-62 'Classic': North Korea, Russia
Il-76 'Candid'/-78 'Midas': Algeria, Belarus, China, Cuba, India, Iran, Iraq, North Korea, Libya, Russia, Syria, Ukraine
Il-86 'Camber': Russia
IPTN (EUROCOPTER)
NAS 330 Puma: Indonesia
NAS 332 Super Puma: Indonesia
NBO-105: Indonesia

JODEL
D.140: France

KAMAN
SH-2 Seasprite: Australia, Egypt, New Zealand, US Navy
KAMOV
Ka-25 'Hormone': India, Russia, Syria, Ukraine, Vietnam
Ka-27/28/29/32 'Helix': India, South Korea, Russia, Ukraine, US Army, Vietnam
Ka-50 'Hokum': Russia*
KAWASAKI
C-1: Japan
OH-1: Japan*
T-4: Japan
KAWASAKI-BOEING VERTOL
CH-47J Chinook: Japan
KV-107: Japan, Saudi Arabia, Sweden
KAWASAKI-LOCKHEED
P-3C Orion: Japan
KAWASAKI/MCDONNELL DOUGLAS HELICOPTERS
OH-6D/OH-6J: Japan

LEARJET (GATES) (see also ShinMaywa)
Models 24/25/35/36/C-21: Argentina, Bolivia, Brazil, Chile, Finland, India, Iraq, Macedonia, Peru, Saudi Arabia, Switzerland, Thailand, USAF, US Army, Venezuela
LET
L-410 Turbolet: Bulgaria, Czech Republic, Estonia, Germany, Hungary, Latvia, Libya, Lithuania, Russia, Slovakia, Slovenia, Tunisia
L-610: Czech Republic
LOCKHEED MARTIN (see also Alenia)
C-5 Galaxy: USAF
C-130/L-100 Hercules: Algeria, Argentina, Australia, Belgium, Bolivia, Bosnia-Herzegovina, Botswana, Brazil, Cameroon, Canada, Chad, Chile, Colombia, Congo (Zaïre), Denmark, Ecuador, Egypt, Ethiopia, France, Gabon, Greece, Honduras, Indonesia, Iran, Israel, Italy, Japan, Jordan, South Korea, Kuwait, Libya, Malaysia, Mexico, Morocco, Netherlands, New Zealand, Niger, Nigeria, Norway, Oman, Pakistan, Peru, Philippines, Portugal, Romania, Saudi Arabia, Singapore, South Africa, Spain, Sri Lanka*, Sudan, Sweden, Taiwan, Thailand, Tunisia, Turkey, United Arab Emirates, United Kingdom, Uruguay, USAF, USCG, USMC, US Navy, Venezuela, Yemen
C-140 JetStar: Libya, Palestine, Saudi Arabia
C-141 StarLifter: USAF
CP-140 Aurora/Arcturus: Canada
F-16 Fighting Falcon: Bahrain, Belgium, Denmark, Egypt, Greece, Indonesia, Israel, Jordan, South Korea, Netherlands, New Zealand*,

Norway, Pakistan, Portugal, Singapore, Taiwan, Thailand, Turkey, United Arab Emirates*, USAF, Venezuela
F-22 Raptor: USAF*
F-117A Nighthawk: USAF
L-188 Electra: Argentina, Bolivia, Honduras
L-1011 TriStar: Jordan, Saudi Arabia, United Kingdom
P-3 Orion: Argentina, Australia, Chile, Greece, Iran, South Korea, Netherlands, New Zealand, Norway, Pakistan, Portugal, Spain, Thailand, US Navy
S-3A/B Viking: US Navy
T-33: Bolivia, Canada, Greece, Iran, Japan, South Korea, Mexico, Pakistan
U-2: USAF

MAULE
M-4 to M-7: Mexico, Thailand, Turkey
MBB (see Eurocopter)
MCDONNELL DOUGLAS (also see Boeing, Mitsubishi)
A-4 Skyhawk: Argentina, Brazil, Indonesia, Israel, New Zealand, Singapore, US Navy
DC-9/C-9 Nightingale/Skytrain II/MD-80: Italy, Kuwait, USAF, USMC, US Navy
KC-10 Extender/KDC-10: Netherlands, USAF
F-4 Phantom II: Egypt, Germany, Greece, Iran, Israel, South Korea, Spain, Turkey, USAF, US Navy
MD-11: Saudi Arabia
MCDONNELL DOUGLAS HELICOPTERS (HUGHES) (see also Kawasaki, Nardi)
369/MD500/H-6: Algeria, Argentina, Colombia, Costa Rica, Croatia, Denmark, Dominican Republic, Finland, Haiti, Honduras, Indonesia, Iraq, Israel, Jordan, Kenya, North Korea, South Korea, Malaysia, Mexico, Peru, El Salvador, Spain, Taiwan, Thailand, US Army, US Navy
MD520: Philippines
MD530: Argentina, Chile, Colombia, Iraq, Mexico, US Army
MD900 Explorer: Belgium, Slovakia*
MIKOYAN (see also Chengdu, Guizhou, Shenyang)
MiG-15 'Fagot': Albania, Angola, Egypt, Guinea Republic, North Korea, Mali, Mozambique, Syria, Yemen
MiG-17 'Fresco': Angola, Cuba, Madagascar, Mali, Mozambique, Syria
MiG-21 'Fishbed': Afghanistan, Algeria, Angola, Azerbaijan, Bulgaria, Cambodia, Congo, Croatia, Cuba, Czech Republic, Egypt, Ethiopia, Guinea-Bissau, Guinea Republic, Hungary, India, Iraq, Libya, Kazakhstan, North Korea, Kyrgyzstan, Laos, Madagascar, Mali, Mongolia, Mozambique, Nigeria, Poland, Romania, Russia, Slovakia, Syria, Turkmenistan, Vietnam, Yemen, Yugoslavia, Zambia
MiG-23 'Flogger': Afghanistan, Algeria, Angola, Belarus, Bulgaria, Cuba, Ethiopia, India, Iraq, Kazakhstan, North Korea, Libya, Poland, Romania, Russia, Sudan, Syria, Turkmenistan, Ukraine, Yemen
MiG-25 'Foxbat': Algeria, Azerbaijan, Belarus, India, Iraq, Kazakhstan, Libya, Russia, Syria, Turkmenistan
MiG-27 'Flogger': India, Kazakhstan, Russia
MiG-29 'Fulcrum': Belarus, Bulgaria, Cuba, Eritrea, Germany, Hungary, India, Iran, Iraq, Kazakhstan, North Korea, Malaysia, Peru, Poland, Romania, Russia, Slovakia, Ukraine, Uzbekistan, Yugoslavia
MiG-31 'Foxhound': Kazakhstan, Russia

MIL (see also PZL)
Mi-4 'Hound': Albania,
Mi-6/22 'Hook': Belarus, China, Egypt, Iraq, Kazakhstan, Laos, Russia, Syria, Ukraine, Uzbekistan, Vietnam
Mi-8/17 'Hip': Afghanistan, Albania, Algeria, Angola, Armenia, Azerbaijan, Bangladesh, Belarus, Bhutan, Bosnia-Herzegovina, Bosnia (Serb), Bulgaria, Burkina Faso, Cambodia, China, Congo, Costa Rica, Croatia, Cuba, Cyprus, Czech Republic, Djibouti, Egypt, Eritrea, Estonia, Ethiopia, Finland, Georgia, Georgia (Abkhazia), Guinea-Bissau, Hungary, India, Indonesia*, Iran, Iraq, Kazakhstan, North Korea, Kyrgyzstan, Laos, Latvia, Libya, Lithuania, Macedonia, Madagascar, Mali, Mexico, Moldova, Mongolia, Mozambique, Myanmar, Nicaragua, Pakistan, Palestine, Papua New Guinea, Peru, Poland, Romania, Russia, Slovakia, Sri Lanka, Sudan, Syria, Tajikistan, Turkey, Uganda, Ukraine, US Army, Uzbekistan, Vietnam, Yemen, Yugoslavia, Zambia
Mi-14 'Haze': Bulgaria, Cuba, Ethiopia, North Korea, Libya, Poland, Romania, Russia, Ukraine, US Army
Mi-24/25 'Hind': Afghanistan, Algeria, Angola, Armenia, Azerbaijan, Belarus, Bulgaria, Cambodia, Croatia, Cuba, Czech Republic, Eritrea, Ethiopia, Georgia, Germany, Hungary, India, Kazakhstan, North Korea, Kyrgyzstan, Libya, Mongolia, Mozambique, Peru, Poland, Russia, Slovakia, Sri Lanka, Sudan, Syria, Tajikistan, Ukraine, US Army, Uzbekistan, Vietnam, Yugoslavia
Mi-26 'Halo': Belarus, Cambodia, India, Kazakhstan, Peru, Russia, Ukraine, Uzbekistan
Mi-28 'Havoc': Russia*
MITSUBISHI
F-1: Japan
F-2: Japan*
MU-2/LR-1: Congo (Zaïre), Dominican Republic, Japan
T-2: Japan
MITSUBISHI (MCDONNELL DOUGLAS)
F-4EJ: Japan
F-15J/DJ: Japan
MORANE-SAULNIER
MS.760 Paris: Argentina, France
MUDRY
CAP 10/10B: France, South Korea, Morocco
CAP 20/21/230: Morocco
MYASISHCHEV
M-17/55 'Mystic': Russia?

NAMC
YS-11: Greece, Japan
NANCHANG
CJ-6/BT-6/PT-6: Albania, Bangladesh, China, North Korea, Zambia
Q-5/A-5 'Fantan': Bangladesh, China, North Korea, Myanmar, Pakistan
NARDI (HUGHES)
NH-300: Greece
NH-500: Italy
NEICO
Lancair 235/320: Bolivia
NEIVA
C-42/L-42/U-42 Regente: Brazil
NH INDUSTRIES
NH 90: France*, Germany*, Italy*, Netherlands*
NORD AVIATION
N2501 Noratlas: Congo
NORTH AMERICAN (see also Rockwell, Sud Aviation)
T-6 Texan/Harvard: Dominican Republic, Paraguay, United Kingdom
NORTHROP GRUMMAN
B-2A Spirit: USAF
C-2A Greyhound: US Navy
E-2 Hawkeye: Egypt, France, Japan,

Singapore, Taiwan, US Navy
EA-6B Prowler: USMC, US Navy
F-5 Freedom Fighter/Tiger II: Bahrain, Botswana, Brazil, Chile, Greece, Honduras, Indonesia, Iran, Jordan, Kenya, South Korea, Malaysia, Mexico, Morocco, Norway, Philippines, Saudi Arabia, Singapore, Spain, Switzerland, Taiwan, Thailand, Tunisia, Turkey, USMC, US Navy, Venezuela, Yemen
F-14 Tomcat: Iran, US Navy
T-38 Talon: Germany, South Korea, Turkey, USAF, US Navy

ORLANDO/SIKORSKY
QS-55: US Army

PAC (AMF/SAAB)
Mushshak: Iran, Oman, Pakistan, Syria
PACIFIC AEROSPACE CORPORATION (AEROSPACE/NZAI)
CT-4 Airtrainer: New Zealand, Thailand
PANAVIA
Tornado: Germany, Italy, Saudi Arabia, United Kingdom
PIAGGIO
P.166: Italy
P.180 Avanti: Italy
PIAGGIO-DOUGLAS
PD-808: Italy
PILATUS (see also Fairchild, Raytheon)
PC-6 Porter/Turbo Porter: Angola, Argentina, Austria, Chad, Colombia, Ecuador, France, Indonesia, Iran, Mexico, Myanmar, Oman, Peru, Slovenia, South Africa, Switzerland, Thailand, United Arab Emirates, US Army
PC-7 Turbo Trainer: Angola, Austria, Bolivia, Botswana, Chad, Chile, Guatemala, Iran, Malaysia, Mexico, Myanmar, Netherlands, Nigeria, South Africa, Surinam, Switzerland, United Arab Emirates, Uruguay
PC-9: Angola, Australia, Croatia, Cyprus, Myanmar, Saudi Arabia, Slovenia, Switzerland, Thailand
PC-12: South Africa
PILATUS BRITTEN-NORMAN
BN-2 Islander/Defender: Angola, Belgium, Belize, Botswana, Cambodia, Central African Republic, Cyprus, Ghana, Guyana, Haiti, India, Indonesia, Ireland, Jamaica, Jordan, Madagascar, Malta, Mauritania, Mauritius, Morocco, Oman, Pakistan, Panama, Peru, Philippines, Seychelles, Surinam, Thailand, United Arab Emirates, United Kingdom, US Army, Venezuela, Zimbabwe
PIPER
PA-18 Super Cub: Argentina, Bosnia (Serb), Israel, Nicaragua, Uruguay
PA-23 Apache/Aztec: Argentina, Bolivia, Cameroon, Costa Rica, Madagascar, Mexico, Venezuela
PA-25/36 Pawnee: Argentina, Colombia, United Arab Emirates
PA-28/-32 Cherokee/Arrow/Dakota/Cherokee Six/Warrior: Argentina, Chile, Colombia, Costa Rica, Finland, Mozambique, Nicaragua, Paraguay
PA-31/T Navajo/Navajo Chieftain/Cheyenne: Argentina, Bahamas, Chile, Colombia, Costa Rica, Dominican Republic, Finland, Guatemala, Honduras, Mauritania, Peru, Sweden, Syria, United Kingdom, US Army
PA-34 Seneca/PA-44 Seminole/PZL M-20: Argentina, Bolivia, Brazil, Costa Rica, Guatemala, Haiti, Honduras, Indonesia, Pakistan, Peru, Poland, Uruguay
PA-38-112 Tomahawk: Indonesia, Lithuania
PZL MIELEC
I-22 Iryda: Poland

M-18 Dromader: Greece
TS-11 Iskra: India, Poland
PZL MIELEC (ANTONOV)
An-28 'Cash': Djibouti, Poland, Venezuela
PZL SWIDNIK
Mi-2 'Hoplite': Algeria, Cyprus, Czech Republic, Djibouti, Estonia, Hungary, North Korea, Latvia, Libya, Lithuania, Myanmar, Nicaragua, Poland, Russia, Slovakia, Syria, Ukraine, US Army
W-3 Huzar/Sokol/Anakonda: Czech Republic, Hungary, Poland
PZL WARSZAWA-OKECIE
PZL-104 Wilga: Egypt, Estonia, Latvia, Lithuania, Ukraine
PZL-130 Orlik: Poland

RAYTHEON (see Beech)
Hawker 800: South Korea*
T-6 Texan II (PC-9): Canada*, Greece*, USAF*, US Navy*
REIMS/CESSNA
F150/152 Aerobat: Burundi, Congo (Zaïre)
F406 Caravan II: France, South Korea, Seychelles
FR172 Rocket: Saudi Arabia
FR337/FTB-337G Milirole: Chad, Congo (Zaïre), Mauritania, Portugal, Togo
ROBIN
HR.100: France
ROBINSON
R22 Beta: Turkey
ROCKWELL (see also Sabreliner)
B-1B Lancer: USAF
Darter Commander 100: Indonesia
Model 114 Commander: Honduras, El Salvador
OV-10 Bronco: Colombia, Indonesia, South Korea, Morocco, Philippines, Thailand, Venezuela
Shrike Commander: Argentina, Iran, Mexico
T-2 Buckeye: Greece, US Navy, Venezuela
Thrush Commander S-2R: Morocco
Turbo Commander: Angola, Bolivia, Colombia, Dominican Republic, Indonesia, Iran, South Korea, Mexico, Pakistan, Peru, Turkey, Uruguay, Venezuela
RUTAN
VariEze: US Army

SAAB (see also PAC)
32 Lansen: Sweden
35 Draken: Austria, Finland, Sweden
37 Viggen: Sweden
39 Gripen: South Africa*, Sweden
105/Sk 60: Austria, Sweden
340: Sweden
MFI-15 Safari: Norway
MFI-17 Supporter: Denmark, Norway, Uganda, Zimbabwe
SABRELINER CORP. (North American/Rockwell)
Sabreliner/CT-39: Argentina, Bolivia, Ecuador, Mexico, USAF, USMC, US Navy
SCHEIBE
SF 25B/C Falke: Oman, Pakistan
SCHWEIZER
RU-38A Condor: USCG
SCHWEIZER (HUGHES)
Model 269: Haiti, Honduras, Pakistan, Spain, Turkey
Model 300: Argentina, Indonesia, Iran, Iraq, North Korea, Nigeria, El Salvador, Sweden, Taiwan, Thailand, Turkey
SCOTTISH AVIATION (see British Aerospace)
SEPECAT
Jaguar: Ecuador, France, India, Nigeria, Oman, United Kingdom
SHAANXI
Y-8: China, Myanmar, Sri Lanka, Sudan
SHENYANG
JJ-2/FT-2: Albania, North Korea, Tanzania
J-5/F-5: Albania, China, North Korea,

Sudan, Tanzania
J-6/F-6 'Farmer': Albania, China, Egypt, Iran, North Korea, Pakistan, Sudan, Tanzania, Zambia
J-8 'Finback': China
JJ-6/FT-6: Bangladesh, China, Egypt, North Korea, Pakistan, Sudan, Zambia
SHIJIAZHUANG (ANTONOV)
Y-5: Albania, China, North Korea
SHINMAYWA
US-1: Japan
SHINMAYWA (LEARJET)
U-36A: Japan
SHORTS
330/C-23: Thailand, United Arab Emirates, US Army
SC-7 Skyvan: Austria, Ghana, Guyana, Oman, United Arab Emirates
Tucano: Kenya, Kuwait, United Kingdom
SIAI-MARCHETTI (see also Agusta)
S.208M: Ethiopia, Italy, Tunisia
SF.260: Belgium, Brunei, Burkina Faso, Burundi, Chad, Congo (Zaïre), Ethiopia, Haiti, Ireland, Italy, Libya, Philippines, Singapore, Sri Lanka, Thailand, Tunisia, Turkey, Uganda, United Arab Emirates, Venezuela, Zambia, Zimbabwe
SIAT
223 Flamingo: Syria
SIKORSKY (see also Agusta, Westland)
S-58/H-34 Choctaw: Haiti, Indonesia, Thailand
S-61/H-3 Sea King: Argentina, Brazil, Canada, Denmark, Japan, Malaysia, Spain, USAF, US Navy
S-65/S-80/H-53 Sea Stallion etc.: Germany, Iran, Israel, Japan, USAF, USMC, US Navy
S-70/H-60 Black Hawk/Seahawk etc.: Argentina, Australia, Bahrain, Chile, Colombia, Egypt, Greece, Israel, Japan, Jordan, South Korea, Malaysia, Mexico, Morocco, Philippines, Saudi Arabia, Spain, Taiwan, Thailand, Turkey, USAF, US Army, USCG, USMC, US Navy
S-76/AUH-76: Argentina, Honduras, Peru, Philippines, Spain, Thailand, Trinidad & Tobago, United Kingdom
SLINGSBY
T-67 Firefly/T-3: Belize, Canada, United Kingdom, USAF
SME
MD3-160 Aerotiga: Indonesia*, Malaysia
SOCATA (AÉROSPATIALE)
Rallye/Guerrier: Central African Republic, France, Morocco, El Salvador, Senegal
TB 9 Tampico/TB 10 Tobago/TB 20 Trinidad: Cambodia, Greece, Indonesia, Israel, Turkey
TBM700: France
SOKO
G-2A Galeb: Bosnia (Serb), Libya, Yugoslavia, Zambia
G-4 Super Galeb: Bosnia (Serb), Myanmar, Yugoslavia
J-21 Jastreb: Bosnia (Serb), Libya, Zambia
SOKO/IAV CRAIOVA (JUROM)
IAR-93/J-22 Orao: Bosnia (Serb), Romania, Yugoslavia
SUD AVIATION (NORTH AMERICAN) (see also Aérospatiale)
Fennec: Uruguay
SUKHOI
Su-7 'Fitter': Afghanistan, Algeria, North Korea, Turkmenistan
Su-15 'Flagon': Azerbaijan
Su-17/-20/22: Afghanistan, Algeria, Angola, Bulgaria, Czech Republic, Egypt, Iraq, Libya, Peru, Poland, Russia, Slovakia, Syria, Turkmenistan, Ukraine, Uzbekistan, Vietnam, Yemen
Su-24 'Fencer': Algeria, Azerbaijan, Belarus, Iran, Kazakhstan, Libya, Russia, Syria, Ukraine, Uzbekistan

Su-25 'Frogfoot': Angola, Armenia, Azerbaijan, Belarus, Bulgaria, Czech Republic, Georgia, Georgia (Abkhazia), Iraq, Kazakhstan, North Korea, Peru, Russia, Slovakia, Turkmenistan, Ukraine, Uzbekistan
Su-27/30/33 'Flanker': Belarus, China, Ethiopia, India, Kazakhstan, Russia, Syria, Ukraine, Uzbekistan, Vietnam
Su-29: Argentina
SUMMIT AVIATION
Sentry 02-337: Haiti, Thailand
SWEARINGEN (see Fairchild)

TECHNOAVIA
SM 92 Finist: Russia
TECNAM
P92 Echo: Cambodia
TONATIUH
Mexico
TRANSALL
C.160: France, Germany, Indonesia, Turkey
TUPOLEV (see also Xian)
Tu-16 'Badger': Russia
Tu-22 'Blinder': Libya, Russia, Ukraine
Tu-22M 'Backfire': Russia, Ukraine
Tu-95/142 'Bear': India, Russia, Ukraine
Tu-134 'Crusty': Belarus, Bulgaria, Cambodia, Czech Republic, Georgia, Kazakhstan, Moldova, Russia, Syria, Ukraine, Uzbekistan
Tu-154 'Careless': China, Czech Republic, Georgia, Germany, North Korea, Poland, Romania, Russia, Slovakia
Tu-160 'Blackjack': Russia

UTVA
UTVA-66: Bosnia-Herzegovina, Bosnia (Serb)
UTVA-75: Bosnia-Herzegovina, Bosnia (Serb), Yugoslavia

VALMET
L-70 Miltrainer/Vinka: Finland
L-90TP Redigo: Eritrea, Finland, Mexico
VOUGHT
A-7 Corsair II: Greece, Portugal, Thailand
F-8 Crusader: France

WACO
VPF-7: Guatemala
WASSMER (CERVA)
CE.43 Guépard: France
WESTLAND
Lynx: Brazil, Denmark, France, Germany, South Korea, Netherlands, Nigeria, Norway, Pakistan, Portugal, South Africa*, United Kingdom
Sea King/Commando: Australia, Belgium, Egypt, Germany, India, Norway, Pakistan, Qatar, United Kingdom
Wasp: Indonesia, Malaysia
Wessex: United Kingdom, Uruguay

XIAN
B-6/H-6: China, Iraq
JH-7: China
Y-7: Cambodia, Laos

YAKOVLEV (see also Aerostar)
Yak-11 'Moose': Yemen
Yak-18 'Max': Afghanistan, Laos, Lithuania, Mali, Russia
Yak-28 'Brewer-D/E': Turkmenistan
Yak-38 'Forger': Ukraine
Yak-40 'Codling': Bulgaria, Cuba, Equatorial Guinea, Ethiopia, Guinea-Bissau, Laos, Poland, Russia, Slovakia, Syria, Yugoslavia, Zambia
Yak-50/52/55: Georgia, Georgia (Abkhazia), Lithuania, Russia

ZLIN
Z 42/43/142/143/242 family: Algeria, Czech Republic, Hungary, Slovenia
Z 326: Cuba, Iraq, Mozambique
Z 526 Trener-Master: Bosnia-Herzegovina, Egypt

Picture acknowledgments

The publishers gratefully acknowledge the assistance of the following individuals and organisations in supplying illustrations for this publication.

Front cover: British Aerospace, Greg L. Davis, Stefan Petersen, Dassault, Zoltan Buza, Patrick Laureau. **6:** John A. Bradley, Yves Debay, MATRA. **7:** Boeing, Dassault. **8:** Jeff Rankin-Lowe, Northrop Grumman. **9:** US Air Force. **11:** Sikorsky, Randy Jolly, Northrop Grumman. **12:** US Air Force, Randy Jolly (three). **13:** Randy Jolly (two). **14:** Boeing, Randy Jolly, US Air Force, Graham Robson. **15:** Gulfstream, Robert F. Dorr. **16:** G.R. Stockle, Jeff Rankin-Lowe. **17:** Teledyne Ryan, Lockheed Martin. **18:** Northrop Grumman, Graham Robson, Lockheed Martin, Jeff Rankin-Lowe, B. Redfern. **19:** Randy Jolly (two), US Air Force, Northrop Grumman. **20:** Randy Jolly (two), Dave Willis, Bob Archer. **21:** Boeing, Jeff Rankin-Lowe. **22:** Randy Jolly (two), Jeff Rankin-Lowe, A. Moulton. **23:** Randy Jolly, Graham Robson, US Air Force, Boeing. **24:** Randy Jolly (two). **25:** Chris Schmidt, Greg L. Davis, US Air Force. **26:** Jeff Rankin-Lowe (two), US Air Force. **27:** Randy Jolly, G.R. Stockle. **28:** Randy Jolly (three). **29:** Randy Jolly (two), Chris Schmidt, MAP. **30:** Randy Jolly, Dave Willis. **31:** Lockheed Martin (two). **32:** Luigino Caliaro, Nate Leong, Lockheed Martin, Jeff Rankin-Lowe. **33:** Gilles Auliard, Jim Winchester, Chris Schmidt, Bob Archer. **34:** Joe Cupido, Lockheed Martin, Randy Jolly. **35:** Dave Willis. **36:** Yves Debay, Kevin Wills, Bob Archer, Dylan Eklund. **37:** MAP, US Air Force. **38:** US Navy, Northrop Grumman. **39:** Yves Debay, Boeing, US Navy. **40:** Yves Debay (three), Stephen J. Brennan. **41:** Yves Debay, Jose M. Ramos, via Jim Winchester. **42:** Jeff Rankin-Lowe, Graham Robson. **43:** Jeff Rankin-Lowe (two), Cdr Dave Baranek, Kevin Wills, Vance Vasquez/US Navy. **44:** Boeing, Mark Munzel, Jamie Hunter, Jose M. Ramos. **45:** Yves Debay, US Navy, B. Redfern, Northrop Grumman. **46:** Jeff Rankin-Lowe (two), Mark Munzel, G.R. Stockle, B. Redfern. **47:** US Navy, Lockheed Martin, Tim Senior. **48:** Boeing, G.R. Stockle. **49:** Jeff Rankin-Lowe, Yves Debay (five). **50:** Graham Robson, Yves Debay. **51:** B. Redfern, Gulfstream, Nigel Pittaway. **54:** Kaman, US Navy. **55:** G.R. Stockle, René J. Francillon (two), Jeff Rankin-Lowe. **56:** Peter B. Lewis, Henry B. Ham. **57:** US Navy, Sabreliner Corp, Raytheon. **58:** Boeing, Jeff Rankin-Lowe. **59:** G.R. Stockle, Jeff Rankin-Lowe (three), Henry B. Ham. **60:** B. Redfern, Alex Hrapunov. **61:** US Navy, Jeff Rankin-Lowe, G.R. Stockle. **62:** Boeing, Jonandrea Gaioni. **63:** Bell, Jeff Rankin-Lowe. **64:** Boeing (two), Robert Sant, Jeff Rankin-Lowe, B. Redfern. **65:** Tieme Festner (two), Sikorsky. **66:** Jonandrea Gaioni, Robert Sant. **67:** Raytheon, Gulfstream, USMC, Carl L. Richards. **70:** Lockheed Martin, B. Redfern, Jeff Rankin-Lowe. **71:** USMC, Jeff Rankin-Lowe (two). **72:** USCG (three), Chris Schmidt. **73:** USCG, Peter J. Cooper. **74:** Peter Steinemann, Graham Robson. **75:** Robert Hewson. **76:** René J. Francillon (two). **77:** Dave Willis, Chris Schmidt. **78:** Robert Hewson, Graham Robson. **79:** Peter J. Cooper. **80:** US Army, David Donald, Robert Hewson. **81:** Robin Polderman. **82:** Peter Steinemann (three). **83:** M. Ogawa via René J. Francillon. **84:** MAP, Andrew H. Cline, Raytheon. **85:** Boeing, Robert Hewson, Sikorsky. **86:** Graham Robson. **87:** David F. Brown. **88:** Boeing, Bell. **89:** Robert Hewson (four), René J. Francillon. **90:** Bell, US Army, Robert Hewson. **91:** Mike Reyno, Luigino Caliaro. **92-94:** Mike Reyno. **95:** Mike Reyno, Jeff Rankin-Lowe. **96:** Northrop Grumman, Aerospace (two). **97:** Andrew H. Cline (two), Ellis. **98:** Alan Key, E.A. Sloot. **99:** Peter R. Foster, Graham Robson, Alan Key. **100:** Alan Key, Paolo Poggi (three). **101:** Paolo Poggi (two), Chris Knott. **102:** Chris Knott (two), E.A. Sloot. **103:** E.A. Sloot (three). **104:** Ben J. Ullings, Alan Key, Tim Spearman. **105:** Chris Knott (two). **106:** Jorge F. Nuñez Padin. **107:** P. Laureau (two). **108:** Jorge F. Nuñez Padin (three), Aerospace, P. Laureau. **109:** Angelo Saini via Dairua, Jorge F. Nuñez Padin, Cees Jan van der Ende/Roland van Maarseveen (two), CASA. **110:** Aerospace (two), Alan Key (two). **111-112:** Carlos Lorch. **113:** Carlos Lorch (two), Newman Homrich/Revista Forca Aerea. **114:** via Revista Forca Aerea, Jackson Flores Jr/Revista Forca Aerea, Peter Steinemann, P. Laureau (two). **115:** P. Laureau (two), Peter Steinemann. **116:** P. Laureau (three). **117:** Alan Key (two), Pilatus. **118:** Alan Key, Roberto Yañez (two). **119:** Alan Key (three), Peter Steinemann (two). **120:** Peter Steinemann, BAe, Aerospace, Alan Key. **121:** Peter R. Foster, Aerospace. **122:** Aerospace (three), Dassault. **123:** Aerospace (three), Enstrom, Pilatus, Jackson Flores Jr/Revista Forca Aerea. **125:** Jackson Flores Jr/Revista Forca Aerea (two), Peter R. Foster, Roberto Yañez. **126:** Lockheed Martin, MAP, René L. Uijthoven. **127:** Roberto Yañez, Iain Mackenzie, MAP. **129:** Andrew H. Cline, Kevin Wills. **130:** Aerospace (two), John Kimberley. **131:** MAP (two), Hans Nijhuis. **132:** Yves Debay, Pilatus. **133:** MAP, Aerospace, Eurocopter, Paul Jackson, P. Laureau. **134:** Yves Debay (two), Tieme Festner, US Air Force. **135:** Tieme Festner, Lockheed Martin. **136:** Kaman, Denis Hughes (two), R. Mateboer, Aermacchi. **137:** MAP (two), EMBRAER. **138:** Aermacchi, Yves Debay, Mark Wagner. **139:** BAe via Michael Stroud, Dornier, McDonnell Douglas. **140:** F.G. Rozendaal. **141:** Aerospace. **142:** MAP, Yves Debay (two). **143:** Jean-Jacques Petit, Gert Kromhout, Dave Willis. **144:** Aerospace, Robin Polderman. **145:** BAe, Aerospace. **146:** Yves Debay (three). **147:** Herman Potgieter, Aerospace. **148:** Herman Potgieter, Denel. **149:** Aerospace (two), D.R. Marriott. **150:** Aérospatiale, R. Shaw. **151:** SIAI-Marchetti, Ian Malcolm. **152:** RNZAF (three). **153:** RAAF (three). **154:** Nigel Pittaway (three). **155:** Pilatus, Nigel Pittaway. **156:** Nigel Pittaway (three), Jim Winchester, Eurocopter. **157-158:** RNZAF. **159:** RNZAF, Kaman, Craig P. Justo, Jim Winchester, via Jim Winchester. **161:** Achille Vigna (four), George Kemp. **162:** Erich Strobl, Georg Mader. **163:** Georg Mader (two). **164:** F.G. Rozendaal (five). **165:** A. Roels/BAF (two), Tieme Festner. **166:** A. Roels/BAF, Frits Widdershoven, René L. Uijthoven, McDonnell Douglas. **167:** Timm Ziegenthaler, Robin Polderman, Alexander Mladenov. **168:** Alexander Mladenov (three), Tieme Festner. **169:** Alexander Mladenov, Georg Mader (three). **170:** Georg Mader, Martin E. Sequest (two). **171:** Erich Strobl (two), F.G. Rozendaal, P. Van Weenen. **172:** RDAF (two). **173:** Jan Jørgensen (four), Gulfstream, Dave Willis. **174:** Jan Jørgensen (four). **175:** J. Laukkanen (two), Tieme Festner, BAe. **176:** Aerospace, J. Laukkanen (two), Dornier, Tieme Festner. **177:** Kevin Wills, Yves Debay. **178:** Michel Fournier, Graham Robson, Gert Kromhout. **179:** Yves Debay, Dave Willis. **180:** Aérospatiale, Georges Delorient, Yves Debay, Michel Fournier. **181:** Aérospatiale, Graham Robson, Tieme Festner. **182:** Dassault, Dylan Eklund, Yves Debay. **183:** Graham Robson, Yves Debay (two), Dassault (three). **184:** Yves Debay (two), Aérospatiale, Stuart Lewis. **185:** Georges Delorient, P. Attrill, Aérospatiale. **186:** Aérospatiale, Dassault, Aerospace, MAP. **187:** DASA, Jeff Rankin-Lowe, Graham Robson. **188:** Robin Polderman, DASA (two). **189:** DASA, B. Redfern, Tieme Festner, Jeff Rankin-Lowe. **190:** DASA (two), Dave Willis, Yves Debay, Chris Schmidt, Timm Ziegenthaler. **191:** Aerospace, Timm Ziegenthaler. **192:** Kevin Wills, Dave Willis, Marcus Fulber (two), René L. Uijthoven. **193:** Stuart Lewis, Gert Kromhout, MBB. **194:** MBB, Tieme Festner, Gert Kromhout. **195:** Lockheed Martin, René van Woezik (two). **196:** René van Woezik (five). **197:** René van Woezik (six), Aerospace. **198:** Boeing, René van Woezik, Georg Mader, J. Gal. **199:** Zoltan Buza, M.D. Tabak, Georg Mader (two). **200:** Irish Air Corps via Robert Hewson (two). **201:** Greg L. Davis, Gert Kromhout, G.R. Stockle, Aeritalia. **202:** Dylan Eklund, M. Court, Luigino Caliaro, R. Mancini. **203:** Luigno Caliaro, Boeing, Agusta, Sergio Bottaro. **204:** Claudio Carretta via Luigino Caliaro, MAP, Robert Sant. **205:** Jan Jørgensen (two), E.A. Sloot, Tieme Festner. **206:** Yves Debay, Tieme Festner. **207:** KLu, Gert Kromhout. **208:** Boeing (two), René L. Uijthoven, KLu (two), Fokker. **209:** KLU, Kevin Wills, Jeff Rankin-Lowe. **210:** RNAF, Paul Jackson, Westland. **211:** Tieme Festner (two), Chris Schmidt. **212:** Gert Kromhout, Stephen J. Brennan, Andrzej Rogucki (two). **213:** Andrzej Rogucki, Stephen J. Brennan (two), Gert Kromhout. **214:** Lockheed Martin, Peter Steinemann. **215:** Eurocopter, Tieme Festner, via Michael Stroud, Westland. **216:** Robin Polderman, Hugo Mambour, Gert Kromhout. **217:** Alexander Mladenov, Dylan Eklund, Gert Kromhout (two). **218:** Sergei Skrynnikov, F.G. Rozendaal, M.J. Gerards. **219:** Yefim Gordon, Hans Nijhuis, Sergei Skrynnikov, M.J. Gerards (two), Artur Sarkisyan. **221:** M.J. Gerards (two), Sergei Skrynnikov. **222:** Régent Dansereau, Aerospace, M.J. Gerards, P. Van Weenen. **223:** M.J. Gerards, Sergei Skrynnikov. **224:** Sukhoi. **225:** F.G. Rozendaal. **226:** David Donald, Sergei Skrynnikov, Victor Drushlyakov. **227:** Georg Mader (two), Marcus Fulber. **228:** M. Malec (two), Salvador Mafé Huertas (three). **229-230:** Salvador Mafé Huertas. **231:** Salvador Mafé Huertas (two), Yves Debay. **232:** Katsuhiko Tokunaga (two). **233:** Robert Hewson (two), Anders Nylen. **234:** Gulfstream, Jim Winchester, Saab, Robert Hewson. **235:** Eurocopter, Stuart Lewis (two). **236:** M. Schenk (three). **237:** Erich Strobl (two), Graham Robson, M. Schenk (two). **238:** Lockheed Martin, M.D. Tabak, René van Woezik, Aerospace. **239:** René van Woezik (five), Aermacchi. **240-241:** René van Woezik. **242:** Georg Mader, Tieme Festner, Chris Ryan, Victor Drushlyakov, Robin Polderman. **244:** BAe, DPR RAF. **245:** David Donald. **246:** DPR RAF, Dylan Eklund, Kevin Wills (three), BAe. **247:** Kevin Wills (three), Dylan Eklund, BAe, Tim Senior, DPR RAF (two). **248:** Strike Command (four), Kevin Wills (two), DPR RAF. **249:** Dylan Eklund, Strike Command, Robert Sant, Tim Senior, Patrick Allen, David Donald (three). **250:** DPR RAF, Westland, BAe, Patrick Allen. **251:** David Donald, BAe, Patrick Allen. **252:** Shorts, Jim Winchester, Chris Lofting, Kevin Wills, David Donald, Paul Jackson, DPR RAF. **253:** DPR RAF. **254:** BAe, Kevin Wills, Tim Senior. **255:** Dave Willis, Westland (three), Tim Senior. **256:** BAe, N. Dunridge (two), Grob, Patrick Allen. **257:** Patrick Allen, David Donald. **258:** David Donald (five), Chris Lofting. **259:** Daniel J. March, Jens Schymura, Robin Polderman. **261:** Lockheed Martin, Aerospace, Simon Watson. **262:** Aerospace (three). **263:** Aerospace, Nikola Dimitrijevio. **264:** Aerospace, Peter R. Foster, Shlomo Aloni, G.A. Boymans. **265:** Peter Steinemann (three), G.A. Boymans, Shlomo Aloni. **266:** Peter R. Foster (two), Peter Steinemann (two), Alan Key, CASA. **267:** Peter Steinemann, Boeing, Shorts, BAe. **268:** R. Shaw, BAe, FFA. **269:** BAe (two), Peter Steinemann (three). **270:** BAe (three), MAP, Boeing, René J. Francillon. **271:** BAe, CASA, Bell. **272:** Shorts, Dassault. **273:** BAe, Aermacchi, Boeing. **275:** Chris Knott/Tim Spearman (seven). **276:** Aerospace, Simon Watson. **277:** Wg Cdr A.P. Mote, Simon Watson (three). **278:** Simon Watson (three). **279:** Simon Watson (three), Pushpindar Singh. **280-281:** Simon Watson. **282:** Simon Watson (three), Peter Steinemann. **283:** Lockheed Martin, Peter Steinemann (two). **284:** Peter Steinemann (three). **285:** Peter Steinemann (two), M.J. Gerards. **287:** Gamma/Frank Spooner, Michael Stroud. **288:** IAI, Aerospace (three). **289:** Aerospace, C.J.P. van Oostman. **290-291:** Aerospace. **292:** John Blake, Boeing, R. Shaw, Lockheed Martin. **293:** BAe, FFA, MAP, John Blake (two). **294:** CASA, John Blake (two), Katsuhiko Tokunaga. **295:** Katsuhiko Tokunaga (three), Yoshitomo Aoki, Atsushi Tsubota. **296:** Katsuhiko Tokunaga (three), Yoshitomo Aoki. **297:** Atsushi Tsubota, Katsuhiko Tokunaga (two), Yoshitomo Aoki. **298:** Yoshitomo Aoki, Stefan Verjans, Peter R. Foster. **299:** Kawasaki, Atsushi Tsubota, Peter R. Foster. **300:** MAP. **301:** Cees Jan van der Ende/Roland van Maarseveen (three), A. Moulton. **302:** Lockheed Martin, Peter R. Foster (three). **303:** Peter R. Foster, McDonnell Douglas, Westland, J. Rossino, Alec Moulton, Cees Jan van der Ende/Roland van Maarseveen (three). **304:** Aerospace, Peter Steinemann (two). **305:** Boeing, RAAF via Nigel Pittaway, Peter R. Foster, Northrop Grumman, BAe. **306:** SOKO, Pilatus, Simon Watson. **307:** Northrop Grumman, Fokker, McDonnell Douglas. **308:** Lockheed Martin, Peter Steinemann (two), Wei-Bin Chang, Ke-Hsiu Lin. **309:** Nigel Pittaway, David Donald, Peter Steinemann (two). **310:** Dassault via Paul Jackson, Peter Steinemann, Sikorsky. **311:** Wei-Bin Chang, Peter Steinemann, Peter R. Foster. **312:** Lockheed Martin, Cees Jan van der Ende/Roland van Maarseveen, Rene van Woezik (two), Simon Watson. **313:** Alenia, Simon Watson (four), Cees Jan van der Ende/Roland van Maarseveen (two). **314:** Cees Jan van der Ende/Roland van Maarseveen (three), Shorts. **315:** F.G. Rozendaal (four). **Back cover:** Peter Steinemann, Sergei Skrynnikov, Yves Debay, Lockheed Martin, Jackson Flores Jr.